PHOTONIC POLYMER SYSTEMS

PLASTICS ENGINEERING

Founding Editor

Donald E. Hudgin

Professor
Clemson University
Clemson, South Carolina

PHOTONIC POLYMER SYSTEMS

FUNDAMENTALS, METHODS, AND APPLICATIONS

edited by

DONALD L. WISE
Northeastern University
Boston, Massachusetts

GARY E. WNEK
Virginia Commonwealth University
Richmond, Virginia

DEBRA J. TRANTOLO
Cambridge Scientific, Inc.
Belmont, Massachusetts

THOMAS M. COOPER
Air Force Research Laboratory
Wright-Patterson Air Force Base, Ohio

JOSEPH D. GRESSER
Cambridge Scientific, Inc.
Belmont, Massachusetts

MARCEL DEKKER, INC.   NEW YORK · BASEL · HONG KONG

Library of Congress Cataloging-in-Publication Data

Photonic polymer systems: Fundamentals, methods, and applications / edited by Donald L. Wise. . .
[et al.].
    p. cm. — (Plastics engineering)
    Includes index.
    ISBN 0-8247-0152-6 (alk. paper)
    1. Photonics—Materials. 2. Polymers—Optical properties. 3. Optoelectronic devices—
Materials. I. Wise, Donald L. (Donald Lee). II. Series: Plastics engineering (Marcel Dekker,
Inc.); 49.
    TA1522.P46 1998
    621.36—dc21
                                       98-24469
                                            CIP

This book is printed on acid-free paper.

Headquarters
Marcel Dekker, Inc.
270 Madison Avenue, New York, NY 10016
tel: 212-696-9000; fax: 212-685-4540

Eastern Hemisphere Distribution
Marcel Dekker AG
Hutgasse 4, Postfach 812, CH-4001 Basel, Switzerland
tel: 44-61-261-8482; fax: 44-61-261-8896

World Wide Web
http://www.dekker.com

The publisher offers discounts on this book when ordered in bulk quantities. For more information,
write to Special Sales/Professional Marketing at the headquarters address above.

Current printing (last digit)
10 9 8 7 6 5 4 3 2 1

PRINTED IN THE UNITED STATES OF AMERICA

Preface

This book is intended primarily for scientists, engineers, and technical personnel who are interested in the materials aspects of optics and photonics. The text provides background characterization, and applications of optics and photonics with an emphasis on materials. It also has a special focus on fundamentals and methods. Differing from the style of formal technical journals, the chapters in this text are in tutorial format, with in-depth discussion of optical and photonic materials fundamentals and methodologies. In addition, supporting chapters are offered on specific applications of optical and photonic properties. Chapters focused on state-of-the-art applications are intercalated within the tutorials by practicing technologists in optics and photonic polymers. We have limited the number of chapters and included a select group of tutorial-type chapters to make this reference text useful to all who are in the field. The focused nature of this book will afford substantial directed input, spirited exchange of problem-solving ideas, and exploration of innovative applications.

Because of the emerging interest in and the potential of optics and photonics, chapter authors with experience with new materials have been included with others of established backgrounds. Readers who are optics or photonic specialists will appreciate chapters dealing with presentation of concepts from those with experience. Chapters from authors of varied backgrounds and disciplines will expand upon present technical understanding of materials within the field of optics and photonics fundamentals, methods, and applications.

Donald L. Wise
Gary E. Wnek
Debra J. Trantolo
Thomas M. Cooper
Joseph D. Gresser

Contents

v

Contributors

Hossin A. Abdeldayem   *Universities Space Research Association, NASA Marshall Space Flight Center, Huntsville, Alabama*

J. A. Akkara   *U.S. Army Natick Research Development and Engineering Center, Natick, Massachusetts*

F. J. Aranda   *University of Massachusetts at Boston, Boston, Massachusetts*

Franco Cacialli   *Cavendish Laboratory, Cambridge, England*

Greg Carlson   *University of Connecticut, Storrs, Connecticut*

M. H. Chan   *Hong Kong Baptist University, Kowloon Tong, Hong Kong*

Kyung M. Choi   *University of California, Irvine, California*

V.-E. Choong   *University of Rochester, Rochester, New York*

Thomas M. Cooper   *Air Force Research Laboratory, Wright-Patterson Air Force Base, Ohio*

Frank R. Denton III   *Motorola, Inc., Lawrenceville, Georgia*

Charles A. DiMarzio   *Center for Electromagnetic Research, Northeastern University, Boston, Massachusetts*

M. A. Drobizhev   *Russian Academy of Sciences, Moscow, Russia*

Donald O. Frazier   *NASA Marshall Space Flight Center, Huntsville, Alabama*

H.-J. Gao   *Chinese Academy of Sciences, Beijing, China*

Y. Gao   *University of Rochester, Rochester, New York*

Kenneth E. Gonsalves   *University of Connecticut, Storrs, Connecticut*

Joseph D. Gresser   *Cambridge Scientific, Inc., Belmont, Massachusetts*

B. R. Hsieh   *Xerox Corporation, Webster, New York*

Alex K.-Y. Jen   *Northeastern University, Boston, Massachusetts*

Joby Joseph   *University of Massachusetts at Boston, Boston, Massachusetts*

Kyoji Kaeriyama *Kyoto Institute of Technology, Kyoto, Japan*

Yoshihiko Kanemitsu *Nara Institute of Science and Technology, Ikoma, Nara, Japan*

B. Kippelen *University of Arizona, Tucson, Arizona*

Gregory J. Kowalski *Northeastern University, Boston, Massachusetts*

Th. Kugler *Linköping University, Linköping, Sweden*

Paul M. Lahti *University of Massachusetts, Amherst, Massachusetts*

DeQuan Li *Los Alamos National Laboratory, Los Alamos, New Mexico*

Li-Sheng Li *University of Illinois at Urbana-Champaign, Urbana, Illinois*

Xiao-Chang Li *Northeastern University, Boston, Massachusetts*

M. Lögdlund *Linköping University, Linköping, Sweden*

Zuhong Lu *Southeast University, Nanjing, Jiangsu, People's Republic of China*

Barry Luther-Davies *The Australian National University, Canberra, Australia*

K. Meerholz *University of Munich, Munich, Germany*

Marc Gregory Mogul *Northeastern University, Boston, Massachusetts*

Stephen C. Moratti *University of Cambridge, Cambridge, England*

M. Nakashima *U.S. Army Natick Research Development and Engineering Center, Natick, Massachusetts*

Lalgudi V. Natarajan *Science Applications International Corporation, Dayton, Ohio*

Mark S. Paley *Universities Space Research Association, NASA Marshall Space Flight Center, Huntsville, Alabama*

S. J. Pang *Chinese Academy of Sciences, Beijing, China*

Y. Park *University of Rochester, Rochester, New York*

Benjamin G. Penn *NASA Marshall Space Flight Center, Huntsville, Alabama*

N. Peyghambarian *University of Arizona, Tucson, Arizona*

Octavio Ramos, Jr. *Los Alamos National Laboratory, Los Alamos, New Mexico*

D. Narayana Rao *University of Massachusetts at Boston, Boston, Massachusetts*

D. V. G. L. N. Rao *University of Massachusetts at Boston, Boston, Massachusetts*

W. R. Salaneck *Linköping University, Linköping, Sweden*

Anna Samoc *The Australian National University, Canberra, Australia*

Marek Samoc *The Australian National University, Canberra, Australia*

Kenneth J. Shea *University of California, Irvine, California*

David D. Smith *NASA Marshall Space Flight Center, Huntsville, Alabama*

S. K. So *Hong Kong Baptist University, Kowloon Tong, Hong Kong*

Debra J. Trantolo *Cambridge Scientific, Inc., Belmont, Massachusetts*

Guangming Wang *Southeast University, Nanjing, Jiangsu, People's Republic of China*

Yu Wei *Southeast University, Nanjing, Jiangsu, People's Republic of China*

Donald L. Wise *Northeastern University, Boston, Massachusetts*

William K. Witherow *NASA Marshall Space Flight Center, Huntsville, Alabama*

Gary E. Wnek *Virginia Commonwealth University, Richmond, Virginia*

Maneerat Woodruff *The Australian National University, Canberra, Australia*

Z. Q. Xue *Peking University, Beijing, China*

Chunwei Yuan *Southeast University, Nanjing, Jiangsu, People's Republic of China*

Yue Zhang* *Northeastern University, Boston, Massachusetts*

Current affiliation: Lightwave Microsystems Corporation, Santa Clara, California

PHOTONIC POLYMER SYSTEMS

1

Optical Properties of σ-Conjugated Silicon Polymers and Oligomers

Yoshihiko Kanemitsu
Nara Institute of Science and Technology
Ikoma, Nara, Japan

I. INTRODUCTION

In exploring new optoelectronic materials and devices, a great deal of research effort is focused on reducing the dimensionality of the electronic structures in semiconductors. In low-dimensional semiconductor systems, three categories are usually considered: two-dimensional (2D) quantum wells, one-dimensional (1D) quantum wires, and zero-dimensional (0D) quantum dots [1]. In particular, 1D semiconductors, often called quantum wires, have attracted considerable attention, because they exhibit a wealth of quantum phenomena and have potential as future optoelectronic devices. Chemically synthesized semiconducting polymers are regarded as natural quantum wires. Unique optical and electronic properties of conjugated polymers are due to the electrons delocalized 1D polymer backbone, and they make it important materials for technological applications like light-emitting diodes, nonlinear optical devices, field-effect transistors, and so on [2].

Recently, in silicon (Si) science and technology, the desire for the integration of optoelectronics devices with Si microelectronics has led to the search for Si-based materials and structures that emit light with a high quantum efficiency. Crystalline Si is the dominant material in microelectronics and is one of the best-studied materials. Crystalline Si does not show efficient light emission at room temperature, because of its band structure with an indirect band gap of $\sim$1.1 eV and a small exciton binding energy ($\sim$15 meV). One promising approach to overcoming the indirect nature of optical transition in Si is the relaxation of the k-selection rule due to the spatial confinement in low-dimensional Si nanostructures [3]. Natural analogs of low-dimensional Si structures such as wires and sheets are chemically synthesized oligomers and polymers. Modern organic synthesis techniques allow us to produce Si materials with controlled structures [4,5]. Studies of optical properties of chemically synthesized Si materials (small Si clusters, Si-based polymers, siloxenes, etc.) are needed to understand the electronic structures and the microscopic luminescence mechanisms in Si nanocrystals and other low-dimensional Si materials, because the local Si structures can be controlled in chemically synthesized Si materials. In this chapter, we discuss the optical properties of σ-conjugated Si polymers and oligomeres.

II. SILICON OLIGOMERS

Long-chain polymers have the complexity of the "real" polymer such as solubility, structural defects, the broad distribution of the chain length, and so on. The optical and electronic properties reflect this disordered nature of polymers. The effective conjugation length and disorders of polymers also depend on the polymer film fabrication method. On the other hand, oligomers are well-defined chemical systems: The conjugation chain length can be exactly controlled. Therefore, oligomers of short-chain lengths have received much attention as model compounds for a better understanding of the electronic and optical properties of polymers. Studies of the size dependence of optical and electronic properties of oligomers with well-defined lengths help to check the theoretical predictions [6,7] and to understand of the microscopic mechanism of radiative and nonradiative recombination processes in polymers [8].

A. Shape Dependence of Optical Properties

The study of the structure and properties of small Si clusters has been an extremely active area of current research. The small Si clusters have been usually produced by pulsed-laser-evaporation methods [9,10]. However, as small clusters are short-lived intermediate species and there are few experimental studies of optical properties in small clusters, we have little information on the optical properties of vapor-phase small clusters. We need to contrive new experimental procedures for the study of the detailed structure and properties of Si clusters; for example, vapor-phase Si clusters deposited onto a solid matrix are prepared for the determination of the structure of small Si clusters [11]. Despite the perfection of a pure cluster in the gas phase, it is a highly defective system from the point of view of semiconductor physics. In pure systems, the surface states appear within the band gap and degrade electrical and optical properties of semiconductor nanocrystals and clusters. Passivation of the surface states is very important in semiconductor physics.

Organic synthesis and purification techniques of Si materials have some advantages over the widely used laser-evaporation methods: The size and shape of Si oligomers or clusters can be exactly controlled. The study of the chemically synthesized Si clusters helps in the understanding of the electronic properties of vapor-phase Si clusters. The synthesis procedure of Si_8 clusters is illustrated in Fig. 1. Synthetic, purification, and characterization methods were described in Ref. 12. The dangling bonds are terminated by the methyl and phenyl groups in the chain structure, by the isopropyl groups in the ladder structure, and by 1,1,2-trimethylpropyl groups in the cubic structure. In each structures, the optical properties are not sensitive to the type of organic termination. The optical absorption

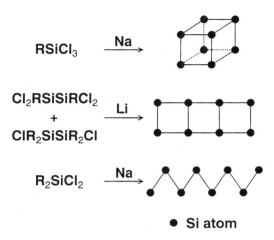

FIGURE 1 Chemically synthesized procedure of Si_8 clusters.

and luminescence properties of these clusters reflect the electronic structures of the central Si skeleton structures. Saturatedly bonded Si_8 clusters terminated by an organic substituent have no dangling bonds and become new model materials for small Si clusters. Here, we discuss optical properties of Si_8 clusters with different structures.

The optical absorption and photoluminescence (PL) spectra of Si_8 clusters with different structures are summarized in Fig. 2. The Si-backbone structures are illus-

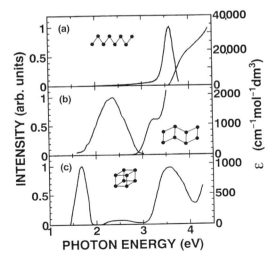

FIGURE 2 Optical absorption and photoluminescence of chemically synthesized Si_8 clusters at room temperature: (a) chain, (b) ladder, and (c) cubic structures. The photoluminescence of the cubic structure is not observed at room temperature. The lowest-absorption peak energy in the cubic structure is low compared with those in the chain and ladder structures.

trated in this figure. In the chain and ladder Si_8 structures, the absorption spectra
are observed in the ultraviolet spectral region. On the other hand, in the cubic
cluster, a small absorption band around 2.5 eV and a large one around 3.5 eV are
observed. The absorption edge energy of the cubic structure is very small compared
with those of the chain and ladder structures [13]. In the chain structure, a sharp,
strong PL band is observed at the absorption edge. In the ladder structure, a broad
PL band was observed with a large Stokes shift in the visible spectral region.
However, in the cubic structure, the efficient PL is not observed at room temperature, but is observed only at low temperatures. Figure 3 shows the temperature
dependence of the PL intensity in Si_8 clusters. The PL intensity in the cubic cluster
abruptly decreases with increasing temperature above 40 K, whereas the PL intensities in the chain and ladder clusters gradually decrease with increasing temperature. The PL decay properties of the cubic Si_8 cluster are entirely different from
those of the Si_8 chain and Si_8 sheet: The lifetimes of the chain and ladder Si_8
structures are 110 ps and 700 ps, respectively, at room temperature. As seen in
Figs. 2 and 3, the optical properties of the very small Si clusters are very sensitive
to the Si backbone structures.

The luminescence characteristics (the slow-decay and temperature-sensitive luminescences) in the cubic structure are different from those in the chain and ladder
structures. The optical absorption edge energy (the lowest-energy gap or the optical
band gap) of the cubic cluster is very small compared with those of the chain and
ladder clusters. The optical properties of the cubic structure are quite different from
those of the chain and ladder structures. It is theoretically pointed out that the

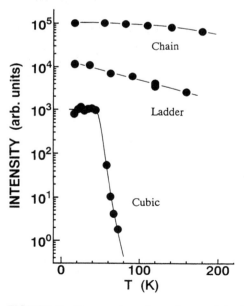

FIGURE 3 Temperature dependence of the luminescence intensity in the chain, ladder, and
cubic Si_8 clusters. The luminescence intensity in the cubic structure abruptly decreases with
increasing temperature above 40 K, whereas the luminescence intensity in the chain and
ladder structures gradually decreases with increasing temperature (From Ref. 13.)

orbital hybridization and the bonding in small clusters are very sensitive to the size and shape of the clusters [14]. The bonding in small clusters is quite different from that in the bulk Si, and the relationship between the band gap and the size of the cluster is not simple in small Si clusters. In our experiments, the Si–Si bond angle in the cubic structure is about 90°, which is different from the bond angle of tetrahedral Si bonds (sp^3-hybridized orbital). On the other hand, the Si–Si bond angle in the chain and ladder structures is nearly equal to that of tetrahedral Si bonds. The molecular orbital characters in the cubic clusters are "p-like" and is different from those in the sp^3-hybridized chain and ladder structures. Therefore, we speculate that the difference of the optical properties between the cubic and the other clusters is caused by that of the orbital hybridization and bonding in the clusters. The study of the electronic properties of Si compounds with different geometries and different orbital hybridizations is very important for the understanding of the presence of the magic numbers of small Si clusters [9] and remarkably size-independent optical properties of small Si clusters [15] and intermediate-sized Si clusters [16].

The luminescence properties of the cubic cluster are briefly summarized here [17]. Figure 4 shows the optical absorption, PL, and PL excitation spectra of the cubic Si_8 cluster. The PL and PL excitation spectra were measured at 20 K. The PL spectrum with a peak of 1.7 eV was observed at temperatures lower than 80 K. Similar PL and PL excitation spectra were reported in octasilacubanes with different organic terminations [18–20]. The organic termination groups do not play a dominant role in luminescence and absorption properties in octasilacubanes. The optical absorption and luminescence properties of the cubic cluster in the visible spectral region are determined by the electronic properties of central Si_8 cubic skeleton.

In the absorption spectrum, a low peak and a high peak are observed at 2.5 eV and 3.5 eV, respectively. The observed peak energy of the strong absorption band is consistent with a theoretical calculation of the lowest optically allowed

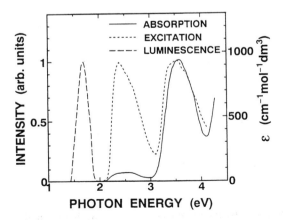

FIGURE 4 Optical absorption, photoluminescence, and photoluminescence excitation spectra in a cubic Si cluster. The photoluminescence and the photoluminescence excitation spectra were measured at 20 K. (From Ref. 17.)

transition energy in a cubic Si_8H_8 cluster (3.4 eV) [21]. On the other hand, two large peaks are observed at 2.5 and 3.5 eV in the PL excitation spectrum. The intensity ratio of PL excitation to the absorption bands at 2.5 eV is much higher than that at 3.5 eV. Although the peak height of the one-photon absorption spectrum at 2.5 eV is very low, the PL spectrum is efficiently observed under 2.5 eV excitation.

The PL decay profile is approximately described by a single-exponential function in the measurement temperatures. The temperature dependence of the PL lifetime is summarized in Fig. 5. The PL lifetime does not depend on the temperature below 40 K. However, at high temperatures above 40 K, the PL lifetime abruptly decreases with increasing temperature. The temperature dependence of the PL lifetime is very similar to that of the PL intensity in Fig. 3. At high temperatures, the PL efficiency and the PL lifetime are limited by nonradiative recombination processes. The above PL experiments strongly suggest that the slow-decay luminescence at low temperatures originates form the "forbidden" radiative recombination such as the singlet–triplet transition or the singlet–singlet transition having the same orbital symmetry. The temperature dependence of both the PL lifetime and the PL intensity can be explained by a simple configuration coordinate model [13].

We attempted to find a light-induced electron-spin resonance (ESR) response in this cubic cluster [17]. In the ESR signal associated with the photoexcited triplet state in the cubic cluster at 30 K, the weak line was observed around 1670 G ($g \sim 4$) and this corresponds to the forbidden $\Delta m_s = \pm 2$ transition between triplet sublevels. This transition is typically seen around half the magnetic field strength at which the $\Delta m_s = \pm 1$ transition occurs and is a signature of the presence of a triple excited state. Moreover, the luminescence properties of the cubic Si cluster are consistent with the characteristics of those of the phosphorescence from the

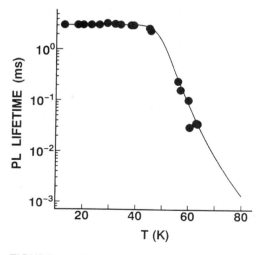

FIGURE 5 The photoluminescence lifetime of a cubic Si cluster as a function of temperature. The solid line is the theoretical calculation based on a single configuration coordinate model. The temperature dependence of the photoluminescence lifetime is very similar to that of the photoluminescence intensity. (From Ref. 13.)

triplet state in chainlike polysilanes. Therefore, it is considered that the luminescence from the cubic Si_8 cluster is attributed to the radiative recombination of the triplet excited state or the singlet–singlet transition having the same orbital symmetry.

The temperature dependence of the PL intensity and the PL lifetime in the cubic cluster are controlled by the nonradiative recombination process. Spectroscopic analysis suggests that the slow-decay and temperature-sensitive luminescences are attributed to the "forbidden" radiative recombination such as the singlet–singlet transition having the same orbital symmetry or the singlet–triplet transition. A large Stokes shift between the PL and absorption peak energies (~0.8 eV) and the luminescence process controlled by nonradiative recombination mean that the cubic Si cluster is a strong electron–phonon coupling system. The luminescence properties of the cubic cluster are different from those of the chain and ladder clusters. The study of the electronic properties of Si compounds with different geometries and different orbital hybridizations becomes very important for the understanding of the structure and properties of small Si clusters.

B. Size Dependence of Optical Properties

Optical studies of oligomers and polymers containing a few to several thousand atoms help to understand how molecules evolve into solids. Here, we report the size-dependent optical properties of confined excitons in chainlike Si oligomers and polymers.

The absorption spectra of chainlike Si-skeleton polymers solved in tetrahydrofuran (THF) were measured in order to eliminate electronic interactions between chains. The average number of Si atoms in chains, N, was varied from 5 to 110. Figure 6 shows extinction coefficients per mole of Si atom (ε_{Si}) spectra and normalized luminescence spectra of chainlike Si polymers. A sharp absorption peak in a chain of Si atoms of $N = 110$ indicates a quasi-1D electronic system, and sharp absorption and emission peaks are due to the lowest 1D exciton state delocalized in Si skeleton. With a decrease in the number of Si atoms in chains, the absorption peak of the lowest exciton state is shifted to the higher energy (blueshift) and the oscillator strength of the lowest exciton state decreases. The size effect of absorption and PL spectra are clearly observed in chains having 20 Si atoms or less. The blueshift of the absorption peak is due to the quantum confinement of the exciton on the Si skeleton [22].

Figure 7 shows the chain-length dependence of the oscillator strength per Si atom for the lowest exciton, f_1, calculated from the absorption spectra. The oscillator strength f_1 is directly proportional to the integrated area of the absorption peak in diluted solutions as follows:

$$f_1 = 10^3 \ln(10) \left(\frac{mc}{\pi h e^2 N_A} \right) \int \varepsilon_{Si} \, dE \tag{1}$$

where m is the mass of electrons, c is the velocity of light, e is the charge of electrons, h is the Planck constant, and N_A is the Avogadro number. Absorption areas of the lowest exciton, $\int \varepsilon_{Si} \, dE$ are those surrounded by the solid and dotted lines plotted in Fig. 6, where the high-energy tail of the lowest peak and the low-energy tail of the second peak are estimated by using Gaussian functions and the

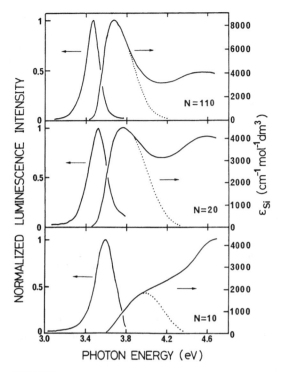

FIGURE 6 Extinction coefficients per Si atom spectra (ϵ_{Si}) and normalized photoluminescence spectra in chainlike Si polymers. N is the averge number of Si atoms in chains. Absorption areas of the lowest excitons are those surrounded by solid and dotted lines. (From Ref. 22.)

dotted lines are optimum tails of Gaussian profiles. Thus, we can estimated the oscillator strength of the lowest exciton, f_1. The oscillator strength per Si atom, f_1, is linearly proportional to the number of Si atoms and then saturates to a constant value. In Si chains of $N < 20$, f_1 is approximately given by $f_1 \propto N$. If chains of Si atoms are solids in nature, the oscillator strength per atom is almost independent of the number of atoms. Therefore, we believe that the size-dependent region is the continuous transition of Si chains from molecule to solid in nature and long chains become solidlike in optical properties.

Figure 8 shows picosecond temporal changes in the exciton luminescence at the peak energies. Picosecond temporal decay of luminescence under 1-ps and 305-nm laser excitation of ~1 pJ per pulse was measured by using a monochromator of subtractive dispersion and a synchroscan streak camera. The temporal resolution of this system was about 30 ps. The luminescence decay was approximately described as a single exponential having time constant τ_{PL}, which increases with decreasing N. A single exponential decay of luminescence implies that the luminescence is determined by a simple relaxation process. Here, we assume that the

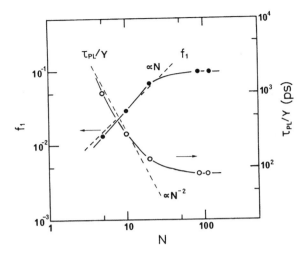

FIGURE 7 The oscillator strength of confined excitons f_1, the lifetime of photoluminescence at the peak energy τ_{PL}, and τ_{PL}/Y as a function of Si atoms in chains, N. Y is the relative quantum yield of photoluminescence. (From Ref. 22.)

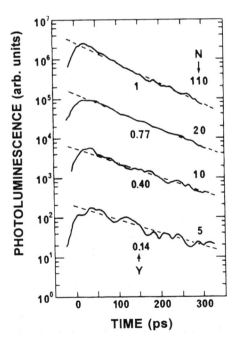

FIGURE 8 Picosecond temporal changes in luminescence due to the exciton at peak energies. The luminescence is fitted by a single exponential. Y is the relative quantum yield of photoluminescence. With increasing N, the luminescence lifetime becomes shorter and the linewidth of luminescence band becomes narrower. (From Ref. 22.)

lifetime of luminescence τ_{PL} is determined by two relaxation channels having the radiative decay rate τ_R^{-1} and the nonradiative decay rate τ_{NR}^{-1}. Because the quantum yield of luminescence η is given by $\eta = \tau_R^{-1}/(\tau_R^{-1} + \tau_{NR}^{-1})$, the radiative decay rate τ_R^{-1} is given by η/τ_{PL}. Although the absolute values of η in Si polymers was not obtained experimentally, we can use the relative yield Y to discuss the size dependence of the radiative decay rate of the exciton and make $Y = 1$ for a chain of $N = 110$. The values of Y for each Si polymer chain are shown in Fig. 8. The PL lifetime τ_{PL} and the radiative decay time τ_{PL}/Y is plotted in Fig. 7. The radiative decay time τ_{PL}/Y increases with the decrease of Si atoms, N. In particular, in chains of $N < 20$, τ_{PL}/Y is approximately given by $\tau_{PL}/Y \propto N^{-2}$. The radiative decay time depends on the oscillator strength per chain. If the exciton is a coherent excitation over the Si chain, the radiative decay time is inversely proportional to Nf_1. Because $f_1 \propto N$ in chains of $N < 20$, the radiative decay time is in proportion to N^{-2}. The size dependence of radiative decay time, τ_{PL}/Y, confirms that of the oscillator strength f_1 of the lowest exciton state. These size dependences imply that in chains having 20 Si atoms or less, the exciton is coherently excited over the chain of Si atoms.

The above considerations and conclusions are supported by the PL quantum yield measurements in polymers [23] and the theoretical calculation of the third nonlinearity $\chi^{(3)}$ [24]. In our work, the PL lifetime is about 76 ps in a Si chain having $N = 110$. Using $\eta = 0.1$ and $\tau_{PL} = 76$ ps, we can estimate the radiative decay rate of excitons delocalized on the Si skeleton, τ_R^{-1}. On the other hand, the oscillator strength per Si atom, f_1, directly gives the radiative decay rate based on one Si atom, τ_{abs}^{-1}. The ratio τ_{abs}/τ_R in a Si chain of $N = 110$ is about 24, which means that the excitons are delocalized over about 24 Si atoms. Therefore, we consider that in chains having Si atoms smaller than about 24, the excitons are confined in chains of Si atoms, and the size dependence of optical properties of excitons can be observed. Moreover, Abe [24] calculated the size dependence of the magnitude of $\chi^{(3)}$ and showed that $\chi^{(3)}$ tends to saturate at 16 Si atoms. The above experiments show that the blueshift of the exciton state energy and the size dependence of the oscillator strength is clearly and experimentally observed in short chains having 20 Si atoms or less. These experimental and theoretical studies shows that the exciton is coherently excited over about 20 Si atoms on the Si-skeleton chain and the oscillator strength of the exciton on the number of Si atoms in chains.

On the other hand, in small Si sheets, the size dependence of the exciton properties is not observed clearly. Figure 9 shows PL and absorption spectra in ladder polysilanes regraded as small Si sheets. Broad PL spectra were observed at the visible region. The absorption spectra in the range below ~4.5 eV were expressed by two absorption bands. The high-energy tail of the lowest peak and the low-energy tail of the second peak were estimated by using Gaussian functions and the broken lines in Fig. 9 are optimum tails of Gaussian profiles. The Stokes shifts between the PL and the lowest absorption peaks are larger than 1 eV in three sheetlike polysilanes. These broad PL spectra with large Stokes shifts were not observed in 1D chainlike polysilanes. A significant size dependence of the lowest exciton energy is clearly observed in small ladder Si polymers. Ab initio calculations suggest that the blueshift of the lowest excitation energy is due to the quantum size effects, and the large Stokes shift is caused by the structure difference between the ground and excited states (i.e., the formation of self-trapped excitons) [25].

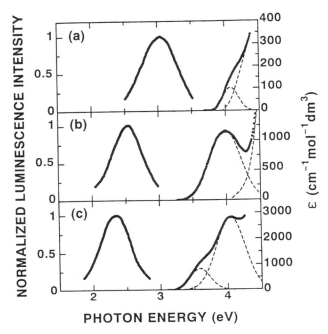

FIGURE 9 Optical absorption and normalized luminescence spectra in small Si sheets: (a) cyclotetrasilane, (b) bicyclohexasilane, and (c) anti-tricyclooctasilane. The broken lines in absorption spectra are optimum tails of two Gaussian functions. (From Ref. 25.)

In ladder polysilanes, the initial luminescence decay was approximately described as a single exponential having time constant τ_{PL}. In ladder polysilanes, the quantum efficiency of PL, η, is very low ($\leq 10^{-3}$) but τ_{PL} is large (≥ 0.8 ns), compared with 1D chainlike polysilanes, where, for example, $\eta \sim 0.1$ and $\tau_{PL} \sim 76$ ps in 1D chainlike polymethylphenylsilanes. Here, if we assume that τ_{PL} and η are only determined by two relaxation channels having a radiative decay rate τ_R^{-1} and a nonradiative decay rate τ_{NR}^{-1}, the radiative decay rate of the excitons is given by η/τ_{PL}. The radiative decay rate of excitons, η/τ_{PL}, is of the order of 10^6 s^{-1} in three sheetlike ladder polysilanes. It is considered that τ_{PL} in Si sheets is mainly limited by nonradiative decay processes. No significant size dependence of the radiative decay rate of excitons was experimentally observed. Small Si sheets have spectroscopic characteristics such as a long τ_{PL}, low quantum efficiencies, and large Stokes shifts.

Figure 10 summarizes the size dependence of the peak energy of the lowest absorption band. The absorption spectrum strongly depends on the size of Si-skeleton chains or sheets. The luminescence properties are very sensitive to the shape of oligomers rather than the size of oligomers. In 1D chainlike polysilanes, a sharp, strong PL band was observed with essentially no Stokes shift: Excitons are delocalized on Si-backbone chains, and luminescence is due to free-exciton recombination. On the other hand, in small Si sheets or ladder polysilanes, a broad, weak PL band was observed with a large Stokes shift. The weak, visible PL in Si sheets is caused by self-trapped exciton recombination and no size dependence of

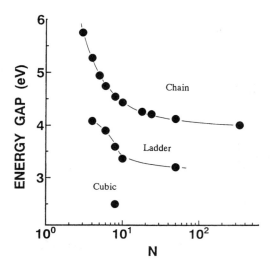

FIGURE 10 The absorption peak energy of the lowest excitation state as a function of the number of Si atom number in chain, ladder, and cubic clusters. The solid lines are guides to the eye.

the radiative decay rate of excitons is caused by the strong localization of excitons in the Si sheets. In Si chains, the oscillator strength per Si atom for the lowest excitons in linearly proportional to the number of Si atoms in chains having 20 Si atoms or less and saturates to a value in long chains. This experimental study is the observation of the continuous transition of a Si chain from molecular to solid form. In Si sheets, broad luminescence spectra with large Stokes shifts were observed at the visible region, and the exciton localization determines the luminescence properties.

C. Comparison Between π- and σ-Conjugated Systems

Chemically synthesized semiconducting polymers are regarded as natural quantum wires. The optical properties of chainlike Si oligomers depend on the chain length. In this section, we discuss optical properties of oligothiophenes in order to compare the optical properties of both σ-conjugated and π-conjugated systems. Polythiophene is a prototype of conjugated polymers having a nondegenerate ground-state structure: Its electronic structure is approximately considered as a polyene in which the conjugated $C\!=\!C$ double bonds are locked in plane. Unique optical and electronic properties of polythiophene are due to the π-conjugated electrons delocalized one-dimensional polymer backbone. We synthesized a new family of thiophene-based oligomers in order to control the optical properties and the backbone rigidity of oligomers [26].

The structures of oligothiophenes (T_n, n is the number of the thiophene ring, $2 \leq n \leq 6$) and thiophene-based oligomers with methylene (T_nMT_n), phenylene (T_nPT_n), and vinylene (T_nVT_n) bridges are shown in Fig. 11. Optical absorption and PL spectra were measured in CH_2Cl_2 solutions (1.0×10^{-5} mol/dm^3). The PL

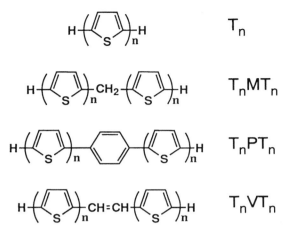

T_n

T_nMT_n

T_nPT_n

T_nVT_n

FIGURE 11 Structures of oligothiophenes (T_n) and thiophene-based oligomers (T_nMT_n, T_nPT_n, and T_nVT_n). (From Ref. 26.)

spectra were measured using a 325-nm excitation light from a He–Cd laser. The calibration for the spectral sensitivity of the entire measuring system was performed by using a tungsten standard lamp. Picosecond PL decay measurements were carried out by using a dye laser (310 nm) with 1-ps pulse duration and a double monochromator of subtractive dispersion and a synchroscan streak camera. The temporal resolution of this system was about 30 ps.

Figure 12 shows absorption and normalized PL spectra of oligothiophenes T_n. With an increase of the chain length of T_n, the shift of the absorption and PL peaks to longer wavelengths is observed. The height of the absorption peak of the lowest excitation state also increases with increasing chain length. The oscillator strength of the lowest excitation state, which is calculated from the integration of the lowest absorption band, linearly increases with the conjugation length. The effective conjugation length of oligothiophenes is estimated to be six thiophene rings or more. The bandwidth of the lowest absorption spectrum narrows with increasing chain length. This result implies that the sharp absorption PL band in oligothiophenes with longer chains is due to the electrons delocalized in the 1D backbone chain. The finite confinement potential for the lowest excitation state V_0 is assumed to be $V_0 = E_S - E_g$, where E_S is the lowest absorption peak energy of the CH_2Cl_2 solvent and E_g is the lowest absorption energy of polythiophene ($n \rightarrow \infty$). Using $E_S = 7$ eV and $E_g = 2.4$ eV in the electrochemically polymerized polythiophene reported in Ref. 27, the calculated results agree well with the observed values of E_L. The blueshift of the lowest absorption peak energy is due to the quantum confinement effect of electrons in the 1D backbone.

The PL decay curves are approximately given by single exponential functions in the picosecond time region. The PL lifetime τ_{PL} increases with increasing chain length. In Fig. 13, the size dependence of the PL lifetime and the PL quantum yields η are summarized. The PL quantum yields η were determined by using quinine sulfate as a standard material. Both the lifetime and the quantum yield increase with increasing chain length. The similar size dependence of τ_{PL} and η

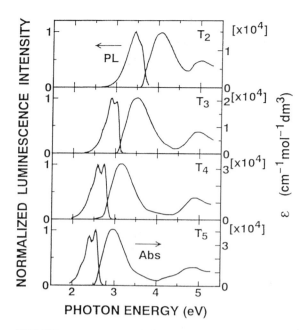

FIGURE 12 Optical absorption and photoluminescence spectra of T_n ($n = 2$–5). The peak energy of the absorption and PL spectra shifts to lower energy with increasing chain length. (From Ref. 26.)

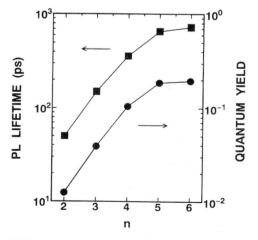

FIGURE 13 Chain-length dependence of the photoluminescence lifetime and the PL quantum yield in T_n. The chain-length dependence of the photoluminescence lifetime is very similar to that of the PL quantum yield. (From Ref. 26.)

implies that the observed PL lifetime is limited by the nonradiative recombination process and that the nonradiative decay rate decreases with increasing chain length.

Here, we assume that the PL decay rate τ_{PL}^{-1} is determined by two relaxation channels having the radiative decay rate τ_{R}^{-1} and the nonradiative decay rate τ_{NR}^{-1} ($\tau_{PL}^{-1} = \tau_{R}^{-1} + \tau_{NR}^{-1}$). The PL quantum yield η is given by $\tau_{R}^{-1}/(\tau_{R}^{-1} + \tau_{NR}^{-1})$ and then τ_{R} is τ_{PL}/η. The chain-length dependence of the estimated radiative lifetime τ_{R} is summarized in Fig. 14. Because both the PL lifetime and the PL quantum yield increase with increasing chain length, the radiative lifetime does not depend on the chain length. In all oligothiophenes, the radiative lifetime is about 4 ns. Therefore, we need to consider the size-independent radiative recombination process. A possible model is that the excitons are spatially localized within two thiophene rings and PL originates from the radiative recombination of localized excitons.

On the other hand, Fig. 13 also indicates that both the efficiency of the nonradiative recombination and the nonradiative decay rate decrease with increasing the chain length. Here, we assume that the end of the chain acts as a nonradiative recombination center and that the exciton migration toward a nonradiative recombination center can be described by the 1D random walks of a neutral particle on a finite chain. The survival probability of excitons at long time t, $S(t)$, is approximately given by [28]

$$S(t) = \left(\frac{8}{\pi}\right) \exp\left(-\frac{\pi^2 D t}{L^2}\right) \tag{2}$$

where D is the diffusion constant and L is the conjugation length. In this model, the PL decay time is determined by the exciton diffusion process, and the PL decay profile is exponential with a time constant of $\tau_{PL} = L^2/\pi^2 D$. Figure 15 shows the size dependence of τ_{PL} as a function of the square of the number of the thiophene ring, n^2, where the chain length L is given by $L = 0.56n$ (nm). A liner relationship

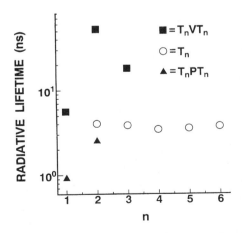

FIGURE 14 Radiative lifetime as a function of the chain length. The radiative recombination rate does not depend on the chain length. The exciton is strongly localized within two thiophene rings. The radiative lifetimes in thiophene-based oligomers, T_nPT_n and T_nVT_n are also shown. (From Ref. 26.)

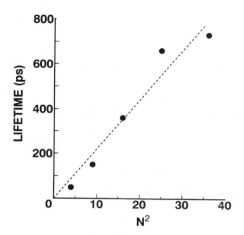

FIGURE 15 Photoluminescence decay time as a function of the square of the chain length. A linear relation suggests that the chain end acts as a nonradiative recombination center and the nonradiative recombination rate is limited by the exciton diffusion to the chain end. (From Ref. 26.)

is observed and the slope gives the diffusion coefficient of about 1.5×10^{-5} cm^2/s, which is small compared with that in an electrochemically polymerized polythiophene [29]. We believe that the difference in the diffusion constant between oligomers and polymers is mainly caused by the difference in the sample structure between the solution and solid films and the difference of the stiffening of the chains between oligomers and polymers. Although the chain length is short ($n = 2$–6), this figure strongly suggests that the nonradiative recombination occurs at the end of the chain and the PL lifetime is controlled by the diffusion of excitons to the chain end.

Figure 16 shows absorption and PL spectra of oligothiphenes and thiophene-based oligomers, T_2, T_2MT_2, T_2PT_2, and T_2VT_2. In these oligomers, the well part is bithiophene, and the barrier part is methylene, phenylene, or vinylene, whose energy gaps are larger than that of bithiophene. Both absorption and luminescence spectra vary with the *barrier* structure. The peak energy of T_2PT_2 and T_2VT_2 shifts to lower energy, compared with that of T_2 and T_2MT_2. The π-conjugated electrons in T_2PT_2 and T_2VT_2 are extended over these oligomers, and their absorption and PL spectra are similar to those of T_4 or T_5. Only methylene effectively acts as a barrier (the electrons are confined to the bithiophene). The electronic properties of the well–barrier–well structures based on conjugated polymers and oligomers are strongly affected by the delocalization nature of the electrons or the excitons.

Picosecond PL decay profiles in T_nMT_n, T_nPT_n, and T_nVT_n were also given by a single-exponential function. However, the PL lifetime and the PL quantum yield strongly depended on the barrier kind of thiophene-based oligomers. The size dependence of the radiative lifetime in T_n, T_nPT_n, and T_nVT_n is also summarized in Fig. 14. In these materials, the τ_R ordering is $T_nPT_n < T_n < T_nVT_n$ and the η ordering is $T_nVT_n < T_n < T_nPT_n$. The dynamics of PL decay, the PL peak energy,

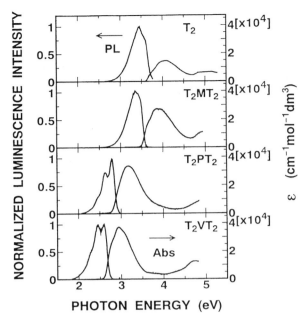

FIGURE 16 Optical absorption and photoluminescence spectra of thiophene-based oligomers, T_2, T_2MT_2, T_2PT_2, and T_2VT_2. The well part in these structures is bithiophene. (From Ref. 26.)

and the PL quantum yield can be controlled by the kind of the barriers in thiophene-based oligomers.

As mentioned earlier, the radiative decay rate in oligothiophenes T_n does not depend on the number of the thiophene ring. On the other hand, the radiative decay rate in thiophene-based oligomers depends on the structure of the barrier part. To understand the PL process in thiophene-based materials, we consider the effect of the rotational motion of thiophene rings on the radiative recombination process in oligothiophenes and thiophene-based oligomers, because the introduction of the barrier part into the polymer backbone changes the backbone rigidity (or chain flexibility). Rotational barriers of the thiophene rings around the interring bond were calculated for T_3, T_1PT_1, and T_1VT_1 using the semiempirical PM3 molecular orbital calculation method. All calculations were done by using MOPAC ver. 6.0 program (QCPE#455). The rotational barrier-height ordering is $T_1VT_1 > T_3 > T_1PT_1$. The radiative lifetime and the PL quantum yield are related to the rotation barrier height: The well–barrier–well structures having lower rotational barrier exhibit the high PL efficiency and the fast radiative decay rate [26]. These results show that the rotation motion plays an important role in the radiative process. The rotational motion of thiophene rings enhances the localization of exciton, causing both the higher PL efficiency and the faster radiative decay.

The size dependence of the optical properties of oligothiophenes indicates that the excitons are spatially localized within two thiophene rings and the luminescence decay time is limited by the exciton diffusion to the chain end of a nonradiative recombination center. Moreover, spectroscopic analysis and molecular orbital calculations suggest that the exciton localization and radiative recombination processes are affected by the rotational motion of thiophene rings. These behaviors of π-conjugated polymers are entirely different from those of σ-conjugated polymers. In σ-conjugated Si polymers, 1D excitons are delocalized 20–30 Si atoms on the backbone chain. We have demonstrated that luminescence properties of quasi-1D thiophene-based oligomers such as luminescence yield, luminescence lifetime, and luminescence wavelength can be controlled by changing the conjugation length of the well part and the barrier structure. The polymeric heterostructures consisting of the well part emitting the light and the barrier part controlling the chain rigidity (the solubility and film-forming properties of the polymers) become unique soluble light-emitting materials.

III. SILICON-BASED POLYMERS

The chainlike σ-conjugated Si polymers are well known as 1D silicon-based materials that have alkyl or aryl groups in their side chains, as mentioned in Section II.B. Reflecting the 1D direct-gap nature, sharp optical absorption bands are observed at 3–4 eV in chainlike Si polymers [30,31]. On the other hand, network polymers show weak luminescence in the visible spectral region [32–35]. In this section, we summarize the optical properties of Si-backbone polymers with different backbone structures and discuss the origin of visible luminescence in Si materials.

We studied the optical properties of various Si-backbone polymers: (a) polymethylphenylsilane, (b) poly(dimethylphenylsilyl)phenylsilane, (c) dodecaisopropyltetracyclodesilane, (d) network polysilynes, and (e) planar siloxene. The Si-backbone structures of these polymers are illustrated in Fig. 17 and, hereafter, we call these structures the chain (Fig. 17a), branch (Fig. 17b), ladder (Fig. 17c), network (Fig. 17d), and siloxene (Fig. 17e) structures. In the chain structure, the organosilicon unit on the polymer backbone has two Si–Si bonds. In the other structures, the polymer backbones are constructed by the organosilicon unit having three Si–Si bonds. Using these samples, we discuss the effects of the "lateral" dimension and backbone geometry on the optical properties of Si polymers.

Figure 18 show luminescence and absorption spectra of the chain, branch, ladder, network, and siloxene structures at room temperature. In the chain structure (Fig. 18a), sharp luminescence and sharp absorption bands are observed. The Stokes shift between luminescence and lowest absorption peaks is relatively small. It is well accepted that the sharp absorption and PL bands are attributed to 1D excitons delocalized on the backbone chain. The structures of chainlike Si polymers take a variety of conformations, depending on organic substituents attached to the polymer backbones [31]. However, the luminescence band is very sharp in all conformations, including disordered forms. Therefore, the sharp luminescence band is the most importnt feature of the optical properties of the chain structure.

In the branch and ladder structures, broad luminescence spectra are only observed in the visible region. The intensities of the lowest absorption peak per organosilicon unit constructing the polymer backbones in the branch and ladder structures are very small compared with that in the chain structure. Time-resolved PL

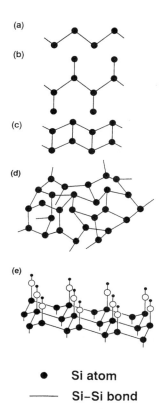

- ●   Si atom
- ——   Si–Si bond

FIGURE 17 Schematics of Si polymers with different backbones: (a) chain, (b) branch, (c) ladder, (d) network, and (e) planar siloxene structures.

measurements show that in the branch and ladder structures, excitons are strongly localized on 2–3 Si atoms, but in the chain structure, excitons are delocalized over about 20–30 Si atoms. Si-backbone polymers consisting of Si atoms with three Si–Si bonds have spectroscopic characteristics that excitons are strongly localized, the PL spectrum is broad, and the quantum efficiency is low. The broad visible PL in Si polymers is caused by the introduction of Si units with three Si–Si bonds into the polymer backbone.

In the network structure, the broad PL is observed in the visible region. The luminescence spectrum and the initial PL decay profiles do not depend on the kind of organic substituents. The PL properties are determined by the disordered network Si-backbone structures. In order to discuss the 2D characteristics of the network polymers, we compare optical absorption and PL spectra in the network structure and planar siloxene. The planar siloxene $[Si_6H_6O_3]N$ is a direct-gap semiconductor because of the orbital mixing of delocalized σ electrons in the 2D Si layer and nonbonding electrons in O atoms, called σ–n mixing [36,37]. In planar siloxene, the sharp PL and absorption bands are observed. A Stokes shift between the PL and absorption peaks is small. The broad PL spectrum in the network structure is different from the sharp PL spectrum due to 2D excitons in planar siloxene.

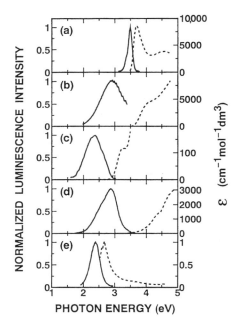

FIGURE 18 Optical absorption and normalized photoluminescence spectra in quasi-1D Si polymers: (a) chain, (b) branch, (c) ladder, (d) network, and (e) planar siloxene structures.

Here, we consider the origin of the broad luminescence spectra in the branch and ladder structures from the electronic band structures of Si polymers. The band calculations were performed by Takeda and Shiraishi with the first-principle local density functional method [36,37]. In theoretical calculations, the organic substitutions without Si atoms in polymers are replaced by H atoms and the trans-planar form is considered as the chain structure, and the branch and ladder structures are constructed using the chain structure of the trans-planar form. The band structures of the chain, planar, and siloxene structures are shown in Fig. 19. We briefly comment on the electronic structures of silicon-based polymers.

Figure 19a shows the energy band structure for trans-planar zigzag polysilanes with D_{2h} symmetry. Both band-edge states, the highest occupied valence band (HOVB) and the lowest unoccupied conduction band (LUCB), are located at point Γ and their parities are different. This chainlike polysilane has a directly-allowed-type band structure with a band gap of 3.89 eV. Both HOVB and LUCB states are formed by the skeletal Si orbital and are delocalized toward the skeleton direction. Therefore, this chainlike polysilane can be describes as "σ-conjugated one-dimensional structure."

Two-dimensional Si-backbone polymers have intermediate electronic properties between 1D chainlike Si polymers of the direct-gap band structure and 3D bulk Si of the indirect-gap band structure. An indirect-to-direct conversion of the optical transition in Si-based materials is considered to be related to the dimensionality of electronic structures. Although the 2D crystalline planar polymer is still a hypothetical polymer, disordered planar and network polymers are of scientific impor-

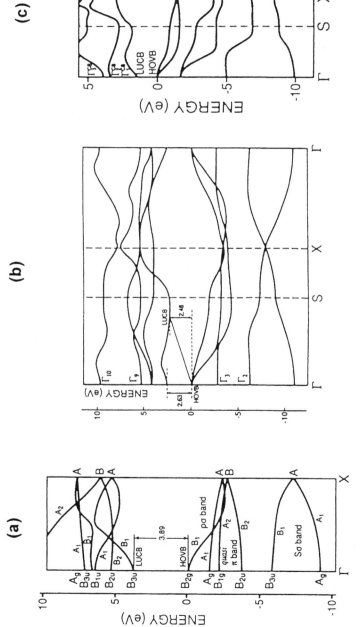

FIGURE 19 Electronic structures of (a) the chain, (b) planar polysilane, and (c) oxidized 2D sheet (planar siloxene) structures. Energy scale is represented from the HOVB in electron volts. The band-gap energy is the energy difference between LUCB and HOVB. (From Ref. 36.)

tance because they serve as a conceptual bridge between the properties of the linear chains and fully coordinated amorphous or crystalline silicon. Figure 19b shows the calculated band structures of the 2D planar structure. The 2D structure has two characteristic band gaps: an indirect one of 2.48 eV and a direct one of 2.63 eV. The 2D structure has an energy structure similar to the indirect-gap one of 3D bulk crystalline Si.

The addition of oxygen atoms to the 2D structure causes the strong modification of the band structure. As shown in Fig. 19c, the oxidized 2D sheet (planar siloxene structure) has a direct-gap structure of the band gap of 1.7 eV at the point Γ. Oxygen atoms position themselves in out-of-plane sites and oxygen atoms do not cut the electron and hole delocalization in 2D sheet. The lifting of the HOVB due to the σ–n mixing and the stabilization of the LUCB due to the oxygen hybridization effectively reduce the band gap at the point Γ. The planar siloxene is the optically allowed direct-gap structure.

Some features of the calculated band structures are summarized as follows: (1) The band-gap energies of the branch and ladder structures are smaller than that of the chain structure. (2) The E–k dispersion of the branch and ladder structures is flat compared with that of the chain structure. The densities of states near the top of valence band and the bottom of the conduction band in the branch and ladder structures are larger than those of the chain structure. (3) The energy of the indirect optical transition is close to that of the direct optical transition in the branch and ladder structure. (4) The 2D structure has an indirect band structure. The oxidized 2D structure has a direct-gap structure because of the σ–n mixing.

It seems that the observed broad PL spectra at the visible region in the branch, ladder, and network structures reflect the above four characteristics of the band structure [(1) luminescence appears in the visible region, (2) and (3) the luminescence spectrum is broad, and (4) the luminescence lifetime is long]. However, it is considered that the band structure of the crystalline state is an oversimplified picture, because the polymers have noncrystalline characteristics, more or less. In particular, the backbone constructed by three Si–Si bonds is more rigid than that by two Si–Si bonds like the chain structure. For example, the ladder structure is a highly strained crystalline one. In the network silicon polymers, their optical properties are determined by exciton localization. The formation of self-tapped excitons and exciton localization at the tail state is very important in optical responses. The dimensional effect on the optical properties of chemically synthesized silicon materials is not clear.

Our study show that the introduction of Si atoms with three Si–Si bonds into the polymer backbone plays a primary role in causing the broad PL spectrum. In amorphous semiconductors and semiconducting glasses, the coordination number is a very important factor for the structure and electronic properties. In Si-based noncrystalline materials, the flexibility of covalent bonds is largest for the twofold coordinated group and least for the tetrahedral coordinated group. In SiO_2 glasses, the oxygen atoms bridging the Si tetrahedra provide the essential flexibility which is needed to form a random covalent network without much strain. Tetrahedral coordinated amorphous silicon without the flexing bridges of Si atoms is highly overconstrained. The Si materials with the three or four Si–Si coordination numbers is locally strained materials: The ladder and network structure are highly strained Si polymers. The disorder and strain of Si-backbone structures constructed by Si units with three or four Si–Si bonds bring about the localized electronic states with amorphous semiconductors and semiconducting glasses.

The branch, ladder, and network Si polymers constructed by the organosilicon unit with three Si–Si bonds exhibit broad luminescence spectra with long decay times. The broad luminescence properties of these structures are entirely different from the sharp luminescence properties of the chain structure constructed by the Si–Si bonds. In the chain structure, the sharp absorption and luminescence bands are mainly observed in the ultraviolet spectral region. These sharp bands due to the quasi-1D excitons are observed even in disordered form. In the disordered network structures, the excitons are strongly localized and the noncrystalline nature dominates the optical properties. The luminescence properties of branch, ladder, and network Si polymers are similar to amorphous semiconductors and semiconducting glasses. The disorder and spatially inhomogeneous strain of Si-backbone structure with Si units of three or four Si–Si bonds bring about the localized electronic states which control the luminescence properties of Si materials.

IV. NONLINEAR OPTICAL RESPONSES

Because chainlike Si polymers show 1D exciton nature, it is expected that nonlinear optical effects due to 1D excitons are clearly observed in chainlike Si polymers. Kajzar et al. [38] first reported third-harmonic generation (THG) measurements on thin polymethylphenylsilane (PMPS) films using a 1.064-μm fundamental. The THG coefficient of PMPS films is large for a material that is transparent in both the visible and infrared regions. Many works have been carried out on $\chi^{(3)}$ measurements at fixed frequencies [39]. Recently, Hasegawa et al. [30,40,41] reported the THG spectrum of chainlike Si polymers. In this section, we discuss the THG spectrum of chainlike Si polymers.

The advantage of nonlinear optical spectroscopy is that we can obtain information on the optical transitions, not only those taking place from the ground states but also those occurring between the excited states, which are forbidden in the ordinary one-photon spectroscopy. We can study the hidden electronic states in Si polymers. The detailed spectra of the complex third-order nonlinear susceptibility $\chi^{(3)}$ ($= |\chi^{(3)}| \, e^{i\phi}$) have been evaluated both for the modulus $|\chi^{(3)}|$ and the phase ϕ for chainlike Si polymers with various backbone conformations, over 0.5–2.2 eV for the fundamental photon energy, by means of THG measurements [40].

Figure 20 shows one-photon absorption and two-photon absorption spectra of a polydihexylsilane (PDHS) film at room temperature [40]. The one-photon-allowed and the two-photon-allowed excited states are located at about 3.30 and 4.19 eV, respectively. The experimental modulus $|\chi^{(3)}|$ and the phase ϕ values of PDHS are plotted in Fig. 21. The linear and nonlinear optical spectra in Figs. 20 and 21 show that the modulus $|\chi^{(3)}|$ spectrum exhibits three peaks in contrast to the single peak in the ordinary one-photon absorption spectrum. The main peak is located at 1.10 eV, a broad peak at 1.50 eV, and a weak peak at 2.10 eV. The phase change also occurs at energies of three peaks. The main peak at 1.10 eV is due to the three-photon resonance ($3\omega = \omega_a$) with the one-photon-allowed excited state of $\omega_a = 3.30$ eV. The 2.20-eV peak is due to the two-photon resonance ($2\omega = \omega_f$) with the one-photon-forbidden excited state of $\omega_f = 4.19$ eV. These peaks are clearly observed in the one-photon and two-photon absorption spectra in Fig. 20. The 1.50-

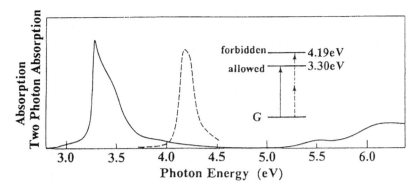

FIGURE 20 One-photon absorption (solid curve) and two-photon absorption (dashed curve) spectra of a polydihexylsilane (PDHS) film at room temperature. (From Ref. 40.)

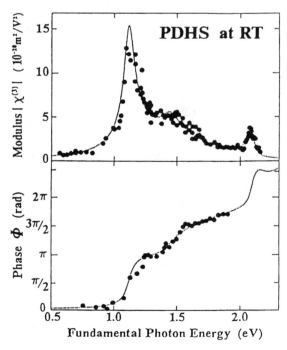

FIGURE 21 Dispersion of modulus $|\chi^{(3)}|$ (upper) and phase ϕ (lower) of third-order non-linear optical susceptibility of polydihexylsilane (PDHS) films at room temperature. (From Ref. 40.)

eV structure in the $|\chi^{(3)}|$ is not clearly observed in the one-photon absorption spectrum. However, Hasegawa et al. considered that the 1.50-eV structure in the $|\chi^{(3)}|$ is due to the three-photon resonance with a one-photon allowed excited states (4.5 eV). These experiments demonstrate that the nonlinear optical spectrum shows the detailed information on the excited states in Si polymers.

Figure 22 shows the $|\chi^{(3)}|$ spectra in polyditetradecylsilane (PDTDS) and polydibutylsilane (PDBS). Spectra of these samples are very similar to each other. Three peaks in the THG spectrum are clearly observed in different Si polymers. Moreover, similar features in the $|\chi^{(3)}|$ spectra are observed even in π-conjugated polythiophenes [42]. The features shown in Fig. 22 are characteristic of the quasi-1D semiconductor polymers in which the σ- and π-conjugated electrons are delocalized on the backbone chains. A common feature in the σ- and π-conjugated polymers may be attributed to the common origin of 1D exciton states. In the case

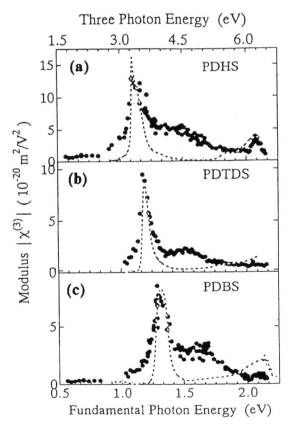

FIGURE 22 Dispersion of modulus $|\chi^{(3)}|$ for the complex third-order nonlinear optical susceptibility of (a) polydihexylsilane (PDHS), (b) polyditetradecylsilane (PDTDS), and (c) polydibutylsilane (PDBS) films at room temperature, plotted against the fundamental photon energy. For comparison, the one-photon absorption spectra are also plotted against the three-photon energy. (From Ref. 40.)

of quasi-1D exciton systems, nonlinear optical spectroscopy is one of the powerful tools used for the understanding on unique feature of 1D excitons in both the σ- and π-conjugated polymers.

V. ELECTROLUMINESCENCE DIODES

Organic electroluminescent (EL) diodes have attracted much interest because of their potential for efficient emission in the visible region and for applications to large-area flat display. Bright EL diodes based on π-conjugated molecules and polymers have been reported [43,44]. Here, we describe EL diodes using σ-conjugated Si polymers.

Polysilanes are hole transport materials [45] and are often used as hole transport layers in organic EL diodes [46]. However, Fuji et al. fabricated EL devices using polymethylphenylsilane (PMPS), polyphenylsilyne (PPS), poly(2-naphtyl-phenylsilane) (PNPS), and polyphenyl-*p*-biphenylsilane (PBPS) [47]. The molecular structures are shown in Fig. 23. The EL diodes consist of an indium–tin–oxide (ITO) electrode, a 100–200-nm-thick polysilane layer, and a silver-containing magnesium (Mg:Ag) electrode. They demonstrated that polysilanes themselves can be emission materials in the ultraviolet and visible spectral regions.

Figure 24 shows PL spectra of PMPS, PSP, PNPS, and PBPS samples [47]. These samples show the sharp luminescence in the ultraviolet and violet spectral regions. This sharp PL is delocalized 1D excitons, and the broad visible PL is caused by the radiative recombination of the localized excitons, as mentioned before. EL spectra under the forward-bias condition at 77 K are shown in Fig. 25. The polysilane diode exhibited rectifying characteristics, and the emission intensity

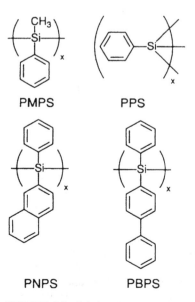

FIGURE 23 Molecular structures of PMPS, PPS, PNPS, and PBPS. (From Ref. 47.)

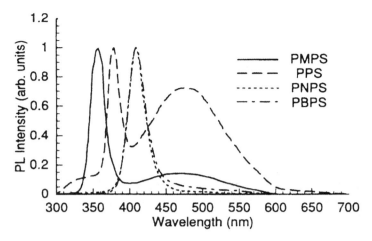

FIGURE 24 Photoluminescence spectra of thin-film PMPS, PPS, PNPS, and PBPS samples at room temperature. (From Ref. 47.)

starts to increase in the range of 10 V. The EL spectra are almost independent of both the applied voltage and the thickness of polysilane films. The EL peak energies of PMPS, PNPS, and PBPS coincide with PL peak energies. The bandwidth of the EL spectrum is narrower in the ultraviolet region. On the other hand, a broad spectrum was only observed in PPS samples. There is a good correlation between the PL and EL spectra. Moreover, the ordering of the EL efficiency is PNPS > PBPS > PMPS > PPS. The localized states control the EL efficiency and is the origin of the broad luminescence in the visible region. For the improvement of the

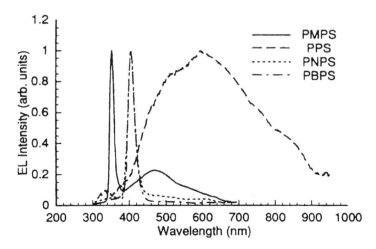

FIGURE 25 Electroluminescence spectra of ITO/polysilane/Mg:Ag diode structures at 77 K. (From Ref. 47).

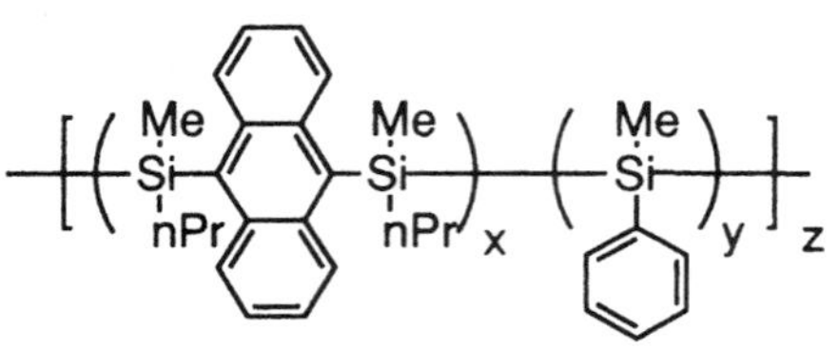

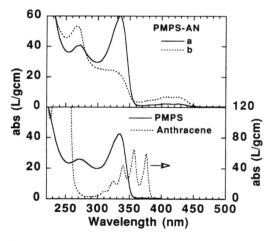

FIGURE 26 Molecular structure of copolymer of polymethylphenylsilane and anthracene units (PMPS–AN). (From Ref. 48.)

EL efficiency, we should develop electron transport materials with large band-gap energies and fabricate multilayer systems consisting a polysilane luminescent layer and an electron transport layer.

Chainlike polysilanes are *p*-type organic semiconductors and show luminescence in the ultraviolet spectra region. For the turning of the luminescence wavelength into the visible region and the improvement of the electron transport in the luminescent layer, copolymers of anthracene and polysilane have been prepared. Anthracene has a high PL efficiency and was studied as a active EL materials in the early stage of organic EL research. Polysilanes containing anthracene units will have possible application as EL materials. Satoh et al. prepared the polymethylphenylsilane (PMPS) containing anthracene (AN) units in the polymer backbone (PMPS–AN) [48]. The chemical structure of PMPS–AN is illustrated in Fig. 26. Two samples of PMPS–AN with different AN contents: (a) 6 wt% AN and (b) 19 wt% AN. The PMPS–AN polymer layer was sandwiched between the positive ITO and the negative Al electrodes.

Figure 27 shows optical absorption spectra of PMPS–AN, PMPS, and anthracene samples. The PMPS–AN samples show a peak at 335 nm due to the 1D

FIGURE 27 Optical absorption spectra in copolymer of polymethylphenylsilane and anthracene units [PMPS–AN, (a) 6 wt% AN and (b) 19 wt% AN] (upper) and polymethylphenylsilane (PMPS) and anthracene (AN) (lower). (From Ref. 48.)

exciton delocalized over PMPS Si chains and a weak peak at 420 nm due to the AN unit. The difference in the absorbance at 335 nm in PMPS, 6 wt% PMPS–AN, and 19 wt% PMPS–AN samples is attributed to the difference in the σ-electron-conjugated length.

Figure 28 shows the EL spectrum in the ITO/PMPS–AN (6 wt%)/Al structure and PL spectrum under 300-nm excitation. In the PL spectrum, a peak due to the PMPS chain is observed at 335 nm and a strong, broad PL peak is due to the AN unit observed at 480 nm. Under ultraviolet (UV) light excitation, the PL efficiency at 480 nm increases with increasing AN concentration. The excitation energy transfer from the PMPS chain to the AN unit efficiently occurs in the PMPS–AN samples with short PMPS chains. The intramolecular energy transfer determines the 480-nm PL efficiency. The PL quantum efficiency in PMPS–AN is about 87% under UV light excitation. On the other hand, in the EL spectrum, the 355-nm EL peak is not observed, only the 480-nm EL is observed. In this EL device, holes move on the PMPS chain and electrons hop between the AN units. The radiative recombination of injected holes and electrons takes place in the AN unit. The EL external efficiency was about 0.01% and the PL was observed at about 6 V bias. These are comparable with those in ITO/poly(*p*-phenylvinylene)/Al devices.

It has been demonstrated that the simple diode structures with thin polysilane films show electroluminescence in the ultraviolet and visible spectral regions. In order to obtain more efficient electroluminescence, we need to clarify the carrier injection and transport mechanisms in organic thin films and develop a good contact between the metal and polysilane film.

VI. SUMMARY

We have discussed briefly the optical properties of chemically synthesized Si oligomers and polymers with different backbone structures. The luminescence properties of Si_8 clusters are very sensitive to the shape of the clusters. In Si polymers, the broad PL originates from the radiative recombination of excitons localized in

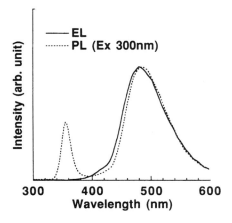

FIGURE 28 Electroluminescence and photoluminescence spectra of ITO/PMPS–AN/Al structures at room temperature. (From Ref. 48.)

the disordered backbone structures with three Si–Si bonds. Synthesis of the crystalline 2D polymers would realize a wavelength-tunable PL with high quantum efficiency. On the other hand, chainlike Si polymers show efficient and sharp PL in the ultraviolet spectral region. Linear and nonlinear optical spectroscopy show that these sharp absorption and PL bands are due to the 1D excitons. The 1D Si polymers would become an important material for technological applications such as light-emitting diodes and nonlinear optical devices. The study of light-emitting Si polymers and oligomers gives guidelines for the design of low-dimensional Si quantum structures.

ACKNOWLEDGMENTS

The author would like to thank Professor H. Matsumoto and Professor S. Kyushin of Gunma University, Professor K. Takeda of Waseda University, Dr. K. Shiraishi of NTT Basic Research Laboratories, K. Suzuki, Y. Shiraishi, and M. Kuroda of Fuji Electric Corporate Research and Development Ltd., and S. Satoh of Toagosei Tsukuba Research Laboratory for useful discussions and collaboration.

REFERENCES

1. Ogawa, T., and Kanemitsu, Y., *Optical Properties of Low Dimensional Materials*, World Scientific, Singapore, 1995.
2. Kanemitsu, Y., Kondo, M., and Takeda, K., *Light Emission from Novel Silicon Materials*, The Physical Society of Japan, Tokyo, 1994.
3. Kanemitsu, Y., Light emission from porous silicon and related materials, *Phys. Rep.*, *263*, 1–93 (1995).
4. Miller, R. D., and Michl, J., Polysilane high polymers, *Chem. Rev.*, *89*, 1359–1410 (1989).
5. Silicon chemistry, *Chem. Rev.*, *95*, 1135–1674 (1995).
6. Abe, S., Exciton versus interband absorption in Peierls insulators, *J. Phys. Soc. Jpn.*, *58*, 62–65 (1989).
7. Ogawa, T., and Takagahara, T., Optical absorption and Sommerfeld factors of one-dimensional semiconductors: An exact treatment of excitonic effects, *Phys. Rev. B*, *44*, 8138–8153 (1991).
8. Kanemitsu, Y., Shimizu, N., Suzuki, K., Shiraishi, Y., and Kuroda, M., Optical and structural properties of oligothiophene crystalline films, *Phys. Rev. B*, *54*, 2198–2204 (1995).
9. Bloomfield, L. A., Freeman, R. R., and Brown, W. L., Photofragmentation of mass-resolved Si^+_{2-12} clusters, *Phys. Rev. Lett.*, *54*, 2246–2249 (1985).
10. Martin, T. P., and Schaber, H., Mass spectra of Si, Ge, and Sn clusters, *J. Chem. Phys.*, *83*, 855–858 (1985).
11. Honea, E. C., Ogura, A., Murray, C. A., Raghavachai, K., Sprenger, W. O., Jarrold, M. F., and Brown, W. L., Raman spectra of size-selected silicon clusters and comparison with calculated structures, *Nature*, *366*, 42–44 (1993).
12. Kyushin, S., Matsumoto, H., Kanemitsu, Y., and Goto, M., Syntheses, structures, properties and photoluminescence of ladder polysilanes and octasilacubanes, in *Light Emission from Novel Silicon Materials*, The Physical Society of Japan, Tokyo, 1994, pp. 46–55.
13. Kanemitsu, Y., Suzuki, K., Kondo, M., Kyushin, S., and Matsumoto, H., Luminescence properties of a cubic silicon cluster octasilacubane, *Phys. Rev. B*, *51*, 10666–10670 (1995).

14. Raghavachari, K., and Logovinsky, V., Structural and bonding in small silicon clusters, *Phys. Rev. Lett.*, *55*, 2853–2856 (1985).
15. Cheshnovsky, O., Yang, S. H., Pettiette, C. L., Crsycrft, M. J., Liu, Y., and Smalley, R. E., Ultraviolet photoelectron spectroscopy of semiconductor clusters: Silicon and germanium, *Chem. Phys. Lett.*, *138*, 119–124 (1987).
16. Rinnen, K. D., and Mandich, M. L., Spectroscopy of neutral silicon clusters, Si_{18}–Si_{41}: Spectra are remarkably size independent, *Phys. Rev. Lett.*, *69*, 1823–1826 (1992).
17. Kanemitsu, Y., Suzuki, K., Kondo, M., and Matsumoto, H., Luminescence from a cubic silicon cluster, *Solid State Commun.*, *89*, 619–621 (1994).
18. Matsumoto, H., Higuchi, K., Kyushin, S., and Goto, M., Octakis(1,1,2-trimethyl-propyl)octasilacubane: Synthesis, molecular structures and unusual properties, *Angrew. Chem. Int. Ed. Engl.*, *31*, 1354–1356 (1992).
19. Furukawa, K., Fujino, M., and Matsumoto, N., Cubic silicon cluster, *Appl. Phys. Lett.*, *60*, 2774–2775 (1992).
20. Sekiguchi, A., Yatabe, T., Kamatani, H., Kabuto, C., and Sakurai, H., Preparation, characterization, and crystal structures of octasilacubanes and octagermacubanes, *J. Am. Chem. Soc.*, *114*, 6260–6262 (1992).
21. Takagahara, T., and Takeda, K., Excitonic exchange splitting and Stokes shift in Si nanocrystals and Si clusters, *Phys. Rev. B*, *53*, R4205–R4208 (1996).
22. Kanemitsu, Y., Suzuki, K., Nakayoshi, Y., and Masumoto, Y., Quantum size effects and enhancement of the oscillator strength of excitons in chains of Si atoms, *Phys. Rev. B*, *46*, 3916–3919 (1992).
23. Thorne, J. R. G., Williams, S. A., Hochstrasser, R. M., and Fagan, P. J., Radiative lifetime of confined excitations in sigma-conjugated silane oligomers, *Chem. Phys.*, *157*, 401–408 (1991).
24. Abe, S., Electronic excitations in polysilane: σ-conjugation, excitons and nonlinear optical response, in *Light Emission from Novel Silicon Materials*, The Physical Society of Japan, Tokyo, 1994, pp. 56–63.
25. Kanemitsu, Y., Suzuki, K., Masumoto, Y., Komatsu, T., Sato, K., Kyushin, S., and Matsumoto, H., Optical properties of small Si skeleton sheets: Ladder polysilanes, *Solid State Commun.*, *86*, 545–548 (1993).
26. Kanemitsu, Y., Suzuki, K., Masumoto, Y., Tomiuchi, Y., Shiraishi, Y., and Kuroda, M., Optical properties of quasi-one-dimensional thiophene-based oligomers, *Phys. Rev. B*, *50*, 2301–2305 (1994).
27. Chung, T. C., Kaufman, J. H., Heeger, A. J., and Wudl, F., Charge storage in doped polythiophene: Optical and electrochemical studies, *Phys. Rev. B*, *30*, 702–710 (1984).
28. Nisoli, M., Cybo-Ottone, A., De Silvestri, S., Magni, V., Tubino, R., Botta, C., and Musco, A., Femtosecond transient absorption saturation in poly(alkyl-thiophene-vinylene)s, *Phys. Rev. B*, *47*, 10881–10884 (1993).
29. Kanner, G. S., Wei, X., Hess, B. C., Chen, L. R., and Vardeny, Z. V., Evolution of excitons and polaritons in polythiophene from femtoseconds to milliseconds, *Phys. Rev. Lett.*, *69*, 538–541 (1992).
30. Hasegawa, T., Iwasa, Y., Sunamura, H., Koda, T., Tokura, Y., Tachibana, H., Matsumoto, M., and Abe, S., Nonlinear optical spectroscopy on one-dimensional excitons in silicon polymer, polysilane, *Phys. Rev. Lett.*, *69*, 668–671 (1992).
31. Tachibana, H., Matsumoto, M., Tokura, Y., Moritomo, Y., Yamaguchi, A., Koshihara, S., Miller, R. D., and Abe, S., Spectra of one-dimensional excitons in polysilanes with various backbone conformations, *Phys. Rev. B*, *47*, 4363–4371 (1993).
32. Bianconi, P. A., and Weidman, T. W., Poly(n-hexylsilyne): Synthesis and properties of the first alkyl silicon $[RSi]_n$ network polymer, *J. Am. Chem. Soc.*, *110*, 2342–2344 (1988).

33. Wilson, W. L., and Weidman, T. W., Excited state dynamics of one- and two-dimensional σ-conjugated silicon frame polymers: Dramatic effects of branching in a series of hexylsilyne-branched poly(hexylmethylsilylene) copolymers, *J. Phys. Chem.*, *95*, 4568–4572 (1991).

34. Wilson, W. L., and Weidman, T. W., Radiative recombination and vibronic relaxation in σ-delocalized silicon-backbone-network polymers: Energy thermalization in poly(n-hexylsilyne), *Phys. Rev. B*, *48*, 2169–2174 (1993).

35. Kanemitsu, Y., Suzuki, K., Kyushin, S., and Matsumoto, H., Visible photoluminescence from silicon backbone polymers, *Phys. Rev. B*, *51*, 13103–13110 (1995).

36. Takeda, K., Si skeleton high-polymers: Their electronic structures and characteristics, in *Light Emission from Novel Silicon Materials*, The Physical Society of Japan, Tokyo, 1994, pp. 1–29.

37. Takeda, K., and Shiraishi, K., Electronic structures of silicon skeleton materials: Toward designing of silicon quantum materials, *Comments Solid State Phys.*, *18*, 91–133 (1997).

38. Kajzar, F., Messier, J., and Rosilio, C., Nonlinear optical properties of thin films of polysilanes, *J. Appl. Phys.*, *60*, 3040–3044 (1986).

39. Kepler, R. G., and Soos, Z. G., Electronic properties of polysilanes: Excitations of σ-conjugated chains, in *Relaxation in Polymers* (T. Kobayashi, ed.), World Scientific, Singapore, 1993, pp. 100–133.

40. Hasegawa, T., Iwasa, Y., Koda, T., Kishida, H., Tokura, Y., Wada, S., Tashiro, H., Tachibana, H., Matsumoto, M., and Miller, R. D., Nonliner optical study of polysilanes, in *Light Emission from Novel Silicon Materials*, The Physical Society of Japan, Tokyo, 1994, pp. 64–81.

41. Hasegawa, T., Iwasa, Y., Kishida, H., Koda, T., Tokura, Y., Tachibana, H., Kawabata, Y., Two-photon resonant third-harmonic generation in polysilanes, *Phys. Rev. B*, *45*, 6317–6320 (1992).

42. Torruellas, W. E., Neher, D., Zanoni, R., Stegeman, G. I., Kajzar, F., and Leclerc, M., Dispersion measurements of the third-order nonlinear susceptibility of polythiophene thin films, *Chem. Phys. Lett.*, 175, 11–16 (1990).

43. Tang, C. W., and VanSlyke, S. A., Organic electroluminescent diodes, *Appl. Phys. Lett.*, *51*, 913–915 (1987).

44. Burroughes, J. H., Bradley, D. D. C., Brown, A. R., Marks, R. N., Mackay, K., Friend, R. H., Burns, P. L., and Holmes, A. B., Light-emitting diodes based on conjugated polymers, *Nature*, *347*, 539–541 (1990).

45. Kepler, R. G., Zeigler, J. M., Harrah, L. A., and Kurtz, S. R., Photocarrier generation and transport in σ-bonded polysilanes, *Phys. Rev. B*, *35*, 2818–2822 (1987).

46. Kido, J., Nagai, K., Okamoto, Y., and Skotheim, T., Polymethylphenylsilane film as a hole transport layer in electroluminescent devices, *Appl. Phys. Lett.*, 59, 2760–2762 (1991).

47. Fujii, A., Yoshimoto, K., Yoshida, M., Ohmori, Y., Yoshino, K., Ueno, H., Kakimoto, M., and Kojima, H., Electroluminescent diodes utilizing polysilanes, *Jpn. J. Appl. Phys.*, *35*, 3914–3917 (1996).

48. Satoh, S., Suzuki, H., Kimata, Y., and Kuriyama, A., Optical and electroluminescence properties of polymethylphenylsilane containing an anthracene unit, *Synth. Met.*, *79*, 97–102 (1996).

2

Synthesis and Properties of Poly(*p*-phenylene) and Its Derivatives

Kyoji Kaeriyama
Kyoto Institute of Technology
Kyoto, Japan

I. INTRODUCTION

Poly(*p*-phenylene) (PPP) is one of the simplest polymers being exclusively composed of benzene rings (Scheme 1). In 1886, it was reported that tridecaphenyl was obtained by the Wurtz–Fittig reaction of *p*-dibromobenzene. In 1936, hexadecaphenyl was prepared by heating *m*-dibromobenzene with methanolic KOH, H_2O, and Pd–CaCO$_3$ at 150°C and 12 atm. Subsequent to this, many scientists studied the preparation of PPP.

In the late 1970s, it was found that polyacetylene exhibited high electrical conductivity, which opened up a new era for conjugated polymers. Polythiophene,

Polyethylene

Polyacetylene

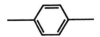

Poly(*p*-phenylene)

SCHEME 1 Simple polymers.

polypyrrole, poly(*p*-phenylenevinylene), and their derivatives were prepared and evaluated as conductive polymers [1–3]. Before this period, PPP was studied as a thermally resistant polymer but is now one of the most promising photoelectronic polymers. It can be doped with electron donor as well as acceptor to form conductive complex. Recently, PPP was shown to be an electroluminescent polymer and should be useful as blue-light emitter. Because thin films are necessary for these applications, new routes have been vigorously searched for; however as yet we do not know the best way to achieve high-quality PPP films.

II. PREPARATION OF POLY(*p*-PHENYLENE)

A. Coupling of Dihalobenzenes

1. Coupling by the Wurtz–Fittig Reaction

Goldschmiedt obtained a substance melting at around 300°C by heating 1,4-dibromobenzene with sodium for 130 h [4]. From elemental analysis, it was assumed to be poly(*p*-phenylene) consisting of 13 phenylene units with terminal bromine atoms.

Metallic sodium and potassium were heated in refluxing xylene for 5 h to form KNa_2 [5]. This liquid alloy was used in preference to either sodium or potassium because small quantities of it could be added to the reaction mixture from a dropping funnel, thus avoiding too violent a reaction. The alloy was added to a mixture of 1,4-dichlorobenzene and dioxane at 95°C. A violent reaction began at once, as indicated by the refluxing of the dioxane. After addition of the alloy and heating for 24–48 h, the polymerization product was collected and dried overnight in a vacuum oven at 85°C (Scheme 2). After thorough extraction with benzene, poly-(*p*-phenylene) was obtained in the yield of 10%. The poly(*p*-phenylene) was practically insoluble in organic solvents.

2. Coupling by the Ullman Reaction

Kern et al. heated 4,4'-diiodo-3,3'-dimethylbiphenyl and copper powder in high-boiling-point solvents such as 1-methylnapthalene for 16 h [6]. The resulting soluble polyphenylenes were fractionated and the highest fraction showed molecular weights up to 300,000. The molecular shape of these polyphenylenes was neither rigid nor rodlike. Side reactions possibly gave meta-linked or branched products.

Activated copper was mixed thoroughly with twice its weight of 1,4-diiodo-2,3,5,6-tetrafluorobenzene and gradually heated to 200°C [7] (Scheme 3). The reaction became exothermic and the temperature rose rapidly to 290°C while the material solidified. The product was extracted with benzene and boiled with pyridine to remove copper salts. The structure was confirmed by mass spectrometry, and elemental analysis indicated $n = 10$. The purified material was a grayish powder which could be heated without decomposition to 500°C in a sealed tube.

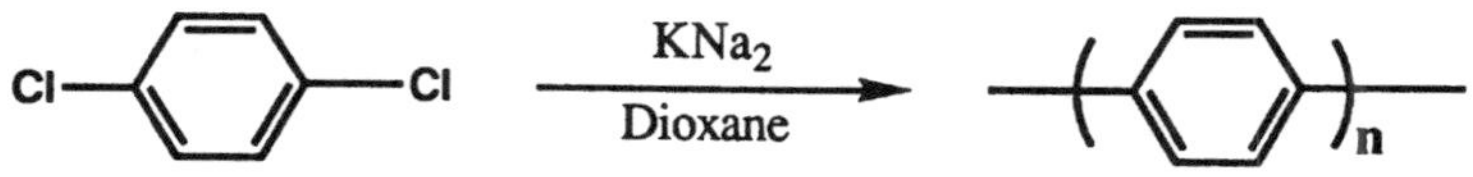

SCHEME 2

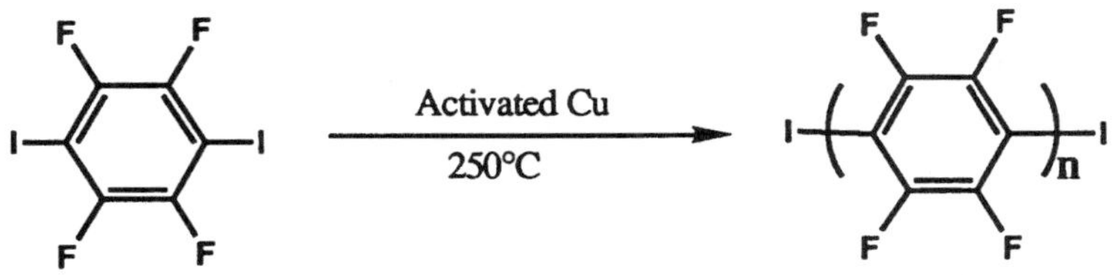

SCHEME 3

3. Grignard Coupling

The above coupling reactions are carried out at high temperatures and accompanied by undesirable side reactions, resulting in branched polymers and a low degree of polymerization. The migration of reaction sites often takes place in these coupling reactions. This is a serious drawback in the preparation of electronic polymers because a meta-linked phenylene unit interrupts conjugation along the PPP backbone.

Kumada et al. found that soluble nickel complexes such as (dppp)NiCl$_2$ [dppp = 1,3-bis(diphenylphosphino)propane] were efficient catalysts for the coupling of Grignard reagents with aryl bromides [8,9]. They clearly showed that no migration of reaction sites took place during the coupling of phenyl and thienyl halides. The mechanism of this catalytic reaction is illustrated in Scheme 4. Here, L$_2$ denotes a bidentate ligand such as dppp. In this catalytic cycle, two organic groups on a nickel complex L$_2$NiRR′ are released by the action of R′Br to undergo coupling, along with the formation of L$_2$NiR′Br. A Grignard reagent reacts with this nickel complex to regenerate L$_2$NiRR′.

An equimolar amount of 1,4-dibromobenzene and magnesium was allowed to react in tetrahydrofuran [10]. When a catalytic amount of (bpy)NiCl$_2$ (bpy = 2,2′-

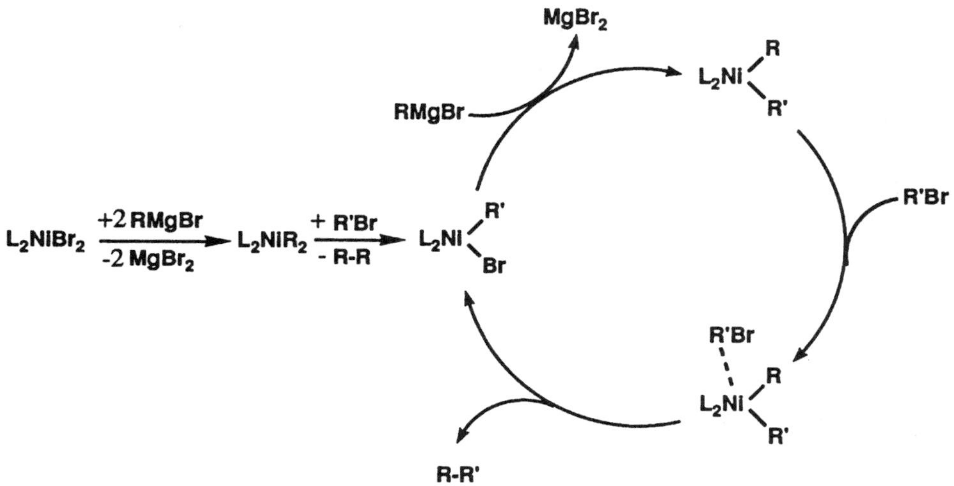

SCHEME 4

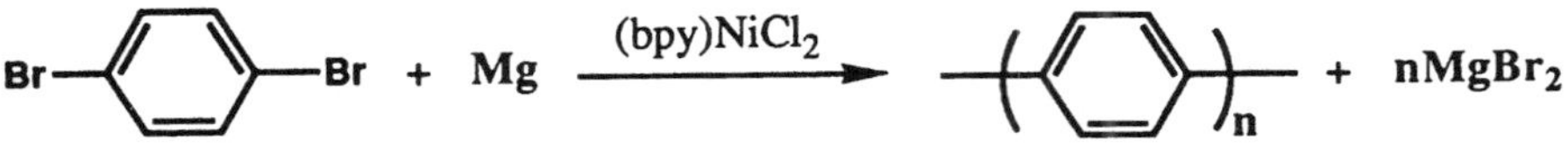

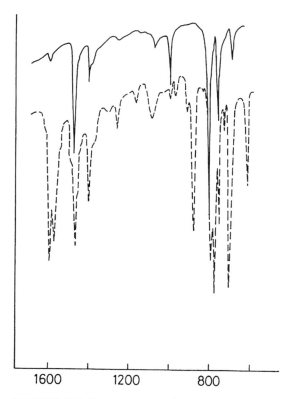

SCHEME 5

bipyridine) was added to the reaction mixture, smooth polymerization started (Scheme 5). The polymerization was almost complete after refluxing the mixture for 1 h. *m*-Dichlorobenzene was polymerized in a similar way. Thus, it was possible to obtain both pure poly(*p*-phenylene) (PPP) and poly(*m*-phenylene) (PMP).

Infrared (IR) spectra of PPP and PMP are shown in Fig. 1. The band due to the *p*-phenylene units occurred at 805 cm^{-1} and out-of-plane bands due to terminal phenyl groups at 760 and 690 cm^{-1}. The ratio of these bands is a measure of the degree of polymerization. Weak C–Br and C–Cl bands are observed at 1065 and 1085 cm^{-1}, respectively.

4. Coupling by the Suzuki Reaction

Grignard coupling gives poly(arylene)s with well-defined structure and can be successfully employed to prepare poly(thienylene)s. It is, however, difficult to prepare

FIGURE 1 Infrared spectra of poly(*p*-phenylene) (solid line) and poly(*m*-phenylene) (dotted line).

$$\text{Ar-Br} \ + \ \text{Ar'-B(OH)}_2 \ \xrightarrow[\text{Na}_2\text{CO}_3]{\text{Pd(PPh}_3)_4} \ \text{Ar-Ar'}$$

SCHEME 6

PPP using this method. Due to poor solubility, PPP oligomers were obtained because premature precipitation prevented further growth. In order to increase solubility, it is necessary to introduce such lateral substituents as an alkyl or alkoxy group. Introduction of these substituents, however, decreases the reactivity of dibromobenzene, which is less reactive than dibromothiophene. Thus, we have to look for another route to functionalized PPP.

The palladium-catalyzed cross-coupling reactions of organic boronic acids or their esters with organic bromides are termed the Suzuki reaction (Scheme 6). As this reaction is tolerant toward a wide variety of functional groups on either coupling partner and involves no migration of reaction sites, it is useful for the synthesis of various functional polyarylenes [11,12].

The reaction mechanism is illustrated in Scheme 7 [13]. The palladium complex **I**, which is in equilibrium with Pd(PPh₃)₄, is attacked by Ar–Br and undergoes oxidative addition. The resulting complex **II** is then allowed to react with Ar′–B(OH)₂. The transmetallation product **III** is converted to a complex **IV** by

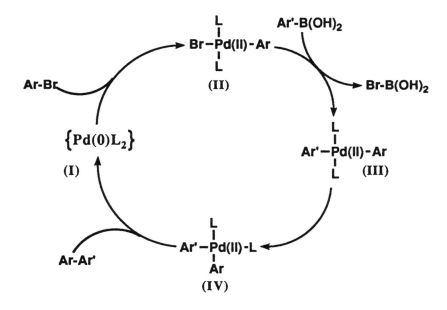

$$\text{L} = \text{P(C}_6\text{H}_5)_3$$

SCHEME 7

trans–cis isomerization. Reductive elimination of Ar–Ar' regenerates the complex **I**. Complexes **II** and **III** are detected by electrospray mass spectrometry. Br–B(OH)$_2$ is decomposed to form B(OH)$_3$ and HBr.

Polycondensation is usually carried out in two-phase systems [14]. One phase is an organic solution of the palladium complex, monomer, and produced polymer; the other phase is an aqueous solution of sodium carbonate to remove acidic by-products from the organic solution (Scheme 8). The reaction mixture is vigorously stirred at refluxing temperatures in the dark under a strictly inert atmosphere.

Poly(2,6-didodecyl-*p*-phenylene) was obtained by the copolymerization of 1,4-dibromo-2,6-didodecylbenzene with 2,6-didodecyl-1,4-bis(boronic acid) or its ester. The acid was a white powder possibly contaminated with boronic anhydride. Stoichiometric copolymerizations could be performed by the gradual addition of 1.03 equivalent of the acid. This copolymerization had to be carried out at elevated temperatures to assure that a stoichiometric balance was obtained before an excess of acid was added. The ester route had the advantage that the ester was highly crystalline and could be readily purified. Stoichiometric copolymerization could be conducted at relatively low temperatures. These copolymerization methods gave polymers with a degree of polymerization of nearly 100. Homopolymerization of 2,6-dihexyl-4-bromobenzenebronic acid also gave poly(2,6-dihexyl-*p*-phenylene), but the degree of polymerization was 30. This low value might result from impurities in the monmer. Copolymerization of diester of benzene-1,4-bis(boronic acid) with 2,5-dibromothiophene also afforded a high-quality copolymer [15].

Benzene-1,4-bis(boronic acid) was copolymerized with 2,5-bis(3-sulfonatopropoxy)-1,4-dibromobenzene to form the copolymer (Scheme 9) [16]. Copolymerization was carried out in a dimethylformamide/water homogeneous solution using a water-soluble palladium complex as catalyst. It is difficult to prepare these copolymers by the other methods.

5. Coupling by Zero-valent Nickel

Bromobenzene can be coupled with zero-valent nickel which is generated in situ by the reduction of nickel bromide with zinc [17].

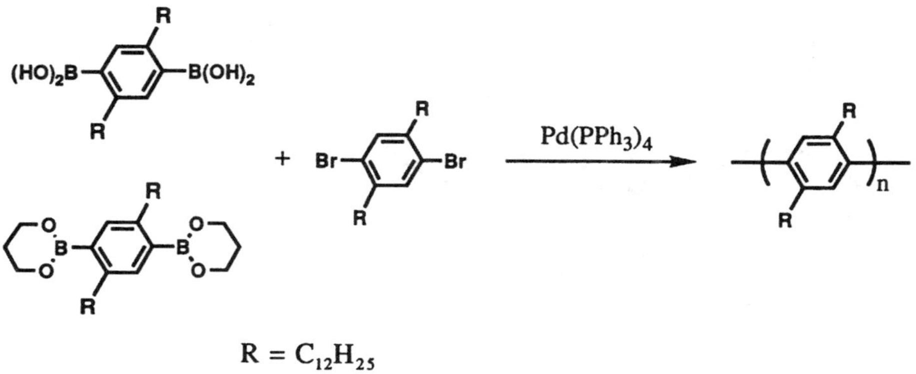

$$R = C_{12}H_{25}$$

SCHEME 8

SCHEME 9

The reaction (Scheme 10) was performed in such polar solvents as *N,N*-dimethylformamide (DMF) and dimethylacetamide in the presence of ligand such as triphenylphosphine. A catalytic amount of nickel bromide is stirred with copious amount of zinc powder and reduced to zero valent as indicated by the red color. The bromobenzene is then added to the reaction mixture. The mechanism of this coupling is shown in Scheme 11.

Because Grignard reagents reacts with ketone and ester groups, Grignard coupling cannot be employed in the coupling of halobenzenes carrying these carbonyl groups. Benzene-1,4-bis(boronic acid) can be coupled with dibromobenzene carrying such substituents as carbonyl and sulfonyl groups; thus, it is possible to prepare copolymers containing these groups by the Suzuki reaction. However, it is impossible to prepare benzene-1,4-bis(boronic acids) carrying these groups.

The coupling by the zero-valent nickel is tolerant to such functional groups as ketone and ester groups. Thus, methyl 2,5-dichlorobenzoate was polymerized by this method [19]. Poly(methyl benzoate) was soluble in organic solvents and its degree of polymerization was about 100. This polymer was hydrolyzed with sodium hydroxide to produce poly(benzoic acid). Decarboxylation with copper oxide afforded PPP with a highly regular structure.

In a similar way, poly(*m*-phenylene) and alkylated PPP were prepared (Scheme 12) [19].

Poly(3-benzoyl-*p*-phenylene) was prepared according to the Scheme 13 [20]. Polymerization of the ketone was performed at a lower temperature. The weight-average degree of polymerization was as high as 300–500. This polymer was soluble in various organic solvents and exhibited high thermal stability as well as a high glass transition temperature. It also showed good mechanical properties.

Electrolysis of 4,4'-dibromobiphenyl in the presence of $NiCl_2$(dppe) [dppe = 1,2-bis(diphenylphosphino)ethane] afforded a homogeneous film on the cathode [21]. The films were brittle and could not be peeled off without damage. This polymer consisted of a regular 1,4-phenylene structure as far as indicated by reflectance IR spectroscopy. The polymerization proceeds via coupling of 4,4'-dibromobiphenyl with zero-valent nickel which is generated by electrochemical reduction of nickel chloride.

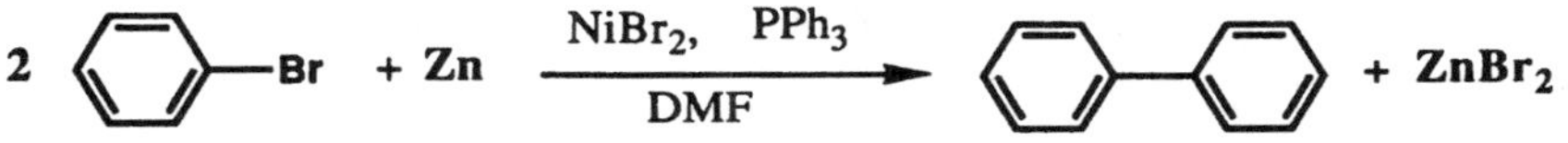

SCHEME 10

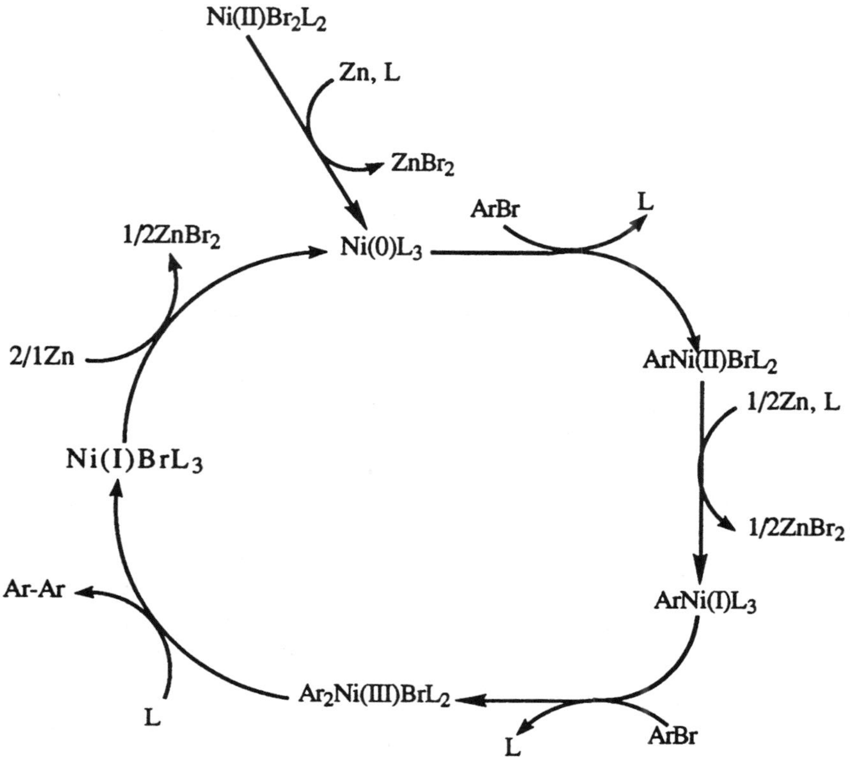

SCHEME 11

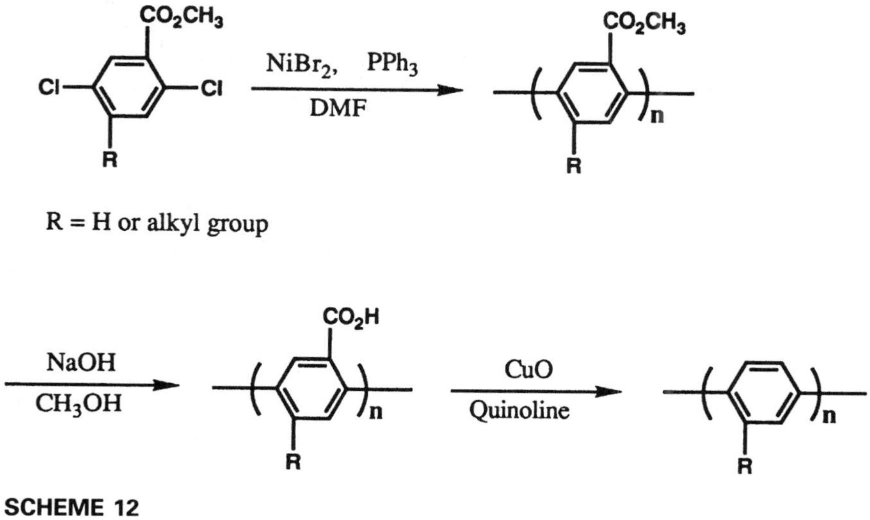

R = H or alkyl group

SCHEME 12

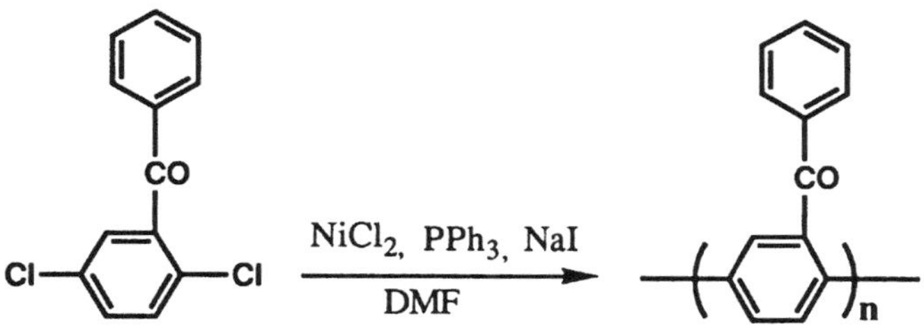

SCHEME 13

Dehalogenative polycondensation of dihalobenzene with bis(1,5-cyclooctadiene)nickel(0) and PPh$_3$ afforded PPP [22].

B. Aromatization of Precursor Polymers

Cyclohexadiene was polymerized with butyllithium or Ziegler–Natta catalyst. The resulting poly(cyclohexadiene) was aromatized with chloranil in a refluxing xylene solution (Scheme 14) [23]. This tan–brown powder showed bands corresponding to aromatic C–H vibration at 3035 cm^{-1} and out-of-plane vibration of para-substituted benzene at 811 cm^{-1} but also bands due to aliphatic C–H stretching at 2925 cm^{-1}, indicating that the dehydrogenated product contained an appreciable amount of nonaromatic units.

This method was further studied by several chemists [24]. The cyclohexadiene was replaced with dicarbonates of 1,2-dihydrocatechol, which was produced by oxidizing benzene with oxygen using a dioxygenase enzyme from a microorganism. The radical polymerization of methyl carbonate gave precursor polymers with Pn = 650–750. After a film was cast·from the precursor polymer, PPP was produced by pyrolysis (Scheme 15). The radical polymerization did not exclusively proceed by a 1,4-addition mechanism and the resulting polymer contained 15% of the 1,2-addition unit. From IR spectra, approximately 15% of absorption due to carbonyl groups remained after pyrolysis, indicating that aromatization did not take place at the 1,2-addition units [25]. Steric hindrance was so large that conjugation along polymeric chains was interrupted at the 1,2-addition units. This is a drawback for electronic polymers.

The procedure was subsequently improved to obtain cis-1,4-addition precursors [26,27]. Dihydrocatechol was allowed to react with trimethyl-chlorosilane. Stereoregular polymerization with the nickel complex afforded the cis-1,4-addition polymer with a Pn of 150. This polymer was hydrolyzed and converted to the acetic ester (Scheme 16). This ester polymer was cast on a NaCl crystal and pyrolyzed

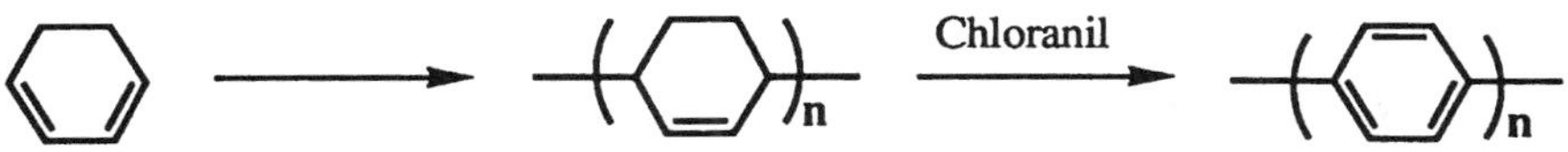

SCHEME 14

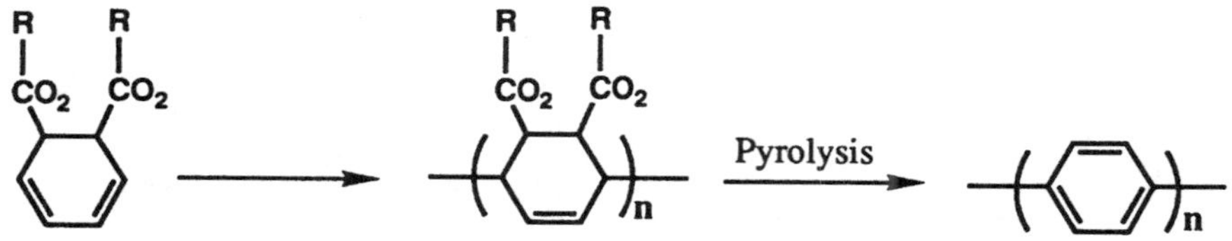
R
CO2
Pyrolysis
n

SCHEME 15

in the presence of various catalysts under argon flush. Pyrolysis with dichlorobenzenesulfonic acid afforded high-molecular-weight PPP with a regular structure.

The stereoregular precursor polymer was aromatized using *o*-phosphoric acid. Flexible, free-standing films of structurally regular PPP were obtained [28]. The PPP was amorphous due to the presence of residual polymetaphosphoric acid even after thorough washing with diluted hydrogen chloride.

The conductivity and absorption maxima of PPP films prepared by these precursor routes are summarized in Table 1. Whereas P1 containing 15% irregular bonding showed 100 S/cm, P3 and P4 exhibited low conductivity in spite of a highly regular structure.

C. Direct Oxidative Coupling of Benzene

1. *Chemical Oxidation of Benzene*

Benzene can be polymerized by oxidation with copper(II) chloride in the presence of aluminum chloride under a nitrogen flow. The resulting polymer is a brown powder [29,30]. This polymerization is termed the Kovacic method. The net reaction for this method is shown in Scheme 17. The polymerization proceeds by an oxidative cationic mechanism according to Scheme 18. This is clearly a chain

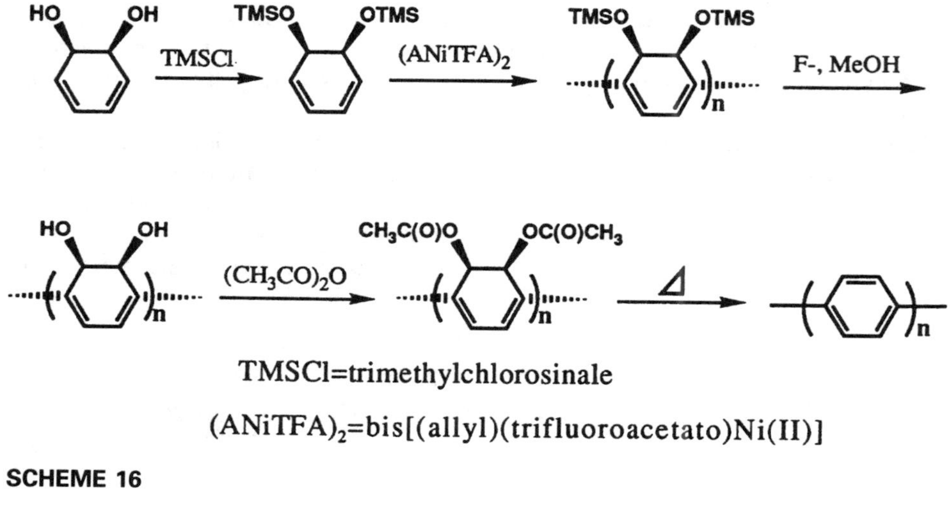
HO
OH
TMSCl
TMSO
OTMS
(ANiTFA)2
F-, MeOH
n
CH3C(O)O
OC(O)CH3
(CH3CO)2O
Δ
TMSCl=trimethylchlorosinale
(ANiTFA)2=bis[(allyl)(trifluoroacetato)Ni(II)]

SCHEME 16

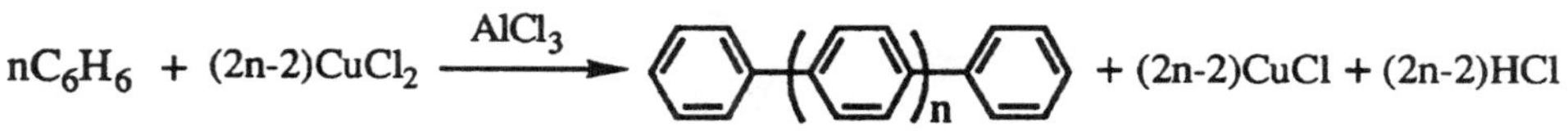

$$nC_6H_6 + (2n-2)CuCl_2 \xrightarrow{\text{AlCl}_3} \text{[structure]} + (2n-2)CuCl + (2n-2)HCl$$

SCHEME 17

polymerization, which propagates by the reaction of monomer with active chain ends.

Poly(p-phenylene) was obtained with a yield of 89–91%. The IR spectra of the PPP exhibited one peak at 803 cm^{-1}, due to disubstituted benzene rings and two small peaks at 763 and 693 cm^{-1} due to terminal phenyl groups in the out-of-plane vibration region. Thus, the PPP was assumed to have a regular 1,4-phenylene structure. Many scientists used this method of preparing PPP in order to investigate its properties. However, the PPP showed more intense electron-spin resonance (ESR) signals than polymers prepared by other methods, indicating that it contained polycyclic defects [31].

PPP was obtained by oxidizing benzene with copper(II) chloride in *N*-butylpyridinium chloride and aluminum chloride solution [32]. The resulting PPP had a degree of polymerization (DP) of 36, higher than PPP prepared by the Kovacic method.

Alkoxy substituents decrease the oxidation potential of benzene. Thus, oxidation of 1,4-dialkoxybenzene with iron(III) chloride gave poly(2,5-dialkoxyphenylene). ^{1}H- and ^{13}C-NMR (nuclear magnetic resonance) spectra revealed that this polymer consisted of almost equal fractions of 1,4- and 1,3-linked units.

2. Electrochemical Oxidation of Benzene

Oxidative electrochemical polymerization is a facile route to prepare conductive polymer films [34]. Furthermore, they are formed on conductive substrates. Thus, this technique is widely employed to prepare polyheterocycles such as polythiophene and polypyrrole. These materials are studied in detail by electrochemical

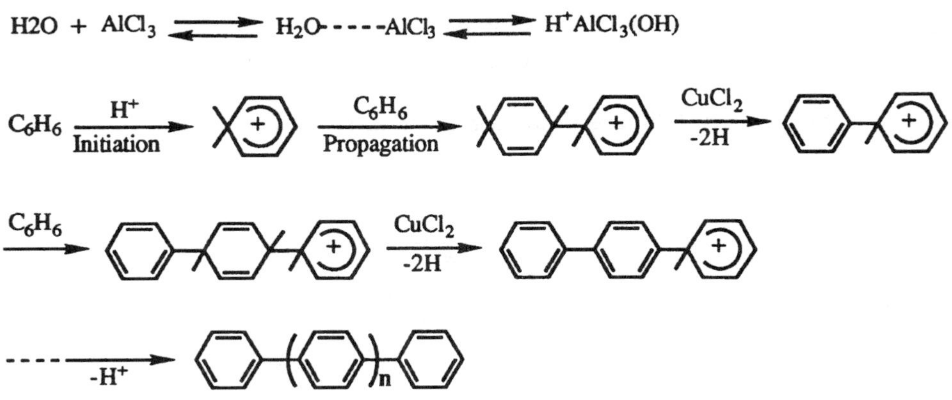

$$H_2O + AlCl_3 \rightleftharpoons H_2O\text{----}AlCl_3 \rightleftharpoons H^+AlCl_3(OH)$$

$$C_6H_6 \xrightarrow[\text{Initiation}]{H^+} \text{[structure]} \xrightarrow[\text{Propagation}]{C_6H_6} \text{[structure]} \xrightarrow[-2H]{CuCl_2} \text{[structure]}$$

$$\xrightarrow{C_6H_6} \text{[structure]} \xrightarrow[-2H]{CuCl_2} \text{[structure]}$$

$$\xrightarrow[-H^+]{\text{----}} \text{[structure]}$$

SCHEME 18

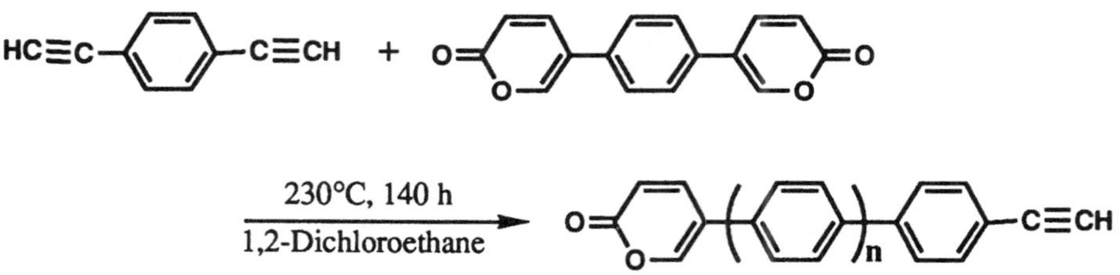

SCHEME 19

and spectroscopic methods. Because the oxidation potential of benzene is so high, polyphenylene films cannot be obtained under the same conditions as those where thiophene and pyrrole are electrochemically polymerized.

Electrolysis of benzene on a 98% hydrogen fluoride layer afforded a polymer film on the platinum electrode [35]. This film was different from chemically prepared PPP. The IR spectra indicated that the film was composed predominantly of 1,4-phenylene units together with an appreciable amount of 1,3-phenylene units. Unlike the chemically prepared PPP, it was amorphous in nature.

Electrochemical polymerization in a two-phase system of benzene–20% fuming sulfuric acid gave a PPP film, which showed an IR spectrum similar to the polymer prepared by the Kovacic method [36].

The addition of a Lewis acid, which lowers the oxidation potential of benzene by complex formation, is necessary to perform electrochemical polymerization of benzene in common organic solvents. Benzene was polymerized in nitrobenzene in the presence of aluminum chloride and a small amount of water or amine [37]. The resulting film contained an appreciable amount of irregular 1,3-phenylene units. Electrochemical polymerization of benzene in the presence of a boron trifluoride–ethyl ether complex afforded a flexible film of polyphenylene [38].

Alkoxy substituents markedly decrease the oxidation potential of benzene. Thus, like thiophene and pyrrole, 2-methoxy-5-alkoxybenzene could be electrochemically polymerized in 0.1 M tetrabutylammonium tetrafluoroborate solution in acetonitrile [39]. The resulting polymers were soluble in common organic solvents. The IR and NMR spectra indicated that these polymers had a linear and well-defined structure. The degree of polymerization was 10–11. Similarly, electrochemical polymerization of 1,2-(methylenedioxy)benzene afforded a soluble polymer, which was partially crystalline and had a regular structure [40].

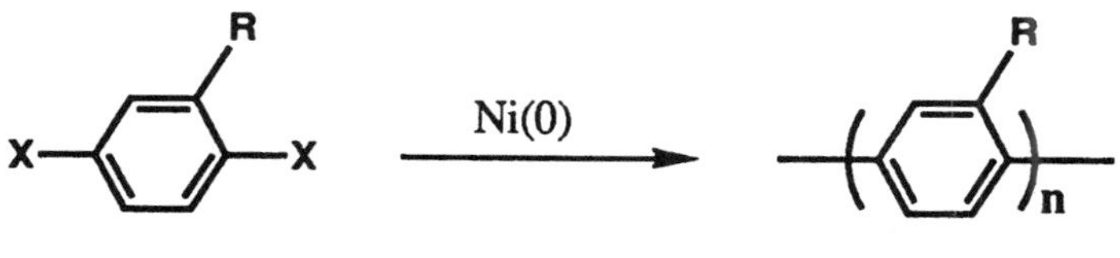

X= CF_3SO_3 or CH_3SO_3

SCHEME 20

It is possible to obtain PPP films by the electrochemical reduction of a Ni(II) complex in the presence of dibromobenzene. This method often affords better PPP film on the cathode than the electrochemical oxidation of benzene [41].

D. Miscellaneous

Poly(*p*-phenylene) has been prepared by the Diels–Alder reaction [42]. The reaction of *p*-diethynylbenzene with 5,5'-*p*-phenylenebis-2-pyrone afforded dark yellow PPP [19]. This showed a similar IR spectrum to the PPP prepared by the Kovacic method. PPP prepared in this way was thermally stable. Thermogravimetric analysis (TGA) indicated a 10% its weight loss at 650°C and 20% at 800°C.

Bistriflates and bismethylates were polymerized with zero-valent nickel in a similar reaction to that described for the dibromides in Section II.A.3 (Scheme 20) [43,44]. This polymerization was also tolerant toward carbonyl groups and is a good method to prepare soluble PPPs carrying a variety of R. A number average degree of polymerization of up to 100 was obtained.

III. PROPERTIES OF POLY(*p*-PHENYLENE)

A. Effect of Doping

The absorption maxima of electronic spectra are a reasonable measure of the effective conjugation length of polymeric chains. Poly(*p*-phenylene) is a conductive polymer but poly(*m*-phenylene) is an insulator. Whereas λ_{max} of *m*-quinqiphenyl is 247 nm, almost identical to 246 nm for biphenyl, λ_{max} of oligomeric PPP increases with the increase in chain length. λ_{max} is 318 nm for *p*-sexiphenylene and calculated to be 339 nm for PPP of infinite length [45].

Except for films doped with metaphosphoric acid, PPP films prepared by aromatization of precursor polymers show λ_{max} ranging from 305 to 320 nm, as shown in Table 1. These values correspond to the length of the *p*-phenylene sequence predicted from the occurrence of *o*-phenylene units and incomplete aromatization.

A PPP film has been prepared by oxidative coupling of benzene with $CuCl_2$/$AlCl_3$ under shear between two rotors [46]. This film showed λ_{max} at about 370 nm. Taking into account a degree of polymerization of about 20 and chlorine content of 3%, this value is too large, probably due to cross-linking and formation of polycyclic units.

TABLE 1 Conductivity and Absorption Maximum of PPP Films Prepared via Precursor Routes

Entry	λ_{max} (nm)	σ (S/cm)	Dopant	Ref.
1		100	AsF_5	24
		1.5×10^{-2}	$FeCl_3$	
		6×10^{-3}	Na	
2	320	2.2	AsF_5	25
3	305–310	10^{-2}–10^{-1}	AsF_5	27
4	336	10^{-1}	AsF_5	28

A PPP film obtained by oxidative electrochemical polymerization of benzene in liquid sulfur dioxide in the presence of Lewis acid shows λ_{max} at 350–360 nm, close to the value of Kovacic PPP [47]. This bathochromic shift is probably due to cross-linking. Because degrees of polymerization estimated from the infrared peak ratio I_{805}/I_{695} are in the range 5–15, this low-molecular-weight oligomer is not expected to form a film without cross-linking.

Because the ionization potential of PPP is high, only a few dopants can be used for polymer doping. AsF_5 is one of the most effective dopants. Ivory et al. have studied the effect of AsF_5 doping on the conductivity of PPP prepared by the Kovacic method [48]. The brown powdery polymer was compressed into 0.1–1.5-mm-thick pellets and annealed at 400°C for 24 h to eliminate chlorine. When the pellets were exposed to 450-Torr AsF_5 vapor, the doping level reached 0.24–0.42 mol of AsF_5 per mol of monomer unit. The conductivity of PPP increased with the increase in doping time as shown in Fig. 2. The doping proceeded by steps. This is similar to the stages observed in the intercalation of graphite. Between one and three stages were observed for the doping of PPP. The highest conductivity was 500 S/cm.

AsF_5-doped PPP is not stable in air. Conductivity decreases by about a factor of 2 after 5 h of exposure to humid air. AsF_5-doped graphite and polyacetylene decay more rapidly. An approximate carrier mobility is calculated to be 10^{-4} m/V s from conductivity and carrier density. The Hall coefficient is about 2.5×10^{-9} m³/A s.

AsF_5-doping causes oxidative couplings of benzene rings, as well as charge transfer. These couplings take place at the *p*-positions of two terminal phenyl groups and possibly cross-links two phenylene groups, thus expanding conjugation systems. This contributes to the high conductivity of AsF_5-doped PPP.

The diameter of PPP fiber prepared by the Kovacic method increased by about 50% after heavy AsF_5-doping, suggesting a penetration of the dopant molecules inside the fiber [49]. This was confirmed by the lowering of crystallinity and the disappearance of a distinct crystalline texture. The new diffraction peaks observed can be explained by an intercalation model.

Cation doping was performed by immersing a PPP pellet in a tetrahydrofuran (THF) solution of potassium naphthalide. The color immediately changed to a gold. Conductivity rapidly increased to 30 S/cm. The doping level was 0.57 mol potassium per monomer unit, higher than that of AsF_5 doping, but conductivity after potassium doping was lower than after AsF_5 doping. This was due to a lower mobility of an electron relative to a hole. The golden luster was lost as soon as the pellet was exposed to air.

Powdery PPP was pressed on a stainless mesh without any plasticizing or conducting additives. PPP was electrochemically doped with Li under galvanostatic conditions. Coulometric measurements indicated that the doping level was 0.07 mol Li per monomeric unit after prolonged electrolysis. The state of Li ions has been studied by the 7 Li NMR free-induction decay technique [50]. Reactions between negatively charged PPP and Li^+, for example, abstraction of hydrogen and formation of the lithium hydride, does not take place, resulting in high thermal and chemical stability of *n*-doped PPP. Clustering of Li is not observed. Li atoms are completely ionized in PPP. Li cations are in two states with different mobility. The mobility of the most mobile Li cations is much less than that of AsF_6 and BF_4

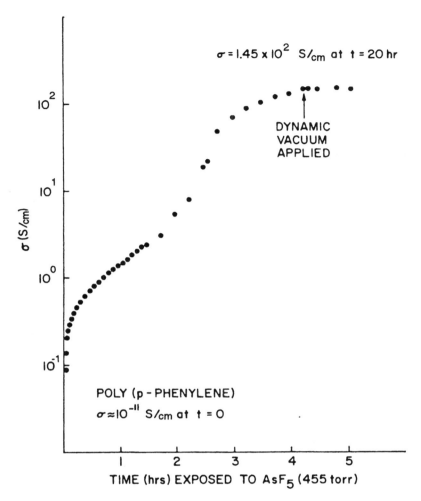

$\sigma = 1.45 \times 10^{2}$ S/cm at $t = 20$ hr

DYNAMIC VACUUM APPLIED

POLY (p - PHENYLENE)
$\sigma \approx 10^{-11}$ S/cm at $t = 0$

TIME (hrs) EXPOSED TO AsF$_5$ (455 torr)

FIGURE 2 Conductivity versus time of exposure of AsF$_5$ at 455 Torr. The conductivity was $\sim 10^{-11}$ S/cm at $t = 0$ and 1.45×10^{2} at $t = 20$ h (4-h exposure to AsF$_5$ and 16 h at dynamic vacuum. (From Ref. 48.)

anions determined by ^{19}F-NMR. For doped PPP, a channel structure with dopants located parallel to the polymer chains is proposed. Significant contraction upon alkali metal doping results in the formation of a close-packed channel. Thus, both intrachannel and interchannel diffusion of Li cations is limited. On the other hand, anion doping leads to a loosely packed polymer structure and the mobility of anions is high.

When PPP prepared by the Kovacic method is doped with K, the K cations aggregate in columns between two parallel polymer chains [51]. The K cations stack over the midpoint of the CH–CH bonds parallel to the polymer chains rather than the centers of the phenyl rings. This structure is compatible with theoretical calculations that indicate that the quinoid structure would have greater electron density over the CH–CH bonds, in contrast to the benzenoid structure with toroids of π-electrons above and below the plane of the phenyl rings [52].

B. Electrochemistry

Poly(p-phenylene) can be doped with cations as well as anions. It undergoes both electrochemical oxidation and reduction. Electrochemical properties of PPP are somewhat dependent on the preparation procedures. For electrochemically prepared PPP, the cyclic voltammogram (CVs) generally shows broad redox peaks. However, PPP prepared by electrochemical oxidation of benzene in a superacid such as CF_3SO_3H shows exceptionally sharp peaks, as shown in Fig. 3 [53]. In these CVs, the potential difference between the anodic and cathodic peaks is large and the peak positions are only slightly dependent on scan rates, being associated with highly energetically stabilized charged species. Clearly, PPP electrochemically prepared in superacid has a well-defined linear structure with narrow molecular-weight

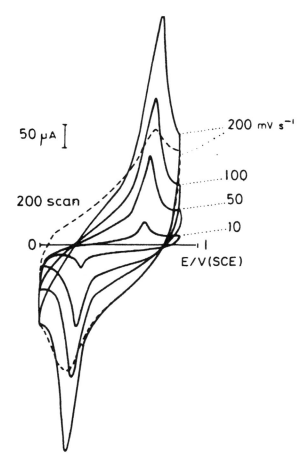

FIGURE 3 Cyclic voltammogram in 95% sulphuric acid for a film obtained in the two-phase benzene/triflic acid system (From Ref. 34.)

distribution and is in an energetically metastable state. After several redox cycles, the peaks are broad, as shown by a dashed curve in Fig. 3. These peaks are very similar to those of PPP prepared in usual electrolyte solutions. A possible explanation for this phenomenon is that PPP prepared in superacid undergoes cross-linking reactions during electrochemical redox cycles, approaching the structure of PPP prepared in usual electrolyte solutions.

It is possible to electrochemically reduce PPP. Although PPP prepared by electrochemical polymerization of biphenyl is irreversibly reduced at -2.0 V (versus Ag/AgCl) in an acetonitrile solution of $(C_4H_9)_4NBF_4$, PPP prepared in the SO_2–SbF_5 system can be reversibly reduced at -1.8 V (versus Ag/AgCl) in tetrahydrofuran containing $(C_4H_9)_4NBF_4$.

Poly(p-phenylene) is obtained on a cathode by polymerizing 4,4′-dibromobiphenyl in the presence of Ni(II) complex. This PPP undergoes both electrochemical oxidation and reduction in a $(C_4H_9)_4NBF_4$ solution in acetonitrile. It is also reduced in a $LiClO_4$ solution. The film changes from yellow to red upon oxidation, and to purple upon reduction [54]. Similar color changes have been reported by Fauvarque et al. [41].

In order to obtain an insight into the electrochemistry of PPP, oliogmeric PPP was studied electrochemically. For a p-sexiphenyl film, the trication is stable at low temperatures and CVs can be cycled repeatedly; however, it is labile at room temperature. When formation of the trication is completely excluded, the p-sexiphenyl can be cycled between oxidized and reduced states more than 500 times without any loss of electrochemical activity (Fig. 4a). When polarity of the electrodes is reversed just before the formation of the trication at room temperature, the initial anodic peak becomes lower and a new peak appears on the negative side (Fig. 4b). A new peak also appears on the negative side of the cathodic peak. These new peaks occur at the same position as the peaks due to dodecaphenyl prepared chemically (Fig. 4c). Clearly, p-sexiphenyl is coupled at terminal phenyl rings to produce dodecaphenyl. When the oxidation potential increases further, charge is lost by proton cleavage and the peaks become broad. Structural deterioration by these terminal coupling and cross-linking reactions is confirmed by Fourier-transform–infrared (FT-IR)[55].

As for substituted PPPs, the electrochemistry of alkoxylated PPP has been studied. In CVs of poly[1,2-(methylenedioxy)benzene], two anodic peaks occur at 0.90 and 1.14 V (versus Ag/AgCl) and two cathoid peaks at 0.72 and 0.86 V [40]. These peaks are due to two successive processes. The electroactivity loss is 21% in the presence of BF_4^- after 60 cycles. Reversibility is dependent on the sizes of anions and cations. The smaller the sizes, the higher the reversibility. Although solubility of poly(3-alkylthiophene) is higher in neutral states than in doped states, the present polymer is more soluble in such polar solvents as dimethyl sulfoxide (DMSO) and dimethylformamide (DMF) after doping with $LiClO_4$ than without doping.

The CV of the copolymer of alkoxylated benzene with bithiophene shows two separate anodic peaks corresponding to two oxidation states [56]. Absorption spectra and magnetic measurements indicate that these peaks are due to the formation of polarons and bipolarons.

The performance of PPP electrodes using PPP prepared by oxidative electrochemical polymerization has been studied [57]. From charge/discharge character-

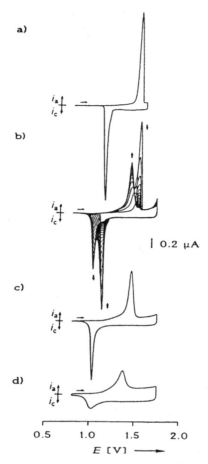

FIGURE 4 Cyclovoltammograms in $CH_2Cl_2/0.1$ M Bu_4NPF_4. (a) *p*-sexiphenyl (thin layer on Pt); (b) *p*-sexiphenyl in the multisweep experiment with $E_\lambda = 1.75$ V; (c) the dodecaphenyl isomer; and (d) polyphenylene. (From Ref. 55.)

istics, this PPP film electrode is estimated to be usable as the active material for a high-energy-density battery.

A composite electrode is prepared from PPP by the Kovacic method. A battery is fabricated using PPP/Na_xPb as a negative electrode and Na_xCoO_2 as a positive electrode [58]. This battery offers a high gravimetric and volumetric energy density. Because dendrite formation does not occur in this battery, cycle life is long and a high charging rate is possible. The cell potential is 2 V.

C. Photoluminescence and Electroluminescene

Many organic molecules emit fluorescence over a wide wavelength range. Aromatic hydrocarbons such as benzene, naphthalene, and pyrene emit fluorescence in high quantum efficiencies. Particularly, the efficiency of 9,10-diphenylanthracene is as high as 90%. In the field of dye lasers, oligo-*p*-phenylenes such as *p*-terphenyl and end-substituted *p*-quarterphenyls are useful for emission of short-wavelength light down to 311 nm [59–61]. Electroluminescence properties were harnessed by Helfrich and Schneider, who fabricated light-emitting devices using single crystals of naphthalene, perylene, and anthracene [62]. However, several hundred volts were required to emit light, because thin films could not be formed from single crystals. In addition, the electroluminescence was faint. Thin films prepared by vacuum-deposition and the Langmuir-Blodgett (LB) technique were also used to fabricate light-emitting devices, but emission efficiency was still low.

Adopting a double-layer structure, with one diamine layer capable of only hole transport, Tang and Vanskyle fabricated a diode emitting green light with a maximum at 550 nm [63]. The external quantum efficiency of this diode is about 1% and the drive voltage 5.5 V. The organic layers are sandwiched between a low-work-function Mg cathode and an indium–tin oxide (ITO) anode. The radiative recombination of electrons and holes takes place in the emitting layer adjacent to the hole transport layer. This study triggered extensive research on organic light-emitting diodes (LED). The conjugated polymer is one of the most promising materials for use in organic LEDs, as it is amorphous in essence and forms thin films without pinholes. Recently, soluble PPP derivatives have been widely used in the fabrication of polymeric LED. PPPs with wide band gaps are suitable for blue-light emission.

Jenekhe and Osaheni studied the effects of excimer and exciplex formation on the luminescence of conjugated polymer thin films [64]. Their study indicated that luminescence originated from excimer emission and that, generally, the quantum efficiency was the result of self-quenching. On the other hand, exciplex formation enhances the solid-state quantum efficiency of conjugated polymers. These are clues that forward the understanding of the photochemical processes of conjugated polymers in terms of supramolecular structure.

Poly(*p*-phenylene) was used to fabricate a polymeric LED [65]. The PPP film was prepared according to the precursor route previously described [24]. This device consisting of a single PPP layer sandwiched between ITO and Al electrodes emitted blue light with a quantum efficiency of 0.01–0.05%.

A film of this polymer exhibits a photoluminescence spectrum with fine structure, tailing down to 700 nm [66]. The main peak is located at 460 nm. The polymer shows an electroluminescence spectrum with a structureless peak at 460 nm and a broad shoulder at around 560 nm.

Yang et al. synthesized LED using poly(2-decyloxy-*p*-phenylene) (DO-PPP) [67]. The fluorescence efficiency was 85% in a 1% solution and 35% in a solid film. A single-layer device consisting of ITO/DO-PPP/Ca showed a maximum external quantum efficiency of 1.8%, but was unstable during measurement due to leakage current. The addition of poly(*N*-vinylcarbazole) (PVK) as a hole transport layer eliminated the leakage problem arising from pinholes. The highest external quantum efficiency obtained for double-layer ITO/PVK/DO-PPP/Ca devices was about 3%, which is equivalent to a 5.4% internal quantum efficiency. Typical quantum efficiencies range between 2.0% and 2.5%. The blue–violet light produced is uniform and bright.

The electronic structure of DO-PPP was investigated by the Fowler–Nordheim's method. A schematic energy diagram is shown in Fig. 5. The single-particle energy gap in DO-PPP is 3.4 ± 0.2 eV.

┬─ 2.3 eV DO-PPP π*

├─ 2.9 eV Ca

ITO 4.7 eV ─┤

├─ 5.2 eV Au

┴─ 5.7 eV DO-PPP π

FIGURE 5   Schematic energy diagram.

The LED turns on at 15 V and reaches a brightness of 490 cd/m² at 30 V. These operating voltages are too high and lead to lower power efficiency. The decrease in the operating voltage improves the power efficiency. The conductivity of the DO-PPP layer was increased by blending DO-PPP with a hole transport material. The operating voltage was lowered from 27 to 18 V by using a blend of DO-PPP with tertiary aromatic amine. The trade-off for using the blend as the light-emitting layer is a reduction in the quantum efficiency from 2.0% to 0.7%.

Jing et al. fabricated a two-layer LED composed of an ITO/poly(*p*-phenylenevinylene) (PPV)/PPP copolymer/Ca [68]. The PPP copolymer was a random 2:1 copolymer of dihexyloxy-*p*-phenylene with *p*-phenylene. Light emission was seen under forward bias at a threshold of about 50 V. The internal quantum efficiency was 0.5%. The electroluminescence spectrum, measured through the semitransparent calcium electrode, showed a peak at approximately 480 nm and a broad emission tail extending to over 800 nm. These features indicated that electroluminescence originated from the copolymer layer.

Grüner et al. prepared oligo(*p*-phenylene) with a rigid planar structure (LPPP) (Scheme 21) [69]. The fluorescence quantum yield of LPPP is 43% in a dilute

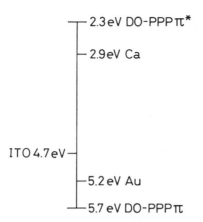

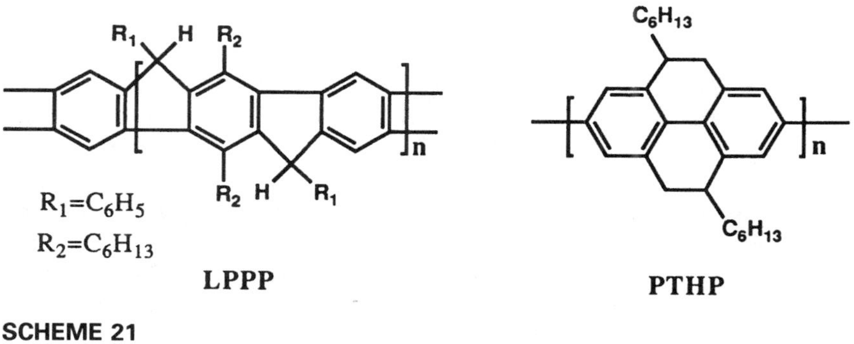

$R_1 = C_6H_5$
$R_2 = C_6H_{13}$

LPPP   **PTHP**

SCHEME 21

dichloromethane solution. A single-layer device with a LPPP layer emits yellow electroluminescence with an internal quantum efficiency of 0.3%. A double-layer ITO/PPV/LPPP/Ca also emits also yellow light with an internal quantum efficiency of 0.6%, without any contribution of PPV to the emission. Blue emission is obtained from a single-layer device consisting of ITO/LPPP: PVK/Ca, where LPPP:PVK denotes a blend of LPPP with PVK. The internal quantum efficiency reduces to 0.16%.

Soluble PPP with tetrahydropyrene units (PTHP) (Scheme 21) was prepared by Ni(O)-catalyzed polycondensation of the corresponding monomeric dibromide [70]. PTHP shows an absorption maximum at 385 nm, which is located at an intermediate wavelength between 330 nm for poly(*n*-dihexyl-*p*-phenylene) and 440 nm for fully planarized LPPP. PTHP shows fluorescence maxima at 425 nm in solution and at 475 nm in a solid film.

A LED has been fabricated using poly(9,9-dihexylfluorene) [71]. This Schottky-type diode is driven at 10 V and emits blue light with a maximum of 470 nm. The emission intensity increases linearly with increasing current density, tending toward slight saturation at higher currents.

Copolymers of diheptyl-*p*-phenylene with aromatic compounds were prepared by the Suzuki reaction [72]. These copolymers emit strong purple or blue fluorescence. Their absorption, excitation, and fluorescence spectra in chloroform solutions are compared in Fig. 6. Although P14NHP and P26NHP show very similar fluorescence and excitation spectra, the absorption spectra of these polymers are remarkably different in that the first absorption of P26NHP is greatly suppressed. The spectra of P910HP shows fine structure, the pattern of which is very similar to that of anthracene. The Stokes shift is as small as 13 nm. These findings suggest that a rigid moiety is responsible for the fluorescence spectrum. The absorption spectrum of P910AHP shifted by only 25 nm relative to that of anthracene.

Remmer et al. studied the optical properties of a series of copolymers of oligo(dipentyloxyphenylene) with vinylene (CPV) (Scheme 22) [73]. The peaks of both the absorption and emission spectra are red-shifted with the increase in molar fraction of vinyl groups. The peak of the emission spectrum can be tuned in the range between 400 and 560 nm. The Stokes shift is also a linear function of the fraction of vinylene groups in the polymer, indicating that geometrical changes in the transition to the excited state becomes smaller with increasing fraction of vinylene groups. Blueshift was observed when vinylene groups were replaced with ethylene groups.

Musfeldt et al. prepared poly(*p*-phenylene) with emitter units separated by *m*-phenylene units and severe steric distortion [74]. Poly(*p*-phenylene) with discrete quater-*p*-phenylene units exhibits absorption and emission maxima at 304 and 406 nm, respectively. The Stokes shift is 1 eV. They also showed that the absorption and emission maxima of poly(*m*-phenylene) were located at 262 and 397 nm, respectively, and that the Stokes shift was as great as 1.6 eV.

D. Supramolecular Structure

The structure of PPP has been analyzed by several authors and the results summarized in Table 2. Although the methods used for polymer preparation were different, the results obtained were very similar, suggesting that the structure of PPP

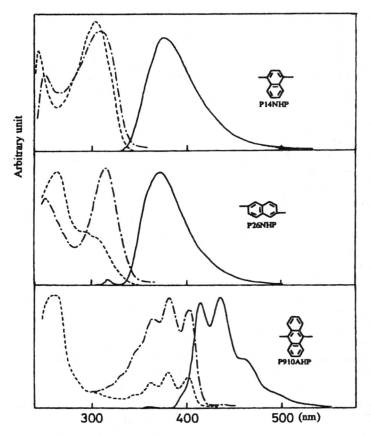

FIGURE 6 Optical spectra of copolymer: absorption (----); excitation (---); fluorescence (—).

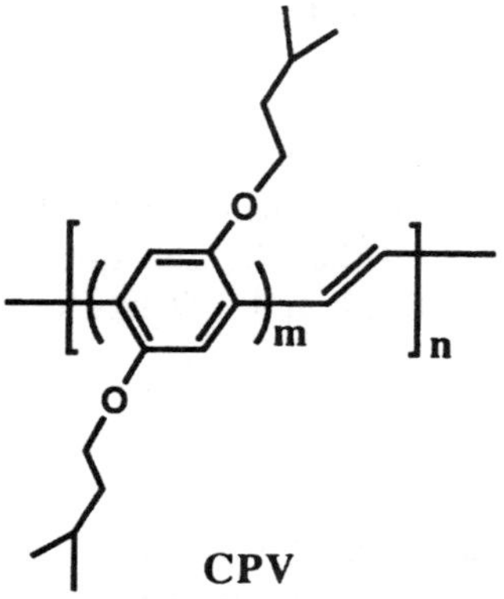

CPV

SCHEME 22

TABLE 2 Results of X-ray Analysis of PPP

Method of Preparation	*d* spacing (Å)			Ref.
Scheme 17	4.53	4.00	3.20	75
Scheme 14	4.54	3.96	3.22	23
Scheme 19	4.60	3.95	3.22	76
Scheme 15	4.5	3.9	3.2	24

is independent of the chemical process of preparation. Comparison with the results of x-ray and neutron-diffraction studies of *p*-terphenyl suggests that the *d* spacing of 4.5 Å corresponds to the intramolecular phenyl group distance in the direction of the principal molecular axis and that the *d* spacing of 4.0 Å corresponds to the distance between similar phenyl group centers in adjacent molecules lying in the same plane. The *d* spacing of 3.2 Å corresponds to the distance between molecular planes [77].

Kawaguchi and Petermann studied the structure of PPP by electron diffraction [78]. A film of the PPP was prepared by the method described previously [46]. It was then annealed in vacuum at 400°C for 24 h. The experimental results indicated that the PPP is paracrystalline, similar to the nematic structure in liquid crystals. Molecular chains are laterally packed in a rather regular way. However, there is remarkable disorder due to the shift of chains in the direction parallel to their axes.

Single chains may have chemically irregular structural units such as the ortho- or meta-configuration. Incorporation of such units into the lattice causes distortion. Chemical irregularity could be one of the possible causes for lattice distortion. The Pn of the PPP is low and there is, thus, the possibility of incorporation of chain ends with random structure into the crystals. This may be the cause of distortion or disorder in the crystal lattice.

An NMR study of deuterated PPP prepared by Scheme 17 indicates that the PPP is approximately 75% crystalline [79]. The predominant mode of motions in the amorphous regions is flipping of the phenylene rings by 180°. This ring flipping has a correlation time of approximately 10^{-7} s at ambient temperature.

Baker et al. studied the crystal structure of *p*-quipuephenyl, *p*-sexiphenyl, and *p*-septiphenyl [80]. These oligomers belong to space group $P2_{1/c}$. The unit cell parameters are very similar but the *a* crystallographic axis increases with the increase in molecular length. The oligomers are centrosymmetric. The angle between the atoms constituting the molecular axis indicate the deviation from linearity. The largest deviation is found to be 3.6° at the center of symmetry for *p*-sexiphenyl. Molecular mechanics calculations predict that whereas isolated oligomers take on nonplanar conformation, intermolecular interactions force a planar conformation in a crystalline environment. Differential scanning calorimetry (DSC) measurements indicate that these oligomers show melting points at 388–420°C and form a liquid-crystalline phase. Sample decomposition above 500°C precludes detailed study.

Wittman and Smith deposited an oriented thin film of poly(tetrafluoroethylene) on a flat substrate by hot-dragging of the polymer [81]. Similarly, Tanigaki et al.

obtained an oriented layer of PPP on the surface of a quartz plate by sliding the polymer at 160–235°C [82]. Polarzied optical spectra, optical micrographs with a crossed polarizer, electron micrographs, and electron-diffraction diagrams clearly show that this friction-deposited PPP is oriented along the direction of friction. The deposited film is very thin, but it is possible to increase the film thickness by epitaxial polymerization of benzene according to Scheme 17.

Wittler et al. studied textures of the liquid-crystalline phase of the alkylated PPPs in Scheme 23 [83]. The polymers show reversible endothermic peaks at temperatures between 160°C and 282°C. Observation under crossed nicols conditions reveals that the melts of 6APPP exhibit a Schlieren texture which is stable up to 300°C. When the polymers are cooled below their melting points, the Schlieren texture remains unchanged. Major disclinations of strength are $\pm 1/2$, and disclinations of strength of ± 1 and $\pm 3/2$ are also found occasionally.

Unusually high disclinations of strength up to -3 were found for the molten mixture of F6APPP with poly(ethylene glycol) (2:8). The core of these disclinations might consist of poly(ethylene glycol) or its mixtures with a low molecular fraction of F6APPP.

Using 12APPP of molecular weight $M_w = 7.9 \times 10^3$ to 1.37×10^5, MacCarthy et al. investigated the structure and properties of the polymer [84]. DSC, rheological measurement, and polarizing optical microscopy indicate that the formation of an anisotropic phase upon melting is dependent on molecular weight: Polymers of $M_w < 30,000$ show only an isotropic phase, polymers of $44,000 < M_w < 73,000$ give coexisting isotropic/anisotropic phases, and polymers of $M_w > 94,000$ show only a single anisotropic phase.

A 12APPP film cast from a *p*-xylene solution was analyzed by x-ray diffraction with the x-ray beams both perpendicular to the film surface and perpendicular to the film edge. When the x-ray beam is perpendicular to the film surface, no reflections indicative of high order are observed. When the x-ray beam is perpendicular to the film edge, the reflections indicate that the film has a periodic structure. These findings lead to the conclusion that whereas main chain layers are parallel to the film surface, the lateral alkyl chains are oriented perpendicular to the film surface. Figure 7 shows that the main chain layers are separated by well-ordered layers of the lateral alkyl chains. A melt-pressed 12APPP film, which was quenched

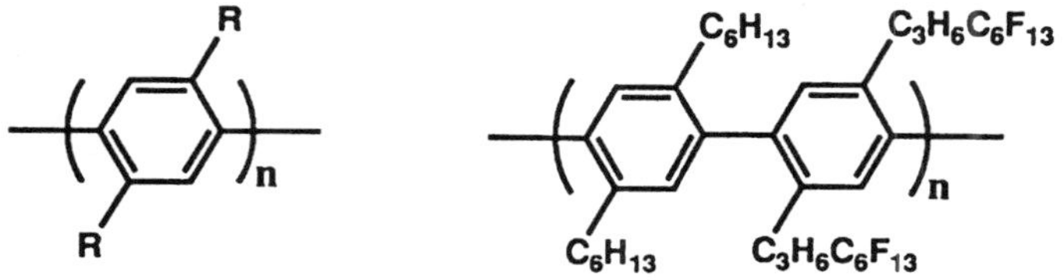

	R
6APPP	C_6H_{13}
12APPP	$C_{12}H_{13}$

SCHEME 23

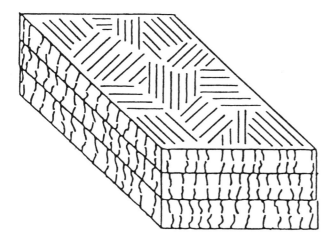

FIGURE 7 Molecular arrangement of 12 APPP.

from the liquid-crystalline phase, exhibits the same type of molecular alignment as the film cast from a solution.

Vahlenkamp et al. studied the structure of poly(2,5-dialkoxy-*p*-phenylene)s containing C_4–C_{12} alkoxy groups [85]. Pn ranges from 10 to 30. The DSC trace of poly(2,5-dibutoxy-*p*-phenylene) (Pn = 30) exhibits two endothermic peaks at 182 and 227°C on heating, and, on cooling, both transitions overlap and occur at 122°C and 116°C. Above 227°C, this polymer forms an anisotropic melt which starts decomposing at 300°C. The same polymer with a lower DP shows one transition at lower temperatures. X-ray diffraction analysis indicates that whereas octyloxy- and dodecyloxy-substituted PPP's exhibit a layered-phase structure, *n*-butoxy- and *i*-pentyloxy-substituted PPPs are packed in a cylindrical structure.

The alternating copolymer of dihexylbenzene with benzenesulfonic acid in a DMSO solution shows birefringence above a concentration of about 5% [86]. After heating a droplet of a solution of the copolymer in DMSO placed between two microscope slides to 70°C and subsequent cooling to room temperature, the formation of parallel aligned dark and bright stripes are observed with crossed polarizers.

An oriented ultrathin film of poly(2,5-diheptyl-*p*-phenylene) was formed by casting the polymer solution on an ordered film of poly(tetrafluoroethylene) [87]. Angle-resolved ultraviolet photoelectron spectroscopy (UPS) indicates that heptyl groups of the poly(2,5-dihepthyl-*p*-phenylene) are aligned parallel to the main chains of poly(tetrafluoroethylene); thus, the main chains of the former lies at an angle of 90° to the latter chains.

A film of poly(2,5-dihepthyl-*p*-phenylene) was prepared by casting from a chloroform solution on an oriented PPP substrate [15]. Contrary to poly(tetrafluoroethylene), in this case the main chains of the alkylated PPP are oriented parallel to the main chains of the PPP substrate.

IV. CONCLUSIONS

Poly(*p*-phenylene) is a very simple polymer consisting of straight array of benzene. PPP and its derivatives exhibit interesting electrical and optical properties and supramolecular structures. Finding a way to high-quality polymers must be an important breakthrough to highly demanding applications.

REFERENCES

1. Skotheim, T. A., *Handbook of Conducting Polymers*, Marcel-Dekker, Inc. New York, 1986.
2. Schlüter, A.-D., and Wegner, G., *Acta Polym*, *44*, 59 (1993).
3. Baker, N. K., Fratini, A. V., Resch T., Knackel, H. C., Adams, W. W., Socci, E. P., and Farmer, B. L., *Polymer, 34*, 1571 (1993).
4. Goldschmiedt, G., *Montash., 7*, 40 (1886).
5. Edwards, G. A., and Goldfinger, G., *J. Polymer Sci., 16*, 589 (1955).
6. Claesson S., Gehm R., and Kern W., *Makromol. Chem., 7*, 46 (1951).
7. Hellman M., Bilbo, A. J., and Pummer, W. J., *J. Am. Chem. Soc., 77*, 3650 (1955).
8. Tamao, K., Sumitani, K., Kiso, Y., Zembayashi, M., Fujioka, A., Kodama, S., Nakajima, I., Minato, A., and Kumada, M., *Bull. Chem. Soc. Jpn.*, *49*, 1958 (1976).
9. Tamao, K., Kodama, S., Nakajima, I., and Kumada, M., *Tetrahedron, 38*, 3347 (1982).
10. Yamamoto, T., Hayashi, Y., and Yamamoto, A., *Bull. Chem. Soc. Jpn.*, *51*, 2091 (1978).
11. Suzuki A., *Acc. Chem. Res., 15*, 178 (1982).
12. Martin, A. R., and Yang, Y., *Acta Chem. Scand., 47*, 221 (1993).
13. Alipantis, A. O., and Canary, J. W., *J. Am. Chem. Soc., 116*, 6985 (1994).
14. McCarthy, T. F., Witteler, H., Pakula, T., and Wegner, G., *Macromolecules, 28*, 8350 (1995).
15. Tanigaki, N., Masuda, H., and Kaeriyama, K., *Polymer, 38*, 1221 (1997).
16. Child, A. D., and Reynolds, J. R., *Macromolecules, 27*, 1975 (1994).
17. Colon, I., and Kelsey, D. R., *J. Org. Chem., 51*, 2627 (1986).
18. Chaturvedi, V., Tanaka, S., and Kaeriyama, K., *Macromolecules, 26*, 2607 (1993).
19. Kaeriyama, K., Mehta, M. A., Chaturvedi, V., and Masuda, H., *Polymer, 36*, 3027 (1995).
20. Wang, Y., and Quirk, R. P., *Macromolecules, 28*, 3495 (1995).
21. Fauvarque, J.-F., Petit, M.-A., Digua, A., and Froyer, G., *Makromol. Chem., 188*, 1833 (1987).
22. Yamamoto, T., Morita, A., Miyazaki, Y., Maruyama, T., Wakayama, H., Zhou, Z., Sasaki, S., and Kubota, K., *Macromolecules, 25*, 1214 (1992).
23. Marvel, C. S., and Hartzell, G. E., *J. Am. Chem. Soc., 81*, 448 (1959).
24. Ballard, D. G., Courtis, A., Shirley, I. M., and Taylor, S. C., *Macromolecules, 21*, 294 (1988).
25. McKean, D. R., and Stille, J. K., *Macromolecules, 20*, 1787 (1987).
26. Gin, D. L., Conticello, V. P., and Grubbs, R. H., *Macromolecules, 116*, 10507 (1993).
27. Gin, D. L., Conticello, V. P., and Grubbs, R. H., *Macromolecules, 116*, 10934 (1993).
28. Gin, D. L., Avlyanov, J. K., and MacDiarmid, A. G., *Synth. Met., 66*, 169 (1994).
29. Kovacic, P., and Oziomek, J., *Macromolecular Synthesis* (J. A. Moore, ed.), John Wiley and Sons, New York, 1977, pp. 109.
30. Kovacic, P., and Oziomek, J., *J. Org. Chem., 29*, 100 (1964).
31. Froyer, G., Maurice, F., Bernier, P., and McAndrew, P., *Polymer, 23*, 1103 (1982).
32. Kobryanskii, V. M., and Arnautov, S. A., *J. Chem. Soc. Chem. Commun., 1992*, 727 (1992).
33. Ueda, M., Abe, T., and Awano, H., *Macromolecules, 25*, 5125 (1992).

34. Goldenberg, L. M., and Lacaze, P. C., *Synth. Met.*, *58*, 271 (1993).
35. Rubinstein, I., *J Polym. Sci., Polym. Chem. Ed.*, *21*, 3035 (1983).
36. Goldenberg, L. M., Roschumpkina, O. S., and Titkov, A. N., *Synth. Met.*, *44*, 107 (1991).
37. Kaeriyama, K., Sato, M., Someno, K., and Tanaka, S., *J. Chem. Soc., Chem. Commun.*, *1984*, 1199 (1984).
38. Ohsawa, T., and Yoshino, K., *Synth. Met.*, *17*, 601 (1987).
39. Moustafid, T. E., Aeiyach, S., Aaron, J. J., Mir-Hedayatullah, H., and Lacaze, P. C., *Polymer*, *32*, 2461 (1991).
40. Taha-Bouamri, K., Aaron, J. J., Aeiyach, S., and Lacaze, P. C., *J. Chem. Soc., Chem. Commun.*, *1994*, 777 (1994).
41. Fauvarque, J. F., Petit, M. A., Digua, A., and Froyer, G., *Makromol. Chem.*, *188*, 1833 (1987).
42. VanKerckhoven, H. F., Gilliams, Y. K., and Stille, J. K., *Macromolecules*, *5*, 541 (1972).
43. Percec, V., Okita, S., and Weiss, R., *Macromolecules*, *25*, 1816 (1992).
44. Percec, V., Base, J. Y., Zhao, M., and Hill, D. H., *Macromolecules*, *28*, 6726 (1995).
45. Stern, E. S., and Timmons, C. J., *Electronic Absorption Spectroscopy in Organic Chemistry*, Edward Arnold Ltd., London, 1970.
46. Tieke, B., Bubeck, C., and Lieser, G., *Makromol. Chem., Rapid Commun.*, *3*, 261 (1982).
47. Aeiyach, S., Soubiran, P., Lacaze, P. C., Froyer, G., and Pelous, Y., *Synth. Met.*, *68*, 213 (1995).
48. Shacklette, L. W., Chance, R. R., Sowa, J. M., Ivory, D. M., Miller, G. G., and Baughman, R. H., *Synth. Met.*, *1*, 307 (1979).
49. Pradere, P., and Boudet, A., *J. Mater. Sci. Lett.*, *7*, 10 (1989).
50. Shteinberg, V. G., Shumm, B. A., Zueva, A. F., and Efimov, O. N., *Synth. Met.*, *66*, 89 (1994).
51. Baughman, R. H., Shacklette, L. W., Murthy, N. S., Miller, G. G., and Elsenbaumer, R. L., *Mol. Cryst. Liq. Cryst.*, *118*, 253 (1985).
52. Bredas, J. L., Chance, R. R., and Baughman, R. H., *J. Chem. Phys.*, *76*, 3673 (1982).
53. Goldenberg, L. M., and Lacaze, P. C., *Synth. Met.*, *53*, 271 (1993).
54. Aboulkassim, A., and Chevort, C., *Polymer*, *34*, 401 (1993).
55. Meerholz, K., and Heinze, J., *Angew. Chem. Int. Ed. Engl.*, *29*, 692 (1990).
56. Child, A. D., Sankaran, B., Larmat, F., and Reynolds, J. R., *Macromolecules*, *28*, 6571 (1995).
57. Morita, M., Komaguchi, K., Tsutsumi, H., and Matsuda, Y., *Electrochim. Acta*, *37*, 1093 (1992).
58. Jow, T. R., and Shacklette, L. W., *Electrochem. Soc.*, *136*, 1 (1989).
59. Zhang, F. G., and Shäfer, F. P., *Appl. Phys.*, *B26*, 211 (1983).
60. Höffer, W., Schieder, R., Telle, H., Raue, R., and Brinkwerth, W., *Opt. Commun.*, *33*, 85 (1980).
61. Furumoto, H. W., and Ceccon, H. L., *IEEE J. Quantum Electron.*, *QE-6*, 262 (1970).
62. Helfirch, W., and Schneider, W. G., *Phys. Rev. Lett.*, *14*, 229 (1965).
63. Tang, C. W., and Vanslyke, S. A., *Appl. Phys. Lett.*, *51*, 913 (1987).
64. Jenekhe, S. A., and Osaheni, J. A., *Nature*, *265*, 765 (1994).
65. Grem, G., Leditzky, G., Ullrich, B., and Leising, G., *Adv. Mater.*, *4*, 36 (1992).
66. Grem, G., and Leising, G., *Synth. Met.*, *55–57*, 4105 (1993).
67. Yang, Y., Pei, Q., and Heeger, A. J., *J. Appl. Phys.*, *79*, 934 (1996).
68. Jing, W. X., Kraft, A., Moratti, S. C., Grüner, J., Cacialli, F., Hamer, P. J., Holmes, A. B., and Friend, R.H., *Synth. Met.*, *67*, 161 (1994).
69. Grüner, J., Wittmann, H. F., Hamer, P. J., Friend, R. H., Huber, J., Scherf, U., Müllen, K., Moratti, S. C., and Holmes, A. B., *Synth. Met.*, *67*, 181 (1994).

70. Kreyenschmidt, M., Uckert, F., and Müllen, K., *Macromolecules*, *28*, 4577 (1995).
71. Ohmori, Y., Uchida, M., Muro, K., and Yoshino, K., *Jpn. J. Appl. Phys.*, *30*, L1941 (1991).
72. Kaeriyama, K., Tsukahara, Y., Negoro, S., Tanigaki, N., and Masuda, H., *Synth. Met.*, *84*, 263 (1997).
73. Remmers, M., Schulze, M., and Wegner, G., *Macromol. Rapid Commun.*, *17*, 239 (1996).
74. Mesfeldt, J. L., Reynolds, J. R., Tanner, D. B., Ruiz, J. P., Wang, J., and Pomerantz, M., *J. Polym. Sci., Part B. Polym Phys.*, *32*, 2395 (1994).
75. Kovacic, P., and Kyriakis, A., *J. Am. Chem. Soc.*, *85*, 454 (1963).
76. Stille, J. K., and Gilliams, Y., *Macromolecules*, *4*, 515 (1971).
77. Rietveld, H. M., Maslen, E. N., and Cleys, C. J. B., *Acta Crystallogr., Sect. B: Struct. Crystallogr. Cryst. Chem.*, *B26*, 693 (1970).
78. Kawaguchi, A., and Petermann, J., *Mol. Cryst. Liq. Cryst.*, *133*, 189 (1986).
79. Dumais, J. J., Jelinski, L. W., Galvin, M. E., Dybowski, C., Brown, C. E., and Kovacic, P., *Macromolecules*, *22*, 612 (1989).
80. Baker, K. B., Fratini, A. V., Resch, T., Kanchel, H. C., Adams, W. W., Socci, E. P., and Farmar, B. L., *Polymer*, *34*, 1571 (1993).
81. Wittmann, J. C., and Smith, P., *Nature*, *352*, 414 (1991).
82. Tanigaki, N., Yase, K., and Kaito, A., *Mol. Cryst. Liq. Cryst.*, *267*, 335 (1995).
83. Wittler, H., Lieser, G., Wegner, G., and Schulze, M., *Makromol. Chem., Rapid Commun.*, *14*, 471 (1993).
84. MacCarthy, T. F., Whittler, H., Pakula, T., and Wegner, G., *Macromolecules*, *28*, 8350 (1995).
85. Vahlenkamp, T., and Wegner, G., *Macromol. Chem. Phys.*, *195*, 1933 (1994).
86. Rulkens, R., Schulze, M., and Wegner, G., *Macromol. Rapid Commun.*, *15*, 669 (1994).
87. Fahlman, M., Rasmusson, J., Kaeriyama, K., Clark, D. T., Beamson, G., and Salaneck, W. R., *Synth. Met.*, *66*, 123 (1994).

3

Synthesis and Properties of Poly(phenylene vinylene)s and Related Poly(arylene vinylene)s

Frank R. Denton III
Motorola, Inc.
Lawrenceville, Georgia

Paul M. Lahti
University of Massachusetts
Amherst, Massachusetts

I. INTRODUCTION AND SCOPE

The poly(arylene vinylene) (PAV) family of conjugated polymers has been of considerable interest for over a century. Computerized literature searches citing poly(1,4-phenylene vinylene) (PPV) and its derivatives alone find thousands of articles. As a result, it is impossible to give a comprehensive review of all PAV and PPV related work in a single book chapter. We shall therefore limit ourselves to discussing the synthetic methodologies which lead to PPV formation, and an overview of some of the properties of PAVs which have made them so promising and widely studied, especially during the past two decades. The focus will be on PPV derivatives, although we will also describe some of the PAVs that have shown considerable promise. In general, we shall also limit discussion to polymers without meta-connectivity. The reader who is interested in specific details is referred to the original literature, and to various specialty reviews which are cited within this chapter.

II. SYNTHESIS OF PAVs

A. Early Synthetic Routes to PAVs

PPV—or, more formally, poly(1,4-phenylene-1,2-ethenediyl)—has apparently been made by various methods for over a century, although not always recognized as such. As early as 1870, Grimaux obtained an insoluble, infusible, yellow precipitate while trying to obtain diethers from α,α'-dichloro- or dibromo-*para*-xylene by

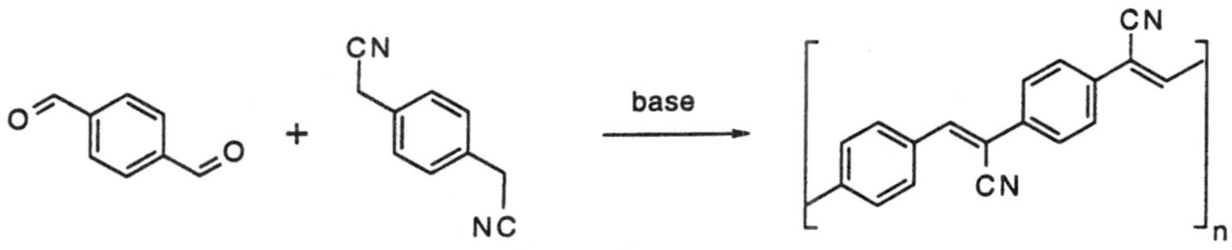

SCHEME 1

treatment with hot aqueous base [1]. In 1904, Thiele and Balhorn obtained yellow insoluble crusts when attempting to reduce $\alpha,\alpha,\alpha,\alpha',\alpha',\alpha'$-hexabromo-*para*-xylene with mercury [2]. In 1931, Quelet repeated Grimaux's work, finding an abundant yield of insoluble yellow condensation product when α,α'-dichloro-*p*-xylene was treated with KOH in 50% aqueous ethanol [3]. Due to a lack of structural analytical methods, these workers did not identify their polymeric products as PPV, but a comparison to modern synthetic methodologies cited below, and the yellow, insoluble nature of the products obtained make highly likely the assumption that somewhat impure forms of PPV were obtained by these workers.

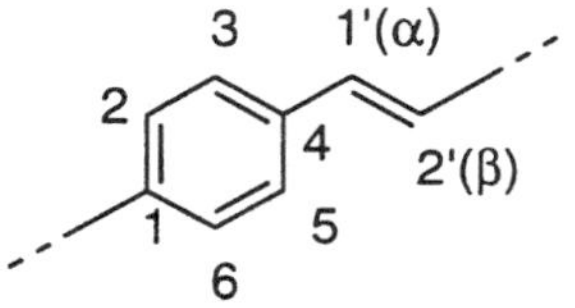

Less serendipitous syntheses of PPV came with deliberate efforts to make polymeric materials. In 1960, Lenz and Handlovitz reported the synthesis of *alt*-PPV-co-poly(phenylene α-cyanovinylene) by a Knoevenagle reaction (Scheme 1) [4]. In the same year, McDonald and Campbell used Wittig chemistry to couple terephthaladehyde and a bis-phosphonium salt of *para*-xylene to obtained intensely yellow, insoluble PPV [5] with a degree of polymerization (DP) of 9 (Scheme 2). These efforts are often considered as ushering in the modern era of PAV syntheses. Some excellent reviews summarize such pioneering synthetic routes to highly conjugated polymers [6–14].

Subsequently, several variants of the Wittig reaction were incorporated into PPV synthesis. Lapitskii et al. prepared a series of poly(phenylene α,ω-polyenes) with up to five double bonds in the polyene units [15]. Interestingly, this group effectively doped the polymers by treating them with concentrated sulfuric acid:

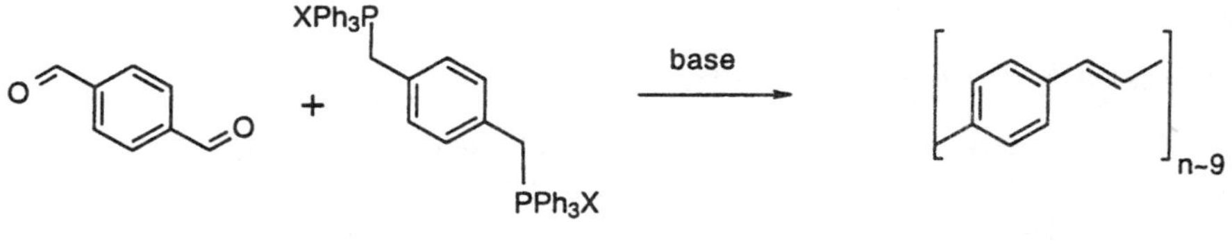

SCHEME 2

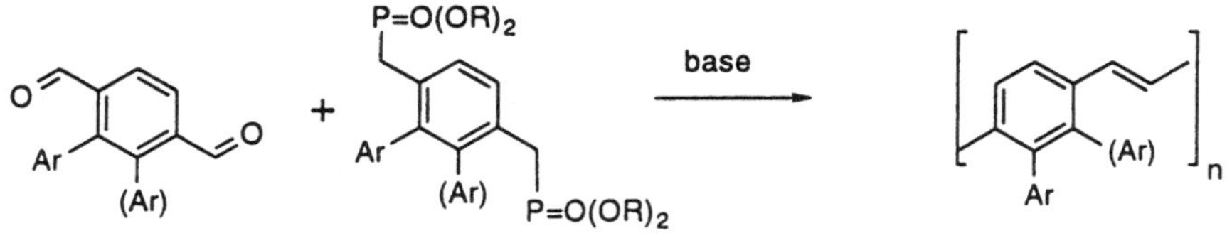

SCHEME 3

They noted the deepening of color, but reported no conductivity measurements. The dramatic benefits of doping were left for polyacetylene workers to discover 10 years later. Others used the Horner–Wittig modification with dialkyl phosphonate esters instead of phosphonium salts to aid solubilization of the growing PPV chain during the reaction. Hörhold's group in East Germany was particularly active in this area, using the synthesis for controlled construction of oligomers and generation of PPVs with aryl-pendant or vinyl-pendant phenyl rings [16–18]: The phenylated PPVs were soluble due to chain deplanarization, an unusual property for the PPVs of that time (Scheme 3). A Wittig variant using diphenyl phosphoxides was also developed by Hörhold and colleagues [19,20].

Using condensation of phosphonium ylides with α,α'-dihydroxyl-α,α'-disulfonato-*para*-xylene salts, Kossmehl claimed yields competitive with the standard Wittig reaction for trans PPVs [7,21]. In addition, this group converted PPV to the all-trans form by boiling in aromatic solvents with iodine, finding that the conductivity of all-trans PPV exceeded that of mixed cis/trans materials (Scheme 4) [22,23]. Subsequently, Kossmehl and Yaridjanian showed that PPV could be made in moderate yield by Siegrist condensation of *para*-tolualdehyde or its anilimino analog with potassium *tert*-butoxide in dimethylformamide; furan and thiophene PAV analogs could be made in somewhat better yields (Scheme 5) [24]. A patent by Thomson similarly showed that phenyl dialdehydes would condense easily with 2,4-dinitro-*meta*-xylene to give polymer with a DP of about 20, giving products that are stable at temperatures up to 400°C (Scheme 6) [25].

In 1965, Moritani et al. described a synthesis of PPV both from α-elimination of bis-halomethylated xylenes, and from bis-diazomethyl arenes (Scheme 7) [26]. The carbenoid conditions gave some PPV, but H-atom abstraction dominated the products of the diazo compound decompositions. Ouchi did similar studies with these carbenoids [27]. Nagai et al. reported reacting the α,α'-diphenyl derivative of bis-diazo-p-xylene in sulfur dioxide [28] to generate phenylated PPV oligomers of DP up to 4. These and related decompositions of diazo compounds [29,30] were

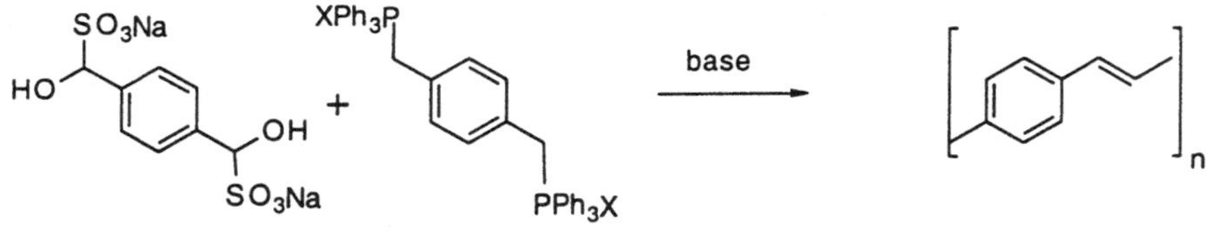

SCHEME 4

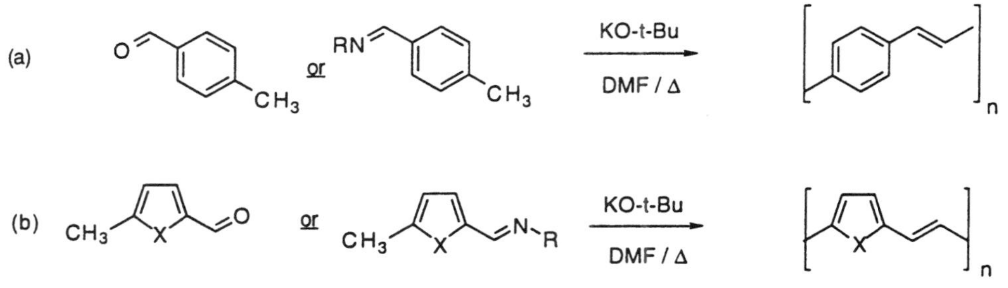

SCHEME 5

intriguing strategies to attempt PPV synthesis, but the resulting material typically suffers from low molecular weight, numerous structural defects such as azine units, and low yields due to competing side reactions.

A useful PPV condensation method has used chromium(II) coupling of chlorides. Hoyt et al. [31] treated α,α'-dichloro-xylenes with $CrCl_2$ to get PPVs with inherent viscosities up to 2.9 (Scheme 8). Hörhold et al. used chromium diacetate in aprotic solvents to obtain soluble phenylated PPVs with apparent molecular weights up to 80,000 [11,32–34]. This chemistry has persisted to more recent usage, as shown by a synthesis of extremely heat-resistant perchlorinated PPVs (Scheme 9) as well as stable perchloroxylylenes via coupling of the appropriate monomer with $SnCl_2$ or $FeCl_2$ [35].

The classification of what we have called "early" syntheses is arbitrary, but in our minds it is constituted by the development of syntheses aimed specifically at soluble, processible PPV derivatives. Hörhold and his co-workers already pursued this strategy in some of the work cited above, but many early workers had to be satisfied with investigating the properties of powdered, insoluble PPV samples, or low-DP oligomeric model systems. This reduced by far the potential utility of the materials for electronic and structural purposes. For this reason, a number of specific methods in current use for synthesis of processible PPVs are separately described in more detail below.

B. Condensation Polymerizations Yielding PAVs

Many (or even most) of the reactions cited above as "early" work involved use of condensation chemistry. For example, modern Wittig methods for PAV synthesis and chromium(II) coupling are essentially the same as those in the above-cited

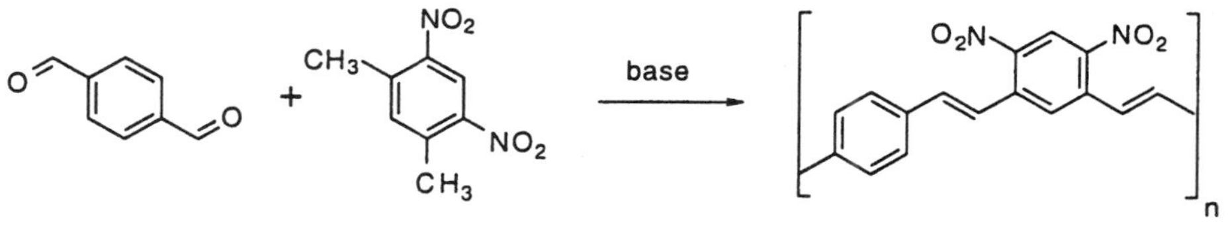

SCHEME 6

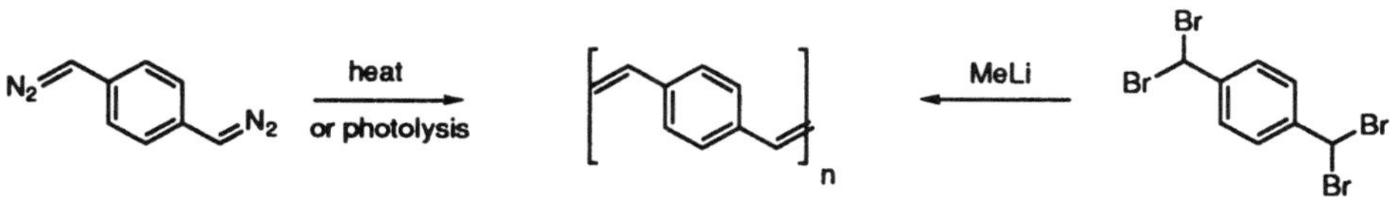

SCHEME 7

work. We, therefore, will focus only on a few, recent examples of the use of this method to make more soluble PAVs of specific interest, or to make PAVs using more recently developed condensation reactions.

Recent Knoevenagel condensations are exemplified by the synthesis of a soluble alternating PPV copolymer incorporating cyano-groups on the ethylenic units (Scheme 10) [36]. This polymer is soluble because of its long alkoxy side chains, as opposed to the insoluble material obtained [4] from the early example shown in Scheme 1. Due to its processibility, this polymer has found use as a component in the engineering of light-emitting diodes [36].

The McMurry condensation of aldehydes and ketones under reducing conditions in the presence of titanium(III) for PPV synthesis apparently was first reported in 1980 [37]. In a generic example, Rajaraman et al. allowed terephthalaldehyde to react with $TiCl_3$ in tetrahydrofuran for 5 h at 70–80°C under nitrogen, and found that the polymer properties obtained were identical to those described in previous preparations of all-trans PPV (Scheme 11). Feast and Millichamp applied this method to the synthesis of poly(4,4′-biphenydiyl 1,2-diphenylvinylene) with considerably longer reaction times, and claimed molecular weights of 75,000 by gel permeation chromatography [38]. Hexyl-substituted PPV was made via the McMurry reaction by Rehaln et al. and was found to be soluble in organic solvents, with a molecular weight of about 8500 [39]. The McMurry synthesis of PAVs has also been scrutinized mechanistically using terephthalaldehyde or 2,5-furandicarboxaldehyde with a $TiCl_4$/Zn coupling system [40].

C.   Soluble Poly(*para*-xylylidene-α-halide) Precursor Synthesis

Soluble precursor routes to conjugated polymers aim at making nonconjugated, flexible precursor polymers which can be processed. Secondary treatment of the precursor polymer—typically by heating or photolysis—yields the final conjugated material as fibers, bubbles, films, or patterns. The precursor polymer may then be

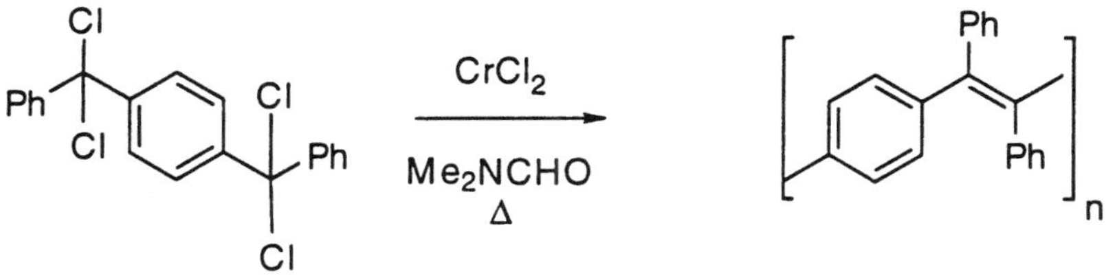

SCHEME 8

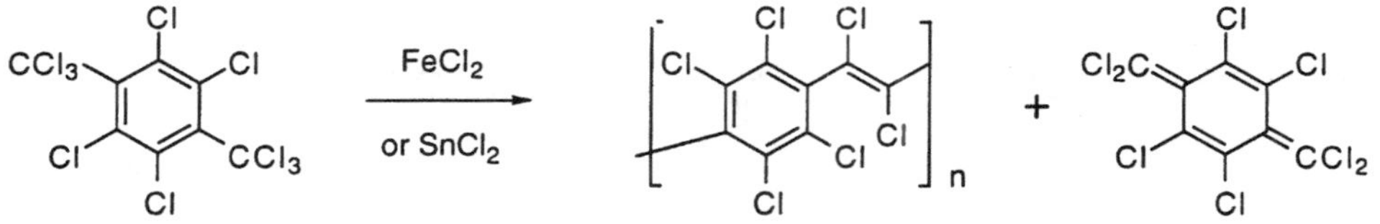

SCHEME 9

dissolved away to leave the conjugated polymer, which in some cases may itself be soluble with appropriate substitution.

A useful soluble precursor route to PAVs is the polymerization of 1,4-bis(chloromethyl)arenes by treatment with about 1 eq. of potassium *tert*-butoxide in nonhydroxylic solvents like tetrahydrofuran. The analogous reaction with 1,4-bis(bromomethyl)arenes can also be used. This methodology was used by Gilch and Wheelwright [41] as one of the more successful early PPV syntheses, elaborated fairly extensively by Hörhold and co-workers [34,42], then recently applied to synthesis of high-molecular-weight, highly phenylated PPVs and dubbed the CPR (chlorine precursor route) or DHCL (dehydrochlorination) method [44–46]. When excess base is avoided, the product of CPR synthesis is a soluble side-chain chloro precursor polymer, which may be thermolytically eliminated to give a conjugated PPV (Scheme 12). As found by early workers, this method suffers from early product precipitation in the absence of solubilizing side chains on the arene ring, but otherwise has an advantage in producing a precursor polymer which is soluble in nonhydroxylic organic solvents and therefore useful for electronic applications that require processing [43]. For example, phenylated PPVs [34,42,44–46] and PPVs with large solubilizing groups on the aryl ring such as cholestanoxy [47] are readily made by CPR methodology.

The mechanism of the CPR method appears not to have been studied up to this point. Presumably, a transient ylid is formed by removal of a benzylic proton by base. Although we are not aware that a *para*-xylylene has been directly observed in this chemistry, it seems plausible that one is formed (Scheme 13), given the reaction's similarity to the Wessling reaction described in the following section. The swift formation of high-molecular-weight material is consistent with radical polymerization of a putative *para*-xylylene, but, at present, this assertion must remain speculative. So long as excess base and raised reaction temperatures are avoided, CPR precursor polymers are produced which are reasonably processible

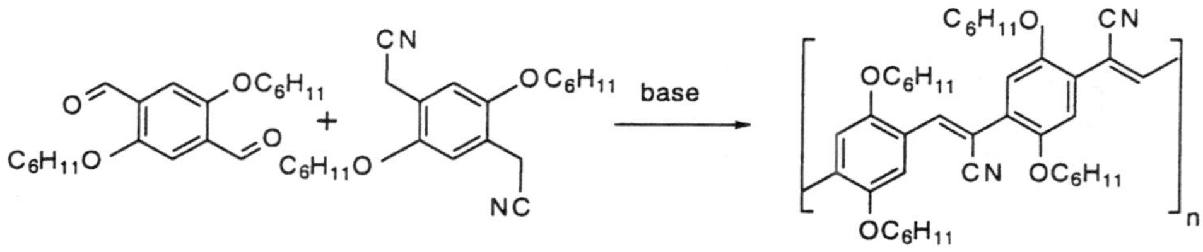

SCHEME 10

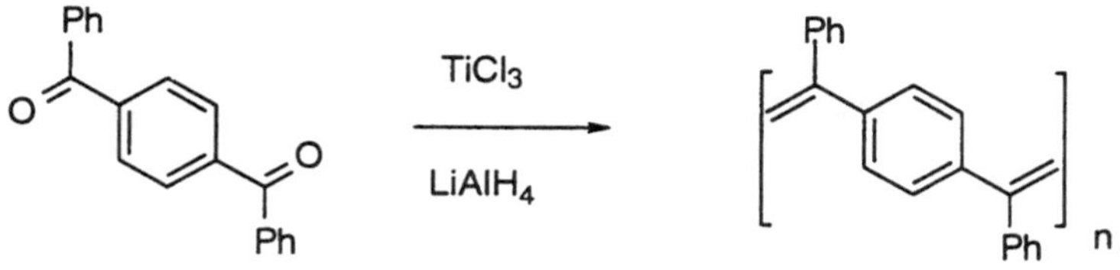

SCHEME 11

in organic solvents and which, upon elimination at 150–300°C, give rise to low-defect, high-molecular-weight PPVs. Although the sulfonium precursor polymer route described in the following section has been more extensively used than the CPR method—in part due to the greater latitude in substituents permitted by the latter method—the substantially defect-free nature of the PPVs produced by the CPR method makes it very attractive in electronic applications [43].

D. Soluble Dialkylsulfonium Polymer Precursor Synthesis

In the 1960s, Wessling and co-workers at Dow Chemical [48,49] developed one of the most important soluble precursor routes to PAVs, based on aqueous solvent synthesis of poly(*para*-xylylene-α-dialkylsulfonium halides) from α,α'-bis(dialkyl sulfonium salts), followed by thermolytic formation of the final conjugated polymer (Scheme 14). Virtually the same process was independently discovered in the same time period by Kambe and Okawara, using tetrafluoroborate sulfonium salts, although the elevated temperature used in this work produced insoluble products [50,51]. Hörhold's group [13] later extended this route to the synthesis of substituted PPVs. In the 1980s, other research groups [52–55] further improved the methodology to allow process-controllable synthesis of various substituted PAV–precursor polymers which could be thermally eliminated to give PPVs and PAVs of quality sufficient to allow them to be used in optical and electronic devices. The result of about two decades of work in this area is a synthesis that can readily give multigram quantities of processible precursor polymers, allowing synthesis of PAVs with a wide variety of substituents under several processing environments. As a result, the sulfonium precursor methodology has been so extensively used in recent times that variants of this methodology constitute the largest portion of the synthetic methods reviewed in this chapter.

The sulfonium precursor synthesis of PPVs involves a reactive, substituted *para*-xylylene (also known as a *para*-benzoquinodimethane) intermediate which

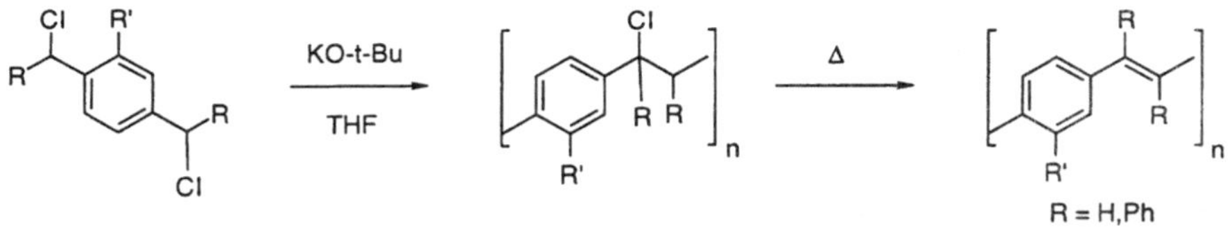

SCHEME 12

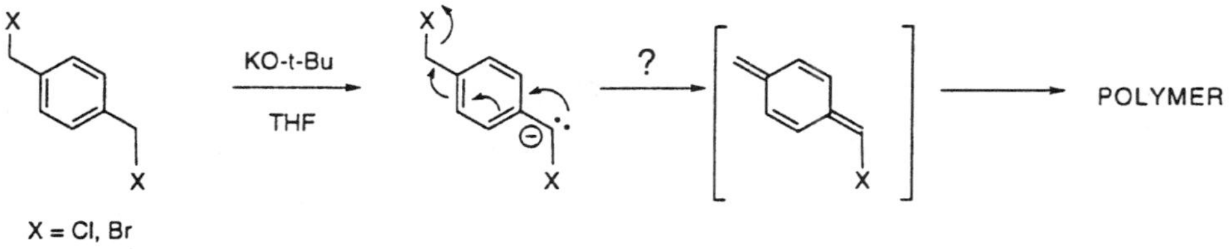

SCHEME 13

polymerizes to give water- or methanol-soluble poly(*para*-xylylene-α-dialkyl-sulfonium halide) (PXD), a polyelectrolyte. In a typical case, a 1,4-bis (dialkylsulfoniomethyl)benzene compound is treated with hydroxide to give 1,6-elimination to an exocyclic sulfonium substituted *para*-xylylene. The xylylene then polymerizes to polyelectrolyte PXD with molecular weights of 10,000 to >1,000,000, which may be precipitated or dialyzed to give typical yields of about 20% high-molecular-weight faction PXD. The polyelectrolyte may also be subjected to anionic exchange, with consequent changes in its solubility properties [49]. Below, we shall return to issues concerning the final elimination step; for now, we will concentrate on the initial, PXD-forming stages of the methodology, as these stages primarily determine the makeup of the final polymer.

To maximize the molecular weight and overall yield of high polymer, it is useful to understand and control this complex mechanistic process. We have carried out a number of mechanistic studies, aimed at improving PAV syntheses. Reviews of the various mechanisms that have been proposed for the Wessling process have been given elsewhere [56,57]. At present, a fairly clear picture has emerged for formation of high-molecular-weight PXDs, which is summarized in Fig. 1.

Overall, a *para*-xylylene-type structure must be formed for this polymerization to work well [58]. Both ultraviolet-visible (UV-vis) and nuclear magnetic resonance (NMR) evidence directly supports the formation of *para*-xylylenes in the Wessling process [57]. Other quinonoidal intermediates appear plausible (Scheme 15), but high-molecular-weight PXDs, in general, do not form without the possibility of 1,4-xylylene formation, with very few possible exceptions [59].* At least one *para*-xylylene—an anthracene analog—does not undergo Wessling polymerization for

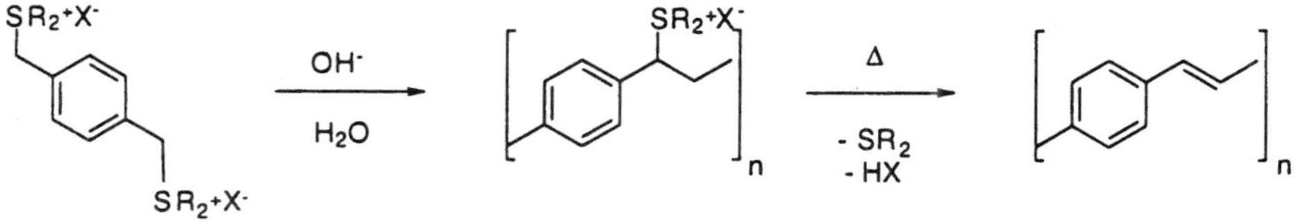

SCHEME 14

---

*For a very unusual case where polymers were claimed when no *para*-xylylene is possible, see Ref. 59. In addition, Ref. 54 reports that 2,6-bis(dimethylsulfoniomethyl)naphthalene polymerizes, whereas Ref. 58 describes failure to polymerize this sort of precursor.

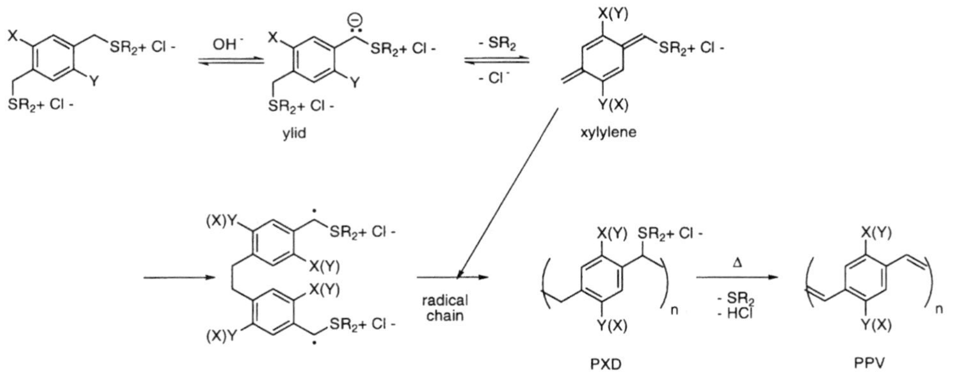

FIGURE 1 Mechanistic processes for sulfonium precursor synthesis of polyphenylenevinylenes (PPVs).

steric reasons (Scheme 15d) [60], although the polymer can be made via an unrelated Heck coupling method [61] that is described in a subsequent section. Manipulation of the equilibria that lead to xylylene formation helps to optimize PXD formation. Both Wessling [49] and Garay and Lenz [58] showed that the use of an aqueous immiscible cosolvent to remove dialkylsulfides from the reaction mixture can increase the yield and DP of PXDs. Denton et al. have shown that adjustment of solvent conditions to maximize xylylene formation in a UV-vis assay allows the identification of solvent systems that will optimize PXD formation [62].

In addition to the PPVs formed from *para*-xylylenes, it is clear that a variety of PAVs can be formed by analogous routes. Related intermediates may be observed as transient species by UV-vis spectroscopy; examples are shown in Fig. 2 [62,63].

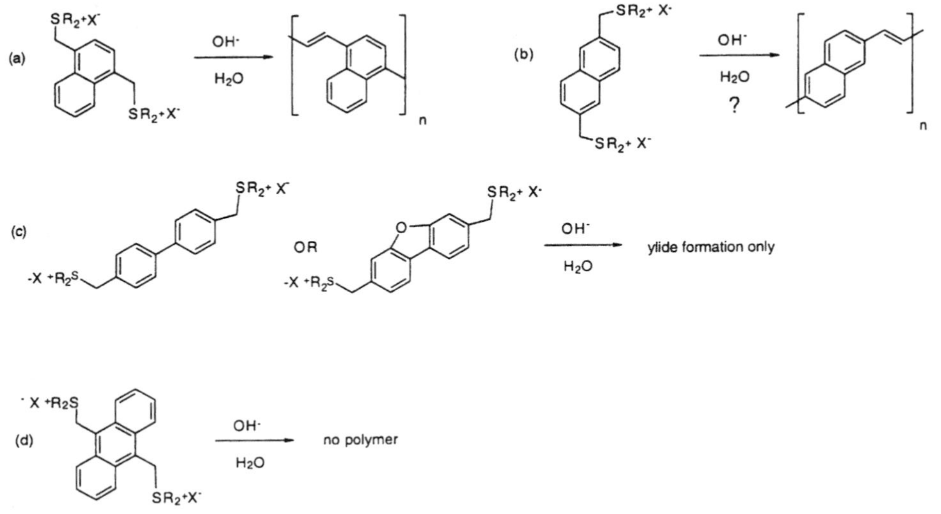

SCHEME 15

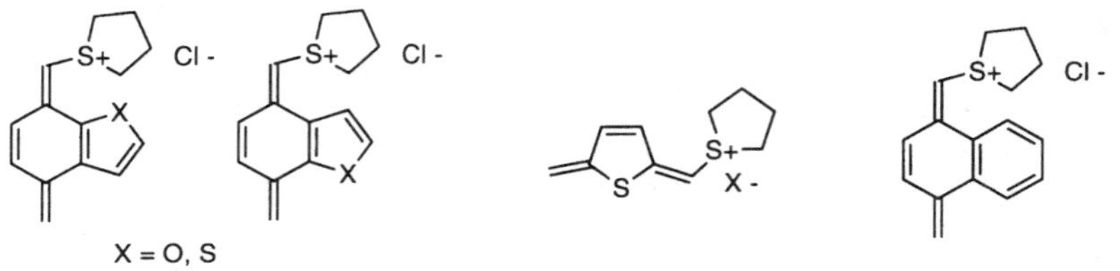

FIGURE 2 Examples of *para*-xylylene analogs generated during PAV synthesis.

The 1,4-relationship of the exocyclic methylene groups appears to be important in these intermediates, as for the xylylenes.

High-molecular-weight material is formed very quickly, within the first few minutes of the reaction. A radical chain process seems most plausible in such a situation. Work by Denton et al. [64] supports the radical chain process in the formation of high-molecular-weight PXD, with the finding that various radical inhibitors limit or prevent the formation of long PXD chains, without affecting formation of the intermediate xylylenes. Denton also found that the rate of xylylene consumption is not significantly altered by addition of the radical inhibitors, ruling out any effect of these additives on the preequilibria that form the xylylene. Analogously, the inhibiting effect of oxygen on the polymerization was noted quite early by Wessling himself [49], in that polymer solution viscosity is much decreased unless reaction solutions are thoroughly purged by inert gases.

It is not clear whether alternative mechanisms may occur in the synthesis substituted PXDs. Both the yield and DP of PXDs can be strongly affected by ring substitution. Whereas ring substitution with electron donors such as alkoxy groups is permissive of high-DP PXDs, substitution with strong electron donors decreases both yields and DPs [63]. It has been argued that an anionic chain mechanism may be an alternative mechanism in the latter cases, such as a case where the aryl ring of the xylylene bears a cyano substituent [65]. Substituents which interfere with the sequence of deprotonation/elimination steps in the overall mechanism can inhibit this polymerization; for example, synthesis of amino-substituted PPVs was problematic for some time, until Stenger-Smith et al. [66] synthesized 2-(*N,N*-dimethylamino)-PPV, **1**.

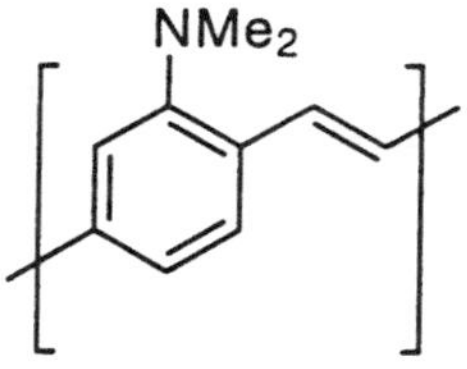

1

Work by Denton et al. supports the radical chain mechanism of Fig. 1, but has not unequivocally identified the initiation process [64]. Wessling [49] suggested head-to-head dimerization of the intermediate xylylene as shown in Fig. 1, followed by regiospecific, radical chain growth. This is similar to the bond-forming initiation polymerization mechanism described by Hall et al. [67–69]. Although there is no direct evidence for this initiation process in the Wessling process, the very small concentration of propagating chains in the reaction is consistent this proposal, and we are unaware of a more compelling argument for alternative initiation processes for PXD formation.

A wide variety of substituents are tolerated by the soluble sulfonium precursor route. Because the polyelectrolyte PXDs are inherently soluble in hydroxylic solvents, one need not use long-chain solubilizing substituents to obtain substantial DPs, as is required for the CPR method. Electron-donor substituents on the aryl ring (X and Y in Fig. 1) of the 1,4-bis(dialkylsulfoniomethyl)benzene reactants yield PPVs with fairly high DPs by this method: Examples are alkoxy [70–72], alkyl [73], and aryl [74] groups. Use of large side chains disrupt the normal PPV crystal packing and give PAVs which are soluble in organic solvent after elimination, as shown by Wudl and co-workers for polymers such as poly(2-methoxy-5-ethylhexyloxy)-1,4-phenylene vinylene) MEH–PPV, which can be made by both CPR and sulfonium methods (Scheme 16) [75,76], although the higher-molecular-weight fraction is insoluble [47] by the CPR method. A number of 2,5-dihalo-PPVs and halo-PPVs have been made [48,63,77]. Strongly electron-deficient substituents tend to give PXDs which do not eliminate easily and which do not give large DPs, but which can still be used to give homopolymers and copolymers such as electron-poor 2,5-dicyano-PPV [78] and copolymers of 2-nitro-PPV with parent PPV [79]. An unusual example is the water-soluble PAV, poly(5-sulfopropoxy-2-methoxy-*para*-phenylene vinylene) sodium salt and its protic analog, **2–3**, which have been described as "self-doping" due to the effect of the counterions associated with the main polymer chains [80,81].

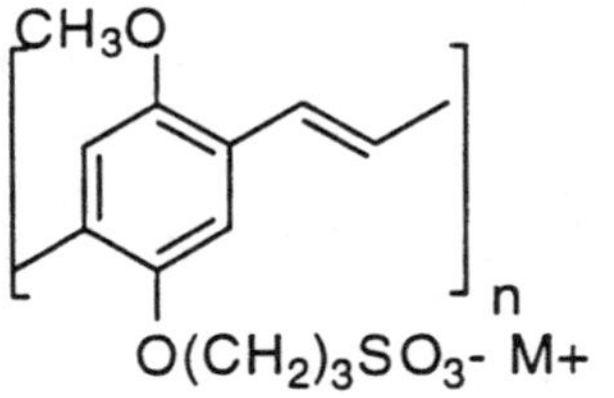

CH_3O

$O(CH_2)_3SO_3^- M^+$

2 (M=Na)
3 (M=H)

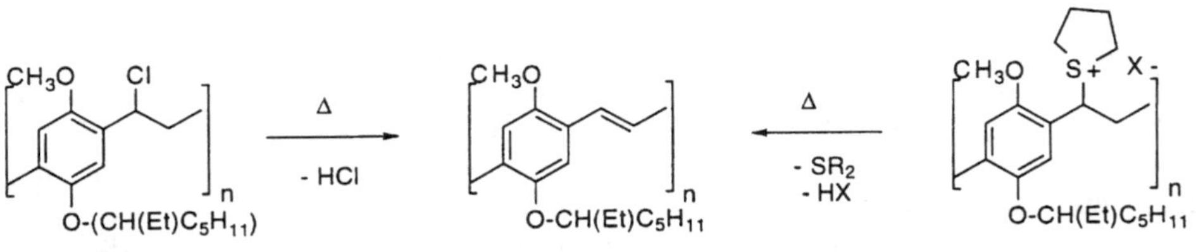

SCHEME 16

Asymmetrically substituted 1,4-bis(dialkylsulfoniumethyl)arenes are likely to yield regiorandomized PPVs, although there may be some preference shown for which benzylic proton is abstracted in the initial ylide-forming step of the Wessling polymerization (Fig. 1). UV-vis evidence is consistent with formation of more than one xylylene in cases where this is possible [62,65]. Small changes in the substituent seem capable of altering the degree of randomization, with a larger substituent apparently favoring greater regioselectivity. 2-Bromo-5-hexyloxy-PPV synthesized by the sulfonium precursor route appears to be largely regiorandom (Scheme 17a) [82], whereas a bromo precursor to 2-bromo-5-dodecyloxy-PPV that was made by a CPR-analogous method was found to be to be largely regiospecific (Scheme 17b) [83]. Numerous studies to date do not explicitly consider this problem of potential regiorandomness, but the matter could be quite important in efforts to control PAV morphology and specific electronic effects. One must therefore recognize that Wessling-based syntheses of these polymers may be subject to considerable regiochemical scrambling [82,84–86] and, therefore, are problematic for applications where regiochemically pure substitution patterns are desired in high-molecular-weight PPVs. We shall see later, fortunately, that other methods can address this problem to some extent.

In addition to the substituted aryl rings that can be incorporated in PPVs by the soluble precursor routes, it is also possible to use other aryl units, so long as they are derivable from *para*-xylylenes or their monocyclic analogs. Poly(2,5-thienylene vinylene) [87,88] poly(2,5-furan vinylene) (Scheme 18) [89] and poly(1,4-naphthalene vinylene) (Scheme 15a) [90] are examples. It also possible to make PAVs consisting of PPV with heterocyclic rings fused onto the arylene units by the precursor route (Schemes 19a and 19b) [91,92].

The strong preference of the Wessling polymerization for xylylene incorporation has rendered virtually nonexistent the copolymerization of PXD monomers with other alkenic monomers. PPV or PAV copolymers with other PPV/PAV-type monomers can be made by treating mixtures of 1,4-bis(dialkylsulfoniomethyl)arenes with base to obtain materials whose compositions seldom reflect the ratio of feed monomers (Scheme 20). For example, Lenz and co-workers published early studies on copolymerization of unsubstituted PPV with dialkoxy derivatives, and found that the alkoxy monomers were incorporated into the chains in approximately triple the proportion present in the feedstock [93–95]. Differential scanning calorimetry, thermogravimetric analysis, UV-vis spectroscopy, and Fourier transform infrared spectroscopy were employed to ascertain that the systems formed were copolymers and not simply blends [96]. Jin et al. have looked at various binary PPV copolymers based on differently substituted PPV monomers, with an eye to varying electronic spectral nature of the final polymers in a controlled fashion [79,97]. A complete survey of all the copolymer studies undertaken using the Wessling process is beyond the scope of this chapter, but the results of these studies concur in finding the product polymers to be a sensitive function of feedstock ratios, reaction solvents and conditions, and the combinations of substituents used in the feedstock mixtures.

Although the literature does not often consider this issue, the production of a regiochemically random distribution of monomers in the final copolymer is not necessarily clear, given the possibility of different reactivity ratios for different

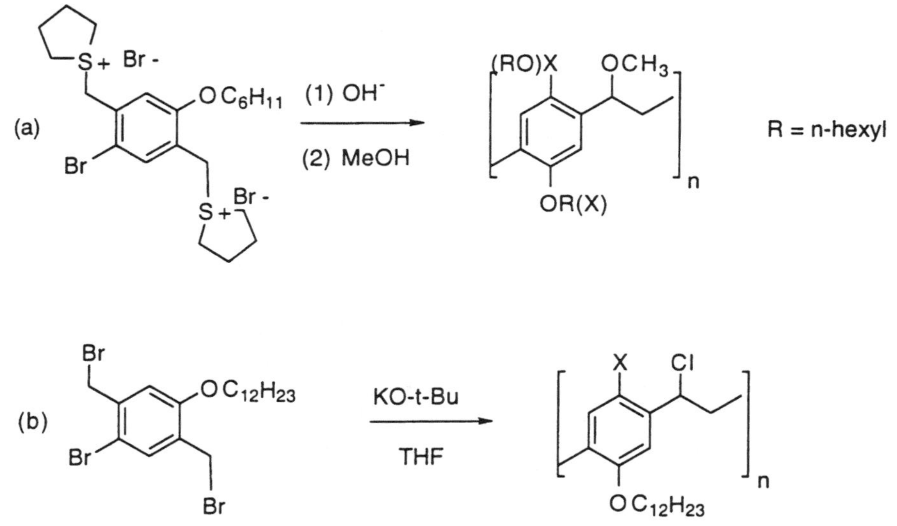

SCHEME 17

polymerizing *para*-xylylenes, and of end-polymerization in such cases. Sanborn and Lahti have found that either block or random copolymer distributions may be obtained in PXD copolymerizations of various 1,4-bis(dialkylsulfoniomethyl)arenes with ^{13}C-labeled, 1,4-bis(tetrahydrothiopheniomethyl)benzene, depending on reaction conditions (Scheme 21) [98]. In another clearly defined case, copolymers of 1,4-phenylene vinylene with 2,5-thienylene vinylene (PTV) showed infrared spectral behavior, clearly indicating the presence of copolymeric links, rather than homopolymer blend formation (Scheme 22a); however, the copolymer of 2,5-dimethyoxy-1,4-phenylene vinylene with 2,5-thienylene vinylene showed apparent blocky behavior (Scheme 22b) [99]. Small changes in mixing rates, concentration, and substituent choice conceivably can render quite difficult the reproducibility of copolymer compositions in this route.

All of the previous discussions in this section may seem to assume the virtually problem-free elimination of precursor PXDs to the final conjugated PPVs. The elimination process has been proposed to occur by both E1cB [100] and by E1 [101] mechanisms. Use of thermogravimetric analysis typically allows one to choose appropriate conditions for the elimination of the various pendant groups in the PXDs under vacuum or inert atmosphere, typically 180–280°C [102]. Films may be cast by heating between plates, or may even be oriented by stretching

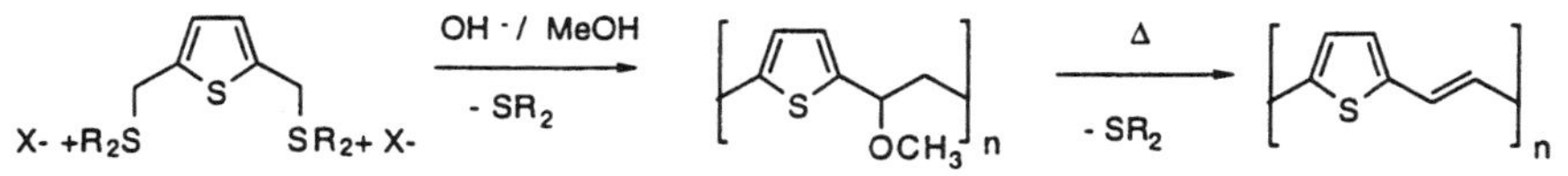

SCHEME 18

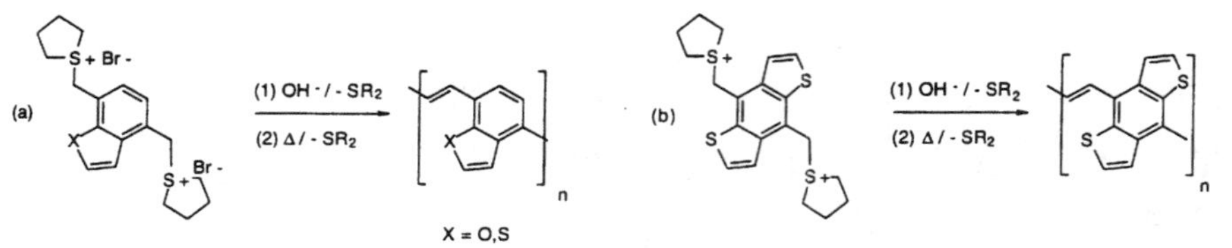

SCHEME 19

during the thermolysis process, when self-plasticization occurs as the elimination proceeds. Fibers may be formed and oriented by extrusion into hot-oil baths. Bubbles may even be blown into the eliminating PXD, leaving free-standing "foams" of PPV. Nominally, the process is quite easy to carry out, with variants for most of this methodology being described by Wessling's seminal paper [49].

However, there are complications with the elimination process. First of all, the by-products are typically both noxious and corrosively acidic, a drawback for large-scale productions. Second, elimination seldom occurs to completion, hence a varying level of deconjugating defects tends to remain in PPVs that are of nominally very high conjugation length (DP >1000). Incomplete elimination is partly a function of the fact that PXDs are fairly defective polymers in which sulfonium pendant groups are partially replaced by hydroxyl and chloride groups during the synthetic process, with the PXD additionally containing a considerable weight fraction in water of hydration for the sulfonium groups (Scheme 23) [102]. Although the chloro and hydroxy groups appear to be readily eliminated by heating, side reactions also produce sulfide pendant chains (Schemes 23 and 24), which appear to extrude much more slowly than the other leaving groups. If any pendant groups are not eliminated during precursor pyrolysis, they represent a source of deconjugating defects in the final PAVs [53].

The side reactions which form sulfide pendant groups (Scheme 24) complicate elimination to a considerable extent when simple dimethylsulfonium groups are used [53]; use of cyclic dialkylsulfonium bis-salts considerably reduces this problem [103–105]. The PPVs derived by thermolysis of PXDs with cyclic sulfonium pendant groups are electronically superior to those derived from thermolysis of dialkylsulfonium pendant groups, as there are fewer defective sites on the PXDs which are sluggish to eliminate [106]. The difficulty with sulfide formation can also be limited in cases where the sulfonium pendant groups may be replaced by

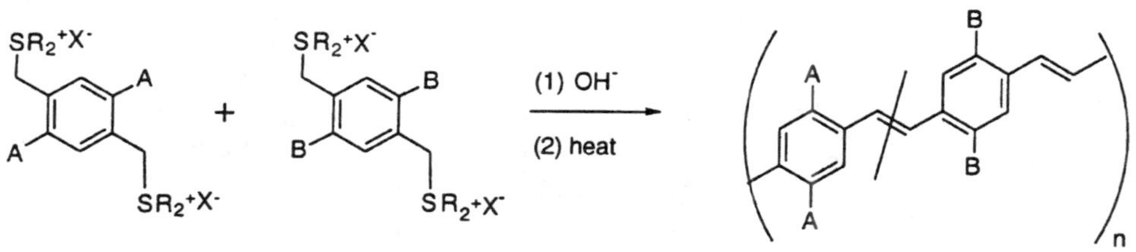

SCHEME 20

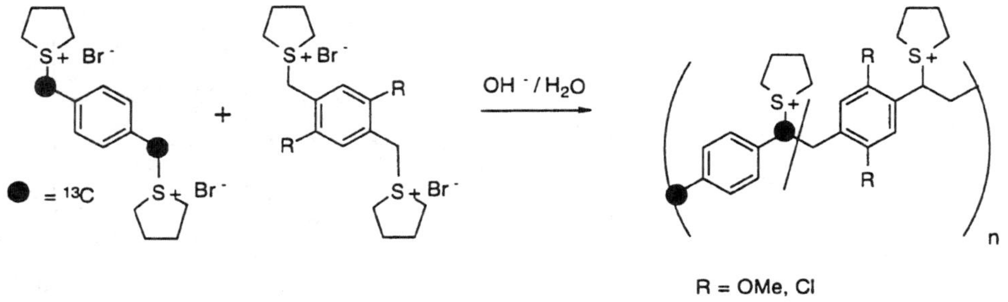

R = OMe, Cl

SCHEME 21

a nonionic group such as an ether (Scheme 25). The sulfonium group has been described as being displaced by methanol to produce polyether precursors which may be thermally eliminated to yield PPV [107], substituted PPVs [108], and PAVs [109]. The polyether precursors are conveniently soluble in nonpolar organic solvents such as chloroform and tetrahydrofuran.

Similar [110–112] neutral precursor polymers with pendant sulfinyl or sulfonyl groups may be thermolyzed to give PPVs and PAVs, although typically under somewhat more forcing conditions than is used for the sulfonium precursor polymers. Such sluggish elimination of pendant groups can actually allow a degree of control over the elimination process, as shown by Burn et al. in varying the conjugation length of a PPV-co-2,5-di-MeOPPV copolymer formed by a partially ether-functionalized precursor polymer (Scheme 26) [113].

The problem of structural defects is of considerable importance in attempting to make reproducible, electronic-quality PPVs and PAVs. PPVs formed via thermolysis of PXDs can have about 0.2–10.0% of remnant sulfur in elemental analysis. Electron-acceptor substituted PPVs tend to show a somewhat higher level of

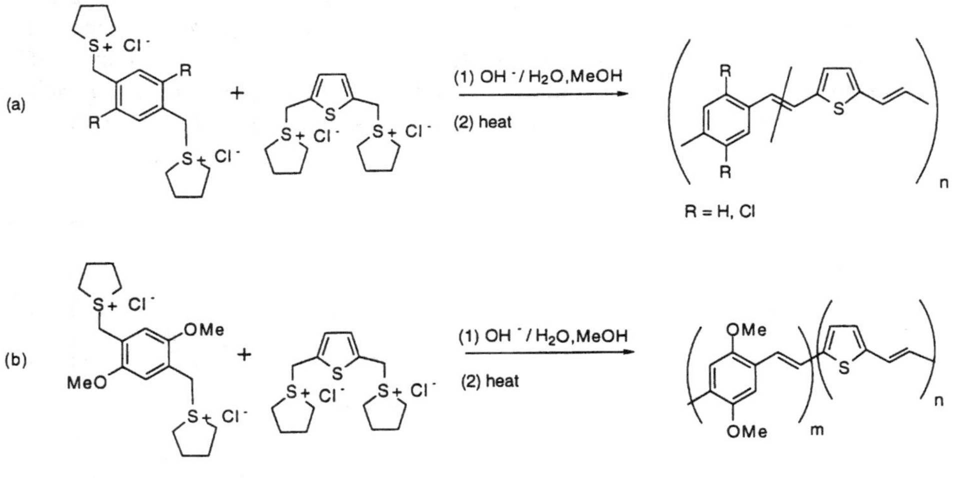

SCHEME 22

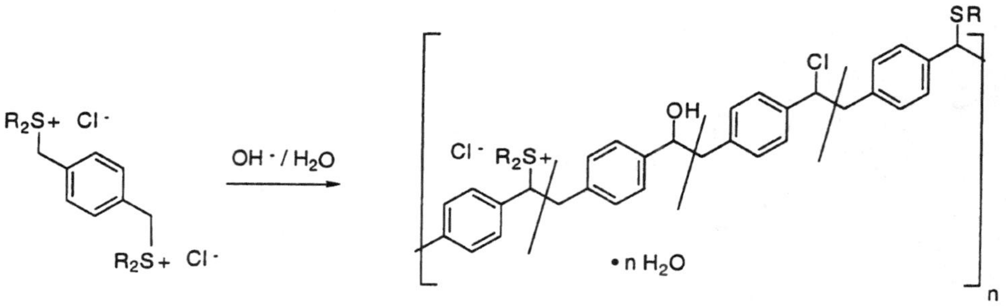

SCHEME 23

defects than electron-donor substituted PPVs [78,85]. The degree of conjugation obtained from PXD elimination is usually quite high, but even minor presence of such defects can be deleterious to electronic device performance in technologies involving conducting devices or production of light-emitting diodes (LEDs). With effort, strategies can be developed to minimize remnant defects during thermolysis of PXDs, such as the acid-atmosphere annealing process of Halliday et al. [101], which is rigorous enough to promote elimination of less reactive ether and sulfide groups found on PXDs and partly functionalized PXDs.

Both temperature and duration of the thermolytic annealing are important. Optimum conditions usually include partial orientation of the sample, which is especially convenient when PPV is stretched during thermolysis of a PXD precursor. Murase et al. obtained uniaxially oriented doped PPV with a conductivity of 2800 S/cm by this method [114]. Machado et al. have demonstrated AsF$_5$-doped PPV film that is stretched with a uniaxial stretching machine, placed under mild tension, and, with optimized annealing conditions, gives a conductivity of >10,000 S/cm [115].

The optimization of counterion choice in the PXDs is also a very important factor in optimizing thermolysis conditions and controlling the electrical properties of product PAVs [116,117]. For example, Patil et al. showed an interesting variant of conjugated polymer conductivity through their "incipient doping" concept. Decomposition products of selected PXD counterions such as AsF$_6$-form dopants (HF, AsF$_5$) during the thermolysis/orientation of the PXD, hence the polymer is self-doped during the process of C=C formation and plasticization [118]. In addition,

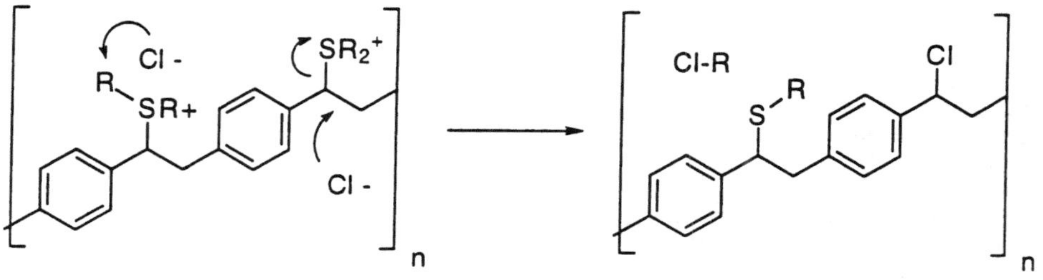

SCHEME 24

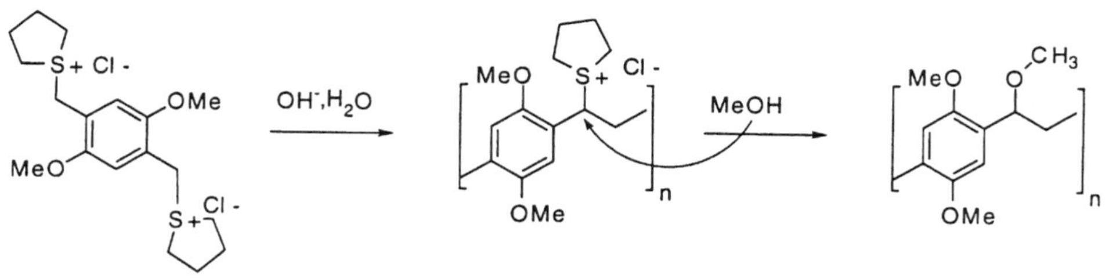

SCHEME 25

the appropriate choice of counteranion and sulfonium pendant groups can allow the elimination temperature for PXDs to be considerably reduced, decreasing the prevalence of the defect-causing side reactions described earlier [119].

Murase has noted that oxidation of PPV films can occur when annealing is carried out in air, with a parallel decrease in conductivity [53]. Papadimitrikopoulos and co-workers have noted that elimination under reducing atmosphere gives PPVs of better luminescent quality than that formed from simple vacuum elimination, and which has fewer of the carbonyl defects formed by reactions with oxygen (Scheme 27) [120–122]. An improvement in electronic properties of PAVs can thus be achieved by appropriate variation of elimination atmosphere [123].

An alternative method for production of PPVs from PXDs is to treat the latter with strong acids at room temperature, a procedure which leads directly to doped PPVs [124] with properties similar to those of doped PPVs from other methods. The method has also been applied to production of poly(2,5-thienylene vinylenes) [125]. The general applicability of this methodology is not clear, but it offers an interesting alternative to thermolytic methods of PXD elimination. Ion irradiation curing has also been claimed as a method for simultaneous elimination of PXDs and doping of the resultant PPV [126] to give material that does not have the typical sulfur or halogen contaminants [127]. Again, it is not clear that this would constitute a generally applicable method.

Overall, the sulfonium precursor route has some great advantages in terms of PPV processability, ease of use, and chain lengths obtained. It also suffers from drawbacks in terms of difficulty in using some substituents, problems in obtaining polymers that are free of structural defects, and control over product regiospecificity. For applications where these drawbacks are not critical, the sulfonium pre-

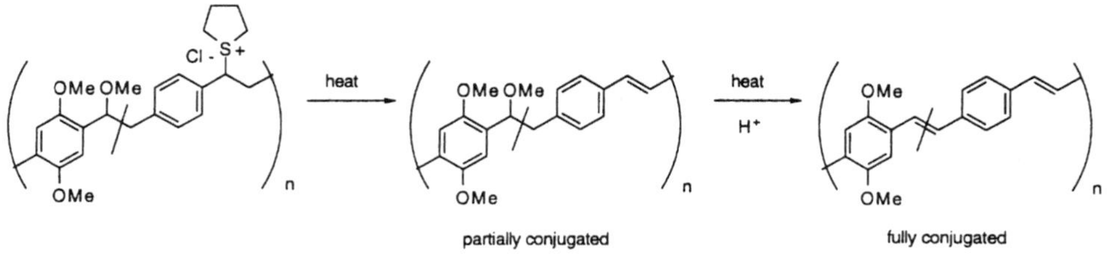

partially conjugated   fully conjugated

SCHEME 26

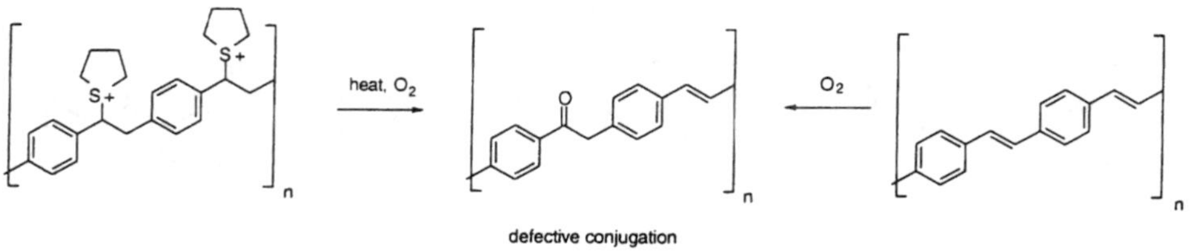

defective conjugation

SCHEME 27

cursor route and the CPR precursor routes are typically the methods of choice for high-molecular-weight, electronic-quality PAVs.

E. PAVs from Carbon–Carbon Coupling Reaction Chemistry

A promising and fairly recent development in PAV synthesis is the use of carbon–carbon bond-forming chemistry. The methodology has limitations due to solubility of the product PAVs, but this can in part be overcome by the use of solubilizing substituents. A wide variety of structure variation can be tolerated by modern methodologies—including the synthesis of soluble precursors—and improvements continue to be reported as experience with the methodology grows for making these polymers.

Simple lithiation has been used in some specialized cases to make PPVs. 4-Bromo-α,β-difluoro-β-chlorostyrene reacts with butyl lithium to give poly-(phenylene α,β-difluorovinylene) [128], obtained primarily as unique cyclo-oligomers ($n = 3$ to 7) (Scheme 28). Lithiation by the Fittig reaction of 1,4-dichlorobenzene in the presence of diphenylacetylene [129,130] gives primarily poly(phenylene α,β-diphenylvinylene) with molecular weights of 5000–20,000. Such methods are of limited use, due to their lack of generality. The use of noble metal catalysis, on the other hand, can extend this type of reaction to permit a wide variety of substitution patterns.

In a series of related publications, Heitz and co-workers have used Heck-type palladium-catalyzed coupling to make a variety of PPV-type polymers containing both electron-donor and electron-acceptor substituents such as fluorine and nitro groups [131–135]. This method can be pursued by three general strategies (Schemes 29a–29c): (1) coupling of dihaloarenes and ethylene to give homo-PAVs, (2) coupling of divinyl arene with dihaloarenes to give alternating PAV copolymers,

SCHEME 28

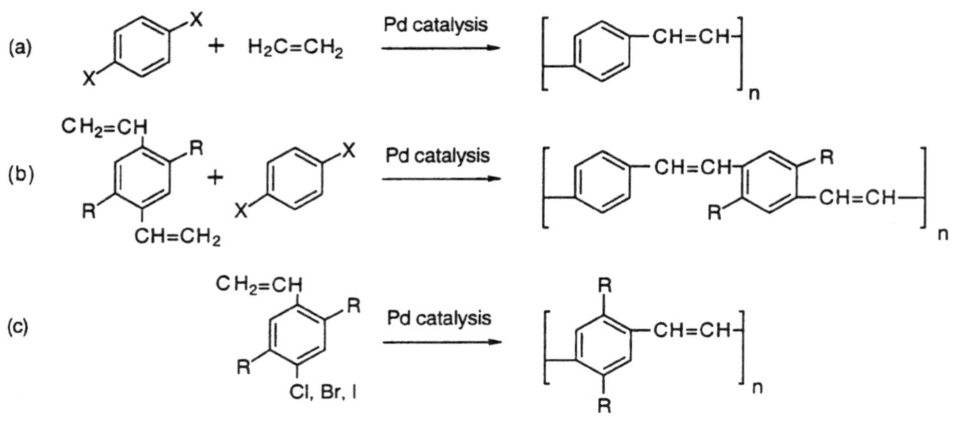

SCHEME 29

and (3) coupling of 1-halo-4-vinylarenes to give homo-PAVs. A wide variety of substituents are tolerated by the methodology. The products of these reactions suffered in many cases from insolubility, as expected when solubilizing substituents are not used. Use of appropriately phenylated monomers allowed synthesis of soluble phenyl-PPVs; related poly(4,4′-biphenyldiyl vinylenes) were also sufficiently flexible to allow solubility [136].

A recent, elegant example of the control achievable in PAV synthesis by the Heck reaction is given by Yu and co-workers [137], who made *alt*-poly(2-dialkylaminophenylene vinylene)-co-poly(2-nitrophenylene vinylene), **4**, for potential use as a nonlinear optical material (Scheme 30) [138]. The monomer for this material was compound **5**, whose synthesis required some effort. But, once **5** was available, catalysis with palladium catalysis led to a polymer with $M_w \approx 35{,}000$, thanks in part to the solubilizing effect of the dialkylamino groups. NMR studies showed the polymer to have been formed regiospecifically, as expected by the synthetic methodology. Such specificity in substituent placement is unusual for a PAV and justifies the effort of making complex monomers to achieve precision in the final, engineered polymer.

In a final (for this chapter), particularly elegant application of Heck methodology, Müllen et al. synthesized *alt*-poly(2,2′-biphenyldiyl 2′-vinylene)-co-(1,4-phenylene vinylene), **6**, from 2,2′-divinylbiphenyl and 1,4-dibromobenzene

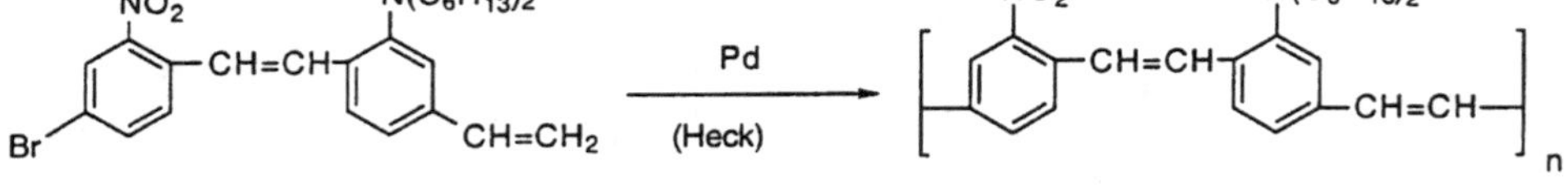

SCHEME 30

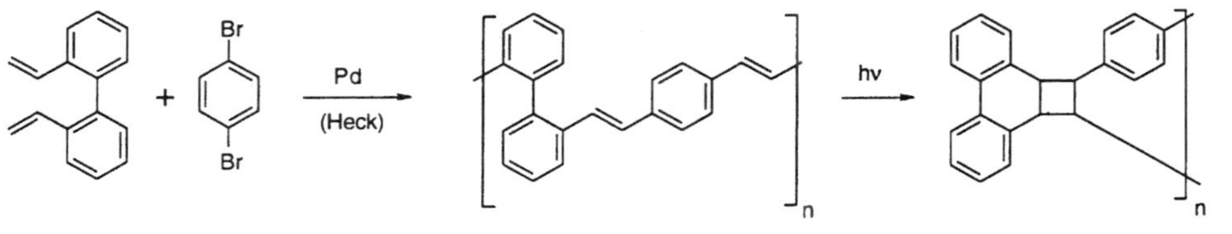

6

SCHEME 31

(Scheme 31). This polymer undergoes solid-state photocyclization to a dihydro-phenanthrene, a fascinating process that is potentially reversible, and so applicable to the synthesis of molecular "switches" [139].

F. PAVs from Metathesis Polymerization Chemistry

Metathesis polymerization has been a very exciting, recent method of making high-molecular-weight polymers with narrow chain-length distributions with a variety of substituents. In pioneering work, 1,4-divinylbenzene was metathesis polymerized by Eichinger and co-workers to give PPV, with ethylene being removed as a side product (Scheme 32) [140]. At about the same time, Grubbs and co-workers developed soluble PPV precursors via ring-opening metathesis polymerization (ROMP) routes, opening the way to PPV derivatives inaccessible from other precursor methods (Scheme 33) [141]. In this method, a bicyclic diene was ROMP polymerized using a molybdenum carbene catalyst to generate an organic solvent-soluble precursor. Pyrolysis of the precursor above 200°C yielded PPV. In related work from the Grubbs group, poly(1,4-naphthalene vinylene) (PNV) precursors were made by ROMP polymerization of substituted benzobarrelenes with molybdenum catalysts (Scheme 33b) [142]. The poly(1,4-dihydronaphthalene) precursor polymers derived therefrom were converted to PNV by oxidation with 2,3-dichloro-5,6-dicyanoquinone (DDQ), resulting in all-trans PNVs or a variable cis-trans ratio, depending on the catalyst used and the reaction stoichiometry. The doped conductivity of these polymers reached 6–15 S/cm, depending on the alkyl substituent –R, well above the conductivity of PNV from the sulfonium [90] precursor route. The use of soluble precursor polymers produced by ROMP synthesis yields advantages similar to those of the sulfonium and CPR precursor methods described earlier, with potential added advantages of fewer defects in the final polymer and greater control of molecular-weight distribution.

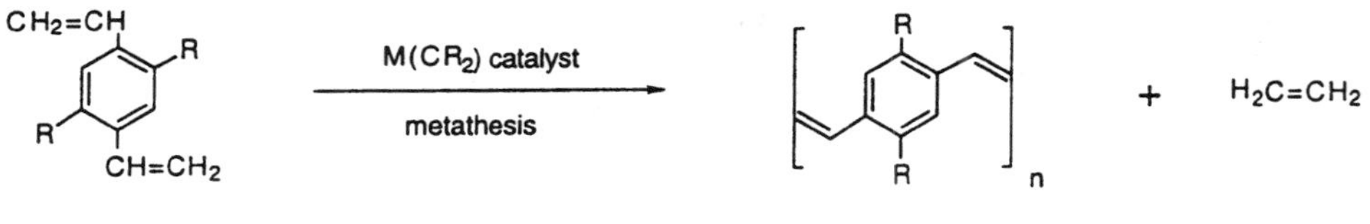

SCHEME 32

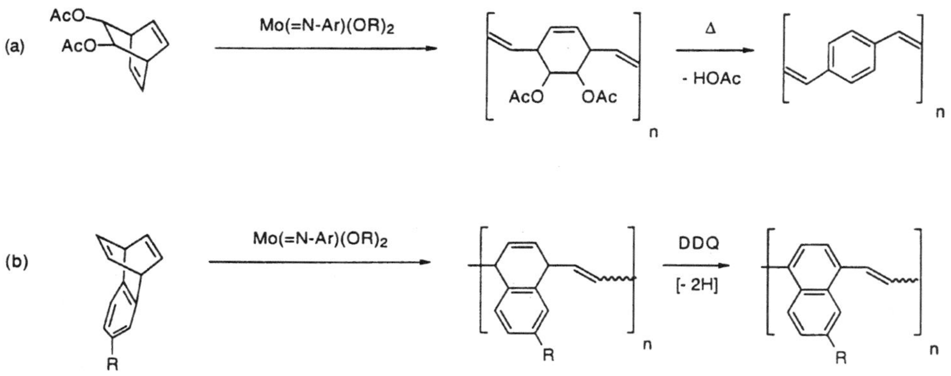

SCHEME 33

Poly(1,4-phenylene vinylene) and some unusual copolymers have also been directly synthesized by ROMP, without formation of soluble intermediate polymers, due to the "living" nature of the reaction mechanism before end-capping. [2.2]Paracyclophane-1,9-diene was reacted with tungsten-based catalysis systems to give PPV directly (Scheme 34). ROMP polymerization of the same monomeric diene in the presence of 1,3,5,7-cyclooctatetraene or cyclopentene allowed synthesis of PPV copolymers incorporating additional vinylene and alkyl spacer units (Scheme 35) [143]. This combination of comonomers is quite unusual for PAV chemistry and holds the promise of yielding interesting electronic and structural polymers, if more generalized synthesis of substituted [2.2]paracyclophane-1,9-dienes can be achieved. For example, a molybdenum catalyst was used to effect living ROMP polymerization of 9-(*tert*-butyldimethylsilyloxy)-[2.2]paracyclophane-1-ene (Scheme 36) [144]. Elimination of the resulting precursor polymer occurred at 190°C under N_2 and HCl atmospheres. Alternatively, the silicon moiety on the precursor could be converted to a hydroxyl group by reaction with NBu_4F. The product polymer yielded PPV under mild dehydration conditions (Scheme 36). Block copolymers suitable for wavelength-tunable photoluminescence could eventually become available due to the living character of this polymerization through incorporation of different chromophoric segments along with the PPV segments.

Studies of divinylbenzene metathesis polymerization using tungsten catalysts have correlated the obtained PPV chain length with various reaction parameters [145]. In a manner similar to the copolymer work with [2.2]paracyclophane-1,9-diene shown in Scheme 34, it has been shown that PPV segments obtained from

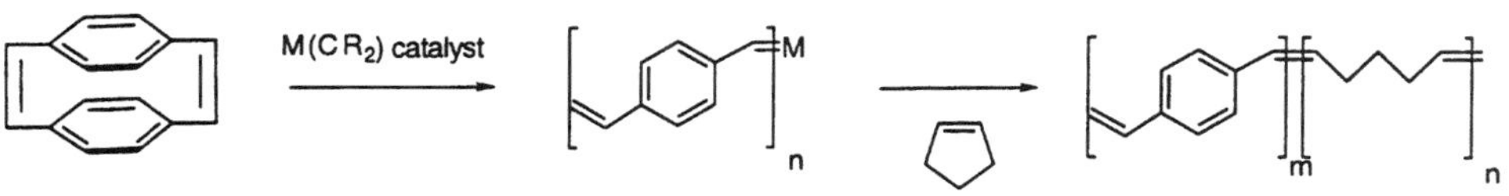

SCHEME 34

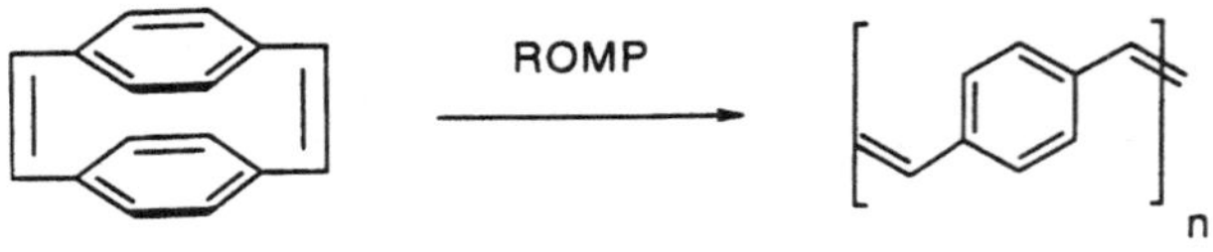

SCHEME 35

an initial living ROMP oligomerization of divinylbenzene can be further reacted with 1,5-cyclooctadiene to yield copolymers containing PPV sequences of fairly well-controlled length [146]. Other PAV methodologies do not allow the sort of block copolymerization which is permitted by these living polymerization methods; hence, ROMP offers unique possibilities for future work.

G. Film-Deposition PAV Syntheses

For some applications, controlled deposition of very thin films is important. In such cases, the methods shown below may be more appropriate than the chemically convenient methods described in earlier sections. Although this area of PAV methodology does not seem to have been extensively explored to date, there is considerable promise for future expansion.

Where feasible, chemical-vapor-deposition (CVD) polymerization is enormously appealing for preparing thin films of PAVs in electronic devices. However, this use has been limited by the dearth of volatile, suitably reactive monomeric substrates. 1,9-Dichloro-[2.2]paracyclophane was reported by Iwatsuki and co-workers to be a suitable CVD precursor monomer for PPV, because it apparently dissociates to the α-chloro-quinodimethane monomers in the gas phase (Scheme 37) [147]. A CPR-type chloro-precursor polymer is deposited, which may be dehydrochlorinated at 300°C under nitrogen in 1 h to yield PPV films. In a similar approach, a group at Philips deposited a precursor polymer by CVD onto various substrates, then thermally converted it into PPV during fabrication of light-emitting diodes [148].

Poly(1,4-phenylene vinylene) has also been electrochemically deposited by reduction of $\alpha,\alpha',\alpha',\alpha'$-tetrabromo-*para*-xyxlene in solution (Scheme 38). This ap-

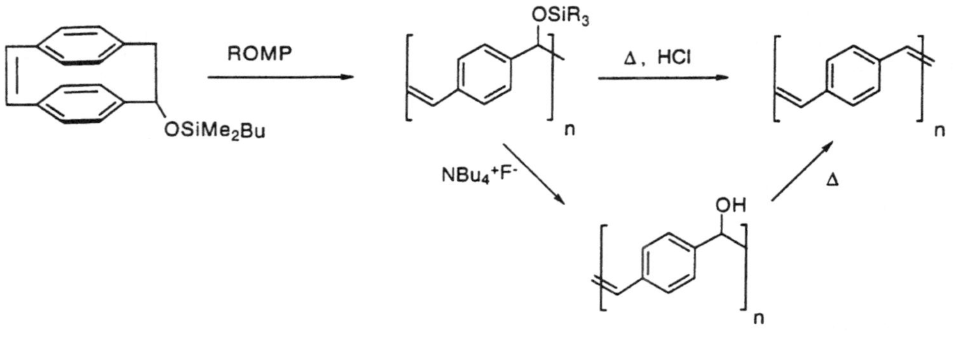

SCHEME 36

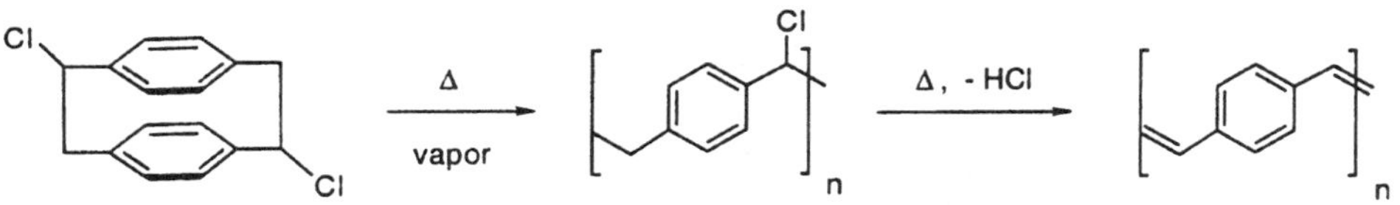

SCHEME 37

proach affords a simple method for depositing alternating layers of PPV and polypyrrole [149]. Although limited in utility for bulk PAV production in most applications, electrochemistry offers a highly controlled method for making PAV films of controlled thickness on very small devices, such as substrate-specific microdetector probes. Given interest in the latter area, it seems plausible that this methodology could be further explored in the future.

H. Modification of PXDs and PAVs

Poly(arylene vinylene) modification is a somewhat limited area, given the typical insolubility of most structural variants and the large DPs in PAVs, which render complete modification of functional groups unlikely. Modification of PXDs from the sulfonium precursor route is more common, although this is usually carried out to alter the PXD solubility and allow processing under conditions different from the usual aqueous or alcoholic media that are used. We briefly summarize some contributions concerning both processes in this section, without attempting to be exhaustive.

The PXDs have been reduced to poly(*para*-xylylene) electrochemically [150] by catalytic hydrogenation [151] and by treatment with sodium naphthalide at −60°C [152]. In the first case, the product was not soluble, and in the second it was of low molecular weight. The modification of PXDs through nucleophilic displacement of the sulfonium side chains by mercaptans and alcohols has been described in a previous section (Scheme 39). Even milder nucleophiles such as carboxylates can be used in very reactive analogs, such as the sulfonium precursor polymer to poly(2,5-thienylene vinylene) [153]. The change in solubility from the hydroxyphilic behavior of PXDs to solubility in neutral solvents such as chloroform and tetrahydrofuran is striking in these cases. This change renders possible multilayer processing of PAVs without mixing of layers at the interfaces, because different PAV precursors can be used that have differing solvent preferences. In addition to the thiophenoxide and methanol substitutions mentioned above, a large variety of inorganic ion displacements have also been tested by Denton [57]. The

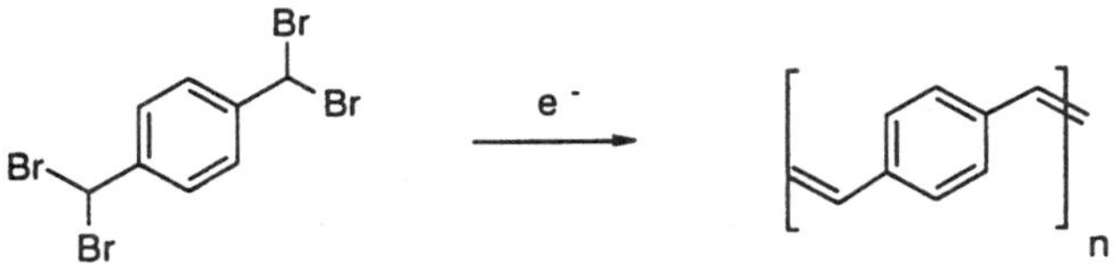

SCHEME 38

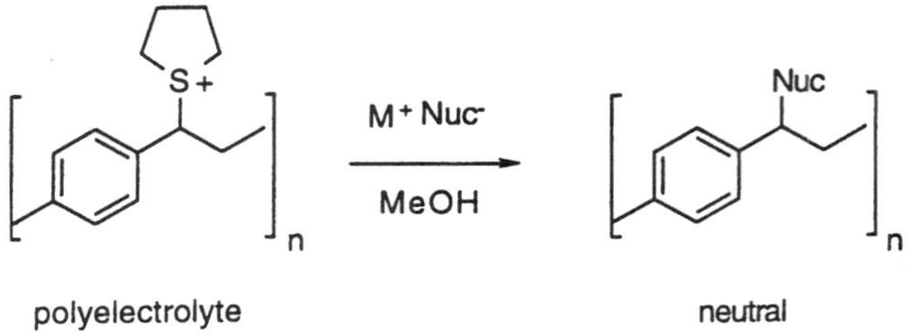

polyelectrolyte   neutral

SCHEME 39

resultant materials in the latter cases sometimes precipitate from the reaction solutions before complete displacement is achieved.

The PXDs have also been cross-linked by heating in the presence of polyfunctional vinyl compounds, which apparently condense with the precursor polymer. This chemistry has been described as an adhesive for epoxy-derived components in circuits [154].

Although some of the early PPV chemists confirmed the presence of vinylene groups by halogenating them, such conversions of PPV have seldom been used en route to related polymers. Hörhold et al. dichlorinated the vinylene units of PPV by treatment with tetra(*n*-butyl)ammonium iodine tetrachloride [12]. McDonald and Campbell characterized their Wittig-derived PPV in part by brominating the double bonds with consequent elimination of color [5]. Hsieh has also described the bromination of PPV and subsequent elimination to obtain poly(*para*-phenylene ethynylene) [155]. The bromination may be complete or partial, depending on conditions selected. Some mechanistic studies on this addition–elimination were reported.

Hörhold found more complex behavior for poly(phenylene α,β-diphenyl vinylene). Whereas the arylated vinylene groups are dichlorinated by Cl_2, equivalent treatment with Br_2 results in bromination of the para positions on the pendant phenyl groups (Scheme 40). Likewise, nitration functionalizes the para sites, as do Friedel–Crafts reactions using $AlCl_3$ and benzoyl chloride, maleic anhydride, or phenyl isocyanate [14,156]. One may also carry out *ipso* substitution on aryl-nitrated PPV derivatives by treatment with KOH and hydrazine in diethylene glycol [157]. Schlenoff and Wang [158] have shown that the surface of PPVs can be modified through BBr_3-mediated cleavage of the ether groups on 2,5-dimethoxy-PPV to give 2,5-dihydroxy-PPV (Scheme 41). This last reaction opens the possibility of attaching a variety of pendant groups onto PPV.

III. PROPERTIES AND APPLICATIONS OF PAVs

Although the number of publications concerning synthesis of PAVs is substantial, the number concerning the properties and applications of PAVs has been even more so, particularly in the past two decades. The promise of PAVs as both structural and electrooptic materials has driven most of this effort. As is true for the synthetic work, it is impossible in a single book chapter to give a comprehensive review of

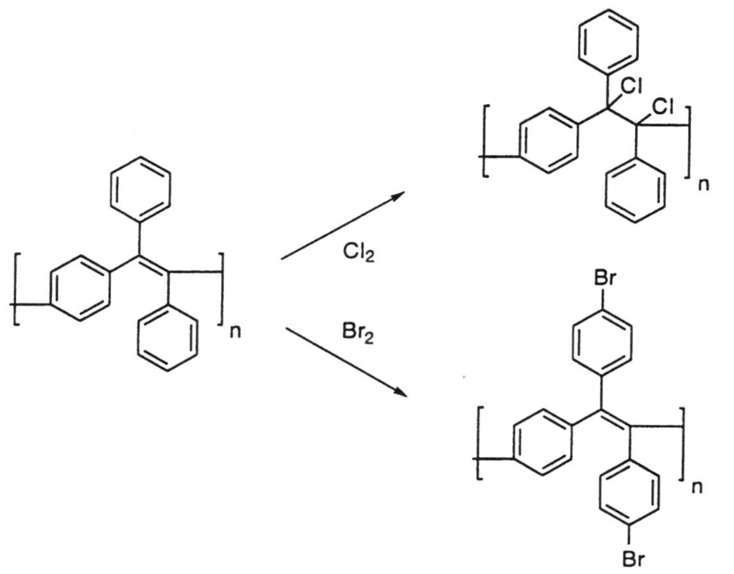

SCHEME 40

PAV property and application studies. We shall therefore strive in this section to give an overview of various interesting properties of PAVs and note some of the more important work concerning major efforts to use PAVs in electronic and other applications.

A. Molecular Weight and Degree of Polymerization

Both the degrees of polymerization and polydispersities of PAVs vary considerably with the method of synthesis. Early condensation methodologies often yielded DPs of only 5–20 before precipitation occurred. The incorporation of solubilizing groups can help considerably. Hörhold's phenylated PPVs made by $CrCl_2$ coupling [12,32,34] had apparent molecular weights up to 80,000. From a Knoevenagle condensation, alternating copolymer **7** was found to have $M_w = 43,000$ and $M_n = 12,000$ [36], a typically polydisperse distribution for a condensation polymerization.

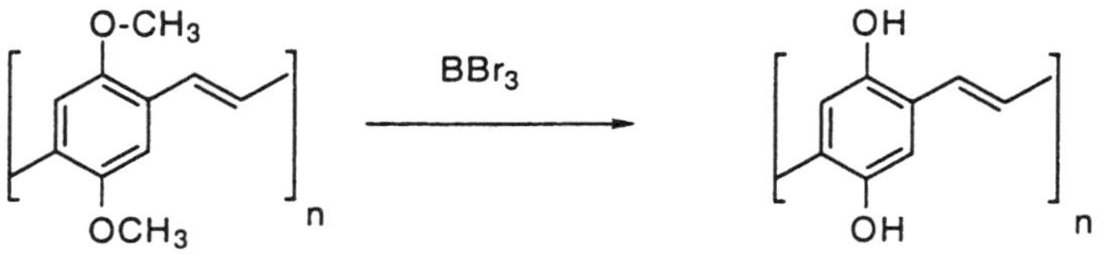

SCHEME 41

HexO   CN

OHex

OHex

CN   OHex

OHex

7

The development of soluble precursor methods led to PAVs with quite large DPs. From the processible sulfonium precursor route, Wessling and co-workers found apparent $M_n = 6 \times 10^5$ (DP ~2800) by gel permeation chromatography (GPC) for PXDs functionalized with mercaptan groups to allow solubility in neutral organic solvents [49], whereas osmometry gave apparent molecular weights of 5.7 $\times 10^4$ to >10^6, depending on reaction conditions [49]. Intrinsic viscosities of the mercapto derivatives ranged from 0.69 to 2.77 dl/g, but had "no clear correlation" with the osmometry results [49]. Hörhold's group similarly found molecular weights of about 10^5 [13].

Machado et al. carried out a GPC study of thiophenoxide-functionalized PXD, and PXD with halogen counterions exchanged for tetrafluoroborate. The results of the GPC work were compared to results of low-angle laser light scattering (LALLS) studies [159]. They found that LALLS gave $M_n = 5 \times 10^5$ and $M_w = 9.9 \times 10^5$ to give a polydispersity index (PDI) of about 2.0. GPC results gave $M_n = 1.9 \times 10^5$ and $M_w = 9.5 \times 10^5$ for thiophenoxide functionalized PXD (PDI = 5.0) and $M_n = 5.5 \times 10^5$ and $M_w = 1.2 \times 10^6$ for the counterion-exchanged PXD (PDI = 2.2), all versus polystyrene standards. The agreement between absolute and GPC-estimated molecular weights is thus reasonably good for these procedures. Wudl's self-doping polyelectrolyte PPV (**2**) (see Section II.D above) had $M_w = 1.12 \times 10^6$ with a PDI of 16. Sarker has examined the dependence of GPC molecular weights for PXDs as functions of various reaction parameters for a number of substituted PPVs and found that the degree of polymerization can depend greatly on both substituent and the purity of the starting bis-sulfonium salt [63].

The polydispersity index of PXDs is variable, depending on synthetic methodology. Derivatized PXD materials can have apparent $M_w/M_n = 2$–10, as shown in the examples above. Larger polydispersities seem to occur in the presence of air and other radical-chain inhibitors. If oxygen is not excluded from a Wessling synthesis, bimodal molecular weight distributions can be obtained [64].

In considering and comparing the results cited above, one must note concerns about using functionalized PXDs to estimate PAV molecular weights. In general, GPC results for derivatized PXDs and for soluble PPVs are somewhat questionable when polystyrene standards are used, due to expected differences between sample and standard with regard to shape and rigidity. Fortunately, GPC results are useful for the order-of-magnitude estimate of DP, which the above examples show to be quite large for precursor-derived PAVs.

Molecular-weight results for PAVs derived from the CPR precursor route are often similar in most respects to results from the sulfonium routes. For example, Hsieh and Feld have found high DPs with $0.5 \times 10^6 < M_w \leq 1.0 \times 10^6$ for phenylated-PPVs (see Scheme 12 above) by the CPR route. The polydispersity of these polymers was fairly large, however, with M_w/M_n = 5–26 under varying re-action conditions [46].

The ROMP methodologies for PAV synthesis have not been explored exten-sively. However, the ROMP-derived processible precursor polymer used by Grubbs et al. to make PPV (Scheme 33) has M_n = 46,000 and M_n = 46,000 (DP >200). The equivalent PDI = 1.23 of this method shows a narrow distribution of chain lengths (PDI = 1.2–1.3) [141] consistent with the living polymerization nature of the ROMP method, giving ROMP a strong advantage in this respect relative to other PAV synthetic methods discussed herein. The Grubbs precursor to PNV showed M_w = 26,000, and M_n = 4900, and so had greater polydispersity (PDI = 5.4) than the PPV precursor [142].

B. PPV Morphology and Orientation

Of the PAVs, PPV has been subjected to the closest morphological investigation. Bradley and Friend have given useful summaries of PPV characterization which include morphological descriptions [160–162]. Our aim herein will be to highlight major features of PPV morphology, and the changes therein upon doping.

Murase et al. reported some of the first x-ray-diffraction studies of PPV from the sulfonium precursor route [53]. Masse et al. subsequently reported a set of wide-angle x-ray-diffraction studies on partially oriented PPV samples to determine crystal structures. Virgin oriented PPV derived by the precursor method is crystal-line with a "herringbone"-type, orthorhombic crystal structure (Fig. 3) [163]. Un-

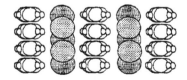

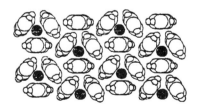

FIGURE 3 Schematic of crystal packing arrays for pristine (top), *p*-doped (middle), and *n*-doped (bottom) poly(1,4-phenylene vinylene).

der scanning tunneling microscopy, semicrystalline domains have been observed with local alignment of chains at the surface [164]. NMR studies give a fairly detailed description of the PPV microstructure and conformation within this morphology [165]. Shifting of the polymer chains relative to one another is found within layers [166]. Interestingly, 2,5-dimethoxy-PPV has better translational ordering than parent PPV, apparently due to the interlocking of pendant moieties on adjacent chains [167]. Although well-defined morphologies are found for PPVs, NMR [168–170] and x-ray [171] studies show constant, temperature-dependent ring rotation of the phenylene ring in the solid state (Fig. 4).

Upon *p*-doping with AsF_5, SbF_5, or H_2SO_4, the morphology changes to give face-to-face stacking of the polymer chains into discrete stacks separated by layers of large dopant counterions (Fig. 3), in which unit-cell dimensions vary with the counterions, but PPV crystallite orientation remains constant [172,173]. Changes in PPV morphology also occur upon *n*-doping by sodium vapor to give roughly cylindrical "cages" of PPV chains, surrounding strings of sodium countercations (Fig. 3) [174]. This morphology begins to lose its integrity at >55 wt% doping levels.

The electrical and optical anisotropy of oriented films, both doped and undoped, has been widely used as a gauge of PPV sample quality. For highly conductive PPV films, Machado was able to increase PPV film anisotropy until conductivity in the axial direction was 80 times greater than transverse conductivity, by use of a uniaxial stretching machine that allowed continuous feed of PXD strips with a high degree of control [115,175]. A more precise measure of ordering is infrared dichroism, which Bradley has discussed in some detail for oriented PPVs [160]. Murase et al. found dichroic ratios up to 51:1 for their highest conductivity materials having a draw ratio of 10 and electrical anisotropy of 100 [176]. In addition, Stenger-Smith et al. have discussed the stress effects of orientation on infrared dichroism [177].

Finally, x-ray- and electron-diffraction methods have been used to probe the polymers derived from various PPV synthetic methods, with an eye to monitoring the degrees of crystallinity and types of crystal packing [178]. These factors are very important in affecting electrooptical properties, so it is no surprise that a considerable volume of work probing PAV microstructure and morphology continues to appear.

C. Conductivity of Doped PAVs

The capability of PAV films and fibers to conduct upon doping with electron acceptors and donors is now well established by a large amount of activity in this

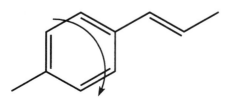

FIGURE 4   Solid-state ring-flipping mode of poly(1,4-phenylene vinylene).

area. The mechanisms by which doped conduction occur in organic systems have been described elsewhere in detail [179]; hence, we shall limit our discussion to the observed conductivities for doped PAVs, not the mechanisms of the conductivity. In initial work, Wnek et al. doped Wittig-derived PPV with AsF_5 to obtain a conductivity of 3 S/cm [180]. This and much of the early work on PAVs describes powder sample conductivities, which are usually quite low (typically 10^{-3}–10^{-1} S/cm). The development of methods to allow processing and orientation of PAVs greatly advanced and accelerated this work; hence, we shall concentrate on conductivities of processible samples.

Considerable effort has been put into use of Lewis acids such as I_2, SbF_5, or AsF_5 to *p*-dope PPVs to a conducting state. The chemistry of the doping process has been described and involves the formation of polaron (radical cations) and bipolaron (spinless dications) defects as charge carriers, depending on the degree of doping (Scheme 42) [179,181]. In one of the more impressive examples, oriented PPV from the soluble precursor route was reported to have a conductivity of about $>10^3$ S/cm along the stretch direction when *p*-doped with AsF_5 [182–184]. A highly doped layer which forms on the polymer during doping seems responsible for much of the conductivity. Rutherford backscattering data found that even after several days of exposure, AsF_5 doping penetrated only 60 nm into the surface of PPV with one dopant counterion for every four to five chain repeat units, yielding an intrinsic conduction $\geq 40,000$ S/cm [185].

Usually, film samples are only evaluated for overall, rather than surface conductivities. As a result, conductivity can vary considerably with dopant, sample orientation, and sample preparation method. SO_3-doped PPV gives conductivity of 1–10 S/cm [186]. The effectiveness of various doping levels of $FeCl_3$ on unoriented PPV films has been studied, with a maximum seen at 35 S/cm [187]. Nonoxidizing Brønsted acids can also be used to induce conductivity [188–190]. A stretched film of PPV doped with H_2SO_4 showed a high conductivity of 27,000 S/cm [191]. In one unusual example, Shi and Wudl showed that salts of poly(5-sulfopropoxy-2-methoxy-*para*-phenylene vinylene) and the corresponding protonated form* (**2–3**) showed self-doping behavior, as being dopable by protic acids. Polymer **2** had a conductivity of 10^{-4}–10^{-2} S/cm—dependent in part on its degree of hydration and association with nonprotic cations—could be compensated (dedoped) with ammonia, and showed IR and UV-vis bands consistent with its self-doped state.

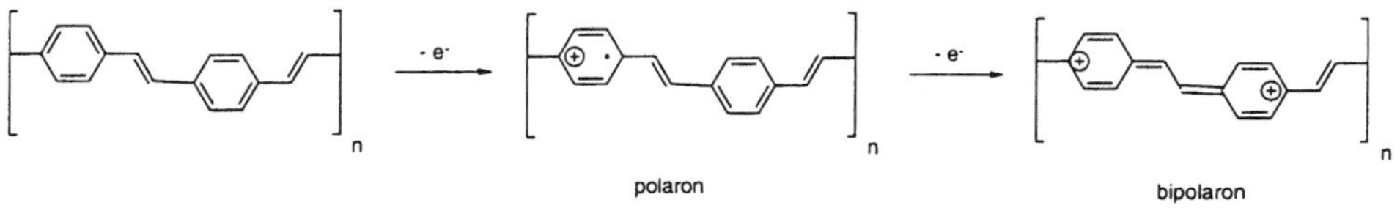

polaron   bipolaron

SCHEME 42

---

*See Ref. 81 for a fuller description of the self-doping phenomenon, which involves the action counterions along the chains of polymers **3–4** to give conductivity, rather than using an added reagent to remove or add electrons into the polymer conduction/valence bands.

2 (M=Na)
3 (M=H)

Iodine has an oxidation potential insufficient to dope PPV itself but readily dopes more electron-rich conjugated polymers. Poly(2,5-furan vinylene) (**8**, X = O) can be doped with I_2 or $FeCl_3$ to give conductivities of ~30 S/cm [192]. For poly(2,5-thienylene vinylene) [PTV, (**8**, X = S)], conductivities of 200 S/cm even without orientation are obtained with iodine doping [87]. 2,5-Dimethoxy-PPV shows a conductivity of 20–430 S/cm upon similar treatment; the final doped states for this and for PTV have considerable long-term stability [87,193]. Oriented dihexyloxy PPV films give a conductivity of up to 3200 S/cm upon doping [194].

8

n-Doping of PPVs with alkali metals has drawn rather less academic interest than *p*-doping in terms of sample stability, but can also give conductivities $>10^3$ S/cm [174,195,196]. Some description of lithium doping of PPVs has been given [197]. Sodium doping of PPV is found to be complete at 55 wt%, past which point morphological decay of the sample occurs [174]. When the series of reducing metals Na, K, Cs, and Rb was investigated, it was determined that Na- and K-doped PPV are dominated by one-dimensional channel structures, whereas cesium produces a tetragonal column structure, and rubidium-doped PPV is unstable toward either structure [198]. Rb *n*-doping of PPV also proved useful in a ultrahigh-vacuum photoelectron spectroscopic study that monitored a polaron to bipolaron transition as the doping level varied [199].

Research is ongoing concerning improvement of doped PAV conductivity, and elucidation of the mechanisms involved. Further data about PAVs as conductors may be found in the reviews cited in Section IV.

D. Nonlinear Optical Behavior of PAVs

The thermal processibility of soluble precursor forms of PAVs can, in principle, be used to pole or orient their morphologies into final form for use as nonlinear optical (NLO) materials. In simplest terms, the NLO effect involves an interaction between light and a transparent material, such that a coherent input light will be transmitted with some component of frequency doubled or tripled energy. PAVs can be rigid and crystalline once processed, characteristics which are important to give long-lived utility in electrooptic devices. At the present time, interest in this aspect of PAVs has abated somewhat in favor of their electroluminescent behavior (see Section III.E), but much as been learned in the past decade. Full discussions of NLO in organic systems have been given elsewhere, to which we refer the reader for background information [200–201]. We will therefore limit ourselves to a brief discussion of some highlight results for NLO in the PAV family of polymers.

The third-harmonic generation (THG) NLO properties of PAVs appear to be more promising than second-harmonic generation (SHG) characteristics. SHG requires noncentrosymmetry of a permanent dipole array, a characteristic which is not readily realized in a solid-state sample. The main requirement for THG is a long conjugation length, which is readily achieved in many PAVs with a large number of substituents. Uniaxially oriented PPV has been reported to have $\chi^{(3)}$ at 602 nm of $\sim 10^{-10}$ esu [202]. Measurements taken for various other oriented and unoriented PPV samples have yielded $\chi^{(3)}$ values of $(0.078–4.0) \times 10^{-10}$ esu [203–206]. 2,5-Dimethoxy-PPV is rather better with a $\chi^{(3)}$ of 4×10^{-9} esu [206,207]. Other alkoxy substituted PPVs have also been reported to have $\chi^{(3)}$ $\sim 10^{-9}$ esu [207,208]. PPV can even be used to fabricate silica sol–gel composites [209] appropriate for use as fiber-optic materials, which could, in principle, make use of the PPV NLO properties for data transmission.

Second-harmonic generation studies have been carried out only for a few PAVs [210,211], although a number of syntheses have been carried out which nominally have the desirable "push–pull" substitution as part of their backbone. Jin et al. have made a number of PPV-based copolymers by the sulfonium precursor route, incorporating both electron-rich and electron-poor monomers (Scheme 43), a number of which have been evaluated for SHG and THF behavior [210]. A "push–pull" substitution pattern incorporating *para*-substituted styryl pendant groups has also been made by the same group (Scheme 44) for similar investigations [211]. The tested SHG d_{33} values of these polymers have ranged from 0.3×10^{-8} to 1.0×10^{-8} esu. Although it is not clear whether PAVs will be economically competitive with other, recently used SHG or THG materials, the potential for such use remains. We therefore refer the reader to some of the reviews cited in the Section IV for further information. Proceedings of a symposium on NLO materials is presented in Ref. 212.

E. Electroluminescent Optical Behavior of PAVs

The fabrication of light-emitting diodes (LED) requires electroluminescent (EL) materials. The basic idea of LED materials is to pump excited states electrically, with consequent emission of light. Because conjugated organic molecules are well known to show strong luminescent properties, it is not surprising that considerable interest in PAV-based EL has developed. Activity in this area has been extensive since the initial report [213] and some efforts have been made to review progress in the area [214–216]. The role of PAVs made by various methods has been increasing steadily in the past 5–10 years, and the level of activity aimed at PAV–LED work is such that it merits a chapter or even a full book by itself. As a result, the work presented below should be considered representative, and not exhaustive.

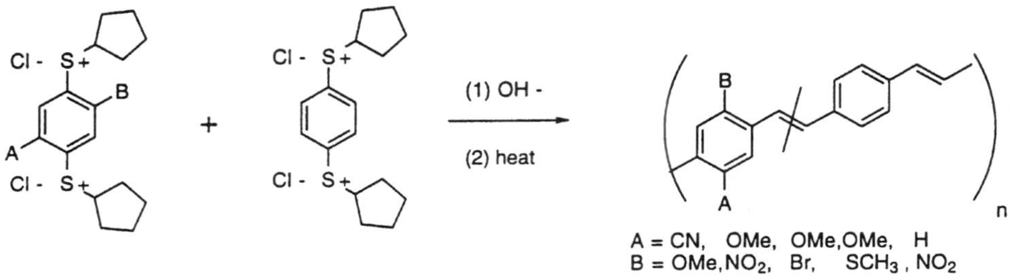

A = CN,   OMe,   OMe,OMe,   H
B = OMe,NO$_2$,   Br,   SCH$_3$, NO$_2$

SCHEME 43

An important desideratum in designing EL behavior is the control of the emission wavelength. This control of emission color can, in principle, be achieved by appropriate substitution, copolymerization, or blending of PAV monomers. Blue-green, yellow, and red emission has been observed for various substituted PAVs, thus constituting the principal colors in a light-emitting display. Computational chemistry suggests rich possibilities for band-gap turning of the PAV through appropriate structural manipulation [217–220].

Photoluminescence (PL) spectra that are very similar to EL spectra in examples studied to date, and these mechanisms of light emission appear to be closely related [221–223]. Because of the initial work in this area, a large number of sophisticated measurements by various methods only changed this view in fairly minor ways [223].

Simple LEDs can be made by spin-casting PAVs onto indium–tin oxide (ITO) coated glass wafers and topped by an electron injection cathode to make single-layer devices, a schematic of which is shown in Fig. 5. The luminescence output can be enhanced by the addition of various electron- and hold-injection layers in multilayer devices, or by variation in the cathodic material (aluminum, calcium, silver) to get better matches to the work function of the emissive layer. Although the majority of devices to date have been tested with pulsed or direct current, some tests have been done with alternating current [224,225] by obvious analogy to the type of current available in a typical household application. Progress has even been

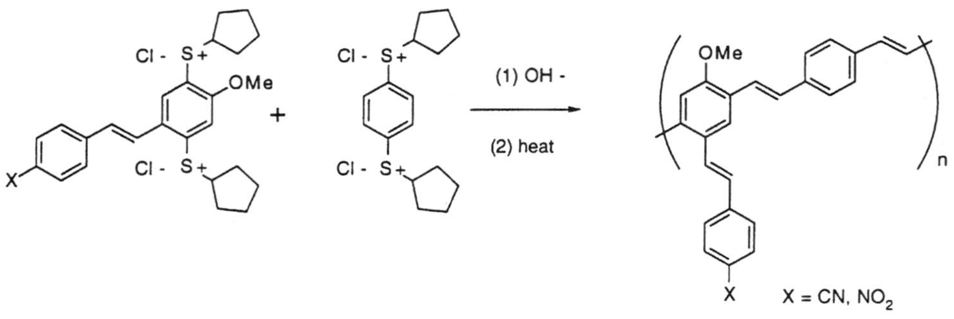

X = CN, NO$_2$

SCHEME 44

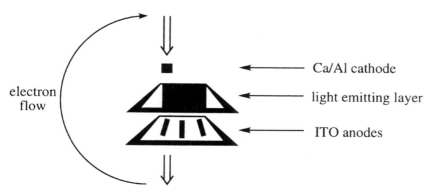

FIGURE 5 Schematic diagram of a simple, single-layer LED device.

made toward making flexible LEDs incorporating PAVs [75,226]. The quantum efficiencies of modern devices have climbed steadily above the 0.01% level achieved in the first ITO/PPV/A1 single-layer LED reported by Burroughes et al. [213].

PAVs synthesized by methods other than the soluble precursor routes can be useful in LED devices, and a number such have been very promising, so long as solubility of the polymer is possible for film casting. For example, the alternating copolymer **7** (see Section III.A above) has one of the most red-shifted EL emission wavelength of the PAVs (710 nm), as well as impressive EL quantum yields [36,227]. Block copolymers incorporating varying arylene unit chromophores with flexible polyethyleen units have been made by Karasz and co-workers as processible LED layers with tunable emission wavelength. The synthesis by this group of polymer **9** (Aryl = *para*-phenylene) was a breakthrough in achieving blue-emitting organic polymer LEDs [228].

$$\left[\begin{array}{c} \text{CH}_3\text{O} \\ -\text{O}- \\ \text{CH}_3\text{O} \end{array} \right. \overset{\text{OCH}_3}{\underset{\text{OCH}_3}{\text{—Aryl—}}} \left. \text{O-(CH}_2)_n- \right]_n$$

9

There has also been much promising work using precursor methods. PPV from the sulfonium precursor route is a useful greenish yellow emissive layer [213]. The use of red-emitting MEH–PPV **(10)** [75] and yellow-emitting derivative **(11)** [229]—among other derivatized homo-PPVs [230]—have been quite successful in device fabrication carried out by Heeger and collaborators. Gurge et al. have noted that 2-alkoxy-5-halo-PPVs **(12)** [231] and ring-fluorinated PPVs **(13–14)** [232] made by the sulfonium precursor route are also promising red emitters. In a PPV-co-2,5-dimethoxy-PPV copolymer made by a modified soluble sulfonium precursor methodology, the final PAV conjugation length can be tuned by choice of elimination conditions, allowing variation of the final EL emission wavelength (Scheme

26) [113]. A similar strategy could be used for copolymers of PPV with other dialkoxy-PPVs made by the CPR method [233]. It has been argued that incomplete, controlled elimination of PAV precursors can be useful to enhance LED emission efficiency [234,235]. However, various sorts of structural and surface defects in the PAV layers can be deleterious to emission efficiency and lifetime, especially carbonyl defects in the PAV chain (see Scheme 27) [120–122,236,237]. Many of the most impressive results are found in the proceedings of topical meetings, and in the patent literature. Details are in some cases difficult to glean due to the rapid pace of progress and the extreme competitiveness of the area at present. As a result, a comprehensive review of the area is not available at present.

The above results are merely representative. Virtually every month seems to bring new possibilities for use of PAV luminescent properties. The possibilities of using PAVs to make arrays of separately addressable diode lasers is extremely exciting. The prospects for the future look bright, and results will surely continue to appear in this area even as this chapter is being published.

IV. SUGGESTIONS FOR FURTHER READING

Space limitations for this chapter preclude a fuller description of the possible applications for which PAVs may be useful. The physical evaluation of electrooptic properties of PAVs is an area of very prolific work at present, with reasonably complete surveys of the literature probably being possible only through computer search methods. However, the brief summaries above are a good indication of the broad range of these possibilities, and the reader is referred to the works mentioned above—and the works cited in those references—for further information. For a broad survey, the doctoral dissertation of one of us was intended to give a comprehensive review of PAV-related literature up to 1989, and to our knowledge contains the largest body of cited work that appears in one place for that topic up to that time [57].

In addition, a variety of topical reviews of PAV work have appeared in English. Reference was made above to reviews of early synthetic work on PAVs [23]. More recent synthetic methodology, including the precursor methods, has also been reviewed [160,238,239]. A review has appeared concerning control of ordering and orientation strategies in PAVs [240]. A review of PAV synthesis and morphology has also appeared [161]. A series of proceedings from the Materials Research

Society is available with details of work presented at autumn colloquia concerning optical and electronic properties of organic compounds, which includes a considerable amount of work involving PAVs [241–244].

Finally, there are a number of compendia which contain considerable information on PAV synthesis, properties, and processing [179,245,246]. These lists are not meant to be comprehensive, but we hope that the reader will find these to be useful references to this staggeringly rich area of applied chemistry.

ACKNOWLEDGMENTS

One of the authors (PML) acknowledges financial support from the Air Force Office of Scientific Research (University Research Initiative), the University of Massachusetts Materials Research Laboratory, and the National Science Foundation during various phases of some of the work cited in this chapter.

REFERENCES

1. Grimaux, E., *Bull. Soc. Chim. (Paris)*, *14*, 133 (1870).
2. Thiele, J., and Balhorn, H., *Ber. Dtsch. Chem. Ges.*, *37*, 1463 (1904).
3. Quelet, R., *Compt. Rend. Hebd. Séances Acad. Sci.*, *192*, 1391 (1931).
4. Lenz, R. W., and Handlovits, C. E., *J. Org. Chem.*, *25*, 813 (1960).
5. McDonald, R. N., and Campbell, T. W., *J. Am. Chem. Soc.*, *82*, 4669 (1960).
6. Feast, W. J., in *Handbook of Conducting Polymers* (T. A. Skotheim, ed.), Marcel Dekker, Inc., New York, 1986, Vol. 1, p. 1.
7. Kossmehl, G. A., in *Handbook of Conducting Polymers* (T. A. Skotheim, ed.), Marcel Dekker, Inc., New York, 1986, Vol. 1, p. 351.
8. Naarman, H., *Angew. Makromol. Chem.*, *109/110*, 295 (1982).
9. Wegner, G., *Angew. Chem. Int. Ed. Engl.*, *20*, 361 (1981).
10. Hörhold, H.-H., and Opfermann, J., *Makromol. Chem.*, *131*, 105 (1970).
11. Hörhold, H.-H., *Zeitschr. Chem.*, *12*, 41 (1972).
12. Hörhold, H.-H., Bergmann, R., Gottschaldt, J., and Drefahl, G., *Acta Chim. Acad. Sci. Hung.*, *81*, 239 (1974).
13. Hörhold, H.-H., Palme, H.-J., and Bergmann, R., *Faserforsch. Tectiltech. Z. Polymerforsch.*, *29*, 299 (1978).
14. Hörhold, H.-H., Helbig, M., Raabe, D., Opfermann, J., Scherf, U., Stockmann, R., and Weiss, D., *Zeitschr. Chem.*, *27*, 126 (1987).
15. Lapitskii, G. A., Makin, S. M., and Berlin, A. A., *Polym. Sci. USSR*, *9*, 1423 (1967).
16. Drefahl, G., Kühmstedt, R., Oswald, H., and Hörhold, H.-H., *Makromol. Chem.*, *131*, 89 (1970).
17. Drefahl, G., Hörhold, H.-H., and Wildner, H., East German Patent 51,436 (1966); *Chem. Abstr.*, *66*, 105346h (1967).
18. Hörhold, H.-H., and Opfermann, J., *Makromol. Chem.*, *176*, 691 (1975).
19. Hörhold, H.-H., and Schön, R., East German Patent 84,272 (1971); *Chem. Abstr.*, *78*, 125164y (1973).
20. Hörhold, H.-H., and Opfermann, J., *Makromol. Chem.*, *178*, 195 (1977).
21. Kossmehl, G., and Bohn, B., *Angew. Chem. Int. Ed. Engl.*, *12*, 237 (1973).
22. Kossmehl, G., Bohn, B., and Braser, W., *Makromol. Chem.*, *177*, 2369 (1976).
23. Kossmehl, G., *Ber. Bunsenges. Phys. Chem.*, *83*, 417 (1979).
24. Kossmehl, G., and Yaridjanian, A., *Makromol. Chem.*, *182*, 3419 (1981).

25. Thomson, D. W., U. S. Patent 3,422,071 (1969); *Chem. Abstr.*, *70*, 58414z (1969).
26. Moritani, I., Nagai, T., and Shirota, Y., *Kogyo Kagaku Zasshi*, *68*, 296 (1965); *Chem. Abstr. 63*, 685f (1965).
27. Ouchi, K., *Aust. J. Chem.*, *19*, 333 (1966).
28. Nagai, T., Namikoshi, H., and Tokura, N., *Tetrahedron*, *24*, 3267 (1968).
29. DeKoninck, L., and Smets, G., *J. Polym. Sci.*, *Polym. Chem. Ed.*, *7*, 3313 (1969).
30. Lee, C. J., and Wunderlich, B., *Makromol. Chem.*, *179*, 561 (1978).
31. Hoyt, J. M., Koch, K., Austin-Sprang, C., Stragevsky, S., and Frank, C. E., *ACS Polym. Prepr.*, *5*, 680 (1964).
32. Hörhold, H.-H., and Gottschaldt, J., East German Patent 104,092 (1974); *Chem. Abstr.*, *81*, 170231r (1974).
33. Hörhold, H.-H., Gottschaldt, J., and Opfermann, J., *J. Prakt. Chem.*, *319*, 611 (1977).
34. Hörhold, H.-H., and Raabe, D., *Acta Polym.*, *30*, 86 (1979).
35. Ballester, M., Castñer, J., and Riera, J., *J. Am. Chem. Soc.*, *88*, 957 (1966).
36. Moratti, S. C., Bradley, D. D. C., Friend, R., Breenham, N. C., and Holmes, A. B., *ACS Polym. Prepr.*, *35*(1), 214 (1994).
37. Rajaraman, L., Balasubramanian, M., and Nanjan, M. J., *Current Sci.*, *49*, 101 (1980).
38. Feast, W. J., and Millichamp, J. S., *Polymer*, *24*, 102 (1983).
39. Rehaln, M., and Schluter, A.-D., *Makromol. Chem. Rapid Commun.*, *11*, 375 (1990).
40. Cooke, A. W., and Wagener, K. B. *Macromolecules*, *24*, 1404 (1991).
41. Gilch, H. G., and Wheelwright, W. L., *J. Polym. Sci. A*, *Polym. Chem.*, *4*, 1337 (1966).
42. Hörhold, H.-H., Ozegowski, J.-H., and Bergmann, R., *J. Prakt. Chem.*, *319*, 622 (1977).
43. Hsieh, B. R., Polyphenylenevinylenes (methods of preparation and properties), in *The Polymeric Materials Encyclopedia, Synthesis, Properties and Applications* (J. C. Salome, ed.), CRC Press, Baca Raton, FL, Vol. 8, pp. 6537–6548.
44. Swatos, W. J., and Gordon, B., III, ACS *Polym. Prepr.*, *31*, 505 (1990).
45. Swatos, W. J., and Gordon, B., III, *Polym. Prep.*, *24*, 143 (1990).
46. Hsieh, B. R., and Feld, W. A., ACS *Polym. Prepr.*, *34*(2), 410 (1993); also a terse update of the 1993 work in Hsieh, B. R., Razafitrimo, H., Gao, Y., Nijakowski, T. R., Feld, W. A., ACS *Polym. Prepr.*, *36*(2), 85 (1995).
47. Wudl, F., Höger, S., Zhang, C., Pakbaz, K., and Heeger, A., ACS *Polym. Prepr.*, *34*(1), 197 (1993).
48. Wessling, R. A., and Zimmerman, R. G., U.S. Patent 3,401,152 (1968); *Chem. Abstr.*, *69*, 87735q (1969).
49. Wessling, R. A., *J. Polym. Sci.*, *Polym. Symp.*, *72*, 55 (1985).
50. Kanbe, M., and Okawara, M., *J. Polym. Sci.*, *Polym. Chem. Ed.*, *6*, 1058 (1968).
51. Kambe, M., and Okawara, M., *Kogyo Kagaku Zasshi*, *71*, 1276 (1968); *Chem. Abstr.*, *70*, 4682s (1969).
52. Kahlert, H., and Leising, G., *ICSM Abstracts*, Abao Terme, Italy, 1984, p. 21; cited in Montaudo, G., Vitalini, D., and Lenz, R. W., *Polymer*, *28*, 837 (1987).
53. Murase, I., Ohnishi, T., Noguchi, T., and Hirooka, M., *Polym. Commun.*, *25*, 327 (1984).
54. Capistran, J. D., Gragnon, D. R., Antoun, S., Lenz, R. W., and Karasz, F. E., ACS *Polym. Prepr.*, *25*, 282 (1984).
55. Gagnon, D. R., Capistran, J. D., Karasz, F. E., and Lenz, R. W., *Polym. Bull.*, *12*, 293 (1984).
56. Lahti, P. M., Poly(arylene vinylenes): Mechanistic control of a soluble precursor method, in *The Polymeric Materials Encyclopedia, Synthesis, Properties and Applications* (J. C. Salome, ed.), CRC Press, Baca Raton, FL, Vol. 7, pp. 5604–5615 (1996).
57. Denton, F. R., III, Ph. D. thesis, University of Massachusetts, Amherst, MA, Univ. Microfilms Int., Order No. DA 9120870; *Diss. Abstr. Int. B*, *52* (2), 833 (1991).

58. Garay, R. O., and Lenz, R. W., *Makromol. Chem.*, 15(Suppl.), 1 (1989).
59. Jin, J-I., Lee, Y-H., Lee, K-S., Kim, S-K., and Park, Y-W., *Synth. Met.*, 29, E47 (1989).
60. Weitzel, H. P., Böhm, A., and Müllen, K., *Makromol. Chem.*, 191, 2185 (1990).
61. Weitzel, H. P., and Müllen, K., *Makromol. Chem.*, 191, 2837 (1990).
62. Denton, F. R., III, Sarker, A., Lahti, P. M., Garay, R. O., and Karasz, F. E., *J. Polym. Sci. A, Polym. Chem.*, 30, 2233 (1992).
63. Sarker, A. M., Ph. D. thesis, University of Massachusetts, Amherst, MA, 1991.
64. Denton, F. R., III, Lahti, P. M., and Karasz, F. E., *J. Polym. Sci. A, Polym. Chem.*, 30, 2223 (1992).
65. Hsieh, B. R., *ACS Polym. Prepr.*, 32(1), 169 (1991); cf. also the original proposal for the anionic mechanism in Lahti, P. M., Modarelli, D. A., Denton, F. R., III, Lenz, R. W., and Karasz, F. E., *J. Am. Chem. Soc.*, 110, 7258 (1988).
66. Stenger-Smith, J. D., Chafin, A. P., and Norris, W. P., *J. Org. Chem.*, 59, 6107 (1994).
67. Hall, H. K., Jr., *Angew. Chem. Int. Ed. Engl.*, 22, 440 (1983).
68. Hall, H. K., Jr., and Padias, A. B., *Acc. Chem. Res.*, 23, 3 (1990).
69. Hall, H. K., Jr., Itoh, T., Iwatsuki, S., Padias, A. B., and Mulvaney, J. E., *Macromolecules*, 23, 913 (1990).
70. Askari, S. H., Rughooputh, S. D., and Wudl, F., *ACS Div. Polym. Mater. Sci. Eng. Prepr.*, 59(2), 1068 (1988).
71. Patil, A. O., Rughooputh, S. D. D. V., Heeger, A. J., and Wudl, F., *ACS Div. Polym. Mater. Sci. Eng. Prep.*, 59, 1068 (1988).
72. Askari, S. H., Rughooputh, S. D., Wudl, F., and Heeger, A. J., *ACS Polym. Prepr.*, 30, 157 (1989).
73. Sonoda, Y., and Kaeriyama, K., *Bull. Chem. Soc. Japan*, 65, 853 (1992).
74. Paulvannan, W. A., and Feld, W. A., *ACS Polym. Prep.*, 32(1), 193 (1991).
75. Braun, D., and Heeger, A. J., *Appl. Phys. Lett.*, 58, 1982 (1991).
76. Wudl, F., Allemand, P. M., Sradanov, G., Ni, Z., and McBranch, D., *ACS Symp. Ser.*, 455, 683 (1991).
77. McCoy, R. K., Karasz, F. E., Sarker, A., and Lahti, P. M., *Chem. Mater.*, 3, 941 (1991).
78. Sarker, A., and Lahti, P. M., *ACS Polym. Prepr.*, 35(1), 790 (1994).
79. Jin, J-I., Park, C-K., Kang, H-J., Yu, S-H., and Kim, J-C., *Synth. Met.*, 41–43, 271 (1991).
80. Shi, S., and Wudl, F., *ACS Div. Polym. Mater. Sci. Eng. Prepr.*, 59(2), 1164 (1988).
81. Shi, S., and Wudl, F., *Macromolecules*, 23, 2119 (1990).
82. Gurge, R. M., Sarker, A., Lahti, P. M., Hu, B., and Karasz, F. E., *Macromolecules*, 29, 4287 (1996).
83. Sarnecki, G. J., Friend, R. H., Holmes, A. B., and Moratti, S. C., *Synth. Met.*, 69, 545 (1995).
84. Kim, J. J., Kang, S. W., Hwang, D. H., and Shim, H. K., *Synth. Met.*, 57, 4024 (1993).
85. Sarker, A., Lahti, P. M., Garay, R. O., Lenz, R. W., and Karasz, F. E., *Polymer*, 35, 1412 (1994).
86. Höger, S., McNamara, J. J., Schricker, S., and Wudl, F., *Chem. Mater.*, 6, 171 (1994).
87. Murase, I., Ohnishi, T., Noguchi, T., and Hirooka, M., *Polym. Commun.*, 28, 229 (1987).
88. Yamada, S., Tokito, S., Tsutsui, T., and Saito, S., *JCS Chem. Commun.*, 1448 (1987).
89. Jen, K.-Y., Jow, T. R., and Elsenbaumer, R. L., *JCS Chem. Commun.*, 113 (1987).
90. Antoun, S., Gagnon, D. R., Karasz, F. E., and Lenz, R. W., *J. Polym. Sci. C, Polym. Lett.*, 24, 503 (1986).
91. Sarker, A., Lahti, P. M., and Karasz, F. E., *J. Polym. Sci. A, Polym. Chem.*, 32, 65 (1994).

92. Pomerantz, M., Wang, J., Seong, S., Starkey, K. P., Nguyen, L., and Marynick, D. S., *Macromolecules*, *27*, 7478 (1994).

93. Lenz, R. W., Han, C. C., and Lux, M., *Polymer*, *30*, 1041 (1989).

94. Jin, J-I., Shim, H-K., and Lenz, R. W., *Synth. Met.*, *29*, E53 (1989).

95. Shim, H-K., Lenz, R. W., and Jin, J-I., *Makromol. Chem.*, *190*, 389 (1989).

96. Lenz, R. W., Han, C. C., and Lux, M., *Polymer*, *30*, 1041 (1989).

97. Jin, J-I., Kim, J-C., and Shim, H-K., *Macromolecules*, *25*, 5519 (1992).

98. Sanborn, J., and Lahti, P. M., unpublished material (1996).

99. Gregorius, R. M., Lahti, P. M., and Karasz, F. E., *Macromolecules*, *25*, 6664 (1992).

100. Karasz, F. E., Capistran, J. D., Gagnon, D. R., and Lenz, R. W., *Mol. Cryst. Liq. Cryst.*, *118*, 327 (1985).

101. Halliday, D. A., Burn, P. L., Friend, R. H., and Holmes, A. B. *JCS Chem. Commun.*, 1685 (1992).

102. Gagnon, D. R., Capistran, J. D., Karasz, F. E., Lenz, R. W., and Antoun, S., *Polymer*, *28*, 567 (1987).

103. Onishi, T., Noguchi, M., and Nakano, T., Japanese Patent JP 02 32, 121 (1990); *Chem. Abstr.*, *113*, 143477y (1990).

104. Lenz, R. W., Han, C.-C., Stenger-Smith, J., and Karasz, F. E., *J. Polym. Sci. A.*, *Polym. Chem.*, *26*, 3241 (1988).

105. Stenger-Smith, J. D., and Lenz, R. W., *Polymer*, *30*, 1048 (1989).

106. Martens, J., Colaneri, N. F., Burn, P., Bradley, D. D. C., Marseglia, E. A., and Friend, R. H., *NATO ASI Ser.*, *Ser. B.*, *248*, 393 (1990).

107. Tokito, S., Momii, T., Murata, H., Tsutui, T., and Saito, S., *Polymer*, *31*, 1137 (1990).

108. Momii, T., Tokito, S., Tsutsui, T., and Saito, S., *Chem. Lett.*, 1201 (1988).

109. Murase, I., Ohnishi, T., Noguchi, T., and Hirooka, M., *Polym. Commun.*, *28*, 229 (1987).

110. Louwet, F., Vanderzande, D., and Gelan, J., *Synth. Met.*, *69*, 509 (1995).

111. Louwet, F., Vanderzande, D., Gelan, J., and Müllens, J., *Macromolecules*, *28*, 1330 (1995).

112. Gelan, J., Vanderzande, D., and Louwet, F., European Patent Appl. EP 644,217 (1995); *Chem. Abstr.*, *123*, 144940 (1995).

113. Burn, P. L., Holmes, a. B., Kraft, A., Bradley, D. D. C., Brown, A. R., Friend, R. H., and Gymer, R. W., *Nature*, *356*, 47 (1992).

114. Murase, I., Ohnishi, T., Noguchi, T., and Hirooka, M., *Polym. Commun.*, *25*, 327 (1984).

115. Machado, J. M., Karasz, F. E., Kovar, R. F., Burnett, J. M., and Druy, M. A., *New Polym. Mater.*, *1*, 189 (1989).

116. Schlenoff, J. B., and Wang, L-J., *Macromolecules*, *24*, 6653 (1991).

117. Beerden, A., Vanderzande, D., and Gelan, J., *Synth. Met.*, *52*, 387 (1992).

118. Patil, A. O., Rughooputh, S. D. D. V., and Wudl, F., *Synth. Met.*, *29*, E115 (1989).

119. Garay, R. O., Baier, U., Bubeck, C., and Müllen, K., *Adv. Mater.*, *5*, 561 (1993).

120. Yan, M., Rothberg, L. J., Papadimitrakopoulos, F., Galvin, M. E., and Miller, T. M., *Phys. Rev. Lett.*, *73*, 744 (1994).

121. Papadimitrakopoulos, F., Konstadinidis, K., Miller, T. M., Pila, R., Chandross, E. A., and Galvin, M. E., *Chem. Mater.*, *6*, 1563 (1994).

122. Papadimitrakopoulos, F., Miller, T. M., Chandross, E. A., and Galvin, M. E., *ACS Polym. Prepr.*, *35*, 215 (1994).

123. Herold, M., Gmeiner, J., and Schowerer, M., *Acta Polym.*, *45*, 392 (1994).

124. Massardier, V., Guyot, A., and Hoang, T. V., *Polymer*, *35*, 1561 (1994).

125. Sagnes, O., Dubois, J. C., Massardier, V., Hoang, T. V., and Guyot, A., European Patent Appl. EP 519,771 (1992); *Chem. Abstr.*, *119*, 215324v (1993).

126. Massardier, V., Michel, P., Dubois, J.-C., Tran Van, H., Guyot, A., and Davenas, J., French Patent FR 92-15383 921221 (1993).

127. Tran, V. H., Nguyen, T. P., Massardier, V., Davenas, J., and Boiteux, G., *Synth. Met.*, *69*, 435 (1995).

128. Panow, E. M., Rybakova, L. F., and Kocheshkov, K. A., *Dokl. Akad. Nauk SSR Ser. Khim.*, *190*, 122 (1970); *Chem. Abstr.*, *72*, 90897 (1970). Panow, E. M., Kocheshkov, K. A., Rybakova, L. F., and Zimin, A. V., British Patent 1,124,913 (1968); *Chem. Abstr.*, *70*, 3503x (1969).

129. Kryazhev, Yu. G., Ermakova, T. G., and Shibanova, E. F., *Bull. Acad. Sci. USSR, Div. Chem. Sci.*, *20*, 2279 (1971).

130. Sinitskii, V. V., Machina, G. F., Kryazhev, Yu, G., and Rozenshtein, L. D., *Bull. Acad. Sci. USSR, Div. Chem. Sci.*, *21*, 932 (1972).

131. Greiner, A., and Heitz, W., *Makromol. Chem. Rapid Commun.*, *9*, 581 (1988).

132. Brenda, M., Greiner, A., and Heitz, W., *Makromol. Chem.*, *191*, 1083 (1990).

133. Greiner, A., Martelock, H., Noll, A., Siegfried, N., and Heitz, W., *Polymer*, *32*, 1857 (1991).

134. Greiner, A., and Heitz, W., *ACS Polym. Prepr.*, *32*, 333 (1991).

135. Martelock, H., Greiner, A., and Heitz, W., *Makromol. Chem.*, *192*, 967 (1991).

136. Heitz, W., *Makromol. Chem.*, *Macromol. Symp.*, *48–49* 15 (1991).

137. Yu, L., and Bao, Z., Poly(phenylenevinylenes) by the Heck coupling reaction, in *The Polymeric Materials Encyclopedia, Synthesis, Properties and Applications* (J. C. Salome, ed.), Baca Raton, FL, Vol. 8, pp. 6532–6537.

138. Pan, M., Bao, Z., and Yu, L., *Macromolecules*, *28*, 5151 (1995).

139. Müllen, K., Böhm, A., Fiesser, G., Garay, R. O., Mauermann, H., and Stein, S., *ACS Polym. Prepr.*, *34*(1), 195 (1994).

140. Kumar, A., and Eichinger, B. E., *Makromol. Chem., Rapid Commun.*, *13*, 311 (1992).

141. Conticello, V. P., Gin, D. L., and Grubbs, R. H., *J. Am. Chem. Soc.*, *114*, 9708 (1992).

142. Pu, L., Wagaman, M. W., and Grubbs, R. H., *Macromolecules*, *29*, 1138 (1996).

143. Thron-Scanyi, E., and Hoehnk, H. D., *J. Molec. Catal.*, *76*, 101 (1992).

144. Miao, Y.-J., and Bazan, G. C., *J. Am. Chem. Soc.*, *116*, 9379 (1994).

145. Thorn-Csanyi, E., and Pflug, K. P., *J. Molec. Catal.*, *90*, 69 (1994).

146. Thorn-Csanyi, E., and Pflug, K. P., *Makromol. Chem., Rapid Commun.*, *14*, 619 (1993).

147. Iwatsuki, S., Kubo, M., and Kumeuchi, T., *Chem. Lett.*, 1071 (1991).

148. Staring, E. G. J., Braun, D., Rikken, G. L. J. A., Demandt, R. J. C. E., Kessener, Y. A. R. R., Bouwmans, M., and Broer, D., *Synth. Met.*, *67*, 71 (1994).

149. Nishihara, H., Akasaka, M., Tateishi, M., and Aramaki, K., *Chem. Lett.*, 2061 (1992).

150. Wessling, R. A., and Settineri, W. J., U.S. Patent 3,480,525 (1969); *Chem. Abstr.*, *72*, 32677d (1970). Wessling, R. A., and Settineri, W. J., U.S. Patent 3,697,398 (1972); *Chem. Abstr.*, *78*, 59879g (1973).

151. Hoeg, D. F., Lusk, D. I., and Goldberg, E. P., *J. Polym. Sci., Polym. Lett.*, *2*, 697 (1964).

152. Wessling, R. A., and Zimmerman, R. G., U. S. Patent 3,706,677 (1972); *Chem. Abstr.*, *78*, 85306n (1973).

153. Murase, I., Ohnishi, T., and Noguchi, T., Japanese Patent 01 79,222 (1989); *Chem. Abstr.*, *111*, 165461z (1989). Murase, I., Ohnishi, T., and Noguchi, T., Japanese Patent 01 79,223 (1989); *Chem. Abstr.*, *111*, 165462a (1989).

154. Sato, M., Kobayashi, A., Ishikawa, H., and Jo, S., Japanese Patent 05,052,756 (1993); *Chem. Abstr.*, *120*, 234225 (1994).

155. Hsieh, B. R., *ACS Polym. Prepr.*, *32*(1), 631 (1991).

156. Hörhold, H.-H., and Helbig, M., *Makromol. Chem., Macromol. Symp.*, *12*, 229 (1987).

157. Thomson, D. W., U.S. Patent 3,422,071 (1969); *Chem. Abstr.*, *70*, 58414z (1969).

158. Schlenoff, J. B., and Wang, L-J., *ACS Polym. Prepr.*, *35*(1), 238 (1994).
159. Machado, J. M., Denton, F. R., III, Schlenoff, J. B., Karasz, F. E., and Lahti, P. M., *J. Polym. Sci.*, *Polym. Phys. Ed.*, *27*, 199 (1989).
160. Bradley, D. D. C., *J. Phys. D: Appl. Phys.*, *20*, 1389 (1987).
161. Friend, R. H., Bradley, D. D. C., and Townsend, P. D., *J. Phys. D: Appl. Phys.*, *20*, 1367 (1987).
162. Bradley, D. D. C., and Friend, R. H., *J. Molec. Electron.*, *5*, 19 (1989).
163. Masse, M. A., Ph. D. thesis, University of Massachusetts, Amherst, MA, (1989), p. 23. Masse, M. A., Martin, D. C., Karasz, F. E., and Thomas, E. L., *ACS Div. Polym. Mater. Sci. Eng. Prepr.*, *57*, 441 (1987).
164. Bond, S. F., Friend, R. H., and Howie, A., *J. Microsc. (Oxford)*, *171*, 199 (1993).
165. Simpson, J. H., Rice, D. M., and Karasz, F. E., *Macromolecules*, *25*, 2099 (1992).
166. Granier, T., Thomas, E. L., and Karasz, F. E., *J. Polym. Sci.*, *Part B: Polym. Phys.*, *27*, 469 (1989).
167. Martens, J. H. F., Bradley, D. D. C., Burn, P. L., Friend, R. H., Holmes, A. B., and Marseglia, E. A., *Synth. Met.*, *41*, 301 (1991).
168. Simpson, J. H., Egger, N., Masse, M. A., Rice, D. M., and Karasz, F. E., *J. Polym. Sci.*, *Part B: Polym. Phys.*, *28*, 1859 (1990).
169. Simpson, J. H., Rice, D. M., and Karasz, F. E., *J. Polym. Sci.*, *Polym. Phys.*, *30*, 11 (1992).
170. Simpson, J. H., Liang, W., Rice, D. M., and Karasz, F. E., *Macromolecules*, *25*, 3068 (1992).
171. Chen, D., Winokur, M. J., Masse, M. A., and Karasz, F. E., *Polymer*, *33*, 3116 (1992).
172. Masse, M. A., Schlenoff, J. B., Karasz, F. E., and Thomas, E. L., *J. Polym. Sci.*, *Polym. Phys.*, *27*, 2045 (1989).
173. Masse, M. A., Martin, D. C., Thomas, E. L., Karasz, F. E., and Petermann, J. H., *J. Mater. Sci.*, *25*, 311 (1990).
174. Chen, D., Winokur, D., Masse, M. A., and Karasz, F. E., *Phys. Rev. B*, *41*, 6759 (1990).
175. Machado, J. M., Ph. D. thesis, University of Massachusetts, Amherst, MA, 1989.
176. Murase, I., Ohnishi, T., Noguchi, T., and Hirooka, M., *Synth. Met.*, *17*, 639 (1987).
177. Stenger-Smith, J. D., Lenz, R. W., and Wegner, G., *Polymer*, *30*, 1048 (1989).
178. Martens, J. H. F., Halliday, D. A., Marseglia, E. A., Bradley, D. D. C., Friend, R. H., Burn, P. L., and Holmes, A. B., *Synth. Met.*, *55*, 434 (1993).
179. Skotheim, T. A. (ed.), *Handbook of Conducting Polymers*, Marcel Dekker, Inc., New York, (1986).
180. Wnek, G. E., Chien, J. C. W., Karasz, F. E., and Lillya, C. P., *Polymer*, *20*, 1441 (1979).
181. Masse, M., Hirsch, J. A., White, V. A., and Karasz, F. E., *New Polym. Mater.*, *2*, 75 (1990).
182. Murase, I., Ohnishi, T., and Noguchi, T., U.S. Patent 4,528,118 (1985).
183. Murase, I., Ohnishi, T., Noguchi, T., Hirooka, M., and Murakama, S., *Mol. Cryst. Liq. Cryst.*, *118*, 333 (1985).
184. Gagnon, D. R., Karasz, F. E., Thomas, E. L., and Lenz, R. W., *Synth. Met.*, *20*, 85 (1987).
185. Masse, M. A., Composto, R. J., Jones, R. A. L., and Karasz, F. E., *Macromolecules*, *23*, 3675 (1990).
186. Sakamoto, A., Furukawa, Y., Tasumi, M., Noguchi, T., and Ohnishi, T., *Synth. Met.*, *69*, 439 (1995).
187. Mertens, R., Nagels, P., Callaerts, R., Van Roy, M., Briers, J., and Geise, H. J., *Synth. Met.*, *51*, 55 (1992).
188. Han, C. C., and Elsenbaumer, R. L., *Synth. Met.*, *30*, 123 (1989).

189. El-Atawy, S., and Davidson, K. J., *J. Chem. Soc., Faraday Trans.*, *90*, 879 (1994).
190. Sakamoto, A., Furukawa, Y., and Tasumi, M., *J. Phys. Chem.*, *98*, 4635 (1994).
191. Ohnishi, T., Noguchi, T., Hirooka, M., and Murase, I., *Synth. Met.*, *41*, 309 (1991).
192. Jen, K-Y., Jow, T. R., and Elsenbaumer, R. L., *JCS Chem. Commun.*, 1113 (1987).
193. Antoun, S., Karasz, F. E., and Lenz, R. W., *J. Polym. Sci, A, Polym. Chem.*, *26*, 1809 (1988).
194. Patil, A. O., *Polym. News*, *14*, 234 (1989).
195. Chen, D., Winokur, M. J., Cao, Y., Heeger A. J., and Karasz, F. E., *Phys. Rev. B*, *45*, 2035 (1992).
196. Simpson, J. H., Rice, D. M., Karasz, F. E., Rossitto, F. C., and Lahti, P. M., *Polymer*, *34*, 4595 (1993).
197. Esteghamatian, M., and Xu, G., *Synth. Met.*, *63*, 195 (1994).
198. Winokur, M. J., Chen, D., and Karasz, F. E., *Synth. Met.*, *41*, 341 (1991).
199. Iucci, G., Xing, K., Loegdlund, M., Fahlman, M., and Salaneck, W. R., *Chem. Phys. Lett.*, *244*, 139 (1995).
200. Messier, J., Kajzar, F., Prasad, P., and Ulrich, D. (eds.), *Nonlinear Optical Effects in Organic Polymers*, Dordrecht, Netherlands, (1989).
201. Prasad, P. N., and Williams, D. J. (eds.), *Introduction to Nonlinear Optical Effects in Molecules and Polymers*, John Wiley & Sons, New York, 1991.
202. Singh, B. P., Prasad, P. N., and Karasz, F. E., *Polymer*, *29*, 1940 (1988).
203. Kaino, T., Kubodera, K-I., Tomaru, S., Kurihara, T., Saito, S., Tsutsui, T., and Tokito, T., *Electron. Lett.*, *23*, 1095 (1987); Kaino, T., Kobayashi, H., Kubodera, T., Kurihara, T., Saito, S., Tsutsui, T., and Tokito, S., *Appl. Phys. Lett.*, *54*, 1619 (1989).
204. Bradley, D. D. C., and Mori, Y., *Jpn. J. Appl. Phys.*, *28*, 174 (1989); McBranch, D., Sinclair, M., Heeger, A. J., Patil, A. O., Shi, S., Askari, S., and Wudl, F., *Synth. Met.*, *29*, E85 (1989).
205. Bubeck, G. A., Kaltbeitzel, A., Lenz, R. W., Neher, D., Stenger-Smith, J. D., and Wegner, G., in *Nonlinear Optical Effects in Organic Polymers* J. Messier, F. Kajzar, P. Prasad, and D. Ulrich (eds.), NATO ASI Ser. E, Kluwer, Dordrecht, p. 143 (1989).
206. Prasad, P. N., and Williams, D. J. (eds.), *Introduction to Nonlinear Optical Effects in Molecules and Polymers*, John Wiley & Sons, New York, 1991, p. 238.
207. Swiatkiewicz, J., Prasad, P. N., Karasz, F. E., Druy, M. A., and Glatkowski, P., *Appl. Phys. Lett.*, *56*, 892 (1990).
208. Wung, C-J., Lee, K-S., Prasad, P. N., Kim, J-J., Jin, J-I., and Shim, H-K., *Polymer*, *33*, 4145 (1992).
209. Wung, C. J., Pang, Y., Prasad, P. N., and Karasz, F. E., *Polymer*, *32*, 605 (1991).
210. Jin, J-I., *Mol. Cryst. Liq. Cryst. Sci. Technol. Sect. A*, *280*, 47 (1996); Jin, J-I., and Shim, H-K., *ACS Symp. Ser.*, *601*, 223 (1995).
211. Jin, J-I., Lee, Y-H., Park, C., and Nam, B., *Macromolecules*, *27*, 5230 (1994).
212. Heeger, A., J., Orenstein, J., and Ulrich, D. R. (eds.), *Materials Research Society Symposium Proceedings*, *Volume 109*, Nonlinear optical properties of polymers, Materials Research Society, Boston, 1987.
213. Burroughes, J. H., Bradley, D. D. C., Brown, A. R., Marks, R. N., MacKay, K., Friend, R. H., Burn, P. L., and Holmes, A. B., *Nature*, *347*, 539 (1990).
214. Friend, R., Bradley, D. D. C., and Holmes, A. B., *Physics World*, *5*, 42 (1992).
215. Holmes, A. B., Bradley, D. D. C., Brown, A. R., Burn, P. L., Burroughes, J. H., Friend, R. H., Greenham, N. C., Gymer, R. W., Halliday, D. A., et al., *Synth. Met.*, *57*, 4031 (1993).
216. Bradley, D. D. C., *Synth. Met.*, *54*, 401 (1993).
217. Lahti, P. M., Obrzut, J., and Karasz, F. E., *Macromolecules*, *20*, 2023 (1987).
218. Brédas, J. L., dos Santos, D. A., Quattrocchi, C., Friend, R. H., and Heeger, A. J., *ACS Polym. Prepr. 35*(1), 185 (1994).

219. Brédas, J. L., and Heeger, A. J., *Chem. Phys. Lett.*, *217*, 507 (1994).
220. Cornil, J., Beljonne, D., and Brédas, J. L., *Polym. Mater. Sci. Eng.*, *72*, 459 (1995).
221. Colaneri, N. F., Bradley, D. D. C., Friend, R. H., Burn, P. L., Holmes, A. B., and Spangler, C. W., *Phys. Rev. B*, *42*, 11670 (1990).
222. Bradley, D. D. C., Brown, A. R., Burn, P. L., Burroughes, J. H., Friend, R. H., Holmes, A. B., Mackay, K. D., and Marks, R. N., *Synth. Met.*, *43*, 3135 (1991).
223. Yan, M., Rothberg, L. J., Papadimitrakopoulos, F., Galvin, M. F., and Miller, T. M., *Phys. Rev. Lett.*, *72*, 1104 (1994).
224. Wang, H. L., Park, J. W., Fu, J. K., Marsella, M. J., Swager, T. M., MacDiarmid, A. G., Wang, Y. Z., Gebler, D. D., and Epstein, A. J., *ACS Polym. Prepr.*, *36*, 45 (1995).
225. Yang, Z., Hu, B., and Karasz, F. E., *Macromolecules*, *28*, 6151 (1995).
226. Herold, M., Gmeiner, J., and Schowerer, M., *Acta Polym.*, *45*, 392 (1994).
227. Greenham, N. C., Moratt, S. C., Bradley, D. D. C., Friend, R. H., and Holmes, A. B., *Nature*, *365*, 628 (1993).
228. Yang, Z., Sokolik, I., and Karasz, F. E., *Macromolecules*, *26*, 1188 (1993); Sokolik, I., Yang, Z., Karasz, F. E., and Morton, D. C., *J. Appl. Phys.*, *74*, 3584 (1993).
229. Zhang, C., Höger, S., Pakbaz, K., Wudl, F., and Heeger, A. J., *ACS Polym. Prepr.*, *35*(1), 329 (1994).
230. Heeger, A. J., *ACS Polym. Prepr.*, *35*(2), 100 (1994).
231. Gurge, R. M., Sarker, A., Lahti, P. M., Hu, B., and Karasz, F. E., *Macromolecules*, *29*, 4287 (1996).
232. Gurge, R. M., Sarker, A., Lahti, P. M., Hu, B., and Karasz, F. E., *Macromolecules*, *30*, 8286 (1997).
233. Burn, P. L., Kraft, A., Baigent, D. R., Bradley, D. D. C., Brown, A. R., Friend, R. H., Gymer, R. W., Holmes, A. B., and Jackson, R. W., *J. Am. Chem. Soc.*, *115*, 10117 (1993).
234. Brown, A. R., Burn, P. L., Bradley, D. D. C., Friend, R. H., Kraft, A., and Holmes, A. B., *Mol. Cryst. Liq. Cryst. Sci. Technol. Sect. A*, *216*, 111 (1992).
235. Zhang, C., Braun, D., and Heeger, A. J., *J. Appl. Phys.*, *73*, 5177 (1993).
236. Yan, M., Rothberg, L. J., Papadimitrakopoulos, F., Galvin, M. E., and Miller, T. M., *Mol. Cryst. Liq. Cryst. Sci. Technol. Sect. A*, *256*, 17 (1994).
237. Hsieh, B. R., Ettedgui, E., Park, K. T., and Gao, Y., *Mol. Cryst. Liq. Cryst. Sci. Technol. Sect. A*, *256*, 71 (1994).
238. Bradley, D. D. C., *Makromol. Chem.*, *Macromol. Symp.*, *37*, 247 (1990).
239. Burn, P. L., Bradley, D. D. C., Friend, R. H., Halliday, D. A., Holmes, A. B., Jackson, R. W., and Kraft, A., *J. Chem. Soc. Perkin Trans. 1*, 3225 (1992).
240. Martens, J. H. F., *Synth. Met.*, *55*, 440 (1993).
241. Chiang, L. Y., Chaikin, P. M., and Cowan, D. O. (eds.), *Materials Research Society Symposium Proceedings Volume 173*, Materials Research Society, Boston, 1989.
242. Chiang, L. Y., Garito, A. F., and Sandman, D. J. (eds.), *Materials Research Society Symposium Proceedings, Volume 247*, Materials Research Society, Boston, 1991.
243. Garito, A. F., Jen, A. K-Y., Lee, C.Y-C., and Dalton, L. R. (eds.), *Materials and Research Society Symposium Proceedings, Volume 328*, Materials Research Society, Boston, 1993.
244. Jen, A. K-Y., Lee, C. Y-C., Dalton, L. R., Rubner, M. F., Wnek, G. E., and Chiang, L. Y., *Materials Research Society Symposium Proceedings, Volume 413*, Materials Research Society, Boston, 1995.
245. Salaneck, W. R., Lundström, I., and Rånby, B., *Conjugated Polymers and Related Materials. The Interconnection of Chemical and Electronic Structure*, Oxford University Press, New York, 1993.
246. Salamone, J. (ed.), *The Polymeric Materials Encyclopedia, Synthesis, Properties and Applications* (12 volumes), CRC Press, Baca Raton, FL, 1996.

4
Conjugated Polymer Light-Emitting Diodes and Microcavities

Franco Cacialli
Cavendish Laboratory
Cambridge, England

I. INTRODUCTION

A. General Remarks

This chapter is concerned with the physics and the technology of light-emitting diodes (LEDs) which use a conjugated, semiconducting polymer as the active material (PLEDs).

The development of polymer-based light-emitting diodes and the study of the related physical and chemical properties have recently become very important areas of interest in the field of advanced materials for electronic applications [1–6].

In addition to the scientific challenge that conjugated polymers and their properties pose to chemists, physicists, and material scientists, an important reason for this activity is the keen interest of the markets for alternative technologies to be used in displays and devices for luminous signaling.

Plastic LEDs offer at least three fundamental advantages over inorganic ones: cheap fabrication technology, possibility of large-area devices, and mechanical flexibility. Ease of chemical tunability, especially of the energy gap, is another plus of organics in general, owing to the great variety of tools available to organic synthesis, but inorganic LEDs are, by now, commercially available in red, green, and blue colors, and tunability is not perceived as a strong advantage anymore.

The market attention for conjugated polymers has offered a unique opportunity for funding further research and, therefore, for developing a real technology out of a scientific achievement. Such a technology must be able to cope with some of the traditionally "weak" properties of these materials, as the limited temporal stability, and its development is still in progress.

In spite of the very advanced state of the art of the organic LEDs (OLEDs) based on low-molecular-weight materials, we will restrict ourselves to polymer LEDs, as these do not suffer from the same recrystallization problems which generally affect short oligomers and because we will draw our examples from our personal research experience or from that of the Cambridge group, and both have proceeded mainly in this direction.

The informed reader will notice that the quoted literature is only a small fraction of what is published in the field. This is because the number of articles and books on the subject is by now huge, and a complete review of the significant work is beyond the purpose of this chapter. Similarly, space constraints do not allow for a full discussion of the general properties and of the theoretical models of semiconducting polymers, for which we refer the reader to the other chapters of this book and to some excellent texts available in the literature [7,8].

Here, we will only provide a brief introduction of the fundamentals and give, instead, an increased attention to some specific issues of semiconducting polymer devices, such as the limitations to the luminescence efficiency or the stability under operation; in addition to discussing the related physical problems, we will try to provide some technological insights by considering proposed and possible strategies to deal with those limitations. A general knowledge of the fundamental concepts of inorganic semiconductor physics [9] and of the general properties of electronic processes in organic materials [10a] will be assumed.

Before starting to consider the details of PLEDs, we would like to draw the reader's attention to a brief consideration of the historical and technological background which hosted the development of organic electroluminescence. The rest of the chapter is organized as follows. In Section II, we introduce the polymer poly (*p*-phenylene vinylene) (PPV), which has been the first to be used and is still one of the best polymers in terms of luminescence efficiency and stability under operation. It will serve as a prototype material for describing the fundamental processes in electroluminescence (EL) diodes, although, in Section III, we will make extensive reference to other conjugated polymers. In Section III, we also point out the unresolved issues in our understanding of the diode's operation principle and the limitations in performance, regarding mainly efficiency and lifetime.

In Section IV, we will present some of the strategies which can be used for raising the efficiency; in Section V, we will be concerned with the stability of the PLEDs and the different aspects of the problem, including considerations on the active material and the interfaces with the electrodes.

In Section VI, we will describe the possibilities of luminescence control accessible through the use of optical resonators, or microcavities, with a cursory mention to the recently reported lasing [11], before presenting our concluding remarks in Section VII.

B. Scientific and Technological Background

The recent interest for conjugated semiconducting polymers follows a more general tendency, in the last 30 years or so, from both the scientific and technical communities to dedicate an increasing attention to the so-called functional organic materials (i.e., organic materials with "complex" functions) relevant to the fields of electronics, photonics, or information technology.

In fact, although the vast majority of conventional polymers are insulators, the possibility that organic substances could have conductivities almost comparable with those of some metals has been suggested nearly 90 years ago [12]. Various historical and technical reasons, and in particular the development of a very successful inorganic semiconductors industry, have since hampered the use and the development of organics in electrical applications.

It was by the 1970s, when the polymer penetration in every-day life was so deep that people were starting to speak in terms of the Polymer Age, that in scientific contexts it was finally realized that organics can be good approximations of reduced dimensionality systems, providing the possibility of model testing and a wealth of new physics. Technologists also appreciated the potential advantages of synthetic compounds with useful electrical properties, such as low weight, tunability of the properties by chemical design, low-cost mass production, possibility of large-area applications, and flexible products. The interest in molecular solids, charge-transfer complexes and conjugated polymers also built up, thanks to the development of xerography and of the related photoconductive films and materials.

This happens in the context of a flourishing, although relatively expensive, inorganic-semiconductors-based microelectronics industry, the evolution of which passes through the reduction of the device physical dimensions. The progress in materials science and thin-film technology produced under market pressure is actually so fast that it is even suggested that intrinsic physical limits to further miniaturization will be reached in a short time. This gives further momentum to the development of an interest in "molecular materials" as potential candidates for applications based on "single-molecule" devices. Although the initial enthusiasm for what was immediately called the field of "Molecular Electronics" quickly overwhelmed the estimation of the difficulties for the development of such ideas, this phase has been fairly important for fostering the development of molecular materials for electronic and electrical applications, if not of a single-molecule electronics.

The origin of electroluminescence in organics can be traced back to the pioneering work on molecular crystals and, in particular, to the experiments of Pope and co-workers on anthracene single crystals in 1963 [10], followed by Helfrich and collaborators in 1965 [13,14]. Significant improvements in terms of driving voltage were then achieved by the use of thin films of similar materials which Roberts and co-workers were able to successfully deposit in 1979 [15] by using a Langmuir–Blodgett technique. The introduction of vacuum sublimation [16], and later of a double-layer structure [17], allowed Tang and co-workers at Kodak to obtain high efficiencies and to eventually set the whole field in motion.

It was only by 1990, however, that EL in a conjugated polymer, namely PPV, was discovered at Cambridge University. The absence of conjugated polymers from the scene of organic EL is, in part, due to the mental attitude which developed soon after the discovery of polyacetylene doping [18], focusing the interest more on the conducting and photoconducting properties obtainable upon doping than on other aspects. This was supported in the first place by the perception that there can be only slight conduction without doping, and in the second place by the fact that doping would have quenched the luminescence more or less irremediably. It is not accidental, then, that device applications concentrated on the transport properties, with several examples of field-effect transistors (FETs), diodes, and photodiodes, and that the only investigation of EL in a polymer, by Partridge, is conducted on poly-vinylcarbazole (PVK), a nonconjugated polymer. It is worth noting at this point that the fundamental discovery of 1990 would have not been possible without a substantial improvement of the synthetic routes of polymer preparation (in this case PPV), which, by then, are refined enough to allow for the deposition of thin films of high optical quality and with relatively high photoluminescence (PL) efficiencies. A similarly important role have had the studies of the surface electronic structure (with techniques such as x-ray or ultraviolet photoelectron spectroscopies—XPS and UPS, respectively) and of their interaction with other materials such as metals. These and the theoretical modeling of intrinsic and doped conjugated polymers have, in fact, provided a crucial element for the fast development of an "organic materials science."

Since 1990, the interest and the number of groups working on polymer and organic electroluminescence has grown very quickly, not only in the universities but also in many companies, and there are now several private research laboratories actively pursuing the optimization of devices, with a view of a commercial product. As a consequence, many further developments and significant additional knowledge has been generated, and we will discuss these in the following sections.

II. PPV AND RELATED MATERIALS

A. Processing

Polymers with extended π-conjugated electron systems are very difficult to process because they are infusible and insoluble in their final form. One way to circumvent this problem is by the use of a nonconjugated soluble or otherwise processible polymeric precursor, to be converted to the conjugated form in situ, by thermal or other treatment. This method was first introduced by Feast and co-workers for polyacetylene [19]. Alternatively, the polymer solubility can be increased by the addition of suitable side chains to the backbone, this second route allowing also for the tuning of the polymer properties through the chemical design of the side chains. Copolymers containing conjugated insoluble units and solubilizing units (conjugated or not) [20] can also be classified in this category. Both approaches have been applied with success to PPV, although the processing from the precursors retains some advantages in the insolubility of the final product. This, in fact, allows for further processing of the film, such as that required in the preparation of multilayer diodes, as discussed in Section IV.

A third method which has been investigated recently is the use of chemical-vapor deposition (CVD) [21,22]. This is potentially very interesting in view of its flexibility, deriving from the large number of controllable parameters (precursors purity, temperature, pressure, irradiation, chemical nature of the precursors), but the state of the art in terms of film quality and LED performance is not as advanced as for the previous methods. Fourth, in situ electropolymerization is also a viable route to conjugated polymers, although it is generally not very flexible compared to CVD [23], and the film quality in the specific case of PPV is relatively poor [24].

Figure 1 shows the chemical structure and one of the most common preparation routes to PPV, the material that more often than others will be used to illustrate general properties and to provide specific examples. The precursor contains a labile tetrahydrothiophenium (THT) group which can be eliminated by thermal treatment at temperatures ranging from 160°C to 350°C for several hours. Note that treatment

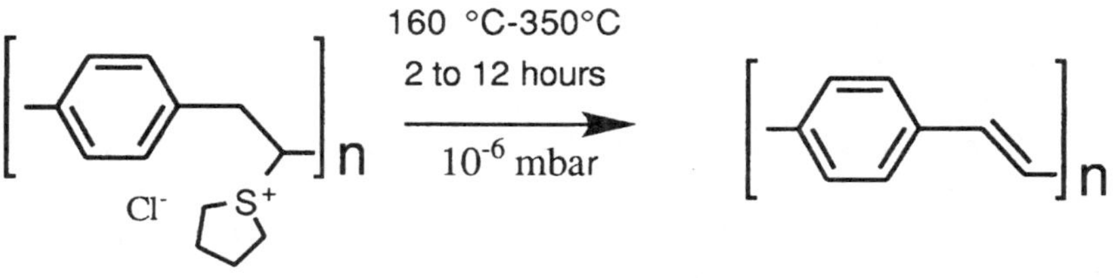

FIGURE 1   A possible precursor route to PPV.

of the precursor and tuning of the conversion conditions allow a wide flexibility in the control of the polymer properties, and in particular of the conjugation length [25], as amply discussed in the literature [26–33]. Conversions at the highest temperatures and for long times will result generally in longer conjugation lengths, red-shifted absorption and emission, and enhanced luminescence quenching, as a consequence of enhanced mobility of the excitons through the conjugated portions of the polymer [34] and of a higher number of oxygen-containing defects [35,36]. Limitation of the conjugation length (e.g., by the introduction of methoxy groups on the benzylic positions adjacent to the phenylene rings [28], which can resist thermal treatment in environments not strongly acidic) has, in fact, demonstrated very effective in controlling the PL and hence the EL efficiency. Attempts to change the optical properties by chemical doping, on the contrary, have generally resulted in thin-film quality not suitable for LED fabrication [37]. Other precursors based on different metastable groups are also possible [32].

Control of the conversion environment is crucial for the final polymer properties and, in particular, for the PL efficiency. As shown by various authors [35,36,38], exposure of PPV to an oxygen-rich environment leads to the formation of a number of defects, which interrupt conjugation and, more importantly, quench the luminescence. These are mainly carbonyl groups which tend to split the exciton by lowering the highest occupied molecular orbital (HOMO) and the lowest unoccupied molecular orbital (LUMO) on the chain segments where they are incorporated. In order to circumvent this problem, the conversion is often conducted in high vacuum (pressure $< 10^{-5}$–10^{-6} mbar) so as to keep the number of defects to a minimum, although conversion in forming gas (15% H_2 in 85% N_2) has been reported to be more effective in the reduction of carbonyl moieties [36].

Accurate study of the elimination reaction and of the consequences on the performance of LEDs have also been reported by Schwoerer and co-workers [38–41]. In particular, by monitoring the ultraviolet–visible (UV-vis) absorption spectrum, they find that the elimination can be completed at temperatures considerably lower if the reaction is carried out in an argon atmosphere [38]. This is very important for the preparation of LEDs on plastic substrates such as poly(ethylene terephtalate). The underlying mechanism is not clear, but it is speculated [38] that the similarity of the masses of Ar, S, and Cl may trigger a kinetic interaction of Ar with Cl and/or THT. Energy-dispersive x-ray analysis (EDX) provide evidence that Cl elimination is much slower than that of THT, and that it is desirable to have long conversion times in order to reduce the content of Cl, which may trigger unwanted polymer reactions and degradation. Quite surprisingly, the Bayreuth group also found that they are not able to detect a significant carbonyl incorporation even for conversion in pure oxygen (160°C, 2 h).

After conversion, the polymer films are uniform and fully dense and basically pinhole-free [42]. They also show a significant degree of crystallinity, even without the application of stretching procedures during the thermal treatment. This can be explained by considering the combined actions of the significant volume reduction (approximately by half) connected with the leaving-group elimination and the adhesion of the precursor to the substrate, which result in a serendipitous stretching of the film and hence in the mentioned crystallinity.

Before passing to illustrate the optical properties of the films so formed we notice that in addition to THT, Cl and HCl are also released during the thermal treatment and that the interaction of these with the local environment, and the substrate, in particular, may become very important for PLEDs.

B. Optical Properties and Energetic Structure of PPV

Figure 2 shows the optical properties of PPV films prepared according to the precursor route of Figure 1. The photoluminescence (PL) has been excited with the 457.9 nm of the Ar-ion laser, whereas the EL has been obtained from a device with indium–tin oxide (ITO) and Ca–Al electrodes. The thickness of the PPV for PL and EL is about 100 nm, whereas the thickness of the film used in absorption is ~125 nm and hence the extinction coefficient at 3 eV is about 1.92×10^5 cm^{-1}.

Note that the absorption edge is located at approximately 2.5 eV and that it overlaps slightly the blue end of the emission spectra. This determines a partial self-absorption of the luminescence which may have implications for the lifetimes

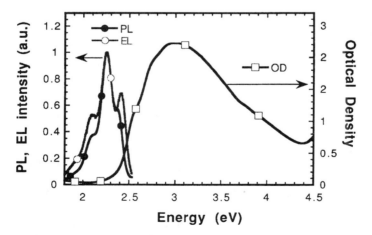

FIGURE 2 The optical properties of films of PPV from the precursor route of Fig. 1. The photoluminescence (PL) has been excited with 457.9 nm radiation from an Ar-ion laser; the EL has been obtained from a device with ITO and Ca–Al electrodes. The optical density (OD) spectrum is from a 125-nm-thick film and is not corrected for reflectivity. The PL and EL spectra are from 125-nm-thick films.

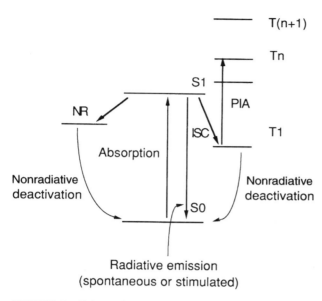

FIGURE 3 Schematic diagram of the electronic transitions in PPV. PIA stands for photo-induced absorption, which can probe an optical transition within the triplet manifold Tx. S0 and S1 are ground and excited singlets. The T1 level is populated by intersystem crossing (ICS). NR and nonradiative species which may be polaron pairs in the case of oxidized PPV.

of PLEDs if they are operated in an oxygen-containing atmosphere, because they can photooxidize by self-absorption of the electroluminescence [43].

Cyclic voltammetry (CV) and differential pulse polarography data place the reduction and oxidation potentials at -1.74 and $+0.76$ V, respectively, versus Ag/AgCl [44], giving an energy gap which matches very closely the onset of electronic absorption. The ionization potential (IP) has been calculated to be at about 5.0 eV below the vacuum level by valence effective Hamiltonian (VEH) methods [45], a value which seems to provide a reasonable explanation of charge-injection processes in PLEDs, consistent with the published values of the work function of common electrodes [2]. Direct measurements of the IP by UPS and by the Kelvin probe are not in good agreement [8], possibly because of different polymer preparation details and surface conditions during the experiment.

The PL emission (Fig. 2) is thought to be due to the radiative decay of cou-lombically bound singlet intrachain excitons [46–49] photogenerated with high quantum yield (owing to the dipole-allowed character of the transition), after ex-citation at energies above the absorption edge, although this model is not univer-sally accepted [50]. The spectrum shows significant structure with several well-resolved phonon lines, spaced of about 0.16 eV, which is attributed to the coupling of the electronic excitations to stretching vibrations of the carbon backbone in the electronic ground state [46].

The electroluminescence spectrum is a very good replica of the PL one and, hence, is assigned to the same emitting species responsible for PL, the small differences being attributable to different interference effects depending on the extent and location of the recombination zone [51].

Note that the inhomogeneous broadening is much smaller in the emission with respect to the absorption, as a result of diffusion processes of the mobile, photo-generated excitons. This diffusion, which can either be a Förster transfer or a tunneling, reduces the set of emitting chromophores quite considerably with respect to those active in absorption [34]. In fact, time-resolved PL measurements [52,53] provide more detailed evidence of the migration process, showing that a small proportion of the excitons decays at "early" times (1 ps or less) with an energy higher than that measured in continuous-wave (CW) experiments or at times longer than 1 ns. Additional evidence supporting this interpretation is provided by site-selective spectroscopy experiments [54], where the shape of the emission spectrum is found to be constant over a large range of excitation energies.

The time constant for the nearly exponential decay of the luminescence is found to be of about 300 ± 30 ps [53] in PPV thin films, showing an absolute quantum efficiency of about 0.27 [55], and hence consistent with a radiative lifetime of ~1.2 ns (the same as for the model oligomer *trans-trans*-distyrylbenzene), and a non-radiative lifetime of about ~450 ps for PPV on glass, in a classical Strickler–Berg analysis [56] where the quantum efficiency is related to the radiative (τ_{rad}^{-1}) and nonradiative (τ_{nonrad}^{-1}) rates by the following equations:

$$\Phi = b \, \frac{\tau_{rad}^{-1}}{\tau_{lum}^{-1}} \tag{1a}$$

$$\tau_{lum}^{-1} = \tau_{rad}^{-1} + \tau_{nonrad}^{-1} \tag{1b}$$

Here, b is the so-called branching ratio (i.e., the portion of the photogenerated excitation which produces emissive excitons). Harrison et al. [57] have recently shown that b is substantially independent of the excitation energy and is almost unity for freshly prepared, nonoxidized PPV from the precursor route, in agreement with Greenham et al. [55] and in disagreement with Yan et al. [58], who had estimated $b \sim 0.1$ for fresh PPV. However, in Ref. 57, it is shown that photooxidation of the polymer introduces a dependence on the excitation energy of b, which is then less than unity at high energies. This may be due to the formation of coulombically bound and nonradiative geminate polaron pairs, as suggested by Yan et al. [58].

The latter is yet another demonstration of the rapid and continuous evolution of the quality of the materials over the last years, which is related to the control of both the polymer purity and the morphological order of the polymer [59,60].

Intersystem crossing and subsequent formation of triplets which couple to the ground state with a nonradiative, dipole-forbidden transition, can be identified as one of the main nonradiative channels in the pristine PPV (see also the scheme in Fig. 3). These can be detected in photo-induced-absorption experiments (PIA) [59], which probe a transition in the triplet manifold at about 1.45 eV. Contrary to the results of early observations [61], there is no trace of charged excitations (bipola-rons) in high-quality, pristine PPV [59]. Other nonradiative processes can be attributed to vibrational deactivation processes and bimolecular recombination of singlets. Although the latter mechanism is likely to be significant only at very high excitation densities, this may be relevant in lasing structures [11,62,63]. Stimulated emission, in fact, is indeed possible in PPV and similar polymers and has been reported by several authors, both in solution (for side-chain polymers) and in solid state [64–67], although early observation had not detected it [34].

The optical properties of PPV, as described above, provide a general reference frame for other polymers as well, although specific features of the optical response may be different, in dependence of the chemical structures and of the polymer morphology. Space constraints prevent a full discussion of many effects discussed in the literature, but we will give some examples relevant to PLEDs in Section II.D.

C. Transport Properties

1. General Concepts

The conductive properties of conjugated polymers were the first to attract considerable attention and hence have been the focus of many studies. Nevertheless, there is no unitary and satisfactory description, to date, of electronic transport. There are several reasons for this, such as the intrinsic disorder of the materials [68], the complexity of their morphological and electronic structures, and the limited purity and reproducibility of the electrical characteristics as compared to inorganic, crystalline semiconductors. The reproducibility problem is probably being addressed with the beginning of industrial interest and subsequent production of large quantities according to strict protocols.

The classical approaches to conventional inorganic and molecular semiconductors are not immediately applicable but offer a reference frame for the elaboration of more specific models and some theoretical tools, which is worth analyzing before considering the specific properties of PPV.

In principle, the disordered character of the charge motion would suggest that the problem can be treated effectively with a general approach based on percolation theory [69], but the particular nature of the electron–electron and electron–phonon interactions of the carriers (polarons, bipolarons, and their combinations) can have a substantial role and need to be examined in some more detail.

Because the conduction mechanism and the dependence of the current on the applied electric field will be determined by the microscopic process which most effectively limits the charge motion through the medium, we can classify the different models according to the type of the active "bottleneck."

Formally, the current density J can always be defined according to Eqs. (2), where n is the number density (concentration) of carriers, q is the electron charge, v is the drift velocity of the carrier, μ is the mobility of the carrier, P is the pressure, T is the absolute temperature, B is the magnetic field, and E is the electric field:

$$J = nq\mu E \tag{2a}$$

$$v = \mu E \tag{2b}$$

$$\mu = \mu(E, T, P, B) \tag{2c}$$

The differences among the various models can then be expressed in the dependence of the mobility on the relevant parameters (E, T, P, B).

The description in terms of a *band model*, for example, puts a condition onto the mobility as in Eq. (3) [70], for the consequences of the uncertainty principle and the relation between bandwidth and the relaxation time of the carriers between scatterings:

$$\mu > \left(\frac{qa^2}{\hbar}\right) \left(\frac{W}{kT}\right) \qquad (3)$$

where a is the lattice constant, $h = 2\pi\hbar$ is the Planck's constant, k is the Boltzmann constant, and W is the width of the energy band.

This condition proves useful enough to exclude the validity of the energy band model in the description of conduction in most molecular solids, where the measured mobilities are too small [70]. Failure to satisfy this condition means that the electron–lattice interaction is too large to be considered only a perturbation, a prerequisite for the applicability of the model. For undoped conjugated polymers, the mobility may be a few orders of magnitude higher than in molecular crystals in relation to a smaller band-gap and unintentional doping, but the use of such a model would be much harder to justify on the ground of symmetry properties.

The other important models which need to be discussed can be grouped in those related to quantum-mechanical tunneling (QMT), hopping, and space-charge-limited conduction (SCLC). In the first two cases, the conduction-limiting factor is the energy barrier between the relevant electron states, whereas, for SCLC, this is the buildup of a charged region which opposes conduction. The specific form of the predicted conductivities (or mobilities) will depend on the details of the model.

Eley and co-workers, for example, have proposed a model for *tunneling* between π-orbitals which can predict the magnitude of the mobility μ and accounts for the anisotropy but fails to explain the negative temperature dependence and the difference between electron and hole mobilities [71].

Alternatively, Sheng et al. suggest considering a heavily doped conjugated polymer in the same way as a granular metal [i.e., as constituted of highly conductive (metallic) islands separated by energy barriers with QMT as the dominant process] a model clearly elaborated for heavily doped conjugated polymers [72–75]. The analysis accounts for both the charging energy of the conducting grains associated with the electron transfer [72] and the influence of the thermal energy of the carriers on both sides of each tunnel junction [74]. In the small island regime (<200 Å), the charging energy is particularly relevant, whereas thermal activation is the principal factor when the potential difference between neighboring islands is comparable with or smaller than kT/q (with the same symbol conventions as above). Equations (4) and (5) summarize the results for the expression of the conductivity σ in the limit of low and high electric fields, respectively:

$$\sigma = \sigma_0 \exp\left[-\left(\frac{T_0}{T}\right)^{1/2}\right] \qquad (4)$$

$$\sigma = \sigma_0 \exp\left[-\left(\frac{E_0}{E}\right)\right] \qquad (5)$$

The separation between the two regimes can be defined by the comparison of the thermal energy kT, with that transferred by the electric field qEd. With an average island dimension d of about 100 Å, the separation is ~1 kV/cm. E_0 and T_0 are constants related to the material and to the spatial extent of the islands; kT_0 can be in the range from a few tens of meV to several eV, whereas E_0 varies between a few tens of kV/cm and several MV/cm. The charging energy is much less important in the case of islands much larger than the critical dimension (200 Å), and the model can be modified by considering the alternated voltage generated across the energy barriers by the random thermal agitation of the carriers inside the islands. The expected dependence of the current is reported in Eq. (6) [73],

$$J = J_1 \exp\left[-\frac{T_1/(T + T_2)}{(E/E_1 - 1)^2} \right] \tag{6}$$

where $E_1 = 4V_b/qw$. T_1 and T_2 are again constants related to the material. V_b is the barrier height, w is the length, and J_1 is only weakly dependent on temperature and electric field.

Note that also models primarily derived for doped conjugated polymers, such as the one discussed above, can be quite relevant for PLEDs because in the normal mode of operation, there is substantial charge injection and it has been shown that electrically induced charges produce effects analogous to chemical doping [76,77].

Poole–Frenkel (PF) emission is also a tunneling-based process [78,79], although the barrier to overcome in this case is the one separating the (intragap) energy level of a trapped electron (hole) and the conduction (valence) band. It is a mechanism relevant to high-field regimes which has been found useful in the description of several molecular materials and of poly(vinylcarbazole) (PVK) in particular [10b]. The predicted dependence of the current density and mobility is reported in Eq. (7):

$$J \approx E \exp\left(-\frac{q[\Phi_b - (qE/\pi\varepsilon_i)^{1/2}]}{kT} \right) \tag{7}$$

where ε_i is the material permittivity.

A minor modification of this model can be introduced to account for observed properties, as in

$$\mu = \mu_0 \exp\left(-\frac{\Delta}{kT} \right) \exp[\gamma\sqrt{E}] \tag{8a}$$

$$\gamma = B\left(\frac{1}{kT} - \frac{1}{kT_0} \right) \tag{8b}$$

This is relevant to the specific case of PLED as will be discussed in the next subsection. μ_0 is a temperature-independent prefactor, Δ is the activation energy, and γ is an empirical modification of the Poole–Frenkel factor first proposed by Gill [80].

Hopping conduction is basically a phonon-assisted transport mechanism between strongly localized states and, as such, is heavily temperature dependent. In particular, the conductivity (mobility) should tend to zero with temperature, whereas it should not for pure tunneling, which is independent of the temperature. In the case of strong localization (fixed range hopping—FRH), hopping is only possible between first-neighbor states with a boltzmannian probability $\exp(-\Delta/kT)$, where Δ is the energy barrier between the two states; for decreasing localization transitions to states progressively more distant (in energy or space), hopping becomes possible with a distribution of energies and/or spatial ranges (variable range hopping—VRH). The temperature dependence is implicity in the Boltzmann factor but also influences the conductivity through a modulation of the localized sites accessible to the hopping. This, in turn, depends on the dimensionality of the process [83,84] and yield

$$\sigma = \sigma_m(T) \exp\left[- \left(\frac{T_m}{T} \right)^{1/(\mathrm{dim}+1)} \right] \tag{9}$$

with $T_m = a^{\mathrm{dim}}[N(E_F)]^{-1}$, where a is the delocalization length, dim is the dimensionality of the material/process, $N(E_F)$ is the density of states at the Fermi level, and $\sigma_m(T)$ is a function of temperature weaker than the exponential. Coulombic electron–electron interaction have been accounted for by Efros and Shkloyskii [84] in a subsequent refinement of the model, which introduces a modification of the exponent from $\frac{1}{4}$ to $\frac{1}{2}$ for three-dimensional (3D) materials. A less severe localization of the carriers along the chain can also be taken into account [85].

In more recent adaptations of hopping theory to the case of charge-transporting molecules dispersed in a polymer matrix, two main approaches can be identified. The first, by Schein [86], stresses the importance of structural relaxation around the excitation (polaron model) and predicts a mobility dependence as in Eqs. (10a) and (10b) for constant field and temperature, respectively:

$$\mu = \mu_0 \exp\left[- \left(\frac{T_0}{T} \right)^2 \right] \tag{10a}$$

$$\mu = \mu_0 \exp[E^{-1} \sinh(aE)] \tag{10b}$$

where T_0, a, and μ_0 are constants.

The second approach, developed by Bässler [87], focuses on the disordered nature of the materials instead and predicts a dependence as in Eqs. (11a) and (11b) for constant field and temperature, respectively:

$$\mu = \mu_0 \exp\left[- \left(\frac{T_0}{T} \right)^2 \right] \tag{11a}$$

$$\mu = \mu_0 \exp(E^{1/2}) \tag{11b}$$

where T_1 and μ_0 are constants. In practice none of these hopping models are of general validity and capable of accounting for all the experimental evidence so far collected.

Space charge limited (SCL) current mechanisms need also to be considered. Independently of the particular tunneling or hopping mechanism which governs the single-transport event, if the mobility is low enough with respect to carrier generation or injection, the charge conservation condition will result in the formation of a charged region which will influence further carrier generation, injection, and the transport itself. The predicted dependence of the current density J on the applied voltage is

$$J = \frac{9}{8} \varepsilon \mu \frac{V^2}{d^3} \tag{12}$$

where V is the voltage applied to the electrodes, d is the thickness of the film, and the other symbols are as previously defined.

Whenever the electrodes impedance is not negligible with respect to that of the semiconductor, the voltage drop over the interface region will be substantial and the contact is said to be non-Ohmic. In this case, *charge injection* can become the dominant limiting mechanism of the current. The problem has been treated very well and extensively for inorganic and organic semiconductors in a number of excellent texts and textbooks [88–90], and we do not consider it meaningful to repeat an analysis which is better conducted elsewhere. We only list the two main processes in Eq. (13) (i.e., Schottky emission) and Eq. (14) (i.e., Fowler–Nordheim emission) [88]:

$$J = A^*T^2 \exp\left(- \frac{q[\Phi_b - (qE/4\pi\varepsilon_i)^{1/2}]}{kT}\right) \tag{13}$$

$$J \propto E^2 \exp\left(- \frac{4\sqrt{2m^*}\,(q\Phi_b)^{3/2}}{3q\hbar E}\right) \tag{14}$$

where A^* is the effective Richardson constant, Φ_b is the barrier height, and m^* is the effective mass of the charge carrier.

2. The Case of PPV

The transport properties of PPV have been studied both in the doped and in the undoped states; the latter are particularly relevant to PLEDs.

A variable-temperature study of conduction with electrodes having an average work-function value (Cr ~ 4.5 eV, Al ~ 4.2–4.3 eV) is reported by Nguyen et al. [91] and is relative to the sandwich structures metal/PPV/metal. The average value of the work function helps to minimize charge injection (holes and electrons), at least at low fields, and hence to extract information about the intrinsic carriers and charge transport. By studying the current/voltage (IV) characteristics as a function of temperature and the thermally stimulated currents, they find a Poole–Frenkel emission in the high temperature range ($T > 200$ K) with a mean activation energy of about 0.3 eV. They also find a less pronounced dependence of the current at low temperature which is attributed to the onset of a hopping regime with probable interactions between localized states. Interestingly, the work conducted recently at Philips Laboratories, on LEDs prepared with a soluble derivative of PPV, found a SCLC behavior for the hole conduction and a Poole–Frenkel temperature dependence of the type expressed in Eqs. (8). The electron conduction is instead found to be trap limited below a certain bias and then SCL again [92,93].

Note also that this is substantially in agreement with the findings of Marks et al. [94], who studied early PLEDs fabricated with PPV from the precursor route and concluded that currents are space charge limited although with significant trapping and, possibly, filamentary injection.

Field-effect measurements on PPV have also been reported [95], but only on the doped material, and much caution must be used in using these data because transport in field-effect transistors (FETs) requires a carrier flow in a very thin layer of the semiconductor close to an interface—clearly, a condition very different to that realized in the bulk of a PLED.

D. Other Polymers

Although PPV has played a prototypical role in the development of PLEDs, because of the initial discovery and of the good performance in terms of lifetime of operation [48,96], many other polymers have been designed, synthesized and used in

LEDs, either as emissive layers or in order to help charge injection and/or balance. These can be summarized as follows:

1. PPV derivatives

These are mainly side-chain, soluble polymers, although there have been examples of precursor-route derivatives [97]. They can be subdivided in smaller classes depending on the nature of the substituents.

Alkyl and Alkoxy Derivatives. Alkyl or alkoxy chains are added in the 2- or 5-position or both on the benzene ring. The alkoxy derivatives are generally affected by a lower band gap owing to the π-electron-donating character of oxygen.

CN Groups Containing Polymers. The electron-withdrawing CN group is used in order to increase the electrode affinity (EA) and, hence, the electron-injection efficiency from the cathode. The polymer shown in Fig. 4a (CN–PPV in brief) proved

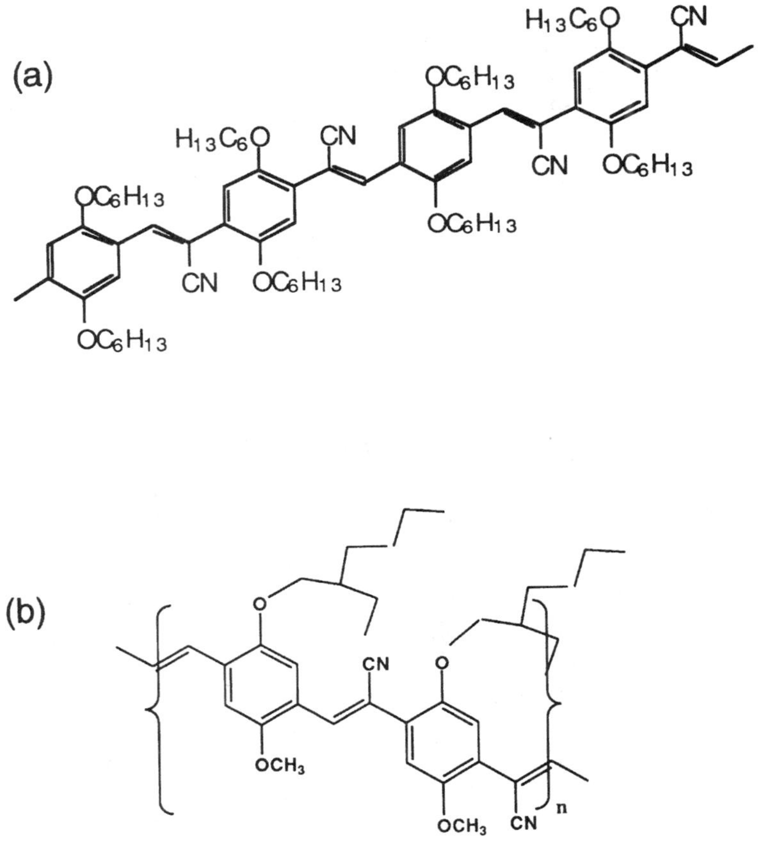

FIGURE 4 The chemical structures of two of the CN-substituted PPVs: (a) poly(cyanoterephtalidene) (CN–PPV); (b) poly[2,5-bis(nitrile methyl)-1-methoxy-4(2′-ethylhexyloxy)benzene-co-2,5-dialdehyde-1-methoxy-4-(2′-ethylhexyloxy)benzene] or MEH–CN–PPV.

particularly effective in helping electron injection and in raising the efficiency of heterojunction LEDs when PPV was used as a hole injector [98]. This type of device will be discussed in more detail in Section V. Quite interestingly, a detailed theoretical study of the electronic structure of several polymers, with CN substitutions at different locations in the repeat unit of PPV, showed that the structure of Fig. 4a is optimal for the maximization of the EA. Asymmetric chain substitution on the 2–5-position is also possible (Fig. 4b MEH–CN–PPV) and yields a polymer with better film-forming properties [99].

F Substitutions. Fluorine is the most electronegative element in the periodic table and its electron-withdrawing effect has been investigated both as a hydrogen-substituting unit or in CF_3-substituted groups. The CF_3-substituted PPVs and related oligomers reported by Lux et al. [100] proved suitable as electron transport materials and also showed reduced tendency to photooxidation, in photobleaching experiments, possibly in connection with steric protection of the vinylic double bond by the bulky groups CF_3. Son et al. also reported the use of CF_3-containing polymers for electron transport [101].

Phenylated PPVs. Attachments of a phenyl ring on either the benzene ring of PPV or on the vinylic carbons results in soluble, green polymers which are electroluminescent. Poly(4,4'-diphenylene diphenylvinylene), for example (Fig. 5), is characterized by high fluorescence yield (0.45 in solid state [102]) and good film-forming properties.

2. N-Containing Polymers for Electron Transport

A large variety of these has been synthesized and a large amount of literature produced.

Oxadiazole-Containing Polymers. The oxadiaxole unit and the relative oligomers and polymers have a high IP [~6 eV for 2-phenyl-5-biphenyl-1,3,4-oxadiazole (PBD)] and, hence, can function as effective hole blockers (see Section IV) in PLEDs. Both side-chain [103,104] and main-chain polymers [105,106] have been reported, although the stability is generally poor, especially under operation [103].

Azomethine-Containing Polymers. This is a relatively recent approach and although it looks promising, not much data are available on the temporal stability [22,107].

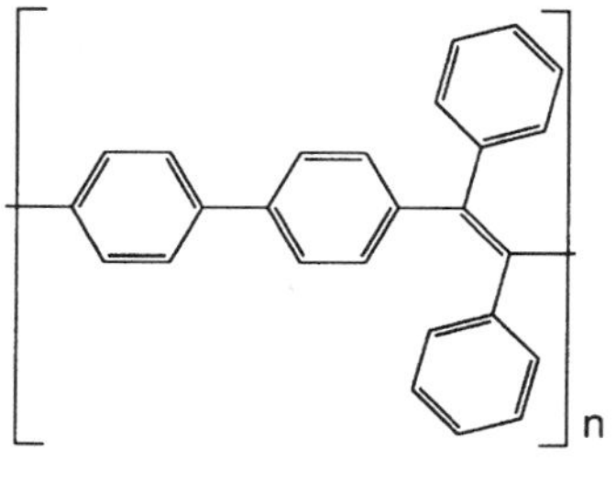

FIGURE 5 The chemical structure of poly(4,4'-diphenylene diphenylvinylene) (PDPV), a phenylated PPV.

Polyquinoxaline-Containing Polymers.   The isolated polyquinoxaline moiety has a reported IP of 9.02 eV and that of polyquinoxaline-containing polymers (PPQ) is approximately 6.7 eV [108] and is therefore of considerable interest as a hole-blocking/electron transport layer to be used on top of an emissive polymer such as PPV.

Polyimide-Containing Polymers.   Polyimide- (PI) containing polymers are still in development [109] and are particularly interesting in view of the high thermal stability of the imide moiety.

3.   Thiophene-Containing Polymers

There is a large variety of these and it is not possible even to attempt to summarize the published work. They are generally attractive for the stability of the thiophene ring, although the PL quantum yield is intrinsically low, in general, owing to the presence of triplet levels almost resonant with the first excited singlet (S1) and therefore of favored intersystem crossing [110].

4.   Si-Containing Polymers

Silicon atoms have been used by the group at Gröningen in order to limit the conjugation length and hence obtain blue-shifted emission [111,112]. PPV with dimethyloctylsilyl side chains has also been reported by Hewang and co-workers [113].

5.   poly(p-phenylene) and Ladder PPP

Poly(p-phenylene) (PPP) has been widely used in order to obtain blue emission. The band gap in this material is increased with respect to PPV as a consequence of a substantial torsion of the plane of the generic benzene with respect to that of the neighbors. A variety of ladder structures has also been proposed, although some of them suffer from aggregation effects [114].

Although the above list is only representative and is selective and not meant to be exhaustive, it provides a sample of the polymeric systems currently under investigation in polymeric EL-related research. The physical properties of these polymers can be considered in the general frame provided by those discussed for PPV, although specific features of the optical or electrical response may be different in their dependence of the chemical structures and of the polymer morphology. This is, for example, the case of the nature of the emissive species. Although in many polymers these can be identified with a reasonable degree of confidence to be intrachain singlet excitons, this is not true for some cases. Studies on CN-containing PPVs, for example [115–117], have found that the photoexcitations are very different, with much longer decay times than in PPV, and it has been suggested that they could be interchain excited dimers (excimers) originating from the influence of the CN groups on the chain packing in the solid state.

III.   SINGLE-LAYER PLEDs

A.   General Remarks

A polymer LED consists, in its simplest form, of a single layer of an electroluminescent polymer sandwiched between two electrodes, as schematically presented in Fig. 6.

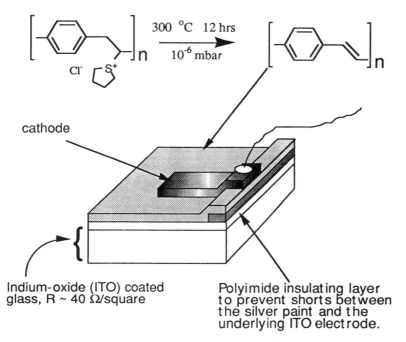

cathode

Indium-oxide (ITO) coated
glass, R ~ 40 Ω/square

Polyimide insulating layer
to prevent shorts between
the silver paint and the
underlying ITO electrode.

FIGURE 6 Schematic structure of a PPV-based LED. Typical polymer thicknesses are in the range 70–200 nm. The cathode is a low-work-function metal or alloy, in order to help electron injection. ITO is typically 100 nm thick and the substrate incorporates a silicon dioxide thin film between the glass and the ITO in order to prevent sodium diffusion inside the active polymer.

At least one of the electrodes has to be transparent in order to guarantee an efficient light output from the structure, and indium–tin oxide (ITO) has by now become a conventional choice, although it is not the ideal material for various reasons, which are discussed in more detail in Section V. The polymer, or its precursor in the case of PPV, is generally spin-coated or blade-cast, whereas the second electrode can be deposited with a variety of techniques ranging from thermal and electron-beam evaporation to radio frequency and DC sputtering.

The electroluminescence process (i.e., the light emission from these structures under the action of an electric field) can be divided further in at least four subprocesses: (1) charge injection, (2) transport, (3) opposite charges capture and exciton formation, and (4) radiative recombination of a fraction of the formed excitons [99]; these can be schematically represented in an energy band diagram as in Fig. 7.

Upon application of an electric field to a structure as in Fig. 6, charge carriers of opposite sign can be injected from the electrodes and transported across the polymer. Low-work-function materials are better suited as injector of electronlike carriers (negative polarons), and high-work-function material, such as ITO (work function ~4.4–5 eV), are more efficient as injector of holelike carriers (positive polarons). Although, for brevity, in what follows we will simply speak of holes

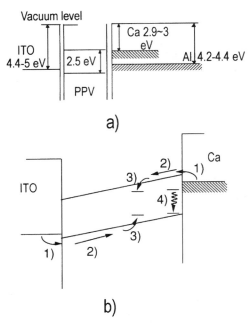

FIGURE 7 (a) Energy band diagram of the (isolated) materials of a PLED; (b) schematic representation of the four subprocesses of EL for a device in forward bias: (1) charge injection, (2) charge transport, (3) opposite charge capture to form an exciton, (4) recombination. (From Ref. 99.)

and electrons, it is important to remember that the addition of a charge carrier to a conjugated molecule determines a substantial perturbation of the local environment (both in terms of the geometric configuration of the chain and, for this reason, of the allowed energy levels), as suggested by the term polaron [118], initially introduced for the description of electrical transport in polar media, where the presence of a carrier determines a *polarization* of the environment in its close proximity [119,120].

Once injected, each carrier can either interact with an oppositely charged particle and form a coulombically bound exciton, or can be swept across the whole polymer film and finally be ejected into the other electrode.

In the previous section, we presented the properties of the photoexcitations and of the charge transport; in what follows, we will present the electrooptical properties of LEDs and discuss them in view of the processes exposed in Section II.B and II.C.

B. Characteristics of Single-Layer PPV LEDs

Figures 8 and 9 show the current–voltage (IV) and luminance–voltage (LV) characteristics of a single-layer PPV device as in Fig. 6 (thickness 280 nm, active area ~4 mm²) and represent an example of the best results obtainable also when varying the conversion temperature of the polymer precursor and the cathode-deposition

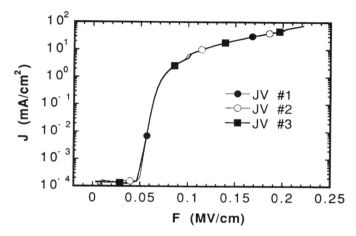

FIGURE 8 Current density versus voltage characteristics of a single-layer PPV device with a Ca cathode, as in Fig. 6 (thickness 280 nm, active area ~4 mm²). Bias is forward for the ITO positive with respect to the Ca–Al electrode.

parameters. Bias is forward for the ITO positive with respect to the Ca–Al electrode.

In spite of a relatively high thickness, the current density reaches 75 mA/cm² at ~6 V with a differential impedance (dV/dI) of about 625 Ω and a static impedance (V/I) of about 2 kΩ. Note that by considering the device dimensions, these values are equivalent to a conductivity of ~1.12 × 10⁻⁶ S/cm and 3.6 × 10⁻⁶ S/cm, respectively, which is greater than expected from the result of earlier observations on undoped, semiconducting polymers such as PPV [8,121].

Currents in reverse bias (i.e., with the ITO negative) are 10⁻³–10⁻⁴ times the maximum currents in forward bias. The rectification ratio [$I(+V)/I(-V)$] increases with the absolute value of the bias V until dielectric breakdown takes place, in

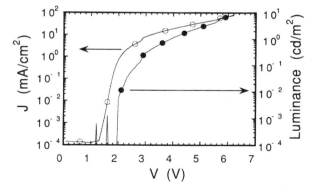

FIGURE 9 Current density of one of the ramps of Fig. 8 and the associated luminance signal. Note that the luminance turn-on voltage is ~2 V, and the current turn-on voltage is ~1.4–1.6 V.

reverse bias, typically for fields exceeding 4 MV/cm. This is very often irreversible, different from the case of inorganic semiconductors, and leads to irreparable damage (short or open circuit). This may be a minor problem for applications, as reverse bias is very ineffective for EL, although emission in this condition has also been reported [122,123].

The three characteristics in Fig. 8 are relative to three consecutive voltage ramps, with nearly perfect reproducibility. Note that the small anomaly at about 0.1 MV/cm (4.8 mA/cm^2 and ~2 mA) is an artifact due to the automatic change of the range of the digital multimeter used for the measurement.

Figure 9 reports both the current of one of the ramps of Fig. 8 and the associated luminance signal. Note that the luminance turn-on voltage (i.e., the point where the signal becomes unambiguously larger than the background noise) is at ~2 V, which corresponds approximately at the lowest emission energy of PPV as derived from PL spectra. Note also that the current turn-on voltage is at a much smaller voltage, i.e. at about 1.4–1.6 V. These differences will become clearer after discussing the various model of LEDs operation which have been put forward and the one by Parker in particular.

Correlation of the JVL (current/voltage/luminance) characteristics with the relevant transport mechanisms are generally complicated by both the uncertainties in the measurement of the film thickness and, most importantly, by the difficulty in providing a unique interpretation to best-fitting studies. These may yield very similar results, in spite of the difference of the models owing to the presence of fitting parameters which can be wildly variable depending on the sample quality. In some cases, the models are not very different as well, as happens with the Fowler–Nordheim (FN) injection and SCLC conduction which have both a square field (F^2) dependence at the highest field values (if the field is high, $1/F$ will tend to 0 and the exponential factor in FN may be less important than the F^2 prefactor). Extension of the set of the data points, which could provide more detailed evidence, is not always feasible without entering a different operation regime or even without altering irreversibly the device characteristics. These considerations should explain, at least partially, the relatively confusing literature. Parker, for example, reports a very detailed study [124] of the current density versus voltage and field (JV and JF) characteristics for a range of single-layer LEDs fabricated with poly(2-methoxy, 5-(2'-ethyl-hexosy)-1,4-phenylene-vinylene) or MEH-PPV, with different thicknesses and several electrode materials. In particular, he uses devices with two very high- (low-) work-function electrodes in order to study the injection/transport of holes (electrons) in the absence of the opposite type of carrier and so extract a "unipolar" injection process. A Fowler–Nordheim analysis is then applied and leads to the conclusion that the relevant injection mechanism is quantum-mechanical tunneling (QMT) through the barriers at the polymer–electrode interfaces. The barriers are thought to be caused mainly by the mismatch between the electrode work functions and the relevant energy levels for hole and electron transport in the polymer. These levels are taken to coincide with the HOMO and LUMO level, but in the context of a rigid-band model instead of in an exciton picture. The barriers are estimated to be 0.2 eV for the hole injection into MEH–PPV and 0.1 eV for electrons. Corroborating evidence in the form of temperature dependence of the current at a particular voltage is also provided, although it is noted that for very low barrier heights (0.1 eV for Ca–MEH–PPV), there is some deviation from

QMT which may be ascribed to thermionic emission. Because the efficiency is dominated by the minority carriers, and the current by the majority carriers, the model predicts that the maximum efficiency is reached when the barrier to injection of the minority carrier is 0 eV and that the minimum luminance turn-on voltage is given by the flat-band condition [124] (i.e., at a voltage which equals the difference of the electrode work functions or, alternatively, the band gap minus the barrier offsets). This is not really consequential, because even in a flat-band condition, the barrier thickness would equal the whole polymer thickness and would make QMT extremely unlikely. The only way in which the experimentally observed turn-on voltage at about the flat-band condition can be reconciled with this model is, in our opinion, by taking into account inhomogeneous broadening in the polymer, and to consider that there will probably be very long conjugated segments for which the barrier heights are virtually 0 eV. Interestingly, it is argued that because the hole barrier is ~0.2 eV and the electron barrier only ~0.1 eV, the efficiency is controlled by the holes, and supporting evidence is provided in the form of an efficiency increase upon replacement of ITO (work function ~4.7 eV) with Au (work function ~5.2 eV). Although the predictions on the dependence of two-carrier device efficiency on the work function of the cathode are also shown to agree very well with experiment, the model fails to account for the JV character-istics in terms of the sum of the single-carrier currents, as noted by Parker who suggests the occurrence of space-charge effects in order to recover the discrepancy.

Indeed, later studies reported by the group at Philips [92,93] challenge the applicability of the FN injection on the basis of quantitative disagreement between the measured and predicted values of the currents, even when corrections are made in order to account for thermionic emission [125], space-charge effects [126], and band-bending [127].

Interestingly, the Philips group found that they can quantitatively account for the JV characteristics of their ITO/PDAPP/Ca diodes [where PDAPPV is a poly(dialkoxy *p*-phenylene vinylene)] with a SCL hole conduction and a field-dependent mobility with Poole–Frenkel behavior [93]. Electron transport is instead found to be characterized by a trap-filling process with linear dependence on the voltage up to a limit (trap-filling limit: TFL) beyond which the transport becomes again space-charge-limited. Note that this trap-filling mechanism is able to provide a good explanation of the increase of efficiency with voltage.

We observe that earlier work by Marks [121] on PPV had also concluded the presence of SCL currents, although in the presence of substantial trapping and, possibly, filamentary injection.

As for the data that we have presented in Figs. 8 and 9, we observe that a straightforward interpretation is not possible in neither model, although the analysis would be possible in terms of a modified SCLC mechanism [128], but that a more complete set of experimental data is necessary before presenting these results. In general terms, we note that we expect higher mobilities for PPV than those found for PDAPPV ($\sim$0.5 10^{-6} cm^2/V s for holes) [92] because of the higher density of conjugated material and of the stronger polymer interaction with the conductive substrate during the precursor treatment. A treatment of the data in Figs. 8 and 9 within a zeroth-order approximation suggests that, as the current density is about 2 orders of magnitude higher than for a similar device thickness [92], a similar relation is expected between the mobilities.

As for the luminance–current characteristics, the slope of which is proportional to the internal quantum efficiency, we observe that both the FN model and the SCLC account for the experimental observation of an efficiency increase with bias [92,124].

The fundamental discrepancy between the two models (FN and SCLC) could be due to the intrinsic difference of the polymer properties (HOMO–LUMO, density of impurities) and different work functions of electrodes which are identical only in principle (this is a problem especially for the ITO, the properties of which can vary significantly depending on the processing parameters). Although the "bulk" (SCLC) model does not fail to account for the behavior of devices with very small barriers, the increase of the device efficiency and luminance with decreasing-work-function cathodes and, in general, the possibility of predicting the qualitative properties of PLEDs considering the electrodes work functions and the polymers frontier levels (as in Fig. 7) suggest that the situation may be more complicated than explained by a SCLC model, and that more experimental evidence is needed in order to reach a definite conclusion.

C. Limitations

The luminance of the device in Figs. 8 and 9 is relatively poor with respect to both the values obtainable with the sublimed molecular films and the PLEDs fabricated with red-orange solution processible polymers. In fact, the latter can easily achieve over 400 cd/m^2 at less than 5 V with an efficiency of a few percent [129], whereas the best results for single-layer PPV devices indicate a value of internal quantum efficiency (photons per electron) well below 1% (namely 0.1% for Ca cathodes and 0.01% for Al ones). This is partly due to the higher band gap that makes double carrier and electron injection more difficult and to the lower PL efficiency of the material [55]. This is low also because of the ITO etching produced by the HCl released during conversion, which creates quenching species.

Brightnesses well above 100 cd/m^2 with low driving voltages and currents and lifetimes (time to reach 50% of the initial luminous emission) of the order of 10^4 h with at least 100 cd/m^2 initial brightness are needed for application; this raises the scientific question of how to control efficiency and lifetime of PLEDs.

We will consider the problems related to the efficiency in Section IV and discuss lifetime-related issues in Section V.

IV. EL EFFICIENCY OF PLEDs

A. General Concepts

The EL efficiency can be expressed for a generic electroluminescent structure as in Eq. (15) [8]:

$$\eta_{EL} = \eta_{PL} r_{st} \gamma_{cap} \tag{15}$$

where η_{EL} is the EL efficiency, η_{PL} is the PL efficiency, r_{st} is the number of singlets over the total number of excitons formed, and γ_{cap} is the number of excitons formed per unitary charge (electron) flowing in the circuit. From Eq. (15), we appreciate the importance of both having a high PL efficiency η_{PL} and a balanced injection of carriers, otherwise γ_{cap} (and therefore η_{EL}) will be much less than unity. Note also that in a simple model, we expect only majority carriers to be ejected from the polymer if we assume a relatively high probability of exciton formation upon collision of two charge carriers with opposite sign and a large imbalance in the hole–electron concentrations. The fate of the formed excitons will depend instead on the available decay channels which will be mainly nonradiative for triplets and radiative and nonradiative for singlets. In a zeroth-order approximation, we can also assume the same capture efficiency for formation of singlets and triplets and hence an upper limit to the electroluminescence efficiency η_{EL} of $\eta_{PL}/4$ (as the number of triplets will be three times the number of singlets for random spin distribution of the injected carriers).

The relation between the cathode work function and the efficiency has been the focus of early investigations [130], and it has been clearly shown in the case of Ca, Mg, and Al that an increase of the former leads to a decrease of the efficiency. This is often taken as an indication of the presence of a potential barrier at the cathode interface, although it should be stressed that the nature of the interfaces is complex and the higher or lower reactivity of the metals with the polymer and the oxygen contaminations left in the polymer or present during the cathode deposition can play a significant role.

The reactivity of Ca and Al with conjugated polymers or short, oligomeric model compounds of the EL polymers has been studied in detail by Salaneck and co-workers [131–134], who have used UPS and XPS techniques, in conjunction with quantum-chemical calculations, in order to study the modifications of the electronic structure of the conjugated materials during the early stage of the metal evaporation in a ultra high-vacuum (UHV) chamber. Whereas Al is found to disrupt the conjugation of the polymer by forming covalent bonds with it, Ca does not alter the chemical structure of the polymer (oligomer), but instead dopes the polymer by charge transfer, which results in the formation of two bipolaronic bands. This will have to be accounted for in future device modeling of PLEDs. In addition, Ca is found to diffuse rapidly into the polymer and to lead to short-circuit formation if evaporated in a UHV environment ($\sim 10^{-10}$ mbar), but this problem can be eliminated by evaporating in a less clean environment ($\sim 10^{-7}$ mbar) so as to let the Ca be partially oxidized and form an oxide layer which seems to act as a diffusion barrier. A thin oxide layer has also been found in other studies [135] and is likely to play a significant role in the limitation of the hole current and, hence, in the efficiency increase. Supporting evidence comes from recent studies of LEDs incorporating an intentional insulating barrier between the cathode and the polymer [136,137].

Because holes are majority carriers and recombination happens very close to the cathode, other significant limitations of the efficiency come from physical interactions of the excited molecules with the metal; of these, the most important are the excitation of plasmons in the cathode and the reduced intensity of the electromagnetic field close to a mirror, which, in a first approximation, imposes a node of the field on its surface [138]. PLEDs are complex optical structures with significant reflection and refraction taking place at interfaces (they are, in fact, poor optical resonators) [51] and it is therefore important to control the position of the recombination region in order to maximize constructive interference.

Here, we will only give two representative examples of how it is possible to control the efficiency by using several polymer layers incorporated in the same

LED structure. This is an old strategy in the field of organic electroluminescence and was first reported by Tang and co-workers [17] for LEDs made with low-molecular-weight materials. The main issues which can be addressed in this way are as follows:

1. Limitation of the excess majority carrier which limits the efficiency
2. Control of the recombination region position inside the device
3. Minimization of the work function–frontier level offsets (note, however, that the operating voltage of the LEDs may not be reduced, owing to possible presence of energy barriers at the polymer/polymer interface)

B. Two-Layer Devices (PPV/CN–PPV)

The technological problem is clearly to find a suitable material in terms of processing, frontier levels, band gap, and stability. This may not be easy in view of the many requirements.

Poly(p-phenylene vinylene) is a good hole conductor, owing to the relatively high (~5 eV) IP, which matches relatively well the work function of ITO (4.4–5 eV, depending on the processing conditions) and because of the unintentional doping by oxygen, which is thought to act as a trap for electrons. Therefore, good candidates for fabrication of heterojunctions (HJ) are materials which show electron transport or electron-injection polymers such as those carrying electron-withdrawing elements, which are often responsible of a lowering of the frontier levels [45].

One of these polymers is CN–PPV (Fig. 4a). It is characterized by an IP of about 5.6 eV, and it proved to be very effective [98] for increasing the EL efficiency of PLEDs in a two-layer configuration ITO/PPV/CN–PPV/metal. This behavior has been interpreted [98] as due to the limitation of the hole current and to the spatial control of the recombination, which is forced to take place at the polymer/polymer HJ rather than close to the quenching cathode. This happens because the IP of CN–PPV is ~0.6 eV lower than that of PPV, so that there is a substantial barrier to hole injection from PPV to CN–PPV. In addition, the EA of CN–PPV is ~0.9 eV lower than in PPV and this has two major consequences:

1. Electron injection from air-stable cathodes but of relatively high work function such as Al (work function ~4.2 eV) is easier, and we therefore expect the difference between the efficiencies obtainable with Ca and Al cathodes to be much smaller.
2. There is a substantial barrier for electron injection into PPV from CN–PPV, and the recombination is expected to take place in CN–PPV for both this reason and because of the lower energy gap.

Both predictions agree with experiment [98].

Figure 10 shows a band diagram of a device under forward bias with a schematic indication of the expected charge accumulation regions at the HJ and subsequent band bending.

Maximum internal quantum efficiencies up to 5% and brightnesses up to 10,000 cd/m^2 have been reported for similar structure but with the MEH–CN–PPV polymer of Fig. 4b [99].

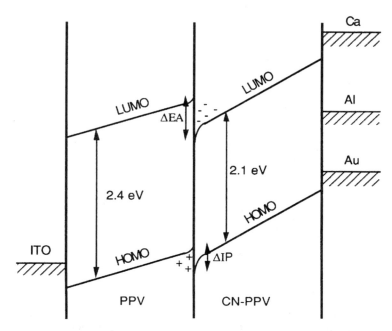

FIGURE 10 Energy-level diagram for a PPV/CN–PPV PLED. ΔEA is estimated to be ~0.9 eV and ΔIP ~0.6 EV. (From Ref. 98.)

IV and luminance versus voltage characteristics are reported in Fig. 11 for a structure as in Fig. 10, for which we measured luminous outputs of 5.2 mW/sr/A normal to the plane of the device, giving a brightness of 2.6 W/sr/m² at a current density of 50 mA/cm² [139].

The efficiency versus current characteristics of a similar device are reported in Fig. 12. They show a good degree of reproduciblity (A to G are relative to consecutive voltage ramps) after an initial period of "burn in" obtained by the application of ~15 voltage ramps of the same width as the one shown. Note that the efficiency is substantially constant for currents higher than 2 mA, which corresponds to 50 mA/cm², and that the absolute value of the efficiency is slightly lower than the maximum values reported in the literature [139].

Although a large part of the work on the CN–PPVs has been carried out on polymers emitting in the orange–red region of the visible spectrum, owing to the red shift caused by the substitution of the solubilising alkoxy chains, chemical tuning of the polymer in order to shift the emission in the green or in the near-infrared is also possible [99,140].

C. Three-Layer Devices (PPV/PVK/PDPV)

Instead of focusing on the many variations and applications of the two-layer device concept, we prefer to discuss an alternative three-layer architecture [102], which we recently found useful for raising the efficiency of a phenylated derivative

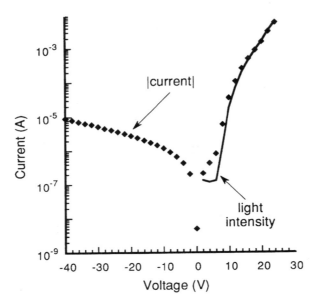

FIGURE 11 Modulus of current and light intensity versus bias voltage for a heterostructure LED device, formed as indium–tin oxide/PPV/CN–PPV/aluminum. Device area is~4 mm². Outputs of 5.2 mW/sr/A were obtained normal to the plane of the device, giving a brightness of 2.6 W/sr/m² at a current density of 50 mA/cm².

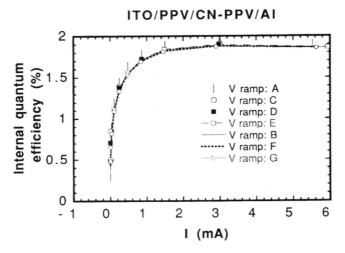

FIGURE 12 Efficiency versus current characteristics for a similar device as in Fig. 11 but with efficiency slightly lower than usual. A to G denote subsequent voltage ramps.

of PPV, namely poly(4,4′-diphenylene diphenylvinylene) (PDPV) [141] (see also Fig. 5).

PDPV is an attractive polymer for electroluminescence (EL) because it has a very high photoluminescence (PL) efficiency in solid state (0.45) along with good solubility in common organic solvents, such as chloroform or toluene, due to the phenyl substitution at vinylenes and disorder induced by cis/trans isomerism.

Polyvinylcarbazole (PVK) is a nonconjugated polymer, but shows good conductive and photoconductive properties due to the close packing of the conjugated chromophores pendant from the olefinic chain [142,143]. For this reason, it found considerable application in electrophotography, whereas more recent work has focused on the use as a hole-injection layer in a variety of organic LEDs [144,145]. Other studies have shown the possibility of raising the PL and EL efficiencies of organic chromophores by blending with PVK, although it is not clear whether this is due to a specific function of the PVK or to the general decrease of quenching with increasing chromophore dilution [146,147].

Our approach is different from the latter because we used PVK as a separate layer (not in blends with the chromophore) and not in close contact with the ITO anode, but as an intermediate layer between the hole-injection material, PPV, and the emissive polymer, PDPV. The idea is basically to make use of PPV as a good hole-injecting layer, capable of forming relatively stable interfaces with ITO and to limit the excess of hole currents with the intermediate PVK layer, which is characterized by a relatively high ionization potential of about 5.8 eV [148], and, therefore, with the formation of a significant energy barrier for holes at the PPV/ PVK interface. Figure 13 shows a schematic energy-level diagram of the structure when the materials are isolated and the Fermi levels are flat. The current and luminance versus voltage characteristics for a structure 300 nm thick (with each layer ~100 nm) are reported in Fig. 14. The internal quantum efficiencies were found to be in the range 0.4–0.55% for Al cathodes and 0.6–0.8% for Ca cathodes. Note that with an Al cathode, we obtain a luminance in excess of 1300 cd/m² at 33 V and at a current density of ~400 mA/cm². We also note that the difference

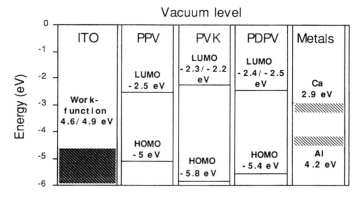

FIGURE 13 The position (with respect to the vacuum) of the HOMOs and the LUMOs for the isolated materials with no applied field. The work functions of indium–tin oxide, Ca and Al are also reported. (From Ref. 102.)

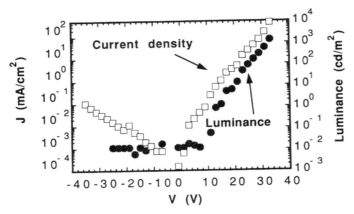

FIGURE 14 Current density (open squares) and luminance (closed circles) versus voltage characteristics for a three-layer ITO/PPV/PVK/PDPV/Al LED. The total thickness of the structure is about 300 nm for an electrode area of about 2 mm^2. (From Ref. 102.)

in efficiency between Ca and Al is only a factor 1.5 instead of 10, as for single layers.

The role of the PVK is evident when comparing diodes which incorporate a PVK layer with single-layer PDPV and double-layer PPV/PDPV diodes. We observed internal efficiencies up to 0.04% for ITO/PDPV/Ca and 0.25% in the ITO/PPV/PDPV/Ca diodes, with emission characteristic of PDPV. In particular, a parameter to classify the degree to which the luminescent properties of a material are exploited in a particular device configuration can be calculated as the ratio of the PL efficiency of the emitting material to the EL efficiency of the diode [103] ($f \equiv \eta_{PL}/\eta_{EL}$), and it is ~$10^4$ (Al) or 10^3 (Ca) for single-layer PDPV LEDs, poorer than for PPV [f(Al) = 2500, f(Ca) = 250] because of the higher PL efficiency and the larger current imbalance. The trilayer structure scores f(Al) $\leq$ 100 and f(Ca) $\leq$ 70. Although the absolute values of the efficiencies are not as high as those achievable with the PPV/CN–PPV structure, we obtain an efficiency improvement of about two orders of magnitude with respect to single-layer devices with Al and more than one order for Ca cathode.

Note that the use of PVK in direct contact with the ITO is possible, and yields efficiencies in the range 0.6–1% but is not recommendable in view of the observation by Adachi et al. that the lifetime of organic LEDs correlates with the barrier at the ITO interface (the lower the barrier, the longer the lifetime) [149], and this justifies further use of PPV as a hole-injection layer, although other materials may actually turn out to be more appropriate. We will discuss this problem in the next section.

V. STABILITY OF PLEDs

The temporal stability of polymer LEDs, both under operation and at rest, is fundamental to the development of a technology with a commercial value and a potential for applications. The target figures which will decide the future of this

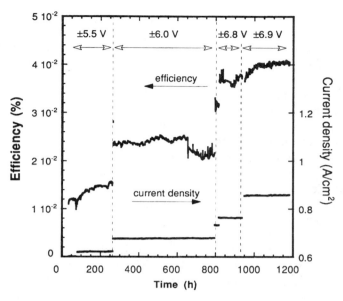

FIGURE 15 Current and internal quantum efficiency versus time for a PPV LED driven with a square-wave voltage which was switched between 5.5 V and 6.9 V during operation, as indicated at the top of the figure. The device had an area of ~4 mm^2 and was maintained at ambient temperature. (From Ref. 48.)

technology are summarized by an initial brightness of at least 100 cd/m^2 (equivalent to that of a CRT) which reduces to 50% of the initial value ("lifetime") in 10^4 h of operation.

Figures 15 and 16 [48] show some results of early investigations that we carried out on a range of single-layer ITO/PPV/Ca devices with an overlayer of Al, aimed at decelerating the degradation of the very reactive Ca electrode (ITO/PPV/Ca–Al

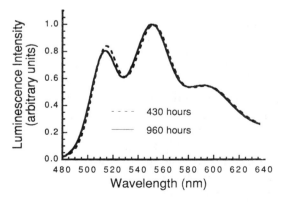

FIGURE 16 Comparison of the emission spectra after 430 and 960 h from the beginning of the lifetime characterization of Fig. 15. (From Ref. 48.)

devices). We tested these devices at both constant current and voltage driving, and in the latter case, both in DC and AC conditions. Although our best result was obtained with pulsed driving (square wave at 1070 Hz, 50% duty cycle, and variable voltage as indicated in Fig. 15), we could not find a clear correlation between the mode of operation (pulsed or DC) and the lifetime.

We were able to operate the device for Figs. 15 and 16 for ~1200 h with a total charge density passed through the device of more than 10^6 C/cm^2 and to check that the nature of the emissive species was not altered, as demonstrated by the nearly identical emission spectra in Fig. 16. Note that these devices were not encapsulated in order not to alter the device structure with the sealant; hence, they were operated in a low vacuum ($\sim 10^{-2}$ mbar) kept by a rotary pump in order to prolong the life of the cathode. At those pressures, oxygen contamination is considerable and the slow increase of the efficiency versus time can be attributed to the formation of a more and more extended CaO barrier at the polymer/Ca interface, although there is no direct evidence of this process. Interruptions of the main supply (and degradation of the vacuum) and experimental accidents shortened the lifetime of the devices, but it was possible to conclude that Ca-ion diffusion inside the polymer was extremely slow. Unfortunately, we were not able to analyze further the stressed devices to get insights into the physical or chemical mechanisms of the aging process.

Although the most recent results [96,129] are very encouraging, with devices having reached the threshold of a few thousand hours, the achievement of the target figures mentioned above is not trivial and requires control over a few processes. The related problems have been addressed in a number of recent studies [43,123,126,150–162] which have focused on the degradation of all the major constituents of the LED structure, namely

1. Metal–polymer interface
2. Polymer
3. Anode–polymer interface

Note that also the results relative to organic LEDs (OLEDs) fabricated with low-molecular-weight materials are relevant to PLEDs.

Several studies [154,159,160] have considered the formation of black or dark nonemissive regions (spots) as a result of operation and have attributed it to the poor metal–polymer interface. In addition to the reactions reported by Salaneck and Bredas [131], the electrode can undergo delamination from the polymer by localized heating and possible bubble formation [163]. Another mechanism, reported by Cumpston and Jensen [159], is the in-plane electromigration of Al cathodes, ascribable to the presence of pinholes or low-impedance regions in the active layer which drain a considerable amount of current and hence cause electromigration of the Al in the electrode with concomitant growth of Al needles above the low-resistivity spot and depletion of Al in the surrounding area. The problem can be addressed by exercising a better control of the deposition of the active layer and of the cathode. In particular, a thicker Al layer and deposition of the latter by sputtering techniques, instead of by thermal evaporation [156,164], should guarantee stronger adhesion and possibly reduced electromigration.

Secondary ion mass spectroscopy (SIMS) studies [151] have reported that a well-defined oxide interlayer can be found at the PPV interface with thermally evaporated Al together with a Cl enrichment at both interfaces. Interestingly, In is also found to penetrate in the film and to affect its characteristics. This is further confirmed by both luminescence spectroscopy and SIMS by Schlatmann and co-workers [123], who were able to detect luminous emission from In contaminations in reverse-biased LEDs.

The group at IBM identified ITO as the possible source of oxygen leading to formation of an aromatic aldehyde* defects in MEH–PPV PLEDs [154], and they proposed the use of an interlayer of a conducting polymer in order to limit the interaction [162]. This approach is found to be effective and it should be noted that it may also serve the function of minimizing the energy barrier between the anode and the emissive layer, as suggested by the work of Adachi et al. [149]. The latter found that lower anodic barriers correlated to longer lifetimes better than other parameters, such as the glass transition temperature or the melting point of the active material, possibly because of a reduction of the effects of electrochemical reactions taking place at the interface [157].

In fact, oxidation, and even self-oxidation, of the polymer during operation [43] is an important problem, and the use of reactive metals such as the alkaline earths or alkali metals could result in a forced choice for their oxygen-gettering virtues. In that case, the availability of a good packaging technology will then become crucial, but this aspect is beyond our present scope.

As far as the chemical stability of the polymeric emitter is concerned, we note that PPV is very stable both morphologically and thermally as a result of the polymeric nature, which reduces the tendency to crystallization (especially with respect to low-molecular-weight materials) and the high content of benzene units, respectively.

To the best of our knowledge, no recrystallization process as a function of storage time has been observed so far, whereas thermogravimetric analysis (TGA) shows that the first significant weight losses are well above 300°C.

Interest has then been focused mainly on the oxidation stability because of the serious consequences on the optical properties which have been discussed in Section II. An excellent review of the understanding of the relevant mechanisms is the one by Cumpston and Jensen [161].

Although there can be several routes along which chemical modifications of the polymer can take place and lead to formation of carbonyl and hydroxyl species, the primary mechanism is thought to be by way of transfer of the excitation energy to molecular oxygen via the triplet excited states of the polymer. This mechanism has been studied in particular for the case of BCHA–PPV [poly(2,5-bis(5,6-cholestanoxy-1,4 phenylene vinylene)] by Scurlock and co-workers [150].

Because the weak point of the PPV molecule is the vinylic bond, substitution of the hydrogens on the aliphatic carbons with other species is a viable method for increasing oxidative stability. PDPV (Fig. 5), for example, shows significant im-

---

*The effect of these defects and, in general, of photooxydation products on the optical properties of PPV have been discussed in Section II.

provements over PPV, owing to the phenyl substitution on the vinylenes [165], although it may undergo a cyclization reaction which leads to the formation of phenanthrene units [44]. Substitution of electron-withdrawing groups is also effective [100].

Alternatively, Scurlock et al. [150] have also suggested the introduction of triplet quenchers in order to avoid excitation of the molecular oxygen and the subsequent oxidation processes. As noted by Cumpston and Jensen [161] however, this may have a limited effect owing to possible "static" oxygen activation (i.e., before the triplet diffuses away from the site of formation, and hence before any effective quenching can take place). This observation is quite relevant, especially in view of the strongly localized character of the triplets [110], and of the fact that a significant migration process (i.e., Förster transfer) is not accessible to triplets.

Note also that in the case of BCHA–PPV, the quantum yield of oxygen activation from the triplets in solid state is reduced to 0.2 from ~1 in solution. It is also worth considering that recent data on PPV [96] report no change in absorption and photoluminescence exposed to white light in air and protected by a layer of transparent Al oxide. This provides a strong indication that interfacial phenomena may actually be more important than bulk degradation at the present state of development of this technology.

VI. CONJUGATED POLYMER MICROCAVITIES

A polymer microcavity can be defined as an optical resonator with a mirror spacing comparable to the (optical) wavelength dimensions, which allows manipulation of the radiative properties of a generic emitter placed in it [166,167].

For low-finesse cavities, the control of spontaneous emission can be viewed either in terms of classical interference effects or within a quantum-mechanical treatment, whereas some of the properties observed in very high-Q cavities are only expected in a quantum-mechanical treatment [168].

Cavity quantum electrodynamic (CQED) effects [166,167] relate to the details of the field modes coupled to the optical transition, and to the orientation of the optical dipole moment. The optical mode structure can be controlled by varying the thickness of the polymer layer and the reflectivity of the mirrors that form the microcavity, enabling enhancement of the emission and narrowing of the linewidth [169]. This increases both the external quantum efficiency in the forward direction and the spectral purity of the device luminescence.

It is this ability of microcavities to control the process of radiative recombination which has made them interesting for both inorganic [166,167,170,171] and organic light-emitting structures [169,172–181].

Although early studies were focused on microcavities with two metallic mirrors, these are always affected by a significant absorption in the visible, which reduces the Q of the cavity and, hence, the intensity of the emission. The problem can be circumvented by using dielectric mirrors (DBR: distributed Bragg reflector) which can be fabricated with a very high reflectivity (~99%) and virtually no loss. ITO electrodes can then be deposited onto the DBR to integrate a PLED into such a microcavity, as schematically shown in Fig. 17, whereas in Fig. 18, we report the optical properties of typical mirrors which can be used for the fabrication of such a microcavity.

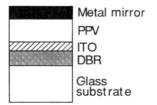

FIGURE 17 Structure of a possible conjugated polymer microcavity device.

The optimization of the emissive properties of a structure as in Fig. 17 is clearly relevant to the improvement of PLED performance and to the realization of optically and electrically driven lasing structures [11,63,182].

Note that because the nature of the radiative excitation involved in electroluminescence (EL) and photoluminescence (PL) is thought to be the same singlet exciton, the results of PL investigations are of general significance to the working mechanism of organic LEDs, and this is important because, in some cases, photoluminescence spectroscopy is much more flexible than electroluminescence.

In Fig. 20, for example, we report a comparison of the PL emission intensity, in the forward direction, of a microcavity with the actual dimensions and parameters indicated in Fig. 19, with both Al and an Ag mirror, together with the PL of a reference sample, consisting of a PPV film of the same thickness on quartz.

We observe that a drastic spectral narrowing in the cavity emission is accompanied by a change of color from green to orange. Note that the cavity with a silver mirror gives a peak emission 17 times more intense than the PPV reference film on glass and almost 11 times more intense than the Al cavity. These figures express the measured *external emission enhancement*, G_{ext}, defined as the intensity

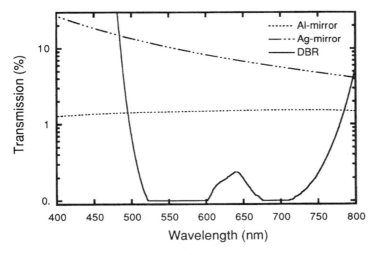

FIGURE 18 Transmission spectra of Al, Ag, and DBR mirrors, with thickness of about 35 nm for the silver and about 15 nm for the aluminum. (From Ref. 169.)

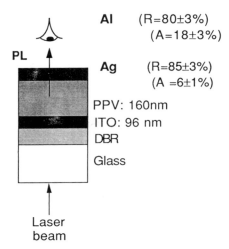

Al   (R=80±3%)
        (A=18±3%)

PL

Ag   (R=85±3%)
        (A =6±1%)

PPV: 160nm

ITO: 96 nm

DBR

Glass

Laser
beam

FIGURE 19 The structure and the parameters of the cavities used for the PL intensity study in Fig. 20.

in the forward direction at the peak maximum, normalized to the intensity of the reference at the same wavelength. After removing the ITO layer, which contributes to the cavity absorption, we obtain further improvements, with a maximum measured gain of 23 and a full width at half-maximum of about 4 nm, corresponding to a quality factor of about 150.

By taking into account the residual absorption in the silver mirror and the nonperfect matching between the antinodes of the electromagnetic field in the cavity and the spatial distribution of the chromophores excited in the PL experiment

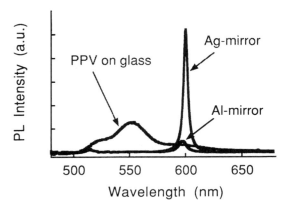

FIGURE 20 The PL spectra observed in the forward direction for a Ag and Al microcavities with the parameters indicated in Fig. 19. The reference spectrum of a PPV film on quartz with the same thickness and measured with the same excitation is also shown for comparison. Excitation is provided by the 457.9 nm of an Ar-ion laser, entering the microcavity from the DBR side (see also Fig. 19). Note that the DBR is transparent at this wavelength.

(determined by the Lambert–Beer law), it is also possible to calculate the ideal enhancement in the forward direction. This turns out to be ~60 for a structure without an ITO electrode. This corresponds to spectrally integrated enhancement factors of 0.7 (uncorrected) and 2.5 ± 0.5 (ideal case). Similarly, spatial integration of the emission at the peak wavelength yields enhancements at 0.36 (uncorrected) and 1.3 ± 0.2 (ideal) with respect to the reference PPV on glass. This figures are found to be in agreement with the values of the absolute PL efficiency (integrated over all angles) and with the temporal evolution of the PL on a subnanosecond scale [182].

A detailed characterization of the role of the chromophore position with respect to the nodes of the field has been given recently by Burns et al., who varied the position of a chromophore layer between two inert spacers, so as to keep constant the overall optical length of the cavity, by applying a Langmuir–Blodgett deposition [183] and by the technique of Lidzey [184], who used a CVD technique for a similar experiment.

We conclude by noting that although the spectrally and spatially integrated enhancement of the emission is limited for a relatively large number of reasons, in the study reported above, the most important are a restricted angular stop band of the DBR and the planar structure of the resonator, which does not confine the modes as effectively as a two-dimensional or a three-dimensional cavity would do (i.e., a cavity of a size comparable with the optical wavelength in two or three dimensions instead of just one). The introduction of the latter and/or mirrors with a larger angular stop band could then bring about substantial improvements.

Note, however, that owing to the very high efficiency of some of the available polymers, also planar structures as those reported above can lase [11].

VII. CONCLUSIONS

Polymeric light emitters have represented a very active and interdisciplinary area of science, which has been growing steadily in the 8 years since the discovery of electroluminescence in PPV in 1990. One of the strengths of this sector is the concomitance of the scientific interests in the universities with the economic interests of the markets for a new and cheap "display technology." Recent advances have been produced in the chemistry of the materials, the technology of fabrication, and the physics of the conjugated systems, in general. This has led to the remarkable achievement of optically excited lasing in a conjugated polymer microcavity structure in the summer of 1996, which seems to suggest the possibility that even more exciting achievements will develop.

ACKNOWLEDGMENTS

I would like to thank all those who have made this work possible, and in particular Professor Wise, who invited me to write the chapter and patiently waited for my manuscript, Professor Friend, who provided me with the possibility of working in the field, and Professor Zerbi, who introduced me to conjugated polymer electroluminescence.

I would also like to thank the Royal Society for the award of a University Research Fellowship.

REFERENCES

1. Burroughes, J. H., Bradley, D. D. C., Brown, A. R., Marks, R. N., McKay, K., Friend, R. H., Burn, P. L., and Holmes, A. B., Light-emitting diodes based on conjugated polymers, *Nature*, *347*, 539–541 (1990).
2. Brown, A. R., Burroughes, J. H., Greenham, N. C., Friend, R. H., Bradley, D. D. C., Burn, P. L., Kraft, A., and Holmes, A. B., Poly(*p*-phenylene vinylene) light-emitting diodes: enhanced electroluminescence efficiency through charge carrier confinement, *Appl. Phys. Lett.*, *61*, 2793–2795 (1992).
3. Ohmori, Y., Uchida, M., Muro, K., and Yoshino, K., Blue electroluminescent diodes utilising poly(alkylfluorene), *Jpn. J. Appl. Phys.*, *30*, L1941–L1943 (1991).
4. Ohmori, Y., Morishima, C., Uchida, M., and Yoshino, K., Time-resolved pulse response of electroluminescence in poly(3-alkylthiophene) diodes, *Jpn. J. Appl. Phys.*, *Pt. 2*, *31*, L568–L569 (1992).
5. Gustafsson, G., Cao, Y., Treacy, G. M., Klavetter, F., Colaneri, N., and Heeger, A. J., Flexible light-emitting diodes made from soluble conducting polymers, *Nature*, *357*, 477–479 (1992).
6. Grem, G., Leditzky, G., Ullrich, B., and Leising, G., Realisation of a blue light emitting device using poly (*para*-phenylene), *Adv. Mater.*, *4*, 36–48 (1992).
7. Skotheim, T. J., *Handbook of Conducting Polymers*, Marcel Dekker, Inc., New York, 1986, pp. 727.
8. Greenham, N. C., and Friend, R. H., Semiconductor device physics of conjugated polymers, *Solid State Phys.*, *49*, 1–149 (1995).
9. Sze, S. M., *Physics of Semiconductor Devices*, John Wiley & Sons, New York, 1981, pp. 868.
10a. Pope, M., Kallmann, H., and Magnante, P., Electroluminescence in organic crystals, *J. Chem. Phys.*, *38*, 2042–2047 (1963).
10b. Pope, M., and Swenberg, C. E., *Electronic Processes in Organic Crystals*, Clarendon Press, Oxford, 1982.
11. Tessler, N., Denton, G. J., and Friend, R. H., Lasing from conjugated polymer microcavities, *Nature*, *382*, 695–697 (1996).
12. Schlenk, W., Ueber chinoide byphenylderivative, *J. Liebigs Annal. Chem.*, *368*, 271–280 (1909).
13. Helfrich, W., and Schneider, W. G., Recombination radiation in anthracene crystals, *Phys. Rev. Lett.*, *14*, 229–233 (1965).
14. Helfrich, W., and Schneider, W. G., Transients of volume-controlled current and of recombination radiation in anthracene, *J. Chem. Phys.*, *44*, 2902–2907 (1966).
15. Roberts, G. G., McGinnity, M., Barlow, W. A., and Vincett, P. S., Electroluminescence, photoluminescence and electroabsorption of a lightly substituted anthracene langmuir film, *Solid State Commun.*, *32*, 683–686 (1979).
16. Vincett, P. S., Barlow, W. A., Hann, R. A., and Roberts, G. G., Electrical conduction and low voltage blue electroluminescence in vacuum-deposited organic films, *Thin Solid Films*, *94*, 476–480 (1982).
17. Tang, C. W., Two-layer organic photovoltaic cell, *Appl. Phys. Lett.*, *48*, 183–185 (1986).
18. Chiang, C. K., Fincher, C. R., Park, Y. W., Heeger, A. J., Shirakawa, H., Louis, E. J., Gau, S. C., and MacDiarmid, A. G., Electrical conductivity in doped polyacetylene, *Phys. Rev. Lett.*, *39*, 1098–1101 (1977).
19. Edwards, J. H., and Feast, W. J., A precursor route to polyacetylene, *Polym. Commun.*, *21*, 595–597 (1980).
20. Yang, Z., Sokolik, I., and Karasz, F. E., A soluble blue-light-emitting polymer, *Macromolecules*, *26*, 1188–1193 (1993).

21. Weaver, M. S., Lidzey, D. G., Fisher, T. A., Pate, M. A., Obrien, D., Bleyer, A., Tajbakhsh, A., Bradley, D. D. C., Skolnick, M. S., and Hill, G., Recent progress in polymers for electroluminescence—Microcavity devices and electron-transport polymers, *Thin Solid Films*, *273*, 39–44 (1996).

22. Fischer, W., Stelzer, F., Meghdadi, F., and Leising, G., Self-assembled aromatic oligoazomethines on polar surfaces, *Synth. Met.*, *76*, 201–204 (1996).

23. Cacialli, F., and Bruschi, P., Site-selective chemical-vapour-deposition of sub-micron wide conducting polypyrrole films: Morphological investigations with the scanning electron and the atomic force microscopic, *J. Appl. Phys.*, *80*, 70–75 (1996).

24. Chang, W. P., Whang, W. T., and Lin, P. W., Characteristics of an electropolymerized PPV and its light-emitting diode, *Polymer*, *37*, 1513–1518 (1996).

25. Heeger, A. J., Kivelson, S., Schrieffer, J. R., and Su, W.-P., Solitons in conducting polymers, *Rev. Mod. Phys.*, *60*, 781–850 (1988).

26. Burn, P. L., Bradley, D. D. C., Brown, A. R., Holmes, A. B., and Friend, R. H., Studies on the efficient synthesis of poly(phenylene vinylene) (PPV) and poly(dimethoxy phenylene vinylene), *Synth. Met.*, *41–43*, 261–264 (1991).

27. Burn, P. L., Holmes, A. B., Kraft, A., Bradley, D. D. C., Brown, A. R., and Friend, R. H., Synthesis of a segmented conjugated polymer chain giving a blue-shifted electroluminescence and improved efficiency, *J. Chem. Soc., Chem. Commun.*, 32–34 (1992).

28. Burn, P. L., Holmes, A. B., Kraft, A., Bradley, D. D. C., Brown, A. R., Friend, R. H., and Gymer, R. W., Chemical tuning of electroluminescent copolymers to improve emission efficiencies and allow patterning, *Nature*, *356*, 47–49 (1992).

29. Burn, P. L., Bradley, D. D. C., Friend, R. H., Halliday, D. A., Holmes, A. B., Jackson, R. W., and Kraft, A. M., Precursor route chemistry and electronic properties of poly(1,4-phenylenevinylene), poly(2,5-dimethyl-1,4-phenylenevinylene), and poly(2,5-dimethoxy-1,4-phenylenevinylene), *J. Chem. Soc., Perkin Trans 1*, *23*, 3225–3231 (1992).

30. Burn, P. L., Kraft, A., Baigent, D. R., Bradley, D. D. C., Brown, A. R., Friend, R. H., Gymer, R. W., Holmes, A. B., and Jackson, R. W., Chemical tuning of the electronic properties of poly(*p*-phenylenevinylene)-based copolymers, *J. Am. Chem. Soc.*, *115*, 10117–10124 (1993).

31. Halliday, D. A., Burn, P. L., Bradley, D. D. C., Friend, R. H., Gelsen, O., Holmes, A. B., Kraft, A., Martens, J. H. F., and Pichler, K., Large changes in optical response through chemical pre-ordering of poly(*p*-phenylenevinylene), *Adv. Mater.*, *5*, 40–42 (1993).

32. Halliday, D. A., Burn, P. L., Friend, R. H., Bradley, D. D. C., Holmes, A. B., and Kraft, A., Extended π-conjugation in poly(*p*-phenylenevinylene) from a chemically modified precursor polymer, *Synth. Met.*, *55–57*, 954–959 (1993).

33. Zhang, C., Braun, D., and Heeger, A. J., Light-emitting diodes from partially conjugated poly(*p*-phenylene vinylene), *J. Appl. Phys.* *73*, 5177–5180 (1993).

34. Samuel, I. D. W., Crystall, B., Rumbles, G., Burn, P. L., Holmes, A. B., and Friend, R. H., Time-resolved luminescence measurements in poly(*P*-phenylenevinylene), *Synth. Met.*, *54*, 281–286 (1993).

35. Yan, M., Rothberg, L. J., Papadimitrakopoulos, F., Galvin, M. E., and Miller, T. M., Defect quenching of conjugated polymer luminescence, *Phys. Rev. Lett.*, *74*, 744–747 (1994).

36. Papadimitrakopoulos, F., Konstadinidis, K., Miller, T. M., Opila, R., Chandross, E. A., and Galvin, M.E., The role of carbonyl groups in the photoluminescence of poly(*p*-phenylenevinylene), *Chem. Mater.*, *6*, 1563–1568 (1994).

37. Esteghamatian, M., and Xu, G., Colour changes in photoluminescence by doped, unconverted and partially converted poly(p-phenylene vinylene), *Appl. Phys. Lett.*, *65*, 1877–1879 (1994).

38. Herold, M., Gmeiner, J., and Schwoerer, M., Preparation of light-emitting diodes on flexible substrates: Elimination reaction of poly(p-phenylene vinylene) at moderate temperatures, *Acta Polym.*, *45*, 392–396 (1994).

39. Gmeiner, J., Karg, S., Meier, M., Riess, W., Strohriegl, P., and Schwoerer, M., Synthesis, electrical conductivity and electroluminescence of poly(p-phenylene vinylene) prepared by the precursor route, *Acta Polym.*, *44*, 201–207 (1993).

40. Herold, M., Gmeiner, J., Riess, W., and Schwoerer, M., Tailoring of the electrical and optical-properties of poly(p-phenylene vinylene), *Synth. Met.*, *76*, 109–114 (1996).

41. Herold, M., Gmeiner, J., and Schwoerer, M., Influence of the elimination temperature on light emitting devices prepared from poly(p-phenylene vinylene), *Acta Polym.*, *47*, 436–441 (1996).

42. Bradley, D. D. C., Precursor-route poly(p-phenylenevinylene): Polymer characterisation and control of electronic properties, *J. Phys. D: Appl. Phys.*, *20*, 1389–1410 (1987).

43. Zyung, T., and Kim, J., Photodegradation of poly(p-phenylenevinylene) by laser-light at the peak wavelength of electroluminescence, *Appl. Phys. Lett.*, *67*, 3420–3422 (1995).

44. Helbig, M., and Hörhold, H. H., Investigations of poly(arylenevinylenes), 40: Electrochemical studies on poly(p-phenylenevinylenes), *Makromol. Chem.*, *194*, 1607–1618 (1993).

45. Brédas, J. L., and Heeger, A. J., Influence of donor and acceptor substituents on the electronic characteristics of poly(*para* phenylene vinylene) and poly(*para* phenylene), *Chem. Phys. Lett.*, *217*, 507–512 (1994).

46. Friend, R. H., Bradley, D. D. C., and Townsend, P. D., Photoexcitation in conjugated polymers, *J. Phys. D: Appl. Phys.*, *20*, 1367–1384 (1987).

47. Rauscher, U., Bässler, H., Bradley, D. D. C., and Hennecke, M., Exciton versus band description of the absorption and luminescence spectra in poly(p-phenylenevinylene), *Phys. Rev. B*, *42*, 9830–9837 (1990).

48. Cacialli, F., Friend, R. H., Moratti, S. C., and Holmes, A. B., Characterisation of properties of polymeric light-emitting-diodes over extended periods, *Synth. Met.*, *67*, 157–160 (1994).

49. Chandross, M., Mazumdar, S., Jeglinski, S., Wei, X., Vardeny, Z. V., Kwock, E. W., and Miller, T. M., Excitons in poly(*para*-phenylenevinylene), *Phys. Rev. B*, *50*, 14702–14705 (1994).

50. Pakbaz, K., Lee, C. H., Heeger, A. J., Hagler, T. W., and Mcbranch, D., Nature of the primary photoexcitations in poly(arylene-vinylenes), *Synth. Met.*, *64*, 295–306 (1994).

51. Burns, S. E., Greenham, N. C., Friend, R. H., Modelling of optical interference effects in conjugated polymer-films and devices, *Synth. Met.*, *76*, 205–208 (1996).

52. Kersting, R., Lemmer, U., Mahrt, R. F., Leo, K., Kurz, H., Bässler, H., and Göbel, E. O., Femtosecond energy relaxation in π-conjugated polymers, *Phys. Rev. Lett.*, *70*, 3820–3823 (1993).

53. Hayes, G. R., Samuel, I. D. W., and Phillips, R. T., Exciton dynamics in electroluminescent polymers studied by femtosecond time-resolved photoluminescence spectroscopy, *Phys. Rev. B—Cond. Matter*, *52*, 11569–11573 (1995).

54. Bässler, H., Site-selective fluorescence spectroscopy of polymers, in *Optical Techniques to Characterise Polymer Systems* (H. Bässler, ed.), Elsevier Science, Amsterdam, 1989.

55. Greenham, N. C., Samuel, I. D. W., Hayes, G. R., Phillips, R. T., Kessener, Y. A. R. R., Moratti, S. C., Holmes, A. B., and Friend, R. H., Measurement of absolute photoluminescence quantum efficiencies in conjugated polymers, *Chem. Phys. Lett.*, *241*, 89–96 (1995).

56. Strickler, S. J., and Berg, R. A., Relationship between absorption intensity and fluorescence lifetimes of molecules, *J. Chem. Phys.*, *37*, 814 (1962).

57. Harrison, N. T., Hayes, G. R., Phillips, R. T., and Friend, R. H., Singlet intrachain exciton generation and decay in poly(p-phenylene vinylene), *Phys. Rev. Lett.*, *77*, 1881–1884 (1996).

58. Yan, M., Rothberg, L. J., Papadimitrakopolous, F., Galvin, M. E., and Miller, T. M., Spatially indirect excitons as primary photoexcitations in conjugated polymers, *Phys. Rev. Lett.*, *72*, 1104–1108 (1994).

59. Pichler, K., Halliday, D. A., Bradley, D. D. C., Burn, P. L., Friend, R. H., and Holmes, A. B., Optical spectroscopy of highly ordered poly(p-phenylene vinylene), *J. Phys. Cond. Matter*, *5*, 7155–7172 (1993).

60. Son, S., Dodabalapur, A., Lovinger, A. J., and Galvin, M. E., Luminescence enhancement by the introduction of disorder into poly(p-phenylene vinylene), *Science*, *269*, 376–378 (1995).

61. Colaneri, N., Friend, R. H., Schaffer, H. E., and Heeger, A. J., Spectroscopy of photoinduced solitons in cis-rich and trans polyacetylene, *Solid State Sci.*, *76*, 118–125 (1987).

62. Schwartz, B. J., DiazGarcia, M. A., and Heeger, A. J., Laser-emission from solutions and films containing a semiconducting polymer and titanium-dioxide nanocrystals, *Chem. Phys. Lett.*, *256*, 424–430 (1996).

63. Hide, F., Diazgarcia, M. A., Schwartz, B. J., Andersson, M. R., Pei, Q. B., and Heeger, A. J., Semiconducting polymers—A new class of solid-state laser materials, *Science*, *273*, 1833–1836 (1996).

64. Moses, D., High quantum efficiency luminescence from a conducting polymer in solution: A novel dye laser, *Appl. Phys. Lett.*, *60*, 3215–3217 (1992).

65. Yan, M., Rothberg, L., Hsieh, B. R., and Alfano, R. R., Exciton formation and decay dynamics in electroluminescent polymers observed by subpicosecond stimulated-emission, *Phys. Rev. B—Cond. Matter*, *49*, 9419–9422 (1994).

66. Brouwer, H. J., Krasnikov, V. V., Hilberer, A., Wildeman, J., and Hadziioannou, G., Novel high efficiency copolymer laser dye in the blue wavelength region, *Appl. Phys. Lett.*, *66*, 3404–3407 (1995).

67. Denton, G. J., Tessler, N., Harrison, N. T., and Friend, R. H., Factors influencing stimulated emission from poly(p-phenylene vinylene), *Phys. Rev. Lett.*, *78*, 733–736 (1997).

68. Stubb, H., Punkka, E., and Paloheimo, J., Electronic and optical properties of conducting polymer thin films, *Mater. Sci. Eng. Rept.*, *10*, 85–140 (1993).

69. Stauffer, D., *Introduction to Percolation Theory*, Taylor & Francis, London, 1987, pp. 87.

70. Kao, K. C., and Hwang, W., *Electrical Transport in Solids*, International Series in the Science of Solid State Vol. 14, Pergamon Press, Oxford, 1981, p. 7.

71. Keller, R. A., and Rast, H. E., Tunnelling model for electron transport and its temperature dependence in crystals of low carrier mobility, *J. Chem. Phys.*, *36*, 2640 (1962).

72. Sheng, P., Hopping conductivity in granular metals, *Phys. Rev. Lett.*, *31*, 44–49 (1973).

73. Sheng, P., Sichel, E. K., and Gittleman, J. I., Fluctuation induced tunnelling conduction in carbon polyvinylchloride composites, *Phys. Rev. Lett.*, *40*, 1197–1201 (1978).

74. Sheng, P., Fluctuation induced tunnelling conduction in disordered materials, *Phys. Rev. B*, *21*, 2180–2187 (1980).

75. Sheng, P., and Klafter, J., Hopping conductivity in granular disordered systems, *Phys. Rev. B*, *27*, 2583–2590 (1983).

76. Ziemelis, K. E., Hussain, A. T., Bradley, D. D. C., Friend, R. H., Rühe, J., and Wegner, G., Optical spectroscopy of field-induced charges in poly(3-hexyl thienylene) MIS structures; evidence for polarons, *Phys. Rev. Lett.*, *66*, 2213–2234 (1991).

77. Harrison, M. G., Ziemelis, K. E., Friend, R. H., Burn, P. L., and Holmes, A. B., Optical spectroscopy of field-induced charge in poly(2,5-dimethoxy-*p*-phenylene vinylene) metal–insulator–semiconductor structures, *Synth. Met.*, *55–57*, 218–222 (1993).

78. Frenkel, J., On pre-breakdown phenomena in insulators and electronic semiconductors, *Phys. Rev.*, *54*, 647–648 (1938).

79. Frenkel, J., On the theory of electric breakdown of dielectrics and electronic semiconductors, *Tech. Phys. USSR*, *5*, 685–686 (1938).

80. Gill, W. D., Drift mobilities in amorphous charge-transfer complexes of trinitrofluorenone and polyvinylcarbazole, *J. Appl. Phys.*, *43*, 5033–5041 (1972).

81. Mott, N. F., and Davis, E. A., *Electronic Processes in Non-Crystalline Materials*, Clarendon Press, Oxford, 1979.

82. Mott, N. F., Electronic processes in non-crystalline materials, *Conduction in Non-Crystalline Materials*, Clarendon Press, Oxford, 1979.

83. Mott, N. F., *Conduction in Non-Crystalline Materials*, Clarendon Press, Oxford, 1987.

84. Efros, A. L., and Shklovskii, Coulomb gap and low temperature conductivity of disordered systems, *J. Phys. C*, *8*, 407 (1975).

85. Roth, S., Bleier, H., and Pukacki, W., Anisotropic variable range hopping transport, *Faraday Discuss. Chem. Soc.*, *29*, 4491–4495 (1989).

86. Schein, L. B., Comparison of charge transport models in molecularly doped polymers, *Phil. Mag. B*, *65*, 795–799 (1992).

87. Bässler, H., Charge transport in molecular doped polymers, *Phil. Mag. B*, *50*, 347–362 (1984).

88. Sze, S. M., *Physics of Semiconductor Devices*, John Wiley & Sons, New York, 1981, p. 403.

89. Henisch, H. K., *Semiconductor Contacts*, Clarendon Press, Oxford, 1984, p. 377.

90. Kao, K. C., Hwang, W., *Electrical Transport in Solids*, International Series in the Science of Solid State Vol. 14, Pergamon Press, Oxford, 1981.

91. Nguyen, T. P., Tran, V. H., Massardier, V., Electrical conduction in poly(phenylenevinylene) thin films, *J. Phys.: Conds. Matter*, *5*, 6243–6252 (1993).

92. Blom, P. W. M., Dejong, M. J. M., and Vleggaar, J. J. M., Electron and hole transport in poly(*p*-phenylene vinylene) devices, *Appl. Phys. Lett.*, *68*, 3308–3310 (1996).

93. Blom, P. W. M., Dejong, M. J. M., and van Munster, M. G., Electric field and temperature dependence on the hole-mobility in poly(*p*-phenylene vinylene), *Phys. Rev. B*, *55*, R656–R659 (1997).

94. Marks, R. N., Bradley, D. D. C., Jackson, R. W., Burn, P. L., and Holmes, A. B., Charge injection and transport in poly(*p*-phenylene vinylene) light emitting diodes, *Synth. Met.*, *55–57*, 4128–4133 (1993).

95. Jarrett, C. P., Friend, R. H., Brown, A. R., and Deleeuw, D. M., Field-effect measurements in doped conjugated polymer-films—Assessment of charge-carrier mobilities, *J. Appl. Phys.*, *77*, 6289–6294 (1995).

96. Pichler, K., Conjugated polymer electroluminescence: Technical aspects from basic devices to commercial products, *Roy. Soc. Phil. Transact. A*, *1725*, 842 (1997).

97. Grimsdale, A. C., Li, X. C., Holmes, A. B., Kraft, A., Moratti, S. C., Cacialli, F., and Friend, R. H., Novel poly(arylene vinylene)s carrying donor and acceptor substituents, *Synth. Met.*, *76*, 165–167 (1997).

98. Greenham, N. C., Moratti, S. C., Bradley, D. D. C., Friend, R. H., and Holmes, A. B., Efficient light-emitting diodes based on polymers with high electron affinities, *Nature*, *365*, 628–630 (1993).

99. Baigent, D. R., Greenham, N. C., Grüner, J., Marks, R. N., Friend, R. H., Moratti, S. C., and Holmes, A. B. Light-emitting diodes fabricated with conjugated polymers—Recent progress, *Synth. Met.*, *67*, 3–10 (1994).

100. Lux, A., Holmes, A. B., Cervini, R., Davies, J. E., Moratti, S. C., Grüner, J., Cacialli, F., and Friend, R. H., New Cf$_3$-substituted PPV-type oligomers and polymers for use as hole-blocking layers in LEDs, *Synth. Metals*, in press.

101. Strukelj, M., Papadimitrakopoulos, F., Miller, T., and Rothberg, L. J., Design and application of electron-transporting organic materials, *Science*, *267*, 1969 (1995).

102. Cacialli, F., Friend, R. H., Haylett, N., Daik, R., Feast, W. J., dos Santos, D., and Brèdas, J. L., Efficient green light emitting diodes from a phenylated derivative of poly(*p*-phenylene-vinylene), *Appl. Phys. Lett.*, *69*, 3794–3796 (1996).

103. Cacialli, F., Li, X., Friend, R. H., Moratti, S. C., and Holmes, A. B., Light emitting diodes from distrylbenzene based polymethacrylates, *Synth. Met.*, *75*, 161–168 (1996).

104. Li, X. C., Yong, T. M., Grüner, J., Holmes, A. B., Moratti, S. C., Cacialli, F., and Friend, R. H., A blue light-emitting copolymer with charge transporting and photocrosslinkable functional units, *Synth. Met.*, *84*, 437–438 (1997).

105. Pei, Q., and Yang, Y., 1,3,4-oxadiazole-containing polymers as electron-injection and blue electroluminescent materials in polymer light-emitting-diodes, *Chem. Mater.*, *7*, 1568–1575 (1995).

106. Pei, Q. I., and Yang, Y., Bright blue electroluminescence from an oxadiazole-containing copolymer, *Adv. Mater.*, *7*, 559–561 (1995).

107. Weaver, M., and Bradley, D. D. C., Organic electroluminescence devices fabricated with chemical vapour deposited polyazomethine films, *Synth. Met.*, *83*, 61–66 (1996).

108. Obrien, D., Weaver, M. S., Lidzey, D. G., and Bradley, D. D. C., Use of poly(phenyl quinoxaline) as an electron-transport material in polymer light-emitting-diodes, *Appl. Phys. Lett.*, *69*, 881–883 (1996).

109. Bouche, C. M., Berdague, P., Facoetti, H., Robin, P., Lebarny, P., and Schott, M., Side-chain electroluminescent polymers, *Synth. Met.*, *81*, 191–195 (1996).

110. Beljonne, D., Cornil, J., Bredas, J. L., and Friend, R. H., Theoretical investigation of the lowest singlet and triplet excited-states in oligo(phenylene vinylene)s and oligothiophenes, *Synth. Met.*, *76*, 61–65 (1996).

111. Hilberer, A., Brouwer, H. J., Vanderscheer, B. J., Wildeman, J., and Hadziioannou, G., Synthesis and characterisation of a new efficient blue-light-emitting copolymer, *Macromolecules*, *28*, 4525–4529 (1995).

112. Garten, F., Hilberer, A., Cacialli, F., Esselink, E., van Dam, Y., Schlatman, B., Friend, R. H., Klapvijk, T., and Hadziioannou, G., blue light emitting diodes based on a partially conjugated Si-containing PPV copolymer in a multi-layer configuration, *Synth. Met.*, *85*, 1253–1254 (1997).

113. Kim, S., Hwang, D., Li, X., Grüner, J., Friend, R., Holmes, A., and Shim, H., Efficient green electroluminescent diodes based on poly(2-dimethyloctylsilyl-1,4-phenylenevinylene), *Adv. Mater.*, *8*, 1996–1998 (1996).

114. Grüner, J., Wittmann, H. F., Hamer, P. J., Friend, R. H., Huber, J., Scherf, U., Müllen, K., Moratti, S. C., and Holmes, A. B., Electroluminescence and photoluminescence investigations of the yellow emission of devices based on ladder-type oligo(para-phenylene)s, *Synth. Met.*, *67*, 181–184 (1994).

115. Hayes, G. R., Samuel, I. D. W., and Phillips, R. T., Ultrafast high energy luminescence in a high-electro-affinity conjugated polymer, *Phys. Rev. B—Cond. Matter*, *54*, R8301–R8304 (1995).

116. Samuel, I. D. W., Rumbles, G., Collison, C. J., Crystall, B., Moratti, S. C., and Holmes, A. B., Luminescence efficiency and time-dependence in a high electron-affinity conjugated polymer, *Synth. Met.*, *76*, 15–19 (1996).

117. Harrison, N. T., Baigent, D. R., Samuel, I. D. W., Friend, R. H., Grimsdale, A. C., Moratti, S. C., and Holmes, A. B., Site-selective fluorescence studies of poly(*p*-phenylene vinylene) and its derivatives, *Phys. Rev. B—Cond. Matter*, *53*, 15815–15822 (1996).

118. Yu, L., *Solitons and Polarons in Conducting Polymers*, World Scientific, Singapore, 1988.

119. Frölich, H., On the theory of electric breakdown in solids, *Proc. Roy. Soc. A*, *118*, 521 (1947).

120. Holstein, T., Polaron motion. I: Molecular crystal model. II: Small polaron, *Ann. Phys. New York*, *8*, 325–342 (1959).

121. Marks, R. N., Ph.D. thesis, Cambridge University, 1993.

122. Garten, F., Vrijmoeth, J., Schlatmann, A. R., Gill, R. E., Klapwijk, T. M., and Hadziioannou, G., Light-emitting-diodes based on polythiophene–Influence of the metal workfunction on rectification properties, *Synth. Met.*, *76*, 85–91 (1996).

123. Schlatmann, A. R., Floet, D. W., Hilberer, A., Garten, F., Smulders, P. J. M., Klapwijk, T. M., and Hadziioannou, G., Indium contamination from the indium–tin-oxide electrode in polymer light-emitting-diodes, *Appl. Phys. Lett.*, *69*, 1764–1766 (1996).

124. Parker, I. D., Carrier tunnelling and device characteristics in polymer light-emitting-diodes, *J. Appl. Phys.*, *75*, 1656–1666 (1994).

125. Davids, P. S., Kogan, S. M., Parker, I. D., and Smith, D. L., Charge injection in organic light-emitting-diodes—tunnelling into low mobility materials, *Appl. Phys. Lett.*, *69*, 2270–2272 (1996).

126. Burrows, P. E., and Forrest, S. R., Electroluminescence from trap-limited current transport in vacuum-deposited organic light-emitting devices, *Appl. Phys. Lett.*, *64*, 2285–2287 (1994).

127. Ettedgui, E., Razafitrimo, H., Gao, Y., and Hsieh, B. R., Band bending modified tunnelling at metal/conjugated polymer interfaces, *Appl. Phys. Lett.*, *67*, 2705–2707 (1995).

128. Cacialli, F., unpublished 1997.

129. Staring, E. G. J., On the photochemical stability of diolkoxy PPV, A quantitative approach. *Roy. Soc. Phil. Transact. A*, *355*, 695–706 (1997).

130. Braun, D., and Heeger, A. J., Visible light emission from semiconducting polymer diodes, *Appl. Phys. Lett.*, *58*, 1982–1984 (1991).

131. Salaneck, W. R., and Bredas, J. L., The metal-on-polymer interface in polymer light-emitting-diodes, *Adv. Mater.*, *8*, 48–50 (1996).

132. Fahlman, M., Bredas, J. L., and Salaneck, W. R., Experimental and theoretical-studies of the pi-electronic structure of conjugated polymers and the low work function metal/conjugated polymer interaction, *Synth. Met.*, *78*, 237–246 (1996).

133. Birgerson, J., Fahlman, M., Broms, P., and Salaneck, W. R., Conjugated polymer surfaces and interfaces—A mini-review and some new results, *Synth. Met.*, *80*, 125–130 (1996).

134. Broms, P., Birgersson, J., Johansson, N., Logdlund, M., and Salaneck, W. R., Calcium electrodes in polymer LEDs, *Synth. Met.*, *74*, 179–181 (1995).

135. Gao, Y., Park, K. T., and Hsieh, B. R., X-ray photoemission investigations of the interface formation of Ca and poly(*p*-phenylene vinylene), *J. Chem. Phys.*, *97*, 6991–6993 (1992).

136. Taliani, C., Marks, R., Muccini, M., and Zamboni, R., The growth and characterization of alpha-sexythiend-based light-emitting diodes, *Roy. Soc. Phil. Transact. A*, *355*, 763–773 (1997).

137. Hung, L. S., Tang, C. W., and Mason, M. G., Electroluminescence of doped organic thin films, *Appl. Phys. Lett.*, *70*, 152–154 (1989).
138. Drexhage, K. H., Interaction of light with monomolecular dye layers, in E. Wolf (ed.), *Progress in Optics*, North-Holland, Amsterdam, 1974, pp. 163–232.
139. Greenham, N. C., Cacialli, F., Bradley, D. D. C., Friend, R. H., Moratti, S. C., and Holmes, A. B., Cyano-derivatives of poly(*p*-phenylene vinylene) for use in thin-film light-emitting diodes, in *Electrical, Optical, and Magnetic Properties of Organic Solid State Materials, Materials Research Society Fall Meeting, Boston, November 29–December 3 1993, Symposium Q, Boston*, A. F. Garito, A. K.-Y. Jen, C. Y.-C. Lee, L. R. Dalton (eds.), Materials Research Society, Pittsburgh, 1993, pp. 351–360.
140. Baigent, D. R., Hamer, P. J., Friend, R. H., Moratti, S. C., and Holmes, A. B., Polymer electroluminescence in the near-infrared, *Synth. Met.*, *71*, 2175–2176 (1995).
141. Feast, W. J., Millichamp, I. S., Friend, R. H., Horton, M. E., Phillips, D., Rughooputh, S. D. D. V., and Rumbles, G., Optical absorption and luminescence in poly(4,4'-diphenylenediphenylvinylene), *Synth. Met.*, *10*, 181–191 (1985).
142. Pope, M., and Swenberg, C. E., Electronic processes in organic crystals, Clarendon Press, Oxford, 1982, pp. 702–703.
143. Tsuchida, A., Nagata, A., Yamamoto, M., Fukui, H., Sawamoto, M., and Higashimura, T., Hole resonance among more than 2 carbazole chromophores in poly(*n*-vinylcarbazole), *Macromolecules*, *28*, 1285–1289 (1995).
144. Parker, I. D., Pei, Q., and Marrocco, M., Efficient blue electroluminescence from a fluorinated polyquinoline, *Appl. Phys. Lett.*, *65*, 1272–1275 (1994).
145. Yang, Y., Pei, Q., and Heeger, A. J., Efficient blue polymer light-emitting diodes from a series of soluble poly(paraphenylenes), *J. Appl. Phys.*, *79*, 934–939 (1996).
146. Vestweber, H., Sander, R., Greiner, A., Heitz, W., Mahrt, R. F., and Bässler, H., Electroluminescence from polymer blends and molecularly doped polymers, *Synth. Met.*, *64*, 141–145 (1994).
147. Zhang, C., von Seggern, H., Pakbaz, K., Kraabel, B., Schmidt, H. W., and Heeger, A. J., Blue electroluminescent diodes utilising blends of poly(*p*-phenylphenylene vinylene) in poly(9-carbazole), *Synth. Met.*, *62*, 35–40 (1994).
148. Kido, J., Kimura, M., and Nagai, K., Multilayer white light-emitting organic electroluminescent device, *Science*, *267*, 1332–1334 (1995).
149. Adachi, C., Nagai, K., and Tamoto, N., Molecular design of hole transport materials for obtaining high durability in organic electroluminescent diodes, *Appl. Phys. Lett.*, *66*, 2679–2681 (1995).
150. Scurlock, R. D., Wang, B. J., Ogilby, P. R., Sheats, J. R., and Clough, R. L., Singlet oxygen as a reactive intermediate in the photodegradation of an electroluminescent polymer, *J. Am. Chem. Soc.*, *117*, 10194–10202 (1995).
151. Sauer, G., Kilo, M., Hund, M., Wokaun, A., Karg, S., Meier, M., Riess, W., Schwoerer, M., Suzuki, H., Simmerer, J., Meyer, H., and Haarer, D., Characterisation of polymeric light-emitting-diodes by SIMS depth profiling analysis, *Fresenius J. Anal. Chem.*, *353*, 642–646 (1995).
152. Hamada, Y., Sano, T., Shibata, K., and Kuroki, K., Influence of the emission site on the running durability of organic electroluminescent devices, *Jpn. J. Appl. Phys. Part 2—Lett.*, *34*, L824–L826 (1995).
153. Cumpston, B. H., and Jensen, K. F., Photooxidation of polymers used in electroluminescent devices, *Synth. Met.*, *73*, 195–199 (1995).
154. Scott, J. C., Kaufman, J. H., Brock, P. J., Dipietro, R., Salem, J., and Goitia, J. A., Degradation and failure of MEH–PPV light-emitting-diodes, *J. Appl. Phys.*, *79*, 2745–2751 (1996).

155. Scott, J. C., Kaufman, J. H., Salem, J., Goitia, J. A., Brock, P. J., and Dipietro, R., MEH–PPV light-emitting-diodes—Mechanisms of failure. *Mol. Cryst. Liq. Cryst.*, *283*, 57–62 (1996).

156. Suzuki, H., and Hikita, M., Organic light-emitting-diodes with radio-frequency sputter-deposited electron injecting electrodes, *Appl. Phys. Lett.*, *68*, 2276–2278 (1996).

157. Aziz, H., and Xu, G., A degradation mechanism of organic light-emitting devices, *Synth. Met.*, *80*, 7–10 (1996).

158. Chao, C. I., Chuang, K. R., and Chen, S. A., Failure phenomena and mechanisms of polymeric light-emitting-diodes—Indium–tin-oxide-damage, *Appl. Phys. Lett.*, *69*, 2894–2896 (1996).

159. Cumpston, B. H., and Jensen, K. F., Electromigration of aluminum cathodes in polymer-based electroluminescent devices, *Appl. Phys. Lett.*, *69*, 3941–3943 (1996).

160. McElvain, J., Antoniadis, H., Hueschen, M. R., Miller, J. N., Roitman, D. M., Sheats, J. R., and Moon, R. L., Formation and growth of black spots in organic light emitting diodes, *J. Appl. Phys.*, *80*, 6002–6007 (1996).

161. Cumpston, B. H., and Jensen, K. F., Photooxidation of electroluminescent polymers, *Trends Polym. Sci.*, *4*, 151–157 (1996).

162. Karg, S., Scott, J. C., Salem, J. R., and Angelopoulos, M., Increased brightness and lifetime of polymer light-emitting-diodes with polyaniline anodes, *Synth. Met.*, *80*, 111–117 (1996).

163. Bijnens, W., Manca, J., and Wu, T.-D., Imaging of the ageing on organic electroluminescent diodes under different atmospheres by impedance spectroscopy, scanning electron microscopy and SIMS depth profiling analysis, *Synth. Met.*, *83*, 261–265 (1996).

164. Suzuki, H., Fabrication of electron injecting Mg–Ag alloy electrodes for organic light-emitting-diodes with radio-frequency magnetron sputter-deposition, *Appl. Phys. Lett.*, *69*, 1611–1614 (1996).

165. Koch, A., M. Phil. thesis, Cambridge University, 1996.

166. Yokoyama, H., Physics and device applications of optical microcavities, *Science*, *256*, 66–70 (1992).

167. Schubert, E. F., Hunt, N. E., Micovic, M., Malik, R. J., Sivco, D. L., Cho, A. Y., and Zydzik, J., Highly efficient light-emitting diodes with microcavities, *Science*, *265*, 943–945 (1994).

168. Berman, P. R., *Cavity Quantum Electrodynamics*, Academic Press, Boston, 1994.

169. Grüner, J., Cacialli, F., and Friend, R. H., Emission enhancement in single-layer polymer microcavities, *J. Appl. Phys.*, *80*, 207–215 (1996).

170. Vredenberg, A. M., Hunt, N. E., Schubert, E. F., Jacobson, D. C., Poate, J. M., and Zydzik, J., Controlled atomic spontaneous emission from Er^{3+} in a transparent Si/SiO_2 microcavity, *Phys. Rev. Lett.*, *71*, 517–520 (1993).

171. Hunt, N. E., Schubert, E. F., Kopf, R. F., Sivco, D. L., Cho, A. Y., and Zydzik, J., Increased fibre communications bandwidths from a resonant cavity light emitting at $\lambda = 940$ nm, *Appl. Phys. Lett.*, *63*, 2600 (1993).

172. Takada, N., Tsutsui, T., and Saito, S., Strongly directed emission from controlled-spontaneous-emission electroluminscent diodes with europium complex as an emitter, *Jpn. J. Appl. Phys.*, *33*, L863–L866 (1994).

173. Dodabalapur, A., Rothberg, L. J., Miller, T. M., and Kwock, E. W., Microcavity effects in organic semiconductors, *Appl. Phys. Lett.*, *64*, 2486–2487 (1994).

174. Lemmer, U., Hennig, R., Guss, W., Ochse, A., Pommerehne, J., Sander, R., Greiner, A., Mahrt, R. F., Bässler, H., Feldmann, J., and Gobel, E. O., Microcavity effects in a spin-coated polymer 2-layer system, *Appl. Phys. Lett.*, *66*, 1301–1304 (1995).

175. Tsutsui, T., Takada, N., Saito, S., and Ogino, E., Sharply directed emission in organic electroluminescent diodes with an optical-microcavity structure, *Appl. Phys. Lett.*, *65*, 1868–1870 (1994).
176. Tsutsui, T., Takada, N., and Saito, S., Control of spontaneous emission using microcavity structures in organic electroluminescent devices, *Synth. Met.*, *71*, 2001–2004 (1995).
177. Wittmann, H. F., Grüner, J., Friend, R. H., Spencer, G. W. C., Moratti, S. C., and Holmes, A. B., Microcavity effect in a single-layer polymer light-emitting diode, *Adv. Mater.*, *7*, 541–543 (1995).
178. Fisher, T. A., Lidzey, D. G., Pate, M. A., Weaver, M. S., Whittaker, D. M., Skolnick, M. S., and Bradley, D. D. C., Electroluminescence from a conjugated polymer microcavity structure, *Appl. Phys. Lett.*, *67*, 1355–1357 (1995).
179. Berggren, M., Inganas, O., Granlund, T., Guo, S., Gustafsson, G., and Andersson, M. R., Polymer light-emitting-diodes placed in microcavities, *Synth. Met.*, *76*, 121–123 (1996).
180. Grüner, J., Cacialli, F., Samuel, I. D., and Friend, R. H., Optical mode structure in a single-layer polymer microcavity, *Synth. Met.*, *76*, 137–140 (1995).
181. Grüner, J., Cacialli, F., Samuel, I. D., and Friend, R. H., Using microcavities to manipulate luminescence in conjugated polymers, in *Microcavities and Photonics Bandgaps*, J. Rarity and C. Weisbuch (eds.), Kluwer Academic, Dordrecht, 1996. pp. 407–417.
182. Cacialli, F., Hayes, G. R., Grüner, J., Friend, R. H., and Phillips, R. T., Light-emitting conjugated polymers in optical microcavities, *Synth. Met.*, in press.
183. Burns, S. E., Pfeffer, N., Grüner, J., Remmers, M., Javoreck, T., Neher, D., and Friend, R. H., Measurements of optical electric field intensities using thin emissive polymer films, *Adv. Mater*, in press.
184. Lidzey, D. G., Pate, M. A., Bradley, D. D. C., Whittaker, D. M., Fisher, T. A., and Skolnick, M. S., Mapping the confined optical field in a microcavity via the emission from a conjugated polymer, *Appl. Phys. Lett.*, *71*, 744–746 (1997).

5

Interface Formation in Organic Semiconductor Devices

V.-E. Choong, Y. Park, and Y. Gao
University of Rochester
Rochester, New York

B. R. Hsieh
Xerox Corporation
Webster, New York

I. INTRODUCTION

Optoelectronics technology, which has a direct impact on modern information processing systems and optical communication, represents a major sector of the world economy. Although conventional inorganic optoelectronic devices based primarily on III–V and II–VI semiconductors are maturing, research advancement in novel materials and device configurations are still needed to reduce cost and to further improve device performance and manufacturability. It turns out that organic materials have the potential to meet these needs [1–5]. Some organic materials have been found to possess the electrical and optical properties traditionally associated with metals and semiconductors, while retaining their mechanical properties. Among these, conjugated polymers are one of the more promising. Thus, the dream of using these materials in a variety of commercial display applications was born.

The first step to the realization of this dream materialized with the discovery of electroluminescence (EL) in poly(p-phenylene vinylene) (PPV), a conjugated polymer, and tris-(8-hydroxyquinoline)aluminum (Alq_3), a sublimable organic compound [6,7]. This discovery attracted a great deal of attention from the scientific and engineering communities because of the novel interactions responsible for light emission, in addition to the application potential. Due to inherent flexibility of organics and relative ease of processing, as well as easy tunability of electric properties by means of chemical substitutions, these materials make possible new venues in the development of electronic devices [6–15]. Since the first literature report on organic devices [6], there has been a rapidly increasing number of publications reporting important and continuing improvements in all features of device performance [12–24]. The color range accessible already includes the three primary colors, red, green, and blue, required for full-color displays [12,13,17–19,21–24].

Just as in their inorganic counterparts, these reports indicate the importance of the charge transport process across interfaces of dissimilar materials in organic semiconductor devices, and the importance of surface preparation [10,11,13,14, 25–33]. Transport and EL measurements show that metals of low work function, such as Ca, inject electrons more efficiently than do those of higher work function, giving rise to brighter EL [10,11,13,14,26,34]. Recent work suggests that charge injection into organic materials proceeds via tunneling across interfacial barriers [26]. Improvement of device performance through interface modification has also been reported [20,34]. Because of the strong dependence of the device performance on interfaces and surface preparation, new techniques have been developed to improve our control in device fabrication. Some of these techniques make it possible to control the fabrication of multilayer devices at the molecular level. Among the more innovative is the self-assembled monolayer technique [35,36], where monolayers of oppositely charged polymers adsorbed from dilute aqueous solutions are stacked onto surfaces to create multilayer thin films. These films comprise of semi-interpenetrated bilayers of polyanions and polycations.

Despite these advances, the detailed properties and roles of interfaces in organic semiconductor devices remain elusive. This chapter is intended to introduce the reader to the field by providing some insights into some of the theoretical and experimental approaches that have been employed thus far, although it is not meant to be an exhaustive study of all the research work and techniques employed in the study of interface formation in organic semiconducting devices. The topic is too broad and is beyond the scope of this work. In writing this chapter, we have naturally chosen to draw on our expertise, which is surface analytical techniques. However, this in no way implies that these tools are the only ones used, or are better than others in the study of organic semiconductor devices.

II. THEORETICAL MODELING

A. Introduction

The theoretical aspects of interface formation in polymeric LED devices closely resemble those of inorganic ones because the active materials in both cases are semiconductors. This allows one to use the extensive knowledge accumulated from many years of studies on the electronic properties of inorganic semiconductors. However, the electronic properties of organic materials, such as conjugated polymers, differ qualitatively from those of crystalline inorganic semiconductors in several important respects. In crystalline materials, the three-dimensional arrangement of the atoms and their resulting interactions ultimately determine the properties of the materials, but conjugated polymers can be treated as quasi-one-dimensional entities [37]. In this approximation, the polymer chains that form a sample are assumed to behave independently of one another, and their electronic properties depend on interactions within the single chains. Interchain interactions, important in some cases, can be treated as a second-order effect. Furthermore, charge transport within a single chain involves new states within the previously forbidden semiconductor band gap as a result of easy deformability of the polymer lattice when charge carriers exist. This property of conjugated polymers demonstrates that many ideas used to explain inorganic semiconductors do not directly carry over to conjugated polymers.

In the study of the metal–conjugated polymer interface formation, two approaches have been developed to understand the interactions that occur at the metal–conjugated polymer interface [33,38–41]. The first one treats conjugated polymers as bulk materials [39]. This model considers charge distribution at and near the metal–polymer interface based on the relative energy levels of the metal Fermi energy, the π and π^* energy levels of the polymer, and the energy of formation of excited states [39]. This model does not, however, take into account specific chemical reactions that may occur between the metal atoms and the polymer. A different treatment of the metal–polymer interface is based on understanding the molecular properties of conjugated polymers [33,38,40,41]. In this case, the interaction between the metal atoms and the polymer is inferred from quantum-chemical calculations, which attempt to minimize the energy of the system. The calculations are usually carried out for small molecules that are presumed to adequately represent the electronic properties of the conjugated polymer. The predictions are verified experimentally for the small molecules and then extrapolated to the polymer. This approach offers the advantage of considering the interaction between the metal atoms and specific functional groups of the conjugated polymer. It does not, however, address charge transport across the interface. Together, these models offer insight into the processes that occur during the metal–polymer interface formation, as well as during charge transport across the interface. In addition, based on the new information available from these two models, it is possible to begin to understand the mechanism of charge injection across the metal–conjugated polymer interface during the operation of actual devices. In order to understand these models better, it is useful to first examine some basic properties of conjugated polymers.

B. Electronic Properties of Conjugated Polymers

Trans-polyacetylene (*t*-PA) is the simplest conjugated polymer. It differs in many respects from PPV, which is the most widely used class of polymers for light-emitting-diode (LED) application. Yet, its chemical simplicity facilitates the understanding of some essential properties of conjugated polymers. The shorthand representation of *t*-PA is shown along with its chemical structure in Fig. 1. The backbone of *t*-PA consists of carbon atoms joined to their nearest neighbors by alternating single and double bonds. The bonding configuration of the carbon atoms along the backbone of *t*-PA results in sp^2 hybridization of the molecular orbitals, so that the molecule tends to lie in a plane defined by the H and C atoms [37]. In addition, due to the double bonds, the p_z orbitals of the carbon atoms extend perpendicular to the backbone of the polymer. Because every carbon atom contributes an electron to its p_z orbital, the functions of the p_z orbitals overlap and these electrons become delocalized, forming an energy band structure.

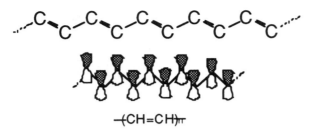

$-(CH=CH)_{\overline{n}}$

FIGURE 1 The chemical structure of *t*-PA. The alternating single- and double-bond structure along its backbone, which gives rise to its novel electronic structure, is shown.

In crystalline undoped semiconductors, the excitation of an electron with sufficient energy moves it to the conduction band. In contrast, the excitation of electrons in *t*-PA results in the formation of new states within the band gap. To understand why this happens, consider the photoexcitation of *t*-PA by a photon equal in energy to the band gap [37]. Instead of promoting a single delocalized electron from the π (valence band maximum) to the π^* (conduction band minimum) level, photoexcitation in *t*-PA excites two electrons. This follows from the quasi-one-dimensional nature of *t*-PA, which easily permits deformations along the polymer chain. Thus, the promotion of two electrons is favored. Due to the resonance structure of *t*-PA, where the positions of the single and double bonds are easily interchanged, the two excited electrons are equivalent, so they both lie exactly mid-gap and photoexcitation results in the formation of new intragap states, as shown in Fig. 2. The mid-gap states are known as solitons and extend over several lattice sites [37].

In conjugated polymers with nondegenerate ground states, such as PPV, single and double bonds are not as readily interchanged as in *t*-PA, so that photoexcitation gives rise to a pair of polarons rather than a pair of solitons, as shown in Fig. 3 [37]. Like solitons, polarons extend over several lattice sites. Unlike solitons, polarons can jump from one chain to the next. Polarons, which also exist in *t*-PA,

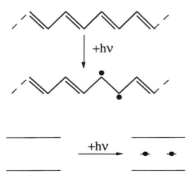

FIGURE 2 Photoexcitation of charge carriers in *t*-PA results in the promotion of two electrons. In addition, rather than bridging the band gap, the electrons sit in new states within the formerly empty gap. These states are solitons.

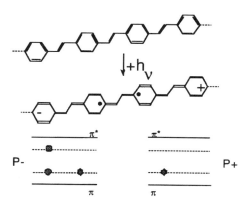

FIGURE 3 Because PPV does not support solitons, photoexcitation results in the formation of a pair of polarons. The charge carriers also sit inside the gap, but they are not mid-gap, as in the case of *t*-PA.

can therefore serve to carry charge in conjugated polymers. In order to understand the polaron structure of *t*-PA, consider the addition of a single electron to *t*-PA. This disturbs the alternating single- and double-bond configuration of the backbone, so that it creates a pattern of bond deformation about 20 sites long. In the deformation process, a level is pulled out of the valence band with its two electrons and a level is pulled out of the conduction band. The stability of the polaron is due to the energy gained when the electron moves from the conduction band into a lower level, exceeding the elastic energy required to form this level. Although its energy levels are in the gap, the polaron can move freely on a single chain by propagating the deformation. Finally, when polarons of opposite sign meet, they form excitons which can then emissively decay.

C. The Continuum Model of the Metal–Polymer Interface

The nondegenerate continuum model of polymer LED by Davids et al. [39] refines the work of Brazovskii and Kirova [42], by specifically considering charge carriers in conjugated polymers with a nondegenerate ground state, such as PPV. The model by Brazovskii and Kirova is itself an extension of a discrete model initially presented by Heeger et al. [43] to understand the interactions at metal–*t*-PA interfaces. In this model, the metal contact and the polymer exchange charges while maintaining quasi-equilibrium in the region of the polymer near each contact [44]. Based on the relative energies of the Fermi level of the metal and the formation energy of polarons and bipolarons within the polymer, conclusions about the extent of charge transfer under static conditions can be made. Thus, the transfer of a large negative charge density into the polymer occurs if the chemical potential of a contact is higher in energy than the negative bipolaron formation energy per particle, and the transfer of a large positive charge density into the polymer occurs if the chemical potential of a contact is lower in energy than the positive bipolaron formation per particle. The transferred charge remains close to the metal–polymer interface. As a result, the transferred charge pins the Fermi energy at the polymer surface near the bipolaron formation energy. If, on the other hand, the chemical potential of a contact lies between the formation energy per particle of the two kinds of bipolarons, there is essentially no charge transfer at the metal–polymer interface, so that the Fermi energy is not pinned.

More specifically, with the atomic positions fixed at the equilibrium ground state, the minimum energy to add an electron to the polymer is defined as Δ. Similarly, the minimum energy to remove an electron (to add a hole) from the polymer is $-\Delta$. Because of lattice relaxation, the addition of an electron or hole to the polymer induces the formation of polarons of energy E_p and $-E_p$, respectively. The polaron relaxation energy therefore equals $\Delta - E_p$. In addition, two polarons can condense to form a bipolaron having a lower energy, $\pm E_b$, than the two separated polarons, so that the bipolaron binding energy equals $2E_p - E_b$. The polymer, assumed to be in local equilibrium with the metal at the interface, is characterized by the values of the polaron formation energy and bipolaron formation energy per particle. After assigning values to the polaron and bipolaron formation energies, as well as to the trap energies and densities, it is possible to model the potential and charge density at the metal–polymer interface for a metal of given chemical potential with respect to the center of the π–π^* gap by solving the Poisson equation. Figure 4 shows the resulting charge density near the metal–polymer interface. In proposing this model, Davids et al. [39] suggest that charge transfer, in fact, occurs at the metal–polymer interface when either the formation energy of polarons or bipolarons is less than the chemical potential of the metal. In addition, it is suggested that the presence of traps in the polymer will accommodate charges as well. As will be seen in later sections, this model appears to agree well with the behavior observed during metal deposition on PPV and its derivatives.

D. Molecular Approach to the Metal–Polymer Interface

Conjugated polymers are essentially one dimensional, and little interaction is assumed between neighboring chains. As a result, the metal–conjugated polymer interface lends itself to molecular modeling [33,38,40,41]. In this case, one con-

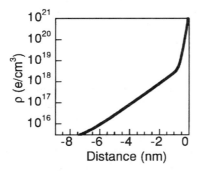

FIGURE 4 A typical charge distribution near the metal–polymer interface following the nondegenerate continuum model calculation of charge transfer at the interface. The distance is from the metal–polymer interface. See Ref. 39 for more detail.

siders metal–polymer interactions from microscopic point of view of quantum interactions. As a result of the difference in chemical potentials between the Fermi energy of the metal and various excited states of the conjugated polymer, it is expected that charge transfer may occur between the metal and polymer. The quantum calculation approach proceeds further in understanding the interactions between the metal and polymer by considering the result of bringing several metal atoms in the vicinity of a conjugated system and observing changes in the electronic structure of the resulting system. In this approach, it is possible to learn more about the metal–polymer interface by considering specific chemical reactions or even charge donation that may occur between metal atoms and the polymer chains. The comparison between experimental results of model conjugated systems and quantum calculations of the same systems provides the means to verify the validity of the calculations. These results are then extrapolated to polymers. The numerical results that serve as the foundation for this derivation are based on the Hartree–Fock (HF) method [45]. Algorithms based on the HF method have been used by other investigators to model the interactions between metal atoms and conjugated polymers [33,38,40,41]. One such algorithm is known as the valence effective Hamiltonian. It is an approximate one-electron pseudopotential method which relies on results obtained for model molecules in order to minimize the number of calculations necessary to evaluate the electronic band structure of a metal–conjugated molecule system [33,38,40,41]. Based on such calculations, it has been postulated that the metal plays an important role in determining the extent of charge transfer that occurs in metal–conjugated polymer interfaces: Ca and Na appear to transfer charge at the interface without significantly altering its chemistry [33,46]. Al, on the other hand, disrupts the chemical structure of the surface of the conjugated polymer sample [38,47]. The results obtained in the case of Ca and Na agree with the general behavior postulated by Davids et al. [39], whereas Al deposition introduces the additional complication of surface chemical reactions, which were not considered by Davids et al. [39]. The results obtained from the molecular modeling approach may therefore point to additional effects which the continuum model overlooks. In the later sections, we employ a simpler version of this quantum-chemical calculation, the semiempirical AM1 method, and calculate changes in valence energy levels upon metal deposition. The results of the calculation is compared with experimental data. We will also illustrate the evolution of electronic structure when Rb was deposited on PPV. Although this system does not represent a typical metal–polymer interface due to Rb diffusion into PPV, it serves a good example of evolution of polaron and bipolaron states.

E. Model of Carrier Transport at the Metal–Polymer Interface

The theoretical models presented in Sections II.C and II.D are extremely valuable in critical assessment of recent studies of the EL efficiency in conjugated-polymer-based LEDs. From these efficiency studies, a hypothesis has emerged, which states that tunneling across interfacial barriers is the primary mode of charge injection in the LEDs [26,44]. In this model, the polymer is assumed to be fully depleted, so that charge transfer does not occur at the interfaces in the absence of an applied bias, in contradiction with the theoretical models described in previous sections. The heights of the interfacial barriers, therefore, equal the differences in work functions between the electrode materials and the conjugated polymer. In addition, in the absence of an external bias, the Fermi levels of the two electrodes are taken to be at the same level for a device under static equilibrium. These conditions lead to a construction of the band diagram of a device shown in Fig. 5. The polymer is sandwiched between two electrodes of different work functions, the Fermi levels of which line up under equilibrium conditions, so that the bands of the polymer tilt.

Under bias, the Fermi levels shift relative to each other so that triangular barriers form at the interface. The tunneling current that results from charge injection through a triangular barrier is given by

$$I \propto F^2 \exp\left(\frac{-\kappa}{F}\right) \tag{1}$$

where I corresponds to the tunneling current, F is the applied electric field, and κ is a parameter that depends on the barrier shape. For a triangular barrier, the factor κ is given by

$$\kappa = \frac{8\pi\sqrt{2m^*}\phi^{3/2}}{3qh} \tag{2}$$

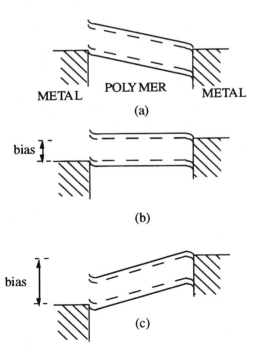

FIGURE 5 Energy diagram of the metal–polymer–metal structure: (a) at zero bias, (b) at forward bias, giving the flat band condition, and (c) at forward bias during the device operation.

where m^* is the effective mass of charge carriers tunneling through the interface, ϕ is the barrier height, q is the charge of the charge carriers and h is Planck's constant. This model considers such properties of LEDs a turn-on voltage and efficiency. The efficiency was reported over a range of barrier heights ranging from 0.1 to 2.4 eV. The agreement between experimental results and the proposed tunneling through rigid barriers appears quite good [26]. Nevertheless, the tunneling current deviates from the predicted behavior for low applied electric fields. This is due to the charge transfer and the formation of a built-in electric field near the metal–polymer interface, as discussed in previous sections. When the built-in electric field, E_{in} is included in Eq. (1), the relation between $\ln(I/F^2)$ and $1/F$ becomes [48]

$$\ln\left(\frac{I}{F^2}\right) = C_1 + \frac{C_2}{F + E_{in}} \tag{3}$$

where C_1 gives the limit of I/F^2 as F becomes infinitely large, and C_2 is proportional to the barrier height at the metal–polymer interface. The value of E_{in} found in experiment is close to the value predicted. Also, the addition of a built-in electric field for $\ln(I/F^2)$ makes this modified tunneling model describe the I–V characteristic throughout the range of the applied electric field.

III. EXPERIMENTAL STUDIES

A. Requirements for Interface Analysis

In a variety of technologies, the importance of the properties of surfaces has been increasingly felt. The technological success of devices whose performance is dictated by the surfaces has driven the development of increasingly sophisticated instrumentation designed to probe better the surface properties. Currently, there exists a large number of surface and interface [49] analytical tools capable of providing complementary information. The usefulness of these tools has been proven by their successful application in the study of inorganic semiconductor devices where the interfaces were found to dictate the performance of the device. Furthermore, several of these tools, such as x-ray and ultraviolet photoemission spectroscopies (XPS and UPS), and near-edge x-ray absorption fine structure spectroscopy (NEXAFS) have already been applied successfully to the study on organic interfaces. In the following sections, we will briefly discuss some of these more frequently used tools.

However, in any systematic surface study of materials, it is essential to recognize basic requirements dictated by the relatively low number of atoms of interest. For a given number of atoms, the number of surface atoms scales as $N_A^{2/3}$. Whereas a solid contains $\sim 10^{23}$ atoms/cm^3, the surface has $\sim 10^{15}$ atoms/cm^2. The relative ratio of surface to bulk atoms presents a handicap when the probing depth exceeds several atomic layers, as the signal due to bulk atoms can quickly overwhelm that originating from the surface. In addition, sample cleanliness is exceedingly important, because the interaction between the surface and contaminants will interfere with a clear characterization of surface properties [49] as illustrated in Section III.C.1.

A simple model of the interaction between the surface of a sample and the surrounding air molecules highlights the importance of a clean environment. Kinetic theory states that the rate ρ of impact of atoms on a surface is given by

$$\rho = \frac{P}{\sqrt{2\pi mkT}} \tag{4}$$

where P is the ambient pressure, m is the mass of the atoms, k is Boltzmann's constant, and T is temperature. If $T = 300$ K and $P = 10^{-6}$ Torr, then $\rho = 5 \times 10^{14}$ cm^{-2}/s for nitrogen. If every atom that strikes the surface sticks to it, a monolayer of nitrogen will grow in 2 s. Because surface contamination may result in interactions between the adsorbate and the substrate, it is necessary to minimize the exposure of the sample to contaminants. Consequently, detailed surface studies require a highly controlled environment with a pressure on the order of 10^{-10} Torr [49]. This range of pressure is usually referred to as ultrahigh vacuum (UHV).

B. Photoelectron Spectroscopy

Photoelectron spectroscopy is based on the photoelectric effect explained by Einstein in 1905. It relies on the creation of photoelectrons via interaction between the irradiating photons and the sample. These electrons are ejected from a wide range of energy levels, ranging from near the Fermi level to the drop core levels of an atom. Because total energy must be conserved in this process, the kinetic energy, E_k, imparted to an electron satisfies

$$E_k = h\nu - E_B - \phi \tag{5}$$

where ϕ is the work function of the sample and E_B is the binding energy of the initial state of the electron with respect to the Fermi level [49]. Figure 6 summarizes this phenomenon schematically. As indicated in Fig. 7, photoelectrons originate from the valence band as well as from core-level states, which correspond to closed atomic shells. An x-ray photon source is used in core-level spectroscopy, whereas an ultraviolet source is used in valence-level spectroscopy. Because these techniques are customarily named according to the type of photon source used, the former is known as x-ray photoelectron spectroscopy (XPS) and the latter is named ultraviolet photoelectron spectroscopy (UPS). These two tools will be discussed further in Sections III.B.1 and III.B.2. In addition to XPS and UPS, synchrotron radiation, which covers photon energies from infrared up to hard x-rays, is ideal for experiments that require the continuous variation of the exciting photon energy, such as near-edge x-ray absorption fine structure spectroscopy (NEXAFS). This type of excitation source also makes constant final-state spectroscopy (CFS), where the electron energy is kept constant while the energy of the exciting photon is varied, possible.

1. X-Ray Photoelectron Spectroscopy

The two most commonly used photon sources for XPS are the unmonochromatized K_α radiation from magnesium or aluminum targets. The Al K_α and the Mg K_α lines peak at 1486.6 eV and 1254.6 eV, respectively. Among the applications of XPS, electron spectroscopy for chemical analysis (ESCA) is one of the best known. It

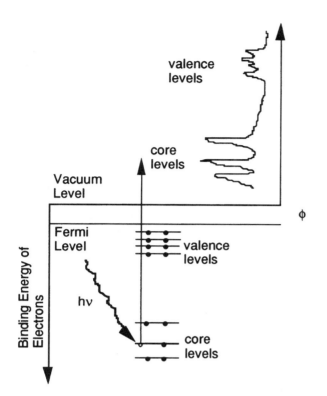

FIGURE 6 A schematic representation of photoelectron emission spectroscopy. Photoelectrons are excited from the core or valence levels depending on the energy of the irradiating photons. The densities of the electronic states are superimposed on a background of secondary electrons.

works on the principle that the binding energy of a given photoelectron is highly dependent on the element from which it originated, so much so, that any changes in the chemical environment of the atom to which it was bound would cause a small chemical shift in the binding energy of a core-level electron [50]. The core-level peaks are easily identified from a handbook of photoemission spectroscopy, although their exact positions depend on the chemical environments of the atoms [50]. Examples of studies that illustrate the capability of ESCA to determine chemical species are presented in Section III.C.

In principle, the underlying physical interactions responsible for changes in the binding energies of different chemical species can be expressed in a straightforward fashion. In a first-order approximation which neglects many-body effects, the energy of a core-level electron depends on the Coulombic attraction of the nuclei and the repulsion of all the other electrons in the system. The change in binding energy, ΔE, resulting from a change in the chemical environment will cause the valence charges to redistribute. $\Delta E(A, B)$ of a particular core-level in two different compounds, A and B, was originally described by Gelius [51] as follows:

$$\Delta E(A, B) = K_c(q_A - q_B) + V_A - V_B) \tag{6}$$

The first term, $K_c(q_A - q_B)$, describes the difference in the electron–electron interaction between core orbital c and the valence charges q_A and q_B, respectively [49]. The coupling constant K_c is the two electron integrals between core and valence electrons. The second term has the character of a Madelung potential, which in the point-charge approximation is

$$V_i = \sum_{i \neq j} \frac{q_j}{R_{ij}} \tag{7}$$

where the summation is over potentials arising from all the other ionic charges q_j centered at positions R_{ij} relative to the atom i in the material. Both terms in Eq. (6) are usually on the order of 10 eV. However, the summation of the two terms are on the order of a few electron volts or less because the ions attract the electron while the valence electrons repel it. As a result of this partial cancellation of the two terms, observed chemical shifts in solids are usually on the order of a few electron volts or less.

Although this straightforward relation has been quite successful in correlating observed chemical shifts in many solids, discrepancies have arisen in a number of systematic studies. Fortunately, these discrepancies are not due to some fundamental error in the above model, but to the fact that the rearrangement of valence charges upon the sudden appearance of a core hole in the atom was neglected [52]. This is because when an atom is ionized in the photoemission process, the remaining electrons will spatially redistribute so as to screen the suddenly created core hole. This screening lowers the energy of the hole state left behind and decreases the measured binding energy. This binding energy effect is commonly referred to as the relaxation energy, E_R. The correction of the relaxation energy becomes particularly important for neutral atoms in a metallic environment, as it can amount to as much as 10 eV when the binding energy is compared to the value obtained from the gaseous atom [49].

2. Ultraviolet Photoelectron Spectroscopy

Although XPS can also be used to obtain a valence band spectrum, the photoionization cross section for the valence photoelectrons is much lower than that of the core electrons. To increase the photoionization cross section, the energy of the photon source used should be of the same order of magnitude as the binding energy of the valence electrons. Therefore, for valence band spectroscopy, an ultraviolet photon source is most commonly used. In UPS, the He resonance lamp is most commonly used to produce the He I (21.2 eV) and He II (40.8 eV) lines. Higher-energy resolution, made possible by the small linewidths (of the order of milli-electron-volts) of the He I and He II lines, is another advantage that UPS has over XPS for valence-level spectroscopy.

3. Synchrotron Radiation Photoelectron Spectroscopies

Synchrotron radiation photoelectron spectroscopy experiments, as the name suggests, are carried out in particle beam accelerators or synchrotrons (e.g., the 2-GeV ring at the Brookhaven National Synchrotron Light Source). From relativistic electrodynamics, we know that when a relativistic charged particle is accelerated, it emits electromagnetic radiation whose angular distribution is given by [53]

$$\frac{dP(t')}{dt} \cong \frac{2}{\pi} \frac{e^2 \gamma^6}{c^3} \frac{|\dot{\mathbf{v}}|}{(1 + \gamma^2 \theta^2)^3} \left(1 - \frac{4 \gamma^2 \theta^2 \cos^2 \phi}{(1 + \gamma^2 \theta^2)^2} \right) \tag{8}$$

where P is the total power radiated, $\dot{\mathbf{v}}$ is the derivative of the velocity of the charge with respect to time, θ and ϕ are the polar and azimuthal angles, respectively, and γ is defined as

$$\gamma = \frac{1}{\sqrt{1 - (v/c)^2}} \tag{9}$$

Its intensity, high degree of polarization, and tunability over a large spectral region makes this electromagnetic radiation an ideal choice for synchrotron radiation photoelectron spectroscopy experiments, reflectivity measurements, and a host of other experiments in related areas. The high intensity facilitates the careful examination of a relatively small number of atoms, and the continuous tunability makes possible many additional techniques, such as NEXAFS, that could not be carried out with conventional photoelectron spectroscopic techniques. This is especially true in the vacuum ultraviolet region of the spectrum ($10 \leq h\nu \leq 100$ eV) because photoelectron spectroscopy is quite surface sensitive at these energies. By measuring the angular distribution of the emitted photoelectrons, one can also gain information about the orbital shapes and bonding symmetries (e.g., a direct determination of the dispersion between the energy E and the wave vector $\mathbf{k}$ for the valence band in a sample).

Near-edge x-ray absorption fine structure spectroscopy is a powerful technique to provide information about the unoccupied electronic structure of surfaces, because it directly probes the evolution of unoccupied states and thus complements the information on occupied states available from photoelectron spectroscopies [54]. NEXAFS spectra are collected by exposing the sample to a monochromatic photon beam of varying energy, and monitoring transitions from a fixed core-level energy to unoccupied states. These transitions, of course, obey the dipole selection rule. Following the excitation of the core-level electron, the core hole decays usually via an Auger process. The photon beam thus induces a current that changes as a function of excitation energy, allowing the determination of the electronic character of the unoccupied electronic states and interatomic bonding [54]. Section III.C will present an application of this technique in the study of metal–polymer interfaces.

C. Examples of Interface Analysis

The first step in a meaningful investigation of interfaces is to identify the surface compositions of one's sample and to establish a procedure which will ensure the reproducible preparation of such a surface. Because we are currently investigating the conjugated polymer–metal interfaces, we will use it to illustrate the usefulness of standard surface analytical tools and the information they provide. The importance of surface states on interface formation in polymer-based LEDs is well known. Therefore, we will first discuss the surface species of poly(p-phenylene

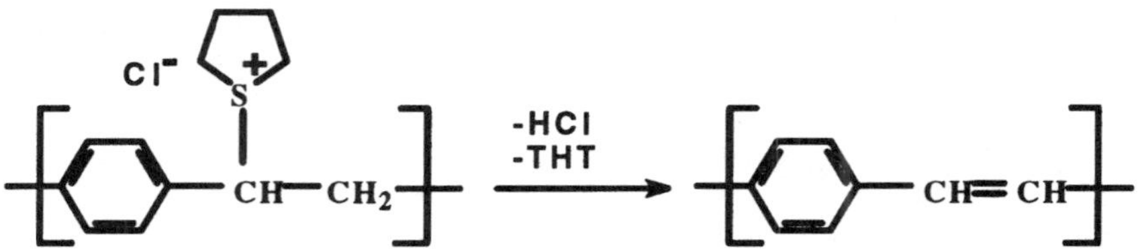

FIGURE 7 The sulfonium-leaving group is eliminated from the precursor polymer upon thermal conversion.

vinylene) (PPV), a conjugated polymer, and then the interface formation of metals with PPV.

1. Surface Species

We consider initially the effects of exposing a sulfonium precursor polymer, shown in Fig. 7, to x-rays prior to thermal conversion. The surface compositions and the thermal conversion conditions for samples 1–6 are given in Table 1. Samples 1–4 were exposed to x-rays prior to thermal conversion. Samples 5 and 6 are precursor polymer films. Although we could make samples with O content as low as 5–6% (samples 3 and 4), 13% O is more representative for the precursor polymer. The XPS survey scans obtained for sample 4 before and after thermal conversion are shown in Fig. 8 (bottom and middle traces, respectively). As expected from the stoichiometry of the precursor polymer, S and Cl were present in approximately equal quantities. Although Cl could be completely eliminated by heating at

TABLE 1 Surface Atomic Compositions for the Sulfonium Polymer Films and the Corresponding PPV Films

Sample no.	Conversion conditions	% C[a]	% O[a]	% S[a]	% Cl[a]
1[b]		(82.6)	(13.0)	(2.6)	(1.8)
	150°C/1 h in UHV	85.8	12.0	1.2	1.0
	200°C/2 h in UHV	89.6	9.5	0.8	0
2[b]		(75.0)	(22.0)	(1.5)	(1.5)
	200°C/2 h in UHV	82.0	17.0	1.0	0
3[c]		(89.8)	(6.0)	(2.1)	(2.2)
	250°C/2 h in UHV	94.8	4.1	1.1	0
4[c]		(91.0)	(5.3)	(2.3)	(1.2)
	320°C/2 h in UHV	95.0	4.0	1.0	0
5[c]		(83.1)	(13.2)	(2.0)	(1.7)
6[c]		(84.0)	(13.0)	(1.6)	(1.4)

[a]Data for the as-cast precursor polymer samples are given in parentheses.
[b]Cast from water solution.
[c]Cast from methanol solution.

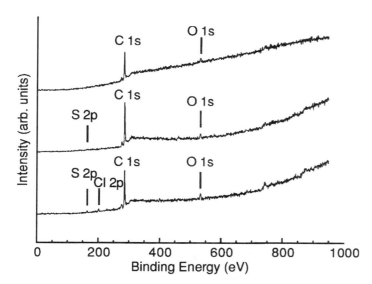

FIGURE 8 Survey scans for selected PPV samples. Bottom scan shows the precursor sulfonium polymer. The sample contains S and Cl. Middle scan shows partially converted PPV with exposure to x-rays prior to conversion. The sample contains S. Top scan shows converted PPV, which was not exposed to x-rays prior to conversion. S is completely eliminated from this sample.

150–200°C, about 1% S remained after conversion at 320°C. This stresses the importance of fully converting the samples before doing any kind of spectroscopy on them.

We were able to completely eliminate sulfur impurities from precursor polymers that had not been exposed to x-rays prior to thermal conversion, as listed in Table 2 and illustrated by the top trace of Fig. 8 for sample 9. The surface compositions for samples converted under argon (samples 7 and 8) or in UHV (samples 9–11) at 320°C are essentially identical. Thermal conversion under argon at 250°C gave sample 12a, which was further heated in UHV at 320°C to give sample 12b. The second heating reduced the O content from 11.5% to 5.0%. By converting at 320°C, samples 7–11 and 12b showed similar surface compositions indicating a high degree of reproducibility associated with our conversion processes. By aging sample 12a in ambient atmosphere for 6 weeks, we obtained sample 13a, which showed increased surface O (14.7%), indicating the oxidation of the PPV surface. The O content decreased to 10.3% after heating at 320°C for 2 h. We found that prolonged heating at 320°C did not further reduce the O content, suggesting that higher temperatures may be required to do so.

The surface O content of 5–13% measured by XPS for PPV exceeds that of the bulk (<1%) determined by elemental analysis [55]. This is likely due to surface-segregated oxygen impurities, as surface oxygen segregation is a well-known phenomenon in polymer systems and can be due to factors such as polymer chain ends, oxygen defects, surface contamination, and surface oxidation [56].

In order to identify the oxygen species, the broad raw oxygen peaks of samples 12a and 12b was decomposed into three Gaussian peak components, as shown in

TABLE 2 Conversion Conditions and Surface Atomic Compositions of PPV Films

Sample no.	Conversion conditions	% C[a]	% O[a]
7	320°C/2 h under argon	94.0	6.0
8	320°C/2 h under argon	95.0	5.0
9	320°C/2 h in UHV	94.9	5.1
10	320°C/2 h in UHV	94.7	5.3
11	320°C/2 h in UHV	94.3	5.7
12a	250°C/2 h under argon	88.5	11.5
12b	320°C/2 h in UHV	95.0	5.0
13a	Sample 12, 6 weeks later	85.3	14.7
13b	320°C in UHV	89.7	10.3

[a]Data given in atomic percent and samples were prepared by spin cast using a methanol solution of the precursor polymer.

Fig. 9. Table 3 summarizes the peak positions and their abundance for several samples. Based on the data of López et al., the components are assigned as follows: carbonyl oxygen (C=O), ~ 531.0 eV, hydroxy/ether oxygen (C–OH/C–O–C), ~532.0 eV, and the carboxylic acid groups (HO–C=O), ~533.0 eV [50,57]. In all cases, component 2, corresponding to the hydroxyl and ether groups, is the dominant species. This is understandable based on the presence of hydroxyl groups in the sulfonium precursor polymer [58,59]. The decrease in the hydroxy/ether component upon heating at 320°C shown in Fig. 9 may be due to dehydration through the thermal elimination of the hydroxy groups and thermal oxidation of

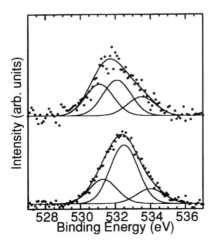

FIGURE 9 Resolved O 1*s* EDCs for samples 12a (bottom) and 12b (top).

TABLE 3 XPS Peak Positions and Relative Areas of Resolved Oxygen Components for Selected PPV Samples

	Component 1		Component 2		Component 3	
Sample no.	Position (eV)	Area (%)	Position (eV)	Area (%)	Position (eV)	Area (%)
7	531.1	18.5	532.0	63.0	533.6	18.5
8	531.1	26.3	532.2	53.5	533.8	20.2
12a	531.3	24.8	532.5	60.0	534.0	15.2
12b	531.0	38.9	532.1	43.6	533.5	17.5
13a	531.6	6.4	532.7	78.2	534.2	15.4
13b	531.5	27.0	532.6	53.1	533.7	19.9
Peak assignment[a]	$C=O$ and $\overline{O-C=O}$ 532.2[a]		$C-OH$ and $\overline{C-O-C}$ 532.8[a]		$O=\overline{C-O}$ 533.7[a]	

[a]Reported values from Refs. 50 and 57.

the hydroxy groups to ketone groups. Such thermal oxidation can be reduced or avoided by performing the conversion under a reducing atmosphere such as forming gas [60]. The presence of oxygen functionalities can also be detected by the analysis of the C 1s peak. For example, the C 1s peak of sample 7 can be resolved into three components with binding energies of 284.5, 285.6, and 287.4 eV. These components can be assigned to C–C, C–O, and C=O, respectively [50,57]. Thus, a "clean" surface is highly desirable because it can simplify the chemistry of interface formation and can thus lead to a better understanding of the underlying interactions.

2. Ca–PPV Interface Study Via XPS and NEXAFS

In our study of the interface formation of Ca with PPV, we made sure that the PPV sample was fully converted before XPS measurements were taken. Studying the behavior of core-level spectra during the deposition of the metal on the polymer allows us to construct a model of the interactions at the Ca–PPV interface. The evolution of the XPS C 1s energy distribution curves (EDC) following the deposition of Ca are shown in Fig. 10. In the XPS data presented, the binding energies of core-level electrons are referenced to the Fermi level of the sample. Because no changes were observed in the shapes of the spectra during metal deposition, we concluded that the metal atoms do not react with the polymer to create new chemical species. The core level EDCs ranged from a full width at half-maximum of 1.8 eV for the pristine surface of 2.1 eV at the final Ca coverage indicating that the interaction of Ca with PPV, if there is any, must be very limited. Also, the exponential decay of the C 1s signal indicates that the Ca atoms do not diffuse significantly into PPV during deposition, and confirms that the deposited Ca does

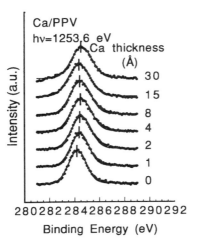

FIGURE 10 Summary of the movements of the major C 1*s* EDCs in PPV during Ca deposition. The rigid shift clearly shows charge transfer across the Ca–PPV interface.

not significantly disrupt C in the PPV substrate. However, a rigid shift of the peak of the C 1*s* core level, which represents band bending, to higher binding energies by about 0.4–0.5 eV was observed. The peak shift of the C 1*s* core level is attributed to the creation of new interface states, evidence for which can be seen in the NEXAFS spectra, resulting in charge transfer at the Ca–PPV interface. The effect of charge at the interface was the formation of a Schottky barrier.

Near-edge x-ray absorption fine structure spectroscopy is a unique technique that provides a more complete understanding of the metal–polymer interface formation by directly probing the evolution of unoccupied states in the polymer. Because of the higher mobilities of holes measured in polymer-based LEDs, they serve as the primary charge carriers in these systems [59,61,62]. Changes in their states during metal/polymer interface formation are therefore expected to play an important role in the interface formation. In addition, the availability of unoccupied states in the vicinity of the metal–polymer interface should contribute strongly to the injection efficiency of electrons across this region. Results from this study thus complement the data gathered from XPS, providing a view of both occupied and unoccupied states at the interface.

The evolution of the NEXAFS spectra following the deposition of Ca, shown in Fig. 11, reveals significant changes in the surface unoccupied levels. The assignment of the resonances is based on labeling features of complex molecules by considering the contributions of their constituent groups—in this case, benzene and ethylene [54]. A comparison of our data to the NEXAFS spectra of polyethylene, polystyrene and polyphenylene shows that the spectrum of clean PPV reflects primarily contributions from the phenylene portion of the molecules [54,63,64]. We ascribe a sharp resonance near 287 eV to the transition of the C 1*s* core-level electron to π^* directly above the band gap. The following features at ~290 eV are the C–H* structures and those at 293–315 eV are the σ_1^* and σ_2^* regions. Using

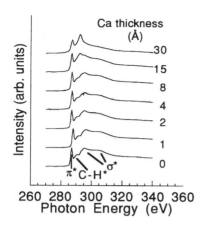

FIGURE 11 Evolution of the NEXAFS spectrum during the deposition of Ca on PPV. The extensive changes measured after as little as 4 Å Ca reflect charge transfer across the metal–polymer interface.

the designation employed to describe semiconductors, the π^* level corresponds to the conduction band minimum.

In Fig. 12, we show a detailed comparison of the NEXAFS signal of the clean PPV surface as well as that obtained following the deposition of 30 Å Ca. The traces have been scaled according to the mesh current and the upper one has been multiplied by 2. Following the deposition of 30 Å Ca, we find an increase in photocurrent within the leading edge of the π^* resonance, indicating the formation of new intragap states, seen in the inset of Fig. 12. Overall, the spectrum loses the structure arising from phenylene and vinylene groups, which supports the notion

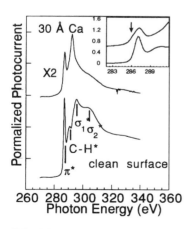

FIGURE 12 Detailed comparison of the NEXAFS spectra obtained for a clean sample of PPV as well as following the deposition of 30 Å Ca. The deposition of Ca dramatically alters the population of unoccupied states in PPV.

of charge transfer at the Ca–PPV interface. This is in accordance with previous studies showing that charge transfer into oligomers results in the deformation of their molecular geometries and gives rise to a quinoid structure [65,66]. This conclusion agrees with theoretical modeling of the Ca–PPV interface, discussed in Section III.C.5, as well as photoelectron spectroscopy indicating new valence states within the band gap of the pristine material as a result of electron donation from Ca [33,38,40,41,64]. The changes observed by NEXAFS are consistent with the formation of polarons or bipolarons, implying the creation of new unoccupied states inside the band gap as well as the isomerization of PPV [46,67]. The appearance of new states within the leading edge of the NEXAFS spectra suggests that the electronic interactions as the Ca–PPV interface may enhance charge transfer at the interface by the creation of unoccupied states at energies very near the Fermi energy of the metal.

3. Metal–DP-PPV NEXAFS Study

In our NEXAFS study of the interface formation between metals and conjugated polymers, the first direct evidence for differences in the evolution of unoccupied states induced by the deposition of Ca and Al on poly(2,3-diphenylphenyl vinylene) (DP-PPV) was observed. In Fig. 13, we show the evolution of the C_K-edge NEXAFS spectrum of DP-PPV during the course of Al deposition. The structure of DP-PPV is shown in the inset. As with PPV, the spectrum of clean DP-PPV reflects contributions primarily from the phenylene portion of the molecules [54,63,68], and the features near 287, ~290, and 293–315 eV are attributed to the π_1^*, π_2^* structures, and the σ_1^* and σ_2^* regions, respectively. A detailed comparison of the NEXAFS spectra obtained for clean DP-PPV as well as after the deposition of 30 Å Al shows a loss in signal intensity within the C–H* and π_2^* region due to Al deposition.

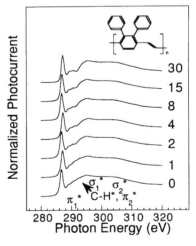

FIGURE 13 Evolution of the NEXAFS spectrum during the deposition of Al on DP-PPV. The spectra change little as the Al layer increases to 30 Å.

We compare our results with our XPS investigations of the effects of Al deposition on PPV. They indicate that, although the polymer does not appear to react with Al, a rigid shift associated with band bending results from metal deposition [32,69]. Molecular calculations on the interfacial reaction of Al with PPV predict that the metal forms covalent bonds which disrupt the backbone of PPV [38]. Conjugation of the system is maintained, however, because the metal atoms create a bridge between conjugated regions [47]. Our NEXAFS data do not clearly point to the disruption of the backbone of the polymer in the case of Al on DP-PPV, although the changes in the C–H* and π_2^* regions indicate some interaction between metal and polymer.

The evolution in NEXAFS spectra following the deposition of Ca differ markedly from those observed for Al. We present in Fig. 14 the evolution of the NEXAFS spectrum of DP-PPV during the course of Ca deposition, which behaves very much like the NEXAFS spectrum of PPV. This is not surprising because the structure of DP-PPV does not differ too much from that of PPV. Just like in PPV, following the deposition of 30 Å Ca, we find within the leading edge of the π_1^* resonance an increase in the photocurrent compared to that of the clean sample, indicating the formation of new intragap states, seen in the inset of Fig. 14. The changes in the unoccupied states of DP-PPV following the deposition of Ca may result from charge transfer across the Ca–DP-PPV interface. Evidence for this process comes from our previous XPS studies which reveals a rigid shift in core-level binding energies with respect to the Fermi level following the deposition of

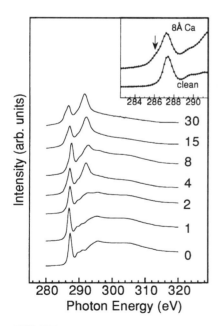

FIGURE 14 Evolution of the NEXAFS spectrum during the deposition of Ca on DP-PPV. Within the leading edge of the π_1^* resonance, new states are found following metal deposition (inset).

Ca on conjugated polymers [32,69,70]. Furthermore, Dannetun et al. have reported the presence of new states at the lower-binding-energy edge of the valence band, ascribing these to the transfer of electrons from Ca into the polymer [33]. A more recent calculation of the evolution of the electronic structure of a four-ring oligomer of poly(p-phenylene vinylene) (3PV) reveals that the interaction of Ca with the conjugated system results in the formation of a new occupied state within the originally forbidden gap. In addition, new unoccupied states appear at the leading edge of the conduction band, as well as at higher energies [41]. The evolution of the NEXAFS spectrum of DP-PPV following Ca deposition clearly reveals the presence of new unoccupied states within the leading edge of the conduction band and at higher energies, in agreement with the calculations. The changes observed by NEXAFS are reminiscent of those reported in poly(3-methylthiophene), where ClO_4^- doping caused the formation of new unoccupied states [44]. The creation of these additional unoccupied states at the Ca–DP-PPV interface may enhance charge injection for EL by providing additional sites for electron transfer into the polymer near the Fermi level of the metal. Based on reported valence band studies, as well as the calculated change in the conduction band of 3PV, the NEXAFS data presented in Fig. 14 therefore suggest charge transfer from Ca to DP-PPV [33,41]. Finally, our previous XPS measurements of the attenuation of the C 1s signal during Ca deposition reveal little if any diffusion at the Ca–polymer interface [28,32]. We note, however, that metal diffusion at the Ca–polymer interface has been reported based on angle-resolved XPS [33]. Even so, this effect is limited in extent, so that much of the interaction between Ca and DP-PPV appears to occur at or very near the interface.

Theoretical modeling of the interaction between Ca and Al and PPV, discussed in Section III.C.5, suggests that Ca gives rise to new valence states resulting from charge transfer, whereas Al reacts with the polymer to form covalent bonds. Photoelectron spectroscopy reveals the presence of new valence states in the case of Ca, but not for Al [33,38,46,47,65]. The NEXAFS data presented here provide an important complement to these studies, as they clearly indicate the creation of new conduction states following the deposition of Ca. By comparison, Al deposition results in relatively minor changes to the conduction band structure.

4. Interface Formation Between Ca and a PPV Oligomer

In the interface formation studies we have considered so far, the oligomers have several advantages over its polymer counterparts. First, it is possible to prepare a clean sample in ultrahigh vacuum (UHV) condition, eliminating the effects of extrinsic impurities. Second, the results obtained from the metal–oligomer system can be directly compared with the theoretical calculations. Third, the use of the oligomer enables the study of the interface while the organic materials are deposited on metal substrates, which is not possible considering typical polymer film preparation methods. Here, we report an XPS and UPS study of interface formation between Ca and a phenylene vinylene oligomer, 4,4'-bis[4-(3,5-di-t-butylstyryl)-styryl]stilbene (5PV) by depositing Ca on 5PV (Ca–5PV) and 5PV on Ca (Ca–5PV).

Figure 15 is the evolution of XPS C 1s peak with increasing Ca coverage. The peaks are normalized to the same height after linear background subtraction. The C 1s peaks move toward higher binding energies (BE) up to $\Theta = 4.0$ Å and a

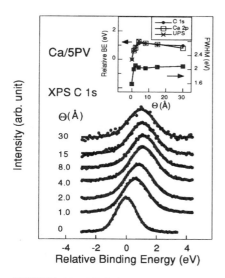

FIGURE 15 XPS C 1*s* peak evolution for Ca–5PV. The inset summarizes these findings along with the movements of Ca 2*p*, and 3.8-eV UPS peaks.

substantial peak broadening is observed at $\Theta = 1.0$ Å. The inset summarizes the movement of C 1*s*, Ca 2*p*, and 3.8-eV UPS peaks as well as the evolution of full width at half-maximum (FWHM) of the C 1*s* peak as a function of Ca coverage. The synchronized movement of the peaks indicates that the movement is from a common origin, which is likely to be Ca-induced energy level bending at the interface (equivalent to band bending in a metal–inorganic semiconductors interface). As observed in the previous UPS spectra [71], gap states are created near the surface due to charge transfer from Ca to 5PV, and the Fermi level is pinned at the top of the filled gap states. This causes a level shift near the surface that results in level bending. Because the Fermi level is the BE reference, the downward level bending increases the distance from the Fermi level to a given core level and results in the peak shift toward higher BEs. As the peak movement at $\Theta = 1.0$ Å indicates that the level bending happens even at a coverage where Ca is not metallic, the individual Ca atoms seem to act as electron donors at this stage. The gap states in conjugated polymers like PPV have been attributed to (bi)polarons [42,72–75]. We expect the situation to be similar in 5PV. The minor peak shifts at $\Theta = 1$–4 Å may be caused by diffusion-induced bending of energy levels, which will be discussed below. The interface formation process is complete after 4 Å Ca have been deposited as indicated by the stabilized peak shape and position at $\Theta > 4.0$ Å. Previous reports on the metal–PPV interface indicated that 15–30 Å of Ca were needed to complete the interface formation process [28,69,70,76]. The relatively quick interface formation process for 5PV is consistent with our argument that the surface impurities of PPV could impede the process. Compared with inorganic semiconductors where the process is completed within a fraction of a monolayer of metal [77–79], the interface formation process for 5PV is still much slower. The gradual interface formation process of Ca–5PV could be due to the Ca atom diffusion and the evolution of gap states.

For 5PV and Ca, the work function ϕ is defined as the distance from the vacuum level, E_{VL} to the Fermi level, E_F. We determined the work functions ϕ_{Ca} and ϕ_{5PV} using the Fermi edge and secondary electron cutoff in UPS spectra [80], respectively, while 5PV was deposited on Ca. We obtained $\phi_{\mathrm{Ca}} = 2.8$ eV, which is in good agreement with the reference value. The relative positions of the C 1s peak as a function of Θ are shown in Fig. 16 after corrected with respect to substrate Ca 2p position. It also shows the relative values of ϕ_{5PV} with respect to ϕ_{Ca} ($\Delta\phi_{\mathrm{5PV}} = \phi_{\mathrm{5PV}} - \phi_{\mathrm{Ca}}$). It is obvious that these two curves are essentially mirror images with a constant offset. The total change (~0.5 eV) is the same for each curve. This is strong evidence that the charge transfer during interface formation results in a shift in binding energy in a manner similar to band bending. It is also clear that the level bending extends up to about 100 Å inside the 5PV layer. The level bending extending approximately 100 Å inside the 5PV is consistent with the results of Lemmer et al. [81] obtained with field-induced photoluminescence quenching.

An energy level diagram deduced from the above observation is illustrated in Fig. 17. We used the lowest BE shoulder in the UPS spectrum at $\Theta = 50$ Å to determine the highest occupied molecular orbital (HOMO) level. Using the Fermi level as the reference, all the relevant energy levels near the interface are specified except for the position of the lowest unoccupied molecular orbital (LUMO), which was not possible to determine by photoemission data alone. The position of LUMO may be directly measured using inverse photoemission, although it can also be deduced from HOMO–LUMO energy-gap measurements.

5. Theoretical Modeling of Metal–PPV Oligomer Interface Formation

All the simulations were done utilizing a commercial program, Spartan V.3.0, developed by Wavefunction, Inc. For this work, a Silicon Graphics computer on the IRIX Release 5.3 platform was used, courtesy of the Center for Photoinduced Charge Transfer. The polymer we plan to simulate is PPV, which has styrene as its repeat units. Ca, Mg, and Al are the three metals that have been chosen to form an interface with PPV. However, it would take a prohibitively long time to model

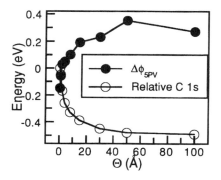

FIGURE 16 Relative position of C 1s peak and the difference between the work functions of Ca and 5PV ($\Delta\phi_{\mathrm{5PV}} = \phi_{\mathrm{5PV}} - \phi_{\mathrm{Ca}}$) with increasing Q. Adjusted for a constant offset, the two curves are essentially mirror images of each other.

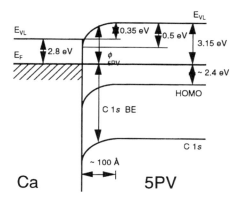

FIGURE 17 Energy level diagram for 5PV–Ca interface constructed from data obtained by XPS and UPS. Positions of relevant levels relative to Fermi level are indicated except for LUMO level.

a polymeric system. Therefore, to simulate the polymeric system, we utilized model systems based on oligomers of PPV. We believe that this is a viable alternative for three reasons. First, light absorption and emission characteristics of an oligomer are not much different from that of PPV [82]. Second, the effective conjugation length of PPV is about four to five repeat units. Finally, the computational cost is much lower for an oligomer.

In modeling work, it is necessary to check if one's software produces results consistent with those available in the literature before applying it to more complicated problems. The optimized geometries of the pristine and Al-doped PPV monomer, or stilbene that we obtained, compared well with theoretical ones reported by Fredriksson et al. [83] and experimental ones reported by Mao et al. [84]. Also, Al atoms interacted with the π conjugated system by coordinating with the vinylene double bond, consistent with Salaneck's results. Having confirmed the validity of Spartan's output, we utilize it to study the effects of different metals, namely Ca, Mg, and Al, on 3PV.

Although at ab initio or semiempirical level, both pristine and metal-doped stilbenes exhibit parallel phenylene rings, the twist angle between the phenylene ring and the vinylene chain differs significantly depending on the type of metal. Al was found to disrupt the conjugation of the model a lot more than Mg or Ca, which results in less efficient charge carrier transportation across the interface. Noting that the device efficiency is highest for Ca systems and lowest for Al systems [85], our results indicate a correlation between the degree of charge transfer across the interface and the performance of the LED, and an inverse relationship between the twist angle and the performance of the LED.

Energy diagrams of the pristine and metal-doped stilbenes, calculated at the ab initio level, are shown in Fig. 18. In these diagrams, the vacuum level is used as the reference. Although some rescaling of the vertical axis would have to be done to fit these calculated energy values with experimental data [86], the untreated values are good enough for our purpose, as we are only interested in the relative changes to these levels caused by the introduction of different metals. For both Ca

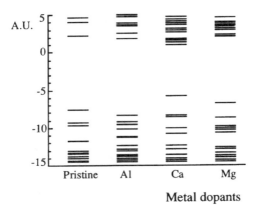

FIGURE 18 Molecular energy levels of pristine, Al-doped, Ca-doped, and Mg-doped stilbene. The shifts of the LUMO and HOMO of the Ca- and Mg-doped systems into the energy gap may be evidence of polaron formation, whereas the rigid shift exhibited by the Al-doped system may be attributed to the formation of covalent bonds.

and Mg, the energy gap was reduced compared with that of the pristine system. The reduction of the gap is caused by the formation of states inside the gap due to charge transfer from metal atoms to stilbene, consistent with the XPS and NEXAFS data presented in the previous section. The reduction is greater for the Ca-doped system, reflecting the higher efficiencies seen in LEDs with Ca electrodes. In the Al-doped system, on the other hand, a rigid shift to lower binding energies of the gap was observed. No intragap states were formed, and the device efficiency was the lowest of the three systems. This further proves that charge transfer across the interface is an important factor in device performance. The same phenomenon was observed in the larger 3PV systems.

6. Polaron Formation in PPV Upon Metal Deposition

The deposition of metal on light-emitting-diode polymers such as PPV induces charge transfer from metal to polymer and dopes the polymer. As discussed in Section II, the charge-transfer-induced state in nondegenerate ground-state polymers, such as PPV, is polaron or bipolaron. UPS has been commonly employed in studying doping-induced changes in the electronic structure of conjugated polymers and molecules, for both p-type [87–90] and n-type [90–92] doping. We use the rubidium–PPV system studied by Iucci et al. [93] as an illustrative example of the polaron formation and its evolution to bipolaron. The interface formed by Rb deposition on PPV is less sharp than that typically formed by Ca or Al deposition [94], although this does not matter in UPS studies because it only samples a less than 10-Å-thick layer at the surface.

Figure 19 shows the He I (21.2 eV) UPS spectra of PPV near the Fermi level while rubidium was deposited. The binding energy in this figure is relative to the vacuum level. The two peaks labeled A and B originated from different electronic states [95]. The stronger peak A is from a flat π band localized to the carbons in the phenyl rings, whereas the weaker peak B is from the top of the valence π band delocalized along the polymer backbone. The deposition of Rb drastically changes

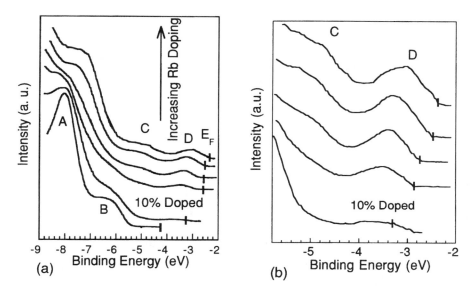

FIGURE 19 (a) UPS spectra of rubidium-doped PPV adapted from Ref. 93. (Courtesy of M. Fahlman). The spectrum of pristine PPV is at the bottom with increasing doping toward the top curve. An expanded view of the lower binding energy region is shown in (b). The energy scale is relative to the vacuum level and the vertical lines indicate the position of the Fermi level.

the electronic structure near the top of valence band, with the formation of new peaks in the previously forbidden energy gap. As more Rb is deposited and the doping level increases from pristine PPV to saturation doping (about two Rb atoms per monomer repeat unit), two different regimes can be identified. At high doping levels, two new peaks, labeled C and D in Fig. 19, appear. These two peaks correspond to the formation of bipolarons, as suggested by experimental and theoretical results on sodium-doped PPV [91]. This is expected for a nondegenerate ground-state system such as PPV. The peak B, which originated from the valence band edge, loses intensity because the formation of bipolaron states leads to the removal of these states to form states within the band gap.

The importance of Fig. 19 is that there is only one new peak in the previously forbidden band gap and it has a finite density of states at the Fermi level. This is expected for a system with a half-filled gap state. In the case of sodium doping of PPV, however, bipolaron bands are observed at all doping levels [91]. The bipolaron bands are well separated from each other, as well as from the valence and conduction bands, leading to two well-resolved peaks in the UPS spectrum [91]. In this case, both bipolaron bands are completely occupied and the Fermi level lies in the energy gap between the upper occupied bipolaron band and the conduction band, leading to no density of states at the Fermi level [96]. However, the upper polaron band is half-filled and a finite density of states at the Fermi level is expected. Therefore, the new peak at low doping (bottom spectrum in Fig. 19b) can be assigned to the upper, singly occupied polaron band. This is a clear illustration of gap state formation due to the metal deposition and the evolution of the gap states that was detected by UPS. Similar results have been observed for a Rb doping of an oligomer of PPV [93]. Obviously, it needs further study both theoretically and experimentally to clarify such issues as why polaron states could be detected in the case of sodium doping of PPV, whereas the evolution of polaron states to bipolaron states is clearly observed in Rb doping.

D. Final Remarks

The above is not intended to be an exhaustive review at all the possible uses of these surface analytical tools for investigations of organic semiconductor devices. Certainly, there are many other examples that illustrate this. Among them is the recent report of a dramatic photoluminescence (PL) quenching of a PPV oligomer due to the deposition of Ca atoms [97] and PL quenching due to electron-acceptor doping [98,99]. In the case of electron-acceptor doping, PL quenching occurs through the dissociation of singlet excitons via electron transfer to the impurity or the dopant electronic levels that reside in the PPV band gap. The Ca-induced PL quenching may be due to a similar mechanism involving the formation of gap states (i.e., polaron states) through charge transfer from Ca to the oligomer [100]. Thus, the electronic properties of these materials are the key to the understanding of their luminescence characteristics under various bulk and interface conditions. Therefore, more tools that probe the electronic structure of materials will be applied to study this phenomenon in the future, and a more complete picture of the interface formation in organic semiconductor devices can be expected to emerge as a result.

ACKNOWLEDGMENTS

This work was supported in part by the National Science Foundation under Grant No. DMR-9612370 and by DARPA DAAL-0196R9133.

REFERENCES

1. Mort, J., *Phys. Today*, *47*, 32 (1994).
2. Pai, D., and Springett, B. E., *Rev. Mod. Phys.*, *65*, 163 (1993).
3. *Flat Information Displays, Market and Technology Trends*, 6th ed., Stanford Resources, Inc., San Jose, 1994.
4. Gooch, C. H., *Injection Electroluminescent Devices*, Wiley, New York, 1973.
5. Craford, M. G., *Inform. Display*, Feb *12* (1993).
6. Burroughes, J. H., Bradley, D. D. C., Brown, A. R., Marks, R. N., MaKay, K., Friend, R. H., Burn, P. L., and Holmes, A. B., *Nature*, *347*, 539 (1990).
7. Tang, C. W., and VanSlyke, S. A., *Appl. Phys. Lett.*, *51*, 913 (1987); Tang, C. W., VanSlyke, S. A., and Chen, C. H., *J. Appl. Phys.*, *65*, 3610 (1989).
8. Aldissi, M. (ed.), *Intrinsically Conducting Polymers: An Emerging Technology*, NATO ASI Series E, Vol. 246, Kluwer, Dordrecht, 1993.
9. Burn, P. L., Holmes, A. B., Kraft, A., Bradley, D. D. C., Brown, A. R., Friend, R. H., and Gymer, R. W., *Nature*, *356*, 47 (1992).
10. Bradley, D. D. C., *Adv. Mater.*, *4*, 756 (1992).
11. Brown, A. R., Bradley, D. D. C., Burroughes, J. H., Friend, R. H., Greenham, N. C., Burn, P. L., Holmes, A. B., and Kraft, A., *Appl. Phys. Lett.*, *61*, 2793 (1992).
12. Ohmori, Y., Uchida, M., Muro, K., and Yoshino, K., *Jpn. J. Appl. Phys.*, *30*, L1938 (1991).

13. Braun, D., and Heeger, A. J., *Appl. Phys. Lett.*, *58*, 1982 (1991).
14. Braun, D., Heeger, A. J., and Kroemer, H., *J. Electron. Mater.*, *20*, 945 (1991).
15. Gustafsson, G., Cao, Y., Treacy, G. M., Klavetter, F., Colaneri, N., and Heeger, A. J., *Nature*, *357*, 447 (1992).
16. Kraabel, B., McBranch, D., Sariciftci, N. S., Moses, D., and Heeger, A. J., *Phys. Rev. B*, *50*, 18543 (1994).
17. Hu, B., Yang, Z., and Karasz, F. E., *J. Appl. Phys.*, *76*, 2419 (1994).
18. Aguiar, M., Akcelrud, L., and Karasz, F. E., *Synth. Met.*, *71*, 2187 (1995).
19. Aguiar, M., Akcelrud, L., and Karasz, F. E., *Synth. Met.*, *71*, 2189 (1995).
20. Wang, H. L., MacDiarmid, A. G., Wang, Y. Z., Gebler, D. D., and Epstein, A. J., *Synth. Met.*, *78*, 33 (1996).
21. Grem, G., Leditzky, G., Ullrich, B., and Leising, G., *Adv. Mater.*, *4*, 36 (1992).
22. Ohmori, Y., Uchida, M., Muro, K., and Yoshino, K., *Jpn. J. Appl. Phys.*, *30*, L1941 (1991).
23. Burn, P. L., Holmes, A. B., Kraft, A., Bradley, D. D. C., Brown, A. R., and Friend, R. H., *J. Chem. Soc., Chem. Commun.*, 32 (1992).
24. Brown, A. R., Burn, P. L., Bradley, D. D. C., Friend, R. H., Kraft, A., and Holmes, A. B., *Mol. Cryst. Liq. Cryst.*, *216*, 111 (1992).
25. Brillson, L. J., in *Handbook on Semiconductors* (P. T. Landsberg, ed.), Elsevier Science Publishers, Amsterdam, 1992, Vol. 1, p. 281.
26. Parker, I. D., *J. Appl. Phys.*, *75*, 1656 (1994).
27. Bröms, P., Birgersson, J., Johansson, N., Lögdlund, M., and Salaneck, W. R., *Synth. Met.*, *74*, 179 (1995).
28. Gao, Y., Park, K. T., and Hsieh, B. R., *J. Appl. Phys.*, *73*, 7894 (1993).
29. Kanicki, J., in *Handbook of Conducting Polymers* (T. A. Skotheim, eds.), Marcel Dekker, Inc., New York, 1986, Vol. 1, p. 543.
30. Horowitz, G., *Adv. Mater.*, *2*, 287 (1990).
31. Tomozawa, H., Braun, D., Phillips, S., Heeger, A. J., and Kroemer, H., *Synth. Met.*, *22*, 63 (1987).
32. Ettedgui, E., Razafitrimo, H., Park, K. T., Gao, Y., and Hsieh, B. R., *Surf. Interf. Anal.*, *23*, 89 (1995).
33. Dannetun, P., Fahlman, M., Fauquet, C., Kaerijama, K., Sonoda, Y., Lazzaroni, R., Brédas, J. L., and Salaneck, W. R., *Synth. Met.*, *67*, 133 (1994).
34. Yang, Y., and Heeger, A. J., *Nature*, *372*, 344 (1994); Yang, Y., Westerweele, E., Zhang, C., Smith, P., and Heeger, A. J., *J. Appl. Phys.*, *77*, 694 (1995).
35. Cheung, J. H., Fou, A. C., and Rubner, M. F., *Thin Solid Films*, *244*, 985 (1994).
36. Fou, A. C., Onitsuka, O., Ferreira, M., Rubner, M. F., and Hsieh, B. R., *J. Appl. Phys.*, *79*, 7501 (1996).
37. Conwell, E. M., and Mizes, H. A., *Handbook on Semiconductors*, Elsevier Science Publishers B.V., Amsterdam, 1992, Vol. 1.
38. Dannetun, P., Löglund, M., Salaneck, W. R., Fredriksson, C., Stafström, S., Holmes, A. B., Brown, A. R., Graham, S., Friend, R. H., and Lhost, O., *Mol. Cryst. Liq. Cryst.*, *228*, 43 (1993).
39. Davids, P. S., Saxena, A., and Smith, D. L., *J. Appl. Phys.*, *78*, 4244 (1995).
40. Brédas, J. L., *Adv. Mater.*, *7*, 263 (1995).
41. Choong, V., Park, Y., Hsieh, B. R., and Gao, Y., *J. Phys. D30*, 1421 (1997).
42. Brazovskii, S. A., and Kirova, N. N., *Synth. Met.*, *55–57*, 4385 (1993).
43. Heeger, A. J., Kivelson, S., Schrieffer, J. R., and Su, W.-P., *Rev. Mod. Phys.*, *60*, 781 (1988).
44. Heeger, A. J., Parker, I. D., and Yang, Y., *Synth. Met.*, *67*, 23 (1994).
45. Ashcroft, N. W., and Mermin, N. D., *Solid State Physics*, W. B. Saunders, Fort Worth, TX, 1976.

46. Dannetun, P., Löglund, M., Fahlman, M., Fauquet, C., Beljonne, D., Brédas, J. L., Bässler, H., and Salaneck, W. R., *Synth. Met.*, *67*, 81 (1994).

47. Dannetun, P., Löglund, M., Fahlman, M., Boman, M., Stafström, S., Salaneck, W. R., Lazzaroni, R., Fredriksson, C., Brédas, J. L., Graham, S., Friend, R. H., Holmes, A. B., Zamboni, R., and Taliani, C., *Synth. Met.*, *55*, 212 (1993).

48. Ettedgui, E., Razafitrimo, H., Gao, Y., and Hsieh, B. R., *Appl. Phys. Lett.*, *67*, 2705 (1995).

49. Park, R. L., and Lagally, M. G. (eds.), *Solid State Physics: Surfaces*, Methods of Experimental Physics 22, Academic Press, Orlando, FL, 1985.

50. Wagner, C. D., Riggs, W. M., Davis, L. E., Moulder, J. F., and Muilenberg, G. E., *Handbook of X-ray Photoemission Spectroscopy*, Perkin-Elmer Corporation, Minneapolis, (1978).

51. Gelius, U., *Phys. Scr.*, *9*, 133 (1974).

52. Barrie, A., *Chem. Phys. Lett.*, *19*, 109 (1973).

53. Jackson, J. D., *Classical Electrodynamics*, Wiley, New York, 1962.

54. Stöhr, J., *NEXAFS Spectroscopy*, Springer Series in Surface Sciences 25, Springer-Verlag, New York, 1992.

55. Hsieh, B. R., *Polym. Bull.*, *25*, 177 (1991).

56. Bhatia, Q. S., Pan, D. H., and Koberstein, J. T., *Macromolecules*, *21*, 2166 (1988), and references cited therein.

57. López, G. P., Castner, D. G., and Ratner, B. D., *Surf. Interf. Anal.*, *17*, 267 (1991).

58. Wessling, R. A., *J. Polym. Sci. Polym. Symp.*, *72*, 55 (1985).

59. Hsieh, B. R., Antoniadis, H., Abkowitz, M. A., and Stolka, M., *Polym. Prepr.*, *33*(2), 414 (1992).

60. Papadimitrakopoulos, F., Konstadinidis, K., Miller, T. M., Opila, R., Chandross, E. A., and Galvin, M. E., *Chem. Mater.*, *6*, 1563 (1994).

61. Antoniadis, H., Hsieh, B. R., Abkowitz, M. A., Jenekhe, S. A., and Stolka, M., *Mol. Cryst. Liq. Cryst.*, *256*, 381 (1994).

62. Antoniadis, H., Abkowitz, M. A., and Hsieh, B. R., *Appl. Phys. Lett.*, *65*, 2030 (1994).

63. Ohta, T., Seki, K., Yokoyama, T., Morisada, I., and Edamatsu, K., *Phys. Scr.*, *41*, 150 (1990).

64. Yokoyama, T., Seki, K., Morisada, I., Edamatsu, K., and Ohta, T., *Phys. Scr.*, *41*, 189 (1990).

65. Lögdlund, M., Dannetun, P., Fredriksson, C., Salaneck, W. R., and Brédas, J. L., *Synth. Met.*, *67*, 141 (1994).

66. Tian, B., Zerbi, G., and Müllen, K., *J. Chem. Phys.*, *95*, 3198 (1991).

67. Bradley, D. D. C., Brown, A. R., Burn, P. L., Friend, R. H., Holmes, A. B., and Kraft, A., *Electronic Properties of Polymers*, Springer Series in Solid-State Sciences 107, Springer-Verlag, New York, 1992, p. 304.

68. Hsieh, B. R., Antoniadis, H., Bland, D. C., and Feld, W. A., *Adv. Mater.*, *7*, 36 (1995).

69. Ettedgui, E., Razafitrimo, H., Park, K. T., and Gao, Y., *J. Appl. Phys.*, *75*, 7526 (1994).

70. Razafitrimo, H., Park, K. T., Ettedgui, E., Gao, Y., and Hsieh, B. R., *Polym. Int.*, *36*, 147 (1995).

71. Park, Y., Ettedgui, E., Choong, V., Hsieh, B. R., Wehrmeister, T., Müllen, K., and Gao, Y., *J. Phys. D* (submitted).

72. Campbell, D. K., and Bishop, A. R., *Phys. Rev. B*, *24*, 4859 (1981).

73. Brazovskii, S. A., and Kirova, N. N., *Mol. Cryst. Liq. Cryst.*, *216*, 151 (1992).

74. Mizes, H. A., and Conwell, E. M., *Phys. Rev. Lett.*, *70*, 1505 (1993).

75. Mizes, H. A., and Conwell, E. M., *Synth. Met.*, *68*, 145 (1995).

76. Hsieh, B. R., Ettedgui, E., Park, K. T., and Gao, Y., *Mol. Cryst. Liq. Cryst.*, *256*, 71 (1994).

77. Sze, S. M., *Physics of Semiconductor Devices*, Wiley, New York, 1981.

78. Spicer, W. E., Lindau, I., Skeath, P., Su, C. Y., and Chye, P., *Phys. Rev. Lett.*, *44*, 420 (1980).
79. Ley, L., Cardona, M., and Pollak, R. A., *Photoemission in Solids I* (M. Cardona and L. Ley, eds.), Springer-Verlag, New York, 1978.
80. Park, Y., Choong, V., Ettedgui, E., Gao, Y., Hsieh, B. R., Wehrmeister, T., and Müllen, K., *Appl. Phys. Lett.*, *69*, 1080 (1996).
81. Lemmer, U., Karg, S., Scheidler, M., Deussen, M., Riess, W., Cleve, B., Thomas, P., Bässler, H., Schwoerer, M., and Göbel, E. O., *Synth. Met.*, *67*, 169 (1994).
82. Woo, H. S., Lee, J. G., Min, H. K., Oh, E. J., Park, S. J., Lee, K. W., Lee, J. H., Cho, S. H., Kim, T. W., and Park, C. H., *Synth. Met.*, *71*, 2173 (1995).
83. Fredriksson, C., Lazzaroni, R., Brédas, J. L., Dannetun, P., Lögdlund, M., and Salaneck, W. R., *Synth. Met.*, *55–57*, 4632 (1993).
84. Mao, G., Fischer, J. E., Karasz, F. E., and Winokur, M. J., *J. Chem. Phys.*, *98*, 712 (1993).
85. Greenham, N. C., Friend, R. H., Brown, A. R., Bradley, D. D. C., Pichler, K., Burn, P. L., Kraft, A., and Holmes, A. B., *Proc. SPIE*, *1910*, 84 (1994).
86. Fahlman, M., Birgersson, J., Kaeriyama, K., and Salaneck, W. R., *Synth. Met.*, *75*, 223 (1995).
87. Salaneck, W. R., Thomas, H. R., Duke, C. B., Paton, A., Plummer, E. W., Heeger, A. J., and MacDiarmid, A. G., *J. Chem. Phys.*, *71*, 2044 (1979).
88. Kamiya, K., Inokuchi, H., Oku, M., Hasegawa, S., Tanaka, C., Tanaka, J., and Seki, K., *Synth. Met.*, *41–43*, 155 (1991).
89. Lazzaroni, R., Lögdlund, M., Stafstrom, S., Salaneck, W. R., and Brédas, J. L., *J. Chem. Phys.*, *93*, 4433 (1990).
90. Tanaka, J., Tanaka, C., Miyamae, T., Kamiya, K., Shimizu, M., Oku, M., Seki, K., Tsukamoto, J., Hasegawa, S., and Inokuchi, H., *Synth. Met.*, *55–57*, 121 (1993).
91. Fahlman, M., Beljonne, D., Lögdlund, M., Friend, R. H., Holmes, A. B., Brédas, J. L., and Salaneck, W. R., *Chem. Phys. Lett.*, *214*, 327 (1993).
92. Fahlman, M., Bröms, P., dos Santos, D. A., Moratti, S. C., Johansson, N., Xing, K., Friend, R. H., Holmes, A. B., Brédas, J. L., and Salaneck, W. R., *J. Chem. Phys.*, *102*, 8167 (1995).
93. Iucci, G., Xing, K., Löglund, M., Fahlman, M., and Salaneck, W. R., *Chem. Phys. Lett.*, *244*, 139 (1995).
94. Salaneck, W. R., and Brédas, J. L., *Adv. Mater.*, *8*, 48 (1996).
95. Lögdlund, M., Salaneck, W. R., Meyers, F., Brédas, J. L., Arbuckle, G. A., Friend, R. H., Holmes, A. B., and Froyer, G., *Macromolecules*, *26*, 3815 (1993).
96. Chance, R. R., Boudreax, D. S., Brédas, J. L., and Silbey, R., in *Handbook of Conducting Polymer* (T. A. Stokheim, ed.), Marcel Dekker, Inc., New York, 1986.
97. Choong, V., Park, Y., Gao, Y., Hsieh, B. R., Wehrmeister, T., Müllen, K., and Tang, C. W., *Appl. Phys. Lett.*, *69*, 1492 (1996).
98. Sariciftci, N. S., Smilowitz, L., Heeger, A. J., and Wudl, F., *Science*, *258*, 1474 (1992).
99. Antoniadis, H., and Hsieh, B. R., *Chem. Phys. Lett.*, *238*, 301 (1995).
100. Rice, M. J., and Garstein, Y. N., *Phys. Rev.*, *B53*, 10764 (1996).

6

Electronic and Chemical Structure of Conjugated Polymers and Model Molecular Systems

M. Lögdlund, Th. Kugler, and W. R. Salaneck
Linköping University
Linköping, Sweden

I. INTRODUCTION

A. Background

A description of the chemical and electronic structure of polymeric and condensed molecule solid systems is an essential ingredient in the establishment of an information data base which could provide for the eventual use of organic molecular and polymeric materials in modern electronics applications. The single most useful experiment tool for these combined purposes has turned out to be photoelectron spectroscopy (PES) [1]. Analysis of PES data often involves the help of the results of appropriate quantum-chemical calculations [2]. A combined experimental–theoretical approach provides a level of information output which is larger than the sum of the two [3]. Although x-ray and other structural data sometimes exist, often even the specific polymeric or molecular-geometrical parameters (atomic coordinates) are provided through theoretical estimates obtained from semiempirical quantum-chemical models [4], the fine-tuning of which enables the interpretation of PES data in terms of certain geometrical parameters. In this chapter, the principles involved in the study of the chemical and electronic structure of conjugated polymers and corresponding model molecular solids using a combination of photoelectron spectroscopy and quantum-chemical calculations are reviewed. Although certain important issues are mentioned, equipment is not discussed. The theoretical models most commonly used are listed, but no comprehensive list or discussions are attempted. Instead, the emphasis is on the illustration of the type of information which may be obtained and some of the pitfalls which may occur in the process. For convenience, the examples present are taken mainly, but not exclusively, from the authors' own works, but without compromise of the subject matter.

The chapter is organized based on materials. First, conjugated materials are introduced, and then a series of illustrated examples are given. Ample references are provided for the interested reader to obtain further information. A basic knowledge of certain principles of organic chemistry and solid-state physics are presumed; but, even here, references are given for more detailed information.

B. Materials

Only linear polymers need be mentioned here. Linear polymers are conceptually simple polymer chains which are composed of a regular repeating sequence of molecular units. The (intermolecular) forces which determine most of the physical properties of simple linear polymers are van der Waals and hydrogen-bonding forces. More complex polymers, branched, cross-linked, and so forth are left for the interested reader to explore [5]. In linear polymer molecules which are *not* conjugated, the electronic structure of the chain of atoms or chemical groups which comprises the backbone of the macromolecule consists of only σ bands (possibly with localized π electronic levels). The large highest occupied molecular orbital–lowest unoccupied molecular orbital (HOMO–LUMO) gap, or electron energy gap, E_g, in the σ system renders these polymer materials electrically insulating and transparent to visible light (although light scattering may occur depending on the morphology of samples studied). π-Conjugated polymers, on the other hand, consist of a regularly alternating system of single (C–C) and double (C=C) bonds; a condition which leads to a lower E_g in the delocalized π system.

Molecular solids of interest here; those used as model systems for conjugated polymers [3] are composed of discrete units which retain their identities as molecules in the solid phase, for example, when condensed from the gas phase into thin films. Molecular solids may be single crystals, polycrystalline, or amorphous in structure [6]. Bonding within the molecules is usually covalent and is localized to within the molecular units. There are usually no *inter*molecular covalent bonds, only weak van der Waals forces [7], at least in the materials of interest here.

The geometrical structures of several common conjugated polymers are shown in Fig. 1, where the necessary (but not sufficient) condition for conjugation, the alternation of the single and double bonds along the polymer backbone, can be seen. The unique electronic properties of conjugated polymers derive from the

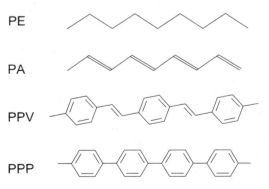

FIGURE 1 The chemical structure of some common conjugated polymers: Polyacetylene (PA), poly(*p*-phenyl vinylene), and poly(*p*-phenylene). At the top is also the nonconjugated polyethylene (PA) displayed for comparison.

presence of π electrons, the wave functions of which are delocalized over long portions of (if not the entire) the polymer chain, when the molecular structure of the backbone is (at least approximately) planar. It is necessary, therefore, that there are no large torsion angles at the bonds, for example, joining aromatic groups of the lower two polymers in Fig. 1, which would decrease the delocalization of the π electrons [2]. The essential properties of the delocalized π-electron system of a typical conjugated polymer, which are different than those for a conventional polymer with σ bands, are as follows: (i) The electronic (π) band gap, E_g, is relatively small ($\sim 1-3\frac{1}{2}$ eV), with corresponding low-energy electronic excitations and semiconductor behavior; (ii) the polymer molecules can be rather easily oxidized or reduced, usually through charge transfer with atomic or molecular dopant species, to produce *conducting polymers*; (iii) net charge carrier mobilities in the conducting state are large enough that high electrical conductivities are realized; and (iv) the charge-carrying species are not free electrons or holes, but self-localized "quasiparticles," which, under certain conditions, may move relatively freely through the materials [8–10].

C. Electronic Structure of Linear Conjugated Polymers

A wide variety of literature exists discussing energy bands in solids from many different points of view [2,11,12]. The essential feature involves extending the molecular orbital eigenvalue problem to systems with one-dimensional periodic boundary conditions, (i.e., over regularly repeating monomeric units). The translational symmetry of the polymer chain implies that the solutions of the Schrödinger equation must be of the Bloch form, $\Psi(k, r) = u_k(r) \exp(ikx)$, where $u_k(r)$ has the period of the lattice (the unit repeating along only one dimension, say x) and k is the crystal momentum along the direction of x. Periodicity in reciprocal space implies that $\Psi(k, r) = \Psi(k + K, r)$, where K is the reciprocal lattice vector. The first Brillouin zone is defined by the region between $-\pi/a \leq k \leq \pi/a$, where a is the magnitude of the real space lattice vector. The first Bragg reflections and the first forbidden electron energy gap occur at the first Brillouin zone. The electronic energy band structure of an arbitrary one-dimensional system will contain many overlapping bands, usually of different symmetry, and spread out over different binding energies. The N electrons of the system will occupy the lowest (deepest binding energy) bands.

The geometrical structure of *trans*-polyacetylene, which is the stable polyacetylene stereoisomer at room temperature, is illustrated in Fig. 1. Of interest is the regular pattern of alternating single and double bonds. Because the carbon atoms are sp^2 hybridized (each carbon atom has three nearest neighbors), the remaining p_z atomic orbitals, one on each carbon atom and oriented perpendicular to the planar zigzag chain, result in a so-called π-band electronic energy band. The calculated energy band structure of *trans*-polyacetylene, calculated using the valence effective Hamiltonian (VEH) method [2,13], is compared with the measured ultraviolet photoelectron spectroscopic (UPS) spectrum [14] in Fig. 2.

Normally, a band structure is presented with the energy vertically and the momentum vector k horizontally. The band structure in the figure is rotated by 90°, to enable a more direct comparison with the experimental spectrum. To facilitate the comparison between the UPS data and the calculated band structure, the density

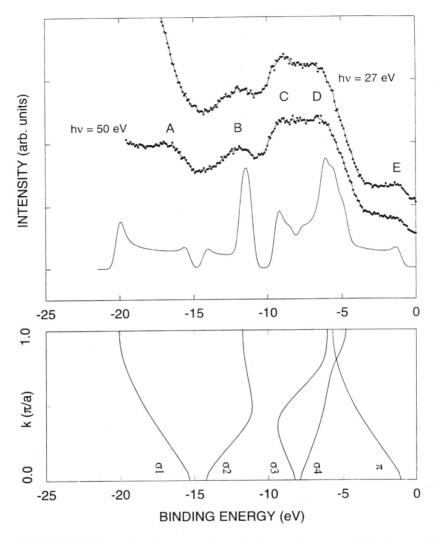

FIGURE 2 Valence band spectra of *trans*-polyacetylene, recorded using synchrotron radiation at 27 eV and 50 eV photon energy, and the corresponding DOVS derived from VEH calculations. The VEH band structure is shown in the lower part of the figure.

of valence states (DOVS), $\rho(E)$, is included. The highest occupied energy band is the π band, which can be seen clearly as the band edge in the UPS spectra. The position of the (π) band edge and changes in the electronic structure in the region of the edge are of central importance in studies of conjugated polymer surfaces and the early stages of formation of the polymer–metal interface. Clearly, the UPS spectra, sensitive to only the topmost molecules of the film, are appropriate in such studies.

D. Charge-Bearing Species in Conjugated Polymers

The intrinsic low-dimensional geometrical nature of linear polymer chains as well as the general property of conjugated organic molecules that the geometric structure is dependent on the ionic state of the molecule [strong electron (hole)–lattice interactions] lead to the existence of unusual charge-carrying species. These species manifest themselves, through either optical absorption in the neutral system or charge transfer doping, as self-localized electronic states with energy levels within the otherwise forbidden electron energy gap [8,15–18]. For example, in *n*-type doping, although the very first single electron on the (any) isolated *trans*-polyacetylene polymer chain goes in as a polaron [8,17], polarons readily combine to form a lower energy state, that of the soliton, named after the mathematics which describes its behavior [15,16]. Although solitons must be formed in pairs [19], usually a single species is diagrammed for convenience. An isolated soliton, containing an excess electron, leads to the formation of a new electronic state within the energy gap. In Fig. 3a are shown schematic representations of the neutral soliton (top), the negatively charged soliton (middle), and the positively charged soliton (bottom). The bond alternation associated with these states has been abbreviated for convenience [8,19]. In nondegenerate ground-state polymers, however, the individual solitons interact (combine) to form spinless bipolarons, a lower-energy configuration. Thus, generally, depending on the symmetry of the ground state, the charge-carrying species are charged polarons, spinless charged solitons, or spinless charged bipolarons [8,15–17,19,20]. The negatively charged bipolaron states for poly(*p*-phenylenevinylene) (PPV), a nondegenerate conjugated polymer, are sketched in Fig. 3b. These species, solitons, polarons, and bipolarons, represent the lowest-energy eigenstates of the coupled electron (hole)–lattice systems [21] and are responsible for the unusual electrical, magnetic, and optical properties of conjugated polymers.

II. CHARGE-STORAGE STATES IN CONJUGATED SYSTEMS

A. Charge-Storage States in Short Polyenes

Since the discovery in the late seventies that the polyacetylene exhibited a relatively high electric conductivity upon reduction/oxidation, many studies has been devoted to the understanding of the charge-storage configuration that occurs upon "doping" [22,23]. Nowadays, it is well established that excess electronic charges, introduced via redox chemistry either by electron acceptors or donors, are accommodated as charged solitons in *trans*-polyacetylene, due to the degeneracy of the ground state [8,15,16]. In the process of forming the solitons, a breaking of the C–C π bonds will occur in pairs and induce a change in the bond-length alternation phase. The two bond-alternating defects (i.e., the soliton and antisoliton) will act as charge-storage states for the charges transferred to/from the polymer chain during the doping process; these solitons typically extends over 15–20 carbons in long polyene chains [8].

More recently, efforts have been devoted [24–27] to trying to relate the phenomena occurring in long polyacetylene chains [8] to those in polyene molecules, which can be viewed as short polyacetylene oligomers. On the basis of theoretical

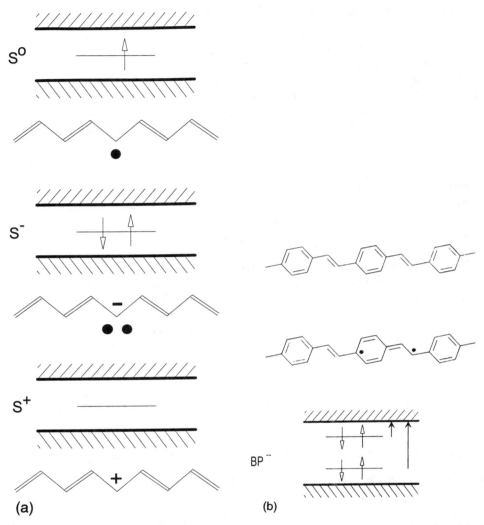

FIGURE 3 Band diagram of solitons (a) with a neutral, negatively charged and positively charged systems, from the top to bottom, and (b) negatively charged bipolarons (BP⁻) in PPV.

considerations [25], it has been proposed that as the length of the chain decreases, the soliton width shrinks [e.g., to reach an extent of about three to five carbon sites for the tetradecaheptaene molecule (seven double-bond polyene)]. Compared with polyacetylene, the finite nature of the short polyenes leads to confinement of the solitons (i.e., to overlap of the soliton–antisoliton wave functions).

Lately, the approach of using conjugated molecules or oligomers of conjugated polymers as models for the conjugated polymers themselves has been extensively used [28–34]. For example, many studies have been devoted to the investigation of the interaction between metals and conjugated molecules, including oligomers of conjugated polymers, using metals such as aluminum, calcium, various alkaline metals, and so forth [30,35–43]. These studies are helpful for the understanding of the possible charge storage configurations induced by charge transfer, as well as for the detailed information about the chemistry involved at the interface upon interface formation between metals and conjugated systems; from a spectroscopic point of view, one of the most important advantages of using molecular materials is the possibility of preparing samples *in situ*, in ultrahigh vacuum (UHV). For example, in the case of the early stages of interface formation between aluminum deposited on various conjugated molecules and polymers, in general, it was found that the interaction caused covalently bonded aluminum atoms, bonded in an sp^3 hybridized configuration, which results in a reduction of the conjugation length.

In this section, the focus is on the experimental and theoretical investigation of the charges-storage configurations in finite polyenes. The molecular systems discussed are β-carotene, α,ω-diphenyltetradecaheptaene, and hexatriene, three polyenelike molecules with polyene segments containing 11, 7, and 3 carbon–carbon double bonds, respectively. The molecular structures are displayed in Fig. 4.

1. *trans-β-carotene*

In a conjugated polymer with a degenerate ground state, there are three expected signatures for the doped state: a mid-gap electronic transition between the soliton gap state and either the HOMO or LUMO; the lack of spins due to the reversed spin-charge relation; and infrared absorption due to the charged solitons. However, for doubly-charged states on a single molecule in short polyenes, the Coulomb repulsion between the defect states is expected to be more important, and thus the nature of the charge storage states might be different.

In a combined experimental and theoretical study of the *p*-type doping of β-carotene in the solid state, a single in-gap absorption band was detected, at 1.35 eV and 1.6 eV, in the cases of I_2 and AsF_5 doping, respectively [32,33]. Room-temperature electron-spin resonance (ESR) measurements of the samples before

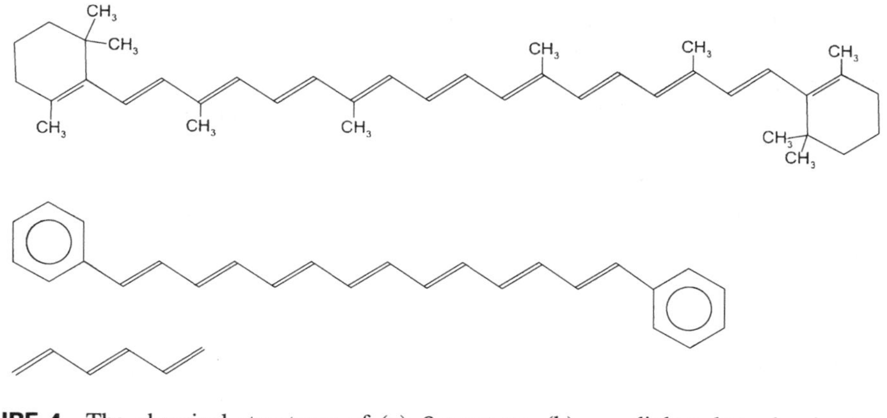

FIGURE 4 The chemical structures of (a) β-carotene, (b) α,ω-diphenyltetradecaheptaene, and (c) 1,3,5-hexatriene.

and after doping showed no signal prior to doping and only a very weak signal after doping; the number of spins per doped β-carotene molecule was estimated to be about 2×10^{-2} for both the I_2 and AsF_5 doping.

Theoretical modeling of the geometrical relaxation, performed using the Hartree–Fock semiempirical Austin Model 1 (AM1) technique, showed that for a singly-charged molecule, a relaxation occurred in the middle of the molecule, extending over about 10–12 carbon atoms, corresponding to the formation of a radical cation, a so-called polaron. In the case of a doubly-charged molecule, two relaxation regions occur, corresponding to the formation of a soliton–antisoliton pair. The solitons are centered about eight carbons sites away from the center of the molecule and extend over about seven carbon atoms. In the case of a long polyene, the soliton–antisoliton pair would have repelled each other such that no overlap between the wave functions would have occurred, resulting in a single in gap state. In this case, however, the finite length of the polyene forces the soliton wave functions to overlap, resulting in the formation of two in-gap states with a separation of about 0.4 eV.

Despite the fact that the calculations indicate the existence of two in-gap states, only a single in-gap band was detected experimentally. However, a simulation of the absorption spectra of the double-charged β-carotene molecule, using the intermediate neglect of differential overlap/configuration interaction (INDO/CI) method, showed that this follows directly from symmetry of the electronic states. The π-electronic states of β-carotene and the doubly-charged molecule, β-carotene^{2+}, are alternatively gerade (g) and ungerade (u) as a function of increasing energy. Thus, the promotion of an electron from the HOMO level (g) of β-carotene^{2+} to the lower soliton level (u) is one-photon allowed (B_u transition), whereas that of the upper soliton level (g) is one-electron forbidden (A_g transition); these results support the experimental results (e.g., the creation of a soliton–antisoliton pair upon doping of β-carotene results in a single in-gap absorption band).

These results imply that the β-carotene^{2+} cation would be the lowest-energy species, which agrees with electrochemical studies, where it was found that the dications are the primary product of electrochemical oxidation [44].

However, in the theoretical study of the defect states in β-carotene and some lower homologs, on the complete neglect of differential overlap (CNDO) level, it was found that the total energy of two isolated singly-charged molecules would be lower than the total energy of a neutral plus a doubly-charged molecule, leading to the conclusion that the singly-charged molecule would be the most stable product [45]. In these calculations, neither the interaction between the molecules nor electron correlation effects are included; the first would be expected to increase the energy of the singly-charged species, whereas the latter would be expected to lower the energy of the doubly-charged species.

2. Diphenylpolyenes

The diphenylenes have been extensively studied both by various experimental and theoretical methods [24,35,46–50]. It has been shown that these molecules, at least the longer ones, are suitable to use in order to mimic the properties of short polyenes; the capping phenyl groups enhance the environmental stability with respect to the unsubstituted polyene molecules, but they do not contribute significantly to the first few frontier orbitals [46,47,49]. The frontier orbitals are, of course, the most important orbitals in, for example, interface formation or doping studies. The separation of the frontier orbitals in terms of polyene- and phenyl-dominated molecular orbitals can be deduced from a comparison between experimental UPS spectra and theoretical quantum-chemical calculations [46,47,49], as well as from experimental "fingerprints" in the x-ray photoelectron spectroscopic (XPS) C(1s) core-level shake-up spectra, as demonstrated below for a diphenyl-polyene with seven carbon–carbon double bonds in the polyene part of the molecule [i.e., α,ω-diphenyl-tetradecaheptaene (DP7)] as displayed in Fig. 4 [49].

The UPS He I and He II valence band spectra of DP7 are compared with the DOVS as convoluted from valence effective Hamiltonian (VEH) calculations in Fig. 5; the corresponding eigenvalues are displayed at the bottom. The initial states of the photoelectrons, in terms of molecular orbitals, can be obtained from calculations. The electrons contributing to the low binding energy part (i.e., from about -5 eV to the valence band edge near -2 eV) originate from the six highest (i.e., lowest binding energy) occupied molecular orbitals. The two peak at lowest binding energy, peaks B and C, correspond to electrons originating from the highest and second highest occupied molecular orbital, respectively, which are mainly localized to the polyene part of the molecule. Peak A corresponds to electrons in four π-orbitals out of which two are doubly degenerate and completely localized to the phenyl groups of the molecule, whereas the other two are more evenly spread out over the molecule. In total, however, peak A is dominated by electrons corresponding to π-orbitals mostly localized to the phenyl groups.

In Fig. 6 are the weak shake-up satellite features, seen on the high-binding-energy side of the XPS C(1s) main peaks, compared for three systems: *trans*-polyacetylene (top), DP7 (middle), and condensed benzene (bottom) [51]. Note the intense peak near 292 eV for benzene and the rather intense peak (not totally

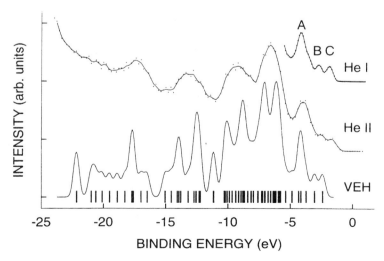

FIGURE 5 UPS He I and He II valence band spectra of DP7 are compared with the DOVS as convoluted from VEH calculations.

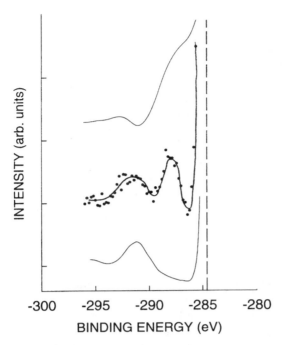

FIGURE 6 Shake-up satellite features in the XPS C(1*s*) spectra are shown for *trans*-polyacetylene (top), DP7 (middle), and benzene (bottom).

resolved) at 288 eV for *trans*-polyacetylene. The two shake-up features in the DP7 spectrum (middle), at about 289 and 292 eV, are at about the same energies as the dominant features in the separate *trans*-polyacetylene and benzene spectra. However, in the case of the 288-eV feature of *trans*-polyacetylene and the 289-eV peak in DP7, a slight difference in energy arises because the π system is more delocalized in real *trans*-polyacetylene and thus this feature lies closer to the main C(1*s*) in this case. Thus, by this "fingerprint" method, the shake-up spectrum for DP7 appears, to a very good approximation, to be a superposition of that of *trans*-polyacetylene and benzene. This also agrees well with simulations of the shake-up feature using the INDO/CI method; it was found that for the important one-electron transitions, in the energy region associated with the polyene part from the experimental "fingerprint" method, the charge redistribution (in response to creation of the core hole) is confined to within the polyene unit [49]. When the core hole is situated on the phenyl groups, there are two regions of shake-up features: one higher-binding-energy region with the most important transition essentially the same as the $\pi-\pi^*$ transition in the benzene molecule and one low-binding-energy region, associated with polyene part. The important contributions in this region, however, involve transitions from orbitals very much localized to the polyene part, to the lowest unoccupied molecular orbital (LUMO), which is almost totally localized on the phenyl groups.

   In the case of the interaction between DP7 and alkali metals, some differences in the evolution of the electronic structure as a function of doping level can be found, although, in all cases, new electronic states within the originally forbidden energy gap can be detected. Here, the focus will be on the evolution of the low-binding-energy region of the valence band spectra (e.g., the region covered by the He I spectrum in Fig. 6) in the cases of interaction between DP7 and Na [50], Rb [52], or Ca [42], upon gradually deposition onto the molecular solid.

Because of the possible use in polymer light-emitting diodes (LEDs), many studies have been performed on the interface formation between aluminum and various conjugated systems [3,53,54]. In general, Al atoms form covalent bonds upon interaction with any conjugated systems; see, for example, Refs. 3, 53, and 54. In the cases of Na, Rb, and Ca, the XPS core-level binding energies of the metal atoms indicate that the atoms are in ionic form (i.e., a charge transfer from the atoms to the DP7 molecules takes place). Also, in contrast to aluminum, where a metal layer is formed almost immediately, it is possible to deposit both Na, Rb, and Ca on the DP7 system without building a metal overlayer (i.e., the metal atoms diffuse into the bulk), although the time to reach equilibrium varies from case to case, as will be discussed in more detail below.

The evolution of low-binding-energy portion of the UPS He I valence band spectra for DP7 at different degrees of doping with (a) sodium [50], (b) rubidium [52], and (c) calcium [42] are shown in Fig. 7, with the curve for the neutral DP7 being at the bottom and with increasing doping upward. The middle curve in each case corresponds to a "low-doping" regime defined as the doping level where new, detectable features first appear in the previously forbidden energy gap.

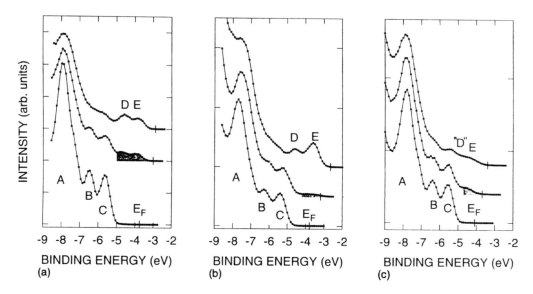

FIGURE 7 The low-binding-energy parts of the UPS spectra recorded for DP7 during successive doping with (a) sodium, (b) rubidium, and (c) calcium. The bottom spectra correspond to the pristine system, the top spectra to saturation doping, and the spectra in the middle correspond to the low doping level where the new doping-induced states first are detected.

In the case of the sodium doping of the DP7 system, the evolution of the spectra upon doping shows that two new features, peaks D and E, appear above the valence π-band edge of the pristine molecule; these two features appear at the very first low-doping step where any detectable features are to be found in the originally empty energy gap [50]. By comparing the experimental results with results from quantum-chemical calculations, these peaks can, as in the case of β-carotene, be assigned to the formation of soliton–antisoliton pairs confined to the polyene part of the molecule, resulting in new states in the otherwise forbidden energy gap between the HOMO and the LUMO of the pristine system. The polyene part in DP7 is, however, shorter than in β-carotene, resulting in a larger overlap between the soliton wave functions, and thus a larger splitting of the energy levels occurs in DP7 (about 0.8 eV) compared to β-carotene (about 0.4 eV). The intensities of peaks B and C decrease upon increased doping, in good agreement with the fact that the formation of a soliton–antisoliton pair leads to the removal of states from the valence band edge (and conduction band edge) of the neutral molecule to form states within the band gap. From the relative intensities of the XPS C(1s) and Na(1s) core levels, it can be deduced that the doping level at which the two new peaks first can be detected is about 0.4 Na per DP7. The saturation doping, defined as the doping level where no further changes in the valence band spectrum upon occurs, corresponds to two Na atoms per DP7, and the binding energy of Na(1s) is ionic, indicating that the sodium atoms donate two electrons to each DP7 molecule.

The deposition of rubidium onto surface of DP7 also results in n-type doping of the material, as the XPS C(1s) and Rb(3$d_{5/2,3/2}$) core-level binding energies indicate that the bond between DP7 and a rubidium atom is ionic. Upon altering the doping level, going from the undoped pristine system to the fully doped system, a shift of the XPS C(1s) core level, and a decrease of the work function of about 0.6 and 1.1. eV, respectively, occur. The latter is consistent with a filling of the previous energy gap with new electronic states, which will shift the position of the Fermi level.

As expected, the doping process is again found to affect essentially the polyene part of the molecule. In the UPS spectra of Rb-doped DP7, the phenyl-group-related peak (A) shows only a broadening upon increased doping, and the electronic structure related to the polyene system is strongly modified. The polyene peaks (B and C) gradually decrease in intensity, and new electronic states appear in the previously forbidden energy gap.

At low doping levels (middle curve in Fig. 7b), a new peak, displaying a finite density of states (DOS) at the Fermi energy, can be detected at about -3.3 eV. Peak C appears to be slightly destabilized and an upward shift of the Fermi level, of about 0.6–0.8 eV, takes place. This can be interpreted in terms of an initial formation of singly-charged molecules, DP7$^-$ (i.e., the formation of radical anions), or polarons. The upper polaron level is half-filled, which results in a finite DOS at the Fermi level. In the case of polaron formation (as well as for bipolarons), however, two new states in the previous energy gap are expected. Quantum-chemical calculations of the electronic structure of singly-charged DP7 molecules show that the highest occupied molecular orbital (HOMO) in the pristine DP7 molecule is very close in energy to the HOMO-1 level (i.e., the lower polaron level). These two contributions cannot be resolved in the UPS spectra and thus, it appears as only one new electronic state is induced in the energy gap at low doping levels.

As the doping level increases, the DOS at the Fermi energy decreases to zero, and two peaks labeled D and E in Fig. 7b, can clearly be observed in the previous energy gap. Thus, at higher doping levels, the situation again becomes similar to the *p*-type doping of β-carotene and the sodium-doping of DP7 (i.e., soliton–antisoliton pairs are formed on the polyene part of the molecules).

The binding-energy difference between peaks D and E in the UPS spectrum (e.g., in the high-doping regime), is about 0.9 eV, whereas in the low-doping regime, the difference between the peak at the lowest binding energy, corresponding to the upper polaron level, and the peak C is about 1.6 eV. The energy difference between the gap states were calculated to be about 0.8 and 1.5 eV for the soliton–antisoliton pair and polaron situation, respectively, in good agreement with the experimental values.

The UPS valence band spectra of DP7, as a function of increasing Ca deposition, are shown in Fig. 7c [42]. Again, in contrast to the sodium doping of DP7, it appears that only one new state is induced within the original forbidden energy gap. However, two new states appear within the band gap, as will be discussed in more detail below. As before, a decrease in the Fermi energy occurs upon doping, consistent with a filling of the band gap with new occupied electronic states. From the estimated saturation doping level, as determined by XPS, it is observed that DP7 can accommodate one Ca atom per DP7 molecule. Also, the binding energy of the Ca($2p$) core-level corresponds to the formation of Ca^{2+} ions upon doping, which indicates that two electrons are transferred from each Ca atom to each DP7 molecule.

Theoretical modeling of the interaction between Ca atoms and DP7, performed using the local spin density (LSD) approximation, revealed that the charge-storage states, accommodating two excess electrons per molecule, can be described as bipolaronlike states [42]. When modeling the situation without taking the counterion into account (i.e., using an isolated molecule with two negative charges), two new occupied states appear above the HOMO of the pristine molecule. However, when the Ca counterion is included, the electrostatic interaction of the large Ca^{2+} ion with the $DP7^{2-}$ leads to a stabilization of the two new states, such that the energy of the lower state (HOMO-1 in Ca/DP7) coincides with the energy of the HOMO of the pristine system. Thus, only one new state (the HOMO of Ca/DP7) appear as a new feature in the energy gap of the pristine molecule, in agreement with the experimental UPS spectra presented in Fig. 7c.

In summary, the metal atoms Na, Rb, and Ca interacts with the polyene part of DP7, such that new doping-induced states appear in the originally empty band gap. Some differences can be pointed out: (i) In the case of sodium doping, only the sign of soliton–antisoliton pairs can be experimentally detected (i.e., dications are formed even at very low doping level); (ii) in the case of rubidium doping, at low and intermediate doping levels it is possible to detect features corresponding to polaron formation (i.e., radical cation), and upon further doping, two new states, corresponding to the formation of soliton–antisoliton pairs, appear; and (iii) in the case of calcium doping, the saturation doping level occurs at a doping level corresponding to one Ca atom per DP7 molecule, where it is suggested that each Ca atom donates two charges (i.e., the same number of charges per molecule as in the cases of Na and Rb saturation doping levels). Only one new state appears to be formed, but the lower soliton level coincides with the old HOMO level in the pristine system.

In the case of *p*-type doping (in solid state and from electrochemically measurements), it has been suggested that the formation of cations should be the most stable configuration. In the case of DP7, however, it appears to be dependent on the actual doping agent used. The reason why polaronlike defects can be detected in the case of Rb doping but not for Na doping might be explained by the different sizes of the ions and by differences in diffusion coefficients; it appears that Rb diffuses much slower than Na [52].

3. *Hexatriene*

The UPS He I (21.2-eV photons) and He II (40.8-eV photons) spectra of thin films of 1,3,5-hexatriene condensed onto sputter-cleaned gold substrates and the density of valence states (DOVS), obtained from VEH calculations, are compared in Fig. 8. The gold substrates were held at $-180°C$ during the condensation of the thin films as well as during the measurements. The DOVS have been obtained by broadening each VEH energy eigenvalue with a Gaussian function with a full width at half-maximum taken to be 0.7 eV. The theoretical and experimental spectra agrees very well concerning the position of the peaks. The intensities, however, are in less good agreement because there are no cross-section effects included in the calculations.

The evolution of the low-binding-energy part of hexatriene UPS valence band, with doping, is shown in Fig. 9. The spectrum for the neutral molecule is shown

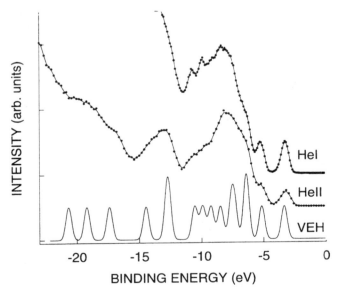

FIGURE 8 UPS He I and He II valence band spectra of hexatriene are compared with the DOVS as convoluted from VEH calculations.

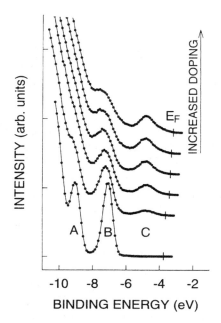

FIGURE 9 Evolution of the electronic structure of hexatriene as a function of sodium doping as recorded by UPS. The spectrum corresponding to the pristine system is displayed at the bottom and for increased doping upward.

at the bottom with spectra for increased doping upward. The binding-energy scale is relative the experimental vacuum level.

Upon doping with sodium, only one new feature appears in the originally forbidden band gap. The new peak, labeled C in Fig. 9, is at the same position, at about -4.8 eV relative to the vacuum level, independent of the doping level. The new feature is first visible in the UPS spectra at about 0.2 Na per hexatriene molecule, as determined from the XPS C(1s) and Na(1s) spectra. Additional doping was done until the growth of the new peak ceased, and beyond until the intensity of the new peak decreased. The spectrum at the top of Fig. 9 is taken at a doping level when the intensity of the new peak was at maximum. This maximum in the intensity occurs at a doping level of about one sodium atom per molecule. Additional doping results in a decrease in the intensity of the new feature. A possible explanation for the decrease is that the peak becomes partially obscured due to the presence of excess sodium on the surface. A slight heating of the sample recovers the spectrum.

The binding-energy difference between the new doping-induced state, peak C, and the HOMO in the pristine system, peak B, is calculated to be 2.74 eV, using the VEH approach based on AM1 optimized geometry for the singly-charged molecule. This is in good agreement with the experimental value of 2.7 ± 0.1 eV. The distortion of the molecule interacting with a single sodium atom resembles the geometrical structure of a polaron on *trans*-polyacetylene.

It has been argued that upon treatment with sodium, hexatriene is likely to dimerize into $(C_6H_8Na)_2$ [55]. A detailed theoretical study of the Na–hexatriene as well as the Na–decapentaene systems have been performed using *ab initio* Hartree–Fock level calculations using the 6-31G* basis set, including electron correlation effects through the use of Möller–Plesset second-order pertubation theory (MP2) [56]. Also, possible dimerization effects have been investigated using semiempirical Hartree–Fock calculations using the PM3 [57] method. The net atomic charges were estimated using neutral population analysis [58].

The neutral charges on the Na atom for the $(C_6H_8)Na$ and $(C_{10}H_{12})Na$ were calculated to be $+94|q_e|$ and $+96|q_e|$, respectively (i.e., the polyene chains have accepted about one electronic charge and thus correspond radical anions). In the case of complexation with two Na atoms [i.e., the $(C_6H_8)Na_2$ and $(C_{10}H_{12})Na_2$ complexes], the neutral charges per sodium atom were calculated to be $+39|q_e|$ for $(C_6H_8)Na_2$ and $+93|q_e|$ for $(C_{10}H_{12})Na_2$. Thus, decapentaene is a much better electron acceptor than hexatriene. In the case of decapentaene, the doubly-charged systems, again, the structural relaxation is consistent with the formation of a soliton–antisoliton pair. It seems that the polyene is not able to accept two electrons unless the system is long enough to support a soliton–antisoliton pair.

The geometry optimization using the PM3 method, including configuration interaction, has been performed on a system of two sodium atoms and two hexatriene molecules, in order to examine the possibility of a dimerization. The calculations were performed such that two $(C_6H_8)Na$ complexes could interact with each other, with all geometrical parameters optimized. Indeed, the geometry converges to a dimerized system. The dimerization process is illustrated in Fig. 10. As can be seen, two sp^3 defects occur at the position where the two molecules are joined. In principle, there are still two separate conjugated polyene segments, each with a single charge.

B. Poly(*p*-phenylene Vinylene)

In connection with the progresses made in the fabrication of light-emitting diodes (LEDs) from conjugated polymers, some of the most promising results have been obtained with poly(*p*-phenylene vinylene) as the light-emission medium, with (hole-injecting) indium-tin oxide (ITO)–glass as the transparent substrate and with a calcium or aluminum electrode as the (electron-injecting) metallic contact. Here, the results from studies of doping-induced effects on doping PPV with sodium and

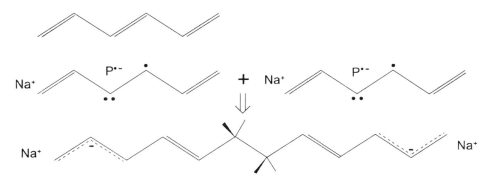

FIGURE 10 Schematic drawing of the dimerization process of two C_6H_8Na complexes.

rubidium, as well as some results from studies of the interaction during the initial stages of interface formation between calcium and a derivative of PPV, poly(2,5 diheptyl-*p*-phenylene vinylene) (DHPPV) [59], will be discussed. The chemical structure of DHPPV is displayed in Fig. 11.

The experimental UPS HeI and HeII valence band spectra of PPV, prepared by the tetrahydrothiophenium precursor route, are compared with the DOVS as calculated from the VEH band structure in Fig. 12 [14]. The calculated density of states have been derived from the energy versus momentum curves (i.e., the band structure displayed at the bottom of Fig. 12). The energy scale is fixed relative to the experimental Fermi level. It has been shown from neutron-diffraction measurements on oriented PPV at room temperature that the ring torsion angles (i.e., the twist of the phenyl rings out of the vinylene plane) for PPV are in the order of 7° ± 6° [60]. Such small torsion angles result in negligible effects on the calculated electronic band structure compared with that for the fully coplanar conformation, and thus the PPV has been taken to be planar in the VEH calculations.

The different peaks in the UPS spectra of PPV can be assigned to different bands from the details in the VEH calculations: Peaks A, B, and C originate from electrons in σ bands; Peak D is built up from contributions from the four highest σ bands, the lowest π band, and a small portion from the relatively flat part of the second π band; peak E is derived from the next highest π band, which is extremely flat, because it corresponds to electronic levels fully localized on the bonds between ortho carbons within the phenyl rings. In general, a flat band results in a high-intensity peak in the DOVS, as there are many states per unit energy just at the flat band. There also are small contributions to peak E from the second and fourth π bands. Finally, peak F is derived from the top part of the highest π band. The larger dispersion of the top π band results in lower intensity in the UPS data.

Some UPS results from *p*-type doping of polyacetylene, using potassium as a doping species, have been reported by Tanaka et al. [61]. The first direct measure of multiple, resolved gap states in a doped conjugated polymer were reported by Fahlman et al., in the case of sodium-doped poly(*p*-phenylene vinylene) [62]. The UPS spectra recorded upon increasing the doping level indicated a slight decrease in the work function at the first doping steps. At about 40% doping, defined as the Na/monomer ratio, a large change of about 1.2 eV occurred, followed by a slight decrease as the doping level approaches 100% (i.e., one sodium atom per "monomer" repeat unit).

Simultaneously with the 1.2-eV change in the work function (i.e., at intermediate doping levels), two new states appear in the previously empty energy gap, with a uniform increase of the intensities of the new gap states as shown in Fig.

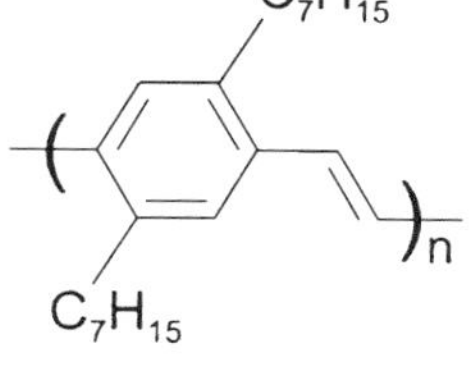

FIGURE 11   Chemical structure of poly(2,5 diheptyl-*p*-phenylene vinylene) (DHPPV).

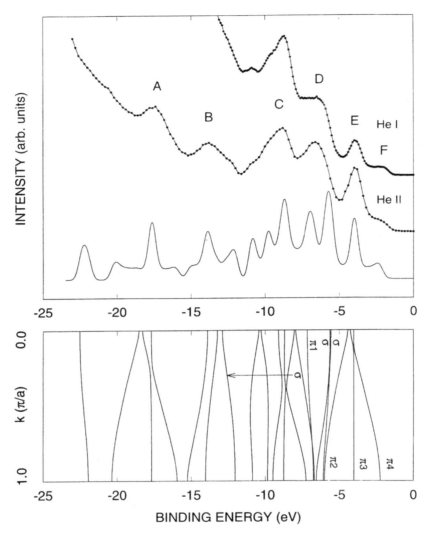

FIGURE 12 He I and He II UPS valence band spectra of PPV compared with DOVS derived from VEH calculations. The VEH band structure is shown in the lower part of the figure.

13. The separation between the two new peaks is about 2.0 eV at the maximum doping level (i.e., near 100%), with the lower-binding-energy peak positioned at about −3.2 eV. At this doping level, the charges in this nondegenerate ground-state polymer can be accommodated in either two polaron bands or in two bipolaron bands.

From model calculations performed using the VEH technique for a 100% doping level, the new states appearing in the previously forbidden energy gap are assigned to two doping-induced bipolaron bands. Also, in the case of polaron formation, a finite density of states would be expected to appear at the Fermi-level

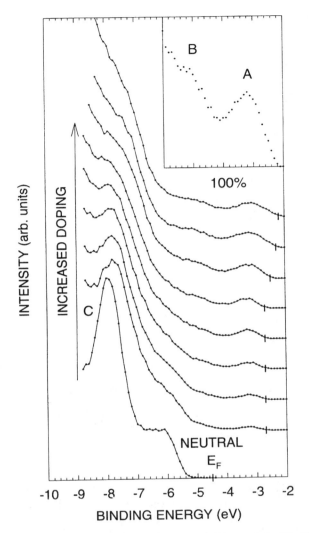

FIGURE 13 The low-binding-energy part of the He I UPS spectra of PPV recorded during successive exposure to sodium.

energy. Thus, the lack of significant density of states at the Fermi level also indicates that a formation of bipolaron bands is most likely.

The low-binding-energy regions of the HeI (21.2 eV) UPS spectra of PPV, as recorded during stepwise exposure to rubidium, are shown in Fig. 14. As discussed above, the two lowest binding energy peaks, labeled A and B, correspond to electrons originating from a flat π-band essentially localized to the ortho-carbons, peak A at about -8 eV, and the highest occupied π-band, peak B at about -6.5 eV, which is delocalized along the polymer backbone [14]

As in the case of the sodium doping, the rubidium doping leads to changes in the structure of the upper π system, with the formation of new features in the

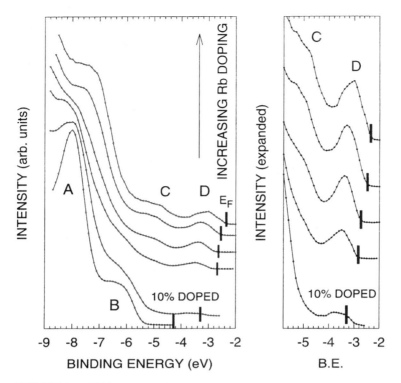

FIGURE 14 UPS spectra of rubidium-doped PPV. The spectrum of pristine PPV is at the bottom, and increasing doping upward. An expansion of the lowest-binding-energy region of the spectra for the doped polymer is shown to the right. The energy scale is referred to the vacuum level. The black bars indicate the position of the Fermi energy.

previously forbidden energy gap. Following successive doping from the pristine PPV up to saturation doping (about two Rb atoms per monomer repeat unit), two different doping regimes can be distinguished. At high doping levels, two new features, labeled C and D in Fig. 14, can be detected in the previously empty energy gap. As in the case of the sodium doping of PPV [62], these two peaks can be related to the formation of bipolarons, as expected for a nondegenerate ground-state system. Note that as the doping preceeds, peak B loses intensity, because the formation of bipolaron states leads to the removal of states from the valence band edge (and conduction band edge) of the neutral molecule to form states within the band gap.

Different from the case of the sodium doping, however, is the observation that, at the very first low-doping steps, only one new feature can be detected in the previously forbidden energy gap. Furthermore, a finite density of states can be detected at the Fermi energy (indicated by the black bars in Fig. 14), which is expected for a system with a half-filled HOMO. On the other hand, bipolaron bands are expected to be well separated from each other, as well as from the valence and conduction bands, leading to two well-resolved peaks in the UPS spectrum [62]. In the bipolaron band model for an *n*-doped polymer, both bands are completely occupied and the Fermi level is found in the energy gap between the upper occupied bipolaron band and the conduction band (i.e., no density of states at the Fermi energy is expected) [63]. In the case of polaron bands, however, the bands can fuse with the conduction and valence bands, leading to a large broadening of the corresponding peaks in the UPS spectrum. In addition, as the upper polaron band is half-filled, a finite density of states at the Fermi level is expected. Thus, the new feature appearing at low doping (bottom spectrum on the right part of Fig. 14) can be assigned to the upper, singly-occupied polaron band. Also, in the case of polaron formation, it is expected that two new bands are induced; however, in this case the energy position of the lower polaron band is very close to the HOMO in the pristine system, preventing this feature from being resolved in the spectrum.

Although it is not clear why no polaron states could be detected in the case of sodium doping of PPV, some differences in the behavior of the doping species are detected. In the case of rubidium doping, new doping-induced states can be detected at about 10%, whereas the doping level had to reach about 40% before any new features at all could be detected in the case of sodium-doping of PPV [62]. Also, the shift of the work function of the samples, as a function of the doping level, differs for the two cases, as illustrated in Fig. 15, where a plot of the work functions of the sodium-doped and rubidium-doped PPV samples are displayed as a function of the dopant concentration. Because new occupied states are created in the previous energy gap upon *n*-type doping, there will be a decrease of the work function (i.e., a shift of the Fermi energy to lower binding energy). When PPV is doped with rubidium, the work function decreases about 1 eV at about 10% doping, corresponding to the appearance of polaron states. An increase from about 10% to 20% doping results in an additional shift of about 0.6 eV. Further doping results in small changes of the work function. A different evolution appears in the case of sodium doping, as discussed above—that is, a steep decline of the work function

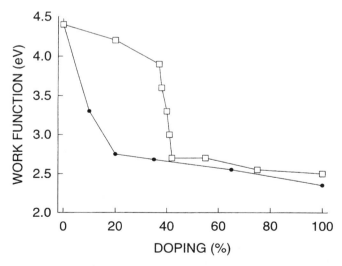

FIGURE 15 Plot of the work function of rubidium-doped (●) and sodium-doped (□) PPV versus the doping level.

of PPV can be seen around 40% doping (i.e., at the lower limit for the detection of the doping-induced states with only minor changes detected at higher and lower doping levels) [62].

Thus, the evolution of the doping-induced states in polyenes and PPV is rather similar in the respect that polaronlike charge-storage states can be detected by UPS at low rubidium doping levels, whereas upon sodium doping, the first detectable doping-induced features correspond to doubly-charged species.

The low-binding-energy parts of the UPS He I valence band spectra of DHPPV, recorded during several stages of deposition of approximately a monolayer of calcium in each step, are shown in Fig. 16. The heptyl side chains of the polymer do not contribute to any intensity below −8 eV [59]; this region is thus very reminiscent of the low-binding-energy part of PPV itself. The spectra are shown with the energy scale relative to the vacuum level; the Fermi energy level for each spectrum is indicated by the bars. The bottommost spectrum corresponds to pristine DHPPV, whereas the uppermost curve corresponds to the saturation doping level, which occurs at about one Ca atom per DHPPV monomer unit, as determined from the relative intensities of the XPS $Ca(2p)$ and $C(1s)$ core levels. From the binding

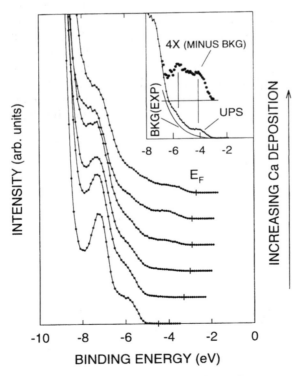

FIGURE 16 The low-binding-energy part of the He I UPS spectra of DHPPV recorded during successive depositions of calcium. The inset shows the fully doped (one Ca atom per monomer) DHPPV with a simple estimate of the inelastic electron background to emphasize the calcium-induced structures.

energy of the Ca (2p) level, it can be deduced that the Ca atoms are in the form of Ca^{2+} ions.

At the initial doping step, the Ca atoms deposited on the surface lead to the complete doping of the top layer, as probed by UPS. Upon further deposition, the amount of Ca at the immediate surface (seen by UPS) does not continue to increase (i.e., saturation doping in the surface region has been reached). Additional Ca atoms then diffuse somewhat into the bulk. Finally, the C(1s) spectrum broadens and shifts slightly toward lower binding energy as the deposition of calcium continues, indicating charge transfer from the calcium to the carbon atoms. As a result of the deposition of Ca, the work function of the sample changes from 4.0 (± 0.1) to 3.0 eV, which is close to that of metallic calcium.

Upon the Ca doping of DHPPV, two new electronic states above the π-band edge of the pristine polymer, are induced: near -4.0 eV and -5.8 eV relative to the vacuum level. This can be enhanced by a simple subtraction of the inelastic electron background signal as shown in the inset of Fig. 16. The position of these new states within the otherwise forbidden energy gap, as well as the relative splitting of the two new states, are similar to those in the case of sodium doping of PPV [62], where two new (bipolaron) states within the gap were observed, as discussed above.

Also, information indicating a limited diffusion of the Ca atoms into the polymer can be extracted from XPS data. The Ca atoms are not localized at the immediate surface, but are rather diffuse into the near-surface region. Although it would be extremely difficult to determine an exact diffusion depth profile, it is obvious from angle-dependent XPS data that the Ca atoms are confined to within the depth observable by XPS. Therefore, the upper and lower limits to the diffusion distance (which is undoubtedly not a sharp diffusion depth) are defined. In this way, it can be estimated that the Ca atoms are distributed in some way within roughly 25 Å of the surface of the DHPPV.

III. ORIENTED CONJUGATED POLYMERIC SYSTEMS

The formation of highly oriented structures such as single crystals, single-domain liquid crystals, and systems comprising uniaxially oriented polymer chains is important in many applications of thin films and interfaces, ranging from materials reinforcement to molecular electronics. Due to the orientational anisotropy of the active molecular units, drastic changes of both the mechanical and physical properties are encountered.

Soon after the discovery of high electrical conductivity in polyacetylene, studies were made on oriented samples. In these experiments, the polymer chains were oriented by stretching the material. This procedure resulted in a distortion of the random coil conformation and an elongation of the polymer chains. As a result, the quasi-one-dimensional delocalized π system was then preferentially oriented in the stretching direction, which gave rise to anisotropic electronic properties. In addition, the experimentally observed macroscopic electronic properties were then close to the intrinsic properties of the individual polymer chains. Accordingly, stretched polyacetylene was shown to display a larger conductivity parallel to the orientation direction than perpendicular to it [64,65].

Processability largely enhances the orientation capability of conjugated polymers. An example is polythiophene, which is rendered soluble by introducing alkyl substituents into the 3-position of the thiophene ring system. Ultrathin films of poly(3-alkylthiophene) can be stretch-oriented even at room temperature. To do so, the film has to be supported by a stretchable substrate like polyethylene foil. The resulting anisotropy of the charge carrier mobility was then determined by coating the polymer film onto a field-effect transistor [66].

In the case of thicker, free-standing conjugated polymer films, elevated temperatures have to be applied in order to facilitate the orientation process. Polyaniline in the insulating emeraldine-base (EB) form is soluble in N-methylpyrrolidinone (NMP). This allows the casting of free-standing, flexible, coppery-colored films, which can be uniaxially oriented by a simultaneous heat treatment and mechanical stretching (the films have to contain about 15% by weight of NMP, which acts as a plasticizer). X-ray diffraction spectra of the stretched films as a function of draw ratio reveal that increased stretching results in a progressive increase in crystallinity. Parallel to that, the solubility in NMP decreases. This increased crystallinity persists even after doping the oriented films with aqueous HCl. The resulting doping stretch-oriented films display a remarkable increase in conductivity both parallel and perpendicular to the stretching direction as compared to doped nonoriented films. This effect is attributed to the increased crystallinity which promotes the intermolecular component of the bulk conductivity [67].

Anisotropic conductivity (σ), thermopower, and dielectric constant studies show that oriented polyaniline salt is representative of a class of quasi-one-dimensional disordered conductors, where coupled parallel chains form "metallic" bundles in which electron wave functions are extended three dimensionally. This is in contrast to isolated conducting chains in conventional one-dimensional conductors. The bundles correspond to crystalline regions of the polymer [68]. Magnetic susceptibility measurements of uniaxially oriented solvent-cast polyaniline films confirms that the polymer behaves as a highly one-dimensional metallic system [69].

Polarized reflectance measurements for doped stretch-oriented EB films (400% elongation) show significant anisotropy in the reflectance and the other optical constants. The frequency-dependent conductivity exhibits a large shift in oscillator strength to lower energies, compared to that of undoped stretch-oriented EB films and compared to that of unstretched films of doped EB, which demonstrates delocalization of conduction electrons in the "polaron band." The dielectric constant of doped stretch-oriented EB indicates that EB is metallike, with maximum conductivity greater than 10 times the DC conductivity. These results indicate that the DC conductivity is likely dominated by interchain processes [70].

A very efficient orientation can be obtained in polymer blends containing the conjugated polymer in an electronically nonactive polymer matrix. An example is substituted poly(p-phenylene vinylene) (PPV) in ultrahigh-molecular-weight polyethylene. The resulting gel can be highly elongated, resulting in a high degree of orientational order and very strongly anisotropic electronic properties [71].

A. Highly Oriented Transfer Films of Poly(tetrafluoroethylene) as Substrates for Oriented Growth of Materials

Of the methods that exist for forming oriented macromolecular structures, few have sufficient generality to make them applicable to materials of differing chemical composition or physical properties. A simple and surprisingly versatile method for orienting a wide variety of crystalline and liquid-crystalline materials, including polymers, monomers, and small organic and inorganic molecules was introduced by Wittman et al. In their technique, a thin, single-crystal-like film of poly(tetrafluoroethylene) (PTFE) is deposited mechanically on a smooth substrate such as glass. Materials grown on this coated surface from solution, melt, or vapor phases show a remarkable degree of alignment [72].

In addition to PTFE, a variety of other polymers were investigated as candidates for the formation of such oriented layers by friction transfer. Even polyethylene, the liquid-crystalline Vectra and fluorinated ethylene–propylene copolymer were found to yield oriented transfer layers. Although these layers induce the oriented growth of a variety of species deposited onto them from the melt, solution, or vapor phase, they were found to be inferior to those of PTFE [73].

B. Preparation of Highly Oriented PTFE Films

When rubbed or slid against a hard surface, PTFE exhibits a low coefficient of friction but a high rate of wear. Deposition occurs as the PTFE chains undergoes scission, creating active groups which chemically react with the counterface. This results in strong adhesion and a coherent transfer film. Further interaction between the bulk polymer and the transfer film gives rise to anisotropic deformation of the unit cell, which results in closeness of adjacent chains and easy shear between chains. Sliding brings about growth as well as reorientation of crystallites situated in a very thin subsurface region of the bulk polymer. Such structural rearrangement facilitates the joining of adjacent aligned crystallites to form films and ribbons, which emerge as debris [74].

In practice, the PTFE films are deposited onto smooth surfaces by slowly sliding a piece of bulk PTFE under load on them at elevated temperatures. They consist of very long, straight, and crystalline ribbons, which may almost completely cover the substrate [75]. The resulting film morphology is strongly influenced by the temperature of the substrate. (High-molecular-weight PTFE powder has a glass transition temperature of about 125 degrees C and an ambient melting point at 334°C [76].)

The temperature influence on friction transfer is to be compared to the temperature influence on the solid-state extrusion-drawing behavior of PTFE powder: Draw is attainable only above 100°C. The maximum achievable extrusion draw ratio (EDR) is almost constant ($\sim$10) from 100°C to 280°C, yet increased rapidly with further increasing temperature and reaches a maximum ($\sim$60) at 330–340°C. At yet higher temperatures, the drawability is lost due to melting. Both the structure and properties of drawn products are found to be completely affected by extrusion temperature and EDR. For extrusion at 330–340°C, near the melting point, an effective and high draw is achieved. The crystalline chain-orientation function, crystallite sizes, both along and perpendicular to the chain axis, differential scanning calorimetry heat of fusion, and flexural modulus increase with EDR and approach a maximum at an EDR of 30–40, depending on the extrusion temperatures. Above a specific EDR, the efficiency of draw decreases due to the formation of flaws. The highly oriented PTFE consisted of microfibrils of a significantly large lateral dimension (~45 nm), as compared to those (6–20 nm) generally found in oriented polymers [76].

C. Characterization of Highly Oriented PTFE Transfer Films

Scanning probe microscopy techniques allow one to directly image the morphological features within oriented PTFE transfer films: Atomic force microscopy (AFM) not only reveals details of the molecular structure, but it can also provide a direct measurement of the absolute thickness and continuity of PTFE-on-glass films. High-magnification images show individual rodlike molecules with an intermolecular spacing of 0.6 nm. The thickness of the films varies from 7 to 32 nm, depending on deposition temperature and mechanical pressure. The continuity of the films strongly decreases at lower temperatures. The remaining single fibers are not stable and can be modified by the imaging tip [77]. PTFE-on-silicon wafer films were shown to consist of narrow ribbons of PTFE, <1 μm wide and <10 nm high, aligned with the film draw direction. Within the ribbons, the PTFE molecular chains are also aligned with the draw direction [78].

Despite the fact that PTFE is electrically insulating, scanning tunneling microscopy (STM) has been applied to examine details of oriented ultrathin PTFE films deposited on various substrates, in particular Pt and highly ordered pyrolytic graphite (HOPG). Ordered structures originating from aligned chains were observed, and reproducible images were obtained at the submicron level. Nanometric molecular details are more difficult to resolve, but indicate interesting features. In particular, they have observed regions of parallel zigzag-shaped molecules, which are separated by approximately 6 Å [79].

X-ray photoelectron spectroscopy (XPS) investigations of PTFE transfer films revealed that the amount of PTFE put down during rubbing increases with deposition temperature and pressure. The XPS azimuthal angle dependence at a low electron take-off angle is consistent with an aligned ribbon morphology. Azimuthal and polar angle dependence and charging studies show that hydrocarbon contamination observed in the XPS spectrum is located on top of the silicon wafer substrate, in the channels between the PTFE ribbons. Imaging XPS confirmed the pressure dependence of PTFE coverage, and at a low deposition pressure, it reveals tracks of PTFE ~100 μm wide, aligned with the film draw direction [78].

Polarization-dependent x-ray absorption near-edge structure (XANES) spectra were obtained for oriented polyethylene, fluorinated polyethylenes, and oligomeric model compounds. Pronounced polarization dependence was observed for each compound, which allowed an unambiguous assignments of the XANES spectrum [80].

The structural change of the evaporated perfluorotetracosane (PFT) n-$CF_3(CF_2)_{22}CF_3$ and poly(tetrafluoroethylene) $(CF_2)_n$ films by the mechanical rubbing process was examined with use of near-edge x-ray absorption fine structure (NEXAFS) spectroscopy. In the PTFE film, the chains were almost parallel to the surface and were uniaxially realigned along the rubbing direction, whereas the chains in the PFT film after evaporation were oriented normal to the surface, and the orientation was hardly affected by the rubbing process. The chains of the PFT evaporated on the rubbed PTFE film aligned along the rubbing direction [81].

The possibility for orientational probing of polymeric thin films by means of computer simulations of NEXAFS spectra was demonstrated using an algorithm for static exchange calculations of core-excited states to study the near-edge carbon and fluorine K absorption in PTFE. Unpolarized and polarization-selected results for oligomers including up to five fluoroethylene subunits were presented. Through experimental comparison, it was confirmed that the PTFE samples used in previous experimental studies are indeed highly oriented, that the zigzag structure prevails over the helical structure, and that the polymer chains are directed along the assumed slide direction [82].

Polarized vacuum–UV absorption and reflection spectra were measured for oriented films of PTFE and its model compound, perfluorotetracosane (PFT) n-$C_{24}F_{50}$. The lowest energy excitation leading to an intense peak was found to be polarized along the molecular chain, and its energy decreases with the chain length (8.05 eV for PFT and 7.66 eV for PTFE) [83].

D. Oriented PTFE Films as Ordered Substrates

A large variety of crystalline and liquid-crystalline organic molecular and polymeric materials have been used to grow oriented films onto highly ordered PTFE films [84]. In the following, a few selected systems will be discussed in more detail.

Highly oriented polysilane layers have potential applications in electrophotography, nonlinear optics, display fabrication, and microlithography. In this context, poly(dimethylsilane) (PDMS), poly(methylphenylsilane) (PMPS), and poly(cyclohexylmethylsilane) (PCHMS) were evaporated on substrates coated with highly oriented PTFE films. The orientation characteristics for the polysilane films were estimated by polarizing microscope images, x-ray diffraction (XRD) patterns, and polarized absorption spectra. Only the PDMS films on PTFE displayed orientational characteristics. X-ray diffraction indicates that the silicon backbone chains lie in a plane parallel to the substrate and that a particular crystalline plain is also parallel to the substrate [85,86].

A lot of effort has been dedicated to the investigation of diacetylenes and their derivatives. The important feature in that context is the solid-state polymerizability of the diacetylene system upon irradiation with long-wavelength UV light. 2-Methyl-4-nitroaniline (MNA) is a well-known material for second-harmonic generation (SHG). The corresponding monomeric diacetylene derivative (diacetylene methylnitroaniline, DAMNA) was used to grow thin crystalline films onto quartz, Teflon, and Kapton substrates by means of vapor deposition in vacuum. These films were then UV polymerized to give crystalline polydiacetylene thin films (PDAMNA). Interestingly, films grown onto Teflon exhibit greater orientation and significantly greater SHG than those grown onto a quartz and Kapton, indicating a high degree of orientational order of the MNA units [87]. Other diacetylenes were also shown to form uniaxially oriented thin layers, which, upon photoirradiation, change to polymer films having extremely high crystallinity and considerable in-plane dichroism. The third-order susceptibility $\chi^{(3)}$ of these uniaxially oriented polydiacetylene films were several times larger than those of the randomly oriented films. Crystal growth could be observed by electron microscopy [88].

In the context of conducting polymers, both the polythiophene and the poly(*p*-phenylene vinylene) (PPV) systems have been investigated: The crystal structure of dimethyl-oligothiophenes grown by epitaxial crystallization on highly oriented PTFE thin film was studied by transmission electron microscopy. The dimethyl-oligothiophenes crystallize principally in a monoclinic phase with common lattice parameters a, b, and β, but a different c for different numbers of thiophene rings: $a = 0.598$ nm, $b = 0.789$ nm, and $\beta = 98°$; and $c = 1.866$ nm for the tetramer, $c = 2.234$ nm for the pentamer, and $c = 2.596$ nm for the hexamer. The space group is $C2/m$. Polymorphism is found, with the coexistence of monoclinic and orthorhombic crystal structures [89].

For PPV, it was demonstrated that conjugated polymers processed via soluble precursor polymers can be oriented by forming them on highly oriented PTFE. The PTFE induces preferential orientation of the precursor polymers, which is preserved and possibly enhanced during the thermal conversion to the final conjugated polymer. Orientation is substantially better if precursor polymers are used which have some content of stiff segments in their backbone. The state of alignment of the polymer was characterized by polarized optical absorption measurements. The orientation is considered to be confined to a thin layer adjacent to the substrate, and from the form of the interband absorption, it is concluded that the extent of conjugation is enhanced in this oriented region [90].

As a second example, aligned poly(2,5-diheptyl-*p*-phenylene) (DHPPP) films were prepared by solution casting onto aligned PTFE substrates [91]. The chemical structure of DHPPP is displayed in Fig. 17. In Fig. 18 are shown three valence band spectra of DHPPP compared with the DOVS derived from VEH calculation for an interring torsion angle of 42° [91]. The inset shows the low-binding-energy part in more detail. The spectra have been recorded with 4-eV photon energy for three different orientations of the sample relative to the polarization direction of the incoming radiation: the incoming light (A) parallel, (B) 45°, and (C) perpendicular to the aligned chains of the PTFE substrate. From the calculations, the peak labeled 1 can be assigned to electrons removed from the two highest lying π bands: $\pi3$ and $\pi2$. The $\pi3$ band is derived from molecular orbitals delocalized along the polymer backbone and has a significant dispersion. The $\pi2$ band is a flat band derived from molecular orbitals localized on the ortho carbons in the phenyl rings. These molecular orbitals correspond to one of the doubly-degenerated e1g molecular orbital in benzene, with linear combination of atomic orbital (LCAO) coefficients close to zero on the carbons connecting to the neighboring phenyl ring. As seen in inset of Fig. 18, the intensity of peak 1 is at maximum for configuration C and completely disappears for configuration A. Configuration B shows something

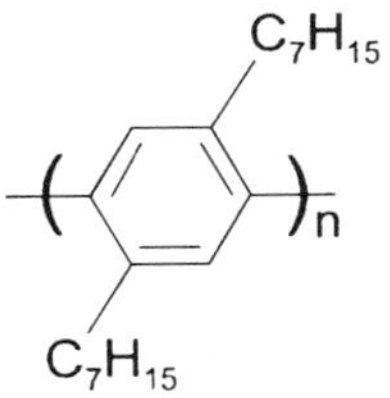

FIGURE 17   Chemical structure of poly(2,5-diheptyl-*p*-phenylene) (DHPPP).

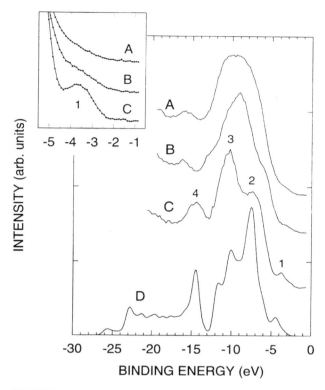

FIGURE 18 UPS spectra of DHPPV, recorded with 40-eV photons, compared with DOVS derived from VEH calculations. Spectra A to C correspond to different experimental configurations as defined in the text.

in between A and C, which would be expected if the polymer chains are almost completely aligned. These results indicate that (i) the DHPPP chains are ordered (parallel to one another), which also have been confirmed by measurements using polarized microscopy [92], and (ii) the DHPPP chains are aligned perpendicular to the PTFE chains (in configuration C, the polarization of the incoming light is perpendicular to the PTFE chains but couples strongest to the dispersed band along the polymer backbone).

E. Oriented PTFE Films as Ordered Substrates for the Adsorption of Biological Macromolecules

Adsorption of protein molecules on solid surfaces is involved in many phenomena of practical and theoretical interest. An important example is the interaction between implants and living tissue. The adsorption of a protein to a solid surface is a complex process influenced by the properties of the protein as well as the substrate surface. In the past, a lot of effort has been dedicated to the investigation of how the chemical and morphological properties of the substrate surface affect the interaction. A large range of microscopic techniques have been applied to obtain images of adsorbed protein molecules. More recently, atomic force microscopy has made it possible to obtain images without a special sample preparation prior to the imaging.

Within this context of protein–substrate interaction, oriented PTFE films offer the possibility to study protein adsorption on ordered surface structures with dimensions similar to the protein molecules themselves. In order to do that, highly oriented fibers of PTFE have been fabricated on silicone substrates. An appropriate choice of substrate temperature, load, and drawing speed allowed one to control the morphology of the resulting PTFE films, resulting in fibers displaying widths in the range 30–600 nm and heights in the 5–20-nm range. Fibrinogen was chosen as a model molecule for the adsorption, as it displays axial dimensions (45 nm long and 6–8 nm in diameter) similar to the widths of the smallest fibers.

It was found that the fibrinogen molecules adsorb with their long axis perpendicular to the fiber direction for fibers having widths of less than 100 nm. The protein molecules appear to form close-packed bands of clusters, consisting of small integer numbers of molecules arranged parallel to each other. In contrast to that, broader fibers are covered by a two-dimensional fibrinogen network displaying a random orientation of the molecules [93].

IV. CONCLUSIONS

The usefulness of photoelectron spectroscopy of polymer condensed molecular surfaces and interfaces has been outlined. A variety of examples has been supplied, mostly, but not exclusively, from our own work. The main points developed through out this review are as follows:

*   The usefulness of surface-sensitive techniques in the study of surfaces and interfaces of conjugated polymers
*   The significant advantages of employing a combined experimental–theoretical approach to the study of conjugated polymer surfaces and interfaces, whereby more information may be obtained than by using either alone

Finally, various methods for achieving aligned polymeric systems and the improved physical properties have been discussed.

ACKNOWLEDGMENTS

Research on conjugated polymers and molecules in Linköping is supported by grants from the Swedish Natural Science Research Council (NFR), the Swedish Research Council for Engineering Sciences (TFR), the Swedish National Board for Industrial and Technical Development (NUTEK), the Neste Corporation, Finland, and Philips Research, NL (within Brite/EURAM Poly Research). The authors also like to acknowledge the important input from all of our many collaborators, for which there is not space enough here for them to be mentioned by name. We do acknowledge specifically, however, the long-time theory–experimental collaborations with S. Stafström and co-workers in Linköping, and, in particular, J. L. Brédas, R. Lazzaroni, and co-workers in Mons, Belgium. The Linköping–Mons collaboration is supported by the ESPRIT Network of Excellence, NEOME, and the ESPRIT Basic Research Action LEDFOS 8013.

REFERENCES

1. Clark, D. T., and Feast, W. J., *Polymer Surfaces*, John Wiley & Sons, Chichester, 1978.
2. André, J. M., Delhalle, J., and Brédas, J. L., *Quantum Chemistry Aided Design of Organic Polymers*, World Scientific, Singapore, 1991.
3. Salaneck, W. R., Stafström, S., and Brédas, J. L., *Conjugated Polymer Surfaces and Interfaces*, Cambridge University Press, Cambridge, 1996.
4. Brédas, J. L., Conjugated polymers and oligomers: designing novel materials using a quantum-chemical approach, *Adv. Mater.*, 7, 263 (1995).
5. Seymour, R. B., and Carraher, C. E., *Polymer Chemistry*, Marcel Dekker, Inc., New York, 1992.
6. Zallen, R., *The Physics of Amorphous Solids*, Wiley–Interscience, New York, 1983.
7. Duke, C. B., Salaneck, W. R., Paton, A., Liang, K. S., Lipari, N. O., and Zallen, R., Equivalence of the electronic structure of molecular glasses, gases and crystals, in *Structure and Excitations of Amorphous Solids* G. Lucovsky and F. L. Galeener, eds., American Institute of Physics, New York, 1976, p. 23.
8. Heeger, A. J., Kivelson, S., Schrieffer, J. R., and Su, W.-P., Solitons in conducting polymers, *Rev. Mod. Phys.*, 60, 781 (1988).
9. Salaneck, W. R., Lundström, I., and Rånby, B., *Conjugated Polymers and Related Materials: The Interconnection of Chemical and Electronic Structure*, Oxford University Press, Oxford, 1993.
10. Salaneck, W. R., and Brédas, J. L., Conjugated polymers, *Solid State Commun.*, 92, 31 (1994); special issue on "Highlights in Condensed Matter Physics and Materials Science."
11. Kittel, C., *Introduction to Solid State Physics*, Wiley, New York, 1986.
12. Hoffman, R., *Solids and Surfaces: A Chemist's View of Bonding in Extended Structures*, VCH, Weinheim, 1988.
13. Brédas, J. L., Chance, R. R., Silbey, R., Nicolas, G., and Durand, P., A nonempirical effective Hamiltonian technique for polymers: Application to polyacetylene and polydiacetylene, *J. Chem. Phys.*, 75, 255 (1981).
14. Lögdlund, M., Salaneck, W. R., Meyers, F., Brédas, J. L., Arbuckle, G. A., Friend, R., Holmes, A. B., and Froyer, G., The evolution of the electronic structure in a conjugated polymer series: Polyacetylene, poly(*p*-phenylene), and poly(*p*-phenylenevinylene), *Macromolecules*, 26, 3815 (1993).
15. Rice, M. J., Charged π-phase kinks in lightly doped polyacetylen, *Phys. Rev. Lett.*, 71, 152 (1979).
16. Su, W. P., Schieffer, J. R., and Heeger, A. J., *Phys. Rev. Lett.*, 42, 1698 (1979).
17. Fesser, K., Bishop, A. R., and Campbell, D. K., Optical absorption from polarons in a model of polyacetylene, *Phys. Rev. B*, 27, 4804 (1983).
18. Schott, M., Undoped (semiconducting) conjugated polymers, in *Organic Conductors: Fundamentals and Applications* (J. P. Farges, ed.), Marcel Dekker, Inc., New York, 1994, p. 539.
19. Brédas, J. L., Chance, R. R., and Silbey, R., Comparative theoretical study of the doping of conjugated polymers: polarons in polyacetylene and polyparaphenylene, *Phys. Rev. B*, 26, 5843 (1982).
20. Brakovskii, S. A., and Kirova, N., Excitons, polarons, and bipolarons in conducting polymers, *JETP Lett.*, 33, 4–8 (1981).

21. Baeriswyl, D., Polarons and bipolarons in conducting polymers: the role of disorder and Coulomb interaction, in *Electronic Properties of Conjugated Polymers III* (H. Kuzmany, M. Mehring, and S. Roth, eds.), Springer-Verlag, Berlin, 1989, p. 54.

22. Skotheim, T. A., *Handbook of Conducting Polymers*, Marcel Dekker, Inc., New York, 1986.

23. Brédas, J. L., and Silbey, R., *Conjugated Polymers: The Novel Science and Technology of Highly Conducting and Nonlinear Optically Active Materials*, Kluwer, Dordrecht, 1991.

24. Spangler, C. W., Nickel, E. G., and Hall, T. J., α,ω-Dipenylpolyenes as model compounds for polymers related to poly(p-phenylenevinylene), *Polym. Prepr.*, *28*, 219 (1987).

25. Brédas, J. L., and Heeger, A. J., The role of solitons in the first B_u excited state of polyene chains: from short polyenes to polyacetylene, *Chem. Phys. Lett.*, *154*, 56 (1989).

26. Soos, Z. G., and Ramasesha, S., Valence bond approach to exact nonlinear, optical properties of conjugated systems, *J. Chem. Phys.*, *90*, 1067 (1989).

27. Stafström, S., Brédas, J. L., Lögdlund, M., and Salaneck, W. R., Charge storage states in polyenes, *J. Chem. Phys.*, *99*, 7938 (1993).

28. Duke, C. B., Paton, A., Salaneck, W. R., Thomas, H. R., Plummer, E. W., Heeger, A. J., and MacDiarmid, A. G., Electronic structure of polyenes and polyacetylene, *Chem. Phys. Lett.*, *59*, 146 (1978).

29. Fichou, D., Horowitz, G., and Garnier, F., Polaron and bipolaron formation on isolated model thiophene oligomers in solution, *Synth. Met.*, *39*, 125 (1990).

30. Ramsey, M. G., Steinmüller, D., and Netzer, F. P., Explicit evidence for bipolaron formation: Cs-doped biphenyl, *Phys. Rev. B*, *42*, 5902 (1990).

31. Schenk, R., Gregorius, H., and Müllen, K., Absorption spectra of charged oligo(phenylenevinylene)s: On the detection of polaronic and bipolaronic states, *Adv. Mater.*, *3*, 492 (1991).

32. Ehrenfreund, E., Hagler, T. W., Moses, D., Wudl, F., and Heeger, A. J., Gap states of iodine-doped β-carotene, *Synth. Met.*, *49–50*, 77 (1992).

33. Ehrenfreund, E., Moses, D., Heeger, A. J., Cornil, J., and Brédas, J. L., Doped β-carotene films: Spinless charge storage stabilized by structural relaxation, *Chem. Phys. Lett.*, *192*, 84 (1992).

34. Chance, R. R., Schaffer, H., Knoll, K., Schrock, R., and Silbey, R., Linear optical properties of a series of linear polyenes: Implications for polyacetylene, *Synth. Met.*, *49–50*, 271 (1992).

35. Tolbert, L. M., and Schomaker, J. A., Alkali-metal doping of polyacetylene model compounds: α,ω-diphenylpolyenes. *Synth. Met.*, *41–43*, 169 (1991).

36. Dannetun, P., Boman, M., Stafström, S., Salaneck, W. R., Lazzaroni, R., Fredriksson, C., Brédas, J. L., Zamboni, R., and Taliani, C., The chemical and electronic structure of the interface between aluminum and polythiophene semiconductors, *J. Chem. Phys.*, *99*, 664 (1993).

37. Dannetun, P., Lögdlund, M., Salaneck, W. R., Fredriksson, C., Stafström, S., Holmes, A. B., Brown, A., Graham, S., Friend, R. H., and Lhost, O., New results on metal–polymer interfaces, *Mol. Cryst. Liq. Cryst.*, *228*, 43 (1993).

38. Fredriksson, C., and Brédas, J. L., Metal/conjugated polymer interfaces: A theoretical investigation of the interaction between aluminum and *trans*-polyacetylene oligomers, *J. Chem. Phys.*, *98*, 4253 (1993).

39. Ramsey, M. G., Steinmüller, D., Schatzmayr, M., Kiskinova, M., and Netzer, F. P., n-Phenyls on transition and alkali metal surfaces: from benzene to hexaphenyl, *Chem. Phys.*, *177*, 349 (1993).

40. Steinmüller, D., Ramsey, M. G., and Netzer, F. P., Polaron and bipolaronlike states in *n*-doped bithiophene, *Phys. Rev. B*, *47*, 13323 (1993).

41. Tanaka, C., and Tanaka, J., Molecular and electronic structure of model compounds of doped polyacetylene, *Bull. Chem. Soc. Jpn.*, *66*, 357 (1993).

42. Dannetun, P., Lögdlund, M., Lazzaroni, R., Fauquet, C., Fredriksson, C., Stafström, S., Spangler, C. W., Brédas, J. L., and Salaneck, W. R., Reactions of low workfunction metals, Na, Al, and Ca, on α,ω-diphenyltetradecaheptaene: Implications for metal/polymer interfaces, *J. Chem. Phys.*, *100*, 6765 (1994).

43. Dannetun, P., Lögdlund, M., Spangler, C. W., Brédas, J. L., and Salaneck, W. R., Evolution of charge-induced gap states in short diphenylpolyenes as studied by photoelectron spectroscopy, *J. Phys. Chem.*, *98*, 2853 (1994).

44. Khaled, M., Hadjipetrou, A., and Kispert, L., Electrochemical and electron paramagnetic resonance studies of carotenoid cation radicals and dications: effect of deuteration, *J. Phys. Chem.*, *94*, 5164 (1990).

45. Valladeres, R. M., Hayes, W., Fisher, A. J., and Stoneham, A. M., Defect electronic states in β-carotene and lower homologues, *J. Phys. Condens. Matter*, *5*, 7049 (1993).

46. Hudson, B. S., Ridyard, J. N. A., and Diamond, J., Polyene spectroscopy. Photoelectron spectroscopy of the diphenylpolyenes, *J. Am. Chem. Soc.*, *98*, 1126 (1976).

47. Yip, K. L., Lipari, N. O., Duke, C. B., Hudson, B. S., and Diamond, J., The electronic structure of bond-alternating and nonalternant conjugated hydrocarbons: Diphenylpolyenes and azulene, *J. Chem. Phys.*, *64*, 4020 (1976).

48. Kohler, B. E., and J. A. Pescatore, J., Transition dipoles of polyacetylene oligomers, in *Conjugated Polymeric Materials: Opportunities in Electronics, Optoelectronics and Molecular Electronics* (J. L. Brédas and R. R. Chance, eds.), Kluwer, Dordrecht, 1990, p. 353.

49. Lögdlund, M., Dannetun, P., Sjögren, B., Boman, M., Fredriksson, C., Stafström, S., and Salaneck, W. R., The electronic structure of α,ω-diphenyltetradecaheptaene, a model molecule for polyacetylene, as studied by photoelectron spectroscopy, *Synth. Met.*, *51*, 187 (1992).

50. Lögdlund, M., Dannetun, P., Stafström, S., Salaneck, W. R., Ramsey, M. G., Spangler, C. W., Fredriksson, C., and Brédas, J. L., Soliton pair charge storage in doped polyene molecules: Evidence from photoelectron spectroscopy studies, *Phys. Rev. Lett.*, *70*, 970 (1993).

51. Dannetun, P., Lögdlund, M., Fredriksson, C., Boman, M., Stafström, S., Salaneck, W. R., Kohler, B. E., and Spangler, C., Chemical and electronic structure of the early stages of interface formation between aluminum and α,ω-diphenyltetradecaheptaene, in *Polymer–Solid Interfaces* (J. J. Pireaux, P. Bertrand, and J. L. Brédas, eds.), IOP Publishing, Bristol, 1992, p. 201.

52. Iucci, G., Lögdlund, M., Spangler, C. W., and Salaneck, W. R., The interaction of rubidium atoms with a model molecule for polyacetylene: A photoelectron spectroscopy investigation, *Synth. Met.*, *76*, 209 (1996).

53. Salaneck, W. R., and Brédas, J. L., The metal-on-polymer interface in polymer light emitting diodes, *Adv. Mater.*, *8*, 48 (1996).

54. Lögdlund, M., Dannetun, P., and Salaneck, W. R., Electronic and chemical structure of conjugated polymers and interfaces as studied by photoelectron spectroscopy, in *Handbook of Conducting Polymers II* (T. Skotheim, R. L. Elsenbaumer, and J. R. Reynolds, eds.), Marcel Dekker, Inc., New York, 1997.

55. Scwarc, M., *Ions and Ion Pairs in Organic Reactions*, Wiley–Interscience, New York, Vol. 2, 1972–1974.

56. Stafström, S., Brédas, J. L., Lögdlund, M., and Salaneck, W. R., Charge storage states in polyenes, *J. Chem. Phys.*, *99*, 7938 (1993).

57. Stewart, J. J. P., *J. Comp. Chem.*, Optimization of parameters for semiempirical methods. II. Applications, *10*, 221 (1989).

58. Reed, A. E., Weistock, R. B., and Wienhold, F., Natural population analysis, *J. Chem. Phys.*, *83*, 735 (1985).

59. Dannetun, P., Fahlman, M., Fauquet, C., Kaerijama, K., Sonoda, Y., Lazzaroni, R., Brédas, J. L., and Salaneck, W. R., Interface formation between poly(2,5-diheptyl-*p*-phenylenevinylene) and calcium: Implications for light emitting diodes, *Synth. Met.*, *67*, 113 (1994).

60. Mao, G., Fischer, J. E., Karasz, F. E., and Winokur, M. J., Nonplanarity and ring torsion in poly(*p*-phenylenevinylene). A neutron-diffraction study, *J. Chem. Phys.*, *98*, 712 (1993).

61. Tanaka, J., Tanaka, C., Miyamae, T., Kamiya, K., Shimizu, M., Oku, M., Seki, K., Tsukamoto, J., Hasegawa, S., and Inokuchi, H., Spectral characteristic of metallic state of polyacetylene, *Synth. Met.*, *55–57*, 121 (1993).

62. Fahlman, M., Beljonne, D., Lögdlund, M., Friend, R. H., Holmes, A. B., Brédas, J. L., and Salaneck, W. R., Experimental and theoretical studies of the electronic structure of Na-doped poly(*p*-phenylenevinylene), *Chem. Phys. Lett.*, *214*, 327 (1993).

63. Chance, R. R., Boudreax, D. S., Brédas, J. L., and Silbey, R., in *Handbook of Conducting Polymers* (T. A. Skotheim, ed.), Marcel Dekker, Inc., New York, 1986.

64. Schimmel, T., Gläser, D., Schwoerer, M., and Naarmann, H., Properties of highly conducting polyacetylene, in *Conjugated Polymers* (J. L. Brédas and R. Silbey, eds.), Kluwer Academic Publishers, Dordrecht, 1991, p. 49.

65. Yan, M., Rothberg, L. J., Papadimitrikopoulos, F., Galvin, M. E., and Miller, T. M., Defect quenching of conjugated polymer luminescence, *Phys. Rev. Lett.*, *73*, 744 (1994).

66. Dyreklev, P., Gustafsson, G., Inganäs, O., and Stubb, H., Aligned polymer chain field effect transistors, *Solid State Commun.*, *82*, 317 (1992).

67. MacDiarmid, A. G., and Epstein, A. J., The concept of secondary doping as applied to polyaniline, *Synth. Met.*, *65*, 103 (1994).

68. Wang, Z. H., Li, C., Scherr, E. M., MacDiarmid, A. G., and Epstein, A. J., Three dimensionality of 'metallic' states in conducting polymers: polyaniline, *Phys. Rev. Lett.*, *66*, 1745 (1991).

69. Adams, P. N., Laughlin, P. J., Monkman, A. P., and Bernhoeft, N., A further step towards stable organic metals. Oriented films of polyaniline with high electrical conductivity and anisotropy, *Solid State Commun.*, *91*, 875 (1994).

70. McCall, R. P., Scherr, E. M., MacDiarmid, A. G., and Epstein, A. J., Anisotropic optical properties of an oriented-emeraldine-base polymer and an emeraldine–hydrochloride–salt polymer, *Phys. Rev. B*, *50*, 5094 (1994).

71. Hagler, T. W., Pakbaz, K., Voss, K. F., and Heeger, A. J., Enhanced order and electron delocalization in conjugated polymers oriented by gel processing in polyethylene, *Phys. Rev. B*, *44*, 8652 (1991).

72. Wittmann, J. C., and Smith, P., Highly oriented thin films of poly(tetrafluoroethylene) as a substrate for oriented growth of materials, *Nature*, *352*, 414 (1991).

73. Motamedi, F., Kyo Jin, I., Fenwick, D., Wittmann, J. C., and Smith, P., Polymer friction-transfer layers as orienting substrates, *J. Polym. Sci., Part B*, *32*, 453 (1994).

74. Biswas, S. K., and Vijayan, K., Friction and wear of PTFE—a review, *Wear*, *158*, 1 (1992).

75. Schott, M., Preparation and properties of highly oriented polytetrafluoroethylene films, *Synth. Met.*, *67*, 1 (1994).

76. Okuyama, H., Kanamoto, T., and Porter, R. S., Solid-state deformation of polytetrafluoroethylene powder. Part I. Extrusion drawing, *J. Mater. Sci.*, *29*, 6485 (1994).

77. Dietz, P., Hansma, P. K., Ihn, K. J., Motamedi, F., and Smith, P., Molecular structure and thickness of highly oriented poly(tetrafluoroethylene) films measured by atomic force microscopy, *J. Mater. Sci.*, *28*, 1372 (1993).

78. Beamson, G., Clark, D. T., Deegan, D. E., Hayes, N. W., Law, D. L., Rasmusson, J. R., and Salaneck, W. R., Characterization of PTFE on silicon wafer tribological transfer films by XPS, imaging XPS and AFM, *Surf. Interf. Anal.*, *24*, 204 (1996).

79. Bodö, P., Ziegler, C., Rasmusson, J. R., Salaneck, W. R., and Clark, D. T., Scanning tunneling microscopy study of oriented poly(tetrafluoroethylene) substrates, *Synth. Met.*, *55*, 329 (1993).

80. Ohta, T., Seki, K., Yokoyama, T., Morisada, I., and Edamatsu, K., Polarized XANES studies of oriented polyethylene and fluorinated polyethylenes, *Phys. Scr.*, *41*, 150 (1990).

81. Nagayama, K., Sei, M., Mitsumoto, R., Ito, E., Araki, T., Ishii, H., Ouchi, Y., Seki, K., and Kondo, K., Polarized NEXAFS studies on the mechanical rubbing effect of poly(tetrafluoroethylene) oligomer and its model compound, *J. Electron Spectrosc. Related Phenom.*, *78*, 375 (1996).

82. Ågren, H., Carravetta, V., Vahtras, O., and Pettersson, L. G. M., Orientational probing of polymeric thin films by NAXAFS: Calculations on polytetrafluoroethylene, *Phys. Rev. B*, *51*, 17848 (1995).

83. Nagayama, K., Miyamae, T., Mitsumoto, R., Ishii, H., Ouchi, Y., and Seki, K., Polarized VUV absorption and reflection spectra or oriented films of poly(tetrafluoroethylene) $(CF_2)_n$ and its model compound, *J. Electron Spectrosc. Related Phenom.*, *78*, 407 (1996).

84. Motamedi, F., Kyo Jin, I., Fenwick, D., Wittmann, J. C., and Smith, P., Orientation of materials onto thin highly ordered PTFE films, *Proc. SPIE*, *1665*, 194 (1992).

85. Aoki, Y., Hattori, R., and Shirafuji, J., Oriented polysilane films evaporated on poly(tetrafluoroethylene) layer, *Technol. Rept. Osaka Univ.*, *45*, 2217 (1995).

86. Hattori, R., Shirafuji, J., Aoki, Y., Fujiki, T., Kawasaki, S., and Nishida, R., Preparation and optical properties of doubly-oriented poly-(di-methyl-silane) films, *J. Non-Crystal. Solids*, *198*, 649 (1996).

87. Paley, M. S., and Frazier, D. O., Diacetylene and polydiacetylene derivatives of 2-methyl-4-nitroaniline for second-harmonic generation, *Chem. Mater.*, *5*, 1641 (1993).

88. Ueda, Y., Kuriyama, T., Hari, T., Watanabe, M., Jingping, N., Hattori, Y., Uenishi, N., and Uemiya, T., Structure and nonlinear optical properties of diacetylene thin films vapor-deposited on highly oriented polytetrafluoroethylene thin films, *Jpn. J. Appl. Phys.*, *1*, 3876 (1995).

89. Yang, C. Y., Yang, Y., and Hotta, S., Crystal structure and polymorphism of dimethyl-oligothiophenes crystallized epitaxially on highly oriented PTFE thin films, *Synth. Met.*, *69*, 1 (1995).

90. Pichler, K., Friend, R. H., Burn, P. L., and Holmes, A. B., Chain alignment in poly(*p*-phenylene vinylene) on oriented substrates, *Synth. Met.*, *55*, 454 (1993).

91. Fahlman, M., Rasmusson, J., Kaeriyama, K., Clark, D. T., Beamson, G., and Salaneck, W. R., Epitaxy of poly(2,5-diheptyl-*p*-phenylene) on ordered polytetrafluoroethylene, *Synth. Met.*, *66*, 123 (1994).

92. Witteler, H., Substituierte poly(*p*-phenylene): Synthese, struktur und phasenverhalten, Ph.D. thesis, University of Mainz, 1993.

93. Rasmusson, J. R., Erlandsson, R., Salaneck, W. R., Schott, M., Clark, D. T., and Lundström, I., Adsorption of fibrinogen on thin oriented poly(tetrafluoroethylene) (PTFE) fibers studied by scanning force microscopy, *Scanning Microsc.*, *8*, 481 (1994).

7

Application of Polymers to Electroluminescence

Guangming Wang, Chunwei Yuan, Zuhong Lu, and Yu Wei
Southeast University
Nanjing, Jiangsu, People's Republic of China

I. INTRODUCTION

Although development of light-emitting devices and lasers using inorganic compound semiconductors has progressed substantially, problems in these devices, especially electroluminescent (EL) diodes emitting blue light, still exist. On the other hand, electroluminescence in organic semiconductors has been known for some time [1–3]. Since Tang et al. reported their findings on the multilayered organic EL diode [4], the EL devices based on organic thin layers have attracted much attention [5–24] because of their possible application as the most promising next-generation flat-panel display systems and their fascinating potential to produce emissions of all colors ranging from blue to red, particularly in the blue band, in EL displays with the aid of the molecular design of organic materials possessing high fluorescent efficiency and semiconducting properties, which are difficult when using inorganic light-emitting diodes only, despite substantial progress achieved in inorganic EL diodes development.

A. Organic Electroluminescence

Organic light emission is generated by the recombination of holes and electrons injected from electrodes of the organic devices in the emitting layer. In organic EL devices, high luminance and high efficiency, and low-voltage drive were realized in organic thin-film EL devices with multilayer structures consisting of the light-emitting layers (LELs) and the carrier transport layers (CTLs) formed by utilizing organic heterostructures [4–7,9,19,20]. The adoption of organic heterostructures in the EL devices is very effective in the improvement of carrier injection from electrodes and carrier confinement in the emission region. With the recent widespread interest in making light-emitting devices from organic materials possessing photoluminescence characteristics, there has been significant effort to understand the organic semiconducting materials and the electronic properties of the layered diode structures [25], as well as the influence of the different metal electrodes on the organic EL efficiency [26,27].

In general, electrodes have a great influence on EL intensity and efficiency for organic EL devices because the dependence of EL intensity on the injection current for the organic EL devices exists. Hsieh and his co-workers [26] have reported the influence of different metallic electrodes on organic EL intensity and efficiency by using Mg/DP-PPV/indium/tin oxide (ITO) and Al/DP-PPV/ITO devices, where DP-PPV is poly(2,3-diphenyl-1,4-phenylene vinylene). The effect of metal work functions (W_{Al} = 4.28 eV and W_{Mg} = 3.66 eV [28]) on the EL efficiency is indicated at 50 times the increase in EL intensity for electrodes of Mg compared with Al. The external quantum efficiency, defined as the number of photons emitted in the forward direction per injected electron, was found to be 0.1% photons/electron and 0.002% photon/electron for the Mg/DP-PPV/ITO and Al/DP-PPV/ITO devices, respectively.

For the metallic electrodes of calcium or indium used in the organic EL devices, experiments were also conducted with poly(2-methoxy, 5-(2'-ethyl-hexoxy)-1,4-phenylene vinylene) (MEH-PPV) used as the emitting layer, and calcium or indium as the cathode [27]. The EL intensity emitted by the polymer EL devices with calcium electrodes as the rectifying contact exceeds by almost an order of magnitudes that emitted by the EL devices with indium electrodes, and the quantum efficiency at 1 mA is about 5×10^{-4} and $\sim 10^{-2}$ photons per electron for indium and calcium electrodes, respectively.

Recently, the development of organic EL devices has focused on the optimization of EL device structures for obtaining high luminance [4–6,9,19,29] and changing the emission characteristics [30–32], and the exploitation of organic materials for obtaining a wide variety of emission colors and enhancing EL efficiency [33–36]. Emitting materials for organic EL devices can be classified into two classes according to their molecular structure: (1) organic molecules and (2) polymers.

B. Application of Polymers to Electroluminescence

Light-emitting devices (LEDs) are fabricated by vacuum sublimation of the organic compounds, and although the efficiencies and selection of color of the emission are a success, there arise, in general, problems associated with the long-term stability of the sublimed organic layer against recrystallization and other structural changes, specially for multilayer-structure organic EL devices, where the weak van der Walls force acts between the layers. Such an interface may not be physically strong enough for practical EL devices. Such problems could be improved by using the polymeric materials in organic EL devices [14]. Conducting polymers with highly extended π-electron systems in their main chains has been of much interest from both fundamental and practical viewpoints [37]. Various new concepts such as polarons and bipolarons have been introduced to explain their unique characteristics [38], and many new functional applications such as the EL devices [10–12,16–18,36], field-effect transistors [39], and photovoltaic cells [40] by utilizing these types of polymer, and their novel characteristics have been proposed.

Polymers were often thought of as being mechanically weak and so their mechanical properties have been somewhat ignored in the past. Because polymers have potential advantages of low cost, ease of fabrication, and relative indifference to low levels of chemical impurities, it seems advisable to investigate if they could be used as the basis for an EL device of reasonable efficiency. The results of these investigations, some of which have been outlined briefly [41], are described in this chapter. This increase in use has been due to several factors: First, the fabrication costs of polymeric components are usually considerably lower than for other materials. Second, polymers melt at relatively low temperature and can be readily molded into quite intricate components using a single molding. Third, one polymer that has been used for years is elastomeric, which can be stretched to very high extensions and will snap back immediately. Because of their remarkable toughness and flexibility and ease of fabrication, they can be used as emitting materials to fabricate flexible EL devices [17]. Finally, polymers are employed as adhesives to simplify techniques for fabricating organic EL layers [14,42,43].

With increasingly widespread use of polymers, the understanding of their EL properties is becoming essential, because the polymeric active layers used in the EL devices might have unique advantages over those that use evaporated films. These advantages in fabricating EL devices include high thermal stability, simplicity of manufacture, ability to prepare a large area, and good mechanical properties.

Since the earliest reports on electroluminescence from conjugated semiconducting polymers, a new field of the research has opened up. In view of the practical applications, the polymer electroluminescence, compared with the organic molecular electroluminescence, is a relatively new discipline which was developed rapidly, following recognition only several years ago. The study of the EL properties of polymers is a subject of recent interest. They are now accepted as EL materials in their own right along with other more conventional EL materials in EL characteristics such as the EL intensity and efficiency, as well as the EL color. Although much progress has been made in understanding the basic principle of operation for these devices, many questions remain unanswered, especially regarding the transport properties and their relation to the amorphous, disordered nature of polymers. One of the fundamental questions to be addressed is whether these materials can be modeled in a standard semiconductor bandlike picture to explain the mechanism of organic electroluminescence.

Light-emitting diodes fabricated from conjugated polymers [10–12] have attracted attention as a result of their potential for use in display technology. In addition to relatively simple device fabrication and the ability to make such devices on flexible substrates [17], the color of the emission can be varied by using different conjugated structures, copolymers, and polymer blends. Red, yellow, green, and blue have been demonstrated with impressive efficiency, brightness, and uniformity.

The color of the EL emission can be changed by using different main-chain molecular structures or through modification of the side-chain structures, as in the poly(p-phenylene vinylene) (PPV) derivatives [26,27]. The EL color can also be tuned by utilizing polymer blends [44]. It was reported that the blue LEDs were made of blends of poly(p-phenylphenylene vinylene) and poly(N-vinylcarbazole), hole-transporting materials [45]. By blending luminescent polymer into the host polymer matrix, the emission peak can be tuned toward a shorter wavelength. When both components of the blend emit light, the polymer blend can be used to fabricate polymer LEDs with desired colors.

Polymers used in the electroluminescence have been greatly developed over the past decades. With focus on the topic of the application of polymers to electroluminescence, this chapter is concerned principally with the relationship between EL materials and their properties.

II. LIGHT EMISSION IN POLYMERS

One of the important advantages for polymer LEDs over other EL devices is their ease of fabrication because the polymer active layers (or their precursor polymers) are soluble and can be deposited on a substrate from solution. In addition, the polymer materials have better sticking properties to substrates and have greater mechanical strength than organic molecules. In this section, several polymers such as poly(*p*-phenylene vinylene) (PPV), poly(*N*-vinylcarbazole) (PVK), and poly(3-alkylthiphene) (P3AT) are introduced as the EL materials.

A. Electroluminescence in Poly(3-alkylthiophene)

Conducting polymers have been a major concern from both fundamental and practical viewpoints. Among various polymers, P3AT has been found soluble in some solvents, and the optical properties of solution for the P3AT change drastically with temperature and concentration [46]. P3AT has been of much interest due to its solubility and novel characteristics, such as thermochromism [47], solvatochromism [48], and the voltage-controlled EL colors from the polymer blends of P3AT [44]. The anomalous dependence of the photoluminescence of P3AT on temperature and alkyl chain length was given in detail in Ref. 49. The photoluminescence of P3AT is first introduced in the following because the photoluminescence bears a close relation to the electroluminescence for organic materials.

1. Characteristics of Photoluminescence in P3AT

Both photoluminescence and electroluminescence originate from the radiative recombination of the exciton state; the former case formed by photoexcitation, and the latter by the combination of oppositely charged polarons (radical ions) generated by injection of electrons and holes from electrodes of the organic EL devices. It is necessary to introduce photoluminescence characteristics of P3AT for studying its electroluminescence.

In the solution of P3ATs with an alkyl chain length longer than 4, the optical transmission intensity and absorption spectrum have been found to change drastically at some critical temperature which depends on the solvent, concentration, and alkyl chain length [50]. Such transition was also observed in polymer solutions [51].

Poly(3-alkylthiophene) can be prepared from a 3-alkylthiophene monomer by utilizing $FeCl_3$ as a catalyst. The molecular structure of P3AT is shown in Fig. 1. The peak positions of absorption spectra for their films are at about 490 nm at room temperature. Figure 2 indicates absorption and photoluminescence spectra of poly(3-dodecylthophene) at various temperatures. It should be noted in this figure that with increasing temperature, the emission peak shifts more or less to a shorter wavelength.

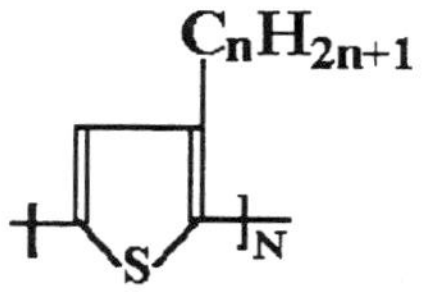

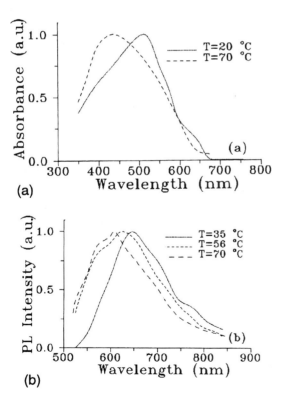

FIGURE 1 Molecular structure of poly(3-alkylthiophene).

The temperature dependence of luminescence intensity is more clearly depicted in Fig. 3; that is, with increasing temperature, the emission intensity increases in the solid phase. These temperature dependencies of photoluminescence are quite anomalous compared with those of ordinary inorganic and organic semiconductors and insulators in which luminescence intensity decreases remarkably with rising temperature due to the increasing probability of nonradiative recombination. This anomalous behavior in P3ATs can be explained qualitatively in terms of conformation change of the polymer main chain as a function of temperature and their influence on the dynamics of photoexcited species. This is done as follows.

Electrons and holes are excited in conduction and valence bands by the excitation with the Ar-ion laser of photon energy 2.54 eV, and some of the electrons

FIGURE 2 Change of normalized absorption (a) and photoluminescence (b) spectra of poly(3-alkylthiophene) film with temperature. (From Ref. 47.)

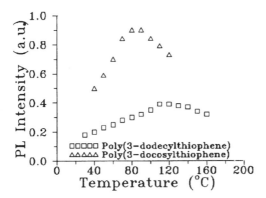

FIGURE 3 Temperature dependence of photoluminescence intensity for poly(3-alkyl-thiophene). (From Ref. 49.)

and holes should turn into negative and positive polarons, respectively, in a very short time [52]. The luminescence in these conducting polymers is considered to be due to the radiative recombination of these excited species. Therefore, when positive and negative excited species moved out from the excitation region and separated, thus escaping from the recombination, the luminescence intensity should be suppressed.

At a higher temperature due to the decreased effective conjugation length, the mobility of the excited species and their escape probability from the excitation region should decrease, which can result in the enhancement of the probability of the recombination and the observed emission of stronger photoluminescence. When the effect of enhancement of recombination probability is larger than the increase of nonradiative recombination at a higher temperature, enhancement of photoluminescence should be observed as in Fig. 3.

The activation energy of luminescence intensity in the solid phase was estimated to be around 0.1–0.2 eV, and this corresponds to the energy necessary for the introduction of the torsion angle substracted with the activation energy of nonradiative recombination [49]. The torsion angle does not change substantially with temperature in the liquid phase compared to the solid state, although it is larger than that in the solid phase. Therefore, the effective conjugation length should not change considerably with temperature in the liquid state, and the probability of confinement of excited species in the excitation region should not change drastically with temperature. In this case, the increased nonradiative recombination at higher temperature will lead to the suppression of luminescence intensity in the liquid phase. In addition, the P3AT with longer alkyl chains indicates stronger luminescence. Figure 4 shows the dependence of peak photoluminescence intensity on the alkyl chain length. This can also be tentatively interpreted in terms of the dynamics of photoexcited species as follows.

The interaction of neighboring polymer main chains should be reduced and the probability of the transfer of photoexcited species from a polymer chain to the neighboring one should be suppressed because of the bulky side-chain effect of the

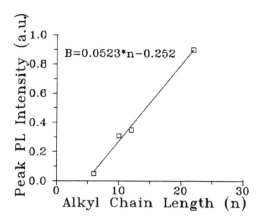

FIGURE 4 Dependence of peak photoluminescence intensity on the alkyl chain length in poly(3-alkylthiophene) films. (From Ref. 49.)

P3ATs with longer alkyl chains. In such a case, the density of positive and negative excited species in a chain will increase, resulting in the enhancement of recombination luminescence; that is, the interchain transfer of excited species should reduce the quantum efficiency of photoluminescence. This also supports the idea that the photoluminescence comes from the recombination of excited species on the same polymer chain.

2. Relationship Between EL Intensity and Alkyl Chain Length in P3AT

Because there is the anomalous dependence of photoluminescence on the alkyl chain length in the P3AT, the alkyl chain length should remarkably affect the EL intensity of P3AT. The light-emitting diodes consist of an ITO-coated glass substrate, an emitting layer of P3AT, and a metal electrode. A thin layer of P3AT could be fabricated by spin-coating onto the surface of an ITO-coated glass substrate using chloroform as the solvent, the thickness of the emitting layer being about 100 nm. Because the quality of the ITO–polymer contact is very important to homogeneous and stable devices, special attention should be paid to the cleaning procedure for the ITO surface prior to spin-coating, based on ultrasonification in polar and nonpolar solvents. Then the metal electrode such as aluminum or magnesium containing indium or Ag is vacuum deposited on the surface of the spin-coating polymer film ($\sim 10^{-6}$ Torr) and at evaporation rates below 1.0 nm/s to avoid unnecessary heating of the polymer's top layer. All the measurements are carried out at room temperature under DC bias conditions.

The emission intensity dependence on injected currents characteristics was obtained when the ITO electrode is positively biased and the metal electrode negatively biased (see Fig. 5). As is evident in Fig. 5, the emission intensity increases superlinearly with increasing injected current, but the EL intensity of the P3AT possessing a longer side chain is stronger than that of the P3AT with a shorter alkyl side chain at the same current density [11]. In Fig. 6, comparison of the EL

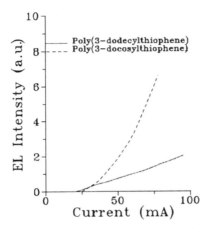

FIGURE 5 Dependence of electroluminescent intensity on injection current of the poly(3-alkylthiophene)-film EL diodes. (From Ref. 11.)

intensity of the P3AT diodes at the same injected current (70 mA) is illustrated. The optimal-fit linear equation can be obtained for the dependence of EL intensity (B) on the alkyl chain length (n):

$$B = 0.0733n - 0.613 \tag{1}$$

suggesting that there is no output light if $n > 8$. The longer the alkyl chain length, the stronger the emission intensity. This fact corresponds to the photoluminescence data of P3AT with different alkyl chain lengths, because the higher photoluminescence intensity is obtained with an increasing alkyl chain length. The emission intensity is enhanced by a confinement of carriers on a main chain with a long interchain distance caused by a long alkyl side chain.

The diodes of P3AT emit red–orange light at room temperature. The emission and photoluminescence spectra are shown in Fig. 7, and the EL spectrum is similar

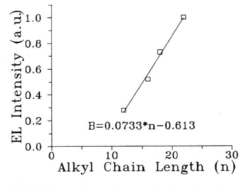

FIGURE 6 Dependence of electroluminescent intensity on alkyl chain length of poly(3-alkylthiophene) light-emitting diodes. (From Ref. 11.)

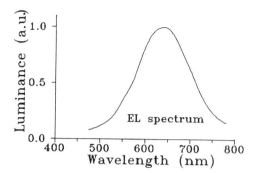

FIGURE 7 Electroluminescent spectrum of the poly(3-alkylthiophene)-film EL diodes. (From Ref. 11.)

to the photoluminescence spectrum, indicating that the EL emitting light originates from the P3AT. Further, the EL spectrum shows the peak intensity at 640 nm corresponding to the photon energy of 1.9 eV, which coincides with the band gap of the P3AT estimated from the absorption spectrum. This suggests that the EL emission occurs from the recombination of the electron–hole pairs in the bands.

3. Relationship Between EL Intensity and Schottky Contact in the EL Diode

In general, ITO is used as positive electrode in the organic EL diode and metals possessing a low work function as a cathode. However, the higher EL efficiency in the reverse-bias operation from P3AT-based light-emitting diodes could be reached [53]. For the contact between polymer film and electrode, the model seems to work quite well to describe most electrical properties and light emission of light-emitting diodes in spite of the basically "simple" approaches underlying the presently popular band model (clean band gap without traps in the semiconductor, no Fermi-level pinning at the interfaces with metal contacts) [4–9,15,54].

In forward-bias conditions, injection of electrons into the conduction band (CB) from a low-work-function electrode and injection of holes into the valence band (VB) from a high-work-function electrode would allow the recombination of excitons formed from opposite charge carriers, which cause the electroluminescence. However, under extreme reverse-bias conditions, electrons would then tunnel from the high-work-function electrode into the CB, and holes from the low-work-function electrode into the VB. Whereas in the forward mode of operations, various mechanisms of charge injection (thermionic emission, tunneling) can play a substantial role, tunneling of charge carriers should be the dominant injection mechanism in the reverse mode of operation.

Figure 8 shows the change of the EL intensity and current crossing an P3AT diode versus the voltages, as well as the relationship between the intensity and the current [here, poly(3-octylthiophene) (P3OT)] as emitting material, with the thickness of polymer thin film about 100 nm and the total area of the device or light emitting equal to 8 mm², indicating that the light output for the P3OT diode under the reverse-bias condition is higher at the same current density than under forward-bias operation. The device acts as a diode, showing large forward currents above

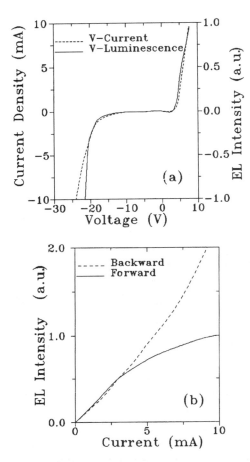

FIGURE 8 Change (a) of electroluminescent intensity and current of an ITO/P3OT/Al EL device with forward and reverse biases, as well as the relationship (b) between the electroluminescent intensity and the current through the EL device. (From Ref. 53.)

3 V (forward is defined as the positive voltage on ITO) and very small currents at the corresponding reverse voltages. Rectification ratios as high as 10^4 have been obtained. In forward and reverse biases, light emission is observed as soon as the current is higher than 6.25×10^{-4} A/cm^2, and this condition is met at entirely different electric fields in both modes of operations [53]. This suggests that the onset of light emission depends on the amount of current flowing through the device and not on the value of the electric field. The data taken under forward bias suggest (thermionic) emission over a Schottky-like barrier of 0.69 eV to be the dominant injection mechanism for electrons at the polymer–aluminum contact.

For voltages corresponding to current densities below 100 mA/cm^2, the luminescence is proportional to the current through the device, whereas at higher voltages, reproducible deviations (superlinear in reverse mode and sublinear in forward mode) from the linear behavior are observed. From Fig. 8a, the recombi-

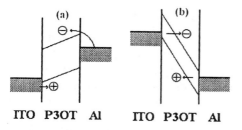

FIGURE 9 Energy diagram of an ITO/P3OT/Al EL diode forward bias (a) [(+)ITO/P3OT/Al(−)] and reverse bias (b) [(−)ITO/P3OT/Al(+)]. (From Ref. 53.)

nation efficiencies for charge carriers with species of the opposite charges depend strongly on the density of available charge carriers, which is determined largely by the heights of the two injection barriers. This result is consistent with the assumption that in the reverse mode of operation, two barriers dominate the injection mechanism, whereas in the forward mode of operation, only the electron-injecting contact limits the total light output. The band picture of an ITO/P3OT/Al device is shown in Fig. 9, based on well-known values for the metal work functions of ITO and Al (4.7 and 4.3 eV, respectively) and the P3OT energy gap of 2.1 eV estimated by the absorption spectrum [53]. In reverse-bias operation, the only possible mechanism for carrier injection into the valence and conduction bands of the polymer is by direct tunneling through the triangular barrier formed at the polymer–metal interface. For a current to flow in the reverse mode, the ITO has to be the electron injector, and Al takes over the role of hole injector.

The EL spectra of the devices in both modes of operation are quite similar (see Fig. 10), implying that the origin of electric-field-driven luminescence is the same. This is in agreement with the concept that two types of charge carrier are needed to form an excited state that can decay radiatively, independent of the method of injecting the charges.

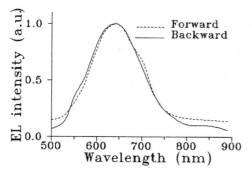

FIGURE 10 Electroluminescent spectra of the poly(3-alkylthiophene)-film EL diodes under forward and reverse biases taken at +8 V and −20 V, respectively. (From Ref. 53.)

B. Electroluminescence in Poly(*p*-phenylene vinylene)

The recent discovery that conjugated polymers with extensive π-electron delocalization can be highly conducting has stimulated great interest among chemists, physicists, and material scientists, and some noteworthy application of these novel materials has been reported [55–57]. Among various conjugated polymers, poly(*p*-phenylene vinylene) (PPV) is the most typical example synthesized through the use of a precursor [58], with the molecular structure shown in Fig. 11.

Especially since Burroughes et al. [10] first demonstrated a greenish yellow light-emitting diode using a polymer, the investigation of elements and devices using PPV derivatives such as poly(2,5-dialkoxy-*p*-phenylene vinylene) and other conjugated polymers have been actively conducted [27,59,60]. The PPV precursor polymer is a positively charged polyelectrolyte. This polymer can be fabricated into thin self-assembled films using alternating layers of positively and negatively charged polymers. On the other hand, PPV can conveniently made into high-quality films and shows strong photoluminescence in a band centered near 2.2 eV, just below the threshold for π to π^* interband transitions [61–63]. PPV is an intractable material with a rigid-rod microcrystalline structure, so that it is infusible and insoluble in common solvents. This gives it excellent mechanical properties, with high elastic moduli and thermal stability. These are excellent properties for a film of polymer once formed, but processing cannot be carried out directly with this material. PPV can, however, be obtained conveniently by in situ chemical conversion of films of a suitable precursor polymer which is processed from solution by spin-coating.

1. Syntheses of Poly(p-phenylene vinylene) and Its Derivatives

Electroluminescence in conjugated polymers was first reported for PPV [10], and a variety of conjugated polymers have been shown to exhibit electroluminescence. Since this discovery, efforts have been directed to increase the range of wavelengths over which polymer electroluminescence can be achieved. Electroluminescence in PPV shows an emission peak around 540 nm appearing as green. Subsequently, polymers based on PPV, such as poly(2-methoxy-5(2'-ethyloxy)-*p*-phenylene vinylene) (MEH-PPV) [27] and cyano derivatives of PPV (CN-PPV) [64], have been reported, which show emission peaks in their EL spectra at wavelengths longer than 600 nm, giving their electroluminescence a red appearance, suggesting that the color of the EL emission can be changed through modification of the side chain in PPV derivatives. PPV is a typical electroluminescence material of polymers. Methods for synthesizing PPV and its derivatives are given below.

Poly(*p*-phenylene vinylene) is synthesized by using a solution-processible precursor polymer. The route most commonly used is the sulfonium precursor, which is conveniently processed from solution in methanol and is converted to PPV by thermal treatment at temperatures of 200°C and 300°C; the synthesization is shown

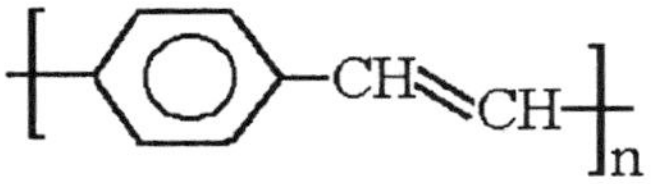

FIGURE 11   Molecular structure of poly(*p*-phenylene vinylene). (From Ref. 10.)

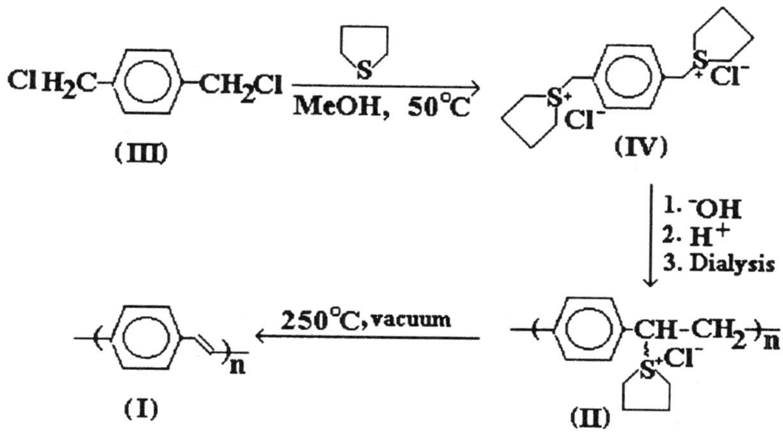

FIGURE 12 Synthetic route of poly(*p*-phenylene vinylene). (From Ref. 10.)

in Fig. 12. This precursor polymer is conveniently prepared from α,α'-dichloro-*p*-xylene through polymerization of the sulphonium salt intermediate [65–67]. In fact, PPV can also be obtained by another method from the sulphonium polyelectrolyte(II) (see Fig. 13) [68]. On the other hand, an alternative strategy for polymer processing is to attach flexible side groups to the polymer chain so that the polymer is directly processible from solution [27]. For a high-EL-efficiency device, the emitting material possesses a good quantum yield for photoluminescence. However, the quantum yield for photoluminescence in organic conjugated molecules can be very high, but the yield in PPV is reduced as the extent of π-electron conjugation is increased [69]. It was found that photoluminescence efficiency can be increased in a polymer by separating the polymer chains [70]. Considering that amorphous PPV would have better chain separation than crystalline PPV would, some polymers for PPV derivatives with high photoluminescence efficiency are given below.

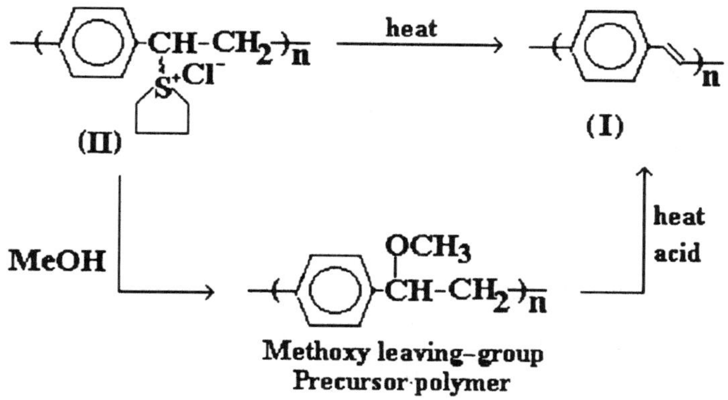

FIGURE 13 Precursor route to poly(*p*-phenylene vinylene). (From Ref. 68.)

One of the PPV copolymers for synthetic pathway is illustrated in Fig. 14. The symbols *m* and *n* represent the molar equivalents of the monomeric salts ① and ② in the reaction, respectively. A small number (*o*) of benzylic positions adjacent to the phenylene rings in ③ are substituted by methoxy groups. Thermal elimination of all the sulphonium groups and a few methoxy groups from ③ produce the polymer ④, which has an arrangement of conjugated units randomly interrupted by saturated units (where *a*, *b*, *c*, and *d* stand for the total number of each structural type). Further acid-promoted conversion of ⑤ produces the conjugated polymer ⑥; the EL efficiency can be raised to above 1% by using the copolymers discussed above [16].

Figure 15 shows another PPV copolymer for a synthetic pathway. Monomer ① is synthesized in one step and found to be stable enough to be stored as a solid at room temperature without any sign of degradation or hydrolysis. The monomer is polymerized by dissolving it in tetrahydrofuran (THF) and then reacting it with potassium *tert*-butoxide at 0°C. The reaction appears to proceed via the quinoid intermediate, because the solution became red almost immediately upon addition of a base. The resulting precursor polymer ② was completely soluble in common organic solvents such as THF, chloroform, 1,4-dioxane, toluene, and cyclohexanone. In the end, a PPV ③ with characteristics of disorder for the structure could be obtained. For this method, the internal quantum efficiency for a ITO/PPV/Al device is 0.22% [71], considerably higher than the 0.01% reported by Burroughes et al. [10] for a device prepared with Wessling's PPV [65,66].

2. Electroluminescent Characteristics for Poly(p-phenylene vinylene) and Its Derivatives

For the EL characteristics of PPV, it could be found that the EL spectrum is identical to its photoluminescence spectrum. Both electroluminescence and photolum-

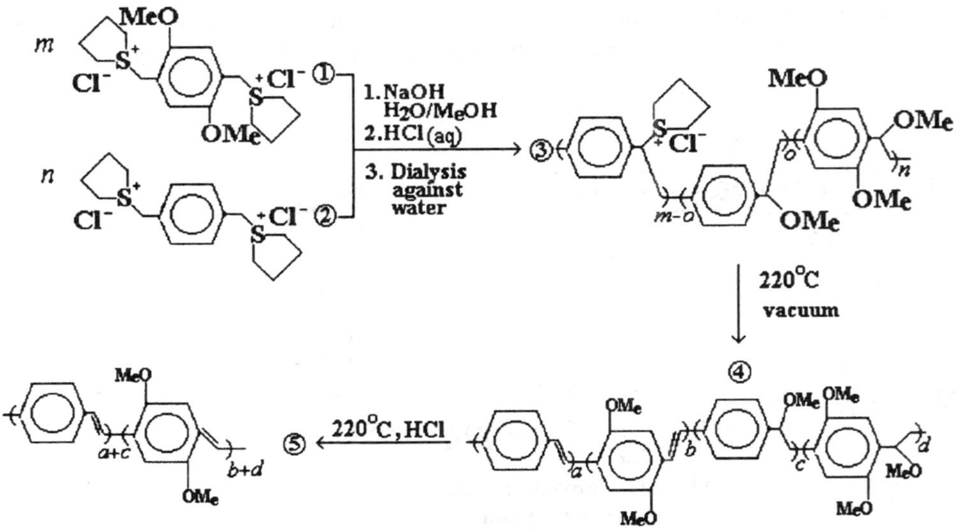

FIGURE 14 Synthetic route of the conjugated copolymers. (From Ref. 16.)

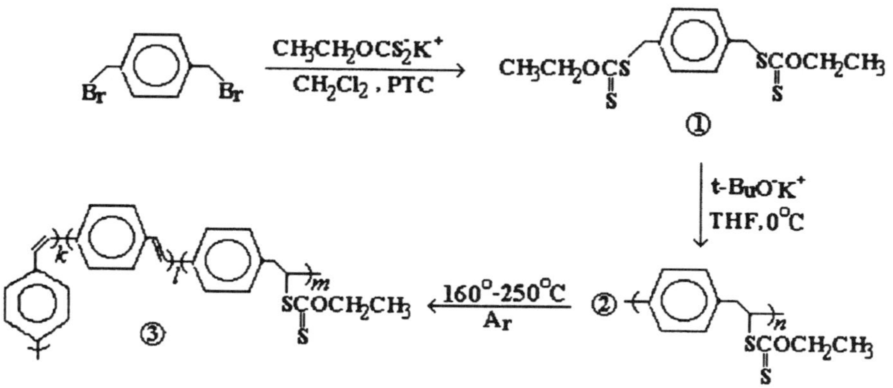

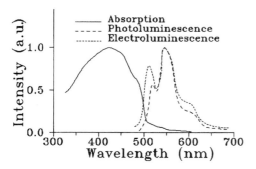

FIGURE 15 The method devised to synthesize an amorphous poly(*p*-phenylene vinylene) having a mixture of *cis*- and *trans*-poly(*p*-phenylene vinylene). (From Ref. 71.)

inescence spectra are shown in Fig. 16, together with the absorption spectrum of PPV. From the absorption spectrum, the band gap can be estimated as 2.5 eV. However, the spectrum changes with ambient temperature for PPV; the EL device, therefore, emits in the green–yellow part of these spectra which are very similar to that measured in photoluminescence, with a peak near 2.2 eV, as shown in Fig. 16. Figure 17 presents typical *V–I* characteristics and the relationship between EL intensity and current for a device having indium oxide as the bottom contact and aluminum as the top contact, indicating that the integrated light output is approximately linear with current.

The EL characteristics of PPV derivatives are greatly affected by the side chain in them, showing that the organic photoluminescence materials have a major advantage of color. Their electroluminescences for several PPV derivatives possessing different side chains are introduced in the following.

One such derivative is poly(2,3-diphenyl-1,4-phenylene vinylene) (DP-PPV), which was also reported as emitting materials to fabricate an EL device [72]. Figure

FIGURE 16 Optical absorption, photoluminescence, and electroluminescent emission spectra for poly(*p*-phenylene vinylene). (From Ref. 68).

FIGURE 17 Change of current (mA) and EL intensity (a.u.) with forward-bias voltage for a ITO/PPV/calcium light-emitting diode. The thickness of the PPV polymer layer is about 100 nm, and the emitting area is about 4 mm². (From Ref. 68.)

18 presents the molecular structure of DP-PPV and the synthetic scheme for DP-PPV. The chlorine precursor polymer of DP-PPV was prepared from the polymerization of 1,4-bis(chloromethyl)-2,3-diphenylbenzene. For the fabrication of the EL thin film, a solution of the precursor polymer in toluene was prepared and used for spin-casting onto an ITO-coated glass substrate. The film was then thermally converted at 290°C with argon gas underflowing for 2 h to give a DP-PPV film with a thickness of about 500°C. Metallic contacts (Al or Mg, 50 nm thick) were deposited on top of the DP-PPV film by vacuum evaporation at a pressure below

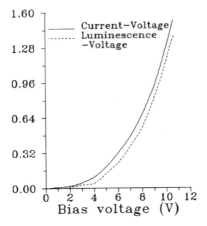

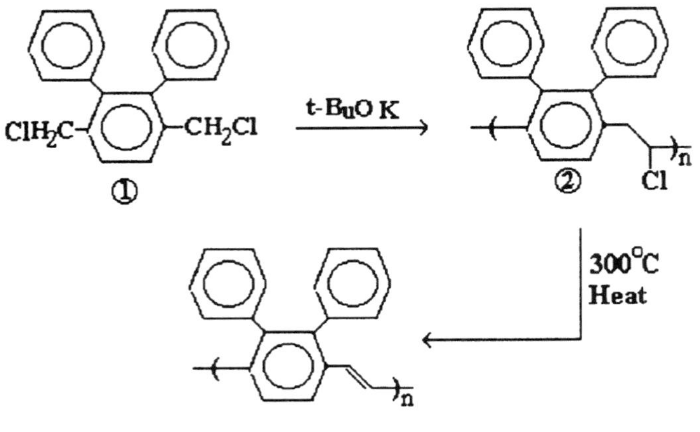

(DP-PPV)

FIGURE 18 Molecular structure of DP-PPV and its synthetic scheme. (From Ref. 72.)

10^{-6} Torr. The Mg electrodes were further passivated by Ag (50 nm). The deposition rate of Mg and Al was 5 nm/s and 0.5–1.0 nm/s, respectively.

The optical absorption, photoluminescence, and EL spectra of DP-PPV are shown in Fig. 19, indicating that the onset of the π–π^* transition of DP-PPV is at about 490 nm, which is very close to the peak at about 500 nm of photoluminescence and EL spectra. This suggests that the emission takes place from the recombination of electron–hole pairs in the band. The maximum of absorption occurs at 400 nm, which is 30–40 nm blue-shifted with respect to that of PPV, and the great difference exists between their photoluminescence and electroluminescence spectra. They may be a result of ineffective conjugation due to the strong steric hindrance between adjacent phenylene rings, caused by the two bulky phenyl substituents. The coincidence in the two luminescence spectra supports the notion that the same kinds of excitations are involved in the two cases, which has been attributed mainly to a singlet polaron–exciton.

In addition, the efficiencies for the DP-PPV EL device are greatly affected by the metallic electrode. As shown in Fig. 20, the EL intensity changes with the injection current for Mg/DP-PPV/ITO and Al/DP-PPV/ITO devices. A more efficient electron injection from the Mg electrode into the CB of DP-PPV accounts for the observed improvement. The linear dependence of the EL intensity with the injection current observed for the Mg device suggests the proportionality between the number of emitted photons and the number of injected charges. On the other hand, injection of electrons from aluminum is very inefficient, particularly at weak electric fields where the EL intensity is a superlinear function of the injection current. This is a result of imbalanced electron and hole currents caused by inefficient electron injection from the aluminum electrode.

Another PPV derivative is poly(2-methoxy,5-(2′-ethylhexoxy)-1,4-phenylene vinylene) (MEH-PPV) [27], which offers the advantage of being soluble in the conjugated form in organic solvents. The mechanical properties of polymers suggest that the structures of the light-emitting diodes that can be attained are more flexible than their inorganic counterparts [17], indicating that the ability to apply

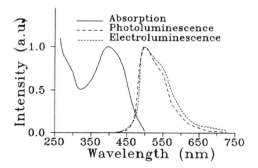

FIGURE 19 Absorption, photoluminescence, and electroluminescent spectra of poly(2,3-diphenyl-1,4-phenylenevinylene) (DP-PPV) (photoluminescence spectrum for DP-PPV excited at a wavelength of 390 nm and electroluminescent spectrum obtained at 7 V forward bias for an ITO/DP-PPV/Mg device). (From Ref. 72.)

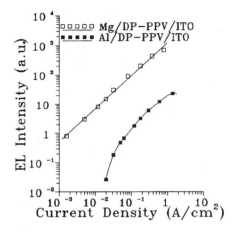

FIGURE 20 Electroluminescent intensity versus injection current density for an ITO/DP-PPV/Mg and a ITO/DP-PPV/Al device. The thickness of DP-PPV in both is 50 nm. (From Ref. 72.)

the active luminescent polymer layer from solution makes it possible to fabricate large-area devices on a flexible substrate.

C. Electroluminescence in Other Polymers

Several polymers such as P3AT, PPV, and the derivatives of PPV have been introduced, and, here, the EL characteristics of others only are given as follows.

The wide-band-gap conducting polymer poly(alkylfluorene) has been documented [73]. Among various poly(alkylfluorene) derivatives, the molecular structure of poly(9,9-dihexylfluorene) is portrayed in Fig. 21. Light-emitting devices consist of an ITO-coated glass substrate, a magnesium-containing indium (Mg:In) electrode, and an emitting layer of poly(9,9-dihexylfluorene). A thin layer of poly(9,9-dihexylfluorene) was prepared by spin-coating with chloroform as a solvent on an ITO glass substrate. The emitting layer was about 100–200 nm thick. The EL device emits blue light at room temperature. The emission spectrum is shown in Fig. 22 at a driven voltage of 14 V forward-bias condition, indicating that the peak intensity at about 470 nm corresponds to the photon energy of 2.6 eV [74]. On the other hand, the band gap of the poly(9,9-dihexylfluorene) has been

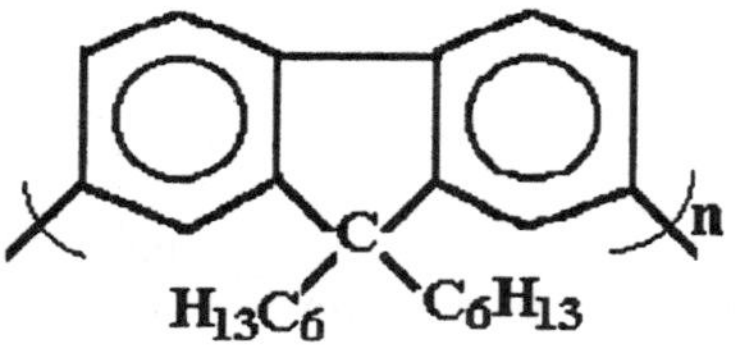

FIGURE 21 Molecular structure of poly(9,9-dihexylfluorene).

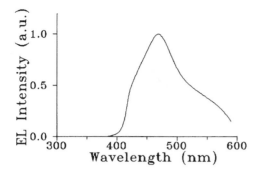

FIGURE 22 Electroluminescent spectrum of the poly(9,9-dihexylfluorene) light-emitting diode at room temperature. (From Ref. 74.)

estimated to be 2.9 eV (420 nm) from the absorption spectrum, showing that the emission is not only a single band-edge emission. However, the emission spectrum shows a small plateau at 420 nm on the shoulder of the main peak. This also seems to suggest that the band-gap emission also induced the recombination to some extent and that the exciton state corresponding to the photon energy of 2.6 eV exists as well.

Another wide-band-gap semiconductor polymer, poly(N-vinylcarbazole) (PVK), has also been reported [75]; the molecular structures of PVK is shown in Fig. 23. PVK has charge-transporting carbazole pendant groups and can be prepared simply by a radical polymerization of N-vinylcarbazole. The polymer has photoconductivity [76] and has an emission peak in the violet–blue region [77]. The polymer transports positive charges through its carbazole side groups, and its hole transport properties were fully investigated [76]. Application of the polymer as the charge transport layer in xerography was also attempted.

The PVK has been made into EL devices [75], but the brightness was, in fact, not high, owing to the lower recombination efficiency between holes and electrons, as organic EL devices are of an injection type which requires injection of both holes and electrons to the emitting layer. Kido et al. [78] reported that EL devices were fabricated by using PVK as a hole-transporting emitter layer and a set of 1,2,4-triazole derivative (TAZ) and 8-hydroxyquinoline aluminum (Alq3) as an electron transport layer. Figure 24 shows these molecular structures of materials, and the configuration of the EL device used in this study. The blue emission peaking at 410 nm and a luminance of 700 cd/m^2 were obtained at a voltage of 14 V.

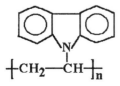

FIGURE 23 Molecular structure of poly(N-vinylcarbazole).

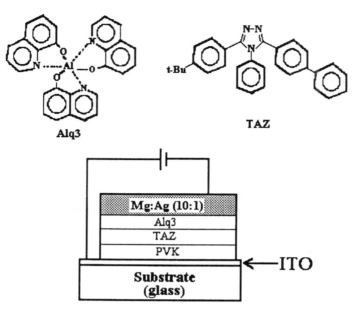

FIGURE 24 Some molecular structures of materials, and the configuration of the EL device used in this study. (From Ref. 78.)

Figure 25 shows the EL spectrum from the EL device with a structure of ITO/PVK/TAZ/Alq3/Mg:Ag, as well as photoluminescence spectrum of PVK film, suggesting that the electroluminescence comes mainly from the PVK.

III. ELECTROLUMINESCENCE BASED ON MOLECULARLY DOPED POLYMERS

Although Tang and his co-workers [4] demonstrated low-voltage-driven EL devices using a very efficient fluorescent material, Alq3, as the emitting layer, there still

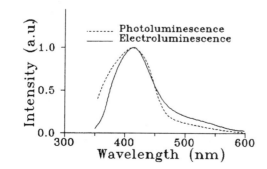

FIGURE 25 Normalized photoluminescence spectrum of poly(*N*-vinylcarbazole) film and electroluminescent spectrum of the EL device. (From Ref. 78.)

remains one problem in these devices—the lifetime of the diode. One reason for the shorter lifetime for EL devices may be the degradation of the device, which is partly caused by crystallization of organic layers, and such a defect may be improved by using less-crystalline-polymer materials. One-layer-type EL devices with intrinsically conductive polymers have been used in the EL devices to study the organic electroluminescence, avoiding the weak van der Waals force between two organic layers. In this section, polymers such as poly(methylmethacrylate) (PMMA), PVK, and P3AT are used as dopants in organic light-emitting devices, respectively. It will be stated that polymers are of great importance to simplifying the techniques for preparing the organic emitting-light devices, to changing the EL characteristics, or to reinforcing organic molecular EL intensity and efficiency.

A. Improvement of Film Quality by Doping with Poly(methylmethacrylate)

In organic EL devices, the microstructure of organic thin films should in general, have a great influence on the organic electroluminescence because the thickness of organic film is a few tens of nanometers, and the higher tunneling current moving through the device could break down the devices under the effect of a strong electrical field. It is very important to examine how the quality of the organic thin-film layers affects the current density crossing the devices, the EL intensity, and the EL efficiency for the EL devices [79,80].

Scanning tunneling microscopy (STM) and atomic force microscopy (AFM) are especially useful approaches to investigating the surface with high resolution, and AFM was used for revealing the surface of organic thin films. It has been obtained that the density and uniformity of organic thin films affect the current density crossing the organic thin films [79,80].

1. Influence of Doped-PMMA Quantities on both Surface Structure and I–V Characteristics for Mixed Films

The ITO glass substrate was used as an organic-device positive electrode of the organic EL devices and was cleaned according to Tang et al. [4]. Thin mixed layers of the organic molecule 1,1-bis(*p*-diethyl-aminophenyl)-4,4-diphenyl-1,3-butadiene derivative (DEAB) as the organic semiconductor material doped with various quantities of PMMA (DEAB/PMMA) by spin-coating onto ITO using chloroform as the solvent, sandwiched between ITO and aluminum electrodes, were fabricated, with the molecular structures of DEAB and PMMA shown in Fig. 26. The weight ratios (WR) of DEAB to PMMA were 10:1, 1:1, and 1:10, and the thickness of the DEAB/PMMA mixed layer was about 80 nm [80]. The sheet resistance of ITO used in this experiment was about 80 $\Omega/\square$ without reference to the influence of it on the EL mechanism. A 500-nm-thick layer of aluminum was also deposited on the organic layer surface as the top electrode at 2×10^{-5} Torr. The substrate was kept at room temperature during the deposition at the rate of 0.8–1.0 nm/s for the aluminum layer. The conducting area of the devices was about 0.2×0.2 cm^2. All measurements were done at room temperature in air under a DC bias condition.

In general, the polymer has better film-forming characteristics than the organic molecules; the DEAB and PMMA thin films were fabricated at the same rotation speed of 4×10^3 rev/min. Figure 27 shows that the three-dimensional AFM images

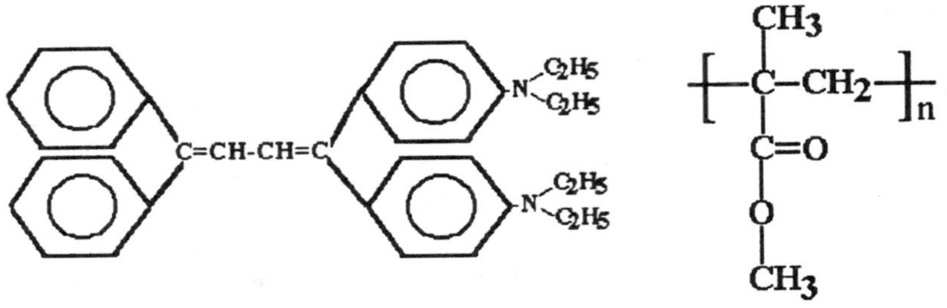

DEAB   **PMMA**

FIGURE 26 Molecular structures of DEAB and PMMA.

of the films of the DEAB and PMMA were obtained with a scan rate of 10.2 Hz by AFM, indicating that the organic molecule film quality could be improved by doping with the PMMA polymers. PMMA is an optically and electrically inert polymer and has good film-forming properties with a high glass transition temperature of 105°C [81]. If there is higher doping-quantity PMMA in DEAB/PMMA thin film, the conductivity of DEAB/PMMA thin film would be reduced greatly. On the other hand, the quality of DEAB film could not be improved by doping with less PMMA. The three-dimensional AFM image of DEAB/PMMA (WR = 1:1) is shown in Fig. 28 [80]. In studying the DEAB/PMMA thin-film conductivity and their electroluminescence, it is very important to discuss the surface structure of the DEAB/PMMA film, which is affected by PMMA-doping quantities [79], because the EL intensity is proportional to the carrier concentrations and carrier recombination efficiencies.

The current–voltage characteristics of the different doping-quantity DEAB/PMMA EL devices are shown in Fig. 29. Because the DEAB is capable of transporting holes [82,83], the electronic current comes mainly from hole transportation of DEAB. However, the I–V characteristics for the EL devices are determined by several factors such as thickness and roughness for organic films, and the contact characteristics between organic film and electrodes in the EL devices.

After eliminating the heterojunctions of the ITO/mixed organic layer (DEAB/PMMA) and the mixed organic layer/Al to affect the I–V characteristics of the DEAB/PMMA film, the behavior of the space-charge-limited (SCL) current for the organic thin films is expressed as $I \propto V^2$ [84]. For the DEAB/PMMA EL devices, because the PMMA is an optically and electrically inert polymer, the I–V characteristics could be described by $I \propto pV^q$, considering the influence of these heterojunctions, where p and q depend on the doped PMMA quantities. The best fits of $I \propto pV^q$ to experimental data give p and q corresponding to different WRs (Table 1). It is shown that the power-law dependence of the I–V data (p, q) was affected by the doping quantities of PMMA, which influence the microstructures (i.e., the roughness of the organic thin film) and the electrical conduction for the DEAB/PMMA film. The relationship between the doping quantities of PMMA and the roughness (RN) for the different mixed (DEAB/PMMA) organic films is summarized in Table 2.

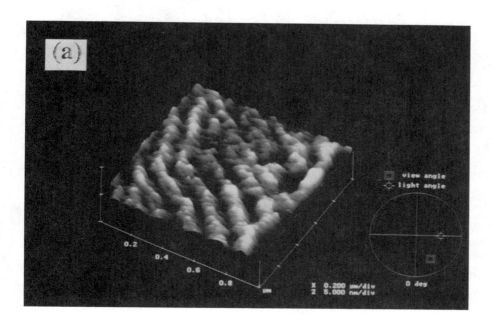

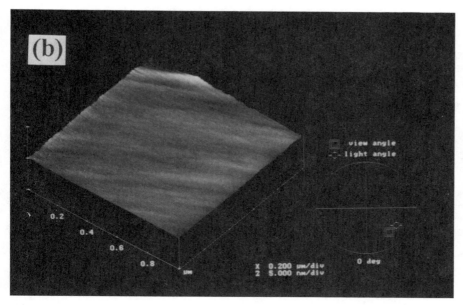

FIGURE 27 Three-dimensional AFM images of the films for DEAB (a) and PMMA (b). (From Ref. 80.)

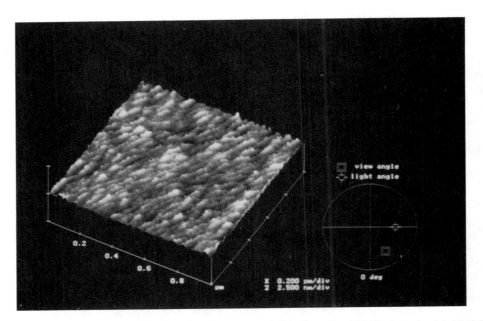

FIGURE 28 Three-dimensional AFM image of the films for DEAB doped with PMMA. (From Ref. 80.)

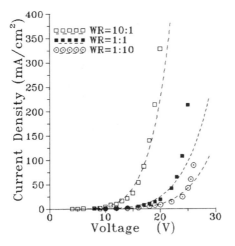

FIGURE 29 Current–voltage characteristics of the doped various quantities of PMMA in DEAB/PMMA thin-film EL devices. (From Ref. 85.)

TABLE 1   Coefficients (p and q) of I–V Nonlinear
Functions Corresponding to Various Doped PMMA
Quanties in DEAB/PMMA Films

	WR		
	10 : 1	1 : 1	1 : 10
p	6.8×10^{-6}	3.8×10^{-7}	2.9×10^{-8}
q	5.79	6.02	6.05

Source: Ref. 80.

In comparison to Tables 1 and 2, one can see that the coefficients (q) of I–V nonlinearity change little if the PMMA in the mixed films is in excess of one doped quantity, indicating that the values of q are mainly determined not by the roughness of the film but by the optically and electrically inert characteristics for the PMMA in this case. On the other hand, the values of q depend on both the optical and electrical characteristics for the PMMA and the roughness of the mixed films if there are fewer doping quantities of PMMA in the mixed film because the uniformity of the organic thin film has strong influence on I–V characteristics [79].

The explanation for the results in Fig. 29 and Table 1 could be stated as follows. For the lower doping-quantity PMMA in DEAB/PMMA films, if the density and uniformity of the organic film is not very high, there are many traps or pinholes in the organic thin-film layer, so that the higher tunneling current acts in this area, which produces the higher current density moving through the device, breaking down the organic film. On the other hand, the current density crossing the DEAB/PMMA film increases with the decrease of PMMA, leading to reducing the I–V nonlinearity for the organic DEAB/PMMA films. For the function of the doped PMMA in the DEAB, there exists an elevation of space-charge density in the DEAB/PMMA mixed organic films with increasing PMMA quantity; the organic thin film is unable to transmit all the space charges so that the current density crossing the devices has higher nonlinear I–V characteristics.

TABLE 2   Relationship Between the Doped PMMA Quantity
and the Roughness (RN) for the Different Mixed (DEAB/
PMMA) Organic Films

	Films				
		DEAB/PMMA (WR)			
	DEAB	10 : 1	1 : 1	1 : 10	PMMA
RN	1.08	0.83	0.46	0.21	0.11

Source: Ref. 80.

2. Relationship Between EL Characteristics and Different Doped Quantities

The radiative recombination luminance B is proportional to electron–hole radiative recombination probability, electron concentration, and hole concentration. The different weight ratios of DEAB to PMMA in the emitting layer should influence the enhancement of DEAB/PMMA EL intensity and EL efficiency.

The luminance (B)–voltage (V) curves for DEAB/PMMA devices with a continuous DC for a forward-bias ITO at positive polarity are shown in Fig. 30. Let the EL efficiency (η) and the normalized η be equal to unity at 20 V for the DEAB/PMMA (WR = 1:1) device. Then η can be defined by

$$\eta = \frac{B}{P} \qquad (2)$$

where B and P represent EL intensity and input power ($P = IV$), respectively. According to Figs. 29 and 30, η can be easily calculated. Figure 31 shows the relationship between the EL efficiency and bias voltages applied to different DEAB/PMMA devices, suggesting that the doping quantities of PMMA has a strong influence on the DEAB electroluminescence, and that the same DEAB/PMMA device has different η's at various applied voltages.

The EL luminance is enhanced and the current density increases with bias voltage elevation. The DEAB/PMMA (WR = 1:1) device has higher EL intensity and EL efficiency than other DEAB/PMMA (WR = 10:1 and 1:10) devices, indicating that an optimum weight ratio of DEAB to PMMA must exist according to the experimental results [85].

For the observed behaviors, organic molecule (DEAB) EL efficiency could be reinforced by doping with PMMA. However, it was found that an optimum weight ratio should exist among these various weight ratios of DEAB to PMMA in the emitting layer to enhance the one-layer DEAB/PMMA EL intensity and EL effi-

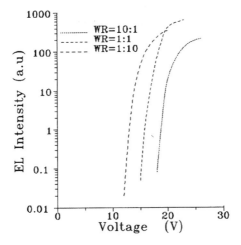

FIGURE 30 Dependence of luminescence (B) on the voltage applied (V) for DEAB/PMMA thin-film EL devices. (From Ref. 85.)

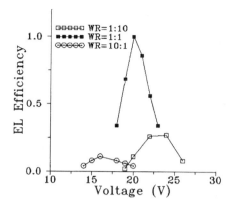

FIGURE 31 Relationship between the EL efficiency and the bias voltages for different DEAB/PMMA thin-film EL devices. (From Ref. 85.)

ciency. The thin-film quality of the organic molecules DEAB by doping with PMMA could be improved, and both the microstructure of organic film and the PMMA-doping quantity have a great influence on the electroluminescence, the organic-film conductivity, and the *I–V* nonlinear characteristics of the DEAB/PMMA EL devices.

Of course, one-layer organic EL devices could be fabricated by doping with several kinds of organic molecules in PMMA to prepare the organic emitting layer. Kido and Kohda [14] have reported *N,N*-diphenyl-*N,N'*-bis(3-methylphenyl)-1,1'-biphenyl-4,4'-diamne (TPD) and tria(8-quinolinolato)-aluminum (Alq3) complex in PMMA. TPD has a high hole drift mobility of 10^{-3} cm²/V s, and its hole transport properties in the polymer matrix were published [86]. Alq3 is a luminescent metal complex possessing electron transport properties and has been a luminescent layer in EL devices [4,7].

The EL device was a single-layer structure with doped PMMA sandwiched between electrodes. The doping concentration was kept constant at 50 wt%, and the molar ratio of TPD to Alq3 (TPD/Alq3) was varied from 0.43 to 1 [14]. A dichloroethane solution containing proper amounts of TPD, Alq3, and PMMA was dip-coated onto an ITO glass substrate with a sheet resistance of 10 Ω/□. The thickness of the doped PMMA layer was about 100 nm. Then, 200 nm of magnesium and silver (10:1) was codeposited on the PMMA layer surface as the top electrode at 3×10^{-5} Torr. The substrate was kept at room temperature during the deposition at a rate of 1.1 nm/s for Mg:Ag and emitting area about 5×5 mm². Green emission was observed from the EL device when operated in a continuous DC mode for a forward-bias ITO at positive polarity. It was found that the EL devices with the molar ratio TPD/Alq3 of 1 and 0.43 exhibited lower EL intensity compared to the device with 067 at the same drive voltages. The EL spectrum is shown in Fig. 32, together with the photoluminescence spectra of Alq3 and TPD molecularly dispersed in PMMA. The EL spectrum is identical to the photoluminescence spectrum of Alq3, indicating that the electroluminescence originates from Alq3. This indicates that TPD plays a role in transporting injected holes from ITO, thus enabling electrons and holes to recombine in the PMMA layer.

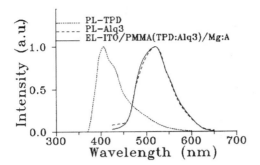

FIGURE 32 Normalized photoluminescence spectra (broken lines) of PMMA doped with TPD or Alq3 and electroluminescent spectrum of ITO/PMMA device doped with TPD and Alq3/Mg:Ag (TPD/Alq3 = 1:1). (From Ref. 14.)

Although the mechanism involved in the excitation of Alq3 is uncertain at present, the carrier transport mechanism in mixed film is believed to be due to hopping between the dopant molecules, which can be assumed to be an oxidation–reduction process. Thus, electrons and holes are transported in the polymer layer through Alq3 and TPD, as shown below [14]:

$$Alq3^- + Alq3 \rightarrow Alq3 + Alq3^- \tag{3a}$$

$$TPD^+ + TPD \rightarrow TPD + TPD^+ \tag{3b}$$

Therefore, the following two mechanisms for the excitation of Alq3 are possible. One is the direct excitation of Alq3 by the reaction between a radical anion of Alq3 and a radical cation of TPD:

$$Alq3^- + TPD^+ \rightarrow Alq^* + TPD \tag{4}$$

The other mechanism involves the excitation of TPD by the reaction of the two charge carriers. It follows that the excited energy is transferred to an adjacent Alq3 molecule via Forster-type energy transfer:

$$TPD^+ + Alq3^- \rightarrow Alq3 + TPD^* \tag{5a}$$

$$TPD^* + Alq3 \rightarrow Alq3^* + TPD \tag{5b}$$

In the photoluminescence of PMMA film containing both Alq3 and TPD, because luminescence from Alq3 was observed by the excitation of TPD, the latter mechanism of energy transfer from TPD to Alq3 cannot be excluded.

Organic molecule EL devices could be simply fabricated by doping with poly(methylmethacrylate). However, it was found that an optimum weight (or molar) ratio must exist among these various weight (or molar) ratios of organic molecules to PMMA in the emitting layer for enhancing the one-layer devices EL intensity and EL efficiency. The thin-film quality of the organic molecules by doping with PMMA could be improved, and both the microstructure of organic film and PMMA-doping quantity have a great influence on the organic-film conductivity and the *I–V* nonlinear characteristics, as well as its EL characteristics.

B. Generation of Charge Transfer by Doping with Poly(3-alkylthiphene)

The photoluminescence of organic crystals has been used to study the energy transfer processes in the solids [87]; it is primarily a random walk to nearest-neighbor molecules, keeping the walk until luminescence does take place or until it steps onto an absorbing site, in which case it is completely absorbed without luminescence. However, the charge transfer can also take place in many fields, such as in molecular heterojunctions at the molecular level [88], in quantum wells [89–91], and in the band-to-band transition approach in polymers to enhance organic electroluminescent intensity and efficiency [24]. It was found very effective that the operating mechanism of optic and electronic devices was studied by analyzing the photoluminescence of the mixed materials.

In order to study the EL characteristics and to simply the fabrication technology for EL devices, the organic molecules by doping with a polymer has been used in the devices. However, studies show that the EL color could be chosen through charge transfer from polymers (guest) to emitting material (host) in our recent work. Here, the DEAB as the emitting material and poly(3-octylthiophene) (P3OT) as the dopant were used; the single-layer film of DEAB doped with P3OT were fabricated by spin-coating onto the ITO glass substrate. The mixed layer was sandwiched between ITO and aluminum electrodes. The weight ratios (WR) of DEAB to P3OT is 8:1 in the mixed layer, and the thickness of the mixed film was about 80 nm. The sheet resistance of ITO used in this experiment is about 80 $\Omega/\square$. In addition, the DEAB thin film, deposited onto ITO by vacuum deposition ($\sim 10^{-5}$ Torr), was about 80 nm. Five hundred nanometers of aluminum was also deposited on the organic layer surface as the top electrode at 2×10^{-5} Torr. The substrate was kept at room temperature during the deposition at the rate of $0.8 \sim 1.0$ nm/s for aluminum. The emitting areas of devices were about 2×2 mm^2. All measurements were done at room temperature in air under DC conditions.

Green and yellow emissions were observed from the single DEAB and DEAB/P3OT EL devices with a continuous DC for a forward-bias ITO at positive polarity, respectively (see Fig. 33). The EL spectra of DEAB and DEAB/P3OT devices at the bias voltage applied and the photoluminescence spectrum of DEAB at room temperature all shown in Fig. 34. It was found that there was a great difference between the positions of peaks in the normalized EL spectra for the EL devices of only DEAB and DEAB doped with P3OT, suggesting that the action

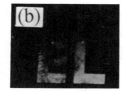

FIGURE 33 Emitting photographs of single DEAB (a) and DEAB doped with P3OT (b) EL devices.

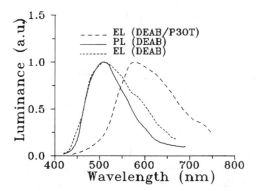

FIGURE 34 Normalized EL spectra of only DEAB device at 15 V and DEAB doped with P3OT device at 12 V, as well as PL spectrum of DEAB film excited at 410 nm. (From Ref. 11.)

between DEAB and P3OT should take place in the mixed film. One may act on another by way of energy transfer or charge transfer.

The typical current–voltage curves for the devices and the luminescence–voltage curves have been obtained in our experiment (see Figs. 35 and 36), indicating that the *I–V* characteristics for a DEAB doped with P3OT device is more nonlinear than that for the DEAB device, and also that the EL intensity for the DEAB/P3OT EL device is at least one order of magnitude higher than that for the DEAB EL device. For the observed behavior, although the EL mechanism of DEAB doped with P3OT is uncertain, we could explain this phenomenon in the following manner: Via energy transfer through light action, the EL process is detailed as follows.

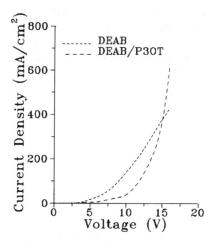

FIGURE 35 Current–voltage characteristics of organic EL devices of only DEAB and DEAB doped with P3OT as emitting layers.

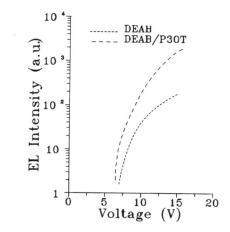

FIGURE 36 Luminance–voltage characteristics of organic EL devices of only DEAB and DEAB doped with P3OT as emitting layers.

$$DEAB^* \rightarrow DEAB + h\nu_{\lambda=507 \text{ nm}} \tag{6a}$$

$$h\nu_{\lambda=507 \text{ nm}} + P3OT \rightarrow P3OT^* \tag{6b}$$

$$P3OT^* \rightarrow P3OT + h\nu_{\lambda=650 \text{ nm}} \tag{6c}$$

where DEAB* and P3OT* denote the DEAB and P3OT excited states, respectively, indicating that the EL highest peak should not be at 575 nm according to Eqs. (6a)–(6c). Evidence suggests that it is impossible to explain the phenomenon by this action.

However, it could be believed that for recombination to occur, a charge transfer state could be formed in both neighboring molecules of DEAB and P3OT. The valence band (VB) for the DEAB molecule (solid state) is 5.1 eV [83], and that for P3OT is about 5.2 eV [16]. Certainly, the latter is only an approximate value to easily explain the phenomena. The values of conductor bands (CB) calculated for DEAB and P3OT are about 2.4 eV and 3.3 eV, respectively, according to their absorption spectra. There could be two exciton energy-level state in the DEAB; the exciton energy levels may be 2.6 eV and 3.0 eV, corresponding to light wavelengths of 507 nm and 575 nm, respectively.

Although the difference between VBs of P3OT and DEAB is small, the hole cannot transfer into the neighbor P3OT from DEAB or in the opposite direction. If this happened, the photoluminescence peak should be at 507 nm or 642 nm, a result that is not identical with the experiment.

The EL peak at 575 nm in Fig. 34 may result from the recombination between holes in VB and electrons in the exciton state 3.0 eV for the DEAB by the charge transfer from P3OT to DEAB in the CBs. The diagram of energy levels for DEAB and P3OT and the EL process are shown in Fig. 37. It can be also expressed as

$$P3OT^- + DEAB \rightarrow P3OT + DEAB^-_{(Eex2=3.0 \text{ eV})} \tag{7a}$$

$$DEAB^-_{(Eex1=2.6 \text{ eV})} + DEAB^+ \rightarrow 2 \text{ DEAB} + h\nu_{\lambda=507 \text{ nm}} \tag{7b}$$

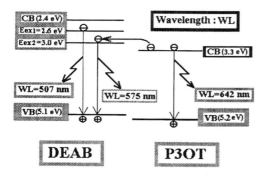

FIGURE 37 Diagram of energy levels for DEAB and P3OT films and electroluminescent progress of the DEAB/P3OT film.

$$\text{DEAB}^-_{(Eex2=3.0\ eV)} + \text{DEAB}^+ \rightarrow 2\ \text{DEAB} + h\nu_{\lambda=575\ nm} \tag{7c}$$

It is obvious that there is the EL peak at 575 nm for the DEAB/P3OT device.

Figure 38 shows the photoluminescence spectra of P3OT excited at 507 nm, and DEAB and DEAB doped with P3OT films excited at 410 nm, separately. The peak of the P3OT photoluminescence spectrum excited at 507 nm is at about 650 nm. Via energy transfer, the process is expressed in detail as

$$h\nu_{\lambda=410\ nm} + \text{DEAB} \rightarrow \text{DEAB}^* \tag{8a}$$

$$h\nu_{\lambda=410\ nm} + \text{P3OT} \rightarrow \text{P3OT}^*_{\lambda=642\ nm} \tag{8b}$$

$$\text{DEAB}^* \rightarrow \text{DEAB} + h\nu_{\lambda=507\ nm} \tag{8c}$$

$$\text{P3OT}^*_{\lambda=642\ nm} \rightarrow \text{P3OT} + h\nu_{\lambda=642\ nm} \tag{8d}$$

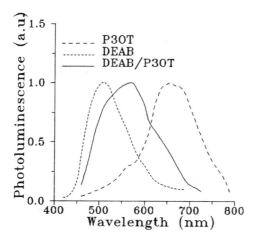

FIGURE 38 Photoluminescence (PL) spectra of DEAB and DEAB doped with P3OT films excited at 410 nm and the P3OT film excited at 507 nm.

$$h\nu_{\lambda=507 \text{ nm}} + \text{P3OT} \rightarrow \text{P3OT*}_{\lambda=650 \text{ nm}} \tag{8e}$$

$$\text{P3OT*}_{\lambda=650 \text{ nm}} \rightarrow \text{P3OT} + h\nu_{\lambda=650 \text{ nm}} \tag{8f}$$

where DEAB* and P3OT* denote the DEAB and P3OT excited states, respectively, showing that the photoluminescence highest peak of DEAB doped with P3OT must not be at 575 nm. It is impossible for energy transfer through light action to explain the experimental results.

The photoluminescence peak 575 nm in Fig. 38 may result from the charge transfer from P3OT to DEAB in the conductor bands; the energy levels of DEAB and P3OT and the photoluminescence radiation of the mixed film are shown in Fig. 39. The process can be expressed as

$$h\nu_{\lambda=410 \text{ nm}} + \text{P3OT} \rightarrow \text{P3OT*} (\text{P3OT}^- + \text{P3OT}^+) \tag{9a}$$

$$h\nu_{\lambda=410 \text{ nm}} + \text{DEAB} \rightarrow \text{DEAB*} (\text{DEAB}^-_{(\text{Eex1}=2.6 \text{ eV})} + \text{DEAB}^+) \tag{9b}$$

$$\text{P3OT}^- + \text{DEAB} \rightarrow \text{P3OT} + \text{DEAB}^-_{(\text{Eex2}=3.0 \text{ eV})} \tag{9c}$$

$$\text{P3OT*} \rightarrow \text{P3OT} + h\nu_{\lambda=642 \text{ nm}} \tag{9d}$$

$$\text{DEAB}^-_{(\text{Eex1}=2.6 \text{ eV})} + \text{DEAB}^+ \rightarrow 2 \text{ DEAB} + h\nu_{\lambda=507 \text{ nm}} \tag{9e}$$

$$\text{DEAB}^-_{(\text{Eex2}=3.0 \text{ eV})} + \text{DEAB}^+ \rightarrow 2 \text{ DEAB} + h\nu_{\lambda=575 \text{ nm}} \tag{9f}$$

There is no doubt the peak of photoluminescence and electroluminescence spectra should be at 575 nm for the mixed film of DEAB and P3OT, according to Eqs. (7a)–(7c) and (9a)–(9f).

Although the difference between the P3OT and DEAB valence bands is small, the hole cannot transfer into the neighbor of P3OT from that of DEAB or in the opposite direction. If this happened, the peak of the photoluminescence should be at 507 nm or 642 nm.

Based on the present experimental results and the above considerations, the findings are summarized as follows. It was found that the charge transfer from P3OT (guest) to DEAB (host) takes place in the DEAB/P3OT mixed film, leading to the fact that the yellow peak (λ = 575 nm) from the DEAB was intensively

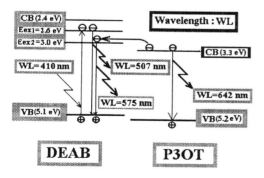

FIGURE 39 Diagram of energy levels for DEAB and P3OT films and photoluminescence progress of the DEAB/P3OT film.

carried out, and that the green peak ($\lambda = 507$ nm) for the strong peak from DEAB was reduced greatly. The doping method has a variety of applications, such as choosing organic EL color and enhancing the EL intensity for the organic EL devices.

C. Enhancement of Carrier Concentration by Doping with Polymers

Because PMMA is optically and electronically inert, the carrier concentrations in the emitting layer must be reduced so that the EL intensity decreases [92], which could be improved by doping with a semiconductor polymer such as poly(*N*-vinylcarbazole) (PVK) and the derivatives of PPV. The role of these polymers includes both enhancement of carrier concentration and simplification of the fabricating organic thin-film technique [93].

White-light-emitting EL devices were fabricated using PVK as a hole-transporting emitter layer and a double layer of 1,2,4-triazole derivative (TAZ) and Alq3 as an electron transport layer [43], where the PVK layer was doped with fluorescent dyes such as blue-emitting 1,1,4,4-tetraphenyl-1,3-butadiene (TPB), green-emitting coumarin 6, and orange-emitting 4,4-dicyano-6-methyl pyvanone derivative (DCM1). The device configuration and the molecular structures of the materials are shown in Fig. 40. Dichloroethane solutions containing PVK were prepared and dip-coated onto an ITO-coated glass substrate with a sheet resistance

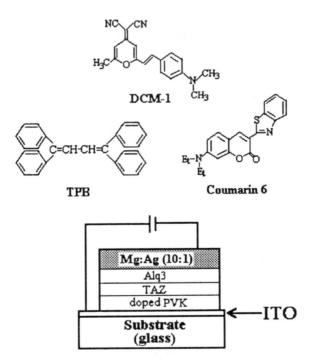

FIGURE 40 Molecular structures of dyes doped in PVK and the diagram of the EL device used. (From Ref. 43.)

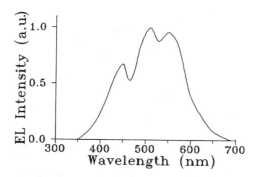

FIGURE 41 Electroluminescent spectrum of an ITO/PVK/TAZ/Alq3/Mg:Ag device. PVK is doped with TPB (5 mol%), coumarin 6 (0.3 mol%), and DCM 1 (0.2 mol%). (From Ref. 43.)

of 15 $\Omega/\square$. The thickness of the PVK layer was 40 nm. Then a 20-nm-thick TAZ layer and a 30-nm-thick Alq3 layer were successively vacuum deposited at 2×10^{-5} Torr onto the polymer layer. The TAZ layer functions as a hole-blocking electron transport layer and an exciton confinement layer, and the Alq3 layer is an electron-injecting layer which transports and injects electrons to the TAZ layer [93,94].

White emission is observed from EL devices when operated in a continuous DC mode with ITO at positive polarity. As shown in Fig. 41, the EL spectrum covers a wide range of the visible region. Three peaks at 450, 510, and 550 nm correspond to the emission from TPB, coumarin 6, and DCM 1, respectively. The intensity of each peak can be altered by changing the concentration of the corresponding fluorescent dye, which makes possible the fine-tuning of the EL color. The excitation mechanism of the dopants are not fully understood, but the excited energy may be transferred from the host matrix, PVK, to the dopants.

Organic EL characteristics could be changed with voltages applied by doping with polymers in the organic molecules. A color-variable light-emitting device has been made possible by using a conducting polymer, poly(2,5-dioctyloxy-*p*-phenylene vinylene) (ROPPV-8), mixed with fluorescent dye, Alq3 [95]. The structure of ROPPV-8 is shown in Fig. 42. The electroluminescence of the diode changes from orange to greenish yellow with increasing applied voltage. The mixed layer of ROPPV-8 to Alq3 in the one-layer device was fabricated by the spin-coating method onto a sufficiently cleaned ITO using a solution with appropriate molar

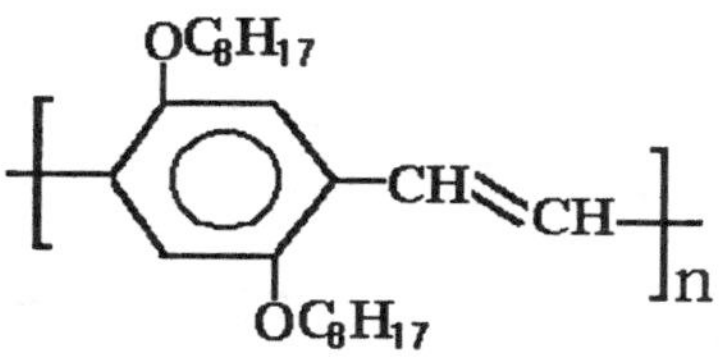

FIGURE 42 Molecular structure of ROPPV-8.

ratios of ROPPV-8 and Alq3 in chloroform. The film is approximately 100 nm thick. Finally, a Mg:In electrode was vacuum-evaporated onto surface of the organic mixed film.

The absorption and the photoluminescence spectra are shown in Fig. 43 for the molar ratio of ROPPV to Aq3 of 3:7. Figure 44a shows the EL spectra of the devices with a ratio of 3:7 in liquid nitrogen. The emission intensity of spectra increase with increasing applied voltage. To compare the spectra at various voltages, EL spectra normalized at the peak intensity are shown in Fig. 44b. It is clear that the emission peak around 530 nm from Alq3 is enhanced with increasing applied voltage. Corresponding to this spectral change, the emission color changes from orange to green–yellow. Similar characteristics were obtained from the devices at other concentration ratios [95].

IV. ELECTROLUMINESCENCE OF POLYMER BLENDS

Using soluble semiconducting and less crystallizable polymers, the fabrication technique for the EL devices are relatively simple. Moreover, the color of the emitted light could be tuned over the visible spectrum with relative ease through changes in main-chain molecular structure, through side-chain modification, or by blending with other polymers. Several techniques have been used to improve the quantum efficiency of polymer EL devices. For example, by using low-work-function metals such as Mg and Ca as the electron-injection electrodes, Braun and Heeger demonstrated the EL devices fabricated with MEH-PPV, with an external EL efficiency of about 1% photon/electron. The organic EL devices of multilayer structures have also been utilized to improve the carrier injection and confine the injected carrier within the emitting layer [64,96,97]. Polymer blends represent an alternative approach to using new materials with improved performance in polymer EL devices. By carefully selecting the component polymers for blending and by adjusting their fractions in the blend organic films in polymer EL devices, one could improve both carrier injection and carrier transport to enhance origin EL intensity and efficiency. Here, the electroluminescence of semiconducting polymer blended with another polymer is examined.

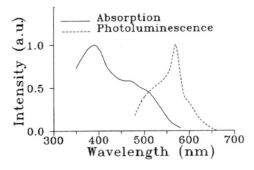

FIGURE 43 Absorption and photoluminescence of the ROPPV/Alq3 mixed film. (From Ref. 95.)

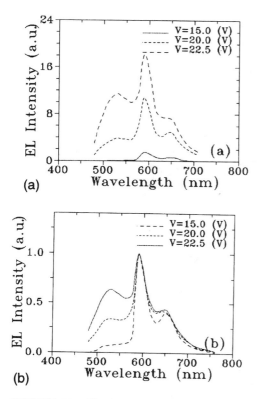

(a)

(b)

FIGURE 44 Change of electroluminescent spectra with applied voltage. (From Ref. 95.)

A. Electroluminescence of Poly(3-alkylthiphene) Blended with Poly(*N*-vinylcarbazole)

The anomalous dependence of the photoluminescence of P3AT on temperature and alkyl chain length has been reported [49], and considerable enhancement of electroluminescent intensity was obtained for a P3AT diode with a longer alkyl side-chain length [11]. According to the comparison of emission intensities of diodes at the same fixed injection current, a higher EL intensity was observed with the longer alkyl side-chain length of P3AT.

In order to enhance the shorter alkyl side-chain P3AT EL intensity, the effect of blended P3AT with another polymer which could produce excited-state charge transfer from guest to host has been documented [98]. The visible red–orange light emission from the EL diode was realized at low voltage by blending poly(3-octylthiophene) (P3OT) with poly(*N*-vinylcarbazole) (PVK) possessing *p*-type semiconductor characteristics [76]. The discussion of P3OT (host) blended with a small amount of PVK (guest) is distinguished from that of PVK blended with poly(3-hexylthiophene (P3HT) [99]. On the other hand, the P3OT/PVK emitting layer by blending the polymer (guest) into the luminescent polymer (host) is differentiated not merely from the mixed layer of organic dyes (no metal element) and chelate–metal complexes [14], but from the mixed layer of polymer and chelate metal complexes as well [42], because crystallization of the EL diodes occurs easily for organic molecules in their diodes.

The EL devices were fabricated, consisting of a mixed layer of P3OT blended with different amounts of PVK and a sample containing only a P3OT layer, sandwiched between the electrodes separately [98]. As a criterion for the molar ratio (MR) of P3OT to PVK, MR=C was expressed by the P3OT/PVK blend absorption spectrum in Fig. 45. The EL layers including different quantities of P3OT were fabricated by spin-coating onto an ITO glass substrate using chloroform as a solvent yielding a thickness of about 100 nm. The ITO sheet resistance used in this experiment was on the order of 120 $\Omega/\square$, insomuch as highly complex technology is not needed for fabricating high-resistance ITO. A 500-nm-thick layer of aluminum was finally deposited on the organic layer surface as the top electrode at 2×10^{-5} Torr. The substrate was kept at room temperature during the deposition at the rate of 1.0 nm/s for aluminum. The emitting areas of the diodes were in the vicinity of 0.3×0.3 cm^2. All measurements were done at room temperature in air under DC conditions.

Red–orange emission was observed from the EL diodes for a forward-bias ITO at positive polarity. Figure 46 shows the normalized EL spectra of P3OT/PVK and only P3OT EL diodes, indicating that the electroluminescence originates mainly from P3OT, except the emission around 411 nm.

The typical current–voltage curves for diodes are shown in Fig. 47, and the luminance–voltage variation is shown in Fig. 48. The EL luminance and the current density increase with bias voltage elevation. The red–orange light can be easily seen at 12 V in a lighted room. Of course, if magnesium and silver (10:1) were used as the cathode instead of aluminum, the EL intensity should be enhanced [4,7,26,27]. The current density for the P3OT/PVK diode has stronger nonlinear current–voltage characteristics compared to that for the P3OT diode, which has higher current density, especially at low voltage (see Fig. 47), but the luminance for the P3OT diode is weak in comparison to that for the P3OT/PVK diode.

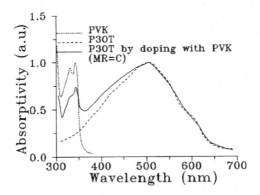

FIGURE 45 Absorption spectra of the films for the P3OT, PVK, and P3OT/PVK (MR=C) blend. (From Ref. 98.)

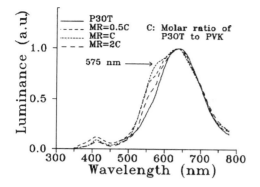

FIGURE 46 Normalized EL spectra of the P3OT diode at 12 V and the P3OT/PVK diode at 10 V. (From Ref. 24.)

To explain the observed behavior, it should be noted that the EL spectra show the highest peak intensity at about 642 nm, corresponding to the band gap of P3OT, found from the absorption spectrum of P3OT. This suggests that the emission comes from recombination of the electron–hole pairs. Compared to the P3OT diode from the EL spectra (see Fig. 46), it is of interest to note that the EL intensity at the 575-nm peak for the PVK/P3OT diode does change relative to that at the 642-nm peak, which results from the blended PVK action. The photoluminance spectrum of PVK excited at 250 nm substantially overlaps the excitation spectrum of P3OT (see Fig. 49), which provides the conditions for energy or charge transfer

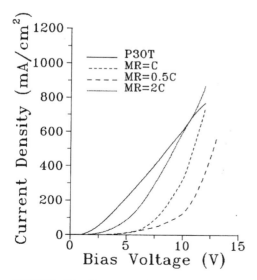

FIGURE 47 Current–voltage characteristics of organic EL diodes of the P3OT and the P3OT/PVK blend as emitting layers. (From Ref. 24.)

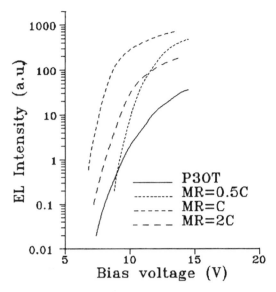

FIGURE 48 Luminance–voltage characteristics of organic EL diodes of the P3OT and the P3OT/PVK blend as emitting layers. (From Ref. 24.)

between the molecules of PVK and P3OT. Figure 50 shows the photoluminance spectrum of the P3OT excited at 411 nm and the normalized photoluminance spectra of the P3OT/PVK blends excited at 250 nm. These spectra suggest that the PVK concentration has a great influence on the peak at 575 nm, but not on the peak at 642 nm from P3OT. Regarding the EL mechanism of the mixed-layer diode, the PVK may act on the P3OT by way of either energy or charge transfer.

For energy transfer through light action, the EL process can be described by

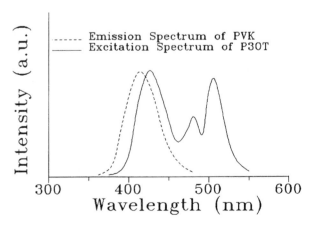

FIGURE 49 Photoluminescence (PL) spectrum of PVK film excited at 250 nm and excitation spectrum of P3OT film. (From Ref. 24.)

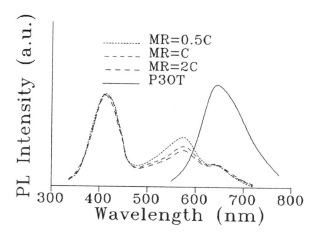

FIGURE 50 Normalized PL spectra of the blend of P3OT and various amounts of PVK films excited at 250 nm and only P3OT film excited at 411 nm. (From Ref. 24.)

$$\text{PVK}^+ + \text{PVK}^- \rightarrow \text{PVK}* \tag{10a}$$

$$\text{PVK}* \rightarrow h\nu_{\lambda=411\ nm} + \text{PVK} \tag{10b}$$

$$h\nu_{\lambda=411\ nm} + \text{P3OT} \rightarrow \text{P3OT}* \tag{10c}$$

$$\text{P3OT}* \rightarrow h\nu_{\lambda=411\ nm} + \text{P3OT} \tag{10d}$$

where P3OT* and PVK* describe the P3OT and PVK excited states, respectively, consistent with the results in Fig. 46. On the other hand, the photoluminance peak at 642 nm from the P3OT/PVK mixed layer (excited at 250 nm) is assumed to be higher than that as 575 nm, according to the energy transfer. In contrast, the photoluminescence peak at 575 nm from P3OT/PVK blends increases with MR elevation, but the one at 642 nm is weaker and small changes are observed at different P3OT/PVK blends (see Fig. 50). Therefore, it is impossible by means of energy transfer from guest to host to explain the phenomenon of enhancing P3OT electroluminescence in association with PVK.

The extra peak at 575 nm of the P3OT/PVK EL spectrum, relative to the P3OT EL spectrum, comes from the action of PVK on P3OT. It could be believed that recombination occurs when an excited-state charge transfer is formed in which an electron–hole pair is situated on neighboring PVK and P3OT molecules. The valence band for PVK (solid state) is 6.1 eV [84] and that for P3OT is approximately 5.2 eV, suggesting that the hole could be easily transferred to a neighboring P3OT from PVK. The photoluminance process is described in Fig. 51, where E_{ex} is the PVK exciton energy level. This process can also be described by

$$h\nu_{\lambda=250\ nm} + \text{PVK} \rightarrow \text{PVK}* \ (\text{PVK}^+ + \text{PVK}^-) \tag{11a}$$

$$\text{PVK}^+ + \text{PVK}^- \rightarrow h\nu_{\lambda=411\ nm} + 2\text{PVK} \tag{11b}$$

$$\text{P3OT} + \text{PVK}^+ \rightarrow \text{P3OT}^+ + \text{PVK} \tag{11c}$$

$$\text{P3OT}^+ + \text{PVK}^- \rightarrow \text{P3OT} + \text{PVK} + h\nu_{\lambda=575\ nm} \tag{11d}$$

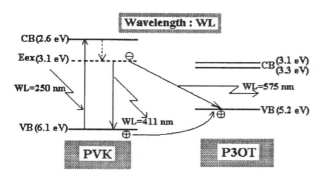

FIGURE 51 Photoluminescent process generated with energy diagram used in this study, where E_{ex} is the electron energy level of the PVK exciton. (From Ref. 98.)

where the PVK* is the PVK exciton state. The peak at 575 nm in Fig. 49 can be explained with the aid of Eq. (11d).

The EL process is indicated in Fig. 52. The electroluminescence comes mainly from the recombination of the P3OT excitons as follows:

$$P3OT^+ + P3OT^- \rightarrow P3OT^* \tag{12a}$$

$$P3OT^- + PVK^+ \rightarrow P3OT^* + PVK \tag{12b}$$

$$nP3OT^* \rightarrow nP3OT + (n - m)\, h\nu_{\lambda=575\ nm} + mh\nu_{\lambda=642\ nm} \tag{12c}$$

assuming that the P3OT exciton (P3OT*) recombines completely. The EL peak at 575 nm increases due to the charge transfer from the PVK dopant to P3OT (see Fig. 46), where the peak at 411 nm originates from the PVK because it is identical with the photoluminescence peak of PVK. The electrons of the P3OT recombine with holes transferred from the PVK in the emitting layer. This may explain why the low current density is accompanied by a higher EL intensity and efficiency.

Let the efficiency of the P3OT/PVK (MR=C) diode be η and the normalized η be unity at 10 V. According to Figs. 47 and 48 and Eq. (1), η can be easily

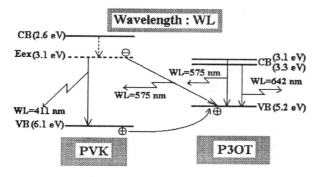

FIGURE 52 Electroluminescent process generated with energy diagram used in this study, where E_{ex} is the electron energy level of the PVK exciton. (From Ref. 98.)

TABLE 3 EL Efficiencies of the P3OT/PVK
Diode and P3OT Diodes at Different Voltages

Voltage (V)		EL efficiencies		
			MR	
	P3OT	2C	C	0.5C
10	0.003	0.095	1.000	0.189
11	0.011	0.118	0.925	0.394
12	0.019	0.054	0.746	0.458

Source: Data from Refs. 24 and 98.

calculated for the P3OT/PVK diode and the P3OT diode at 10, 11, and 12 V, respectively (see Table 3). The results suggest that the PVK has a strong influence on the P3OT electroluminescence and that the organic EL diodes have different η values at various applied voltages.

The EL intensity for the shorter alkyl side-chain P3AT is enhanced by doping with the *p*-type material PVK. The major reason for the lower EL intensity of the P3OT EL diode is the lower carrier radiative recombination probability, which can be reinforced by change transfer from the PVK dopant to P3OT.

B. Electroluminescence Based on Other Polymers Blend

Influence was also found of the blend concentration on the polymer EL efficiency [100,101]. The polymer EL devices were fabricated by using poly(*p*-phenylphenylene vinylene) (PPPV) blended with PVK, a commercially available hole-transporting polymer. The molecular structure of PPPV is shown in Fig. 53. As a result of the phenyl side groups, PPPV is soluble in the conjugated form. The emitting layer of the blended PPPV/PVK can be spin-cast from chloroform solution with excellent reproducibility, and the films are light–green, homogenous, and dense in structure.

To determine the optimum concentration of PPPV in PVK, the PPPV/PVK ratio was varied from 1:1000 to 1:10 (w/w) [100]. Electron-injecting calcium contacts were deposited onto the surface of the blend by vacuum evaporation at pressures below 4×10^{-7} Torr, yielding active areas of 0.1 cm². The efficiencies increase sharply with increasing injected current and then tend to saturate. It was found that the EL efficiency also depends on the PPPV/PVK ratio and that the

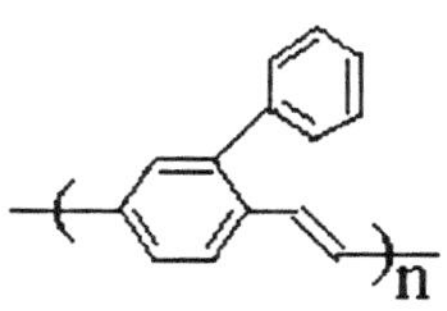

FIGURE 53 Molecular structure of poly(*p*-phenylphenylene vinylene). (From Ref. 100.)

PPPV/PVK EL device with the ratio of 1:50 possesses the highest efficiency among these ratios. The emission efficiency dependence on PPPV content for the EL devices with different PPPV/PVK ratios (w/w) is shown in Fig. 54. The efficiency increases with increasing PPPV/PVK ratio while the ratio is below 1:50 and then decreases slightly for ratios above 1:50. The quantum efficiencies are as high as 0.16% photons per electron for the devices in which the PPPV/PVK ratio is 1:50.

The PVK in the blend is believed to perform the following functions [100]: (1) The polymer PVK helps to form thin, dense, uniform films. Because the solubility of high-molecular-weight PPPV in chloroform is limited, a thin film cast from saturated PPPV–chloroform is not thick enough to make EL devices; (2) as a hole-transporting polymer, PVK carries charge to the PPPV in the blend film. Of course, the turn-on voltage of EL devices made with blends of PPPV and PMMA is much higher than for devices made with the PPPV/PVK blends.

For the poly(3-hexylthiphene) blended with poly(2-methoxy-5-(2′-ethyl-hexoxy)-1,4-phenylene vinylene) (MEH-PPV), the solution of P3HT:MEH-PPV blends in p-xylene with the P3HT/(P3HT/MEH-PPV) ratio (R) varying from 0 to 100 wt% were prepared from two master solutions of 1 wt% P3HT and 0.5 wt% MEH-PPV [101]. The EL devices were fabricated by spin-casting the polymer blend from solution onto a glass substrate partially coated with ITO, followed by Ca metal evaporation to form the electron-injection electrode. The mixed films were uniform with thickness ranging from 150 to 300 nm. The external EL quantum efficiency η of Ca/P3HT:MEH-PPV/ITO devices initially increases with P3HT content and goes through a maximum at $\eta = 1.7\%$ photon/electron with only 1 wt% P3HT, which is greater than in Ca/MEH-PPV/ITO by a factor of 2–3 and greater than in Ca/P3HT/ITO by more than three orders of magnitude.

The external quantum efficiencies, measured at 4 V, are plotted in Fig. 55 against P3HT concentration, indicating that blending dilute concentrations of P3HT into MEH-PPV improves the EL efficiency significantly without any sacrifice on operating voltage.

The EL intensity for the polymers could be enhanced by blending with other polymers. The major reason for enhancing the EL intensity of the polymer EL diode is higher carrier radiative recombination probability, which can be completed by charge transfer from one polymer into another. On the other hand, blending of

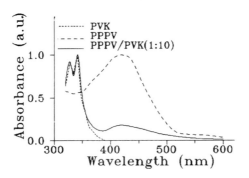

FIGURE 54 Absorption spectra of the PPPV, PVK, and PPPV/PVK blend (1:10) film. (From Ref. 100.)

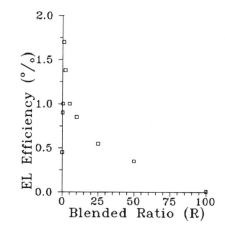

FIGURE 55 External quantum efficiency η at $V = 4$ V as a function of P3HT concentration R. (From Ref. 101.)

semiconducting polymers can also simplify the technique for preparing the EL device.

V. CONCLUSIONS

Polymers have better sticking properties to substrates, and greater mechanical strength than organic molecules. In view of physical, chemical, and electroluminescent characteristics for the polymers, the methods of synthesizing poly(*p*-phenylene vinylene) and its derivatives polymers are introduced. The EL characteristics are also discussed for several typical polymers such as PPV, P3AT, and PVK. The anomalous dependence of the photoluminescence and electroluminescence on temperature and alkyl chain length exists in the P3AT. The EL characteristics of PPV derivatives are greatly affected by the side chain in them, indicating that the polymers possessing photoluminescence characteristics have a major advantage of the color.

Considering the characteristics of single-layer EL devices, one-layer organic EL devices are fabricated by using organic molecules doped with polymers (e.g., PMMA, P3AT, and PVK) by the spin-coating method. The EL characteristics and V–I features for these EL devices are also investigated, suggesting that the doping quantity has a great influence on EL intensity, EL efficiencies, and EL characteristics. It was found that an optimum weight ratio must exist among these various weight ratios of organic molecules to PMMA in the emitting layer for enhancing the one-layer devices EL intensity and EL efficiency, and that the charge transfer from P3AT (guest) to DEAB (host) takes place in the DEAB/P3AT mixed film, leading to the fact that the yellow peak ($\lambda = 575$ nm) from the DEAB was intensively carried out and that the green peak ($\lambda = 507$ nm) for the strong peak from DEAB was reduced greatly. The doping method has a variety of applications through polymers doped in organic molecules, such as choosing organic EL color and enhancing the EL intensity for the organic EL devices. These results are partially based on the improvement of the EL thin film, the generation of charge transfer from guest to host, or the enhancement of carrier concentration in the emitting layer.

The poly(3-alkylthiophene) EL intensity and efficiency could be enhanced by blending with the *p*-type polymer PVK, resulting from the charge transfer from guest to host. For the other polymer blends, it also found that the EL color can be tuned and that the EL efficiencies can be reinforced by the blending method.

REFERENCES

1. Pope, M., Kallmann, H. P., and Magnante, P., Electroluminescence in organic crystals, *J. Chem. Phys.*, *38*, 2042–2043 (1963).
2. Helfrich, W., and Schneider, W. G., Recombination radiation in anthracene crystals, *Phys. Rev. Lett.*, *14*, 229–231 (1965).
3. Helfrich, W., and Schneider, W. G., Transients of volume-controlled current and of recombination radiation in anthracene, *J. Chem. Phys.*, *44*, 2902–2909 (1966).
4. Tang, C. W., and Vanslyke, S. A., Organic electroluminescence diodes, *Appl. Phys. Lett.*, *51*, 913–915 (1987).
5. Adachi, C., Tokito, S., Tsutsui, T., and Saito, S., Electroluminescence in organic films with three-layer structure, *Jpn. J. Appl. Phys.*, *27*, L269–L271 (1988).
6. Adachi, C., Tokito, S., Tsutsui, T., and Saito, S., Organic electroluminescent device with a three-layer structure, *Jpn. J. Appl. Phys.*, *27*, L713–L715 (1988).
7. Tang, C. W., Vanslyke, S. A., and Chen, C. H., Electroluminescence of doped organic thin films, *J. Appl. Phys.*, *65*, 3610–3616 (1989).
8. Adachi, C., Tsutsui, T., and Saito, S., Blue light-emitting organic electroluminescent devices, *Appl. Phys. Lett.*, *56*, 799–801 (1990).
9. Adachi, C., Tsutsui, T., and Saito, S., Confinement of charge carriers and molecular excitons with 5-nm-thick emitter layer in organic electroluminescent devices with a double heterostructure, *Appl. Phys. Lett.*, *57*, 531–533 (1990).
10. Burroughes, J. H., Bradley, D. D. C., Brown, A. R., Marks, R. N., Mackay, K., Frien, R. H., Burn, P. L., and Holmes, A. B., Light-emitting diodes based on conjugated polymers, *Nature*, *347*, 539–541 (1990).
11. Ohmori, Y., Uchida, M., Muro, K., and Yoshino, K., Visible-light electroluminescent diodes utilizing poly(3-alkylthiophene), *Jpn. J. Appl. Phys.*, *30*, L1938–L1940 (1991).
12. Ohmori, Y., Uchida, M., Muro, K., and Yoshino, K., Blue electroluminescent diodes utilizing poly(alkylfluorene), *Jpn. J. Appl. Phys.*, *30*, L1941–L1943 (1991).
13. Era, M., Adachi, C., Tsutsui, T., and Saito, S., Organic electroluminescent device with cranine dye Langmuir–Blodgett as an emitter, *Thin Solid Films*, *210/211*, 468–470 (1992).
14. Kido, J., and Kohda, M., Organic electroluminescent devices based on molecular doped polymers, *Appl. Phys. Lett.*, *61*, 761–763 (1992).
15. Brown, A. R., Bradley, D. D. C., Burroughes, J. H., Friend, R. H., Grennham, N. C., Burn, P. L., Homes, A. B., and Kraft, A., Poly(*p*-phenylenevinylene) light-emitting: Enhanced electroluminescent efficiency through charge carrier confinement, *Appl. Phys. Lett.*, *61*, 2793–2795 (1992).
16. Burn, P. L., Holmes, A. B., Kraft, A., Bradley, D. D. C., Brown, A. R., Friend, R. H., and Gymer, R. W., Chemical tuning of electroluminescent copolymers to improve emission efficiencies and allow pattering, *Nature*, *356*, 47–49 (1992).
17. Gustafsson, G., Cao, Y., Treacy, G. M., Klavetter, F., Colaneri, A., and Heeger, A. J., Flexible light-emitting diodes made from soluble conducting polymers, *Nature*, *357*, 477–479 (1992).

18. Kido, J., Hougawa, K., Okuyawa, K., and Nagai, K., Bright blue electroluminescence from poly(*N*-vinylcarbazole), *Appl. Phys. Lett.*, *63*, 2627–2629 (1993).

19. Hosokawa, C., Higashi, H., and Kusumoto, T., Novel structure organic electroluminescent cell with conjugated oligomers, *Appl. Phys. Lett.*, *62*, 3238–3240 (1993).

20. Kido, J., Hougawa, K., Okuyawa, K., and Nagai, K., White light-emitting electroluminescence devices using the poly(*N*-vinylcarbazole) emitter layer doped with three fluorescent dyes, *Appl. Phys. Lett.*, *64*, 815–817 (1994).

21. Berggren, M., Inganas, O., Gustafsson, G., Rasmusson, J., Andorsson, M. R., Hjertberg, T., and Wennerstron, O., Light-emitting diodes with variable colours from polymer blends, *Nature*, *372*, 444–446 (1994).

22. Zhang, C., Seggern, H. V., Kraabel, B., Schmiclt, H. W., and Heeger, A. J., Blue emission from polymer light-emitting diodes using non-conjugated polymer blends with air-stable electrode, *Synth. Met.*, *72*, 185–188 (1995).

23. Wang, G.-M., Yuan, C.-W., Wu, H.-M., and Wei, Y., Influence of doped poly(*N*-vinylcarbazole) on poly(3-octylthiphene) electroluminescence, *Jpn. J. Appl. Phys.*, *34*, L182–L184 (1995).

24. Wang, G.-M., Yuan, C.-W., Wu, H.-M., and Wei, Y., Importance of poly(*N*-vinylcarbazole) dopant to poly(3-octylthiophene) electroluminescence, *J. Appl. Phys.*, *78*, 2679–2683 (1995).

25. Colvin, V. L., Light-emitting diodes made from cadmium selenide nanocrystals and semiconducting polymers, *Nature*, *370*, 354–357 (1994).

26. Hsieh, B. R., Antoniadis, H., Bland, D. C., and Field, W. A., Chlorine precursor route chemistry to poly(*p*-phenylene vinylene)-based light-emitting diodes, *Adv. Mat.*, *7*, 36–38 (1995).

27. Braun, D., and Heeger, A. J., Visible light emission from semiconducting polymer diodes, *Appl. Phys. Lett.*, *58*, 1982–1984 (1991).

28. Michaelson, H. B., The work function of the elements and its periodicity, *J. Appl. Phys.*, *48*, 4729–4733 (1977).

29. Kido, J., Kimura, M., and Nagai, K., Multilayer white light-emitting organic electroluminescent devices, *Science*, *267*, 1332–1334 (1995).

30. Nakayama, T., Itoh, Y., and Kakuta, A., Organic photo- and electroluminescent devices with double mirrors, *Appl. Phys. Lett.*, *63*, 594–595 (1993).

31. Takada, T., Tsutsui, T., and Saito, S., Control of emission characteristics in organic thin-film electroluminescent diodes using an optical-microcavity structure, *Appl. Phys. Lett.*, *63*, 2032–2034 (1993).

32. Dedabalapur, A., Rothberg, L. J., and Miller, T. M., Color variation with electroluminescent organic semiconductors in multimode resonant cavities, *Appl. Phys. Lett.*, *65*, 2308–2310 (1994).

33. Hamada, Y., Sano, T., Fujita, M., Fujii, T., Nishio, Y., and Shibata, K., Organic electroluminescent device with 8-hydroxyquinoline derivatives metal complexes as an emitters, *Jpn. J. Appl. Phys.*, *32*, L524–L515 (1993).

34. Adachi, C., Nagai, K., and Tamoto, N., Molecular design of hole transport materials for obtaining high durability organic electroluminescent diodes, *Appl. Phys. Lett.*, *66*, 2679–2681 (1995).

35. Strukelj, M., Papadimitrakopoulos, F., Miller, T. M., and Rotherg, L. J., Design and application of electron-transporting organic materials, *Science*, *267*, 1969–1972 (1995).

36. Son, S., Dodabalapur, A., Lovinger, A. J., and Galvin, M. E., Luminescence enhancement by the introduction of disorder into poly(*p*-phenylene vinylene), *Science*, *269*, 376–378 (1995).

37. Su, W. P., Schrieffer, J. R., and Heeger, A. J., Solitons in polyacethlene, *Phys. Rev. Lett.*, *42*, 1698–1701 (1979).

38. Tian, C., Jin, G., Chao, F., and Costa, M., Optical spectra of a conducting polymer (polymethyl-3-thiophene) at several stages of the electrode position process, *Thin Solid Films*, *233*, 91–95 (1993).

39. Ohmori, Y., Takahashi, H., Muro, K., Uchida, M., Kawai, T., and Yashino, K., Fabrication and characteristics of Schottky gated poly(3-alkylthiophene) field effect transistors, *Jpn. J. Appl. Phys.*, *30*, L610–L611 (1991).

40. Yu, G., Gao, J., Hummelen, J. C., Wudi, F., and Heeger, A. J., Polymer photovoltaic cells: Enhanced efficiencies via a network of internal donor–acceptor heterojunctions, *Science*, *270*, 1789–1791 (1995).

41. Partridge, R. H., U.K. Patent 74/44704 (1974).

42. Uchida, M., Ohmori, Y., Noguchi, T., Ohnishi, T., and Yoshino, K., Color-variable light-emitting diode utilizing conducting polymer containing fluorescent dye, *Jpn. J. Appl. Phys.*, *32*, L921–L924 (1993).

43. Kido, J., Hongawa, K., Okuyawa, K., and Nagai, K., White light-emitting organic electroluminescent devices using the poly(*N*-vinylcarbazole) emitter layer doped with three fluorescent dyes, *Appl. Phys. Lett.*, *64*, 815–817 (1994).

44. Berggren, M., Inganas, O., Gustafsson, G., Rasmusson, J., Andorsson, M. R., Hjertberg, T., and Wennerstrom, O., Light-emitting diodes with variable colours from polymer blends, *Nature*, *372*, 444–446 (1994).

45. Zhang, C., Seggern, H. V., Pakbaz, K., Kraabel, B., Schmidt, H. W., and Heeger, A. J., Blue electroluminescent diodes utilizing blends of poly(*p*-phenylphenylene vinylene) in poly(9-vinylcarbazole), *Synth. Met.*, *62*, 35–40 (1994).

46. Yoshino, K., Nakjima, S., Gu, H. B., and Sugimoto, R., Optical properties of solution of polythiophene derivatives as function of alkyl chain length, concentration and temperature, *Jpn. J. Appl. Phys.*, *26*, L1371–L1373 (1987).

47. Yoshino, K., Nakjima, S., Park, D. H., and Sugimoto, R., Thermochromism, photochromism, and anomalous temperature dependence of luminescence in poly(3-alkylthiophene) film, *Jpn. J. Appl. Phys.*, *27*, L716–L718 (1988).

48. Yoshino, K., Love, P., Onoda, M., and Sugimoto, R., Dependence of absorption spectra and solubility of poly(3-alkylthiophene) on molecular structure of solvent, *Jpn. J. Appl. Phys.*, *27*, L2388–L2391 (1988).

49. Yoshino, K., Manda, Y., Sawada, K., Onoda, M., and Sagimoto, R. I., Anomalous dependence of luminescence of poly(3-alkylthiophene) on temperature and alkyl chain length, *Solid State Commun.*, *69*, 143–146 (1989).

50. Yoshino, K., Nakjima, S., and Sugimoto, R., Absorption and emission spectral changes in a poly(3-alkylthiophene) solution with solvent and temperature, *Jpn. J. Appl. Phys.*, *26*, L2046–L2048 (1987).

51. Yoshino, K., Nakjima, S., Park, D. H., and Sugimoto, R., Spectra change of polymer film containing poly(3-alkylthiophene) with temperature and its application as optical recording media, *Jpn. J. Appl. Phys.*, *27*, L454–L456 (1988).

52. Kaneto, K., Uesugi, F., and Yoshino, K., Optical anisotropies of photo- and doping-induced absorptions in stretched polythiophene films, *J. Phys. Soc. Jpn.*, *57*, 747–749 (1988).

53. Garten, F., Schlatmann, A. R., Gill, R. E., Vrijmoeth, J., Klapwijk, T. M., and Hadziioannou, G., Light emission in reverse bias operation from poly(3-octylthiophene)-based light emitting diodes, *Appl. Phys. Lett.*, *66*, 2540–2542 (1995).

54. Parker, I. D., Carrier tunneling and device characteristics in polymer light-emitting diodes, *J. Appl. Phys.*, *75*, 1656–1670 (1994).

55. Skotheim, T. A., *Handbook of Conducting Polymers*, Marcel Dekker, Inc., New York, 1986, Vols. 1 and 2.

56. Yoshino, K., *Dodenseikobunshi no Kiso to Oyo*, IPC, Tokyo, 1988 (in Japanese).

57. Roncali, J., Conjugated poly(thiophene): Synthesis, functionalization, and applications, *Chem. Rev.*, *92*, 711–738 (1992).

58. Onoda, M., Nakayama, H., Amakawa, K., and Yoshino, K., Electronic properties of polyarylene–vinylene conducting polymers, *IEEE Trans. Electron Insul.*, *EI-27*, 636–646 (1992).

59. Ohmori, Y., Uchida, M., Muro, K., and Yoshino, K., Effects of alkyl chain length and carrier confinement layer on characteristics of poly(3-alkylthiophene) electroluminescence diodes, *Solid State Commun.*, *80*, 605–608 (1991).

60. Onoda, M., Uchida, M., Ohmori, Y., and Yoshino, K., Organic electroluminescence devices using poly(arylene vinylene) conducting polymers, *Jpn. J. Appl. Phys.*, *32*, 3895–3899 (1993).

61. Friend, R. H., Bradley, D. D. C., and Townsend, P. D., Photo-excitation in conjugated polymers, *J. Phys. D*, *20*, 1367–1384 (1987).

62. Bradley, D. D. C. and Friend, R. H., Light-induced luminescence quenching in precursor-route poly(*p*-phenylene vinylene), *J. Phys. Condens. Matter*, *1*, 3671–3678 (1989).

63. Yoshino, K., Takiguchi, T., Hayashi, S., Park, D. H., and Sugimoto, R. I., Electrical and optical properties of poly(*p*-phenylene vinylene) and effects of electrochemical doping, *Jpn. J. Appl. Phys.*, *25*, 881–884 (1986).

64. Greenham, N. C., Moratti, S. C., Bradley, D. D. C., Friend, R. H., and Holmes, A. B., Efficient light-emitting diodes based on polymers with high electron affinities, *Nature*, *365*, 628–630 (1993).

65. Wessling, R. A., and Zimmerman, R. G., U.S. Patent No. 3401152 (1968).

66. Wessling, R. A., and Zimmerman, R. G., U.S. Patent No. 3706677 (1972).

67. Bradley, D. D. C., Precursor-route poly(*p*-phenylenevinylene): polymer characterisation and control of electronic properties, *J. Phys.*, *D20*, 1389–1410 (1987).

68. Baigent, D. R., Greenham, N. C., Gruner, J., Marks, R. N., Friend, R. H., Moratti, S. C., and Homes, A. B., Light-emitting diodes fabricated with conjugated polymers—recent progress, *Synth. Met.*, *67*, 3–10 (1994).

69. Wong, K. S., Bradley, D. D. C., Hayes, W., Ryan, J. F., Friend, R. H., Lindenberger, H., and Roth S., Correlation between conjugation length and non-radiative relaxation rate in poly(*p*-phenylene vinylene): a picosecond photoluminescence study, *J. Phys. C. (Solid State Phys.)*, *20*, L187–L194 (1987).

70. Yan, M., Rothberg, L. J., Papadimitrakopoulos, F., Galvin, M. E., and Miller, T. M., Defect quenching of conjugated polymer luminescence, *Phys. Rev. Lett.*, *73*, 744–747 (1994).

71. Son, S., Dodabalapur, A., Lovinger, A. J., and Galvin, M. E., Luminescence enhancement by the introduction of disorder into poly(*p*-phenylene vinylene), *Science*, *269*, 376–378 (1995).

72. Hsieh, B. R., Antoniadis, H., Bland, D. C., and Feld, W. A., Chlorine precursor route chemistry to poly(*p*-phenylene vinylene)-based light emitting diodes, *Adv. Mater.*, *7*, 36–38 (1995).

73. Fukuda, M., Sawada, K., and Yoshino, K., Fusible conducting poly(9-alkylfluorene) and poly(9,9-dialkylfluorene) and their characteristics, *Jpn. J. Appl. Phys.*, *28*, L1433–L1435 (1989).

74. Ohmori, Y., Uchida, M., Muro, K., and Yoshino, K., Blue electroluminescent diodes utilizing poly(alkylfluorene), *Jpn. J. Appl. Phys.*, *30*, L1941–L1943 (1991).

75. Partridge, R. H., Electroluminescence from polyvinylcarbazole films: 3. Electroluminescent devices, *Polymer*, *24*, 748–754 (1983).

76. Seanor, D. A., *Electrical Properties of Polymers*, Academic Press, New York, 1982, pp. 116–122.

77. Mort, J. and Pfister, G., *Electronic Properties of Polymers*, Wiley, New York, 1982, pp. 187–188.
78. Kido, J., Hongawa, K., Okuyama, K., and Nagai, K., Bright blue electroluminescence from poly(N-vinylcarbazole), *Appl. Phys. Lett.*, *63*, 2627–2629 (1993).
79. Wang, G-M., Ding, Y., and Wei, Y., Relationship between organic thin film uniformity and its electroluminescence, *Appl. Surf. Sci.*, *93*, 281–283 (1996).
80. Wang, G.-M., Wu, J.-W., Wei, Y., and Wang, G.-M., Study on the conductivity of organic film by doping with polymers, *Phys. Lett. (A)*, *224*, 116–120 (1996).
81. Brandrup, J., and Immergut, E. H., *Polymer Handbook*, 3 rd ed., Wiley, New York, 1989.
82. Hagiwara, T., U.S. Patent 4,751,163 (1988).
83. Enokida, T., and Hirohashi, R., Morphology and hole transport of butadiene derivative, *J. Appl. Phys.*, *70*, 6908–6914 (1991).
84. Seanor, D. A., *Electrical Properties of Polymers*, Academic Press, New York, 1982, pp. 32–35.
85. Wang, G.-M., Wang, G.-M., Lu, Z.-H., Yuan, C.-W., and Wei, Y., Influence of poly(methylmethacrylate) dopant on butadiene derivative electroluminescence, *Chin. J. Southeast Univ.*, *13*, 87–90 (1997).
86. Stolka, M., Yanus, J. F., and Pai, D. M., Hole transport in solid solutions of a diamine in polycarbonate, *J. Phys. Chem.*, *88*, 4707–4711 (1984).
87. Rosenstoke, H. B., Energy transfer in organic solids, *J. Chem. Phys.*, *48*, 532–533 (1968).
88. Isoda, S., Nishikawa, S., Ueyama, S., Hanazato, Y., Kawakubo, H., and Maeda, M., Photo-induced electron transfer in molecular heterojunction using flavin-prophyrin Langmuir–Blodgett multilayers, *Thin Solid Films*, *210/211*, 290–292 (1992).
89. Hutchings, D. C., Park, C. B., and Miller, A., Modeling of cross-well carrier transport in a multiple quantum well modular, *Appl. Phys. Lett.*, *59*, 3009–3011 (1991).
90. Donovan, K. J., Paradiso, R., Wilkins, R. F., et al., Electron transfer across organic quantum well: New results on amphiphilic phthalocyanines, *Thin Solid Films*, *210/211*, 253–256 (1992).
91. Ohmori K., Fujii, A., Uchida, M., Moishima, C., and Yoshino, K., Fabrication and characteristics of 8-hydroxyquinoline aluminum/aromate diamine organic multiple quantum well and its use for electroluminescent diode, *Appl. Phys. Lett.*, *62*, 3250–3252 (1993).
92. Kido, J., Hongawa, K., Okuyama, K., and Nagai, K., White light-emitting organic EL devices, in *International Workshop on Electroluminescence Digest of Technical Papers*, 1994, p. 100.
93. Kido, J., Shionoya, H., and Nagai, K., Single-layer white light-emitting organic electroluminescent devices based on dye-dispersed poly(N-vinylcarbazole), *Appl. Phys. Lett.*, *67*, 2281–2283 (1995).
94. Kido, J., Ohtaki, C., Hongawa, K., Okuyama, K., and Nagai, K., 1,2,4-triazole derivatives as an electron transport layer in organic electroluminescent diodes, *Jpn. J. Appl. Phys.*, *32*, L917–L920 (1993).
95. Uchida, M., Ohmori, Y., Noguchi, T., Ohnishi, T., and Yoshino, K., Color-variable light-emitting diode utilizing conducting polymer containing fluorescent dye, *Jpn. J. Appl. Phys.*, *32*, L921–L924 (1993).
96. Aratani, S., Zhang, C., Pakbaz, K., Hoger, S., Wudl, F., and Heeger, A. J., Improved efficiency in polymer light-emitting diodes using air-stable electrodes, *J. Electron Matter*, *22*, 745–749 (1993).
97. Parker, I. D., Pei, Q., and Marrocco, M., Efficient blue electroluminescence from a fluorinated polyquinoline, *Appl. Phys. Lett.*, *65*, 1272–1274 (1994).

98. Wang, G.-M., Yuan, C.-W., Lu, Z.-H., and Wei, Y., Enhancement of organic electroluminescent intensity by charge transfer from guest to host, *J. Luminesc.*, *68*, 49–54 (1996).

99. Nishino, H., Yu, G., Heeger, A. J., Chen, T. A., and Rieke, R. D., Electroluminescence from blend films of poly(3-hexylthiophene) and poly(*N*-vinylcarbazole), *Synth. Met.*, *68*, 243–247 (1995).

100. Zhang, C., Seggern, H. V., Pakbaz, K., Kraabel, B., and Schmidt, H. W., Blue electroluminescent diodes utilizing blends of poly(*p*-phenylphenylene vinylene) in poly(9-vinylcarbazole), *Synth. Met.*, *62*, 35–40 (1994).

101. Yu, G., Nishino, H., Heeger, A. J., Chen, T. A., and Rieke, R. D., Enhanced electroluminescence from semiconducting polymer blends, *Synth. Met.*, *72*, 249–252 (1995).

8
Optical Absorption of Polymers

S. K. So and M. H. Chan
Hong Kong Baptist University
Kowloon Tong, Hong Kong

I. INTRODUCTION

Polymers are now routinely used as materials for fabricating optical fibers or waveguides because they are flexible, processable, and are available at low cost compared to glass [1]. In these applications, the polymer involved should be exceptionally transparent for high performance. Polymer optical fiber fabricated from high-purity polycarbonate or poly(methyl methacrylate) (PMMA) can have attenuation loss as low as tens of decibels per kilometer. Although the attenuation is very small, it must be measured accurately in order to access the quality of a polymeric material for a specific application. Sometimes, it is sufficient to determine the value of light attenuation at a given wavelength. However, if the origin of light attenuation is to be understood in any polymeric material, spectroscopy should be performed. Light attenuation in a polymer can be attributed to the following two factors: absorption losses and scattering losses. The origins of each loss mechanism will be differentiated below. The purpose of this chapter is to introduce a highly sensitive technique, known as photothermal deflection spectroscopy (PDS), for evaluating the optical absorption of polymers. We will only concentrate on absorption measurements of polymeric materials in the visible and near-ultraviolet (UV) regions (0.5–2.2 μm). Scattering loss measurement will be discussed briefly.

II. BASICS OF OPTICAL ABSORPTION

When light is incident onto a medium, energy conservation demands that

$$T + R + A + S = 1 \tag{1}$$

where T is the transmittance, R is the reflectance, A is the absorbance of the sample, and S is the scattered portion of the incident light. The relative magnitudes of the quantities in Eq. (1) can be calculated if the complex refractive index, $N \equiv n - i\kappa$, of the material is known. The meaning of the complex refractive index can be understood as follows. Suppose the incident light (of frequency f and free-space wavelength λ) is in the form of a plane wave traveling in the z-direction from left to right, as indicated in Fig. 1. At $z = 0$, a material of complex refractive index, N, and thickness z_0 is inserted normal to the direction of light propagation. The wave amplitude at any time t for $z < 0$ can be described by

$$E(z, t) = E_0 \exp[i(\omega t - kz)] \tag{2}$$

Here, $\omega = 2\pi f$ and $k = 2\pi/\lambda$ are the angular frequency and the wave number, respectively. Inside the medium ($z_0 > z > 0$), the plane wave will have a different phase and amplitude and can be written as

$$E(z, t) = E_0 \exp(-iknz) \exp(-k\kappa z) \exp(i\omega t) \tag{3}$$

The intensity, I, of the light wave is equal to $|E|^2$. Thus,

$$I = |E_0|^2 \exp(-2k\kappa z) \tag{4}$$

which decays exponentially with z for $z > 0$. The real part of the refractive index, n, is responsible for the phase change introduced by the medium, and the imaginary part, κ, is responsible for the attenuation of the incident light.

The attenuation of the incident light can be attributed to the following two factors. First, light attenuation can originate from scattering. The effect of scattering is to change the direction of the incident light. For small scatterers (dimensions comparable to the wavelength of the incident light), the scattered light is spread out in all direction. For large scatterers (dimensions much larger than the wavelength), the scattered light is concentrated in the forward direction (i.e., the direction of propagation of the incident light). Second, light attenuation can be the consequence of absorption of light by the materials. In semiconductors, strong absorption of light occurs when the incident photons have energies larger than the

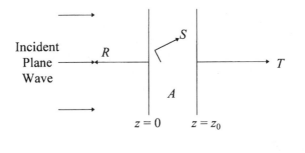

FIGURE 1 Incident plane wave travels from left to right along the z-direction and is reflected (R), absorbed (A), and scattered (S) by a medium of thickness z_0, with complex index of refraction N. The fraction of the transmitted light is denoted by T.

fundamental energy band gap. Polymer materials are generally transparent for visible light. However, strong absorption can occur in the infrared region if the incident photons have energies that match those of the vibrational energies of the polymer backbone (e.g., C—C bond) or its functional groups (e.g., C=O bond). Depending on the nature of the material, the absorbed photon energy may be converted into other forms (e.g., electrical or chemical energies). If the absorbed photon energy is released radiatively, fluorescence occurs. The absorbed photon energy can also be released nonradiatively, in the form of heat. The detection of heat released, as we will discuss below, turns out to be an exceptionally sensitive method of measuring the optical absorption of the polymers.

When both absorption and scattering effects are considered together, they constitute the attenuation of light through a medium. The absorption coefficient, α, and the scattering coefficient, σ, can be defined by

$$T = \exp[-(\alpha + \sigma)z] \tag{5}$$

Hence,

$$(\alpha + \sigma) = \frac{4\pi\kappa}{\lambda} \tag{6}$$

The quantity $\alpha + \sigma$ is known as the attenuation or the extinction coefficient. For materials with $\alpha \gg \sigma$, the absorption coefficient can be expressed in the form

$$\alpha = \frac{4\pi\kappa}{\lambda} \tag{7}$$

Many techniques are available to measure the attenuation of light by a material. However, it is often difficult to differentiate the portion of the absorbed light from the scattered light, especially when one or both of these loss mechanisms are small. For polymer materials, absorption and scattering can be separated only by very careful measurements [2]. We will concentrate our discussion on measuring the optical absorption by materials. Techniques for measuring scattering coefficient have been reviewed elsewhere [3].

III. LIGHT ATTENUATION IN POLYMERS

Absorption loss of a polymer can be classified into intrinsic and extrinsic losses. Intrinsic absorption loss in the UV region is associated with electronic excitations of chromophores in the repeating units of a polymer [4]. Examples of strong UV absorbing chromophores include carbonyl, nitro, or ethylenic double bonds. These electronic excitations are typically in the energy range 4–6 eV (200–310 nm) which may correspond to optical transitions of $n \rightarrow \sigma^*$, $n \rightarrow \pi^*$, and $\pi \rightarrow \pi^*$, where n represents a nonbonding orbital. For example, in the carbonyl group, the $n \rightarrow \pi^*$ involves the transfer of the electron density from the oxygen to the carbon atom. In principle, this transition is dipole-forbidden, but it can be observed. The coupling of the electronic and vibrational modes of the carbonyl group relaxes the dipole selection rule and gives rise to an observable electronic transition which is commonly called the vibronic transition. In contrast, the $\pi \rightarrow \pi^*$ transition is dipole allowed. Therefore, the transition is relatively strong. The UV absorption of a chromophore is not completely restricted in the UV region but decays exponentially with increasing wavelength [1]. Although this UV absorption tail is small, it may dominate the absorption loss of a polymer in the visible region where other loss mechanisms are absent.

Intrinsic absorption loss for a polymer in the near-IR (infrared) region (0.8–2.5 μm) is usually due to vibrational overtone absorption of molecular C–H, its bending vibration, or their combination bands. These vibrational overtone absorptions are forbidden optical transitions in the parabolic approximation of the C–H binding potential. However, the actual molecular binding potential is anharmonic. Thus, these overtone absorptions can be observed, although they are usually weak in intensities, especially for the higher harmonics which have absorption energies in the visible or the near-infrared range. To observe the overtones using transmission measurements, the polymer samples are usually cast into long rods in order to enhance the absorption loss features [1]. Using this method, Kaino has observed the fourth to seventh harmonics of C–H vibrations in PMMA. Extrinsic absorption loss in a polymer can be controlled during the preparation process. Extrinsic absorption is associated with impurities inside the polymer. Notable impurity in a polymer is absorbed water which has O–H vibration overtones in roughly the same energy range as the C–H overtones. For example, the third O–H overtone is at 748 nm, whereas the corresponding C–H overtone is at 930 nm.

In the absence of absorption, polymers may appear to be hazy or milky. Such appearance is due to scattering of light by inhomogeneities in a polymer. As in absorption loss, scattering loss can also be intrinsic or extrinsic. Intrinsic scattering loss originates from Rayleigh scattering which has the well-known $1/\lambda^4$ dependence. Rayleigh scattering in a polymer is due to the presence of inhomogeneities of sizes much smaller than the wavelength of the scattered light. These inhomogeneities are present in a polymer even under ideal preparation conditions. In contrast, extrinsic scattering loss in a polymer originates from inhomogeneities of sizes comparable or larger than the wavelength. These inhomogeneities exist in forms of small impurity particles, air bubbles, or concentration (or refractive index) fluctuations which are introduced during the synthesis of the polymer sample. Density fluctuations have been observed in some polymers such as PMMA or polystyrene which have been used in optical fibers. Concentration fluctuations can occur in some high polymers due to the presence of low-molecular-weight species.

So far, very few data are available for the attenuation of light by a transparent polymer in the visible and the near-infrared range. The primary reason is that in this spectral region, the absorption of a polymer is extremely weak. Few techniques are available for such measurements, especially when the polymer is in the form of thin films which one often encounters in a compact electrooptic device. Our main goal here is to introduce the technique of photothermal deflection spectroscopy (PDS) and to apply PDS to measure absorption of polymers in form of thin films of thickness between 1 and 100 μm. We think that such spectra will be very useful for scientists working in electrooptic polymers. For example, industrial scientists who are developing polymers for fiber-optic communication or waveguide applications may find such a technique useful in evaluating loss mechanisms in these polymers.

IV. TECHNIQUES FOR MEASURING ABSORPTION OF POLYMERS

Techniques for measuring small optical absorptions have been reviewed previously [3,5]. We will restrict our discussion to those few techniques having special relevance to polymers in the visible and near-infrared range (200–2500 nm). This wavelength region is of interest for most electrooptic applications. Below, more conventional techniques will be considered first and their limitations will be outlined. Then, the special technique of photothermal deflection spectroscopy will be introduced.

A. Absorption Measurements Based on Transmission and Reflection

The most common method of measuring optical absorption is to use a UV-vis spectrophotometer. In essence, a spectrophotometer measures the reflected and the transmitted light from the material of interest. The results are analyzed to obtain the optical constants. We will concentrate on the fundamentals principles and highlight some of the common concerns when we apply this technique to polymers. Experimentally, a simple spectrophotometer consists of a broadband light source which emits photons between 0.2 and 2 μm. In practice, two light sources are often used: one emits in the UV range (e.g., a mercury lamp) and the other emits in the visible and the IR (e.g., a tungsten lamp). The broadband light source, together with a grating monochromator, produces tunable photons for transmittance and reflectance measurements. If the polymer under investigation has small optical absorption, the sample should be cast into a thick disk or even a long rod for enhanced signals. Otherwise, the sample can be prepared in the form of a thin film on a transparent substrate (e.g., a quartz disk for ease of handling).

For a homogeneous medium of complex refractive index $N = n - i\kappa$, of uniform thickness d, the intensities of the reflected and the transmitted light can be deduced from the Fresnel equations. In the case of normal incidence, the Fresnel equations can be reduced to the following forms:

$$R = \frac{(n-1)^2 + \kappa^2}{(n+1)^2 + \kappa^2} \left(1 + \frac{16(n^2 + \kappa^2)\exp(-2\alpha d)}{[(n+1)^2 + \kappa^2]^2 - [(n-1)^2 + \kappa^2]^2 \exp(-2\alpha d)}\right) \tag{8}$$

$$T = \frac{16(n^2 + \kappa^2)\exp(-\alpha d)}{[(n+1)^2 + \kappa^2]^2}$$
$$\left(1 + \frac{[(n-1)^2 + \kappa^2]^2}{[(n+1)^2 + \kappa^2]^2 - [(n-1)^2 + \kappa^2]^2 \exp(-2\alpha d)}\right) \tag{9}$$

In both Eqs. (8) and (9), the second term inside the parentheses is due to multiple reflections. For simplicity, we have assumed that scattering is zero. Experimentally, R and T can be measured. Hence, the optical constants n, κ, and α can be determined from Eqs. (7)–(9). These equations are quite cumbersome. For many polymers in the visible range, the absorption is quite small. In the limit of small $\alpha d << 1$ and $\kappa << n$, Eqs. (8) and (9) can be further simplified to

$$R = \frac{(n-1)^2}{(n+1)^2} \left\{ 1 + \frac{2n}{n^2+1} \left[1 - \left(2 + \frac{(n-1)^4}{4n(n^2+1)} \right) \alpha d \right] \right\} \qquad (10)$$

$$\begin{aligned} T = &\frac{16n^2}{(n+1)^4} + \frac{2n(n-1)^4}{(n^2+1)(n+1)^4} \\ &- \alpha d \left(\frac{16n^2}{(n+1)^4} + \frac{2n(n-1)^4}{(n+1)^4(n^2+1)} + \frac{2(n-1)^8}{(n+1)^4(n^2+1)^2} \right) \end{aligned} \qquad (11)$$

Let us consider how effective these equations are when they are applied to polymers in the visible range, bearing in mind that we have ignored scattering effects. Suppose the transmittance can be measured very accurately and the index of refraction is only known to within 1%. For example, let $T = 0.90$ and $n = 1.5000$. Then, $\alpha d = 0.025$. Now, if $n = 1.5100$, then for the same T, $\alpha d = 0.022$. So a 0.7% error in the index of refractive measurement will affect the value of the absorption by 12%. Hence, n must be known very accurately if we want to obtain a sensible measurement for the absorption. Conversely, suppose the refractive index is known precisely and T can be measured to within 1%. For example, let $T = 0.920$ and $n = 1.5000$. Then, $\alpha d = 3.3 \times 10^{-3}$. Now, if $T = 0.922$, then for the same n, $\alpha d = 1.2 \times 10^{-3}$. So a 0.2% error in the transmission measurement will affect the value of the absorption by a factor of 3. Note that we have already ignored scattering losses in the calculations. If scattering is included, the uncertainty will be even higher. These numerical examples show that spectrophotometry is not a very reliable technique for small optical absorption measurements.

In order to derive absorption loss from spectrophotometry described above, scattering loss of the sample must be found separately. The fundamental physics of light scattering is a complex subject [6]. Experimentally, the scattering absorption coefficient can be expressed as [7]

$$\sigma = A + \frac{B}{\lambda^2} + \frac{C}{\lambda^4} \qquad (12)$$

The first term above represents the portion of scattering loss due to inhomogeneities of sizes much larger than the wavelength λ. The second term is due to inhomogeneities of sizes comparable to λ (Mie scattering). The last term is due to Rayleigh scattering. Scattering losses can be measured directly or indirectly. Direct method for measuring scattering can be achieved by placing an integrating sphere around the sample, or a detector at 90° off the forward-scattering position. Indirectly, scattering losses can be deduced by subtracting the absorption loss from the attenuation loss. A polynomial fit of the light attenuation coefficient with Eq. (12) will enable us to find the coefficients A, B, and C. A practical way of evaluating these coefficients is to identify a region where the polymer is known to have "zero" or very small absorption, fit the attenuation loss data with Eq. (12), and then extrapolate the scattering loss to the region of interest to deduce the absorption loss. For a newly developed polymer with little known optical properties, this procedure is not feasible. Direct measurement of scattering is always preferred.

B. Waveguide Loss Spectroscopy

Unlike inorganic glassy materials, polymers can be easily cast into smooth thin films. Waveguides based on polymer thin films have been demonstrated to be very useful as a means of coupling optical components for integrated optoelectronic devices [8]. Although polymer waveguides tend to be more lossy (10–100 dB/km for PMMA fibers) than their inorganic counterparts (about 0.3 dB/km for SiO_2 fiber), they are easy to process. Moreover, they can be conveniently integrated to current integrated circuit (IC) technologies.

In waveguide loss measurement, the polymer under investigation is first cast into a thin film on a smooth substrate, as shown in Fig. 2. The refractive index of the polymer film should be higher than the substrate. Incident light is launched into the polymer film via a prism coupler. Once entered into the prism, the light undergoes multiple internal reflections in both the air–polymer and polymer–substrate interfaces. The outgoing light can be detected by a second coupler or an optical fiber which can be scanned precisely along the surface by a stepper motor. The waveguide loss measurement, in essence, is a transmission technique, except that the length of interaction between the light and the absorbing medium is lengthened many times inside the waveguide. The waveguide measures the extinction of light, which is the sum of scattering and absorption, rather than absorption alone. For a new electrooptic polymer whose physical properties are not well characterized, significant scattering losses (which may be due to a large amount of inhomogeneities) can severely affect a reliable determination of absorption. So, this technique can only provide an upper bound for the intrinsic absorption loss of the sample. Furthermore, the waveguide technique is not very well adapted for performing spectroscopy. Lasers (e.g., a tunable laser) at many discrete wavelengths must be used as the incident light for full spectral information. It is clear that to measure the absorption of polymer in a wide spectral range, an alternative technique should be sought.

C. Photothermal Methods

In the last decade, a class of experimental methods based on the photothermal effect have been successfully developed to measure absorption coefficients of materials. Photothermal effects refer to the generation of heat by materials upon absorption of light. A variety of techniques have been developed to detect the heat generated by the photothermal effects. Some techniques, such as the photopyroelectric spectroscopy (PPS) [9], directly measure the temperature rise of the sample

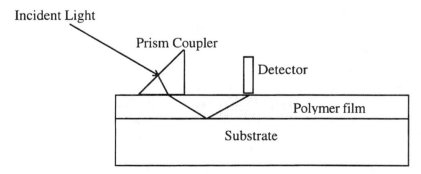

FIGURE 2 Waveguide loss measurement.

or its surrounding medium. The detection of the heat generated can also be measured indirectly. In photoacoustic spectroscopy (PAS), for example, the photothermal effect is detected in the form of sound waves by a gas microphone or piezoelectric sensor [10]. For maximum sensitivity, photothermal techniques may require the attachment of a sensor to the sample. However, the physical properties of the sample can be perturbed irreversibly by this process.

In contrast, photothermal deflection spectroscopy (PDS) is a sensitive technique which does not require the adhesion of any sensor. The principle of operation of PDS is based on the photothermal deflection of a laser beam, in a manner analogous to the mirage effect. This technique has become quite popular in the 1980s and has been often used to study the optical absorption of semiconductors, especially at a wavelength region where the absorption is very low. For example, the infrared optical absorption spectra of amorphous silicon can be measured very effectively by this technique [11]. From the PDS-derived absorption spectra, the amount of defects present in the semiconductor can be calculated [12,13]. Owing to its non-contact detection scheme, PDS is particularly useful when the sample under investigation is in the form of a thin film. Absorption spectra of organic and polymer thin films, with special relevance to optoelectronics, have been recently reported by us and others [14]. The absorption spectra, as revealed by PDS in the near-IR region, are usually very rich and complex. A typical absorption spectrum in this region consists of a number of peaks which can be assigned to the vibrational overtones of the C–H stretching mode, C–H bending vibrations, or combination band absorption. Thermal conductivities of polymer crystals can also be determined by PDS [15]. Below, we will concentrate our discussion on the basic physics of this particular technique and its experimental implementation, and highlight its special merits. Rather than giving sketchy outlines of the instrumentation and its operation, we will present, below, the detailed implementation of PDS in our laboratory. We hope that interested readers will find this section a useful reference for constructing a practical PDS system.

1. *Theoretical Background of Photothermal Methods*

The physics of PDS is based on the mirage effect. A modulated light source (the pump beam) is allowed to impinge perpendicularly on the sample under investigation. The sample usually has a flat surface and is assumed to be surrounded by a transparent fluid. After absorption of the modulated incident light, the sample releases a portion of the absorbed energy in the form of heat, which produces a periodic temperature rise in the surrounding fluid. As a result, the refractive index of the fluid changes in the vicinity of the irradiated region. This refractive index change can be detected by a probe laser. There are two ways of probing the change. In *collinear PDS*, the gradient of the refractive index is detected by a laser probe beam parallel to the pump beam. In *transverse PDS*, the gradient of the refractive index is detected by a probe laser perpendicular to the pump beam. The two versions of the PDS experiment are shown in Fig. 3.

To see how the measurement of the deflection of the probe beam can be related to the absorption of the sample, let us consider the temperature rise produced by the pump beam. For simplicity, we assume that the sample is irradiated by the pump beam, of the form of plane waves with intensity I_0 (in mW/cm^2), and modulated by a chopper with frequency ω (Fig. 4). The thermal-diffusion equations are

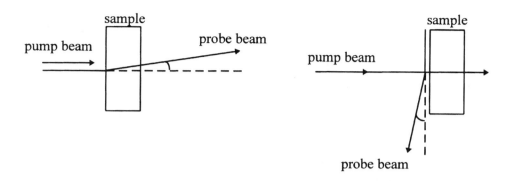

Collinear PDS   Transverse PDS

FIGURE 3 Geometry for collinear and transverse PDS.

one dimensional. We assumed that the sample has a thickness d, mounted on a transparent backing material with thickness l_b. The front of the sample is in thermal contact with a transparent fluid which extends a distance l_f. We also assume that l_f and l_b are much larger than the lengths through which the periodic heat diffuses. The heat-diffusion equations in the three regions can be written in the following standard forms:

$$\frac{\partial^2 T_f}{\partial z^2} = \frac{1}{D_f}\frac{\partial T_f}{\partial t} \tag{13a}$$

$$\frac{\partial^2 T_s}{\partial z^2} = \frac{1}{D_s}\frac{\partial T_s}{\partial t} - \frac{Q(z,t)}{k_s} \tag{13b}$$

$$\frac{\partial^2 T_b}{\partial z^2} = \frac{1}{D_b}\frac{\partial T_b}{\partial t} \tag{13c}$$

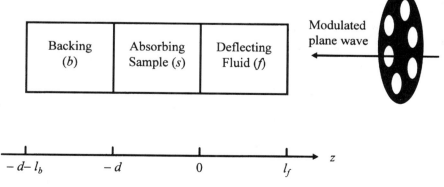

FIGURE 4 Geometry for the heat-diffusion equations.

In Eq. (13), $D_i = k_i/(\rho_i C_i)$ denotes the heat diffusivity in the region i where k_i ($i = f, s, b$) is the corresponding thermal conductivity, ρ_i is the density, and C_i is the specific heat. The source term $Q(z, t)$ in Eq. (13b) can be written as

$$Q(z, t) = \frac{\alpha I_0 \eta}{2} \exp(\alpha z)[1 + \exp(i\omega t)] \qquad (14)$$

where η is the conversion efficiency of the nonradiative deexcitation and α is the absorption coefficient of the sample. The thermal-diffusion equations can be solved by imposing the appropriate boundary conditions, including the continuity of temperature and heat flux at each interface.

We will only concentrate on transverse PDS, which is the most commonly adopted experimental configuration due to easy alignment. For transverse PDS, the probing of the photothermal effect is performed entirely inside the deflecting fluid. Thus, knowledge of the optical properties of the sample is not required. The magnitude of the deflection angle θ can be shown to be given by

$$\theta = L_1 \left(\frac{1}{n_0} \frac{dn}{dT} \right) \frac{dT_f}{dz} \qquad (15)$$

where L_1 is the interaction length (which is the distance between the two intercepts of the probe laser and the pump beam), n_0 is the refraction index of the deflecting fluid at equilibrium, and dn/dT is the temperature coefficient of the refractive index of the deflection fluid. When the thickness d of the sample is much smaller than the thermal diffusion length $\mu \equiv (2D_f/\omega)^{1/2}$ of the fluid (the so-called thermally thin regime), the magnitude of the measured signal S can be simplified to [11]

$$S \propto \left(\frac{1}{n_0} \right) \left(\frac{dn}{dT} \right) I_0[1 - \exp(-\alpha d)] \exp \left(-\frac{z}{\mu} \right) \qquad (16)$$

Equation (16) succinctly summarizes the few factors affecting the PDS signal: the intensity of the pump beam (I_0), the distance of the probe beam from the sample surface (z), $(1/n_0)(dn/dT)$, and the absorption coefficient of the sample. The first three factors are independent of the sample. For $\alpha d \ll 1$, the deflection signal is proportional to the absorption coefficient α. For large αd, the deflection signal saturates. Denoting the saturation signal by S_{sat}, we can also write the PDS signal in a reduced form:

$$\frac{S}{S_{sat}} \approx 1 - \exp(-\alpha d) \qquad (17)$$

Figure 5 is the log–log plot of the PDS signal versus the absorbance (αd) showing, clearly, the linear regime for αd less than about 0.1. Hence, PDS is easy to interpret when the sample has small optical absorption. According to Eq. (17), the PDS signal should be normalized to the saturated signal. In practice, the PDS signal can be normalized to I_0, or to the saturated deflection signal provided independently by a sample with large absorption.

In PDS, the heat released by optical absorption is probed directly. With reasonable precautions, a sensitivity factor of $\alpha d \sim 10^{-5}$ can be easily obtained for a transverse PDS setup. In contrast, conventional techniques rely on the accurate measurements of the incident, the reflected, and the transmitted light from which

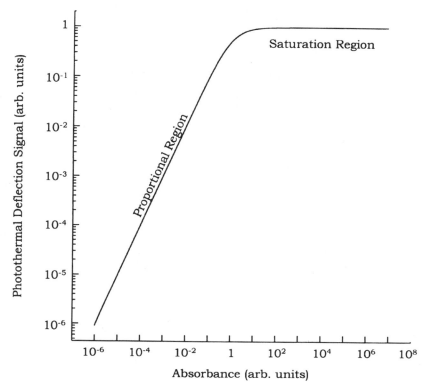

FIGURE 5 Log–log plot of the photothermal deflection signal versus absorbance.

the absorption is calculated. The sensitivity is restricted to about $\alpha d \sim 10^{-2}$. For polymers in the visible or the near-infrared region, the absorption is very small. As a result, scattering may be the dominant loss mechanism. The great merits of PDS is not only its high sensitivity to absorption but also its relative insensitivity to scattering losses. Elastically scattered light will affect the PDS signal only when the light is scattered along the direction of the probe beam and subsequently detected [11]. Such an event is highly unlikely. Nevertheless, careful analysis revealed that if the sample scatters strongly, the photothermal signals can be significantly affected [16].

2. Experimental Aspects of Photothermal Deflection Spectroscopy

The instrumentation for PDS is quite simple (Fig. 6). In our laboratory, we use a 1-kW Xe arc lamp and a ¼-m grating monochromator with selective long-pass filters as a tunable light source (the pump beam). If the UV region is not the spectral range of interest, a tungsten filament lamp should be used because it has a smooth and broad power spectrum in the infrared region and may be less susceptible to normalization problems. For measurements in a wide spectral range, selective gratings must be used to enhance the power of the pump beam, and suitable long-pass filters should be chosen in order to block higher-order light from the monochromator. The pump beam is modulated by a mechanical chopper before irradiating

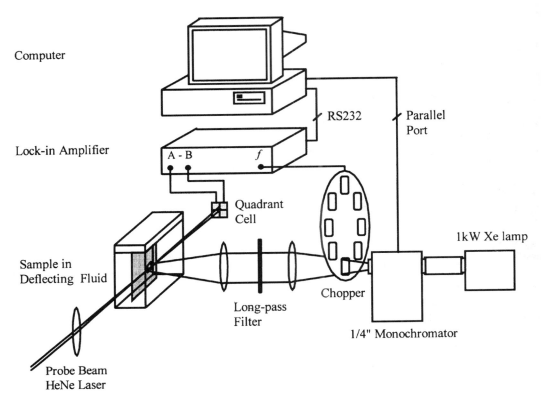

Computer

Lock-in Amplifier

Sample in
Deflecting Fluid

Probe Beam
HeNe Laser

FIGURE 6   Experimental setup of photothermal deflection spectroscopy.

on the sample. Typical chopping frequencies vary between 10 and 120 Hz, depending on the signal-to-noise ratio and the thermal-diffusion length of the sample. A collimating plano-convex lens is placed near the exit of the monochromator for collimation. Another lens is used to tightly focus the pump beam onto the sample. The dimension of the focused pump beam is about 1.5 mm × 4 mm. Typical incident power ranges from a few to 100 mW. The sample is inserted inside a rectangular cell constructed from pyrex or quartz. The cell is filled with a suitable deflecting fluid which must satisfy some simple criteria [17]. First, the deflecting fluid must be entirely transparent over the spectral range of investigation. Second, it should have a large value of temperature coefficients of the refractive index. Table 1 lists the values of n and dn/dT for some common fluids at room temperature [18]. Carbon tetrachloride is one of the most common and useful deflecting fluids because of its relatively large $(1/n_0)(dn/dT)$ and it is transparent over a very large spectral range ($\lambda > 350$ nm). Many of the fluids listed also have large temperature coefficients, but absorb strongly in the infrared. For PDS measurements extending to the UV region (down to about 250 nm), hexane is also very useful. However, hexane cannot be used in the infrared range because it starts to absorb at wavelengths of 800 nm or above. Many polymers or organic crystals are soluble in carbon tetrachloride. An alternative deflecting fluid must be used. Fluorinated hy-

TABLE 1 Physical Properties of Selected Deflecting Fluids

Deflecting fluids	Boiling point (°C)	Density (g/cm^3)	Refractive index		
			n	$dn/dT \times 10^3$ (°C^{-1})	$(1/n)(dn/dT) \times 10^3$ (°C^{-1})
n-Pentane	36	0.6214	1.3547	−0.530	−0.391
n-Hexane	69	0.6548	1.3723	−0.542	−0.395
n-Heptane	98	0.6795	1.3851	−0.508	−0.367
Cyclohexane	81	0.7739	1.4325	−0.540	−0.377
Benzene	80	0.8737	1.4979	−0.640	−0.427
Dichoromethane	40	1.3168	1.4214	−0.632	−0.445
Chloroform	61	1.4799	1.4426	−0.598	−0.415
Tetrachloromethane	77	1.5842	1.4644	−0.571	−0.390
Bromoform	150	2.8775	1.5948	−0.570	−0.357
Carbon disulphide	46	1.2556	1.6240	−0.780	−0.480
Acetonitrile	82	0.7768	1.3416	−0.450	−0.335
Nitromethane	101	1.1312	1.3795	−0.480	−0.348
Acetone	56	0.7851	1.3561	−0.500	−0.369
Nitrobenzene	211	1.1984	1.5506	−0.400	−0.258
Pyridine	116	0.9778	1.5067	−0.500	−0.332

From Ref. 18.

drocarbons, such as perfluorohexane, perfluornonane, and Flourinert FC-75 [2] have been used as the deflecting fluids for polymer samples.

The probe laser used in the PDS experiment should have a stable power (of the order of a few milliwatts is sufficient) and small angular divergence. In addition, the probe laser must have a good pointing stability, which can be a major noise source, especially at low chopping frequencies. Few commercial laser manufacturers specify the pointing stability of their products. Currently, we use a He–Ne laser (Uniphase, model 1103P) which is ideal for our applications. An inexpensive laser diode can also be used as a substitute, although it is more difficult to focus and align. During alignment, the probe beam is focused so that it is parallel to and above the sample surface. The distance between the probe beam and the sample surface should be minimized because the photothermal signal decays exponentially from the sample surface. To facilitate such adjustment, the sample cell can be mounted on a three-dimensional minitranslation stage (Edmund Scientific). The photothermal deflection is detected by a quadrant cell (United Detector Technology) under reverse bias. A lock-in amplifier (Standard Research Model SR830), operated under differential inputs, is used for phase-sensitive detection. The entire experiment was fully computer controlled.

Optical alignment is a critical step to improve the signal-to-noise ratio. The distance between the probe laser and the sample surface should be minimized. The power density of the pump beam should be maximized on the sample. Spurious noise, especially at low frequencies, may originate from the scattering of probe beam by small particles within the deflection fluid. Therefore, proper filtering (e.g., by means of a membrane filter) of the deflecting fluid should be performed. All PDS experiments should be done on a vibrationally isolated platform (e.g., on an air-damped optical table). In addition, a shielding box covering the entire sample cell may greatly suppress temperature drift due to undesirable air movement.

Normally, the PDS experiment should be operated in the linear regime in order to obtain meaningful absorption data. To check linearity, complementary transmission measurements should be performed. Therefore, two absorption spectra, one derived from PDS and the other from transmission measurement, should be obtained for any sample (Fig. 7). For large absorbance, the PDS signal tends to saturate, and therefore the PDS signal does not correspond to the absorbance of the sample. But for small absorbance, transmission measurement has low sensitivity. So, calibration of absorption coefficients can only be performed in the intermediate regime where both techniques can be applied for reasonably accurate absorption measurements. The best way to calibrate the PDS spectra and to ensure the linearity is to match the PDS absorption spectra to transmission measurement. For a semiconductor, this procedure can be done at the Urbach slope near the

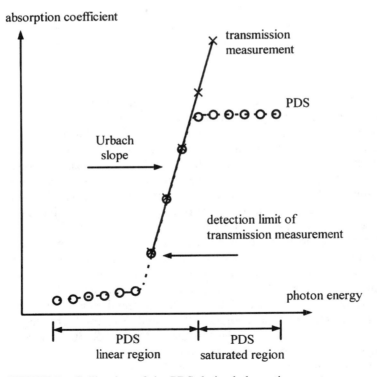

FIGURE 7   Calibration of the PDS-derived absorption spectrum.

fundamental band gap. In the case of a polymer, calibration can be done at the overtone absorption peaks. For self-consistency, the product of α and d should always be checked to ensure linearity.

In general, PDS raw data can be normalized to the incident power of the pump beam, which can be detected by a power meter. In our laboratory, we use a pyrolytic power meter which has a flat spectral response. PDS spectra can also be normalized to the spectrum of a thick carbon film which has a large αd so that the PDS signal saturates. According to Eq. (17), the deflection signal should be proportional to the power spectrum of the pump beam. Figure 8 shows the power spectrum of the pump beam and the PDS spectrum of a carbon film. The two spectra are very well correlated. In principle, normalization to the power spectrum derived from a carbon sample should be more reliable.

A newly constructed PDS setup should always be checked using standard samples in order to access its performance and sensitivity. Semiconductors such as silicon, semi-insulating (SI) GaAs, and SI-InP can be used because their optical properties are very well defined and well known. The optical absorption spectra of some boron-doped silicon derived from PDS are shown in Fig. 9. PDS can only be applied to deduce the subgap optical absorption of silicon according to Eq. (17). For these semiconductors, the PDS signals are so large that for photon energies above their fundamental band gaps, the deflection signals can be observed by the naked eye. If semiconductors are not readily available, thin films of polymers (e.g.,

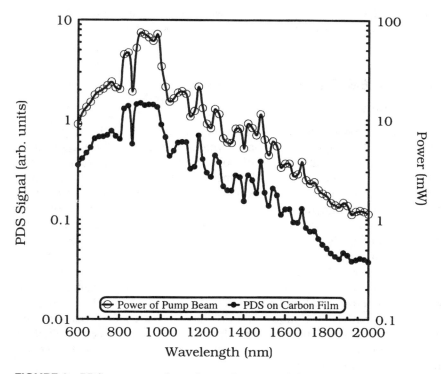

FIGURE 8 PDS spectrum of a carbon reference and the power spectrum of the corresponding pump beam.

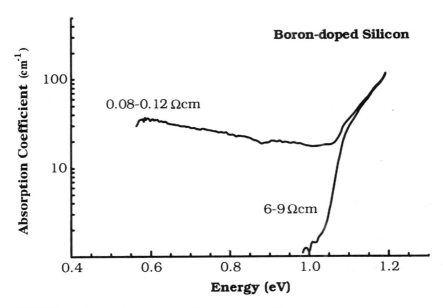

FIGURE 9   Absorption spectra of boron-doped silicon.

polycarbonate, PMMA) can also be used. As we indicate in the next section, the PDS spectra of these polymers have very distinct peaks which can be used as calibration standards for the photothermal spectrometer.

V.   PDS MEASUREMENTS ON POLYMERS

A.   Spectroscopies

The studies of optical absorption by photothermal techniques have been well documented [19,20], but their applications on polymers only emerged quite recently. An early study on the optical absorption of polyacetylene appeared soon after the invention of PDS [21]. The first detailed PDS measurement of the optical absorption on polymer thin films was reported by Skumanich [22], who obtained absorption spectra of polycarbonate thin film. PDS has also been applied to study photon-induced absorption in conjugated polymers [23]. In addition, PDS was found to be very useful in elucidating the mechanisms of light attenuation in nonlinear optical polymer systems [24] and polymer waveguides [2]. Recently, we have obtained high-quality PDS-derived absorption spectra for various polymer systems [14]. Other than PDS, a number of researchers are beginning to take advantage of the high sensitivity of photothermal techniques for studying the optical and thermal properties of polymers. The technique of PPS has been used to investigate the optical gap of poly(3-butylthiophene) thin films [25]. PAS has been applied to understand the changes in optical absorption of polyaniline thin films treated by different preparation conditions [26].

Sample preparation of polymer thin films for PDS is straightforward. Spin-coating appears to be the most common method. During preparation, the sample polymer is dissolved in a solvent which is then cast onto a flat surface (usually a small glass or a quartz disk). Depending on the viscosity of the solution and the spinning speed, uniform thin films of polymer between 1 and 10 μm can be obtained after the solvent evaporates. If the solvent is involatile, mild baking of the sample is necessary. Thicker films (>20 μm) can be obtained by just spreading a thin layer of the polymer solution over a small container and letting the solvent evaporate. Thin films of polycarbonate, for example, can be formed by dissolving chips of the polymer in chloroform. After the solvent evaporates, a free film of the polymer is formed which can be cut into desirable sizes for optical investigation. The chemical structures of several polymers are shown in Fig. 10, and their associated PDS-derived absorption spectra are shown in Figs. 11 and 12.

As transverse PDS only provides relative absorption measurements, these spectra have been calibrated by transmission measurements at regions of the spectra where they have relatively strong absorptions. The sensitivity of PDS is clearly demonstrated in Fig. 11. In these spectra, the y axes are plotted in terms of the dimensionless quantity αd (i.e., the absorbance) which can also be interpreted as the optical density (OD). From the relation $I/I_0 = 10^{-OD}$, $\alpha d = 2.303$ OD. Of the three absorption spectra shown in Fig. 11, the spectrum for the LDPE is entirely below the detection limit of an ordinary UV-visible spectrophotometer. The dotted lines in Fig. 11 represent such a detection limit. For scientists who are developing new polymers for optical fibers, the absorption minima are the regions of interest. The locations and widths of the minima decide what kind of laser diode should be used for optical communication. The optical densities at the minima will be useful for evaluating the loss of the polymer optical fiber.

The PDS-derived absorption can be divided into two regions: the UV/visible and the IR region. In the UV-visible region, the optical absorption of a polymer is, in general, featureless and increases toward the UV region. The absorption in this region primarily arises from the absorption tail of the electronic transitions in the UV region. For example, for both polyetherether ketone (PEEK) and polyimide (PI), strong absorptions begin to appear for $\lambda < 400$ nm. The absorption can be attributed to the optical excitation of the aromatic groups and is basically a $\pi \rightarrow \pi^*$ transition. The $n\pi^*$ chromophores (C=O) present in PEEK and PI are also expected to contribute to the absorption in this region. For low density polyethylene (LDPE), the absorption in the visible region is extremely weak because of the absence of π bonds. Electronic excitations of the σ bonds are expected to occur for $\lambda < 250$ nm.

In the near-IR region of the absorption spectra, well-defined absorption peaks are observed for all the polymer films. These absorption peaks can be grouped into two categories which are both associated with the vibrational overtones of the C–H stretching modes. The first group of absorption peaks can be assigned directly to C–H overtone vibrations. For example, in PEEK, absorption peaks can be observed at 1670, 1140, and 850 nm which can be assigned to the first, second, and third aromatic C–H overtone vibration energies, respectively. Similarly, in LDPE, absorption peaks occur at 1730, 1210, and 930 nm, and they can be assigned to the first, second, and third aliphatic C–H stretching vibrations, respectively. In polyimide, both aromatic and aliphatic stretches are present. Their overtone vibrations

Low-Density Polyethylene (LDPE)

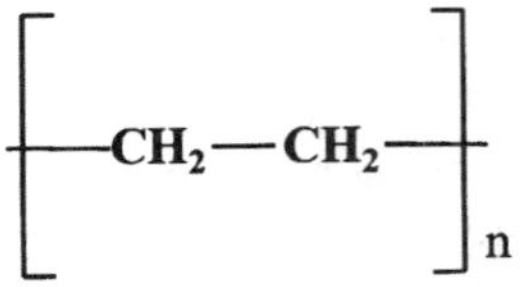

Polyetherether ketone (PEEK, ICI)

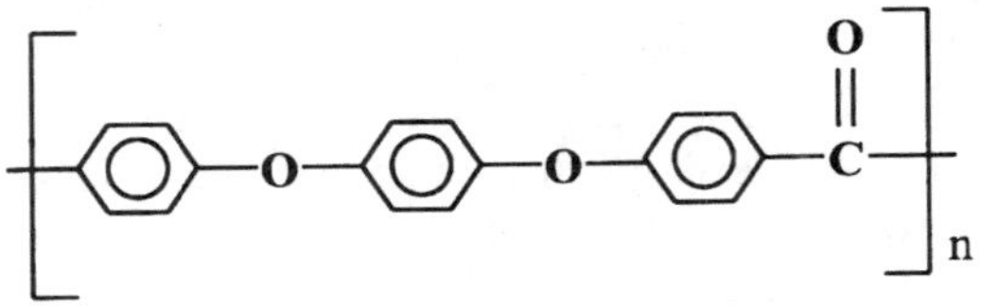

Polyimide (XU 218)

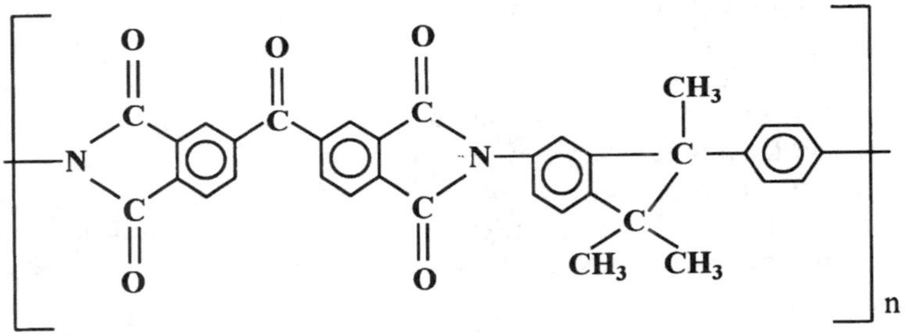

Mylar

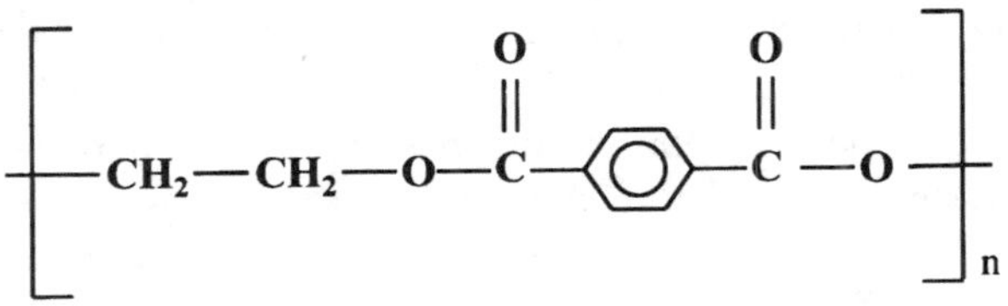

Polycarbonate (PC)

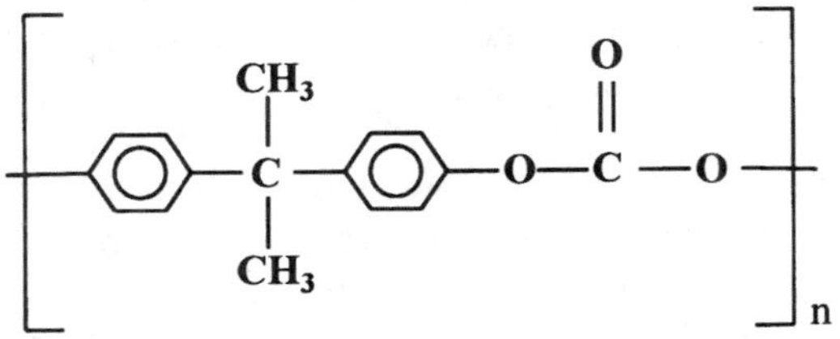

FIGURE 10 Chemical structures of selected polymers.

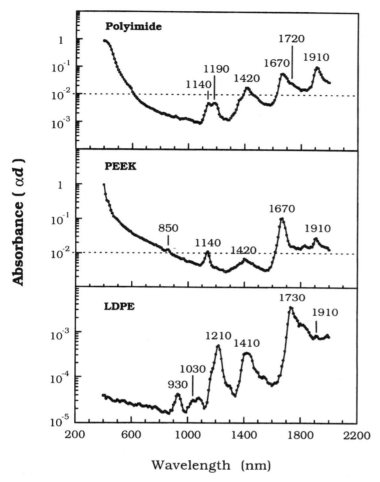

FIGURE 11 Absorption spectra of polyimide, PEEK, and LDPE. The dotted lines represent the approximate detection limit of a common spectrophotometer.

can be observed simultaneously as doublets, although their peak positions are not completely resolved (e.g., in the second C–H overtones).

Apart from the overtone vibrations discussed above, another set of absorption peaks located at about 1910 and 1420 nm can be observed in all of the absorption spectra. An analysis of the fundamental IR data of the polymer spectra reveals that these peaks are very likely to be associated with combination losses. In general, combination losses are weak in intensities, but if the excitations are strongly coupled, they can be observed. Absorption spectra of very high-purity PMMA rod have been reported by Kaino [1] between 0.5 and 2.5 μm. Besides C–H overtone vibrations, they also observed peaks associated with combination losses. These losses have been assigned to the combination band of the C–H stretch (or its overtones) and its bending vibration, δ. In Figs. 11 and 12, the absorption peaks at about 1910 and 1420 nm match those in PMMA. The peak at 1910 nm, for

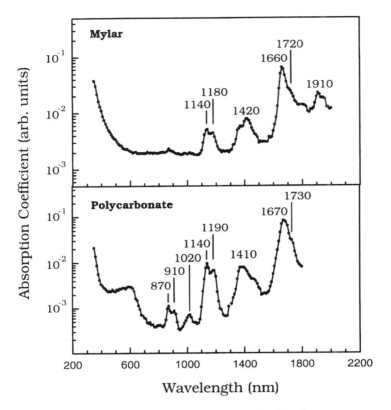

FIGURE 12   Absorption spectra of mylar and polycarbonate.

example, can be assigned to the sum of a C–H stretch and the first overtone of the δ vibration. Using the notation adopted by Skumanich [22], this peak is labeled |1, 2⟩. Similarly, the peak at 1410 nm can be assigned to the combination mode of the first overtone of C–H stretch and a δ vibration, or |2, 1⟩. The detail assignments of the peaks in Fig. 11 are summarized in Table 2.

The overtone absorption peaks in Figs. 11 and 12 can be described by transitions between anharmonic vibrational levels in a Morse potential, according to the expression [27]

$$\frac{E_n}{hc} = (n + \tfrac{1}{2})\nu_h - (n + \tfrac{1}{2})^2 x\nu_h \qquad (18)$$

where E_n is the vibrational energy level with quantum number n and ν_h (cm^{-1}) is the frequency corresponding to harmonic oscillator (in the ground state). The parameter x is the anharmonicity coefficient which describes how far the binding potential deviates from a parabolic potential well. For a $0 \rightarrow n$ transition, the absorption frequency is $\nu_{0 \rightarrow n} = n\nu_h - n(n + 1)x\nu_h$ which can be rearranged as [22]

$$\frac{\nu_{0 \rightarrow n}}{n} = \nu_h(1 - x) - \nu_h xn = a - bn \qquad (19)$$

TABLE 2 Peak Assignments for the PDS-Derived Absorption Spectra of LDPE, PEEK, and PI

	LDPE		PEEK		PI	
	(nm)	(cm^{-1})	(nm)	(cm^{-1})	(nm)	(cm^{-1})
Fundamental frequencies						
Aliphatic C–H stretching mode	3356	2980	—	—	3319	3013
Aromatic C–H stretching mode	—	—	3264	3064	3264	3064
Peak assignments						
$\|1, 2\rangle$	1910	5236	1910	5236	1910	5236
First aliphatic C–H overtone	1730	5780	—	—	1720	5814
First aromatic C–H overtone	—	—	1670	5988	1670	5988
$\|2, 1\rangle$	1410	7092	1420	7042	1420	7042
Second aliphatic overtone	1210	8264	—	—	1190	8403
Second aromatic overtone	—	—	1140	8772	1140	8772
$\|3, 1\rangle$	1030	9709	—	—	—	—
Third aliphatic overtone	930	10753	—	—	—	—
Third aromatic overtone	—	—	850	11765	—	—
Anharmonicity coefficient x						
Aliphatic C–H stretching mode	3.4%		—		3.5%	
Aromatic C–-H stretching mode	—		2.3%		3.5%	

Note: Combination peaks are represented by the notation $\|m, n\rangle$ where m and n represent the number of quanta in the C–H stretching and δ vibrations, respectively. The fundamental frequencies are calculated from Eq. (19).

Here, a and b can be used as fitting parameters in a linear plot of $v_{0\to n}/n$ against n. The results of plots of $v_{0\to n}/n$ against n for polyimide are shown in Fig. 13. From the linear regression, the fundamental absorption frequencies can be obtained. The extrapolated fundamental absorption frequencies are around 3.3 μm (3030 cm^{-1}) for the aliphatic C–H stretching vibration and 3.2 μm (3125 cm^{-1}) for aromatic C–H stretching vibration. These frequencies are consistent with the results obtained from standard infrared spectroscopy. The anharmonicity coefficients are typically less than 4%. The fitting parameters obtained for different polymer films are shown in Table 2.

The intensities of the vibration overtones fall rapidly with increasing nth harmonics, roughly a factor-of-10 drop for each increase in n. In the red region of the spectra, the overtone losses are very weak. The most promising polymer materials for optical fiber (e.g., PMMA or polystyrene) have attenuation of about 10–100 dB/km in this region [1]. For such small attenuation, both scattering and absorption losses must be considered. To lower absorption losses, the positions of the overtone losses can be red-shifted by replacing hydrogen in C–H bonds with more massive atoms. Systematic research has been under progress to study the effect of substituting hydrogen by its isotope, deuterium, or even by more massive atoms such as fluorine [1]. To lower extrinsic scattering losses, the homogeneities of the polymers must be improved.

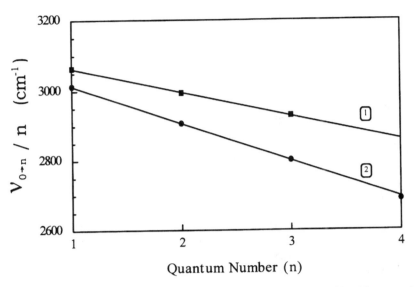

FIGURE 13   Linear fits to the vibrational overtones of polyimide: (1) aromatic C–H stretch; (2) aliphatic C–H stretch.

B.   Applications to Electrooptic Polymers

A class of conducting polymers, known as poly(*p*-phenylene vinylene) (PPV) and its derivatives, have recently generated a lot of research activities. PPVs are conjugated polymers because of the alternation of single and double bonds along the polymer chain. Delocalization of the π electrons along the polymer chain leads to significant electrical conductivity [28]. These polymers are highly fluorescent in the visible range. When a DC voltage is applied across a thin film of PPV, visible light is emitted which can be observed under ordinary room lights [29]. Many efforts are now devoted to fabricating light-emitting diodes using PPV because of its potentials in making large-area flat-panel displays. Applications of the electrooptical properties of this material are also discussed by several authors in this monograph. Despite the excitement generated by PPV, it is still far from commercial use because of its chemical instability and therefore limited lifetimes when used for display purposes. When exposed to visible light and in the presence of oxygen, PPV tends to degrade. Infrared spectroscopy indicates that substantial amounts of carbon–carbon π bonds are broken in this process. Simultaneously, there is strong growth in C=O mode, suggesting that the degradation process is related to photooxidation. PDS can be used to elucidate this process. Absorption spectra of thin films of PPV (derived from PDS) before and after laser irradiation (at 514 nm) are shown in Fig. 14 [30]. Before laser irradiation, a sharp increase in the optical absorption occurs at about 590 nm (2.1 eV), which corresponds to the well-known $\pi \rightarrow \pi^*$ electronic transition associated with the polymer chain. In addition, a series of absorption peaks occur in the range 1.1–2.2 μm, corresponding to the C–H vibrational overtones. After prolong laser treatment (about 20 h), the absorption edge at 590 nm is blue-shifted to about 400 nm. The change of the

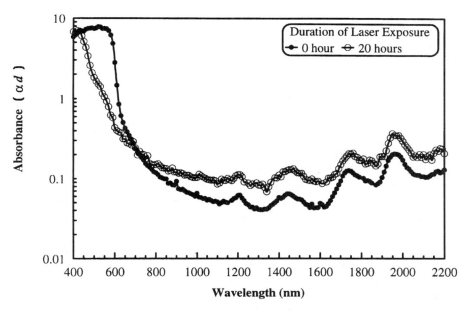

FIGURE 14 Absorption spectra of PPV before and after laser irradiation.

absorption edge can be understood in terms of the reduction of the conjugation length due to the opening of the vinyl double bonds along the polymer backbone. Simultaneously, there is a growth of the optical absorption below the absorption edge. The rise of absorption in this region can be associated with increased defect absorptions.

Other than conducting polymers, electrooptic polymers are often prepared in the form of guest–host systems which are used to fabricate devices for various applied areas (e.g., in second-harmonic generation, light-emitting diodes, and light sensors). The guest–host system can be formed by simply dispersing an optically active material (the chromophore) inside an optically inactive polymer matrix which acts as the host. Common host materials are highly transparent and stable polymers (e.g., polycarbonate or PMMA). Alternatively, the chromophore can be attached to a polymer backbone, via chemical means, and form an optically active side chain. The resulting guest–host system is then subjected to optical or electrical stimulation for the appropriate responses. Many guest–host systems have the following characteristics: (1) The system is in the form of thin film and (2) the optically active material is present in a small concentration and forms minute clusters which act as scattering centers. It is expected that the optical properties of the guest material may be perturbed by the host. The technique of PDS is best adapted for measuring small optical absorption for such system. Guest–host systems developed for nonlinear optical purposes have been studied by PDS and by waveguide loss measurements [24].

VI. CONCLUSIONS

We have shown that PDS can be a valuable tool for studying the absorption of polymers, especially in the wavelength region between 0.7 and 2.2 μm where very few nondestructive techniques have comparable sensitivities. To further explore the potential of PDS as a spectroscopic tool, the resolution of the existing spectra should be improved. The instrumental resolution in a PDS experiment is defined by the slit width of the monochromator. For the spectra shown in Figs. 11 and 12, the instrumental resolution is of the order of 10 nm. On the other hand, the narrowest peak in the data appears to be at least three times broader. Thus, it appears that the intrinsic linewidths of the overtone vibrations may have the major contribution to the experimental linewidths. The intrinsic linewidth of a overtone vibration should be related to the lifetime of the vibrational excited state which should increase with energy due to the increase in the density of states [22]. Such results have indeed been obtained for the C–H overtones of benzene in the gas phase [27]. In Figs. 11 and 12, there are quite a few spectral peaks that appear to have some fine structures, but cannot be well resolved. For example, in the LDPE spectrum, the peak at 1210 nm has a shoulder at the short-wavelength region. Further improvement in the instrumental resolution (e.g., by using a more narrow-band light source such as a tunable laser) may help us to decide if the resolution in PDS is instrumental limited, or if the intrinsic resolution has already been achieved.

Another subject of interest is to use PDS to study the absorptions of oriented polymer films versus the polarization of the pump beam. In the previous discussion, we have assumed that the polymer samples are isotropic. However, polymers chains can be aligned with one another by poling or more sophisticated methods. In an oriented polymer sample, it is expected that the optical absorption will be highly anisotropic. It is natural that PDS measurements should be extended to include polarization dependent measurements on such systems. In one report, the optical absorption of perfectly aligned chains of polydiacetylenes was measured using polarized light with an electric field parallel to and perpendicular to the chain axis [31]. Their results indicate that the polymer absorbs more strongly when the light polarization is parallel to the chain in the region of nonresonant absorption. Recent polarization measurements have been reported for a guest–host system consisting of dye Disperse Red 1 in a matrix of PMMA [32]. More systematic studies of this kind should be performed in future.

ACKNOWLEDGMENTS

Support of this work by the Research Grant Council of Hong Kong is gratefully acknowledged. The authors would like to thank Dr. Louis M. Leung for introducing to us the subject of electrooptic polymers.

REFERENCES

1. Kaino, T., Polymer optical fibers, in *Polymers for Lightwave and Integrated Optics*, (L. A. Hornak, ed.), Marcel Dekker, Inc., New York, 1992, Chap. 1.
2. Kowalczyk, T. C., Kosc, T., and Singer, K. D., Loss mechanisms in polymer waveguides, *J. Appl. Phys.*, 76, 2505 (1994).

3. Meeten, G. H., *Optical Properties of Polymers*, Elsevier, London, 1989, Chap. 1.
4. Klopffer, W., *Introduction to Polymer Spectroscopy*, Springer-Verlag, Berlin, 1984, pp. 30–32.
5. Hordvik, A., Measurement techniques for small absorption coefficients: recent advances, *Appl. Opt.*, *16*, 2827 (1977).
6. Munk, P., *Introduction to Macromolecular Science*, Wiley, Singapore, 1989, Chap. 3.
7. Busse, L. E., McCabe, G. H., and Aggarwal, I. D., Wavelength dependence of the scattering loss in fluoride optical fibers, *Opt. Lett.*, *15*, 423 (1990).
8. Kenny, J. T., Nurse, J. C., Chon, J. C., Binkley, E. S., Stiller, M., Ball, D. W., and Jen, A. K-Y., NLO polymer material systems for electro-optic devices, *Mater. Res. Soc. Symp. Proc.*, *413*, 159 (1996).
9. Christofides, C., Thermal wave photopyroelectric characterization of advanced materials: state of the art, *Crit. Rev. Solid State Mater. Sci.*, *18*, 113 (1993).
10. Mandelis, A. (ed.), *Photoacoustic and Thermal Wave Phenomena in Semiconductors*, North-Holland, New York, 1987.
11. Amer, N. M., and Jackson, W. B., Optical properties of defect states in a-Si:H, in *Semiconductors and Semimetals* (J. I. Pankove, ed.), Academic Press, New York, 1984, Vol. 21B, pp. 83–112.
12. Chan, M. H., So, S. K., Chan, K. T., and Kellert, F., Defect density measurements of low temperature grown molecular beam epitaxial GaAs by photothermal deflection spectroscopy, *Appl. Phys. Lett.*, *67*, 834 (1995).
13. Chan, M. H., So, S. K., and Cheah, K. W., Optical absorption of free-standing porous silicon films, *J. Appl. Phys.*, *79*, 3273 (1996).
14. So, S. K., Chan, M. H., and Leung, L. M., Photothermal deflection spectroscopy of polymer thin films, *Appl. Phys.*, *A61*, 159 (1995).
15. Quelin, X., Perrin, B., Louis, G., and Peretti, P., Three-dimensional thermal-conductivity-tensor measurement of a polymer crystal by photothermal probe-beam deflection, *Phys. Rev. B*, *48*, 3677 (1993).
16. Tam, A. C., Applications of photoacoustic sensing techniques, *Rev. Mod. Phys.*, *58*, 381 (1986).
17. Montecchi, M., and Masetti, E., Characterization of some suitable deflecting liquids in photothermal deflection spectroscopy, *Appl. Opt.*, *29*, 3989 (1990).
18. Milan, H., and Antonin, V., *Interpretation and Processing of Vibrational Spectra*, John Wiley & Sons, Chichester, 1978.
19. Bialkowski, S. E., *Photothermal Spectroscopy Methods for Chemical Analysis*, Wiley, New York, 1996.
20. Welsch, E., Absorption measurements, in *Handbook of Optical Properties, Vol. 1: Thin Films for Optical Coatings*, (R. E. Hummel and K. H. Guenther, eds.), CRC Press. Boca Raton, FL, 1995, Chap. 9.
21. Weinberger, B. R., Roxlo, C. B., Etemad, S., Baker, G. L., and Orenstein, J., Optical absorption in polydiacetylene: A direct measurement using photothermal deflection spectroscopy, *Phys. Rev. Lett.*, *53*, 86 (1984).
22. Skumanich, A., and Moylan, C. R., The vibrational overtone spectrum of a thin polymer film, *Chem. Phys. Lett.*, *174*, 139 (1990).
23. Seager, C. H., Sinclair, M., McBranch, D., Heeger, A. J., and Baker, G. L., Photothermal deflection spectroscopy of conjugated polymers, *Synth. Met.*, *49–50*, 91 (1992).
24. Skumanich, A., Jurich, M., and Swalen, J. D., Absorption and scattering in nonlinear optical polymeric systems, *Appl. Phys. Lett.*, *62*, 446 (1993).
25. Melo, W. L. B., Pawlicka, A., Sanches, R., Mascarenhas, S., and Faria, R. M., Determination of thermal parameters and the optical gap of poly(3-butylthiophene) films by photopyroelectric spectroscopy, *J. Appl. Phys.*, *74*, 979 (1993).

26. Toyoda, T., and Nakamura, H., Photoacoustic spectroscopy of polyanniline films, *Jpn. J. Appl. Phys.*, *34*, 2907 (1995).
27. Child, M. S., Local mode overtone spectra, *Acc. Chem. Res.*, *18*, 45 (1985).
28. Skotheim, A. T. (ed.), *Handbook of Conducting Polymers*, Marcek Dekker, Inc., New York, 1986.
29. Burroughes, J. H., Bradley, D. D. C., Brown, A. R., Marks, R. N., Mackay, K., Friend, R. H., Burns, P. L., and Holems, A. B., Light emitting diode based on conjugated polymers, *Nature*, *347*, 539 (1990).
30. So, S. K., Chan, M. H., Hon, C. S., and Leung, L. M., Optical degradation of poly(p-phenylene vinylene), *Mater. Res. Soc. Symp. Proc.*, *413*, 159 (1996).
31. Thakur, M., Frye, R. C., and Greene, B. I., Nonresonant absorption coefficient of single-crystal films of polydiaacetylene measured by photothermal deflection spectroscopy, *Appl. Phys. Lett.*, *56*, 1187 (1990).
32. Einsiedel, H., and Mittler-Neher, S., Polarization-dependent photothermal beam deflection for the determination of the order parameter C_2 in a laser-oriented polymer system, *Appl. Opt.*, *35*, 5406 (1996).

9

Relaxation Processes in Organic Polymer Systems as Probed by Energy-Selective Laser Spectroscopy

M. A. Drobizhev
Russian Academy of Sciences
Moscow, Russia

I. INTRODUCTION

Since the late eighties the methods of energy-selective (or otherwise site-selective) laser spectroscopy (SLS) initially developed for the investigation of homogeneous spectra of isolated impurity centers found use in the study of molecular extended systems. Usually, one recognizes two versions of SLS: fluorescence line narrowing (FLN) and spectral hole burning (HB). The application of these methods to isolated noninteracting molecules has been extensively reviewed; see Refs. 1–8. However, to our knowledge, there is only one review devoted to the applications of FLN to conjugated polymers [9].

The aim of this chapter is to consider a wider set of systems with intermolecular interaction in an excited state from the standpoint of what possibilities the SLS can provide to study relaxation pathways in them. We shall consider the relaxation of electronic excitation with an incomplete loss of its energy. At least three physically different mechanisms of such relaxation can be regarded:

1. Nonradiative transfer of excitation energy among separate molecules or different segments of molecular aggregates or polymers.
2. Relaxation over a system of levels of the exciton (electron conductance) band. This implies the possibility of initial excitation of the states above the bottom of the band. This can take place in the case of two-particle excitation (electron + hole) or in the case of a Frenkel exciton band in a strongly disordered system.
3. Trapping or self-trapping of a moving exciton (electron) due to its coupling to impurities, defects, or phonons of the lattice. The self-trapping of the exciton resembles the electron-vibrational (or electron–phonon) relaxation in molecules.

Then, excitation can relax down to the ground state with the emission of a lower-energy photon (Stokes-shifted luminescence) or nonradiatively. The listed processes depend crucially on the dimensionality of the system, electron–photon coupling strength, and other parameters.

A concentrated solution of identical molecules placed and oriented chaotically where an energy migration can take place serves as the simplest example of systems with intermolecular interaction. In such systems, the interaction can be so weak that it does not affect the homogeneous spectrum of each molecule.

More complex systems are the physical dimers (bound by Coulomb or van der Waals interactions in the ground state), oligomers, and molecular aggregates (chain-like J-aggregates, H-aggregates, and others). These systems exhibit a strong resonance interaction in the excited state and, in the latter case, the translational symmetry along the chain direction. This results in a dramatic change in the spectrum: The excited level of a single molecule splits into a number (two for the dimer and N for the N-molecular aggregate) of levels giving rise to an excitonic band. The excitation (electron + hole pair) in such aggregates is confined to a single molecule but can move coherently along the chain and is, therefore, described by the Frenkel exciton [10]. Any of the three above-mentioned mechanisms of relaxation can take place here. As it will be shown later, the SLS can help to distinguish among them.

Finally, we shall consider systems consisting of a number of conjugated monomers, such as chemical dimers, oligomers, and polymers. These arrays also have translational symmetry. However, as compared to molecular aggregates, where the electron excitation is confined to a single molecule (Frenkel exciton), an excited electron in conjugated polymers is delocalized at least over some segment of several monomer units. However, the delocalization length is debatable in the literature [11]. If this length is very large, the electron-hole correlation can be neglected and the polymer can be described by a semiconductor model. Optical transitions from the valence band to the conductance band occur [12]. The free charge carrier in this model is assumed to relax with polarization of the medium, forming the self-localized quasi-particle of polaron type [12]. Another approach is based on the model with a high electron-hole correlation [9]. In this approach, an excitation can be described by the Frenkel or Wannier excitons. Due to a high intrinsic disorder and (or) nonrigidity of a one-dimensional (1D) polymer chain, the initially created nearly free exciton quickly localizes on a particular segment of the chain. Its subsequent motion can be described in terms of a random walk similar to nonradiative energy transfer in concentrated solutions. If such a localized exciton has sufficiently high energy, the probability of finding a segment alongside with lower energy is high enough and the exciton can migrate over the bulk of polymer. Upon excitation in the low-energy tail of the density of states (DOS), the probability of energy transfer is low and the exciton dies on the parent segment. In this case, only the intrasegment vibrational relaxation is possible. Therefore, the physical picture of relaxation in bulk polymers is most complex and interesting.

The SLS is the most suitable method for establishing the particular mechanism of relaxation. Its advantage is in the selection of a group of centers with a particular transition energy from an inhomogeneous ensemble and monitoring of its relaxation. It is not necessary to include any chemical impurities which are usually used as donors or acceptors of energy, disturbing, however, the intrinsic structure of the host. Thus, the SLS can provide the possibility of clarifying the nature of excitations in conjugated polymers [9,13,14].

The chapter is organized as follows: In Section II, the principles of SLS (luminescence line narrowing and spectral hole burning) are outlined. A few examples showing how intermolecular interactions can influence the spectra obtained using selective excitation are presented in Section III. In Section IV we describe the relaxation processes in molecular J-aggregates in isotropic solutions. These systems can be regarded as a model of exciton dynamics in well-organized 1D systems without energy transfer among different chains. In Section V, the results obtained with conjugated polymers using SLS are reviewed.

II. PRINCIPLES OF ENERGY-SELECTIVE SPECTROSCOPY

We consider here only the basic principles of SLS which will be inevitably applied to extended systems. More pronounced reviews can be found in Refs. 1–8.

In disordered solid materials, the energies of optical transitions of chromophores are scattered due to inhomogeneous surrounding of each chromophore. This leads to the inhomogeneous broadening of optical spectra which obscures a true homogeneous spectrum. Energy-selective spectroscopy consists of monochromatic excitation of chromophores whose transition energies are in resonance with excitation frequency. A narrow hole in the absorption spectrum appears due to the selective photochemical (or photophysical) elimination of centers from resonance with laser radiation. A spectrally broad antihole appears in another region. Selective excitation results also in appreciable narrowing of the luminescence spectrum because only a subset of centers of the entire ensemble is excited. If the inhomogeneous distribution of energies is not changed during the time of the experiment, its influence can be eliminated and the true homogeneous spectrum can be recovered either from the shape of the hole in the absorption spectrum or from the selectively excited luminescence spectrum.

The homogeneous spectrum of a molecule in a solid matrix at low temperatures consists of a very narrow (10^{-4}–1 cm^{-1}) and intense zero-phonon line (ZPL) and a sufficiently broad (10–10^2 cm^{-1}) structureless phonon wing (PW); see Fig. 1. The ZPL corresponds to electronic transition without the participation of phonons. The PWs in absorption and luminescence spectra correspond to electronic transitions with the creation of phonons. In absorption spectrum, the PW lies at a higher-energy side of the ZPL. In luminescence spectrum, it lies at a lower-energy side of the ZPL, in accordance with the mirror symmetry principle; see Fig. 1. This feature (ZPL + PW) is, as a rule, replicated by intermolecular (or local) dispersionless vibrations. Thus, in a homogeneous spectrum, a number of narrow ZPLs (corresponding to 0–0 and a series of vibronic transitions) each with its own PWs are observed.

Most studies are concentrated on the two aspects of selectively obtained spectra: (1) the spectral width of the narrow ZPL and (2) the shape of the entire homogeneous spectrum. The width of the ZPL for 0–0 transition (coincident in absorption and emission) is usually studied by the HB version of the SLS because of the problem of scattered light in recording the fluorescence spectrum. The entire spectrum (ZPL + PW) can be obtained by both the HB and FLN techniques (for

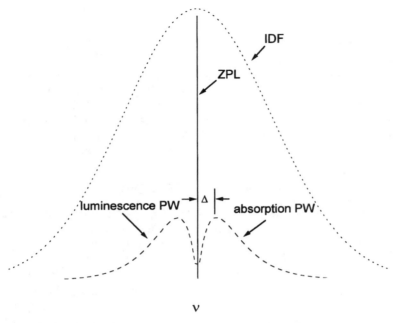

FIGURE 1 Schematic view of homogeneous absorption and emission spectra with a narrow zero-phonon line (ZPL), phonon wings (PW), and an inhomogeneous site distribution function (IDF).

0–0 or vibronic transitions, respectively). However, the FLN is the choice for systems with low hole-burning quantum yield.

The effect of selection in the systems with noninteracting molecules reveals itself primarily in the narrowing of the luminescence spectrum and its shifting with excitation frequency. It should be noted here that these effects can be observed only if excitation and emission belong to the same electronic transition. If, for example, we excite the $S_0 \rightarrow S_2$ transition monochromatically and record the $S_1 \rightarrow S_0$ luminescence, the selectivity is lost after relaxation [1]. On the other hand, vibronic relaxation within the same electronic term does not lead to the loss of selectivity. This fact can be used for the determination of the nature of relaxation in extended systems.

Let us consider now the two above-mentioned aspects of the SLS in detail.

A. Broadening of the Zero-Phonon Hole

A narrow zero-phonon hole (ZPH) arises in absorption spectrum as a result of resonant selective excitation of centers via their ZPLs. A subsequent probing photon will not be absorbed by these centers via their ZPLs. The ZPL has, as a rule, a Lorentzian shape, reflecting homogeneous broadening. If all experimental conditions (temperature, pressure, etc.) are equal upon burning and recording of the hole and burning fluences and the time of the burning-reading cycle are very small, the hole width Γ_H is determined by a convolution of two homogeneous ZPLs of width γ:

$$\Gamma_H = 2\gamma \tag{1}$$

According to the uncertainty principle, γ (in cm^{-1}) is equal to

$$\gamma = \frac{1}{\pi c T_2} = \frac{1}{2\pi c T_1} + \frac{1}{\pi c T_2^*} \tag{2}$$

where T_2 is the total dephasing time, T_2^* is the pure dephasing time, and T_1 is the population decay time of the excited state. At low (but nonzero) temperatures, the temperature dependence of γ is completely determined by the pure dephasing contribution, which vanishes at $T \rightarrow 0$ K. Note that T_1 coincides (for singlet–singlet transitions) with the fluorescence decay time τ_F only for the isolated two-level molecule.

The temperature dependence of dephasing rate of isolated centers in amorphous glasses and polymers is determined by coupling of electronic excitation to phonons and tunneling two-level systems (TLSs) of a glass. (Here, the notion of two-level systems implies that in a number of asymmetric double-well potentials, inherent in the amorphous matrix, only two levels either in the excited or ground state of the nearest wells are considered [15,16]. This notion should not be confused with a two-level system involving electronic excited and ground states.) At very low temperatures ($T < 4$ K), the role of phonons is small and the broadening is determined by the interaction with TLSs and described by the power law $1/T_2 \propto T^\mu$ with $\mu \approx 1.3$ [5]. At higher temperatures, the role of phonons becomes appreciable. The broadening law of the Arrenius type (exchange model [17]), implying the coupling to an effective dispersionless phonon of frequency ω, is widely used now. Thus, the homogeneous width is described by a combination [6,18]:

$$\gamma(T) = \gamma(0) + AT^{1.3} + B \exp\left(-\frac{\hbar\omega}{kT}\right) \tag{3}$$

A persistent ZPH can broaden itself in an amorphous matrix during the time of experiment [19,20]. This is explained by the nonequilibrium state of amorphous matter resulting in continuous redistribution of transition energies within an inhomogeneous absorption band. Such broadening is called the spectral diffusion and depends near linearly on temperature (at constant experimental time) [6]. This functional dependence deviates only slightly from the TLS-coupling term in the dephasing rate and, therefore, these two contributions can be combined in a single term. Therefore, the temperature dependence of the hole width (at low fluences and intensities of burning light) is described by

$$\Gamma_H(T) = 2\gamma(0) + A'T^\mu + B' \exp\left(-\frac{\hbar\omega}{kT}\right) \tag{4}$$

with $\mu \approx 1.3$.

B. The Entire Shape of Selective Spectra

Consider now the selectively excited luminescence spectrum and its relation with the homogeneous one. Similar to HB, the FLN is a two-quantum method. The primary quanta of light with frequency ν_{ex} excite selectively a number of centers via their absorption ZPLs and PWs and then the secondary quanta are emitted with a distribution of intensity reflecting the homogeneous luminescence spectra. Therefore, the selectively excited spectrum does not coincide with the homogeneous one.

Let $g(\nu_0)$ be the inhomogeneous distribution function (IDF) of centers (a number of centers with ZPLs maxima at ν_0), $a(\nu_{ex} - \nu_0)$ the homogeneous absorption spectrum, and $f(\nu - \nu_0)$ the homogeneous luminescence spectrum. As has been mentioned earlier, the homogeneous spectrum consists of a narrow ZPL and a broad PW:

$$a(\nu_{ex} - \nu_0) = \alpha z(\nu_{ex} - \nu_0) + (1 - \alpha)p(\nu_{ex} - \nu_0) \tag{5a}$$

$$f(\nu - \nu_0) = \alpha z(\nu - \nu_0) + (1 - \alpha)p(\nu - \nu_0) \tag{5b}$$

Here, α is the Debye–Waller factor, which characterizes the strength of electron–phonon coupling, $z(\nu - \nu_0)$ is the ZPL (normalized to unity), and $p(\nu - \nu_0)$ is the corresponding PW. We assume in Eqs. (5a) and (5b) that the rule of mirror symmetry of spectra takes place (the excited- and ground-state potentials are not distributed but only shifted along a configurational coordinate) and, therefore, $a(\nu_{ex} - \nu_0) = f(\nu_0 - \nu)$. The observed absorption spectrum is described by the convolution

$$A(\nu_{ez}) = \int_{-\infty}^{\infty} a(\nu_{ex} - \nu_0)g(\nu_0)\,d\nu_0 \tag{6}$$

Thus, absorption spectrum is structureless and broader than the IDF. The selectively excited luminescence spectrum is described by the following convolution (the laser line is assumed to be the delta function):

$$F(\nu, \nu_{ex}) = \int_{-\infty}^{\infty} f(\nu - \nu_0)a(\nu_{ex} - \nu_0)g(\nu_0)\,d\nu_0 \tag{7}$$

where the excitation frequency ν_{ex} can be regarded as a parameter. After substituting Eqs. (5a) and (5b) into Eq. (7), one can see that the observed luminescence spectrum consists of four terms [21]: F_{zz}, F_{zp}, F_{pz}, and F_{pp}. The first corresponds to excitation via ZPL of the homogeneous absorption spectrum and the emission of a photon also within the ZPL (0–0 or vibronic) of the homogeneous luminescence spectrum. The second corresponds to excitation via the ZPL and emission within the PW of the luminescence spectrum. The third and fourth terms are responsible for excitation via PWs of the homogeneous absorption spectrum and emission within the ZPL and PW of luminescence spectrum, respectively. Consequently, the observed luminescence spectrum consists of a narrow ZPL (F_{zz}) and a broad phonon wing ($F_{zp} + F_{pz} + F_{pp}$). This feature is replicated by vibronic transitions. The contrast of the luminescence spectrum (relative intensities of ZPLs and a broad phonon background) depends on the electron–phonon coupling strength (Debye–Waller factor) and on excitation frequency [22]. It follows from Eq. (7) and the properties of convolution that if the homogeneous absorption spectrum is much narrower than the IDF, the observed shape of the luminescence spectrum does not depend on the excitation frequency, and the entire spectrum shifts resonantly (or quasi-resonantly for vibronic replica) with excitation frequency. Such behavior, indeed, takes place at least for vibronic ZPL (the 0–0 ZPL is obscured by scattered light) of molecules in glasses and polymers used as matrices. As far as the PW of the homogeneous spectrum is concerned, it can be of comparable width with the IDF. Model calculations of the observed luminescence spectrum and its dependence on excitation frequency in this case were carried out in Refs. 23–25. The shapes of the IDF and PWs in absorption and luminescence were taken to be Gaussians:

$$p(\nu - \nu_0) - (2\pi\sigma_p^2)^{-1/2} \exp\left(-\frac{(\nu - \nu_0 \pm \Delta)^2}{2\sigma_p^2}\right)$$

$$g(\nu_0) = (2\pi\sigma^2)^{-1/2} \exp\left(-\frac{\nu_0^2}{2\sigma^2}\right)$$

where σ_p and σ are standard deviations of PW and IDF, respectively, the maximum of IDF is considered to be at $\nu_0 = 0$, the upper sign in the first equation corresponds to luminescence and the lower to absorption, and Δ is the separation between the ZPL and PW maxima, called the Stokes losses. In this case, the spectrum [Eq. (7)] can be calculated analytically. It has been shown [23–25] that the spectrum [Eq. (7)] shifts with the excitation frequency tuned in the entire range of IDF. In addition, as the excitation frequency increases, the following occur:

1. The separation between ν_{ex} and the observed phonon side-band maximum increases.
2. The integral intensity of the ZPL decreases in relation to the phonon side band.
3. The observed phonon side band broadens.

Let us consider here an interesting specific case of strong exciton–phonon coupling (small Debye–Waller factor) when the ZPL is vanishing in homogeneous spectra. As will be shown below, this is precisely the case in π-conjugated polymers. Thus, the selectively excited (via PWs) luminescence spectrum has the only contribution, F_{pp}. Let us write it explicitly as [see Eqs. (5a), (5b), and (7)]

$$F_{pp}(\nu, \nu_{ex}) = (1 - \alpha)^2 \int_{-\infty}^{\infty} p(\nu - \nu_0)p(\nu_0 - \nu_{ex})g(\nu_0)\, d\nu_0 \tag{8}$$

The dependence of this function on excitation frequency for the Gaussian shapes of IDF and PW has been considered in Refs. 24 and 25. The F_{pp} has a Gaussian shape with standard deviation

$$\sigma_{pp} = [(\sigma^{-2} + \sigma_p^{-2})^{-1} + \sigma_p^2]^{1/2} \tag{9}$$

At the same time, the absorption spectrum [Eq. (6)] is also Gaussian with standard deviation

$$\sigma_a = (\sigma^2 + \sigma_p^2)^{1/2} \tag{10}$$

The frequency ν_{em} of maximum intensity of luminescence PW in vibronic replica with intermolecular frequency Ω shifts with excitation frequency ν_{ex} as [24,25]

$$\nu_{em} = \frac{\nu_{ex}}{n^2 + 1} - \frac{\Delta(n^2 + 2)}{n^2 + 1} - \Omega, \tag{11}$$

where $n = \sigma_p/\sigma$.

The last three equations are very important in the following analysis of data obtained with conjugated polymers and, therefore, we shall consider them and conditions of their derivation in detail. First, the dependence [Eq. (11)] of ν_{em} on ν_{ex} is linear. The slope k of this dependence is equal to $(n^2 + 1)^{-1}$ and is less than unity. Thus, if the IDF is Gaussian in shape, the maximum of the luminescence band will shift nonresonantly and will intersect the resonance line $\nu_{em} = \nu_{ex}$ at

$$\nu^* = -\frac{\Delta(n^2 + 2)}{n^2} - \frac{\Omega(n^2 + 1)}{n^2} \tag{12}$$

Therefore, the true Stokes losses Δ can be obtained from this frequency if n and Ω (for vibronic transition) are known. Note that in the case of a narrow PW ($n \ll 1$), Eq. (11) gives a quasi-resonant behavior: $\nu_{em} = \nu_{ex} - 2\Delta - \Omega$ with constant shift $2\Delta + \Omega$.

It is obvious from Eqs. (9) and (10) that the measurable ratio σ_{pp}/σ_a can also be presented in terms of n:

$$\frac{\sigma_{pp}}{\sigma_a} = \frac{n(n^2 + 2)^{1/2}}{n^2 + 1} \tag{13}$$

Therefore, we can relate the two observables σ_{pp}/σ_a and k by

$$\frac{\sigma_{pp}}{\sigma_a} = (1 - k^2)^{1/2} \tag{14}$$

and obtain the shift Δ for 0–0 transition from measured values of k and ν^*:

$$\Delta = -\frac{\nu^*(1 - k)}{1 + k} \tag{15}$$

In the model under consideration, the luminescence spectrum will shift linearly with ν_{ex} in the entire range of IDF with the slope less than unity. This simple "two-Gaussians" model (Gaussian IDF and PWs) should, however, be used with care.

Consider, for example, a more natural PW, described by the asymmetric function

$$\begin{aligned} p(\nu - \nu_0) &= \beta^{-2}(\nu - \nu_0) \exp\left(-\frac{\nu - \nu_0}{\beta}\right), & \nu - \nu_0 > 0 \\ p(\nu - \nu_0) &= 0, & \nu - \nu_0 < 0 \end{aligned} \tag{16}$$

This type of PW simulates the $T \to 0$ limit more adequately because it is free from the anti-Stokes part. First, let the IDF to be a Guassian with standard deviation σ and the homogeneous luminescence spectrum have a mirror symmetry with respect to the absorption one. In this case, the luminescence spectrum will (with an accuracy of constant factor) be

$$\begin{aligned} F_{pp}(\nu, \nu_{ex}) = \exp\left(\frac{\nu - \nu_{ex}}{\beta}\right) \int_\nu^{\nu_{ex}} [\nu_0(\nu + \nu_{ex}) - \nu_0^2 - \nu\nu_{ex}] \\ \exp\left(-\frac{\nu_0^2}{2\sigma^2}\right) d\nu_0 \end{aligned} \tag{17}$$

The integration can be performed numerically. The corresponding dependence of the maximum position ν_{em} on ν_{ex} is shown in Fig. 2. It is interesting that if β and σ are comparable, ν_{em} depends linearly on ν_{ex} only in the low-energy side of IDF. On the other hand, ν_{em} does not depend on ν_{ex} in the high-energy side of IDF. As a result, one can approximate the entire dependence by two straight lines in the

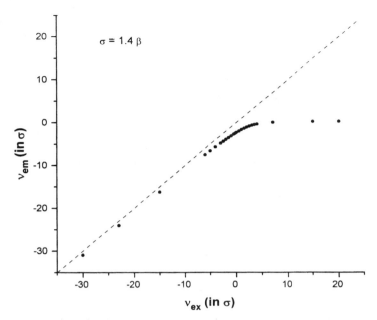

FIGURE 2 Dependence of ν_{em} on ν_{ex} in the case of the Gaussian shape of IDF and the Gamma distribution [Eq. (16)] for PW, as presented in Fig. 1.

low- and high-energy sides of IDF. Our calculations show that the intersection point of two asymptotic lines lies in the high-energy side of IDF at about 1σ–2σ for β/σ varying from 2.9 to 0.7. It is interesting that the slope k of the low-energy part of dependence is nearly constant when one varies the β/σ from 2.9 to 0.7 and it is equal to 0.85 for ν_{ex} in the range -4σ–0. (Note, for comparison, that in the "two Gaussians" model, the slope changes from 0.0625 to 0.5 under these conditions.)

Another deviation from the model of "two Gaussians" can be an exponential dependence of the density of states (DOS) on the frequency in the red tail (Urbach tail). Such behavior is inherent for the density of excitonic states of systems with disorder [26–28]. Let us consider, therefore, the homogeneous spectra determined by Eq. (16) and obeying the rule of mirror symmetry and

$$g(\nu_0) = C \exp(\eta\nu_0) \tag{18}$$

where C and η are the constants. Substitution of Eqs. (16) and (18) into Eq. (8) gives, with an accuracy of constant,

$$F_{pp}(\nu,\nu_{ex}) = \exp\left(\frac{\nu - \nu_{ex}}{\beta}\right) \int_{\nu}^{\nu_{ex}} \exp(\eta\nu_0) (\nu + \nu_{ex} - \nu_0^2 - \nu\nu_{ex}) \, d\nu_0 \tag{19}$$

The integral in Eq. (19) can be calculated analytically. The resulting luminescence spectrum is

$$F(\nu,\nu_{ex}) = \exp\left(\frac{\nu - \nu_{ex}}{\beta}\right) \left(\frac{\nu_{ex} - \nu)[\exp(\eta\nu_{ex}) + \exp(\eta\nu)]}{\eta^2}\right.$$
$$\left. - \frac{2[\exp(\eta\nu_{ex}) - \exp(\eta\nu)]}{\eta^3}\right)$$

Its maximum is determined by the transcendental equation

$$\frac{x(e^x + 2)}{e^x - 1} = 2 + \eta\beta \qquad (20)$$

where $x = \eta(\nu_{ex} - \nu_{em})$. For the vibronic luminescence transition observed at $\nu_{vib} = \nu_{em} - \Omega$, one must set $\nu_{em} = \nu_{vib} + \Omega$ to determine x in Eq. (20). It is evident that the result depends only on the difference $\nu_{ex} - \nu_{em}$ and, therefore the dependence ν_{em} versus ν_{ex} (or ν_{vib} versus ν_{ex} for vibronic transitions) has the slope $k = 1$. Investigation of Eq. (20) shows that it has no solutions if $\eta\beta < 0.74$, a single solution if $\eta\beta > 1$, and two solutions in a narrow range of parameter values if $0.74 < \eta\beta < 1$. The shift $\nu_{em} - \nu_{ex}$ can be obtained from the plot of Fig. 3. Note that in this particular case, the shift is not directly related to the Stokes losses (equal to β) but depends on the inhomogeneous Urbach parameter η as well.

III. THE INFLUENCE OF INTERMOLECULAR INTERACTION ON SELECTIVE SPECTRA

Let us now consider some prominent examples of the influence of intermolecular interaction on selectively obtained spectra. The systems considered here are simpler than the true polymers, but they can model different relaxation processes observed in polymers successfully. The intermolecular coupling value gradually increases in this set of systems.

A. Intermolecular Energy Transfer

The first example concerns the energy transfer of Förster type between equivalent molecules having different energies of ZPLs but with overlapping PWs. The dipole–dipole interaction between molecules is not strong enough to change the structure of absorption spectrum of noninteracting molecules. Thus, in these systems, the intermolecular interaction is hidden and cannot be unraveled using standard nonselective spectroscopy.

In Ref. 29, the effect of the dye concentration on the zero-phonon hole width was studied for free-base chlorin in polystyrene. First, it was found that the hole width at 1.2 K depends on excitation wavelength for concentrations greater than 10^{-5} M. It decreases with increasing wavelength and reaches a minimum value (equal to the hole width of a low-concentration sample) only in the long-wavelength edge of inhomogeneous band; see Fig. 4. The hole width is found to be proportional to the fraction of integrated absorption with energies lower than the laser frequency (equal to the error function for the Gaussian IDF). This means that the hole width is a measure of the energy transfer rate, which is proportional to the number of acceptor states. At the lowest concentration (10^{-5} M), the hole width does not depend on the wavelength. Second, the coefficient A' in temperature dependence [Eq. (4)] of the hole width is almost the same for all concentrations. On the other

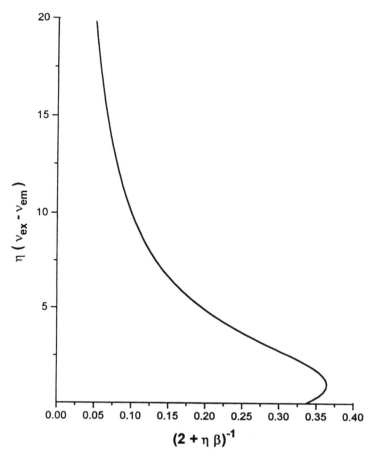

FIGURE 3 Dependence of the observed normalized shift $\eta(\nu_{ex} - \nu_{em})$ on the model parameter in the case of exponential IDF [Eq. (18)] and Gamma distribution for PW [Eq. (16)].

hand, the limiting value of Γ_H at $T \to 0$ is determined by the fluorescence lifetime τ_F, measured after spectral relaxation [$\gamma(0) = 1/(2\pi c \tau_F)$] only for the lowest concentration. At higher concentrations, this value is greater. Qualitatively, similar results were obtained in Refs. 30 and 31 for porphyrin-doped Langmuir–Blodgett films and in Ref. 32 for chlorin and tetraphenylporphyrin in polystyrene.

These results were explained by the energy transfer from centers with high energy to those with lower energy. The transfer probability per unit time from the molecule with energy E is proportional to the number of centers with energies smaller than E. Excitation created in the low-energy site cannot be transferred and dies in the same site. It is interesting that the energy transfer process contributes only to the temperature-independent part of the linewidth. This suggests that the observed relaxation is a T_1 process.

Another approach was employed in Ref. 33. The authors used selective excitation of luminescence to study the energy transfer in chlorophyll-a and pheophytin-a frozen solutions. For concentrated solutions, it was found that increasing of ex-

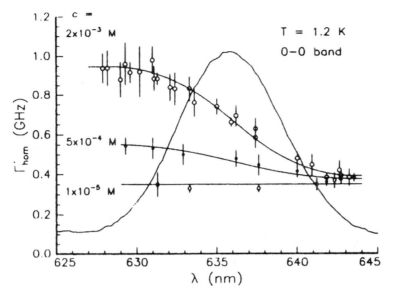

FIGURE 4 The homogeneous linewidth γ versus excitation wavelength for free-base chlorin in polystyrene at different concentrations at $T = 1.2$ K. The $S_1 \leftarrow S_0$ absorption band is also plotted. (From Ref. 29.)

citation frequency results, as compared to diluted solutions, in smoothing of the ZPLs intensity (against the phonon background) and red-shifting of the whole spectrum. It should be noted that the ZPLs are present in the luminescence spectrum even after the energy transfer. These observations are in qualitative agreement with the model calculation accomplished in this work [33].

Jankowiak et al. [34] observed the dependence of the phosphorescence spectrum of glassy 2-bromonaphthalene films on the excitation frequency at low temperature. The film under study was a bulk nondiluted disordered system of chromophores where triplet excitons can migrate by hopping. The peaks of the spectrum shifted with excitation frequency if the latter was scanned across the red side of the inhomogeneously broadened absorption band; see Fig. 5. On the other hand, the positions of these peaks became independent of excitation frequency in the blue side of the band, at approximately $\nu_{ex} > 2\sigma_a$. The authors explained this dependence by the presence of energy transfer at high-energy excitation and the absence of it at low-energy excitation. The low-energy part of the dependence obviously implies that, in this range, the energy transfer is not significant. But the saturation in the blue side can be due to either the energy transfer or to the specific shape of the homogeneous spectrum with the PW width comparable to that of IDF in the system without energy transfer (see Fig. 2).

B. Relaxation in Physical Dimers and Oligomers

In physical dimers (trimers, oligomers), the constituent molecules are tightly bound to each other by Coulomb or (and) van der Waals interactions. The resonance

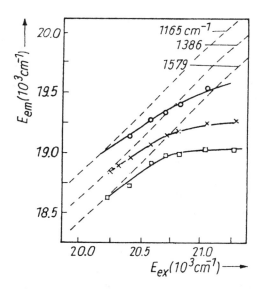

FIGURE 5 Variation of the energy of the 1165-, 1368-, and 1579-cm^{-1} vibronic phosphorescence bands with excitation energy. Dashed curves indicate the resonance case $\nu_{em} = \nu_{ex}$, where Ω is the energy of the vibrational model. (From Ref. 34.)

dipole–dipole interaction of the excited molecule with its neighbors gives rise to an excitonic splitting of the excited-state level of a single molecule into a number (equal to the number of molecules in the complex) of levels.

Dimers of porphyrins were studied by hole burning in Refs.. 35 and 36. It was established that the hole width of a dimer is larger than that of the corresponding monomer in the temperature ranges from 1.2 to 4.2 K [35] and from 10 to 20 K [36]. Two reasons for this difference were discussed in Ref. 36. First is the phonon scattering between two excitonic levels E_+ and E_- of a dimer. This process leading to a dephasing of the probed state (e.g., E_-) is important when excitonic splitting $\Delta E = E_+ - E_-$ is smaller than kT. In dimers of etioporphyrin studied in Ref. 36, the excitonic splitting ΔE amounts to some tens of wave numbers [37] and, therefore, the above condition is not fulfilled in the temperature range studied. The second possibility is a non-radiative transitions $E_+ \rightarrow E_-$ between two excitonic states. These transitions can occur if (a) both excitonic states are optically accessible (the molecular planes are nonparallel in the dimer) and (b) the splitting ΔE is smaller than the inhomogeneous broadening. The latter situation is probably typical for dimers of etioporphyrin [37]. The relaxation of such a type can be called the "intraband" excitonic relaxation. In our opinion, this type of relaxation can, however, also have a minor influence on the hole width because (1) the absorption to a higher component (E_+) is weak (this is supported by an efficient fluorescence of the dimer from the E_- state [37]) and (2) the relaxation of this state is faster than that of the E_- state. Both of these effects reduce the hole-burning efficiency and, therefore, the contribution of the hole burnt in the $S_0 \rightarrow E_+$ transition is expected to be smaller than that burnt in the $S_0 \rightarrow E_-$ transition.

There is, however, a third possible mechanism of relaxation, not considered in Ref. 36 in the context of ZPH broadening. It is related to a change of intermolecular geometry of a dimer after excitation. Rashba [38] showed that if the parameter of electron–phonon coupling is small, the probabilities (average in time) of finding an excitation in either molecule of a dimer are equal. (At each given moment of time, the electronic density is concentrated on a single molecule, but it oscillates with the frequency $f = 2\pi\Delta E/h$). This state is similar to a free exciton in a crystal. But if the electron–phonon coupling exceeds some critical value, the structure of a dimer changes, the excitation localizes on one single molecule, and the energy of this state lowers with respect to the lowest E_- state of an unperturbed dimer. This relaxation is analogous to the self-trapping of an exciton in crystals. Obviously, such a relaxation process can result in additional broadening of the homogeneous line in the dimer as compared to the monomer. Note that independent experimental suggestions of structural rearrangement in etioporphyrin dimers can be found in Refs. 36 and 37.

Another very interesting observation is the absence of a fine structure (ZPLs) in the fluorescence spectrum of dimers upon selective laser excitation at low temperature [36]. A similar effect as well as the independence of the observed fluorescence spectrum on excitation wavelength was observed in trimers of *C*-phycocyanine [39]. These observations together with the possibility of hole burning in absorption bands of these systems can be explained by the absence of correlation between absorption and emission excited states, which, therefore, belong to different electronic transitions (see Section II). This agrees well with the model [38] of structural rearrangement in the complex (self-trapping of exciton).

C. Selective Spectra of Chemical Oligomers

If the resonance intermolecular interaction in physical dimers, oligomers, and aggregates amounts from some tens to hundreds of wave numbers, this interaction in chemically bound molecular subunits with conjugated π bonds can amount to thousands of wave numbers. In contrast to physical oligomers, the electronic density is delocalized in this case over the whole assembly. Therefore, chemical oligomers can be regarded as long organic molecules with a system of conjugated bonds.

Selectively excited fluorescence spectra of a set of oligo(*p*-phenylenevinylene)s (OPV) were obtained in Ref. 40; see Fig. 6. An important observation is that the sharpness of vibronic ZPLs decreases with the increase of oligomer length. The ZPLs are observed for $n = 2$ and 3 and are almost blurred for $n = 4$, where n is the number of monomer units. This effect can be due to the increase of vibrational degrees of freedom and (or) to the increase of electron–phonon coupling with oligomer size. In pertinent polymer poly(*p*-phenylenevinylene) (PPV), the ZPLs are also absent under the same conditions. This is a crucial point for the analysis of the site-selective spectroscopy of conjugated polymers presented in Section V.

IV. SELECTIVE SPECTROSCOPY OF MOLECULAR J-AGGREGATES (SCHEIBE POLYMERS)

It had been discovered 60 years ago that cyanine dyes can form specific J-aggregates with a very narrow absorption band shifted to the red relative to that of

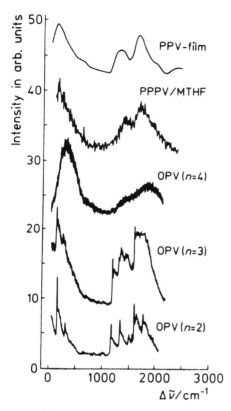

FIGURE 6 Site-selective fluorescence spectra of oligo(*p*-phenylenevinylene)s (OPV), poly(*p*-phenyl-phenylenevinylene) (PPPV) in MTHF, and PPV films, measured at 6 K. The abscissa scale is relative to the excitation energy ν_{ex}. (From Ref. 40.)

monomer [41,42]. The molecules in aggregate are coupled in the ground state by ionic and van der Waals interactions building a linear chain [43]. In this respect, a J-aggregate is a polymer of the same nature as physical dimers and oligomers considered earlier. The optical properties suggest the fundamental excitations in these systems to be the Frenkel excitons [10]. Indeed, only if the excitation (electron + hole) is concentrated on a single molecule, the optical line will be narrow due to a specific selection rule based on the wave-vector conservation law [10,44]. This implies that only an exciton with a wave vector close to zero can be produced by a photon, because the wavelength of the photon is much greater than the intermolecular distance in the aggregate.

The upper electronic level of a molecule is split into N levels in the aggregate due to a resonance dipole–dipole interaction making the excitonic band. Here, N is the number of coherently coupled molecules (coherence length). In an ideal chain, the exciton wave vector k is a quantum number, labeling each level, and due to the wave-vector conservation law, only the lowest (for negative resonance interaction) level with $k = 0$ is optically accessible. The striking feature of J-aggregates is that the oscillator strength is proportional to the coherence length N [45].

J-Aggregates are usually embedded in amorphous media that possess, as compared to molecular crystals, a greater structural disorder. This disorder can be described by an inhomogeneous distribution of site energies (diagonal disorder) and a distribution of resonance coupling constants between sites (nondiagonal disorder); see Refs. 46 and 47, and references therein. In addition, in reality some defects can break the chain into relatively decoupled segments. Furthermore, a real chain is deformable, and molecular (or nuclear) displacements can affect the exciton dynamics via the exciton–phonon coupling. In perturbed chain, the states with different k values, belonging to an unperturbed system, are mixed and, therefore, the selection rule ($k = 0$) is not too severe. Other levels become accessible for absorption. Consequently, the disorder will result in inhomogeneous line broadening, although the stationary states are still delocalized over particular segment lengths.

A. Relaxation in Isotropic Solutions of J-Aggregates

In spite of considerable attention paid to molecular J-aggregates for many years, some of their optical properties remain unclear. Among them is such an important problem as the origin of relaxed states. Are they the final states of the energy transfer process between rigid subunits (segments) of the long physical chain or localized (trapped, self-trapped) excitations of the deformed chain? In the first case, the longest segments with the most delocalized states are the luminescent ones, whereas in the second case, the emitting states are the localized ones (relative to free-exciton states) being connected with the local lattice distortions. The third possibility is a relaxation within the density of states (DOS) of the excitonic band. It implies absorption not only by the band bottom states, which is facilitated due to the above-mentioned mixing of excitonic states in perturbed lattice.

Little attention has been paid in the literature to the possibility of photo-induced chain deformation as a possible mechanism of exciton relaxation in J-aggregates. In particular, exciton self-trapping by chain distortion can be considered as a channel of exciton relaxation in a deformable lattice [48]. This process crucially depends on the system dimensionality and size restrictions [48–51]. It was found theoretically for three-dimensional (3D) systems that if the parameter of exciton–phonon coupling exceeds some critical value, the stable self-trapped state (ST) appears below the bottom of the band of free-exciton states (F) [38,52]. In this case, two energy minima are separated by a potential barrier. Contrary to 3D systems, as has been shown by Rashba [38], in infinite, strictly 1D chains, self-trapping occurs for any nonvanishing exciton–phonon coupling and the barrier between ST and F states is absent.

The first experiment on persistent hole burning in J-aggregates was performed by De Boer and coauthors [53]. They studied the J-aggregates of pseudoisocyanine (PIC) bromide in a frozen glass of water/ethylene glycol (WEG) (1/1) mixture. J-Aggregates of PIC is the most widely studied system based on the pioneering works of Jelley and Scheibe [42,43]. In this system, there exist at low temperatures, two J bands ("blue" and "red"), assigned to aggregates with different structure. They have extremely small widths of absorption bands (of ~30 cm^{-1}) and no measurable Stokes shift of fluorescence in WEG at liquid-helium temperatures [45,53,54].

Subsequent in-depth hole-burning studies of J-aggregates of PIC chloride and iodide in WEG glass was accomplished in Refs. 55–58. Note that the possibility of hole burning in the absorption band of J-aggregates of PIC and other dyes (see below) suggests that this band, even being very narrow, is inhomogeneously broadened at low temperatures. The authors [55] observed the antihole (increasing of optical density) distributed over the inhomogeneous band rather symmetrically on both sides of the resonant hole. The antihole area was equal to the hole area. This led the authors to the conclusion that the hole-burning mechanism is a light-induced conformational change of the aggregate chain. However, the redistribution of surrounding solvent molecules (nonphotochemical HB) was not excluded.

Unusual behavior of spectral holes in J-aggregates, as compared to a single molecule in a glassy matrix, has been observed for PIC iodide red aggregates [56]. First, the hole width remained practically constant at 4 K for a week. For monomers in alcohols, the hole broadens by a factor of 2 in these conditions due to the spectral diffusion. Also, the broadening of the hole after excursion of temperatures up to 70 K is quite small in J-aggregates. The temperature broadening of the hole is also very intriguing. The hole width remains constant and equal to 0.36 cm^{-1} full width at half-maximum (FWHM) if one increases temperature up to ~10 K. The broadening at higher temperatures is well described by a sum of activation exponential terms without a power-law contribution employed for isolated molecules in glasses [see Eq. (3)]. All these facts suggest that the coupling to TLS of glass is rather small and the spectral diffusion plays a minor role in the hole broadening; in other words, the exciton in the J-aggregate chain is completely decoupled from the host glass [56]. Another evidence of this fact is the similarities of the low-temperature limits of homogeneous linewidths, obtained by accumulated photon echo and HB. Photo-echo decay in PIC Br red aggregates is nonexponential with T_2 from 12 to 54 ps [53]. The effective T_2 time has been obtained recently to be 40 ps [59]. The hole width of 0.36 cm^{-1} of PIC J-aggregates [56] corresponds to $T_2 = 60$ ps [see Eqs. (1) and (2)].

The PIC J-aggregates in WEG glass seems exceptional among other J-aggregate systems. Indeed, J-aggregates of other dyes in WEG or even J-aggregates of PIC in other matrices (like Langmuir–Blodgett films or polymers) possess measurable Stokes shifts and greater inhomogeneous absorption widths. The well-studied examples are the aggregates of PIC iodide in a Langmuir–Blodgett (LB) film [60], and the aggregates of 6,6′-dimethoxy-3,3′-disulfopropyl-9-ethylthiacarbocyanine (TC) [61] and 5,5′,6,6′–tetrachloro-1,1′-diethyl-3,3′-di(4-sulfobutyl)-benzimidazolocarbocyanine (TDBC) [62] in WEG. In these systems, the dephasing time T_2 measured by the accumulated photon echo at low temperature (1.5 K) is disperse; it increases with increasing wavelength within the inhomogeneous absorption band. In addition, $T_2 \neq 2\tau_F$ (τ_F is the fluorescence lifetime) even at the lowest temperature. These facts in conjunction with the Stokes shift observed at low temperatures suggest that some relaxation processes occur before relaxation to the ground state. As has been mentioned before, there are at least three possible mechanisms of this relaxation: (a) energy transfer between segments of chains, (b) intraband relaxation, and (c) trapping or self-trapping of exciton via exciton–phonon interaction to the states lower than the bottom of the band. Wiersma and coauthors [60–62] supposed the intraband relaxation mechanism (b) to be responsible for the observed Stokes shift and disperse dephasing.

The self-trapping mechanism (c) is inherent in 1D systems, where the free state (F) is unstable, whereas the self-trapped one (ST) provides the energy minimum at any exciton–phonon coupling strength. We should emphasize that the intrinsic feature of this mechanism is the direct connection of the Stokes shift to the exciton–phonon coupling strength that is not so evident for two other possible relaxation mechanisms (a) and (b). Within the framework of the self-trapping mechanism, the Stokes shift would gradually increase from zero with increasing exciton–phonon coupling in 1D systems [63,64]. One can imagine that in this case the optical excitation creates a nearly free exciton (due to the Franck–Condon principle) at the bottom of the exciton band and that this exciton disturbs a lattice around it, making the self-consistent ST state with lower energy. Therefore, the fluorescent states are qualitatively different from the absorbing ones.

Going to the particular results of SLS, we shall consider, first, J-aggregates of 3,3′,9-triethyl-5,5′-dichlorothiacarbocyanine iodide (TDC) in WEG [65]. The fluorescence spectrum of this system possessing a Stokes shift of 100 cm^{-1} at low temperature is presented in Fig. 7. One can see that it does not depend on excitation wavelength. It is structureless (does not contain ZPLs) upon selective laser exci-

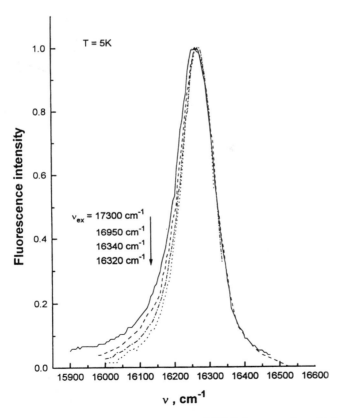

FIGURE 7 Fluorescence spectra of TDC J-aggregates obtained upon selective laser excitation at different excitation frequencies at 5 K. The excitation frequencies, indicated from top to bottom, correspond to the spectra with decreasing intensity in the red tail.

tation. The same features have been observed for other J-aggregates in Langmuir–Blodgett film [66] and as was mentioned earlier (Section III.B) for physical oligomers of other dyes. These facts together with the possibility of hole burning in absorption (see below) suggest the presence of electronic relaxation with the lack of correlation (such as a self-trapping or intraband relaxation) rather than the energy transfer between various segments. In the latter case, the fluorescence spectrum must shift with excitation frequency tuned in the red side of absorption spectrum (see Ref. 33 and Section V) that is not observed in experiment.

To relate the Stokes shift (SS) value with the exciton–phonon coupling strength, we shall consider the J-aggregates of two close thiacarbocyanine dyes: TDC and THIATS (triethyl ammonium salt of 3,3'-disulfopropyl-9-ethyl-5,5'-dichloro-thiacarbocyanines) [67]. They possess the Stokes shifts differing by four times at 6 K: SS = 100 cm^{-1} for TDC and SS = 25 cm^{-1} for THIATS J-aggregates.

Figure 8 shows the temperature dependence of the fluorescence linewidths (HWHM) for J-aggregates of TDC and THIATS [67,68]. Note the negative second derivative of the experimental curve even at low temperatures, which is rather unusual for excitonic states in molecular crystals. This result can be understood if one remembers that the frequency of the phonon bound to the ST exciton in a 1D lattice tends to zero [69,70]. One can see also from Fig. 8 that the line broadens faster for TDC than for THIATS, which correlates with the difference in the Stokes shifts. To describe quantitatively the temperature dependences of the linewidths δ (in cm^{-1}), we used the well-known relation for localized excitations in the classic approximation [71,72]:

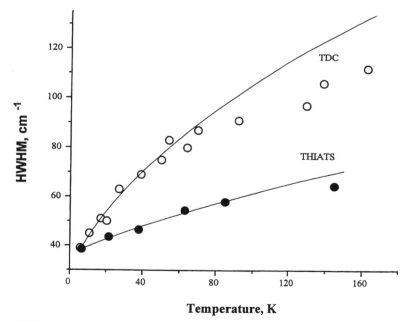

FIGURE 8 Temperature dependence of fluorescence linewidth for TDC and THIATS J-aggregates.

$$\delta = (a + bT)^{1/2} \tag{21}$$

where the parameter a describes the static disorder (inhomogeneous) contribution and $b = 4(\ln 2)k_B E_{LR}/h$. E_{LR} is the lattice relaxation energy released after phototransition. The broadening of the TDC line is well described by Eq. (21) with $E_{LR} = 53 \pm 4$ cm^{-1} in the restricted low-temperature range from 6 to 54 K. Extending the fitting range to higher temperatures gives higher mean square deviations. In the case of THIATS, the fitting of experimental points by Eq. (21) gives $E_{LR} = 13 \pm 1$ cm^{-1} in the same temperature range. We should emphasize here that the Stokes shift is proportional to E_{LR}. This unambiguously suggests the essential role of the electron–phonon coupling in creation of the Stokes shift. Another interesting feature is the quite small values of E_{LR} as compared to that usually observed for isolated molecules (100–1000 cm^{-1}).

Apart from the relative agreement of the Stokes shifts and the parameter of temperature line broadening, we attempted to relate the absolute values of these observables in the case of TDC aggregates. Let us consider the specific form of the adiabatic potential of the ST exciton in the 1D lattice. This potential is given by the sum [63,64]

$$P(Q) = E(Q) + \tfrac{1}{2}Q^2$$

where Q is the average lattice distortion in the area of a ST exciton and $E(Q)$ is the binding energy of an exciton in the lattice deformed by the exciton itself. By definition [63,64], a direction of Q is decided so that $E(Q) < 0$ for $Q > 0$ and $E(Q) = 0$ for $Q < 0$. The function $E(Q)$ can be calculated if the exciton–phonon coupling parameter g is known [63]. The constant g is defined as a ratio of the lattice relaxation energy Δ gained by the complete localization of an exciton on a single site and the half-width B of the exciton band: $g = \Delta/B$. Higai and Sumi [64] calculated the dependence of the Stokes shift on g in the discrete 1D lattice. Using their calculations, we obtained $g = 0.35$ for TDC. (The value $B = 1400$ cm^{-1} was estimated as a difference between the absorption peaks of J-aggregates and monomers at low temperature.) As a result, we can calculate the adiabatic potential $P(Q)$ for our system (Fig. 9). It is evident that this potential is asymmetric. In the energy range from -34 to 0 cm^{-1} (0–49 K), it can be well described by a parabola with a minimum at $Q_{min} = 10.7$. Therefore, we can obtain, in this temperature range, $E_{LR} = Q_{min}^2/2 = 57$ cm^{-1}. The same calculations for THIATS give $g = 0.17$ and $E_{LR} = 14$ cm^{-1}. These values are in good agreement with those obtained from the fluorescence temperature broadening that proves the ST model considered.

It is evident from Fig. 8 that at $T > 60$ K, the linewidth of TDC J-aggregates increases slowly than the model function (21) with $E_{LR} = 53$ cm^{-1}. The high-temperature behavior of the linewidth can be explained by the following considerations [65]. The optical transitions that occur during the time when $Q < 0$ does not contribute to the linewidth because, in this area, the potential curves of the ground and excited states are not shifted; see Fig. 9. Therefore, the temperature broadening slows down.

The unexpectedly low values of the Stokes shifts and lattice relaxation energies of the systems studied can be explained if one takes into consideration the dimensionality of the system. Indeed, the large Stokes shifts are inherent only in 3D

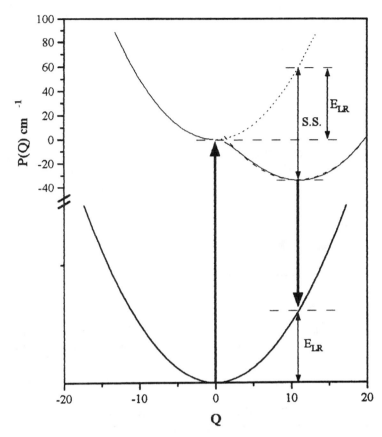

FIGURE 9 Adiabatic potentials (appropriate to TDC J-aggregates) calculated according to Ref. 64 with $B = 1400$ cm^{-1} and $g = 0.35$; E_{LR} is the energy of lattice relaxation. The bold arrow represents the optical transitions. The broken curve is the parabolic approximation of the low-energy region of the excited-state potential.

systems, where localized excitation tends to occupy only a single site [73,74]. In this case, the lattice relaxation energy $E_{LR} = \Delta$ must be greater than B for self-trapping to occur. It is easy to show that for a 3D lattice SS $= 2\Delta - B > B$ and, therefore, the Stokes shift must be at least greater than the bandwidth B. On the other hand, the stable ST state in a 1D lattice occupies a great number of sites if the parameter $g = \Delta/B$ is small enough [64]. In the 1D case, the Stokes shift of Davydov solitons (involving only longitudinal displacements of molecules) can be expressed as SS $= 1.5E_{LR}$ [75]. Nearly the same relation is true for the ST excitons (see Fig. 9 and Ref. 64) and the SS can have any value, which vanishes when the coupling constant g tends to zero [63,64,75].

The hole burning in the absorption bands of J-aggregates of TDC and THIATS is found to be possible [67,76]. Temperature dependences of the hole width are shown in Fig. 10. It is evident that at $T < 10$ K, the hole width depends only slightly on temperature (cf. PIC J-aggregates [56]). In addition, the hole width of

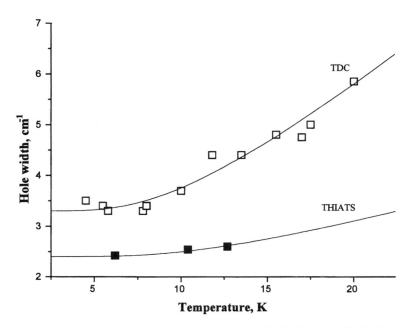

FIGURE 10 Temperature dependence of the width of spectral holes burnt near the absorption band maximums of TDC and THIATS J-aggregates. Full curves are the approximations by Eq. (22).

TDC rises faster than that of THIATS and the low-temperature limit of the width is greater for TDC. This can imply that the exciton–phonon coupling as measured in the absorption transition is also greater for TDC.

To describe the hole broadening, we used the simple model of exciton–phonon scattering [72] describing the homogeneous absorption line of nearly free excitons. This model assumes a single mode of electronic excitation per site, the Franck–Condon approximation for absorption transition, a single and harmonic mode ω of lattice vibration per site (Einstein model), and linear and site-diagonal electron–phonon interaction of contact (short range) type. Within this framework and in the case of weak exciton–phonon intraband scattering, the linewidth is determined as follows [72]:

$$\gamma^*(T) = \pi\hbar\omega\Delta \quad \coth\left(\frac{\hbar\omega}{2kT}\right)\rho(E_e) \tag{22}$$

Here, $\rho(E_e)$ is the density of excitonic states (DOS) assumed to be constant near the absorption maximum E_0.

It should be stressed that Eq. (22) is similar to the expression for the homogeneous line broadening due to the scattering in course of the so-called "direct processes," that is, the transitions with absorption and emission of a phonon by the exciton. In this case [77–79],

$$\gamma^*(T) = c\rho(E_e + \hbar\omega)n + c\rho(E_e - \hbar\omega)(n + 1) \tag{23}$$

where $n = [\exp(\hbar\omega/kT) - 1]^{-1}$ is the Bose–Einstein occupation number for phonons and c is the parameter proportional to exciton–phonon coupling. If the DOS can be considered nearly constant in the range from $E_0 - \hbar\omega$ to $E_0 + \hbar\omega$, one arrives at Eq. (22). Our important finding is that the best fitting of experimental hole-broadening data on TDC J-aggregates to Eq. (23) gives $\rho(E_0 + \hbar\omega) = \rho(E_0 - \hbar\omega)$ with a good accuracy [76]. This is possible due to a high inhomogeneous broadening (static disorder) of the absorption band of the system, as compared to the PIC J-aggregates. Note that the processes with phonon emission [second term in Eq. (23)] have been omitted in the analysis of homogeneous line broadening of PIC J-aggregates [54] because the inhomogeneous broadening is comparable with the effective phonon frequency in this case and the DOS has an abrupt edge on the red side. Thus, the temperature dependences of the hole widths of TDC and THIATS J-aggregates can well be described by Eq. (22) with effective phonon frequencies $\omega = 20$ and 25 cm^{-1}, respectively. The temperature-independent line-width is determined by spontaneous one-phonon emission to a greater extent than by relaxation to the ground state. This process can be regarded as the first step of self-trapping [76]. Whether this step involves the intraband relaxation (without displacement of nuclei) or direct transition to the ST state (with nuclear displacements) is probably determined by the ratio of phonon frequency and the depth of the ST state potential.

Further significant evidence in favor of self-trapping in J-aggregates can be obtained from the consideration of the dependence of the SS on exciton–phonon coupling constant $g = \Delta/B$ [68]. The dimensionality of the lattice is of crucial importance for this dependence. It was found theoretically for 1D Davydov solitons that the binding energy of resting soliton and, therefore, SS [75] $\sim g^2$ for small values of g [80,81] and $\sim g$ for large g [81]. Qualitatively, the similar dependence was obtained in Refs. 63 and 64 for ST excitons in a 1D lattice. In contrast, in 2D and 3D cases, SS does not depend on g for small g, and $\sim g$ for large g [63].

Assuming the model of exciton–phonon scattering described above, we can obtain the g values from the low-temperature hole-burning and photon-echo data for a number of J-aggregates. At very low temperatures, Eq. (22) transforms into

$$\gamma^*(0) = \gamma^* = \pi\hbar\omega\Delta\rho(E_e). \tag{24}$$

Besides our hole-burning results, we analyzed a number of the photon-echo and hole-burning data [54,56,59,60–62] in order to obtain a relationship between the parameter g and the Stokes shift in a number of different J-aggregates. As a first approximation, we take the ω value equal for all the systems. Then, $\Delta \propto \gamma^*/\rho(E_0)$ and $g \propto \gamma^*/\rho(E_0)B$. To obtain γ^*, we subtracted the fluorescence lifetime contribution from the low-temperature limit of homogeneous linewidth (as in Ref. 61) [see Eq. (2)]:

$$\gamma^* = (\pi c T_2)^{-1} - (2\pi c \tau_F)^{-1} \tag{25}$$

An amplitude of the density of states $\rho(E_e)$ near the absorption maximum E_0 can be estimated from the inhomogeneous absorption linewidth W. It was shown [72] that in the case of diagonal disorder with the standard deviation D of energies, $W \propto D^{4/3}$ for the 1D lattice. On the other hand, an amplitude of the normalized one-dimensional DOS near the absorption maximum $\rho(E_0) \propto D^{-2/3}$ [72,82]. Therefore,

$\rho(E_0) \propto W^{0.5}$ and $g \propto \gamma*W^{0.5}/B$. All the parameters used for this combination are summarized in Table 1.

Figure 11 shows the dependence of the Stokes shift measured in units of B on the parameter $\gamma*W^{0.5}/B$. It is evident that this dependence is close to quadratic, which is an intrinsic feature of self-trapped excitons of large radius in 1D systems. The radii of these states N_b are calculated with the help of Ref. 64 and are presented in Table 1. Note that the precise value of N_b for PIC J-aggregates cannot be obtained because of vanishing Stokes shift in this system. But, some nonvanishing defect between T_2 and $2T_1$ exists (see Table 1); therefore, we can obtain from the relative $\gamma*W^{0.5}/B$ value that $N_b \approx 200$ for this system. This large number corresponds to a very weak exciton–phonon interaction in PIC J-aggregates.

Therefore, our analysis suggests that the relaxed states of J-aggregates are described by the self-trapped excitons of large radius.

B. Relaxation in Macroscopic Complexes of J-Aggregates

Another type of relaxation which occurs in macroscopically arranged J-aggregates has been studied recently [83,84] by hole-burning spectroscopy. Wavelength and polarization dependence of the hole-burning efficiency in PIC J-aggregates oriented by the spin-coating method in polyvinylalcohol matrix has been investigated. It has been found that the spectral hole is asymmetric in shape with the antihole on the higher-energy side. Furthermore, the efficiency of burning is higher on the lower-energy side of the absorption spectrum and falls drastically (by three orders of magnitude) on the higher-energy side in the case of burning (and probing) light polarization parallel to the orientation of aggregates. The wavelength dependence of the burning efficiency is smoother for perpendicular polarization. The authors [83,84] believe that the absorption of light with polarization parallel to the orientation axis is due to macroscopic complexes of J-aggregates (macroaggregates) which are aligned during the spin, and the absorption with perpendicular polarization of light is due to usual single-chain J-aggregates (mesoaggregates). A drastic decrease of the hole-burning efficiency in the high-energy side of the spectrum of macroaggregates has been explained by a fast relaxation process, probably by bi-excitonic annihilation [84] in these closely packed systems. A fast energy transfer among different neighboring chains has not been excluded.

V. SELECTIVE SPECTROSCOPY OF CONJUGATED POLYMERS

A. π-Conjugated Polymers

π-Conjugated polymers attract much interest today from the point of view of the nature and dynamics of fundamental excitations in them and from a technological point of view as well [85]. The most widely studied are the *cis* or *trans*-poly(acetylene) (PA), poly(diacetylene) (PDA), poly(*p*-phenylene) (PPP), poly(thiphene) (PT), poly(3-dodecylthiophene) (PDT), poly(phenylenevinylene) (PPV) and its derivatives, and others. The nature of electronic excitations in such one-dimensional systems is in the center of debates now [9,11,13,86].

There are two different approaches to the description of optical transitions in π-conjugated polymers. The first one is based on neglecting the electron–hole and electron–electron correlation in the polymer chain, which possesses low disorder.

TABLE 1 Parameters of a Number of J-Aggregates[a]

	T_2^b (ps) γ_2 (cm^{-1})	τ_F^b (ps) γ_F (cm^{-1})	γ^* (cm^{-1})	W (cm^{-1})	B^c (cm^{-1})	$\gamma^* W^{1/2}/B$	SS (cm^{-1})	SS/B	N_b^d	Ref.
1. PIC	40 0.265	70 0.076	0.19	30	1300	0.0007	<1	<10^{-3}	200	54, 59
2. THIATS	1.2	18^e 0.29	0.9	120	1550	0.0064	25	0.016	17	67
3. TDBC	1.5	18 0.29	1.2	160	2080	0.0074	40	0.019	15	62
4. TC	6.4 1.66	18 0.29	1.37	240	2500	0.0085	80	0.032	11	61
5. TDC	1.65	18^e 0.29	1.36	220	1400	0.0144	100	0.071	7	67
6. PIC in LB film	5.4 1.97	10.5 0.53	1.44	130	1400	0.0163	130	0.093	6	60

Note: All the data are presented for J-aggregates in water/ethylene glycol frozen matrices, except for the last row, where a Langmuir–Blodgett film was used as a matrix. All the data refer to liquid-helium temperatures.

[a]Dephasing time (T_2), fluorescence lifetime τ_F, scattering rate γ^*, obtained according to Eq. (25), inhomogeneous linewidths (W), excitonic half-width (B), Stokes shift (SS), and the number of molecules bound in the self-trapped state of large radius (N_b).

[b]Obtained at absorption maximum (or interpolated to it).

[c]Half-width of exciton band obtained as the difference between the absorption maxima of monomer and J peak (measured at liquid-helium temperatures).

[d]Estimated using calculations [64] from the measured values of SS/B.

[e]Assumed to be equal to 18 ps as for TDBC and TC [61,62].

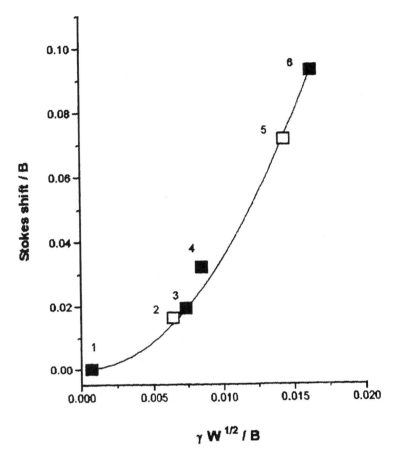

FIGURE 11 The dependence of SS/B on a combination proportional to exciton–phonon coupling constant g. Filled squares are obtained from photon-echo data and open squares are obtained from hole-burning data. The numbers 1–6 correspond to systems, as summarized in Table 1.

It states that the absorption of light is due to the transition from valence band to conduction band, as in semiconductors. Initially produced nearly free electrons and holes then become self-trapped due to electron–phonon interaction to give polarons or bipolarons. These quasi-particles recombine to emit the luminescence with a large red shift. In polymers with a degenerate ground state (like PA), the relaxed states are considered to be domain wells (or solitons) [12]. The band-to-band model is based on the similarity of the polymer steady-state absorption spectrum and the DOS of disordered 1D system [12]. Other evidences result from photoconduction action spectra [87] and photo-induced absorption experiments [88].

Another (molecular) model argues that elementary excitations are the bound electron–phonon pairs; that is, excitons of Frenkel or Wannier type localized on some restricted segment of the strongly disordered chain. The disorder explains why the absorption band is not as narrow as in nearly free-exciton transition (as is observed in J-aggregates). The molecular model implies that the polymer is described by a set of practically decoupled segments, among which incoherent energy transfer can occur, as in the case of the concentrated solution of molecules. Absorption spectrum in this case consists of inhomogeneously broadened vibronic progressions of different segments and is similar to that of oligomers, but somewhat red-shifted [89]. The conclusions derived from photoconduction data have been subjected to question because of the possibility of thermal degradation of excitons [90]. Site-selective and polarization spectroscopy support the molecular approach [89,91]. The red shift of absorption spectra of polymers with respect to those of oligomers suggests, on the other hand, the greater (in some cases much greater) delocalization of the wave function in polymers. A high degree of delocalization is also established recently in a well-oriented and well-ordered solution of PPV by using polarized-electroabsorption spectroscopy [11]. This could suggest that the nature of excitation depends on the morphology of the particular sample [11]. In addition, the nature of excitation could depend on the type of polymer and could be different for different polymers. For example, there is no doubt that in PDA, the lowest allowed transition is of an excitonic nature [92], the band-to-band transition being much weaker and shifted by 0.4 eV to the blue [93].

Energy-selective spectroscopy is, probably, the most apparent and appealing in the establishment of the nature of fundamental excitations in a particular polymer [9]. This statement is based on the observation that if one excites a polymer above some frequency ν_{loc}, called the localization threshold, the emission spectrum does not depend on frequency, but if $\nu_{ex} < \nu_{loc}$, the emission maximum ν_{em} shifts with ν_{ex}, just as in the case of noninteracting molecules (see Section II.B). This is explained as follows [9]: Upon exciting above ν_{loc}, the initially generated excitonic states relax fast, down to the tail states of the DOS, from which the luminescence occurs. This is confirmed also by time-resolved energy-selective spectroscopy [94]. Upon low-energy excitation, the tail states are populated directly, giving rise to luminescence from the parent center. If the region where the shift is observed covers some peak of inhomogeneous absorption spectrum, this peak can be attributed to an immobilized bound exciton. The energy region where ν_{em} does not depend on ν_{ex} can be associated with a fast (compared to the luminescence rate) electron thermalization in the band-to-band model or with a fast incoherent downhill energy transfer in the course of a random walk of a localized exciton in the molecular model.

Now, we shall consider our results obtained for PPP thin film as an example of application of SLS to conjugated polymers. We shall include in the consideration a very low-energy region of excitation. The PPP currently attracts attention as a material with blue luminescence for applications in polymer-based opto-electronic devices. The SLS results obtained for other polymers by other authors will be briefly considered within the framework of the model calculations (presented in Section II.B).

1. Selective Spectroscopy of the PPP Film [95]

The position of the room-temperature luminescence spectrum of the PPP film is independent of excitation frequency for $\nu_{ex} > 23,500$ cm^{-1}. The Gaussian widths of the vibronic bands of this spectrum constitute about 500 cm^{-1}.

The luminescence excitation spectrum is independent of recording wavelength and is virtually structureless (Fig. 12). Its FWHM amounts 7700 cm^{-1} and maximum is located at $\nu_m = 25,500$ cm^{-1} with a low-energy shoulder, which possibly can be assigned to the 0–0 transition. A fitting by a sum of Gaussians gives 24,000 cm^{-1} for the maximum position and $\sigma_a = 600$ cm^{-1} for the Gaussian width of this lowest energy transition.

Excitation by eight argon laser lines in the red tail of the absorption band results in the shift of the luminescence spectrum even at room temperature (Fig. 13). In the $21,155 < \nu_{ex} < 21,839$-cm^{-1} frequency range, the luminescence spectrum consists of two peaks (which we shall call the blue and red peaks) and a long-wavelength shoulder (Fig. 13). Fluorescence spectra at low-frequency excitation were fitted by three Gaussians. The blue and red peaks are formed by Gaussians with approximately equal widths $\sigma_{pp} \approx 460$ and 430 cm^{-1}, respectively, which do not depend on excitation frequency.

Upon decreasing excitation frequency, the blue peak gradually approaches the laser line and eventually becomes resonant with it. The shift between the blue peak and the laser line cannot be precisely determined in this case because of the scattered laser light.

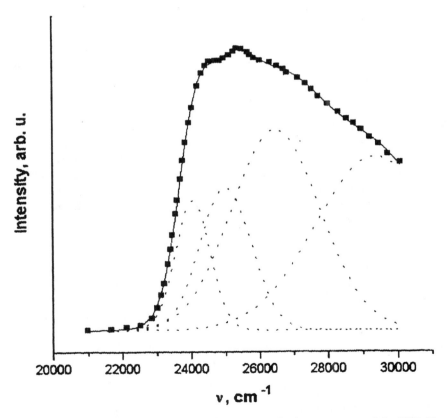

FIGURE 12 Room-temperature luminescence excitation spectrum of the PPP film recorded at 485 nm. Fitting by a sum of four Gaussians is shown.

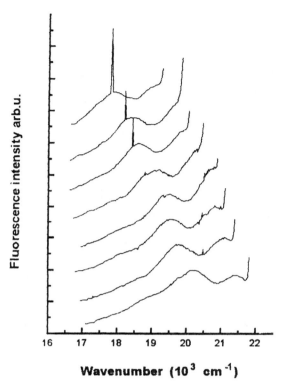

FIGURE 13 Fluorescence spectra of the PPP film, obtained for different excitation frequencies in the low-energy tail of excitation spectrum at room temperature. All spectra are normalized to unity and shifted along the ordinate axes. The narrow lines in the upper three spectra are Raman scattering from the quartz substrate.

Cooling the sample to 5 K results in some resolution of vibronic bands and their narrowing. Note that ZPLs are not observed at low temperature upon selective laser excitation. This is consistent with previous observations for PPV, PPPV, and their derivatives [40,89,91] and is due to rich vibronic structure and (or) strong electron–phonon coupling in polymers (see Section III.C).

Let us now consider in detail the shifts of luminescence peaks of the PPP with ν_{ex} at room temperature. These dependences are shown in Fig. 14. The dependence of ν_{em} on ν_{ex} for the blue peak is well described by a straight line with the slope $k_b = 0.75 \pm 0.05$ in the 21,155–21,839-cm^{-1} range. As for the red peak, the slope was measured in two independent experiments and was equal to $k_r = 0.87 \pm 0.05$ and 0.92 ± 0.03, but in a more extended spectral range from 19,436 to 21,839 cm^{-1}. The intersection point of the ν_{em} versus ν_{ex} line and constant level in the high-frequency region is located at $\nu_{loc} \approx 22,400$ cm^{-1} ($-2.7\sigma_a$ from the 0–0 transition) for both emission peaks (see Fig. 14). The difference in k_b and k_r suggests that either the two peaks belong to different species or the slope depends in general on the spectral region. The first possibility is evidently ruled out because the excitation spectrum does not change upon recording fluorescence in either of

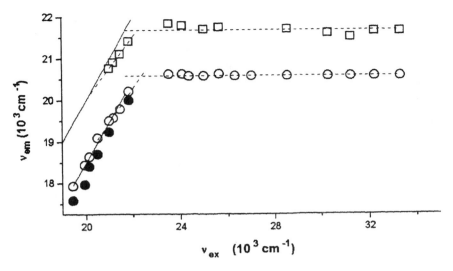

FIGURE 14   The dependence of ν_{em} on ν_{ex} for the blue (open squares) and red (open circles) peaks. The solid lines represent best fits by $\nu_{em} = $ const. for the high-frequency region and by $\nu_{em} = k\nu_{ex} + c$ for the low frequency region with $k_b = 0.75$ and $k_r = 0.92$ (see text). Filled circles represent the red peak position at 5 K.

these peaks. Therefore, we suppose that there are two different spectral regions with different slopes. As the crossover point between the two regions is unknown a priori, we fitted six low-frequency points and five high-frequency points independently and obtained the following dependencies: $\nu_{em} = 0.98\nu_{ex} - 1065$ cm^{-1} for the low-frequency region and $\nu_{em} = 0.80\nu_{ex} + 2580$ cm^{-1} for the high-frequency region. We, therefore, can conclude that in the low-energy region, the red peak shifts in parallel with excitation, and in the high-frequency region, it shifts slower, with the slope $k_r = 0.80 \pm 0.05$ (i.e., just as the blue peak). To check this behavior for the blue peak we determined the frequency of its long-wavelength inflection point. This point shifts linearly with excitation frequency with the slope 0.99 ± 0.05. This proves once more a presence of the crossover frequency for both peaks. Below this frequency, the fluorescence peaks shift with the slope $k = 1.0$, and above it, they shift with the slope $k < 1$ (~0.8 in our case).

Thus, one can conclude that in the red tail of the excitation spectrum, there exist at least two different regions, marked by $k = 1$ and $k < 1$, respectively, with a crossover point at $\nu^* = 20,900$ cm^{-1} ($-5.2\sigma_a$).

Note that the ν_{em} values for the red peak at low temperatures (5 K) do not differ substantially from the room-temperature results (see filled circles in Fig. 14).

The observed phenomena can be explained (cf. Ref. 9) assuming that the excitation, created above the localized threshold ν_{loc}, relaxes to the lowest bound excitonic states very fast as compared to the luminescence decay rate. These states then emit the relaxed fluorescence, the spectrum of which is independent of excitation frequency. The relative density of states decreases drastically in the red tail of the excitation spectrum and the excitation remains at the center where it was generated. Monochromatic laser excitation below ν_{loc} selects the (practically de-coupled) segments, whose homogeneous spectra occur near the excitation frequency. In our case, a quasi-selective excitation via phonon bands takes place because sharp zero-phonon lines are absent both at room and liquid-helium temperatures. (This fact explains why the zero-phonon hole burning is not possible in π-conjugated polymers; see Ref. 40.) After a fast intramolecular vibronic relaxation, these centers emit fluorescence at ν_{em} determined by Eq. (8). It should be noted that in a solid sample, the local environment remains unchanged during the fluorescence lifetime; therefore, it is possible to observe selective luminescence even at room temperature.

The entire spectral range can be divided into four regions: (1) $\nu_{ex} > \nu_{loc}$—the region of independence of ν_{em} on ν_{ex} where the fast relaxation processes dominate; (2) $\nu_i < \nu_{ex} < \nu_{loc}$—the region of nonlinear dependence of ν_{em} on ν_{ex}, where the energy transfer between centers (or other fast relaxation) still occurs [9]; (3) $\nu^* < \nu_{ex} < \nu_i$—the region of linear dependence with the slope $k < 1$; (4) $\nu_{ex} < \nu^*$—the region of linear dependence with the slope $k = 1$.

In region (3), the model of "two Gaussians" described in Section II.B can be applied. From the measured value of $\sigma_{pp}/\sigma_a = 460/600 = 0.72$ (for the red peak), we can estimate, using Eq. (14), $k = 0.7$. This value corresponds well to that ($k = 0.75$) obtained from the ν_{em} versus ν_{ex} dependence. Using Eqs. (9), (10), and (14) and assuming $k = 0.8$, we can obtain $\sigma_{pb} = 340$ cm^{-1} and $\sigma = 670$ cm^{-1}. The Stokes shift $\Delta = 365$ cm^{-1} is obtained from the ν^* and $k = 0.75$ values, using Eq. (15).

On the other hand, a quasi-resonant behavior of both peaks in region (4) excludes the applicability of the model of "two Gaussians" because it predicts that $k < 1$ in the entire spectral range. In addition, this model predicts the intersection of the ν_{em} versus ν_{ex} dependence with the resonance line ($\nu_{em} = \nu_{ex}$) even for vibronic transitions [see Eq. (12)], which is not observed for the red peak in our case. This suggests some deviations from the model of "two Gaussians" in the far red tail of the IDF to take place. Both other models, described in Section II.B are consistent with the observed dependence. The model assuming the Gamma distribution for PW [Eq. (16)] and Gaussian IDF gives $k \approx 0.9$ in the vicinity of $\nu_{ex} = -5\sigma$ and a smooth increase of the slope to the red (see Fig. 2). The model assuming a Urbach tail in the IDF predicts strictly $k = 1$ in the spectral range of the tail.

2. Selective Spectroscopy of Other Polymers

Let us now compare our results with those obtained by Bassler and co-authors with other polymers. The main result obtained with poly(p-phenyl-phenylenevinylene) (PPPV) isolated in a matrix of methyltetrahydrofuran (MTHF), is that the luminescence spectrum shifts even at excitation frequencies higher than the onset of absorption ($\nu_{loc} > 0$) [89]. This unambiguously supports the conclusion about the absence of a fast relaxation process and calls for the localized exciton description of the lowest electronic transition in the isolated polymer chain. The same picture is observed in poly-2-isopropenylnaphtalene (IPN), where the chromophore is the naphthalene substituent. Some results obtained with various polymer systems are presented in Table 2.

Note that region (3) with $k < 1$ is observed for a number of polymers. This suggests the comparable width of IDF and PW and immobilized nature of excita-

TABLE 2 Parameters of the Dependence of ν_{em} on ν_{ex} for a Number of Polymer and Related Systems.

System	σ_{pp}/σ_a [a]	k [b]	Range [c]	ν^{*} [d] (cm^{-1})	Δ [e] (cm^{-1})	ν_{loc} [d]	Ref.
PPV film	120/650 = 0.18	1	−2.2 to −2.1	—	—	−2	91
PPPV in MTHF	190/690 = 0.28	0.96	−2.7 to −1.3	—	—	−1.3	40
PPPV in MTHF	390/625 = 0.62	0.78	−1.6 to +2.1	−2200	270 (0–0)	>0	89
PPPV in PC (1%) stretched	—	0.5	—			<0	96
PPPV in PC (1%) unstretch.	—	0.6	—			<0	96
PDA in MTHF	155/550 = 0.28	0.98	−1.3 to −0.2	—	—		99
PDT in MTHF	240/875 = 0.27	1	−2.1 to −1.9	—	—	−1.8	97
PPP film	460/600 = 0.77	0.75 (0–0)	−5.2 to −3.7	−3300	365 (0–0)	−2.7	95
	430/600 = 0.72	0.8 (0–1)					
		1 (0–1)	−7.9 to −5.2				
IPN[f] in MTHF	40/120 = 0.33	0.97	−0.5 to +1.1	—	—	>0	99
Bromonaphthalene film	105/160 = 0.65	0.65 (0–0)	−2.2 to +1.1	−330	70 (0–0)	+2	34

[a] σ_a and σ_{pp} are measured in cm^{-1}.
[b] The slope of the ν_{em} versus ν_{ex} dependence.
[c] The studied range of the ν_{em} versus ν_{ex} dependence measured from absorption maximum in units of σ_a.
[d] Measured from absorption maximum.
[e] Calculated using Eq. (15).
[f] Poly-2-isopropenylnaphtalene.

tions in this spectral region. The dependence of σ_{pp}/σ_a on k, taken from Table 2, is shown in Fig. 15. The line presents relation (14) connecting these parameters in the model of "two Gaussians." One can see that it qualitatively describes the experimental data. Therefore, in the region not very far from the IDF maximum, this model seems reasonable. We obtained also the values of Stokes losses Δ from Eq. (15), using the position of the intersection point ν^{*} for the 0–0 transition and presented them in Table 2. Note also that as in the PPP film, the crossover from $k < 1$ to $k = 1$ behavior is also observed in PPPV dispersed in polycarbonate [96].

The localization threshold ν_{loc} is obtained to be $(2–3)\sigma_a$ below absorption maximum for 3D polymer films and $(1–2)\sigma_a$ below or in some cases even above ab-

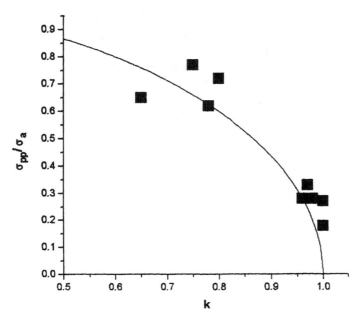

FIGURE 15 Dependence of σ_{pp}/σ_a on k for a number of polymer systems presented in Table 2.

sorption maximum for 1D polymer chains diluted in inert matrix (see Table 2). This observation is consistent with Monte Carlo simulations of an incoherent random walk in 3D [98] and 1D [99] systems and is another indirect support for molecular model.

In conclusion, note that the results obtained by selective laser spectroscopy show only that below some threshold energy ν_{loc}, excitations in conjugated polymers are bound, thus being neutral localized excitons. In some cases, this threshold lies above the absorption peak and, therefore, this peak belongs to localized excitons. This results from the statement that during the thermalization process, free electrons and holes lose the memory on their initial excitation energy very quickly [100]. The semiconductor model is not excluded in other cases (when ν_{loc} is in the red tail of absorption) because the sub-band-gap transitions can lead to a formation of bound excitons.

B. σ-Conjugated Polymers

Another class of 1D polymers constitute the σ-conjugated polysilanes. Excitation nature and relaxation mechanisms in poly(di-n-hexyl-silane) (PDHS) has been studied by selective spectroscopy in Ref. 14. Absorption peak is narrow (250 cm^{-1} at low temperatures). The fluorescence spectrum is structureless and red-shifted by 200 cm^{-1} unless the excitation is not at the low-energy side. Upon low-energy excitation, narrow vibronic ZPLs appear in the fluorescence spectrum. Electronic excitations in PDHS are presumably described by Frenkel excitons [14]. Two different models of disorder have been considered to describe absorption spectrum of

PDHS: the independent segment model and the continuous energy disorder model. The first implies that the polymer chain consists of a set of virtually decoupled segments separated by some local imperfections (cf. molecular model for π-conjugated polymers). The second considers a continuous chain with each site subject to dispersion of transition energy (diagonal disorder) or resonance transfer energy (nondiagonal disorder) with Gaussian distribution (cf. the model of disorder without correlation in J-aggregates). It has been shown that the second model describes the absorption spectrum of PDHS better than the first one [14].

A persistent spectral hole can be burned in the PDHS absorption spectrum [14]. Its width is much larger than the fluorescence lifetime-limited value and increases linearly with the fraction of absorption integrated over all energies lower than the laser frequency. The authors have concluded that this width is governed by the energy transfer process (cf. Section III.A). The temperature dependence of the hole width is well described by an activation exponential term without a power-law contribution like in J-aggregates (see Section III.A), suggesting that the coupling to TLS of glass is rather small.

The Stokes shift and the shape of the luminescence spectrum has been discussed in the framework of different relaxation processes: self-trapping (polaron formation), intraband relaxation, and energy transfer between spatially separated chromophores. Polaron formation has been ruled out because of the loss of luminescence polarization in the low-energy side of the spectrum. The intraband relaxation model gives a very narrow fluorescence spectrum and very low Stokes shift at reasonable chain lengths as compared to experimental ones. On the other hand, the energy transfer between different localized states can account for the observed fluorescence spectrum [14].

Therefore, the main relaxation process in PDHS is assumed to be the energy transfer between localized states.

VI. CONCLUSIONS

Different relaxation mechanisms in extended molecular systems are considered. Relaxation in model systems such as concentrated solutions of chromophores, dimers, and oligomers studied by energy selective spectroscopy is discussed.

Relevant parameters of exciton–phonon coupling in molecular J-aggregates have been estimated from independent measurements of the temperature dependence of fluorescence linewidths and spectral hole widths. The results of experiments are interpreted in the framework of exciton self-trapping (ST) in 1D molecular chain. The measured Stokes shift is found to be proportional to the exciton–phonon coupling strength squared, which is in accordance with the model of large-radius ST excitons.

The dependence of selectively excited luminescence spectra of π-conjugated polymers on excitation frequency is discussed in terms of different models for inhomogeneous distribution function and homogeneous vibronic spectra of segments. The very far red tail of inhomogeneous distribution of the PPP film is studied in detail. The crossover from the nonresonance linear dependence of ν_{em} on ν_{ex} (with $k < 1$) to a quasi-resonance one (with $k = 1$) is discovered in this region. The quasi-resonant behavior can be due to the exponential (Urbach) tail in the IDF. The results obtained in the literature with other polymers are well described in the nonresonance region by the model involving Gaussian shapes of IDF and PW. Upon exciting above the localization threshold, ν_{em} is independent of ν_{ex}, suggesting that the excitation relaxes to lower-lying states of chromophores within the IDF.

ACKNOWLEDGMENTS

I wish to thank Dr. M. N. Sapozhnikov, Dr. O. P. Varnavsky, Dr. A. G. Vitukhnovsky, and Dr. V. M. Kobryanskii for valuable help and discussions, and my wife for her stimulation of this work and patience. The work was supported, in part, by the Russian Foundation for Basic Research, projects Nos. 96-02-16527 and 96-03-32568.

REFERENCES

1. Personov, R. I., Site selection spectroscopy of complex molecules in solutions and its applications, in *Spectroscopy and Excitation Dynamics of Condensed Molecular Systems* (V. M. Agranovich and R. M. Hochstrasser, eds.), North-Holland, Amsterdam, 1983, pp. 555–619.
2. Small, G. J., Persistent nonphotochemical hole burning and the dephasing of impurity electronic transitions in organic glasses, in *Spectroscopy and Excitation Dynamics of Condensed Molecular Systems* (V. M. Agranovich and R. M. Hochstrasser, eds.), North-Holland, Amsterdam, 1983, pp. 515–554.
3. Friedrich, J., and Haarer, D., Photochemisches Lochbrennen und Optische Relaxationsspektroskopie in Polymeren und Glasern, *Angew. Chem.*, 96(2), 96–123 (1984). [English transl. in *Angew. Chem. Int. Ed.*, *23*, 113 (1984).]
4. Moerner, W. E. (ed.), *Persistent Spectral Hole Burning: Science and Applications*, Topics in Current Physics, Vol. 44, Springer-Verlag, Berlin, 1988.
5. Volker, S., Spectral hole-burning in crystalline and amorphous organic solids, in *Relaxation Processes in Molecular Excited States*, (J. Fünfschilling, ed.) Kluwer, Dordrecht, 1989, pp. 113–242.
6. Narasimhan, L. R., Littau, K. A., Pack, D. W., Bai, Y. S., Elschner, A., and Fayer, M. D., Probing organic glasses at low temperature with variable time scale optical dephasing measurements, *Chem. Rev.*, *90*, 439–457 (1990).
7. Jankowiak, R., Hayes, J. M., and Small, G. J., Spectral hole-burning spectroscopy in amorphous molecular solids and proteins, *Chem. Rev.*, *93*, 1471–1502 (1993).
8. Osad'ko, I. S., Selective spectroscopy of chromophore doped polymers and glasses, in *Advances in Polymer Science, Vol. 114*, Springer-Verlag, Berlin, 1994, pp. 123–186.
9. Bässler, H., Site-selective fluorescence spectroscopy of polymers, in *Optical Techniques to Characterize Polymer Systems* (H. Bassler, ed.), Studies in Polymer Science 5, Elsevier, Amsterdam, 1989, pp. 181–225.
10. Franck, J., and Teller, E., Migration and photochemical action of excitation energy in crystals, *J. Chem. Phys.*, *6*, 861–872 (1938).
11. Hagler, T. W., Pakbaz, K., and Heeger, A. J., Polarized-electroabsorption spectroscopy of a soluble derivative of poly(p-phenylenevinylene) oriented by gel processing in polyethylene: Polarization anisotropy, the off-axis dipole moment, and excited-state delocalization, *Phys. Rev. B*, *49*, 10968–10975 (1994).
12. Heeger, A. J., Kivelson, S., Schrieffer, J. R., and Su, W. P., Solitons in conducting polymers, *Rev. Mod. Phys.*, *60*, 781–850 (1988).

13. Bässler, H., Exciton and charge transport in random organic solids, in *Disorder Effects on Relaxation Processes. Glasses, Polymers, Proteins* (R. Richert and A. Blumen, eds.), Springer-Verlag, Berlin, 1994, pp. 485–507.

14. Tilgner, A., Trommsdorf, H. P., Zeigler, J. M., and Hochstrasser, R. M., Poly(di-*n*-hexyl-silane) in solid solutions: Experimental and theoretical studies of electronic excitations of a disordered linear chain, *J. Chem. Phys.*, 96, 781–796 (1992).

15. Anderson, P. W., Halperin, B. I., and Varma, C. M., Anomalous low-temperature thermal properties of glasses and spin glasses, *Phil. Mag.*, 25, 1–9 (1972).

16. Phillips, W. A., Tunneling states in amorphous solids, *J. Low Temp. Phys.*, 7, 351–360 (1972).

17. De Bree, P., and Wiersma, D. A., Application of Redfield theory to optical dephasing and line shape of electronic transitions in molecular mixed crystals, *J. Chem. Phys.*, 70, 790–801 (1979).

18. Jackson, B., and Silbey, R., Theoretical description of photochemical hole burning in soft glasses, *Chem. Phys. Lett.*, 99, 331–334 (1983).

19. Breinl, W., Friedrich, J., and Haarer, D., Logarithmic decay of photochemically induced two-level systems in an organic glass, *Chem. Phys. Lett.*, 106, 487–490 (1984).

20. Breinl, W., Friedrich, J., and Haarer, D., Spectral diffusion of a photochemical proton transfer system in an amorphous organic host: Quinizarin in alcohol glass, *J. Chem. Phys.*, 81, 3915–3921 (1984).

21. Abram, I. I., Auerbach, R. A., Birge, R. R., Köhler, B. E., and Stevenson, J. M., Narrow-line fluorescence spectra of perylene as a function of excitation wavelength, *J. Chem. Phys.*, 63, 2473–2478 (1975).

22. Kohler, B. E., Site selection spectroscopy, in *Chemical and Biochemical Applications of Lasers*, 4 (C. B. Moor, ed.), Academic Press, New York, 1979, pp. 31–53.

23. Kikas, J., Effects of inhomogeneity and site selective impurity–phonon coupling in solid solutions, *Chem. Phys. Lett.*, 57, 511–513 (1978).

24. Sapozhnikov, M. N., Selective laser spectroscopy of complex molecules in inhomogeneous matrices: model calculations, *Sov. Phys.–Dokl.*, 31, 323–328 (1986).

25. Sapozhnikov, M. N., Model calculations of luminescence spectra of impurity molecules in solid state at monochromatic excitation, *J. Opt. Spektrosk.*, 61, 21–27 (1986).

26. Abe, S., and Toyozawa, Y., Interband absorption spectra of disordered semiconductors in the coherent potential approximation, *J. Phys. Soc. Jpn.*, 50, 2185–2194 (1981).

27. Schreiber, M., and Toyozawa, Y., Numerical experiments on the absorption lineshape of the exciton under lattice vibrations. 3. The Urbach rule, *J. Phys. Soc. Jpn.*, 51, 1544–1549 (1982).

28. Economou, E. N., Soukoulis, C. M., Cohen, M. H., and Zdetsis, A. D., Quantitative results near band edges of disordered systems, *Phys. Rev. B*, 31, 6172–6183 (1995).

29. Koedijk, J. M. A., Creemers, T. M. H., den Hartog, F. T. H., Bakker, M. P., and Völker, S., The effect of pressure, time and concentration in hole-burning studies, *J. Lumin.*, 64, 55–61 (1995).

30. Bernard, J., Orrit, M., Personov, R. I., and Samoilenko, A. D., Hole burning on porphyrin centers in Langmuir–Blodgett films, *Chem. Phys. Lett.*, 164, 377–382 (1989).

31. Romanovskii, Yu, V., Personov, R. I., Samoilenko, A. D., Hollidai, K., and Wild, U. P., Concentration effects on spectra and persistent hole-burning of porphyrin-doped Langmuir–Blodgett films, *Chem. Phys. Lett.*, 197, 373–379 (1992).

32. Kulikov, S., and Galaup, J.-P., Influence of concentration and energy transfer on permanent spectral hole widths, *J. Lumin.*, 53, 239–243 (1992).

33. Avarmaa, R., Jaaniso, R., Mauring, K., Renge, I., and Tamkivi, R., Influence of energy transfer on the structure of site-selective spectra of molecules, *Molec. Phys.*, 57, 605–621 (1986).

34. Jankowiak, R., Ries, B., and Bassler, H., Spectral diffusion and triplet exciton localization in an organic glass, *Phys. Stat. Sol. (b)*, *124*, 363–371 (1984).

35. Thijssen, H. P. H., Dicker, A. I. M., and Volker, S., Optical dephasing in free-base porphin in organic glasses: A study by photochemical hole-burning, *Chem. Phys. Lett.*, *92*, 7–12 (1982).

36. Pahapill, Yu, Photoburning of the hole in absorption spectra of monomer and dimer of ethioporphyrin I in organic glasses, *Proc. Acad. Sci. Estonian SSR. Phys. Math.*, *35*, 416–423 (1986) (in Russian).

37. Dvornikov, S. S., Soloviyov, K. N., and Tsvirko, M. P., The spectroscopic display of etioporhyrin I dimerization in hydrocarbon solvents, *Zh. Prikl. Spektrosk.*, *28*, 798–803 (1983) (in Russian).

38. Rashba, E. I., A theory of strong coupling of electronic excitations with lattice vibrations in molecular crystals. II. *Opt. Spektrosk.*, *2*, 88–98 (1957) (in Russian).

39. Köhler, W., Friedrich, J., Fischer, R., and Scheer, H., Site-selective spectroscopy and level ordering in *C*-phycocyanine, *Chem. Phys. Lett.*, *143*, 169–173 (1988).

40. Mahrt, R., Yang, J.-P., Greiner, A., Bässler, H., and Bradley, D. D. C., Site-selective spectroscopy of poly(*p*-phenylenevinylene)s and oligomeric model compounds, *Makromol. Chem., Rapid Commun.*, *11*, 415–421 (1990).

41. Scheibe, G., *Angew. Chem.*, *49*, 563 (1936).

42. Jelley, E. E., Spectral absorption and fluorescence of dyes in the molecular state, *Nature*, *138*, 1009–1010 (1936).

43. Scheibe, G., Lage, Intesitat und Struktur von Absorptionbanden, in *Optische Anregungen organischer Systeme* (W. Foerst, ed.), Verlag Chemie, Weinheim, 1966, p. 109.

44. Elliott, R. J., Intensity of optical absorption by excitons, *Phys. Rev.*, *108*, 1384–1389 (1957).

45. De Boer, S., and Wiersma, D. A., Dephasing-induced damping of superradiant emission in J-aggregates, *Chem. Phys. Lett.*, *165*, 45–53 (1990).

46. Klafter, J., and Jortner, J., Effects of structural disorder on optical properties of molecular crystals, *J. Chem. Phys.*, *68*, 1513–1522 (1978).

47. Fidder, H., Knoester, J., and Wiersma, D. A., Optical properties of disordered molecular aggregates—a numerical study, *J. Chem. Phys.*, *95*, 7880–7890 (1991).

48. Rashba, E. I., Self-trapping of excitons, in *Excitons* (E. I. Rashba and M. D. Sturge, eds.), North-Holland, Amsterdam, 1982, p. 543.

49. Ueta, M., Kanzaki, H., Kobayashi, K., Toyozawa, Y., and Hanamura, E., *Excitonic Processes in Solids*, Springer-Verlag, Berlin, 1986.

50. Song, K. S., and Williams, R. T., *Self-Trapped Excitons*, Springer-Verlag, Berlin, 1993.

51. Rashba, E. I., Critical length for self-trapping of excitons, *Synth. Met.*, *64*, 255–257 (1994).

52. Toyozawa, Y., Self-trapping of an electron by the acoustical mode of lattice vibration. 1, *Prog. Theoret. Phys.*, *26*, 29–44 (1961).

53. De Boer, S., Vink, K. J., and Wiersma, D. A., Optical dynamics of condensed molecular aggregates: An accumulated photon-echo and hole-burning study of the J-aggregates, *Chem. Phys. Lett.*, *137*, 99–106 (1987).

54. Fidder, H., Terpstra, J., and Wiersma, D. A., Dynamics of Frenkel excitons in disordered molecular aggregates, *J. Chem. Phys.*, *94*, 6895–6907 (1991).

55. Hirschmann, R., Köhler, W., Friedrich, J., and Daltrozzo, E., Hole burning in excitonic states of long-chain molecular aggregates, *Chem. Phys. Lett.*, *151*, 60–64 (1988).

56. Hirschmann, R., and Friedrich, J., A hole burning study of excitonic states of chain molecules in glasses, *J. Chem. Phys.*, *91*, 7988–7993 (1989).

57. Hirschmann, R., and Friedrich, J., Hole burning of long-chain molecular aggregates: Homogeneous line broadening, spectral-diffusion broadening, and pressure broadening, *J. Opt. Soc. Am. B*, *9*, 811–815 (1992).

58. Pschierer, H., and Friedrich, J., Pressure broadening and motional narrowing in excitonic states of J-aggregates, *Phys. Stat. Sol. (b)*, *188*, 43–49 (1995).

59. Fidder, H., and Wiersma, D. A., Collective optical response of molecular aggregates, *Phys. Stat. Sol. (b)*, *188*, 285–295 (1995).

60. Terpstra, J., Fidder, H., and Wiersma, D. A., A nonlinear optical study of Frenkel excitons in Langmuir–Blodgett films, *Chem. Phys. Lett.*, *179*, 349–354 (1991).

61. Fidder, H., and Wiersma, D. A., Exciton dynamics in disordered molecular aggregates: Dispersive dephasing probed by photon echo and Rayleigh scattering. *J. Phys. Chem.*, *97*, 11603–11610 (1993).

62. Moll, J., Daehne, S., Durrant, J. R., and Wiersma, D. A., Optical dynamics of excitons in J-aggregates of a carbocyanine dye, *J. Chem. Phys.*, *102*, 6362–6370 (1995).

63. Sumi, H., and Sumi, A., Dimensionality dependence in self-trapping of excitons, *J. Phys. Soc. Jpn.*, *63*, 637–657 (1994).

64. Higai, S., and Sumi, H., Two types of self-trapped states for excitons in one dimension, *J. Phys. Soc. Jpn.*, *63*, 4489–4498 (1994).

65. Drobizhev, M. A., Sapozhnikov, M. N., Scheblykin, I. G., Varnavsky, O. P., Van der Auweraer, M., and Vitukhnovsky, A. G., Relaxation and trapping of excitons in J-aggregates of a thiacarbocyanine dye, *Chem. Phys.*, *211*, 455–468 (1996).

66. Nabetani, A., Tomioka, A., Tamaru, H., and Miyano, K., Optical properties of two-dimensional dye aggregates, *J. Chem. Phys.*, *102*, 5109–5117 (1995).

67. Drobizhev, M. A., Sapozhnikov, M. N., Scheblykin, I. G., Varnavsky, O. P., Van der Auweraer, M., and Vitukhnovsky, A. G., Exciton dynamics and trapping in J-aggregates of carbocyanine dyes, *Pure Appl. Oct.*, *5*, 1–13 (1996).

68. Drobizhev, M. A., Sapozhnikov, M. N., Varnavsky, O. P., Vitukhnovsky, A. G., and Schreiber, M., Self-trapping of excitons in one-dimensional molecular J-aggregates, in *Excitonic Processes in Condensed Matter. Proc. 2nd Intern. Conf. on Excitonic Processes in Condensed Matter, 14–17 August, 1996, Kurort Gohrisch, Germany* (M. Schreiber, ed.), Dresden University Press, Dresden, 1996, pp. 23–26.

69. Melnikov, V. I., Coupled electron and phonon states in a one-dimensional system, *Zh. Eksp. Teor. Fiz.*, *72*, 2345–2349 (1977) (in Russian).

70. Shaw, P. B., and Whitfield, G., Vibrational excitations of a one-dimensional electron–phonon system in strong coupling, *Phys. Rev.*, *B17*, 1495–1505 (1978).

71. Pekar, S. I., *Research in Electron Theory of Crystals*, AEC, Division of Technical Information, Washington, DC, 1963, translation series No. AEC-tr-5575 Physics.

72. Schreiber, M., and Toyozawa, Y., Numerical experiments on the absorption lineshape of the exciton under lattice vibrations. 1. The overall lineshape, *J. Phys. Soc. Jpn.*, *51*, 1528–1536 (1982).

73. Sumi, H., and Toyozawa, Y., Urbach–Martienssen rule and exciton trapped momentarily by lattice vibrations, *J. Phys. Soc. Jpn.*, *31*, 342–358 (1971).

74. Toyozawa, Y., and Shinozuka, Y., Stability of an electron in deformable lattice—force range, dimensionality and potential barrier, *J. Phys. Soc. Jpn.*, *48*, 472 (1980).

75. Eremko, A. A., Gaididei, Yu.B., and Vakhnenko, A. A., Dissociation-accompanied Raman scattering by Davydov solitons, *Phys. Stat. Sol. (b)*, *127*, 703–713 (1985).

76. Drobizhev, M. A., Novkiov, A. V., Sapozhnikov, M. N., Varnavsky, O. P., Vitukhnovsky, A. G., and Ushomirsky, M. N., A comparative study of temperature broadening of spectral holes burnt in absorption bands of monomers and J-aggregates of a thiacarbocyanine dye, *Chem. Phys. Lett.*, *234*, 425–430 (1995).

77. Davydov, A. S., *Theory of Molecular Excitons*, Plenum Press, New York, 1971.

78. Toyozawa, Y., Theory of line shapes of the exciton absorption bands, *Prog. Theoret. Phys.*, *20*, 53–81 (1958).

79. Craig, D. P., and Dissado, L. A., Dispersion and resonance terms in exciton–phonon coupling: Absorption band profiles in molecular crystals, *Chem. Phys.*, *14*, 89–110 (1976).

80. Davydov, A. S., and Kislukha, N. I., Solitary excitons in one-dimensional molecular chains, *Phys. Stat. Sol. (b)*, *59*, 465–470 (1973).

81. Kuprievich, V. A., On autolocalization of the stationary states in a finite molecular chain, *Physica*, *14D*, 395–402 (1985).

82. Halperin, B. I., Green's functions for a particle in a one-dimensional random potential, *Phys. Rev.*, *A139*, 104–117 (1965).

83. Misawa, K., Machida, S., Horie, K., and Kobayashi, T., Wavelength and polarization dependence of spectral hole-burning efficiency in highly oriented J-aggregates, *Chem. Phys. Lett.*, *240*, 210–215 (1995).

84. Kobayashi, T., and Misawa, K., Excitons in J-aggregates with hierarchical structure, in *Excitonic Processes in Condensed Matter. Proc. 2nd Intern. Conf. on Excitonic Processes in Condensed Matter, 14–17 August, 1996, Kurort Gohrisch, Germany* (M. Schreiber, ed.), Dresden University Press, Dresden, 1996, pp. 23–26.

85. Burroughes, J. H., Bradley, D. D. C., Brown, A. R., Burn, P. L., Friend, R. H., Holmes, A. B., Mackay, K. D., and Marks, R. H., Light-emitting diodes based on conjugated polymers, *Nature*, *347*, 539–541 (1990).

86. Lee, C. H., Moses, G., Yu, D., and Heeger, A. J., Picosecond transient photoconductivity in poly(*p*-phenylenevinylene), *Phys. Rev. B.*, *49*, 2396–2407 (1994).

87. Lee, C. H., Yu, D., and Heeger, A. J., Persistent photoconductivity in poly(*p*-phenylenevinylene): Spectral response and slow relaxation, *Phys. Rev.*, *B, 47*, 15543–15553 (1993).

88. Vardeny, Z., Ehrenfreund, E., Brafman, O., Nowak, M., Schaffer, H., Heeger, A. J., and Wudl, F., Photogeneration of confined soliton pairs (bipolarons) in polythiophene, *Phys. Rev. Lett.*, *56*, 671–674 (1986).

89. Rauscher, U., Schutz, L., Greiner, A., and Bässler, H., Site-selective spectroscopy of matrix-isolated conjugated polymers, *J. Phys.: Condens. Matter*, *1*, 9751–9763 (1989).

90. Gailberger, M., and Bässler, H., dc and transient photoconductivity of poly(2-phenyl-1,4-henylenevinylene), *Phys. Rev.*, *B, 44*, 8643–8651 (1991).

91. Rauscher, U., Bässler, H., Bradley, D. D. C., and Hennecke, M., Exciton versus band description of the absorption and luminescence spectra in poly(*p*-phenylenevinylene), *Phys. Rev. B*, *42*, 9830–9836 (1990).

92. Lochner, K., Bässler, H., Tieke, B., and Wegner, G., Photoconduction in polydiacetylene multilayer structures and single crystals. Evidence for band-to-band excitation, *Phys. Stat. Sol. (b)*, *88*, 653–661 (1978).

93. Sebastian, L., and Weiser, G., One-dimensional wide energy bands in a polydiacetylene revealed by electroreflectance, *Phys. Rev. Lett.*, *46*, 1156–1159 (1981).

94. Kersting, R., Lemmer, U., Mahrt, R. F., Leo, K., Kurz, H., Bässler, H., and Göbel, E. O., Femtosecond energy relaxation in π-conjugated polymers, *Phys. Rev. Lett.*, *70*, 3820–3823 (1993).

95. Drobizhev, M. A., Sapozhnikov, M. N., and Kobryanskii, V. M., Selective excitation of the luminescence of a polymer film at room temperature, *JETP Lett.*, *63*, 182–186 (1996).

96. Heun, S., Mahrt, R. F., Greiner, A., Lemmer, U., Bässler, H., Halliday, D. A., Bradley, D. D. C., Burn, P. L., and Holmes, A. B., Conformational effects in poly(*p*-phenylene vinylene)s revealed by low-temperature site-selective fluorescence, *J. Phys.: Condens. Matter*, *5*, 247–259 (1993).

97. Mahrt, R., and Bässler, H., Light and heavy excitonic polarons in conjugated polymers, *Synth. Met.*, *45*, 107–117 (1991).

98. Elschner, A., Mahrt, R. F., Pautmeier, L., Bässler, H., Stolka, M., and McGrane, K., Site-selective fluorescence studies on polysilylenes, *Chem. Phys.*, *150*, 81–91 (1991).

99. Pautmeier, L., Rausher, U., and Bässler, H., Spectral diffusion in 1D systems: Simulation and experimental results for matrix-isolated polydiacetylene and polyisopropenylnaphthalene, *Chem. Phys.*, *146*, 291–301 (1990).

100. Su, W. P., and Schrieffer, J. R., Soliton dynamics in polyacetylene, *Proc. Natl. Acad. Sci. USA*, *77*, 5626–5629 (1980).

10
Semiconducting Polymers as Light-Emitting Materials

Xiao-Chang Li
Northeastern University
Boston, Massachusetts

Stephen C. Moratti
University of Cambridge
Cambridge, England

I. INTRODUCTION

Semiconducting polymers usually possess conjugated backbones which allow π electrons to significantly delocalize along the chain. Bulk electrical conductivity is very poor unless these materials are "doped" by adding or removing charges from the conjugated backbone.* A full range of conductivities, ranging from insulating to metallic (10^{-16}–10^5 S/cm), have been achieved by the judicious selection of polymer backbone units and proper dopants. The conjugated backbone in these polymers can also be regarded as an extreme example of a long-chain chromophore. Indeed, most conjugated polymers appear colored and show interesting photophysical phenomena, such as photoconductivity [1], nonlinear optical properties [2], and photoluminescence [3].

Electroluminescence through charge injection under a high applied field was known as early as the 1960s [4,5]. Progress in organic electroluminescence was slow until Tang et al. reported improved performance by the use of multilayer structures [6–8]. These were fabricated by the vacuum sublimation of organic chromophores or charge transport materials to form amorphous films. These devices showed high brightness and good efficiency, and the use of organic compounds meant that it was easy to choose the color of emission. In 1990, Burroughs et al. [9] discovered that poly(p-phenylene vinylene) (PPV) can emit green light under

---

*For up-to-date reviews, see relevant chapters.

an applied electric field when it is sandwiched between indium–tin oxide (ITO) and aluminum. Initially, the light emitted was very faint—yet it was bright enough to point to a new promising application for conducting polymers. Very soon afterward, visible red light emission was reported by Braun and Heeger using a poly(2,5-dialkoxy phenylene vinylene) as the emissive layer [10]. Following these two pioneering works, ever-increasing numbers of reports of electroluminescence in conjugated polymers have been published by scores of research groups from all over the world.

Conjugated polymers used for applications requiring good electrical conductivity usually need to be "doped" with strong acids or oxidants, which can lead to long-term stability problems even on storage. In contrast, an electroluminescent (EL) polymer needs only a low intrinsic conductivity, although the electric field strengths are so high that current densities can be quite appreciable. These EL polymers can be deposited from solution by simple spin-coating or dip-coating techniques over larger areas than are feasible by sublimed film techniques. Semiconducting polymers also usually have excellent thermal stability and good mechanical properties. These characteristics, in combination with control of color by the use of chemical design, make these materials serious competitors to existing light-emitting devices (LEDs) and flat panel technology.

Research into light-emitting polymers is still very new and is in its developing stage. Nevertheless, striking progress has been achieved over the past 7 years in many aspects; in improving processibility, in the tuning of color and color purity, in increasing device efficiencies and brightness, and in the construction of complicated device architecture. As a result, there are several companies with products now close to commercialization.

In this chapter, we will present the most recent state-of-the-art developments in this field as well as the basic principles behind the technology. Major issues discussed will include the following: (i) construction and operating principles of polymeric light-emitting devices (PLEDs); (ii) structures and synthetic strategies for light-emitting polymers; (iii) the control of the band gap and the tuning of color emission; (iv) improvement of device quantum efficiency; (v) multilayered PLEDs and charge-transporting polymers; (vi) potential applications of PLEDs.

II. CONSTRUCTION AND OPERATING PRINCIPLES OF PLEDS

A. Various Designs of PLEDs

Electroluminescence is the term used to describe the overall process that occurs when light is emitted upon application of an electric field to a semiconducting material. In inorganic materials, both alternating bias (AC) and direct bias (DC) modes are common. In organic devices, the major emphasis has been on DC applications for reasons explained later. The most common device architecture employed is shown in Figure 1. A fluorescent polymer is sandwiched between two metallic or conductive electrodes. At least one of these metallic contacts must be semitransparent for the light to escape normal to the plane to the device. The most widely used transparent conductor is indium–tin oxide (ITO), which is usually bought precoated on glass. On top of the ITO, an emissive polymer layer is coated directly from solution to a thickness of about 20–200 nm. To avoid shortcut circuits

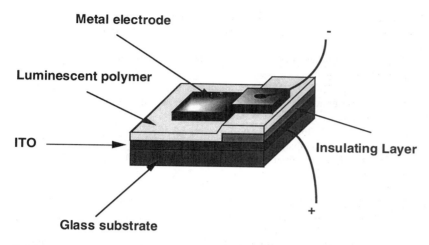

Metal electrode

Luminescent polymer

ITO

Insulating Layer

Glass substrate

FIGURE 1 Schematic diagram of a typical single-layer-polymer-based LED.

between the positive and negative contacts through the extremely thin polymer layer, the ITO electrode is often patterned by etching, or an insulating layer can be used, as shown in Fig. 1. A charge-injecting metal contact such as calcium or aluminum is then typically deposited by vacuum evaporation. This simple, single-layer device can be constructed in a matter of hours in a standard laboratory and is expected to form the basis of a profitable new flat-panel industry.

To improve stability and to enhance device efficiency (photons of light emitted per injected charge), a multilayered device is usually employed. This involves the use of various electron and/or hole (positive charge carrier) transporting materials (Fig. 2). The use of one or two charge transporting layers is used to balance the injection of positive and negative charges into the emissive polymer, which leads to enhanced light emission. Detailed discussion on the use of charge transporting materials will be found in Section VI.

Whereas rigid, flat-screen devices are probably the most important application for emissive materials, the special properties of polymers allow for many other interesting applications. Polymers can be deposited by a number of techniques over a variety of large flat and curved surfaces, including spin-coating, dip-coating, electropolymerization, draw-coating, and spray-coating. With the likely exception of the last technique, all the others have been used to construct PLEDs. The pliable nature of most polymers also offers new possibilities to fabricate totally flexible LEDs. Gustafsson et al. realized a fully polymer LED [11] by using poly(ethylene terephthalate) as the substrate (Fig. 3), soluble polyaniline (doped with camphor sulfonic acid) as the hole-injecting and conducting electrode (~0.5 μm), a red fluorescent PPV derivative as the electroluminescent layer (100–150 nm), and calcium (<100 nm) as the electron-injecting top contact. This PLED was found to be mechanically robust and could be sharply bent without failure. In addition, it shows high efficiency and a low turn-on voltage (2–3 V).

Similarly, a kind of reversed cylindrical PLED can also be constructed. Figure 4 shows the structure used, in which a simple aluminum rod is used as the electron-

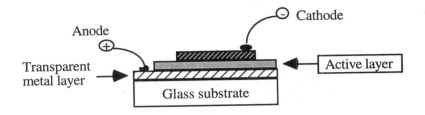

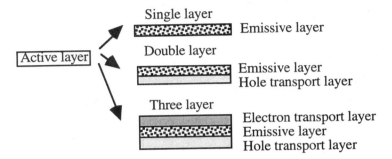

FIGURE 2   Cross-sectional view of a multilayered LED.

injecting substrate. Emissive polymer and hole-injecting polyaniline are subsequently deposited by dip-coating. Light emission was observed when the aluminum rod was biased negatively with respect to the polyaniline layer.

Discovered by Pei et al. [12,13], an unusual device configuration is the use of an electroluminescent polymer blended with an ionic conductive polymer. When this blend is sandwiched between two electrodes and sufficient potential is applied, a light-emitting *p-n* junction diode is created in situ through simultaneous *p*-type and *n*-type doping of the polymer on opposite sides of the device. Red, green, and blue light emission have been achieved by varying the electroluminescent polymer. The fabrication of this so-called light-emitting electrochemical cell (LEC) is principally the same as the above-mentioned PLEDs, but both the thickness of active layer and the nature of electrodes used are less critical than LEDs, due to the greatly

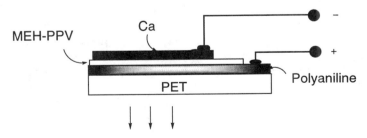

FIGURE 3   Cross-sectional view of the structure of a flexible PLED.

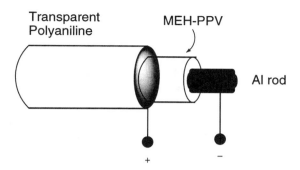

FIGURE 4   Structure of an inverted cylindrical PLED.

enhanced charge injection and transportation. Although the response time for LECs (around 1s to several minutes depending on diffusion speed) are usually greater than PLEDs, their turn-on voltages are lower than that of LEDs and almost equal to the band gap of the emissive material.

The ability to create anisotropic materials by simple stretch orientation is a unique feature of polymers. Anisotropic electronic conductivity has been achieved as a result of orientation of random-coil polymer chains. Optical absorption, as well as photoluminescence, will be stronger parallel to the orientation direction as well. In highly oriented polymers, such as gel-processed polyethylene/dialkoxy-PPV blends, it was found possible to achieve a very large anisotropy in photoluminescence intensity (60:1) [14]. With this method, a polarized, red electroluminescent device was achieved by Dyreklev et al. [14] with an intensity ratio between light emitted parallel and perpendicular to the stretching direction of up to 3.1 (Fig. 5). Using Langmuir–Blodgett techniques, Cimrova et al. [15] have succeeded in obtaining highly aligned poly(2,5-diisopentoxy-*p*-phenylene) and thus achieving polarized blue emission with an intensity ratio of up to 3.5.

It is even possible to construct a planar configuration PLED, in which two metal electrodes (separated by ~20 μm) are deposited on a luminescent polymer layer, so that the electrical field and current are in the plane of the film. These planar PLEDs are compatible with standard photolithography and silicon technology, and therefore hold promising technological potential [16,17].

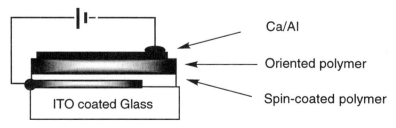

FIGURE 5   Cross section of the structure of a polarized PLED.

B. Basic Principles

Although PLEDs can be constructed in a wide variety of ways, the basic operating mechanism is the same in all cases. In order to understand the principles behind light emission in PLEDs, it is instructive to start with the simpler process of photoluminescence (PL) [18]. The great similarity between the EL and PL emission spectra of PPV [19] and other materials suggests that similar mechanisms are at play.

When a molecule is irradiated by light, photoexcitation of an electron from the highest occupied molecular orbital (HOMO) (or ground state S_0) to the lowest unoccupied molecular orbital (LUMO) generates an excited state (S_1) which can lose the absorbed energy in the following ways:

(a) Radiationless transitions, such as internal conversion or intersystem crossing (macroscopically observable by heat formation)
(b) Emission of radiation (fluorescence and phosphorescence)
(c) Photochemical reactions (e.g., rearrangements, dissociations, dimerizations, photo-additions, reactions with neighboring particles, etc.).

Process (a) and (b) are represented schematically in Fig. 6.

Therefore, much of the light energy absorbed by a molecule may be lost by processes other than fluorescence. Indeed, it is rare for an organic compound to emit all of its absorbed energy back as light (i.e., to have a quantum efficiency of unity). As shown in Fig. 6, in most organic molecules the emitted light (hv_f) is of

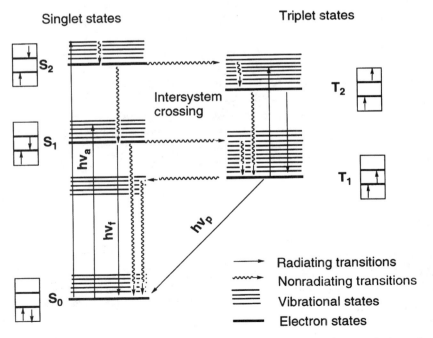

FIGURE 6 Relationship between absorption, emission, and nonradiative vibrational processes (Jablonski scheme).

a lower energy than of that originally absorbed (hv_a). This difference between absorbed and emitted light is termed the Stokes shift. Phosphorescence (hv_p) is usually very inefficient for organic materials at room temperature, and thus inter-system crossing does not lead to useful levels of light emission.

A simplified band scheme for the photoluminescence of a conjugated polymer system is shown in Fig. 7. Excitation of an electron from the polymers HOMO (or valence band) into the LUMO (or conduction band) is followed by a relaxation process to give an excited state known as a singlet exciton. This relaxation process, which involves both spatial reorganization of the conjugated backbone and a shift of the conjugated electrons, also results in a large Stokes shift. Nonradiative decay processes are not shown; for example, those involving the formation of charged species or the formation of triplet excitons (detected by photo-induced absorption) [20] which have been found to be the main channels to diminish emission efficiency in conjugated PPV.

The similarity between the PL and EL spectra in luminescent materials suggests that the same emitting species (a singlet exciton) is involved in each case. However, the mechanism of formation is much more complicated in the latter case. Injection of electrons into the LUMO of the polymer can occur at one electrode and removal of electrons from the HOMO can occur at the other [the process of removal of a negative charge from the HOMO leaves a positive charge (or hole) in the band]—this process is more usually referred to as hole injection. Under the influence of the applied electric field, the oppositely charged species tend to drift toward each other. The combination of the charge carriers occurs on a segment of polymer chain to form a singlet exciton, which can then radiatively decay with the emission of visible light.

Photoluminescence

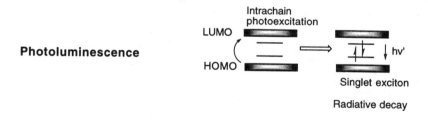

Electroluminescence

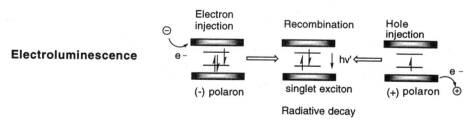

FIGURE 7 Comparison between photoluminescence and electroluminescence in conjugated polymers. (Adapted from Ref. 18.)

Emission spectra from polymeric LEDs are generally very broad due to vibronic side bands and the inhomogeneity of the material (defects, kinks in the structure, etc.). This can be partially overcome by the use of Fabry–Perot microcavity structures. Striking modifications to both EL and PL emission involving narrowing/strengthening of the main transition have been achieved [21,22].

Photoluminescent efficiencies are usually defined simply as (photons emitted/photons absorbed) × 100%. However, there is a wide range of ways that can be used to define EL efficiencies. One of the most common is device quantum efficiency, defined as the ratio of photons emitted/charge injected (both holes and electrons) × 100%. Much of the light (up to 75%) of any light generated within a simple LED can be lost through absorption and wave-guiding out the edges instead of going in the useful forward direction. Thus, a further distinction must be made between external quantum efficiency, which measures the light leaving the device, and internal quantum efficiency, which is based on the calculated total amount of light generated within the film. Device brightness is often reported in the units of candela per meter squared. A figure of 100 cd/m^2 is considered bright enough for a flat panel display, whereas for lighting applications up to 5000 cd/m^2 may be required. The best organic devices can often achieve the latter figure (up to 90,000 cd/m^2 has been reported [23]), although not necessarily for prolonged periods.

Internal efficiency measurements can be very useful as, in principle, they give information on how much further a device can be improved. Unless otherwise stated, efficiencies quoted in this chapter should be taken to be internal. As nonemitting triplet states are also generated when two charge carriers combine to form an exciton, the maximum internal efficiency possible is believed to be 25%, although this figure has been challenged as possibly being too high by some workers, as well as being too low by others.

The close relationship between the PL and EL implies that increasing the photoluminescence yield will result in equal improvements in electroluminescence efficiency, all other factors being equal. Photoluminescence efficiencies of conjugated polymers used in PLEDs range from below 1% to about 80% in solid state and can reach almost up to 100% in solution. However, EL efficiencies are rarely above 5%. This indicates that many other factors have to be considered in EL devices, such as charge-injection ratios, charge mobilities, and extra quenching processes due to the electrodes and charge carriers. These factors will be discussed in more detail in Section V. First, the effect of polymer design on color and fluorescence efficiency as well as general synthetic methodology will be covered.

III. CHEMICAL STRUCTURES AND SYNTHETIC STRATEGIES FOR LIGHT-EMITTING POLYMERS

A. Requirements for Electroluminescence on a Polymer

Very little research has been done on the overall structural requirements for a polymer to exhibit electroluminescence. For a given polymer, it is still hard to say in advance whether it would prove useful in a PLED. This is due to the complicated mechanism for singlet exciton formation in these devices, which involves double charge injection, charge transportation, recombination, and emission. It is quite common for a given material to give poor results with one electrode or charge transport material, but excellent performance with others. The most important prerequisite for the emissive layer is that polymer must be fluorescent.

Several electronic considerations are believed to be important in order to favor photoluminescence in organic compounds. In photoluminescent compounds generally, the longest wavelength of absorption corresponds to a $\pi \rightarrow \pi^*$ excitation. On the other hand, species in which the longest absorption wavelength corresponds to an $n \rightarrow \pi^*$ transition (common in molecules containing heteroatoms or heterocyclic aromatic molecules) are seldom fluorescent [24]. Thus, it is not strange that both polypyrrole and polyaniline are not fluorescent, although they are good semiconducting polymers. It is observed that the excited singlet state normally has an half-life of about 10^{-8} s. If the half-life is more than 10^{-8} s, intersystem crossing will be favored; a shorter half-life implies rapid deactivation of the molecule by other processes.

The fluorescence behavior of an organic compound is dependent on three main factors, namely the nature of the carbon skeleton, the geometrical arrangement of the molecule, and the type and position of any substituents. Increasing the extent of conjugation and therefore increasing the mobility of the π electrons often results in an increase in fluorescence intensity. For instance, anthracene has a higher efficiency (0.32 in hexane) than naphthalene (0.1 in hexane), which, in turn, is more fluorescent than benzene (0.04). Increasing planarity and rigidity also helps to increase fluorescence, as both will also enhance the free mobility of the π electrons and charge transportation. In solution, increased rigidity can also help by reducing vibrational–rotational interactions in the excited state, which can lead to intersystem crossing. Biphenyl and fluorene possess the same degree of conjugation, but the aromatic rings in the latter compound are held rigidly in a planar configuration, whereas those in biphenyl are not. Fluorene (Fig. 8) has a quantum efficiency of about 0.54 in hexane, whereas that of biphenyl is only 0.23 [25].

However, if the extent of conjugation is very large, other effects can come into play. The mobility of excited states tends to increase as the conjugation increases, thus they are more likely to encounter a quenching centre before decaying radiatively, and so reduce emission intensity. As well, the band gap can decrease to an extent (<1.5 eV) where thermal deactivation of excited states can become important. Time-resolved photoluminescence of poly(p-phenylene vinylene) in the picosecond regime shows [26] that the nonradiative decay rate increases with increasing polymer chain conjugation length and that the luminescence decay is much faster for excitation parallel to the polymer chain than for perpendicular excitation.

The substituents on a conjugated system may have a very profound effect on the fluorescence properties. Substituents which enhance π electron mobility will

Fluorene   Biphenyl

FIGURE 8 The structural comparison between fluorene and biphenyl.

normally increase fluorescence. Often, a combination of electron-donating (positively mesomeric) substituents, such as $-NH_2$, $-N(CH_3)_2$, OCH_3, and OH, with electron-withdrawing substituents, such as $C=O$, CN, SO_3H, and COOH, is used to enhance fluorescence. Alkyl groups generally have little effect on fluorescence except on steric grounds. Large atoms such as bromine and iodine will reduce fluorescence wherever they occur as substituents, due to enhancement of intersystem crossing.

Photoluminescence can often be greatly enhanced by increasing the intrinsic stiffness of a polymer backbone or by inducing large bulky side groups to weaken intermolecular interactions [27]. This is shown in Fig. 9, where the use of larger side groups in PPV derivatives lead to higher photoluminescent efficiencies. However, very large aliphatic side chains may cause problems in EL devices by lowering the overall charge carrier mobility.

The fluorescence–structure relationships of heterocyclic compounds are poorly understood at present. However, it appears that heterocyclics with a longest absorption wavelength corresponding to an $n \rightarrow \pi^*$ transition are likely to be nonfluorescent, whereas those corresponding to a $\pi \rightarrow \pi^*$ transition are more likely to be fluorescent. For instance, pyridine is nonfluorescent, whereas quinoline is weakly fluorescent. However, both polypyridine and polyquinoline are much more fluorescent as the $\pi \rightarrow \pi^*$ transition becomes more dominant.

B. Synthetic Strategies for Light-Emitting Polymers

One of the main problem with highly conjugated polymers is their lack of processibility, as most of them are neither fusible nor soluble due to their rigid backbones and strong intermolecular interactions. In order to get around this problem, several synthetic strategies have been developed during last several years. These include (a) soluble precursor routes, (b) conjugated polymers with long flexible solubilizing groups, and (c) copolymers containing fixed conjugated segments and flexible or angular spacers.

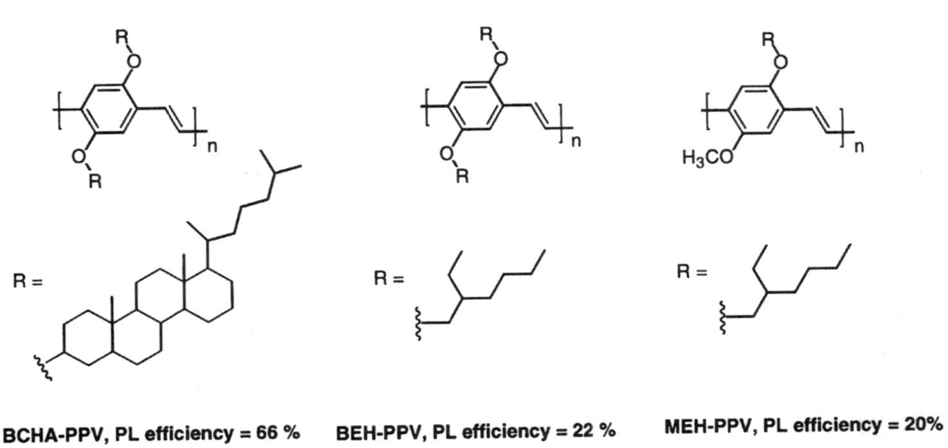

BCHA-PPV, PL efficiency = 66 %   **BEH-PPV, PL efficiency = 22 %**   **MEH-PPV, PL efficiency = 20%**

FIGURE 9 The influence of side-chain structures on PL efficiency in PPV derivatives.

Soluble Precursor Routes Towards Conjugated Polymers

The first reported electroluminescent polymer was poly(p-phenylene vinylene) which consists of alternating phenyl and double-bond units. This polymer is bright yellow with a HOMO–LUMO energy gap of around 2.4 eV (517 nm) and a photoluminescent emission peak at 551 nm. It is neither soluble nor fusible because of its rigid backbone. In order to prepare good quality films, the Cambridge group used poly(p-xylidene sulfonium chloride) as a precursor for PPV in LEDs, synthesized by a route originally described by Wessling [28] and later modified by Lenz et al. [29].

Scheme 1 outlines the standard preparation of PPV (**6**). Treatment of 1,4-bis(chloromethyl)-benzene (**1**) with tetrahydrothiophene results in formation of bis-sulfonium salt (**2**). The polymerization of (**2**) is carried out in aqueous solution at low temperature in order to minimize any subsequent elimination reactions of the polymer which might lead to insoluble products such as (**5**). Although the mechanism of the polymerization reaction is ambiguous, it has been presumed that the initial reaction is a proton abstraction by the base on the benzylic position of the monomer to form a sulfonium ylid, followed by a 1,6-elimination reaction to form a quinoidal intermediate [29], which undergoes rapid radical-promoted polymerization (Scheme 2).

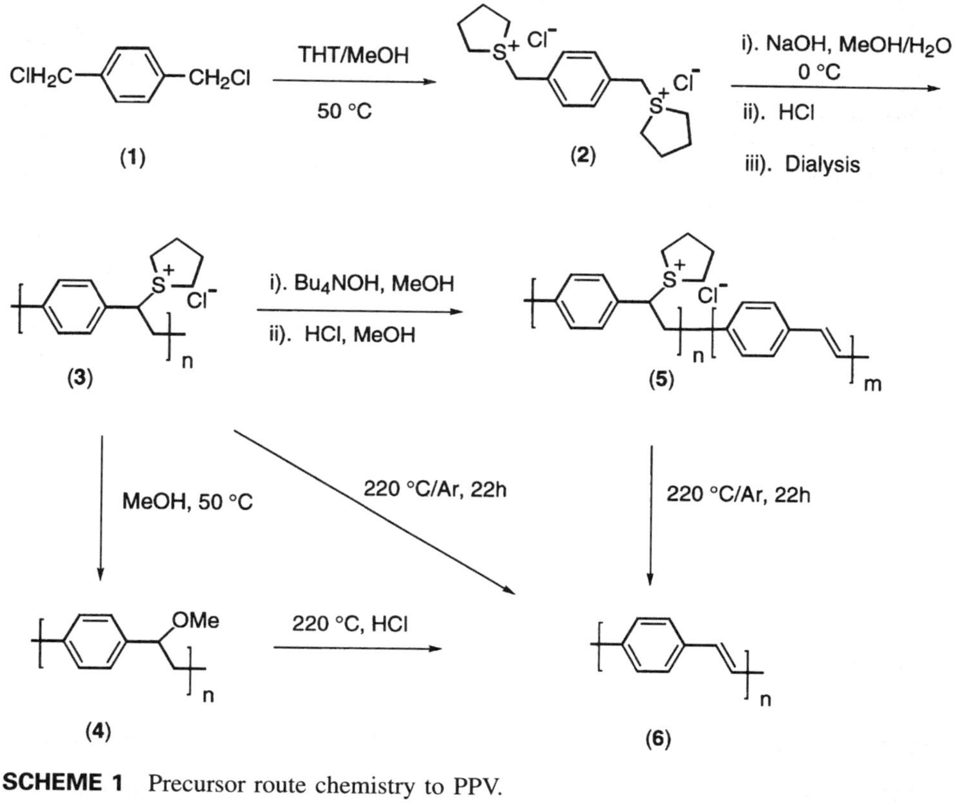

SCHEME 1 Precursor route chemistry to PPV.

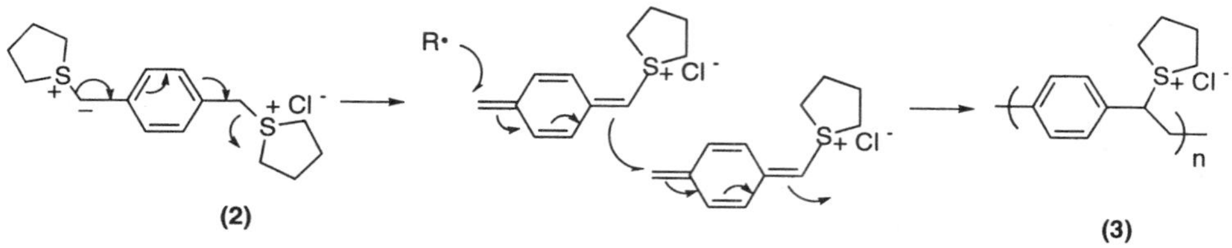

SCHEME 2 Proposed mechanism for the preparation of the sulfonium precursor polymer to PPV.

The polymerization mixture is purified by dialysis against water to remove low-molecular-weight oligomer and other impurities such as unreacted monomer, tetrahydrothiophene, and inorganic salts. Good quality PPV precursor [poly(*p*-xylidene sulfonium chloride)] is lightly colored and fully soluble in polar solvents such as water and methanol. Thin polymer film can be obtained by spin-coating or casting the precursor polymer solution on glass, followed by heat treatment at 220°C under inert atmosphere or vacuum. The thermal conversion of the sulfonium polymer can be prone to side reactions, and some carbonyl peaks are usually seen in the infrared (IR) spectra of the final conjugated polymer. It has been suggested that if the conversion is carried out in a reducing atmosphere (e.g., 15% hydrogen in argon), the measured amount of the carbonyl groups is substantially reduced. As aromatic ketones are well-known triplet sensitizers, reduction in the overall carbonyl level can result in the photoluminescence intensity of the polymer increasing by as much as five-fold [30].

The sulfonium precursor polymer is rather unstable and odorous, and a more stable and environmentally friendly precursor is the methoxy polymer (**4**). This can be produced by simply heating the sulfonium polymer (**3**) in methanol. The increased stability of the methoxy precursor polymer allows better characterization by nuclear magnetic resonance (NMR) and standard molecular-weight determinations by gel permeation chromatography (GPC). High molecular masses, often well over 10^5 Da, have been measured for this methoxy precursor polymer.

Another route to poly(arylene vinylene)s is via a halide precursor, as a modification to the Gilch route [31] to PPV-type materials [32,33]. Similarly, poly(heteroaromatic vinylene) can also be synthesized using halide precursor route. As shown in Scheme 3, bis(halomethyl)pyridine (**7**) in the presence of one equivalent of base will undergo a Wessling-type polymerization as above to form an intermediate polymer (**8**) which is soluble in formic acid and is readily converted

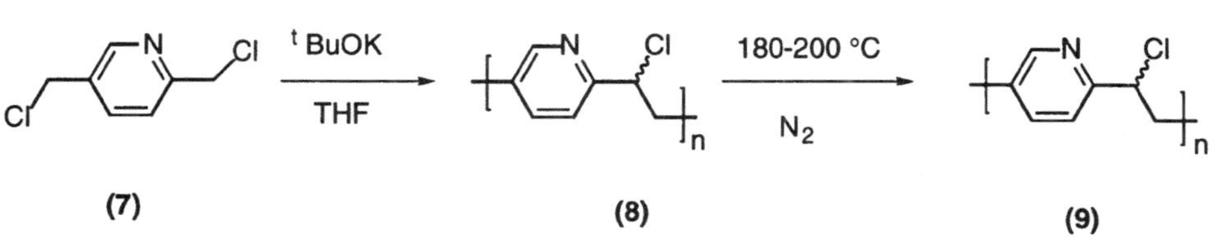

SCHEME 3 Chlorine precursor route to polypyridine vinylene (PPyV).

into conjugated poly(pyridine vinylene) (PPyV). Excess base will produce the fully conjugated polymer (**9**) directly. Thermal treatment of the precursor polymer will produce elimination of the respective hydrogen halide to form the required double bonds [34]. Poly(pyridine vinylene) fluoresces orange red and has a HOMO–LUMO band gap of 2.2 eV. PPyV has a lower luminescent quantum efficiency compared with its analog PPV, but a higher electron affinity according to cyclic voltammogram measurements (with reduction potential of -1.3 V against ferocene/ferocene$^+$, 0.42 V). PPyV was used as an orange–red emissive layer in a LED using aluminum as the cathode.

Recently, other new precursor routes have been developed. A metal carbene-catalyzed "living" ring-opening metathesis polymerization (ROMP) was used to produce PPV [35]; see Scheme 4. In this process, a barrelene derivative (**10**) was synthesized and polymerized by a molybdenum bis(hexafluoro-*tert*-butoxyl) carbene catalyst (**11**) to generate a soluble precursor polymer (**12**). Thermal elimination of carboxylic acids from (**12**) at temperatures above 200°C led to the formation of PPV. Using this method, poly(1,4-naphthalene vinylene) and other derivatives can also be made [36].

A xanthate-precursor route has been developed by a team in Bell AT&T Laboratories [37]; see Scheme 5. The monomer is synthesized in one step and is stable enough to store as a solid at room temperature without any sign of degradation or hydrolysis. The polymerization is carried out in the presence of base at 0°C. The resulting precursor polymer has very high molecular mass (1.05×10^5 Da) and is soluble in common organic solvents such as chloroform and toluene. The elimination of xanthic acid or conversion of the precursor polymer starts at 180°C from thermal gravimetric analysis. Fourier-transform infrared (FTIR) revealed that the precursor polymer is not completely converted to PPV even at 250°C and the

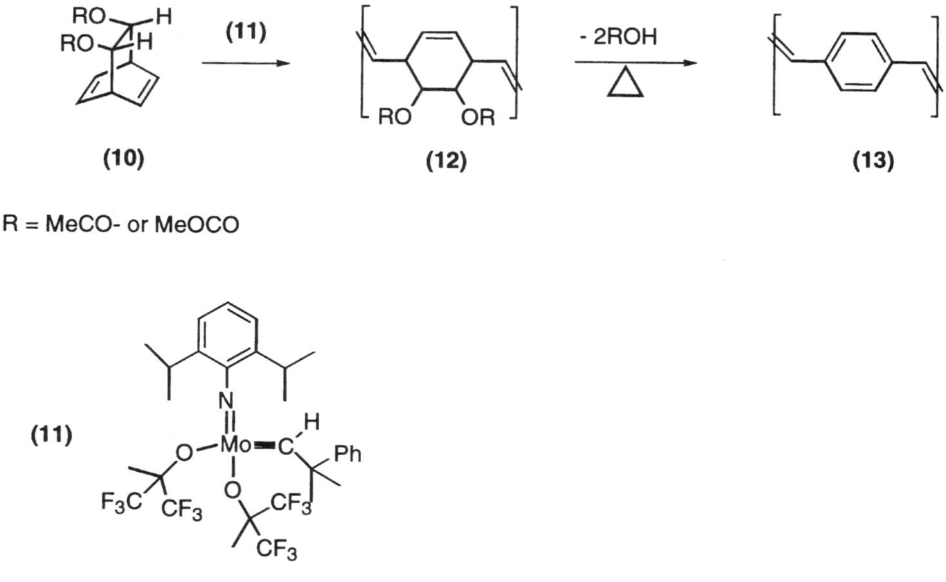

(10)   **(12)**   **(13)**

R = MeCO- or MeOCO

(11)

SCHEME 4 A ring-opening polymerization precursor route for PPV.

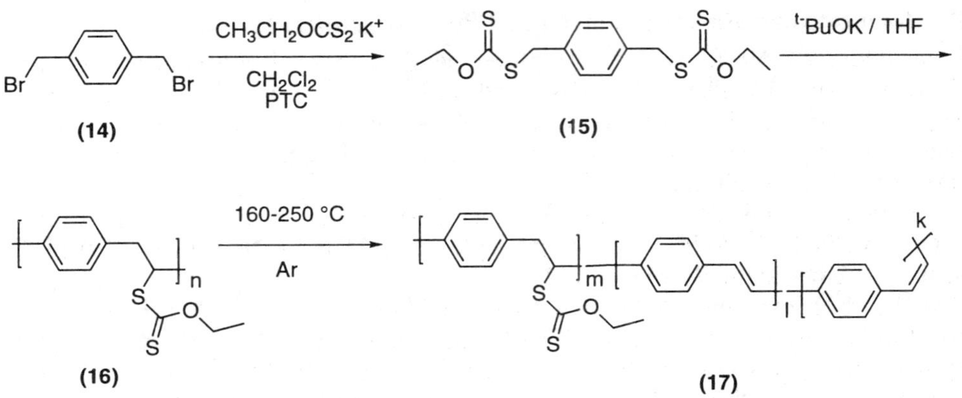

SCHEME 5 A xanthate group precursor route for PPV.

converted PPV contains cis linkages. The unconverted and the *cis*-PPV segments interrupt conjugation and result in the formation of an amorphous material. This is different from PPV produced via the sulfonium method, which is normally semi-crystalline, and with the double bonds largely in a trans configuration. The more amorphous nature of the xanthate-derived films was reported to produce a much more stable device.

Kim and Ober have developed a spin-coatable polymeric precursor route for the preparation of polyphenylene (PPP) films [40]; see Scheme 6. Radical polymerization of cyclohexadiene-1,2-diol derivatives gave a largely 1,4-linked soluble polymer (**19**). This precursor could be aromatized to poly(phenylene) either by heating or by ultraviolet (UV) exposure in the presence of a photo-acid generator. This unique photoaromatization capability suggests that poly(phenylene) precursors can be imaged by deep-UV microlithography to produce patterned devices.

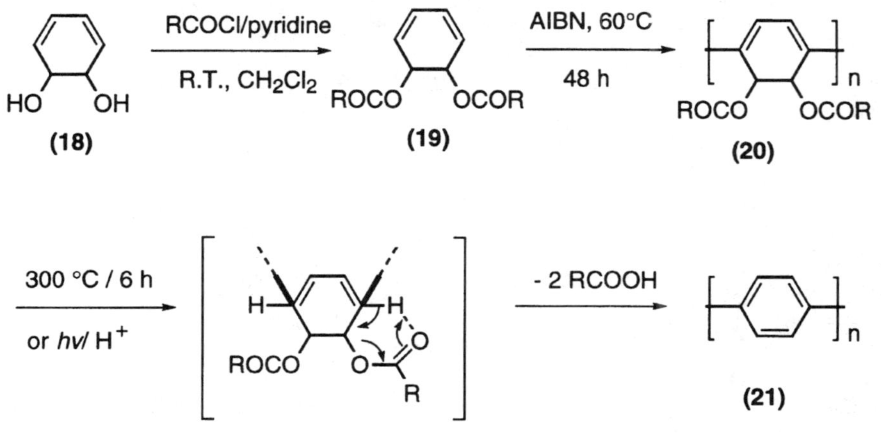

SCHEME 6 A soluble precursor route for PPP.

2. Conjugated Polymers with Flexible Solubilizing Groups

It is often desirable to avoid any high-temperature thermal conversion reactions. The conjugated polymer then needs to be directly soluble and the most common way of arranging this is the use of long, flexible groups. These are usually alkyl or alkoxy substituents directly attached to the aromatic ring, although they can be part of the main chain in semiconjugated polymers.

Scheme 7 shows the synthesis of the most widely studied PPV derivative, poly[(2-methoxy-5-(2-ethylhexyloxy)phenylene) vinylene] (MEH–PPV) (**24**) [27]. In this scheme, methoxy phenol (**22**) is converted into the 2-ethylhexyloxy ether (**23**) by base-promoted alkylation. Chloromethylation of (**23**) produces the bis(chloromethyl) benzene (**24**), which is polymerized directly to MEH–PPV (**25**) by base (i.e., potassium *tert*-butoxide). The polymer is soluble in chlorinated solvent, pyridine, and tetrahydrofuran. It usually has very high molecular weight (M_n normally larger than 10^5 Da).

The polymer has strong photoluminescence in solution, whereas in solid state, it is a bit weaker (PL efficiency of between 10% and 20% as a thin film). Unlike PPV, MEH–PPV is red colored because of the two alkoxy substituents. An electroluminescent device using this polymer (ITO/MEH–PPV/Ca) achieved the quite respectable efficiency of 1.0% with red light emission (λ_{max} = 630 nm).

In order to get soluble blue LEPs, Huber and Scherf [39] have synthesized a polyphenylene through a Pd-catalyzed cross-coupling reaction. The polymer obtained had very good solubility in chlorinated solvents such as chloroform as a result of the cyclophane substitution; see Scheme 8. The polymer was found to fluoresce with a purple–blue color (λ_{max} = 415 nm).

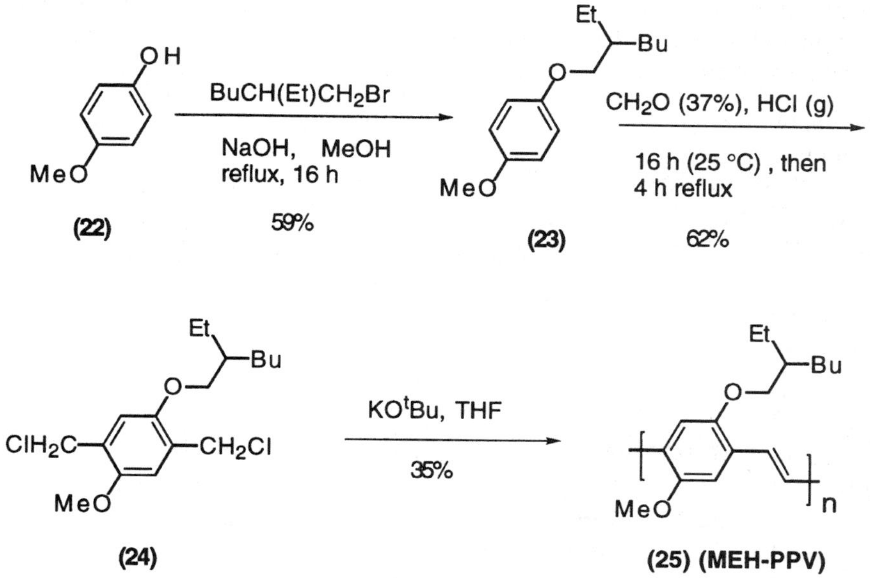

SCHEME 7 Synthesis of MEH–PPV.

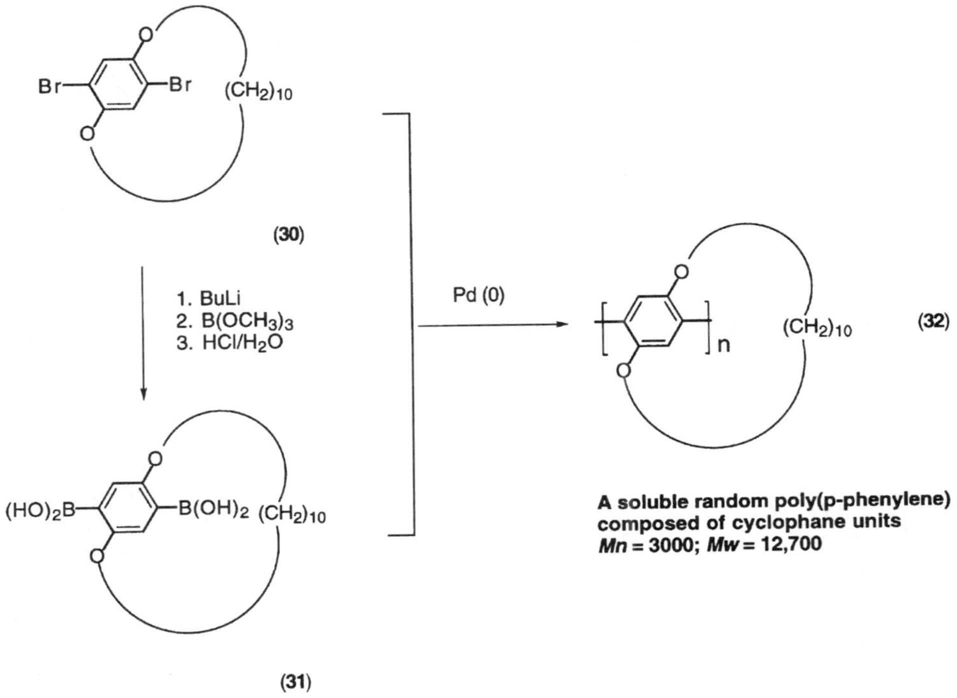

SCHEME 8 Synthesis of soluble PPP with cyclophane groups.

Toward the achievement of functional solubilizing side group, oxygen-rich side groups have been introduced into conjugated polyphenylene [40] and into polyfluorene [41]. As shown in Scheme 9, the side methoxyethoxy substituents function as both a solubilizing group and a potential group to solvate such salt as lithium triflate. The polyfluorene (29) has a band gap of 2.8 eV and as high a photoluminescent quantum efficiency as 73% (solid film) with an emission peak around 450 nm. Sky-blue light has been achieved both in a simple LED and in a light-emitting electrochemical cell in which lithium triflate salt was added.

There are many other soluble conjugated polymers or copolymers that contain alkoxy, alkyl, and silyl substituents including PPV [42–45] and PPP derivatives [46] and copolymers [47,48]. Solubilizing alkoxy or alkyl groups are also used in such heteroaromatic polymers as polythiophenes [49], polypyridines [50–52], and others [53,54]. Recently, electroactive polymers possessing crown ether sites [55] have been synthesized and used in electrochemical light-emitting diodes.

3. Polymers Containing Fixed Conjugated Segments and Flexible Spacers

Another strategy toward EL polymers is the use of covalently attached fluorescent chromophores. One problem when designing new fully conjugated polymers is that one can never be sure what the emission color will be, or even if it will be fluorescent at all. However, there are a large number of highly fluorescent chromophores reported in the literature. By attaching these groups into a polymer chain,

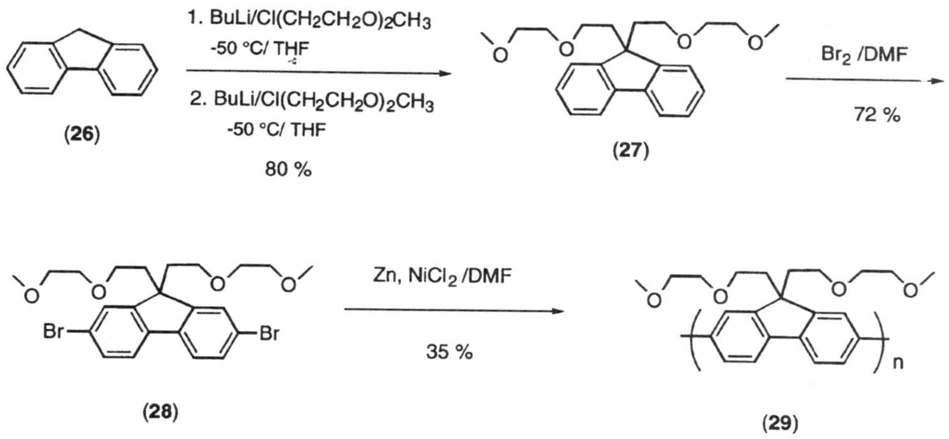

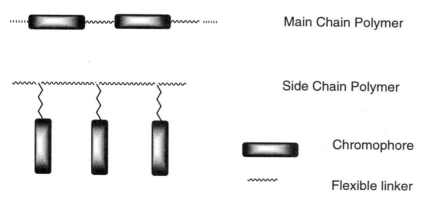

SCHEME 9   Synthesis of a polyfluorene with oxygen-rich side chains.

it is possible, in many cases, to combine the stability advantages of polymers with the high fluorescence and known properties of the small chromophore. Good purity blue light (450 nm) is especially desirable but can be quite difficult to achieve in fully conjugated polymers due to exciplex formation and other effects. In contrast, there are a large number of small molecules that show high PL efficiency in this region. Therefore, a straightforward way to get luminescent polymers is to covalently bond these chromophores onto suitable polymers. The chromophoric segments can either be in the main chain or be present in polymer side chain (Fig. 10).

Using a Wittig reaction, a main-chain blue-light-emitting polymer (λ_{max} = 465 nm) has been synthesized as shown in Scheme 10 [56].

A similar method has been used to synthesize main-chain blue electroluminescent polymers with well-defined conjugation lengths and silane-containing flexible side groups [57,58]. Yellow–green polyesters containing polyphenylene vinylene segments have also been reported [59].

Main Chain Polymer

Side Chain Polymer

Chromophore

Flexible linker

FIGURE 10   Polymers containing fixed conjugated segments and flexible spacers.

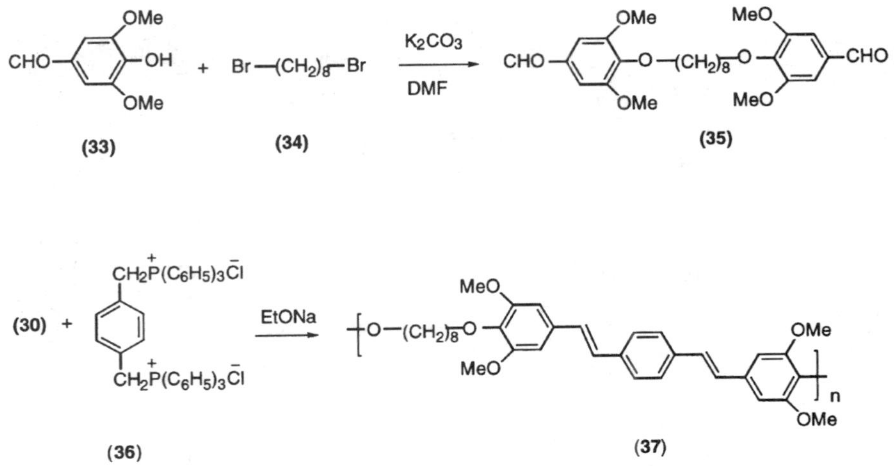

SCHEME 10 Synthesis of a main-chain blue LEP.

Blue-emitting distyrylbenzene derivatives has also been chemically bonded to such flexible polymer chains as polystyrene [60,61] and polynorbornene [62]. Using radical polymerization, it is possible to synthesize copolymers such as (**39**) containing both blue-light-emitting chromophores and charge-transporting chromophores as side chains [63,64], or even with a photo-cross-linkable blue light-emitting polymer (LEP) [65]; see Scheme 11.

For achieving light emission shorter than blue light, it is necessary to chose chromophore with shorter conjugation length. In a particular case, poly-(methylphenylsilane) was used as emissive layer sandwiched between ITO and aluminum, and near-UV light emission (353 nm) was emitted under 6 V [66].

4. Other Methods

There are a few reports using other film-forming techniques for polymer deposition. Ma et al. have reported a plasma polymerization method for the formation of polynaphthalene thin film on substrate such as ITO-coated glass and was able to construct a blue-light-emitting diode [68]. PPV films can also be polymerized electrochemically onto an ITO-coated glass substrate and used in green-light-emitting devices [68].

IV. CONTROL OF BAND GAP AND EMISSION COLOR OF EMISSIVE POLYMERS

Shortly after the first report of green emission from a polymer-based light-emitting diode based on PPV, red light emission was reported using a close derivative, MEH–PPV. A blue-light-emitting diode has also been reported under another conjugated polymer—poly(p-phenylene) (PPP). This ability to produce a large number of different colors and hues gives organic LEDs a big advantage in flat-panel displays.

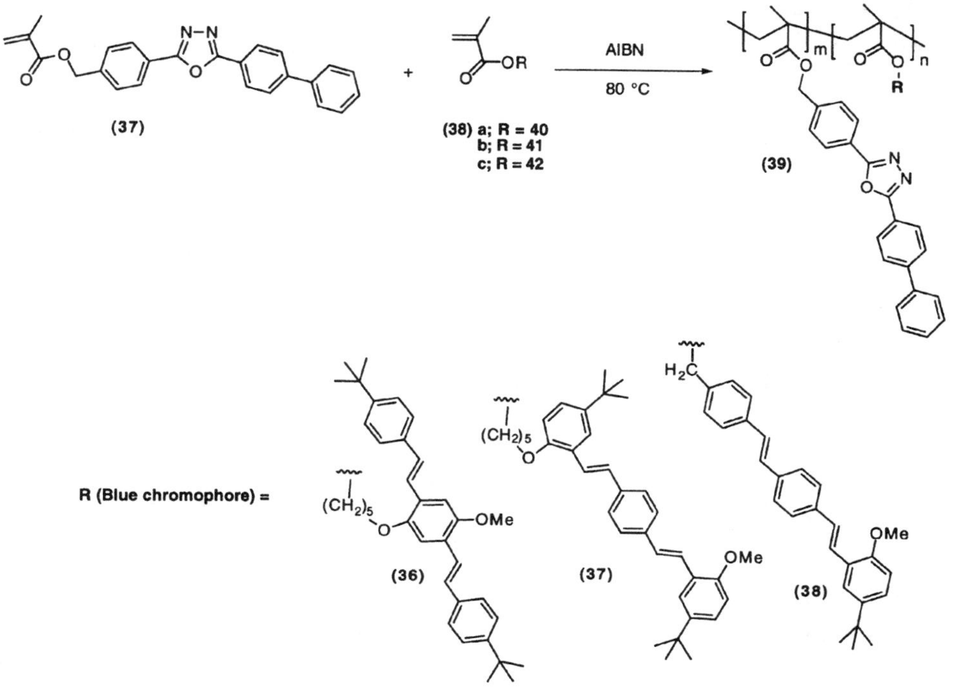

SCHEME 11 Synthesis of PMMA bearing with aromatic oxadiazoles and blue chromophores.

There are several approaches that can be used to vary the band gap and emission of a conjugated polymer. The nature of any substituents can have a strong influence, either through electronic or steric effects. Table 1 shows the calculated variation of electronic properties of several PPVs with various substituents [69]. For instance, by using alkoxy substitution on the benzene ring of PPV, the band gap (E_g) can be reduced significantly, leading to red LEPs. Ionization potentials can be tuned by the inclusion of electron-donating groups such as alkoxy or amino substituents, or the presence of electron-withdrawing substituents such as the cyano group.

The band gap of many thiophene-based polymers can be tuned by changing the size of the substituents. In polythiophenes, enhanced steric hindrance from side groups leads to increasing planarity of the main chain and an increasing band gap. A range of thiophene-based polymers have been synthesized with emissive colors spanning the entire visible spectrum and beyond (Table 2) [70].

Another way of blue-shifting the emission of conjugated polymers is to incorporate conjugation-breaking units in the main chain. This can be achieved by deliberate incorporation of nonconjugated comonomers, or by controlling the polymerization or postpolymerization conversion conditions [71]. Various conjugation-breaking linkages may be used, including flexible alkyl or alkoxyl spacers, meta-benzene, and the elements O, S, N, and Si. Limiting the conjugation

TABLE 1 Influence of Substituents on the Electronic Properties of PPVs

Polymer	E_g	I_p	E_a
(6)	2.32	5.05	2.73
(43)	2.07	4.72	2.65
(44)	2.17	5.27	3.10
(45)	2.24	5.15	2.91
(46)	1.57	5.12	3.15
(47)	1.74	5.06	3.24

Note: I_p = ionization potential; E_a = electron affinity; E_g = band gap.

of a polymer in such a way often has a useful effect of increasing the fluorescent efficiency, in part due to lowering exciton diffusion rates. This was achieved in a PPV precursor polymer by limiting the degree of final conversion into the fully conjugated polymer [72]. In another example, incorporation of a small amount of a comonomer into a normal PPV–sulfonium precursor polymer results on conjugation-breaking segments and improved EL efficiency [73]. However, this resulted in a statistical incorporation of a comonomer, and, hence, a wide range of conjugation lengths was produced. For more effective control of color, one can use a nonconjugated polymer containing small chromophores either in the main chain or the side chain. As an example, several block copolymers containing distyrylbenzene

TABLE 2 Relationship of Polymer Structure and Electroluminescence Emission of Polythiophenes

Polymer structure	(48)	(49)	(50)	(51)
EL peak (nm)	460	570	625	685

or its substituted derivatives and solubilizing flexible spacers have been synthesized with blue emission [57,59,74]. Polyquinoline with a hexafluoropropylene linkage or spacer has been used as a blue PLED with high efficiency (internal quantum efficiency of 4% at 450 nm) but low brightness (30 cd/m²) [75].

Polyphenylene and its derivatives represent a typical blue-light-emitting polymer, although the exact emission may vary from violet to green depending on substituents. Because the first blue PLED was reported using poly(*p*-phenylene) as the emissive layer [76,77], a whole series of soluble blue-light-emitting PPPs have been synthesized [15,46,78–81]. The phenylene units can be bridged to form "ladder"-type PPPs, with the aim of obtaining more stable and efficient blue emission [82,83]. Device efficiencies as high as 4% have been achieved using these materials [84].

Microcavity structures have received a great deal of attention for their ability to tailor the emission properties of LEDs. In particular, the broad-emission characteristics of most organic materials can be transformed into a very narrow output spectrum [21]. By using such a structure, lasing pumped optically in a high-*Q* cavity has been recently demonstrated [85,86].

V. IMPROVEMENT OF DEVICE QUANTUM EFFICIENCY

Initial reports of conjugated polymer electroluminescence involved poly(phenylene vinylene) derivatives, with device quantum efficiencies of only 0.01–0.05% [9]. Soon after, up to 1% quantum efficiency was achieved using a PPV copolymer containing nonconjugated segments [87]. The most striking improvement of quantum efficiency is perhaps a double-layer red PLED using a cyano-substituted PPV as the emissive layer and PPV as the hole transport layer with a 4% internal quantum yield (as detailed below) [88]. Efficiencies of close to 11% have been achieved in another cyano-PPV derivative [82], which given the PL efficiency of the polymer of ~40%, may be close to the maximum theoretically possible.

The quantum efficiency in PLEDs is affected not only by the nature of the emissive polymer but also by the device structure. This is because other important aspects include charge carrier transport to and within the material and unwanted exciton quenching reactions—especially at the metal electrode interfaces.

TABLE 3 Work Function of Several Metals

Metal	Ca	Mg	ITO	Al	In	Ag
Work function	2.9 eV	3.7 eV	4.9 eV	4.3 eV	4.2 eV	4.4 eV

Therefore, for optimum device performance, the following three factors should be considered: (1) carrier injection and transport; (2) formation of excitons; and (3) radiative recombination of excitons [89]. This can be written as

$$\phi_{el} = \phi_{pl}\phi_i\phi_e$$

where ϕ_{pl} is the photoluminescence efficiency, ϕ_i is the charge balance efficiency, and ϕ_e is the singlet exciton quantum yield (often taken to be 25%).

Thus, it is very important to equalize the number of electrons and holes reaching the emissive zone to maximize ϕ_i. PPV is a widely used material in PLEDs but has a low oxidation potential, as do many other simple conjugated polymers. Under normal circumstances, mainly holes are injected into the polymer, leading to poor EL efficiencies. It is therefore necessary to enhance electron injection into many such polymers.

The most direct method is to use a low work-function metal to improve electron injection. Many luminescent polymers show much higher EL efficiencies when calcium is used as the electron-injecting contact as compared to most other metals. Unfortunately, calcium is well known to oxidize very easily when in contact with air or moisture, and, hence, may not be useful in a commercial environment. Table 3 lists some common metals and their work functions. Often, mixtures of metals are used to give a combination of low work function and increased environmental stability, such as Mg/Ag and Li/Al alloys.

Another way is through the use of polymers with a much higher electron affinity. This has been achieved by the Cambridge group by synthesizing a PPV-derivative (**54**) with electron-stabilizing cyano-substituents [90]; see Scheme 12.

This red polymer has an optical energy–gap of 2.05 eV (as measured by the onset of absorbance), which is very close to MEH–PPV but with a much higher PL efficiency. A single-layer device both using aluminum and calcium gave devices with 0.15% efficiency. Furthermore, the improved electron affinity also allowed the

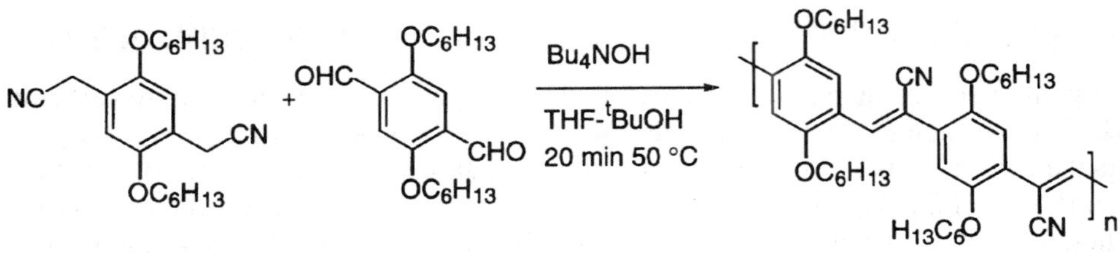

SCHEME 12 Synthesis of cyano-substituted PPV.

use of the more stable aluminum metal as cathode without a loss of efficiency. Remarkably, an internal efficiency of 4% was found with a double-layer device (ITO/PPV/CN–PPV/Al). Its emission proved to be solely from the cyano polymer with a peak wavelength of 710 nm. The fields needed to drive the devices were also significantly reduced (a current of 5 mA/cm^2 required a field of 4×10^5 V/cm, about half that required for a single-layer device) [88].

There are increasing number of reports of other high-electron-affinity luminescent polymers being synthesized [91]. Other cyano-substituted polymers have been reported [90] and heterocyclic electron-deficient rings (such as pyridine or oxadiazole) have also been used.

Alternatively, electron injection can also be improved through the use of an extra electron-transporting layer. Examples include high-band-gap materials such as oxadiazoles or triazines that also have a high oxidation potential. Thus, they often function by blocking hole transport, rather than necessarily having a high natural electron affinity. (2-(4-Biphenyl)-5-(4-*tert*-butyl-phenyl)-1,3,4-oxadiazole) (PBD) is a very commonly used compound in such layers [92,93]. More details about charge-transporting polymers will be discussed in the next section.

A prime consideration when improving efficiencies is to enhance the fluorescence quantum yields of the materials involved. This can often be achieved through elimination of defects or impurities, or prevention of crystallization. However, many stable and potentially attractive materials such as perylene derivatives show poor fluorescence in the solid state, but high fluorescence in dilute solution. Thus, by blending different polymers to form solid-state solutions, often much higher PL and EL efficiencies are possible than by using either material alone. General considerations for improving PL are discussed in previous sections. However, it is not uncommon for EL efficiency to decrease with increasing PL efficiency due to other factors [42].

In electroluminescence, the existence of a bound triplet excited state can severely limit the quantum efficiency. If the triplet binding energy and the corresponding cross section for forming a triplet from a pair of injected carriers were large, the singlet to triplet ratio would be determined by spin statistics (3:1). That means the maximum EL quantum efficiency would be limited to 25%. Although this limitation has been generally accepted, Pakbaz et al. [94] pointed out that if the dynamics are such that the cross section for triplet formation from a pair of oppositely charged free carriers is relatively small, the limiting quantum efficiency could approach 100%. This is very encouraging, but the state of art is still far from even 25%.

VI. CHARGE TRANSPORT POLYMERS AND MULTILAYERS PLEDs

As mentioned above, electron injection and transport is often much more difficult than it is for holes. It is very common to use electron-transporting layers to try to equalize the two kinds of charge injection and transportation. In one of the first examples in PLEDs, Burn et al. [87] reported a 10-fold enhancement of internal efficiency by using (2-(4-biphenyl)-5-(4-*tert*-butyl-phenyl)-1,3,4-oxadiazole) (PBD)/poly(methyl methacrylate) (PMMA) blend on top of PPV. It was considered that the role of the electron transport layer is to confine holes to the emissive layer, as shown in Figure 11.

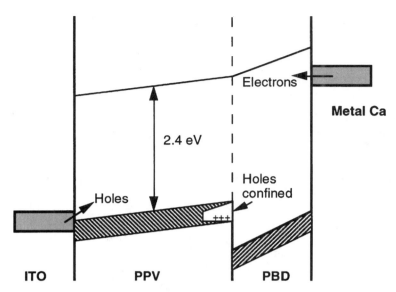

FIGURE 11   Schematic band diagram of a double-layer device under a forward bias utilizing PBD/PMMA as electron-transporting/hole-blocking layer.

A polymer blend with PMMA was used in this case, as good quality PBD films cannot be cast directly from solution due to crystallization. Although such polymer-dispersed films have enjoyed wide application in multilayered PLEDs, such devices tend to suffer from short lifetimes. It is believed that this may be due to problems such as aggregation and recrystallization of the PBD under operation. To address this issue, many polymers containing PBD or other electron-deficient aromatic heterocycles have been synthesized. In a typical example, Li et al. [63] reported a polymethacrylate bearing aromatic oxadiazoles side chains; see Scheme 13.

These polymers were used in various multilayered devices with PPV as the emissive layer. Compared with single-layer PPV devices, the internal efficiency of the two-layer PPV device with oxadiazole-bonded PMMA (**60**) as the charge-transporting layer was increased by a factor of 4 to 0.04% (at a current density of 0.5 mA/cm^2). It was also observed that the turn-on voltage for the two-layer device was significantly lower than that of an analogous single-layer PPV device, as shown in Fig. 12. The downturn in efficiency shown in the inset of Fig. 12 on increased current may be due to a number of effects, such as heating or charge-carrier-induced quenching.

A similar polymethacrylate–PPD derivative and several polyethers containing PPD units have also been reported by Bell AT&T laboratories [95]. LEDs containing these electron-transporting (ET) polymers were found to be much more stable than those without an ET layer. However, increased efficiencies were not always found with these EL layers [96].

Processable main-chain oxadiazole polymers have also been synthesized via incorporation of flexible spacers [97–99], or by a precursor route [100]. Oxadiazole-containing polymers are usually wide-band-gap materials and are,

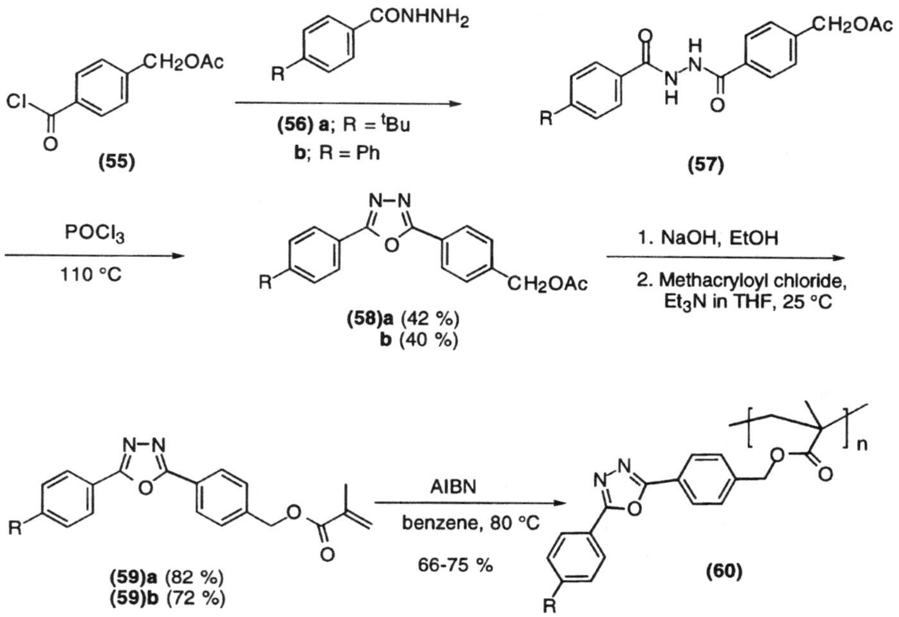

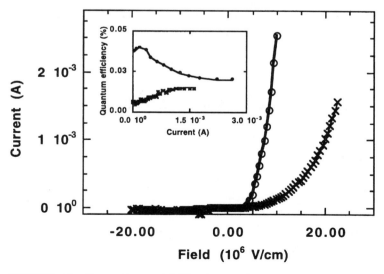

SCHEME 13 Synthesis of polymethacrylate polymers bearing PPD and PBD.

FIGURE 12 Comparison of LED characteristics by using PMA–PPD as electron transporting (×: ITO/PPV/Ca; ○: ITO/PPV/PMA–PPD/Ca).

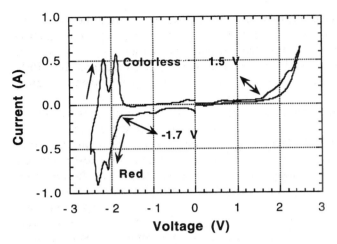

FIGURE 13 Cyclic voltammetry of an oxadiazole polymer film on Pt electrode in MeCN solution containing Bu_4NClO_4 (0.1 M); sweep rate 10 mV/s at 20°C [Reference against Ag/Ag^+ (0.1 M)]

therefore, not very colored. As they are also generally quite photoluminescent, they can also be used as the emissive layer in blue devices [101,102].

The electron affinity of oxadiazole materials is not particularly large and may actually be less than some of emissive polymers. However, the oxidation potential is very high and irreversible ($\sim$1.5 V versus ferrocene/ferrocene$^+$). Thus, oxadiazoles, may act not so much by accepting electrons, but by blocking the passage of holes through the device. Figure 13 shows the cyclic voltammogram of an oxadiazole polymer.

These oxadiazole compounds or polymers show quasi-reversible *n*-doping properties. It can be seen from Fig. 13 that there are two apparent cathodic reduction steps beginning from -1.7 eV. This may be explained by the respective formation of radical anions and dianions as shown in Fig. 14.

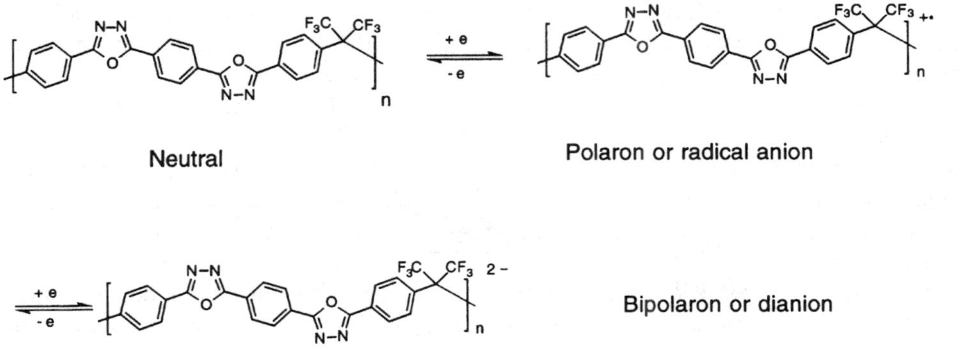

Neutral   Polaron or radical anion

Bipolaron or dianion

FIGURE 14 The formation of a radical anion and dianion in an oxadiazole polymer.

Table 4 lists the electronic properties of some oxadiazole polymers in comparison to some emissive polymers [103]. Some polymers (e.g., **54**) act as electron transport materials because of a low reduction potential. Other polymers, such as the oxadiazole series **62–64,** actually are worse at accepting electrons than PPV, but have such a high oxidation potential that holes are blocked from traveling through [63,101,104].

These charge-transporting polymers have a variety of useful roles to play in the operation of the device. These include (1) assisting effective carrier injection from the electrode to the emitting layer, (2) confining the carriers within the emitting layer and, thus, increasing the probability of recombination processes leading to radiative decay, and (3) preventing the quenching of excitons at the electrode interface by shifting the recombination zone away from the metal.

There have been a few reports that have shown that charge-transporting polymers can be blended or mixed with emissive polymers or compounds to construct single-layered PLEDs. In some cases, good results were obtained upon mixing a hole transport polymer, an emissive material, and an electron-transporting material into a single blend.

Because aromatic oxadiazole polymers show good photoluminescence as well, light emission from these charge-transporting polymers has also been demonstrated if a good hole-transporting layer, such as polyaniline, was used. Berggren et al. obtained white light emission [105] and violet emission [106], respectively, by blending geometrically restricted polythiophenes with an oxadiazole compound (PBD), which played both a charge-transporting and electroluminescent role.

In some cases, the use of hole transport polymers can also improve the PLEDs performance. This is especially effective when emissive polymers with higher electron affinity are employed. For instance, the efficiency of double-layer devices (ITO/PPV/CN–PPV/Al) exhibit a 20-fold increase in efficiency over single-layer cyano-PPV devices, the PPV layer acting solely as a hole-transporting material. Polyoxadiazole polymers show poor light-emitting performance in a single-layer device, whereas bright-blue light emission was achieved by incorporating a polyaniline hole transport layer [97]. By using electropolymerized poly(3-*n*-octylthiophene) as a hole-transporting layer, light emission from an organic chromophore could also be enhanced [107]. An increase of 100-fold in internal quantum efficiency was achieved by dispersing a nonconjugated blue-luminescent copolyester containing 1,2-dinaphthylene vinylene units, PBD (electron-transporting compound) and poly(9-vinylcarbazole) (hole-transporting polymer) [108].

By careful use of charge-transporting materials, Wang et al. have constructed an alternating-current PLED [109]. This device has more symmetry than is usual, with an electron-transporting material sandwiched between two hole-transporting layers. The actual structure of the device was Al/EB/PPy/EB/ITO-glass, in which EB stands for emeraldine base form of polyaniline and PPy for polypyridine. Holes can be injected from either electrode, and due to the relatively fast dynamic response, the device can also be operated in an alternating current mode. It is suggested that the nearly Ohmic behavior of the electrode–redox polymer contact and the presence of a large density of redox polymer–emissive polymer interface states have a central role in the operation of the device.

TABLE 4 Charge-Injection and Band-Gap Data for Different Conjugated Polymers

Polymer[a]	E^{Ox}	E^{Red}	E_{gec}	E_{gopt}
	(V versus Ag/Ag$^+$)		(eV)	
(6)	0.85	-1.7	2.55	2.5
(25)	0.34	-1.95	2.29	2.1
(54)	1.05	-1.6	2.65	2.1
(61)	1.0	-1.4	2.4	2.3
(62)	1.6	-1.8	3.4	3.73
(63)	1.45	-1.7	3.15	3.24
(64)	1.54	-1.6	3.14	3.19

Note: Electrochemical data were recorded in MeCN solution containing Bu$_4$NClO$_4$ (0.1 M), sweep rate 10 mV/s at 20°C versus Ag/Ag$^+$ (0.1 M).
[a]Hex = hexyl; EH = 2-ethylhexyl

VII. POTENTIAL APPLICATIONS OF POLYMERIC LEDs

The driving force for the fast development of light-emitting polymers is closely linked with their potential for applications in various fields [110]. It has been shown that PLEDs can emit different colors with good efficiency (1–4%) at low voltages. They are light and flexible and can be easily processed to form large-area, robust films. The response time for PLEDs is fast (usually submicrosecond), and the intensity of light is proportional to the current. It is obvious that PLEDs have great potential as a new lighting source, for small-scale instrumental displays, and ultimately in large-area, full-color flat-panel displays. The potential market for flat-panel displays is at least $10 billion/year.

For commercial application, a PLED should meet at least the following requirements:

1. A distinct threshold in the brightness–voltage relationship which allows for multiplexing of passive arrays
2. A high efficiency to avoid heating effects
3. A long working life

Polymer LEDs based on PPV are being commercialized in the United Kingdom by Cambridge Display Technology Ltd (CDT Ltd). A dot matrix has already been successfully made on a 6×2-cm display with a 16×60-pixel format (as shown in Fig. 15). Working devices have already been shown about 10,000 h working life and a 2-year shelf life [111]. CDT hopes to begin pilot production for commercial application. Commercial development using LEP technology is also engaged actively in UNIAX, Hoechst, and Philips corporations. Products using PLEDs as back lights for mobile phones and personal digital assistants (PDAs) are expected to come into being within 2 years.

The ability to create an anisotropic material by simple stretch orientation is an unique feature of polymers. When such a stretched light-emitting polymers are used in PLEDs, polarized light emission can be achieved. This kind of LED could be very suitable for background illumination in a liquid-crystal display, which normally requires a polarizing filter.

Once established, it is hoped that this new LEP technology will replace existing cathode-ray tubes in television and computer monitors. Present limitations are lifetimes, color purity, and power efficiencies. Specifications for commercial devices include a working lifetime of 20,000 h, a shelf life of 5 years, a light-emission efficiency of ~5%, and operating voltages of 5 V or less. Obviously, there is still a gap between the present technology and the more vigorous industrial requirements. In addition, blue emission is still a problem, as such larger-band-gap materials usually require larger threshold voltages for charge injection. Breakdown mechanisms of devices are not known for sure but are often due to the formation of nonemissive regions, or dark spots, which seem to be related to moisture ingress into the electrode interface region [112], or related to possible electrochemical dissolving of metal electrode under a high electric field [113].

VII. CONCLUSION

Almost all semiconducting polymers that contain luminescent conjugated backbones, or chromophores in the main chain or as side groups, may be considered

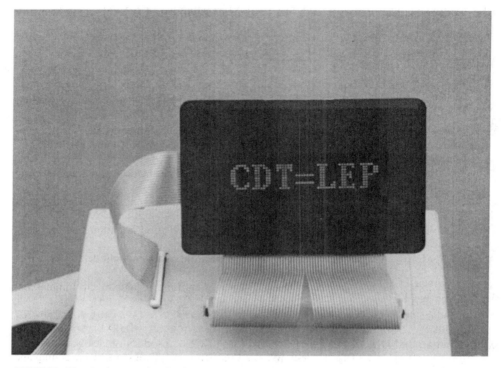

FIGURE 15   A dot-matrix display device utilizing PLEDs technology. (Courtesy of CDT Ltd.)

as potential electroluminescent materials. Electroluminescent emission resembles its photoluminescent emission because of the same excited-state formation in both cases. The maximum emission peak position is governed by the band gap of the luminescent polymer, and so the emission color can be tuned by careful design of polymers with suitable band gap. By careful use of substituents, the solubility, color, and charge-accepting properties of these materials can be significantly varied. The promise of cheap, large-area, multicolored displays has spurred a large amount of research into optimization of the properties of such polymers. Commercial applications of PLEDs may first be realized in backlighting and simple monochromic displays in the near future. For high information displays and full-colour flat-panel displays, it remains to be seen whether the particular advantages of polymer-based displays can win out over the large number of competing technologies.

ACKNOWLEDGMENTS

We are indebted to Dr. Andrew Holmes and Professor Richard H. Friend for their invaluable discussion and support. Part of the work that we have reviewed in this chapter has been carried out in Cambridge, involving our colleagues. These include Dr. Raoul Cervini, Dr. George Spencer, Dr. Andrew Grimsdale, and Ms. Shu-Mai Chang (the Chemical Laboratory), Dr. Johannes Grüner and Dr. Franco Cacialli (the Cavendish Laboratory), and Dr. Paul May (Cambridge Display Technology Ltd.)

REFERENCES

1. Halls, J. J. M., Walsh, C. A., Greenham, N. C., Marseglia, E. A., Friend, R. H., Moratti, S. C., and Holmes, A. B., Efficient photodiodes from interpenetrating polymer networks, *Nature*, *376*, 498–500 (1995).
2. Feast, W. J., Tsibouklis, J., Pouwer, K. L., Groenendaal, L., and Meijer, E. W., Synthesis, processing and material properties of conjugated polymers, *Polymer*, *37*, 5017–5047 (1996).
3. Yu, G., and Heeger, A. J., High efficiency photonic devices made with semiconducting polymers, *Synth. Met.*, *85*, 1183–1186 (1997).
4. Pope, M., Kallmann, H., and Magnante, P., Electroluminescence in organic crystals, *J. Chem. Phys.*, *38*, 2042–2043 (1963).
5. Helfrich, W., and Schneider, W. G., Recombination radiation in anthrancene crystals, *Phys. Rev. Lett.*, *14*, 229 (1965).
6. Tang, C. W., and VanSlyke, S. A., Organic electroluminescence diodes, *Appl. Phys. Lett.*, *51*, 913–915 (1987).
7. Adachi, C., Tsutsui, T., and Saito, S., Blue light-emitting organic electroluminescent device, *Appl. Phys. Lett.*, *56*, 799–801 (1990).
8. Dodabalapur, A., Organic light emitting diodes, *Solid State Commun.*, *102*, 259–267 (1997).
9. Burroughes, J. H., Bradley, D. D. C., Brown, A. R., Marks, R. N., Mackay, K., Friend, R. H., Burn, P. L., and Holmes, A. B., Light-emitting diodes based on conjugated polymers, *Nature*, *347*, 539–541 (1990).
10. Braun, D., and Heeger, A. J., Visible light emission from semiconducting polymer diodes, *Appl. Phys. Lett.*, *58*, 1982–1984 (1991).
11. Gustafsson, G., Cao, Y., Treacy, G. M., Klavetter, F., Colaneri, N., and Heeger, A. J., Flexible light-emitting diodes made from soluble conducting polymers, *Nature*, *357*, 477–479 (1992).
12. Pei, Q., Yu, G., Zhang, C., Yang, Y., and Heeger, A. J., Polymer light-emitting electrochemical cells, *Science*, *269*, 1086–1088 (1995).
13. Pei, Q., Yang, Y., Yu, G., Cao, Y., and Heeger, A. J., Solid state polymer light-emitting electrochemical cells: Recent developments, *Synth. Met.*, *85*, 1229–1232 (1997).
14. Dyreklev, P., Berggren, M., Inganäs, O., Andersson, M. R., Wennerstrom, O., and Hjertberg, T., Polarised electroluminescence from an oriented substituted polythiophene in a LED, *Adv. Mater.*, *7*, 43–45 (1995).
15. Cimrova, V., Remmers, M., Nether, D., and Wegner, G., Polarised light emission from LEDs prepared by the LB technique, *Adv. Mater.*, *8*, 146–149 (1996).
16. Lemmer, U., Vacar, D., Moses, D., Heeger, A. J., Ohnishi, T., and Noguchi, T., Electroluminescence from PPV in a planar metal–polymer–metal structure, *Appl. Phys. Lett.*, *68*, 3007–3009 (1996).
17. Yu, G., Pei, Q., and Heeger, A. J., Planar light-emitting devices fabricated with luminescent electrochemical polyblends, *Appl. Phys. Lett.*, *70*, 934–936 (1997).
18. Holmes, A. B., Bradley, D. D. C., Brown, A. R., Burn, P. L., Burroughes, J. H., Friend, R. H., Greenham, N. C., Gymer, R. W., Hallicay, D. A., Jackson, R. W., Kraft, A., Martens, J. H. F., Pichler, K., and Samuel, I. D. W., Photoluminescence and electroluminescence in conjugated polymeric systems, *Synth. Met.*, *55–57*, 4031–4040 (1993).
19. Brown, A. R., Greenham, N. C., Burroughes, J. H., Bradley, D. D. C., Friend, R. H., Burn, P. L., Kraft, A., and Holmes, A. B., Electroluminescence from multilayer con-

jugated polymer devices: Spatial control of exciton formation and emission, *Chem. Phys. Lett.*, *200*, 46 (1992).

20. Colaneri, N. F., Bradley, D. D. C., Friend, R. H., Burns, P. L., Holmes, A. B., and Spangler, C. W., Photoexcited states in poly(*p*-phenylene vinylene): Comparison with trans, trans-distyrylbenzene, a model oligomer, *Phys. Rev. B*, *42*, 11670 (1990).

21. Wittmann, H. F., Gruner, J., Friend, R. H., Spencer, G. C. W., Moratti, S. C., and Holmes, A. B., Microcavity effect in a single-layer polymer light-emitting diode, *Adv. Mater.*, *7*, 541–544 (1995).

22. Fisher, T. A., Lidzey, D. G., Pate, M. A., Weaver, M. S., Whittaker, D. M., Skolnick, M. S., and Bradley, D. D. C., Electroluminescence from a conjugated polymer microcavity structure, *Appl. Phys. Lett.*, *67*, 1355–1357 (1995).

23. Friend, R. H., Denton, G. J., Halls, J. J. M., Harrison, N. T., Holmes, A. B., Kohler, A., Lux, A., Moratti, S. C., Pichler, K., Tessler, N., Towns, K., and Wittmann, H. F., Electronic excitations in luminescent conjugated polymers, *Solid State Commun.*, *102*, 249–258 (1997).

24. Kasha, M., Characterization of electronic transitions in complex molecules, *Faraday Discuss. Chem. Soc.*, *9*, 14–19 (1950).

25. Bowen, E. J., *Advances in Photochemistry*, Interscience, New York, 1963, p. 32.

26. Wong, K. S., Bradley, D. D. C., Hayes, W., Ryan, J. F., Friend, R. H., Lindenberger, H., and Roth, S., Correlation between conjugation length and non-radiative relaxation rate in PPV: A picosecond photoluminescence study, *J. Phys. C: Solid State Phys.*, *20*, L187–L194 (1987).

27. Wudle, F., Allemand, P. M., Srdanov, G., Ni, Z., and McBranch, D., Materials for non-linear optics: Chemical perspectives, *ACS Symp. Serv.*, *455*, 683–686 (1991).

28. Wessling, R. A., The polymerization of xylyene bisdialkyl sulfonium salts, *J. Polym. Sci. Polym. Symp.*, *72*, 55–66 (1986).

29. Lenz, R. W., Han, C.-C., Stenger-Smith, J., and Karasz, F. E., Preparation of poly(phenylene vinylene) from cycloalkylene sulfonium salt monomers and polymers, *J. Polym. Sci. A: Polym. Chem.*, *26*, 3241–3249 (1988).

30. Papadimitrakopoulos, F., Konstadinidis, K., Miller, T. M., Opila, R., Chandross, E. A., and Galvin, M. E., The role of carbonyl groups in the photoluminescence of poly(*p*-phenylenevinylene). *Chem. Mater.*, *6*, 1563–1568 (1994).

31. Gilch, H. G., and Wheelwright, W. L., Polymerisation of bis(halomethyl)benzenes, *J. Polym. Sci. A-1*, *4*, 1337 (1966).

32. Swatos, W. J., and Gordon III, B., Chlorine precursor route for making PPV derivative conjugated polymers, *Polym. Prep.*, *31*, 505 (1990).

33. Hsieh, B. R., Antoniadis, H., Bland, D. C., and Feld, W. A., Chlorine precursor route (CPR) chemistry to poly(*p*-phenylene vinylene)-based light emitting diodes, *Adv. Mater.*, *7*, 36–38 (1995).

34. Li, X. C., Cacialli, F., Cervini, R., Holmes, A. B., Moratti, S. C., Grimsdale, A. C., and Friend, R. H., Precursor route chemistry and optoelectronic properties of poly(pyridine vinylene), *Synth. Met.*, *84*, 159–160 (1997).

35. Conticello, V. P., Gin, D. L., and Grubbs, R. H., Ring opening polymerisation for precursor of poly(*p*-phenylene vinylene), *J. Am. Chem. Soc.*, *114*, 9708 (1992).

36. Pu, L., Wagaman, W., and Grubbs, R. H., Synthesis of poly(1,4-naphthylenevinylenes): Methathesis polymerization of benzobarrelenes, *Macromolecules*, *29*, 1138–1143 (1996).

37. Son, S., Dodabalapur, A., Lovinger, A. J., and Galvin, M. E., Luminescence enhancement by the introduction of disorder into poly(*p*-phenylene vinylene), *Science*, *269*, 376–378 (1995).

38. Kim, H. K., and Ober, C. K., Development of poly(phenylene)-based materials for thin-film applications—optical wave-guides and low dielectric materials, *J. Macromol. Sci. Pure Appl. Chem.*, *18*, 877–897 (1993).

39. Huber, J., and Scherf, U., A soluble poly(para-phenylene) composed of cyclophane units: Poly(2,5-(oxydecanoxy)-1,4-phenylene), *Macromol. Rapid Commun.*, *15*, 897–902 (1994).

40. Balanda, P. B., and Reynolds, J. R., Methoxyethoxy- and triethoxy-substituted PPP via Suziki cross-coupling, *Polym. Prep.*, *37*, 828–829 (1996).

41. Pei, Q., and Yang, Y., Efficient photoluminescence and electroluminescence from a soluble polyfluorene, *J. Am. Chem. Soc.*, *118*, 7416–7417 (1996).

42. Staring, E. G. J., Demandt, C. J. E., Braun, D., Rikken, L. J., Kessener, Y. A. R. R., Venhuizen, T. H. J., Wynberg, H., Hoeve, W. T., and Spoelstra, K. J., Photo- and electroluminescence efficiency in soluble poly(dialkyl-*p*-phenylenevinylene), *Adv. Mater.*, *6*, 934–937 (1994).

43. Claussen, W., Schulte, N., and Schlueter, A.-D., A poly(*P*-phenylene) decorated with Frechet-type dendritic fragments of the first generation, *Macromol. Rapid Commun.*, *16*, 89–94 (1995).

44. Kim, S. T., Hwang, D. H., Li, X. C., Gruner, J., Friend, R. H., Holmes, A. B., and Shim, H. K., Efficient green electroluminescent diodes based on poly(2-dimethyloctylsilyl-1,4-phenylenevinylene), *Adv. Mater.*, *8*, 979–982 (1996).

45. Grimsdale, A. C., Cacialli, F., Grüner, J., Li, X. C., Holmes, A. B., Moratti, S. C., and Friend, R. H., Novel poly(arylene vinylene)s carrying donor and acceptor substituents, *Synth. Met.*, *76*, 165–167 (1996).

46. Yang, Y., Pei, Q., and Heeger, A. J., Efficient blue polymer light-emitting-diodes from a series of soluble poly(paraphenylene)s, *J. Appl. Phys.*, *79*, 934–939 (1996).

47. Davey, A. P., Elliott, S., and O'Connor, O. B., W., New rigid backbone conjugated organic polymers with large fluorescence quantum yields, *J. Chem. Soc., Chem. Commun.*, 1433–1434 (1995).

48. Bao, Z., Chan, W., and Yu, L., Synthesis of conjugated polymer by the Stille coupling reaction, *Chem. Mater.*, *5*, 2–3 (1993).

49. Cheng, H., and Elsenbaumer, R. L., New precursors and polymerisation route tor the preparation of high molecular mass poly(3,4-dialkoxy-2,5-thienylenevinylene)s, *J. Chem. Soc., Chem. Commun.*, 1451 (1995).

50. Yamamoto, T., Maruyama, T., Zhou, Z. H., Ito, T., Fukuda, T., Yoneda, Y., Begum, F., Ikeda, T., Sasaki, S., Takezoe, H., Fukuda, A., and Kubota, K., Pi-Conjugated poly(pyridine-2,5-diyl), poly(2,2'-bipyridine-5,5'-diyl), and their alkyl derivatives—preparation, linear structure, function as a ligand to form their transition-metal complexes, catalytic reactions, *N*-type electrically conducting properties, optical-properties, and alignment on substrates, *J. Am. Chem. Soc.*, *116*, 4832–4845 (1994).

51. Tian, J., Wu, C.-C., Thompson, M. E., Sturn, J. C., Register, R. A., Marsella, M. J., and Swager, T. M., Electroluminescent properties of self-assembled polymer thin films, *Adv. Mater.*, *7*, 395–398 (1995).

52. Marsella, M. J., Fu, D.-K., and Swager, T. M., Synthesis of regioregular poly(methyl pyridinium vinylene): An isoelectronic analogue to poly(phenylene vinylene), *Adv. Mater.*,, *7*, 145–147 (1995).

53. Yamamoto, T., Suganuma, H., Maruyama, T., and Kubota, K., Poly(4,4'-dialky-2,2'-bithiazole-5,5'-diyl). New electro-withdrawing pi-conjugated polymers, *J. Chem. Soc., Chem. Commun.*, 1613 (1995).

54. Kanbara, T., Miyazaki, Y., and Yamamoto, T., New pi-conjugated heteroaromatic alternative copolymers with electron-donating thiophene and electron-withdrawing quinoxaline units, *J. Polym. Sci., Polym. Chem.*, *33*, 999 (1995).

55. Simonet, J., Patillon, H., Belloncle, C., Simonet-Gueguen, N., and Cauliez, P., Electroactive, polymers possessing crown ether sites: Anodic polymerisation, *Synth. Met.,* *75,* 103 (1995).

56. Yang, Z., Sokolik, I., and Karasz, F. E., A soluble blue light emitting polymers, *Macromolecules, 26,* 1188 (1993).

57. Zyung, T., Hwang, D.-H., Kang, I.-N., Shim, H.-K., Hwang, W.-Y., and Kim, J.-J., Novel blue electroluminescent polymers with well-defined conjugation length, *Chem. Mater., 7,* 1499–1503 (1995).

58. Pohl, A., and Bredas, J. L., Influence of silicon atoms on the pi-conjugation in electroluminescent polymers, *Int. J. Quantum Chem., 63,* 437–440 (1997).

59. Hay, M., and Klavetter, F. L., Aliphatic phenylene vinylene copolymers: Tuning the colour of luminescence through co-monomer feed ratios, *J. Am. Chem. Soc., 117,* 7112–7118 (1995).

60. Aguiar, M., Hu, B., Karasz, F. E., and Akcelrud, L., Light-emitting polymers with pendant chromophoric groups: 2. Poly[styrene-co-(*p*-(stilbenylmethoxy)styrene)], *Macromolecules, 29,* 3161–3166 (1996).

61. Hesemann, P., Vestweber, H., Pommerhne, J., Mahrt, R. F., and Greiner, A., A blue light emitting polymer with phenylenevinylene segments in the side-chain, *Adv. Mater., 7,* 388–390 (1995).

62. Lee, J.-K., Schrock, R. R., Baigent, D. R., and Friend, R. H., A new type of blue-light-emitting electroluminescent polymer, *Macromolecules, 28,* 1966–1971 (1995).

63. Li, X.-C., Cacialli, F., Giles, M., Grüner, J., Friend, R. H., Holmes, A. B., Moratti, S. C., and Yong, T. M., Charge-transport polymers for light-emitting-diodes, *Adv. Mater., 7,* 898–900 (1995).

64. Cacialli, F., Li, X. C., Friend, R. H., Moratti, S. C., and Holmes, A. B., Light-emitting-diodes based on poly(methacrylates) with distyrylbenzene and oxadiazole side-chain, *Synth. Met., 75,* 161–168 (1995).

65. Li, X. C., Yong, T. M., Grüner, J., Holmes, A. B., Moratti, S. C., Cacialli, F., and Friend, R. H., A blue light emitting copolymer with charge transporting and photo-crosslinkable functional units, *Synth. Met., 84,* 437–438 (1997).

66. Suzuki, H., Temperature-dependence of the electroluminescent characteristics of light-emitting-diodes made from poly(methylphenylsilane), *Adv. Mater., 8,* 657 (1996).

67. Ma, Y.-G., Zhang, H.-F., Shen, J.-C., Liu, S.-Y., and Liu, X.-D., Blue LEDs using plasma polymerised naphthalene as emitting layer, *Chem. J. Chinese Univ., 7,* 1095–1096 (1995).

68. Chang, W. P., Whang, W. T., and Lin, P. W., Characteristics of an electropolymerized PPV and its light-emitting diode, *Polymer, 37,* 1513–1518 (1996).

69. Bredas, J. L., Santos, D. A. D., Quattrocchi, C., Friend, R. H., and Heeger, A. J., Electronic structure of poly(paraphenylene vinylene): Influence of copolymerisation and derivaization on light-emitting characterisation, *Polym. Prep., 35,* 185–186 (1994).

70. Berggren, M., Inganäs, O. G., G., Rasmusson, J., Anderson, M. R., Hjertberg, T., and Wennerstrom, O., Light-emitting diodes with variable colours from polymer blends, *Nature, 372,* 444–446 (1994).

71. Brouwer, H. J., Hilberer, A., Krasnikov, V. V., Werts, M., Wildeman, J., and Hadziioannou, G., Leds based on conjugated ppv block copolymers, *Synth. Met., 84,* 881–882 (1997).

72. Burn, P. L., Kraft, A., Baigent, D. R., Bradley, D. D. C., Brown, A. R., Friend, R. H., Gymer, R. W., Holmes, A. B., and Jackson, T. W., Chemical tuning of the electronic properties of PPV based copolymers, *J. Am. Chem. Soc., 115,* 10117–10124 (1993).

73. Burn, P. L., Holmes, A. B., Kraft, A., Bradley, D. D. C., Brown, A. R., and Friend, R. H., Synthesis of a segmented conjugated polymer chain giving a blue-shifted electroluminescence and improved efficiency, *J. Chem. Soc. Chem. Commun.*, 32 (1992).

74. Luessem, G., Festag, R., Greiner, A., Schmidt, C., Unterlechner, C., Heitz, W., Wendorff, J. H., Hopmeier, M., and Feldmann, J., Polarized photoluminescence of liquid-crystalline polymers with isolated arylenevinylene segments in the main chain, *Adv. Mater.*, 7, 923–925 (1995).

75. Paker, I. D., Pei, Q., and Marrocco, M., Efficient blue electroluminescence from a fluorinated polyquinoline, *Appl. Phys. Lett.*, 65, 1272 (1994).

76. Ohmori, Y., Uchida, M., Muro, K., and Yoshino, K., Blue polymeric LED, *Jpn. J. Appl. Phys.*, 2, 20, L1938 (1991).

77. Grem, G., Leditzky, G., Ullrich, B., and Leising, G., Realization of a blue-light-emitting device using poly(p-phenylene), *Adv. Mater.*, 4, 36–37 (1992).

78. Stampfl, J., Tasch, S., Leising, G., and Scherf, U., Quantum efficiency of electroluminescent poly(para-phenylene), *Synth. Met.*, 71, 2125 (1995).

79. Grem, G. M. V., Meghdadi, F., Paar, C., Stampfl, J., Tasch, S., and Leising, G., Stable poly(para-phenylene)s and their application in organic light-emitting devices, *Synth. Met.*, 71, 2193–2194 (1995).

80. Hilberer, A., Brouwer, H.-J., van der Scheer, B.-J., Wildeman, J., and Hadziioannou, G., Synthesis and characterization of a new efficient blue-light emitting copolymer, *Macromolecules*, 28, 4525–4529 (1995).

81. Jing, W. X., Kraft, A., Moratti, S. C., Gruener, J., Cacialli, F., Hamer, P. J., Holmes, A. B., and Friend, R. H., Synthesis of a polyphenylene light emitting copolymer, *Synth. Met.*, 67, 161–163 (1994).

82. Huber, J., Mullen, K., Salbeck, J., Schenk, H., Scherf, U., Stehlin, T., and Stern, R., Blue-light-emitting diodes based on ladder polymers of the polyphenylene type, *Acta Polym.*, 45, 244–247 (1994).

83. Grüner, J., Hamer, P. J., Friend, R. H., Huber, H.-J., Scherf, U., and Holmes, A. B., A high efficient blue-light-emitting diode based on novel ladder poly(p-phenylene)s, *Adv. Mater.*, 6, 749–752 (1994).

84. Tasch, S., Niko, A., Leising, G., and Scherf, U., Highly efficiency electroluminescence of new wide-band gap ladder-type poly(para-phenylenes), *Appl. Phys. Lett.*, 68, 1090–1092 (1996).

85. Tessler, N., Denton, G. J., and Friend, R. H., Lasing from conjugated-polymer microcavities, *Nature*, 382, 695–697 (1996).

86. Weaver, M. S., Lidzey, D. G., Fisher, T. A., Pate, M. A., Obrien, D., Bleyer, A., Tajbakhsh, A., Bradley, D. D. C., Skolnick, M. S., and Hill, G., Recent progress in polymers for electroluminescence—Microcavity devices and electron-transport polymers, *Thin Solid Films,*, 273, 39–47 (1996).

87. Burn, P. L., Holmes, A. B., Kraft, A., Brown, A. R., Bradley, D. D. C., and Friend, R., Light-emitting diodes based on conjugated polymers: Control of colour and efficiency, *Mater. Res. Soc. Symp. Proc.*, 247, 647–654 (1992).

88. Greenham, N. C., Moratti, S. C., Bradley, D. D. C., Friend, R. H., and Holmes, A. B., Efficient light-emitting-diodes based on polymers with high electron-affinities, *Nature*, 365, 628–630 (1993).

89. Bradley, D. D. C., Electroluminescence: A bright future for conjugated polymers? *Adv. Mater.*, 4, 756–758 (1992).

90. Moratti, S. C., Cervini, R., Holmes, A. B., Baigent, D. R., Friend, R. H., Greenham, N. C., Gruner, J., and Hamer, P. J., High electron affinity polymers for LEDs, *Synth. Met.*, 71, 2117–2120 (1995).

91. Li, X. C., Grimsdale, A. C., Cervini, R., Holmes, A. B., Moratti, S. C., Yong, T. M., Gruner, J., and Friend, R. H., *Synthesis and Properties of Novel High Electron Affinity*

Polymers for Electroluminescent Devices (S. A. Jenekhe and K. J. Wynne, ed.), American Chemical Society, *672*, 322–344 (1997).

92. Yu, G., Nishino, H., Heeger, A. J., Chen, T., and Rieke, R. D., Enhanced electroluminescence from semiconducting polymer blends, *Synth. Met.*, *72*, 249–252 (1995).

93. Pommerehne, J., Vestweber, H., Guss, W., Mahrt, R. F., Bässler, H., Porsch, M., and Daub, J., Efficient two layer LEDs on a polymer blend basis, *Adv. Mater.*, *7*, 551–554 (1995).

94. Pakbaz, K., Lee, C. H., Heeger, A. J., Hagler, W. W., and McBranch, D., Nature of the primary photoexcitations in poly(arylene-vinylenes), *Synth. Met.*, *64*, 295–306 (1994).

95. Strukelj, M., Miller, T. M., Papadimitrakopoulos, F., and Son, S., Effects of polymeric electron transporters and the structure of poly(p-phenylenevinylene) on the performance of light-emitting-diodes, *J. Am. Chem. Soc.*, *117*, 11976–11983 (1995).

96. Strukelj, M., Papadimistrakoppoulos, F., Miller, T. M., and Rothberg, L. J., Design and application of charge transporting polymers, *Science*, *267*, 1969 (1995).

97. Pei, Q., and Yang, Y., Bright blue electroluminescence from an oxadiazole-containing copolymer, *Adv. Mater.*, *7*, 559–561 (1995).

98. Li, X. C., Holmes, A. B., Kraft, A., Moratti, S. C., Spencer, G. C. W., Cacialli, F., Grüner, J., and Friend, R. H., Synthesis and optoelectronic properties of aromatic oxadiazole polymers, *J. Chem. Soc., Chem. Commun.*, 2211–2212 (1995).

99. Buchwald, E., Meier, M., Karg, S., Posch, P., Schmidt, H. W., Strohriegl, P., Reiss, W., and Schwoerer, M., Enhanced efficiency of polymer light-emitting-diodes utilizing oxadiazole polymers, *Adv. Mater.*, *7*, 839 (1995).

100. Li, X. C., Spencer, G. C. W., Holmes, A. B., Moratti, S. C., Cacialli, F., and Friend, R. H., The synthesis, optical and charge-transport properties of poly(aromatic oxadiazole)s, *Synth. Met.*, *76*, 153–156 (1996).

101. Yang, Y., and Pei, Q., Electron injection polymer for polymer light emitting diode, *J. Appl. Phys.*, *77*, 4807 (1995).

102. Schulz, B., Kaminorz, Y., and Brehmer, L., New aromatic poly(1,3,4-oxadiazole)s for light emitting diodes, *Synth. Met.*, *84*, 449–450 (1997).

103. Li, X. C., Kraft, A., Cervini, R., Spencer, G. C. W., Cacialli, F., Friend, R. H., Grüner, J., Holmes, A. B., De Mello, J. C., and Moratti, S. C., The synthesis and optoelectronic properties of oxadiazole-based polymers, *Mater. Res. Soc. Symp. Proc.*, *413*, 13–22 (1996).

104. Brown, A. R., Bradley, D. D. C., Burroughes, J. H., Friend, R. H., Greenham, N. C., Burn, P. L., Holmes, A. B., and Kraft, A., Poly(p-phenylenevinylene) light-emitting diodes: Enhanced electroluminescent efficiency through charge carrier confinement, *Appl. Phys. Lett.*, *61*, 2793–2795 (1992).

105. Berggren, M., Gustafsson, G., Inganas, O., Andersson, M. R., Hjertberg, T., and Wennerstrom, O., White light from an electroluminescent diode made from poly[3(4-octylphenyl)-2,2'-bithiophene] and an oxadiazole derivative, *J. Appl. Phys.*, *76*, 7530–7534 (1994).

106. Berggren, M., Granstrom, M., Inganäs, O., and Andersson, M., Ultraviolet electroluminescence from an organic light-emitting diode, *Adv. Mater.*, *7*, 900 (1995).

107. Osaka, T., Komaba, S.-I., Fujihana, K., Okamoto, N., and Kaneko, N., Enhancement properties of organic electroluminescence device using electropolymerised poly(3-n-octylthiophene) thin films, *Chem. Lett.*, *11*, 1023–1024 (1995).

108. Zhang, C., von Seggern, H., Kraabel, B., Schmidt, H.-W., and Hegger, A. J., Blue emission from polymer light-emitting diodes using non-conjugated polymer blends with air-stable electrodes, *Synth. Met.*, *72*, 185–188 (1995).

109. Wang, Y. Z., Gebler, D. D., Lin, L. B., Blatchford, J. W., Jessen, S. W., Wang, H. L., and Epstein, A. J., Alternating-current light-emitting devices based on conjugated polymers, *Appl. Phys. Lett.*, *68*, 894–896 (1996).
110. Rothberg, L. J., and Lovinger, A. J., Status of and prospects for organic electroluminescence, *J. Mater. Res.*, *11*, 3174–3187 (1996).
111. Service, R. F., Organic light emitters gain longevity, *Science*, *273*, 878–880 (1996).
112. Burrows, P. E., Bulovic, V., Forrest, S. R., Sapochak, L. S., McCarty, D. M., and Thompson, M. E., Reliability and degradation of organic light emitting devices, *Appl. Phys. Lett.*, *65*, 2922–2924 (1995).
113. Aziz, H., and Xu, G., Electric-field-induced degradation of poly(*p*-phenylenevinylene) electroluminescent devices, *J. Phys. Chem. B*, *101*, 4009–4012 (1997).

11
Poly(p-phenylenevinylene): An Attractive Material for Photonic Applications

Anna Samoc, Marek Samoc, Maneerat Woodruff, and Barry Luther-Davies
The Australian National University
Canberra, Australia

I. INTRODUCTION

Nonlinear optical effects were virtually unknown before the advent of lasers in the sixties. However, the appearance of the laser and especially the possibility of generating short light pulses brought forward a plethora of effects which rely on the nonlinear relation between the amplitude of the electromagnetic field of the light wave and the material response of the medium in which the light propagates (see Refs. 1–4). The technological revolution caused by the availability of coherent high-intensity light is still in the making. One of the more spectacular achievements in the recent years has been the easy availability of extremely short laser pulses from solid-state lasers (such as the titanium–sapphire laser). The application of nonlinear optical effects in the technologies using lasers is closely related to the development of materials exhibiting desired properties. We describe in this chapter the results of studies of third-order nonlinear optical properties of poly(p-phenylenevinylene) (PPV).

A. Definitions of Nonlinear Optical Properties

The nonlinear optical properties of materials can be conveniently described by using a power-series expansion of the relation between the electrical polarization **P** and the electrical field **E**:

$$\mathbf{P} = \chi^{(1)} \mathbf{E} + \chi^{(2)} \mathbf{EE} + \chi^{(3)} \mathbf{EEE} + \cdots \tag{1}$$

The above equation relates a vector (**P**) to various powers of another vector (**E**). This means that the susceptibilities of various orders $\chi^{(n)}$ appearing in the power expansion have to be tensors of appropriate ranks. The linear susceptibility $\chi^{(1)}$ describes the relation between three cartesian components of the **P** vector and the three components of the **E** vector and therefore can be presented as a 3×3 matrix or a second-rank tensor. The second-order susceptibility $\chi^{(2)}$ has to relate three components of **P** to products of cartesian components of **E** (e.g., E_1^2, E_1E_2, E_1E_3, etc.). There will be nine products of E_iE_j type but only six unique ones (because $E_iE_j = E_jE_i$). Therefore, $\chi^{(2)}$, which is a third-rank tensor, can be alternatively expressed as a $3 \times 3 \times 3$ matrix which relates the **P** vector components to the 3×3 matrix of E_iE_j products or, using a reduced number of components, as a 3×6 matrix relating **P** to a column vector composed of six unique E_iE_j products. In the case of the third-order susceptibility $\chi^{(3)}$, the **P** vector components are related to the 27 combinations of $E_iE_jE_k$ products. The third-rank $\chi^{(3)}$ tensor is therefore a $3 \times 3 \times 3 \times 3$ matrix—the number of independent components again being reduced by the possibility of permuting the indices in the $E_iE_jE_k$ products.

The susceptibilities of various orders $\chi^{(n)}$ have to fulfill the Neumann principle [5] stating that any material property must be invariant with respect to symmetry operations constituting the point group which describes the medium symmetry. This principle has far-reaching implications for the susceptibilities. A well-known implication is that all tensor components of even-order susceptibilities (including $\chi^{(2)}$) have to vanish for media belonging to centrosymmetric point groups. In general, the number of independent tensor components is strongly reduced by the presence of symmetry operations, some of the components being equal to zero.

Equation (1) deals essentially with a static field and the resulting static polarization. The introduction of optical frequency AC fields brings some complication. The optical field may contain components at various frequencies ω_i, in general:

$$E = \sum E(\omega_i) \exp(i\omega_i t) + \text{c.c} \tag{2}$$

Substitution of such an expression for the electrical field into Eq. (1) leads to the conclusion that the nonlinear polarization will contain components not only at the original frequencies ω_i but also at combinations of the input frequencies. For example, the presence of nonzero $\chi^{(2)}$ leads to the appearance of sum and difference frequency generation as well as to the second-harmonic generation (SHG) [i.e., the appearance of the polarization (and the electromagnetic wave) at $2\omega_i$]. In the case of $\chi^{(3)}$, the frequency mixing will lead, among others, to the third-harmonic generation (THG) [i.e. the generation of $3\omega_i$]. An important phenomenon is, however, the generation of nonlinear polarization at the same frequencies as those corresponding to the input. Such nonlinear components in the polarization can lead to the generation of new light beams propagating in directions different from those of the input light beams. They can also effectively speed up or slow down the propagation of the existing light beams or modify their propagation by imposing focusing or defocusing effects. Many (but not all) of these effects can be very simply described by introducing the concept of the nonlinear refractive index n_2. The refractive index of a medium can be treated as a function of the light intensity (the intensity being related to the square of the electromagnetic wave amplitude, $I = 2\varepsilon_0 nc \, |E^2|$, ε_0 being the permittivity of free space), as

$$n = n(0) + n_2 I \tag{3}$$

where $n(0)$ is the zero-intensity refractive index. Equation (3) represents so-called Kerr-type nonlinearity in which the change of the refractive index Δn is proportional to the light intensity.

The assumption made in Eq. (1) that the polarization of the medium can be expressed as a simple function of the electric field acting on it is, in general, not applicable to time-dependent fields. Even in the case of the linear polarization of a dielectric, the time-dependent polarization vector should be, in fact, treated as a *nonlocal* function of the field acting on it. For the third-order polarization term, this nonlocality can be formally expressed as

$$P^{(3)}(t) = \iiint \chi^{(3)} (t - t_1, t - t_2, t - t_3) E(t_1) E(t_2) E(t_3) \, dt_1 \, dt_2 \, dt_3 \qquad (4)$$

where $P^{(3)}$ is the third-order component of the polarization.

It is customary to deal with Fourier components of the input electric fields and those of the polarization, rather than with time dependencies; therefore, one usually prefers to consider nonlinear optical interactions separately for each frequency combination, writing the nonlinear relation as

$$P^{(3)}(\omega_4) = D\chi^{(3)}(-\omega_4; \omega_1, \omega_2, \omega_3) E(\omega_1) E(\omega_2) E(\omega_3) \qquad (5)$$

where D is a degeneracy factor which may be necessary depending on the type of interaction considered [1–4].

There are several important consequences of the nonlocal character of the material response. If the polarization component of a certain order at the frequency ω lags behind the driving field at the same frequency, then the phase shift between the field and the polarization can be simply treated as the result of the complex character of the appropriate susceptibility $\chi^{(i)}$. The presence of an imaginary part of a susceptibility is equivalent to loss or gain seen by a certain combination of input fields. For example, an imaginary part of the degenerate third-order susceptibility $\chi^{(3)}(-\omega; \omega, -\omega, \omega)$ is responsible for two-photon absorption and, in the case of the negative sign of this material property, for the induced transmission. The nonlocal character of the polarization may, however, be manifested in a way more complicated than just the phase shift between the field product and the polarization component oscillating at the same frequency. The polarization envelope function can itself be time dependent. There are many examples of physical phenomena that would lead to such a behavior. One common example is that of absorption of light, leading to the generation of a population of excited species whose presence changes the optical properties (refractive index and absorption coefficient or the real and imaginary parts of the linear susceptibility) of the medium. The growth and relaxation of the population of the species excited by a short light pulse will be reflected in the temporal dependence of the third-order nonlinear polarization.

B. Material Requirements for Photonic Switching

One of the most exciting prospects of utilizing nonlinear optical effects is that of all-optical switching for extremely fast signal processing in telecommunication, computing, and so forth. Because the third-order nonlinear optical properties lead, among others, to the appearance of refractive-index changes occurring on the time scale of ultrashort femtosecond laser pulses, it has been postulated that switches can be built that would exploit these effects. Examples of such devices (e.g., nonlinear coupler, nonlinear Fabry–Perot etalon) have been discussed, for example, in Ref. 6. Very interesting devices may be envisaged also if it becomes possible to use waveguides induced by optical solitons for guiding light with light [7–17]. However, in order to build photonic devices using this concept and operating under reasonable technological conditions, it is necessary to develop materials of suitable optical properties. These properties have been enumerated in many publications (see Refs. 6, 18, and 19).

We concentrate in this chapter on the determination of only one type of a nonlinear coefficient for PPV and its derivatives, namely the nonlinear refractive index n_2, defined by Eq. (3) and related to the degenerate third-order susceptibility $\chi^{(3)}(-\omega;\omega,-\omega,\omega)$ by

$$n_2 = \frac{3}{4\varepsilon_0 n^2 c}\chi^{(3)}(-\omega;\omega,-\omega,\omega) \tag{6}$$

SI units are used for both n_2 and $\chi^{(3)}$. Among many possible scenarios for using third-order nonlinearities for photonic switching, most of them use the nonlinear phase generated by the refractive-index change. Such a nonlinear phase change is related to the real part of $\chi^{(3)}$ or, treating n_2 as complex, to the real part of n_2. This "refractive" nonlinearity has to compete with linear and nonlinear light-intensity losses encountered by the propagating light beams. In quantitative terms, the suitability of a nonlinear material for an application as a photonic switch is related to its ability to generate a reasonably high (let us say, of the order of π or higher) nonlinear phase change on the propagation distance corresponding to a loss of intensity by a factor of $1/e$. This notion leads to the definition of merit factors, the two most important ones being the W factor and the T factor [6,18].

The W factor is related to the maximum phase change available for a single "absorption length" [i.e., α^{-1}, α being the absorption coefficient, derived from $\ln(I_0/I) = 1$] and is defined as

$$W = \frac{\mathrm{Re}(n_2)I_{\mathrm{sat}}}{\lambda\alpha} \tag{7}$$

I_{sat} is the light intensity at which the relation between the refractive-index change and the intensity saturates. Alternatively, one can compare various materials assuming I_{sat} to be the same operating light intensity (e.g., 1 GW/cm²). It may be interesting to mention that, from the point of view of the value of the W factor, silica (SiO₂) remains one of the best nonlinear materials. The low value of n_2 (about 3 × 10^{-16} cm²/W) is compensated by an extremely low absorption coefficient, on the order of dB/km which corresponds to α on the order of 10^{-6} cm⁻¹ and W of the order 10^4.

In the case of materials with one-photon absorption at shorter wavelengths and higher energies, the limiting factor in an application for photonic switching becomes the nonlinear loss due to multiple-photon absorption, mostly, the two-photon absorption. Assuming that the absorption of light is described by

$$\frac{dI}{dz} = -\alpha I - \beta I^2 \tag{8}$$

where α is the one-photon absorption coefficient and β is the two-photon absorption coefficient, one can relate β to the imaginary part of the nonlinear refractive index or third-order susceptibility by

$$\beta = \frac{3\pi}{\varepsilon_0 n^2 c\lambda}\, \mathrm{Im}[\chi^{(3)}(-\omega;\omega,-\omega,\omega)] \tag{9}$$

or

$$\beta = \frac{4\pi}{\lambda}\, \mathrm{Im}(n_2) \tag{10}$$

(all parameters in SI units). In the absence of one-photon absorption, the intensity of light varies as

$$I(z) = \frac{I(0)}{1 + \beta I(0)z} \tag{11}$$

One can define the two-photon absorption length as $1/\beta I(0)$. The nonlinear phase change acquired by a beam undergoing two-photon absorption is given by

$$\Delta\varphi = \int_0^L \Delta K(z)\, dz = \int_0^L \frac{2\pi \Delta n(z)\, dz}{\lambda} \tag{12}$$

Assuming Kerr-type nonlinearity [i.e., $\Delta n = n_2 I(z)$], we get

$$\Delta\varphi = \frac{2\pi n_2 I(0)}{\lambda} \int_0^L \frac{dz}{1 + \beta I(0)z} = \frac{2\pi n_2}{\lambda\beta}\, \ln[1 + \beta I(0)L] \tag{13}$$

Thus, for the propagation on the distance of one absorption length, one obtains the nonlinear phase change which is on the order of $2\pi/T$, where T is the two-photon absorption merit factor defined by

$$T = \frac{\beta\lambda}{\mathrm{Re}(n_2)} = \frac{4\pi\, \mathrm{Im}(n_2)}{\mathrm{Re}(n_2)} = \frac{4\pi\, \mathrm{Im}(\chi^{(3)})}{\mathrm{Re}(\chi^{(3)})} \tag{14}$$

In order for the nonlinear phase change to be close to 2π or larger, T has to be on the order of unity or smaller.

Therefore, the requirements for photonic switching material are usually taken as comprising, among others, two simple criteria: $W \gg 1$ and $T \ll 1$. The assessment of the suitability of a material must therefore involve precise determination not only of the magnitude of the real part of the nonlinearity itself but also of the ratio of the nonlinear coefficient to the absorption coefficient and of the real and imaginary components of the nonlinearity. Unfortunately, literature information about the merit factors of various nonlinear optical materials suggested for photonic applications is rather scarce.

From the practical point of view, two parameters are also of utmost significance: the time needed for the switch to operate and the energy spent in a single switching event. These two parameters are related to properties of a nonlinear material and also to the design of the switch. However, it is generally agreed that it is the solving of the problem of obtaining the suitable material properties which is the key to achieving a successful photonic switch. The problem of achieving a high speed of operation can essentially be related to the necessity of using only those kinds of nonlinear optical response mechanisms which do not lead to the generation of relatively long-lived species nor other long-living effects changing the optical properties of a medium such as thermal effects. Minimizing the energy needed for the switch to operate involves increasing the nonlinearity, but improvements also can be achieved by increasing the propagation path of the switching beam and reducing its active area by confining it in a waveguide. Other technological requirements like environmental and photochemical stability of a nonlinear material and its compatibility with fiber-optic technologies can also be specified.

II. CONJUGATED POLYMERS AS THIRD-ORDER NONLINEAR OPTICAL MATERIALS

Conjugated polymers have been suggested as potentially attractive media for photonic switching (see, for example, Refs. 19–25) for several reasons. The most immediate reason is that high third-order nonlinearities are possible in the systems having highly delocalized π-electron orbitals. Considering the third-order nonlinear optical (NLO) response due to the electrons moving in a one-dimensional anharmonic potential well or in a system composed of several wells which can be treated as coupled anharmonic oscillators (e.g., Refs. 26 and 27), one finds that the nonlinear response increases rapidly with the length of a linear π-electron system. Models have been suggested to describe such a dependence, either in terms of the length of a π-electron system (the conjugation length) or in terms of the "band gap" of the system (see, for example, Refs. 28–31). Quantum-chemical methods can be used to obtain information on such dependencies (see, for example, the review in Ref. 24). Quantitatively, different results can be obtained from various theoretical approaches; the qualitative picture seems, however, to be well established.

Most approaches treat conjugated polymers in terms of an oriented gas model in which the electronic contribution to the macroscopic nonlinear response (described in terms of the macroscopic susceptibilities $\chi^{(n)}$) is treated as deriving from the electronic nonlinear response of the molecules. Therefore, on the microscopic scale, one considers the dipole moment of a molecule to be given by

$$\mu = \mu(0) + \alpha\mathbf{E}_{loc} + \beta\mathbf{E}_{loc}\mathbf{E}_{loc} + \gamma\mathbf{E}_{loc}\mathbf{E}_{loc}\mathbf{E}_{loc} + \cdots \tag{15}$$

where α, β, and γ are the polarizabilities of consecutive orders which are microscopic analogs of the susceptibilities $\chi^{(n)}$, and $\mathbf{E}_{loc}$ stands for the local electric field. The polarizabilities are tensors of the same ranks as the nonlinear susceptibilities of the same order; for example, the third-order polarizability γ (also called the second hyperpolarizability) is the fourth-rank tensor. The relation between the microscopic and macroscopic quantities has to take into account the density of the microscopic linear entities, their orientation, and the local field correction [28].

In order to find what correlations exist between the nonlinearity and the electronic structure of conjugated polymers such as poly(p-phenylenevinylene), (abbreviated PPV), one can perform quantum-chemical simulations of oligomeric structures similar to those which can be expected for partly converted PPV chains (see Section III). We used the Austin Model 1 (AM1) method in the molecular orbital package (MOPAC) quantum-chemical package to optimize molecular geometries and to perform computations of third-order optical nonlinearity as a function of frequency and of the number of conjugated units in a polymer chain. The electronic spectra of the oligomers were simulated using the intermediate neglect of differential overlap/spectroscopic (INDO/S) method. With the increase of the number of conjugated mer units, the polarizabilities of various orders increase in a superlinear manner (the linear increase would, of course, be equivalent to simple additivity). Figure 1 shows an example of calculated third-order polarizabilities as a function of the number of conjugated entities in a system of poly(*p*-phenylenevinylene) oligomers.

The rapid increase in the hyperpolarizability gives way to saturation for about 8–12 units of phenylenevinylene moieties. At the same time, the energy of the lowest absorption transition $E_{0 \to 1}$ (often referred to as the "band gap") also stabilizes. In qualitative terms, the system has reached its "maximum conjugation length" and further improvement in the nonlinear properties is not possible. Because both the lowest absorption energy and the polarizabilities depend on the degree of π-electron delocalization along the conjugated chain, there is often postulated an interrelation between these two quantities. Bubeck and co-workers [30,31] has provided ample experimental evidence that there exists a simple power-law dependence between $\chi^{(3)}(-3\omega; \omega, \omega, \omega)$ and the energy of the lowest absorption transition. They presented this relation in the form of $\chi^{(3)}/\alpha_{max}$ versus λ_{max} double logarithmic plots and found that the exponent of the power-law relationship between these parameters is about 10, which is larger than what could be expected from simple models.

In practical terms, obtaining a high nonlinearity involves achieving a high value of the conjugation length. This is possible in several aromatic polymers, including the archetypal conducting polymer: polyacetylene. It has been postulated that polyacetylene can be the "ideal" material which satisfies the important criterion for conjugate polymers that large third-order NLO response should be achievable in conjugated polymers with degenerate ground states [29]. Polyacetylene, is, however, quite difficult to handle because of its rather high chemical reactivity. The

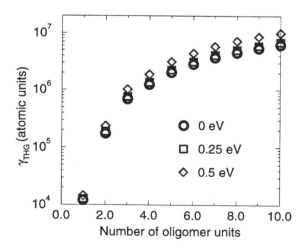

FIGURE 1 The orientationally averaged third-order hyperpolarizability γ_{THG} [i.e., $\gamma(-3\omega; \omega, \omega, \omega)$ calculated by the MOPAC93/AM1 quantum-chemical program for a series of poly(*p*-phenylenevinylene) oligomers at three input frequencies ω corresponding to photon energies of 0, 0.25, and 0.5 eV.

low stability of polyacetylene is assumed to be a consequence of high concentration of conformational defects formed at the formation of the *trans*-(CH)$_x$ polymer in solid phase. One should mention that there has been intensive research effort to stabilize this polymer through encapsulation, using antioxidants or blending with other polymers [32,33]. On the other hand, another archetype, poly(*p*-phenylene) which is very stable, is not a promising candidate for high nonlinearity because of a relatively short conjugation length limited by the tendency of neighboring benzene rings to lose coplanarity due to the steric hindrance effect of closely lying hydrogen atoms. Poly(*p*-phenylenevinylene) is therefore a natural choice for at least model studies of the nonlinear effects. It is a conjugated nondegenerate ground-state polymer. However, PPV satisfies all the other following criteria specified for the "ideal" NLO material in Ref. 29:

- Energy gap greater than 2 eV (for good transparency)
- No side chains (no dilution to ensure high π-electron density)
- Processable (optical quality thin films)
- Oriented and ordered (anisotropy)

PPV combines a good chemical and thermal stability with a high π-electron conjugation caused by the fact that, as shown in Fig. 2, the molecule should be essentially planar and rigid-rod-like.

Poly(*p*-phenylenevinylene) is certainly an extremely interesting chemical substance because of its electrical and optical properties. Especially the possibility of obtaining efficient electroluminescence from PPV cells [34–37] has stimulated numerous vigorous research programs. The PPV backbone can also be substituted (e.g., in the 2,5-positions of the phenyl ring or at the vinylene units) and thereby modified to obtain materials of various properties. We concentrate in the following mostly on the properties of unsubstituted PPV and its composites, with only a limited amount of discussion of the properties of substituted PPVs.

III. SYNTHETIC ROUTE TO POLY(*p*-PHENYLENEVINYLENE)

Poly(*p*-phenylenevinylene) and some of its derivatives can be obtained by a soluble precursor route which facilitates processing of PPV in the form of thin films. It proceeds through the synthesis of a sulfonium polyelectrolyte precursor solution which is then used for casting film of the precursor polymer. The film is converted

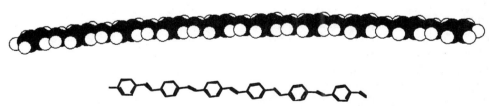

FIGURE 2 A view of a long (10 units) PPV oligomer structure optimized by a AM1 (MOPAC93) geometry optimization routine. Energy minimization of a rough structure built using molecular modeling tools leads to structures with little deviations from planarity.

to conjugated polymer by the elimination reaction. This synthetic route [38–74] follows the pioneering work of Kanbe and Okawara [38], Wessling and Zimmerman [39–41], Hoerhold and Opfermann [42] and the research performed by Murase et al. [43,44], Karasz and co-workers at the University of Massachusetts [45–61], and the Cambridge group [63–74] and is currently widely exploited by many research groups around the world. The advantages of the soluble precursor route include the possibility of purification and handling of the precursor polymer and the production of fully dense, high-molecular-weight films of the conjugated polymer in a variety of forms. In particular, the polymer can be stretch-oriented during thermal transformation to yield highly anisotropic samples which allow characterization of the intrinsic anisotropic properties.

The preparation of a soluble nonconjugated polyelectrolyte precursor polymer proceeds via the base-promoted polymerization of a bis-sulfonium salt monomer. In this reaction, half of the sulfonium groups of the di-sulfonium salt is released; the other half remains on the polymer backbone. Due to the charge on these groups, they solubilize the polymer in protic solvents such as water and methanol. The sulfonium groups are eliminated from the polymer during the precursor transformation into the conjugated polymer by thermal decomposition or other methods. A second strategy has been developed [62,65–73] in which the solubilizing and leaving group is an alkoxy group. The precursor polymer containing methoxy groups can be dissolved in polar aprotic solvents such as chloroform, dichloromethane, and tetrahydrofuran. The properties of the polyelectrolyte precusor depend on the nature of the leaving group, the type and concentration of ions and counterions, and the solvent.

A direct approach in the synthesis of PPV can be realized via a Wittig reaction [75–78] of *p*-xylene bis(triphenylphosphonium chloride) with terephthalaldehyde or the dehydrohalogenation of *p*-xylylidene dihalides. Such an approach was reported in the literature to result in the formation of powders of insoluble oligomers which could be processed only in the form of pressed pellets. Lack of solubility of highly conjugated polymers primarily due to the rigid nature of the polymer backbone makes morphological modification and fabrication into a useful physical forms difficult. The versatility of chemical synthetic approaches [79] leads one to believe that future research will help with this problem and make it possible to fully explore the potential of the electronic structures offered by conjugated polymers. Recently, successful syntheses leading to soluble poly(phenylenevinylene)s have been accomplished via the Heck coupling reaction [80–81].

The soluble sulfonium precursor route, often called the Wessling route, has the inherent advantage of an increased flexibility in the design of syntheses of various PPV derivatives (e.g., Refs. 57–62, 65–68, and 82–85). The polymerization of the bis-sulfonium salt monomer substituted on the phenylene ring permits the synthesis of many other substituted derivatives and analogs of PPV of different electronic properties. In this synthetic strategy, various groups of different size and electronegativity can be added to the phenyl ring in the main chain of the polymer or to the backbone, changing the electron density, the intramolecular and intermolecular packing and the chemical reactivity of the polymer. Careful planning may yield precursor polymers which are soluble and stable, thus allowing both purification and processing by conventional techniques prior to the final conversion to the fully conjugated polymer. The important advantage of the precursor route is that it allows the soft film of the precursor to stretch during thermolysis, which induces a high degree of molecular orientation of the polymer chains in the conjugated polymer.

The synthetic procedure in the soluble sulfonium precursor route consists of several steps which are given here for the case of unsubstituted PPV, and the tetrahydrothiophenium group chosen as the active group to be eliminated in the step of obtaining the conjugated polymer from the precursor. The synthesis steps are as follows:

1. Synthesis of the bis-sulphonium salt monomer in the reaction in Scheme 1: α,α'-Dichloro-p-xylene is reacted with tetrahydrothiophene to give p-xylylene-bis(tetrahydrothiophenium chloride), [52547-07-6]. The monomer salt carries various names in the literature; for example, 1,1'-[1,4-phenylenebis(methylene)]-bis[tetrahydrothiophenium] dichloride; p-xylylene bis(tetramethylenesulphonium chloride); 1,4-phenylenedimethylene-bis(tetramethylene sulphonium chloride); α,α'-bis(tetrahydrothiophenio)-p-xylene dichloride. Its *Chemical Abstracts* index name is thiophenium, 1,1'-[1,4-phenylenebis(methylene)]bis[tetrahydro-dichloride].

The solvent for the reaction is methanol or methanol mixed with water. The bis-sulfonium salt can be prepared following the procedure described in, for example, Refs. 48 and 61.

2. Isolation of the bis-sulphonium salt monomer by crystallization. The salt being less soluble than the other reagents can be precipitated in cold acetone, filtered, dried, and purified by reprecipitation. The salt is very hygroscopic. The important property of the monomer salt is its low stability. The salt can decompose during several weeks, especially being in contact with air. Partial degradation with the loss of the sulfide indicates a slow reversal of the reaction [48,61].

3. Polymerization of the bis-sulphonium salt monomer to form the precursor polymer poly(p-xylylene tetrahydrothiophenium chloride) [110866-77-8] is outlined in Scheme 2.

The solvent in this reaction can be methanol, aqueous methanol, or water; the base may be organic or inorganic (e.g., tetrabutylammonium hydroxide n-Bu$_4$NOH or NaOH). The polymerization proceeds via the elimination of the sulfonium group according to the E1cb mechanism in which a reactive quinomethane intermediate is formed [38,41,50–54,86] or E1 mechanism [64]. In the standard procedure (see, for example, Refs. 47 and 65–67), the 0.4 M/L methanolic solution of the monomer salt is polymerized with equimolar quantity of the 0.4 M/L solution of the base in methanol (1 h, 0°C, N$_2$) subsequently being neutralized to pH = 5–7. Other procedures [48,66,87] use less concentrated solutions (e.g., 0.2 M mixed with the equivalent or less than stoichiometric amount of the base). The resulting solution is a polyelectrolyte which contains many species, namely the precursor polymer, the unreacted monomer salt, tetrabutylammonium chloride or NaCl, tetrahydrothiophene dissolved in the solution, and possibly the impurities formed during polymerization or from decomposition of the monomer salt need to be separated from the precursor polymer in the purification step.

The high- molecular-weight water-soluble precursor polymer, *p*-xylylene-bis(tetrahydrothiophenium chloride) homopolymer, (*Chemical Abstracts* index name: thiophenium, 1,1′-[1,4-phenylenebis(methylene)]bis[tetrahydro-, dichloride, homopolymer]) can be obtained as a colorless substance when the polymerization reaction is carried out in the absence of light. Yellow coloration of the polymer solution indicating partial conversion to the conjugated polymer through the base-promoted elimination of tetrahydrothiophenium group can be observed if an excess of base is used in the polymerization.

4. Purification of the precursor polymer by dialysis. The solution of the polyelectrolyte polymer is transported into the dialysis tube, which is immersed in the solvent (e.g., water, methanol, or methanolic solution) to remove low-molecular-weight (MW) oligomers, the residual monomer, and by-products. The purification proceeds by means of diffusion through the selectively permeable membrane of the MW cutoff (e.g., 3500). Usually, the polymer solution is dialyzed for 3 days with frequent change of the solvent.

The methanol or water solution of the precursor polymer is used for casting films. Alternatively, the precursor solution can be mixed with a compatible solution of another polymer to form a composite material. The dried, almost colorless film of the precursor polymer can be converted into the conjugated yellow polymer, PPV, by pyrolysis. Storage at room temperature and the exposure of the films to light results in a fluorescent greenish-to-yellow coloration of the polymer, indicating again partial conversion to the conjugated polymer, PPV.

5. Conversion of the precursor polymer into the π-conjugated polymer: poly(*p*-phenylenevinylene), poly-*p*-xylidene (PPV) [*26009-24-5P*] can be accomplished either by base-induced or thermal elimination of the sulfonium group as it is shown in Scheme 3.

The reaction can be carried out under vacuum or in an inert gas atmosphere (argon, nitrogen), to prevent oxidation, in the temperature range 200–300°C. The thermally induced conversion of the precursor polymer into the conjugated polymer occurs in a complex sequence of reactions rather than in a single step where the gaseous products, tetrahydrothiophene and the corresponding acid of the counterion, are eliminated. The degree of conversion to PPV depends on the temperature and duration of pyrolysis. Thermal gravimetric analysis (TGA), differential scanning calorimetry (DSC), and mass spectrometry show several characteristic thresholds in thermolysis (e.g., around 100°C extending to 120–130°C, which was attributed to the elimination of water, and 230–240°C attributed to the elimination of the sulfonium groups [49]). The elimination efficiency was found to be dependent on the type of the leaving group and the nature of the counterion present in the polyelectrolyte [48,49,56,88–91]. Four major thermal transition maxima were observed [e.g., in the poly(xylylenetetrahydrothiophenium chloride) with the counterion exchanged to iodine], they were at 60, 101, 173, and 564°C [89]. The weight loss at 60°C was ascribed to loss of the hydration water and at 564°C to polymer decomposition. The effectiveness of pyrolysis was correlated with the amount of sulfur and chlorine left in the polymer. The polymers containing the cycloalkylene sulfonium (tetrahydrothiophenium) groups were eliminated more effectively when thermally treated at 180, 210, or 300°C (less residual S and Cl was found in elemental analysis) than the polymers with dialkyl sulfonium chloride [49]. This agrees well with observations made by Murase et al. [44] that films derived from dimethylsulfonium salt can contain several percent of sulfur (e.g., 2.7% if treated at 200°C and <0.5% after heating at 300°C). The loss of weight in poly(p-xylylenetetrahydrothiophenium chloride) in Schlenoff and Wang's experiment [88] was completed at 200°C. Similar observations were also made with DSC experiments [90]. It has been found that the intensity and position of the low-temperature endothermic peak depends on the amount of water entrapped in a film, which accounts for the differences between samples of different batches. In the bromide sulfonium salt polyelectrolyte, the main loss of weight due to elimination of sulfonium groups occurred below 100°C [91].

These experiments also prove that PPV is heat resistant. The decomposition point of PPV observed in Refs. 47, 48, and 89 was in the temperature range 536°–600°C.

An improvement in optical properties of PPV was observed when the elimination of the sulfonium groups was performed at elevated temperature under a flowing hydrogen chloride–argon mixture. The addition of the HCl catalyst to the reaction environment serves to help in the efficiency of elimination of the sulfonium groups [62,64–73]. The conversion of the sulfonium precursor using strong acids could be performed also at room temperature, giving a less dense polymer of optical properties similar to standard PPV [92].

There have been attempts for using a photochemical method of conversion using, for example, UV irradiation of a 200-W XBO xenon lamp (190 nm–infrared) [93]. The product showed only partial conversion. However, the light-assisted conversion method might be successfully applied to microstructuring of PPV by UV interferometry [94].

The details of the synthesis route are crucial for obtaining highly nonlinear films of good optical quality. There are several issues of importance. Optical properties of PPV depend on the extent of π-electron delocalization; that is, on the number of phenylenevinylene units which are formed during precursor conversion in such a way that there are no defects of chemical or structural type interrupting the conjugation.

It seems that the thermal pyrolysis is a very successful way of obtaining highly conjugated PPV. High temperature, however, may activate additional processes of decomposition, nucleophilic substitution, and incipient doping [41,48,49,56,95,96].

Chemical defects may be present in PPV chains, for example, due to the sulphonium groups not eliminated or due to other substituents displacing the sulfonium group and the carbonyl groups generated during conversion in the presence of oxygen [63,97,98]. The presence of forming gas (mixture of H_2 and N_2) during the conversion reaction may be helpful in reducing the number of defects caused by the oxidation of PPV [97].

The solvent, particularly methanol, may react with the precursor polymer displacing the sulfonium group and introducing a methoxy group on the olefinic carbon [62,65–73]. Scheme 4 shows an example of such a methoxy defect caused by a substitution reaction occurring spontaneously on storage of the PPV precursor solution in methanol even at room temperature. However, the substitution proceeds more effectively at elevated temperature (e.g., 55°C [65–67]) or at room temperature following the exchange of the counterion to *p*-toluenesulfonate [62,66]. As shown above, the methoxy group is more difficult to eliminate than the sulfonium group under the standard conditions (under vacuum) and therefore may remain in partially converted PPV causing the blue shift of absorption spectra, reduction of the refractive index and the third-order nonlinearity.

The methoxy substitution of the sulfonium groups was exploited in a modified precursor route inducing extended π-conjugation in PPV [70–72]. The method introduces an additional step of the intrachain preordering of poly(*p*-phenylenevinylene) by the base-induced partial elimination of the sulfonium groups. This leads to the formation of the relatively long conjugated rigid-rod segments (on a 10-nm scale) intersected by flexible spacers of the precursor polymer having not eliminated methoxy groups. The subsequent more effective elimination of the methoxy groups via thermolysis at 220°C for, for example, 12 h in the presence of hydrochloric acid under Ar results in highly ordered conjugated chains.

At this point, we address the technologically oriented issues related to the requirements for PPV as a candidate for photonic devices:

- Optical properties of PPV are dependent on conjugation. The extent of conjugation is determined by the number of defects. Good control over the process of formation of the conjugated structures on the intramolecular level is required.
- Highly converted PPV is air-stable. The presence of impurities and reactive regions of the residual precursor moieties may be responsible for the occasional lower stability of the polymer. Procedures to improve purity in the converted PPV polymers are required.
- PPV has a tendency for crystallization. Crystallinity was observed in the stretch-oriented [87,99–103] and the nonoriented samples [63,90]. In some applications,

as for waveguiding structures, very low light-scattering levels are required, therefore amorphous rather than partly crystalline structure is essential. The noncrystalline character of the polymer should thus be combined with extended intramolecular ordering. A good degree of control over the morphology and homogenity of such films is needed.

*   The precursor polymer is a polyelectrolyte whose composition and structure result from the procedures applied in the precursor route. The precursor solution exhibits a tendency for gelation depending on the solvent and the details of the precursor route.

Optical experiments performed by us and described in later sections of this chapter have been done on PPV films prepared to have low light scattering and hence providing a low background for the measurements. The homogeneous films were prepared by thermal conversion of precursor films cast from methanolic and aqueous solutions or by spin-coating. The precursor polymers were prepared according to the procedure described by Gagnon et al. [48] using tetrahydrothiophene in the preparation of the bis-sulfonium salt. The polymerization was carried at the ratio of 1 : 0.9 mol equiv. of the salt to base. Methanol mixed with water in various proportions was used during preparation of the monomer salt, polymerization, and dialysis. PPV films were obtained by spinning of the solution of the precursor polymer onto silica or glass microscope slides. Some batches of the material were prepared in air. In other batches, elaborate precautions against oxygen contamination were taken. The conversion was performed under various heating times, temperatures, and inert atmosphere schemes. The pyrolysis was carried out either under the dynamic vacuum (about 3×10^{-3} mm Hg) or the flow of nitrogen, or the forming gas (the mixture of 95% of nitrogen and 5% of hydrogen) or the mixture of nitrogen and hydrochloric acid. The heating speed was programmed (e.g., the heating ramp was either $(1 + 2)$ h or has been extended to $(10 + 5)$ h. The numbers in the parentheses show the time of ramping the temperature and the time of keeping the temperature constant. Thin films of PPV < 1 μm, were optically clear, yellow, or yellow-greenish; thicker films were often opaque to some degree. The light scattering could be reduced under the influence of methanol. We applied the procedure in which the precursor solution prior to deposition was diluted with methanol (e.g., in the proportion 2 : 1), stored at room temperature for at least 1 day, then filtered (0.5 μm) and concentrated with a rotary evaporator. The influence of methanol on the precursor polymer at slightly elevated temperature has already been stated [62,65–73]. We believe that the procedure we used led us to the same effect of the substitution of the tetrahydrothiophenium groups in the precursor polymer. As a consequence, the precursor was partially substituted with methoxy groups possibly acting as the defects breaking the conjugation length of the rigid-rod molecules of poly(*p*-phenylenevinylene) and had only a limited tendency for crystallization.

IV. LINEAR OPTICAL PROPERTIES OF PPV

To obtain information on the electronic and molecular structure of the polymer in the pristine and doped states, spectroscopic techniques such as ultraviolet-visible

(UV-vis), infrared, and Raman spectroscopy are useful tools. These techniques have been used in the studies of the evolution of molecular structure of the sulfonium polyelectrolyte precursor during transformation to the conjugated polymer [63,104,105]. Figure 3 shows examples of absorption spectra obtained on gradual conversion of a spin-coated film of a poly(*p*-xylylene tetrahydrothiophenium chloride) precursor to poly(*p*-phenylenevinylene).

The absorption in the visible is the result of a relatively strong $\pi-\pi^*$ transition involving the conjugated electron backbone. The electronic spectra of poly(*p*-phenylenevinylene) depend on the extent of delocalization of π electrons along the polymer backbone. The chain segments with longer conjugation have their electronic transitions at a longer wavelength range of the absorption spectrum. Studies on phenylenevinylene oligomers (e.g., Refs. 106 and 107) have shown that the absorption maximum shifts to the red with the elongation of the conjugated π-electron system. The energy of $\pi-\pi^*$ transitions of the oligomers starts approaching a plateau with increasing number of monomeric units for $n > 5$. Using the value of the $\pi-\pi^*$ transition for PPV and the plot of the energies of the highest occupied molecular orbital–lowest unoccupied molecular orbital (HOMO–LUMO) transition in oligomers versus $1/n$, Tian et al. [106,107] concluded that the "effective conjugation length" in the polymer was not larger than 5–10 units. They have also shown from the analysis of the vibrational spectra that it is less than 7–9 units.

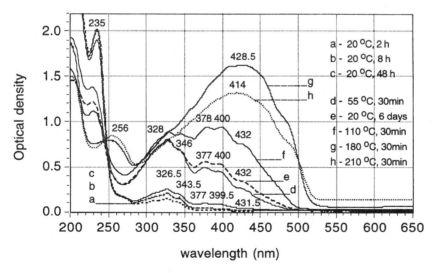

FIGURE 3 Experimentally observed changes of intensity of the vibronic transitions in electronic spectra under stepwise thermal conversion of the precursor polymer to PPV. The increase in the absorption reflects the formation of more conjugated chains in the polymer. Curves a, b, and c show absorption in the film of the precursor polymer stored at room temperature under nitrogen for 2, 8, and 48 h since the film deposition, respectively. Curves d, e, f, g, and h stand for the spectra of the film heated at 55°C (30 min), then stored at room temperature for a week, and heated again at 110, 180, and 210°C for a period of 30 min each. The resulting polymer film was ~0.11 μ thick.

In Sakamoto et al.'s work [108], the monomer (dimethylstilbene), dimer, and trimer of PPV were found to have the main absorption peaks at 317, 360, and 386 nm, respectively, and to have well-defined vibrational structure.

The optical spectra in Fig. 3 are similar to the absorption spectra of thin films of the diethyl sulfonium bromide precursor at various stages of thermal conversion to PPV, published in Refs. 63, 104, and 105. The experiment shown in Fig. 3 was performed with a sample converted and stored under nitrogen, but the sample was not protected against oxygen during the spectroscopic measurements. This might have some influence on the later stages of conversion leading to the observed reduction of absorptivity and less structured spectra of the film converted at the highest temperature, as indicated in curve h of Fig. 3.

An interesting feature in the absorption spectra of PPV is the presence of a vibronic structure clearly observed at the early stages of conversion of the precursor film. Figure 3 shows examples of the electronic spectra of a partly converted precursor with marked positions of absorption maxima and the shoulders. Their positions were derived from the analysis of the spectra with the help of the second derivative of the absorption curves. The energy difference calculated from the positions of the peaks show a rather regular pattern in which the value $\sim 1510 \pm 50$ cm^{-1} appears repeatedly. This value is characteristic of the frequency separation between the double maxima in segmentlike structures built in the electronic spectrum upon partial conversion. Absorptivity changes observed upon progressing conversion indicate the increase of the population of conjugated chains giving contributions to characteristic vibronic transitions. The repeated frequency difference detected in the spectra in Fig. 3 is close to the characteristic C–C ring stretch modes observed in the Raman and infrared spectra of PPV (cf. e.g., Refs. 105–109). The comparison can only be approximate, because vibronic transitions in electronic spectra relate to the vibrational modes of the molecule in the excited state which may have geometry and the force constants different from the molecule in the ground state, whereas the infrared and Raman spectra relate to the molecule in the ground state. The vibronic structure in Fig. 3 is less visible for relatively well-converted PPV sample. As will be shown later, its appearance in a well-converted polymer depends on the conditions applied during conversion (i.e., the intrachain ordering introduced into the polymer).

Optical studies are useful in investigation of the conditions needed for the effective transformation of the precursor to PPV. As the reaction proceeds, the conjugation length increases, and there is a shift of spectral features to a longer wavelength. The intrinsic absorption can be tuned by adjusting the conversion conditions. For example, high temperature and vacuum give material of different optical properties than the material converted at high temperature under the flow of a gas HCl catalyst [64]. A substantial batochromic shift to longer wavelengths is observed if HCl is used for conversion. The spectral tuning can also be realized in a chemical way by using precursors with different leaving groups during the elimination reaction. The experiments reported in Ref. 110 gave different spectra for PPV made from the acyclic and the cyclic sulfonium salt precursor polymer. The polymer prepared from the tetrahydrothiophenium precursor absorbed at a longer wavelength. A new approach in the precursor route inspired by the work of Tokito et al. [62] and developed by the research of the Cambridge group shows the way to a highly conjugated structure of the "improved" PPV. The Cambridge group developed a strategy based on the methanol-modified precursor and the efficient conversion conditions described in Refs. 71 and 72 (see Section III).

The optical spectra of the improved PPV [71,72] show a sharp red-shifted peak of the (0–0) electronic transition at ~2.45 eV (506 nm) with the vibronic progression toward higher energy. The presence of this structure has been attributed to high molecular order in polymer chains [111,112]. In the case of PPV, the driving force leading to an improved intrachain order is the elimination reaction occurring in the preordered structure; in the case of a similar structure in poly(2-methoxy-5-(2′-ethylhexyloxy)-1,4-phenylenevinylene) (MEH–PPV) in polyethylene [112], the order is induced by stretching.

Earlier studies of the conversion of the stretch-oriented precursor polymer films to PPV have been performed using x-ray techniques [90,100]. The results indicated growth of microcrystalline domains during the thermal treatment. It was accompanied by an increase of the orientational ordering, a decrease in the interchain distances, and an increase in the unit length along the chain direction.

The optical spectra play an essential role in our research of nonlinear optical properties of PPV which we performed on spin-coated (unoriented) films. We used UV spectra as a basic spectroscopic technique for the characterization of PPV and its conjugation in the material used in the NLO measurements. We prepared a set of precursors using the procedure described in Section III. The synthesis was performed with various modifications of the precursor synthetic route using a combination of solvents and bases as specified in Table 1. Several batches have been prepared and characterized. Figure 4 shows examples of the spectra of PPV prepared from the batches described as the batch Nos. 6A, 11, and 7A in Table 1.

The spectra in Figure 4 have not been corrected for reflection losses. Curve 3 shows the spectrum of PPV prepared from the precursor synthesized in methanol (batch 6A); curve 2 shows the spectrum of PPV from precursor No. 11 (synthesis performed in the methanol–water solution), and Curve 1 shows the spectrum of PPV from the precursor No. 7A (precursor prepared in water). The blue shift of the spectra of the films prepared from the methanolic solutions can be interpreted as being caused by the influence of methanol on the precursor polymer which leads to partial substitution with methoxy groups possibly acting as the defects breaking the conjugation in PPV. The difference in the spectra reflects different conjugation achieved in the films. It was caused by differences in the precursors not fully eliminated because of the unsatisfactory conditions (vacuum, 200°C) applied during conversion. Some amount of sulfur and chlorine was detected in the films proving this explanation.

Conjugated polymers (like many organics) may be susceptible to degradation when exposed to light and air. Indeed, we have observed some changes in the optical properties of PPV; however, they were more pronounced in the less converted films.

Figure 5 shows the spectra observed in the film well converted at high temperature, stored in an ambient atmosphere for a year, then again treated thermally under the forming gas. The changes in absorption were about 10%.

Figure 6 shows a more pronounced aging effect in the sample which was intentionally converted to a lesser degree using a lower temperature. The aging effect might be inhibited by repeated heating in the presence of HCl.

TABLE 1 Optical Data and Conditions of Preparation (Polymerization and Purification) of the Precursor Polymer Poly(p-xylylene tetrahydrothiophenium chloride) and Poly(p-phenylenevinylene)

Precurs. batch	Polymerization conditions — Solvent or solvent mixture (vol%), type of the base	Dialysis — Solvent in dialysis, Membrane MW cutoff	Refractive indices in precursor films measured at 632.8 and 810 nm		Refractive indices of PPV films measured at 632.8 and 810 nm		UV Spectra of PPV	
			n_{TE} precursor/ 633 nm 810 nm	n_{TM} precursor/ 633 nm 810 nm	n_{TE} PPV 633 nm 810 nm	n_{TM} PPV 633 nm 810 nm	λ_{max} (nm)	ϵ_{max} (10^5 cm^{-1})
6A	Anhydrous methanol, organic base	Methanol 12,000	1.616 1.606	1.610 1.600	1.950 1.876	1.606 1.593	380–400	1.1–1.2
10	Anhydrous methanol, organic base	Methanol 2,000	1.623 1.612	1.614 1.605	1.989 1.903	1.614 1.597	390–400	1.1–1.3
11	80 : 20 methanol : water, organic base	80 : 20 methanol : water 3,500	1.581 1.571	1.552 1.544	2.228 2.070	1.607 1.586	428–430	1.8–2.0
9	50 : 50 methanol : water, organic base	Water 3,500	1.639 1.628	1.596 1.587	2.215 2.049	1.603 —	426–430	1.6–1.8
8	50 : 50 methanol : water, inorganic base	Water 3,500	1.648 1.636	1.602 1.592	2.217 2.072	1.611 1.590	420–430	1.5–1.7
7A	Water, inorganic base	Water 3,500	1.628 1.618	1.586 1.578	2.300 2.128	1.614 1.596	453–458	1.8–1.9
5	Water, inorganic base	Water 12,000	1.648 1.633	1.589 1.580	2.282 2.115	1.596 1.579	427–456	1.8–2.3

Note: The spin-coated films of the precursor polymer of various batches were converted to PPV by programmed heating at 200°C for (1 + 2) h under vacuum. ϵ_{max} is the decadic absorption coefficient at the maximum absorption wavelength, λ_{max}.

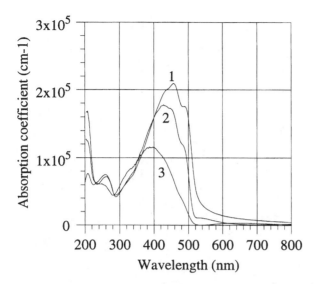

FIGURE 4 Examples of spectra of PPV obtained using various modifications of the precursor route showing the changes in λ_{max} and the maximum absorption coefficient. (From Ref. 113.)

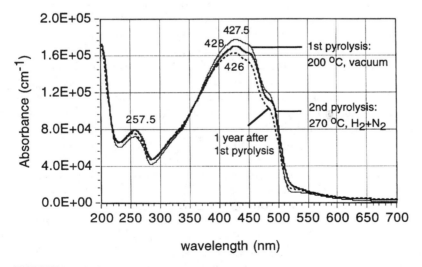

FIGURE 5 Aging observed in the film of PPV initially converted at 200°C in vacuum for (1 + 2) h. The film was stored at ambient conditions for a year, then it was heated again at 270°C for (1 + 2) h in the flow of gas mixture of nitrogen with hydrogen (5%).

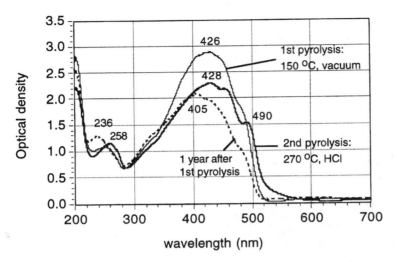

FIGURE 6   Aging in the film of PPV initially converted at 150°C in vacuum for (1 + 2) h. After a year of keeping it in ambient conditions, the sample was heated again at 270°C for (1 + 2) h under the flow of nitrogen mixed with vapors of hydrochloric acid.

A characteristic feature of PPV films obtained by spinning or doctor-blading of the precursor and subsequent thermal conversion is a high anisotropy of the optical parameters [114,115]. A convenient way of investigating these properties is by prism coupling [116,117]. Figure 7 shows the principal setup for coupling light into a thin-film waveguide using a prism.

Figure 8 shows the typical results obtained with a prism coupler (Metricon) at two wavelengths. Figure 8 shows the dependence of the intensity of the light reflected from the prism–PPV interface on the propagation constant which is defined as $\beta = n \sin \Theta$, where β is the propagation constant, n is the refractive index, and Θ is the light beam incidence angle at the interface. The sharp minima in the above dependencies correspond to guided mode resonances [i.e., to β values characterizing waveguide modes possible in the thin film of PPV of a given thickness (1 μ in the case of the film investigated here)]. For TE (transverse electric) modes, the polarization of light incident on the prism–PPV interface is selected to be perpendicular to the plane of incidence (the s polarization); therefore, the guided wave that is generated will have its electric field vector **E** oscillating within the plane of the film. On the other hand, selecting the polarization of the incident beam in the incidence plane (the p polarization) leads to the generation of TM (transverse magnetic) waveguide modes, the electric field oscillating essentially perpendicular to the film plane. Knowledge of β values for several TE and TM modes allows one to determine the corresponding values of the refractive indices.

Table 1 lists typical refractive-index values determined by the prism-coupling technique in spin-coated and doctor-bladed films of unsubstituted PPV and the precursor polymers prepared at various synthetic conditions. The values of refractive indices of the precursor polymers are typically in the range 1.61–1.64 for the TE modes and 1.59–1.61 for the TM modes at 633 nm. At 810 nm, they are 1.60–1.63 for the TE modes and 1.58–1.60 for the TM modes. PPV films converted

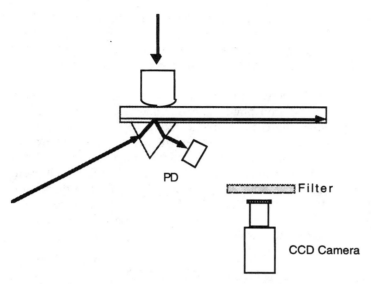

FIGURE 7 A measurement of waveguiding by prism coupling. The film on a substrate (glass, silica, or silicon) is pressed against a high-refractive-index prism and the whole assembly is rotated to find angular positions at which coupling into waveguide modes is possible. For low-loss samples, it is also possible to observe the propagation of guided light by detecting the light scattered from the sample with a suitably positioned charge coupled device (CCD) camera and the loss factor can be determined. PD = photodiode.

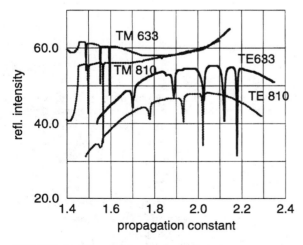

FIGURE 8 Waveguide mode resonances for a 1-μm-thick PPV film obtained by spin-coating.

under vacuum at 200°C had the values of refractive index in the range 1.92–2.30 for TE and 1.60–1.61 for TM at 633 nm, and 1.84–2.13 for TE and 1.57–1.60 for TM at 810 nm. These values agree reasonably well with the data published in, for example, Refs. 114, 115, and 118, although the differences in the materials can be detected.

The huge anisotropy of the refractive indices indicates that the conjugated chain molecules lie preferentially in the film plane. The orientation of polymer chains can be additionally increased in the case of free-standing PPV films by stretch-aligning them during the conversion. The anisotropy of the refractive index becomes then extremely high. Swiatkiewicz et al. [119] determined $n_{\text{parall}} = 2.89$ and $n_{\text{perp}} = 1.63$ (at 602 nm) as the principal values of the refractive-index ellipsoid along the stretch direction of a PPV film and perpendicular to it. The dispersion of the refractive index is also highly anisotropic, due to the dominance of the lowest-energy absorptive transition in the dispersion terms describing the dependence of the TE refractive index on the wavelength.

V. NONLINEAR OPTICAL MEASUREMENTS ON PPV

Several studies have been performed on the third harmonic generation by PPV [30,31,118,120,121] and its derivatives [30,31,122,123]. The nonlinear susceptibilities $\chi^{(3)}(-3\omega; \omega, \omega, \omega)$ determined by these studies are important for understanding the relations between structure and third-order nonlinearities of molecules and polymers. An assessment of technological suitability of PPV requires, however, that the *degenerate* third-order susceptibility $\chi^{(3)}(-\omega; \omega, -\omega, \omega)$ is known. This can be obtained by several techniques, and, in fact, there is a number of determinations of $\chi^{(3)}$ and n_2 in the literature [113,119,124–128].

Experimental techniques used for the determination of the material nonlinearity can be characterized according to the type of information provided. We have used several techniques which are summarized in Table 2. One can select several techniques from Table 2 in order to obtain relatively complete information about the

TABLE 2 Comparison of Experimental Techniques for Measuring the Degenerate Third-Order Nonlinearity

Technique	Temporal resolution	Real part of n_2	Imaginary part of n_2	Modulus of n_2
DFWM	Yes	No	No	Yes
Kerr gate	Yes	No	No	Yes
Heterodyne Kerr gate	Yes	Yes	Yes	
Single beam power-dependent transmission	No	No		Yes
Closed-aperture Z-scan	No	Yes	Yes	
Open-aperture Z-scan	No	No		Yes

nonlinear optical properties of a material at a given wavelength and it is possible to use the redundancy to cross-check the reliability of various techniques.

A. Degenerate Four-Wave Mixing

The term "four-wave mixing" relates to nonlinear optical processes in which three optical fields interact to create a fourth optical field, usually each of the fields belonging to a different light beam. If all the beams have the same frequency, the process is called degenerate four-wave mixing (DFWM). For thin films of conjugated polymers and very short laser pulses, the usual geometry of the experiment is the so-called BOXCARS geometry or forward-scattering geometry [4]. Figures 9 and 10 show a scheme of an experiment in which such a geometry is used.

The principle of this experiment is as follows (see Refs. 4 and 129]). When two beams of the same frequency are crossed in space, the electromagnetic field exhibits interference structure which results in the spatial variation of the light intensity. If the crossing takes place in a nonlinear medium, its optical properties can also be modified according to the spatial structure of the electromagnetic field and the light intensity. The resulting pattern of the refractive index and/or absorption coefficient variation formed in a material has a character of a volume diffraction grating. One should note here that, in fact, the presence of intensity modulation is not always necessary to modulate the properties of a medium. For two beams

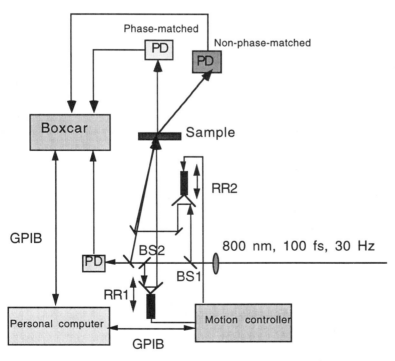

FIGURE 9 A scheme of a forward DFWM setup using the BOXCARS geometry showing the phase-matched and non-phase-matched signal detection. PD—photodiode; RR—retroreflector; BS—beam splitter; GPIB—data collection bus.

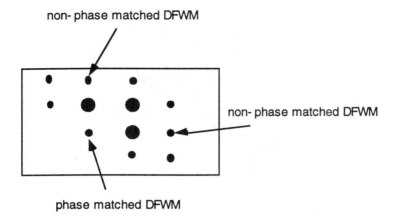

FIGURE 10 A view on the screen after the sample. The three strong beams are the original beams. Other beams are due to self-diffraction of pairs of beams and the four-wave mixing.

of mutually perpendicular light polarizations crossing in a medium, there is no intensity modulation but only a spatially periodic variation in the polarization of the summed electromagnetic field. This may also be sufficient to create a grating-like structure.

The grating has a three-dimensional character. For thin films, the depth modulation can be disregarded and the grating can be treated as a periodic surface array of reflective/absorptive elements. Such an array should act in a way similar to an ordinary difraction grating and diffract any beams incident on it. On the other hand, for a thicker sample, the grating must be treated as a volume diffractive element. For efficient diffraction from a volume grating, the waves diffracted at different depths of the grating have to add constructively. This is equivalent to requiring that the Bragg condition is fulfilled for the diffraction geometry. The condition can be written in the vector form as

$$\mathbf{k}_4 = \mathbf{k}_3 + \mathbf{b}$$

where $\mathbf{k}_3$ is the wave vector of the incident beam, $\mathbf{k}_4$ is the wave vector of the diffracted beam, and $\mathbf{b}$ is the grating vector (see Fig. 11). Since $|\mathbf{k}_3| = |\mathbf{k}_4|$, the Bragg condition essentially determines an angle between $\mathbf{k}_4$ and $\mathbf{k}_3$ (or between $\mathbf{k}_3$ and $\mathbf{b}$) at which efficient (phase-matched) diffraction is possible. The above condition is often referred to as the phase-matching condition. It can also be recast into a more general form, knowing that the $\mathbf{b}$ grating vector is a result of the interference of two incident beams with wave vectors $\mathbf{k}_1$ and $\mathbf{k}_2$; therefore,

$$\mathbf{b} = \mathbf{k}_1 - \mathbf{k}_2$$

and the phase-matching condition assumes the general form

$$\sum \mathbf{k}_i = 0$$

The experimental geometry shown in Figs. 9–11 fulfill the phase-matching condition for the diffracted beam traveling toward the fourth corner of the rectangle

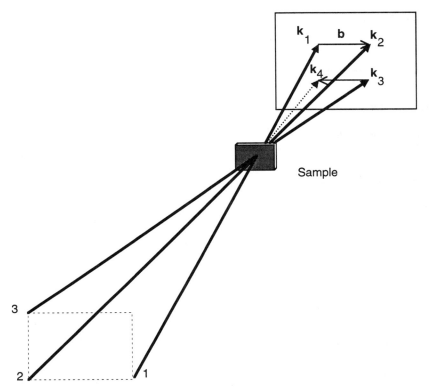

FIGURE 11 BOXCARS geometry of the interaction of the beams in the DFWM experiment.

formed by the exit points of the three beams. On the other hand, the diffraction spots marked in Fig. 10 as "non-phase-matched" are the result of non-Bragg diffractions from a relatively thin volume grating. Both the phase-matched and non-phase-matched diffraction signals are useful for the determination of the nonlinearity of conjugated polymers such as PPV. The non-phase-matched signal is especially useful if one needs to discriminate between a relatively weak signal originating from a very thin film of a polymer (e.g., thinner that 20 nm) and that from a thick glass plate on which the film is deposited.

The DFWM experiment is usually performed in such a way that two beams are timed to arrive at the sample at exactly the same moment and the third beam has a variable time delay to read the evolution of the changes in the sample during the laser pulse and after it.

B. DFWM Measurements in PPV Films at 800 nm

Figure 12 shows examples of time-resolved DFWM signals obtained from a film of PPV [127]. Figure 13 shows the power dependence of the peak of the DFWM signal. It is evident that the signal has at least two components: the fast component corresponding to the laser pulse duration and a delayed component responsible for

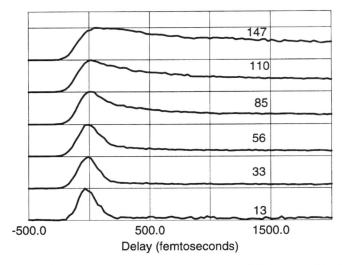

FIGURE 12 DFWM traces taken at the total light intensities in GW/cm² marked in the figure for a 2-μm-thick film of PPV. (From Ref. 127.)

the "tail" in the temporal dependence of the signal. It is evident that the "tail" depends on the light intensity of the beams used to write the transient grating. Such a behavior has been explained in Ref. 128 as the result of the action of two components of the nonlinearity: the instantaneous response due to the electronic non-linearity and the delayed response due to the excited states generated mostly by two-photon absorption of the pump beams. The formalism used in Ref. 128 can

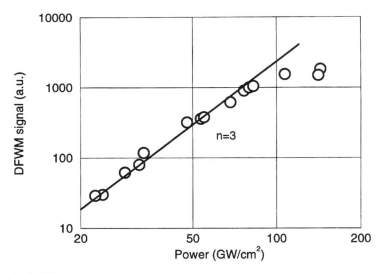

FIGURE 13 Power dependence of the maximum of the DFWM signal for the same sample of PPV as in Fig. 12. (From Ref. 127.)

be employed for the fitting of a DFWM temporal profile of the kind observed in PPV. We assume that the polarization of the medium can be written as

$$\mathbf{P} = \chi_g^{(1)} \mathbf{E} + \chi_g^{(3)} \mathbf{EEE} + \frac{N_e}{N_0}(\chi_e^{(1)} - \chi_g^{(1)})\mathbf{E} \tag{16}$$

where the subscript g refers to ground state and the subscript e to the excited state, and the indices (1) and (3) denote the linear and the third-order susceptibilities, respectively. N_0 is the total concentration of molecules. The nonlinear polarization is therefore considered to be the result of instantaneous third-order response and the response of excited molecules of a concentration N_e which are generated by two-photon absorption according to

$$\frac{dN_e}{dt} = \kappa_2 N_0 I(t)^2 - \frac{N_e}{\tau} \tag{17}$$

where κ_2 is related to the two-photon absorption coefficient [$\kappa_2 = \beta/(h\nu)$] and τ is the lifetime of the excited species. It is assumed that $N_e \ll N_0$. Using the polarization defined as above in the coupling equation describing the four-wave mixing process, one can derive approximate relations describing the temporal evolution of the DFWM signal. It may be useful to note that, in general, the DFWM signal derives from mixing of various combinations of the amplitudes of electromagnetic fields constituting the three incident beams. These combinations can be treated as resulting from diffraction of a given beam from a transient grating formed by the two remaining beams. For a given combination of such fields, one approximates the intensity of the DFWM signal as the function of the delay of one of the beams as

$$I_{\text{DFWM}}(t_D) \propto \left(\int_{-\infty}^{\infty} \Delta\chi(t)E(t - t_D)\, dt \right)^2 \tag{18}$$

where $\Delta\chi(t)$ is the amplitude of the transient susceptibility grating containing the instantaneous and delayed terms:

$$\Delta\chi(t) = \chi^{(3)}E^2(t) + \frac{N_e(t)}{N_0}(\chi_e^{(1)} - \chi_g^{(1)}) \tag{19}$$

$N_e(t)$ in the above equation is given by the solution of Eq. (17). It can be noted that because the intensity of the DFWM signal depends on the square of the total susceptibility change, the decay time of a delayed part of the DFWM as the function of the delay time should correspond to a half of the decay time of the excited-state population (i.e., I_{DFWM} is proportional to $e^{-2t/\tau}$ in the tail region). Figure 14 shows an example of a DFWM trace fitted with a theoretical calculation based on Eqs. (16)–(19). It has been assumed that the shape of the laser pulses used for the DFWM measurements can be approximated by a sech^2 function [i.e., $I(t) \propto E(t)^2 \propto$ sech$^2(t/t_p)$, and laser pulse intensity FWHM being 120 fs.

Figure 14 shows that the approximations used here allow one to determine the characteristic decay time τ for the tail of the DFWM curve ($\tau = 4$ ps in Fig. 14). We find that this parameter varies with the input power and, therefore, is a function of the density of the excited states generated by the two-photon absorption process. This is not unexpected because at the high concentration of excited species, their

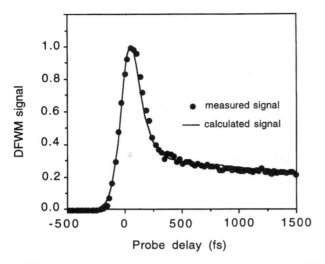

FIGURE 14 An example of a numerical fit for a DFWM transient signal with a strong contribution of a delayed response (a 2-μm-thick sample at the total intensity of about 45 GW/cm^2).

decay may involve bimolecular pathways (e.g., the bimolecular annihilation of excitons) [130].

The identity of the excited species contributing to the DFWM signal cannot be established from the DFWM measurements alone. Relatively long lifetimes are involved at lower powers (on the order of 100 ps and more); it can, therefore, be inferred that the states involved are excitons, which may or may not be identical to those states which contribute to the two-photon-induced fluorescence of PPV. It is necessary to note here that the exact identity of the excited states in PPV has been a subject of a prolonged discussion (see, for example, Refs. 131–141). Generation of free carriers has been considered [131–133] as an alternative to generation of neutral Frenkel–Wannier excitons [134,135]. It has been argued that both the intrachain excitons (which are fluorescent species) and interchain excitons (spatially indirect excitons) are generated [136–138] upon photoexcitation of PPV. Recent photoconductivity and electric field fluorescence quenching experiments [142] conclude that bound rather than free charge carrier pairs are generated in PPV.

As mentioned earlier, the DFWM signal intensity is proportional to the square of the transient susceptibility change. In fact, as both the real ("refractive") and imaginary ("absorptive") parts of the susceptibility may be involved, the intensity of the instantaneous part of the DFWM signal can be written as

$$I_{\mathrm{DFWM}} \propto [\mathrm{Re}(\chi^{(3)})]^2 + [\mathrm{Im}(\chi^{(3)})]^2 \tag{20}$$

This relation allows one to determine the modulus of the third-order susceptibility, or, rewriting the equations in terms of a complex nonlinear refractive index n_2, the modulus of n_2. A convenient way of determining the nonlinearity of a polymer film is by comparing the DFWM signal from the film with that from a well-known standard (e.g., a thin plate of silica). The nonlinearity can then be calculated as

$$\ln_2^{\text{polymer}}| = C_{\text{refl}} \, C_{\text{abs}} \, n_2^{\text{silica}} \, \frac{L_{\text{silica}}}{L_{\text{polymer}}} \left(\frac{I_{\text{DFWM}}^{\text{polymer}}}{I_{\text{DFWM}}^{\text{silica}}} \right)^{1/2} \tag{21}$$

where C_{refl} and C_{abs} are correction factors taking into account the differences in the reflection and absorption losses for the polymer sample and the silica sample, respectively.

Figure 15 shows results of DFWM determination of the modulus of n_2 in a wide selection of PPV films obtained using various modifications of the precursor synthetic route. The large spread of the results has its origin in the differences in the material. Electronic spectra of the films prepared from the precursors synthesized at various combinations of the solvent and base during the polymerization show a strong λ_{max} dependence as seen in Fig. 16. It appears that the details of the modified synthetic route described in Table 1 are essential both for the linear and the nonlinear properties of these films. We found it useful to identify PPV samples with the positions of λ_{max} in absorption spectrum measured in the same sample (if it was not too thick) or in a thinner sample prepared from the same material and converted under the same conditions and used in DFWM experiments. The changes of the decadic absorption coefficients versus λ_{max} in different materials in Fig. 17 show a similar trend as the change of nonlinear refractive indices versus λ_{max} in Fig. 18. Figure 18 shows the results of the DFWM measurements in the same small selection of PPV films characterized with absorption data in Figure 16 and Figure 17. It indicates that linear and nonlinear effects depend strongly on the length and the number density of conjugated chains in the material. Depending on the degree of π-electron conjugation, manifested as the red shift on the main absorption maximum and an increase of absorption coefficient as shown in Figs. 16 and 17, the nonlinearity at 800 nm increases quite rapidly, reaching values of $|n_2|$ on the order of 10^{-11} cm^2/W.

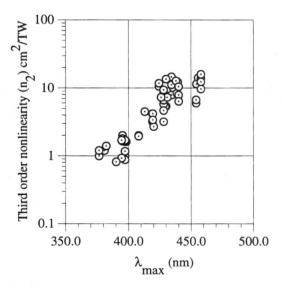

FIGURE 15 The dependence of n_2 determined at 800 nm by the DFWM technique on the λ_{max} for PPV films. (From Ref. 113.)

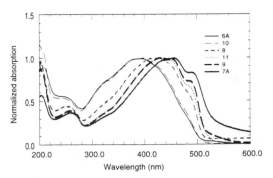

FIGURE 16 Examples of the absorption spectra of PPV films obtained by the precursor route showing the changes in the position of λ_{max}. The spectra were measured in PPV films prepared from the sulfonium precursor prepared in different solvents: in methanol—curves 6A and 10; in water–methanol solution—curves 8, 9, and 11; in water—curve 7A. All films were converted at the same conditions (under vacuum at 200°C). (From Ref. 143.)

The differences between materials converted at the same conditions but prepared from different precursors may be accounted for by various number of defects induced either by the conditions of preparation of the precursor solutions or during preparation of the films. Stronger conversion conditions with the help of HCl markedly improve both the linear and the nonlinear properties of PPV. The value $|n_2| = 1.7 \times 10^{-11}$ cm^2/W (this is equivalent to $\chi^{(3)} = 2.2 \times 10^{-9}$ esu) was measured by us at 800 nm in the highly converted PPV films.

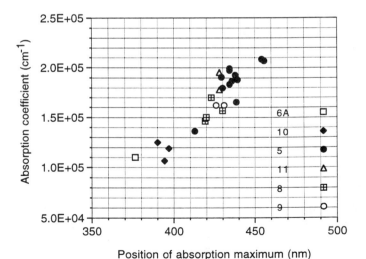

FIGURE 17 Absorption coefficients versus position of the absorption maximum λ_{max} in a selection of PPV films prepared using modified conditions in the precursor route described in Table 1 and Fig. 16. (From Ref. 143.)

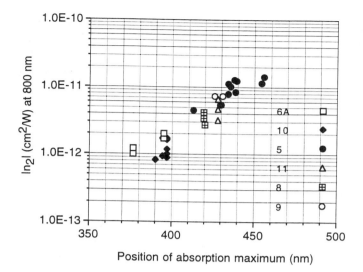

FIGURE 18 Modulus of nonlinear refractive index $|n_2|$ determined by femtosecond DFWM at 800 nm versus position of the absorption maximum in PPV films prepared from a number of precursors (cf. Table 1 and Fig. 16) synthesized under various conditions and converted in vacuum at 200°C. (From Ref. 143.)

The $|n_2|$ data found in PPV are quite respectable values of the nonlinearity. A device having a cross section of $A = 10\mu m^2$ and a length of $L = 10$ mm operating with a nonlinear phase change of 2π would need an input light intensity of $I = 2\pi\lambda/n_2L = 5 \times 10^7$ W/cm² and the peak power of only 5 W. For 1-ps pulses, this results in a very modest requirement of only 5 pJ per switching event. Such a prediction makes it very exciting to investigate the possibility of using PPV in a photonic switch. The major problem in building such a switch is, of course, obtaining long enough propagation of a beam in the nonlinear material. Unfortunately, simple DFWM experiments performed in transmission on very thin films of the polymer do not provide any information on the degree of linear and nonlinear losses in the film. It is also not known if the value of $|n_2|$ determined from DFWM is dominated by the real or by the imaginary part of the nonlinearity and if the sign of the real part of the nonlinearity is positive (self-focusing) or negative (self-defocusing). Additional experiments are necessary, especially to determine the losses.

C. DFWM Measurements of PPV Nonlinearity at Other Wavelengths

Among the literature data on the nonlinearity of PPV at various wavelengths, there are several determinations of $\chi^{(3)}$ from the DFWM technique. However, it is not entirely reliable to compare these data directly because of the different pulse durations used in the measurements, different standards used for the determination of $\chi^{(3)}$ of PPV, and, finally, differences in the material itself used by different authors.

Because the differences in the degree of π-electron conjugation of the material prepared by varying synthetic routes can lead to an order-of-magnitude spread of the n_2 values at 800 nm (Figs. 15 and 18), one can expect that some of the differences among various literature data can be due to this particular factor. Generally, values of $\chi^{(3)}$ between 10^{-10} esu and 10^{-9} esu have been reported, which correspond roughly to the n_2 range 10^{-12}–10^{-11} cm^2/W. The Buffalo group reported measurements of PPV at 602 and 620 nm with 400-fs and 80-fs laser pulses [119,124,128]. They initially reported $\chi^{(3)} = 4 \times 10^{-10}$ esu at 602 nm for uniaxially stretch-oriented PPV [124]. A value given in Ref. 119 for $|\chi^{(3)}_{1111}|$ is, however, as high as 1.1×10^{-9} esu. The negative sign of Re($\chi^{(3)}$) has been deduced on the basis of the anisotropy of the nonlinearity. Measurements on PPV in the visible range have also been reported by Bubeck et al. [125,126].

Our own measurements of DFWM at 1.054 μm and at 527 nm on some samples have provided the values of 2×10^{-12} and 1×10^{-11} cm^2/W, respectively, for these two wavelengths. Figure 19 shows an example of a DFWM signal obtained with 7-ps laser pulses at 1.054 μm. Evident is the absence of a tail. The two-photon and multiphoton absorption at 1.054 μm is apparently too weak to produce a high concentration of excited states responsible for the long-lived nonlinear response. The theoretical curve in Fig. 19 corresponds to the DFWM signal expected for a sech^2 laser pulse with a FWHM of 6.6 ps.

D. Two-Photon Absorption

Earlier studies of PPV [119,124,125,128], performed mostly using wavelengths between 570 and 620 nm, have shown that linear and nonlinear absorption in PPV in that wavelength range may be too high for device applications. Although the problem of linear absorption becomes less critical at longer wavelengths, the two-photon absorption coefficient (β) may still be too high for all-optical switching using PPV because it is essential that the two-photon figure of merit, $T = \beta\lambda/n_2$, which characterizes the nonlinear phase accumulation over a two-photon absorption length, is greater than m where m is of the order of unity but whose exact value depends on the device structure [6].

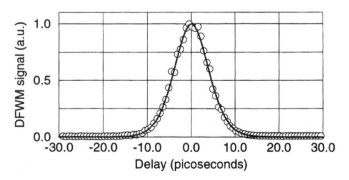

FIGURE 19 Experimental (circles) and calculated (solid line) DFWM curves for a PPV film obtained at 1.054 μm.

The two-photon absorption coefficient β can be measured in a number of ways. Conceptually, the simplest technique is the measurement of a sample transmittance as a function of the input power. Figure 20 shows an example of such an experiment for a 3-μm-thick PPV film. It is evident that the transmission decreases with the incident light intensity, indicating strong two-photon absorption. Considering the transmission of a beam to be controlled by a two-photon process, one may prefer to plot the inverse of the transmittance as a function of the input light intensity using an inverted form of Eq. (11):

$$\frac{1}{T} = \frac{I_0}{I_{\text{tr}}} = 1 + \beta I_0 L \tag{22}$$

This equation predicts a linear dependence between the inverse of the transmittance and the input light intensity (Fig. 21). Two-photon absorption coefficients (β) may be simply determined from the slope of the inverse transmittance of the sample as a function of the incident beam intensity. However, corrections are needed because of the fact that one deals with spatial and temporal dependencies of the light intensity (see, for example, Ref. 4). We used a correction factor of 2.2 for the β values presented in Ref. 113.

The result of the determination of the two-photon absorption coefficient at 800 nm by this procedure in films of PPV obtained under varying conditions is that the β value can vary in a range of about 20 to 80 cm/GW [113]. As shown in Fig. 22, this variation in the values of the two-photon absorption coefficient parallels the variation in the $|n_2|$ values determined by the DFWM technique. The reason for the changes of the both quantities is the same; that is, the more conjugated polymer samples have both a higher $|n_2|$ and a higher imaginary part of n_2. As shown in Fig. 11, the ratio of these two quantities is, however, not constant but becomes

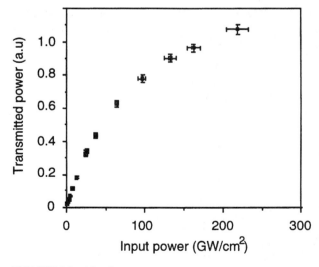

FIGURE 20 Nonlinear transmission of 800-nm light in a 3-μm-thick film of PPV plotted as the transmitted power against the incident power.

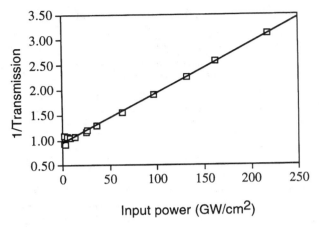

FIGURE 21 Data from Fig. 20 replotted as the inverse transmission versus the input light intensity.

smaller as the nonlinearity increases. This leads to the two-photon loss merit factor T at 800 nm becoming smaller and finally being lower than unity which is considered a threshold value for photonic switching applications. Nevertheless, in absolute terms, the two-photon absorption coefficients are very high and therefore may be of interest themselves, for applications unrelated to photonic switching (e.g., for laser pulse diagnostics by two-photon-induced fluorescence) [144].

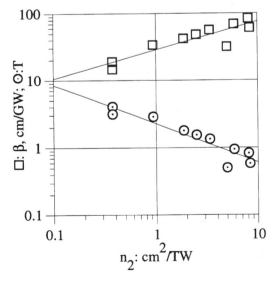

FIGURE 22 The two-photon absorption coefficient β and the merit factor T plotted as the function of the nonlinearity measured by DFWM in a series of PPV samples. (From Ref. 113.)

In the considerations above, the ratio $T = 4\pi\,\text{Im}(n_2)/\text{Re}(n_2)$ was taken approximately equal to $4\pi\,\text{Im}(n_2)/|n_2|$. This is justified a posteriori because of relatively low values of T. $T = 1$ corresponds to $\text{Im}(n_2) = 1/4\pi\,\text{Re}(n_2)$, thus the approximation that $|n_2| = |\text{Re}(n_2)|$ is justified.

Another possibility of measuring the two-photon absorption coefficient in a single-beam setup is by using the Z-scan technique described in further sections of this chapter. However, both the single-beam transmission measurements and Z-scan do not provide any information on the temporal dependencies of the nonlinear losses. More detailed information about these losses can be obtained from pump–probe experiments in which transient (induced) absorption or transient (induced) transmission is probed with a variable time delay.

Figure 23 shows an example of a pump–probe experiment in which the pump beam and the probe beam are both at the same wavelength of 800 nm. The effect consists of two components, both of them being of absorptive character (i.e., showing as negative signals on the pump intensity background). The fast component is related to the two-photon absorption. The appearance of a minimum can be rationalized as the influence of the pump beam on two-photon absorption of the probe beam in the sense that one photon comes from the pump beam and the other one comes from the probe beam. The depth of the minimum is related to the two-photon absorption coefficient [128,145]. The β values obtained from the pump–probe experiments are in good agreement with the two-photon absorption coefficients measured by the power-dependent transmission method.

The instantaneous effect is accompanied by a longer tail which is also of absorptive character and which can be attributed to the absorption induced by the excited states generated by two-photon absorption. As from transient absorption studies, [139,141,145], the induced change of transmission in PPV can be positive or negative depending on the wavelength. By using two different wavelengths for the pump and the probe, there is a possibility of inducing a change of transmitted probe intensity which actually corresponds to a gain. Such a gain [146–148] is a result of stimulated emission from excited states generated by the pump beam.

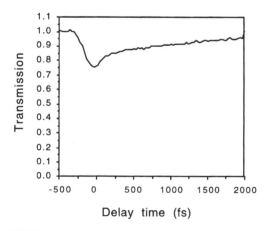

FIGURE 23 Time-resolved pump–probe measurement of induced absorption in a PPV film.

Combining DFWM measurements with the determination of β by a single beam or pump–probe technique allows one, in principle, to determine the T merit factor. When the one-photon absorption coefficient at the wavelength of interest is known, the W factor can be determined, too, and one can decide on the suitability of a material for photonic switching. We discuss the one-photon losses of PPV in more detail in Section VI.B. We note here that the usual one-photon loss factor in pure PPV at 800 nm is of the order of 100 dB/cm; therefore, n_2 of the order 1×10^{-11} cm^2/W leads to a W factor of about 5 at 1 GW/cm^2. Thus, the material has a realistic application potential, especially if the one-photon losses can be reduced [149].

E. DFWM Measurements of Dimethoxy-Substituted Poly(*p*-phenylenevinylene)

An example of a substituted poly(*p*-phenylenevinylene) is poly(2,5-dimethoxy-1,4-phenylenevinylene), abbreviated DMOPPV, and having the following structure:

We present here a determination of nonlinear optical parameters at 800 nm for this polymer: The motivation of this research has been to check whether substitution of the benzene rings in PPV has a positive effect on the material nonlinearity and whether the material parameters are superior to these of unsubstituted PPV. Previous work on DMOPPV seemed to indicate that the third-order nonlinearity might be larger than that in PPV when the third-harmonic generation results were compared [122]. Four-wave mixing results obtained at 600 nm also suggested the same pattern [150]. Swiatkiewicz et al. [150] quoted a $\chi^{(3)}$ of 4×10^{-9} esu for a stretch-oriented film of DMOPPV at 602 nm, whereas the same group gave 1.1×10^{-9} esu [119] for a stretch-oriented film of unsubstituted PPV. The methoxy substitution certainly has an effect of reducing the band gap: DMOPPV is red, whereas unsubstituted PPV is usually yellow–orange. The absorption maximum for films of DMOPPV is about 490 nm; see Fig. 24. Therefore, one can argue that a higher nonlinearity of DMOPPV measured at 602 nm compared to PPV measured at the same wavelength may be partly due to a stronger dispersion effect. Figure 25 shows a time-resolved DFWM signal from a 0.6 μm DMOPPV film at 800 nm. The signal is similar to those obtainable from unsubstituted PPV under similar conditions.

We have verified that, at relatively low intensities, the height of the peak in the DFWM signal scales with the cube of the input intensity as expected for the electronic nonlinearity far from saturation. The signal starts to saturate at the input intensity of about 50 GW/cm^2. The average $|n_2|$ value derived from measurements on several DMOPPV samples is 4×10^{-12} cm^2/W. One can note that this value falls in the middle of the range of n_2 values determined by us for PPV. In fact, this value corresponds roughly to the nonlinearity of PPV with an absorption maximum

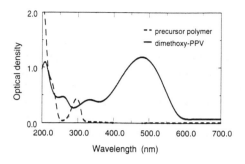

FIGURE 24 Absorption spectra of a precursor of DMOPPV and a converted film.

at about 430 nm. Therefore, a red shift of DMOPPV spectrum compared to that of unsubstituted PPV does not result in an increase of the absolute value of the degenerate nonlinearity at 800 nm.

The two-photon absorption coefficient β determined for the DMOPPV films was about 5×10^{-8} cm/W. A comparison with PPV shows, again, that this value corresponds to the middle of the range of values obtained for the unsubstituted polymer. Defining the merit factor T as $T = \beta\lambda/n_2$, we obtain $T = 1.0$ for DMOPPV. This value of the merit factor is on the borderline of applicability. It is, however, possible that DMOPPV might have a better merit factor at wavelengths other than 800 nm.

F. Optical Kerr Ellipsometry in Dimethoxy-Substituted PPV

Because only the modulus of n_2 is available from the DFWM measurements, one cannot determine the sign of the nonlinearity from these measurements alone. The good optical quality of DMOPPV films allowed for an application of a technique

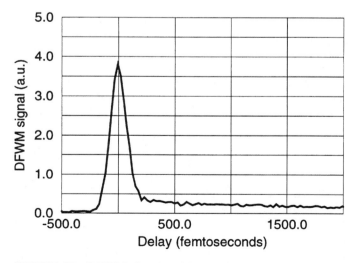

FIGURE 25 DFWM signal at 800 nm for a dimethoxy-PPV film.

which can provide this information, namely the phase-tuned heterodyne Kerr gate (Kerr ellipsometry) [151–154]. In a simplest setting of an optical Kerr effect measurement, a sample is exposed to a linearly polarized pump beam which induces birefringence and dichroism through the real part and the imaginary part of the third-order nonlinearity, respectively. The induced changes are probed by a probe beam polarized at 45° to the pump. The birefringence and dichroism result in the probe beam acquiring a perpendicularly polarized field component. It can be shown that the intensity of the pump beam transmitted through a crossed polarizer is proportional to the square of the modulus of the complex nonlinear phase change.

The principle of the phase-tuned heterodyne Kerr gate measurement is the observation of a Kerr gate signal as a function of the position of the analyzer which is rotated from the crossed position by a few degrees to admit some contribution of the other polarization field component, which then acts as a local oscillator. When this local oscillator is additionally phase shifted by $\pi/2$ by use of a suitably oriented quarter-wave plate, the Kerr gate signal is linearly dependent on the angle of rotation of the analyzer, the slope of the relation being proportional to the real part of the nonlinearity. Thus, the Kerr ellipsometry can be used to provide information about the content of refractive and absorptive (real and imaginary) nonlinearity components as a function of time. Figure 26 shows a scheme of the setup for time-resolved Kerr gate measurements. A set of time-resolved Kerr gate measurements performed on a 0.6-μm-thick film of DMOPPV [17] is shown in Fig. 27. It can be seen that the Kerr signal depends on the angular position of the analyzer in such a way that the signal changes from being negative at negative angles to being positive at positive angles. In a reference experiment, a silica plate gave an opposite behavior; that is, with the same position of the quarter-wave plate, the silica signal changed from being positive at negative angles to being negative at positive angles. This clearly shows that the real part of the nonlinearity in DMOPPV is negative because the n_2 of silica is known to be positive. One can note here that the behavior of the heterodyne signal reverses when the quarter-wave plate is rotated by 90° (i.e., the slow and the fast axes of the plate are exchanged). In fact, the measurement of a reference silica sample is, in principle, unnecessary, as the sign of the phase bias introduced by the quarter wave plate determines the sign of the slope of the dependence of the heterodyne Kerr signal on the angle. In the example shown in Fig. 27, the "tail" of the signal apparently contains a larger proportion of the refractive component than the "instantaneous" part of the signal because it is more sensitive to the rotation of the analyzer (cf. for example, curves at 0° and −2° in Fig. 27). Some contribution to the "instan-

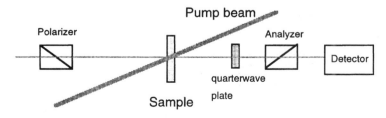

FIGURE 26 The principal setup for Kerr ellipsometry.

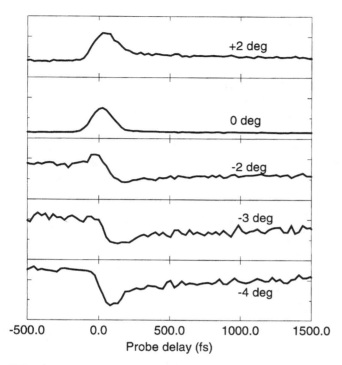

FIGURE 27 Example of a series of measurements of heterodyne Kerr effect in a 0.6-μm-thick film of dimethoxy-PPV. The angle of the analyzer is quoted for consecutive scans. (From Ref. 17.)

taneous" portion of the signal from the silica plate may influence this behavior. A 0.6-μm-thick DMOPPV film with a nonlinearity of about -4×10^{-12} cm^2/W should, in principle, produce about an eight times larger contribution to the nonlinear phase change than a 1-mm-thick silica plate with a nonlinear index of 3×10^{-16} cm^2/W (assuming that the off-diagonal, e.g., $\chi^{(3)}_{1122}$, nonlinearity terms of DMOPPV and silica responsible for the Kerr effect are in the same relation as the diagonal $\chi^{(3)}_{1111}$ terms responsible for the n_2 values determined by DFWM). However, it is also possible that the effective T factor should, indeed, be considered a function of time.

We conclude that the NLO properties of poly(2,5-dimethoxy-p-phenylene-vinylene) measured at 800 nm with femtosecond pulses are very similar to the values of those of medium conjugated unsubstituted PPV. It is likely that larger differences in the nonlinearities observed earlier at shorter wavelengths [150] might have been due to a stronger dispersion effect in the case of the dimethoxy-substituted polymer.

G. Z-scan Measurements in PPV Films and Solutions of Substituted PPV

The Z-scan [155,156] is a technique of measuring the third-order nonlinearity in which a sample is scanned along the path of a laser beam which is being focused by a lens and the far-field on-axis light intensity is monitored with a detector with an aperture in front of it. The nonlinear behavior of the sample is equivalent to the formation of an induced positive or negative lens in the case of positive (self-focusing) or negative (self-defocusing) third-order nonlinearity, respectively, changing beam intensity at the aperture plane; see Fig. 28.

Figure 28 presents the principle of the measurement for the case of $n_2 > 0$ (self-focusing). Because the light intensity varies due to the focusing action of the linear lens, the induced (nonlinear) lens strength induced in the sample varies, too, with the position of the sample with respect to the focused beam waist. Placing a positive induced lens before the waist will bring the waist closer and, therefore, will reduce the far-field on-axis intensity at the aperture plane, whereas placing the same induced lens after the waist will induce some focusing on the divergent beam and, therefore, increase the on-axis intensity. The S shape resulting from scanning the sample can be analyzed using calculation techniques described by Sheikh-bahae et al. [156], and nonlinear properties of the sample can be derived.

Characteristic lowering and increase of the on-axis intensity shows the focusing and defocusing effect of the sample and is thus a measure of the refractive component of the nonlinearity. At the same time, one can also determine the absorptive (i.e., related to nonlinear absorption or induced transmission) part of the nonlinearity. This can be done either by running an "open-aperture" Z-scan (i.e., collecting all the light transmitted through the sample) or by investigating the detailed shape of a "closed-aperture" Z-scan.

Figure 29 shows shapes of Z-scans expected for the case of purely refractive nonlinearity, the case of positive imaginary part of the nonlinearity (induced absorption case), and the case of negative imaginary part of the nonlinearity (induced transmission case). In practice, the closed- and open-aperture scans can be performed simultaneously; see the setup in Fig. 30. The Z-scans curves can be ana-

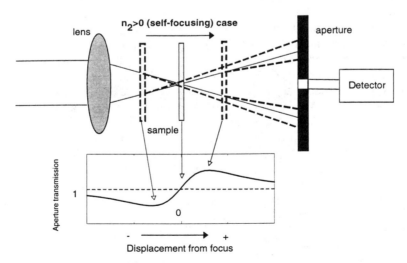

FIGURE 28 Principle of the Z-scan. Three positions of the sample are shown: before the focus, where focusing of light results in lowering of the aperture transmission; at the focus; and after the focus, where the focusing results in an increase of the aperture transmission.

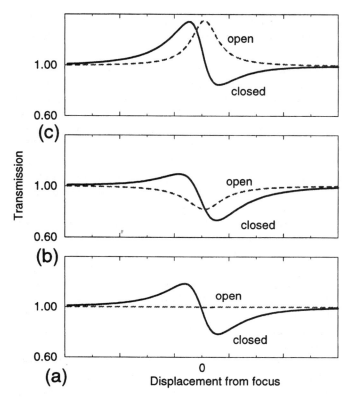

FIGURE 29 Theoretical shapes of open- and closed-aperture Z-scans for the cases of $n_{2,real} < 0$ and no nonlinear absorption, $n_{2,imag} = 0$ (a), $n_{2,real} < 0$ and induced absorption, $n_{2,imag} > 0$ (b), and $n_{2,real} < 0$ and induced transmission (absorption bleaching), $n_{2,imag} < 0$ (c).

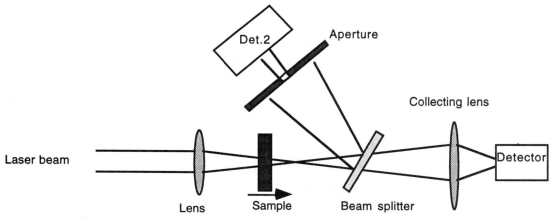

FIGURE 30 A simple setup for simultaneous measurement of the open- and closed-aperture Z-scans.

lyzed by numerical fitting using expressions derived by Sheikh-bahae et al. [156] to obtain the nonlinear phase change $\Delta\phi_{real}$ induced by the third-order nonlinearity and the T factor (defined here as $T = 4\pi\Delta\phi_{imag}/\Delta\phi_{real}$) for a given sample. In practice, the amplitude of a closed-aperture Z-scan is approximately proportional to the real part of the nonlinear phase change $\Delta\phi_{real}$. The asymmetry of a closed-aperture scan gives the T factor (for $T = 0$, the scan is essentially S shaped and symmetric). The depth of a dip in the open-aperture scan, on the other hand, is related to the value of the imaginary part of the nonlinear phase change $\Delta\phi_{imag}$ (i.e., to the product $T\Delta\phi_{real}/4\pi$). One can also use a procedure consisting in dividing the closed-aperture scan by the open-aperture scan in order to resolve information on the real part of the phase change. Such a procedure gives a scan which is essentially free from the influence of the imaginary part of the phase change (i.e., free from the two-photon absorption) [156].

An example of a set of Z-scans on a 2-μm-thick film of PPV is given in Fig. 31. These data were obtained with 30-Hz repetition frequency 35-ps pulses at 527 nm (about 0.5 μJ/pulse, about 2.9 GW/cm^2). The Z-scans are dominated by a strong induced absorption effect. The numerical fitting of the shapes of the Z-scans and a comparison with a Z-scan obtained for a silica plate leads to $n_{2,real} = -1.0 \times 10^{-11}$ cm^2/W and $\beta = 1.1 \times 10^{-6}$ cm/W and the T factor of 5.5 for PPV at 527 nm. These parameters are not very attractive from the practical point of view, however; it has to be realized that they are obtained with relatively long pulses at the wavelength where one-photon absorption becomes already important and therefore a part of the induced absorption may come from one-photon generated and relatively long-lived excited states.

It can be remarked here that the real part of the nonlinear refractive index of PPV is found to be negative also at 800 nm. The Z-scan technique is a convenient

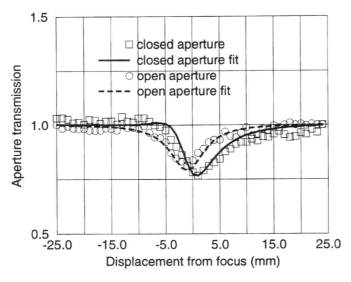

FIGURE 31 Open- and closed-aperture Z-scans performed on a 2-μm-thick PPV sample with 35-ps laser pulses at 527 nm.

tool for verifying the sign of the nonlinearity which is unavailable from DFWM measurements. A negative sign of the nonlinearity is, however, always rather suspect because of the fact that the *thermal nonlinearity* effect is usually negative. Therefore, it is important that care is taken to avoid thermal effects which may be interfering in the Z-scan determination of the fast (electronic) nonlinearity sign. One important factor in any Z-scan measurement is the total average power used in the measurement and absorbed by the sample. The total power used in the experiment described above is only 15 μW and the total absorbed power is only about 3 μW. A typical focal length of the lenses used in the experiments is 15–25 cm.

We performed Z-scan measurements on solutions of several soluble poly(*p*-phenylenevinylene)s [157]. The presence of aliphatic chains or aromatic substituents on the PPV chain can render the resulting polymers soluble in common organic solvents. The solubility of these polymers makes it easy to determine their nonlinear optical properties by investigating concentration changes of the NLO response of solutions. A typical concentration range is about 0–0.2 wt% of a substance in solution if its nonlinearity n_2 is on the order 10^{-12} cm^2/W. Figure 32 shows examples of closed-aperture scans for a series of solutions of one of soluble derivatives of PPV [157]. It can be seen that the amplitudes of the scans decrease with the increasing concentration, implying that the real part of the nonlinearity of the solute is opposite in sign to that of the solvent. At the same time, there is an

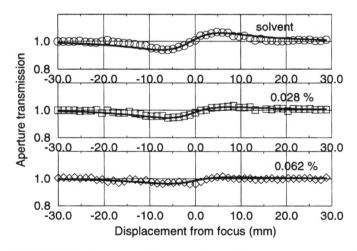

FIGURE 32 Changes of Z-scans ($f = 175$ mm lens was used) with concentration of a soluble PPV in 1,1,2,2-tetrachloroethane. (From Ref. 157.) Closed-aperture Z-scans on a 1-mm cell filled with the solvent and solutions of a soluble PPV polymer with the formula

increasing asymmetry of the scans, indicating that the imaginary part of the non-linearity is present and leads to nonlinear absorption losses.

Concentration dependencies of $\Delta\phi_{real}$ and $\Delta\phi_{imag}$ can be analyzed to extract the information on the nonlinear optical properties of the solute. In this analysis, we assume that the nonlinearity of a solution can be approximated by the expression

$$n_2 = g n_2^{(1)} + (1 - g) n_2^{(2)} \tag{23}$$

where g is the weight fraction of the solute, $n_2^{(1)}$ is its nonlinear refractive index, and $n_2^{(2)}$ is the nonlinear refractive index of the solvent. Figure 33 shows an example of such concentration dependencies. The nonlinear phase change is related to the nonlinear refractive index by

$$\Delta\phi = \frac{2\pi n_2 I L}{\lambda} \tag{24}$$

where I is the light intensity and λ is the wavelength. Knowledge of the absolute light intensity may be used for conversion from phase-change values to the non-linearity values. It is, however, more convenient to perform the measurements in a relative manner. The values of the nonlinear optical parameters were calibrated by performing measurements of the nonlinear phase change for a silica plate for which $n_{2real} = 3 \times 10^{-16}$ cm^2/W was assumed. From the magnitude of the phase change, it has been determined that the peak on-axis intensity in these measurements was 12 GW/cm^2. Using the nonlinearity for silica, the n_2 for the solvent and for the solutions could be determined.

In the above example illustrating the use of the Z-scan technique for the determination of nonlinear properties of substances in solutions, the results show the real (refractive) part of the nonlinearity to be negative (self-defocusing) and the imaginary part of the nonlinearity highly positive for the investigated polymer. This leads to a T factor larger than unity. It should be mentioned that a markedly smaller

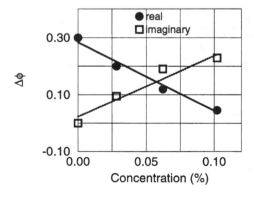

FIGURE 33 The dependence of the nonlinear phase change on the concentration of the polymer in 1,1,2,2-tetrachloroethane solution gives the negative value of $n_{2real} = -2.6 \times 10^{-12}$ and the positive value of $n_{2imag} = 1.1 \times 10^{-12}$ cm^2/W for the real and imaginary parts of n_2, respectively, of the polymer.

T value has been obtained in the case of measurements in tetrahydrofuran (THF) compared to the results obtained for tetrachloroethane.

The extrapolated n_2 results obtained from Z-scan measurements on solutions are generally in good agreement with the $|n_2|$ data obtained from DFWM measurements on thin films. This can be considered to be partly fortuitous, as there are several factors which influence the nonlinearity of solutions and solid films in a different way. Extrapolation of the nonlinearity of dilute solutions to the 100% concentration of a polymer involves an assumption that the hyperpolarizability of polymer chains is not modified by intermolecular interactions at a high concentration of the polymer and that the local field factor is constant. The problem of intermolecular interactions is difficult to treat, but the local field factor is certainly different for pure polymer because of a different refractive index of the polymer compared to the solution. Additionally, the polymer films exhibit a certain degree of anisotropy, and the measurement performed with the electric field in the plane of the polymer film will probe an enhanced nonlinearity because of the preferential orientation of the chains on the surface. Finally, the simple linear extrapolation should take into account the density difference between the solutions and the pure polymer. Apparently, all these differences do not account for more than a factor of 2, as the agreement of the thin film DFWM results with those on solutions was found to be better than within such a factor. Admittedly, substitution of PPV chains with bulky side groups isolates, to a certain degree, the chains from mutual interactions and, therefore, one may expect that the difference between the nonlinear optical properties of a substituted PPV molecule in the solid film and the same molecule in a solution may be not very pronounced.

VI. NONLINEAR AND WAVEGUIDING PROPERTIES OF PPV COMPOSITES

The major problem with using the nonlinear optical properties of PPV in a device is in obtaining good quality, highly transparent films suitable for waveguiding applications. Optical waveguide devices may involve propagation of light through distances of many centimeters. Depending on the way in which the PPV films are prepared, the linear losses of the waveguides can vary within a certain range of values; however, they are usually much higher than those required for making waveguide devices with lengths on the order of 1 cm. Gymer et al. [115] quoted losses of about 40–60 dB/cm for copolymer (PPV–DMOPPV) waveguides, but the films used were so thin (e.g., $0.15 \mu m$) that a substantial correction might be necessary to account for the large extent of the modal field being actually outside of the PPV film. Therefore, the losses in thicker films can be expected to be even higher. Michelotti et al. [158] observed much lower losses in a PPV derivative polymer.

Efficient interaction of the guided light with the nonlinear material forming a slab waveguide can be achieved if a single-mode transverse electrical (TE) field and single-mode transverse magnetic (TM) field is well confined within the thickness of the film [159,160]. The relationship among the waveguide parameter *V*, the number of modes of the waveguide, and the material parameters is described by [159, Table 12-2, p. 244]

$$V = k\rho \, (n_{co}^2 - n_{cl}^2)^{1/2} \tag{25}$$

where k is the wave vector ($k = 2\pi/\lambda$), ρ is the half-width of the waveguide, and $(n_{co}^2 - n_{cl}^2)^{1/2}$ is the numerical aperture describing the difference of refractive indices between the core n_{co} (the film) and the cladding n_{cl}. The waveguide is single mode if $0 < V < \pi/2$. For example, the structure composed of a waveguiding film of PPV of thickness equal to 0.15 μm, refractive index equal to 2 at 0.63 μm, and using silica ($n = 1.457$) as a cladding will be single mode ($V = 0.33\pi$); however, the field will be largely distributed outside the film. Equation (25) provides a tool for designing the modal characteristics of a device by tuning the parameters like thickness and wavelength having assumed, for example, a constant difference of refractive indices between film and the substrate. Another scenario might be adjusting the difference of refractive indices. The control over refractive indices can be, in principle, realized by preparing the polymer with a different degree of polymerization. In the case of PPV, its refractive indices depend on the degree of conversion into a conjugated structure. Table 1 shows a large difference between refractive indices of the films of precursor polymer and the converted PPV. The gradual increase of refractive indices upon conversion has been observed in a separate experiment.

We observed good waveguiding properties of the films of the precursor polymer before conversion. The propagation losses measured with a He–Ne laser and a 810 nm laser diode in the films of the precursor before conversion were in the range 3–7 dB/cm, being slightly anisotropic. The losses depend on the way in which the precursor was prepared and on the actual level of conversion in the film. After pyrolysis, the losses become highly anisotropic; only TM modes can propagate even in visually quite transparent film of PPV showing relatively little light scattering. The optical losses of TM modes in the best films were at the level of 30 dB/cm at 633 and ~20 dB/cm at 810 nm.

One possible solution for controlling optical properties of waveguiding structures is to prepare composites of PPV with suitably chosen host materials. Composites offer tunability of refractive indices by the adjustment of the concentration of PPV in a matrix. Diluting of PPV by mixing it into a matrix may be thought to lead to the proportional decrease of the nonlinearity (assuming that the matrix nonlinearity can be neglected compared to that of PPV) and also to the proportional decrease of the loss factors. Such behavior would produce composite materials with the T and W factors equal to those of pure PPV. One can hope, however, that the additional tunability of PPV properties available by modifying the matrix and the conversion procedures may actually result in the improvement of the merit factors.

We describe here two types of such composites: ormosils containing unsubstituted PPV and the composite of poly(N-vinylpyrrolidone) (PVP) with unsubstituted PPV. The crucial problem in obtaining composites of unsubstituted PPV is that the material has to be prepared by mixing the precursor of PPV with other components and then the conversion reaction has to be carried out on the film cast from the obtained solution. Thus, there has to be a solvent compatibility of the PPV precursor with the other components to achieve miscibility. The heat treatment has to be adjusted in such a way that the composites have a good degree of conversion of the PPV precursor to the conjugated structure while the host material has the required optical and mechanical properties. This has already been suggested by

Karasz et al. [161]. Karasz and Prasad and co-workers described the preparation and properties of the PPV/sol–gel composite [162–165]. We describe the preparation of PPV composites and their linear and nonlinear optical properties, with special emphasis on the properties of thin film waveguides obtained from them.

A. Sol–Gel and Ormosil Composites of PPV

Sol–gel processing of optical glass materials [166] is a very interesting technique allowing one to obtain glasses containing additives which are impossible to incorporate into conventional glass materials prepared by melt-driven high-temperature routes. The basic steps in preparation of sol–gel silica glass are the hydrolysis of alkoxysilanes, usually tetramethoxysilicate (TMOS) or tetraethoxysilicate (TEOS) with the general formula $Si(OR)_4$ and the condensation of the resulting silanols followed by drying and a low-temperature sintering. The reaction of hydrolysis can be acid catalyzed and temperature dependent with hydrolysis time inversely proportional to the acid concentration [167]. A shorter hydrolysis time is observed with an increase of temperature. The intermediate stage of the sol–gel process involves formation of OH groups of the precursor (sol). Polycondensation reaction results in a wet silica gel. The summary reaction is

$$Si(OR)_4 + 2H_2O = SiO_2 + 4\ ROH$$

The time of the formation of silica gel is a function of the hydrolysis time which, in turn, depends on the concentration of water, an acid, and alcohol. An important point is that the sol obtained from TMOS or TEOS can be combined with various water- or alcohol-soluble chemicals and, if necessary, an additional solvent to dope the resulting silica glass. In such a way, it is possible to prepare an optically clear mixture of two very unlike components: PPV and silica. The mixing of TMOS (or TEOS) with a water/alcohol pristine solution of the PPV precursor leads to the formation of a wet gel doped with the unconverted precursor. After forming a film of such a mixture, the thermal conversion can be carried out, resulting in drying (and further condensation) of the gel and, at the same time, conversion of the PPV into the final conjugated form. The PPV conversion reaction releases HCl, which acts as a catalyst for the sol–gel hydrolysis reaction; the process is therefore quite complicated and not easy to control. However, good-optical-quality PPV/sol–gel composites have been obtained and characterized although cracking of thicker films is often a problem.

Our studies on the preparation and linear and nonlinear optical properties of PPV/sol–gel [168] have confirmed the observation of Prasad's group that there is usually a blue shift of the main absorption maximum of the PPV incorporated into the silica sol–gel matrix compared to the analogous material converted in the pristine form. Figure 34 shows a spectrum of a sol–gel/PPV composite obtained by hydrolysis of TMOS. Depending on the conditions of the reaction and the thermal treatment involved, the material may have varying properties; generally, however, the resulting composites have the main absorption maximum below 400 nm.

Unfortunately, such a blue shift is accompanied by a drastic lowering of the material nonlinearity in the 800-nm region. This property of PPV/sol–gel composites is quite unfavorable. There may be several reasons for such a behavior of the PPV chains in the matrix, two major issues being the possibility of substitution

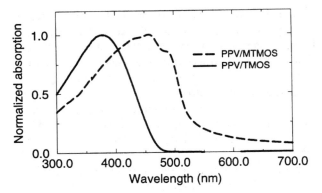

FIGURE 34 Comparison of examples of absorption spectra obtained for PPV composites in a sol–gel matrix and in an ormosil matrix. The precursors used for preparation of the films are TMOS = $(CH_3O)_4Si$ and MTMOS = $(CH_3O)_3SiCH_3$, respectively. The sol–gel film was converted at 200°C in vacuum and the ormosil film was converted at 220°C under the N_2/HCl flow. (After Ref. 168.)

of the tetrahydrothiophene groups by groups more difficult to remove thermally, thus leading to an increased number of π-conjugation defects and the effect of the pore size on the structure of PPV chains contained within the pores [168].

Among the materials which we considered as an alternative to sol–gel silica were PPV/ormosil composites. One usually defines ormosil as a sol–gel analog in which some of the Si–O bonds have been replaced by Si–R bonds, R being an organic group. Ormosils can be prepared in a way analogous to the sol–gel preparation, from precursors of the $RSi(OR')_3$ type. The hydrolysis leads to formation of silicalike glassy structure in which each silicon atom forms three bonds to oxygen and one bond to an organic group.

Mixing an ormosil precursor with an ordinary sol–gel precursor (e.g., TMOS) leads to sol–gel glasses with a different ormosil content.

We studied optical properties of PPV/ormosil composites (Fig. 34) and found that it is possible to obtain composites which, being converted under HCl, have the position of an absorption maximum of PPV very similar to that in pristine PPV. Therefore, one should expect that the nonlinear optical properties of such composites should also be similar to those of pristine PPV. In general, these films were prepared as follows. Alkoxysilanes of the general formula $Si(R)_a(OR)_{4-a}$ ($a = 0$, 1) were reacted with water in the presence of a cosolvent ROH. The catalyst used was aqueous HCl. Films were cast by spin-coating and converted thermally under vacuum or under the N_2/HCl flow. Films thus prepared were clear and of good optical quality, with the absorption maximum controllable in the range 350–450 nm. The factors which permit tuning of the absorption maximum include the reactivity of the sol–gel precursor, the water to silicon ratio, the temperature cycle used in processing of the composite, and the presence or absence of HCl as the catalyst in the conversion process.

Figure 35 shows a comparison of the nonlinear refracive index n_2 determined from DFWM for a series of sol–gel and ormosil composites of 20% content of PPV plotted against the position of the absorption maximum of the sample. For comparison, we show a similar collection of data for pristine PPV. The difference between the nonlinearities of the composites and those of pure PPV can be ascribed to the difference in the density of molecules and the wavelength dependence in both types of samples.

B. PPV/PVP Composites

Waveguiding can be observed in another type of nonlinear composites investigated by us, the guest–host polymer blend of PPV with poly(*N*-vinylpyrrolidone) (PVP):

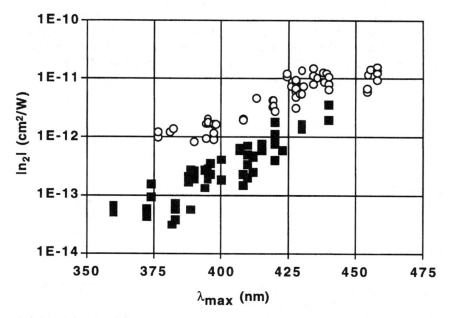

FIGURE 35 Dependence of the $|n_2|$ determined by DFWM for a series of PPV composites obtained by the sol–gel route on the position of the absorption maximum. The black squares are the 20% wt/wt composites and open circles are results for a series of neat PPV films. (From Ref. 168.)

$$\left(-\underset{\underset{\displaystyle N}{|}}{CH}-CH_2-\right)_n$$

Polyvinylpyrrolidone [169] is an inert polymer soluble in many solvents, among them water, alcohol, and, in an exceptionally good solvent for spin-casting of films, *N*-methylpyrrolidone. Waveguiding properties of the films of pure PVP have been found to be quite outstanding. High-quality films of PVP can be obtained by spin-casting of the solutions followed by drying at 150–200°C under vacuum. The waveguide loss factor of PVP is found to be generally below 0.5 dB/cm. Low optical losses and the advantage of being dissolved in a solvent common with the precursor of PPV put PVP in a range of convenient polymer matrices for PPV composites. PVP is transparent up to ~220 nm. Refractive indices of spin-cast PVP films are 1.522 for the TE mode and 1.524 for the TM mode at 633 nm. These values vary slightly with humidity level, especially if the polymer is processed at lower temperatures. PVP indices are not much different from the refractive index of glass, thus several-micrometer-thick-film cast on a glass substrate can be single mode. This can be important if one thinks about the compatibility of the waveguide core dimensions needed in a device based on a silica fiber attached to a planar waveguide structure.

We investigated the linear and nonlinear properties of PPV/PVP films as a function of concentration of PPV [170,171]. The composites were prepared in a two-step procedure under nitrogen atmosphere. First, the solutions were prepared by mixing a required content of PVP powder (40,000 or 360,000 Da) to the pristine solution of the precursor containing usually ~0.5–1% of the precursor polymer. Optically, clear homogeneous solutions were used for the film spin-casting. In the second step, the dried films were thermally converted under vacuum or in the flow of N_2/HCl mixture. Temperature of conversion influenced optical and waveguiding properties of PPV–PVP composites. Our experiments were performed for the films converted at 130, 150, 200, and 270°C.

The spin-coated and solution-coated films of the composites of PVP with the precursor polymer prepared by slow evaporation of the solvent under an inert-gas atmosphere were homogeneous, rather soft. After conversion, the films were optically clear and fully dense without voids. They were yellow-greenish and highly fluorescent if converted under vacuum or in an inert-gas atmosphere. Films thicker than ~10 μm had some orange-skin-type pattern. Films thicker than ~5–7 μm and containing a low concentration of PPV frequently suffered cracking if conversion proceeded under vacuum at temperatures of 150–200°C.

The films of the PPV/PVP composites were studied in a wide range of concentrations with UV-vis spectroscopy and prism coupling. We found that they are strongly anisotropic, proving that the PVP matrix allows for the preferential orientation of PPV polymer chains parallel to the substrate in a similar way as was observed in the neat PPV films. Positions of absorption maxima are found to be labile, depending on the PPV content and temperature of conversion. The films showed a characteristic shift of the position of absorption maximum with increasing PPV concentration. There was a range of the concentrations (~5–40%) in which a well-pronounced red shift of λ_{max}, accompanied by a substantial enhancement in absorption coefficients, has been observed. Similar enhancement was observed in the measurements of the nonlinear refractive index $|n_2|$ with degenerate four-wave mixing at 800 nm in this range of concentrations in PPV/PVP composites.

Figure 36 shows changes of the nonlinear refractive index $|n_2|$ versus PPV concentration in the films of PPV/PVP composites converted at 150°C for (2 + 3) h in vacuum. The precursor solution used here for blending with PVP was synthesized in the conditions described for the precursor No. 11 in Table 1. As can be seen from Fig. 36, the nonlinear properties of PPV in PVP composite cannot be approximated with a linear dependence on concentration. The enhancement of the absorptive, refractive, and nonlinear properties can be interpreted as an evidence for the positive influence of PVP on the conversion process allowing for a larger number of well-conjugated chains to be formed if the conversion proceeds in the matrix of PVP in comparison to neat PPV. The composites converted at other temperatures like 130°C for (2 + 6) h and 200°C for (1 + 2) h under vacuum exhibited different values of the linear and nonlinear optical parameters; however, the trend in the concentration dependence was very much similar.

We found that the nonlinearity of composites increases with increasing temperature of conversion and with the action of HCl. For example, the values of the decadic absorption coefficient and nonlinear refractive index measured in the low concentrated PPV/PVP samples, when extrapolated to 100% PPV, were found to be $\varepsilon = 4.1 \times 10^5$ cm^{-1} and $|n_2| = 3.8 \times 10^{-11}$ cm^2/W, respectively, whereas pure PPV film converted at the same conditions had the values $\varepsilon = 2.1 \times 10^5$ cm^{-1} and

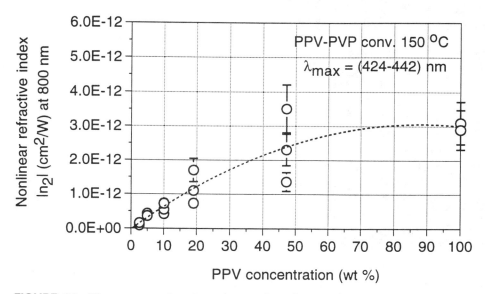

FIGURE 36 The concentration dependence of nonlinear refractive index measured with femtosecond DFWM at 800 nm in the films of PPV/PVP composites (each concentration was represented by two films prepared at the same conditions). (After Ref. 170.)

$|n_2| = 1.7 \times 10^{-11}$ cm²/W. Absorption maxima in these two materials occurred at 456 nm and 454 nm, respectively.

The PPV/PVP composites were found to have waveguiding properties strongly dependent on PPV concentration and conditions of conversion. Figure 37 shows a series of a CCD camera images of waveguiding of the 810-nm diode laser beam in the films of PPV/PVP composites from Fig. 36. The streak of light propagating

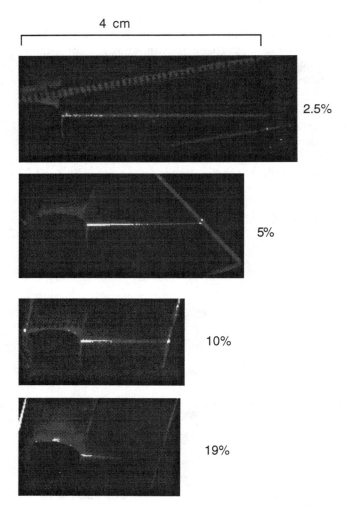

FIGURE 37 The side-view image of the waveguiding in films of PPV/PVP composites of various PPV concentrations captured with a CCD camera. Waveguide losses were determined through the analysis of the decay of intensity of the streak of light scattered on imperfections along the path of the laser beam (810 nm in the above experiment) propagating inside the film. (From Ref. 170.)

within the film containing various concentrations of PPV was visible because of light scattered out of the waveguide.

Due to the intrinsic (material) losses, the intensity of guided light decays in proportion to the propagation distance. The waveguide linear loss factor is described by Lambert–Beer Law (see, for example, Ref. 160) $I = I_0 \times 10^{-\varepsilon L}$, where I_0 is the incident intensity, I is the transmitted intensity through the distance L, and ε is the decadic absorption coefficient expressed in the units of cm^{-1}. It is common to express the loss factor as ε' in decibels (dB/cm^{-1}) ($\varepsilon' = 10$ dB/cm is equivalent to $\varepsilon = 1$ cm^{-1}).

If there is a reduction of the intensity of propagating beam, then a reduction of the intensity of scattered light is observed, too. We determine waveguide losses with the measurements of the decay of intensity of scattered light measured along the distance of the propagation. We used in our determination the logarithmic form of the Lambert–Beer Law: $\log_{10} I = \log_{10} I_0 - (\varepsilon'/10)L$. This method provided us with information on the total intrinsic losses arising both from absorption and scattering.

The total linear losses of TE and TM modes propagating in PPV/PVP composites were found to be highly anisotropic. The TE mode loss data changed linearly with PPV concentration within the detectable range (0–20% PPV) [174]. The TM mode loss data, however, after initial increase showed a saturationlike behavior in the films containing more than 20% PPV. The optical losses for TE were found to be larger than for TM modes. The value for the TE mode extrapolated to 100% PPV was about 180 dB/cm at 810 nm [173,174] in the samples converted at 150°C. The films of the composites converted at 130°C had TE losses of about 90 dB/cm if extrapolated to 100% PPV. Generally, the total losses in PPV/PVP composites are high.

Optical losses measured with a CCD camera technique depend on both the absorption and scattering contributions. The increase of total losses with concentration and their dependence on the temperature of conversion may have its origin in an increase of absorption due to, for example, doping. It may also be caused by light scattering which increases with the formation of imperfect PPV semicrystalline domains embedded in the amorphous matrix of the precursor in PVP, similar to that observed in x-ray studies of conversion of the precursor polymer [63,90,100]. Photodeflection spectroscopy provides an independent source of information on the absorption losses (see, for example, Refs. 170 and 171). Our preliminary experiments show that total absorption losses in the composites are markedly influenced by scattering [172]. Scattering might be induced by defects (e.g., pores formed in the self-shrinking film by the by-products): tetrahydrothiophene and HCl escaping during the elimination reaction. We observed that it is possible to reduce the total losses using lower speed of heating during thermolysis. It is believed that selection of the type of precursor polymer and careful control of the conversion conditions for the precursor in the composites would bring useful waveguiding characteristics and third-order nonlinearity.

VII. CONCLUSIONS

Several polymer systems, poly(*p*-phenylenevinylene) (PPV), poly(2,5-dimethoxy-1,4-phenylenevinylene and other derivatives of PPV, and a series of PPV composites, were investigated. The following issues were addressed: the suitability of the π-conjugated polymers for all-optical switching and their processibility and stability. We conclude that π-conjugated polymers remain attractive candidates as materials for all-optical switching technology because their optical properties are susceptible to tuning through the compositional and structural changes that can be achieved during chemical processing. The material properties required for all-optical switching are defined both in terms of the various figures of merit (T, W) which relate linear and nonlinear properties of material to the switching behavior and in terms of the thermal, mechanical, and optical stability of the materials. Poly(p-phenylenevinylene) is the most promising material which has better third-order nonlinear optical properties than other organic systems that we know.

In the conjugated form, PPV molecules have nearly planar rigid-rod structures showing a strong vibronic absorption in UV-visible spectral range. The molecules become oriented in spin-cast PPV films obtained by the precursor route showing a large anisotropy of refractive indices. We have demonstrated that the chemical and physical environment used to synthesize and process the precursor polymer has a marked effect on the effective conjugation length of the resulting PPV chains, which directly affects the third-order nonlinearity. A remarkable change of the position of absorption maximum, absorption coefficient, and refractive indices upon conversion has been demonstrated. We have used femtosecond degenerate four-wave mixing, Z-scan, optical Kerr ellipsometry, and power-dependent transmission measurements to determine the sign and amplitude of the real and imaginary components of the third-order nonlinearity at 800 nm in PPV films with different conjugation length and in PPV derivatives in the form of films and solutions. We found that PPV with the highest conjugation length has sufficient nonlinearity ($n_2 > 10^{-11}$ cm^2/W) to provide the potential for all-optical switching at picojoule energy levels, while the linear and nonlinear absorption is at the level where the various switching figures of merit can be satisfied.

We demonstrated the suitability of the above-mentioned techniques in cross-correlated studies determining reliable data on nonlinear optical parameters in PPV and its poly(2,5-disubstituted phenylenevinylene) derivatives. Possessing good thermal and mechanical stability and resistance to photodegradation when irradiated in the infrared, PPV appears, therefore, to be a very interesting candidate for all-optical switching. However, some material properties must still be improved before the application can be achieved. We showed that the aging effect in PPV can be stabilized with more efficient processing. The conjugation length achieved in our material can also be improved with a more effective synthetic and conversion approach.

Nonlinear optical properties of PPV can be utilized in a most convenient way in waveguiding structures of composites in which the nonlinear phase change induced by the presence of PPV is built over a relatively long propagation distance. The progress toward waveguiding in PPV was achieved in the polymer system of PPV blended with poly(N-vinylpyrrolidone) (PVP) and in the sol–gel ormosil composites. The waveguiding losses depend on PPV concentration. The nonlinearity of the composites depend on the condition of conversion. The linear and nonlinear properties of PPV/PVP system were investigated as a function of PPV concentration. An enhancement of the linear and nonlinear optical properties, due to better conjugation, is observed in composites PPV/PVP at low concentrations ($\sim$5–40%).

ACKNOWLEDGMENTS

This research was partly supported in its early stages by Harry Triguboff AM Research Syndicate and recently by the Australian Photonics Cooperative Research Centre. We greatly appreciate the support of Dr. S. B. Wild and Professor J. W. White of Research School of Chemistry and the technical help of Mrs. R. M. Krolikowska. We also acknowledge the Technical Staff of the Laser Physics Centre, Research School of Physical Sciences and Engineering at The Australian National University.

REFERENCES

1. Shen, Y.R., *The Principles of Nonlinear Optics*, Wiley, New York, 1984.
2. Butcher, P. N., and Cotter, D., *The Elements of Nonlinear Optics*, Cambridge University Press, New York, 1990.
3. Boyd, R. W., *Nonlinear Optics*, Academic Press, New York, 1992.
4. Sutherland, R. L., *Handbook of Nonlinear Optics*, Marcel Dekker, Inc., New York, 1996.
5. Nye, J. F., *Physical Properties of Crystals*, Oxford University Press, Oxford, 1957.
6. Stegeman, G. I., Nonlinear guided wave optics, in *Contemporary Nonlinear Optics*, Academic Press, New York, 1992, pp. 1–40.
7. Snyder, A. W., Mitchell, D. J., Poladian, L., and Ladouceur, F., Self-induced optical fibers: Spatial solitary waves, *Opt. Lett.*, *16*, 21–23 (1991).
8. Snyder, A. W., Poladian, L., and Mitchell, D. J., Stable black self-guided beams of circular symmetry in a bulk Kerr medium, *Opt. Lett.*, *17*, 789–791 (1992).
9. Snyder, A. W., and Mitchell, D. J., Spatial solitons of the power-law nonlinearity, *Opt. Lett.*, *18*, 101–103 (1993).
10. De la Fuente, R., Barthelemy, A., and Froehly, C., Spatial-soliton-induced guided waves in a homogeneous nonlinear Kerr medium, *Opt. Lett.*, *16*, 793–795 (1991).
11. Allan, G. R., Skinner, S. R., Andersen, D. R., and Smirl, A. L., Observation of fundamental dark spatial solitons in semiconductors using picosecond pulses, *Opt. Lett.*, *16*, 156–158 (1991).
12. Swartzlander, Jr., G. A., Andersen, D. R., Regan, J. J., Yin, H., and Kaplan, A. E., Spatial dark-soliton stripes and grids in self-defocusing materials, *Phys. Rev. Lett*, *66*, 1583–1586 (1991).
13. Luther-Davies, B., and Xiaoping, Y., Waveguides and Y junctions formed in bulk media by using dark spatial solitons, *Opt. Lett.*, *17*, 496–498 (1992).
14. Luther-Davies, B., and Xiaoping, Y., Steerable optical waveguides formed in self-defocusing media by using dark spatial solutions, *Opt. Lett.*, *17*, 1755–1757 (1992).
15. Krolikowski, W., and Kivshar, Y. S., Soliton-based optical switching in waveguide arrays, *J. Opt. Soc. Am. B*, *13*, 876–887 (1996).
16. Samoc, A., Samoc, M., Woodruff, M., and Luther-Davies, B., Photophysical processes involved in creation of dark spatial solitons in composite photonic media, *Molec. Cryst. Liq. Cryst.*, *253*, 133–142 (1994).
17. Samoc, M., Samoc, A., Luther-Davies, B., and Woodruff, M., The concept of guiding light with light and negative third-order optical nonlinearities of organics, *Pure Appl. Opt.*, *5*, 681–687 (1996).
18. Mizrahi, V., DeLong, K. W., Stegeman, G. I., Saifi, M. A., and Andrejco, M. J., Two-photon absorption as a limitation to all-optical switching, *Opt. Lett.*, *14*, 1140–1142 (1989).

19. Kuzyk, M., All-optical materials and devices, in *Organic Thin Films for Waveguiding Nonlinear Optics* (F. Kajzar and J. D. Swalen, eds.), Gordon and Breach, Amsterdam, 1996, pp. 759–820.

20. Heeger, A. J., Moses, D., and Sinclair, M., Semiconducting polymers: Fast response non-linear optical materials, *Synth. Met. 15*, 95–104 (1986).

21. Prasad, P. N., and Williams, D. J., *Introduction to Nonlinear Optical Effects in Molecules and Polymers*, Wiley, New York, 1991.

22. Boyd, G. T., Polymers for nonlinear optics, in *Polymers for Electronic and Photonic Applications* (C. P. Wong, ed.), Academic Press, Boston, 1993, pp. 467–505.

23. Kuzyk, M. G., Polymers as third-order nonlinear-optical materials, in *Polymers for Electronic and Photonic Applications* (C. P. Wong, ed.), Academic Press, Boston, 1993, pp. 507–548.

24. Bredas, J. L., Adant, C., Tackx, P., Persoons, A., and Pierce, B. M., Third-order nonlinear optical response in organic materials: Theoretical and experimental aspects, *Chem. Rev.*, *94*, 243–278 (1994).

25. Lee, K.-S., Samoc, M., and Prasad, P. N., Polymers for photonic applications, in *Comprehensive Polymer Science, First Supplement* (S. L. Aggarwal and S. Russo, eds.), Pergamon Press, Oxford, 1992.

26. Prasad, P. N., Perrin, E., and Samoc, M., A coupled anharmonic oscillator model for optical nonlinearities of conjugated organic structures, *J. Chem. Phys.*, *91*, 2360–2365 (1989).

27. Kavanaugh, T. C., and Silbey, R. J., A simple model for polarizabilities of organic polymers, *J. Chem. Phys.*, *95*, 6924–6931 (1991).

28. Chemla, D. S., and Zyss, J. (eds.), *Nonlinear Optical Properties of Organic Molecules and Crystals*, Academic Press, New York, 1987, Vols. I and II.

29. Hagler, T. W., and Heeger, A. J., The role of soliton–antisoliton configurations with Ag symmetry in the nonlinear optical response of polyacetylene, *Chem. Phys. Lett*, *189*, 333–339 (1992).

30. Bubeck, C., Relations between structure and third-order nonlinearities of molecules and polymers, in *Organic Thin Films for Waveguiding Nonlinear Optics* (F. Kajzar and J. D. Swalen, ed.), Gordon and Breach, Amsterdam, 1996, pp. 137–161.

31. Mathy, A., Ueberhofen, K., Schenk, R., Gregorius, H., Garay, R., Muellen, K., and Bubeck, C., Third-harmonic generation spectroscopy of poly(p-phenylenevinylene): A comparison with oligomers and scaling laws for conjugated polymers, *Phys. Rev. B*, *53*, 4367–4376 (1996).

32. Baker, G. L., Progress towards processable, environmentally stable conducting polymers, in *Electronic and Photonic Applications of Polymers* (M. J. Bowden and S. R. Turner, eds.), Advances in Chemistry Series 218, American Chemical Society, Washington, DC, 1988.

33. Kobryanskii, V. M., Preparation and optical properties of trans-polyacetylene with low content of defects, *Synth. Met.*, *69*, 41–42 (1995).

34. Burroughes, J. H., Bradley, D. D. C., Brown, A. R., Marks, R. N., MacKay, K., Friend, R. H., Burn, P. L., and Holmes, A. B., Light-emitting diodes based on conjugated polymers, *Nature*, *347*, 539–541 (1990).

35. Braun, D., and Heeger, A. J., Visible light emission from semiconducting polymer diodes, *Appl. Phys. Lett.*, *58*, 1982–1984 (1991).

36. Gustaffson, G., Cao, Y., Treacy, G. M., Klavetter, R., Colaneri, N., and Heeger, A. J., Flexible light-emitting diodes made from soluble conducting polymers, *Nature*, *357*, 477–479 (1992).

37. Bradley, D. D. C., Electroluminescence: A bright future for conjugated polymers? *Adv. Mater.*, *4*, 756–758 (1992).

38. Kanbe, M., and Okawara, M., Synthesis of poly-*p*-xylylidene from *p*-xylylene-bis(dimethylsolfonium) tetrafluoroborate, *J. Polym. Sci., A-1, 6,* 1058–1060 (1968).

39. Wessling, R. A., and Zimmerman, R. G., Polyelectrolytes from bis sulfonium salts, U.S. Patent 3,401,152 (1968); *Chem. Abstr. 69,* 87735q (1968).

40. Wessling, R. A., and Zimmerman, R. G., Polyxylylidene articles, U.S. Patent 3,706,677 (1972); *Chem. Abstr. 78,* 85306n (1973).

41. Wessling, R. A., The polymerization of xylylene bisdialkylsulfonium salts, *J. Polym. Sci., Polym. Symp., 72,* 55–66 (1985).

42. Hoerhold, H.-H., and Opfermann, J., Poly-*p*-xyliden. Synthesen und Beziehungen zwischen Struktur und electrophysikalischen Eigenschaften, *Makromol. Chem., 131,* 105–132 (1970).

43. Murase, I., Ohnishi, T., Noguchi, T., and Hirooka, M., Highly conducting poly(*p*-phenylene vinylene) prepared from a sulphonium salt, *Polym. Commun., 25,* 327–329 (1984).

44. Murase, I., Ohnishi, T., Noguchi, T., Hirooka, M., and Murakami, S., Highly conducting poly(*p*-phenylene vinylene) prepared from a sulphonium salt, *Molec. Cryst. Liq. Cryst., 118,* 333–336 (1985).

45. Capistran, J. D., Gagnon, D. R., Antoun, S., Lenz, R. W., and Karasz, F. E., Synthesis and electrical conductivity of high molecular weight poly(arylene vinylenes), *ACS Polym. Prepr., 25,* 282–283 (1984).

46. Gagnon, D. R., Capistran, J. D., Karasz, F. E., and Lenz, R. W., Conductivity anisotropy of doped poly(*p*-phenylene vinylene) films, *ACS Polym. Prepr., 25,* 284–285 (1984).

47. Karasz, F. E., Capistran, J. D., Gagnon, D. R., and Lenz, R. W., High molecular weight polyphenylene vinylene, *Molec. Cryst. Liq. Cryst., 118,* 327–332.

48. Gagnon, D. R., Capistran, J. D., Karasz, F. E., Lenz, R. W., and Antoun, S., Synthesis, doping, and electrical conductivity of high molecular weight poly(*p*-phenylene vinylene), *Polymer, 28,* 567–573 (1987).

49. Lenz, R. W., Han, C. C., Stenger-Smith, J., and Karasz, F. E., Preparation of poly(phenylene vinylene) from cycloalkylene sulfonium salt monomers and polymers, *J. Polym. Sci., A 26,* 3241–3249 (1988).

50. Lahti, P. M., Modarelli, D. A., Denton III, F. R., Lenz, R. W., and Karasz, F. E., Polymerization of α,α'-bis(dialkylsulfonio)-*p*-xylene dihalides via *p*-xylylene intermediates: Evidence for a nonradical mechanism, *J. Am. Chem. Soc., 110,* 7258–7259 (1988).

51. Garay, R., and Lenz, R. W., Anionic polymerization of *p*-xylenesulfonium salts, *Macromol. Chem., 15* (Suppl.): 1–7 (1989).

52. Garay, R. O., and Lenz, R. W., Effect of the reaction state on the polymerization of *p*-xylenesulfonium salts, *J. Polym. Sci., Part A: Polym. Chem., 30,* 977–982 (1992).

53. Liang, W., Rice, D. M., Karasz, F. E., Denton III, F. R., and Lahti, P. M., Preparation of poly(*p*-phenylene vinylene) deuterium labelled in the vinylene position, *Polymer, 33,* 1780–1782 (1992).

54. Denton III, F. R., Lahti, P. M., and Karasz, F. E., The effect of radical trapping reagents upon formation of poly(α-tetrahydrothiophenio *para*-xylene) polyelectrolytes by the Wessling soluble precursor method, *J. Polym. Sci., Part A: Polym. Chem., 30,* 2223–2231 (1992).

55. Machado, J. M., Denton III, F. R., Schlenoff, J. B., Karasz, F. E., and Lahti, P. M., Analytical methods for molecular weight determination of poly(*p*-xylylidene dialkyl sulfonium halide): Degree of polymerization of poly(*p*-phenylene vinylene) precursors, *J. Polym. Sci., Part B: Polym. Phys., 27,* 199–205 (1989).

56. Montuado, G., Vitalini, D., and Lenz, R. W., Mechanism of thermal generation of poly(p-phenylene vinylene) from poly(p-xylene-α-dimethylsulphonium halides), *Polymer*, *28*, 837–842 (1987).

57. Han, C. C., Lenz, R. W., and Karasz, F. E., Highly conducting, iodine-doped copoly(phenylene vinylene)s, *Polym. Commun.*, *28*, 261–262 (1987).

58. Antoun, S., Karasz, F. E., and Lenz, R. W., Synthesis and electrical conductivity of poly(arylene vinylene). I. Poly(2,5-dimethoxyphenylene vinylene) and poly(2,5-dimethylphenylene vinylene), *J. Polym. Sci., Part A: Polym. Chem.*, *26*, 1809–1817 (1988).

59. Lenz, R. W., Han, C. C., and Lux, M., Highly conducting, iodine-doped arylene vinylene copolymers with dialkoxyphenylene units, *Polymer*, *30*, 1041–1047 (1989).

60. Denton III, F. R., Sarker, A., Lahti, P. M., Garay, R. O., and Karasz, F. E., *Para*-xylenes and analogues by base-induced elimination from 1,4-bis-(dialkylsulfoniomethyl)arene salts in poly(1,4-arylene vinylene) synthesis by the Wessling soluble precursor method, *J. Polym. Sci., Part A: Polym. Chem.*, *30*, 2233–2240 (1992).

61. Gregorius, R. M., Lahti, P. M., and Karasz, F. E., Preparation and characterization of poly(arylenevinylene) copolymers and their blends, *Macromolecules*, *25*, 6664–6669 (1992).

62. Tokito, S., Momii, T., Murata, H., Tsutsui, T., and Saito, S., Polyarylenevinylene films prepared from precursor polymers soluble in organic solvents, *Polymer*, *31*, 1137–1141 (1990).

63. Bradley, D. D. C., Precursor-route poly(p-phenylenevinylene): Polymer characterisation and control of electronic properties, *J. Phys. D, Appl. Phys.*, *20*, 1389–1410 (1987).

64. Halliday, D. A., Burn, P. L., Friend, R. H., and Holmes, A. B., A study on the elimination reaction of sulfonium polyelectrolyte precursor polymers to poly(p-phenylenevinylene), *J. Chem. Soc., Chem. Commun.*, 1685–1687 (1992).

65. Burn, P. L., Bradley, D. D. C., Brown, A. R., Friend, R. H., and Holmes, A. B., Studies on the efficient synthesis of poly(phenylenevinylene) (PPV) and poly(dimethoxy phenylenevinylene) (dimethoxy-PPV), *Synth. Met.*, *41–43*, 261–264 (1991).

66. Burn, P. L., Bradley, D. D. C., Friend, R. H., Halliday, D. A., Holmes, A. B., Jackson, R. W., and Kraft, A., Precursor route chemistry and electronic properties of poly(p-phenylenevinylene), poly[(2,5-dimethyl-p-phenylene)vinylene] and poly[(2,5-dimethoxy-p-phenylene)vinylene], *J. Chem. Soc., Perkin Trans. 1*, 3225–3231 (1992).

67. Burn, P. L., Kraft, A., Baigent, D. R., Bradley, D. D. C., Brown, A. R., Friend, R. H., Gymer, R. W., Holmes, A. B., and Jackson, R. W., Chemical tuning of the electronic properties of poly(p-phenylenevinylene)-based copolymers, *J. Am. Chem. Soc.*, *115*, 10117–10124 (1993).

68. Burn, P. L., Holmes, A. B., Kraft, A., Bradley, D. D. C., Brown, A. R., Friend, R. H., and Gymer, R. W., Chemical tuning of electroluminescent copolymers to improve emission efficiencies and allow patterning, *Nature*, 356, 47–49 (1992).

69. Burn, P. L., Holmes, A. B., Kraft, A., Bradley, D. D. C., Brown, A. R., and Friend, R. H., Synthesis of a segmented conjugated polymer chain giving a blue-shifted electroluminescence and improved efficiency, *J. Chem. Soc. Chem. Commun.*, 32–34 (1992).

70. Burn, P. L., Bradley, D. D. C., Brown, A. R., Friend, R. H., Halliday, D. A., Holmes, A. B., Kraft, A., and Martens, J. H. F., Control of electronic and physical structure through modification of the synthesis and processing in precursor-route poly(arylenevinylene) polymers, in *Electronic Properties of Polymers* (H. Kuzmany,

M. Mehring, and S. Roth, eds.), Springer Series in Solid-State Sciences No. 107, Springer-Verlag, Berlin, 1992, pp. 293–297.

71. Halliday, D. A., Burn, P. L., Bradley, D. D. C., Friend, R. H., Gelsen, O. M., Holmes, A. B., Kraft, A., Martens, J. H. F., and Pichler, K., Large changes in optical response through chemical pre-ordering of poly(*p*-phenylenevinylene), *Adv. Mater.*, *5*, 40–43 (1993).

72. Halliday, D. A., Burn, P. L., Friend, R. H., Bradley, D. D. C., Holmes, A. B., and Kraft, A., Extended π-conjugation in poly(*p*-phenylenevinylene) from a chemically modified precursor polymer, *Synth. Met.*, *55–57*, 954–959 (1993).

73. Kraft, A., Burn, P. L., Holmes, A. B., Bradley, D. D. C., Brown, A. R., Friend, R. H., and Gymer, R. W., Chemical control of colour and electroluminescent device efficiency in copolymeric poly(arylenevinylenes), *Synth. Met.*, *55–57*, 936–941 (1993).

74. Cherry, M. J., Moratti, S. C., Holmes, A. B., Taylor, P. L., Gruener, J., and Friend, R. H., The dispersion polymerisation of poly(*p*-phenylenevinylene), *Synth. Met.*, *69*, 493–494 (1995).

75. McDonald, R. N., and Campbell, T. W., The Wittig reaction as a polymerization method, *J. Am. Chem. Soc.*, *82*, 4664–4671 (1960).

76. Wnek, G. E., Chien, J. C. W., Karasz, F. E., and Lillya, C. P., Electrically conducting derivative of poly(*p*-phenylene vinylene), *Polymer*, *20*, 1441–1443 (1979).

77. Gourley, K., Lillya, C. P., Reynolds, J. R., and Chien, J. C. W., Electrically conducting polymers: AsF$_5$-doped poly(phenylenevinylene) and its analogues, *Macromolecules*, *17*, 1025–1033 (1984).

78. Hoerhold, H. H., Helbig, M., Raabe, D., Opfermann, J., Scherf, U., Stockmann, R., and Weis, D., Poly(phenylenevinylen); Entwicklung eines electroactiven Polymermaterials vom unschmelzbaren Pulver zum transparenten Film, *Z. Chem.*, *27*, 126–137 (1987).

79. Meier, H., The photochemistry of stilbenoid compounds and their role in material technology, *Angew. Chem. Int. Ed. Engl.*, *31*, 1399–1420 (1992).

80. Bao, Z., Chen, Y., Cai, R., and Yu, L., Conjugated liquid crystalline polymers-soluble and fusible poly(phenylenevinylene) by the Heck Coupling reaction, *Macromolecules*, *26*, 5281–5286 (1993).

81. Bao, Z., Chen, Y., and Yu, L., New metalloporphyrin containing polymers from the Heck coupling reaction, *Macromolecules*, *27*, 4629–4631 (1994).

82. Askari, S. H., Rughooputh, S. D., and Wudl, F., Soluble substituted-PPV conducting polymers: Spectroscopic studies, *Synth. Met*, *29*, E129–E134 (1989).

83. Wudl, F, Allemand, P.-M., Srdanov, G., Ni, Z., and McBranch, D., Polymers and an unusual molecular crystal with nonlinear optical properties, in *Materials for Nonlinear Optics: Chemical Perspectives* (S. R. Marder, J. E. Sohn, and D. Galen, eds.), ACS Symposium Series 455, American Chemical Society, Washington, DC, 1991, pp. 683–686.

84. Yang, Z., Geise, H. J., Nouwen, J., Adriaensens, P., Franco, D., Vandrzande, D., Martens, H., Gelan, J., and Mehbod, M., Modification of poly(*para*-phenylene vinylene) by introduction of aromatic groups on the olefinic carbons, *Synth. Met.*, *47*, 111–132 (1992).

85. Shim, H. K., Hwang, D. H., and Lee, J. I., Synthesis, electrical and optical properties of asymmetrically monoalkoxy-substituted PPV derivatives, *Synth. Met.*, *55–57*, 908–913 (1993).

86. Cho, B. R., Han, M. S., Suh, Y. S., Oh, K. J., and Jeon, S. J., Mechanism of polymerization of α,α'-bis(tetrahydrothiophenio)-*p*-xylene dichloride, *J. Chem. Soc., Chem. Commun.*, 564–566 (1993).

87. Zhang, X. B., Van Tendeloo, G., Van Landuyt, J., Van Dijck, D., Bries, J., Bao, Y., and Geise, H. J., An electron microscopic study of highly oriented undoped and FeCl₃-doped poly(p-phenylenevinylene), *Macromolecules*, 29, 1554–1561 (1996).

88. Schlenoff, B. J., and Wang, L. J., Elimination of ion-exchanged precursors to poly(phenylenevinylene), *Macromolecules*, 24, 6653–6659 (1991).

89. Beerden, A., Vanderzande, D., and Gelan, J., The effect of anions on the solution behaviour of poly(xylylene tetrathiophenium chloride) and on the elimination to poly(p-phenylene vinylene), *Synth. Met.*, 52, 387–394 (1992).

90. Ezquerra, T. A., Lopez-Cabarcos, E., Balta-Calleja, F. J., Stenger-Smith, J. D., and Lenz, R. W., Real-time X-ray scattering study during the thermal conversion of a precursor polymer to poly(p-phenylene vinylene), *Polymer*, 32, 781–785 (1991).

91. Garay, R. O., Baier, U., Bubeck, C., and Muellen, K., Low-temperature synthesis of poly(p-phenylene vinylene) by the sulfonium salt route, *Adv. Mater.*, 5, 561–564 (1993).

92. Massardier, V., Guyot, A., and Tran, V. H., Direct conversion of sulfonium precursors into poly(p-phenylenevinylene) by acids, *Polymer*, 35, 1561–1563 (1994).

93. Bullot, J., Bulieu, B., and Lefrant, S., Photochemical conversion of polyphenylene-vinylene, *Synth. Met.*, 61, 211–215 (1993).

94. Schmid, W., Dankesreiter, R., Gmeiner, J., Vogtmann, Th., and Schwoerer, M., Photolithography with poly-(p-phenylene vinylene) (PPV) prepared by the precursor route, *Acta Polymer.*, 44, 208–210 (1993).

95. Tokito, S., Tsutsui, T., Saito, S., and Tanak, R., Optical and electrical properties of pristine poly(p-phenylenevinylene) film, *Polym. Commun.*, 27, 333–335 (1986).

96. Patil, A. O., Rughooputh, S. D. D. V., and Wudl, F., Poly(p-phenylene vinylene): incipient doping in conducting polymers, *Synth. Metals*, 29, E115–E119 (1989).

97. Papadimitrakopoulos, F., Yan, M., Rothberg, L. J., Katz, H. E., Chandross, E. A., and Galvin, M. E., Thermal and photochemical origin of carbonyl group defects in poly(p-phenylenevinylene), *Molec. Cryst. Liq. Cryst.*, 256, 663–669 (1994).

98. Hsieh, B. R., Ettedgui, E., and Gao, Y., The surface species of poly(p-phenylene vinylene) and their effects on metal interface formation, *Synth. Met.*, 78, 269–275 (1996).

99. Granier, T., Thomas, E. L., Gagnon, D. R., Karasz, F. E., and Lenz, R. W., Structure investigation of poly(p-phenylene vinylene), *J. Polym. Sci., Part B, Polym. Phys.*, 24, 2793–2804 (1986).

100. Moon, Y. B., Rughooputh, S. D. D. V., Heeger, A. J., Patil, A. O., and Wudl, F., X-ray scattering study of the conversion of poly(p-phenylene vinylene precursor to the conjugated polymer, *Synth. Met.*, 29, E79–E84 (1989).

101. Masse, M. A., Martin, D. C., Thomas, E. L., Karasz, F. E., and Petermann, J. H., Crystal morphology in pristine and doped films of poly(p-phenylene vinylene), *J. Mater. Sci.*, 25, 311–320 (1990).

102. Martens, J. H. F., Bradley, D. D. C., Burn, P. L., Friend, R. H., Holmes, A. B., and Marseglia, E. A., Control of order in poly(arylene vinylene) conjugated polymers, *Synth. Met.*, 41–43, 301–304 (1991).

103. Briers, J., Eevers, W., Cos, P., Geise, H. J., Mertens, R., Nagels, P., Zhang, X. B., Van Tendeloo, G., Herrebout, W., and Van der Veken, B., Molecular orientation and conductivity in highly oriented poly(p-phenylene vinylene), *Polymer*, 35, 4569–4572 (1994).

104. Obrzut, J., and Karasz, F. E., Ultraviolet and visible spectroscopy of poly(paraphenylene vinylene), *J. Chem. Phys.*, 87, 2349–2358 (1987).

105. Bradley, D. D. C., Evans, G. P., and Friend, R. H., Characterisation of poly(phenylenevinylene) by infrared and optical absorption, *Synth. Met.*, 17, 651–656 (1987).

106. Tian, B., Zerbi, G., Schenk, R., and Muellen, K., Optical spectra and structure of oligomeric models of polyparaphenylenevinylene, *J. Chem. Phys.*, *95*, 3191–3197 (1991).

107. Tian, B., Zerbi, G., and Muellen, K., Electronic and structural properties of polyparaphenylenevinylene from the vibrational spectra, *J. Chem. Phys.*, *95*, 3198–3207 (1991).

108. Sakamoto, A., Furukawa, Y., and Tasumi, M., Infrared and Raman studies of poly(*p*-phenylenevinylene) and its model compounds, *J. Phys. Chem.*, *96*, 1490–1494 (1992).

109. Bradley, D. D. C., Friend, R. H., Lindenberger, H., and Roth, S., Infra-red characterization of oriented poly(phenylene vinylene), *Polymer*, *27*, 1709–1713 (1987).

110. Stenger-Smith, J. D., Lenz, R. W., and Wegner, G., Spectroscopic and cyclic voltammetric studies of poly(*p*-phenylene vinylene) prepared from two different sulphonium salt precursor polymers, *Polymer*, *30*, 1048–1053 (1989).

111. Pichler, K., Halliday, D. A., Bradley, D. D. C., Friend, R. H., Burn, P. L., and Holmes, A. B., Photoinduced absorption of structurally improved poly(*p*-phenylene vinylene)—No evidence for bipolarons, *Synth. Met.*, *55–57*, 230–234 (1993).

112. Hagler, T. W., Pakbaz, K., Voss, K. F., and Heeger, A. J., Enhanced order and electronic delocalization in conjugated polymers oriented by gel processing in polyethylene, *Phys. Rev. B*, *44*, 8652–8666 (1994).

113. Samoc, A., Samoc, M., Woodruff M., and Luther-Davies, B., Tuning the properties of poly(*p*-phenylenevinylene) for use in all-optical switching, *Opt. Lett. 20*, 1241–1243 (1995).

114. Burzynski, R., Prasad, P. N., and Karasz, F. E., Large optical birefringence in poly(*p*-phenylene vinylene) films measured by optical waveguide techniques, *Polymer*, *31*, 627–630 (1990).

115. Gymer, R. W., Friend, R. H., Ahmed, H., Burn, P. L., Kraft, A. M., and Holmes, A. B., The fabrication and assessment of optical waveguides in poly(*p*-phenylenevinylene/poly(2,5-dimethoxy-p-phenylenevinylene) copolymer, *Synth. Met.*, *55*, 3683–3688 (1993).

116. Tien, P. K., Ulrich, R., and Martin, R. J., Modes of propagating light waves in thin deposited semiconductor films, *Appl. Phys. Lett.*, *14*, 291–294 (1969).

117. Ulrich, R., and Torge, R., Measurement of thin film parameters with a prism coupler, *Appl. Opt.*, *12*, 2901–2908 (1973).

118. McBranch, D., Sinclair, M. Heeger, A. J., Patil, A. O., Shi, S., Askari, S., and Wudl, F., Linear and nonlinear optical studies of poly(*p*-phenylene vinylene) derivatives and polydiacetylene–4BCMU, *Synth. Met.*, *29*, E85–E90 (1989).

119. Swiatkiewicz, J., Prasad, P. N., and Karasz, F. E., Anisotropy in the complex refractive index and the third-order nonlinear optical susceptibility of a stretch-oriented film of poly(*p*-phenylene vinylene), *J. Appl. Phys.*, *74*, 525–530 (1993).

120. Bradley, D. D. C., and Mori, Y., Third harmonic generation in precursor route poly(*p*-phenylene vinylene), *Jpn. J. Appl. Phys.*, *28*, 174–177 (1989).

121. Kaino, T., Kubodera, K. I., Tomaru, S., Kurihara, T., Saito, S., Tsutsui T., and Tokito, S., Optical third-harmonic generation from poly(*p*-phenylenevinylene) thin films, *Electron. Lett.*, *23*, 1095–1097 (1987).

122. Kurihara, T., Mori, Y., Kaino, T., Murata, H., Takada, N., Tsutsui, T., and Saito, S., Spectra of $\chi^{(3)}(-3\omega;\omega,\omega,\omega)$ in poly(2,5-dimethoxy *p*-phenylene vinylene) (MO–PPV) for various conversion levels, *Chem. Phys. Lett.*, *183*, 534–538 (1991).

123. Murata, H., Takada, N., Tsutsui, T., Saito, S., Kurihara, T., and Kaino, T., Frequency dependence of third-order nonlinear susceptibilities in polyarylenevinylene thin films, *Synth. Met.*, *49*, 131–139 (1992).

124. Singh, B. P., Prasad, P. N., and Karasz, F. E., Third-order non-linear optical properties of oriented films of poly(*p*-phenylene vinylene) investigated by femtosecond degenerate four wave mixing, *Polymer*, 29, 1940–1942 (1988).

125. Bubeck, C., Kaltbeitzel, A., Grund, A., and LeClerc, M., Resonant degenerate four wave mixing and scaling laws for saturable absorption in thin films of conjugated polymers and Rhodamine 6G, *Chem. Phys.*, 154, 343–348 (1991).

126. Bubeck, C., Kaltbeitzel, A., Lenz, R. W., Neher, D., Stenger-Smith, J. D., and Wegner, G., Nonlinear optical properties of poly(*p*-phenylene vinylene) thin films, in *Nonlinear Optical Effects in Organic Polymers* (J. Messier, F. Kajzar, P. Prasad, and D. Ulrich, eds.), NATO ASI Series E: Appl. Sci., Vol. 162, Kluwer Academic Publishers, Dordrecht, 1989, pp. 143–147.

127. Luther-Davies, B., Samoc, M., Samoc, A., and Woodruff, M., Third-order nonlinearity of poly(*p*-phenylenevinylene) at 800 nm, *Nonlinear Opt.*, 14, 161–167 (1995).

128. Pang, Y., Samoc, M., and Prasad, P. N., Third-order nonlinearity and two-photon-induced molecular dynamics: Femtosecond time-resolved transient absorption, Kerr gate and degenerate four-wave mixing studies in poly(*p*-phenylene vinylene)/sol–gel silica film, *J. Chem. Phys.*, 94, 5282–5290 (1991).

129. Eichler, H. J., Guenther, P., and Pohl, D. W., *Laser-Induced Dynamic Gratings*. Springer Series in Optical Science Vol. 50, Springer-Verlag, Berlin, 1986.

130. Kepler, R. G., Valencia, V. S., Jacobs, S. J., and McNamara, J. J., Exciton–exciton annihilation in poly(*p*-phenylenevinylene) films, *Synth. Met.*, 78, 227–230 (1996).

131. Pakbaz, K., Lee, C. H., Heeger, A. J., Hagler, T. W., and McBranch, D., Nature of the primary photoexcitations in poly(arylene-vinylenes), *Synth. Met.*, 64, 295–306 (1994).

132. Lee, C. H., Yu, G., Moses, D., and Heeger, A. J., Picosecond transient photoconductivity in poly(*p*-phenylenevinylene), *Phys. Rev. B*, 49, 2396–2407 (1994).

133. Lee, C. H., Yu, G., and Heeger, A. J., Persistent photoconductivity in poly(*p*-phenylenevinylene): Spectral response and slow relaxation, *Phys. Rev. B*, 47, 15543–15553 (1993).

134. Rauscher, U., Baessler, H., Bradley, D. D. C., and Hennecke, M., Exciton versus band description of the absorption and luminescence spectra in poly(*p*-phenylenevinylene), *Phys. Rev. B*, 42, 9830–9836 (1990).

135. Kersting, R., Lemmer, U., Deussen, M., Bakker, H. J., Mahrt, R. F., Kurz, H., Arkhipov, V. I., Baessler, H., and Goebel, E. O., Ultrafast field-induced dissociation of excitons in conjugated polymers, *Phys. Rev. Lett.*, 73, 1440–1443 (1994).

136. Rothberg, L. J., Yan, M., Son, S., Galvin, M. E., Kwock, E. W., Miller, T. M., Katz, H. E., Haddon, R. C., and Papadimitrakopoulos, F., Intrinsic and extrinsic constraints on phenylenevinylene polymer electroluminescence, *Synth. Met.*, 78, 231–236 (1996).

137. Rothberg, L. J., Yan, M., Papadimitrakopoulos, F., Galvin, M. E., Kwock, E. W., and Miller, T. M., Photophysics of phenylenevinylene polymers, *Synth. Met.*, 80, 41–58 (1996).

138. Conwell, E. M., and Mizes, H. A., Does photogeneration produce bipolarons in poly(para-phenylene vinylene)? *Synth. Met.*, 78, 201–207 (1966).

139. Yan, M., Rothberg, L. J., Kwock, E. W., and Miller, T. M., Interchain excitations in conjugated polymers, *Phys. Rev. Letters*, 75, 1992–1995 (1995).

140. Leng, J. M., Jeglinski, S., Wei, X., Benner, R. E., Vardeny, Z. V., Gou, F., and Mazumdar, S., Optical probes of excited states in poly(*p*-phenylenevinylene), *Phys. Rev. Lett.*, 72, 156–159 (1994).

141. Hsu, J. W. P., Yan, M., Jedju, T. M., Rothberg, L. J., and Hsieh, B. R., Assignment of the picosecond photoinduced absorption in phenylene vinylene polymers, *Phys. Rev. B*, 49, 712–715 (1994).

142. Esteghamatian, M., Popovic, Z. D., and Xu, G., Carrier generation process in poly(*p*-phenylene vinylene) by fluorescent quenching and delayed-collection-field techniques, *J. Phys. Chem.*, *100*, 13716–13719 (1996).

143. Samoc, A., Samoc, M., Woodruff, M., and Luther-Davies, B., The dependence of the third order nonlinear optical properties on the linear optical properties of poly(*p*-phenylene vinylene)–PPV, *ACS Polym. Mater. Sci. Eng.*, *72*, 224–225 (1995).

144. Luther-Davies, B., Samoc, M., Swiatkiewicz, J., Samoc, A., Woodruff, M., Trebino, R., and Delong, K. W., Diagnostics of femtosecond laser pulses using films of poly(*p*-phenylenevinylene), *Opt. Commun.* *131*, 301–306 (1996).

145. Lemmer, U., Fischer, R., Feldmann, J., Mahrt, R. F., Yang, J., Greiner, A., Baessler, H., Goebel, E. O., Heesel, H., and Kurz, H., Time-resolved studies of two-photon absorption processes in poly(*p*-phenylenevinylene)s, *Chem. Phys. Lett.*, *203*, 28–32 (1993).

146. Yan, M., Rothberg, L., Hsieh, B. R., and Alfano, R. R., Exciton formation and decay dynamics in electroluminescent polymers observed by subpicosecond stimulated emission, *Phys. Rev. B*, *49*, 9419–9422 (1994).

147. Yan, M., Rothberg, L. J., Papadimitrakopoulos, F., Galvin, M. E., and Miller, T. M., Spatially indirect excitons as primary photoexcitations in conjugated polymers, *Phys. Rev. Lett.;* *72*, 1104–1107 (1994).

148. Moses, D., High quantum efficiency luminescence from a conducting polymer in solution: A novel polymer laser dye, *Appl. Phys. Lett.*, *60*, 3215–3216 (1992).

149. Luther-Davis, B., Samoc, M., Samoc, A., Woodruff, M., Fotheringham, D., White, J., and Jin, C-Q., Ultrafast measurements of third order nonlinear optical properties of organic materials, in *International Quantum Electronics Conference, 14–19 July 1996, Sydney, Australia*. Optical Society of America, Washington, DC, 1996, p. 17/133.

150. Swiatkiewicz, J., Prasad, P. N., Karasz, F. E., Druy, M. A., and Glatkowski, P., Anisotropy of the linear and third-order nonlinear optical properties of a stretch-oriented polymer film of poly[2,5-dimethoxy paraphenylenevinylene], *Appl. Phys. Lett.*, *56*, 892–894 (1990).

151. Orczyk, M. E., Samoc, M., Swiatkiewicz, J., Manickam, N., Tomoaia-Cotisel, M., and Prasad, P. N., Optical heterodyning of the phase-tuned femtosecond optical Kerr gate signal for the determination of complex third-order susceptibilities, *Appl. Phys. Lett.*, *60*, 2837–2839 (1992).

152. Orczyk, M. E., Samoc, M., Swiatkiewicz, J., and Prasad, P. N., Dynamics of third-order nonlinearity of canthaxanthin carotenoid by the optically heterodyned phase-tuned femtosecond optical Kerr gate, *J. Chem. Phys.*, *98*, 2524–2533 (1993).

153. Pfeffer, N., Charra, F., and Nunzi, J. M., Phase and frequency resolution of picosecond optical Kerr nonlinearities, *Opt. Lett.*, *16*, 1987–1989 (1991).

154. Philippart, V., Dumont, M., Nunzi, J. M., and Charra, F., Femtosecond Kerr ellipsometry in polydiacetylene solutions: Two photon effects, *Appl. Phys. A56*, 29–34 (1993).

155. Sheik-bahae, M., Said, A. A., and Van Stryland, E. W., High-sensitivity, single-beam n_2 measurements, *Opt. Lett.*, *14*, 955–957 (1989).

156. Sheikh-bahae, M., Said, A. A., Wei, T., Hagan, D. J., and van Stryland, E. W., Sensitive measurement of optical nonlinearities using a single beam, *IEEE J. Quantum Electron.* *26*, 760–769 (1990).

157. Samoc, M., Samoc, A., Luther-Davies, B., Bao, Z., Yu, L., Hsieh, B., and Scherf, U., Femtosecond Z-scan and degenerate four-wave mixing measurements of real and imaginary parts of the third-order nonlinearity of soluble conjugated polymers, *J. Opt. Soc. Am. B*, *15*, 817–825 (1998).

158. Michelotti, F., Gabler, T., Hoerhold, H., Waldhausl, R., and Brauer, A., Prism coupling in DMOP–PPV optical waveguides, *Opt. Commun.*, *114*, 247–254 (1995).

159. Snyder, A. W., and Love, J. D., *Optical Waveguide Theory*, Chapman & Hall, London, 1983.

160. Yardley, J. T., Design and characterization of organic waveguides for passive and active optical devices, in *Organic Thin Films for Waveguiding Nonlinear Optics* (F. Kajzar and J. D. Swalen, eds.), Gordon and Breach, Amsterdam, 1996, pp. 607–688.

161. Karasz, F. E., Williams, G., and Attard, G. S., Functional polymers and guest–hosts polymer blends for optical and electronic applications, *SPIE*, *878*, 123–127 (1988).

162. Wung, C. J., Pang, Y., Prasad, P. N., and Karasz, F. E., Poly(*p*-phenylene vinylene)–silica composite: A novel sol–gel processed non-linear optical material for optical waveguides, *Polymer*, *32*, 605–608 (1991).

163. Wung, C. J., Wijekoon, W. M. K. P., and Prasad, P. N., Characterization of sol–gel processed poly(*p*-phenylenevinylene) silica and V_2O_5 composites using waveguide Raman, Raman and FTi.r. spectroscopy, *Polymer*, *34*, 1174–1178 (1993).

164. Embs, F. W., Thomas, E. L., Wung, C. J., and Prasad, P. N., Structure and morphology of sol–gel prepared polymer-ceramic composite thin films, *Polymer*, *34*, 4607–4612 (1993).

165. Prasad, P. N., Karasz, F. E., Pang, Y., and Wung, C. J., U.S. Patent 5,130,362 (1992).

166. Zarzycki, J., *Glasses and the Vitreous State*, Cambridge Solid State Science Series, Cambridge University Press, Cambridge, 1991.

167. Boonstra, A. H., and Bernards, T. N. M., The dependence of the gelation time on the hydrolysis time in a two-step SiO_2 sol–gel process, *J. Non-Cryst. Solids*, 105, 207–213 (1988).

168. Woodruff, M., Samoc, M., and Luther-Davies, B., Comparison of the linear and non-linear optical properties of poly(*p*-phenylenevinylene)/sol–gel composites derived from tetramethoxysilane and methyltrimethoxysilane, *Chem. Mater.*, *8*, 2586–2594 (1996).

169. Hort, E. V., and Waxman, B. H., N-vinyl monomers and polymers, in *Kirk-Othmer Encyclopedia of Chemical Technology*, *Vol. 23*, 3rd ed., Wiley, New York, 1983, pp. 960–979.

170. Seager, C. H., Sinclair, M., McBranch, D., Heeger, A. J., and Baker, G. L., Photothermal deflection spectroscopy of conjugated polymers, *Synth. Met.*, *49*, 91–97 (1992).

171. Skumanich, A., and Scott, J. C., Photothermal deflection spectroscopy: A sensitive absorption technique for organic thin films, *Molec. Cryst. Liq. Cryst.*, *183*, 365–370 (1990).

172. Fotheringham, D., Master thesis (in preparation).

173. Samoc, A., Samoc, M., and Luther-Davies, B., Linear and nonlinear optical properties of poly(*p*-phenylenevinylene)-poly(*N*-vinylpyrrolidone) (PPV-PVP) composites, *SPIE*, *3147*, 166–177 (1997).

174. Samoc, A., Samoc, M., and Luther-Davies, B., Nonlinear optical and waveguiding properties of poly(*p*-phenylenevinylene)-poly(1-vinyl-2-pyrrolidone) (PPV-PVP) blends, *Eur. Opt. Soc. Topical Mtg. Dig. Series*, *Vol. 11*, 16–17 (1997).

12
Hybrid Materials for Electrical and Optical Applications, Sol–Gel Synthesis of Bridged Polysilsesquioxanes

Kyung M. Choi and Kenneth J. Shea
University of California
Irvine, California

I. INTRODUCTION

The increasing demand for new materials that satisfy multiple roles, for example, electrical and optical functions, has provided the impetus for an important new area of research in materials science. Traditional techniques which utilize multiple components, including inorganics, organics, metals, polymers, and ceramics, have produced advanced polymeric or ceramic materials that span many desired functions. However, because simple physical mixing of two or more components often results in a phase-separation problem, efforts to develop new *hybrid materials* based on molecular building blocks have intensified. These efforts have required the collaborative research of chemists, physicists, engineers, and materials scientists.

Usually, physical mixtures will maintain many of the original properties of the individual components. Current research is directed toward improving the interface between components by using additives, coupling agents, and fillers to solve interface problems. One potential advantage of hybrid materials is that the molecular-level mixing results in the loss of each individual component's identity and creates new materials without phase-separation and interface problems. In this chapter, we describe the preparation of new hybrid materials assembled from molecular building blocks for use in electrical and optical applications.

The properties of inorganic oxides can be modified by incorporation of organics to produce hybrids of inorganic oxides and organic polymers (Fig. 1) [1–5]. These hybrid materials can range from physical mixtures of inorganic oxides and organic compounds including *blends and composites* to *molecular composites* that utilize formal chemical linkages between the organic and inorganic domains. The resulting properties of molecular composites are influenced by individual components and their respective domain sizes.

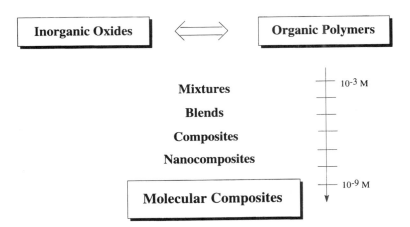

FIGURE 1 Illustration of the range of domain sizes in hybrid materials.

The physical and optical properties of these composites are often enhanced by a reduction of the domain size. As the domain size is reduced, the interfacial surface area increases. Substantial improvement in performance is achieved with domains on the nanometer scale. These materials have been referred to as *nanocomposites*. Further reduction in the domain size results in *molecular composites*. By incorporating organic fragments between inorganic oxides in these molecular composites, it is possible to create new hybrid materials with properties that can differ from those of the single component. This approach allows two inherently incompatible substances (i.e., organic polymers and inorganic oxides) to be made compatible by *premixing* at the molecular level before conversion to a new material.

The success of this approach is due in part to the fact that the silicon atom forms stable bonds to both oxygen and carbon. This permits the design and synthesis of advanced hybrid materials with specific physico-chemical properties. With molecular composites, no phase separation between the organic and inorganic phases will occur; thus, hybrid glasses containing organic functionalities will show good homogeneity and high transparency so that these will be free from significant light scattering. This concept is important for fabrication of optical materials based on hybrid materials.

Porosity is a factor to consider for optical applications. It can be controlled in inorganic oxides by the choice of sol–gel reaction and processing conditions. Hybrid materials prepared from molecular building blocks have an additional "handle" for porosity control, namely the organic fragment which can also influence the porosity of the final material.

A. Sol–Gel Chemistry

Sol–gel chemistry is a mild method for producing inorganic/organic hybrid glasses or ceramics [6]. The basic reactions consist of hydrolysis and condensation to produce Si–O–Si networks under sol–gel conditions compatible with silicon–carbon bonds.

Simple inorganic oxides such as silicates are synthesized by sol–gel polymerization from tetraethoxysilane (TEOS). A silicate-like lattice that incorporates organic spacers at regular intervals can be prepared by sol–gel polymerization of bis(triethoxysilyl)-bridged organic spacers (Fig. 2). The properties of these modified silica materials, bridged polysilsesquioxanes, can be modulated by the choice of molecular building blocks.

The chemical connectivity of bridged polysilsesquioxanes is shown in Fig. 2. Sol–gel polymerization of bis(triethoxysilyl)alkyl or aryl monomers results in the formation of this three-dimensional network which is distinguished by incorporation of an organic fragment as an integral component of the lattice.

Both silica and modified silica such as bridged polysilsesquioxane can be prepared by *sol–gel chemistry*. Sol–gel polymerization can be catalyzed by either an acid or a base. The chemistry involves hydrolysis of ethoxysilyl groups to yield silanols and their subsequent condensation of either other silanols or ethoxysilanes to form siloxane linkages (Si–O–Si).

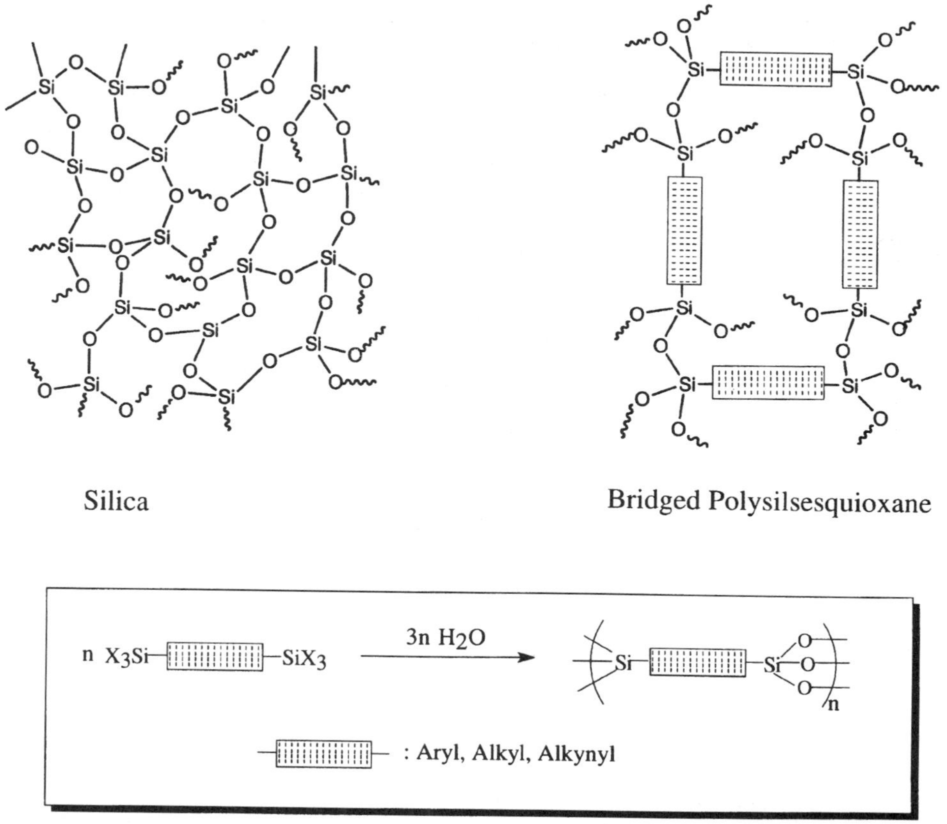

Silica   Bridged Polysilsesquioxane

FIGURE 2 Schematic diagram of silica and bridged polysilsesquioxane networks. The shaded rectangles correspond to the variable organic fragments.

B. Sol–Gel Processing

Bridged polysilsesquioxanes can be fabricated in a variety of forms. Figure 3 illustrates several of these including coated films, fibers, *xerogels*, and *aerogels*. All are prepared from sol–gel solutions containing monomer, solvent, water, and catalyst. Prior to gellation, fibers can be drawn and thin films can be cast. In time, an infinite network (wet gel) is produced. Following aging, the wet gel is dried by solvent evaporation to yield a *xerogel*. Significant volume shrinkage occurs at this stage. *Monoliths* can be produced by controlling the rate of drying and minimizing capillary stress. The wet gel can also be dried under supercritical conditions. The resulting dried gels are referred to as *aerogels*. These materials show less volume shrinkage and higher porosity than xerogels [6].

Aerogels are prepared by drying the wet gel containing alcohol at a temperature and pressure above the supercritical point of the alcohol. Alternatively, aerogels are prepared by supercritical carbon dioxide extraction, which involves replacing the original solvent in the gel with supercritical carbon dioxide, then slowly venting the carbon dioxide. Aerogels are often white opaque materials because of light scattering due to micron-sized pores [7].

II. HYBRID MATERIALS BASED ON POLYSILSESQUIOXANES

A. Preparation of Arylene-Bridged Polysilsesquioxanes

The arylene-bridged silsesquioxanes contain relatively stiff organic fragments. A number of bis(triethoxysilyl)aryl monomers have been synthesized; representative structures of these monomers are shown in Fig. 4 [8–10]. Porosities of these arylene-bridged polysilsesquioxane xerogels have been measured by the N_2 adsorption/desorption method and selected results are summarized in Table 1.

B. Preparation of Alkylene-Bridged Polysilsesquioxanes

Bis(triethoxysilyl)alkyl monomers have also been synthesized [11] (Fig. 5) by hydrosilylation of the corresponding α-ω-alkyldienes with triethoxysilane employing chloroplatinic acid (H_2PtCl_6) [12] or a divinyldisiloxane platinum complex (Karstedt's catalyst [13]).

The alkyl spacers introduce variable flexibility into the resulting xerogels. The alkyl group length has been varied in an attempt to change pore size and overall porosity. The porosity analyses of these alkylene-bridged polysilsesquioxanes have been measured by either the N_2 or Ar adsorption/desorption method and the results are summarized in Table 2. As noted in the table, surface areas of the resulting xerogels decrease with increasing length of the alkyl chain in the sol–gel monomers. A qualitative relationship between porosity and the organic bridging group exists, indicating that the microstructure of bridged polysilsesquioxanes can be influenced by molecular building blocks.

C. Preparation of Alkynylene- and Alkenylene-Bridged Polysilsesquioxanes

Bis(triethoxysilyl)alkynyl and alkenyl monomers have also been synthesized and polymerized by sol–gel chemistry [8,14]. The polymerized structures are shown in

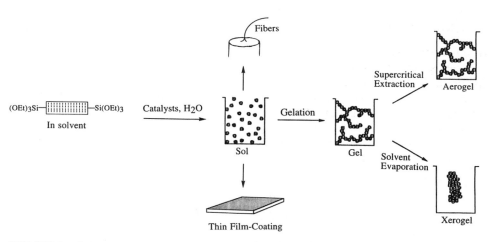

FIGURE 3 Sol–gel processing for formation of fibers, thin films, xerogels, and aerogels.

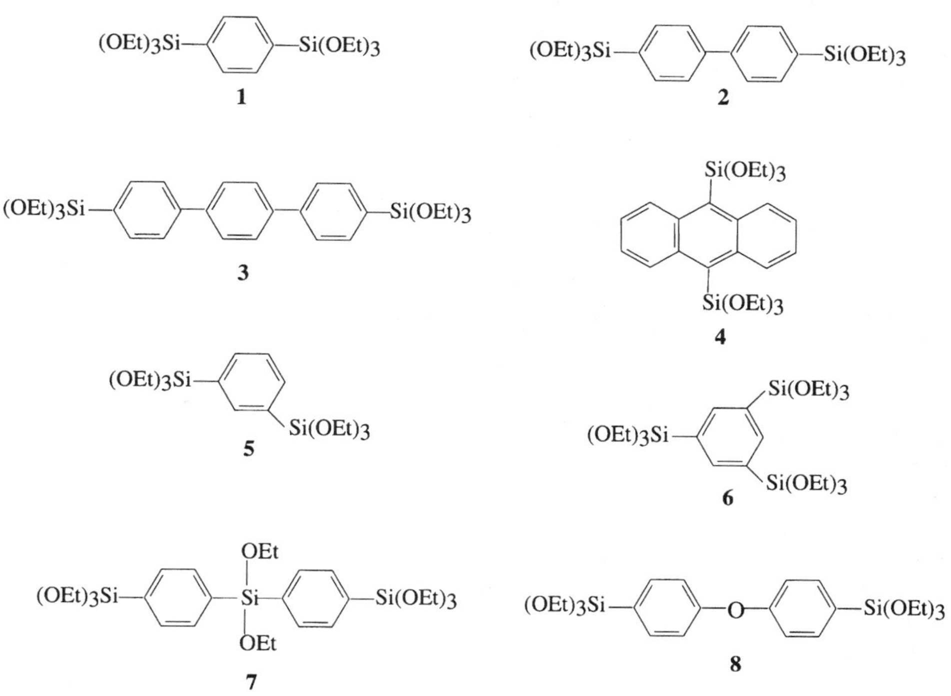

FIGURE 4 Aryl-bridged sol–gel processable monomers. **1:** 1,4-bis(triethoxysilyl)benzene; **2:** 4,4′-bis(triethoxysilyl)biphenyl; **3:** 4,4″-bis(triethoxysilyl)terphenyl, **4:** 9,10-bis(triethoxysilyl)anthracene; **5:** 1,3-bis(triethoxysilyl)benzene; **6:** 1,3,5-tris(triethoxysilyl)benzene; **7:** bis(4-triethoxysilylphenyl)diethoxysilane; **8:** bis-(triethoxysilyl)diphenyl ether.

TABLE 1 Representative Surface Areas and Average Pore Sizes for Arylene-Bridged Polysilsesquioxane Xerogels

Monomer	Sol–gel conditions	Surface area (BET: m²/g)	Mean pore diameter (Å)
1	0.2M Monomer/THF[a] 570 mol% NH₄OH	1180	45.6
2	0.2 M Monomer/THF 570 mol% NH₄OH	1150	—
6	0.2 M Monomer/THF 570 mol% NH₄OH	756	33.8
7	0.2 M Monomer/THF 570 mol% NH₄OH	956	24.0

[a]THF: tetrahydrofuran.

Fig. 6. In these polymerizations, the silicon–alkynyl bond undergoes cleavage under alkaline conditions or with the fluoride ion. These monomers can also be used for secondary modification of the polymeric architecture by subsequent polymerization of the carbon–carbon triple bond. For example, thermal polymerization of the 1,3-

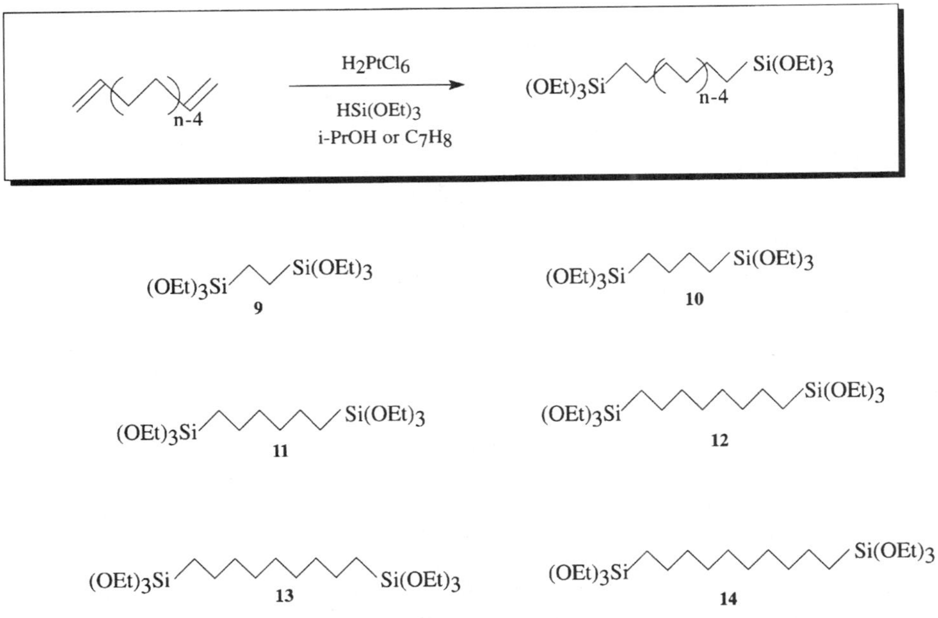

FIGURE 5 Synthesis of alkyl-bridged sol–gel processable monomers. **9**: ethyl-; **10**: butyl-; **11**: hexyl-; **12**: octyl-; **13**: nonyl-; and **14**: decyl-bridged monomers.

TABLE 2 Surface Areas and Pore Diameters for Alkylene-Bridged Polysilsesquioxane Xerogels

Monomer	Sol–gel conditions	Surface area (BET: m^2/g)	Mean pore diameter (Å)
9	0.2M Monomer/EtOH 10% NaOH	747	37.8
10	0.2 M Monomer/EtOH 10% NaOH	730	35.0
11	0.2 M Monomer/EtOH 10% NaOH	582	35.4
12	0.2 M Monomer/EtOH 10% NaOH	459	55.3
13	0.2 M Monomer/EtOH 10% NaOH	269	105
14	0.2 M Monomer/EtOH 10% NaOH	94	—

butadiynylene bridge in the bridged alkynyl silsesquioxane forms an interpenetrating network of silsesquioxane and butadiyne polymers [14].

As for alkenylene-bridged polysilsesquioxanes, the unsaturated bonds can be used for formation of π-complexes by metallic groups. π-Bond formation between the metal and the olefin increases the effective size of the polymer chain growing in solution and increases the rate of gelation [15]. This metal coordination offers the potential for an additional method of controlling xerogel morphology.

III. CHARACTERIZATION

Following sol–gel polymerization, the chemical connectivity and the degree of condensation of the gel can be determined by several spectroscopic techniques including solid-state nuclear magnetic resonance (NMR) [8,11], infrared (IR) [16], and Raman [17] spectroscopies.

The degree of hydrolysis can be determined from the amount of residual ethoxysilyl groups in the materials. An IR study for analysis of residual ethoxysilyl

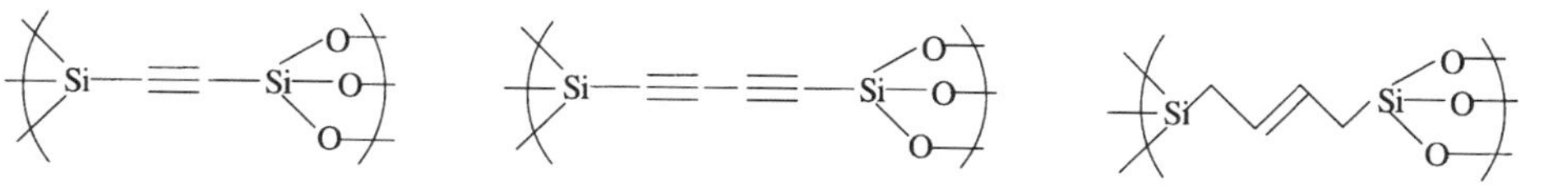

FIGURE 6 Alkynylene- and alkenylene-bridged polysilsesquioxanes. **15:** 1,2-ethynylene-; **16:** 1,3-butadiynylene-; **17:** *trans*-1,4,-2-butenyl-bridged polysilsesquioxanes.

groups in silica has been reported [16] and a Raman study for monitoring hydrolysis and condensation in sol–gel polymerization of TEOS has been reported [17].

A. Solid-State NMR Spectroscopy

^{13}C and ^{29}Si solid-state NMR have been extensively employed to characterize bridged polysilsesquioxanes. ^{13}C-NMR can be used to identify residual ethoxysilyl groups and ^{29}Si-NMR can be used to identify a formation of the "Si–O–Si" linkages. Cross-polarized magic angle spinning (CP/MAS) NMR and single-pulse magic angle spinning (SP/MAS) NMR techniques have been used for the characterization of bridged polysilsesquioxanes.

Figure 7 shows representative ^{29}Si-NMR chemical shifts for silicate and silsesquioxane materials. The chemical shifts of ^{29}Si-NMR spectra for silicate materials have been described using the Q^n, T^n, D^n, and M^n notations [18]. According to these notations, Q refers to the NMR shift of a silicon connected to four oxygens, T refers to the NMR shift of a silicon connected to three oxygens and one alkyl group, D refers to the NMR shift of a silicon connected to two oxygens and two alkyl groups, and M refers to the NMR shift of a silicon connected to one oxygen and three alkyl groups.

For example, the chemical shifts of a silicate silicon connected to one, two, three, and four siloxanes is referred to Q^1, Q^2, Q^3, and Q^4, respectively. Chemical shifts of silicon in arylene- and alkylene-bridged polysilsesquioxanes connected to

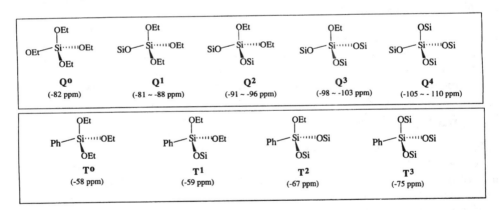

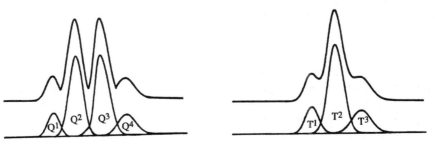

FIGURE 7   ^{29}Si-NMR chemical shifts of silicons in silicates (Q notation) and phenylene-bridged polysilsesquioxanes (T notation).

one, two, and three siloxanes are referred to as T^1, T^2, and T^3, respectively. Figure 7 shows chemical shifts of silicons in a typical silicate-bridged xerogel. The chemical shift moves upfield with increasing numbers of siloxane bond. The degree of condensation (DC) in silsesquioxanes can be described by

$$DC = \frac{(0.5)(\text{Area } T_1) + (1.0)(\text{Area } T_2) + (1.5)(\text{Area } T_3)}{1.5} \tag{1}$$

Deconvolution of the spectrum permits quantification of each peak.

The degree of hydrolysis can be determined by ^{13}C-NMR and is calculated from the ratio of the integrals of residual ethoxysilyl groups to that of the aromatic resonances. The degree of hydrolysis (DH) can be calculated by using the following equations:

$$DH = \frac{600 - M_{\text{ethoxy}} (\%)}{600} \tag{2}$$

$$M_{\text{ethoxy}} (\%) = \left(\frac{\text{IEM}}{\text{IAC}}\right) \times 100 \tag{3}$$

where M_{ethoxy} is the mol% of residual ethoxy groups; each mole of bis(triethoxysilyl) monomer has 600 mol% or six equivalents of ethoxy groups. IEM is the integral of ethoxy methylene and IAC is the integral of one aromatic carbon.

Figure 8 shows ^{13}C and ^{29}Si solid-state NMR spectra of phenylene-bridged polysilsesquioxanes prepared under basic conditions. In Fig. 8, a single peak at 134 ppm is observed in ^{13}C-NMR of the phenylene-bridged xerogel prepared under basic conditions and the three ^{29}Si peaks corresponded to the T^1, T^2, and T^3 environments. The degree of condensation of the phenylene-bridged polysilsesquioxane is calculated using Eq. (1).

Figure 9 shows the ^{13}C and ^{29}Si solid state NMR spectra of a hexylene-bridged polysilsesquioxane prepared under basic conditions. In the ^{13}C-NMR spectra, three peaks are formed from the carbon atoms in different chemical environments. In the ^{29}Si-NMR, two different siloxane peaks (T^2 and T^3) are also observed. In general, a higher degree of condensation is observed under basic conditions.

The degree of condensation for arylene-bridged polysilsesquioxanes introduced in Fig. 4 formed under acidic or basic conditions is summarized in Table 3. The degree of condensation for many different types of arylene-bridged polysilsesquioxanes range from 65% to 80%. Higher degrees of condensation would result in fewer Si–OH groups and a greater hydrophobicity.

B. Porosity Analysis

Porosity is an important concern for the development of hybrid materials for electrical and optical applications. For example, nano-sized particle fabrication in xerogels requires a high porosity of materials and homogeneous pore distribution. Porous hybrid glasses can provide various domain sizes for these purposes.

Pores can be divided into three groups based on the mathematical models used to calculate pore sizes from gas sorption data. *Micropores*, with pore diameters less than 20 Å, can be measured by Ar adsorption. *Mesopores*, with pore diameters in

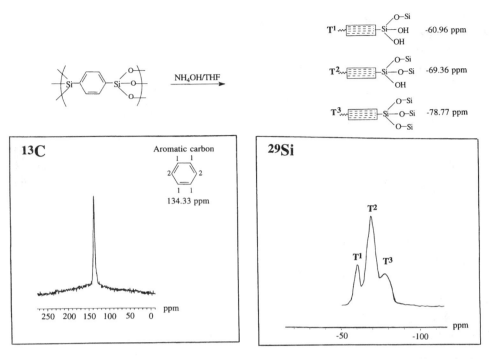

FIGURE 8 ^{13}C and ^{29}Si solid-state NMR spectra for phenylene-bridged polysilsesquioxane prepared under basic conditions.

the region of 20–500 Å, can be determined by N_2 adsorption porosimetry. Pore diameters larger than 500 Å are designated as *macropores* and can be evaluated by mercury porosimetry [19].

Studies of bridged polysilsesquioxanes have permitted development of qualitative empirical relationships between the porosity of materials and the chemical structure of sol–gel monomers. The porosity analysis of xerogels (surface areas between 20 and 1000 m^2/g) have been determined by either the N_2 or Ar adsorption/desorption method. Surface areas and pore diameters can be determined by Brunauer, Emmett, Teller (BET) and Barrett, Joyner, Halenda (BJH) methods, respectively, in N_2 adsorption/desorption porosimetry [20].

Figure 10 shows typical adsorption/desorption isotherm and pore volume distribution plots for a phenylene-bridged polysilsesquioxane prepared under basic conditions. The surface area was calculated to be 1156 m^2/g (BET). The pore diameter of the xerogel was calculated to be 43 Å (BJH). The surface area of a phenylene-bridged xerogel prepared under *acidic* conditions was determined to be 531 m^2/g with an average pore diameter of 50 Å.

Figure 11 shows a porosity analysis of hexylene-bridged xerogel prepared under basic conditions. The xerogel shows the bulk of the porosity resides in the mesopore region. Surface area and pore diameter of the xerogel was found to be 571 m^2/g and 43 Å, respectively. Nonporous materials are produced from 1,6-bis(triethoxysilyl)hexane under acidic conditions. The change in porosity of a series

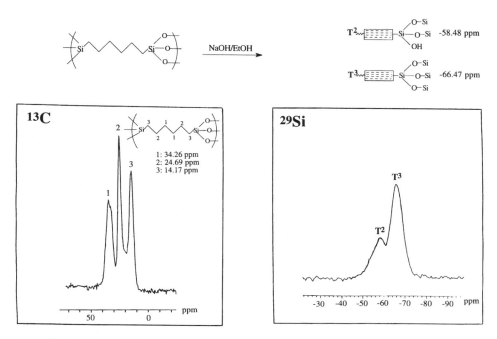

FIGURE 9 ^{13}C and ^{29}Si solid-state NMR spectra for hexylene-bridged polysilsesquioxane prepared under basic conditions.

of alkylene-bridged xerogels with increasing alkyl-chain length is shown in Fig. 12 and Table 2. The surface areas of these materials decrease with increasing alkyl length. In summary, the porosities and pore structures of silsesquioxanes strongly rely on the sol–gel conditions and molecular structures of the monomers, so that their porosities can be influenced by choice of sol–gel conditions and molecular tailoring. However, with some monomers, such as arylene-bridged xerogels, no significant relationship between the average pore sizes and the length of the arylene spacers was found.

The factors influencing porosity are still under investigation in an effort to control the porosities. For example, optically transparent polysilsesquioxanes can be achieved by using an acidic catalyst rather than base. In general, alkylene-bridged polysilsesquioxanes show better optical transparency than arylene-bridged xerogels.

C. Thermal Stability

Thermal stability of bridged polysilsesquioxanes has been evaluated by thermogravimetric analysis (TGA) and differential scanning calorimetry (DSC) analysis. The onset of decomposition for bridged polysilsesquioxanes occurs at temperatures intermediate between that of organic network polymers and the melting range of silicate materials. Figure 13 shows a TGA curve for phenylene-bridged polysilsesquioxane. A slight weight loss ($\approx$5%) at 120°C due to loss of adsorbed solvent is observed followed by little change until the temperature approaches 500°C.

TABLE 3 Corrected Areas for T^1, T^2, and T^3 in ^{29}Si-NMR and Degree of Condensation for Arylene-Bridged Polysilsesquioxanes

Xerogel[a]	Monomer conc. solvent	Catalyst (mol%)	H_2O (equiv.)	Drying condition[b]	Area T^1 (%)	Area T^2 (%)	Area T^3 (%)	Degree of condensation (%)
Xerogel (1)	0.2 M/THF	HCl (10.8)	6.0	A	20.4	61.7	17.9	65.8
Xerogel (1′)	0.2 M/THF	NH_3 (570)	>10	B	17.4	60.8	21.8	68.1
Xerogel (2)	0.2 M/THF	HCl (10.8)	6.0	A	23.6	62.1	14.4	63.7
Xerogel (2′)	0.4 M/EtOH	HCl (10.8)	6.0	A	20.2	61.8	18.0	65.9
Xerogel (3)	0.18 M/THF	HCl (10.8)	6.0	A	42.4	49.6	8.0	55.2
Xerogel (3′)	0.2 M/THF	NH_3 (570)	>10	B	10.0	40.6	49.4	79.8
Xerogel (4)	0.2 M/THF	NaOH (5.4)	3.0	A	16.2	42.4	41.4	75.1

[a]Xerogels 1, 2, 3, and 4 were made from monomers 1, 2, 3, and 4 shown in Fig. 4.
[b]Drying conditions: A—vacuum dry at room temperature; B—vacuum dry at 100°C.

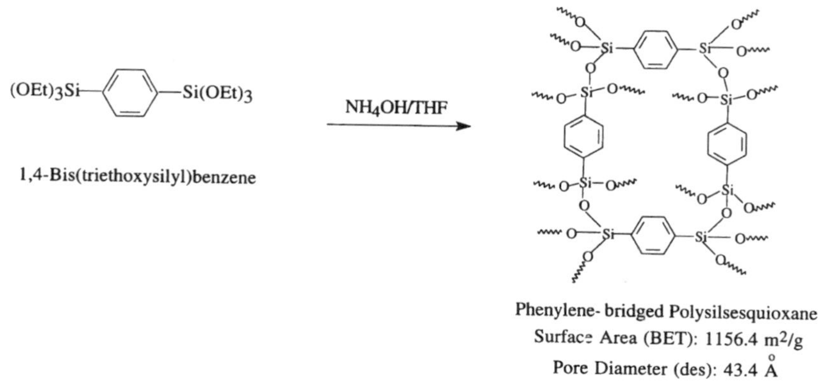
(OEt)3Si
Si(OEt)3
1,4-Bis(triethoxysilyl)benzene
NH4OH/THF
Phenylene- bridged Polysilsesquioxane
Surface Area (BET): 1156.4 m2/g
Pore Diameter (des): 43.4 Å

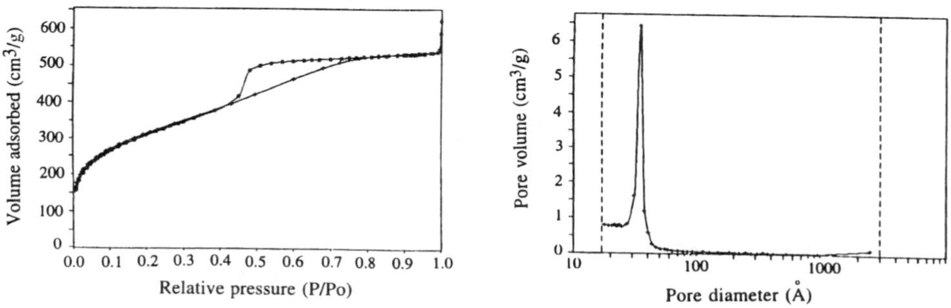
Volume adsorbed (cm3/g)
Relative pressure (P/Po)
Pore volume (cm3/g)
Pore diameter (Å)

FIGURE 10 Isotherm plot (bottom, left) and desorption pore volume distribution plot (bottom, right) for a phenylene-bridged polysilsesquioxane prepared under basic conditions and measured by the N_2 adsorption/desorption method.

At temperatures above 500°C, silicon–carbon linkages are broken and the xerogel loses hydrocarbon components. After heating at 1200°C, a black, carbon-rich residue remains. There is a significant difference between the calculated weight loss for pyrolysis to silica and the experimentally observed weight loss for pyrolysis of arylene-bridged xerogel at 800°C.

Ceramic materials have been prepared from sesquioxanes; for example, silicon carbide has been prepared by high-temperature pyrolysis at 1400°C of sawdust and sand by the *Acheson process*. Silicon carbide is also prepared from preceramic polymers such as polycarbosilanes or polysilanes [21]. Arylene-bridged polysilsesquioxane may also be a suitable preceramic polymer for the preparation of silicon carbide (SiC) by pyrolysis xerogels containing aryl or alkyl spacers [22].

D. Mechanical Properties

The physical strength of bridged polysilsesquioxanes has been measured to evaluate their hardness and toughness. Biaxial flexure measurements of bridged polysilsesquioxanes have been performed using a universal strength-testing machine and a sample holder that is specially designed for the measurements.

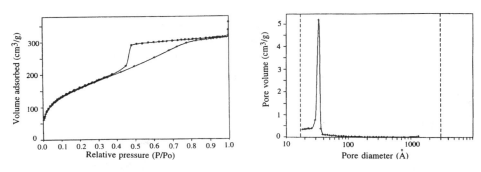

Hexylene-bridged Polysilsesquioxane
Surface Area (BET): 571.2 m^2/g
Pore Diameter (des.): 42.6 Å

FIGURE 11   Isotherm plot (bottom, left) and desorption pore volume distribution plot (bottom, right) for a hexylene-bridged polysilsesquioxane prepared under basic conditions and measured by the N$_2$ adsorption/desorption method.

For biaxial flexure measurements, a disk-shaped hexylene-bridged polysilsesquioxane xerogel prepared using formic acid in diethyl ether was employed. The glass disk was then heated at 180°C under a high vacuum (<1 mm Hg) for 2 h. The glass sample was then placed in the sample holder and biaxial flexure of the hybrid glass was calculated using Eq. (4) [23]. The tensile stress (S) in the center of the surface in tension can be written as

$$S = -\frac{3P(X-Y)}{4\pi d^2}$$

$$X = (1 + v)\ln\left(\frac{B}{C}\right)^2 + \left(\frac{1-v}{2}\right)\left(\frac{B}{C}\right)^2$$

$$Y = (1 + v)\left[1 + \ln\left(\frac{A}{C}\right)^2\right] + (1 - v)\left(\frac{A}{C}\right)^2 \tag{4}$$

where v is Poisson's ratio, B is the radius of the loaded area, A is the radius of the support circle, C is the radius of the specimen, P is the loaded force, and d is the sample thickness.

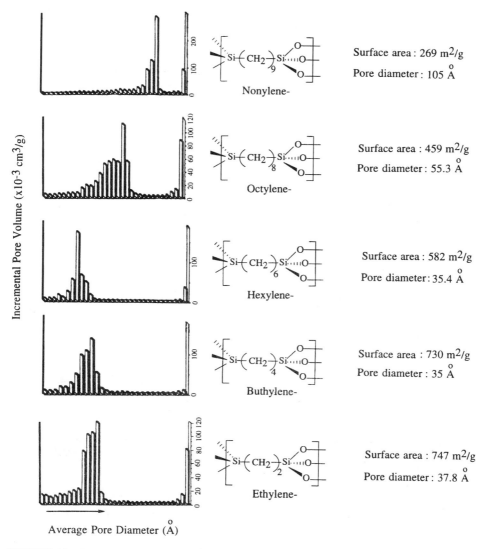

FIGURE 12 Porosities of base-catalyzed alkylene-bridged polysilsesquioxanes.

The biaxial strength of hexylene-bridged polysilsesquioxane (HCOOH catalyst) was calculated to be 34.5 MPa from Eq. (4). This value is lower than that of normal inorganic silicates, but novel applications for these hybrid glasses might be suggested. For example, inorganic glass and quartz materials have been used as a filler for the preparation of composites to improve their physical strength; however, the composites filled with inorganic or quartz glass show a poor toughness and flexibility. In order to improve the toughness and flexibility of composites, a small amount of hybrid glasses could be added as a filler. Hybrid glasses can be a good candidate for a specific filler material encapsulating inorganic salts or particles.

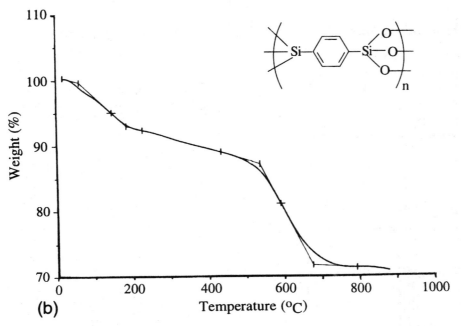

FIGURE 13 (a) Pyrolysis of arylene-bridged polysilsesquioxanes to silica. (b) TGA analysis of phenylene-bridged polysilsesquioxane. The sample was heated at a 10°C/min rate from 25°C to 1000°C under nitrogen.

IV. APPLICATIONS

As the demand for advanced materials increase, there is a growing interest in new approaches for fabrication of novel optical devices based on hybrid materials. Organofunctional silsesquioxanes are highly condensed *molecular composites* which show interesting physical and optical properties that can be modified by varying the organic functionality in the molecular building blocks. As a result, these hybrid materials have potential applications in optics, physics, chemistry, materials science, and medical areas.

This *molecular-level mixing* results a loss of individual identity of inorganic oxides, but important new functionalities from the organic fragments can be introduced. For example, the refractive index of hybrid glasses can be altered by inserting different molecular building blocks in the network for optical-device applications.

Hybrid-glass materials also have found a number of applications in optical technology, including optical fibers [24], microlens arrays [25], waveguides, interconnects, laser devices including couplers and nonlinear optical devices, laser writers, photochromic glasses [26], photoconductive glasses [27], ferroelectric glasses [28], highly flexible glasses [29], optical memory devices [30], fluorescent glasses, and fluorinated glasses for waterproof optical-coating materials. Bridged polysilsesquioxanes may offer improved optical properties compared with those of inorganic glasses.

A. Nano-Sized Particles in Porous Bridged Polysilsesquioxanes

The *quantum-size effect* of nano-sized particles (particle size <10 nm) has become an important area of research with the potential for optical-device applications [31]. Nano-sized particles show interesting optical properties and their utilization in devices is receiving attention. A number of nano-sized semiconductors or transition metal particles have been prepared in polymers [32], glasses [33,34], zeolites [35], and micelles [36].

A potential application of nano-sized semiconductor particles is as a photo-catalyst of electron transfer reactions. These nano-sized semiconductor particles can be likened to microelectrodes because of the quantum-size effect. For example, Hoffman et al. reported an efficiency of photoinitiated polymerization for vinyl monomers using either bulk or nano-sized ZnO particles [37]. The result established that nano-sized semiconductors can result in higher quantum yields for the photo-initiated polymerization. The fluorescence of nano-sized CdS has been found to be stronger than that of the bulk phase [38]. Wang et al. reported that polymers embodying nano-sized PbS and CdS have potential use as third-order nonlinear optical devices [39].

Sol–gel technology offers a practical chemical approach to produce nano-sized particles doped in transparent hybrid-glass materials for electro-optical devices. We have investigated the preparation of nano-sized particles in polysilsesquioxanes. Nano-sized CdS, Cr, Fe, Co, and Pt particles have been prepared in xerogels by either an external or internal doping method [40–47]. In addition, heterogeneous systems, such as Cr°/CdS or Fe°/CdS clusters have also been prepared by a combination of internal and external doping methods. The size and identity of these particles have been characterized by ultraviolet (UV) absorption, fluorescence, transmission electron microscopy (TEM), energy dispersive analysis by X-ray detection (EDAX), electron spectroscopy for chemical analysis (ESCA), and electron-diffraction techniques.

1. Preparation of Nano-Sized CdS by the External Doping Method

Nano-sized CdS particles have been incorporated into transparent arylene- or alkylene-bridged polysilsesquioxanes by external doping. When dried xerogel prepared from 1,4-bis(triethoxysilyl)benzene or 1,6-bis(triethoxysilyl)hexane was treated successively with an aqueous solution of $CdCl_2$ and Na_2S, nano-sized CdS particles (58 or 90 Å) were deposited uniformly in the glassy matrices. The procedures are shown in Figs. 14 and 15 [40].

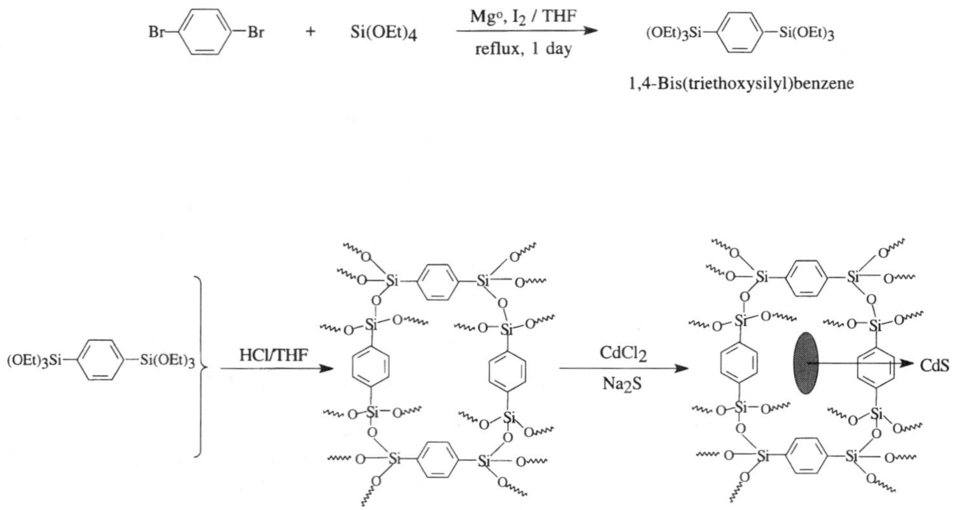

FIGURE 14 Synthesis of bis(triethoxysilyl)benzene (top) and preparation of CdS particles into phenylene-bridged polysilsesquioxane.

The CdS phases prepared in the xerogels were analyzed by TEM, EDAX, and electron-diffraction techniques. Figure 16 shows TEM images of phenylene-bridged xerogel containing CdS particles; dark features dispersed in TEM images were identified by the electron-diffraction pattern as a cubic crystalline phase of CdS (Fig. 17). The EDAX pattern of the dark spots in the TEM images shows clear Cd

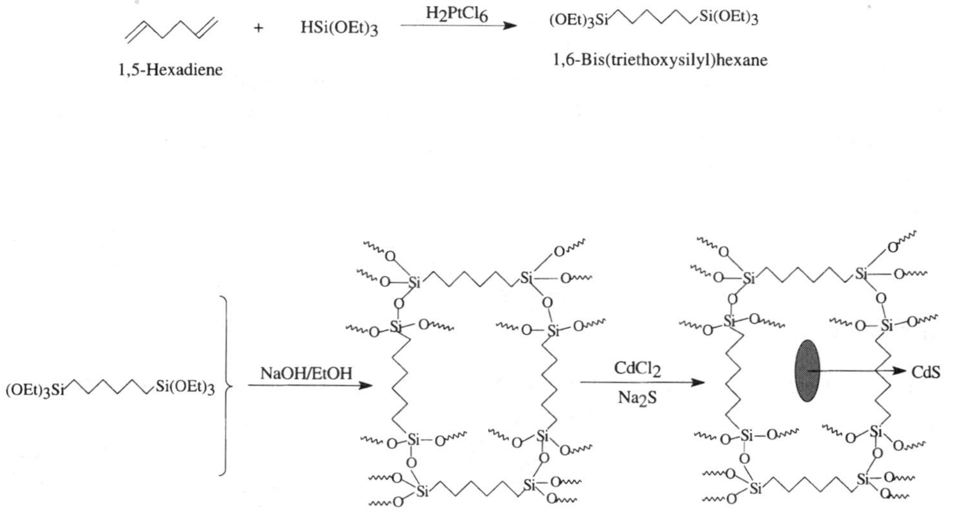

FIGURE 15 Synthesis of bis(triethoxysilyl)hexane (top) and preparation of CdS particles into hexylene-bridged polysilsesquioxane.

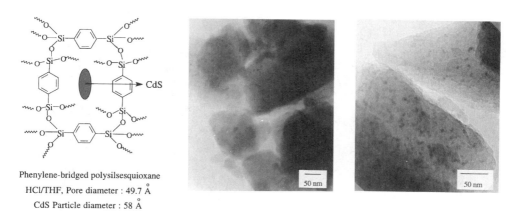

Phenylene-bridged polysilsesquioxane

HCl/THF, Pore diameter : 49.7 Å

CdS Particle diameter : 58 Å

FIGURE 16 Transmission electron microscopic images of phenylene-bridged polysilsesquioxane doped with CdS.

and S peaks. An average diameter for the CdS particles prepared in the phenylene-bridged xerogel (58 ± 12 Å) was directly measured from the TEM images.

When polysilsesquioxane xerogels with different bridging groups were employed, CdS particles with different average particle sizes were produced. In the TEM image, dispersion of CdS particles prepared in hexylene-bridged xerogels is also shown (Fig. 18). The average diameter of CdS particles prepared in the xerogel

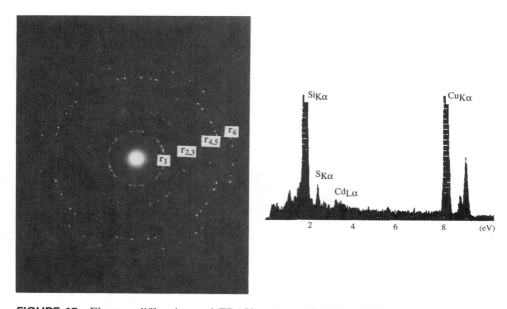

FIGURE 17 Electron-diffraction and EDAX patterns of CdS particles prepared into phenylene-bridged polysilsesquioxane; in electron diffraction pattern, [r_1(111); r_2(220); r_3(311); r_4(420); r_5(333, 511); r_6(531)].

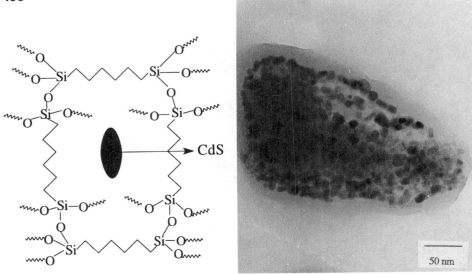

FIGURE 18 Transmission electron microscopic image of hexylene-bridged polysilsesquiox-ane doped with CdS.

was 90 Å. A rough correlation between average pore size of the xerogel matrices and the particle size of the semiconductor was observed. This result suggests some influence of the porous matrix on particle growth.

In order to achieve a homogeneous dispersion containing large quantities of colloidal particles, additional metal ions and chelating agents have been used in previous studies to achieve higher colloidal particle concentration [41]. In this work, the Cd^{2+} source was included in a sol–gel polymerization solution to distribute the CdS seeds more uniformly in the matrix. The wet gel containing Cd^{2+} ions was subsequently contacted with a sulfide source to achieve a high concentration of colloidal CdS dispersed uniformly in hybrid glasses [42].

Alteration of semiconductor and xerogel morphology could be achieved by modifying the doping procedures. The procedures for this experiment and pore analysis of the resulting xerogels are shown in Fig. 19. The porosity of the gel prepared from sol–gel mixture containing a Cd^{2+} source was significantly lower compared to that of the gel prepared without a Cd^{2+} source in the sol–gel mixture (surface areas: 754 and 571 m^2/g). A possible explanation for this observation is that divalent Cd^{2+} ions can coordinate several SiOH groups, facilitating their condensation or drawing polymer silsesquioxane chains closer together. This effect could result in a more highly condensed and tighter xerogel with a smaller average pore size (pore diameters: 79 and 47 Å, respectively). When the sulfide source is added to the *dried* gel containing Cd^{2+} ions, a xerogel with a surface area and pore diameter of 502 m^2/g and 50 Å was obtained. The results indicated that the surface areas have been gradually decreased by CdS doping due to a coverage of matrix surfaces by Cd and S phases. However, when the sulfide source is added to the *wet* gel containing Cd^{2+} ions, this process would remove most of the Cd^{2+} ions from the wet gel and produce CdS precipitation on the surface; thus, the surface area was not significantly reduced.

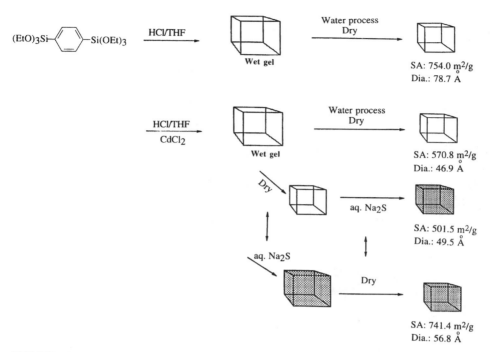

FIGURE 19 Schematic summarizing the influence of added Cd^{2+} ions and CdS doping on the porosities of phenylene-bridged xerogels.

Figure 20 shows a TEM image of the resulting xerogel containing colloidal CdS phases; the dark features dispersed in the matrix were identified as nano-sized CdS. The particle size of the colloidal CdS has been determined by spectroscopic techniques including UV, ESCA, and fluorescence spectroscopies. Based on the "size quantized effect," a blue shift of the UV band edge of the CdS doped in the xerogel permitted calculation of the average diameter of CdS to be 45 Å (Fig. 21) [42,43].

The ESCA measurements of the $S(2p)$ peak in CdS shifts with decreasing CdS particle sizes. In previous studies of bulk CdS, the $S(2p)$ peak was observed at 162 eV and an oxidized sulfur (SO_4^{2-}) peak was observed at 168 eV [44]. In this study, the ESCA peak of colloidal CdS phases doped in a bridged polysilsesquioxane shows a single sulfur peak $S(2p)$ at 164.6 eV. This blue shift is due to the quantum size effect (Fig. 22).

The fluorescent spectra of both bulk CdS single crystal and colloidal nano-sized CdS phases doped in bridged polysilsesquioxane xerogels were obtained (Fig. 23). A maximum peak for bulk CdS was observed at 705.64 nm and a maximum peak of CdS prepared in the xerogel was observed at 667.94 nm. The blue shift in fluorescence emission for CdS [45] is consistent with nano-scale particles.

In summary, polysilsesquioxane xerogels can be used as a matrix for growth of nano-sized particles. These transparent hybrid glasses have potential applications in nano-optic devices.

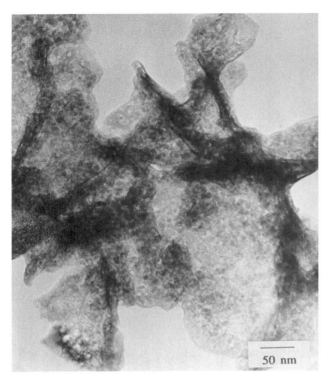

FIGURE 20 Transmission electron microscopic image of phenylene-bridged xerogel containing colloidal CdS; dark features dispersed in the matrix were identified as CdS with a cubic structure by the electron-diffraction pattern.

2. Preparation of Nano-Sized Cr and Fe Metal Particles by the Internal Doping Method

Nano-sized transition metal particles have been doped in bridged polysilsesquioxanes by incorporating sol–gel processable metal precursors in the polymerization reaction (Fig. 24). A zero-valent transition metal π complex is cocondensed with other sol–gel monomers to prepare a porous dried xerogel containing metal carbonyl groups. Metal clusters were liberated by thermolysis or photolysis from those metal precursors.

Figure 25 shows a procedure for doping chromium metal particles in a phenylene-bridged xerogel. After thermolysis of the xerogel containing a chromium precursor at 120°C under <1 mm Hg, chromium metal was liberated. Figure 26 shows TEM and electron-diffraction patterns of xerogel containing Cr metal particles. Irregularly shaped dark features in TEM image were identified as chromium metal particles with a cubic structure by electron diffraction pattern [46].

Preparation of iron metal particles in phenylene-bridged xerogel has been also achieved from sol–gel processible iron metal precursor by the internal doping method. Figure 27 shows a similar preparation of iron metal particles in porous xerogel [47]. Both thermolysis and photolysis have been utilized to liberate iron.

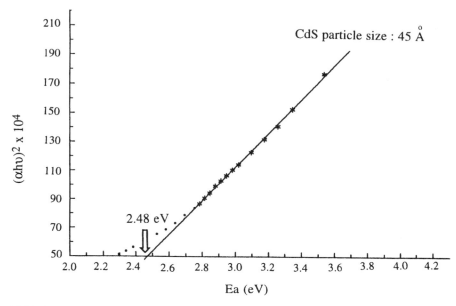

FIGURE 21 Blue shift of the UV band edge for colloidal CdS doped into the phenylene-bridged xerogel.

Figure 28 shows an IR spectrum of a disappearance of carbonyl group peaks by photolysis. Figure 29 shows TEM image of the xerogel doped with iron metal particles. The dark features in TEM image were identified as iron metal with an average particle diameter of 40 ± 5 Å. The EDAX and electron-diffraction patterns of the dark features in the TEM image were identified as iron metal with a cubic structure (Fig. 30).

Nano-sized metal fabrication in transparent media has been the subject of numerous investigations for possible optical-device fabrication as well as conducting glasses because a silica doped with nano-sized iron metal shows a conductivity in the semiconductor region [48].

3. Preparation of Heterogeneous Cr°/CdS and Fe°/CdS Phases in Xerogels by a Combination of Internal and External Doping Methods

Mixed nanophases of semiconductor/metal clusters, such as Cr°/CdS or Fe°/CdS have been realized by a combination of internal and external doping methods. The combined doping method results in porous materials containing intimate *nanocomposites* of metal and cadmium sulfide phases. A xerogel containing Cr metal was prepared by internal doping with the Cr metal precursor. The undoped xerogel (X-1) and xerogels containing the Cr metal precursors (X-2), Cr metal particles (X-2a), and Cr°/CdS clusters (X-3) have been analyzed by N_2 adsorption/desorption methods (Fig. 31, Table 4). Doping a porous matrix with chromium metal and with CdS can result in changes in the final morphology of the xerogel.

Figures 32 and 33 show the TEM and EDAX images and electron-diffraction patterns of the heterogeneous Cr°/CdS system doped in phenylene-bridged xerogel.

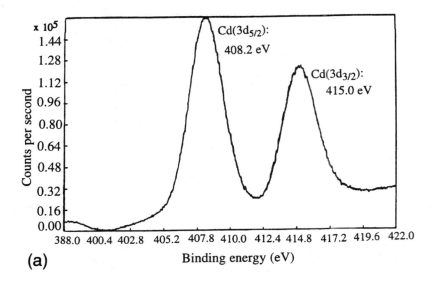

(a)

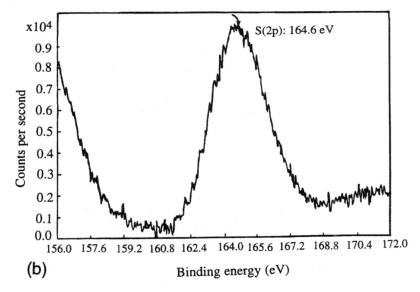

(b)

FIGURE 22 ESCA spectra of colloidal CdS-doped phenylene-bridged xerogel.

In the TEM image (Fig. 32), the features of this material differ significantly from xerogels doped individually with either CdS or Cr. There are numerous features not observed in a blank xerogel, X-1. The dark features are characterized by two distinct phases, darker phases surrounded by lighter phases. These features are

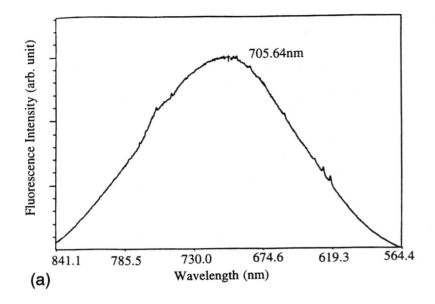

(a)

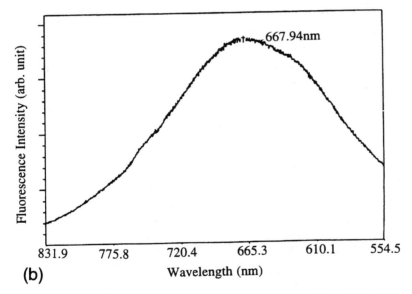

(b)

FIGURE 23 Laser fluorescence spectra of (a) single crystalline CdS and (b) nano-sized CdS doped into the phenylene-bridged xerogel.

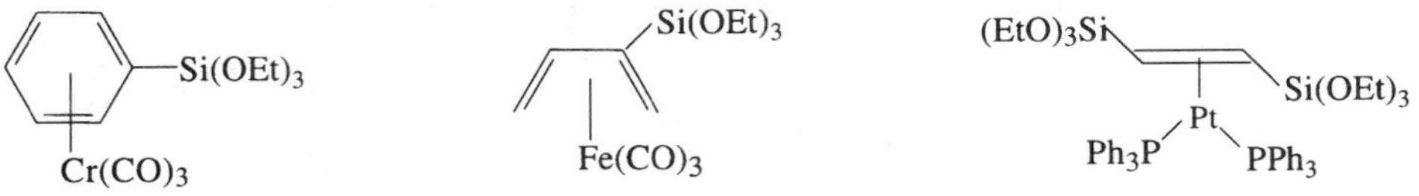

FIGURE 24 Sol–gel processible Cr, Fe, Co, and Pt metal precursors.

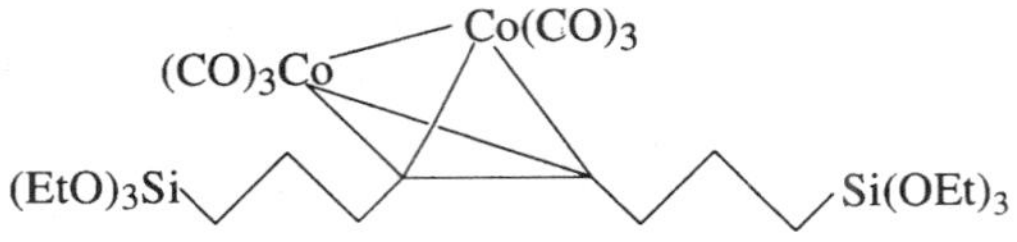

distributed throughout the matrix. To establish these two distinct phases, electron-diffraction techniques have been employed. Upon focusing on a small portion of an individual darker phase, the domain was found to consist largely of microcrystals of Cr metal (EDAX) with a cubic structure (electron diffraction), but when the fringes of the darker phases were focused with the electron beam, they were found to contain microcrystals of both CdS and Cr metal.

Figure 33 shows an interesting comparison of those electron-diffraction patterns arising from the darker and lighter phases. An electron-diffraction pattern focused

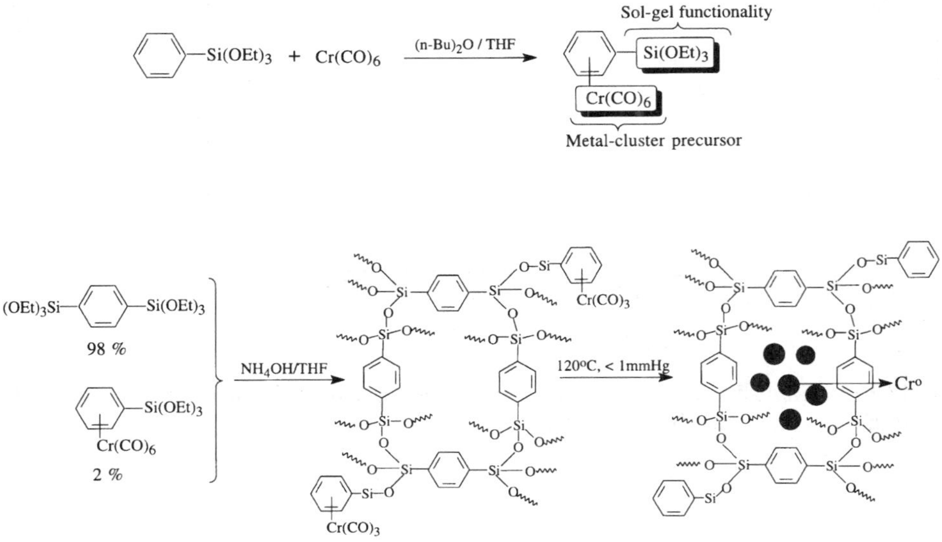

FIGURE 25 Synthesis of sol–gel processible chromium metal precursor (top). Preparation of chromium metal particle-doped phenylene-bridged xerogel.

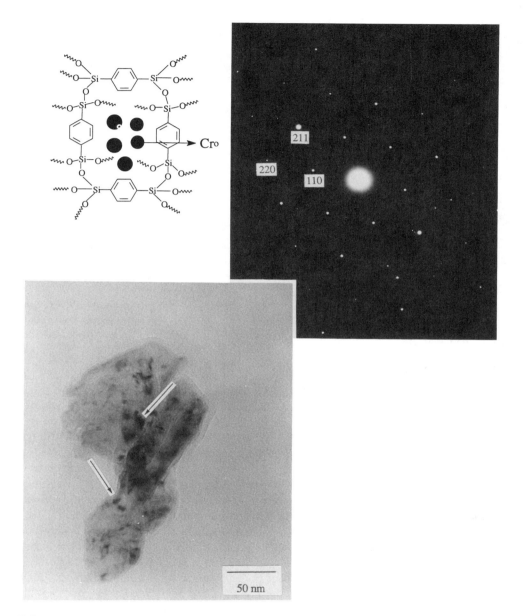

FIGURE 26 Transmission electron microscopy image and an electron-diffraction pattern of chromium metal-doped phenylene-bridged xerogel. The dark features in TEM image were identified as chromium metal with a cubic structure by the electron-diffraction technique.

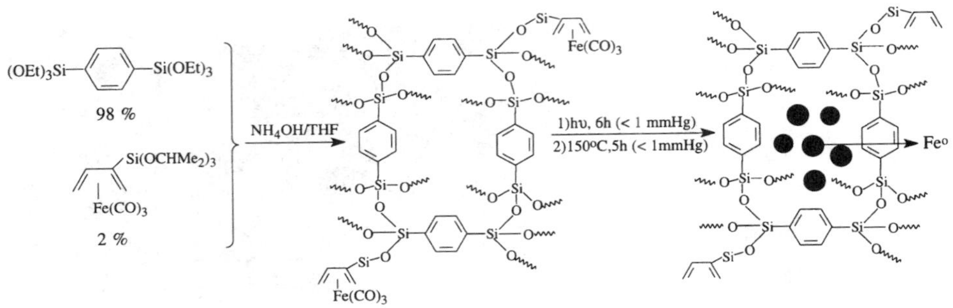

FIGURE 27 Preparation of iron metal particles into phenylene-bridged polysilsesquioxane.

on an individual darker phase shows only a Cr metal pattern (Fig. 33a), whereas an electron-diffraction pattern focused on the fringe shows both Cr and CdS patterns (Fig. 33b). This is in good agreement with the EDAX result. Thus, a possible structure for the heterogeneous clusters, darker Cr metal particles surrounded by lighter CdS phases, was suggested.

In summary, polysilsesquioxanes are good candidates for a *quantum confinement matrix* for the growth of nano-sized homogeneous or heterogeneous phases for electrical and optical applications.

B. Optical Devices Based on Bridged Polysilsesquioxanes

Nonlinear optical (NLO) materials must possess a variety of opto-electronic properties including high optical damage threshold, and high and stable optical responses, as well as fast response time. Currently, NLO devices are based on ferroelectric inorganic crystals such as $LiNbO_3$, KNb_3, $Ba_2NaNb_5O_{1.5}$, $Ba_2LiNb_5O_{1.5}$, and $BaTiO_3$. More recently, it was discovered that laser devices based on organics can have a faster response time (femtoseconds) than that of inorganic-based devices (picoseconds). A number of NLO devices based on organic polymers and polymeric materials, especially π-conjugated systems such as polyacetylene, polythiophene, polyphenylene, and polypyrrole, have been investigated for that purpose [49,50].

For second-order NLO process, a stringent symmetry requirement is imposed upon the materials, the active molecular units must be arranged in a noncentrosymmetric fashion to generate the nonzero second-order NLO coefficient. Single crystals with proper symmetry grown from both inorganic and organic materials have been shown to perform effectively in second-order NLO processes. For organic polymers fabricated in the form of thin films, the noncentrosymmetry is achieved by the *electric field* (EF) *poling process* [51]. In this process, a strong electric field is applied across the material to orient molecules with nonzero dipole moments in the direction of the electric field. Optically active groups have been attached to polymer backbones; however, pure polymers doped with these organic NLO functionalities often do not form good quality films [52].

For this reason, different procedures have been suggested to enhance the optical quality of these polymeric films and to achieve a stable NLO coefficient. Ye et al. reported that hydrogen-bonding can lead to a considerable increase in stability by establishing a weak cross-linked network [53]. Robello et al. reported that cross-

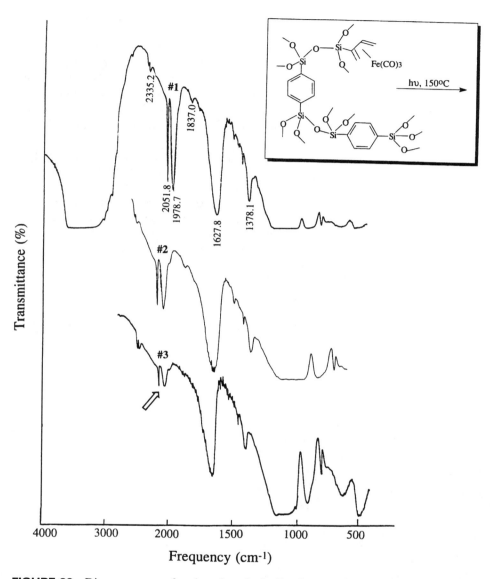

FIGURE 28 Disappearance of carbonyl peaks in Fourier-transform infrared (FTIR) spectra of the phenylene-bridged xerogel doped with iron–metal precursor during thermolysis and photolysis.

links in multifunctional acrylic systems can lead to considerable increase in the stability of poled structures [54]. In addition, inorganic/organic hybrid NLO materials with organic NLO functionalities between two inorganic oxides were used to produce new NLO materials. This new NLO material combines advantages from inorganic and organic polymeric materials. Sol–gel technology offers many advantages for material processing, and in the resulting hybrid material based devices,

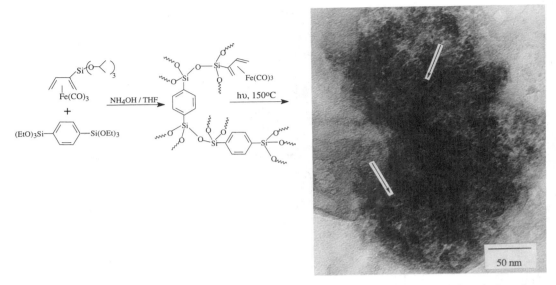

FIGURE 29 Transmission electron microscopic image of the ion metal-doped phenylene-bridged xerogel.

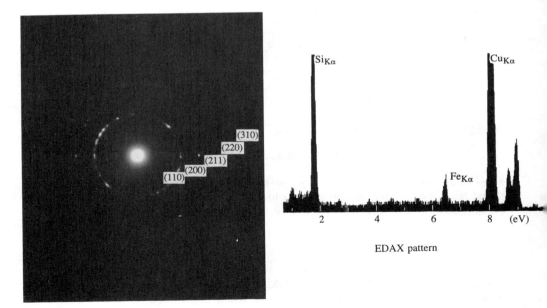

EDAX pattern

FIGURE 30 Electron-diffraction and EDAX patterns of the iron metal-doped phenylene-bridged xerogel; the dark features shown in the TEM image were identified as iron metal with a cubic structure (α-Fe).

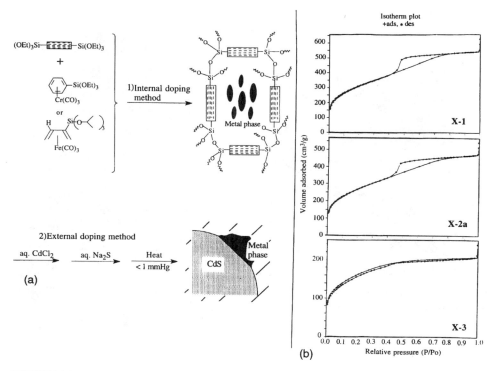

FIGURE 31 (a) Preparation of heterogeneous Cr/CdS or Fe/CdS clusters into phenylene-bridged xerogel. (b) Isotherm plots of undoped xerogel (X-1) and xerogels doped with Cr metal (X-2a) and Cr/CdS clusters (X-3).

TABLE 4 Pore Structure Analysis of X-1, X-2a, and X-3 as Determined by N_2 Adsorption/Desorption Method

Xerogel	Parameters[a]	BET	BJH
X-1	Surface area (m^2/g)	1121.56	
	V_m (cm^3/g) at STP	257.64	
	d (des.) (Å)		48.98
X-2a	Surface area (m^2/g)	963.97	
	V_m (cm^3/g) at STP	221.44	
	d (des.) (Å)		52.15
X-3	Surface area (m^2/g)	574.45	
	V_m (cm^3/g) at STP	131.96	
	d (des.) (Å)		57.78

[a]V_m: volume of sorbent; d: (des.); pore diameter calculated from desorption leg of the isotherm.

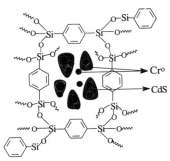

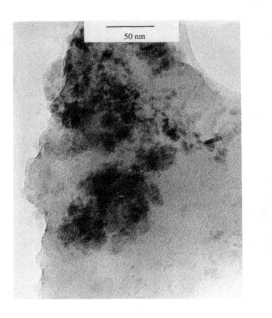

TEM image

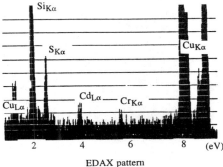

EDAX pattern

FIGURE 32 Transmission electron microscopic image and electron-diffraction pattern of Cr/CdS cluster-doped phenylene-bridged xerogel.

we would expect successful incorporation of organic NLO functionalities into modified glassy matrices [55].

High optical quality such as good microhomogeneity and low light scattering can be achieved in hybrid glasses because no phase-separation or microscopic defects are obtained. Furthermore, the hybrid glasses show a high physical strength and excellent thermal stability compared with that of devices based on polymeric films. For this reason, the potential of inserting organic functionalities into glasses by sol–gel technology has been recognized as a new approach for NLO devices which would be provided by "molecular tailoring" in order to impart specific functions into glasses. *Ormosils* (ORganically MOdified SILicates) inserted with nonlinear optical functionalities are of interest.

1. Second-Order NLO Devices Based on Alkylene-Bridged Polysilsesquioxanes Containing Side-Chain Chromophores

In this work, sol–gel processible monomers containing side-chain chromophores (so called *optiphores*) which result in a large molecular *hyperpolarizability* have been synthesized.

Nitroaniline-derived sol–gel processible monomers such as 4-nitro-*N*,*N*-bis[(3-triethoxysilyl)propyl]aniline (NTPA monomer **22**), 2,4-dinitro-*N*,*N*-bis[3-(triethoxysilyl)propyl]aniline (DNTPA monomer **23**) and *N*′,*N*′-bis[(3-triethoxy-

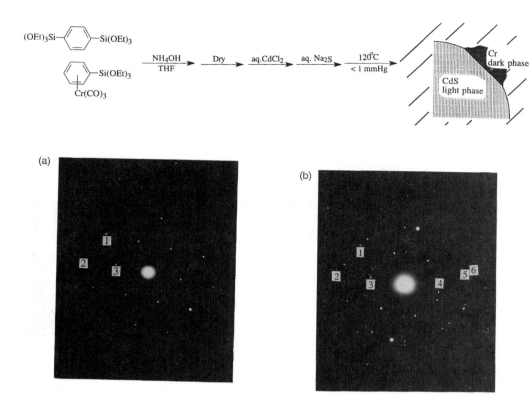

FIGURE 33 Electron-diffraction patterns focused on (a) an individual dark phase of cluster and (b) a hole cluster; part (a) shows diffraction pattern of Cr metal with a cubic structure and part (b) shows diffraction patterns of both Cr metal single crystalline and CdS polycrystalline.

silyl)propyl]dansyl sulfonamide (TPDS monomer **24**) have been prepared and cast as optically transparent sol–gel film [55]. The sol–gel polymerizable monomers containing optiphores, as shown in Fig. 34, were polymerized using formic acid. The NLO side-chain chromophores will be arranged in a direction perpendicular to the polymer backbone after polymerization. After the thin film was spin-coated (1000 rpm for 20 s) on an indium–tin–oxide (ITO) -coated glass slide, it was dried (Fig. 35).

Poling was performed using a needle-and-plane configuration with a 5-kV field and an approximate needle-to-plane distance of 1 cm. The second-order NLO signals of these films were measured by the electric-field-induced second-harmonic generation (EFISH) method [56,57].

For the experiments, nitroanaline films were cast in formic acid, followed by heating, and electric field (EF) poling for the second-harmonic generation (SHG) measurements. EF poling of sol–gel films was performed before significant cross-linking of the sol–gel matrix occurred. The sol–gel films were prepared immediately prior to EF poling and concurrent with thermal curing of the samples. Poling was conducted while monitoring the signal; the signal intensity began to appear,

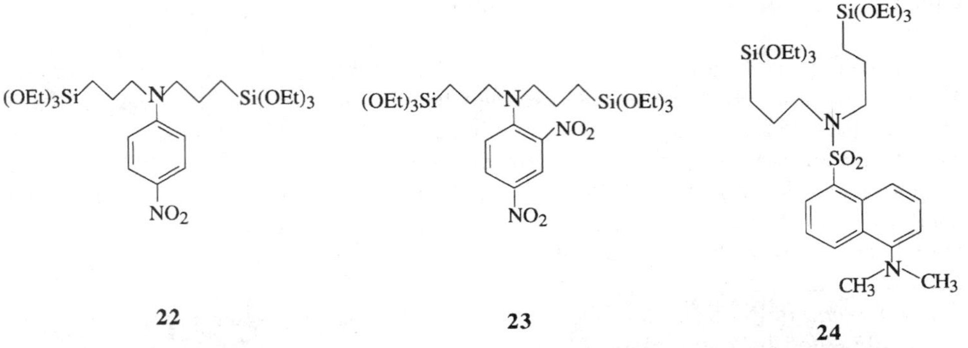

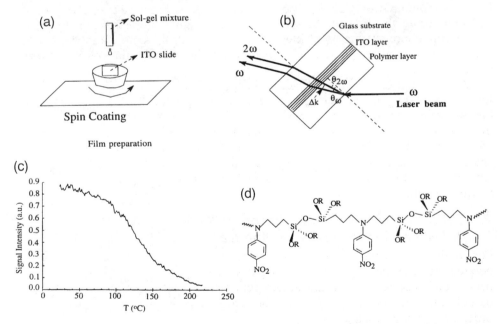

FIGURE 34 Sol–gel processible optiphores for preparation of second-order nonlinear optical devices.

FIGURE 35 Sol–gel film formation by the spin-coating method (a) and a schematic geometry of the film by beam propagation (ω: frequency of the initial laser beam; Δk: phase mismatch between incident fundamental and generated second-harmonic signals, θ_ω and $\theta_{2\omega}$: the angles of refraction of the two frequencies) (b). Second-harmonic signal loss of a poled sol–gel film derived from DNTPA versus temperature in the absence of a poling field (c) and proposed structure of the NTPA-based sol–gel film (d).

which leveled off at 180°C and then continued to increase upon cooling to about 100°C.

A YAG laser operating a 1064-nm wavelength was used with a pulse delay of 10 ns. The laser beam was split into two beams, one as a signal beam and the other as a reference beam. A 3-mm-thick Y-cut quartz crystal was used as reference. The reference was at a 6° angle to the incident laser beam for maximum signal output. The poled film was at 45° to the incident beam. A neutral density filter was used to control the second-harmonic signal intensities to accommodate a large signal range. From this experiment, the *bulk second-order nonlinear susceptibility* (d_{33}) were calculated from the intensity of reference quartz crystal using the *Maker fringe technique* [57,58], and the *electrooptic coefficients* (r_{33}) at $\lambda = 1.3$ μm have been calculated by extrapolation from the d_{33} values [59].

The SHG signals of sol–gel films derived from NTPA and DNTPA have been obtained from the experiments, and their d_{33} values were calculated. From the measurements, the second-order susceptibility (d_{33}) and electrooptical coefficient (r_{33}) for NTPA were calculated as 35–37 and 9–10 pm/V, respectively. The d_{33} and r_{33} values for DNTPA were also calculated as 9–10 and 2–2.5 pm/V, respectively. NTPA film shows comparable magnitude of the second-harmonic susceptibility as lithium niobate (LiNbO$_3$, d_{33}; 44 pm/V).

In the sol–gel films, the side-chain optiphores are oriented by poling. Upon cooling, the films exhibited thermal stability. The thermal stability of poled sol–gel film was monitored by heating a poled sample in the absence of a poling voltage and monitoring the loss of the second-harmonic signal. Linear polymers doped with an optiphore show a relatively stable signal until the temperature approaches the glass transition temperature (T_g), followed by a catastrophic loss of the SHG signal.

As shown in Fig. 35, the sol–gel film derived from DNTPA exhibited a gradual loss of signal over the temperature range 25–200°C when heated at 10°C/min, with the maximum loss occurring at about 100–150°C. The shape of the loss curve of the SHG signal versus temperature is indicative of a distribution of relaxation times of the oriented optiphores in the sol–gel matrix. The current sol–gel systems are analogous to highly cross-linked organic polymers possessing optiphores as short side chains. These side chains are oriented by poling, and upon cooling, exhibit thermal stability.

2. Third-Order NLO Devices Based on Alkylene-Bridged Polysilsesquioxanes Containing Nano-Sized Metals

For applications of third-order NLO devices, nano-sized particles have been incorporated into transparent media [32–38,60–64]. Because polysilsesquioxanes have been established as a suitable quantum confinement matrix for growth nano-sized particles, transition metal particles can be incorporated into hybrid glasses for third-order NLO device applications.

Nano-sized metal particles which are trapped in either polymeric films or inorganic oxides are expected to have optical properties similar to metal quantum wells. For phase-conjugation applications, the selection of appropriate particles for which would be expected a high *degenerate four-wave mixing* (DFWM) reflection coefficient is often complicated by the difficulty of growing crystals of suitable dimensions with the precisely desired band edge, absorption coefficient, and overall optical quality.

Semiconductor microcrystals have also been doped into transparent media to achieve third-order NLO responsive materials with significant χ^3 responses. For example, CuCl microcrystals in NaCl host crystals were prepared by the Transverse Bridgman method followed by heat treatment [61]. The mixed semiconductor CdS_xSe_{1-x} was doped into borosilicate glasses and the glass filters exhibited third-order nonlinearities of $\sim 10^{-9}$–10^{-8} esu for DFWM with short laser pulses (~ 10 ns) [64].

In this work, nano-sized Cr° and $Cr^\circ/CrOx$ systems doped into bridged polysilsesquioxanes were prepared by sol–gel polymerization for a third NLO material (Fig. 36). A different procedure than that for preparation of either homogeneous Cr metals or mixed $Cr^\circ/CrOx$ phases in xerogels from sol–gel mixture are shown. As mentioned earlier, under basic conditions, Cr metals have been liberated by thermolysis [46]. However in acidic solution, the simultaneous sol–gel copolymerization of 1,6-bis(triethoxysilyl)hexane and decomposition of an η^6-aryl chromium complex coupled with subsequent drying at 120°C in a high vacuum results in formation of $Cr^\circ/CrOx$-rich domains [65]:

$$RCr(CO)_3 + HCl \rightarrow \text{``CrO''} + CO + H_2 \qquad (5)$$

A thin sol–gel film (~ 1 mm) containing a mixed $Cr^\circ/CrOx$ phase was prepared

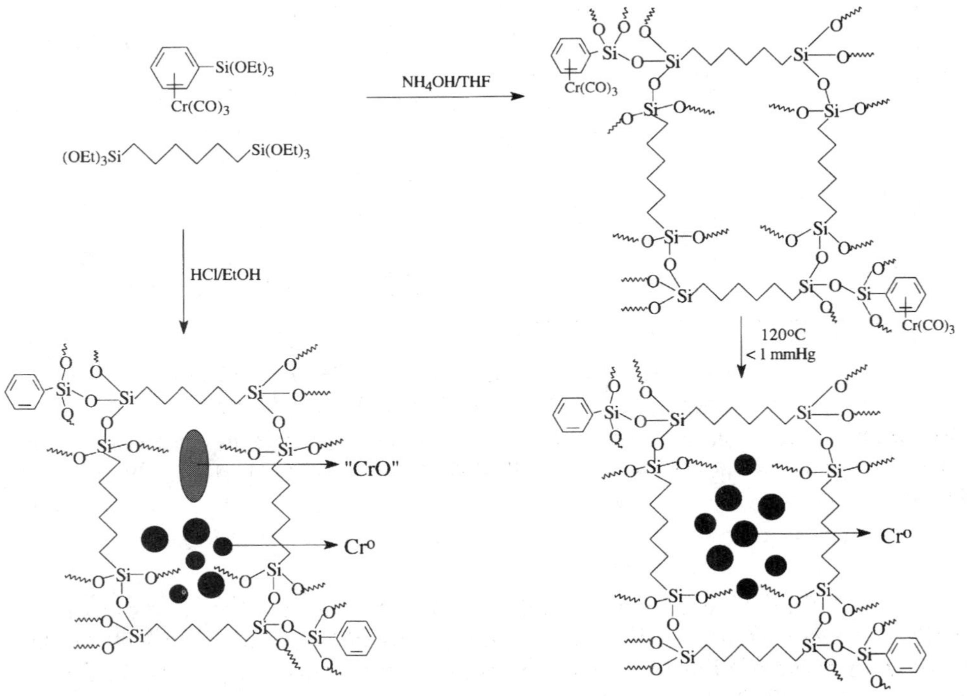

FIGURE 36 Schematic of the sol–gel conditions for preparations of either Cr metal or Cr /CrOx clusters into hexylene-bridged polysilsesquioxanes.

by copolymerization of 1,6-bis(triethoxysilyl)hexane and chromium tricarbonyl η^6-(triethoxysilyl)benzene in ethanol, catalyzed by HCl. The sol–gel polymerization was run in a flat-bottom quartz container and gelation occurred within 1 h. Final drying of the film was achieved by heating at 120°C in vacuum. A transparent green-colored sol–gel film was obtained.

Also, the heating process could result a formation of small amount of Cr metal, which means that most chromium tricarbonyl in the chromium precursor could be oxidized by acid in the earlier step to produce CrOx; however, a small amount of chromium tricarbonyl remained in nonoxidized form. Subsequently, nonoxidized chromium tricarbonyl produced small amounts of Cr metals. From the result, formation of the mixed Cr°/CrOx system in the xerogel can be expected; however, the exact chemical composition of the "CrOx" is not readily established, but it is comprised of a mixture of chromium oxide and Cr° forms. The third-order nonlinear optical properties of the film containing Cr°/CrOx have been measured by the degenerated four wave mixing technique. The sol–gel film containing mixed Cr°/CrOx system exhibits an interesting optical property showing multiple responses of *electronic*, *population*, and *density* χ^3.

3. Hybrid Materials for Optical Fibers

In optical communications, optical fibers composed of *core* and *clad* materials with different refractive indices have been used to transfer light through glass media. As shown in Fig. 37, an input electronic pulse is modulated by a laser. The input pulse must travel through the medium without significant light loss. The light pulse is demodulated to an output electronic pulse at the end of the communication system. Optimally, the optical pulses will be transferred without electronic interference. In order to reduce the optical loss that arises from light scattering or the presence of impurities, a number of engineering techniques have been employed.

Sol–gel processable monomers can be prepared with high chemical purity. With a suitable choice of organic fragment, the material can exhibit extremely low light absorption and scattering. In order to minimize light scattering, high-purity and extremely low porous xerogels have been prepared from specifically designed sol–gel monomers under acidic conditions. In addition, sol–gel technology allows us copolymerization with TEOS; thus, desired structures can be fabricated for coatings. Coating process of these sol–gel monomers on the surface of inorganic silica is also available for new structures of optical fibers with good homogeneity, low light scattering, and controllable refractive indices.

Sol–gel processable monomers have been synthesized for possible applications as optical coatings and as fiber-optic materials. Examples include monomers such as *N,N'*-bis(3-triethoxysilylpropyl)carbonate, 1,6-bis(triethoxysilyl)hexane, and 1,8-bis(triethoxysilyl)oxane. The refractive indices of bis(triethoxysilyl)hexane and bis(triethoxysilyl)oxane were measured to be 1.421 and 1.424 at 23.5°C, respectively.

In order to eliminate significant light-scattering sources, nonporous hybrid glasses were prepared. Figure 37 shows the transparency of sol–gel glasses prepared from *N,N'*-bis(3-triethoxysilylpropyl)carbonate and bis(triethoxysilyl)hexane and transmittances of these hybrid glasses. Optical fibers based on selected bridged polysilsesquioxanes could find applications as flexible replacements to inorganic fibers.

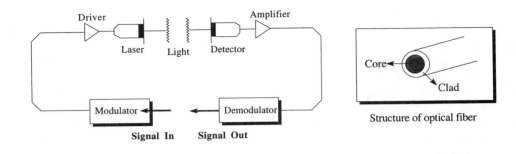

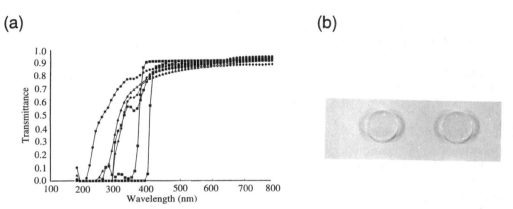

(a)   **(b)**

FIGURE 37   New optical fiber materials based on hybrid glasses for optical communication. (a) UV transmittance obtained from a variety of hybrid glasses with extremely low porosities shown in (b).

4. Fluorescent Hybrid Glasses

Fluorescent materials have been used as light-generation sources as well as analytical microenvironment probes. Their applications have extended to emergency-light devices [66], scintillation counters [67], and solid-state dye lasers [68].

Sol–gel processible monomers containing fluorescent functionalities have been synthesized for fluorescent hybrid-glass application. Two fluorescent sol–gel monomers of 9,10-bis(triethoxysilyl)anthracene and 5-(dimethylamino)-*N*-(phenyl-methyl)-1-napthalene-sulfonamide have been introduced in this work.

Figure 38 shows fluorescent excitation and emission spectra of 9,10-bis (triethoxysilyl)anthracene (monomer **4**), a xerogel copolymerized from monomer **4** and TEOS, and a xerogel copolymerized from monomer **4** and bis-(triethoxysilyl)benzene (monomer **1**). The excitation spectrum of monomer **4** (10^{-4} *M* in tetrahydrofuran) revealed an absorption maximum at 395 nm with vibrational fine structure and a mirror-image emission band with a maximum at 440 nm (Fig. 38a). While the dilute solution of monomer **4** fluoresced violet, the neat monomer or concentrated solutions appeared yellow–green (λ_{max} = 550–600 nm) due to excimer formation [69].

The excitation and emission spectra of dried xerogels copolymerized from either TEOS and monomer **4** or from bis(triethoxysilyl)benzene and monomer **4** are

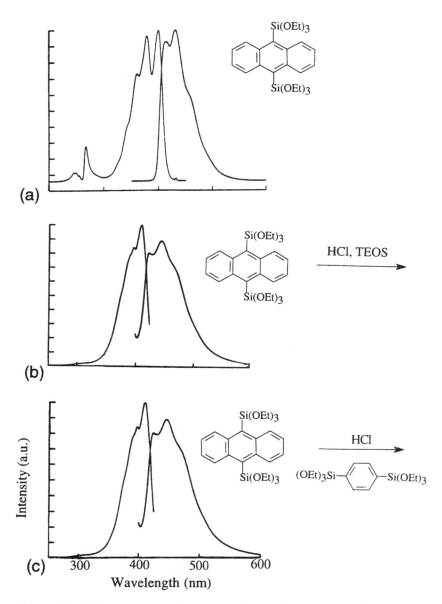

FIGURE 38 Fluorescent excitation and emission spectra of (a) 9,10-bis(triethoxy-silyl)anthracene (monomer **4**), (b) a xerogel copolymerized with monomer **4** and TEOS, and (c) a xerogel copolymerized with monomer **4** and bis(triethoxysilyl)benzene (monomer **1**).

shown in Figs. 38b and 38c, respectively. The excitation and emission spectra of these copolymerized xerogels exhibit no significant changes in emission bands, which proves that organic portions inserted in hybrid glasses maintain their optical properties in the dried xerogel.

TABLE 5 Classification of Ion-Containing Polymers

Type	Composition	Mobile species	Examples[a]
Gel polymer electrolyte	Polymer, salt, solvent	Cations anions, solvent	PVF$_2$, PC + LiClO$_4$
Ionomer	Polymeric salt	—	Nafion
Solvating polymer	Polymer complexed with salt	Cations, anions	PEO + LiClO$_4$
Solvating ionomer	Solvation polymer + ionomer	Cations, anions	

[a]PVF$_2$: poly(vinylidine fluoride); PC: propylene carbonate; PEO: poly(ethylene oxide).

C. Conducting Hybrid Glasses

Conducting polymers have been widely studied with regard to their conduction mechanisms and applications as a substitute for metals. For example, conducting polymers including doped poly(acetylene), polypyrrole, and polyaniline have been prepared by either chemical or electrochemical methods for use in, for example, polymeric batteries [70,71]. The conduction mechanisms of these organic polymer–ion conductors involve primarily *solitons*, *polarons*, or *bipolarons* and their conductivities approach 10^3–10^4 S/cm [72]. Also, a number of polymer–salt complexes have been prepared as a polymer–ion electrolyte; they are classified into four classes: *gel polymer electrolyte*, *ionomer* (*polyelectrolyte*), *solvating polymer*, and *solvating ionomer*, as listed in Table 5; the conduction mechanisms of these polymers involve ion conduction and their conductivities are close to the semiconductor region [73].

Conducting xerogels based on polysilsesquioxanes have also been achieved. Because a glass material doped with nano-sized iron particles has been shown to exhibit conductivity in the semiconductor region [48], conducting xerogels can be prepared by doping metal particles into porous matrices. In our earlier work, several procedures for preparation of nano-sized metal clusters in xerogels have been introduced and success of this approach would be depend on increasing the dopant level of metal particles. Conducting xerogels also can be achieved by doping ions from salts in porous media.

V. CONCLUSION

Bridged polysilsesquioxanes are a new family of hybrid organic–inorganic materials prepared by sol–gel chemistry. Trialkoxysilyl groups bridged by organic spacers permit the formation of network polymers capable of forming gels with a wide variety of organic functionalities, including arylene, alkylene, alkenylene, and alkynylene groups. The properties of these materials can be readily manipulated by changing the sol–gel reaction or processing conditions to afford either high-surface-area aerogels and xerogels or nonporous materials. The capability for molecular engineering based on selection of the organic spacer distinguishes bridged polysil-sesquioxanes from other sol–gel processed materials.

REFERENCES

1. (a) Schmidt, H. K., *J. Sol–Gel Sci. Technol.*, *1*, 217 (1994). (b) Sanchez, C., and Ribot F. (eds.), *First European Workshop on Hybrid Organic–Inorganic Materials*, 1993, Ch. 9. (c) Schmidt, H. K., *J. Non-Crystal Solids*, *100*, 51 (1988). (d) Wilkes, G. L., Huang, H. H., and Glaser, R. H., *Adv. Chem. Ser.*, *224*, 207 (1990). (e) Schmidt, H. K., *J. Non-Crystal Solids*, *112*, 419 (1989). (f) Wu, E., Cheu, K. C., and Mackenzie, J. D., *Res. Soc. Symp. Proc.*, *32*, 169 (1984). (g) Schmidt, H. K., *Sol–Gel Science and Technol. Mater.*, World Scientific, Singapore, 1989, p. 432.
2. (a) Sperling, L. H., *Interpenetrating Polymer Networks and Related Materials*, Plenum Press, New York, 1981. (b) Novak, B. M., *Adv. Mater.*, *5*, 422 (1993). (c) Mark, J. E., *CHEMTECH*, *19*, 230 (1989). (d) Huang, H., Orler, B., and Wikes, G. L., *Macromolecules*, *20*, 1322 (1987). (e) Avnir, D., Levy, D., and Reisfeld, R., *J. Phys. Chem.*, *88*, 5954 (1984). (f) Voronokov, M. G., and Laurentyev, V. I., *Topics Current Chem.*, *102*, 199 (1982). (g) Agaskar, P. A., Day, V. W., and Klemperer, W. G., *J. Am. Chem. Soc.*, *109*, 5554 (1987).
3. (a) Schmidt, H. K., *J. Non-Cryst. Solids*, *73*, 681 (1985). (b) Wilkes, G. L., Orler, B., and Huang, H. H., *Polym. Prepr.*, *26*, 300 (1985). (c) Wolter, H., Glaubitt, W., and Rose, K., *Mater. Res. Soc. Symp. Proc.*, *271*, 719 (1992). (d) Sur, G. S., and Mark, J. E., *Eur. Polym. J.*, *21*, 1051 (1985). (e) Wilkes, G. L., Orler, B., and Huang, H. H., *Polym. Prepr.*, *26*, 300 (1985).
4. (a) Huang, H. H., Orler, B., and Wikes, G. L., *Macromolecules*, *20*, 1322 (1987). (b) Hu, Y., and Mackenzie, J. D., *Mater. Res. Soc. Symp. Proc.*, *271*, 681 (1992). (c) Chung, Y. J., Ting, S., and Mackenzie, J. D., *Better Ceramics Through Chemistry IV*, *Vol. 180*, (B. J. J. Zelinski, C. J. Brinker, D. E. Clark, and D. R. Ulvich, eds.), Materials Research Society, 1990, p. 981.
5. (a) Morikawa, A., Iyoku, Y., Kahimoto, M., and Imai, Y., *J. Mater. Chem.*, *2*, 679 (1992). (b) Boulton, J. M., Fox, H. H., Neilson, G. F., and Uhlmann, D. R., *Better Ceramics Through Chemistry IV*, *Vol. 180*, (B. J. J. Zelinski, C. J. Brinker, D. E. Clark, and D. R. Ulvich, eds.), Materials Research Society, 1990, p. 773.
6. Brinker, C. J., and Scherer, G. W., in *Sol–Gel Science*, Academic Press, New York, 1990.
7. Fricke, J., *J. Non-Crystal. Solids*, *100*, 169 (1988).
8. (a) Shea, K. J., Loy, D. A., and Webster, O. W., *Polym. Mater. Sci. Eng.*, *63*, 281 (1990). (b) Shea, K. J., Webster, O. W., and Loy, D. A., *Mater. Res. Soc. Symp. Proc.*, *180*, 975 (1990). (c) Shea, K. J., Loy, D. A., and Webster, O. W., *J. Am. Chem. Soc.*, *114*(17), 6700 (1992). (d) Shea, K. J., and Loy, D. A., *Chem. Rev.*, *95*, 1431 (1995).
9. (a) Shea, K. J., Loy, D. A., and Webster, O. W., *Chem. Mater.*, *1*, 572 (1989). (b) Loy, D. A., Shea, K. J., and Russick, E. M., *Mater. Res. Soc. Symp. Proc.*, *271*, 699 (1992).
10. (a) Corriu, R. J. P., Moreru, J. J. E., Thepot, P., and Wong, C. M. M., *Chem. Mater.*, *4*, 1217 (1992). (b) Corriu, R. J. P., Moreau, J. J. E., Thepot, P., Man, M. W. C., Chorro, C., Lere-Porte, J. P., and Sauvajol, J. L., *Chem. Mater.*, *6*, 640 (1994).
11. Oviatt, H. W., Jr., Shea, K. J., and Small, J. H., *Chem. Mater.*, *5*(7), 943 (1993).
12. Speier, J. L., Webster, J. A., and Barnes, G. H., *J. Am. Chem. Soc.*, *79*, 974 (1957).
13. Karstedt, B. D., U.S. Patent 3,775,452 (1973).
14. Corriu, R. J. P., Moreau, J. E. E., Thepot, P., and Man, M. W. C., *J. Mater. Chem.*, *4*, 987 (1994).

15. Cerveau, G., Corriu, R., and Costa, N., *J. Non-Crystal Solids*, *163*, 226 (1993).
16. Smith, A. L., *Spectrochim. Acta*, *16*, 87 (1960).
17. (a) Lippert, J. L., Melpolder, S. B., and Kelts, L. W., *J. Non-Crystal. Solids*, *83*, 353 (1986). (b) Mulder, C. A. M., and Damen, A. A. J. M., *J. Non-Crystal. Solids*, *93*, 169 (1987). (c) Bertoluzza, A., Fagnano, C., Morelli, M. A., Gottardi, M., and Guglielmi, M., *J. Non-Crystal. Solids*, *48*, 117 (1982). (d) O'Keeffe, M., and Gibbs, G. V., *J. Chem. Phys.*, *81*, 876 (1984).
18. Eaborn, C., *Organosilicon Compounds*, Butterworth's Scientific, London, 1960, p. 19.
19. Gregg, S. J., and Sing, K. S. W., *Adsorption, Surface Area, and Porosity*, 2nd ed., Academic Press, London, 1982, p. 25.
20. (a) Brunauer, S., Emmett, P. H., and Teller, E. J., *J. Am. Chem. Soc.*, *60*, 309 (1938). (b) Barrett, E. P., Joyner, L. G., and Halenda, P. P., *J. Am. Chem. Soc.*, *73*, 373 (1951).
21. (a) Yajima, S., Hayashi, J., and Omori, M., *Chem. Lett.*, 931 (1975). (b) West, R. C., in *Ultrastructure Processing of Ceramics, Glasses, and Composites* (L. L. Hench and D. R. Ulrich, eds.), Wiley, New York, 1984, p. 235.
22. Ishida, H., Shick, R., and Hurwitz, F., *J. Polym. Sci. Polym. Phys. Ed.*, *29*, 1095 (1991).
23. Wachtman, J. B., Jr., Capps, W., and Mandel, J., *J. Mater.*, *7*(2), 188 (1972).
24. (a) Sakka, S., in *Sol–Gel Technology for Thin Films, Fibers, Preforms, Electronics, and Specialty Shapes* (L. C. Klein, ed.), Noyes Publications, Park Ridge, NY, 1988, pp. 140–161. (b) Sakka, S., In *Treatise on Materials Science and Technology*, (M. Tomozawa and R. H. Doremus, eds.), Academic Press, New York, 1982, Vol. 22, pp. 129–167.
25. Chia, T., and Hench, L. L., in *Sol–Gel Optics, Processing and Applications* (C. K. Lisa, ed.), Kluwer Academic, Boston, 1994, Chap. 22, pp. 511–538.
26. (a) Kaufman, V. R., Levy, D., and Avnir, D., *J. Non-Crystal. Solids*, *82*, 103 (1986). (b) Levy, D., and Avnir, D., *J. Phys. Chem.*, *92*, 4734 (1988).
27. (a) Domes, H., Fischer, R., Haarer, D., and Strohriegel, P., *Makromol. Chem.*, *190*, 165 (1989). (b) Tieke, B., and Chard, M. O., *Polymer*, *30*, 1150 (1989).
28. Michael, S. W., in *Siloxane Polymers* (S. J. Clarson and J. A. Semlyen, eds.), PTR Prentice-Hall, Inc., Englewood Cliffs, NJ, 1993, Chap. 6, p. 287.
29. (a) Huang, H. H., Orler, B., and Wilkes, G. L., *Macromolecules*, *20*, 1322 (1987). (b) Kohjiya, S., Ochial, K., and Yamashita, S., *J. Non-Crystal. Solids*, *119*, 132 (1990).
30. Levy, D., Einhorn, S., and Avnir, D., *J. Non-Crystal. Solids*, *113*, 137 (1989).
31. (a) Brus, L. E., *J. Chem. Phys.*, *79*(11), 5566 (1983). (b) Brus, L. E., *J. Chem. Phys.*, *80*(9), 4403 (1984).
32. (a) Kuczynski, J. P., Milosavljevic, B. H., and Thomas, J. K., *J. Phys. Chem.*, *88*(5), 980 (1984). (b) Meissner, D., Memming, R., and Kastening, B., *Chem. Phys. Lett.*, *96*, 34 (1983). (c) Wang, Y., Suna, A., Mahler, W., and Kasowski, R., *J. Chem. Phys.*, *87*(12), 7315 (1987).
33. (a) Kuczynski, J. P., and Thomas, J. K., *J. Phys. Chem.*, *89*, 2720 (1985). (b) Weller, H., *Angew. Chem., Int. Ed. Engl.*, *32*, 1 (1993).
34. (a) Takada, T., Yano, T., Yasumori, A., Yamane, M., and Mackenzie, J. D., *J. Non-Crystal. Solids*, *147*, 631 (1992). (b) Lukehart, C. M., Carpenter, J. P., Milne, S. B., and Burnam, K. J., *Chemtech*, *23*, 29 (1993).
35. Wang, Y., and Herron, N., *J. Phys. Chem.*, *91*, 257 (1987).
36. Tricot, Y. M., and Fendler, J. H., *J. Phys. Chem.*, *90*, 3369 (1986).
37. (a) Hoffman, A. J., Yee, H., Mills, G., and Hoffman, M. R., *J. Phys. Chem.*, *96*, 5540 (1992). (b) Hoffman, A. J., Mills, G., Yee, H., and Hoffman, M. R., *J. Phys. Chem.*, *96*, 5546 (1992).
38. Weller, H., Fojtik, A., and Henglein, A., *Chem. Phys. Lett.*, *117*(5), 485 (1985).
39. Wang, Y., Suna, A., and Mahler, W., *Mater. Res. Soc. Symp. Proc.*, *109*, 187 (1988).

40. (a) Choi, K. M., and Shea, K. J., *Chem. Mater.*, *5*, 1067 (1993). (b) Choi, K. M., and Shea, K. J., *J. Phys. Chem.*, *98*(12), 3207 (1994).

41. (a) Sanchez, C., and Ribot, F., in *First European Workshop on Hybrid Organic Inorganic Materials*, 1993. (b) Fukumi, K., Chayahara, A., Kadono, K., Sakaguchi, T., Horino, Y., Miya, M., Fujii, K., Hayakawa, J., and Satou, M., *J. Appl. Phys.*, *75*(6), 3075 (1994).

42. Choi, K. M., Hemminger, J. C., and Shea, K. J., *J. Phys. Chem.*, *99*, 4720 (1995).

43. (a) Pankove, J. I., in *Optical Processes in Semiconductors*, Dover, New York, 1971, p. 36. (b) Dutton, D., *Phys. Rev.*, *88*, 1861 (1988).

44. (a) Colvin, V. L., Goldstein, A. N., and Alivisatos, A. P., *J. Am. Chem. Soc.*, *114*, 5221 (1992). (b) Lichtensteiger, M., Webb, C., and Lagowski, J., *Surf. Sci.*, *97*, L375 (1980). (c) Nuzzo, R. G., Zegarski, B. R., and Dubois, L. H., *J. Am. Chem. Soc.*, *109*, 733 (1987).

45. O'Neil, M., Marohn, J., and McLendon, G., *J. Phys. Chem.*, *94*, 4356 (1990).

46. (a) Choi, K. M., and Shea, K. J., *J. Am. Chem. Soc.*, *116*, 9052 (1994). (b) Choi, K. M., and Shea, K. J., *Mater. Res. Soc. Symp. Proc.*, *346*, 763 (1994).

47. Choi, K. M., and Shea, K. J., *J. Sol–Gel Tech.*, *5*, 143 (1995).

48. Roy, S., and Chakravorty, D., *J. Mater. Res.*, *9*(9), 2314 (1994).

49. (a) Meredith, G. R., Buchalter, B., and Hanzlik, C., *J. Chem. Phys.*, *78*, 1533 (1983). (b) Borrelli, N. F., *Phys. Chem. Glass*, *12*, 93 (1971). (c) Dory, M., Bodart, V. P., Delhalle, J., Andre, J. M., and Bredas, J. L., *Mater. Res. Soc. Symp. Proc.*, *109*, 239 (1988).

50. (a) Marder, S. R., Beratan, D. N., and Cheng, L. T., *Science*, *252*, 103 (1991). (b) Robello, D. R., Dao, P. T., Phelan, J., Revelli, J., Schildkraut, J. S., Scozzafava, M., Ulman, A., and Willand, C. S., *Chem. Mater.*, *4*, 425 (1992). (c) Osaheni, J. A., Jenekhe, S. A., Vanherzeele, H., Meth, J. S., Sun, Y., and MacDiarmid, A. G., *J. Phys. Chem.*, *96*, 2830 (1992).

51. Singer, K. D., Kuzyk, M. G., and Sohn, S. E., in *Nonlinear Optical and Electroactive Polymers* (P. N. Prasad and D. R. Ulrich, eds.), Plenum Press, New York, 1988.

52. (a) Haruvy, Y., and Webber, S. E., *Chem. Mater.*, *4*, 89 (1992). (b) Griesmar, P., Sanchez, C., Puccetti, G., Ledoux, I., and Zyss, J., *Molec. Eng.*, *1*, 205 (1991). (c) Toussaere, E., Zyss, J., Griesmar, P., and Sanchez, C., *Nonlinear Opt.*, *1*, 349 (1991). (d) Zhang, Y., Prasad, P. N., and Burzynski, R., *Chem. Mater.*, *4*, 851 (1992).

53. Ye, C., Marks, T. J., Yang, J., and Wong, G. K., *Macromolecules*, *20*, 2324 (1987).

54. Robello, D. R., Ulman, A. Willand, C. S., and Williams, D. J., U.S. Patent 4,796,971 (1989).

55. Oviatt, H. W., Jr., Shea, K. J., Kalluri, S., Shi, Y., Steier, W., and Dalton, L. R., *Chem. Mater.*, *7*(3), 493 (1995).

56. Prasad, P. N., and Williams, D. J., in *Introduction to Nonlinear Optical Effects in Molecules and Polymers*, Wiley, New York, 1991.

57. Mortazavi, M. A., Knoesen, A., Kowel, S. T., Higgins, B. G., and Dienes, A., *J. Opt. Soc. Am. B*, *6*, 733 (1988).

58. Jerphagnon, J., and Kurtz, S. K., *J. Appl. Phys.*, *41*(4), 1667 (1970).

59. Shi, Y., Ranon, P. M., Steier, W. H., Xu, C., Wu, B., and Dalton, L. R., *Appl. Phys. Lett.*, *63*(16), 2168 (1993).

60. Wang, Y., *J. Am. Chem. Soc.*, *116*, 397 (1994).

61. Masumoto, Y., Yamazaki, M., and Sugawara, H., *Appl. Phys. Lett.*, *53*(16), 1527 (1988).

62. (a) Schmid, G., *Chem. Rev.*, *92*, 1709 (1992). (b) Brust, M., Walker, M., Bethell, D., Schiffrin, D. J., and Whyman, R., *J. Chem. Soc., Chem. Commun.*, 801 (1994).

63. Akers, K. L., Cousins, L. M., and Moskovits, M., *Chem. Phys. Lett.*, *190*(6), 614 (1992).

64. (a) Hinsch, A., and Zastrow, A., *J. Non-Crystal. Solids*, *147/148*, 579 (1992). (b) Jain, R. K., and Lind, R. C., *J. Opt. Soc. Am.*, *73*(5), 647 (1983). (c) Schmidt, H. K., *J. Sol–Gel Technol.*, *1*, 217 (1994).

65. (a) Gribov, B. G., Mozzhukhin, D. D., Kozyrkin, B. I., and Strizhkova, A. S., *J. Gen. Chem. USSR* (Engl. Transl.), *42*, 2521 (1972). (b) Fischer, E. U., and Brunner, H., *Chem. Ber.*, *95*, 1949 (1962).

66. Renschler, C. L., Clough, R. L., and Shepodd, T. J., *J. Appl. Phys.*, *66*, 4542 (1989).

67. Neame, K. D., and Homewood, C. A., in *Liquid Scintillation Counting*, Wiley, New York, 1974.

68. Peterson, O. G., and Snavely, B. B., *Appl. Phys. Lett.*, *12*, 238 (1968).

69. (a) Coffey, S., and Van Alphen, J., in *Chemistry of Carbon Compounds* (E. H. Rodd, ed.), Elsevier, Amsterdam, (1956). Vol. IIIB, p. 1361. (b) Guillet, J., in *Polymer Photophysics and Photochemistry*, Cambridge University Press, Cambridge, 1985, pp. 141–194.

70. Chiang, C. K., Druy, M. A., Gau, S. C., Hegger, A. J., Louis, E. J., MacDiarmid, A. G., Park, Y. W., and Shirakawa, H., *J. Am. Chem. Soc.*, *100*, 1013 (1978).

71. (a) Mermilliod, N., Tanguy, J., and Petiot, F., *J. Electrochem. Soc.*, *133*(6), 1073 (1986). (b) Pfluger, P., Krounbi, M., Street, G. B., and Weiser, G., *J. Chem. Phys.*, *78*(6), 3212 (1983). (c) Kitani, A., Kaya, M., and Sasaki, K., *J. Electrochem. Soc.*, *133*(6), 1069 (1986).

72. Skotheim, T. A., in *Handbook of Conducting Polymers*, *Vol. 2*, Marcel Dekker, Inc., New York, 1986.

73. Ferraro, J. R., and Williams, J. M., in *Introduction to Synthetic Electrical Conductors*, Academic Press, New York, 1987.

13

Morphology and Structure of Bilayered Polymer Assemblies with Nonlinear Optical Properties

Li-Sheng Li

University of Illinois at Urbana-Champaign
Urbana, Illinois

Polymers are a new class of promising nonlinear optical materials. The most important step in developing this class of materials is to optimize their optical nonlinearity at the molecular level; then a detailed understanding of the relationship between the optical nonlinearities and morphology as well as structure of bulk materials will be of fundamental importance. In this chapter, two types of polymer will be discussed in detail. One is the chiral polymer with either one-dimensional (comb or comb-ladder) or two-dimensional (molecular sheet) molecular architectures which self-assemble into bilayered structures in liquid-crystalline or crystalline phases. These polymers exhibit second-order nonlinear optical properties. The other type of polymer is the bilayered polydiacetylene single crystal with third-order nonlinear optical properties and chromic transitions.

I. ONE-DIMENSIONAL AND TWO-DIMENSIONAL CHIRAL POLYMERS

A. Molecular Precursors of the Polymers

The molecular precursors of these polymers are enantiomerically enriched oligomers (R and S enantiomers) containing an acrylate at one terminus and a highly dipolar nitrile group at the stereocenter of each molecule. The chemical structures of the oligomers R-1 and S-1 are [1,2]

R-1

S-1

These oligomers easily self-assemble into a layered structure. The morphology and structure of films of the oligomers R-1 and S-1 are similar. Figure 1a is an electron micrograph of a film of the oligomer R-1 [obtained by solution-casting (0.1% by weight in chloroform) at room temperature], showing self-assembled layers stacked in parallel. The small-angle electron-diffraction pattern inset in Fig. 1a reveals several orders of small-angle reflections, indicating that the thickness of the layers is very regular and is equal to 303 Å. Figure 1b is an electron-diffraction pattern of the oligomer R-1 film, oriented properly with respect to the image shown in Fig. 1a. This pattern contains both small-angle and wide-angle reflections and is similar to a fiber pattern with c^* (the layer normal) as the fiber axis. The reflections in the pattern can be indexed by the following monoclinic unit cell parameters: $a = 11.62$ Å, $b = 5.41$ Å, and $c = 369$ Å, with $\beta = 124.8°$. The c axis (oligomer chain axis) in the layer is tilted by $34.8°$ with respect to the layer normal (c^* axis). Because the length of an extended oligomer is ~46 Å, there must be eight molecular sublayers in each layer and possibly with head-to-head and tail-to-tail packing of the oligomers. According to the fiber-like electron-diffraction pattern shown in Fig. 1b, the molecular tilt directions within each layer form a helical distribution on moving from the first sublayer to the eighth sublayer, completing $360°$ rotation of the c axis about the c^* axis. This is a chiral smectic H phase (S_H^*) [2,3].

When the oligomer film is poled in an external electric field of 67 KV/cm at 80°C to create a macroscopic noncentrosymmetric structure, the second-order susceptibility $\chi^{(2)} = 1.25 \times 10^{-9}$ esu (the fundamental wavelength = 1064 nm) [4]. If the oligomer film is doped with 2% by weight of disperse red 1 (4-[ethyl(2-hydroxyethyl) amino]-4'-nitroazobenzene), $\chi^{(2)}$ increases to 4.75×10^{-9} esu [4]. However, the large second-harmonic signals of the film generated by poling decay gradually due to formation of the chiral smectic H phase of the oligomers, in which a cancellation of the second-order nonlinear effect occurs.

B. The One-Dimensional Polymer

The one-dimensional (comb or comb-ladder) side-chain polymers R-2 or S-2 can be synthesized through dilute-solution free-radical polymerization of R-1 or S-1 [3,5]. The chemical structure of the side-chain polymer S-2 is shown by

S-2

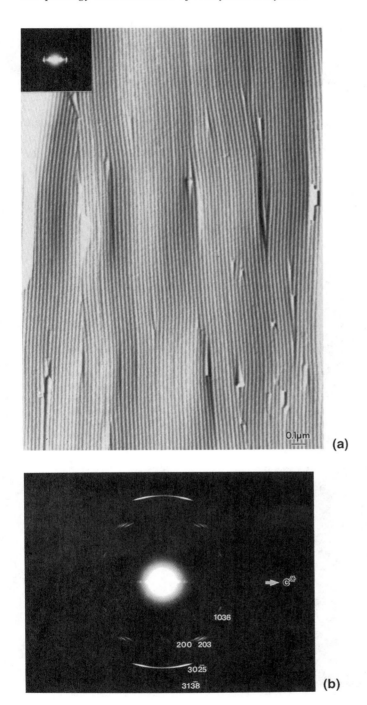

(a)

(b)

FIGURE 1 (A) Electron micrograph of a film of the oligomer R-1 with an inset of a small-angle electron-diffraction pattern oriented properly with respect to the image, showing self-assembled layers stacked in parallel. (B) Electron-diffraction pattern of the film shown in (A), oriented properly with respect to the image.

This side-chain polymer exhibits two different types of liquid-crystalline structures (the chiral smectic E and the chiral smectic H_c phases) depending on the temperature at which the sample is prepared [3].

1. The Chiral Smectic E Phase (S_E^*) of the Polymer

Thin films of the polymer S-2 can be prepared by casting the polymer solution (0.1% by weight in chloroform) on a glycerin surface preheated to 110°C and subsequently cooled to room temperature. Figure 2a is an electron micrograph of the film, showing that the film is composed of layers stacked in parallel with the normal to the layer planes parallel to the film's normal. Figure 2b is a selected-

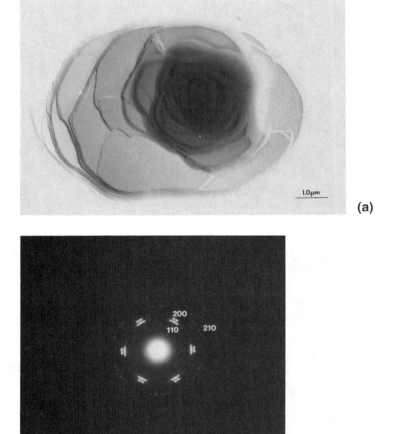

(a)

(b)

FIGURE 2 (A) Electron micrograph of a film of the one-dimensional side-chain polymer S-2 in the S_E^* phase. (B) Selected-area electron-diffraction pattern taken from a thin region of the film shown in (A).

area electron-diffraction pattern taken from a thin region (~50 Å thick) of the film, revealing reflection spots distributed with a hexagonal geometry. However, this pattern cannot be indexed by hexagonal unit-cell parameters, requiring instead the following orthorhombic unit-cell parameters: $a = 8.24$ Å, and $b = 5.34$ Å, with $\gamma = 90°$. Because this pattern contains only $hk0$ reflections, the c axis (side-chain axis) must be parallel to the beam direction and, therefore, perpendicular to the layer planes. The hexagonal geometry of the distribution of the $hk0$ reflections in the pattern can be explained by three different orientations of identical orthorhombic reciprocal lattices related to one another by rotations of $\pm 60°$ about the c axis, as shown schematically in Fig. 3a and 3b for a single layer. However, the pattern in Fig. 3b contains only two 110, one 200, and two 210 reciprocal lattice points. The experimental pattern shown in Fig. 2b reveals three 110 reflections, two 200 reflections, and four 210 reflections. It is very important to see that in the pattern of Fig. 2b, the angle between every two neighboring [110]* reciprocal vectors and between the two [200]* reciprocal vectors is 5.9°. If the reciprocal lattices of two neighboring layers are rotated by 5.9° relative to each other about the c axis, one observes exactly the pattern shown in Fig. 2b. Therefore, the pattern shown in Fig. 2b is produced by a bilayer structure in which the two sublayers are twisted by 5.9° relative to each other about the c axis. Figure 3c shows the superposition of the orthorhombic $a*b*$ reciprocal lattice planes in two sublayers rotated by 5.9° relative to each other about the c axis, with each sublayer consisting of three different orientations of the orthorhombic $a*b*$ reciprocal lattice planes related to one another by rotations of $\pm 60°$ about the c axis. The simulated pattern shown in Fig. 3c matches the electron-diffraction pattern (Fig. 2b) very well, indicating that each layer is really composed of two sublayers rotated by 5.9° relative to each other about the c axis. Figure 4 is another selected-area electron-diffraction pattern taken from a different region of the thin film of the polymer, showing five 110 and four 200 reflections. Again, the angle between every two neighboring [110]* or every two neighboring [200]* is 5.9°. Therefore, this pattern can be interpreted as one resulting from two bilayers (four sublayers). The 5.9° rotation angle between sublayers is related to the reciprocal lattice parameters and is characteristic of this polymer. It is important to point out that the number of 110 reflections in the electron-diffraction patterns is always odd and larger than or equal to 3, and the number of 200 reflections is always even and larger than or equal to 2. This implies, of course, the presence of a bilayer structure. Furthermore, x-ray diffraction data shown in Fig. 5 reveal four sharp reflections (001, 002, 003, and 004), indicating that the thickness of the layers is 52.6 Å. This thickness is larger than the length of one fully extended side chain and shorter than that of the two. This also indicates that a bilayer structure exists in the polymer film. In addition, the second-order reflection of the side-chain polymer in the x-ray diffractogram is much stronger than the first one. This implies that the electron-density profile along the c axis within each bilayer is not a single sinusoidal function [6], which suggests that the backbones and flexible spacers are confined in the central region of the bilayer with lower electron density. Figure 6 is a schematic representation of the bilayer structure of the polymer. The observed $\pm 60°$ rotations of the orthorhombic lattices about the c axis in each sublayer are dynamic phenomena, which originate from the $\pm 60°$ rotations of side chains about the c axis. If there were no twisting between sublayers in the film, the structure would be typical of that smectic E phase. How-

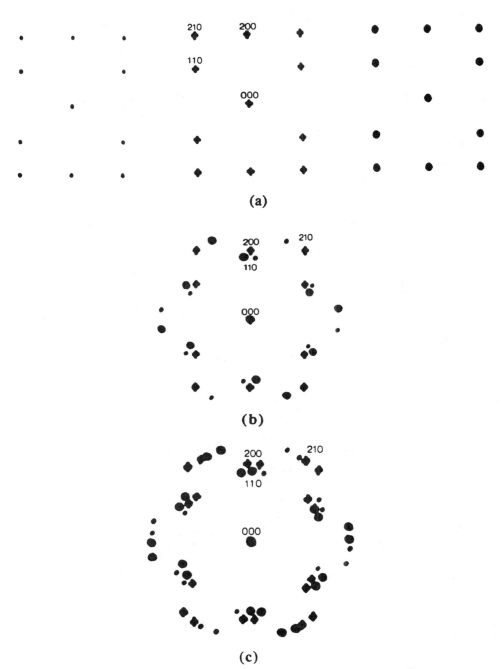

FIGURE 3 (a) Schematic representation of three identical orthorhombic $a*b*$ reciprocal lattice planes. (b) Three different orientations of the orthorhombic $a*b*$ reciprocal lattice planes related to one another by rotations of $\pm 60°$ about the c axis. (c) Schematic representation of $a*b*$ reciprocal lattice planes corresponding to two sublayers rotated by $5.9°$ relative to each other about the c axis.

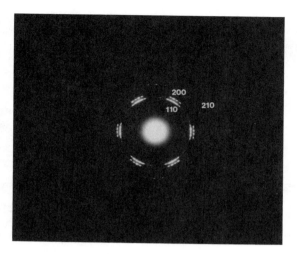

FIGURE 4   Selected-area electron-diffraction pattern of two bilayers (four sublayers) of the side-chain polymer S-2 in the S_E^* phase.

ever, there is a 5.9° rotation about the c axis between adjacent sublayers, which indicates that the side-chain polymers are in a liquid-crystalline chiral smectic E phase (S_E^*) [3].

Usually in a chiral smectic phase, molecules are tilted with respect to the smectic layer normal and the molecular tilt directions twist in a helical fashion as layers are traversed [7]. Now, in the S_E^* phase, the side chains are parallel to the layer normal. The absence of the tilt of the side-chain axes in the S_E^* phase is

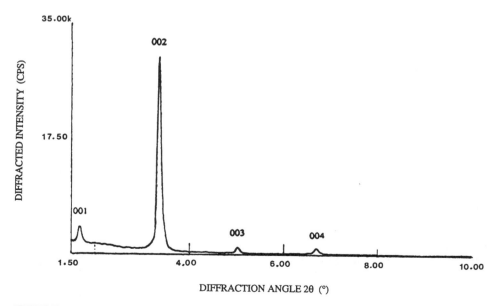

FIGURE 5   X-ray diffractogram of the side-chain polymer S-2 film in the S_E^* phase.

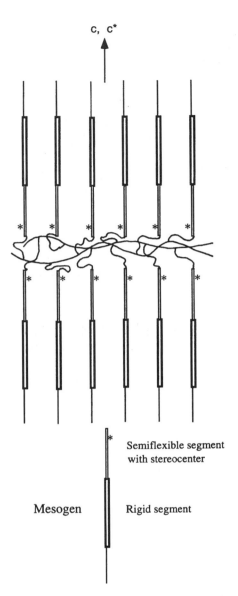

c, c*

* * * * * *

* * * * * *

*

Semiflexible segment
with stereocenter

Mesogen   Rigid segment

FIGURE 6 Schematic representation of a bilayer structure of the side-chain polymer in the S_E^* phase, showing side chains packed in parallel with polymer backbones and spacers confined in a thin disordered region between two mesogenic sublayers.

probably the consequence of partial reaction among nitrile groups at 110°C during sample preparation. Therefore, in some local regions of the film, the molecular architecture could be comb-ladder-like.

2. The Chiral Smectic H_c Phase (S^*_{Hc}) of the Polymer

When films of the side-chain polymer S-2 are prepared by casting a solution (0.1% by weight in chloroform) on a glycerin surface at 77°C and subsequently cooling them to room temperature, the films experience no reaction among nitrile groups and thus have a comb-shaped molecule. Under such preparation conditions, a very different morphology from that of the S^*_E phase is observed. Figure 7a is an electron micrograph taken from the thinnest region of such a film, showing tilted layers in fan-shaped domains. There are also some defects shown in the micrograph, such as edge dislocations and curved layers. Figure 7b is a selected-area electron-diffraction pattern taken from one of the domains. This pattern is similar to that of the chiral smectic E phase. The 110 and 200 reflection spots are separated and both of them are distributed with a hexagonal geometry. This indicates that there are also orthogonal a^*b^* reciprocal lattice planes with the c axis (side-chain axis) parallel to the beam direction. Therefore, the c axis (side-chain axis) is tilted with respect to the layer normal (c^* axis). Because the $hk0$ reflection spots are distributed with a hexagonal geometry, there must be three different orientations of the orthogonal a^*b^* reciprocal lattice planes related to one another by rotations of $\pm60°$ about the c axis. There are many 110 and 200 reflection spots, and the angle between every two neighboring $[110]^*$ or every two neighboring $[200]^*$ is 5.9°. Therefore, each tilted layer is also a bilayer which is composed of two sublayers rotated by 5.9° relative to each other about the c axis, and in each sublayer there are three different orientations of the orthogonal ab lattice planes related to one another by rotations of $\pm60°$ about the c axis. Figure 8a is an x-ray diffractogram of the polymer film, revealing four sharp Bragg reflections (001, 002, 003, and 004) from the tilted layers, corresponding to a period of 51.4 Å. Here, again, the second-order reflection is the strongest one, indicating that the electron-density profile along the c^* direction within the bilayer is not a single sinusoidal function [6]. This suggests, again, that the backbones and flexible spacers are confined in the central region of the bilayer. Figure 8b shows the same x-ray-diffraction results, but with a magnified intensity scale. In addition to the four sharp peaks, other broader peaks appear, corresponding to the diffraction from reciprocal planes perpendicular to the side-chain axes. Because in the side-chain axis direction there are displacements of parallel side chains which are not correlated from one to another in the tilted layers, the peaks are broad. The d-spacings of the broader peaks indicate that the period in the side-chain axis direction is 62.6 Å. Therefore, the side-chain axes are tilted by 34.8° with respect to the layer normal.

According to the electron-diffraction and x-ray diffraction results, the unit cell of the polymer film has monoclinic symmetry, and the unit-cell parameters are as $a = 10.06$ Å, $b = 5.35$ Å, and $c = 62.6$ Å, with $\beta = 124.8°$. Figure 9 is a schematic representation of the tilted bilayer structure of the polymer.

Based on the electron microscopy, electron-diffraction, and x-ray-diffraction results, it is clear that the c-axis (side-chain axis) direction is fixed and is perpendicular to the polymer film plane. The rotations of $\pm60°$ of the monoclinic lattices in each sublayer and the 5.9° rotation between sublayers are about the c axis. The c^* directions (the layer normal directions) change as layers are traversed. This can even be seen in the image shown in Fig. 7a, in which the layer normals are rotated

(A)

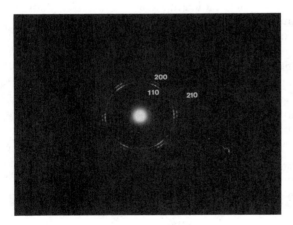

(B)

FIGURE 7 (A) Electron micrograph of a thin film of the side-chain polymer in the S_{Hc}^* phase, showing tilted layers in fan-shaped domains. (B) Selected-area electron-diffraction pattern taken from one of the domains shown in (A), revealing reflection spots distributed with a hexagonal geometry.

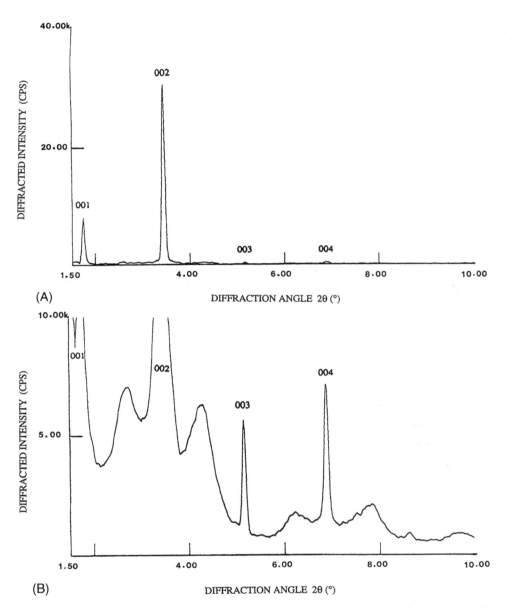

FIGURE 8 (A) X-ray diffractogram of a film of the side-chain polymer S-2 in the S_{Hc}^* phase. (B) The x-ray diffractogram shown in (A) with a magnified intensity scale.

about the c axis from layer to layer forming fan-shaped domains with edge dislocations to fill the space. This is a chiral smectic H_c phase (S_{Hc}^*) [3]. This phase is different from the chiral smectic H phase (S_H^*) of the oligomer R-1, in which the c^* direction (the layer normal direction) remains constant and the c-axis directions

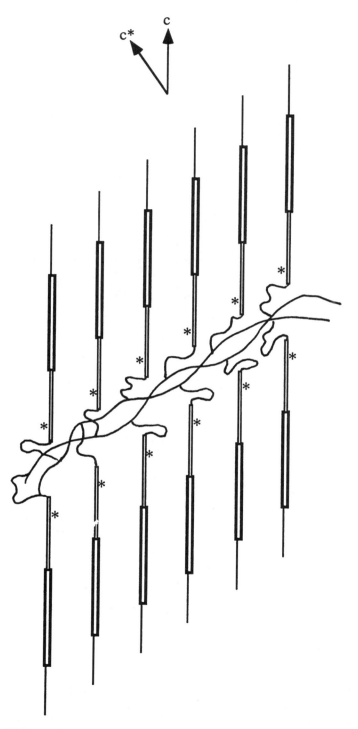

FIGURE 9 Schematic representation of a tilted bilayer structure of the side-chain polymer S-2 in the S_{Hc}^* phase, showing side chains packed in parallel with polymer backbones and spacers confined in a thin disordered region between two mesogenic sublayers.

(molecular chain axis directions) twist in a helical fashion while moving from sublayer to sublayer.

The $\chi^{(2)}$ of the pure polymer film after poling (67 kV/cm) at 80°C is equal to 3.0×10^{-10} esu (the fundamental wavelength = 1064 nm) [4]. When the polymer is doped with 2% disperse red 1 and poled at 80°C, the $\chi^{(2)}$ increases to 2.2×10^{-9} esu [4]. However, the strong second-harmonic signals of the film generated by poling decay even faster than those of the poled oligomers.

C. The Two-Dimensional Polymer

Thin films of the oligomer R-1 or S-1 can be polymerized in bulk simply by heating at temperatures above 100°C for 20 h and subsequently cooling to room temperature [1,2]. It is important to see that the morphology and structure of the heated films change dramatically. Figure 10a is an electron micrograph of the oligomer R-1 film after heating at 125°C for 20 h. This figure no longer shows edge-on layers with layer normals perpendicular to the film's normal (Fig. 1a). Instead, flat and plate-like layers stacked in parallel, and with the layer normal parallel to the film's normal, appear. Figure 10b is an electron-diffraction pattern of the heated film. This pattern is completely different from that of the oligomer film shown in Fig. 1b. This pattern shows an orthorhombic $a*b*$ reciprocal lattice plane, indicating that the layered structure shown in Fig. 10a is a single crystal. X-ray diffraction of the film reveals four reflections (002, 003, 004, and 005), indicating that the thickness of each layer is 50.2 Å (Fig. 11). Based on the electron-diffraction and x-ray-diffraction results, the orthorhombic unit-cell parameters of the single crystal are $a = 8.38$Å, $b = 10.56$ Å, and $c = 50.2$ Å, with $\alpha = \beta = \gamma = 90°$. It is important to point out that the thickness of each layer (50.2 Å) is larger than the length of one fully extended oligomer and is significantly smaller than that of the two oligomers (~92 Å). This indicates that each layer is a bilayer which is similar to that of the comb-ladder polymer S-2 in the S_E^* phase. This also implies that the acrylates of the oligomers have been reacted among each other during annealing at 125°C. It can be seen from the x-ray diffractogram (Fig. 11) that the second-order reflection is the strongest one and the first-order reflection cannot even be seen. This implies again that the electron-density profile along the c axis within each layer is not a single sinusoidal function [6]. Therefore, it is possible that the polyacrylate backbones and flexible spacers are confined in the central region of the bilayer. Furthermore, the unit-cell parameter b of the single crystal is about twice as large as that of the oligomers in the S_H^* phase or the side-chain polymers in the S_E^* and S_{Hc}^* phases. This suggests that chemical reactions among nitrile groups have also taken place forming imine bonds that connect two or more oligomers in a plane in the sublayer near the central region of each bilayer. The chemical reactions among acrylate groups and among nitrile groups catenate the oligomers through two types of stitching into a bilayer two-dimensional (2D) polymer which results in dramatical changes in the morphology and structure of the oligomer film. The evidence for such reactions are further provided with differential scanning calorimetry, ^1H and ^{13}C nuclear magnetic resonance (NMR), infrared spectra, as well as gel permeation chromatography [1,2]. It is important to note that the $\pm 60°$ and 5.9° rotations of the ab lattice planes about the c axis disappear in the 2D-polymer single crystal, indicating a much lower mobility of the 2D polymer in

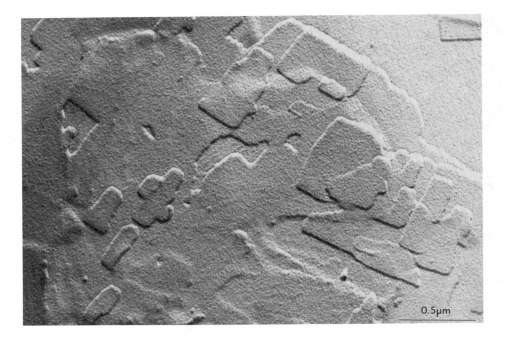

(A)

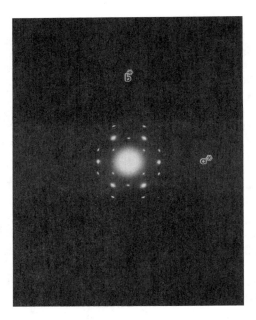

(B)

FIGURE 10 (A) Electron micrograph of a film of the oligomer R-1 after heating at 125°C for 20 h and subsequently cooling to room temperature. (B) Selected area electron diffraction pattern of the heated film shown in (A).

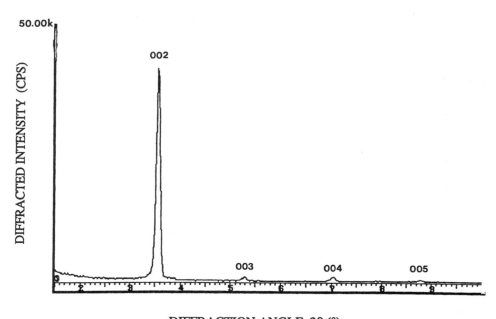

FIGURE 11 X-ray diffractogram of a heated film of the oligomer R-1 (two-dimensional polymer).

comparison with its one-dimensional (1D) analogs (comb or comb-ladder polymers). This also implies that a high degree of nitrile reaction has taken place in the 2D-polymer single crystal.

The $\chi^{(2)}$ of the 2D-polymer film (obtained by heating the oligomer R-1 film at 110°C in a DC electric field and subsequently poling in the electric field of 67 kV/cm at 80°C) is equal to 1.07×10^{-9} esu) (the fundamental wavelength = 1064 nm) [4]. If the film of the 2D polymer is doped with 2% of disperse red 1, the $\chi^{(2)}$ increases to 3.5×10^{-9} esu [4].

Now, it is very interesting to note the extent to which the architecture of the chiral polymers and the morphology as well as structure of their films affect their nonlinear optical properties. Figure 12 shows the time dependence of the second-harmonic laser beam intensity (532 nm) in films of the doped (2% of disperse red 1) 2D polymer and 1D-comb polymer after poling in the electric field of 67 kV/cm at 80°C [1,4]. It can be seen from Fig. 12 that the intensity of the second-harmonic generation (SHG) of the doped 2D-polymer film remains unchanged for a period exceeding 1 year. In contrast, the film of the doped 1D-comb polymer exhibits a rapid decay of the second-harmonic signal with time. The difference between the 2D and 1D polymers is that the 1D-comb polymers form a bilayer chiral smectic H_c phase (S_{Hc}^*), in which the side chains are highly mobile and rotate by $\pm 60°$ about the c axis (side-chain axis) with high frequency and, at the same time, the lattices in each sublayer are regularly rotated by 5.9° about the c axis relative to those in the adjacent sublayer. Therefore, the cancellation of the second-order nonlinear effect occurs, resulting in a rapid decay of the SHG intensity of

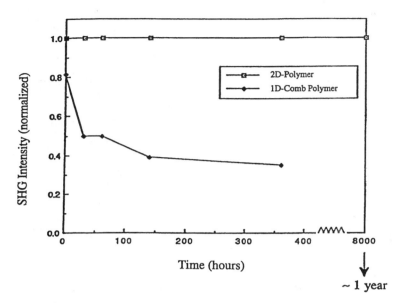

FIGURE 12 Time dependence of the second-harmonic laser beam intensity (532 nm) in doped (2% disperse red 1) films of the 2D polymer and 1D-comb polymer after poling in the electric field of 67 kV/cm at 80°C. (Courtesy of H. C. Lin.)

the 1D-polymer film after poling. In contrast, the 2D polymers form single crystals in which the ±60° and 5.9° rotations of the *ab* lattice planes about the *c* axis vanish and the side chains are much less mobile than the 1D polymers in the S_{Hc}^* phase. Therefore, the SHG intensity of the 2D-polymer film remains unchanged for a long time after poling. The 2D polymers might be a promising nonlinear optical material.

II. THE BILAYERED POLYDIACETYLENE SINGLE CRYSTAL

It is well known that large polydiacetylene single crystals can be obtained directly from their diacetylene monomer single crystals by topochemical polymerization which converts the monomers with conjugated triple bonds into polydiacetylenes with polyconjugated single–triple–single–double bonds along each polymer backbone [8–11]. Polydiacetylene single crystals have a large third-order nonlinear susceptibility $\chi^{(3)}$ due to their extended π-electron conjugation length [12,13], and many polydiacetylenes exhibit chromic transitions [12–26]. In this section, a bilayered polydiacetylene single crystal will be discussed in detail. The monomer of the polydiacetylene is a linear diacetylene which has the following chemical structure [25]:

There is an –OH group at one terminus and a mesogenic fragment at the other. The conjugated triple bonds are the reactive element for the topochemical polymerization. The monomer single crystals obtained by solution casting (0.1% by weight in CH_2Cl_2) can be polymerized directly into polymer single crystals by ultraviolet (UV) irradiation. The polydiacetylene single crystals exhibit third-order nonlinear optical properties, with the third-order susceptibility $\chi^{(3)} \sim 10^{-11}$ esu at 25°C at a fundamental wavelength of 1064 nm [25]. The polydiacetylene single crystals are blue at room temperature and exhibit chromic transitions during heating or immersing in CH_2Cl_2 [25,26]. In order to understand the origin of the chromic transitions, the first important step is to reveal the morphology and structure of the polydiacetylene single crystals [26].

A. The Morphology and Structure of the Polydiacetylene Single Crystal

Figure 13a is an electron micrograph of the polydiacetylene sample, showing large plate-like crystals with sharp edges. Figure 13b is another electron micrograph of the polydiacetylene sample, showing that the crystals are composed of multilayers stacked in parallel with the layer normals parallel to the film's normal. Figure 14 is an x-ray diffractogram of the polydiacetylene single crystals, revealing four sharp peaks (002, 003, 004, and 005). It can be seen from the diffractogram that the second-order reflection is the strongest one, indicating that the electron-density profile along the layer normal is not a single sinusoidal function [6]. This suggests that there is a mosaic structure along the layer normal direction. The *d*-spacings of the reflections indicate that the thickness of each layer is 80.0 Å which is about twice of the length of a fully extended monomer chain (~40 Å). Therefore, each layer must have a bilayer structure which consists of two sublayers linked by hydrogen bonds, as revealed by Fourier-transform infrared spectra [25]. The central region of the bilayer is composed of hydrogen-bonded side chains with –OH terminal groups and outer regions of the bilayer contain mesogenic segments. Figure 15 is a schematic representation of the bilayered single-crystal sheet of the polydiacetylene. Figures 16a and 16b are electron-diffraction patterns obtained from two different single crystals of the polydiacetylene. These two patterns show the same a^*b^* reciprocal lattice plane with an orthorhombic unit cell, indicating that the *c* axis (side-chain axis) is perpendicular to the layer plane. However, the sharp streaks appearing in these two patterns are not in the same direction. They are along the layer lines either parallel to the [110]* reciprocal vector direction or parallel to the [$\bar{1}$10]* reciprocal vector direction. Figure 17 is a schematic representation of the orthorhombic a^*b^* reciprocal lattice plane and the orthorhombic *ab* lattice plane in real space. Lines in the [110]* and [$\bar{1}$10]* directions represent the layer lines. Because the streaks appear on these layer lines, there must be periodic structures in the direction perpendicular to these layer lines in real space. These directions are the two *ab* plane face-diagonal directions, and the periodic structures along these directions are the polyconjugated backbones of the polydiacetylene. The reflections from the polyconjugated backbones appearing to be streaks on the layer lines are due to a displacement of parallel polyconjugated backbones which are not correlated one to another in the *ab* plane face-diagonal direction. The streaks are very sharp, indicating that the molecular backbones in these polydiacetylene single crystals are fairly long. The intensities of the streaks along the layer lines are not uniform, showing some detailed structures of the

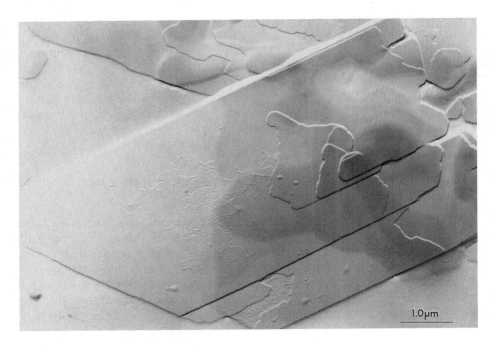

(A)

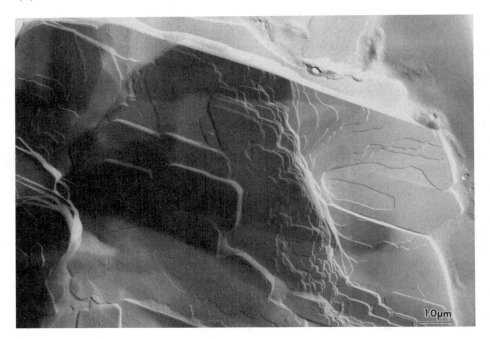

(B)

FIGURE 13 (A) Electron micrograph of the polydiacetylene sample shadowed with Pt/C, showing large, plate-like crystals with sharp edges. (B) Electron micrograph of the polydiacetylene single crystals shadowed with Pt/C, showing multilayers stacked in parallel.

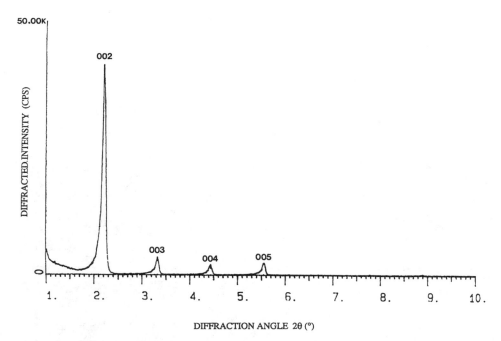

FIGURE 14 X-ray diffractogram of the polydiacetylene single crystals, showing four sharp peaks.

alternating single–triple–single–double bonds along the polyconjugated backbone. According to the electron-diffraction and x-ray-diffraction results, the orthorhombic unit-cell parameters are as $a = 7.63$ Å, $b = 5.94$ Å, and $c = 80.0$ Å, with $\alpha = \beta = \gamma = 90°$. The d-spacing of the layer lines measured from the electron-diffraction patterns is equal to 9.67 Å (the length of the ab plane face diagonal). The length of one chemical repeat unit of the backbone (single–triple–single–double bonds),

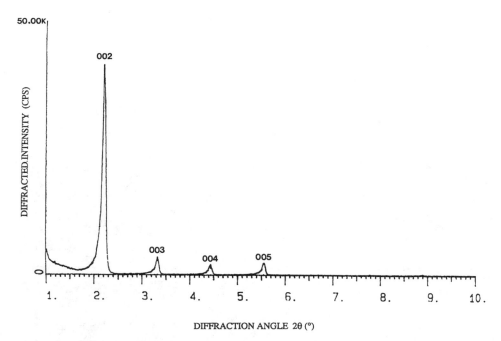

calculated using standard bond length values, is equal to 4.90 Å. Therefore, there must be two chemical repeat units per half-unit cell in each sublayer, and each chemical repeat unit is equal to 4.84 Å. The orthorhombic unit cell of the polydiacetylene single crystal is schematically represented in Fig. 18. This figure shows two half-unit cells linked by hydrogen bonds. The polyconjugated backbones are in either of the two ab plane face-diagonal directions, and the side chains are along the c-axis direction in the corner and in the middle of the unit cell.

B. Chromic Transitions of the Polydiacetylene Single Crystal

The polydiacetylene single crystals are blue at room temperature. During heating or immersing in CH_2Cl_2 (which is a solvent for the monomer, but not for the polymer) chromic transitions take place [25,26]:

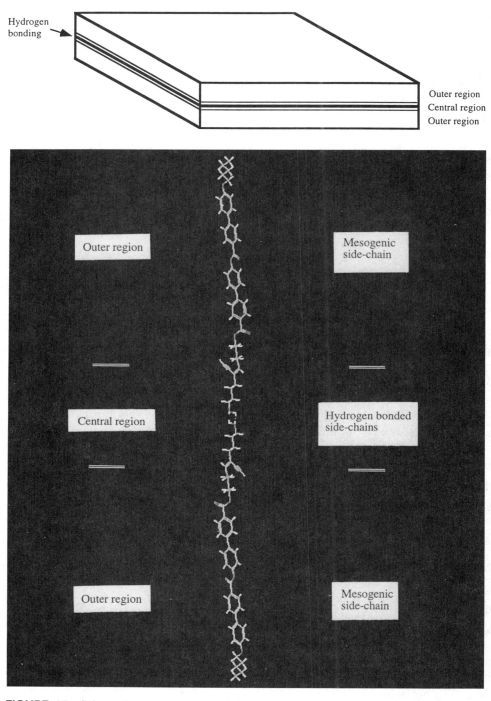

Hydrogen bonding

Outer region
Central region
Outer region

Outer region

Central region

Outer region

Mesogenic side-chain

Hydrogen bonded side-chains

Mesogenic side-chain

FIGURE 15 Schematic representation of a bilayered polydiacetylene single-crystal sheet with two sublayers linked by hydrogen bonds. The central region of the bilayer is composed of hydrogen-bonded side chains with –OH terminal groups and outer regions consist of mesogenic side chains.

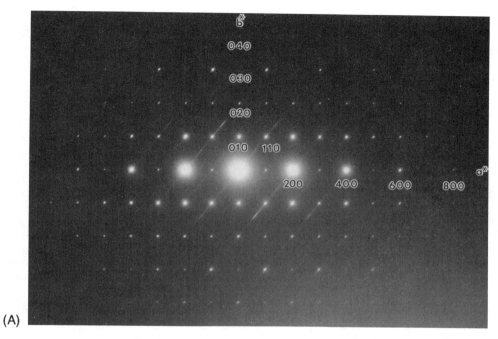

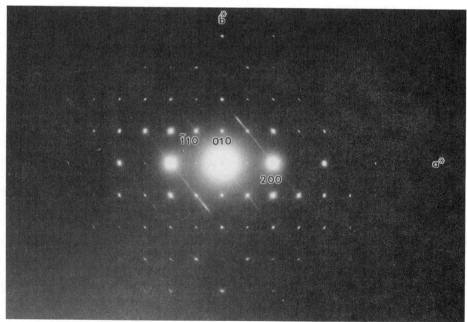

FIGURE 16 (A) Electron-diffraction pattern of a polydiacetylene single crystal, showing an orthorhombic $a*b*$ reciprocal lattice plane and a series of sharp streaks on the layer lines parallel to the [110]* reciprocal vector direction. (B) Electron-diffraction pattern of a polydiacetylene single crystal, showing the same orthorhombic reciprocal lattice planes as that shown in (A). The sharp streaks on layer lines are not parallel to the [110]* direction, but parallel to the [$\bar{1}$10]* direction.

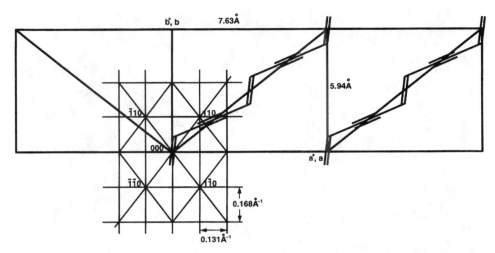

FIGURE 17 Schematic representation of the orthorhombic a^*b^* reciprocal lattice plane and the orthorhombic ab lattice plane in real space. Lines in the $[110]^*$ and $[\bar{1}10]^*$ directions represent the layer lines.

$$\text{Blue form} \underset{\sim 77^\circ C}{\rightleftharpoons} \text{Red form} \xrightarrow{\sim 125\text{-}150^\circ C} \text{Orange form}$$

$$\text{Blue form} \xrightarrow{CH_2Cl_2} \text{Red form}$$

Figure 19 shows the temperature dependence of the orthorhhombic unit-cell parameters a and b of the polydiacetylene single crystal, measured from electron-diffraction patterns taken at temperatures from 18°C to 200°C. It can be seen from the figure that both a and b remain constant with increasing temperature until 110°C. Hence, the length of the chemical repeat unit (single–triple–single–double bonds) of the polyconjugated backbone also remains constant (4.84 Å) until 110°C. However, the color of the single crystals changes from blue to red at ~77° and remains red in the temperature range from ~77°C to ~110°C.

If the red form of the polydiacetylene single crystal has exactly the same crystalline structure as that of the blue form, then it is interesting to consider what the origin of the thermochromic transition might be. Figure 20 is an electron micrograph of a polydiacetylene single crystal taken at 77°C. This figure shows that the original flat and smooth blue-form single crystal is fractured into parallel strips along the backbone direction (Because the crystal is not shadowed with Pt/C, the contrast is low.) Figure 21 is an electron micrograph of a polydiacetylene single crystal taken at 100°C, showing even clearer strips along the backbone direction. Figure 22a is an electron micrograph of a polydiacetylene single crystal heated to 110°C, showing strips parallel to the backbone direction. However, when the single crystals are heated to 110°C and subsequently cooled to room temperature, all the crystals completely recover. There are no longer narrow strips, but large and flat crystals (Fig. 22b), and, at the same time, the color of the single crystals becomes blue again. This indicates that the origin of the thermochromic transition from the blue form to the red form is associated with the planar dimension of the single

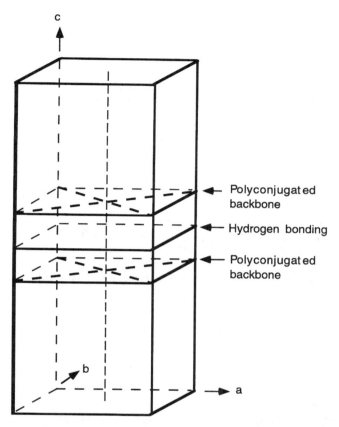

FIGURE 18 Schematic representation of the orthorhombic unit cell of the polydiacetylene single crystal, showing two half-unit cells linked by hydrogen bonds. The polyconjugated backbones are in either of the two *ab* plane face-diagonal directions, and the side chains are in the corner and in the middle of the unit cell along the *c*-axis direction.

crystal in the direction perpendicular to the backbone and this transition is reversible. The healing process associated with the reversible thermochromic transition in the solid state is quite surprising and its mechanism is not clear at this moment.

It is interesting to see that the chromic transition from the blue form to the red form also occurs during the immersion of the blue-form single crystals at room temperature in CH_2Cl_2 and the red color of the crystals remains red at room temperature when removed from the CH_2Cl_2 and after the CH_2Cl_2 has evaporated. Figure 23a is an electron micrograph of the red form of the polydiacetylene single crystals, obtained from the blue-form single crystals by immersion in CH_2Cl_2, showing that the original large, perfect, blue-form single crystals are also fractured into strips along the polyconjugated backbone direction. Figure 23b is an electron-diffraction pattern of a red-form polydiacetylene single crystal, showing sharp streaks and reflections with the same orthorhombic unit-cell parameters as those of the blue-form single crystal. Figure 24 is an electron micrograph of many small, red-form polydiacetylene single crystals, showing that each original blue-form sin-

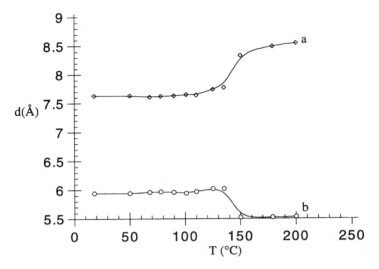

FIGURE 19 Temperature dependence of the orthorhombic unit-cell parameters *a* and *b* in the polydiacetylene single crystals.

FIGURE 20 Electron micrograph of a polydiacetylene single crystal taken at 77°C, showing that the original blue-form single crystal is fractured into strips along the backbone direction.

FIGURE 21 Electron micrograph of a polydiacetylene single crystal taken at 100°C, clearly showing strips along the backbone direction.

gle crystal is fractured into narrow strips along the polyconjugated backbone direction. These results are exactly the same as those obtained by heating. This indicates that the difference between the blue form and the red form of the polydiacetylene single crystals is only the planar dimension in the direction perpendicular to the polyconjugated backbone. Thus, one could speculate that the delocalization of π electrons in the layer plane occurs not only along the polyconjugated backbone direction but also in the direction perpendicular to it. According to electron images, the planar crystal dimensions parallel and perpendicular to the backbone direction are between 0.2 μm and several microns in the blue-form crystals. However, in the red-form single crystals, although the planar dimensions along the backbone direction are the same as those of the blue-form crystals, the dimensions in the direction perpendicular to the backbone are much smaller than those in the blue-form crystals. They are only on the order of 80 to several hundred angstroms. This suggests that in the red-form crystals, the π-electron delocalization in the layer plane along the direction perpendicular to the backbone is reduced.

When the polydiacetylene single crystals are heated through the temperature range from ~125°C to 150°C, the red color of the crystals turns to orange. It can be seen from Fig. 19 that the parameter a increases very rapidly from 7.74 Å at 125°C to 8.33 Å at 150°C. Meanwhile, the parameter b decreases very rapidly from 6.02 Å at 125°C to 5.52 Å at 150°C. Correspondingly, the length of the chemical repeat unit (single–triple–single–double bonds) of the backbone increases from 4.90 Å at 125°C to 5.00 Å at 150°C. When the crystals are heated above 150°C,

(A)

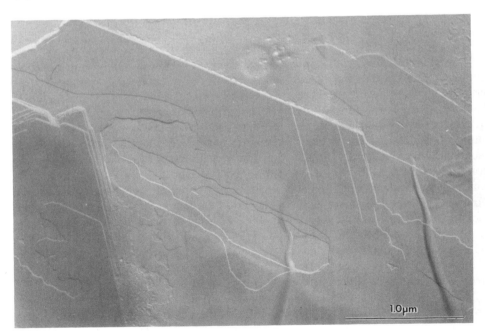

(B)

FIGURE 22 (A) Electron micrograph of polydiacetylene single crystals taken at 110°C, showing strips parallel to the backbone direction. (B) Electron micrograph of polydiacetylene single crystals heated to 110°C and subsequently cooled to room temperature, then shadowed with Pt/C, showing no strips, but large and flat crystals.

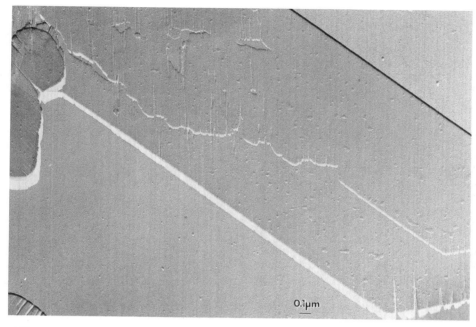

(A)

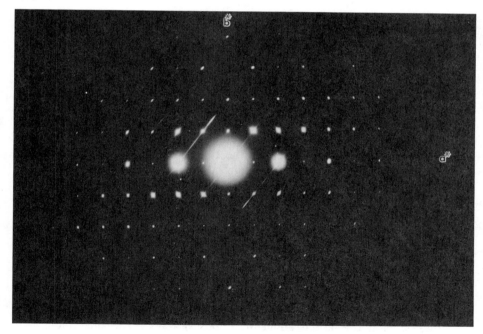

(B)

FIGURE 23 (A) Electron micrograph of red-form polydiacetylene single crystals obtained from the blue-form single crystals after CH_2Cl_2 immersion, showing that the original large perfect blue-form single crystals are fractured into strips along the polyconjugated backbone. (B) Electron-diffraction pattern of a red-form polydiacetylene single crystal, showing sharp reflections and streaks on the layer lines parallel to the [110]* direction.

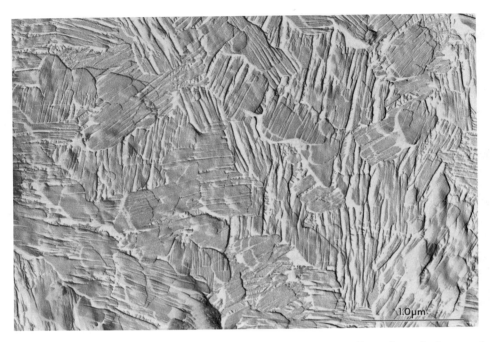

FIGURE 24 Electron micrograph of many small, red-form polydiacetylene single crystals, showing that each original blue-form single crystal is fractured into narrow strips along the polyconjugated backbone.

the parameter b does not change any more, but the parameter a still slowly increases. This indicates that the thermochromic transition of the polydiacetylene single crystals from the red form to the orange form occurs in the temperature range from ~125°C to 150°C. In addition to the unit-cell parameter changes during this transition, extra reflections appear in the electron-diffraction patterns. Figure 25 shows an electron-diffraction pattern of a polydiacetylene single crystal taken at 125°C. This pattern shows six extra reflections with $d = 4.40$ Å arrayed at equal angles around the central spot near the 110, 200, $1\bar{1}0$, $\bar{1}\,\bar{1}0$, $\bar{2}00$, and $\bar{1}10$ reflections. This indicates that there are rotations of side chains about the c axis in the crystal. As there are hydrogen bonds between two sublayers in each single-crystal sheet, the side chains with –OH terminal groups in the central region of the bilayer probably cannot rotate at this temperature. It is very important to note that the d-spacing of the six extra reflections is equal to 4.40 Å, which is exactly equal to the d_{200} of the side-chain polymer S-2 in the crystal smectic E_h phase at 67°C [3]. This indicates that in the sublayers the side chains with mesogenic segments start performing cooperative rotations about the c axis at 125°C. The side chains with mesogenic segments now are also in a crystal smectic E_h phase in which three orthorhombic lattices, with $a = 8.80$ Å and $b = 5.08$ Å ($a = 3^{1/2} b$), are related to one another by rotations of $\pm 60°$ about the c axis. The three orthorhombic a^*b^* reciprocal lattice planes of the smectic E_h phase are shown schematically in Fig. 26 [3], in which the six innermost indexed reciprocal lattice points are equivalent

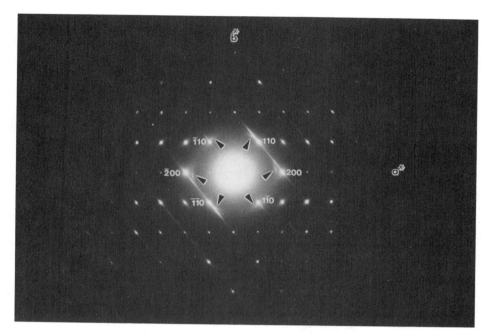

FIGURE 25 Electron-diffraction pattern of a polydiacetylene single crystal taken at 125°C, showing six extra reflections (indicated by arrowheads) arrayed at equal angles around the central spot near the 110, 200, $1\bar{1}0$, $\bar{1}\bar{1}0$, $\bar{2}00$, and $\bar{1}10$ reflections.

to the six extra reflections in the pattern shown in Fig. 25. The side-chain rotations about the c axis would probably twist the polyconjugated backbone, causing a shortening of the conjugation length along the backbone direction. Figure 27 is an electron-diffraction pattern taken at 150°C. This pattern also shows six extra, but diffuse, reflections with $d = 4.49$ Å distributed in a hexagonal geometry, indicating that the side chains with mesogenic segments are probably performing independent rotations about the c axis in the sublayers due to the enlarged d-spacing. Now, the side chains with mesogenic segments are in a hexatic smectic B phase [3]. Figure 28a is an electron-diffraction pattern of a polydiacetylene single crystal taken at 197°C. This pattern no longer shows the six hexagonally distributed extra reflections. Instead, a diffuse ring appears in the pattern, indicating that both hexagonal bond-orientational order and positional order of the side chains with mesogenic segments become short range in the plane perpendicular to the c axis. The side chains with mesogenic segments now are in a smectic A phase. This would significantly twist the polyconjugated backbone. Figure 28b is an electron micrograph of the polydiacetylene single crystal taken at 197°C, showing crystal strips parallel to the polyconjugated backbone direction. The twisting of the backbones still cannot be seen clearly in the low-contrast image. However, as shown in Fig. 29a, when the polydiacetylene single crystal is heated to 238°C the strips are seriously twisted, indicating that the conjugation length along the backbone direction is significantly reduced. Figure 29b is an electron-diffraction pattern of a polydiacetylene single crystal taken at 238°C, showing only one diffuse streak on the second-layer line

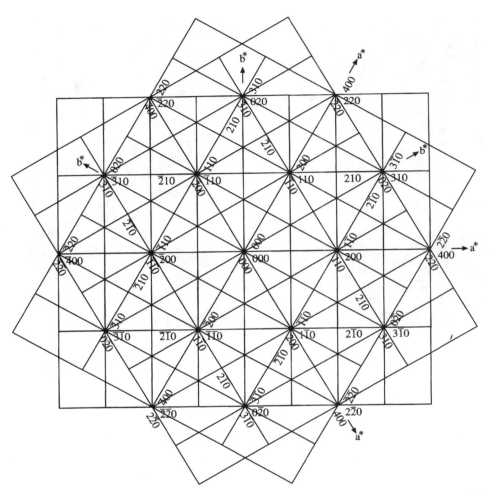

FIGURE 26 Schematic representation of three identical orthorhombic $a*b*$ reciprocal lattice planes with $b* = 3^{1/2}a*$ related to one another by rotations of $\pm60°$ about the c axis (the crystal smectic E_h phase).

parallel to the [110]* direction. This indicates that the conjugation length along the backbone direction has been significantly shortened. The reflections in the pattern are not sharp and the diffuse ring is clearly shown in the pattern. In addition to the shortening of the conjugation length along the backbone, the parameter a increases from 8.33 Å at 150°C to 8.54 Å at 200°C (Fig. 19) and, correspondingly, the length of the chemical repeat unit of the backbone increases from 5.00 Å at 150°C to 5.09 Å at 200°C. According to these electron-diffraction and electron microscopy results, the thermochromic transition from the red form to the orange form can be accounted for by a reduction of electronic delocalization along the backbone direction resulting from the backbone twisting and the lengthening of the chemical repeat unit of the backbone. Therefore, in the orange form, the π electrons in the layer plane are less delocalized in both directions, parallel and perpendicular

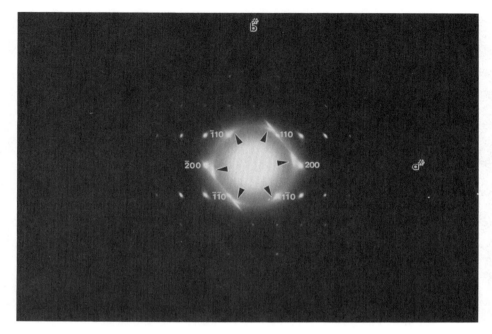

FIGURE 27 Electron-diffraction pattern of a polydiacetylene single crystal taken at 150°C, showing six extra diffuse reflections (indicated by arrowheads) distributed in a hexagonal geometry.

to the backbone, and the length of the single–triple–single–double bonds is larger than or equal to 5.00 Å

Figure 30a is an electron micrograph of a large polydiacetylene single crystal heated to 200°C and subsequently cooled to room temperature and then shadowed with Pt/C, which retained the orange color at room temperature. This micrograph shows twisted narrow strips connected to each other, forming a network-like structure. Figure 30b is the corresponding electron-diffraction pattern. This pattern still shows six innermost reflections distributed in a hexagonal geometry. According to the electron-diffraction pattern and electron micrograph shown in Fig. 30, it is obvious that the thermochromic transition from the red form to the orange form is irreversible.

C. Conclusions

Based on the x-ray-diffraction, electron-diffraction, and electron microscopy results obtained at room and elevated temperatures, the following conclusions can be reached:

1. The polydiacetylene single crystal has a bilayered structure. Each bilayer consists of two sublayers linked by hydrogen bonds. The unit cell of the single crystal is orthorhombic. The polyconjugated backbones are in either of the two *ab* plane face-diagonal directions. There are two repeat units (single–triple–

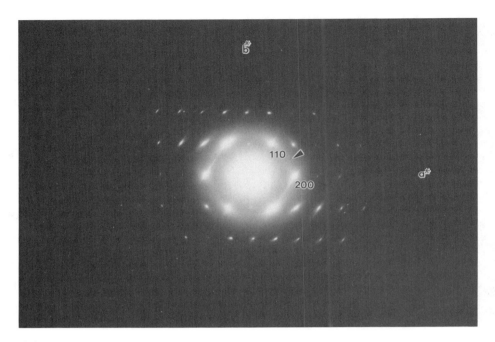

(A)

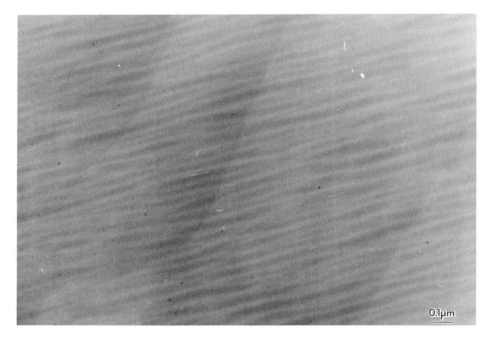

(B)

FIGURE 28 (A) Electron-diffraction pattern of a polydiacetylene single crystal taken at 197°C, showing an a^*b^* reciprocal lattice plane and a diffuse ring (indicated by an arrowhead). (B) Electron micrograph of the polydiacetylene single crystal taken at 197°C, showing crystal strips along the backbone direction.

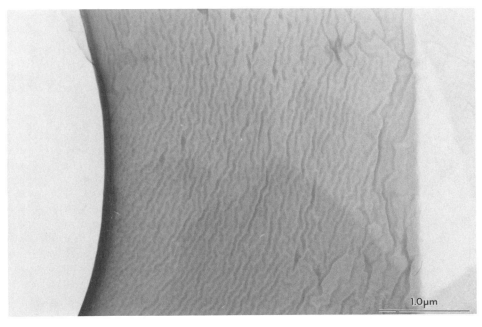

(A)

(B)

FIGURE 29 (A) Electron micrograph of a polydiacetylene single crystal taken at 238°C, showing twisted strips. (B) Electron-diffraction pattern of the polydiacetylene single crystal taken at 238°C, showing only one diffuse streak on the second layer line (indicated by an arrow) and a diffuse ring (indicated by an arrowhead).

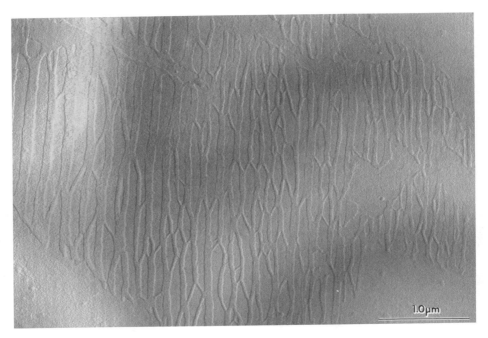

(A)

(B)

FIGURE 30 (A) Electron micrograph of a polydiacetylene single crystal heated to 200°C and subsequently cooled to room temperature and shadowed with Pt/C, showing a network-like structure formed by twisted narrow strips. (B) Electron-diffraction pattern of the polydiacetylene single crystal heated to 200°C and subsequently cooled to room temperature, showing six innermost reflections (indicated by arrowheads) distributed in a hexagonal geometry.

single–double bonds) per half-unit cell in each sublayer and the side chains are perpendicular to the layer plane along the *c*-axis direction.

2. The origin of the chromic transitions of the polydiacetylene single crystals appears to be associated with the extent of delocalization of π electrons in the layer plane along the directions parallel and perpendicular to the conjugated backbone.

3. In the blue-form single crystals, π electrons in the layer plane are more delocalized in both directions, parallel and perpendicular to the backbone, and the length of the single–triple–single–double bonds is equal to 4.84 Å.

4. In the red-form single crystals, the π electrons in the layer plane are less delocalized in the direction perpendicular to the backbone due to the fracture of the blue-form single crystals along the backbone direction. The crystalline structure of the red-form single crystal geometrically is the same as that of the blue form, and the length of the single–triple–single–double bonds is also equal to 4.84 Å.

5. In the orange-form crystals, the π electrons in the layer plane are less delocalized in both directions, parallel and perpendicular to the backbone. The reduction of the π-electron delocalization along the backbone direction is due to the twisting of the backbones and the lengthening of the chemical repeat unit of the backbone (≥ 5.00 Å).

6. The thermochromic transition from the blue form to the red form is reversible. In contrast, the thermochromic transition from the red form to the orange form is irreversible.

ACKNOWLEDGMENTS

The author is grateful to Professor W. C. Bigelow and Dr. L. F. Allard for valuable comments.

REFERENCES

1. Stupp, S. I., Son, S., Lin, H. C., and Li, L. S., *Science*, *259*, 59 (1993).
2. Stupp, S. I., Son, S., Li, L. S., Lin, H. C., and Keser, M., *J. Am. Chem. Soc.*, *117*, 5212 (1995).
3. Li, L. S., Hong, X. J., and Stupp, S. I., *Liquid Crystals*, *21*, 469 (1996).
4. Lin, H. C., Ph.D. thesis, University of Illinois at Urbana-Champaign, 1992.
5. Hong, X. J., and Stupp, S. I., *Polm. Prepr.*, *30*, 469 (1989).
6. Davidson, P., and Levelut, A. M., *Liquid Crystals*, 11, 469 (1992).
7. Gray, G. W., and Goodby, J. W., *Smectic Liquid Crystals—Textures and Structures*, Heyden & Son, Inc., Philadelphia (1984).
8. Wegner, G., *Z. Naturforsch*, *24b*, 824 (1969).
9. Wegner, G., *Faraday Discuss. Chem. Soc.*, *68*, 494 (1980).
10. Baughman, R. H., *J. Polym. Sci., Polym. Phys. Ed.*, *12*, 1511 (1974).
11. Baughman, R. H., *J. Appl. Phys.*, *43*, 4362 (1972).
12. Patel, G. N., Chance, R. R., and Witt, J. D., *J. Chem. Phys.*, *70*, 4387 (1979).
13. Chance, R. R., *Macromolecules*, *13*, 396 (1980).
14. Muller M. A., Schmidt, M., and Wegner, G., *Makromol. Chem., Rapid Commun.*, *5*, 83 (1984).
15. Lim, K. C., and Heeger, A. J., *J. Chem. Phys.*, *82*, 522 (1985).

16. Bloor, D., and Chance, R. R., *Polydiacetylenes*, NATO ASI Ser. E, Martinus Nijhoff Publishers, Dordrecht/Boston/Lancaster, (1985).

17. Chance, R. R., *Encyclopedia of Polymer Science and Engineering*, Wiley, New York, 1986, Vol. 4.

18. Tanaka, H., Thakur, M., Gomez, M. A., and Tonelli, A. E., *Macromolecules*, *20*, 3094 (1987).

19. Tanaka, H., Gomez, M. A., Tonelli, A. E., and Thakur, M., *Macromolecules*, *22*, 1208 (1989).

20. Tanaka, H., Thakur, M., Gomez, M. A., and Tonelli, A. E., *Polymer*, *32*, 1834 (1991).

21. Tanaka, H., Gomez, M. A., Tonelli, A. E., Lovinger, A. J., Davis D. D., and Thakur, M., *Macromolecules*, *22*, 2427 (1989).

22. Nava, A. D., Thakur, M., and Tonelli, A. E., *Macromolecules*, *23*, 3055 (1990).

23. Chance, R. R., Baughman, R. H., Muller, H., and Eckhardt, C. J., *J. Chem. Phys.*, *67*, 3616 (1977).

24. Dobrosavljevic, V., and Stratt, R. M., *Phys. Rev.*, *35B*, 2781 (1987).

25. Huggins, K. E., Son, S., and Stupp, S. I., *Macromolecules*, *30*, 5305 (1997).

26. Li, L. S., and Stupp, S. I., *Macromolecules*, *30*, 5313 (1997).

14

Part I: Polymeric Optical Guided-Wave Devices

Marc Gregory Mogul and Donald L. Wise
Northeastern University
Boston, Massachusetts

Joseph D. Gresser and Debra J. Trantolo
Cambridge Scientific, Inc.
Belmont, Massachusetts

Gary E. Wnek
Virginia Commonwealth University
Richmond, Virginia

Charles A. DiMarzio
Center for Electromagnetic Research
Northeastern University
Boston, Massachusetts

I. INTRODUCTION

There is a growing need in both government and industry for increased data-handling capabilities, including acquisition, processing, transmission, and storage. Optics, including nonlinear optics, will play a vital role in these emerging areas. Moreover, the use of optics in electronic processing systems would allow for parallel processing, faster speeds, and higher data-storage densities, significantly increasing the overall system throughput.

Biopolymers have a controlled molecular structure and morphology and are promising candidates for nonlinear optical materials (NLOM). This project will establish the practicality of using aligned biopolymers as materials for an optical guided-wave device, in conjunction with the inherent advantages of using a nonlinear optical material of subsequent electrooptic applications. This work focuses on the importance of the variables associated with aligned thin-film processing and the subsequent determination of its electrooptic capabilities. The goal of this biopolymer-based modulator is to demonstrate electrooptic coefficients greater than 30 pm/V at 633 nm with a percentage modulation of 20% or better.

In preliminary studies, films of poly(benzyl-L-glutamate), PBLG, prepared by solvent evaporation in a homogeneous electric field showed structural features and NLO properties, both of which were absent in control films prepared at zero-field strength (i.e., without a field). More recent work has investigated the effects of molecular weight, electric field strength, and solution concentration on the degree of orientation of PBLG and further correlates the degree of orientation with enhancement of second-harmonic generation (SHG) as measured by the second-order susceptibility. This chapter focuses on the efforts to optimize film processing conditions to maximize order and overall film quality with respect to a polymeric optical guided-wave device.

II. TECHNICAL BACKGROUND

A waveguide acts as a propagation medium for guided optical waves. Waveguides can be roughly classified into optical waveguides for optical integrated circuits and optical fibers. The fiber type can support guided waves for large distances (hundreds of kilometers) with low loss. It is generally manufactured from a variety of silica glasses arranged in a cylindrical geometry, with the core region of a higher refractive index than the cladding. This index relationship is a necessary condition for total internal reflection, the phenomenon which is harnessed for low-loss transmission. The overall diameter of a standard communication fiber, core, and cladding is 125 μm, which is comparable to the thickness of a human hair [1]. A multimode fiber has a cylindrical geometry, with core diameter, typically on the order of 50–100 μm and at 125 nm cladding. This fiber can be formed from either glass or plastic. However, this multimode fiber can support more than one characteristic propagation mode and, consequently, suffers from much greater dispersion than the single-mode fiber. In addition, because of dispersion, this type of fiber is extremely lossy and can only be used for medium-bandwidth, short-distance applications. This, in turn, is covered with a plastic "buffer" which can be removed via chemicals or a razor blade. Light actually travels in the core region as is represented in Fig. 1a.

Light will propagate in one or more of a set of modes in which E and H are solutions of Maxwell's equations, subjected to boundary conditions. In general, each mode will have a unique characteristic transverse size, d_i, and a divergence angle, α. These are related by diffraction: $\alpha_i \geq \lambda/d_i$. If $d_i < D$ and $\alpha_i < \alpha_{\text{acceptance}}$, the mode is not lossy. Otherwise, the mode is diminishing as it travels down the propagation length. For single-mode fibers, the relationship is $\alpha_{\text{acceptance}} = \lambda/D$. The aforementioned total internal reflection can occur when light is incident on an interface between two lossless dielectric media with different refractive indices. This phenomenon can occur provided the light is incident from the high-index medium at a shallow enough angle, at or beyond the critical angle. If the waveguide is made of a polymer, similar to the two-glass examples, the ratio of the indices of the polymer and glass used would determine the possibility of light propagation. This is represented in Fig. 1b.

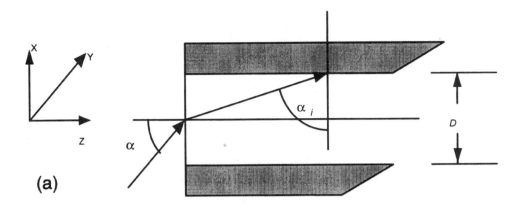

(a)

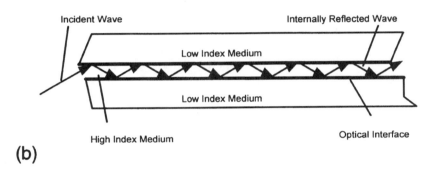

(b)

FIGURE 1 Representation of optical waveguide and wave propagation.

In the above case, there is no transmission loss across the interface, which acts as a highly efficient reflector, The "lossless" media are ideal media that do not really exist and must be regarded as media where the losses are arbitrarily small, but not zero. The fact is that there is a small amount of loss in glass, but no Fresnel loss. The field emitted by any source of finite extent must decay as it travels along, going down to 0 at infinity.

The electromagnetic field is represented by four electromagnetic field vectors that are functions of position, r(m), and time, t(s). The four vectors, electric field **E** (V/m), magnetic field **H** (A/m), electric flux density **D** (C/m²), and magnetic flux density B (Wb/m²), all conform to Maxwell's equations:

$$\nabla \times \mathbf{E} = -\frac{dB}{dt} \quad \text{and} \quad \nabla \times \mathbf{H} = -\frac{dD}{dt} + \mathbf{J} \tag{1}$$

Here, **J** (A/m²) is the electric current density and stems from the conservation of charge. The basic result about energy exchanges in macroscopic electromagnetism is that the power carried by the field into a bounded volume, V, limited by a closed surface, S, is defined as the inward flux of the Poynting vector $\mathbf{E} \times \mathbf{H}$ through S [2]. This Poynting vector is time dependent and it gives the instantaneous power density. For varying fields that are harmonic time dependent, $e^{i\omega t}$, the complex Poynting vector equation is considered, where the real part gives the time-averaged power density. The plane-wave solution of the wave equation in lossy dielectrics is

$$E = E_x + \exp(-k''z) \exp(-jk'z) \tag{2}$$

This relationship has the form of a plane wave, whose amplitude decays exponentially with distance z. The real part of the propagation constant k', defines the phase variation of the wave, whereas the imaginary part, k'', defines the amplitude variation and is known as the absorption constant. This is often given the symbol, α [3]. A medium absorbs electromagnetic energy if ε (permittivity) has a negative imaginary part. In addition, if the beam divergence in a fiber causes the angle of incidence of the orbbing to be $> \theta_c$, Fresnel losses occur at the interface and light propagates out of the fiber. Lossless media correspond to the case when ε and μ are purely real. Because at least the dielectric effects are never negligible, all materials are both dispersive and lossy. This concept is effectively a power-conservation relation, because it relates the rate of change of stored energy to the outward energy flow and the energy dissipated. This power flow is measured in watts per meter squared. However, it should be noted that this vector is an extremely fast varying function, which is not measurable by any practical technique. It is, therefore, necessary to define an alternative quantity related to power that is directly measurable.

A convenient way to measure power is to calculate the associated time-averaged Poynting vector or irradiance. The power is obtained by integrating the Poynting vector, over a time T, large compared with the period of oscillations, and dividing by T for the $E = E_\sigma$ point and so forth. It follows that the time-averaged Poynting vector is

$$\mathbf{S} = \frac{1}{2 \operatorname{Re}[\mathbf{E} \times \mathbf{H}^*]} \tag{3}$$

The asterisk indicates complex conjugation. Irradiance, $f = |s|$, is still not a directly measurable quantity. What can be measured is the time-averaged power P flowing through a given surface. This averaged power is found as the integral of the normal component of $\mathbf{S}$ over the surface. The irradiance carried by a plane wave traveling in the $+z$ direction can be written as

$$I = \frac{dP}{dA} = \frac{1}{2(nE_0^2/Z_0)} \tag{4}$$

where the time-averaged power is very important because it is one of the few parameters of the high-frequency electromagnetic field that can actually be measured. Suitable detectors consist of solar cells and other semiconductor $p–n$ junction photodiodes.

A perfect monochromatic wave, which must exist for all time without changing in frequency or amplitude, cannot carry any information. Only the modulation of such a wave creates the ability to transport data. A framework for the analysis of such a system is provided by the Fourier-transform theory, which states that any signal may be decomposed into an infinite sum of signal frequency terms. The beam propagation method consists of approximate mathematical methods for solving problems of light propagation in dielectric optical waveguides of arbitrary shape. The principle idea behind this method is the realization that the optical wave is subject to two main influences. The first is that because of the wave nature, the light is subject to diffraction. The second is due to the fact that it propagates in a slightly inhomogeneous medium. The light rays of a wave moving from a plane experience different amounts of phase shift depending on their x and y positions in these planes. Thus, a continuous medium is conceptually replaced by a sequence of lenses separated by short sections of homogeneous space. The Fourier-transform algorithm is therefore a convenient way to separate rapid variations of amplitude in the z direction.

An optical field oscillates at an extremely high frequency, so a light wave can act as a high-frequency carrier. The optical fiber provides a medium for which data can be transferred at very high rates. In fact, the optical fiber offers an increase in channel capacity by a factor of 10^5 over the microwave guide [4]. In addition, the raw materials of optical figures consist basically of purified sand. Coupled with the fiber's reduction in size, weight, and cost, this material is replacing copper cables in many applications involving the transfer of low-power signals.

Alternatively, plastic-coated silica (PCS) fibers and all plastic fibers are often used. The all-plastic fibers, for example, are generally manufactured with a polystyrene core with a methyl methacrylate cladding. Although these are cheap, they suffer from considerably higher propagation loss than silica fiber. A PCS fiber exhibits losses on the order of 10 dB/km, whereas an all-plastic fibers losses are approximately 500 dB/km [5]. Therefore, all-plastic fibers are suitable only for transmission over short distances, normally tens of meters.

The semiconductor laser is a small, efficient light source and is a component which has sustained the development of guided-wave optical devices. Because the semiconductor laser is also a guided-wave device and emits light comparable to that of the core of an optical fiber, it immediately finds utility in fiber systems. Furthermore, the semiconductor laser can emit light of a wavelength at which silica fiber shows minimum dispersion and minimum propagation loss, near $\lambda_0 = 1.3$ and 0.5 μm, respectively. With fine-tuning of the semiconductor laser, the output can very nearly consist of a single frequency. This is a very important advantage for communications, because it reduces the effect of dispersion.

Planar waveguide integrated optics is concerned with the manipulation of sheet beams. These can propagate in any direction parallel to the surface of a high-index guiding layer. This layer provides optical confinement in a single direction. By loading a thin film with a higher refractive index than the substrate, the light can be trapped inside this film. Waveguides in which the refractive index changes in stages are called step-index (SI) optical waveguides, whereas those with a gradual refractive-index change are called graded-index (GI) optical waveguides. Figure 2 shows a typical three-layer SI planar waveguide. This is formed by depositing a thin layer of material of high refractive index on a thicker, lower-index substrate. The third layer is usually air, or an additional low-index cover layer can be used. Wave guidance is governed by total internal reflection at the layer interfaces. An

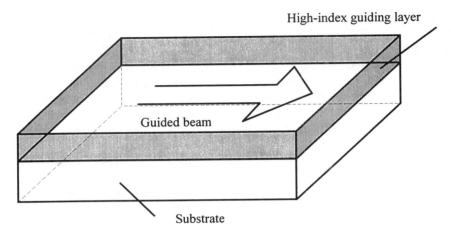

FIGURE 2 Typical three-layer SI planar waveguide.

interesting aspect of this system is that sheet beams allow in two dimensions (2D) many of the operations that are possible using free-space optics in three dimensions (3D); for example, focusing by a lens or beam deflection. Therefore, planar integrated-optical chips can be used for signal processing based on the Fourier-transform properties of a lens.

An optical waveguide that is uniform in the direction of propagation is the most basic type of waveguide, but this alone is not sufficient for construction of an optical integrated circuit. In reality, an appropriate combination of various forms of optical waveguides is placed on a substrate to construct an optical circuit with desired features. A 2D optical waveguide can trap light in the direction of the thickness (y direction) but can allow light to spread in the horizontal direction (x direction). In order to facilitate the construction of optical integrated circuits, various types of 3D optical waveguides, or optical channel waveguides, which trap the light in both x and y directions, have been manufactured. Although waveguides come in various forms and with a variety of functions, such as crossed waveguides which are used for combining, the fact remains that the optical waveguide that is uniform in the direction of propagation is the most basic form. Therefore, this discussion will be limited to optical waveguides in which material constants such as structure and refractive index do not change in the direction of propagation.

Figure 2 is an example of a three-layer optical waveguide, the third layer being air. If c, s, and f represent the cladding, substrate, and film, respectively, then the relationship between the refractive indexes is $n_c \leq n_s \leq n_f$. Assume the critical angles θ follow similar nomenclature; it then follows that there are three separate modes which need to be characterized. Therefore, when $\theta_s < \theta$, total reflection occurs at both interfaces, resulting in the light being transmitted while trapped in the thin film. This mode is called the guided, bound, or trapped mode. On the other hand, when θ is smaller than θ_s and $\theta_c < \theta < \theta_s$, the conditions for total reflection are not met at the lower interface, and part of the light is radiated into the substrate. This mode is called the substrate radiation mode. When q becomes even smaller and $\theta < \theta_c$, total reflection conditions are not met at either the upper or lower interface, and the light is radiated out at both sides. This mode is called the substrate-cladding radiation mode. Moreover, if the film thickness varies along the direction of propagation, this variation will cause the angle of incidence to vary and will produce a relative radiation loss.

If other materials are used to fabricate the waveguide, more functions involving the manipulation of light are possible. For example, using electrooptic substrates like $LiNbO_3$, GaAs, and InP, modulation and switching can be performed at extremely high speeds (tens of gigahertz). Figure 3 is a schematic of an electrooptical directional coupler. This device can be used to switch light between two adjacent parallel waveguides under electrical control. This operation can be used to place information on the carrier wave and then route it around the network. In addition, integration also offers the possibility of combining optical components with their controlling electronics, in the form of integrated optoelectronics. In this way, light can be used as a method of communicating between very high-speed electronic circuits in fast computers.

An important building block element for integrated-optic switches and a variety of other devices is the directional coupler, which is a pair of strip waveguides closely spaced over the interaction length L. The spacing between the waveguides is comparable to the waveguide width, so that the evanescent tail of the mode profile of each waveguide extends into that of the other. As a result, light guided in the two waveguides interacts or is coupled. Light injected into one waveguide gradually "leaks" (or is coupled) into the second as it travels along the waveguide. This is a distributed coupling, and the net effect is strong only if light coupled to the second waveguide at each point along their interaction length is in phase with light that has coupled at an earlier point. To achieve significant coupling between the waveguides, the speed of light in the two waveguides must be the same. In this case, the coupler is referred to as being phase matched, and all of the light injected into one waveguide will be coupled to the second after a characteristic interaction length, referred to as the coupling length. The directional coupler can be easily understood by comparing it to a coupled pair of pendulums [6]. Consider two pendulums having equal periods that are connected or coupled transverse to their

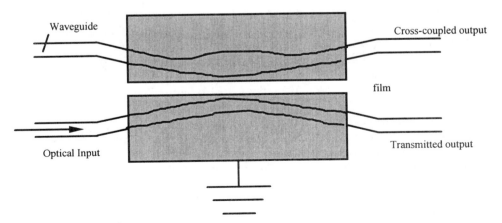

FIGURE 3 Schematic of an electrooptical directional coupler.

common direction of motion. If only one is started swinging, the second one slowly begins to swing. After a characteristic time, all of the energy is transferred to the second pendulum. In an identical time period, the kinetic energy will be transferred back to the initial pendulum while the second stops. This situation is oscillatory until friction damps out all motion. In the directional coupler, given enough interaction length, light will also periodically transfer between the initial and second waveguide along the length of the coupler. Typically, the device is made to be one coupling length long, such that the light injected into one waveguide exits from the other.

In the case of the coupled pendulums, if the characteristic periods are sufficiently different, there is no coupling of energy between them, despite a coupling element. Similarly, there is no net coupling of light between the two waveguides, in spite of their close proximity, if the refractive indices and, therefore, the speed of light in the two waveguides are sufficiently different. In this case, light injected into one waveguide exits from the same waveguide. By using an electrooptic material like lithium niobate for the substrate, one can introduce this refractive-index difference or mismatch dynamically by the application of a voltage to electrodes placed over or along the waveguides. The routing of light between two alternate paths is therefore controlled by voltage.

The directional coupler operation can be described quantitatively by the solution of the well-known coupled-mode equations [7]. For input to one waveguide, and under the assumption of no loss, the crossover frequency, n, to the second waveguide can be written in terms of k, the coupling coefficient per unit length, and Δ, the phase mismatch between the two waveguides. The value of k depends exponentially in the interwaveguide separation normalized by the mode diameter, which, in turn, depends slightly on the wavelength. The coupling length for which complete crossover occurs for no index mismatch ($\Delta = 0$) is $L = \pi/2k$. For the electrooptically controlled switch, Δ depends on the induced index change, which depends, in turn, on the applied electric voltage. The scale of the voltage axis depends on the electrooptic material, device length, and optical wavelength. In lithium niobate directional couplers 1 cm long, the voltage swing from complete to zero interwaveguide transfer is approximately 2 and 8 V for wavelengths of 0.6328 (red) and 1.3 μm, respectively. These voltages are many times smaller than those achievable in bulk (nonwaveguide) devices in the same material. This voltage reduction is a result of the small lateral dimension over which the electric voltage is applied and of the long interaction length, which is not limited by diffractive spreading, as in the bulk case. The switching can be either discrete between the two outputs (cross-talk as low as -30 dB has been achieved) or continuous [8]. Therefore, the switched directional coupler can also be used as a voltage-controlled tap or power splitter.

The full optical-bandwidth performance potential of fiber-optic systems, however, is limited by the lack of a reconfigurable device capable of relaying signals from fiber to fiber without interruption of optical integrity. The ability to transit the switch signals in a purely optical domain provides cost, power, weight, and size advantages.

Using integrated optics, there lies the possibility of fabricating an optical interferometer, using a coil of optical fibers (which provide the sensor element) and a single-channel waveguide integrated chip (which carries a number of beamsplitting and signal-processing components). Such devices can be arranged to sense variations in a wide range of physical parameters.

The development and fabrication of device-quality, thin-film materials for electrooptics and nonlinear-optics applications is being undertaken. The focus of this work is to characterize the organic polymer (PBLG) with respect to future applications in the realm of integrated optical circuitry. The first type of optical circuit relies on the electrooptic effect, in which light is switched from one optical waveguide to an adjacent one by applying a voltage. The second type relies on the optically induced refractive-index change, in which the intensity of the light within the guide determined whether the device switches.

Both the electrooptic effect and the optical Kerr effect can be described using the nonlinear optical susceptibility tensors. The central expression in this quantitative approach is the expansion of the total optical polarization density as a power series in the electric field:

$$P_i = \varepsilon_0(\Sigma\ \chi_{ij}^{(1)}E_j + \Sigma\ \Sigma\ \chi_{ijk}^{(2)}E_jE_k + \Sigma\ \Sigma\ \Sigma\ \chi_{ijkl}^{(3)}E_jE_kE_l + \cdots) \qquad (5)$$

describes the polarization P_i generated in a medium throughout which light is propagating, where the χ's are the nth-order complex optical susceptibilities. The first term is the linear polarization, part of the linear susceptibility tensor. This is responsible for the conventional index of refraction and absorption of a material. Similarly, the second and all subsequent terms are nonlinear optical effects. These play important roles in practical applications. Whereas third-order (and other odd-order) nonlinear polarization occurs in all media, second-order (or other even-order) polarization occurs only in media that lack inversion symmetry. This second-order susceptibility is responsible for such nonlinear effects as second-harmonic generation, sum and difference frequency generation, the rectification of light, and the well-known electrooptic effect or Pockels effect. In addition, the third-order susceptibility tensor is responsible for a number of additional optical effects including nonlinear rectification and absorption, third-harmonic generation, and stimulated Raman scattering.

In order for an organic molecular to exhibit electrooptic behavior, it must possess a large molecular dipole. In order for a material containing such molecules to be electrooptic, all (or at least many) of the individual molecular dipoles must be aligned in the same direction. A large number of organic moleculars now have been shown to have promise for this purpose, although many of these molecules will likely prove unsuitable for device applications because of excessive absorption losses or other problems. However, virtually all of the most promising compounds are "push–pull" donor–acceptor molecules such as *p*-nitroaniline, which has an electron-rich amino group at one end and an electron-deficient group at the other. Similar compounds with larger dipoles (but also with correspondingly greater optical absorption throughout more of the visible spectrum) have been made by several laboratories by connecting two aromatic rings with a carbon–carbon or nitrogen–nitrogen double bond [9]. Again, the amino group is at one end of the long molecular axis and a nitro or tricyanovinyl group at the other.

There are several methods for fabricating these electrooptic polymer systems into thin films for use in integrated optical applications. The most common are spinning, casting, or dipping the polymer (or diffusing it) into a suitable substrate, such as glass. Recently, organic waveguides also have been fabricated on substrates such as silicon and gallium arsenide. Monolithically integrating active devices such as lasers, photodetectors, and transistors with the organic waveguides on a common substrate provides complete processing capability.

Consider a symmetric slab waveguide consisting of a linear core and a nonlinear cladding. Figure 4 shows the optical power dependence of the local electric field strength for the transverse electric mode. The local transverse magnetic field shows a similar relationship. A and C in this figure correspond to positions in the nonlinear cladding and liner core, respectively. At low powers, the field strength is higher at the core center A than at C in the cladding, increasing at both locations as the power increases. At high powers, however, the relationship between the field strength is reversed. In addition, whereas the field strength in the cladding tends to increase monotonically as the power increases, it tends to decrease at the core center. This is an excellent way to take advantage of the nonlinear characteristics of a waveguide. In doing so, one is able to concentrate higher field strengths in either the film or the cladding by varying the optical power utilizing the aforementioned electrooptic effect or the refractive-index change in different parts of the film.

III. POTENTIAL APPLICATIONS

In the past, the use of plastic optics was typically limited to products with lower performance requirements or implied lower quality. These restrictions were the result of the small number of optical-grade plastics available and the ability to mold and manufacture the types of tolerance required for more sophisticated products. Compared with available glasses, the number of optical-grade thermoplastics is still

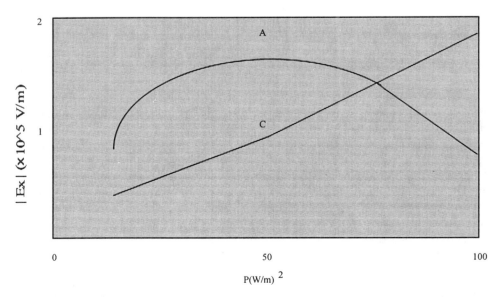

FIGURE 4 Optical power dependence of the local electric field strength for the transverse electric mode.

very limited. The most common are acrylic, styrene, polycarbonate, and various copolymers. However, advances in materials and molding technology affords new efficiencies for lens builders. Clark and Cohen from Eastman Kodak Co. have demonstrated that with strict attention by operators, or the addition of robotic handling to remove parts from the mold, one is able to maintain a consistent molding cycle [10]. Demonstrating a capability to manufacture plastic optics in high volume is critical when dealing with the ever-present world of high-volume life. Furthermore, many of the practices developed by Eastman Kodak are obviously appropriate for plastic-optics manufacture regardless of volume.

Photorefractive polymers represent a new and growing class of optoelectronic materials. Photorefractivity is a specific mechanism for producing a hologram in an optically nonlinear material. Photoexcited charge carriers migrate to produce an internal electric field, and this internal space-charge electric field produces a modulation of the refractive index. This effect has been studied in several known inorganic crystals such as $LiNbO_3$ and $BaTiO_3$. Polymers have several potential advantages over inorganic crystalline photorefractives, primarily because the dielectric constant is far lower. Polymers also possess other advantages over inorganics such as compositional flexibility and ease of sample preparation, doping, and processability. Meerholz et al. suggest that the large nonlinear response of the azo dye 2,5-dimethyl-4-(p-nitrophenylazo)anisol is partly due to the orientational enhancement mechanism which consists of adding a plasticizing agent, N-ethyl-carbazole [11]. Despite the fact that photorefractive polymers were first identified less than 4 years ago, their index modulation and net gain at present are far superior to many of the well-known and extensively studied inorganic crystals. Much remains to be done because other properties such as optical quality and grating storage lifetime have to be optimized. However, the material's leaps in performance over just a few years provide a promise of utility regarding photorefractive polymeric materials.

There is a growing need in both government and industry for increased data-handling capabilities, including acquisition, processing, transmission, and storage. Optics will play a vital role in these areas. The possibility of designing new optical materials at the molecular level has created substantial interest in recent years. Specifically, certain organic polymers, when appropriately designed and fabricated, have many desirable properties for use as electrooptic materials. In order for an organic molecule to exhibit electrooptic behavior, it must possess a large molecular dipole. In order for a material containing such molecules to be electrooptic, a majority of the individual molecular dipoles must be aligned in the same direction. Researchers at Battelle and elsewhere are now working to improve the orientational stability of poled polymer materials [12]. These researchers have done extensive waveguide development with materials such as organic polydiacetylenes. This polymer exhibits a large third-order nonlinear optical response as well as the flexibility they offer in designing molecular structures for specific applications. The fabrication of these materials into device-quality thin films has proven to be difficult, however, largely because of their propensity for forming highly scattering polycrystalline solids. As a result, the promised desirable nonlinear optical properties of the polydiacetylenes have yet to be realized in a useful waveguide device.

If an electrooptic device is to be of practical use, it is desirable to use a structure in which there is a long interaction length between the light and the active polymer. This maximizes the nonlinear effects which occur. Therefore, much work has been carried out in producing an electrooptic modulator in the form of a waveguide, along which light is transmitted for a distance of many millimeters. If polymer waveguides can be deposited onto semiconductor substrates, then there is the possibility of combining both electrooptic and electrical processing functions into a single component. Early work on electrooptic polymers used active monomeric guides in a polymeric host. One of the first reports of such a system is of the azo dye, disperse red I, in a matrix of poly(methylmethacrylate), in which second-harmonic generation was observed [13]. Electrooptic modulators have been made using guest–host systems by several research groups. For example, Haas and Yammamoto made a 10-mm-long channel waveguide using methylnitroaniline in poly(methylmethacrylate) (PMMA), and observed 16% modulation for an applied voltage of 100 V at a wavelength of 633 nm [14]. Two main drawbacks exist for such systems. First, it has been found that the activity drops over a relatively short period of time as the guest molecules return to a random orientation. The second disadvantage of guest–host systems is that the amount of guest that can be added is limited; hence, the electrooptic coefficients are usually only of the order of a few pm/V, compared with 30 pm/V for lithium niobate [14]. In addition, Hoechst Celanese have recently produced polymers with an enhanced electrooptic coefficient of 38 pm/V at 1300 nm [14]. The demonstrator modulator based on the acrylate polymer has also been reported by the same individuals and is estimated to have an electrooptic coefficient of 30 pm/V at 633 nm for a thermopoling field of 20 V/μm. The modulator showed 17% modulation, and there was no observable degradation in its performance over a period of 2 years.

Recently, researchers have further improved the stability of nonlinear polymers by using cross-linked systems. Akzo Research Laboratories have developed high-thermal-stability, high-coefficient cross-linked polymers [15]. Eich et al. have reported on a polymer, which can be cross-linked during poling, which has a final second-harmonic coefficient of 13.5 pm/V, and does not show relaxation even at 85°C [16].

Researchers in Sweden have made the first polymer-based light-emitting diode (LED) that gives off polarized light [17]. In the future, improved versions of this device might be used to provide background illumination in liquid-crystal displays (LCDs), which currently require a polarizing filter to furnish the necessary light for polarization. The researchers made the LED using a substituted polythiophene film that luminesces when an electric current is passed through it. The polymer chains in such material normally have random orientation. As a result, the light they emit is not polarized. However, when the polymer film is stretched, the macromolecules are extended and become aligned. In this way, the light that the polymer film emits is polarized. The Swedish researchers Peter Dyreklev and Olle Inganas of the Laboratory of Applied Physics at Linkoping University and their co-workers have gone an important step further. They have demonstrated that such a stretch-orientated polymer can be made to give off polarized light under electrical stimulation in a thin-film device [17].

A chemiluminescence fiber-optic system coupled to flow-injection analysis (FIA) and ion-exchange chromatography has been developed for determining glucose in blood and urine. Immobilized glucose oxidase acted on β-D-glucose to produce hydrogen peroxide, which was then reacted with luminol in the presence of ferricyanide to produce a light signal [18]. Endogenous ascorbic acid and uric acid present in urine or blood samples were effectively retained by an upstream acetate anion exchanger. In addition, acetaminophen could also be absorbed by this ion exchanger. The detection system exhibited a sensitivity of 1.315 ± 0.044 RU/μm for glucose with a minimum detection level of 1 μm. When applied for the determination of urinary and blood glucose levels, the results obtained compared well with those of the reference hexokinase assay. Immobilized glucose axidase was reused for over 500 analyses without losing its original activity. A conservative estimate for the reuse of the acetate ion-exchange column was about 100 analyses.

Optically induced variations of properties of polymeric materials such as absorption, refractive index, and birefringence have recently been of considerable interest for the development of optical memory, optical data processing, and optical display devices [19]. Polymeric materials have several unique advantages for these applications, such as ease of processing, good mechanical strength, and low cost, compared to other inorganic and organic low-molecular-weight compounds. Recently, it was reported that birefringence could be optically induced in polymers containing azobenzene groups by polarized light. The mechanism involves repeated trans–cis photoisomerization of azobenzene groups and thermal cis–trans relaxation, resulting in the alignment of azobenzene groups in the direction perpendicular to the polarization of the incident beam. Consequently, there is a decrease of absorption in the direction parallel to the polarization of the incident beam and an induced birefringence in the material. Therefore, photoinduced orientation and recording of erasable holographic gratings in a new photoresponsive amorphous polymer containing azobenzene groups were studied [20]. Using linearly, polarized white light and an Ar$^+$ laser beam at 488 nm, orientation of azobenzene groups perpendicular to the polarization direction of the incident light was observed in polymer films. Holographic gratings with high diffraction efficiency were recorded by polarized Ar$^+$ laser beams at 488 nm. These gratings were stable at room temperature. The writing of multiple gratings at the same spot on the polymer films was also investigated. The holographic gratings can be erased by heating the samples above the glass transition temperature and can be rewritten on the same spot. This was performed for 50 cycles.

As nonlinear optical materials continue their journey from being laboratory curiosities to real engineering building blocks, additional properties continue to surface. In the case of second-order electrooptic (EO) materials some of the more recent questions center around alignment and chromophore thermal stability. It has been concluded that organic electrooptic materials must not only have high EO coefficients and long-term alignment stability at room temperature but these materials must also be able to retain their alignment at the elevated temperatures realized during fabrication and integration with commercial state-of-the-art microelectronic systems. Although the upper limit of temperature stability required is still under question, it seems probable that a practical EO NLO material will have to demonstrate stability above 300°C at least for short periods of time [21]. Current property requirements indicate that for second-order guest–host polymeric materials to be both useful in an engineering sense and cost-effective, they must possess much greater thermal stability. In a presentation by Reinhardt et al., a new molecular design and synthetic strategy was presented to produce second-order chromophores with much higher thermal stabilities [22]. Preliminary results of electric-field-induced second-harmonic-generation (EFISH) experiments on these compounds were revealed. These experiments indicate that molecules containing nonstandard aromatic heterocyclic donor and acceptor groups appear to be examples of a totally different class of second-order NLO chromophores. These types of molecule, because of their increased aromatic nature, have the potential for producing second-order active materials with much higher thermal stability than other types of chromophore. A secondary benefit of using heterocyclic rings as donors and acceptors is the possibility of reduced toxicity by eliminating the use of organic functional groups such as nitro. In addition to higher thermal stability and reduced toxicity, these types of chromophore also seem to have reduced absorption in the visible, thus making them usable over a greater wavelength range.

Among the several advantageous properties of molecular polymeric nonlinear optical materials is their ease of fabrication into low-scatter thin films for use in guided-wave devices. The waveguide geometry is the most convenient way to control and enhance the processes most useful for optical computers and other integrated optics devices. As the light in a waveguide can be confined to a very small cross-sectional area over long lengths, more efficient nonlinear interactions can take place. The advantage due to increased energy density in waveguide devices over bulk devices can be as high as the ratio of the length of the waveguide to the wavelength of light, which can be a factor of 10^4 [23]. Of particular importance are frequency conversion devices and devices utilizing field-induced refractive-index changes, including the photorefractive effect. With respect to the former, the most important design consideration is the phase matching of the converted light to the fundamental beam. In addition to the advantages of waveguides for integrated device applications, waveguide measurement techniques are often preferable for the important task of characterizing nonlinear polymer films. In techniques utilizing transmission through thin films, the efficiencies for various processes are limited by the fact that the interaction length cannot exceed the film's thickness and coupling to different dielectric directions may be impossible or require highly oblique incident angles. The waveguide geometry overcomes these difficulties. Glass and co-workers have described a prism-coupled waveguide technique to measure photorefractive properties of slab waveguides which would be especially appropriate for the thin-film polymeric photorefractive materials [24]. Researchers have adapted this technique for measurements of photo-induced diffraction in waveguides [25]. In this case, the photo-induced grating arises not from photorefraction, but from photochromic effects due to cis–trans isomerization in azo dyes. These individuals discuss their recent observations and measurements of photo-induced diffraction in polymer waveguides using single prism coupling in a manner similar to that described by Glass. Figure 5 is a representation of the geometry used for the diffraction efficiency measurements. Two beams with angular separation 2θ are incident at the guided mode angle. Difficulties in determining the energy density in the guide due to uncertain coupling efficiency, interaction length, and beam overlap volume limited the confidence in measurements in single-prism-coupled waveguide geometries. It is noted that other methods for overcoming these limitations such as end-fire coupling, two-prism coupling, and other techniques are available.

Matrix–matrix multiplication is an important operation in many computational and processing applications. These operations include correlation, convolution, Fourier transform of temporal signals, and two-dimensional images, to mention a

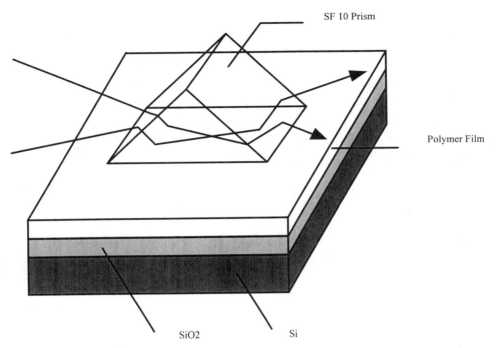

FIGURE 5 Measurement of diffraction efficiency in photo-induced polymer waveguides.

few. In addition, a large number of signal and image processing algorithms can be expressed in terms of matrix operations. Direct matrix–matrix multiplication is often avoided in electronic computers because it is an $O(N^3)$ (where $N \times N$ is the number of elements in each matrix) operation which requires a long computation time for serial machines. However, attempts to avoid matrix–matrix multiplications usually lead to other complicated algorithms. Furthermore, there are situations where direct matrix–matrix multiplications are inevitable. Alternatively, optical computing offers the advantages of parallelism and large capacity. Such capabilities have been successfully demonstrated in parallel vector–matrix multiplication [26]. Recently, nonlinear optical techniques have been employed in the parallel matrix–matrix multiplication [27]. These techniques require complicated alignment and suffer from severe energy loss. Gu et al. proposed and demonstrated a new method which utilizes grating degeneracy in photorefractive media in conjunction with an incoherent laser array to implement parallel optical matrix–matrix multiplication [28]. Specifically, multiplications are implemented by photo-induced index gratings whose amplitudes are determined by the interference between coherent beams, whereas summations are implemented by grating degeneracy. Such a matrix–matrix multiplier is capable of handling large matrices such as those with 1000×1000 elements. Figure 6 shows the schematic diagram which describes the principle of operation of matrix–matrix multiplication. Both matrix A ($N \times N$) and matrix B ($N \times N$) are placed at the front focal plane of lens L_1. At the rear focal plane of lens L_1, a volume holographic medium such as a photorefractive crystal is inserted to record the multiplication of the two matrices. The recorded infor-

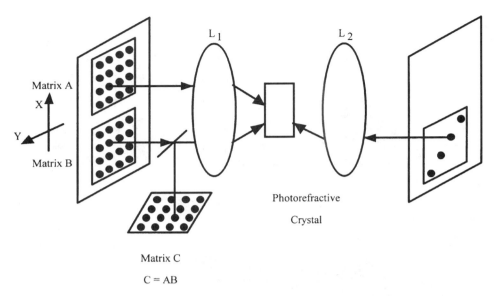

L_1 L_2

Matrix A

X

Y

Matrix B

Photorefractive

Crystal

Matrix C

C = AB

FIGURE 6   Schematic illustrating the principle of operation of matrix–matrix multiplication.

mation is read out by a set of reading beams which consist of N diagonally aligned point sources placed at the front focal plane of lens L_2. The diffracted readout beams are directed by a beamsplitter to the output plane which is located at the focal plane of lens L_1. To realize matrix–matrix multiplication, the illumination of matrices A and B is chosen so that all the pixels within each line along the x direction are mutually coherent, whereas pixels with different y values at the input plane are mutually incoherent. The N reading points are diagonally aligned at the reading plane. Diffraction occurs at different angles among the pixels, which, in turn, forms different columns in matrix C. The experimental results were in excellent agreement with the theoretical predictions. Such a matrix–matrix multiplier, with its large capacity and parallelism (matrices with 1000×1000 elements), can potentially be used in optical computing, photonic switching, and optical neural networks.

VI. SUMMARY

Biopolymers have a controlled molecular structure and morphology which make them promising candidates for nonlinear optical materials. However, many of the material's capabilities are lost if the incident photons are absorbed or dispersed. Improving upon not only the polymers alignment but its transparency to specific wavelengths should produce a film with superior qualities to that of its inorganic counterpart.

During this investigation, an effort to improve the film quality has been attempted by utilizing other helicogenic solvents, including the standard methylene chloride. Specifically, 1,4-dioxane has been employed. However, this substance is a higher-boiling solvent with relatively low vapor pressures under ambient conditions and therefore requires excessive time for evaporation. This information has influenced research efforts to focus on the effects of solvent type and film clarity as well as the electrode design to facilitate alignment and quality of film. Subsequently, films will be analyzed for their waveguide abilities, which is the backbone capability for a whole host of other material uses.

A measured amount of PBLG solution (2.5%) in 1,4-dioxane was placed in a cavity and distributed by manipulating a thin Teflon leaf. The dimensions of the well are 5.5 × 3.5 × 2.5 cm (Figure 7). After distributing the PBLG solution within the cavity between the two platinum electrodes, a glass covering is immediately installed over the cell. The ensuing enclosed system is maintained in a level position throughout the film formation. A horizontal distribution of material is necessary to avoid a nonuniform polymeric film. A Del RHVS Series Detector Supply unit, adjustable from 0 to 60 kV DC, was utilized as a power source. The voltage was then turned on and the solvent inside was allowed to evaporate slowly. The time for film formation remained constant at approximately 20 min. After completion, the film is carefully removed by using tweezers and then stored between two glass coverslips. Although the applied field should not produce any molecular translation, such motion is an indication of an impurity effect and random thermal motion. This condition needs to be rectified in order to create an environment in which the field strength has no effect on the film thickness and uniformity.

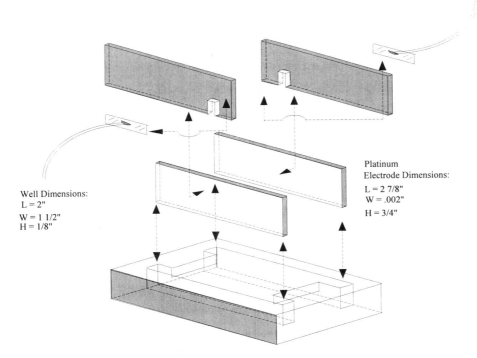

Well Dimensions:
L = 2"
W = 1 1/2"
H = 1/8"

Platinum
Electrode Dimensions:
L = 2 7/8"
W = .002"
H = 3/4"

FIGURE 7 Schematic for in-place poling of polymer films.

REFERENCES

1. Syms, R., and Cozens, J. *Optical Guided Waves and Devices*, McGraw-Hill Book Co., New York, 1992, pp. 97–109.
2. Marcuse, D., in *Theory of Dielectric Optical Waveguides*, 2nd ed., (P. F. Liao, eds.), Academic Press, New York, 1991, pp. 2–23.
3. Vassallo, C., in *Optical Waveguide Concepts*, (H. Hung-chia, ed.), Elsevier, Amsterdam, 1991, pp. 119–144.
4. Okoshi, T., Okamoto, K., and Hotate, K., *Optical Fiber*, Ohm Publishing Co., Tokyo, 1983, pp. 27–29.
5. Kugagami, H., Koshiba, M., and Suzuki, M., Numerical analysis of silicon-clad planar optical waveguides, *IECE Japan, J69-C*, 856–864 (1986).
6. Alferness, R. C., Optical guided-wave devices, *Science, 234*, 825–829, (1986).
7. Lytel, R., and Stegman, G. I., in *Nonlinear Optical Effects in Organic Polymers*, (J. Messier et al., eds.), Kluwer Academic Publishers, Boston, 1989, pp. 277–289.
8. Alferness, R. C., Optical guided-wave devices, *Science, 234*, 825–829 (1986).
9. Kurmer, J. P., and Schwerzel, R. E., Organic materials for integrated optical waveguides, *Photonics Spectra, 23*, 169–173 (1989).
10. Clark, D., and Cohen, L., Plastics optics shine in high-volume production, *Photon. Spectra*, 97–100 (March 1995).
11. Moerner, W. E., and Peyghambarian, N., Advances in photorefractive polymers: Plastics for holography and optical processing, *Optics Photon. News, 6*(3), 24–29 (1995).
12. Carole, J., Polymers for non-linear optical devices, *Chem. Ind. 19*, 600–608, (October 1, 1990).
13. Lytel, R., and Stegman, G. I., in *Nonlinear Optical Effects in Organic Polymers* (J. Messier et al., eds.), Kluwer Academic Publishers, Boston, 1989, pp. 379–381.
14. Haas, D. and Yamamoto, K., Polymeric materials and electro-optic waveguide modulators, in *Proc. Topical Meeting on Integrated Photonics Research*, Hilton Head, SC, 1990.
15. Horsthuis, W. H. G., Van der Horst, P. M., and Mohlmann, G. R., Developments in high temperature, stable nonlinear optical polymers, in *Proc. Topical Meeting on Integrated Photonics Research*, Hilton Head, SC, 1990.
16. Eich, M., et al., Novel second-order nonlinear optical polymers via chemical cross-linking-induced vitrification under electric fields, *J. Appl. Phys., 66*, 3241–3247 (1989).
17. Dyreklev, P., Polarized light-emitting diode based on polymer, *C&EN 73*(4), 28–29 (January 23, 1995).
18. Cattaneo, M. V., and Luong, J. H. T., On-line chemiluminescence assay using FIA and fiber optics for urinary and blood glucose, *Enzyme Microbiol. Technol., 15*, 424–428 (1993).
19. Todorov, T., Nikolova, L., and Tomova, N., A new high-efficiency organic material with reversible photoinduced birefringence, *Appl. Opt., 23*, 4309–4313 (1984).
20. Kim, D. Y., Li Lian, Jeng, R. J., Kumar, J., Fiddy, M. A., and Tripathy, S. K., Nonlinear optical photoresponsive polymer for reversible optical data storage, *SPIE Org. Biol. Optoelectron., 1853*, 23–28 (1993).
21. Wu, J. W., Valley, F., Ermer, S., Binkley, E. S., Kenney, J. T., and Lipscomb, G. F., Thermal stability of electro-optic response in poled polyimide systems, *Appl. Phys. Lett., 58*(3), 225–227 (1991).
22. Reinhardt, B. A., Kannan, R., and Dillard, A. G., Synthetic approaches for novel optoelectronic and polymeric materials, *SPIE Org. Biol. Optoelectron., 1853*, 50–59 (1993).
23. Zyss, J., and Chemla, D. S., Quadratic nonlinear optics and optimization of the second-order nonlinear optical response of molecular crystals, in *Nonlinear Optical Properties*

of Organic Molecules and Crystals (D. S. Chemla and J. Zyss, eds.), Academic, New York, 1987, pp. 23–191.

24. Glass, A. M., Kaminow, I. P., Ballman, A. A., and Olson, D. H., Absorption loss and photorefractive index changes in Ti: $LiNbO_3$ crystals and waveguides, *Appl. Opt.*, *19*, 276–281 (1980).

25. Andrews, J. H., Singer, K. D., and Cahill, P. A., Nonlinear optical interactions in polymer waveguides, *SPIE Org. Biol. Optoelectron.*, *1853*, 221–232 (1993).

26. Marder, S. R., Beratan, D. N., and Chemg, L. T., *Science*, *103*, 252 (1991).

27. Marder, S. R., Chemg, L. T., and Teimann, B. G., *J. Chem. Soc., Chem. Commun. 9*, 672 (1992).

28. Gu, C., Campbell, S., and Yeh, P., Nonlinear optical matrix multiplier, *SPIE Org. Biol. Optoelectron.*, *1853*, 239–252 (1993).

14

Part II: Sensor Protection from Lasers

Debra J. Trantolo and Joseph D. Gresser
Cambridge Scientific, Inc.
Belmont, Massachusetts

Donald L. Wise and Gregory J. Kowalski
Northeastern University
Boston, Massachusetts

D. V. G. L. N. Rao and F. J. Aranda
University of Massachusetts at Boston
Boston, Massachusetts

Gary E. Wnek
Virginia Commonwealth University
Richmond, Virginia

I. OVERVIEW

The overall objective of this project was the development of high χ^3 materials with optical clarity and mechanical strength for sensor and eye protection. One specific end product of the proposed investigations is a material suitable for protection of the human eye against laser irradiation. Such a material must not only have a reasonable optical limiting response but also have optical clarity, thermal stability, and high impact strength. To develop this material, promising χ^2-active materials were processed into films using polycarbonate (PC), relying on its requisite optical and mechanical properties, as the host polymer. Three nonlinear optically (NLO) active components were evaluated: (1) a tetrabenzoporphyrin (TBP), (2) a polyaniline (PANI), and (3) a sol complex. Thus, a full range of both molecular and polymeric NLO-active substances were initially targeted for this survey.

Although materials with high third-order susceptibilities demonstrate optical switching or limitation in response to critical input intensity, application in devices for sensor protection (to include that of the human eye) has not yet been brought to practicable fruition. In addition to the physics requirements for translation of input laser energy, devices such as goggles require an optically clear carrier of high-transparency and good mechanical properties. Thus, our determination of feasibility was directed toward the development of such a material by incorporation of χ^3-active materials into a polymeric carrier which is strong and transparent, focusing almost exclusively on the materials-processing constraints of a valid protection system. In addition to the first-order translation of solution chemistries to solid-phase materials, a supporting theme of our study was the investigation of the alignment of the guest–host systems as a route to the development of protective materials exploiting our patented system for in-plane poling to determine the feasibility of the proposed approach. Although χ^3 properties do not require noncentrosymmetry, the NLO behavior frequently is enhanced by alignment, which removes the symmetry element. In addition, the structural integrity of polymers is frequently improved by alignment.

We report herein on the preparation and characterization of guest–host polymer thin films. Results of our work suggest that solid-phase NLO materials can be prepared, thus demonstrating their potential for sensor protection from lasers. The results are very encouraging. We have obtained excellent optical power limiting results for the films that we manufactured. In summary, (1) we have developed techniques to fabricate reverse saturate absorber (RSA) films, (2) various NLO materials can be processed using these techniques, (3) the techniques are suitable for the development of multilayer limiters, and (4) we have developed a sound agenda for the development and improvement of solid multilayer optical power limiters.

II. PROJECT RATIONALE

The thrust of the work was to demonstrate high χ^3 values in composite films of a host polymer into which χ^3-active materials have been incorporated. It was reasoned that whereas a single material might eventually meet optical, mechanical, and processing criteria, a viable approach would be to explore hybrid, multicomponent, and multifunctional materials with each component serving one or more specific needs. Specifically, doped polymer systems were investigated in which the host or matrix offered the desired optical transparency and mechanical toughness, with molecularly dispersed dopants providing the requisite optical limiting behavior. It was anticipated that the host polymer would serve to "matrix-isolate" χ^3 chromophores, thus reducing the potential for both phase segregation of the guest molecules and photophysical processes among them which might be undesirable.

Our efforts have been guided by the "materials tetrahedron" shown below, which emphasizes that the ultimate goal of performance is realized through the linkage to structure and properties which are, in turn, dictated by synthesis and processing.

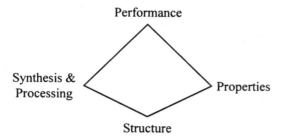

A. The Host Polymer

The host polymer may be χ^3 inactive but must have excellent mechanical and optical properties. Polycarbonate (PC) is the host of choice, having high impact strength, optical clarity, and processability. PC (Lexan®, GE Plastics) is optically clear due to the absence of crystallinity and is rather easily processed from solution (methylene chloride is a commonly used solvent) or the molten state ($T_g \sim 150°C$). Also, the polymer is tough as a result of short-range segmental motions which act as energy absorbers. Lucite® or poly(methyl methacrylate) (PMMA) was also investigated as a host. PMMA is amorphous and has good optical transparency and easy processability into films but lacks the toughness of polycarbonate.

It would be ideal to identify a host which was also χ^3 active. An enzymatically synthesized polyaniline (PANI) was examined in this context both as a free-standing film and as a composite with PC. The initial plan considered that if PANI was film-forming, the χ^3 guests [TBP and a tricyanovinyl aliline (TCVA)-coated silver sol (TCVA/Ag)] would be incorporated directly into the PANI; should film formation not be practicable, PANI was to be incorporated into PC prior to introducing the TBP and sol guests.

B. Guest NLO-Active Materials

Three classes of the guest χ^3 materials were chosen in order to afford flexibility in the design of the guest–host systems. The three materials were (1) polyaniline, (2) a zinc tetrabenzoporphyrin derivative, and (3) a silver sol–organic chromophore complex.

1. Polyanilines:

Conducting polymers are typically characterized by at least modest π-electron delocalization, and this, coupled with the high polarizability of these electrons, makes conjugated polymers interesting candidates for applications relying on third-order nonlinearities. Moreover, the optical and electrical properties of these polymers are readily modulated by redox reactions of the backbones. Polyaniline (R = H, above) is one of the most widely studied conducting polymers due to the easy conversion of neutral to cationic and conducting forms by protonation (with a concomitant change in optical absorption), the excellent environmental stability of both forms, the opportunity to control the processibility via the choice of conjugate base of the acid, and the ability to control the processibility and optical and electronic properties by the choice of the R group. For example, when R = methyl, the polymer (poly-*o*-toluidine) is more soluble in common solvents [1], and it was reasoned that such substituents would also afford enhanced solubility in a polymeric host such as Lexan® polycarbonate.

Polyaniline and derivatives are typically prepared by oxidation of aniline with, for example, persulfate in the presence of HCl. However, it is also possible to polymerize aniline via enzymatic catalysis, and this may be advantageous if the polymer chain structure is more polarizable than that obtained by typical wet chemical conditions.

2. Tetrabenzoporphyrin Derivatives:

Metal-containing tetrabenzoporphyrins were selected as active materials for sensor protection for several reasons. First, they, like related phthalocyanines [2], have strong optical absorptions and, complexed with selected metals, can show efficient intersystem crossing and hence large triplet–triplet absorptions which afford optical limiting. Second, both the metal and the substituent group on the porphyrin ring can be changed, offering tunability of optical properties and enhanced solubility in host polymers. Third, porphyrins are redox active and it should be possible to tune the optical absorption of metal-containing porphyrins by altering the redox state of the metal.

It is, of course, important to achieve a high chromophore density to maximize the optical response, but, as shown in our results, chromophore aggregation has an undesirable effect on the optical limiting capability of zinc tetrabenzoporphyrin. However, we believe that we have the ability to control host–guest interactions and, thus, it should be possible to achieve a good balance of chromophore loading without significant interchromophore interactions.

3. Silver Sol–Chromophore Complexes. Earlier work by Korenowski and Wnek [3,4] demonstrated that silver sols dispersed in a polymer containing *p*-tricyanovinylaniline units exhibited large third-order NLO responses from four-wave degenerate mixing experiments. Noble metal nanoparticle systems are of interest because of their intrinsically large third order nonlinear susceptibilities [3,4]. When combined in a composite, interactions between the polymer host bearing NLO chromophores and metal colloids were shown to yield enhanced third-order nonlinear susceptibilities. A proposed model for the enhancement is shown in Fig. 8.

1

Further work in Wnek's laboratory demonstrated that Ag sols to which *p*-tricyanovinylanilines having a long alkyl tail at the aniline nitrogen are dispersible in dimethylformanide. For example, a gray–black powder of Ag in alcohol, prepared from reduction of silver salts with $NaBH_4$, immediately disperses to form an optically clear magenta solution upon addition of a small amount of compound **1**.

Replacement of ethanol as it evaporates with DMF preserves the Ag colloid. This suggests that the NLO chromophores need not be present as part of a polymer into which the sols might be dispersed. Rather, we believed that it would be possible to disperse homogeneously Ag sols modified with **1** into a variety of polymer hosts. These hosts may confer mechanical stability with the Ag colloid, absorbed **1** being the NLO-active entity. As will be discussed in the following, only a limited amount of work was done on the Ag sol system, principally because of the promising results obtained from the Zn tetrabenzoporphyrin/Lexan® composite.

III. EXPERIMENTAL METHODS AND RESULTS

A. Introduction

The thrust of the work was to demonstrate high χ^3 values in composite films of a host polymer into which χ^3-active materials have been incorporated. The host polymer may be χ^3 inactive but must have excellent mechanical and optical properties.

METAL ADSORBATE

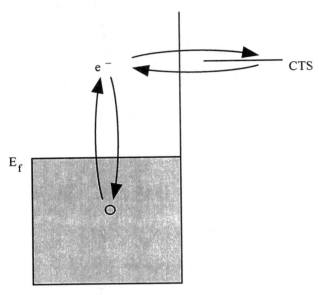

FIGURE 8 Charge transfer scheme. The diagram shows the photo-induced promotion of a valence electron (below the Fermi level) to the conduction band. The promoted electron can then couple to an uncoupled molecular orbital state or charge transfer stage (CTS) of the absorbate–metal complex. The electron is then transferred back to the conduction band, followed by recombining with the hole in the valence band.

Polycarbonate (PC) was the host of choice, having high impact strength, optical clarity, and processibility. It was a supporting goal to identify a host which was also χ^3 active. Polyaniline (PANI) was examined in this context both as a free-standing film and as a composite with PC. If PANI was processible into a film, other χ^3-active materials would be incorporated into it.

The experimental program included preparation of χ^3-active PANI and incorporation of PANI into a PC. This program utilized the most common commercial PC synthesized from bisphenol A, 2,2′-bis(4-hydroxyphenol) propane, known as Lexan®. PC composites were also prepared using two other χ^3 materials, a TBP and an Ag sol, thus surveying a range of molecular offerings to address requirements for both optical efficiency and materials processing.

The tasks of the program are summarized as follows: Preparation and Characterization of Polyaniline, Acquisition of TBP and Preparation of Silver Sols, Preparation of PANI Films and PANI/PC Films, Alignment of PANI and PANI/PC Films, Spectral Characterization: UV/Vis and IR Absorption, Polymer Characterization: DSC and TGA, Mechanical Characterization: Tensile Strength and Modulus, and NLO Characterization: Degenerate Four-Wave Mixing.

B. Materials

The materials used are as follows:

- Aniline (Sigma Chemical Co.)
- 4-*n*-Butylaniline (TCI, Tokyo, Japan)
- 4-*sec*-Butylaniline (TCI, Tokyo, Japan)
- 1,4-Dioxane (Fisher Scientific)
- Dimethylformamide, HPLC grade, used as received (Aldrich Chemical Co.)
- Formaldehyde solution, 37.1% (Fisher Scientific)
- Gelatin (from porcine skin, ~300 bloom) (Sigma Chemical Co.)
- HEPES, *N*-(2-hydroxyethyl)piperazine-*N'*-ethane sulfonic acid (Sigma Chemical Co.)
- Horseradish peroxidase, Type II, 150–200 units/mg solid (Sigma Chemical Co.)
- Hydrogen peroxide, 30% (w/w) (Sigma Chemical Co.)
- Methylene chloride, HPLC grade (Sigma Chemical Co.)
- Polycarbonate (PC), Lexan 9034 Sheets (General Electric)
- Silver nitrate, ACS (Bradford Scientific)
- Zinc meso-tetra-*p*-methoxyphenyl-tetrabenzoporphyrin (ZnOCH$_3$TBP) and zinc meso-*p*-hydroxyphenyl-tetrabenzoporphyrin (ZnOHTBP) were kindly supplied by Dr. Masato Nakashima of the U.S. Army Natick Labs.
- *N*-Methyl-*N*-dodecyl-4-tricyanovinyl aniline (TCVA) was kindly supplied by Professor Gary E. Wnek of Virginia Commonwealth University.

C. Synthesis of Polyanilines

The PANI syntheses were done in collaboration with Dr. Joseph A. Akkara of the U.S. Army NatickLabs; see the Acknowledgment.

1. The HEPES buffer [20% aq. HEPES (250 m*M*)/80% dioxane, v/v] was prepared by dissolving 5.9575 g of HEPES in distilled H$_2$O and bringing the volume to 100 ml. The concentration of this solution is 250 m*M*. Twenty milliliters of this solution was then dissolved in 80 ml of 1, 4-dioxane. The pH was adjusted to 7.5.

   The 30% H$_2$O$_2$ (1.1 g) was dissolved in a small portion of the buffer; 50 mg of horseradish peroxidase (HRP) was dissolved in a small portion of the buffer; and the aniline (0.91 g) was dissolved in the remaining buffer.

   The HRP solution was added to the aniline solution followed by the addition of the H$_2$O$_2$ solution. This mixture was allowed to react at room temperature for up to 24 h with gentle stirring. PANI forms and can be separated by centrifugation at 5000 × **g** for ~10 min.

2. Our Natick collaborators suggested a PANI copolymer based on their observations relating to ease of synthesis and characterization. Two derivatized anilines were used in the syntheses, *sec*-butyl and an *n*-butyl aniline, each being copolymerized with an equimolar ratio of aniline. These syntheses are thus reported below.

   The reaction was carried out at room temperature in a solution of dimethylformamide (DMF) and succinate buffer (0.1 *M* and pH 5.5) in the ration of 1:1 (v/v). The monomers, aniline (5.0 g) and 4-*n*-butylaniline (8.0 g), were dissolved in 200 ml of DMF. Horseradish peroxidase (120 mg) was dissolved in 200 ml of succinate buffer (pH 5.5, 0.1 *M*) and added slowly to the aniline DMF solution with constant stirring via stirring bar. With continued stirring, the polymerization reaction was initiated by the dropwise addition of hydrogen peroxide (30%) to the above reaction solution; 16 ml of hydrogen peroxide was added over an 8-h period. The reaction was allowed to proceed with stirring to a total reaction time of 20 h. Formation of a PANI is evidenced by the formation of a precipitate. The precipitate was isolated via centrifugation (30 min at 5000 × **g**), washed with water, and air-dried. To increase the yield, the reaction supernatant was reduced in volume via rotary evaporation. This viscous residue and the primary precipitate were dried under vacuum at 50°C. (Reaction yields were generally ~60%).

"PsBAA" is the copolymer of a 1:1 mole ratio of 4-*sec*-butylaniline and aniline (sample JA-31). PnBAA is the copolymer of *n*-butylaniline and aniline. The initial PANI experiments were done using the PsBAA. Although this sample was not highly purified, it enabled us to explore its film-forming capacity and to develop techniques for incorporating the polymer into a polycarbonate (PC) support.

D. Preparation of Silver Sols

Previously, we had prepared silver sols by reduction of silver nitrate with sodium borohydride [5]. The silver colloid was prepared via a chemical method (rather than a laser ablation method). Silver nitrate can be reduced to silver metal with an alkali aluminum hydride. Several aspects of the complete system, including the solubility of the reactants used (here, *N*-methyl-*N*-dodecyl-4-tricyanovinyl aniline, TCVA) as well as the reactivities of the two organic compounds toward the reducing agent, had to be understood before choosing the solvent system. A main issue in the preparation of the colloid, its isolation, and subsequent use in the film preparations requires a cognizance of both the colloidal dispersion and solubility. Ultimately, the colloid should be isolable and then suspendable in a solvent which dissolves the polycarbonate (PC) host. The inorganic reagents are soluble in water, methanol, and 95% ethanol but insoluble in most organic solvents (ether, chloroform, etc.). On the other hand, PC is insoluble in those same solvents, whereas TCVA is only partially soluble in ethanol.

In our earlier work, it appeared that a silver colloid was obtained when 96% ethanol was used as a solvent and the excess borohydride was decomposed with acetic acid before adding TCVA to the colloid. A solution of silver nitrate (9 mg in 100 ml of ethanol) was added dropwise to a solution of sodium borohydride (14 mg in 100 ml of ethanol). The borohydride solution immediately becomes yellow, and as more silver nitrate was added, the color progresses to red, then dark gray. At the end of the addition (~15 min), one drop of glacial acetic acid was added followed by 14 mg of TCVA. (If TCVA is added to the colloid without decomposing the excess NaBH$_4$, its color changes from red to yellow, indicating that TCVA gets reduced by hydrides.) However, within a short time, a dark gray suspension precipitates with the formation of a shiny silver layer on the glassware.

At this point, the TCVA/silver mixture is in ethanol solution. In our initial preparations, this mixture was then transferred to a DMF-based medium. DMF is one solvent for film formation. In order to transfer from the ethanolic colloid, 10 g of DMF were added to the suspension and slowly evaporated with a stream of air until no more than 3 g of the total suspension remained. Here, the solvent composition is predominantly DMF.

In another embodiment of the colloid preparation, it appeared that the mixture could be isolated directly from the ethanol solution. The ethanol solution was evaporated to dryness, yielding a dark magenta particulate. Preliminary investigations suggested that this TCVA/silver "colloid" could be resuspended in methylene chloride. This is an attractive solvent for film preparation because it is a good solvent for PC and has a reasonable vapor pressure.

We attempted to determine the "stoichiometry" of this silver/TCVA complexation. The colloid preparation was followed spectrophotometrically, anticipating that the TCVA absorbance would be different than that of the TCVA/silver complex. This had the potential of offering additional direction to the colloid preparation. However, attempts to follow the color progression during colloid preparation (i.e., the progression from yellow to red) were not successful. Even without further addition, the yellow color develops at a rate too fast to allow sampling for spectral analysis.

With minor modification, this is the technique used by Creighton et al. [6]. We repeatedly observed a progression of color from yellow through red to blue gray with final precipitation of particulate silver. Although the addition of a stabilizing agent such as TCVA produces a relatively stable sol in that the particle growth as judged by the color progression was slower, the stability was still judged insufficient for further processing into PC-based films.

We have therefore explored an alternative method which appears to merit continued use. If the reduction is carried out with formaldehyde rather than with the borohydride, the reaction is slower, as is the progression of color changes. Silver nitrate was dissolved in an aqueous gelatin solution. To this was added the formaldehyde. A deep red color developed, which, at this writing, is stable for at least 21 days. The solution appears clear when viewed directly against a light source, but hazy when viewed perpendicular to the light. This qualitative observation indicates light scattering by small particles. A spectrum of the diluted suspension (taken at day 21) has an absorption band at 427 nm which does not shift on further dilution by a factor of 2.5. (The spectra are displayed in Fig. 9.) The diluted solutions were yellow to yellow-brown, and the absorption at 427 nm is characteristic of silver particles substantially smaller than the wavelength of light. Creighton [6] reports an absorption at about 380 nm for a silver sol, but on addition of pyridine, this peak diminished with appearance of longer wavelength absorptions.

(The use of gelatin to stabilize gold sols were reported by Michael Faraday in 1857. The sols were prepared by striking an arc between two gold electrodes immersed in water. The sols are still stable and are on display in the British Museum.)

Although gelatin is a very effective stabilizing agent, its NLO activity is not known to us. Thus, the use of the NLO-active TCVAs was included in the program. To plan on the outcome of the revised sol preparation, TCVA was treated with an excess of formaldehyde with no noticeable change. (This is being confirmed by absorption measurements.) Although gelatin can be cross-linked by dialdehydes (e.g., glutaraldehyde), the reaction is too slow with formaldehyde to permit stabilization of the sol.

Several experiments were conducted using formaldehyde reduction of silver nitrate without gelatin to form the sol. In one series of these experiments, TCVA

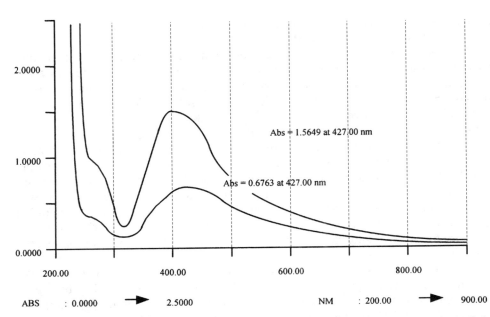

FIGURE 9 Silver sol complexed with gelatin. Upper curve: 2.5 × concentration of the lower curve (sample 82-7).

and silver nitrate were codissolved in aqueous ethanol followed by the addition of formaldehyde. No immediate reaction was observed, although, over the course of several days, silver slowly precipitated. These experiments showed that TCVA is fairly effective in stabilizing the silver ion and may significantly slow the growth of the sol particles.

In another experiment, an aqueous silver nitrate solution was treated with formaldehyde. When the yellow color was beginning to turn red, an alcoholic TCVA solution was added. After 5 min, the solution appeared slightly hazy and red brown in color, but no precipitate of silver had formed. After 4 days, a slight sliver coating formed on the inside of the test tube. The solution was red-brown with a very slight haze.

We tentatively conclude from these experiments that TCVA is fairly effective at stabilizing the silver ion and somewhat effective in stabilizing the sol. However, stability must be improved before further processing into films can be undertaken.

The stoichiometry of the reduction process is

$$2AgNO_3 + HCNO + H_2O \rightarrow HCO_2H + 2HNO_3 + 2Ag$$

Reduction of 50 mg of $AgNO_3$ should require 11 μl of 37.1% HCHO solution. A further advantage of this approach is that the reduction can be stopped by oxidizing excess HCHO with dilute hydrogen peroxide without destroying the sol and without oxidizing the TCVA.

E. PANI Film Preparation and UV/VIS Characterization

1. PANI Copolymers

A solution of PsBAA in methylene chloride (4.3 mg/ml) was cast on a glass microscope slide and allowed to evaporate. The resulting deposit of PsBAA was not a film. Films were formed, however, from cosolutions of PC and PBAA. A solution of PsBAA in methylene chloride (4.3 mg/ml) was mixed with a solution of PC in methylene chloride (8.6 mg/ml) in a volume ratio of 1 : 2. This cosolution was cast on a glass microscope slide and allowed to evaporate to form a film with a 4 : 1 weight ratio of PC : PsBAA. A second film was cast from a cosolution containing a 1 : 4 volume ratio of the two stock solutions to form a film with a PC : PsBAA ratio of 8 : 1. Both films were flexible and transparent although, under optical microscopic examination, each showed minor inclusions of some foreign insoluble material, probably impurities in the original sample.

Visible spectra were taken of a PsBAA solution in methylene chloride over the concentration range 7.9–63.4 μg/ml. The spectra are displayed in Fig. 10. Absorbance versus wavelength data is given as follows:

Concentration (mg/ml)	Absorbance (wavelength, nm)		
0.0634	2.4274 (331)	1.0976 (420)	— (532)
0.0317	1.6871 (318)	0.5468 (422)	0.1709 (532)
0.01585	0.9602 (314)	0.2995 (420)	0.0959 (532)
0.007925	0.4992 (313)	0.1554 (424)	0.0507 (532)

Absorbance versus wavelength is plotted in Fig. 11. Two peaks appear in the visible region of the spectrum. The short-wavelength peak (313–331 nm) exhibits a pronounced hypochromic displacement (blue shift) at increasing concentrations. This effect is much less noticeable at the longer wavelength peak at 420 nm. Although no peak appears at 532 nm, absorbances were noted. Linear regressions

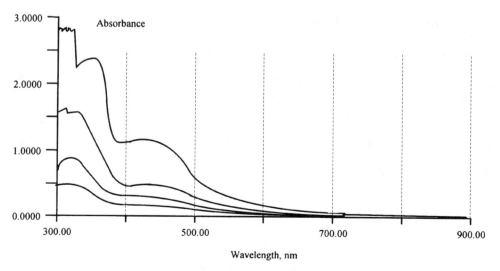

FIGURE 10 Spectra of PBAA in methylene chloride.

Absorbance

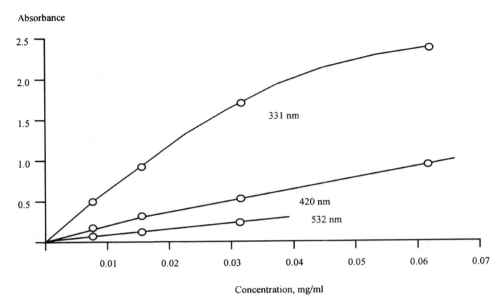

Concentration, mg/ml

FIGURE 11 Beer–Lambert plots for poly(4-*sec*-butylaniline-co-aniline (1:1) (PBAA) in methylene chloride.

in the linear regions are given below, the slopes of which are the extinction coefficients in AU/mg/ml (AU = absorbance units).

Wavelength	Conc. range (mg/m)	Linear regression*	Correlation Coeff.
313–331	0–0.01585	$A = 0.00637 + 60.580C$	0.99974
420	0–0.0634	$A = 0.01302 + 17.112C$	0.99965
532	0–0.0317	$A = 0.00528 + 5.343C$	0.99974

*Includes 0,0 point.

Two spectra were also taken of the 8 : 1 PC : PsBAA film by fixing it to the face of a quartz cuvette in two positions with respect to the beam (Fig. 12). The film thickness was approximately 10 μm. Peaks were observed at 421 nm (Abs = 0.4884, 0.4728) and at 315 nm (Abs = 1.0725, 1.1922). The absorbance at 532 nm was 0.2514 and 0.2267. A control of pure Lexan film of about the same thickness showed no absorbance in the visible region.

It is interesting to compare the ratio of absorbances of the short-wavelength peak to that at 420 nm. The ratio decreases with increasing concentration, and at the highest concentration, the ratio is very close to that observed for the PBAA/PC film.

Conc. (mg/ml)	Ratio
0.0079	3.213
0.0159	3.206
0.0317	3.085
0.0634	2.211
Film	2.196, 2.522* (Average = 2.359)

*Two measurements at different orientations of the film in the cell.

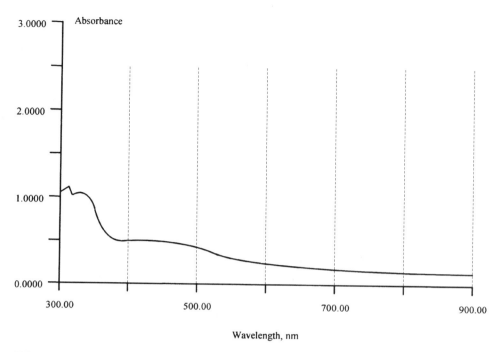

FIGURE 12 Absorbance of a PC/PBAA (8:1) thin film.

Polycarbonate (from bisphenol A, 4,4′-dihydroxy-diphenyl-2,2′-propane) has a heat distortion temperature of 138–143°C at 66 psi [7] and a glass transition temperature reported variously as 144–145°C [8] and 150°C [9]. Odian [9] reports its crystalline melting point as 270°C. Two sheets of PC (General Electric Lexan 9034, 2 mm thick × approx. 1 in.^{2}) were placed between two stainless-steel cylinders with flat polished surfaces. This assembly was placed in a hydraulic press and heated by heating tapes wrapped around the cylinders. At a temperature of 132–142°C and a pressure of approximately 2000 psi, the sheets fused to form a solid transparent composite without significantly compromising its optical clarity. A spectrum of this composite in the visible was taken for baseline data and is shown as the lower curve of Fig. 13. As seen in this figure, PC is transparent to 400 nm; below 400 nm, the absorbance increases rapidly.

The 8 : 1 PC/PsBAA film was sandwiched between two of the 1-in.^{2} Lexan sheets, and a fused composite prepared as described above. A spectrum of this composite is also shown in Fig. 13. The maximum of the 420-nm peak is not clearly defined due to the slight absorbance of the PC. Absorbances of the fused PC plates with and without the PsBAA are shown below. (All spectra of PC/PsBAA film and Lexan plates were taken against air as a reference.)

Absorbance	420 nm	532 nm
Fused PC plates*	0.1089	0.0990
Fused PC plates with PsBAA	0.3409	0.2254
Ratio of Absorbances	0.319	0.4392

*Thickness of the fused Lexan plates is 4 mm.

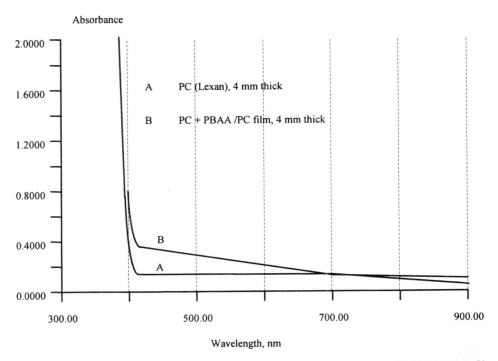

FIGURE 13 Spectra of Lexan plates, 4 mm thick (A); Lexan plates with PC/PBAA film (B).

Spectra were taken of a second thin (10 μm) PC/PsBAA film with an 8 : 1 ratio of PC : PsBAA. Spectra were taken both before and after heat treatment at 132–142°C to look for evidence of thermal degradation (Fig. 14). Absorbances at both wavelength maxima and at 532 nm were recorded.

| | Wavelength (nm) | | | |
	316	421	532	A_{316}/A_{421}
Before heat treatment	1.143	0.462	0.202	2.474
After heat treatment	0.798	0.361	0.168	2.211

The ratios of absorbances at 315 and 421 nm before and after heat treatment were 2.474 and 2.211. Although there is a 10.6% reduction in the ratio, this does not necessarily indicate thermal degradation, although this point must be addressed more fully by infrared spectroscopy. Again, the differences in the spectra are probably due to variation in placing the film against the face of the quartz cuvette rather than to changes in chemical composition. Figure 14 displays the thin-film spectra before and after heat treatment.

These results are very encouraging in that it appears quite possible to easily fabricate structurally strong composites of good optical quality by this technique. The optical density of the PC/PBAA film can be adjusted by controlling the ratio

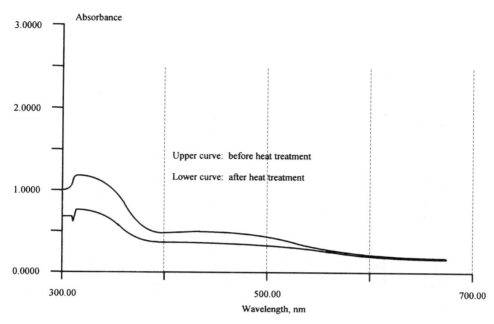

FIGURE 14 Spectra of PC/PBAA thin film before and after heat treatment.

of the two polymers and the film thickness, a function of the solution concentration, and surface area over which the solution is cast.

2. PANI Homopolymers

The UV/visible spectra of the n-PANI in $MeCl_2$ solutions are displayed in Fig. 15. Three peaks were observed at wavelengths of 236, 290, and 338 nm. Absorbance versus concentration data is given in Table 1 and is plotted in Fig. 16. The extinction coefficients of n-PANI at these wavelengths were computed from the slopes of the plots and are reported in Table 2.

It is interesting to compare the extinction coefficient of the n-PANI in $MeCl_2$ to the extinction coefficient in the composite film. UV/visible spectra of films 79-36-5 and 79-43-1 to 79-43-4 were taken by mounting the film in the spectrophotometer perpendicular to the beam path. One peak was observed at 337–342 nm (see Fig. 17). Absorbance versus weight ratio data are reported in Table 3, and the plot of these data is displayed in Fig. 18. (Note that the weight fractions of n-PANI in the films have been converted into units of mg/cm^3; the density of PC is 1.2 g/cm^3.) Again, a linear regression was performed on the points to obtain the slope, from which the extinction coefficient can be calculated. The results of the linear regression are summarized in Table 4.

The extinction coefficient of the composite film is 0.0186 in a unit of AU/ [$(mg/cm^3)(20 \ \mu m)$] (see Table 4), or 9.3 AU/(mg/cm^3 cm). This number was compared to the extinction coefficient of n-PANI in $MeCl_2$ [24.5 AU/(mg/ml cm) reported in Table 2]. The absorbance of n-PANI in PC composite films is 2.63 times lower than the absorbance in $MeCl_2$ at 338 nm wavelength.

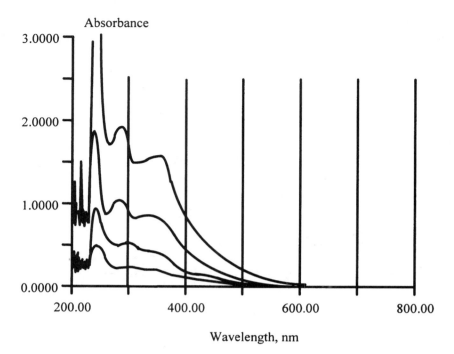

FIGURE 15 Spectra of *n*-PANI in methylene chloride.

F. Alignment of PANI/PC Composites

The PC/PBAA solutions were evaporated in the presence of electric fields of up to approximately 10 kV/cm. It was thought that given a field direction parallel to the film surface, there should be some degree of alignment of the polymers. Although the dipole moment of the PBAA is not known, PC has a reported dipole moment of 0.67 D, which may confer some alignment to the PANI. As pointed out by Prasad and Williams [10], alignment may enhance χ^3 values, and polymer alignment could also potentially add strength to the polymer composite. Alignment can be detected by polarized infrared spectroscopy as changes in the intensity of bands with transition moments aligned parallel or perpendicular to the field direction.

TABLE 1 Absorbance of *n*-PANI in MeCl$_2$

Conc. of *n*-PANI (μg/ml)	Absorbance		
	at 236 nm	at 290 nm	at 338 nm
64.1	3.1802	1.9455	1.5743
32.0	1.9158	1.0396	0.8465
16.0	0.9802	0.5198	0.4307
8.0	0.4901	0.2525	0.2228

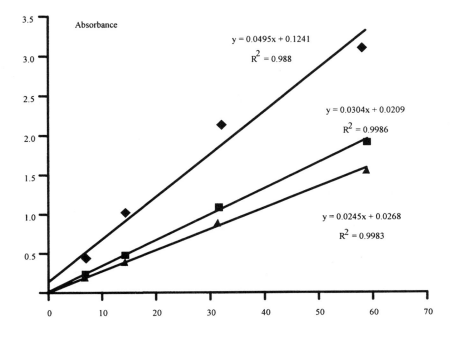

$y = 0.0495x + 0.1241$
$R^2 = 0.988$

$y = 0.0304x + 0.0209$
$R^2 = 0.9986$

$y = 0.0245x + 0.0268$
$R^2 = 0.9983$

Conc. of PANI in Methylene Chloride, ug/ml

FIGURE 16 Beer–Lambert plot for *n*-PANI in methylene chloride.

PANI and PC were codissolved in $MeCl_2$ to form solutions with certain PANI/PC ratios. The solutions were cast on glass embedded in a Teflon well using our cell configuration designed for electric field alignment. An electric field was applied parallel to the solution surface after the solution was cast. Thus, the polymer films were formed on the glass after the solvent flashed off. The field strength was 2.63 kV/cm for the aligned films; the control films were prepared at zero field strength. Typically, the films were strong enough to be recovered from the glass with gentle pulling. The films were placed in desiccators under vacuum to remove the residual solvent. All of the films were flexible and transparent; however, film integrity was proportional to the concentration of PC in the film. The sample information is tabulated in Table 5, which gives the *n*-PANI/PC ratios, the field strengths, and the film thickness.

TABLE 2 Results of Linear Regression for *n*-PANI/$MeCl_2$ Solutions

Wavelength (nm)	Conc. range (μg/ml)	Linear regression	Correlation coeff. (R^2)
236	8.0–64.1	$A = 0.0495C + 0.1241$	0.9880
290	8.0–64.1	$A = 0.0304C + 0.0209$	0.9986
338	8.0–64.1	$A = 0.0245C + 0.0268$	0.9983

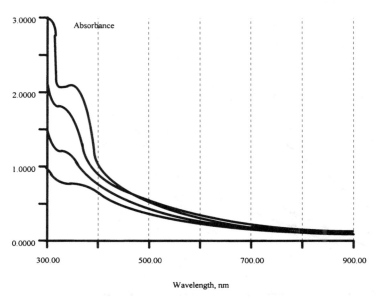

FIGURE 17 Spectra of *n*-PANI composite films.

G. Infrared Characterization of the Aligned *n*-PANI/PC Films

Films 79-36-2 and 79-36-5 were examined for the degree of alignment due to the applied electric field during the film preparation. Figures 19 and 20 show the infrared (IR) spectra of film 79-36-2 polarized of 0° and 90°, respectively. From these spectra, no significant difference was observed in absorbance, suggesting that there was little or no alignment in an electric field. This was our expectation because no strong dipole exists in both PC and *n*-PANI. In addition, the spectrum of films 79-36-2 and 79-36-5 were compared to the spectrum of film 79-43-4 (100% PC film) (see Fig. 21). There was no absorbance from *n*-PANI observed in the spectrum, implying that the content of *n*-PANI in PC matrix is too low to be detected in the IR.

H. Preparation of TBP/PC Composites

In the initial experiments with TBP, composites similar to those in Section III.E were prepared using TBP in PC. The TBP used was $ZnOCH_3TBP$. A film was first formed in PC using solutions of both in methylene chloride. The final TBP/PC film was in the weight ratio of 10.6 : 89.4. This film was then laminated between two 1-in.^{2} square Lexan sheets, again as described previously.

I. Optical Characterization

1. Degenerate Four-Wave Mixing

The degenerate four-wave mixing (DFWM) measurements follow the techniques described by Rao et al. [11]. All measurements were done using a Quantel Nd : YAG frequency-doubled laser at 532 nm with a picosecond pulse. We first report

TABLE 3 Absorbance of *n*-PANI in PC
Composite Films

Sample	Wt. Ratio of *n*-PANI to PC	Absorbance at 337–342 nm
79-36-5	1 : 48.5	1.2103
79-43-1	1 : 20.8	1.7339
79-43-2	1 : 12.5	2.1458
79-43-3	1 : 101	0.7325

on some of our initial measurements. In Sections III.E and III.G, the preparation and characterization of these composites was described. The NLO measurements are presented in Table 6.

These initial results were very encouraging in that it appeared that some NLO activity can be retained in the composite. At this point in the project, it seemed also possible to maintain mechanical resilience. It was quite possible to fabricate with ease structurally strong composites by this technique. Theoretically, the optical density of the PC film can be adjusted by controlling the ratio of the two materials and the film thickness, a function of the solution concentration and surface area over which the solution is cast.

2. Z-Scan Measurements

The Z-scan technique is a single-beam technique which allows the determination of the real and imaginary parts of the third-order susceptibility. In the Z-scan experiments, the transmittance of the nonlinear medium through a finite aperture in the far field is measured as a function of the sample position Z to determine the real part of the susceptibility. A schematic of the Z-scan technique is shown in Fig. 22. On the other hand, the transmittance of the nonlinear medium without an ap-

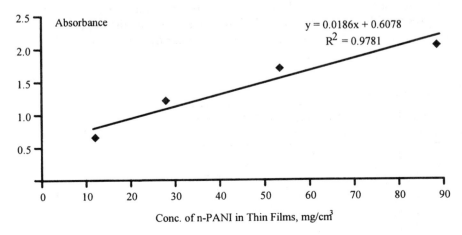

FIGURE 18 Beer–Lambert plot for *n*-PANI/PC composite films.

TABLE 4 Results of Linear Regression for *n*-PANI/PC Composite Films

Wavelength (nm)	Conc. range (μg/ml)	Linear regression	Correlation coeff. (R^2)
337–342	0.98–7.4	$A = 0.0186C + 0.6078$	0.9781

erture is measured as a function of the sample position Z to determine the imaginary part of the susceptibility. The zero position for Z is taken to be that of the focal plane.

Open-aperture Z-scan measurements of solutions of ZnTBP and thin films of PMMA and PC doped with ZnTBP exhibit reverse saturable absorption demonstrating that the idea of doping the polymer with the dye is essentially sound and works. Two traces, one for THF solution and the other for a PMMA film, are shown in Figs. 23 and 24.

3. Power Limiting Measurements

Power limiting measurements were obtained with a standard $f/5$ experimental setup. The center portion of an expanded beam is chosen as the input in order to simulate a distant source. The laser source used in the experiments is a Q-switched Nd : YAG laser frequency doubled to give 10-ns pulses at 532 nm wavelength at a repetition rate of 10 Hz. The output power is plotted versus input power. The power limiting curve is shown in Fig. 25. We obtained a power limiting curve for ZnTBP doped film where the host was PC. The free-standing film was sandwiched between transparent Lexan (PC) sheets. The impact strength for the sandwiches is that of PC (12–16 ft-lb/in.) The tensile strength is 8500–9500 psi. Water (moisture) absorption for a $\frac{1}{8}$-in. bar in 24 h is 3.0%. The linear transmission of the film is 32%. The experimentally observed limiting threshold is 1 μJ, which corresponds to a fluence of about 35 mJ/cm^2. Obviously, there is much room for improvement concerning the limiting threshold if the sensor is a human eye. The experimentally observed damage threshold is 100 μJ, which corresponds to a fluence of about 3.5 J/cm^2. It is noteworthy that beyond the damage threshold, the transmission drops sharply and irreversibly, which is a more desirable outcome than the opposite effect. The ratio of high fluence transmission (1 J/cm^2) to linear transmission is 0.02. The

TABLE 5 Sample Data

Sample	*n*-PANI to PC ratio	Field strength (kV/cm)	Thickness (μm)
79-36-2	1 : 48.5	2.63	20
79-36-5	1 : 48.5	0	20
79-43-1	1 : 20.8	0	20
79-43-2	1 : 12.5	0	20
79-43-3	1 : 101	0	20
79-43-4	0 : 1	0	20

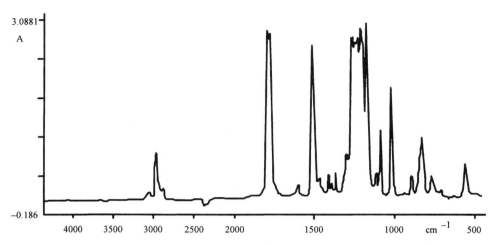

FIGURE 19 Infrared spectrum of *n*-PANI thin film (polarized at 0°) (sample 79-36-2).

presence of the film between a laser source and a sensor or eye would increase the dynamic range of the sensor by at least two orders of magnitude, as evidenced by the width of the limiting region in the graph.

Power limiting curves displaying the dependence of the power limiting on wavelength (Fig. 26) and solvent for ZnTBP (Fig. 27), as well as curves comparing C60, a liquid crystal, and ZnTBP solutions are also included for reference (Fig. 28).

IV. CONCLUSIONS AND DISCUSSION

A. Processing of Solid-Phase χ^3 Materials

In this project, our focus has been on the processing of χ^3 materials in mechanically strong polymeric hosts. We have chosen PC as the host material. The objective is to combine the χ^3 material into a host such that the resulting composite maintains the NLO activity that the starting chromophore displays in solution.

A number of film preparation techniques have been investigated in this project. Again, the objective is to put each of the three χ^3 materials, PANI, TBP, and the silver sol, into a mechanically strong host. Early on, it was clear that the PANI, as received from the U.S. Army Labs in Natick, would not be a candidate host because its low molecular weight precluded its use as a supporting film. PC is a good film-former and has demonstrated adequate performance as a host for the χ^3 materials. Films were made by casting a solution of the chromophore and PC either directly on a 2-mm PC sheet or on Teflon. Those films cast on the PC sheet are then covered with another layer of 2-mm PC and effectively sandwiched by placing two stainless-steel cylinders and heating to 132–142°C at 2000 psi. Those films cast on Teflon are removed from the casting surface and then laminated between the PC sheets, using the same assembly.

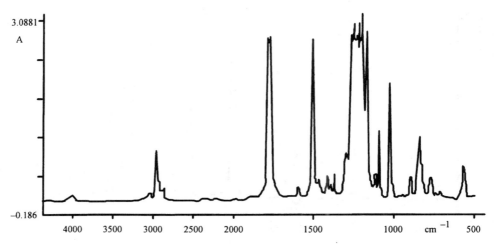

FIGURE 20   Infrared spectrum of *n*-PANI thin film (polarized at 90°) (sample 79-36-2).

1. Solvents

We have observed that the use of appropriate solvents is of utmost importance in the processing of the samples. Samples of TBP in methylene chloride show the highest sensitivity toward photodegradation. Methylene chloride is compatible with both the TBP dyes and PMMA and PC, but the damage threshold is reached very early for both films, as well as solutions (less than 100 laser shots at 532 nm wavelength with 30-mJ and 25-ps pulses at a repetition rate of 7Hz).

Tetrahydrofuran (THF) is a particularly good solvent for TBPs but is incompatible with both PMMA and PC. Methyl ethyl ketone, 1,4-dioxane, and toluene

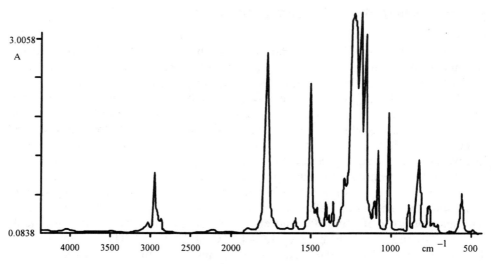

FIGURE 21   Infrared spectum of PC thin film (sample 79-43-4).

TABLE 6 Sample DFWM Measurements

Sample	Sample description	χ^3 (esu)
79-23	PsBAA/PC	4.8×10^{-12}
79-43-1	n-PANI/PC (1 : 20.8)	1.3×10^{-11}
79-36-2	n-PANI/PC (1 : 48.5)	4.8×10^{-12}
79-34-4	ZnTBP/PC (1 : 50)	1.2×10^{-10}
83-FA-1	ZnTBP/PC (1 : 50)	6.6×10^{-14}
83-FA-2	PC only	6.2×10^{-11}

were also tried because they are compatible with both TBPs and PMMA. Of these candidates, the most promising two are 1,4-dioxane and toluene. In these cases, it takes around 10,000 laser shots to photodegrade the TBP, and TBP solutions evaporate slowly enough to form good-surface-quality evaporated films. Furthermore, the evaporated films adhere well to polycarbonate substrates, thus allowing the fabrication of multilayer sandwiches.

2. Aggregation

The TBP dyes form aggregates. We studied the effect of aggregation by measuring Langmuir–Blodgett films of TBP. The films show no reverse saturable absorption. A possible mechanism for this occurrence is that in the aggregates, there are many more pathways for deexcitation compared to the single molecules. The effects of aggregation are evident in the linear absorption spectrum of the samples where a blue-shifted and broadened Soret band is evidenced.

3. Moisture

The effects of moisture (water) in the samples is very dramatic. With increased concentration of water, the otherwise sharp and intense Soret band first broadens

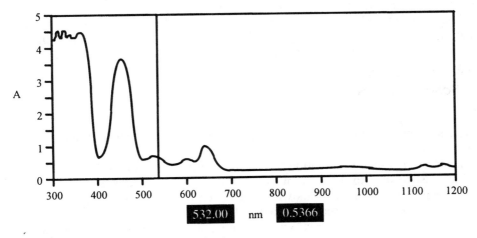

FIGURE 22 ZnTBP/PC in Lexan laminate.

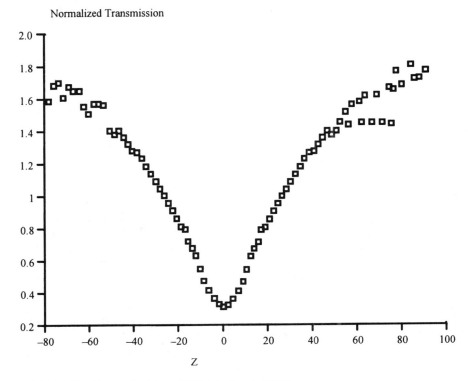

FIGURE 23 Power limiting: TBP solution in THF.

and blue-shifts and eventually splits up. Measurements of intensity-dependent transmission on THF solutions and THF/water (1 : 1 by volume) made at the onset of reverse saturable absorption clearly indicate that the increasing absorption observed in the THF solutions completely disappears in the case of the THF/water mixtures. The reason is that water induces the formation of TBP aggregates.

The above remarks are of serious concern in the choice of a processing technique that is suitable for obtaining multilayer optical limiters. It is useful, then, to summarize the solvent preparation methods en route to recommending solvent-free composite processing.

4. Solvent Processing

To summarize:

- The solvent must be compatible with both the nonlinear material and the polymer host.
- Substrate on which evaporation takes place must have a smooth surface.
- One must be able to detach film from substrate after evaporation.
- Free-standing film is then sandwiched between uniformly thick, optically clear, and transparent Lexan (polycarbonate) sheets.
- Sandwich is placed in a press with a heating sheet surrounding the sandwich.

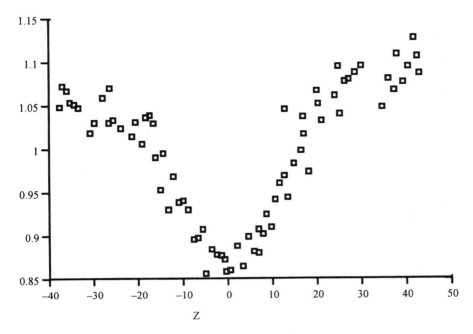

FIGURE 24 Power limiting: TBP in PMMA matrix.

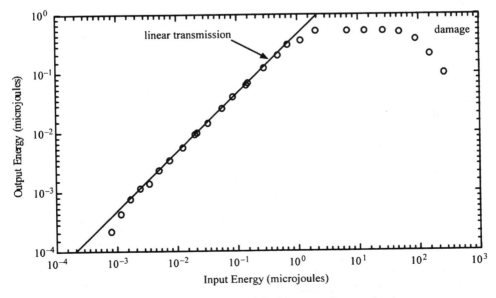

FIGURE 25 ZnTBP in polycarbonate sandwiched between Lexan sheets.

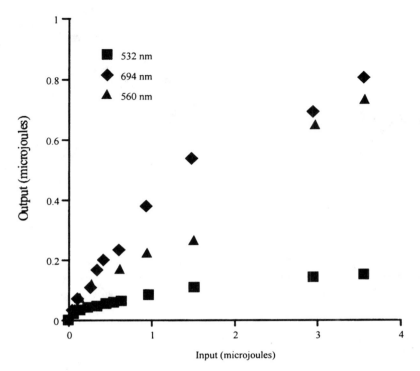

FIGURE 26   Power limiting wavelength effect.

- Temperature is controlled to induce adhesion of the sandwich by going slightly above and below the glass transition temperature of the host.
- Contacts between press and sandwich must be smooth and uniform surfaces to prevent deformation of the sandwich.

Problems

- Control of uniformity of thickness
- Difficulty in attaining an optically smooth surface
- Bubble formation
- Evaporation rate (nonuniform distribution of NLO molecules)

Advantages

- Solutions are very homogeneous blends.

5.   *Recommendations for Solvent-Free Composite Processing*

- Mill polymer host into a fine powder (variation: mill dry solid formed from solution of host and NLO material).
- Dry powder for 48 h in oven at a temperature below the glass transition temperature of the host to remove any moisture or remaining solvent.
- Dust host powder with NLO material to the target proportions.

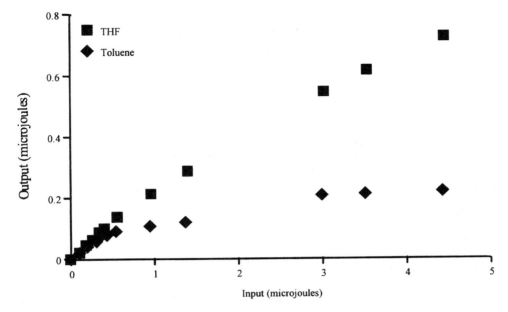

FIGURE 27 Power limiting solvent effect.

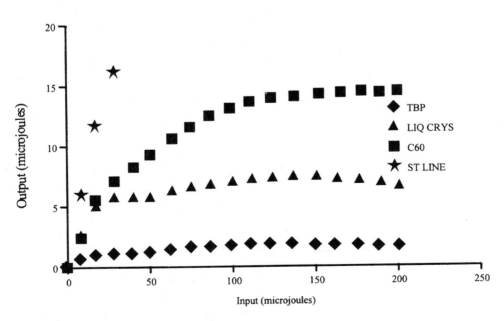

FIGURE 28 Power limiting comparison.

- Shake and coat host powder pellets.
- Extrude mixture through a pellet die to form a blended pellet.
- Extrude as many times as necessary to form uniform blend.
- Add pellet to a laboratory-size injection molder.
- Inject blended pellets into an optically smooth, hot, metal mold (die) to form film samples.

Problems

- Processing temperatures must be compatible with both the host and the NLO material. (PC and PMMA are compatible with TBPs and phthalocyanines but may not be with other NLO materials.)
- Drying and processing may require a gas atmosphere other than air.

Advantages

- Uniform thickness throughout film.
- Ability to control of thickness.
- Optically smooth surfaces.
- Solvent incompatibilities may be avoided.
- Different shapes can be given to the finished material by simply changing the metal mold (die).

B. Nonlinear Absorption

The imaginary part of the third-order susceptibility is responsible for nonlinear absorption. Two-photon absorption, saturable absorption, and excited-state absorption are the most relevant types of nonlinear absorptive processes. Transitions that involve one photon and transitions that involve two photons have different selection rules; two-photon processes involve the simultaneous absorption of two photons to excite a material. Saturable absorption (SA) involves the saturation of a given transition, by populating an excited state of the material so that the material which initially absorbed at that wavelength becomes more transparent. Excited-state absorption involves a sequential process in which a photon is initially absorbed and the molecule remains in an excited state for a finite length of time so that a second photon that arrives during that time is also absorbed to put the molecule into an even higher excited state.

The basic difference between two-photon and excited-state transitions is that the former involve intermediate extremely short-lived virtual states, whereas the later involve intermediate real states whose lifetimes are not determined by the Heisenberg uncertainty relations but instead their lifetimes are determined by the electronic structure of the molecules in the materials. Two-photon absorption (TPA) processes are dependent on the intensity (energy per unit time per unit area) of the incident light, whereas excited-state absorption (ESA) processes are dependent on the fluence (energy per unit area) of the incident light. The two processes can be resolved by obtaining measurements of the samples for different incident pulse widths. For optical power limiters, we are mostly interested in a fluence-dependent material because if the nonlinear transmission characteristics of the material were intensity dependent, that would make the amount of power limiting dependent on the incident laser pulse's width, which is undesirable.

For an ESA material, the ground-state absorption cross section is smaller than the excited-state absorption cross section. Thus, a suitable material for optical power limiting must have as large as possible a ratio of excited-state absorption cross section to ground-state absorption cross section. Materials that meet this requirement are the tetrabenzporphyrins (TBP), the phthalocyanines (Pc), and the buckminsterfullerenes. The largest ratios have been observed for the tetrabenzporphyrins for which σ_{eff}/σ_0 is around 30 at 532 nm wavelength. For the phthalocyanines, it is between 10 and 18 at 532 nm wavelength. For the buckminsterfullerenes (C60), it is one order of magnitude larger, but only at 694 nm wavelength, which is a region of low sensitivity for the human eye.

In summary, we require a material with a large excited- to ground-state cross-section ratio in the regions of the visible spectrum, where the eye is most sensitive. During the day, that is around 560 nm wavelength due to cone cells, and at night, it is around 510 nm wavelength due to the rod cells. The maximum absorption for the excited state of both TBPs and Pcs is centered close to these wavelengths. This justifies our choice of pursuing these materials as our guest NLO system in future work. Nevertheless, the proposed methodology for processing solid films can be used for any other material that may be of interest for power limiting.

Our optical characterization of films which we prepared suggests that solid-phase NLO materials can be prepared, thus demonstrating their potential for sensor protection from lasers. The results are very encouraging. NLO activity, as measured by DFWM, is retained in the composite; open-aperture Z-scan measurements of solutions of ZnTBP and thin films of PMMA and PC doped with ZnTBP exhibit reverse saturable absorption and power limiting measurements yielded a power limiting curve for ZnTBP/PC films. These power limiting experiments are particularly important. The experimentally observed limiting threshold was 1 μJ, which corresponds to a fluence of about 35 mJ/cm^2. Although there is much room for improvement concerning the limiting threshold if the sensor is a human eye (the experimentally observed damage threshold is 100 μJ, which corresponds to a fluence of about 3.5 J/cm^2), it is important to note that beyond the damage threshold, the transmission dropped sharply and irreversibly, a desirable outcome. Our results can be compared with those reported by Kost et al. [12] for optical power limiting at 532 nm wavelength with 8-ns pulses. Although Kost et al. used a different experimental setup and did not report damage thresholds, they report ratios of high-fluence transmission (1 J/cm^2) to linear transmission for films of several RSA materials doped in PMMA. They report 0.28 for C60, 0.40 for chloroaluminum phthalocyanine, 0.67 for *N*-methylthioacridone, and 0.94 for King's complex. Our results represent a one-order-of-magnitude improvement over their best result. This supports the project concept of doping the high impact polymers with NLO dyes.

C. Conclusions

In summary:

- We have developed techniques to fabricate RSA films.
- Various NLO materials can be processed using these techniques.
- The techniques are suitable for the development of multilayer limiters.
- The films have excellent mechanical characteristics.

- We have developed a sound agenda for the development and improvement of solid multilayer optical power limiters.
- We have obtained excellent optical power limiting results for the films that we manufactured.

V. FUTURE DIRECTIONS

A. Modeling of Thermal Responses

For the thermally induced nonlinear optical behavior of the films developed in this work, both the thermally induced defocusing and self-focusing effects and the thermal damage can be predicted from a simulation code that has been developed and verified for nonlinear optical liquids and organic molecules. This code can be used for the solid films once the nonlinear optical properties and mechanical properties of the films have been measured. Although the thermal effects may be small for the picosecond measurement techniques used in this present work, the algorithm provides a means of extrapolating the results to longer pulse widths or to picosecond-pulse trains that extend for milliseconds or nanoseconds. Operating versions of the code are available at Northeastern University and the U.S. Army Natick Research, Development and Engineering Center [13–15].

The simulation code or computer algorithm is a three-dimensional, transient thermal response model that is solved using a finite-difference technique. The algorithm predicts the temperature of the material and the phase change and refraction of the laser radiation as it travels through the material. The inputs to the algorithm are the geometry of the system, characteristics of the incident laser radiation, and thermal and optical properties of the material. The algorithm includes a simultaneous numerical solution for the electromagnetic-wave equation using a ray-trace technique. In the ray-trace technique, the incident radiation is divided into a set of plane waves which travel through the film. This electromagnetic-wave model has been verified to be an accurate approximation in thin liquid cells (thickness up to 2 mm). A modified Euler integration method is used to solve the finite-difference formulation. The modified Euler integration method allows the nonlinear variation of the properties, especially the dependence of the absorption coefficient and the index of refraction on the laser radiation intensity and the density of the material. The density of the material is related to the temperature and pressure fields within the film through the equation of state. The algorithm includes the traditional Fourier conduction and continuum equation of state for pulses that are on the order of nanoseconds and microscale heat transfer and absorption coefficient models for laser pulses in the picosecond range. User-specified models are included in the algorithm for the absorption coefficient and index of refraction that are based on reversed saturable absorption processes or other χ^3 processes reported in the literature. The preprocessor subroutine included in the algorithm allows the user to specify the temporal and spatial variation of the incident radiation. These choices include the Gaussian spatial and temporal laser radiation, ramp, and square-wave inputs.

The algorithm has been verified by comparing the predictions to experimental results reported in literature citations. The algorithm is capable of reproducing thermally induced Z-scan behavior in CS_2 exposed to laser radiation at 532 nm and 10.6-μm wavelengths. The algorithm has also been shown to be able to predict the thermal reflectance effect in nickel films that are exposed to picosecond laser pulses. These successes of the algorithm for predicting thermally induced nonlinear optical behavior suggest that it can be used as a development and design tool for the films being developed in this project. Using the framework that has been established in the algorithm, the effects of the substrate or of the residual solvent on the threshold of thermal damage can be determined. The thermally induced nonlinear behavior can also be calculated based on test data to further supplement the design of the protection system.

B. Multicomponent Multiphase Materials

The promising results obtained for molecularly dispersed ZnTBP in PC suggest that the multicomponent guest-in-host approach is viable. However, there are broader implications from this work. For example, it should be possible to extend the multicomponent approach to multiphase materials, where individual phases confer specialized properties to the composite material (Fig. 29).

For example, two different chromophores with different optical characteristics can be bound in adjacent domains. Alternatively, one domain might contain the optically active molecules and the other may serve a mechanical/structural role. This idea was recently exploited in the preparation of light-emitting polymer thin films based on the comb fluorpolymer Nafion®, where ionic domains bind either luminescent or carrier transport molecules and fluorocarbon domains offer mechanical integrity [17]. It is, of course, important for the phase sizes to be small so as not to scatter light, and this is easily possible with, for example, block copolymers where the average domain size is $\sim$10–30 nm [18]. Moreover, the domain shape (spheres, cylinders, sheets) can be controlled in block copolymer films by the relative block lengths and processing conditions.

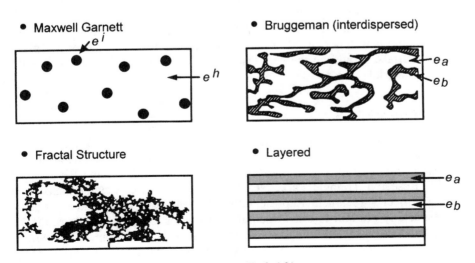

FIGURE 29 Composite geometries. (From Ref. 16.)

C. A Comprehensive Protection System

The achievement of reverse saturable absorber (RSA) properties in a polymer composite needs to be combined with other laser protection materials to provide a total system with increased protection. This combination can be a hybrid construction with layers of different laser protection materials, including the RSA materials of this project and new types of material. It can also be the use of multifunctional materials where each has more than one laser protection property (i.e., RSA, NLO refractive, new types, etc.).

An example of a new type of laser protection mechanism to be combined with the baseline RSA material is the "picosecond photochromic approach." It is based on using an absorber which absorbs protons and forms a colored or photochromic product in a 1-ps time scale. Most photochromic processes take more than 1 s to form the colored product and are, thus, not suited for laser protection. Typical properties of the picosecond photochromic materials in a polymer composite roughly developed for the RSA materials are as follows:

No Laser	Laser Exposure at 1 MW/cm^2 for 0.4 J/cm^2
50% transmission	13% transmission
40% transmission	8% transmission
17% transmission	1% transmission
10% transmission	0.14% transmission

These properties are short of the reduction to the 0.01% transmission that is desired for laser protection. It should be possible to improve the picosecond photochromic properties as a single laser protection system to a higher level, but even greater improvements should be possible by hybrid construction with other types of NLO systems, such as RSA materials of this project, and by selecting materials for the picosecond photochromic system that are multifunctional and simultaneously exhibit other NLO properties, such as RSA.

The optimization and combinations of the RSA materials, of other NLO materials, and of the new types of NLO-like materials, such as the picosecond photochromic materials, should be the best laser protection that can be achieved with NLO-related materials as a whole.

ACKNOWLEDGMENTS

This work of "Sensor Protection from Lasers" is based on a Final STTR Technical Report (August 12, 1996–February 11, 1997) under Contract No. DAAH04-96-C-0073 and sponsored by the U.S. Army Research Office, 4300 S. Miami Blvd., P.O. Box 12211, Research Triangle Park, NC 27709–2211. We especially thank our Contracting Officer's Technical Representative, Dr. Mikael Ciftan, and the coauthors are especially grateful to our Principal Investigator, Debra J. Trantolo, Ph.D. We also acknowledge the technical assistance of Joseph A. Akkara, U.S. Army Natick Labs with respect to polyaniline NLO materials, Masato Nakashima, U.S. Army Natick Labs with respect to porphyrin NLO materials, and Joseph F. Roach, U.S. Army Natick Labs for his overall guidance.

REFERENCES

1. Focke, W. W., Wei, Y., Wnek, G. E., Ray, A., and MacDiarmid, A. G., *J. Phys. Chem.*, *93*, 495 (1989).
2. Miles, P., *Appl. Opt.*, *33*, 6965 (1994).
3. LaPeruta, R., Van Wagenen, E. A., Roche, J. J., Kitipichai, P., Wnek, G. E., and Korenowski, G. M., *SPIE Proc. Nonlinear Opt. Mater. 1497*, 57 (1991).
4. Kitipichai, P., LaPeruta, R., Korenowski, G. M., and Wnek, G. E., *MRS Symp. Proc.*, *247*, 117 (1992).
5. Trantolo, D. J., Mogul, M. G., Wise, D. L., Wnek, G. E., Frazier, D. O., and Gresser, J. D., *SPIE, 2809*, 106 (1996).
6. Creighton, J. A., Blatchford, C. G., and Albrecht, M. G., *Chem. Soc. J., Faraday Trans. II, 75*, 790–798 (1979).
7. Billmeyer, F. W., *Textbook of Polymer Chemistry*, Wiley–Interscience, New York, 1962.
8. Bandrup, J., and Immergut, E. H. (eds.), *Polymer Handbook*, 3rd ed. Wiley, New York, 1989.
9. Odian G., *Principles of Polymerization*, 2nd ed., Wiley, New York, 1981.
10. Prasad, P. N., and Williams, D. J., *Introduction to Nonlinear Optical Effects in Molecules and Polymers*, Wiley, New York, 1991.
11. Rao, D. V. G. L. N., Aranda, F. J., Cheng, C. F., Akkara, J. A., Kaplan, D. L., and Roach, J. F., in *Frontiers of Polymers and Advance Materials* (P. N. Prasad ed.), Plenum Press, New York, 1994, pp. 219–228.
12. Kost, A., Tutt, L., Klein, M. B., Dougherty, T. K., and Elias, W. E., *Opt. Lett.*, *18*, 334 (1993).
13. Kowalski, G. J., Wahl, E. H., and Roach, J. F., Materials for optical limiting, *MRS Symp. Proc.*, *374* (1995).
14. Kowalski, G. J., in *High Heat Flux Engineering III*, SPIE, Denver, CO, 1996a.
15. Kowalski, G. J., in *Nonlinear Optical Liquids*, SPIE, Denver, CO, 1996.
16. Boyd, R. W., Nanocomposite Materials for Nonlinear Optics, in *Army Eye/Sensor Optical Protection Workshop*, Rochester, NY, 1996.
17. Karasz, M. A., and Wnek, G. E., *Electrochim. Acta* (to be published).
18. Bates, F. S., and Fredrickson, G. H., *Ann. Rev. Phys. Chem.*, *41*, 525 (1990).

15

Noncrystalline Organic Photorefractive Materials: Chemistry, Physics, and Applications

K. Meerholz
University of Munich
Munich, Germany

B. Kippelen and N. Peyghambarian
University of Arizona
Tucson, Arizona

I. INTRODUCTION

The photorefractive (PR) effect arises in materials combining photosensitivity, photoconductivity, and a field-dependent refractive index, most commonly through the linear electrooptic or "Pockels" effect. For three decades, photorefractive materials have attracted a lot of attention, because they are promising recording media for a variety of applications related to holography. One of the most promising is holographic storage of entire images. By using different multiplexing schemes, much higher data storage densities compared with conventional sequential bitwise storage mechanisms like on magnetic tapes can be achieved. Other applications include phase conjugation, novelty filtering, and optical correlation. In contrast to more traditional holographic recording media, such as silver halide films or thermoplastic polymers, PR materials do not require any processing, and the holograms are erasable. Thus, PR materials are suitable for reversible storage and real-time applications.

Photorefractivity was first reported in 1966 as a detrimental effect in an electrooptic inorganic ferroelectric crystal, lithium niobate ($LiNbO_3$). Since then, a variety of other PR materials, mostly inorganic crystalline materials, have been discovered and investigated [1]. Despite the proof-of-principle demonstration of a number of applications, none of them has been commercialized on a large scale. One reason for this is the tremendous progress that has been achieved in integrated-circuit technologies, making electronic computing fast, powerful, and rather inexpensive, whereas optical schemes require costly high-precision optics. The other is the difficulty to reproducibly grow crystals with sufficiently high-PR performance.

Recently, due to the progress made in the fields of CCD (charge-coupled device) and SLM (spatial light modulator) technologies, holographic processing using PR crystals has experienced a renaissance because of the highly parallel nature of optical processing. Nevertheless, the relatively high cost for a high-quality crystal prevents their use in widespread applications.

The search for new PR materials has continued for several decades and has mostly centered around finding materials with faster response times, better electrooptic (EO) coefficients, better optical quality, and better cost-effectiveness. Also, several fixing schemes have been investigated in order to achieve long-term storage. In addition to many new inorganic crystalline materials, photorefractivity has been discovered in semiconductors [2], organic molecular crystals [3], and, recently, in noncrystalline organic materials such as polymers, liquid crystals, and amorphous glasses, which will be the topic of this chapter. In particular, the discovery of photorefractivity in polymers by IBM researchers in 1991 [4] has triggered a great deal of research efforts toward improving the performance of this new material class. Within only 3 years after the first report on PR polymers, major improvements were accomplished [5–7], yielding materials with extremely high net gain ($\Gamma > 200$ cm^{-1}) [8], a measure of the PR performance, exceeding that of their inorganic counterparts ($\Gamma \approx 40$–50 cm^{-1} in BaTiO$_4$). Note, however, that the technically important figure is the product of gain coefficient and active layer thickness, Γd, which is much more favorable for the crystals due to their larger thickness of several millimeters compared with typical thickness of "only" 100–200 μm for organic devices.

The improvement of organic amorphous PR materials was mainly achieved by improving the quality and uniformity of polymeric materials, allowing the use of high electric fields. The most important step forward, however, was that researchers have decreased the glass transition temperature of the polymers, which permits reorientation of the nonlinear optical chromophores at room temperature by the photorefractive space-charge field within the viscous matrix. This phenomenon is referred to as the "orientational enhancement mechanism" [9] and produces a large birefringence contribution to the total change of the refractive index in a PR material. Most of the recent high-performance PR polymers owe at least 60%, if not more, of their index modulation amplitude to the orientational birefringence.

Currently, inorganic crystals are still the preferred materials for applications such as holographic storage. Organic PR materials suffer from drawbacks, such as (1) stability for long-time storage, (2) short storage times, (3) high scattering levels leading to poor hologram reconstruction, (4) the need to apply high electric fields, and (5) their slow response times. However, it should be possible to overcome these issues, as organic materials have several inherent advantages over inorganic materials for producing photorefractivity:

• Wide flexibility to vary the chemical functions in the material. One can take advantage of the intensive research efforts and the major improvements that have been achieved over the last decade on organic photoconductors for xerography, on one hand, and organic NLO materials for EO switching and frequency doubling, on the other.

- Low dielectric constants ($\varepsilon < 10\text{–}15$), which minimizes unwanted screening of the internal photorefractive space-charge field. This compares to inorganic crystals, which can have dielectric constants greater than 1000.
- Easy processing; that is, they can be mass produced in any shape and also in large areas. The reproducibility is much better than in inorganic PR crystals.
- Orientational effects which is an advantage to obtain larger photorefractivity.

In this chapter, we will first define "photorefractivity" and describe the PR effect as it is understood in inorganic crystals (Section II). We will then lay out the chemistry and physics of noncrystalline organic PR materials (Section III). Following an experimental Section IV, we will review the results obtained with the different amorphous organic materials that have been investigated for photorefractivity over the last 3 years, with an emphasis on composite materials based on the photoconducting polymer poly(N-vinylcarbazole), (PVK) (Section V). The advantages and disadvantages of the different material classes will be discussed. For earlier work, the reader is referred to the reviews in Refs. 5–7. We will then discuss special effects which were discovered in the new noncrystalline materials (Section VI) and present some possible applications that have been already demonstrated with PR polymers (Section VII). The chapter will conclude with an outlook on future development.

II. PHOTOREFRACTIVITY AND HOLOGRAPHY

When two coherent light beams overlap—one containing information about an arbitrary image or object ("object beam"), the other being a plane wave ("reference beam")— a characteristic interference pattern is produced. By overlapping them in an appropriate medium, the material's absorption and/or index of refraction are altered nonuniformly as a result of the nonuniform illumination. A holographic "grating" (hologram) is formed which can diffract light. This hologram contains intensity and phase information about the object. The information stored in the material can then be read out by a plane wave incident from the original direction of the reference beam. The virtual image of the object is reconstructed as a diffracted signal in the original direction of the object (Fig. 1).

For the evaluation of novel holographic recording materials, a much simpler situation is commonly utilized for screening purposes. Two coherent plane waves are overlapped inside the material. They interfere, generating a spatially modulated sinusoidal pattern of high- and low-light-intensity regions:

$$I(x) = I_0 \left[1 + m \cos\left(\frac{2\pi x}{\Lambda_g} \right) \right] \tag{1}$$

where $I(x)$ is the local light intensity at the coordinate x, $I_0 = I_1 + I_2$ is the total light intensity incident on the material, $m = 2(I_1 I_2)^{1/2}/(I_1 + I_2)$ is the fringe visibility, and Λ_g is the period of the sinusoidal pattern, which is determined by the wavelength λ of the light used to write the hologram and the angle 2θ between the overlapping coherent beams measured inside the material:

$$\Lambda_g = \frac{\lambda}{2n \sin \theta} \tag{2}$$

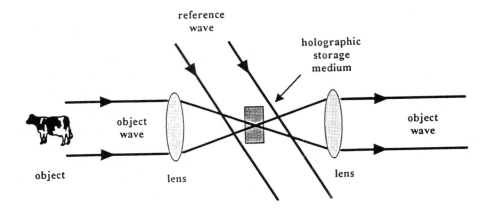

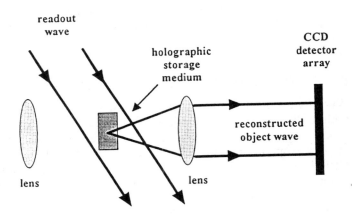

FIGURE 1 Illustration of holographic storage. *Top*: Recording. Two laser beams are used—one carries phase and amplitude information from the object, the other acts as a reference. The two beams intersect in the storage medium. The beams interfere and form a characteristic light-intensity pattern. The refractive index of the storage medium is altered in the light regions, generating a hologram of the object. *Bottom*: Reconstruction. The information is retrieved from the storage medium using a reading beam. It is diffracted by the holographic pattern stored in the medium and produces the identical signal as the original object beam on the detector array.

Technically, one of the most important figures is the angular selectivity for readout of a hologram. For example, sharp angular selectivity is essential for the multiplexing of several holograms in the same volume of a recording material. Two limits are distinguished: in the Raman-Nath regime, which applies to thin films, the grating spacing is much larger than the thickness of the holographic medium ($\Lambda_g \gg d$), leading to multiple diffraction orders during readout and poor angular selectivity; in the other limit which applies to thick holograms, the Bragg regime, the grating spacing is much smaller than the active layer thickness ($\Lambda_g \ll d$), allowing ideally only one diffraction order which can be read out only if the Bragg condition is fulfilled exactly.

A. Holographic Recording Mechanisms— Definition of "Photorefractivity"

At first sight, the term "photorefractive" suggests that one refers to *photo-induced refractive index changes* in a material. This is certainly true; however, there is a large number of mechanisms that would fall under this general definition which are commonly not considered "photorefractive," such as the following:

*   Absorption gratings obtained, for example, through photochromism, photoisomerization, or photobleaching processes (reversible or irreversible)
*   Thermal gratings (reversible)
*   Density gratings through photopolymerization (irreversible)
*   Transient gratings of excited states [e.g. in $\chi^{(3)}$ four-wave mixing experiments (reversible)]

All these mechanisms have been utilized to record holograms. Their common feature is that the recorded index pattern is in phase with the incident light pattern (phase shift $\Theta = 0, \pi$); that is, the maximum index change occurs where the maximum light intensity is incident on the material (Fig. 2a). Note that the refractive index remains unchanged where the material was not illuminated.

By contrast, the term "photorefractive" is only used for those mechanisms where nonuniform illumination leads to an internal space-charge field which then modulates the refractive index of the material. In most cases, the response of the material is nonlocal; that is, the resulting index pattern is *not* in phase with the incident light pattern ($\Theta \neq 0, \pi$). The origin of the PR effect will be explained in the following section.

B. The Photorefractive Effect in Inorganic Crystals

The photorefractive effect arises in materials combining photosensitivity, photoconductivity, and a field-dependent refractive index, most commonly through the linear electrooptic or "Pockels" effect. The different steps of the PR grating formation are illustrated in Fig. 2b. Note, that the PR effect absolutely requires nonuniform illumination in order to be observed, whereas all other mechanisms listed in the previous section may take place even under uniform illumination. In fact, uniform illumination erases the hologram; that is, photorefractive materials are suitable for dynamic (i.e., reversible) and real-time holograpy.

In the regions where the light intensity is high, charge carriers (referred to as "electrons" with negative charge and "holes" with positive charge) are generated inside the material through absorption of photons. By contrast, where the light intensity is low, only few or no charge carriers are formed. Typically, one type of carrier is much more mobile than the other. The mobile carrier type can migrate away from the regions of high light intensity under the influence of an electric field or diffuse due to the concentration gradient following the nonuniform excitation. Left behind in the bright regions of the grating are the oppositely charged, less mobile carriers which are generally treated as practically fixed charges.

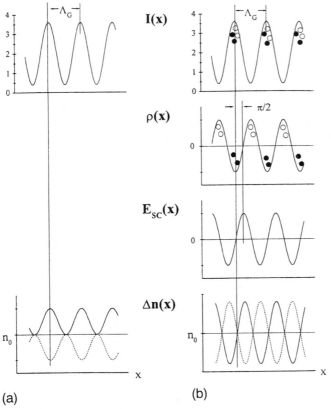

FIGURE 2 Holographic grating formation. (a) In a linear medium, nonuniform illumination leads to maximum changes of the refractive index right where the light intensity is maximum (local grating). The solid (dashed) line in the left bottom figure symbolizes the photo-induced increase (decrease) of the bulk refractive index. (b) In a photorefractive medium, nonuniform illumination leads to a nonlocal grating (for details of the mechanism, see text). The solid (dashed) line in the bottom figure symbolizes the electric-field-induced increase (decrease) of the bulk refractive index.

The charge transport process is limited by traps present in the material. Traps are sites with lower energy for the carriers than the surrounding sites (e.g., impurities, vacant sites in the crystal lattice, or defects such as domain boundaries). As long as there is enough local light intensity, the carriers can be photo-reactivated and drift/diffuse further away from the location where they were originally generated. Typical distances for the carriers are in the micrometer range. Over time, transport of the carriers leads to the separation of the centers or positive and negative charge and, gradually, to the buildup of an internal electric field, the so-called "space-charge field," $E_{SC}(x)$. Note that the space-charge field is $\pi/2$-phase-shifted with respect to the space-charge distribution $\rho(x)$ due to the Poisson relation (the field is strongest between the centers of positive and negative charge):

$$\frac{dE_{SC}(x)}{dx} = \frac{4\pi\rho(x)}{\varepsilon} \tag{3}$$

with ε being the DC dielectric constant of the medium and ε_0 the permittivity (MKS units). The basic physical model to describe photorefractivity in inorganic materials was developed by Kukhtarev [10]. The amplitude of the first-order (Bragg) component of the steady-state space-charge field $E_{SC}(x)$ created by a sinusoidal light distribution is given by

$$E_{SC}(x) = m\left(\frac{E_0^2 + E_D^2}{(1 + E_D/E_q)^2 + (E_0/E_q)^2}\right)^{1/2} \tag{4}$$

where E_0 is the component of the externally applied field along the grating wave vector $K = 2\pi/\Lambda_g$. The diffusion field E_D and the trap-limited field E_q are given by

$$E_D = \frac{Kk_BT}{e} \tag{5a}$$

$$E_q = \frac{eN_T}{K\varepsilon\varepsilon_0} \tag{5b}$$

Here, k_B is the Boltzmann constant, T is the temperature, e is the elementary charge, ε_0 is the permittivity, and N_T is the number density of photorefractive traps. Equation (5b) shows that the maximum value for the space-charge field for a given material depends on the trap density and the dielectric constant. In most of the inorganic crystals, these two parameters are not very favorable for the buildup of strong space-charge fields. High doping (i.e., trap) levels are difficult to achieve in crystals because their incorporation into the crystal lattice leads to the loss of optical quality. Also, the dielectric constant is high due to the large polarizability of inorganic PR crystals (see below).

The last step in the PR grating formation is the modulation of the refractive index of the material by the space-charge field. In PR crystals, this commonly takes place via the linear electrooptic (EO) effect:

$$\Delta n(x) = \frac{1}{2}[n^3 \, r_{\text{eff}} \, E_{SC}(x)] \tag{6}$$

with n the average refractive index of the material and r_{eff} the effective electrooptic coefficient. The basis for the EO effect is the second-order nonlinear susceptibility $\chi^{(2)}_{IJK}(-\omega; 0, \omega)$, which depends on the material's symmetry, the orientation of the material during the interaction process, and the direction of the fields involved in the process. As a prerequisite, photorefractive materials must be noncentrosymmetric in order to exhibit macroscopic second-order nonlinear properties. In inorganic crystals, the nonlinearity is a bulk property mainly driven by the ionic polarizability of the material; large nonlinearities are thus always accompanied by high bulk dielectric constants, resulting in small variations of the commonly used figure of merit $Q = n^3 r_{\text{eff}}/\varepsilon_r$, which basically describes the induced optical nonlinearity (refractive index change for a given field) with respect to the screening properties of the medium (strength of space-charge field for given number of charges).

The refractive-index pattern resulting from the modulation of the bulk index by the space-charge field mimics the original light-intensity pattern and constitutes a grating which can diffract light. However, due to the displacement of the charge carriers, the maximum index change is not exactly in the same location as the maximum light intensity. This phase shift Θ is the unique feature of the PR effect, distinguishing it from any other mechanism that produces a light-induced change of the refractive index. In the Kukhtarev model [10], it is given by

$$\Theta = \text{arctg} \left[\frac{E_D}{E_0} \left(1 + \frac{E_D}{E_q} + \frac{E_0^2}{E_D E_q} \right) \right] \tag{7}$$

If transport takes place by diffusion or drift only, the phase shift is $\Theta = \pi/2$; in all other cases when both mechanisms contribute to the charge transport, Θ depends on the relative importance of the diffusion and drift processes and most importantly on the trap-limited space-charge field E_q (i.e., $0 < \Theta < \pi/2$).

The phase shift between light-intensity pattern and the resulting index grating gives rise to asymmetric energy transfer (also referred to as "two-beam coupling"), whereby one of the writing beams gains energy at the expense of the other—a property that can be exploited in photonic devices for image amplification. The beam coupling occurs, because both writing beams ideally fulfill the Bragg condition for the grating they generate. As a result, they are self-diffracted on that same grating. This applies to any holographic recording mechanism (i.e., also to local gratings). In the latter case, the transmitted light of one beam is $\pi/2$-phase-shifted compared to the diffracted light of the other. After passing through the material, both waves ($I_{1,\text{trans}}$ and $I_{2,\text{diffr}}$, and $I_{2,\text{trans}}$ and $I_{1,\text{diffr}}$, respectively) interfere equally for both beam directions. For nonlocal gratings on the other side ($\Theta \neq 0$), the phase relationship between the transmitted part and the diffracted part is such that the interference is constructive on one beam, whereas it is destructive on the other, leading to energy transfer between the beams. The photorefractive gain Γ is defined as

$$\Gamma = \frac{4\pi}{\lambda} (\hat{e}_1 \hat{e}_2^*) \Delta n \, \sin \Theta \tag{8}$$

where $\hat{e}_i$ are the polarization vectors of the two beams. The dimension of Γ is cm^{-1}; that is, it is length independent. Following Eq. (8), maximum steady-state energy transfer will occur for $\Theta = \pi/2$ (90°). The direction of the energy transfer is determined by the sign of Γ, which can be used to determine the sign of the index modulation amplitude or to identify the type of mobile carrier, if the sign of Δn is known. Thus, one can use a PR material to transfer optical energy from a pump beam (I_1) to a signal beam (I_2). For $I_1 \gg I_2$, the pump beam can be assumed to be undepleted, and only absorption losses have to be taken into account, leading to

$$I_1(d) = I_1(0) \exp(-\alpha l_1) \tag{9a}$$

$$I_2(d) = I_2(0) \exp\left[(\Gamma - \alpha)l_2\right] \tag{9b}$$

where l_i are the interaction lengths of the beams in the sample. The gain coefficient needs to be larger than the absorption coefficient so that beam 2 experiences net optical gain. The maximum achievable amplification γ_{max} of the signal beam is given by

$$\gamma_{max} = \frac{I_2(d)}{I_2(0)} = \exp[(\Gamma - \alpha)l_2] \tag{10}$$

This model, introduced by Kukhtarev and co-workers, has provided a solid framework for the theoretical description of photorefractive gratings in inorganic crystals such as LiNbO$_3$. Already at this early stage of the research on organic photorefractive materials, it became clear that the Kukhtarev model can be used to understand the general schemes; however, to quantitatively describe the performance of the organic materials, the fundamental differences between inorganic and organic materials have to be taken into account (see Section III).

III. PHOTOREFRACTIVITY IN AMORPHOUS ORGANIC PR MATERIALS

A. Material Classes

As was pointed out earlier, a material has to combine photosensitivity and photoconductivity, provide traps for the photogenerated charge carriers, and possess a field-dependent refractive index in order to be photorefractive. Unlike in inorganic crystals, in polymeric systems the individual properties required for photorefractivity can be optimized more or less independently from each other by varying the different functionalities to obtain the best possible combination for a particular application. Note that some of the compounds can be multifunctional, providing, for example, photosensitivity and EO properties at the same time. Two main material classes can be distinguished: monolithic and multicomponent systems. In the first category, three types of materials have been studied—fully functionalized polymers, where all functionalities are covalently bound to the polymer backbone; multifunctional glass-forming low-molecular-weight compounds; and liquid crystals. The second category includes all kinds of composite materials that are obtained by blending several compounds (e.g., low-molecular-weight compounds into a matrix polymer). This so-called "guest–host" approach opens up a large number of opportunities to obtain a PR material and has, therefore, been the most intensively studied class of organic PR materials. The matrix can fulfill one of the required functionalities or just be a binder without any functionality. The latter choice is usually less favorable, because a major part of the material is inactive this way, diminishing the bulk photoconductivity and EO properties. Therefore, in most cases, the host polymer is partially functionalized. One can distinguish between EO polymers mixed with a low-molecular-weight photoconductor (such as the first PR polymer [4]) and photoconductive polymers mixed with different low-molecular weight EO chromophores. A drawback of the guest–host approach is that all components need to be highly compatible with each other. Incompatibility often results in phase separation and ultimately limits the content of one or more components and, thus, the PR figure-of-merit. In this respect, monolithic systems are the materials of choice. The photosensitizer is generally at low concentration and is typically blended into the material.

Photorefractivity has been demonstrated in all of the above-mentioned categories of PR polymers. However, composites based on the polymeric photoconductor poly(N-vinylcarbazole) (PVK) doped with different monomeric EO chromophores [5–7[and recently glassy monomeric materials have shown, by far, the best performance among all of them (see Section V). Note, that the presence of all required properties at the same time does not necessarily result in photorefractivity [11].

B. General Description of the Functional Units and the Resulting Physical Properties

Conventional inorganic PR crystals such as $LiNbO_3$ and amorphous PR polymers are characterized by very different electronic structures and bulk properties, leading to distinctly different physical behaviors of these two material classes. In this section, we will give a quick overview and qualitatively compare the properties of inorganic and organic PR materials. For details we refer the reader to the literature cited in the text.

In inorganic PR materials, photoconductivity takes place via a band transport mechanism. By contrast, in organic amorphous materials, photoconductivity is provided by redox-active molecules that are easily oxidized (reduced). The most commonly used model to describe charge transport in amorphous organic matrices is the hopping model developed by Bässler [12,13]. The hole (electron) transport proceeds via hopping between spatially separated redox sites which undergo a series of oxidation/reduction steps (trapping/detrapping); that is, between radical cations (anions). The photoconductivity depends on the polarity of the matrix, the average distance between two conducting redox sites, the wavelength, and the temperature. Transport in such amorphous organic solids is highly dispersive due to the wide distribution of the energy of each site and the distance between two neighboring sites. Both aspects are modeled assuming Gaussian distributions for the density of states [12,13]. Transport is strongly field dependent:

$$\mu(\overline{\sigma}, \Sigma, E) = \mu_0 \exp\left[-\left(\frac{2\overline{\sigma}}{3}\right)^2\right] \begin{cases} \exp[C(\overline{\sigma}^2 - \Sigma^2)E^{1/2}], & \Sigma \geq 1.5 \\ \exp[C(\overline{\sigma}^2 - 2.25)E^{1/2}], & \Sigma < 1.5 \end{cases} \quad (11)$$

Here, $\overline{\sigma} = \sigma/k_B T$ is a measure for the energetical disorder in the system, σ is the width of the energetical distribution function, Σ is a parameter taking into account the spatial disorder, μ_0 is the carrier mobility without disorder, and $C = 2.9 \times 10^{-4}$ $(cm/V)^{1/2}$.

Most organic photoconductors are p-type (i.e., hole mobility is generally better than electron mobility). Typical examples of hole conductors include N-alkylcarbazoles, triarylamines, N,N-diarylhydrazones, and conjugated polymers such as substituted poly(p-phenylenevinylenes) and polythiophenes (Fig. 3).

Due to the high dispersity of charge transport, the hopping sites with the lowest energy may act as traps. In the limit, every redox site involved in the conducting pathway may act as a "shallow" trap for a charge carrier, as long as it is not photoreactivated; that is, the number of possible traps in PR polymers equals basically the number of conducting moieties, enabling the buildup of strong trap-limited space-charge fields [Eq. (5b)], which may reach values comparable to the external field. By contrast, in PR crystals, only impurities act as traps. Those are naturally

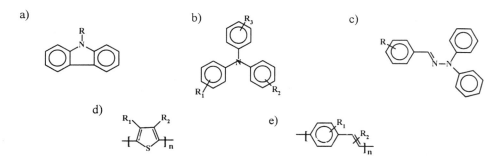

FIGURE 3 Classes of organic hole conductors used in organic photorefractive materials: (a) *N*-alkylcarbazoles, (b) triarylamines, (c) *N,N*-diarylhydrazones, (d) polythiophenes, and (e) poly(*p*-phenylenevinylenes); R = H, alkyl, or alkoxy.

at low concentration, limiting maximum strength of the internal space-charge field. Additional "deep" traps can be added to the organic PR materials. Those are molecules which possess a lower oxidation (reduction) potential than the principal hole (electron) photoconductor.

Most common organic charge transport molecules show absorption only in the ultraviolet (UV) part of the electromagnetic spectrum. In order to generate charge carriers through absorption of photons in the visible or even in the near infrared (NIR), materials are sensitized by adding appropriate functional molecules. Such "photosensitizers" include (intermolecular) charge transfer complexes between the hole (electron) conducting moieties and an electron-accepting (-donating) guest molecule (the best known example is 2,4,7-trinitro-9-fluorenane PVK/TNF), fullerenes such as C_{60}, and commonly used pigments such as thiapryyllium and squaryllium (Fig. 4). Upon photoexcitation, an electron is transferred between the photoconductor and the sensitizer, creating at first a bound electron–hole pair. The separation of this pair to generate free carriers (involving the transfer of an electron between the sensitizer and the photoconductor) competes with its recombination,

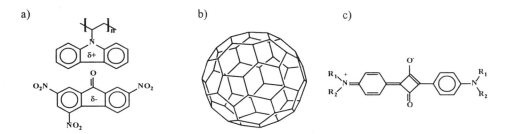

FIGURE 4 Examples for photosensitizers used in organic photorefractive materials: (a) charge transfer complex PVK/TNF, (b) C_{60}, and (c) squaryllium dyes.

and due to the low dielectric constant of organic materials, the geminate charge generation efficiency is poor ($< 10^{-4}$). The generation of free-charge carriers can be enhanced by application of an electric field, leading to a strongly field-dependent Onsager-type expression for the charge generation efficiency in the amorphous organic materials. A good numerical approximation to Onsager's quantum efficiency $\phi(E)$ was given by Mozumder [14] in terms of the infinite sum which was found to converge for $n > 10$:

$$\phi(E) = \phi_0 \left[1 - \zeta^{-1} \sum_{n=0}^{\infty} A_n(\eta) A_n(\zeta) \right] \quad \text{with}$$

$$\eta = \frac{e^2}{4\pi\varepsilon_0 \varepsilon k_B T r_0} \quad \text{and} \quad \zeta = \frac{e r_0 E}{k_B T} \quad (12)$$

$A_n(x)$ is a recursive formula given by

$$A_n(x) = A_{n-1}(x) - \frac{x^n \exp(-x)}{n!} \quad \text{and} \quad A_0(x) = 1 - \exp(-x) \quad (13)$$

The primary quantum yield ϕ_0 (i.e., the fraction of absorbed photons resulting in bound thermalized electron–hole pairs) is considered independent of the applied field r_0 is a parameter describing the thermalization length between the bound electron and hole.

Finally, the field-dependent refractive-index change is achieved by including highly polarizable noncentrosymmetric molecules with a permanent dipole moment into the materials. Such molecules typically possess a rod-like shape, consisting of a π-electron "bridge" end-capped with substituents of disparate electron affinities (electron donors and acceptors, respectively) to induce the asymmetry. The delocalization of the π electrons along the molecular axis leads to an increased electron density at the acceptor site at the expense of electron density at the donor group. In most cases, the resulting permanent dipole moment μ is parallel with the intramolecular charge transfer axis of the molecule. Furthermore, strong asymmetry of the linear polarizability along the molecular axis and perpendicular to it, $\alpha_\parallel$ and $\alpha_\perp$, respectively, as well as first hyperpolarizability, β, are obtained. Some typical classes of EO molecules that have been used in amorphous organic PR materials are shown in Fig. 5. Starting from p-nitroaniline (pNA) and urea, tremendous progress has been achieved in the field of EO molecules over the years [15,16].

Historically, because a lot of work had been done in inorganic PR crystals, where the field-dependent refractive index is provided by the electrooptic effect, research on organic PR materials had originally focused to maximize the EO performance of the materials. Later, it was discovered that the performance of PR polymers was much better than expected from separate investigations of the EO effect and considering the molecular hyperpolarizability β [9]. This was explained by orientational effects giving rise to a birefringence contribution to the total index modulation amplitude.

To observe macroscopic second-order nonlinear optical (NLO) effects in the material, the randomly distributed dipoles need to be oriented (e.g., by poling in an external electric field). In high-T_g polymers, the poling is performed close to the glass transition temperature, and the material is cooled afterward to room temperature with the external field applied, freezing the orientation of the chromo-

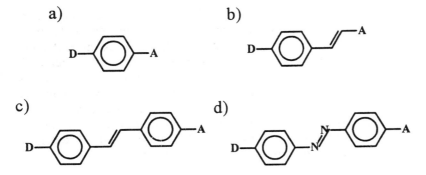

FIGURE 5 Classes of EO chromophores used in organic photorefractive materials: (a) benzene dyes, (b) styrene dyes, (c) stilbene dyes, (d) stilbenoide azo dyes; A = acceptor [e.g., $-NO_2$, $-CCN=C(CN)_2$]; D = donor (e.g., $-OR$, $-NR_2$; R = alkyl).

phores. By contrast, low-T_g polymers can be poled at room temperature. The orientation of dipoles by an electric field is described by the so-called "oriented gas model" [17]. This model assumes a Maxwell–Boltzmann distribution for the dipoles and can be used to calculate the macroscopic changes of the linear and nonlinear polarizabilities based on the microscopic molecular constants:

$$\Delta\chi_{ZZ}^{(1)} = \frac{2F^{(1)}}{45} N(\alpha_\parallel - \alpha_\perp) \left(\frac{\mu^*}{k_B T}\right)^2 E_p^2 \tag{14a}$$

$$\Delta\chi_{XX}^{(1)} = -\frac{1}{2}\Delta\chi_{ZZ}^{(1)} \tag{14b}$$

$$\Delta\chi_{ZZZ}^{(2)} = \frac{1F^{(2)}}{5} N\beta\left(\frac{\mu^*}{k_B T}\right)E_p \tag{15a}$$

$$\Delta\chi_{XXX}^{(2)} = \frac{1}{3}\Delta\chi_{ZZZ}^{(2)} \tag{15b}$$

Here, E_p is the poling field, Z and X denote the directions parallel and perpendicular to the poling direction, respectively, N is the number density of the chromophores, and μ^* is their dipole moment $F^{(1)}$ and $F^{(2)}$ are the local field correction factors. Note that for the linear optical properties $\chi^{(1)}$, the anisotropy of the polarization changes parallel and perpendicular to the poling field direction is negative [Eq. (14b)], whereas it is positive for $\chi^{(2)}$ [Eq. (15b)].

In photorefractive holographic experiments on high-T_g prepoled polymers, the refractive-index changes originate solely from the EO effect due to the structural rigidity which ideally does not allow for rotational diffusion of the chromophores. By contrast, low-T_g polymers are poled at room temperature in situ during recording. The total local electric field, E_{tot}, is the superposition of the uniform external field, E_{dc}, and the nonuniform internal space-charge field, E_{SC}. The result is a periodically (in amplitude *and* direction) varying local poling field, leading to spatial variations of the direction of the local orientation and the degree of poling (Fig.

6). Thus, the modulation of the refractive index through the EO effect is enhanced and, furthermore, the linear polarization of the material (birefringence) is also modulated. As a result, the PR effect in such materials with low T_g is much stronger than in a hypothetical PR polymer, where all chromophores would be oriented in the same direction (as in permanently poled polymers). This phenomenon was first discussed by Moerner et al. as the so-called "orientational enhancement mechanism" [9]. Calculations based on the oriented gas model [17] taking into account the tilted geometry commonly used for the characterization of organic PR polymer (see Section IV) yield the following expressions for the field-induced index changes sensed under Bragg conditions by either s or p polarization:

$$\Delta n_S^{(1)} = -\frac{2\pi}{n} B E_{dc} E_{SC} \cos \varphi \tag{16a}$$

$$\Delta n_P^{(1)} = \frac{2\pi}{n} B E_{dc} E_{SC} [2 \cos \varphi \sin \alpha_1 \sin \alpha_2$$
$$- \cos \varphi \cos \alpha_1 \cos \alpha_2 + \tfrac{3}{2} \sin \varphi \sin(\alpha_1 + \alpha_2)] \tag{16b}$$

$$\Delta n_S^{(2)} = \frac{8\pi}{n} C E_{dc} E_{SC} \cos \varphi \tag{17a}$$

$$\Delta n_P^{(2)} = \frac{8\pi}{n} C E_{dc} E_{SC} [\cos \varphi \cos \alpha_1 \cos \alpha_2 + 3 \cos \varphi \sin \alpha_1 \sin \alpha_2$$
$$+ \sin \varphi \sin(\alpha_1 + \alpha_2)] \tag{17b}$$

with

$$B = \frac{2}{45} N f_\infty (\alpha_\| - \alpha_\perp) \left(\frac{\mu^*}{k_B T}\right)^2 \tag{18a}$$

and

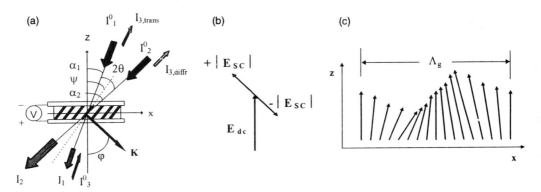

FIGURE 6 (a) Experimental geometry used for holographic two-beam coupling and four-wave mixing experiments; (b) vector diagram illustrating the relative directions of the externally applied field and the evolving photorefractive space-charge field; (c) local total poling field vector.

$$C = \frac{Nf_0 f_\infty f_\infty \beta_{ZZZ}}{15} \left(\frac{\mu^*}{k_B T} \right) \tag{18b}$$

φ is the tilt angle of the device, $\alpha_{1,2}$ are the angles between each beam and the sample normal, respectively (Fig. 6a), and $f_{0,\infty}$ are local field factors taking into account interactions with the surrounding matrix. Thus, a useful figure-of-merit to classify nonlinear optical molecules/moieties used in organic amorphous materials with low/high glass transition temperature is

$$F_{\text{ltg}} = \frac{(A \, \mu^{*2} \, \Delta\alpha)/k_B T + \mu^* \, \beta}{M} \tag{19a}$$

$$F_{\text{htg}} = \frac{\mu^* \, \beta}{M} \tag{19b}$$

where A is prefactor taking into account the different degeneracy of the two contributions and M is the molecular weight of the chromophore. According to Eqs. (19), the use of chromophores with large dipole moments and large anisotropy for the linear polarization and *without* apparent second-order nonlinear optical properties and also liquid crystals in organic PR materials is possible (see Sections V.B.2 and V.B.3) when the orientational mobility is large enough.

Note that according to Eq. (14a), the linear contribution to the refractive-index change is quadratic in electric field. The same quadratic dependence applies to the electrooptic coefficient times the field (Eq. 6), and the electrooptic coeffient scales linear with the electric field (Eq. 15a). Because the total local electric field E_{tot} (the local poling field) is the superposition of the uniform external field E_{dc} and the nonuniform internal space-charge field E_{SC}, this square dependence has important consequences on the development of the PR grating in materials which are poled in situ as illustrated by

$$E_{\text{tot}}(x) = [E_{\text{dc}} + E_{\text{SC}} \sin(Kx)]^2 \tag{20}$$
$$= E_{\text{dc}}^2 + 2E_{\text{dc}}E_{\text{SC}} \sin(Kx) + [E_{\text{SC}} \sin(Kx)]^2$$

Thus, besides the grating with period Λ_g probed under Bragg condition [proportional to the middle term in Eq. (20)], an additional grating with half the grating period is formed inside the material because the last term in Eq. (20) is proportional to $E_{\text{SC}}(Kx)^2$ [$\sin^2(Kx) \propto \sin(2Kx)$; grating vector $2K$]. This behavior is very different from the situation in PR crystals or permanently poled polymers where gratings with higher spatial frequencies can exist only due to the nonsinusoidal shape of the primary index grating.

IV. EXPERIMENTAL

A. Device Preparation

As was pointed out earlier, the electric field plays an important triple role in amorphous organic PR materials: (1) enhance the efficiency of charge carrier photogeneration, (2) increase photoconductivity, and (3) poling of the molecular dipoles to achieve a macroscopic index change. Typical photorefractive devices using organic amorphous materials as the active medium thus consist of two glass slides coated with transparent electrodes (e.g., indium–tin oxide, ITO) for the application of the electric field, which is then perpendicular to the device normal (Fig. 7). To mix the components properly, appropriate amounts of each component of a PR material under investigation are dissolved in a solvent (e.g., methylenechloride, chloroform, or toluene). The solution is filtered to remove particles which may act as scattering centers, and the solvent is then allowed to evaporate. In some cases, a vacuum should be applied in order to remove any residual solvent, which may ease dielectric breakdown later.

Some of the resulting composite material is placed onto an electrode and heated until it softens. Spacers are placed around the material and the second electrode is pressed on top, yielding uniform films of a thickness given by the spacers used. In some cases, commercial liquid-crystal cells can be used which are filed by capillary forces. Typical thicknesses are 100–200 μm, which is a compromise between maximizing the thickness for long interaction lengths and high Bragg selectivity on one hand, and maximizing the external field for high photogeneration efficiency, fast charge transport, and good poling on the other. The devices are finally sealed with epoxy resin.

B. Four-Wave Mixing and Two-Beam Coupling Experiments

The holographic characterization of the new devices is typically done by means of the four-wave mixing (FWM) and the two-beam coupling (TBC) techniques. The experimental geometry is illustrated in Fig 6a. In both cases, two equally polarized coherent laser beams ("writing beams" 1 and 2) of wavelength λ with intensity

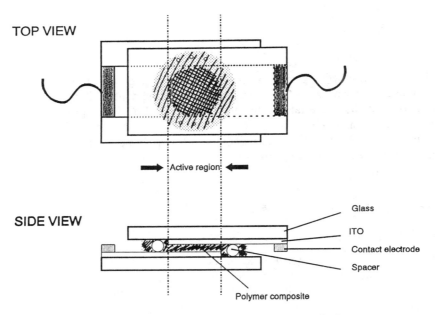

FIGURE 7   General geometry of organic photorefractive devices.

ratio $b = I_1/I_2 \geq 1$ are overlapped in the sample with an angle 2θ to create a fringe pattern with period Λ_g. In order to have a nonzero component of the external field along the grating wave vector, the experiments are performed in a tilted geometry (tilt angle ψ, also defined outside the material) to allow the drift of photogenerated charge carriers. In a TBC experiment, the intensities of the two "writing beams" are monitored after passing through the device. From these data, the optical gain coefficient Γ is calculated according to

$$\Gamma = \frac{1}{d}\left[\cos\alpha_1 \ln\left(\frac{I_1}{I_1^0}\right) - \cos\alpha_2 \ln\left(\frac{I_2}{I_2^0}\right)\right] \tag{21}$$

Here, $I_{1,2}^0$ are the intensities of the two writing beams after passing through the device independently, $I_{1,2}$ are the intensities after passing simultaneously through the device, d is the active layer thickness, and $\alpha_{1,2}$ are the angles of the beams with the sample normal inside the material. The occurrence of beam coupling is considered proof for the PR nature of the grating.

In the FWM experiments, index gratings are probed by a weak "probe" beam $(3, I_3 << I_{1,2})$ in a backward geometry (Fig. 6a). In most cases, the writing beams are simultaneously incident on the devices, because even at low light levels, the grating would be erased when uniformly illuminated by the probe beam. For the degenerate case $(\lambda_{1,2} = \lambda_3)$, the probe beam is counterpropagating with the writing beam 1, whereas for the nondegenerate case $(\lambda_{1,2} \neq \lambda_3)$, the probe beam must be adjusted to the appropriate Bragg angle. The intensities of the transmitted and the diffracted light are monitored to obtain the diffraction efficiency which is defined as the intensity ratio of the diffracted light and the incident probe intensity, $\eta = I_{3,\text{diff}}/I_3^0$. For slanted lossy dielectric gratings, Kogelnik's coupled-wave model [18] can then be used as a good approximation to calculate the total amplitude of the refractive-index modulation Δn from the experimental FWM data for transmission gratings:

$$\eta = \exp(-2D)\frac{\sin^2[(\nu^2 - \xi^2)^{1/2}]}{1 - \xi^2/\nu^2}$$

$$\nu = \delta\frac{\pi d \Delta n}{(\cos^2\theta - \cos^2\varphi)^{1/2}}, \quad \xi = -D\tan\theta\cot\varphi, \quad D = \frac{\alpha}{2}d\frac{\cos\theta\sin\varphi}{\cos^2\theta - \cos^2\varphi}$$

$$\delta = 1 \text{ for } s\text{-polarized light}, \quad \delta = \cos 2\theta_i \text{ for } p\text{-polarized light} \tag{22}$$

where n is the bulk refractive index, $\lambda\ (= \lambda_3)$ is the readout wavelength, and α is the absorption coefficient; the angles θ and φ are specified in Fig. 6a. Note that this model applies strictly only to linear gratings. Thus, s-polarized writing beams are typically used in a FWM experiment for the evaluation of organic PR materials, because the beam coupling is smaller than for p-polarized beams (see Section IV). In this way, the nonuniformity of grating amplitude and phase throughout the device can be reduced to a minimum.

The index grating amplitude Δn and the gain coefficient Γ both strongly depend on the total incident light intensity I, the intensity ratio between the beam I_1, I_2, and I_3, the polarization of the beams, the grating period Λ_g determined by the angle between the writing beams 2θ [Eq. (2)], the tilt angle φ, the active device thickness d, the applied electric field E and so forth.

C. Determination of the Phase Shift

The unique property of PR materials is the phase shift Θ between the original light grating and the resulting index grating Θ, leading to energy transfer between the two writing beams. It can be derived from the index modulation amplitude Δn for a particular polarization obtained from FWM experiments using the coupled-wave model [Eq. (22)] and the gain coefficient Γ in TBC experiments for the identical polarization using Eq. (8). The close relationship between Δn and the gain is illustrated in Fig. 8.

The phase shift Θ can also be determined independently by the so-called "moving-grating technique" [3b,19]. A key feature of this method is the homodyne detection scheme, yielding large amplitude variations even if the diffraction efficiency for the identical material would be small. Here, after a grating has been written in the PR material, the sample is translated with constant speed. Because the recording is reversible in photorefractive materials, the translation must be faster than the materials' response time so that the grating remains stable during the translation. The intensities I_1 and I_2 of the two "writing" beams are modulated when the sample is displaced by x, because Θ is artificially altered. For symmetric geometry, they are given by

$$I_1(x) = I_1^0 \exp\left(-\frac{\alpha d}{\cos \theta}\right)[1 - 2 Ad \cos (\phi_A + Kx) + 2P \sin(\phi_P + Kx)] \quad (23a)$$

$$I_2(x) = I_2^0 \exp\left(-\frac{\alpha d}{\cos \theta}\right)[1 - 2 Ad \cos(\phi_A + Kx) - 2P \sin(\phi_P + Kx)] \quad (23b)$$

where $I_{1,2}^0$ are the initial intensities of the writing beams, α is the absorption coefficient, d is the sample thickness, θ ($= \theta_1 = -\theta_2$) is the angle of the beams with the sample normal (see Fig. 6a), K is the grating wave vector, $\phi_{A,P}$ are the phase shifts between the light intensity pattern and the absorption and the index grating, respectively, and A and P are the amplitudes of the gratings, respectively, given by

$$A = \frac{\Delta \alpha d}{4 \cos \theta} \quad (24a)$$

$$P = \frac{\pi \Delta n d}{\lambda \cos \theta} \quad (24b)$$

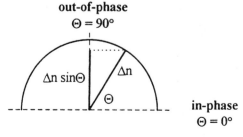

FIGURE 8 Amplitude of the index of refraction tested in TBC (only out-of-phase component $\Delta n \times \sin \Theta$) and FWM experiments (total amplitude Δn).

As can be seen from Eqs. (23a) and (23b), absorption gratings occur with identical signs for both beams, whereas the index gratings have opposite signs, indicating asymmetric energy exchange between the beams. For a pure absorption grating, the beams oscillate in-phase, whereas for a pure index grating, a π-phase shift will occur. For intermediate cases, the sum $I^{(+)} = I_1 + I_2$ and the difference $I^{(-)} = I_1 - I_2$, respectively, of the two signals are calculated to determine the individual contributions to the total index grating amplitude:

$$I^{(+)}(x) = I_0 \exp\left(-\frac{\alpha d}{\cos \theta}\right)[2 - 4A \cos(\phi_A + Kx)] \tag{25a}$$

$$I^{(-)}(x) = I_0 \exp\left(-\frac{\alpha d}{\cos \theta}\right)[-4P \sin(\phi_P + Kx)] \tag{25b}$$

The amplitudes Δn and $\Delta \alpha$ of the refractive index and amplitude gratings, respectively, as well as their phase shifts ϕ_A and ϕ_P can be determined from the Eqs. (24) and (25). As indicated above, this analysis is valid when the two beams impinge on the sample symmetrically with respect to the sample normal. In the tilted configuration, the analysis is more complicated and additional approximations are needed [19].

D. Determination of Photoconductivity, and Electrooptic Activity

Critically important properties of a PR medium include its photoconductivity and its electrooptic activity. The former can be determined via a discharge technique commonly used to characterize xerographic conductive layers or by a time-of-flight experiment [12,13]. These two techniques are typically applied to thin layers of the organic photoconductor. By contrast, an in situ transient diffraction experiment ("holographic time-of-flight", HTOF) can be performed with the identical PR samples used for the holographic experiments [20]. Two picosecond or nanosecond laser pulses create a sinusoidal distribution of charge carriers, which drifts under the influence of an external electric field. As charge separation advances, a space-charge field builds up that can be probed by a continuous-wave (CW) laser beam due to the EO effect. The space-charge field and thus the diffraction efficiency η reach a maximum when the mobile carriers have drifted to a position of anticoincidence with the immobile charge carriers. Further drift will result in a reduction of E_{SC} and η until coincidence is reached, and so on. The diffraction efficiency should thus oscillate with time, yielding the carrier drift mobility according to

$$\mu_{dr} = \frac{L_{dr}}{Et}, \qquad L_{dr} = \frac{\Lambda_g}{2 \cos \varphi} \tag{26}$$

with E the applied field, L_{dr} the drift length of the carriers, and t the transient time to reach the first maximum.

The EO properties can be checked by interferometric techniques or by a ellipsometric measurement first described by Teng and Man [21] and by Schildkraut [22]. The refractive index of the material is modulated by an AC electric field with given frequency Ω. These techniques generally apply to prepoled films, and the effect of the field on the orientation of the molecules can be neglected. However, due to the low glass transition temperature of most amorphous organic photorefractive materials, this has to be taken into consideration, as the modulation field can reorient the molecules depending on the AC frequency used. The experimental setup to investigate the EO properties of actual photorefractive devices is schematically shown in Fig. 9. A DC voltage is applied to the device to achieve some orientation of the chromophores (corresponds to the prepoling mentioned above). An AC field is then additionally applied to modulate the index of refraction. For the low-frequency (LF) limit, modulation will provide insight into the electrooptic *and* birefringence properties of the material, whereas in the high-frequency (HF) modulation limit, only the electrooptic part is probed [23]. Due to the square dependence of poling-induced index changes on the electric field ($E_{tot} = E_{ac} + E_{dc}$), besides a signal at the excitation frequency Ω, one can also detect a signal with the frequency 2Ω [compare Eq. (20)]:

$$R^{LF}(\Omega) \propto \tfrac{3}{2}B + 4C \tag{27a}$$

$$R^{HF}(\Omega) \propto 2C \tag{27b}$$

The constants B and C are given in Eqs. (18). Figure 9 also schematically shows

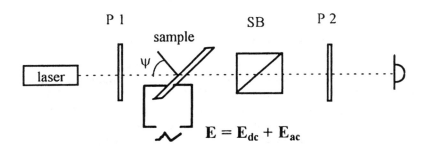

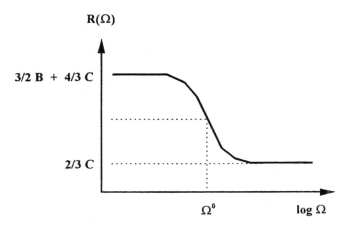

FIGURE 9 *Top*: Experimental setup for transmission ellipsometric measurements: P = polarizer, SB = Soleil–Babinet compensator, D = detector; P1, is polarized at 45°, P2 at −45° (i.e., perpendicular to P1). *Bottom*: General behavior of the response $R(\Omega)$ as a function of $\log(\Omega)$. The constants B and C are given in the text [Eqs. (18)].

the expected frequency dependence of $R(\Omega)$. The frequency Ω^0 (climax of the response function) is characteristic for the orientational mobility of the chromophores in the matrix. $1/\Omega^0$ is considered the mean value of the distribution function describing the different rotational dynamics time constants for a dipole in the system [24].

V. RESULTS AND DISCUSSION

In this section, experimental results on different material classes will be described. The emphasis will be on composites based on the photoconducting polymer poly(N-vinylcarbazole) (PVK) which have been studied most intensively in the field of organic photorefractive materials.

Figure 10 schematically shows the performance of an amorphous organic PR material with a low glass transition temperature in TBC and FWM experiments. Originally ($t < 0$), there is no external field applied to the sample, the two pump beams start out at some given energy (I_1 and I_2, respectively), and diffraction is not observed ($I_3 = 0$). This is because the EO chromophores are isotropically distributed inside the matrix and, thus, there is no macroscopic EO effect, one of the requirements for photorefractivity.

At $t = 0$, the external electric field is switched on, and the photorefractive grating builds up. As a result, the probe beam is diffracted by this grating, and asymmetric energy transfer due to self-diffraction is observed between the two pump beams: Beam 1 is amplified, whereas beam 2 is diminished. If the field is further increased ($t = 1$), more energy is transferred between the pump beams, and the diffraction becomes more efficient. When the field is switched off ($t = 2$), the pump beam intensities relax back to their original values and the diffraction disappears due to the relaxation of the Bragg grating. This clearly demonstrates the need of an electric field for the orientation of the EO chromophores, as was described in section III.

For the opposite field direction but same field strength ($t = 3$), the diffraction efficiency is identical; by contrast, energy transfer occurs in the opposite direction (i.e., beam 2 is then amplified and beam 1 diminished). This is because the direction of carrier transport is reversed. When one of the pump beams is blocked with the electric field still applied ($t = 4$), the other pump beam uniformly illuminates the grating and erases it. When the beam is restored at $t = 5$, a new grating identical to the previous one is written. At time $t = 6$, all beams are blocked and the electric field is switched off. Finally, at time $t = 7$, the electric field and the probe beam are restored. The occurrence of a diffracted signal demonstrates the storage capabilities of the material. Note, however, that the grating is readily erased by the uniform illumination of the probe beam during readout, and the diffraction relaxes back to zero. Erasure by the weak probe beam is much slower than with the strong pump 2 (compare $t = 4$).

A. Composite Materials

The first photorefractive polymer was proposed by Schildkraut at Eastman Kodak [25], but photorefractivity was really demonstrated for the first time in a polymer by Ducharme and co-workers at IBM Almaden [4]. The performance of that first

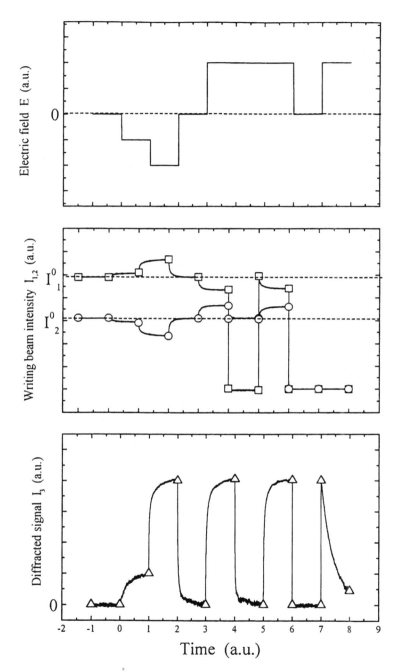

FIGURE 10 Schematic performance of an amorphous PR material with low glass transition temperature in TBC and FWM experiments. (*top*) electric field, (*middle*) writing beam intensities, and (*bottom*) diffracted beam intensity as a function of time (see text for details).

material was limited, but it initiated a very active field of research. In this section, we will review some of the photorefractive polymers that have been proposed since.

1. PVK-Based Systems

The PVK-based composites showed the most promising performance levels to date. Almost simultaneously, photorefractivity was reported in such materials by Zhang et al. [26], Silence et al. [27], and Kippelen et al. [28]. The EO chromophores used were DEANST, F-DEANST, and 2,5DMNPAA, respectively, and a thiapyrillium dye, the fullerene C_{60}, and TNF as the sensitizers (see Fig. 11 for chemical structures and full names of the compounds). Most importantly, an additional plasticizer was used for the efficient orientation of the chromophores even at room temperature, leading to record diffraction efficiencies of up to 6% at the time [28]. The reported response times were approximately 100–200 ms in all cases. The first material with net gain was also a PVK-based composite [29]. A large number of different chromophores have been checked in combination with PVK since.

In 1994, a breakthrough in the young field of PR polymers was reported by Meerholz et al. at the University of Arizona [8]. Improvement of the material's homogeneity and uniformity allowed the application of higher electric fields and yielded excellent photorefractive performance. For the first time, the performance of an organic PR material was brought to levels comparable to, in some respects even better than the long-known inorganic counterparts. The material consisted of 2,5DMNPAA (50 wt%), PVK (33 wt%), and TNF (1 wt%). Additionally, a plas-

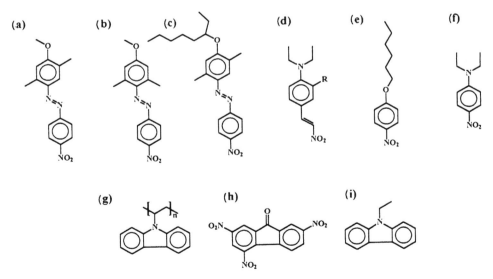

FIGURE 11 Chemical structures: (a) 2,5-dimethyl-4-(*p*-nitrophenylazo)anisole (2,5DMNPAA), (b) 3,5-dimethyl-4-(*p*-nitrophenylazo)anisole (3,5DMNPAA), (c) 1-2'-ethylhexyloxy)-2,5-dimethyl-4-(*p*-nitrophenylazo)benzene (EHDNPB), (d) R = H, diethyl-amino-4-nitrostilbene (DEANST), R = F, 2-fluoro-diethylamino-4-nitrostilbene (F-DEANST), (e) 4-(hexyloxy)nitrobenzene (HONB), (f) 4-(diethylamino)nitrobenzene (EPNA), (g) poly(*N*-vinylcarbazole), (h) 2,4,7-trinitro-9-fluorenone (TNF), and (i) *N*-ethylcarbazole.

ticizer was added to obtain a material with a glass transition temperature T_g close to room temperature and to be able to perform the poling of the chromophores at room temperature. The material of choice was *N*-ethylcarbazole (ECZ, 16 wt%, Fig. 11i) which is assumed to just vary the T_g without reducing the photoconductivity of the material. Later, tricresyl phosphate (TCP) was used for similar reasons [30].

This composite—even though reported more than 2 years ago—is still among the best organic PR materials. Therefore, results obtained with this material (or closely related composites) will be presented here in more detail. Figure 12 shows the linear absorption spectrum of this composite. The absorption peaks at ≈ 350 nm originate from carbazole units; the peak at 400 nm originates from the 2,5DMNPAA. The absorption of the PVK/TNF complex extends over the entire visible and NIR part of the electromagnetic spectrum and does not show any specific peak. For wavelengths $\lambda > 650$ nm, the absorption of the composite is due to PVK/TNF. At the operating wavelength of the experiments described in the following ($\lambda = 675$ nm) and without applied field, the absorption coefficient is $\alpha = 8$ cm^{-1}. It is important that α is not too high in order not to limit the performance of the material, but it must be high enough to ensure sufficient charge carrier generation.

First, the holographic steady-state performance of 105-μm-thick devices was studied as a function of the external electric field. Figure 13 shows two-beam

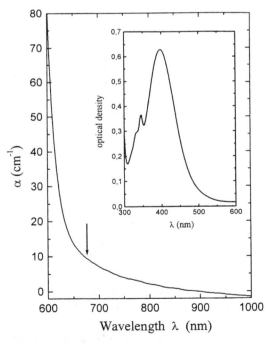

FIGURE 12 Absorption spectrum of a 105-μm-thick photorefractive device of DMNPAA: PVK : ECZ : TNF 50 : 33 : 16 : 1 wt%; inset: absorption spectrum of a thin film ($d \approx 100$ nm).

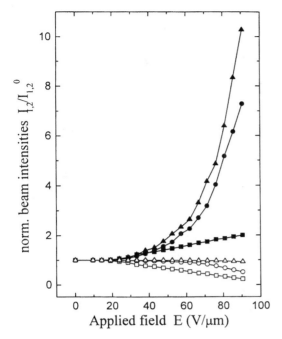

FIGURE 13 TBC experiments on 105-μm-thick devices of DMNPAA : PVK : ECZ : TNF 50 : 33 : 16 : 1 wt%. External field dependence of the normalized beam intensities $I_{1,2}/I_{1,2}^0$ for *p*-polarized writing beams, $\lambda = 675$ nm, $2\theta = 22°$, $\psi = 60°$; one beam was constant at 1 W/cm^2 (open symbols), the other (filled symbols) was varied: 0.77 W/cm^2 (squares), 0.077 W/cm^2 (circles), and 0.0077 W/cm^2 (triangles). The solid lines are guides to the eye.

coupling (TBC) results. With increasing field, more and more energy is transferred from one beam to the other. For a small modulation depth m (large beam intensity ratio $b = I_1/I_2$), the signal beam is amplified more than for large m (small b). For clarity reasons, the normalized intensities are shown. If the field direction is reversed, the beam which gained energy for positive fields loses for negative fields (not shown in Fig. 13; see Fig. 10). This is because the charge carriers drift in the opposite direction, leading to sign inversion of the phase shift Θ between the light-intensity pattern and the written index grating.

As shown in Fig. 14, the steady-state gain coefficient Γ calculated according to Eq. (21) is identical for the three different modulation depths. The gain monotonically increases, reaching $\Gamma_p = 220$ cm^{-1} at $E = 90$ V/μm. The gain by far exceeds the absorption in the sample at this voltage ($\alpha = 13$ cm^{-1}), giving a net optical gain of $\Gamma_{p,\text{net}} = 207$ cm^{-1}. The fact that gain is observed proves the photorefractive (i.e., nonlocal) nature of the index grating recorded in the material. For *s*-polarized beams, on the other hand, the gain is much smaller ($\Gamma_s = -40$ cm^{-1} at 90 V/μm) and occurs in the opposite direction; that is, the beam which gained energy when it was *p*-polarized, now loses energy for the identical field direction. The reduced gain for *s*-polarization compared with *p*-polarization originates from the fact that the poling-induced index changes for both linear and the nonlinear

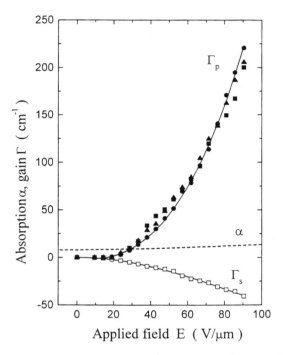

FIGURE 14 TBC experiments on 105-μm-thick devices of DMNPAA : PVK : ECZ : TNF
50 : 33 : 16 : 1 wt%. External field dependence of the gain coefficient for s- (open symbols)
and p-polarized writing beams (filled symbols), $\lambda = 675$ nm, $2\theta = 22°$, $\psi = 60°$; one beam
was constant at 1 W/cm^2, the other was varied: 0.77 W/cm^2 (squares), 0.077 W/cm^2 (circles), and 0.0077 W/cm^2 (triangles); Also shown is the absorption coefficient α (dashed
line). The solid lines are guides to the eye.

polarizabilities are larger for p-polarized light than for s-polarized light [see Section
III, Eqs. (16) and (17)].

According to Eq. (8), the sign change of the gain coefficient Γ when the polarization is changed from p to s reflects opposite signs of the index grating amplitudes sensed by the two light polarizations, as the carrier drift direction (determined by the field direction) is identical. Such a behavior is unknown for inorganic
PR crystals. It can be explained in terms of the orientational enhancement mechanism [9] discussed in Section III, leading to birefringence (BR) as well as EO
contributions to the refractive-index modulation. The anisotropy $\Delta n_p / \Delta n_s$ of the
poling-induced index change sensed by p- and s-polarized light has opposite signs
for the birefringence ($\Delta n_p^{BR} / \Delta n_s^{BR} < 0$) and for the EO effect ($\Delta n_p^{EO} / \Delta n_s^{EO} > 0$)
[Eqs. (16b) and (17b)]. Assuming the superposition of the individual contributions
to the total index change, the sign of the total anisotropy will be determined by
the relative strength of the two contributions. For the material investigated here,
the sign change of the gain coefficient Γ when the polarization is changed from p
to s proves that the electric-field-induced birefringence is dominant in this material.
This holds true for most of the other materials studied to date.

Frequency-dependent ellipsometric measurements were performed on the identical devices using the composite 2,5DMNPAA : PVK : ECZ : TNF in the proportion 50 : 40 : 10 : 1 wt% [23]. This experiment yields information about the *effective* electrooptic coefficient r_{eff} which may have EO as well as BR contributions. For the pure EO effect ($B = 0$, $C \neq 0$), a factor of 2 is expected from theory between the HF and LF limits [compare Eqs. (27)]. In our experiment (Fig. 15), the response function $R(\Omega)$ levels off for $\Omega > 200$ Hz and $\Omega < 0.05$ Hz. The increase going from high to low frequencies is more than a factor of 20, indicating strong enhancement by the poling-induced birefringence at low frequency.

Figure 16 shows the FWM experiment performed with *s*-polarized writing beams on the identical sample. The diffraction efficiency for a *p*-polarized readout beam increases with the electric field and reaches a maximum of $\eta = 86\%$ at $E = 61$ V/μm. At the same time, the transmitted intensity almost vanishes, indicating almost complete internal diffraction of the reading beam. A further increase of the field leads to periodic energy transfer between the diffracted and the transmitted beam; at $E = 81$ V/μm, all light is redirected into the original probe wave even though the grating became twice as efficient as for $E = 61$ V/μm. This oscillatory behavior is in agreement with the coupled-wave model [Eq. (22)]. Similar experiments were performed for *s*-polarized readout (not shown). For the same reasons already explained for the gain coefficient, the diffraction efficiency is smaller for *s*-polarized than for *p*-polarized readout, and no maximum is observed for fields up to 90 V/μm. Using Eq. (22), we calculated the total index modulation amplitude Δn. As shown in Fig. 17, it reaches $\Delta n_p \approx 6.4 \times 10^{-3}$ at 90 V/μm for *p*-polarized readout and $\Delta n_s \approx -1.5 \times 10^{-3}$ for *s*-polarized readout. The negative (opposite) sign for *s*-polarized readout cannot be determined from the data directly, due to

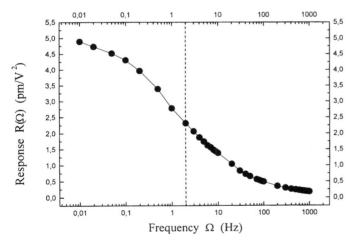

FIGURE 15 Ellipsometric experiment performed on 105-μm-thick devices of DMNPAA : PVK : ECZ : TNF 50 : 40 : 9 : 1 wt%. Response function $R(\Omega)$ as a function of the frequency Ω of applied field at $\lambda = 690$ nm with $E_{dc} = E_{ac} = 500$ V, $\psi = 45°$. The solid line is a guide to the eye. The dashed line indicates the estimated response time $\tau = 1/\Omega^0$ of the material (climax of the function).

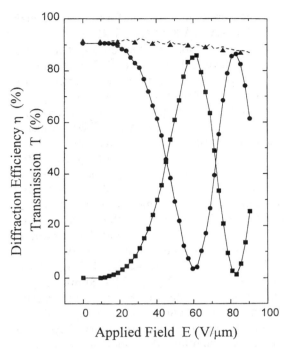

FIGURE 16 Degenerate FWM experiment performed on 105-μm-thick devices of DMNPAA : PVK : ECZ : TNF 50 : 33 : 16 : 1 wt% with s-polarized writing beams (power density $I_1 \approx 1$ W/cm^2, $I_2 \approx 0.77$ W/cm^2) and p-polarized reading beam ($I_3 \approx 0.35$ nW/ cm^2), $\lambda = 675$ nm, $2\theta = 22°$, $\psi = 60°$. External field dependence of the diffraction efficiency (squares) and the transmission in the presence (circles) and in the absence (triangles) of the writing beams. The dashed line is the sum of the transmitted and diffracted intensity in the presence of the writing beams, indicating that the maximum achievable diffraction efficiency is limited by absorption and reflection losses ($\approx 12\%$). The solid lines are guides to the eye.

the sin^2 dependence, but was a result of the TBC experiment. Plotted on a log scale, a slope of 2.1 is obtained (inset Fig. 17), a value close to the exponent 2 expected from the orientational enhancement model [Eqs. (14)–(17)]. Considering the large index modulation amplitude, only a relatively moderate TBC gain constant is observed. This suggests that the phase shift Θ between the light fringe pattern and the index grating is much smaller than the optimum value of $\pi/2$ (90°).

Figure 18 shows the field dependency of the phase shift Θ calculated from the gain coefficient Γ_s and the index modulation amplitude Δn_s, both for s-polarization using Eq. (8). Without external field, a phase shift of 90° is expected because only diffusion processes can take place; however, this was not measurable within the experimental sensitivity of the experiment. With increasing field, the phase shift gradually decreases and levels off to $\Theta \approx 17°$ for $E > 40$ V/μm. These results were confirmed almost quantitatively by moving-grating experiments on a very similar composite where PVK and ECZ were substituted with a carbazole-substituted siloxane with inherently lower glass transition temperature [33]. In this material, a value of $\theta \approx 20°$ was reported for high fields.

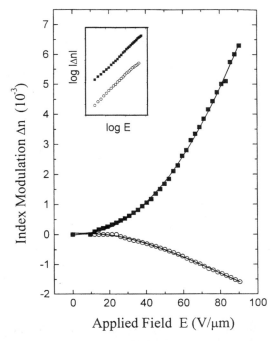

FIGURE 17 External field dependence of the total refractive-index modulation amplitude Δn for s-polarized (circles) and p-polarized (squares) readout of a hologram written with s-polarized beams obtained from the FWM data using the coupled-wave model for slanted lossy dielectric transmission gratings and taking into account the change in absorption, but not the small phase and amplitude variations caused by energy exchange between the writing beams. The solid lines are guides to the eye. Inset: Double logarithmic plot of the external field dependence of the total refractive-index modulation amplitudes $|\Delta n_s|$ (circles) and $|\Delta n_p|$ (squares) for s- and p-polarized readouts, respectively.

The value of the grating spacing of the gratings recorded in the PR materials can be easily varied by changing the opening angle 2θ between the writing beams. Figure 19 shows the results of such variation in a FWM experiment. The diffraction efficiency η_p for p-polarized readout increases with decreasing angle 2θ (i.e., increasing grating spacing). As discussed in Section III, for large Λ_g (> 5 μm) the diffraction is still in the Bragg regime for thick gratings, however, and non-Bragg diffraction into higher orders can be observed, as will be discussed in Section VI.A.

The dynamics of grating formation is very complex. The speed within one single material depends on the applied electric field, the light intensity, and the grating spacing. The analysis of the data is very complicated, as it is known that *all* processes involved in the grating buildup, such as photogeneration, charge transport, and orientation of the EO chromophores, are highly dispersive [i.e., a large number of time constants with a certain (broad) distribution are expected for each of those processes]. For example, holographic time-of-flight experiments were performed by Malliaras et al. on the PR composite PVK : HONB : TNF [32]. Surprisingly, not the expected oscillatory behavior was observed for the time-

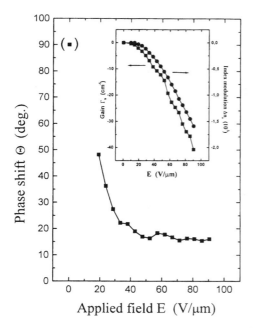

Applied field E (V/μm)

FIGURE 18 External field dependence of the phase shift Θ in holographic experiments performed on 105-μm-thick devices DMNPAA : PVK : ECZ : TNF 50 : 33 : 16 : 1 wt%, λ = 675 nm, 2θ = 22°, ψ = 60°. The data point for E = 0 was not measured, but the phase shift for transport by diffusion only (90°) was used. Inset: Field dependence of the index modulation amplitude and the gain coefficient for s-polarization which were used to calculate Θ by using Eq. (8). The lines are guides to the eye.

dependent diffraction efficiency η (see Section IV.D), but after a fast initial rise, η reaches a maximum at t_{max} and then levels off to a plateau value (Fig. 20, top). This was interpreted in terms of a strongly dispersive charge transport and qualitatively modeled using the Scher–Montroll theory [33]. The best fit was obtained for a disorder parameter a between 0.6 and 0.7 (Fig. 20, bottom). A small value of a is associated with a high degree of disorder, whereas $a = 1$ corresponds to perfect order. The field-dependent drift mobility was found to be $\mu_0 \approx 8 \times 10^{-4}$ cm²/Vs, in agreement with literature values for PVK : TNF composites [34].

As discussed in Section IV.D, the frequency-dependent ellipsometric measurements yield information about the rotational mobility of the EO chromophores. From the experimental data in Fig. 15 [23], we can estimate $\Omega^0 \approx 2$ Hz, corresponding to response times of about 500 ms. This agrees perfectly with the grating buildup time found for the identical material if the data are fitted by an arbitrary exponential function. This result indicates that the holographic response time found in these materials at high electric fields is mostly determined by the orientational mobility of the EO chromophores.

Because of the high dispersity of all the processes involved in the grating buildup, it is physically not meaningful to use (multi-)exponential fits to determine the

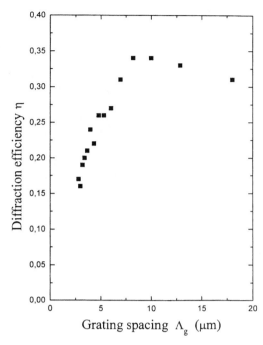

FIGURE 19 Degenerate FWM experiments performed on 105-μm-thick devices of DMNPAA : PVK : ECZ : TNF 50 : 33 : 16 : 1 wt% with *s*-polarized writing beams ($I_1 \approx I_2 \approx 1$ W/cm^2) and *p*-polarized reading beam ($I_3 \approx 0.35$ mW/cm^2), $\lambda = 675$ nm, $\psi = 60°$. Grating spacing dependence of the steady-state diffraction efficiency η_p at an external field of $E = 50$ V/μm.

response time of the materials. Thus, we define the response time as the time which is needed to reach 90% of the final steady-state value. It can be adjusted between 100 ms and several minutes, depending on the applied electric field and the glass-transition temperature of the composite. The storage time of the gratings defined as the time within which the diffraction efficiency relaxes to 10% of its original value varies from hours to days. Note, however, that the grating is readily erased by the uniform probe beam during hologram readout (compare Fig. 10, last step).

Improvement of the Device Stability. The results described so far were obtained with a composite where two-thirds are low-molecular-weight components. At the same time, the glass transition temperature is low, because molecular mobility is needed in order to take advantage of the orientational enhancement mechanism. Such heavily doped multicomponent PR materials were found metastable, tending to phase-separate over time under loss of the optical transparency. This is clearly seen in a typical DSC scan (differential scanning calorimetry) of the above-mentioned composite (Fig. 21). One observes the glass transition at 17°C, an exothermic process at about 80°C, and an endothermic process at 150°C. The energy flux involved in the two latter processes is similar. One can, therefore, conclude that the two peaks are related to each other. If the temperature scan is reversed at 120°C, the exothermic peak is absent during the following heating scan (not

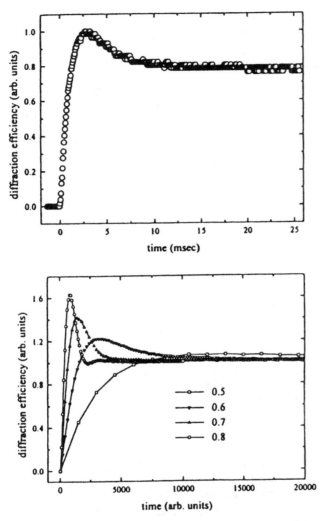

FIGURE 20   *Top*: HTOF experiment on 100-μm-thick devices PVK : TNF : HONB 59.5 : 0.5 : 40 wt%. Transient diffraction efficiency of a grating written at 532 nm with 10-ns pulses from a frequency-doubled Nd : YAG laser (5 mJ/cm^2) and read out by a Bragg-matched *p*-polarized HeNe Laser ($\lambda = 633$ nm) at $E = 55$ V/μm and $2\theta = 30°$, $\psi = 45°$. *Bottom*: Simulated HTOF traces for various values of the disorder parameter *a* for $\Lambda_g = 200$ μm. (From Ref. 32.)

shown). It only reappears if the sample is heated above approximately 150°C and then quenched. If the heating is performed on a hot plate, one can observe that the originally transparent material turns opaque at intermediate temperatures (~100°C) and becomes clear again at elevated temperatures (> 150°C). Therefore, the exothermic and endothermic processes can be identified to be the crystallization and the melting of the 2,5DMNPAA, respectively (melting of the pure 2,5DMNPAA occurs at 165°C, it is expected to be lower in the presence of other compounds).

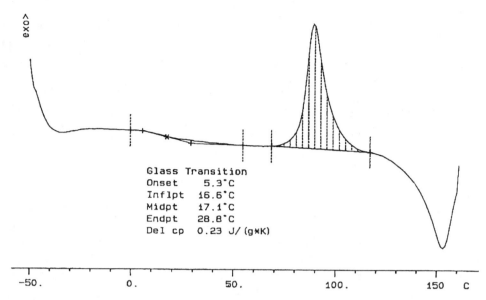

FIGURE 21 DSC trace obtained from the composite DMNPAA : PVK : ECZ : TNF 50 : 33 : 16 : 1 wt% recorded at a speed of 20 K/min.

The occurrence of the two peaks proves that the optically transparent material is thermodynamically unstable at room temperature and will phase-separate over time, depending on the molecular mobility in the matrix and the number of crystallization centers upon preparation.

There are several possibilities to improve the shelf life of the PR polymers. The best choice in this respect is the use of monolithic materials which will be discussed in Section V.B. Other possibilities include the reduction of the chromophore and plasticizer content, the use of chromophores which are highly compatible with the matrix polymer [35,36], or the use of a polymer matrix with an a priori low-T_g matrix polymer [31] to get rid of the additional plasticizer, such as ECZ. In all these cases, the PR performance is somewhat reduced, but devices with acceptable long-term stability (< 12 months) are obtained. Another unique way to improve the shelf life of PR polymer devices is to use composites with mixtures of two isomeric EO chromophores [37]. Because the PVK-based composite using 2,5DMNPAA as the EO chromophore has proven to exhibit excellent PR performance (see above [9]), the isomer with 3,5-dimethyl substitution (see Fig. 11b) was chosen. Mixtures of 2,5DMNPAA and 3,5DMNPAA with varying molar fractions x of 3,5DMNPAA [$x = m_{3.5}/(m_{2.5} + m_{3.5})$, m_i are the weights of each isomer] were prepared and the phase-transition temperatures were measured by DSC. The data indicate the existence of a eutectic mixture with a melting temperature of 110°C for $x = 0.72$ (Fig. 22, top). For the eutectic composition, the melting point is the lowest for any binary mixture of the two DMNPAA isomers, indicating the weakest possible interactions between the dye molecules. Therefore, the eutectic composition is expected to be the best to prevent crystallization in PR devices.

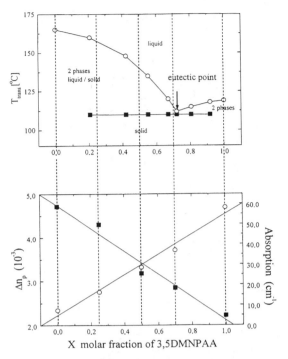

FIGURE 22 *Top*: Phase diagram for binary mixtures of 2,5DMNPAA and 3,5DMNPAA. *Bottom*: Absorption coefficient α (open circles) and index modulation amplitude Δn_p (filled squares) at an electric field of $E = 86$ V/μm calculated from the FWM data in Fig. 23 using Eq. (22) as a function of the molar fraction (x) of 3,5DMNPAA. The lines are guides for the eye.

Then, composites of varying molar fractions x but constant total content of EO chromophore (40 wt%), PVK (45 wt%), ECZ (14 wt%), and TNF (1 wt%) were prepared. Figure 23 shows the results of the FWM experiments performed with 105-μm-thick samples. In all cases except $x = 1$, a maximum is observed for the field-dependent diffraction efficiency. The maximum achievable diffraction efficiency η_{max} is found to decrease with increasing 3,5DMNPAA content x. η_{max} is limited by the absorption and reflection losses as was demonstrated by simultaneously recording the transmitted light intensity (not shown for clarity reasons). The decrease of η_{max} can be thus explained by the increased absorption of 3,5DMNPAA compared with 2,5DMNPAA. As expected, a linear dependence of the absorbance α on the molar fraction was found (Fig. 22, bottom). The FWM data also show that the field necessary to reach η_{max} shifts to higher values with increasing x, indicating that the PR figure-of-merit decreases. The index modulation amplitude Δn at $E = 90$ V/μm was calculated for each mixture from the FWM data using Eq. (22), again yielding a linear relationship between Δn and x (Fig. 22, bottom).

A set of unsealed samples was kept for over a year under random laboratory conditions; that is, random temperature (0°C $< T <$ 45°C), humidity (0–100%) and

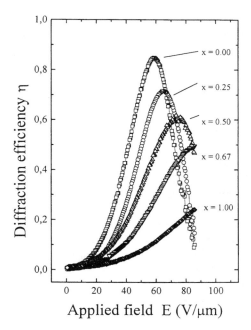

Applied field E (V/μm)

FIGURE 23 FWM experiments performed on 105-μm thick devices of DMNPAA : PVK : ECZ : TNF 40 : 40 : 19 : 1 wt% with s-polarized writing ($I_1 \approx I_2 \approx 1$ W/cm^2, $\lambda = 633$ nm) and p-polarized reading beams ($I_3 \approx 0.35$ nW/cm^2, $\lambda = 675$ nm), $2\theta = 22°$, $\psi = 60°$. Field dependence of the diffraction efficiency for varying molar fractions x of 3,5DMNPAA.

illumination (dark, extreme sunlight). As expected, the sample using the eutectic mixture showed best device stability. It was the only one which had remained optically clear even after that time; all the other had lost at least some of their transparency, and the device using pure 3,5DMNPAA was completely opaque. In order to study the long-term stability of the devices in more detail, they were heated to 60°C to accelerate the phase-separation process. The experiment indicated that the time in which the transmission of devices decreases to some constant value (e.g., 50%) of the initial value strongly depended on the molar fraction x. It was best for samples of the eutectic mix, whereas samples with the pure isomers or any other binary mixture prepared under identical conditions lost their transparency within much shorter times.

2. Other Polymeric Composite Materials

Whereas a lot of PR polymer research has been done on materials based on a photoconducting polymer (e.g., PVK) doped with a low-M_w EO chromophore, very few have reported on the "reverse" material class (i.e., EO polymers as host for low-M_w photoconductors). The first PR polymer composite reported on [4] belonged to this second class of composite materials, consisting of the NLO-active polymer bis-phenol-A-diglycidylether with the covalently bound chromophore NPDA, and DEH was doped into the polymer to obtain photoconductivity. NPDA also acted as the photosensitizer. A similar system also doped with DEH (29 wt%) was reported recently, where the NPDA was substituted by the stilbene dye NAS

[38], which showed improved performance compared with the first PR polymer. (For the chemical structures and explanation of the abbreviations, see Fig. 24.) Diffraction efficiencies of $\eta = 12\%$ (at 35 V/μm) and a gain of $\Gamma = 56$ cm^{-1} (at 65 V/μm) were reported for $\lambda = 650$ nm. However, due to the strong absorption at the operating wavelength ($\alpha \approx 200$ cm^{-1}), no net gain was observed [38]. Similarly, in all cases known to date, the performance of the materials based on NLO polymers is reduced compared with the PVK-based systems. A plausible explanation is the restricted molecular mobility of the EO chromophores when attached to a polymer backbone.

A third class of composite materials is based on inert polymers (Fig. 25). The problem of phase separation can be amplified in this case because the incorporation of high amounts of charge transport agent limits the amount of EO chromophores and vice versa. In order to circumvent this problem, dual-function dopants (Figs. 24c and 24d) have been proposed [39,40]. PMMA was used as an inert binder polymer in both cases. Again, the PR performance levels of PVK-based materials were out of reach. Diffraction efficiencies of 1% and less were only obtained. Silence et al. [39] performed a study using different inert binder polymers together with dual-function dopants. The effect of the binder can be understood qualitatively as being related to the effect that the local polarity of the binder has on the charge

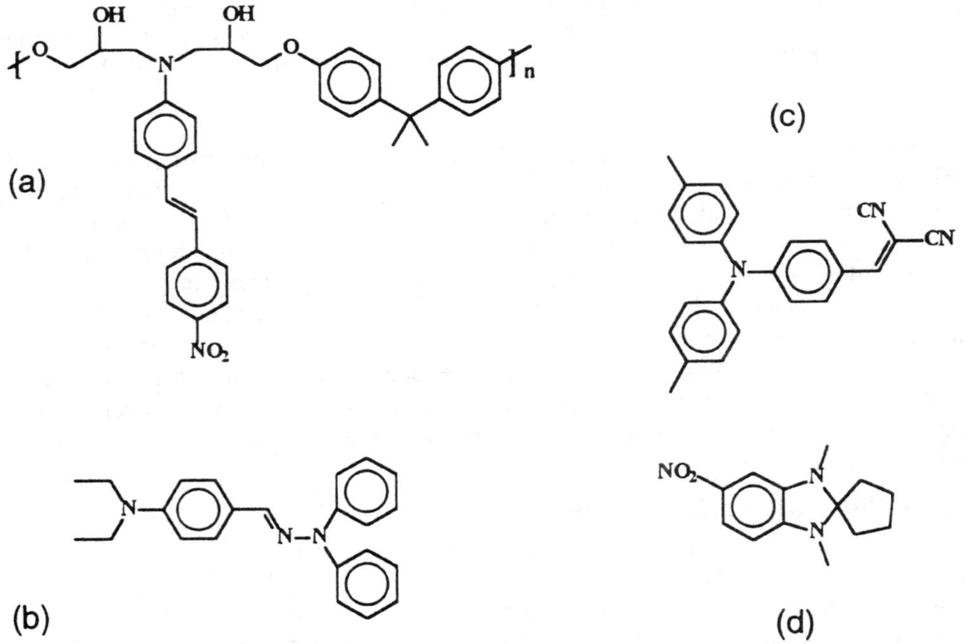

FIGURE 24 Chemical structures: (a) bisphenol A 4.4′-nitroaminostilbene (BisA-NAS), (b) 4-(*N,N*-diethylamino)benzaldehyde diphenylhydrazone (DEH), (c) 4-(*N,N*-di(*p*-tolyl)amino)-$\beta\beta$-dicyanostyrene (DTADCST), (d) 1,3-dimethyl-2.2-tetramethylen-5-nitrobenzimidazoline (DTNBI).

transport mobility. The principal drawback of this material class are their slow response times (1–100 s) presumably due to the reduced carrier mobility.

A class of inorganic/organic hybrid materials are the so-called "sol–gel glasses" (ormosils) combining the excellent optical qualities of inorganic glasses with the wide synthetic flexibility of organic compounds. Furthermore, the chances for permanent poling of the EO chromophores are expected to be much better than in amorphous polymeric materials with a high glass transition temperature. Photorefractive sol-gel glasses have been proposed by Burzynski et al. [41] and by Chaput et al. [42]. The performance of these materials is limited at this stage, but the approach looks promising.

B. Monolithic Systems

The most efficient polymeric PR materials are composite materials based on PVK. The high concentration of the EO chromophore required to produce large refractive-index modulations results in metastable systems in which the chromophores tend to phase-separate over time, a process that seriously compromises the optical quality and thus the device stability and durability. A majority of composite materials share this shortcoming: The PR performance generally increases with increasing chromophore content, reaching an upper limit determined by the compatibility of the chromophore and/or photoconductor and the host polymer. A number of ways to improve on the shelf life of composite devices was described earlier. However, as was pointed out, highly doped composites remain metastable. This can be overcome by using monolithic materials.

The fully functionalized polymer approach has the evident advantage of long-term stability and minimized phase separation. But the time-consuming chemical synthesis and the limited design possibilities are inherent disadvantages for this approach. Another rather new approach to monolithic photorefractive materials is the use of glass-forming compounds and liquid crystals.

1. Fully Functionalized Polymers

A number of fully functionalized PR polymers (i.e., with photosensitive, charge-transporting and electrooptic moieties either in the polymeric main chain or as pending side groups) have been synthesized over the last couple of years in an effort to minimize the phase-separation problem related to highly doped composites. The first example of an electrooptically active photoconducting polymer was given by Tamura et al. [43]. It consisted of poly(methyl methacrylate) with pending carbazole units. Some of them were additionally substituted by dicyanovinyl groups in order to introduce EO chromophores and photosensitivity. (Fig. 26, **P1**). Photorefractivity was demonstrated by field-dependent four-wave mixing experiments [44], but the low performance of the material did not enable two-beam coupling experiments.

Later, work in this field was mostly put forward by Yu et al. [45]. For example, the polymer **P3** contained independent EO, photoconductive, and photosensitive moieties linked to a saturated backbone. **P2** was the first example of a PR polymer using conductive polymers as the backbone. However, the PR performance of polymers **P2** and **P3** was limited because of their high absorbance of the materials due to high sensitizer contents (e.g., 10% in **P2**, Fig. 26). Recently, a very prom-

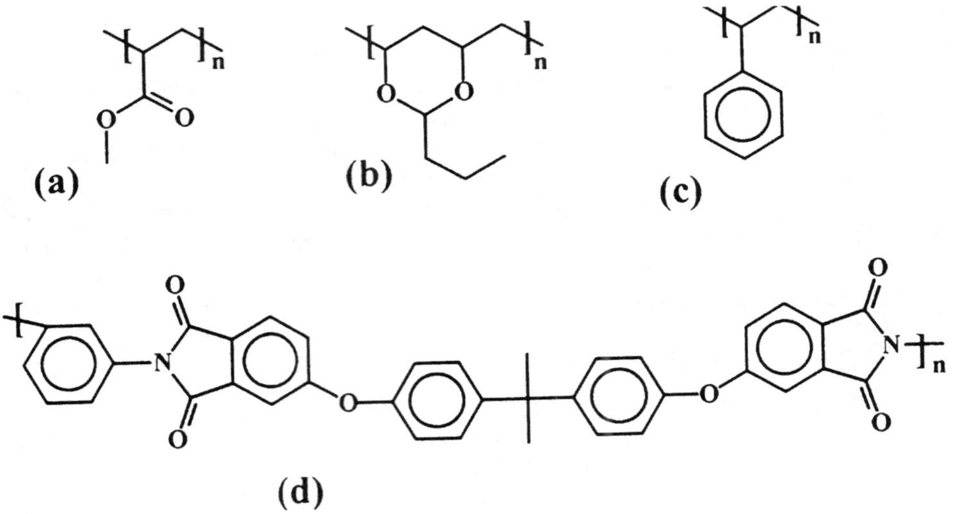

FIGURE 25 Inert polymeric binders: (a)poly(methylmethacrylate) (PMMA), (b) polyvinylbutyral (PVB), (c) polystyrene (PS), (d) polyimide Ultem.

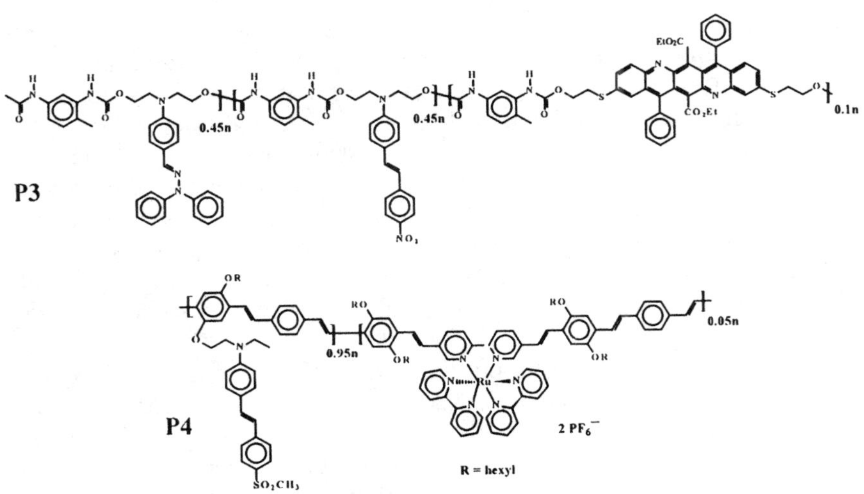

FIGURE 26 Examples of fully functionalized polymers: **P1** [43], **P2** [45a], **P3** [45b,45c], **P4** [45d].

ising polymer, **P4**, showing net optical gain without applied electric field, was reported [45d]. Good photoconductivity is again provided by the conjugated polymer backbone. Efficient charge generation occurs through electron transfer from the Ru complexes onto the backbone. Finally, the EO properties are provided by a stilbene dye. The polymer was poled at its glass transition temperature and then cooled to room temperature; 90% of the original orientation was retained after 1 month as checked by second-harmonic-generation measurements. The response of the material was initially very slow; however, after the initial grating was written, erasure and rewriting of the grating were faster. This phenomenon is commonly known for other PR polymeric materials, and is referred to as "optical trap activation" [46]. Optical gain coefficients as large as 300 cm^{-1} without an applied electric field were observed, clearly exceeding the absorption of the materials ($\alpha \approx 100^{-1}$). As expected for diffusion-limited charge transport, the phase shift was found to be 90°. Surprisingly, the promising photorefractive properties of these materials have never been confirmed by four-wave mixing experiments.

2. Glass-Forming Low-Molecular-Weight Compounds

An important aspect of this material design strategy is the multifunctionality of the chromophore. Dual-function dopants exhibiting photoconductive and EO properties at the same time have been studied in composite materials [39,40]. However, with appropriate chemical substitution, such compounds may form glasses in the pure form, making the use of a binder polymer obsolete.

One approach to glass-forming low-molecular-weight compounds proposed by Wang et al. is based on oligomers of carbazole which has proven to exhibit excellent photoconducting properties. Recently, carbazole starbust dendrimers, cyclic octamers, and conjugated trimers were reported [47] (see Fig. 27a). The appearance of the compounds is glassy at room temperature. The materials were photosensitized by adding TNF. In holographic experiments performed at 633 nm on 130-μm-thick samples of the trimer, gain coefficients Γ_p of up to 80 cm^{-1} and diffraction efficiencies η_p of up to 20% at 30–40 V/μm were observed. The absorption of the materials varied between 4 and 13.5 cm^{-1}, depending on the TNF content (i.e., extremely high net gain was observed considering the low fields that were applied). The phase shift between light-intensity pattern and index grating was found to be approximately $\Theta = 30°$ for low fields ($E < 15$ V/μm), whereas it approached $\Theta = 90°$ for $E > 20$ V/μm. The latter is certainly the origin for the excellent performance of this material.

The second approach was to use glass-forming chromophores with negligible nonlinearity ($\beta \approx 0$) and high dipole moments μ. This approach takes full advantage of the orientational enhancement mechanism by optimizing the term $\Delta\alpha\mu^2$ in the PR figure-of-merit F_{ltg} for organic materials with low T_g [Eq. (19)]. The first examples for this category belonged to the family of 2,6-dialkyl-4H-pyranylidenemolonodinitrile (Fig. 27b [48]); later, a closely related series of compounds was proposed, the *N*-alkyl-2,6-dimethyl-4H-pyridone-4-ylidenecyanoalkyl-acetates (Fig. 27c [49]). The latter materials show a high dynamic range in FWM experiments of $\Delta n \approx 2 \times 10^{-3}$ (10^{-2} at $E = 40$ (90) V/μm–1.5 times the value of the best material reported earlier [8]—and a large net gain of $\Gamma_{\text{net}} \approx 65$ cm^{-1} at $E = 40$ V/μm. However, a major drawback of these materials are the very slow response times compared with most of the other PR polymers reported to date, most probably due to the poor photoconductivity. A significant increase in the hologram growth rates can be achieved by doping the chromophore with a small amount of polymer, such as PMMA; nevertheless, the response times remain slow ($\tau = 83$ s at $E = 40$ V/μm [49]).

3. Liquid Crystals

The observation of photorefractivity in liquid crystals was first reported by Rudenko [50a] and by Khoo et al. [50b]. It was accomplished by doping 4-(n-pentyl)-4-cyanobiphenyl (5CB, Fig. 27d) and similar nematic liquid crystals with Rhodamine 6G and other laser dyes for photosensitivity. The thickness of the samples was $d = 100$ μm, the sample tilt $\psi = 0.5$ radian ($\approx 30°$), and $2\theta = 1.8 \times 10^{-3}$ ($\approx 0.1°$), yielding grating spacings of $\Lambda_g = 278$ μm. Experiments were performed at $\lambda = 514$ nm. Photorefractivity with a gain of $\Gamma \approx 25$ cm^{-1} was observed for applied fields of only 1.5 V. The absorption loss was negligible; however, the scattering losses were about 20 cm^{-1}, leaving only a small net gain in these materials. Under prolonged illumination of the dye-containing liquid-crystal devices, strong permanent gratings evolved. They were explained due to perturbations of the surface director axis alignment by the current flux under prolonged application of the DC electric field [50c, 50d].

Another liquid-crystal system consisting of 5CB and 4-(n-octyloxy)-4-cyanobiphenyl (8OCB, Fig. 27d) sensitized with perylene and N,N'-di(n-octyl)-1,4,5,8-naphthalenetetra-carboxydiimide (DI) was studied by Wiederrecht et al. [51]. An improved performance compared with the laser-dye sensitized devices was obtained. For 37-μm-thick samples and applied voltages up to 2.5 V, response times as fast as 40 ms were reported for a grating spacing $\Lambda_g \approx 2$ μm. However, the diffraction efficiencies for these small values of the grating spacing are very small. For larger Λ_g, the response time increases dramatically, giving 14 s for $\Lambda_g = 57$ μm. Under the latter conditions, a gain coefficient or $\Gamma_p = 640$ cm^{-1} was calculated; the absorption was $\alpha = 0.5$ cm^{-1}. The gain decreased considerably with the grating spacing. Further improvement of this system was achieved by varying the sensitizer [51b].

VI. SPECIAL EFFECTS

A. Dynamic Non-Bragg Diffraction

In experiments performed on thick holographic gratings, in most cases only the Bragg order is considered. Diffraction into higher orders on thick gratings has been addressed in PR crystals and liquid crystals [52], but the correlation between energy exchange in dynamic self-diffraction and the non-Bragg orders has not been described adequately. Due to the very large refractive-index modulation amplitudes achievable in highly efficient organic PR materials ($\Delta n \sim 10^{-2}$), the picture complicates due to the appearance of non-Bragg (i.e., higher) orders even in diffraction on thick holograms. It should be emphasized that this phenomenon is different from Raman–Nath diffraction on thin gratings which also gives rise to multiple diffraction orders.

The material used to study this effect was the composite 2,5DMNPAA : PVK : ECZ : TNF in the proporation 50 : 39 : 10 : 1 wt% [53,54]. The experiment was performed using *s*-polarized beams (marked $+0$ and -0 in Fig. 28) with a diameter of 450 μm originating from a HeNe laser (633 nm, 4 mW). The geometry used was $\psi = 60°$ and $2\theta = 9°$ in air. Inside the sample, the grating period was $\Lambda_g = 7$ μm. The powers of the transmitted writing beams as well as the powers of the two first-order non-Bragg diffracted beams ($+1$ and -1 in Fig. 28) were monitored by photodiodes. The angles of diffraction for the non-Bragg orders $+1$ and -1 were measured to by $\theta_{+1}^{ext} = 11.6° \pm 0.5°$ and $\theta\;_{-1}^{ext} = -17.7° \pm 0.5°$ in air with respect to the bisector of the two writing beams, which is in excellent agreement with theoretical expectations taking into account a bulk refractive index of $n = 1.75$ measured by ellipsometry.

Figure 29 (top) shows the asymmetric energy transfer between the $+0$ and -0 beams as a function of the applied field. Similarly, Fig. 29 (bottom) shows the field dependence of the diffraction efficiency for the $+1$ and -1 non-Bragg orders defined as $\eta_{\pm 1} \equiv I_{\pm 1}/[I_{+0} + I_{-0}]$ for both field polarities. The diffraction efficiency $\eta_{\pm 1}$ increases with the applied field; however, $\eta_{+1} \neq \eta_{-1}$. Furthermore, the non-Bragg diffraction efficiency is distinctively asymmetric with respect to the polarity of the field. From the experimental results, it can be concluded that $\eta_{+1}(+E_0) \approx \eta_{-1}(-E_0)$ and vice versa—the small difference is due to the asymmetry associated with the sample tilt. The polarization of the ± 1 beams was found to be the same as of the ± 0 beams (i.e., *s* in this case). The ± 2 diffraction orders can also be observed under certain conditions (very small intersection angles 2θ, large grating spacings Λ_g). However, they are much weaker than the ± 1 orders and were, therefore, ignored in the analysis. For *p*-polarized writing beams, the diffraction

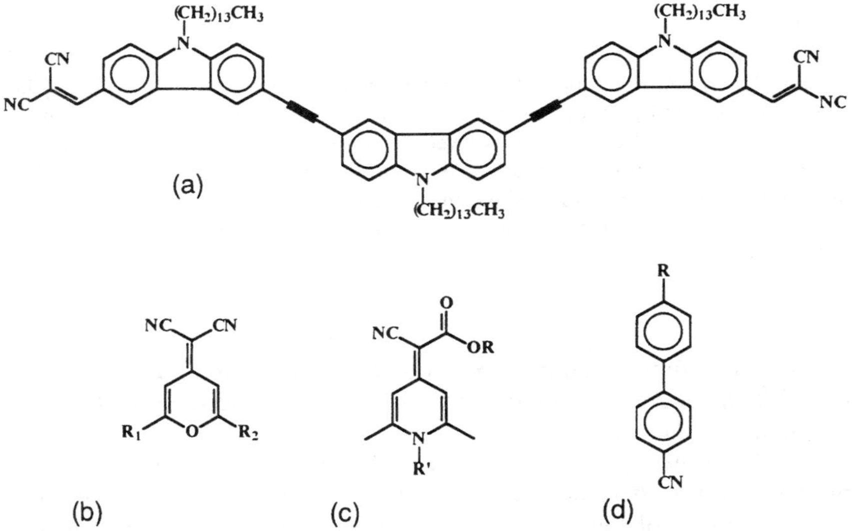

FIGURE 27 Glass-forming low-M_w compounds and liquid crystals used for photorefractive devices: (a) carbazole trimer [47], (b) 2,6-dialkyl-4H-pyranylidenemalono-dinitriles [48], (c) *N*-alkyl-2,6-dimethyl-4H-pyridone-4-ylidenecyano-alkylacetates [49], (d) R = C_5H_{11}: 4-(*n*-pentyl)-4-cyanobiphenyl (5CB); R = OC_8H_{17}: 4-(*n*-octyloxy)-4-cyanobiphenyl (8OCB).

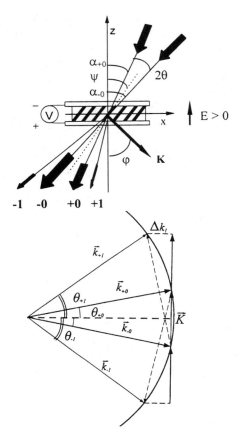

FIGURE 28 *Top*: Experimental geometry for the investigation of non-Bragg diffraction. The beams ± 0 are the transmitted writing beams and the beams ± 1 are the non-Bragg (higher order) diffracted beams. K is the grating wave vector. *Bottom*: Ewald sphere showing the zero- and the first-order diffraction processes on the grating created by the ± 0 beams (with grating wave vector K). The dashed lines show the weaker gratings formed by interference of the ± 1 and the ± 0 beams. These gratings were ignored in the theoretical calculations (see text and Ref. 54).

into the ± 1 orders has similar features, except that it is larger and reaches 10% at $E > 90$ V/μm. When the diffraction into the non-Bragg orders becomes that strong, it cannot be ignored in data analysis.

A model based on the standard PR model of the dynamic self-diffraction in a PR medium [10] has been developed to describe this effect and an analytical solution for the amplitudes of the first non-Bragg orders has been obtained in the approximation of undepleted pump beams. The analytical solution obtained in this approximation describes the main features of the experimental observations quite well. The following scaling rules were obtained for the complex amplitudes $\mathcal{E}_{\pm 1}$ and the diffraction efficiency $\eta_{\pm 1}$:

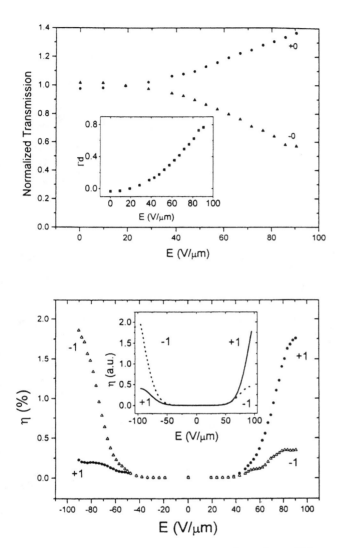

FIGURE 29 *Top*: TBC experiment performed on 105-μm-thick devices of DMNPAA : PVK : ECZ : TNF 50 : 39 : 10 : 1 wt% with *s*-polarized writing beams, power density $\approx$ 1 W/cm^2, λ = 675 nm, 2θ = 9°, ψ = 60°. Transmitted intensities of the ± 0 beams as a function of the applied electric field (positive direction only). The inset shows the gain coefficient calculated according to Eq. (21). *Bottom*: Diffraction efficiency of the $+1$ (filled circles) and the -1 (open triangles) non-Bragg order beams as a function of the applied electric field. The field polarity convention is illustrated in Fig. 28. The inset shows model calculations according to the full theory described in detail in Refs. 53 and 54 [compare Eq. (28)].

$$\mathscr{E}_{\pm 1} \propto \frac{\mathscr{E}_{\pm 0}^2 \mathscr{E}_{\mp 0}^*}{I_0 + I_b}$$

$$\eta_{\pm 1} \equiv \frac{I_{\pm 1}(d)}{I_0} \propto \frac{I_{\pm 0}^2 I_{\mp 0}}{I_0(I_0 + I_b)^2} \tag{28}$$

Here, $I_0 = I_{+0} + I_{-0}$ is the total coherent light intensity, and I_b is the (incoherent) background light intensity. Following Eq. (29), the asymmetry for the diffraction efficiency of the ± 1 orders as a function of the electric field is a result the self-diffraction process during which one of the ± 0 beams is depleted and the other amplified. If, for example, the $+0$ beam is amplified and the -0 beam diminished, then η_{+1} will grow faster with the field than η_{-1} and vice versa as was observed experimentally. Such a behavior will not occur for diffraction on thin gratings in the Raman—Nath regime. Note that for local gratings, the non-Bragg orders may also appear; however, they would be independent of the field polarity.

According to Eq. (28), the two first non-Bragg diffraction orders represent the phase-conjugated (i.e., $\sim \mathscr{E}_{\pm 0}^*$) and the phase-doubled (i.e., $\sim \mathscr{E}_{\pm 0}^2$) replicas of the input waves. To verify these wavefront phase transformations experimentally for the ± 1 orders, an experiment was performed with the two main (writing) beams having different divergences: beam -0 was collimated and beam $+0$ was diverging (Fig. 30). The divergence was introduced by a lens in the beam path. At a certain distance after passing through the sample, a screen was introduced into the beams' path and the image of the ± 0 and ± 1 beams was photographed. The picture clearly shows that the beam -1 is focused and, therefore, is converging, whereas the beam $+1$ diverges approximately two times faster than $+0$. This effect can be used for phase-conjugation applications in a forward configuration.

Following Eq. (29), the incoherent background light intensity I_b can modulate the first non-Bragg order diffraction efficiency η_{+1}. It allows the performance of incoherent-to-coherent image conversion via this process. When the sample was illuminated with a spatially modulated beam (e.g., by placing an image into the beam path) of a separate 675-nm laser diode, the ± 1 order beams carried this image in reverse contrast [54].

B. Trap-Controlled PR Response

In most organic PR materials studied, the nature of the traps and their number are unknown. Due to the hopping mechanism for charge transport in amorphous glasses, basically every charge transporting moiety may act as a shallow trap. However, traps strongly influence the performance of a photorefractive material, in particular the trap-limited saturation field E_q [according to Eqs. (4) and (5)] and the carrier mobility and dark conductivity.

Detailed studies were performed by Malliaras et al. [55] and Zhang et al. [56]. Both groups used PVK-based PR composites and added compounds with a reduced oxidation potential compared to PVK. In both cases, a dramatic increase of the grating growth rate and the decrease of the dark decay was observed when little amounts of traps were added (Fig. 31, top). After a certain number density of traps ($N_t \approx 7 \times 10^{18}$ cm^{-3} [56]) was exceeded, the response and decay times decreased again. This effect has been studied in organic photoconductors previously and is well understood [57]. It can be explained as follows: When there are no deep traps

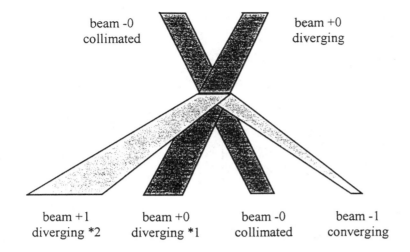

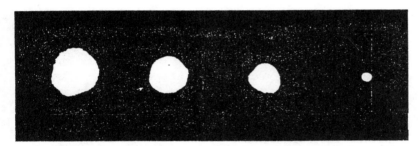

FIGURE 30 (Top) Illustration of the experiment performed to demonstrate the phase-conjugated and the phase-doubling nature of the non-Bragg diffracted beams: The -0 beam is collimated, the $+0$ beam is diverging, the -1 beam is converging (demonstrating the phase-conjugate relationship to the $+0$ beam), and the $+1$ beam is diverging about two times faster than the $+0$ beam (demonstrating phase doubling). (Bottom) Photograph of all beams after passing through the sample. The exposure time was such that all four spots were saturated in intensity regardless of their absolute intensity.

present in the material, charge transport proceeds via the majority charge-transporting units. The addition of a small number of deep traps efficiently localizes the charge carriers, leading to a reduction of the carrier mobility. For higher concentrations of the trapping moieties though, the mean distance between them becomes small enough and transport can take place via the "traps" rather than the original photoconductor (see Fig. 32). The trap density also strongly affects the phase shift Θ between light-intensity pattern and index grating [55]: Starting from $32°$ without traps added, it decreases to almost zero and then increases up to $45°$ for high trap concentrations (Fig. 31, middle). Surprisingly, the index modulation amplitude varies very little when the trap concentration is changed (Fig. 31, bottom).

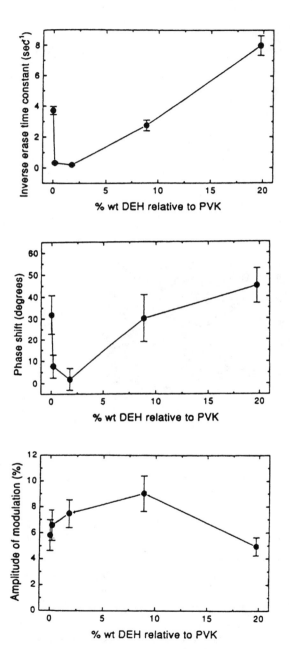

FIGURE 31 Holographic experiments performed on 57-μm-thick devices of PVK : TNF : EPNA with *p*-polarized writing beams ($I = 0.6$ W/cm^2, $\lambda = 633$ nm, $2\theta = 30°$, $\psi = 45°$): inverse erase time for holograms (top), phase shift Θ, (middle) and index modulation amplitude (bottom) as a function of the weight percentage of DEH relative to PVK. The lines are guides to the eye. (From Ref. 55.).

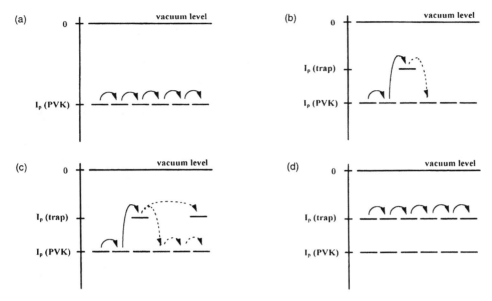

FIGURE 32 Illustration of charge transport in PR polymers with increasing amount of traps: (a) hopping transport via the main photoconductor; (b) trapping in sites with low ionization potential; (c) detrapping; and (d) hopping via the low I_p compound. (From Refs. 6 and 57.)

C. Influence of the Glass Transition Temperature

The orientational enhancement mechanism is the origin of the excellent performance of the organic PR materials. It is thus obvious that the rotational mobility of a chromophore in a viscous matrix will determine the orientational dynamics as well as the degree of orientation in the steady state, two important figures for the performance of an organic PR material. A commonly used measure for the mobility is the glass transition temperature T_g. For $T > T_g$, the rotational mobility is high and the oriented-gas model can be applied; by contrast, for $T < T_g$, the rotational mobility is low and the elastic constants of the matrix have to be taken into account in order to determine the degree of orientation.

There are two ways to study the influence of orientational mobility on the performance of organic PR materials: (1) vary the temperature at which the holographic experiment is performed [58] and (2) vary the glass transition temperature [59]. In both cases, a better performance of the PR material was observed when the reduced temperature $T_r = T - T_g$ was increased. This was explained by higher charge generation efficiency, by higher photoconductivity, and by better and faster alignment of the chromophores.

D. Quasi-Nondestructive Readout

One of the most important issues that need to be addressed in the content of holographic storage is the erasure of holograms during readout due to the uniform generation of charge carriers which smears out the grating. In inorganic crystals, so-called "fixing schemes" have been developed and employed; however, once a hologram is fixed, several processing steps are needed to erase the information (e.g., heating) and the optical quality is degraded. In organic PR polymers, researchers have tried to overcome this inherent problem, but with limited success so far. Silence et al. [60] reported on "quasi-nondestructive readout" of holograms in a particular PR polymer composite consisting of the bifunctional molecule DTNBI (33 wt%, see Fig. 24d) doped into PMMA and sensitized with C_{60} (0.2 wt%). For light intensities higher than 2×10^{-5} W/cm^2, biexponential decay ($\tau_1 \approx 300$ s, $\tau_2 \approx 6000$ s) of the gratings was observed. By contrast for lower intensities, the diffraction efficiency was found to decrease to ~35% of its original value and even increases slightly over time. The authors used a model which assumes two trap levels: a shallow one which can be thermally deactivated and a deep trap level.

VII. APPLICATIONS

The search for holographic storage media, erasable as well as write-once read-many (WORM), has been an active field for more than 30 years since the discovery of photorefractivity in LiNbO$_3$ and has mostly concentrated on inorganic crystals as the storage medium. Since the discovery of PR polymers and the recent mayor advances achieved in this field [5–7], such materials have also been seriously considered as erasable holographic storage media. The potential of the organic materials was demonstrated by recording a gray-scale hologram of a U.S. penny in a photorefractive polymer [61,62]. Likewise, organic PR materials can have a variety of other useful applications, which can be subdivided into two main areas, real-time holographic optical processing , and holographic data storage [49,63]. These examples will be presented in more detail below.

A. Choice of the Material

An important aspect in the design of a well-performing PR device is the choice of the right material for a particular application. Since the discovery of PR polymers in 1991, it became a general rule that materials with fast response times (100 ms in PR polymer composites [26–28], 40 ms in liquid crystals [51]) generally exhibit rather short storage times, whereas materials showing long storage times have a very slow response (several tens of seconds) [55,56,60]. This is due to the fact that in the materials investigated so far, low dark conductivity (required to keep the charges in the places where they were originally stored) resulted in low photoconductivity and vice versa. In the following, we will generalize about the requirements for an organic amorphous PR material in terms of the grating buildup, the steady-state performance, the grating decay, and the erasure.

The *grating buildup* speed (response time) of a PR material limits its use in PR devices for real-time optical processing. It is determined by the charge photo-generation efficiency and the photoconductivity of the material. Furthermore, in low-T_g materials, the reorientation of the chromophores takes place every time a new hologram is written, introducing an additional limiting factor for the grating buildup. The rotational mobility of the chromophores in the matrix is high if the reduced temperature $T_r = (T - T_g)$ is high and decreases dramatically for $T_r < 0$.

Apparently, in the fastest materials reported so far, the response is determined by the rotational mobility of the chromophores in the matrix, thus response times much faster than 40 ms (the limit observed for the switch times in nematic liquid-crystal cells) cannot be expected. At low electric fields, charge transport becomes the time-limiting process.

The *steady-state performance* of a PR composite, in particular the dynamic range Δn and the gain coefficient Γ, are mostly determined by the strength of the internal space-charge field and the nature of the chromophore, its concentration, the degree of orientation, and the rotational mobility of the chromophores. To maximize the macroscopic PR performance of the material, it should withstand high external fields to allow the buildup of strong space-charge fields. Generally speaking, the strength of an organic material toward dielectric breakdown increases with increasing T_g. The EO chromophore should possess a large PR figure-of-merit [Eq. (19)] [i.e., large dipole moment μ (for efficient poling), a large anisotropy of the linear polarizability $\Delta\alpha$, a large first hyperpolarizability β, and a relatively low molecular weight M]. In addition to that, a low absorption at the operating wavelength is beneficial in order not to inherently limit the maximum achievable diffraction efficiency and the net gain. The only compound absorbing at the operating wavelength should be the sensitizer. The large μ and β of an EO chromophore can be achieved by choosing strong acceptors (e.g., nitro or tricyanovinyl groups) and donors (e.g., dialkylamino group). However, the use of chromophores with extremely large first hyperpolarizabilities is most often contradictory with a low absorption in the visible or near-infrared part of the electromagnetic spectrum, because an increased optical nonlinearity of a molecule usually goes along with a red-shift of the wavelength of maximum absorption. The concentration of the EO chromophores should be as high as possible. The degree of orientation depends on the absolute temperature T [oriented-gas model, higher T reduces the degree of orientation, see Eqs. (14) and (15)]. However, for $T_r < 0$, deviations from the oriented-gas model taking into account the repulsive elastic forces of the surrounding matrix for the chromophores have to be considered. Finally, to take full advantage of the orientational enhancement mechanism, the T_g of the PR polymer should be below the operating temperature ($T_r > 0$). T_g (T_r) can be adjusted within a certain range by adding plasticizers. As a compromise, $T_r \approx 1$ is chosen in order to take advantage of the orientational enhancement mechanism, to have reasonably fast orientation times, and to be able to reach high enough electric field strengths.

The lifetime (*storage time*) of a grating written in a PR material mostly depends on the ability of the material to keep the charge carriers in the locations where they were originally trapped during grating formation. The most important processes leading to the gradual decay of the grating are the thermally activated diffusion leading to the recombination of charge carriers and the chemical decomposition of the involved radical cations and anions. The first process is determined by the dark conductivity of the material at a given electric field strength and temperature; the latter is determined by the chemical stability of the radicals in the matrix (i.e., strongly depends on the presence of acidic impurities such as water).

A hologram written in a PR polymer can be *erased* by illuminating the material uniformly. Therefore, the readout of a grating erases it, and low light levels have to be used in order avoid this (see Section VI.D, [60]). The erasure time depends on the photoconductivity of the material for the redistribution of the charge carriers and the rotational mobility of the chromophores in materials where the orientational enhancement mechanism is active. In most materials known to date, dark conductivity and photoconductivity are related to each other; that is, when the photoconductivity is increased, the dark conductivity is increased as well. As a result, materials showing long storage time exhibit slow response times due to their poor photoconductivity.

The previous remarks clearly demonstrate that there is not one single amorphous organic PR material for real-time *and* storage applications. The material requirements for the two areas are quite different; thus, the material for a particular application has to be chosen very carefully. For real-time applications [e.g., optical correlation or interferometry (see below)], fast response times are preferable for achieving high processing rates. Only one hologram is present at a time; therefore, the signal-to-noise ratio is not crucial because the entire dynamic range of a material can be used. By contrast, for storage applications, a large number of holograms will be stored in the same volume of the photorefractive material (e.g., by angular or wavelength multiplexing). Thus, the dynamic range has to be shared between the different holograms, making the need for good signal-to-noise ratio much stronger. As a result, the linear optical properties of the materials, in particular optical clarity, uniformity, and scattering, become at least as important as the nonlinear ones for storage applications. Whereas in typical FWM experiments as described in Section V, the diffraction efficiencies are large and background scattering effects are barely noticeable, in practice the diffraction efficiencies that are of interest are only on the order of a fraction of a percent, and under these conditions even slight scattering becomes detrimental.

For real-time applications, PR composites with low glass transition temperatures near room temperature ($T_r \approx 1$) and good photoconductivity seem the materials of choice, offering large diffraction efficiencies, reasonably fast response times (100 ms), and high spatial resolution (grating period $\Lambda_g \approx 1$ μm). The use of such high-performance PR polymers in an optical pattern recognition system for security verification has been demonstrated recently (see Section VII.B.2). The price to pay is the necessity for high voltages (several thousands of volts); about 2000–3000 V are necessary to achieve 10% diffraction efficiency in 100–200-μm-thick layers of the best-performing organic PR materials. This is more than enough for typical charge-coupled-device (CCD) detection. Note, however, that the current flux ($I < 10^{-6}$ A) and, as a result, the overall power consumption are low. An alternative class of materials for real-time applications are liquid crystals (LCs), in particular because of the low operating voltage. However, LC devices driven by a DC voltage suffer from electrochemically induced damage [50c]. Furthermore, it is a disadvantage that photorefractivity is strong for large grating spacings Λ_g (several hundreds of μm), but much less for small spacings. For $\Lambda_g \approx 2$–5 μm (typical values used with the PR composites), the dynamic range is reduced compared with the composite materials [50,51]. As a result for high-resolution optical computing, the PR composite materials are preferable. On the other hand, the PR polymer PMMA : DTNBI : C_{60} seems particularly interesting for storage applications, because extremely long dark lifetimes and quasi-nondestructive readout have been observed with this material [60]. Multiplexing of 10 holograms has been demonstrated in this material. Also, the glass-forming chromophores (Section V.B.2) seem suitable for storage applications due to their large dynamic range and their excellent optical quality (low scattering [49]). However, in both cases, very slow response times were anticipated, preventing direct use of such materials in actual storage devices at this time. Materials with low dark conductivity and high photoconductivity would be the materials of choice.

B. Real-Time Applications

1. Dynamic Holographic Interferometry

Dynamic holographic interferometry (DHI) has been proven to be an important technique for nondestructive testing of vibrating objects. DHI produces interferograms showing the vibrational mode patterns of objects of arbitrary shape and size (e.g., a car, turbine blade, etc.) in operation. Such interferograms allow the quantitative evaluation of the vibrational frequencies and the amplitudes. In DHI, a plane-wave reference beam and a signal beam which carries information about the vibrating object under test interfere in the holographic medium. Through four-wave mixing, an image of the vibrating object under test can be reconstructed by diffraction of a reading beam from the dynamically recorded hologram. When the recording time of the hologram is longer than the vibration period, a time-average image is obtained in which brighter regions correspond to the nodes of the vibrating object (parts that are not moving), and the darker regions correspond to the moving parts. The number of these fringes is proportional to the vibrational amplitude.

A key component of such a system is the storage medium, which ideally should have high sensitivity and can be used many times without performance reduction. DHI is traditionally performed by a double-shot technique using silver halide plates or thermoplastic polymers. Due to the fact that these materials need processing, real-time monitoring is impossible. Thermoplastic polymers can be recycled approximately 10–15 times via heating of the material before they have to be completely replaced. PR polymers offer a great advantage in this respect by providing reversible recording, so the testing can be performed in real time. Furthermore, sensitivity and response time can be easily adjusted by the external applied field. As a proof-of-principle example, the mode patterns of a circular membrane excited by a loudspeaker at different excitation frequencies and amplitudes (Fig. 33) were recorded in the high-performance PR polymer DMNPAA : PVK : ECZ : TNF (50 : 40 : 10 : 1) [61].

2. Optical Correlation

Another very promising application for PR polymers is their use in an all-optical image correlator [64,65]. There are many possible areas in which an optical correlator proves useful, such as target recognition, fingerprint recognition, and other security applications. Due to rapid technological progress, especially in computers, CCD technology, and color printers and scanners, forgery and counterfeit of valuable documents such as credit cards, IDs (identification documents), or other important objects becomes increasingly simple. Current techniques such as embossed holograms on credit cards are no longer a reliable solution to this problem, as they are visible and may thus be easily copied. There is, therefore, a need to develop new inexpensive optical methods for mass security applications in order to handle the counterfeiting problem better. An example is a low-cost security verification

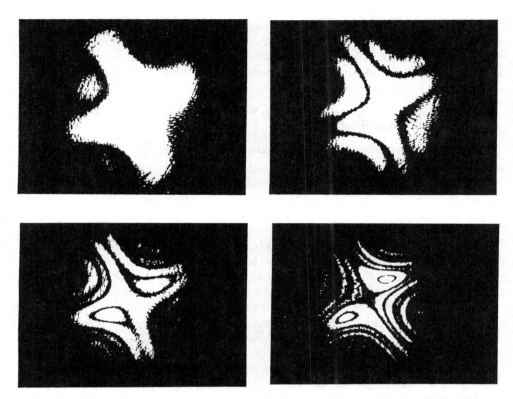

FIGURE 33 Time-average interferograms of the vibrational mode of a circular membrane excited acoustically at 6.15 kHz for different amplitudes of the excitation: (top left) 30, (top right) 60, (bottom left) 120, and (bottom right) 240 mV. Recording was performed with a 633-nm HeNe laser in DMNPAA : PVK : ECZ : TNF 50 : 40 : 10 : 1 wt%; readout was provided by a 675-nm laser diode.

system based on the optical encoding of documents with pseudorandomly generated phase masks, encoding the information in the phase rather than in the amplitude of the light. The inspection of these masks is performed by obtaining the all-optical spatial correlation of two images in a high-performance PR polymer in a four-wave-mixing geometry [65].

Figure 34 shows the experimental setup used. L2 and L3 are the Fourier-transform lenses (focal length $f = 85$ mm). Recording of the hologram was performed in real time with a 633-nm HeNe laser with a total power of 1.5 mW. The writing beams were s-polarized and overlap in the back focal plane of lens L2 inside the photorefractive polymer and form a phase grating. The illuminated area on the sample was ~4 mm in diameter at normal incidence. As in regular FWM experiments, the sample was tilted at $\psi = 60°$. A separate p-polarized 675-nm laser diode with $\ll 1$ mW power was used for the readout. The photorefractive polymer was the composite DMNPAA : PVK : ECZ : TNF with the composition 50 : 37 : 12 : 1 wt%. A field of 50 V/μm was applied during the recording of the holo-

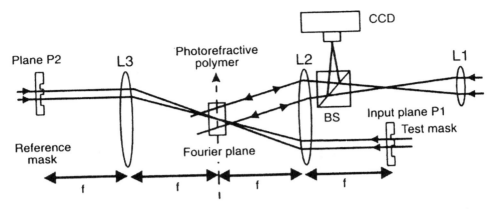

FIGURE 34 Schematic of the four-wave-mixing photorefractive correlator for security applications.

graphic filter and the implementation of the correlation. The input card containing the phase mask to be checked is placed in the plane P1 which is to be correlated with the master phase image in the plane P2. The hologram written by the interference of the reference beam and the beam going through the test mask forms a holographic filter for the master mask. The spatial cross-correlation of the two phase masks was formed inside the photorefractive polymer. Diffraction occurs only where the Fourier transforms of the two masks overlap. The correlation signal emerges, counterpropagating with the second writing beam, and is picked up by the beamsplitter BS and projected onto a CCD camera. Here, the CCD was only used to visualize the correlation peak and to compare it to the background noise. In a practical system, the correlation can be verified with a simple photodiode detector with appropriate threshold (see below).

For demonstration purposes, the phase masks used were binary random patterns of 64 × 64 pixels; the whole array was 5 × 5 mm^2 and was formed in a 2-μm thin photoresist layer ($n = 1.64$) deposited on a glass substrate, corresponding to a phase modulation depth of 4π. In principle, larger codes and gray levels can be used to improve the discrimination power. Figure 35 shows an example of the cross-correlation of the master phase mask with a matching test mask. It clearly shows a sharp correlation peak in the center of the CCD image which indicates the "match." A mask different from the control image yields only a random background signal (not shown) Therefore, discrimination of the original document (secured with a phase mask) and a copy (missing the mask) is achieved by measuring the intensity in the center of the correlation plane and comparing it with a threshold that is adjusted for a chosen security level.

Because the recording process is based on the photorefractive effect, the stored hologram can be erased and a new one can be written in real time. This reversible real-time recording and processing enables the testing of a variety of different documents encoded with different phase masks, and their comparison with a corresponding master mask database. Its response time of $\approx$1s is satisfactory for security verification tasks and does not preclude fast database searches.

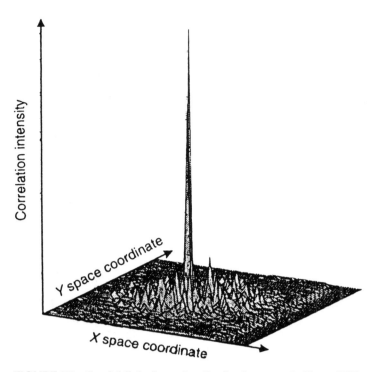

FIGURE 35 Spatial light-intensity distribution recorded by a CCD camera in the correlation plane P1 (see Fig. 34) when the test phase mask matches the master mask.

C. Holographic Storage

Finally, another important application of PR materials is holographic data storage. The storage of a large number of holograms multiplexed (by angle or wavelength) in one single location of a storage medium leads to very high storage densities. The storage density is limited by the dynamic range of the material, the Bragg selectivity, the scattering level, and the dark lifetimes of the stored holograms.

One drawback for the organic materials is that high electric fields are required during operation, limiting the active layer thickness to several hundred μm. As a result, the Bragg selectivity is reduced and the number of holograms that can be multiplexed is smaller than in inorganic crystals which can have thicknesses of several mm. This problem can be overcome by using stratified volume devices consisting of multiple polymer layers alternating with glass spacers and indium–tin oxide (ITO) electrodes [66]. In this way, high electric fields can be applied and, at the same time, longer interaction lengths are achieved, resulting in considerably better Bragg selectivity. Data pages containing 64 kbit were stored in the PR polymer PMMA : DTNBI : C_{60} in a typical 4f–Fourier-plane geometry by using the HOST setup developed at IBM [63]. A chrome-plated quartz plate containing an array of 256×256 "bits" in the form of black or white (transparent) squares served as the object to be stored. The average diffraction efficiency is on the order of 10^{-4} for the entire image and approximately 10^{-9} for each pixel. This illustrates the entremely high demands on optical quality of the materials. Figure 36 shows a typical histogram of a single data page [63]. Obviously, the noise level in the data page is increased for the reconstructed page in comparison with the transmitted image. For error-free readout of such a data page, it is extremely important that the two zones of the histogram corresponding to the "zeros" and "ones" are clearly separated in order to have an unambiguous threshold level for the assignment of each bit. In the example presented here, the performance of the material was excellent, yielding a bit/error rate of 10^{-5} if a threshold comparing the intensities of adjacent bits was used.

The exposure time needed to record efficient holograms was rather long (500 s). This is too long for an industrial application; by contrast, writing times in the order several tens of seconds would be acceptable because of the highly parallel nature of the storage process, allowing the storage of pages of data bits or entire gray-

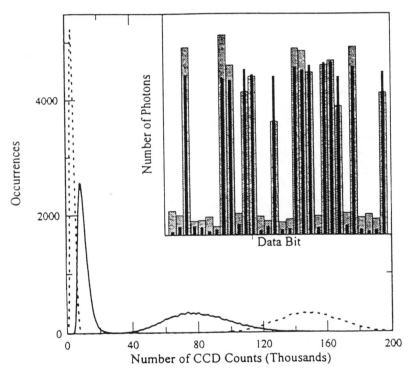

FIGURE 36 Intensity distribution of data bits contained in the original transmitted image (dashed curve) and the reconstructed hologram (solid curve) obtained by averaging the intensities for each four-pixel data bit and ignoring the opaque border regions. The horizontal scales for the histogram are not complete. Inset: Line sections of the pixel counts in a transmitted image (black strips) and the subsequent reconstructed hologram (gray bars). The hologram was recorded in $\sim$ 130-μm-thick devices of PMMA : DTNBI : C_{60} in the proportion 66.5 : 33 : 0.5 wt% using a Kr laser (λ = 676 nm) at E = 76 V/μm, 2θ = 40°, ψ = 40°. The intensity of the object and reference beam were 1 and 10 mW, respectively. The exposure time was 500 s. (From Ref. 63.)

scale images (corresponding to several megabytes) at a time. To engineer materials with faster response and long storage at the same time is one of the future problems to be solved.

VIII. CONCLUSION

In conclusion, amorphous organic PR materials are emerging as attractive materials for optical devices and processing systems. The materials are easy to process compared to time-consuming crystal growth. Very compact devices as well as large-area ones can be fabricated with excellent reproducibility. Organics offer wide flexibility for varying the chemical functions that can be incorporated in the material.

In just 5 years, the performance of this new material class has improved dramatically and became a true alternative to their long-known inorganic counterparts. Starting from diffraction efficiencies of 10^{-4} reported in 1991 on the first PR polymer [4], nowadays complete internal diffraction and optical gains in the order of 200 cm^{-1} and more are the state of the art [8,49]. This breakthrough could be mainly achieved by lowering the glass transition temperature of the composite close to room temperature, allowing the so-called "orientational enhancement mechanism" [9] to come into play whereby the chromophores are in situ poled by the photorefractive space-charge field. This leads to strong modulations of the refractive index through birefringence. Soon it became clear that this was even the dominant contribution in such materials [8,49]. However, a severe problem related to such highly doped composite materials was the loss of optical transparency due to phase separation. Meanwhile, this problem has been overcome, and composite devices with excellent performance and long-term stability of more than a year at the same time have been fabricated.

Only organic noncrystalline PR materials with high rotational mobility of the chromophores enable orientational effects, a strong advantage for obtaining good photorefractivity. Taking advantage of the orientational enhancement mechanism culminated in the report on a glassy low-molecular-weight compound with no apparent second-order nonlinearity, but excellent PR performance due to a strong birefringence [48,49]. The latter material class combines its excellent performance with high optical quality and long-term stability. Recent materials have optimized $\Delta\alpha$, but even better performance can be expected for materials using chromophores with large dipole moments. However, the compatibility with unpolar organic matrics may be the limiting factor here. Photorefractivity was also reported in liquid crystals [50,51], but with reduced spatial resolution compared to the composites and short device stability due to electrolytic damage of the devices.

In contrast to more traditional holographic recording media, such as silver halide films or thermoplastic polymers, information is reversibly stored in photo-refractive materials. Thus, storage, readout, and erasure can be done such that one device can be used for real-time monitoring without any additional intermediate developing steps. This definitely represents a major advantage over existing testing technologies. The proof-of-principle for the use of noncrystalline organic PR materials in optical correlation and storage devices using two-dimensional digital and/or gray-scale images and three-dimensional objects has been successfully demonstrated [49,61–65]. The wavelengths at which devices are typically operated (600–800 nm) are compatible with commercial low-power laser diodes. The sensitivity at these wavelengths is very high and, most importantly, can easily be adjusted by the external electric field. The compatibility with laser diodes is instrumental for widespread applications in order to keep the overall manufacturing cost low and to miniaturize the devices. The latter issue became even more favorable due to recent developments of high-precision plastic optical elements such as lenses, beamsplitters, and waveguide integrated optics.

Quo Vadis? Naturally, after only 5 years of research in this new exciting field, much remains to be done in terms of understanding of the basic physics as well as the development of new materials and their optimization for particular applications. Although the stability problem particularly related to highly doped composites has been overcome, little is known about the long-term performance of the materials under continuous operation at this point. This information is crucial before steps toward commercialization can be taken. The next issues to be addressed are faster response times, a reduction of the voltages necessary to achieve good PR performance, and the extension of the PR performance to wavelengths of technical interest, such as the telecommunications wavelengths 1300 and 1500 nm. A very promising area will be the holographic investigation of biological samples (tissue) with a transparency window around 800 nm.

ACKNOWLEDGMENT

This research has been supported by the U.S. Office of Naval Research through the Center of Advanced Multifunctional Nonlinear Optical Polymers and Molecular Assemblies (CAMP), the U.S. Air Force Office of Scientific Research (AFOSR), the U.S. National Science Foundation (NSF), and a NATO collaboration grant. K. M. acknowledges support from the Volkswagen-Stiftung. The authors wish to thank Dr. B.L. Volodin and Dr. Sandalphon from the Optical Sciences Center of the University of Arizona, Professor N. V. Kukhtarev from Alabama A&M University, Professor B. Javidi from the University of Connecticut, and Prof. C. Bräuchle and R. Bittner from the University of Munich for their fruitful collaboration.

REFERENCES

1. Günter, P., and Huignard, J.-P., *Photorefractive Materials and their Applications*, Springer-Verlag, Berlin, 1988/1989, Vols. 1 and 2.
2. (a) Partovi, A., Glass, A. M., Olson, D. H., Zydzik, G. J., Short, K. T., Feldman, R. D., and Austin, R. F., High sensitivity optical image processing device based on CdZnTe/ZnTe multiple quantum well structures, *Appl. Phys. Lett.*, *59*, 1832 (1992); (b) Partovi, A., Glass, A. M., Olson, D. H., Zydzik, G. J., O'Bryan, H. M., Chin, T. H., and Knox, W. H., Cr-doped GaAs/AlGaAs seim-insulating multiple quantum well photorefractive devices, *Appl. Phys. Lett.*, *62*, 464 (1993); (c) Wang, Q., Nolte, D. D., and Melloch, M. R., Two-wave mixing in photorefractive AlGaAs/GaAs quantum wells, *Appl. Phys. Lett.*, *59*, 256 (1991).
3. (a) Sutter, K., Hulliger, J., and Günter, P., Photorefractive effect observed in the organic crystal 2-cyclooctylamino-5-nitropyridine doped with 7,7,8-tetracyanoquinodimethane, *Solid State Commun.*, *74*, 867 (1990); (b) Sutter, K., and Günter, P., Photorefractive

gratings in the organic crystal 2-cyclooctylamino-5-nitropyridine doped with 7,7,8,8,B-tetracyanoquinodimethane, *J. Opt. Soc. Am, B7*, 2274–2278 (1990); (c) Sutter, K., Hulliger, J., Schlesser, R., and Günter, P., Photorefractive properties of 4'-nitrobenzylidene-3-acetamino-4-methoxyaniline, *Opt. Lett.*, *18*, 778 (1993).

4. Ducharme, S., Scott, J. C., Twieg, R. J., and Moerner, W. E., Observation of the photorefractive effect in a polymer, *Phys. Rev. Lett.*, *66*, 1846 (1991).

5. Moerner, W. E., and Silence, S. M., Polymeric photorefractive materials, *Chem. Rev*, *94*, 127 (1994), and references therein.

6. Zhang, Y., Burzynski, R., Ghosal, S., and Casstevens, M. K., Photorefractive polymers and composites, *Adv. Mater.*, *8*, 111–125 (1996), and references therein.

7. Kippelen, B., Meerholz, K., and Peyghambarian, N., An introduction to photorefractive polymers, in *Nonlinear Optics of Organic Molecules and Polymers* (H. S. Nalwa and S. Miyata, eds.), CRC Press, Boca Raton, FL, 1997, pp. 465–513, and references therein.

8. Meerholz, K., Volodin, B., Sandalphon, Kippelen, B., and Peyghambarian, N., A photorefractive polymer with high optical gain and diffraction efficiency near 100%, *Nature*, *371*, 497–500 (1994).

9. Moerner, W. E., Silence, S. M., Hache, F., and Bjorklund, G. C., Orientationally enhanced photorefractive effect in polymers, *J. Opt. Soc. Am.*, *B11*, 320 (1994).

10. (a) Kukhtarev, N. V., Kinetics of recording and erasing of holograms in electrooptic crystals, *Sov. Tech. Phys. Lett.*, *2*, 1114 (1976); (b) Kukhtarev, N. V., Dynamic holographic gratings and optical activity in photorefractive crystals, in (P. Günter and J.-P., Huignard, eds.), *Photorefractive Materials and their Applications* Springer-Verlag, Berlin, 1988, Vol. 1.

11. Silence, S. M., Donckers, M. C. J. M., Walsh, C. A., Burland, D. M., Moerner, W. E., and Twieg, R. J., Electric field-dependent nonphotorefractive gratings in a nonlinear photoconducting polymer, *Appl. Phys. Lett.*, *64*, (1994).

12. Bässler, H., Charge transport in random organic photoconductors, *Adv. Mater.*, 5, 662 (1993).

13. Borsenberger, P. M., and Weiss, D. S., *Organic Photoreceptors for Imaging Systems*, Marcel Dekker, Inc., New York, 1993.

14. Mozumder, A., Effect of an external electric field on the yield of free ions. I. General results from the Onsager theory, *J. Chem. Phys.*, *60*, 4300 (1974).

15. Marder, S. R., Cheng, L. -T., Tiemann, B. G., Friedle, A. C., Blanchard-Desce, M., Perry, J. W., and Skindhoj, J., Large first hyperpolarizabilities in push–pull polyenes by tuning of the bond length alternation and aromaticity, *Science*, 263, 511–514 (1994).

16. Boldt, P., Bourhill, G., Bräuchle, C., Jin, Y., Kammler, R., Müller, C., Rase, J., and Wichern, J., Tricyanochinodimethane derivatives with extremely large second-order optical nonlinearities, *J. Chem. Soc., Chem. Commun.*, 793–795 (1996).

17. Wu, J. W., Birefringent and electro-optic effects in poled polymer films: Steady-state and transient properties, *J. Opt. Soc. Am.*, *B8*, 142–152 (1991).

18. Kogelnik, H., Coupled wave theory for thick hologram gratings, *Bell Syst. Tech. J.*, *48*, 2909–2943 (1969).

19. Walsh, C. A., and Moerner, W. E., Two-beam coupling measurements of grating phase in a photorefractive polymer, *J. Opt. Soc. Am.*, *B9*, 1642–1647 (1992).

20. (a) Jonathan, J. M. C., Roussignol, P., and Roosen, G., *Opt. Lett.*, *13*, 224 (1988); (b) Partanen, J. P., Jonathan, J. M. C., and Hellwarth, R. W., Direct determination of electron mobility in photorefractive BSO by a holographic time-of-flight technique, *Appl. Phys. Lett.*, *57*, 2404–2406 (1990); (c) Nouchi, P., Partanen, J. P., and Hellwarth, R. W., Simple transient solutions for photoconduction and the space-charge field in a photorefractive material with shallow traps, *Phys. Rev.*, *B47*, 15581–15587 (1993).

21. Teng, C. C., and Man, H. T., Simple reflection technique for measuring the electro-optic coefficient of poled polymers, *Appl. Phys. Lett.*, *56*, 1734 (1990).
22. Schildkraut, J. S., Determination of the electrooptic coefficient of a poled polymer film, *Appl. Opt.*, *29*, 2839 (1990).
23. Sandalphon, Kippelen, B., Meerholz, K., and Peyghambarian, N., Ellipsometric measurements of poling birefringence, the Pockels effect, and the Kerr effect in high-performance photorefractive polymer composites, *Appl. Opt.*, *35*, 2346–2354 (1996).
24. (a) Winkelhahn, H.-J., Servay, T. K., and Neher, D., A novel concept for modeling the time-temperature dependence of polar order in nonlinear optically active polymers, *Ber. Bunsenges.*, *Phys. Chem.*, *100*, 123–133 (1966); (b) Heldmann, C., Brombacher, L., Neher, D., and Graf, M., Dispersion of the electro-optical response in poled polymer films determined by Stark spectroscopy, *Thin Solid Films*, *261*, 241–247 (1994).
25. Schildkraut, J. S., Photoconducting electro-optic polymer films, *Appl. Phys. Lett.*, *58*, 340–342 (1991).
26. Zhang, Y., Cui, Y., and Prasad, P. N., Observation of photorefractivity in a fullerene-doped polymer composite, *Phys. Rev.*, *B46*, 9900–9902 (1992).
27. Silence, S. M., Donckers, M. C. J. M., Walsh, C. A., Burland, D. M., Twieg, R. J., and Moerner, W. E., Optical properties of poly (*N*-vinylcarbazole)-based guest–host photorefractive polymer systems, *Appl. Opt.*, *33*, 2218 (1994).
28. Kippelen, B., Sandalphon, Peyghambarian, N., Lyon, S. R., Padias, A. B., and Hall, H. K. J., New highly efficient photorefractive polymer composite for optical-storage and image-processing applications, *Electron. Lett.*, *29*, 1873–1874 (1993).
29. Donckers, M. C. J. M., Silence, S. M., Walsh, C. A., Hache, F., Burland, D. M., Moerner, W. E., and Twieg, R. J., Net two-beam-coupling gain in a polymeric photo-refractive material, *Opt. Lett.*, *18*, 1044–1046 (1993).
30. Orczyk, M. E., Swedek, B., Zieba, J., and Prasad, P. N., Enhanced photorefractive performance in a photorefractive polymeric composite, *J. Appl. Phys.*, *76*, 4995–4998 (1994).
31. Zobel, O., Eckl, M., Strohriegl, P., and Haarer, D., A polysiloxane-based photorefractive polymer with high optical gain and diffraction efficiency, *Adv. Mater.*, 7, 911–914 (1995).
32. Malliaras, G. G., Krasnikov, V. V., Bolink, H. J., and Hadziioannou, G., Holographic time-of-flight measurements of the hole-drift mobility in a photorefractive polymer, *Phys. Rev.*, *B52*, R14324–R14327 (1995).
33. Scher, H., and Montroll, E. W., Anomalous transit-time dispersion in amorphous solids, *Phys. Rev. B*, *12*, 2455 (1975).
34. Bos, F. C., and Burland, D. M., *Phys. Rev. Lett.*, *58*, 152 (1987).
35. Poga, C., Burland, D. M., Hanemann, T., Jia, Y., Moylan, C. R., Stankus, J. J., Twieg, R. J., and Moerner, W. E., Photorefractivity in new organic polymeric materials, in *Xerographic Photoreceptors and Photorefractive Polymers* (S. Ducharme and P. M. Borsenberger, eds.), SPIE, San Diego, 1995, Vol. 2526.
36. Cox, A. M., Blackburn, R. D., West, D. P., King T. A., Wade, F. A., and Leigh D. A., Crystallization-resistant photorefractive polymer composite with high diffraction efficiency and reproducibility, *Appl. Phys. Lett.*, *68*, 2801–2803 (1996).
37. Meerholz, K., Bittner, R., Bräuchle, C., Volodin, B. L., Sandalphon, Kippelen, B., and Peyghambarian, N., Improved long-term stability of high-performance photorefractive polymer devices, in *Organic Photorefractive Materials and Xerographic Photoreceptors*, (S. Ducharme and J. W. Stasiak, eds.), SPIE, San Diego, 1996, Vol. 2850, pp. 100–107.
38. Liphard, M., Goonesekera, A., Jones, B. E., Ducharme, S., Takacs, J. M., and Zhang, L., High-performance photorefractive polymers, *Science*, *263*, 367–369 (1994).

39. Silence, S. M., Scott, J. C., Stankus, J. J., Moerner, W. E., Moylan, C. R., Bjorklund, G. C., and Twieg, R. J., Photorefractive polymers based on dual-function dopants, *J. Phys. Chem.*, *99*, 4096–4105 (1995).

40. Zhang, Y., Ghosal, S., Casstevens, M. K., and Burzynski, R., Bifunctional chromophore for photorefractive applications, *Appl. Phys. Lett.*, *66*, 256–258, (1995).

41. Burzynski, R., Casstevens, M. K., Zhang, Y., and Ghosal, S., Novel composites: Second-order nonlinear optical and polymeric photorefractive materials for optical information storage and processing applications, *Opt. Eng.*, *35*, 443–451 (1996).

42. Chaput, F., Riehl, D., Boilot, J. P., Cargnelli, K., Canva, M., Levy, Y., Brun, A., New nonlinear sol-gel films exhibiting photorefractivity, *Chem. Mater.*, *8*, 312–314 (1996).

43. Tamura, K., Padias, A. B., Hall, H. K. J., and Peyghambarian, N., A new polymeric material containing the tricyanovinylcarbazole group for photorefractive applications, *Appl. Phys. Lett.*, *60*, 1803–1805 (1992).

44. Kippelen, B., Tamura, K., Peyghambarian, N., Padias, A. B., and Hall, Jr., H. K., Photorefractivity in a functional side-chain polymer, *Phys. Rev.*, *B48*, 10710–10718 (1993).

45. (a) Yu, L., Chan, W., Bao, Z., and Cao, S. X. F., Photorefractive polymers 2. Structure design and property characterization, *Macromolecules*, *26*, 2216 (1993); (b) Yu, L., Chen, Y., Chan, W. K. and Peng, Z., Conjugated photorefractive polymer, *Appl. Phys. Lett.*, *64*, 2489 (1994); (c) Yu, L., Chan, W. K., Peng, Z., and Gharavi, A., Multifunctional polymers exhibiting photorefractive effects, *Acc. Chem. Res. 29*, 13–21 (1996); (d) Peng, Z., Ghavari, A. R., and Yu, L., Rational synthesis of photorefractive polymers, in *Organic Photorefractive Materials and Xerographic Photoreceptors* (S. Ducharme and J. W. Stasiak, eds.), SPIE, San Diego, 1996; Vol. 2850, pp. 63–68.

46. Silence, S. M., Bjorklund, G. C., and Moerner, W. E., Optical trap activation in a photorefractive polymer, *Opt. Lett.*, *19*, 1822–1824 (1994).

47. Wang, L., Zhang, Y., Wada, T., and Sasabe, H., Photorefractive effect in a photoconducting electro-optic carbazole trimer, *Appl. Phys. Lett.*, *69*, 728–730 (1996).

48. Wortmann, R., Poga, C., Twieg, R. J., Geletneky, C., Moylan, C. R., Lundquist, P. M., DeVoe, R. G., Cotts, P. M., Horn, H., Rice, J. E., and Burland, D. M., Design of optimized photorefractive materials: a novel class of chromophores, *J. Chem. Phys.*, *105*, 10637–10647 (1996).

49. Lundquist, P. M., Wortmann, R., Geletneky, C., Twieg, R. J., Jurich, M., Lee, V. Y., Moylan, and Burland, D. M., Organic glasses: a new class of photorefractive materials, *Science*, *274*, 1182–1185 (1996). Table 2 in this reference comparing different organic PR materials contains a typing error. Δn for 2BNCM is only $\approx 1.5 \times 10^{-3}$.

50. (a) Rudenko, E. V., and Sukhov, A. V., Photoinduced electrical conductivity and photorefraction in a nematic liquid crystal, *JETP Lett.*, *59*, 142–146 (1994); (b) Khoo, I. C., Li, H., and Liang, Y., Observation of orientational photorefractive effects in nematic liquid crystals, *Opt. Lett.*, *19*, 1723–1725 (1994); (c) Khoo, I. C., Holographic grating formation in dye- and fullerene C_{60}-doped nematic liquid crystals, *Opt. Lett.*, *20*, 2137–2139 (1995); (d) Khoo, I. C., Liang, Y., and Li, H., Observation of stimulated orientational scattering and cross-polarized self-starting phase conjugation in a nematic liquid-crystal film, *Opt. Lett.*, *20*, 130–132 (1995); Khoo, I. C., Orientational photorefractive effects in nematic liquid crystal films, *IEEE J. Quantum Electron.*, *QE-32*, 525–534 (1996).

51. (a) Wiederrecht, G. P., Yoon, B. A., and Wasielewski, M. R., High photorefractive gain in nematic liquid crystals doped with electron donor and acceptor molecules, *Science*, *270*, 1794–1797 (1996); (b) Wiederrecht, G. P., Yoon, B. A., and Wasielewski, M. R., Photorefractive liquid crystals, *Adv. Mater.*, *8*, 535–539 (1996).

52. Khoo, I. C., and Liu, T. H., Theory and experiment on multi-wave mixing-mediated probe-beam amplification, *Phys. Rev.*, *B38*, 4036–4044 (1989).

53. Volodin, B. L., Kippelen, B., Meerholz, K., Kukhtarev, N. V., Caulfield, H. J., and Peyghambarian, N., Non-Bragg orders in dynamic self-diffraction on thick phase gratings in a photorefractive polymer, *Opt. Lett.*, *21*, 519–521 (1996).
54. Volodin, B. L., Kippelen, B., Meerholz, K., Peyghambarian, N., Kukhtarev, N. V., and Caulfield, H. J., Study of non-Bragg orders in dynamic self-diffraction in a photorefractive polymer: Experiment, theory, and applications *J. Opt. Soc. Am.*, *B13*, 2261–2267 (1996).
55. Malliaras, G. G., Krasnikov, V. V., Bolink, H. J., and Hadziioannou, G., Control of charge trapping in a photorefractive polymer, *Appl. Phys. Lett.*, *66*, 1038–1040 (1995).
56. Zhang, Y., Ghosal, S., Casstevens, M. K., and Burzynski, R., Trap controlled photorefractive response in polymeric composites, *Polym. Prepr.*, *35*, 233 (1994).
57. Pai, D. M., Yanus, J., and Stolka, M., Trap-controlled hopping transport, *J. Phys. Chem.*, *88*, 4714 (1984).
58. Swedek, B., Prasad, P. N., Cui, C., Cheng, N., Zieba, J. W., Winiarz, J., and Kim, K. S., Theoretical and experimental studies of photorefractivity in novel polymeric composites, in *Organic Photorefractive Materials and Xerographic Photoreceptors* (S. Ducharme and J. W. Stasiak, (eds.), SPIE, San Diego, 1996, Vol. 2850, pp. 89–99.
59. Bolink, H. J., Malliaras, G. G., Krasnikov, V. V., and Hadziioannou, G., Effect of plasticization on the performance of a photorefractive polymer, *J. Phys. Chem.*, *100*, 16356–16360 (1996).
60. Silence, S. M., Twieg, R. J., Bjorklund, G. C., and Moerner, W. E., Quasinondestructive readout in a photorefractive polymer, *Phys. Rev. Lett.*, *73*, 2047–2050 (1994).
61. Volodin, B. L., Sandalphon, Meerholz, K., Kippelen, B., Kukhtarev, N. V., and Peyghambrian, N., Highly efficient photorefractive polymers for dynamic holography, *Opt. Eng.*, *34*, 2213–2223 (1995).
62. Levi, B. G., Search and Discovery, *Physics Today*, *48*(1), 17–19 (1995).
63. Lundquist, P. M., Poga, C., DeVoe, R. G., Jia, Y., Moerner, W. E., Bernal, M. -P., Coufal, H., Grygier, R. K., Hoffnagle, J. A., Jefferson, C. M., Macfarlane, R. M., Shelby, R. M., and Sincerbox, G. T., Holographic digital data storage in a photorefractive polymer, *Opt. Lett.*, *21,* 890–892 (1996).
64. Halvorson, C., Kraabel, B., Heeger, A. J., Volodin, B. L., Meerholz, K., Sandalphon, and Peyghambarian, N., Optical correlator using photorefractive polymers, *Opt. Lett.*, *20*, 76 (1995).
65. Volodin, B. L., Kippelen, B., Meerholz, K., Javidi, B., and Peyghambarian, N., A polymeric optical pattern-recognition system for security verification, *Nature*, *383*, 58–60 (1996).
66. Stankus, J. J., Silence, S. M., Moerner, W. E., and Bjorklund, G. C., Electric-field-switchable stratified volume holograms in photorefractive polymers, *Opt. Lett.*, *19*, 1480 (1994).

16

Optical Information Processing with Bacteriorhodopsin

D. V. G. L. N. Rao, F. J. Aranda, D. Narayana Rao, and Joby Joseph
University of Massachusetts at Boston
Boston, Massachusetts

J. A. Akkara and M. Nakashima
U. S. Army Natick Research Development and Engineering Center
Natick, Massachusetts

I. INTRODUCTION

Light-induced processes in nature can be divided into two fundamental groups: visual and photosynthetic. Visual photoprocesses pertain to the acquisition and processing of information by organisms. Photosynthetic photoprocesses pertain to the use of light by organisms for energetic purposes [1]. The retinal chromophore is commonly associated with visual processes. Likewise, chlorophyll is commonly associated with energetic processes. Bacteriorhodopsin (bR), discovered in 1971, is a photodynamic protein complex found in living systems. It is related to the visual pigment rhodopsin contained in the cone cells of the human retina [2]. Subsequently, is was also realized that the bR-rich purple membrane of the *Halobacterium halobium* is responsible for energetic processes in the bacterium [3]. The purple membrane is the only crystalline membrane found in nature. The halophilic *Halobacterium halobium* is found in salt marshes. The protein and lipid composition of the purple membrane of the *Halobacterium halobium* is 0.2 lipid-to-protein ratio of dry weight. The major lipids in the membrane are phosphatidylglycerophosphate, glycolipids, and neutral lipids that make up for 52%, 30%, and 6%, respectively, of the membrane [4].

In environments with high temperature and salt concentration, the oxygen that is needed for the oxidative phosphorilation required to synthesize ATP is not available. The *Halobacterium halobium* is able to synthesize ATP necessary for its phototrophic growth by producing the purple membrane. A halobacterial cell under these extreme conditions covers a large extent of the cell surface (up to 80%) [5]

with the bR protein. In its natural state, the bR molecules perform the biological function in the halobacterial cell of converting light into an electrochemical ion gradient across the membrane. Protons are pumped across the membrane, from the inside (cytoplasmic) to the outside (extracellular) [3]. The pH gradient that results from this proton transfer is responsible for the proton-motive force that allows the bacterium to synthesize ATP from inorganic phosphate and ADP. Thus, the bacterium is able to synthesize ATP via photochemical and respiratory processes. The bR molecule contains seven transmembrane helical segments and consists of a polypeptide chain composed of 248 amino acids. bR is a transmembranous protein that passes completely through the membrane several times. This is its mode of attachment [6]. Bacteriorhodopsin has been crystallized [7]. It can be induced to form two-dimensional crystals when purified and reconstituted with phospholipids in adequate conditions. The two-dimensional arrays have been used [6,8] to obtain structural information. bR was studied [9] with a three-dimensional resolution of 6.5 Å.

The active chromophore in the bR protein is a retinal molecule linked via a protonated Schiff base near the middle of helix G to lysine 216. The relevant location for the proton pumping activity of the bR protein consists of an ion pair which is composed of a protonated Schiff base and an anionic aspartic acid in position 85 (Asp85 residue). Protons transferred from the Schiff base to Asp85 are the primary mechanism in the funneling of protons toward the extracellular side of the membrane and other proton transfers that ensue. This mechanism is also associated with conformational changes of the bR molecule and its chromophoric group during the photocycle [5]. The bR photocycle is shown schematically in Fig. 1. In the initial B state of bR, also called the light-adapted state, the retinal chromophore is in its all-trans molecular configuration. The B state has an absorption peak at 570 nm. It can be excited by means of light in the red, yellow, or green

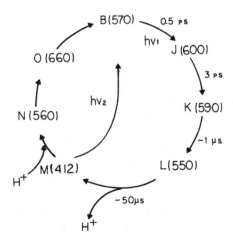

FIGURE 1 bR photocycle. Upon excitation with photon hv1, the molecule goes through several short-lived intermediate states in the long-lived M state. A blue photon hv2 stimulates the photochemical relaxation to the B state. The numbers in parenthesis indicate absorption peaks in nanometers.

parts of the spectrum because a broad absorption band is associated with it. Once a photon is absorbed, the retinal chromophore undergoes conformational changes, isomerizing about a double bond located between positions 13 and 14 in a subpicosecond time frame. The quantum yield of this primary photoreaction [10] is approximately 0.64. The chromophore subsequently relaxes by going through a series of short-lived intermediates to the M state which has an absorption maximum at 410 nm wavelength. The M state is associated with a 13-cis retinal molecular configuration. The M state can revert to the initial state via a thermal relaxation process or via a photochemical process upon excitation with blue light. The thermal relaxation of the chromophore from the blue-shifted M state is initiated by the reprotonation of the aspartic acid in position 96 (Asp96 residue). The retinal molecule is then able to isomerize once again and relax to the all-trans (B) state. The lifetime of the M state depends on the reprotonation process. It can be altered by different means: drying, controlling pH, reducing the temperature, and genetic mutation. It has been shown by Chen et al. [11] that in bR films of high pH, the reprotonation process is inhibited as a result of which the lifetime of the M state is increased from milliseconds to tens of seconds.

Bacteriorhodopsin and photopolymers containing bR have shown great promise as candidate materials for applications in photonics technology. Several applications have been proposed in image and information processing [12–16]. Some of the remarkable nonlinear optical properties of bR and photopolymers containing bR have already been the subject of research [17–23]. bR, like most complex organic materials, lends itself to many manipulations through the use of the techniques of synthetic chemistry and molecular bioengineering which permit modification of its optical properties to suit the requirements of a given technological application without degradation of its inherent mechanical and thermal stability or other physical characteristics. From the point of view of optical engineering, this endows bR with distinct advantages over other optical materials.

A. Advantages of bR over Conventional Materials

Biological molecules can be used to fabricate devices with smaller sizes and faster data handling capabilities than currently available semiconductor devices. Proteins, in particular, have proven to be ideally suited for applications in optical information processing and all-optical computing.

The basic building blocks in semiconductor-based electronics and computing are logic gates. These switches have two stable states often referred to as logic 0 and logic 1. Computers encode all information in terms of these two logic states or bits. Molecules are, in principle, the smallest switches that we can envision. If the atoms in some molecules change position in determined ways, the molecule is said to isomerize. Control of such an isomerization process to attain at least two stable states in one molecule allows us to use one state as 0 and the other state as 1, constituting a molecular switch. This field of research has been named molecular electronics or protein-based electronics. Optics is used to induce the structural changes in the molecule that are associated with the different logic states. Light is used to address the molecule, as each molecular state is associated with distinct optical absorption spectra. Owing to the fact that proteins such as bacteriorhodopsin are more sensitive to light than inorganic crystals, the molecular switch can be realized with very low light levels, resulting in an extremely energy efficient switch. Of considerable significance is the fact that both the spectrum and kinetic aspects of the bR photocycle can readily be modified. This is accomplished by replacing the light-absorbing component of the protein, which is a retinal (vitamin-A-like) chromophore. This chromophore can be replaced by natural and synthetic analogs which can shift the bR spectrum to virtually any color [24]. It has also been reported that genetic mutations of bR that can be attained by biotechnological procedures can alter both the kinetic and spectral properties of bR [25]. Alterations such as genetic manipulation of proteins are plainly impossible with inorganic materials.

The most severe bottleneck in modern computers is transfer of data within the computer, which ultimately limits their performance. This problem can be greatly alleviated through the use of parallel processing and light-based interconnections that employ efficient protein-based switches which allow the implementation of storage, transfer, and manipulation of massive amounts of data.

Furthermore, the cost of producing, the low-power requirements of bR, and its environmental friendliness make it an attractive material. Current estimates make it cost-effective compared to the price of semiconductor-based memories. Cleansing and etching in semiconductor processing involve a substantial amount of pollution, whereas preparation and isolation of the bR is environmentally safe; after all, proteins play a fundamental role in nutrition.

B. Stability

In addition to large quantum yields and the distinct absorption of B and M, bR has several intrinsic properties of importance in optical engineering. First, this molecule exhibits a large absorption cross section. Second, bR is a two-dimensional protein which has the structure of a hexagonal array of trimers with intercalated lipids which confers it a crystal-like architecture [5]. This structure is to some extent the reason why bR is very stable. Furthermore, adaptation to high-salt environments make bR very robust to degradation by environmental perturbations, and, thus, unlike other biological materials, it does not require special storage. Dry films of bR have been stored for several years without degradation [26,27]. In our lab, the film we made 5 years ago exhibits no noticeable changes in its characteristics. Light-sensitive proteins like bR have been optimized by evolution for millions of years; thus, bR also exhibits extremely high stability toward photo degradation. No noticeable change is observed after a bR film is switched between bR and the M state more than a million times with a quartz lamp with appropriate color filters. In addition, it has been reported by Shen et al. [28] that dry films of bR are structurally stable up to a temperature of 140°C.

II. MATERIALS

The following procedure describes the method of growing large batches of *Halobacterium halobium*, preparing growth medium, harvesting the bacteria, and isolating of the purple membranes from the bacteria.

A. Cultures

Halobacterium halobium S-9 was received from S. Subramaniam (MIT, Cambridge, MA). The original culture was received from W. Stoekenius [30–32]. *Halobacterium halobium* mutant cultures D85N, D96N, and D115N/D97N, indicating site mutations at Asp85, Asp96, and Asp115 [33–38], respectively, were received from Richard Needleman (Wayne State University School of Medicine, Detroit, MI).

B. Preparation of Culture Medium

1. Reagents per Liter of the Medium for the Wild Type

The medium of Oesterhelt and Stoeckenius was used for the growth of the *Halobacterium halobium* [30–32]. This growth medium was prepared by mixing basal salts medium and Oxoid Bacteriological Peptone. The composition of the basal salts medium per liter of the medium is given below.

2. Preparation of Basal Salts Medium

NaCl	250.00g	Sodium citrate 2 H_2O	3.00 g
KH_2PO_4	100.0 mg	Glycerol	1.0 ml
$MgSO_4$	9.77 g	$CaCl_2$ (anhydrous)	0.20 g
NH_4Cl	5.00 g	KCl	2.00 g

The above reagents, in the order given above, were added to about 800 ml of water, one ingredient at a time, and were dissolved before the next ingredient was added, to prevent the formation of an insoluble precipitate. Incomplete dissolution of the different ingredients may result in poor bacterial growth. The complete growth medium was prepared by adding 10 g of Oxoid Bacteriological Peptone L-37 (Oxoid Ltd., Basingstoke, Hampshire, UK) to the above basal salts medium. The peptone was dissolved and the pH of the medium was adjusted to 7.0–7.4 by adding sodium hydroxide and the volume adjusted to 1 L. The 1-L medium in a 2.8-L Fernbach flask was covered by aluminum foil and sterilized at 15 psi for 20 min.

3. Reagents per Liter of the medium for the Mutants

The following medium was used and was prepared as per R. Needleman [33,38]:

Deionized water	800.00 ml
NaCl	250.00g
$MgSO_4 \cdot 7H_2O$,	20.00 g
$MnSO_4 \cdot H_2O$, 2 H_2O	3.00 g
KCl	2.00 g
100 Peptone (Oxoid)	10.00 g
Tris.HCl	7.85 g
Trace metals	100.00 ml

[Trace metals consist of $ZnSO_4 \cdot 7H_2O$, 0.65 g; 0.15 g; Fe(NH)$_4$SO$_4 \cdot$6Na citrate H_2O, 0.4 g; $CuSO_4 \cdot 5 H_2O$, 0.075 g; and HCl (0.1 N) ml.]

The pH of the medium was adjusted to 7.4 and volume made up to 1 L. The medium was then sterilized by filtration using a 0.45-μ filter.

C. Culture Conditions

The 1-L medium was inoculated with about 200 ml of bacterial culture reserved from a previous run. Cells were grown at 40°C in a 2.8-L Fernbach flask with a rotary shaker for good aeration in the presence of light. The agitation was set at about 300 rpm with an aeration of the medium set at about 3.0 L/min. It should be emphasized that too much aeration will inhibit the formation of the purple membrane. Illumination was required for the accelerated bacterial growth. The bacterial growth was monitored by measuring the absorbance at 660 nm and the purple membrane production was followed by measuring the absorbance at 570 nm. The cells were harvested while the culture was in the prestationary phase, when an optical density of about 3.0 was observed at 570 nm.

D. Preparation of Purple Membrane

Cells are harvested by centrifugation at 4°C for 15 min at 13,000 × g. The cells are washed with basal salts medium (without peptone) and suspended in 100 ml of basal salts medium [31]. The cell suspension was mixed with 2 mg of deoxyribonuclease I (4000 units, Sigma Chem. Co., St. Louis, MO), incubated at room temperature for 48–72 h, and dialyzed overnight in a 2-L sodium chloride solution (0.1 M) using dialysis tubing of 12,000–14,000 molecular weight cutoff (Spectrum Medical Industries, Inc., Los Angeles, CA). The dialysate was centrifuged for 40 min at 40,900 × **g** in the cold (4°C). After decanting the supernatant, the pellet was suspended in 0.1 M NaCl and centrifuged for 40 min at 40,900 × **g** and 4°C. This washing procedure was repeated until the supernatant was almost colorless. The sediment was suspended in deionized water. Washing was continued with deionized water (two or three times) and the sediment was suspended in 6–10 ml of deionized water and frozen at −20°C.

E. Solutions

Bacteriorhodopsin was isolated from wild-type *Halobacterium halobium* at the U.S. Army Natick R.D.&E. Center. The aprotein (protein minus the chromophore) was prepared by bleaching with hydroxylamine HCl, pH 7.0 as described by Oesterhelt et al. [31,32]. Briefly, the method of preparation was as follows. The hydroxylamine HCl was prepared as a 1.0 M solution in deionized H_2O and the pH adjusted to 7.0 with NaOH. One milliliter of protein suspension received 100 μl of 1.0 M hydroxylamine solution. The suspensions were then irradiated with light filtered through a Zeiss Ikon gelbfilter (yellow) from a Schott KL 1500 cold light source (fiber-optic) Illuminator, equipped with a Xenophot HLX xenon light bulb (OSRAM) and Schott KG1 heat filter (color temperature was approximately 3200 K, and brightness was approximately 10 Mlx). Bleaching was continued for about 15 h before the suspension was completely colorless. The suspensions were then centrifuged, the supernatant removed, and the pellet resuspended. for a total of three washings in H_2O or D_2O to remove the free retinal chromophore and the hydroxylamine.

F. Thin Films

Two types of bR films were used in the experiments. Thin films of wild-type bR dispersed in a polymer matrix were purchased from Wacker Chemical (U.S.A.) Inc.; they are a product of Consortium fur Electrochemische Industrie GmbH. They were 35 μm thick and were sandwiched between glass plates.

The method of preparation for the high-pH films is as follows: The purple membrane was first washed with deionized water and then passed through a 5-μm pore size filter to remove particulate matter. Stock solution of 40% (w/w) acrylamide was made with an acrylamide to N,N'-methylene-bis-acrylamide ratio of 20:1. The concentrated bR solution (3.5 ml) was then mixed with the acrylamide solution (0.5 ml). Two glass plates and three 3-mm-thick spacers were used to form a rectangular gel cassette. The gel solution was prepared by mixing the polymerization catalyst ammonium persulfate (0.03% w/w) and initiator $N,N,N'N'$-tetramethylethylenediamine (1 μl/ml) with bR/acrylamide solution. Immediately after the preparation of the gel solution, it was poured into the cassette. After polymerization, the cast gel was removed from the cassette and rinsed with deionized water. It was then soaked in a borate buffer (pH 10, 10 mM) for 24 h. The buffer-equilibrated gel was then covered with two gel-drying cellulose films and held firmly in a drying cell. The film was dried at room temperature for about 24 h. The dried bR–polymer film was then held between two glass plates to prevent deformation. The final film had an optical density of 2 at 568 nm, 0.14 at 633 nm, and 0.45 at 458 nm. The main advantage of preparing the film with this method is that the lifetime of the M state can be varied by changing the pH of the socking buffer solution.

III. NONLINEAR OPTICS

It was realized as far back as James Clerk Maxwell [39] that nonlinear effects occurred in electromagnetism. The saturation of the magnetization in ferromagnets is a classical example. The *p-n* junction in semiconductor electronic devices so familiar to us today has nonlinear electrical characteristics. The contemporary electronics and computer industries rely heavily on such nonlinear behavior of the electrical and magnetic properties of materials.

With the advent of the laser in 1960 as a source of coherent, monochromatic, and high-intensity light, the study of nonlinear optical phenomena in the visible part of the spectrum of electromagnetic waves became feasible. The next decades of this century have witnessed the flourishing of the science of Nonlinear Optics.

In the sixties and seventies, most experiments undertaken involved relatively simple and well-understood material systems. Gases, simple solvents, and crystals such as quartz were studied. The observation of second-harmonic generation, that is, the partial conversion of a ruby laser beam of 694 nm wavelength into half the wavelength (or twice the frequency) of ultraviolet radiation at 347 nm in a quartz crystal [40] marks the birth of Nonlinear Optics.

In later years, it was realized that nonlinear optical phenomena, much like electrical nonlinearities in electronics, are of fundamental importance in the development of the new technological frontier of photonics. In photonics, photons instead of electrons are utilized for the acquisition, storage, transmission, and processing of information [41].

It is understood that all media are intrinsically nonlinear. Vacuum is itself nonlinear in the sense that photons effectively interact in a vacuum due to the vacuum polarizations [39]. However, nonlinear phenomena are far more effective in media in which the photons interact by inducing a polarization in a material. The ability of a material to respond to interacting photons depends on its physical nature. The electronic structure of its atoms and molecules and their dynamics, the degree of symmetry, the packing, and overall arrangement all influence the nonlinear behavior of materials. It is then understandable that a substantial effort is underway to understand, characterize, and exploit various classes of materials. Recently, there has been an enormous growth in the study of the more complex and rich domain of organic molecular materials. Organic materials are promising candidates for nonlinear optical applications because they posses ultrafast nonlinearities and can respond several orders of magnitude faster than inorganic materials. Some organic molecules found in nature have been optimized by evolution through millions of years. The response to light photons of the natural pigments that are involved in photosynthesis (e.g., chlorophyll), the phycobiliproteins responsible for algal photosynthesis, and the retinal-containing proteins involved in visual processes and anioxyogenic photosynthesis in bacteria (e.g., bR) are a few examples of some of the most elegant and efficient light-driven materials found in biological systems.

Nonlinear optics is the study of optical processes that take place when the optical properties of some material system are altered by its interaction with light. For most material systems, the optical properties will only noticeably change when exposed to very intense light. Lasers provide the light-intensity levels needed to induce significant modifications of the optical properties of a material. In historical perspective, one can thus understand that the advent of the laser was a prerequisite [42] to the birth of Nonlinear Optics. The discovery of second-harmonic generation in quartz by Franken and co-workers [40] in 1961 is generally considered to be the beginning of Nonlinear Optics. It is no coincidence that it happened 1 year after the first successful demonstration of the operation of a laser.

The term "nonlinear" refers to the relation between the response of the material and the incident optical-field strength. If the response of a material is proportional to the square of the strength of the applied optical field, the interaction is said to be a second-order nonlinear optical interaction. If the response is proportional to the third power of the optical field, it is a third-order nonlinear optical interaction, and so on. Higher-order interactions are limited in most materials because the fields needed to observe such higher-order phenomena may exceed the threshold for ionization and, hence, the dielectric will break down.

The polarization induced in the medium by the electric field is the key concept to understand the interaction of a molecular system with optical radiation. Light incident on a medium induces an oscillating dipole moment in the electron or ion distribution. For an incident electromagnetic field, there is an electric polarization generated within the material. The induced polarization in the material can be observed macroscopically. The index of refraction and absorption of a material are determined by the linear polarization.

The interaction of an electric field with a dielectric medium produces a separation of negative and positive charges; this separation of charges produces a polarization in the medium, the direction of which varies subject to the oscillations of the applied field. When the fields applied are weak, regardless of whether the field propagates or not, the polarization induced by the field in the medium has a linear dependence on the applied electric field **E**; the electric displacement in Gaussian units is described by [43]

$$\mathbf{D} = \mathbf{E} + 4\pi\mathbf{P} \tag{1}$$

The linear susceptibility $\chi^{(1)}$ determines the propagation behavior at low intensities. It has real and imaginary parts that can be related to the indices of refraction and absorption loss. In lossless, isotropic media,

$$\chi^{(1)} = \frac{n_2 - 1}{4\pi} \tag{2}$$

where n is the index of refraction of the dielectric medium. In absorptive media, n^2 must be replaced by the complex dielectric constant ε, whose real part is the square of the index of refraction and whose imaginary pat is related to the linear absorption coefficient α. In nonisotropic media, $\chi^{(1)}$ is a three-dimensional or second-rank tensor.

In general, the displacement of electrons in the materials is proportional to the incident electromagnetic field. The linear dependence is described by

$$\mathbf{P} = \chi^{(1)} \mathbf{E} \tag{3}$$

The proportionality factor $\chi^{(1)}$ is, in general, a tensor termed the susceptibility. In a medium where several fields are incident, the induced polarization is a linear superposition of the alterations that the sane fields would separately produce. For small displacements, electrons or ions can be described by harmonic oscillations. When the applied optical field is intense, the displacement becomes large and the potential function becomes anharmonic. The polarization developed in this regime contains components that depend on the second and higher powers of the incident electromagnetic fields. The polarization radiates a second optical field that interferes with the incident field. The high-frequency permittivity, permeability, or conductivity of the medium will become field dependent. Maxwell's equations are no longer linear and the principle of superposition becomes invalid. Thus, at higher intensities, we need to include a nonlinear polarization term $\mathbf{P}_{nonlinear}$ such that the overall polarization becomes

$$\mathbf{P} = \chi^{(1)} \mathbf{E} + \mathbf{P}_{nonlinear} \tag{4}$$

This extra term is labeled nonlinear because its relation with the electric field is not a linear one. The nonlinear polarization has the general form

$$\mathbf{P}_{nonlinear} = \chi^{(2)} \mathbf{E}^2 + \chi^{(3)} \mathbf{E}^3 + \cdots \tag{5}$$

where $\chi^{(2)}$ and $\chi^{(3)}$ represent the second-order and third-order susceptibilities, respectively. They are tentorial, dispersive, and complex in nature. They represent the magnitudes of higher-order interactions among applied fields, and they are determined by features at the molecular level, including electronic delocalization, molecular structure, and overall molecular packing. Altogether, we have for the polarization

$$\mathbf{P} = \mathbf{P}_{linear} + \mathbf{P}_{nonlinear} = \chi^{(1)} \mathbf{E} + \chi^{(2)} \mathbf{E}^2 + \chi^{(3)} \mathbf{E}^3 + \cdots \qquad (6)$$

The important point to note is that at high intensities, the electric fields can interact with each other, be modulated, and be changed in frequency through these higher-order terms in the polarization vector. At microwave frequencies and below, this type of interaction has been known and utilized for many years; at optical frequencies, however, this field of study, namely Nonlinear Optics, has been made possible only with the advent of the laser. The interest in nonlinear optics lies in the research, first, of interactions of light with matter; second, on the understanding of the physical processes responsible for nonlinear phenomena; and, finally, in the utilization of these effects for potential device applications.

Whereas the polarization $\mathbf{P}$ is a macroscopic measure of the dipole moment averaged over the volume of the sample, on a molecular scale the ith component of the dipole moment can be expressed in an analogous manner as

$$P_i = \alpha_{ij} E_j + \beta_{ijk} E_j E_k + \gamma_{ijkl} E_j E_k E_l \qquad (7)$$

Here, the coefficients β_{ijk} and γ_{ijkl} are termed the first and second molecular hyperpolarizabilities, respectively.

IV. DEGENERATE FOUR-WAVE MIXING

The process of four-wave mixing involves nonlinear interaction of three waves to produce a fourth one. When all four waves are of the same frequency, the process is called degenerate four-wave mixing (DFWM). The first observation of DFWM was made by Stepanov et al. [44]. The general theory was worked out by Hellwarth et al. [45] in 1977. Yariv and Pepper showed that the process of DFWM is a special way to characterize the third-order susceptibility [46]. The DFWM geometry is shown schematically in Fig. 2. Several other alternative schemes are also possible. We use the backward wave geometry also called phase conjugate geometry. In DFWM, two counterpropagating waves, called the forward and the backward pump beams, interact with a third wave, called the probe beam, in a nonlinear medium. The probe beam is incident at an angle with respect to the pumps. As a result of the interaction, a fourth wave, called the phase conjugate or signal beam, is generated.

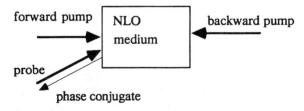

FIGURE 2 DFWM schematic.

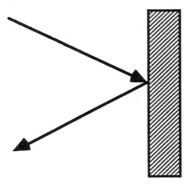

FIGURE 3 Conventional mirror.

The interaction can be explained in two different ways. First, we can argue that the signal beam is the phase conjugate of the probe beam. In such case, the nonlinear medium is acting as a phase conjugate mirror. A phase conjugate mirror is a mirror that reflects the phase conjugate of the incident wave. The qualitative differences between a regular mirror and a phase conjugate mirror are shown in Fig. 3 and 4.

The signal I_{pc} is the phase conjugate of the probe beam I_p. They are two counterpropagating waves which are complex conjugates of each other. The conjugate wave will propagate in the direction opposite the incident field and retraces the original path of the incident wave. The phase fronts of the two waves have identical shapes, except that they move in opposite directions. Figure 5 shows schematically what is meant by the phase reversal of the incident wave front. The phase conjugate wave can be thought of as the time reversal of the probe wave.

Another way to explain the interaction is as follows. We can argue that the interaction of the forward pump beam and the probe is such as to create an interference pattern with a spatially varying intensity distribution. The medium is nonlinear and it has a nonlinear index of refraction and/or nonlinear absorption. The

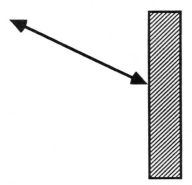

FIGURE 4 Phase conjugate mirror.

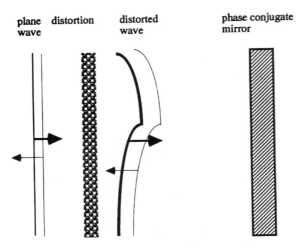

plane distortion   distorted   phase conjugate
wave   wave   mirror

FIGURE 5   The concept of wave front reconstruction by phase conjugation.

result of interference is thus to create a refractive index and/or absorptive grating. The grating is a volume-diffraction grating. When the backward pump beam enters the medium, it is partially diffracted by the grating. In this case, the signal beam can be thought of as the diffracted backward pump beam. Thus, the DFWM process can be though of a real-time holography. Schematics illustrating the processes of writing and reading of conventional holography are shown in Figs. 6 and 7, respectively. Note that if we overlap the two diagrams, we have backward DFWM as shown in Fig. 2.

There are, in fact, three gratings formed in backward DFWM [39,47], each one formed by the interaction of a pair of beams. The three gratings involved in backward DFWM are shown in Figs. 8–10.

A.   Holographic Analogies of DFWM

Holography allows us to record all the information (amplitude and phase) that is carried by a laser beam on some medium. The image can be reconstructed after recording by illuminating the medium with another laser beam. In contrast, a simple

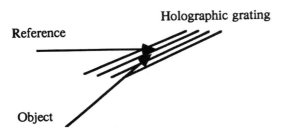

Holographic grating
Reference

Object

FIGURE 6   Writing the hologram in conventional holography.

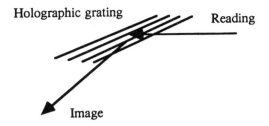

FIGURE 7 Reading the hologram in conventional holography.

photograph of an object can only show the amplitude variation of the recorded image. As a result, far more realistic images with three-dimensional appearance can be recorded by holography. In addition, the information density of a hologram can be extremely high, so that this technique may be used for information storage.

In backward DFWM, the vertically polarized laser beam is split into three beams, two counterpropagating pump radiation fields $E_{fp}(\omega, t)$ and $E_{bp}(\omega, T)$ (forward and backward pumps, respectively) and a third probe beam $E_p(\omega, t)$ incident at a small angle with respect to the direction of the forward pump, which are spatially overlapped in the nonlinear medium to produce a periodic intensity grating which can be represented by [48]

$$I(x) = (I_{fp} + I_p)\left[1 + C \cos\left(\frac{2\pi x}{\Gamma}\right)\right] \tag{8}$$

where I_{fp} and I_p are the intensities of the forward pump and probe, respectively, C is the contrast of the fringe pattern given by

$$C = \frac{2\sqrt{I_{fp}I_p}}{I_{fp} + I_p} \tag{9}$$

and the periodicity Γ of the absorptive or phase grating is given by

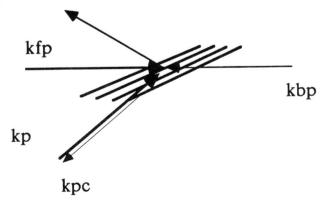

FIGURE 8 Holographic grating formed by the forward pump and the probe.

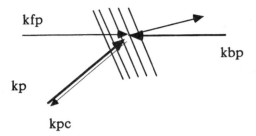

FIGURE 9 Holographic grating formed by the backward pump and the probe.

$$\Gamma = \frac{\lambda}{2 \sin \theta} \tag{10}$$

where λ is the wavelength of the DFWM beams and θ is the half-angle between the forward pump and the probe. This periodic grating is responsible for the diffraction of the backward pump which appears as the fourth beam $E_{pc}(\omega, t)$. The fourth beam is a phase conjugate replica of the probe beam and counterpropagates along the same direction as the probe beam. The diffraction efficiency η for the grating formed is defined as the ratio of the diffracted phase conjugate intensity I_{pc} to readout backward pump intensity I_{bp} and for a thick holographic grating in which both the absorption of the material as well as the refractive index are modulated as a result of the interference given by the above equation; it is given by [48]

$$\eta = \frac{I_{pc}}{I_{bp}} = \left[\sin^2\left(\frac{\pi n_1 d}{\lambda \cos \theta}\right) + \sinh^2\left(\frac{\alpha_1 d}{2 \cos \theta}\right) \right] \exp\left(\frac{2\alpha_0 d}{\cos \theta}\right) \tag{11}$$

where λ is the wavelength of the recording beams, d is the thickness of the sample, θ is the half-angle between the forward pump and probe, α_0 is the average absorption coefficient in the grating, and n_1 and α_1 are the modulation amplitudes of the refractive index and absorption coefficient, respectively.

B. Optical Phase Conjugation and All-Optical Light Modulation

Optical phase conjugation by backward DFWM is one of the methods that can be used to determine the absolute value of the third-order nonlinear optical suscepti-

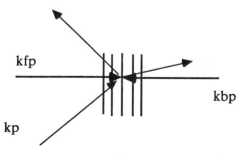

FIGURE 10 Holographic grating formed by the forward pump and the backward pump.

bility. With bR solutions and films, the large intensities that are usually required in DFWM experiments are not necessary. Optical phase conjugation can be attained with very small intensities from low-power continuous wave (CW) lasers. We used two different methods to ascertain the value of $\chi^{(3)}$ for the high-pH chemically stabilized film previously described. Following the procedure of Tompkin et al. [49], we estimated an effective third-order nonlinear susceptibility for the observed saturation intensity I_S for the film, which was 1.3 mW/cm², according to

$$\chi^{(3)} = \frac{n_0^2 c^2 \alpha_0}{24\pi^2 \omega I_S}(\delta + i) \tag{12}$$

n_0 and α_0 are the linear refractive index and absorption, respectively, and $\delta = (\omega - \omega_0)T_2$ is the detuning of the laser frequency ω from the resonance frequency ω_0 normalized to the dipole dephasing time T_2. We obtain the dipole dephasing time experimentally from the width of the absorption peak centered at 570 mm in the linear absorption spectrum of the film. We thus estimate a value of $0.14 + 0.36i$ esu for $\chi^{(3)}$. Experimentally, we calculated $\chi^{(3)}$ using [50]

$$|\chi^{(3)}| = \sqrt{\frac{I_{pc}}{I_{fp}I_{bp}I_p}} \frac{n_0^2 c^2}{16\pi^2 \omega L} \tag{13}$$

with $I_{pc} = 3 \times 10^{-3}$, $I_p = 30$, $I_{fp} = 70$, $I_{bp} = 50$ μW/cm²; we obtain $\chi^{(3)} = 0.6$ esu, which is in line with the estimated value from I_S.

Light modulation plays a fundamental role in the development of optical and optoelectronic systems for future information technologies. Spatial light modulators (SLMs) have been proposed to provide interconnections in parallel information and image processing. The most common SLMs are made from ferroelectric liquid crystals, and there is a substantial effort underway to utilize multiple quantum-well (MQW) structures for this purpose [51].

An all-optical light-intensity modulator has been developed using the chemically stabilized bR thin film. Independent of the slow thermal relaxation of the chromophore from 13-cis (M) to the all-trans (B) configuration, we can stimulate the transition using blue light in a time frame of several nanoseconds. The source of blue light in the experiments had a wavelength of 458 nm. The net result of the presence of blue light is to effectively increase the saturation intensity of the film by several orders of magnitude. We monitored the saturation behavior of the bR film in the presence and absence of a blue beam as a function of the incident red-light intensity and the results are shown in Fig. 11. The source of red light was a 632 nm wavelength. When only the red beam is present, we obtain a saturation intensity of 1.3 mW/cm² by numerically fitting the data in Fig. 11. However, when a blue beam of 1.1 W/cm² is simultaneously present, the saturation intensity increases to 50 mW/cm². If we further increase the blue-light power, the saturation intensity will increase accordingly.

The experimental arrangement for the light-intensity modulator consists of a Spectra Physics He-Ne CW laser with output at 632 nm. A Coherent Innova 70 Spectrum argon–krypton CW laser with an intracavity prism was tuned to give 458-nm-wavelength laser light. The modulator consists of a backward (DFWM) setup [52] with the red He–Ne beams plus a modulating blue beam. A schematic diagram of the modulator is shown in Fig. 12. The wavelength used in the exper-

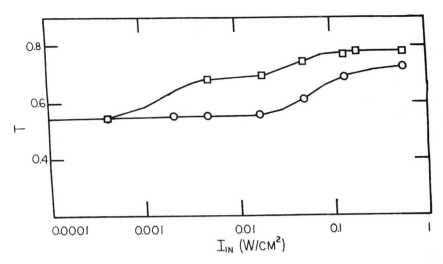

FIGURE 11 Transmission of the chemically stabilized bR film as a function of the incident 632-nm light intensity. The squares represent data obtained in the absence of blue light and the circles represent data obtained when 458-nm light is simultaneously present.

iments has large diffraction efficiency and relatively low absorption. By adjusting the relative blue to red light intensities, the intensity of the output beam can be controlled.

The modulated signal is obtained by physically separating the phase conjugate beam with the aid of a beamsplitter and measuring the beam intensity with a Hamamatsu R928 photomultiplier tube. Energy measurements were done with an

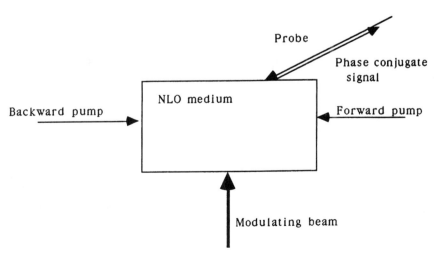

FIGURE 12 Schematic of the light-intensity modulator. The nonlinear medium is a chemically stabilized bR film.

EG&G calibrated silicon photodetector. The detectors were interfaced to a personal computer through an Alpha Products analog-to-digital converter or to an oscilloscope for data acquisition and analysis. The blue light is made to overlap in the region of the film where the DFWM interaction takes place. The diameter of the blue beam spot is slightly larger than that of the red beams. A mechanical chopper with different aperture sizes or a rotating mirror with a fixed aperture are used to modulate the blue-light intensity. It should be noted that the blue light need not be derived from a laser because the photochemical M to B transition can be stimulated with a broad range of wavelengths in the short-wavelength part of the optical spectrum. The experimental arrangement is shown in Fig. 13. The DFWM signal arises as a result of the index grating that is formed from the B to M state transition and the change in the index of refraction is governed by the Kramers–Kronig dispersion relations [53]. As a result of the photochromic transition induced by the red light in the region of the DFWM interaction and the low saturation intensity of the films, the B to M transition will be saturated locally and all of the bR in the region will be switched to the M state. This saturation happens at relatively low powers of the incident light because of the long thermal relaxation time of the M to B transition, which is of the order of tens of seconds. No index grating will exist due to the above mechanism because all the molecules remain in the M state and the dark interference regions in the grating tend to be washed off by light scattered from inhomogeneities in the film. Gratings formed by other mechanisms

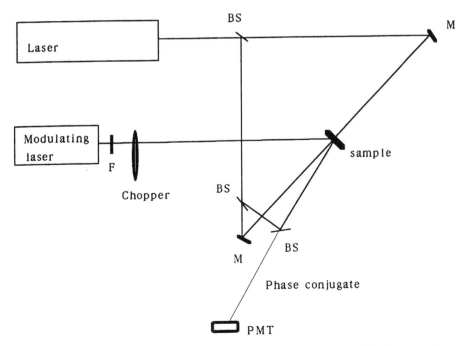

FIGURE 13 Experimental arrangement for the light-intensity modulation experiments. The laser used for DFWM is a Spectra Physics He–Ne and the modulating laser is a Coherent Innova 70 Spectrum Ar–Kr. Bs-beam-splitter; M-mirror; F-filter; PMT-photomultiplier tube.

may be present, but we found their contribution to the signal to be below the noise level and thus negligible at these intensities.

On the other hand, when blue light is simultaneously present, a fast photochemical reaction is induced that reverts the bR molecules in the M state back to the initial B state within less than a microsecond. The effect of the presence of the blue light is thus to trigger the photocycle. An index grating will, therefore, be formed and a phase conjugate beam will appear. The uniform blue-light illumination also prevents the washing out of the interference fringes. The outcome of modulating the blue light results in the phase conjugate beam becoming effectively modulated. We monitored the evolution of the phase conjugate signal as a function of the blue beam power while holding the power of the four-wave-mixing red beams fixed. The forward pump and the probe were set at 67 mW/cm² and the backward pump was set at 55 mW/cm². The data obtained are shown in Fig. 14. Different curves will be obtained if we were to change the intensities of the DFWM beams; the peak will shift toward higher or lower intensities with an increase or decrease of the intensities of the red beams, respectively. The great sensitivity of the chemically stabilized bR film is due to the low saturation intensity. Thus, the beams used in the DFWM setup can be extremely weak. We were able to use 0.8 mW/cm² for the backward pump, 0.75 mW/cm² for the forward pump, and 0.78 mW/cm² for the probe. These powers can easily be achieved with currently available low-power laser diodes. The blue laser is chopped in pulses with average powers in the range 60–500 μW/cm². The use of this type of geometry and the wavelengths chosen make the light modulation process dependent only on the dynamics of the B to M transition and the control of the photochemical transition back to the B state triggered by the blue light. We thus take advantage of the great sensitivity afforded by the low saturation intensity while not being constrained on a time scale by the long relaxation lifetime. We have been able to attain modulation

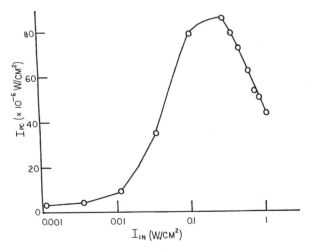

FIGURE 14 Evolution of the phase conjugate signal intensity as a function of the incident 458-nm light intensity. The forward pump and the probe were set at 67 mW/cm² and the backward pump was set at 55 mW/cm².

with 0.4-ms pulses with repetition rates of 250 Hz, limited only by the speed of the rotating mirror used in the experiments. An added advantage is that this technique affords grate signal-to-noise contrast because the phase conjugate beam will be either nonexistent or present. It could also be treated as an all-optical switch where a low-power blue pulse switches on the red signal beam. A picture of the light modulated phase conjugate signal is shown in Fig. 15. We can easily extend the use of this technique to obtain spatial light modulation by the use of a pixelated chemically enhanced bR sample which would allow us to implement optical processing algorithms.

C. All-Optical Logic Gates with Bacteriorhodopsin Films

We demonstrated an all-optical switch using molecular states in a bacteriorhodopsin thin film. All-optical logic gates were implemented with wild-type and chemically stabilized films of bacteriorhodopsin using a two-color backward degenerate four-wave-mixing geometry. Each of the two wavelengths in the experimental system acts as an input to the all-optical gate and the phase conjugate signal beam bears the output of the gate.

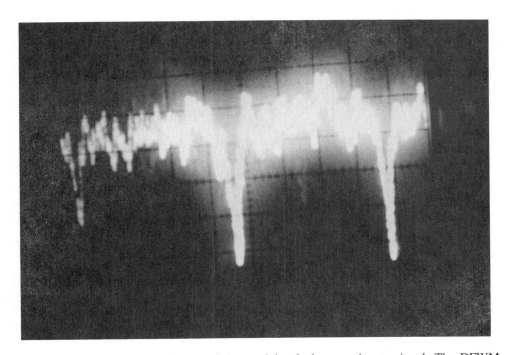

FIGURE 15 Oscilloscope picture of the modulated phase conjugate signal. The DFWM beams were set at 0.8 mW/cm² for the backward pump, 0.75 mW/cm² for the forward pump, and 0.78 mW/cm² for the probe. The blue laser is chopped in pulses with average intensity in the range 60–500 μW/cm². The horizontal scale on the photograph is 1 ms/division. High-frequency noise is due to radio-frequency background from the laser affecting the photomultiplier tube.

Haronian and Lewis [54] demonstrated the microfabrication of electroded bR thin-film pixels of 50 μm $\times$ 50μm size on quartz substrates. They report that the size of the bR micropixels was limited only by the mask used and diffraction issues. So, in principle, micrometer, size bR devices are feasible. Light-addressed (all-optical) logic gates will undoubtedly play a fundamental role in the development of optical and optoelectronic systems for future information technologies. We developed a method of implementing both AND and OR all-optical logic gates. All other gates can be developed using combinations of these two. The nonlinear optical media used in the experiments are wild-type and chemically stabilized films of bacteriorhodopsin. A different approach to implement an AND gate in bR using sequential photoexcitation has been proposed [55]. However, to the best of our knowledge, an OR gate, which is essential to implement optical information processing, cannot be attained with this scheme.

The experimental arrangement for implementing the optical gates is shown in Fig. 16. It consists of a Spectra Physics He–Ne CW laser with output at 632 nm which is used as the source of red light. A Coherent Innova 70 Spectrum argon–krypton CW laser tuned to give 458-nm-wavelength laser light is used a the source of blue light. The geometry is the same as for a one-color DFWM, except in this case we have two wavelengths following the same paths. Both the blue and red beams form holographic gratings in the films. Mechanical shutters are used to control the blue and red beams.

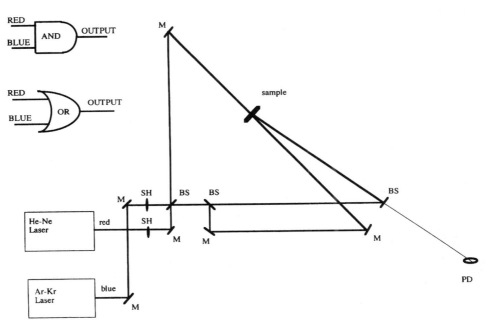

FIGURE 16 Experimental arrangement for all-optical logic gates. M-mirror, SH-mechanical shutter, BS-beamsplitter, PD-photodiode. Red input is 633 nm wavelength and blue input is 458 nm wavelength.

We used a chemically stabilized bR film to implement the function of the AND gate. When red light is only present in the region of interaction, the B-to-M transition will be locally saturated in the regions of constructive interference. Light scattering by inhomogeneities in the film will further saturate the regions where the red beams interfered destructively with the outcome that all the bR molecules in the region will be in the M state. In an analogous fashion, when blue light is only present, all the bR molecules in the region will be in the B state. The absorption of the red or blue light is saturated in each case. Saturation is observed at very low intensities of the incident light because of the long thermal relaxation time of the M-to-B transition, which is of the order of tens of seconds. In both cases, no index grating will exist because all the molecules remain in the M or B state. The observed saturation intensity I_S for our film is 1.3 mW/cm^2 [56]. When both wavelengths are simultaneously present, the two holographic gratings can coexist. The effect of the simultaneous presence of the two wavelengths is an increase in the saturation intensity by at least two orders of magnitude. A complex index grating will therefore be formed and a phase conjugate beam bearing both wavelengths will appear. Light scattered at both wavelengths prevents the local saturation of bR molecules in either B or M states, provided that the relative intensity of blue to red light is appropriate. We are currently using low intensities of the order of 25 mW/cm^2 for the blue and red beams before splitting them into the DFWM arrangement. The system functions as an AND gate because only when both wavelengths are present there is a phase conjugate signal. When only blue is present, all bR molecules are in the B state and there is no signal; similarly, if there is only red light, it will keep the bR molecules in the M state and there is no signal. Finally, we have the trivial case of neither red nor blue input. The output of the gate is shown in Fig. 17.

An OR gate is implemented at the same intensity levels by the use of a wild-type bR film, for which the saturation intensity for red and blue wavelengths is higher than that of the chemically enhanced bR film. We observe phase conjugate signals with this film when both wavelengths are present, as well as when either blue or red is singly present, owing to the fact that at this level of intensity (25 mW/cm^2), we are well below the saturation of bR into either state. When only blue light is on, we observe a phase conjugate signal. The reason is that when both wavelengths are off, not all the molecules are in the B state. If the gate is operated under ambient light conditions, there will always be a small fraction of molecules in the M state due to the ambient light. If the gate is operated in absolute dark again, a fraction of the molecules are in the M state (roughly 50% of them); this is called the dark adapted state. Problems may arise if the gate is operated under ambient light composed spectrally of only deep blue, violet, and near-ultraviolet wavelengths. Thus, the phase conjugate signal will be absent only when both beams are turned off. The output of the gate is shown in Fig. 18.

Several advantages of this technique are noteworthy. First, we must emphasize that both the gates are implemented with the same experimental setup. The only relevant parameter that determines operation as an AND or an OR gate is the saturation intensity of the bR sample. This implies that any of the previously mentioned methods known to affect the lifetime of the M state can be used to tailor the optical gates for a given application. Conversely, by adjusting the intensities of

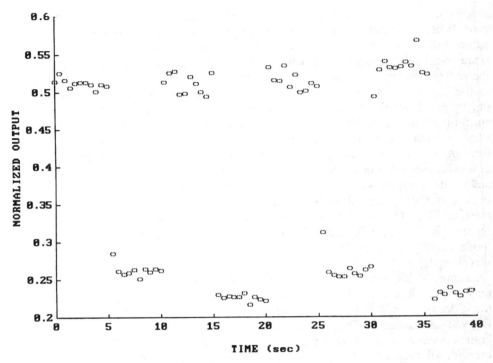

FIGURE 17 AND gate output as a function of time in seconds and input states: 0–5, red and blue on; 5–10, red on, blue off; 10–15, both on; 15–20, both off; 20–25, both on; 25–30, blue on, red off; 30–35, both on; 35–40, both off.

the red and blue beams, either film can be used as both the AND and OR gates. The combination of these factors thus offers remarkable engineering flexibility. The technique affords adequate signal-to-noise contrast because the logic states are separated by three divisions, as shown in Figs. 17 and 18. The phase conjugate beams at both wavelengths travel in the same direction. Care must be taken in the event of cascading gates that the medium in which the phase conjugate output signal is to propagate is transparent to both wavelengths.

The limiting factor for the switching speed of the gates would be the B-to-M transition time and the M-to-B photochemically induced transition time. For water suspensions of bR, these times are known to be of the order of microseconds and nanoseconds, respectively. For our films, these times are unknown and they are the subject of ongoing research. The speed of the gates could be controlled in several ways: first, through the use of pulsed lasers that would deliver sufficient energy to saturate either transition in a short time; second, by optimizing the separation between successive input pulses; and third, by modifying the sensitivity by advanced biotechnology.

With the bR films currently available the limiting speed of operation of the AND gate is of the order of microseconds because transient gratings may exist during the time that it takes to saturate the B-to-M transition. Considering the vigorous activity to develop new bR-related materials for a variety of applications,

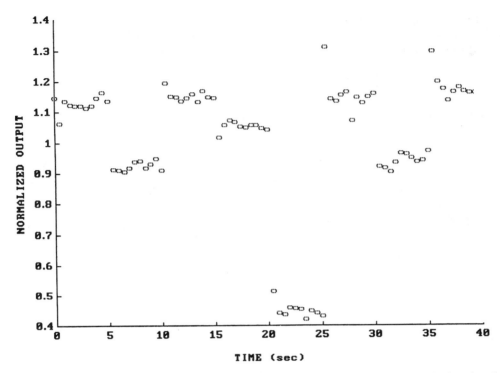

FIGURE 18 OR gate output as a function of time in seconds and input states: 0–5, red and blue on; 5–10, red on, blue off; 10–15, both on; 15–20, blue one, red off; 20–25, both off; 25–30, both on; 30–35, red on, blue off; 35–40, both on.

it is quite likely that advances in chemical synthesis and bioengineering methods will enable the control of the B-to-M transition time and the confinement of any transients in the AND gates to shorter time scales.

A careful choice of wavelengths would permit the adjustment of relative diffraction efficiency so as to enhance the relative amount of one wavelength to another in the phase conjugate output beam. By adjusting the relative blue to red light intensities, we were able to control the intensity of each in the output beam.

V. NONLINEAR REFRACTIVE INDEX

With the advent of lasers, it was observed that the index of refraction of a material varies with the incident intensity. This phenomenon is observed for most types of materials whether isotropic or not. The effective index of refraction of a substance can then be described by [42]

$$n = n_0 + n_2 I \tag{14}$$

where n_0 is the linear index of refraction, I is the intensity of the optical beam, and n_2 is the nonlinear index of refraction or, as it is also called, the second-order index of refraction. The intensity is defined as the square of the time-averaged electric field. The nonlinear index of refraction gives rise to self-focusing and self-defocusing of laser beams. Self-focusing and self-defocusing are shown schematically in Figs. 19 and 20, respectively. If we write the time-averaged electric field as

$$E(t) = E(\omega)e^{-i\omega t} + \text{complex conjugate (c.c)} \tag{15}$$

we can rewrite the effective index of refraction as

$$n = n_0 + 2n_2|E(\omega)|^2 \tag{16}$$

This last equation implies a change in the refractive index $\Delta n = 2n_2|E(\omega)|^2$, which is called the AC Kerr effect. Historically, it was known for a long time that the index of refraction of a material could be changed by applying a DC field across the material, this was known as the electrooptic Kerr effect or DC Kerr effect.

In the polarization expansion, the term which affects the propagation of a beam of light through a nonlinear medium is

$$P(\omega) = 3\chi^{(3)}(\omega = \omega + \omega - \omega)|E(\omega)|^2E(\omega) \tag{17}$$

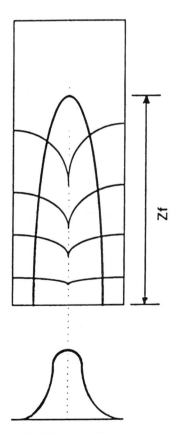

FIGURE 19 Self-focusing on a Gaussian beam in a nonlinear medium.

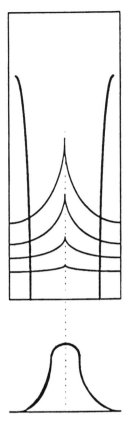

FIGURE 20 Self-defocusing of a Gaussian beam in a nonlinear medium.

for a linearly polarized beam of light. The overall polarization taking into account linear and nonlinear terms is

$$P_{\text{total}} (\omega) = \chi^{(1)} E(\omega) + 3\chi^{(3)} (\omega = \omega + \omega - \omega)|E(\omega)|^2 E(\omega) \equiv \chi_{\text{effective}} E(\omega) \tag{18}$$

where we have defined the effective susceptibility of the medium as

$$\chi_{\text{effective}} = \chi^{(1)} + 3\chi^{(3)}(\omega = \omega + \omega - \omega)|E(\omega)|^2 \tag{19}$$

If we now make use of the general definition of the refractive index

$$n^2 = 1 + 4\pi\chi_{\text{effective}} \tag{20}$$

we obtain, upon substitution, a relation of the form

$$[n_0 + 2n_2|E(\omega)|^2]^2 = 1 + 4\pi\chi^{(1)} + 12\pi\chi^{(3)} (\omega = \omega + \omega - \omega)|E(\omega)|^2 \tag{21}$$

which, if we expand it and keep only terms up to the square of the field, implies that

$$n_0 = (1 + 4\pi\chi^{(1)})^{1/2} \tag{22}$$

and that the relation between n_2 and the third-order susceptibility is [42]

$$n_2 = \frac{3\pi\chi^{(3)}}{n_0} \tag{23}$$

Which is valid for a one-beam type of experiment. If we had two beams (a pump and a probe), the degeneracy factor 3 is changes to 6, whether the probe is at the same frequency as the pump or not.

A more common way of describing the above is by defining the intensity as

$$I = \frac{n_0 c}{2\pi} |E(\omega)|^2 \tag{24}$$

which is the averaged intensity of the optical field. We then have that the relation between n_2 and the third-order susceptibility is [42]

$$n_2 = \frac{12\pi^2\chi^{(3)}}{n_0^2 c} \tag{25}$$

A word of caution concerning the units is in order. In most articles found in the literature, the intensity is measured in W/cm^2, which implies that the mixed cgs, MKS units of n_2 are cm^2/W. The conversion relations then are given by

$$n_2\left[\frac{cm^2}{W}\right] = 10^7 \frac{12\pi^2\chi^{(3)}(esu)}{n_0^2 c} = \frac{0.0395}{n_0^2}\chi^{(3)}(esu) \tag{26}$$

bR has a mixed absorptive–dispersive nonlinearity. It behaves like a saturable Kerr medium; thus its complex index of refraction is better described by [57]

$$n = n_0' + in_0'' + (n_2' + in_2'')\frac{I}{1 + I/I_s} \tag{27}$$

The real part of this expression is the effective index of refraction for a bR type of system.

VI. NONLINEAR ABSORPTION

The imaginary part of the third-order susceptibility is responsible for nonlinear absorption. Two-photon absorption (TPA), saturable absorption (SA), and excited-state absorption (ESA) are the most relevant types of nonlinear absorptive processes [58–63].

Transitions that involve one photon and transitions that involve two photons have different selection rules. Two-photon processes involve the simultaneous absorption of two photons to excite a material.

Saturable absorption involves the saturation of a given transition, by populating an excited state of the material so that the material which initially absorbed at that wavelength becomes more transparent.

Excited-state absorption involves a sequential process in which a photon is initially absorbed and the molecule remains in an excited state for a finite length

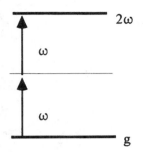

FIGURE 21 Two-photon absorption.

of time so that a second photon that arrives during that time is also absorbed to put the molecule into an even higher excited state. The basic difference between two-photon and excited-state transitions is that the former involve intermediate extremely short-lived virtual states, whereas the latter involve intermediate real states whose lifetime is not determined by the Heisenberg uncertainty relations, but, instead, their lifetimes are determined by the electronic structure of the molecules in the materials. Two-photon-absorption processes are dependent on the intensity of the incident light, whereas excited-state-absorption processes are dependent on the fluence (energy per unit area) of the incident light. A schematic indicating the qualitative aspects of two-photon absorption is shown in Fig. 21. A schematic indicating the process of saturable absorption for a general molecular four-level system is shown in Fig. 22. A schematic indicating the process of excited-state absorption for a general molecular five-level system is shown in Fig. 23. The two processes can be resolved by obtaining measurements of the samples for different incident pulse widths.

Our experiments exploit the dynamics of the B and M states which are the two most stable states in the bR photocycle. These states are not electronic excited states because they correspond to two different conformations of the retinal molecule. B corresponds to the all-trans configuration and *M* to the 13-cis configuration. For all practical purposes, we can neglect the remaining short-lived intermediate states of the photocycle. We may then analyze the saturation dynamics of bR

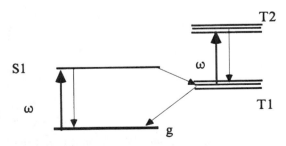

FIGURE 22 Saturable absorption.

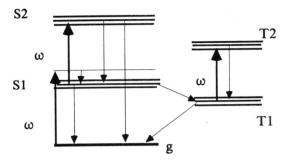

FIGURE 23 Excited-state absorption.

using a simple two-level model [10]. Two-level schematics for the bR states are shown in Fig. 24. The population at M and B states can be described by a rate equation:

$$\frac{dM}{dt} = \sigma_1 FB - \frac{M}{\tau} - \sigma_2 FM \tag{28}$$

where F is the photon density flux of the incident beam, M and B are the populations per unit volume in the M and B levels, respectively, σ_1 and σ_2 are the absorption cross sections for the nonradiative transitions B to M and M to B, respectively, and τ is the relaxation time for the transition M to B. The steady-state solution of Eq. (28) yields the population in the M and B states as

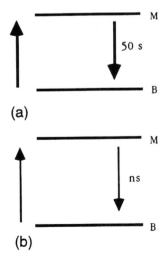

FIGURE 24 Two-level diagrams for bR: (a) in the absence of blue light, (b) in the presence of blue light.

$$B = N \left(\frac{1 + \sigma_2 F \tau}{1 + (\sigma_1 + \sigma_2) F \tau} \right) \tag{29}$$

with $M = N - B$, where N is the density of bR molecules in the sample. The intensity-dependent nonlinear absorption is described by

$$\alpha = N\sigma_1 \left(\frac{1 + 2\sigma_2 F \tau}{1 + (\sigma_1 + \sigma_2) F \tau} \right) = \alpha_0 - \frac{gI}{1 + I/I_s} \tag{30}$$

$$\frac{dI}{dz} = -\alpha(I)I$$

where $\alpha_0 = N\sigma_1$, $g = N\sigma_1(\sigma_1 - \sigma_2)\tau/h\nu$, and the saturation intensity $I_s = h\nu/(\sigma_1 + \sigma_2)\tau$. For light of wavelengths close to or greater than 570 nm, $\sigma_1 \gg \sigma_2$ and $\alpha(I)$ exhibits saturable absorption. The term α_0 includes linear absorption as well as any loss due to scattering.

A. Z-Scan Measurements

The Z-scan technique is a single-beam technique which allows the determination of the real and imaginary parts of the third-order susceptibility. In the Z-scan experiments [60], the transmittance of the nonlinear medium through a finite aperture in the far field is measured as a function of the sample position Z to determine the real part of the susceptibility. A schematic of the Z-scan technique is shown in Fig. 25. On the other hand, the transmittance of the nonlinear medium without an aperture is measured as a function of the sample position Z to determine the imaginary part of the susceptibility. The zero position for Z is taken to be that of the focal plane. A TEM$_{00}$ Gaussian beam traveling in the $+Z$ direction is given by

$$E(z,r,t) = E_0(t) \frac{\omega_0}{\omega(z)} \exp\left[-\left(\frac{r^2}{\omega^2(z)} \right) - \left(\frac{ikr^2}{2R(z)} \right) \right] e^{-i\phi(z,t)} \tag{31}$$

where $\omega^2(z) = \omega_0^2 (1 + z^2/z_0^2)$ is the beam radius, $R(z) = (1 + z_0^2/z^2)$ is the radius of curvature of the wave front at z, $z_0 = k\omega_0^2/2$ is the confocal parameter of the

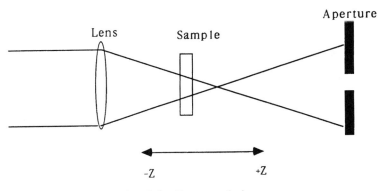

FIGURE 25 Schematic of the Z-scan technique.

beam, $k = 2\pi/\lambda$ is the wave vector, and λ is the laser wavelength, all of these parameters assumed in free space. If the sample length L is smaller than the confocal parameter z_0, the sample can be considered thin. Under such an assumption, the amplitude and phase of the electric field as a function of z' are now governed in the slowly varying envelope approximation by a pair of differential equations:

$$\frac{d\Delta\phi}{dz'} = \Delta n(I)k \tag{32}$$

$$\frac{dI}{dz'} = -\alpha(I)I \tag{33}$$

where z' is the propagation depth in the sample and the absorption $\alpha(I)$ is assumed to comprise linear and nonlinear absorption terms. If we solve Eqs. (32) and (33), we can obtain the complex electric field exiting the sample. The far-field pattern of the beam at the aperture plane can be obtained through the Gaussian decomposition method.

The normalized transmittance $T(z)$ for measurements taken in the far field through a small aperture plotted versus the sample position exhibits a valley–peak trace for the positive refractive nonlinearity as the sample is translated across the focal plane from $-z$ to $+z$ and a peak–valley trace for a negative refractive nonlinearity. The peak and valley separation is experimentally found to remain nearly constant for large phase distortions and is given by

$$\Delta z_{p-v} \cong 1.7z_0 \tag{34}$$

To a first approximation, the magnitude of the phase distortion can be found either by computer simulation or by using the empirical formulas given by Sheik-Bahae et al. [60]:

$$\Delta T_{p-v} \cong 0.406(1 - S)^{0.25} |\Delta\Phi_0| \cong 0.406(1 - S)^{0.25} \left| \frac{\pi n_2}{\lambda} |E|^2 L_{\text{eff}} \right| \tag{35}$$

for $|\Delta\Phi_0| \leq \pi$, where $S = 1 - \exp(-2r_{\text{ap}}^2/\omega_{\text{ap}}^2)$, and r_{ap} and ω_{ap} are the radius of the aperture and the radius of the beam spot at the location of the aperture, respectively.

Experimentally, one can first obtain a closed-aperture Z-scan for a reference sample in order to establish empirically the radius of the focal spot (thus avoiding errors due to the calculation based on the assumption of a diffraction limited spot) and thereby the intensity at the focal plane. Furthermore, the value of the real part of $\chi^{(3)}$ for the unknown sample can be readily obtained by simply comparing the observed ΔT_{p-n} for the reference with that of the sample obtained under identical conditions. They are related by

$$\chi^{(3)}_{\text{sample}} = \chi^{(3)}_{\text{reference}} \frac{(\Delta T_{p-v} n_0)_{\text{sample}}}{(\Delta T_{p-v} n_0)_{\text{reference}}} \tag{36}$$

Alternatively, once $\Delta\phi_0$ is known, we can calculate the nonlinear index of refraction n_2 as

$$n_2 = \frac{\Delta n_0}{I_0} \tag{37}$$

where

$$\Delta n_0 = \frac{\Delta \phi_0}{k L_{\text{eff}}} \tag{38}$$

and

$$L_{\text{eff}} = \frac{1 - e^{-\alpha L}}{\alpha} \tag{39}$$

with k the wave vector, L the sample length, α the linear absorption coefficient, and I_0 the on-axis irradiance at the focus and is given by $I_0 = 2P/\pi\omega_0^2$, where P is the power in watts and ω_0 is the radius of the waist at the focus. Values of n_2 for the pure sample are obtained from the solution measurements by multiplying them by ρ/C, where ρ is the density of the solute and C is the solute concentration in g/cm^3. The values of n^2 and $\chi^{(3)}$ are related in MKSA units by [60]

$$\text{Re } \chi^{(3)} = 2n_0^2 \varepsilon_0 c n_2 \tag{40}$$

where n_0 is the linear index of refraction, ε_0 is the dielectric constant, and c is the speed of light.

Open-aperture Z-scans allow for the determination of the intensity-dependent changes in transmission.

B. Z-Scan Results

We studied the effective nonlinear refractive-index coefficient n_2 of wild-type bR suspensions in water using the Z-scan technique with a low-power continuous-wave laser at 647.1 and 488.0 nm wavelength [64]. We used a wide range of powers at 647.1 nm wavelength, where the absorption is small and the change in refractive index in accordance with the Kramers–Kronig dispersion relations for the changes in absorption due to the B-to-M photochromic transition is expected to be large to establish the basic mechanism responsible for the high nonlinearity. Our results indicate that the magnitude and the sign of n_2 depend strongly on the light intensity and excitation wavelength, respectively. Negative values for n_2 are obtained with 647.1 nm excitation wavelength. Positive values for n_2 are obtained with 488.0 nm excitation wavelength. The observed self-defocusing and self-focusing phenomena can be attributed to the index change due to the light-induced transition between the photochromic states. The results elucidate the origin of n_2. The bR samples used in the experiments were water suspensions of wild-type bR. The concentration of the samples used is 3.84 mg/ml. The samples had a pH of 7 and the measurements were obtained at 20°C. The linear absorption coefficient α at 647.1 nm is 3.01 cm^{-1}. The samples were held in 2-mm path-length spectroscopic cuvettes.

In our experiments, $\omega_0 = 4.5 \times 10^{-3}$ cm. Values of n_2 for the pure sample are obtained from the solution measurements by multiplying them by ρ/C, where ρ is the density of bR and C is the solute concentration in g/cm^3.

Negative values for n_2 in the range -1.0×10^{-4} to -3.0×10^{-3} cm^2/W are obtained for on-axis laser irradiances at the focus in the range 3.7×10^2 to 6.0×10^0 W/cm^2 at 647 nm wavelength. All the measurements are carried out with the same sample and same experimental setup.

Open-aperture and Z-scans allow for the determination of the intensity-dependent changes in transmission. The bR solution exhibits saturable absorption at both wavelengths. The results obtained by open-aperture Z-scan experiments are shown in Figs. 26 and 27. The closed-aperture Z-scan measurements contain contributions from both the intensity-dependent changes in the transmission and the intensity-dependent changes in the refractive index. The phase distortion produced by the change in the index of refraction can be retrieved by dividing the normalized closed-aperture Z-scan data by the normalized open-aperture Z-scan data. The results obtained from the division for both wavelengths are shown in Figs. 28 and 29.

The dependence of n_2 on the light intensity at both wavelengths indicates that the self-defocusing and self-focusing observed are not due to the purely electronic nonlinearity because in that case, the value of n_2 would have been independent of the light intensity. The observed changes in index of refraction arise as a result of the photochromic transitions in the bR photocycle. This changes can be interpreted by the Kramers–Kronig dispersion relation that relates the changes in the index of refraction to the changes in absorption. For a photochromic medium, we can write a Kramers–Kronig type of dispersion relation where if we denote $A_i(\lambda)$ and $A_f(\lambda)$ for the absorption of the initial and final states as a function of wavelength, respectively, then the corresponding change in refractive index is given by

$$\Delta n(\lambda) = \frac{2.3026}{2\pi^2 t} \, \text{P.V.} \int_0^\infty \frac{A_f(\lambda') - A_i(\lambda')}{1 - \frac{\lambda'^2}{\lambda^2}} \, d\lambda' \tag{41}$$

where $\Delta n(\lambda)$ is the change in index of refraction as a function of wavelength, P.V. denotes the Cauchy principal value of the integral, and t is the thickness (nm) of the sample. For bR, the most stable intermediate is the long-lived M state. Gross et al. [53] showed that the change in refractive index associated with the absorption change that results from the B-to-M transition is negative for wavelengths longer than 570 nm in the visible and near-infrared and positive for wavelengths shorter than 570 nm. We thus expect a negative n_2 at 647.1 nm and a positive n_2 at 488 nm. Thus, the observed sign of the n_2 at both wavelengths is consistent with that predicted by the Kramers–Kronig dispersion relation for photochromic transition between B and M states. Effects such as the temperature dependence of the index of refraction governed by the thermooptic coefficient dn/dt of the refractive index that would result in thermal defocusing can be ruled out because we obtain identical results when the experiments are carried out with a chopped beam. Thermal effects were investigated by Werner et al. [19] and found to be negligible compared to the observed changes in index.

The intensity dependence of n_2 can be understood if we take into account the nonlinear saturable absorption. When the bR sample contains a mixture of molecules in i (B) and f (M) states, the refractive index of the material can be more accurately written as

$$n = n_i C_i + n_f C_f \tag{42}$$

in the steady state where n_i and n_f are the refractive indexes of the initial and final states, respectively, and C_i and C_f are the ratios of molecules in each state to the total number of molecules. Using the two-level rate equation, we have

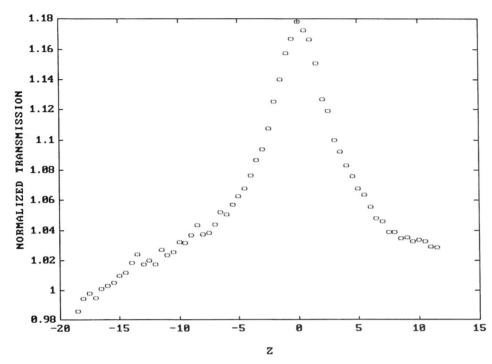

FIGURE 26 Open-aperture Z-scan of bR water solution at 488 nm wavelength indicating saturable absorption.

$$n = n_i + \frac{\Delta n}{I_S} \frac{I}{1 + I/I_S} \tag{43}$$

where I_S is the saturation intensity and Δn is the change in index of refraction given by the Kramers–Kronig relation. From this relation, we can derive n_2 as

$$n_2 = \frac{\Delta n}{I_S} \frac{I}{1 + I/I_S} \tag{44}$$

This last equation implies that n_2 is a function of the intensity when the light intensity is much smaller than I_S, $n_2 = \Delta n/I_S$. However, as the intensity of the probing light exceeds the saturation intensity, n_2 starts to decrease. The discrepancies in the values previously reported in the literature could be accounted for in terms of the intensity at which the measurements were made. In addition, the equation also shows that n_2 is also strongly dependent on the saturation intensity which is a function of the lifetime of the photochromic M state. It has been known that the lifetime of the M state varies depending on pH, humidity, and other environmental factors. These factors could also affect the n_2 value measured. Equation (44) was used to fit the experimental data obtained from the Z-scan measurements at different intensities. The value of Δn used was 0.0049 which was calculated from the Kramers–Kronig formula [Eq. (41)] for our bR sample under the assumption

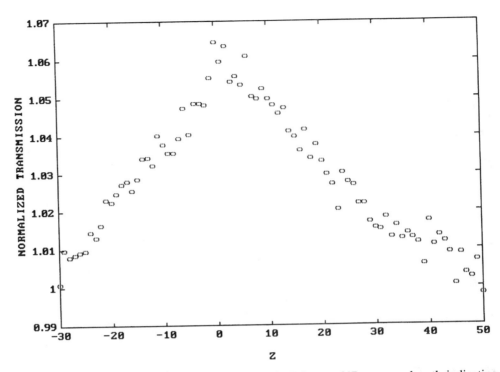

FIGURE 27   Open-aperture Z-scan of bR water solution at 647 nm wavelength indicating saturable absorption.

of 100% photoconversion. The best fit yields a value for the saturation intensity of 2.355 W/cm². The saturation intensity is related to the lifetime of an excited state by $I_S = h\nu/\sigma\tau$ where h is Planck's constant, ν is the photon frequency, and σ is the absorption cross section. For bR, σ is 2.15×10^{-17} cm². The value for τ thus obtained is 6 ms, which is consistent with the values for the lifetime of the M state for water solutions of wild-type bR reported in the literature. Our results also agree with the values previously reported [19,65] for n_2 at the intensities that were used in those experiments.

VII.   OPTICAL BISTABILITY

Optical bistability was first observed in sodium vapor [66] in 1974; it had been previously present in lasers.

Current research in optical bistability focuses on decreasing the size of the bistable devices, improving the switching time, decreasing the operating power requirements, and achieving reliable operation at room temperatures.

Any system that has two possible output states for a given input state is said to be bistable. If the system has multiple output states for a given input state, then it is said to be multistable. In an optical system, bistability implies the possibility

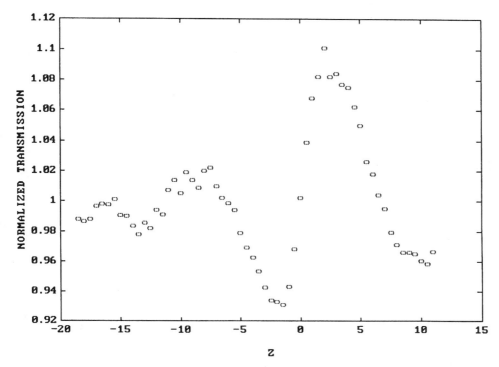

FIGURE 28 *Z*-scan of bR water solution at 488 nm wavelength (result of dividing the closed and open apertures) indicating a positive value for n_2.

of having two different output intensities for one given input intensity through some range of input intensities. Optical bistability thus gives rise to hysteresis loops in the graph with output intensity plotted versus input intensity. A typical bistable loop is shown in Fig. 30. This definition is not meant to comprise irreversible systems (e.g., damaged or burnt material by the action of light) which are, of course, bistable, but the initial state cannot be recovered. We are speaking about reversible systems that can be reset by merely reducing the intensity of the input light.

Nonlinearity alone is not sufficient to assure bistability. Bistable optical systems require feedback. Bistable systems can be subdivided into several groups:

Extrinsic: Feedback is supplied by an external source (e.g., external signal supplied from a detector monitoring the transmitted intensity, external DC field, mirrors, etc.).

Intrinsic: Feedback is supplied by some mechanism inherent to the medium itself; that is, the intensity dependence arises from a direct interaction of light and matter (also referred to as all-optical).

Dispersive: Bistability arises as a result of changes in the refractive index of the material; feedback thus occurs by way of an intensity-dependent refractive index.

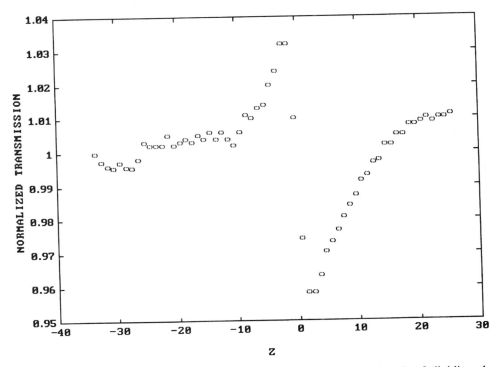

FIGURE 29 Z-scan of bR water solution at 647 nm wavelength (result of dividing the closed and open apertures) indicating a negative value for n_2.

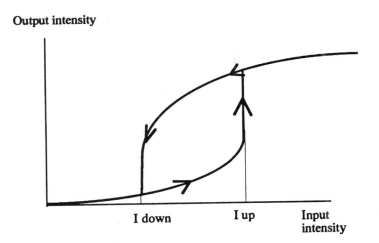

FIGURE 30 Typical bistable loop.

Absorptive: Bistability arises as a result of changes in the absorption of the material; feedback thus occurs by way of an intensity-dependent absorption.

A common example of a bistable optical system is a Fabry–Perot interferometer filled with a saturable absorber with saturation intensity I_S. In such a case, the intensity inside the interferometer can be written as

$$I_{\text{inside cavity}}(L) = e^{-\alpha L} T_{\text{mirror}} I_{\text{input}} \tag{45}$$

where α is the linear absorption coefficient, T is the transmission coefficient of the mirror, and L is the length taken along the optic axis of the interferometer. The transmitted intensity is given by

$$I_{\text{transmitted}} = e^{-\alpha L} T_{\text{mirror}}^2 I_{\text{input}} \tag{46}$$

The second equation holds as long as $I_S > T_{\text{mirror}} I_{\text{input}}$. When the intensity inside the interferometer becomes much larger than the saturation intensity of the medium; that is, the input intensity is larger than the product of the transmission coefficient of the mirror with the saturation intensity of the medium, and the intensity of the light transmitted through the etalon will be approximately equal to the input intensity. This is very similar to what happens when lasers are Q-switched with saturable dyes. At this point, the intensity inside the etalon will equal the ratio of the transmitted intensity divided by the mirror transmission coefficient. The possibility of bistability is then suggested by noting that the following two relations can be satisfied by the same input intensity $I_S = I_{\text{input}}$:

$$I_S > T_{\text{mirror}} I_{\text{input}}$$
$$I_{\text{input}} > T_{\text{mirror}} I_S \tag{47}$$

with the only requirement that the mirror transmission be less than 1 (less than 100%), which is always the case. We therefore have two output states for one given input intensity. Both cases are pictorially represented in Figs. 31 and 32.

A. Optical Bistability Without External Feedback

Optical bistability can be observed in systems lacking external feedback [66–70]; in our previous example, if we removed the mirrors, we would be able to observe mirrorless decreasing absorption optical bistability. In materials for which the absorption increases with increasing intensity [materials that possess reverse saturable absorption (RSA)] such as materials that exhibit two-photon or excited state absorption, we can observe mirrorless increasing absorption optical bistability.

For both types of material, the absorption is a function of some parameter N of the medium:

$$A \equiv A(N) \tag{48}$$

N could be carrier density, as in semiconductor materials, temperature, excited-state population, and so forth. The material parameter N is, in general, proportional to the incident power. The condition for bistable behavior is

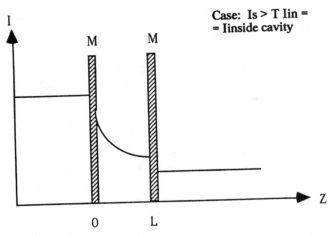

FIGURE 31 Fabry–Perot etalon filled with a saturable absorber in the down state.

$$\frac{dA}{dN} > \frac{A}{N} \tag{49}$$

which just implies that the plot of the absorption as a function of N or input power for that matter must be steeper than a straight line. For this condition to exist, A must be more than linearly proportional to N over some range. The transmission as a function of material excitation for a material exhibiting RSA is shown in Fig. 33.

For the transmission characteristics shown in Fig. 33 where the straight lines labeled A to D represent four different input powers in increasing order, we can plot the corresponding bistable loop as shown in Fig. 34.

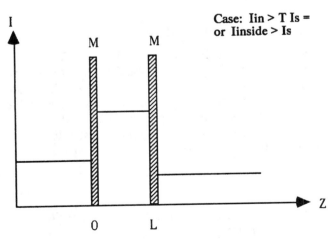

FIGURE 32 Fabry–Perot etalon filled with a saturable absorber in the up state.

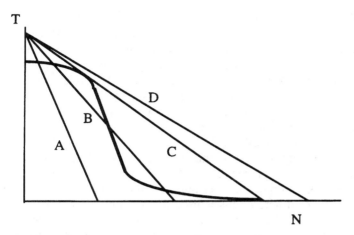

FIGURE 33 Transmission as a function of the material parameter N.

B. Mirrorless All-Optical Bistability

We observed mirrorless all-optical bistability induced by the nonlinear absorption of bR. The nonlinear absorption and the finite lifetime of the excited state permit the existence of two stable output (transmission/absorption) states for a given input [66–69]. Traditionally, Fabry–Perot etalons have been used to provide the feedback necessary in a bistable process [66]. In the bR samples, we observe bistable behavior in the absence of mirrors or any other type of external feedback. When a laser pulse of an appropriate wavelength is incident on a bR sample, initially some of the light may be absorbed due to linear absorption. For sufficiently high inten-

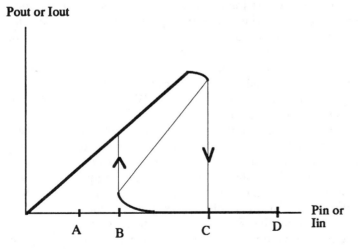

FIGURE 34 Bistable loop resulting from the data in Fig. 33.

sities, nonlinear saturable absorption will commence. A further increase in the intensity will continue to decrease the absorption of the sample, thus providing positive feedback. If the lifetime of the excited state responsible for the nonlinear absorption is sufficiently long, when the intensity of the light begins to decrease, saturable absorption will still be taking place at the same intensity levels at which there was no nonlinear absorption previously. This process gives rise to a second output state at that intensity. The three samples studied were wild-type bR (WT bR) in water solution and in a thin-film dispersed in a polymer matrix, as well as a water solution of the genetic variant bR_{D96N} (Asp96, Asn).

The experimental arrangement for studying intensity-dependent absorption consists of a mode-locked, repetitively pulsed Quantel Nd : YAG laser, frequency doubled to give 532 nm. The duration of the laser pulse is 30 ps. The output energy per pulse is 25 mJ. A 10-cm-focal-length lens is used for the high intensities. The intensity was varied using a standard combination of a half-wave plate retarder and two polarizers. A beamsplitter was used to sample the input beam. Incident intensities are calculated from energy measurements. The uncertainty in the absolute values of the intensity may be +5%, but relative values are more accurate. Total energy transmission is obtained directly from a pair of Molectron pyroelectric energy meters. The meters were interfaced to a personal computer for data acquisition and analysis. The nonlinear sample was placed in the path of the laser beam. Two-millimeter glass cuvettes are used for the water solutions.

All the samples studied displayed bistable behavior. The experimental results are shown in Fig. 35, illustrating hysteresis in the transmission response. The bistability is induced at low incident intensities for dry samples of wild-type bR (thin film) and solutions of the mutant, which for both, the reprotonation process is somewhat inhibited and the lifetime of the M state is long. For water solutions of wild-type bR, on the other hand, the lifetime of the M state is relatively short due to reprotonation and the optical bistability occurs only at intensities two to three orders higher. The hysteresis loops are nonsquare, making the actual switching transitions diffuse. This can attributed to different extinction coefficients for each state causing a differential absorption at the two absorption peaks so that the local absorption is a function of position along the optical path, thus smearing out the sample's response. A similar behavior was observed by Kirkby et al. [71] in an organic material. The experiments were repeated several times with the same set of samples and consistent results are obtained demonstrating the remarkable regenerative properties of bacteriorhodopsin-based materials. In the case of the film, we observe some damage at intensity levels of order 1 GW/cm^2. For the solution samples, we do not see any damage even at these levels.

The experimental points of transmissivity of the samples versus incident intensity are displayed in Fig. 35. The solid lines obtained by the theoretical fit of Eq. (30) closely follow the experimental points and the parameters obtained by best fit are shown in Table 1 for the three samples studied, for both the upper and lower curves. It may be seen from Table 1 that the saturation intensity I_S is about three orders higher for the wild-type bR in water solution compared to the value for thin film and bR-mutant water solution. The ratio of time constants in any two samples can be obtained from the theoretical relations for the parameters at the same wavelength, as the absorption cross sections are the same for all the samples [23].

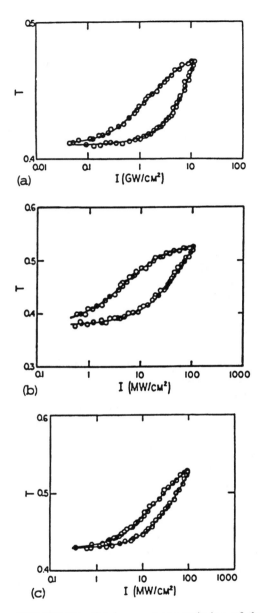

FIGURE 35 Total energy transmission of the bR samples measured as a function of the incident intensity: (a) wild-type bR in water solution, (b) wild-type bR thin film in a polymer matrix, and (c) mutant bR D96N in water solution. The circles represent average values of observed results with a spread of about 3%.

TABLE 1 Values of Best-Fit Parameters for the Samples Studied

Sample	Conc. (mg/ml)	α_0 (cm^{-1})	Trace	I_S (MW/cm^2)	g (cm/GW)
WT bR (water)	0.80	4.45	Upper	1,500	0.105
			Lower	70,000	0.014
WT bR (film)	50.0	277	Upper	4	85.0
			Lower	70	7.7
bR D96N (water)	0.76	4.25	Upper	18	14.0
			Lower	70	5.0

The results indicate that the thin-film sample of wild-type bR and the water solution of the mutant are characterized by relaxation times for the M state about three orders longer relative to the wild-type bR water solution. The increase is attributed to relative values of humidity in the first case and mutagenesis in the second case. This correlates well with the observed behavior of switching at low light intensities (two to three orders) for wild-type bR thin film and water solution of mutant compared to wild-type bR in water solution.

VIII. OPTICAL PROCESSING APPLICATIONS OF PHOTO-INDUCED ANISOTROPY IN BACTERIORHODOPSIN FILMS

We undertook detailed studies on the intensity dependence of photo-induced dichroism in bacteriorhodopsin films in view of applications for optical image processing. Under the illumination of an actinic light of 570 nm wavelength, photo-induced dichroism in a bacteriorhodopsin film induces polarization rotation for a probe beam of the same wavelength. We have successfully exploited the observed behavior of probe polarization for optical Fourier processing. A simple optical image processing system was experimentally implemented, which does not require any precise alignment or vibration isolation systems. Experimental demonstration of image processing applications such as edge enhancement, noise filtering, band-pass filtering, flow visualization, beam profiling, and so forth are presented.

Adaptive spatial filtering in an optical system employing the photo-induced anisotropy of a bacteriorhodopsin film has been used as a conventional Schlieren system. The high-pass filtering characteristics of the system has been applied for flow visualization. The system is self-adaptive with no precise alignment requirements at the filtering plane compared to the conventional system and many other Fourier plane processing systems.

Illumination of bR film at a wavelength near to 570 nm leads to the transition of molecules to the M state, reducing the absorption coefficient of the film at that wavelength, and these molecules are bleached. Upon illumination of the film with a linearly polarized light (actinic beam), the film shows anisotropic properties of photo-induced dichroism and photo-induced birefringence. The magnitude and sign of induced anisotropy are dependent on the actinic light intensity and wavelength. Turning off the actinic light returns the film to its initial isotropic state with a relaxation time equal to that of the intermediate M state. The occurrence of photo-induced anisotropy in a bR film is an effect of the photoselective bleaching of the bR molecules on light illumination. Because the actinic light is linearly polarized and the bR molecules in the film are randomly oriented, only those bR molecules oriented, with their transition dipole moments for absorption, in or near the electric field direction of the light are bleached. Considering the simple case of bR molecule's transition between B and M states, their photostationary state distribution can be written as [20]

$$\sigma_1 IN = \frac{1}{\tau} N_2$$

where σ_1 is the cross section for the B-to-M transition, τ is the lifetime of the M state, I is the light intensity at 570 nm, and N_1 and N_2 are the concentration of bR molecules in the B and M states, respectively.

Taking into account the anisotropy of bR molecules,

$$\sigma_1 = \sigma_\| \cos^2 \phi + \sigma_\perp \sin \phi$$

where ϕ is the angle between the molecule axis and the polarization direction, and $\sigma_\|$ and $\sigma_\perp$ are the molecular cross sections for light polarized parallel and perpendicular to the molecule axis, respectively.

The dichroism shown by photoselective bleaching of bR molecules with polarized light can then be obtained as [20].

$$D_\perp - D_\| = D_0 \frac{4}{a_1(1 + k_1)} \left(\frac{a_1 + k_1 + 2}{2[(1 + a_1)(1 + k_1 a_1)]^{1/2}} - 1 \right)$$

where, D_0 is the optical density, k_1 is the dichroic ratio, and $a_1 = \sigma_\| I\tau$.

When the wavelength of actinic light is ~570 nm, induced dichroism predominates over induced birefringence. The presence of dichroism upon actinic light illumination can produce a rotation of the plane of polarization of a probe beam passing through the dichroic parts of the film. In the following section, we study the dependence of this angle of rotation and hence that of the photo-induced dichroism on the intensities of probe and actinic beams. The actinic-light-induced angular rotation of a probe beam's polarization and its dependence on the probe beam intensity as explained below are the basic principles employed for optical processing described in this chapter.

A. Optical Image Processing

The optical Fourier transform is a powerful tool in optical computing and processing systems [72,73]. Use of nonlinear optical materials for the implementation of Fourier-transform operations such as edge enhancement, bandpass filtering, noise removal, and pattern recognition [5,74] is well established. The parallel processing nature of optics and the real-time characteristics of many nonlinear optical materials contribute to the importance of optical Fourier processing with nonlinear materials. Photorefractive materials are by far the most popular as the nonlinear media for implementation of optical Fourier-transform systems [74,75]. Recently, bR has shown great promise as a nonlinear material for optical computing and processing. Bacteriorhodopsin shows many intrinsic optical and physical properties that make it suitable for use as a real-time spatial light modulator [56,76] and also as an optical storage medium [77]. Its useful optical properties have been used by many researchers to implement applications such as pattern recognition, image subtraction and addition, [5,21] spatial filtering [76], interferometry, holographic correlation, and image transmission with phase conjugation [79]. The spatial filtering performed by Thoma et al. [76] involves a control beam that precisely manipulates spatial frequencies at the Fourier plane. Imam et al. [79] recently demonstrated an incoherent-to-coherent converter, using the photo-induced anisotropic properties of bR thin films. The logarithmic transmission characteristics of bR films were used by Downie [80] to implement optical image processing. Recently Takei and Shimizu [81] used the photo-induced refractive-index change of bR for spatial light modulation.

We use the photo-induced anisotropy in a bR film for achieving a real-time, self-adaptive spatial filtering system for optical Fourier processing. We experimentally demonstrated edge enhancement. The photo-induced anisotropy of bR films depends on the intensities of the illuminating beams. The incoherent-to-coherent converter scheme of Imam et al. [79] used the photo-induced anisotropy that is induced by an actinic beam. We have noticed that at an optimum constant actinic beam intensity, the photo-induced dichroism and the resultant polarization rotation of a probe beam show an intensity dependence. Our system exploits the observed intensity dependence of photoanisotropy in a bR film for Fourier processing applications. The advantages of this system are simplicity and ease of operation, with no requirement for precise alignment at the Fourier plane, vibration isolation, or coherent light. The conventional spatial-filtering technique [72] with selective masking of Fourier frequencies has the disadvantages of requiring mechanical masks to block undesired frequency components at the Fourier plane. We measured the photo-induced dichroism as a function of actinic beam intensity; the results are shown in Fig. 36. The dependence of the induced angular rotation of probe beam polarization as a function of the intensity of the actinic light beam was thus determined using a weak probe beam so that it would not contribute to the photo-induced dichroism. The photo-induced angular rotation of the probe beam polarization reached a maximum for an optimum actinic beam intensity of ~ 10 mW/cm^2 and decreased with a further increase in the actinic beam intensity. Similar behavior was observed by Burykin et al. [82] for photo-induced dichroism. By keeping the actinic beam intensity near 10 mW/cm^2, where the angle of rotation and hence the corresponding dichroism is maximum, we gradually increased the intensity of the probe beam. Figure 37 shows the experimental data on the degree of rotation of probe beam polarization as a function of probe beam intensity under constant actinic beam intensity of 10 mW/cm^2. In our experiment the analyzer was mounted on a rotation stage with a minimum detectable angular rotation of 0.1°. The system is highly sensitive to changes in probe intensity; a more accurate rotation stage could be used. The effect of increasing the probe beam intensity is to decrease the degree of polarization rotation of the probe beam because high probe intensities reduce the anisotropy of the bR film. With intensities of ~ 10 mW/cm^2 for the probe beam, the angular rotation experienced by the probe beam polarization is nearly zero. The result from Fig. 37 that increasing the probe beam's intensity reduces the angle of rotation of the probe beam passing through the bR under the influence of actinic light can be effectively exploited for optical Fourier processing.

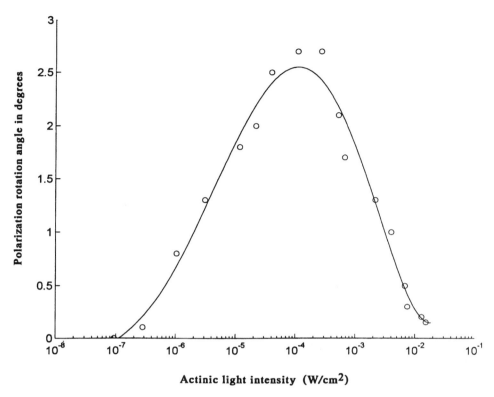

FIGURE 36 Dependence of photo-induced polarization rotation on actinic beam intensity. The solid curve is intended solely as a visual aid.

Figure 38 is a schematic of the optical Fourier processing system. Lens L1 forms the Fourier transform of the object information (O) at the bR film (bR), and lens L2 forms the inverse Fourier transform at the CCD plane to yield the processed image. The actinic beam illuminates the film uniformly. Both the actinic beam and the probe beam that illuminate the object (O) are derived from an Ar–Kr laser tuned to 570 nm wavelength. The actinic beam is linearly polarized at 45° with respect to the plane of polarizer P. Initially, with no actinic beam present, the polarizer (P) and the analyzer (A) are crossed with respect to each other. The probe beam illuminating the object produced an intensity of >10 mW/cm^2 at the center of the Fourier transform of O formed at the bR film plane. On illumination of the bR with an actinic beam of intensity 10 mW/cm^2, only the edges of the object appear at the CCD, yielding high-frequency spatial filtering or edge enhancement. Figure 39 shows the result of edge enhancement; the inner portion of a razor blade was used as the object. This self-adaptive edge enhancement is explained as follows. The Fourier transform of an arbitrary object information formed at the bR plane has an intensity distribution with high intensities for low-frequency components and low intensities for high-frequency components. Figure 37 shows that at optimum actinic beam intensity, the effect of increasing the probe beam's intensity is to decrease its degree of polarization rotation owing to photo-induced dichroism.

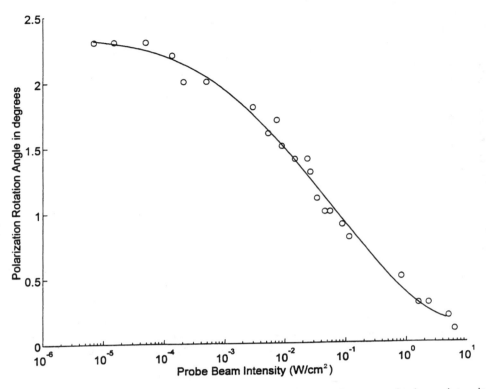

FIGURE 37 Dependence of photo-induced polarization rotation on probe beam intensity with a constant actinic beam intensity of ~10 mw/cm². The solid curve is intended solely as a visual aid.

That means that high-frequency components at the bR plane experience higher degrees of polarization rotation than the zero- and low-frequency components. Hence, if the input object information has a single linear polarization after its passage through the bR under actinic light illumination, it has a range of polari-

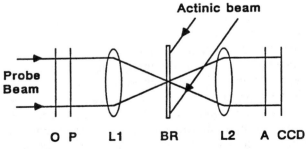

FIGURE 38 Schematic experimental setup for optical Fourier processing using photo-induced dichroism in a bR film.

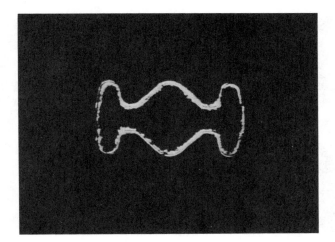

FIGURE 39 Experimental result of edge enhancement with the inner portion of a razor blade as the object.

zations of different orientations. The Fourier processing is accomplished through the analyzer, which blocks specific polarization components and, in turn, blocking the corresponding spatial frequencies. When the analyzer is at a right angles to the input beam polarization, zero- and low-frequency components, which experience almost no polarization rotation because of their high intensities, are blocked by the analyzer. But the high-frequency components that correspond to the edges of the object experience polarization rotation and are thus transmitted through the analyzer to appear at the CCD camera to yield edge enhancement. Rotation of the analyzer serves as a variable spatial filter for Fourier processing.

We extended our investigations on the use of the principle of self-intensity dependence of probe beam polarization rotation for several other optical processing applications. We demonstrated an interesting application of optical Fourier processing by imposing noise on an object to perform low-frequency spatial filtering which amounts to object smoothing and noise removal. A glass plate with black dots, as noise, was placed close to the object, which is the inner portion of a razor blade. Figure 40 shows the noise-embedded object. Now when the bR film is illuminated with actinic light and the analyzer is rotated by ~3° so that the analyzer is crossed to the polarization, rotated high-frequency components of the object yield the result shown in Fig. 41. The appearance of only the low-frequency components at the CCD camera gives image smoothing with the dotted noise removed from the object. Rotating the analyzer to the position where it blocks the low frequencies gives the result shown in Fig. 42, with edge and noise enhancement. Thus, it was possible to get edge enhancement and noise removal operations in a single setup with simple adjustments of the analyzer.

In conclusion, we have successfully demonstrated a self-adaptive optical Fourier-processing system that the photo-induced dichroic characteristics of bacteriorhodopsin film. The dependence of photo-induced dichroism on probe light intensity has effectively been used to imprint continuous polarization variations on the dif-

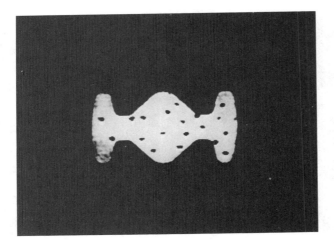

FIGURE 40   Razor blade plus noise object.

ferent spatial-frequency components of an object information. Spatial filtering of desired frequencies is then performed by an analyzer. The simplicity of the present scheme is straightforward, with the processing requiring no alignment at the Fourier plane.

No interference recordings are involved in this experiment; hence, vibration isolation systems are not required. A coherent source is not a requirement for the experiment; a white-light source with an appropriate wavelength filter at ~570 nm can induce photoanisotropy in the bR film. The improved performance of the present system compared with that of other techniques is based on the advantage that a single optical setup can perform low-pass filtering and high-pass filtering oper-

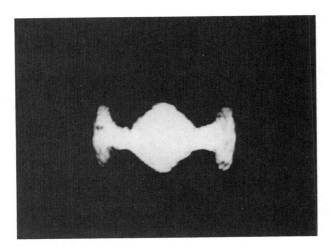

FIGURE 41   Experimental result of noise removal.

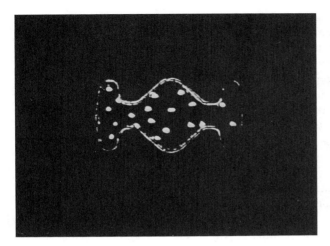

FIGURE 42 Experimental result of edge enhancement of the inner portion of a razor blade and the noise.

ations with simple adjustments of the analyzer. One could also perform the same operation by varying the intensities of the beams.

B. Real-Time Self-Adaptive Schlieren System

Schlieren systems are traditionally used for flow visualization, optical aberration measurements, and optical testing. The basic principle of operation in a Schlieren optical system relies on the detection of light deviated from its nominal path due to optical phase disturbances in the test region. In the conventional Schlieren system, observation of the deviated light is done by blocking of the undeviated light using a knife edge inserted at the Fourier-transform plane of a lens or a concave mirror. In the focusing Schlieren system, a source grid–cutoff grid combination is used to cut off the undeviated light. Recently, Downie [78,80] described a focusing Schlieren system employing the real-time recording nature of bacteriorhodopsin which behaves as a negative image in the plane of the cutoff grid. The system required the illumination bR with appropriate wavelengths of light intermittently for its success, making it only partially real time. The principle of operation of the system described in this chapter is entirely different from the one by Downie [78,80]. Following the same principle of operation of the above-described Fourier processing experiment, we demonstrate a Schlieren system using bacteriorhodopsin, for airflow visualization. Our system used the photoanisotropy on bR for a conventional Schlieren system, whereas Downíe's system used the saturable absorption of bR for a focusing Schlieren system.

The schematic of the Schlieren system with bR as shown in Fig. 38 is very similar to the conventional system, except for the fact that the knife edge is replaced with a bR film illuminated by a uniform beam of 570 nm wavelength and some polarization components. The experimental setup is the same as for the edge enhancement experiments with the object O replaced by an air disturbance. This uniform beam acts as an actinic beam which induces photoanisotropy in the bR film. A monochromatic parallel beam of light at a wavelength near the absorption peak ~570 nm of the bR passes through the test region and the Fourier transform of the test region is formed at the bR film plane. Both the actinic beam and the beam which passes through the test region, the probe beam, are linearly polarized, with actinic beam polarization at 45° with respect to that of the probe beam. The actinic beam induces photoanisotropy and dichroism in the bR film.

It can be understood from the previous sections that with optimum actinic beam intensity, the effect of increasing the probe beam intensity is to decrease its degree of polarization rotation. Hence, the high-frequency components of the Fourier transform formed at the bR experience more degrees of rotation than the zero- and near-frequency components. High-intensity low-frequency components experienced almost no rotation in its polarization and are blocked by the analyzer when the analyzer is kept crossed to the polarizer in the experimental setup. This blockage is similar to the insertion of a knife edge in the conventional Schlieren system. Light diffracted or deviated from its nominal path in the test region appears as high frequencies in the Fourier transform and the detection of only these components in the view plane leads to the visualization of disturbances in the test region.

Hence, as discussed in the previous section, with the bR film inserted between crossed polarizer and anlayzer, the zero and near frequencies with high intensities are cut off at the analyzer. This gives rise to the detection of high frequencies which reveal the details of phase disturbances and flow visualization. A proof of concept experiment for airflow visualization is described below.

The actinic beam polarization is made 45° to the probe beam polarization by the use of polarization rotator. With the presence of the actinic beam, the phase disturbances at the test region become visible at the CCD. Figure 43 is the CCD camera picture showing the visualization of an air jet from the nozzle of a lens cleaning air bottle.

C.   Laser Beam Profiling

The case where the probe beam has a Gaussian intensity profile is described below. The Gaussian beam output from the laser is passed through the polarizer, the bR film, and the analyzer, to be captured by the CCD. Actinic light of 10 mW/cm^2 uniformly illuminates the bR film. Because the Gaussian beam has intensity variation across the beam, the anisotropy and, hence, the angle of rotation experienced by the probe beam is different for different contours across the beam. The center of the beam with maximum intensity experiences the least rotation and the angle of rotation increases as we move away from the center. With the analyzer crossed to the polarizer, the center of the probe beam is blocked and rotating the analyzer blocks different contours across the beam. The pictures of the probe beam captured at the CCD with gradual rotation of the analyzer are shown in Figs. 44a–44e. The incident Gaussian beam transforms to a donut profile and then back to a Gaussian beam with reduced width.

D.   Optical Logic Operations

We constructed an optical system for performing two-dimensional logic operations. The operations performed were OR, NOR, XOR, AND, and so forth. Three input

FIGURE 43   Result of airflow visualization experiment.

logic operations were also implemented. The photo-induced dichroism of bR is the physical mechanism exploited in the optical system.

In its normal state with no light illumination, a bR-doped polymer film is isotropic with random distribution of bR molecules. When excited by a linearly polarized light (actinic light), only those bR molecules whose transition dipole moments for absorption lie in or near the direction of the electric field of the actinic light are bleached at the actinic light wavelength due to the B-to-M transition.

If a probe is incident in the regions illuminated by the actinic light, it will no longer be interacting with an isotropic film but rather the film is now anisotropic and dichroic. Due to the dichroism, the actinic light illumination produces an angular rotation of the plane of polarization of the probe beam.

The experimental setup of the logic operation optical system contains three SLMs as the electronic to optical interface. The two-dimensional inputs for the logic operations are presented through the SLMs in accordance with the experiment described below. The output of a He–Ne laser at 632.8 nm wavelength is used as the actinic light. The output of a Green He–Ne laser at 543 nm wavelength is used as the probe beam. The bR film is kept between two crossed polarizers to get zero output at the screen when no actinic light is incident on the bR film. The two polarizers and the screen are kept in the path of the probe beam. There are two actinic light beams that induce dichroism and they are made orthogonally polarized with respect to each other by the use of a polarization rotator. The polarization rotator can be arranged so that their polarizations are made parallel for some of the logic operations. The plane of polarization of these two actinic beams are at 45° to the plane of polarization of the probe beam. The experimental setup is shown in Fig. 45.

The actinic light through T1 rotates the plane of polarization of the probe beam in the clockwise (cw) direction and the actinic light through T3 rotates the plane of polarization of the probe beam in the counterclockwise (ccw) direction. The

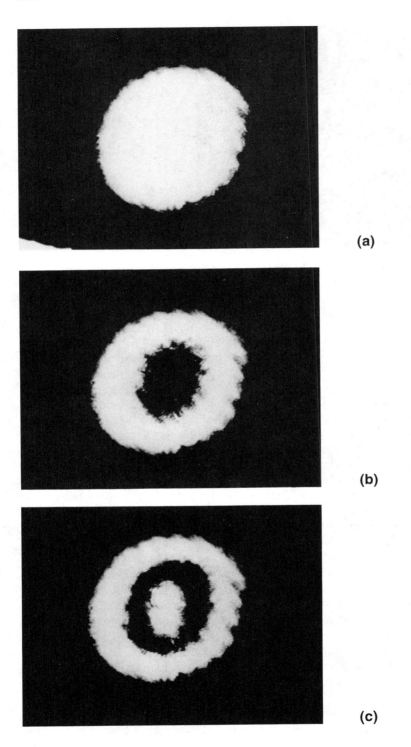

(a)

(b)

(c)

FIGURE 44 Experimental results for a probe with a Gaussian intensity distribution.

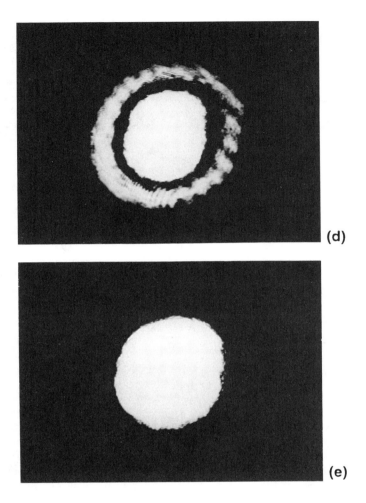

FIGURE 44 Continued

combination of rotation of the analyzer A and the polarization rotation of one of the actinic beams allows for many logic operations to be performed. The output of the logic operations is detected at the screen S. The two-dimensional inputs for the logic operations are presented at the planes T1 and T2 or T3. For simplicity, one of the two-dimensional patterns chosen is a vertical slit which is kept at T1 and the other a horizontal slit which is kept at plane T2 or T3 in accordance with the logic operations that are given below.

1. OR operation; note that this is equivalent to the addition of the two patterns. Both actinic beams have the same polarization. No pattern is presented at T2 (i.e., beam through T2 if fully transmitted). Pattern B is presented at T3. Polarizer P and

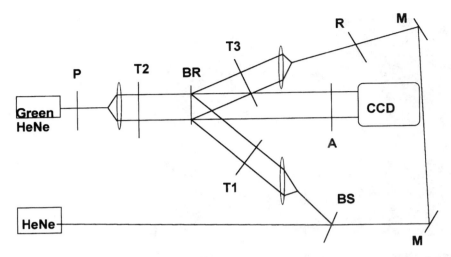

FIGURE 45 Schematic experimental setup for optical logic operations using photo-induced dichroism in a bR film.

analyzer A are crossed. The results are shown in Fig. 46. The truth table for OR operation is

A	B	Output
1	1	1
1	0	1
0	1	1
0	0	0

2.   NOR operation.
The experimental conditions are the same as for OR operation but the analyzer is rotated by ~1° cw in order to block the polarization-rotated probe beam. The results are shown in Fig. 47. The truth table for NOR operation is

A	B	Output
1	1	0
1	0	0
0	1	0
0	0	0

3.   XOR operation; note that his is equivalent to the subtraction of the two patterns. The experimental conditions are same as for OR operation, but the polarization of the actinic light through T3 is rotated to make it orthogonal to the other actinic beam through T1. The results are shown in Fig. 48. The truth table for XOR is

A	B	Output
1	1	0
1	0	1
0	1	1
0	0	0

FIGURE 46 Result of OR logic operation.

4. AND operation.

The actinic beam through T3 is blocked. Pattern B is presented at plane T2 and pattern A is at T1. Polarizer P and analyzer A are crossed. The results are shown in Fig. 49. The truth table for AND is

A	B	Output
1	1	1
1	0	1
0	1	1
0	0	0

5. (NotA)B operation.

The experimental conditions are the same as for AND operation, but the analyzer is rotated $\sim -1°$ cw in order to block the polarization-rotated probe beam. The results are shown in Fig. 50. The truth table for AB is

A	B	Output
1	1	0
1	0	0
0	1	1
0	0	0

We have successfully demonstrated a self-adaptive optical processing system using the photo-induced dichroic characteristics of a bacteriorhodopsin film. The dependence of photo-induced dichroism on probe light intensity has effectively been utilized to imprint continuous polarization variations on the different spatial frequency components of an object information. Spatial filtering of desired frequencies are then performed by an analyzer. The simplicity of the present scheme is straightforward with no alignment required at the Fourier plane to perform the

FIGURE 47 Result of NOR logic operation.

processing. No interference recordings are involved in this experiments, hence vibration isolation systems are not required A coherent source is not a requirement to perform the experiment, a white light source with an appropriate wavelength filter at ~ 570 nm can induce photoanisotropy in the bacteriorhodopsin film. Simple adjustments on the analyzer are employed for processing operations.

FIGURE 48 Result of XOR logic operation.

FIGURE 49 Result of AND logic operation.

FIGURE 50 Result of (notA)B logic operation.

REFERENCES

1. Sharkov, A. V., and Matveets, Y. A., *Laser Picosecond Spectroscopy and Photochemistry of Biomolecules* (V. S. Letokhov, ed.), Adam Hilger, Bristol, UK, 1987, p. 56.
2. Andrews, D. L., *Lasers in Chemistry*, Springer-Verlag, Berlin, 1990.
3. Birge, R. R., and Zhang, C. F., *J. Chem. Phys.*, *92*, 7178 (1990).

4. Gennis, R. B., *Biomembranes Molecular Structure and Function*, Springer-Verlag, New York, 1989, p. 22.

5. Oesterhelt, D., Bräuchle, C., and Hampp, N., *Q. Rev. Biophys.*, *24*, 425 (1991).

6. Henderson, R., and Unwin, P. N. T., Three dimensional model of purple membrane obtained by electron microscopy, *Nature*, *257*, p. 28 (1975).

7. Michel, H., Characterization and crystal packing of three-dimensional bR crystals, *EMBO J.*, *1*, 1267 (1982).

8. Michel, H., Oesterhelt, D., and Henderson R., Orthorhombic Two-dimensional Crystal Form of Purple Membrane, *Proc. Natl. Acad. Sci.*, *USA*, *77*, 338 (1980).

9. Henderson, R., and Unwin, P. N. T., Three dimensional model of Purple membrane obtained by electron microscopy, *Nature*, *257*, 28 (1975).

10. Sasaki, J., Brown, L. S., Chon, Y., Kandori, H., Maeda, A., Needleman, R., and Lanyi, J. K., *Science*, *269*, 73 (1995).

11. Chen, Z., Lewis, A., Takei, H., and Nebenzahl, I., *Appl. Opt. 30*, 5188 (1991).

12. Hampp, N., Bräuchle, C. and Oesterhelt, D., *Mater. Res. Soc. Bull.*, *XVII*, 56 (1992).

13. Birge, R. R., *Annu. Rev. Phys. Chem.*, *41*, 683 (1990).

14. Lewis, A., and Del Piore, V., *Phys. Today*, *41*(1), 38 (1988).

15. Korchemskaya, E. Y., Soskin, M., and Taranenko, V. B., *Sov. J. Quantum Electron.*, *17*, 450 (1987).

16. Hong, F. T., *Biosystems*, *19*, 223 (1986).

17. Huang, J. Y., Chen, Z., and Lewis, A., *J. Phys. Chem.*, *93*, 3314 (1989).

18. Werner, O., Fischer, B., Lewis, A., and Nebenzahl, I., *Opt. Lett.*, *15*, 1117 (1990).

19. Werner, O., Fischer, B., and Lewis, A., *Opt. Lett.*, *17*, 241 (1992).

20. Vselvolodov, N. N., Druzhko, A. B., and Djurkova, T. V., *Molecular Electronics: Biosensors and Biocomputers* (F. T. Hong, ed.), Plenum Press, New York, 1989, p. 381.

21. Bazhenov, V. Y., Soskin, M. S., Taranenko, M. S., and Vasnetsov, M. V., *Optical Processing and Computing* (A. Arsenault, ed.), Academic Press, New York, 1989, p. 103.

22. Taranenko, and V. B., Vasnetsov, M. V., *Optical Memory and Neural Networks*, SPIE, Bellingham, WA, 1991, Vol. 1621, p. 169.

23. Rao, D. V. G. L. N., Aranda, F., Wiley, B. J., Akkara, J. A., Kaplan, D. L., and Roach, J. F., *Appl. Phys. Lett.*, *63*, 1489 (1993).

24. Marcus, M. A., Lewis, A., Racker, E., and Crespi, H., *Biochem. Biophy. Res. Commun.*, *78*, 669 (1977).

25. Gilles-Gonsales, M. A., Hackett, N. R., Jones, S. J., Khorana, H. G., Lee, D. S., Lo, K. M., and McCoy, J. M., *Methods Enzymol*, *125*, 190 (1985).

26. Varo, G., and Keszthelyi, L., *Biophys. J.*, 43, 47 (1983).

27. Vselbolodov, N. N., and Ivanitsky, G. R., *Biofizika*, *30*, 883 (1985).

28. Shen, Y., Safinya, C. R., Liang, K. S., Ruppert, A. F., and Rothschild, K. J., *Nature*, *366*, 48 (1993).

29. Oesterhelt, D., Schuhmann, L., and Gruber, H., *FEBS Lett.*, *44*, 262 (1974).

30. Stoeckenius, W., Lozier, R. H., and Bogomolni, R. A., *Biochim. Biophys. Acta*, *505*, 215–278 (1979).

31. Oesterhelt, D., Schuhmann, L., and Gruber, H., *FEBS Lett.*, *44*, 262–265 (1974).

32. Oesterhelt, D., and Stoeckenius, W., *Methods in Enzymology*, *Vol. 31*, Academic Press, New York, 1982, pp. 667–678.

33. Needleman, R., Chang, M., Ni, B., Varo, G., and Fornes, J., *J. Biol. Chem.*, *266*, 11478–11484 (1991).

34. Khorana, H. G., in *Microbial Energy Transduction: Genetics, Structure, and Function of Membrane Proteins* (D. C. Youvan and F. Daldal, eds.), Cold Spring Habor Laboratory, Cold Spring Harbor, NY, 1986.

35. Huang, K. S., Radhakrishman, R., Bayley, H., and Khorana, H. G., *J. Biol. Chem.*, *257*, 13616 (1982).
36. Liao, X., and Wise, J. A., *Gene*, *88*, 107–111 (1990).
37. Ni, B., Chang, M., Duschl, A., Lanyi, J., and Needleman, R., *Gene*, *90*, 169–172 (1990).
38. Needleman, R., *New Photonic Materials from Genetically Engineered Bacteriorhodopsin*, ARO Proposal 30336-LS. U.S. Army Research Office, Research Triangle Park, NC, 1992.
39. Shen, Y. R., *The Principles of Nonlinear Optics*, Wiley–Interscience, New York, 1984.
40. Franken, P. A., Hill, A. E., Peters, G. W., and Weinreich, G., *Phys. Rev. Lett.*, *7*, 118 (1961).
41. Prasad, P. N., and Williams, D. J., *Introduction to Nonlinear Optical Effects in Molecules and Polymers*, Wiley, New York, 1991.
42. Boyd, R. W., *Nonlinear Optics*, Academic Press, New York, 1993.
43. Jackson, J. D., *Classical Electrodynamics*, 2nd ed., McGraw-Hill, New York, 1975.
44. Stepanov, B. I., Ivakin, E. V., and Rubanov, A. S., *Sov. Phys. Doklady*, *26*, 46 (1971).
45. Hellwarth R. W., Third order susceptibilities of liquids and gases, *Progr. Quant. Elec.*, *5*, 1 (1977).
46. Yariv, A., and Pepper, D. M., *Opt. Lett.*, *16* (1977; Yariv, A., *IEEE Jour. Quant. Electron.*, QE-14, 650 (1978).
47. Fisher, R. A., (ed.), *Optical Phase Conjugation*, Academic, New York, 1983.
48. Kogelnik, H., *Bell Syst. Tech. J.*, *48*, 2909 (1969).
49. Tompkin, W. R., Boyd, R. W., Hall, D. W., and Tick, P. A., *J. Opt. Soc. Am.*, *B4*, 1030 (1987).
50. Acioli, L. H., Gomes, A. S. L., and Rios Leite, J. R., *Appl. Phys. Lett.*, *53*, 1788 (1988).
51. Lentine, A. L., Lee, J. N., Lee, S. H., and Efron, U., *Appl. Opt.*, *33*, 2767 (1994).
52. Rao, D. V. G. L. N., Aranda, F., Roach, J. F., and Remy, D. E., *Appl. Phys. Lett.*, *58*, 1241 (1991).
53. Gross, R. B., Izgi, K. C., and Birge, R. R., *SPIE*, *1662*, 186 (1992).
54. Haronian, D., and Lewis, A., *Appl. Phys. Lett.*, *61*, 2237 (1992).
55. Birge, R. R., *Sci. Am.*, *90*, 272 (March 1995).
56. Aranda, F. J., Garimella, R., McCarthy, N. F., Narayana Rao, D., Rao, D. V. G. L. N., Chen, Z., Akkara, J. A., Kaplan, D. L., and Roach, J. F., *Appl. Phys. Lett.*, *67*, 599 (1995).
57. Gluckstad J., and Saffman, M., *Opt. Lett.*, *20*, 551 (1996).
58. Sperber, P., and Penzkofer, A., *Opt. Quatum Electron.*, *18*, 381 (1986).
59. Guha, S., Frazier, C. C., Porter, P. L., Kang, K., and Finberg, S., *Opt. Lett.*, *14*, 942 (1989).
60. Sheik-Bahae, M., Said, A. A., Wei, T., Hagan, D. J., and Van Stryland, E. W., *IEEE J. Quantum Electron.*, *46*, 760 (1990).
61. Smith, W. L., in *Handbook of Laser Science and Technology* (M. J. Weber, ed.), CRC, Boca Raton, FL, 1986, Vol. III, p. 229.
62. Guha, S., Kang, K., Porter, P., Roach, J. F., Remy, D. E., Aranda, F. J., and Rao, D. V. G. L. N., *Opt. Lett.*, *17*, 264 (1992).
63. Wei, T. H., Hagan, D. J., Sence, M. J., Van Stryland, E. W., Perry, J. W., and Coulter, D. R., *Appl. Phys. B*, *54*, 46 (1992).
64. Aranda, F. J., Rao, D. V. G. L. N., Wong, C. L., Zhou, P., Chen, Z., Akkara, J. A., Kaplan, D. L., and Roach, J. F., *Opt. Rev.*, *2*(3), 204–206 (1995).
65. Wang, Q., Zhang, C., Gross, R. B., and Birge, R. R., *Opt. Lett.*, *18*, 775 (1993).
66. Gibbs, H. M., *Optical Bistability: Controlling Light with Light*, Academic Press, Orlando, FL, 1985.

67. Haus, J. W., Sung, C. C., Bowden, C. M., and Cook, J. M., *J. Opt. Soc. Am. B*, *2*, 1920 (1985).
68. Miller, D. A. B., *J. Opt. Soc. Am. B.*, *1*, 857 (1984).
69. Miller, D. A. B., Gossard, A. C., and Wiegmann, J., *Opt. Soc. Am. B*, *1*, 477 (1984).
70. Mizrahi, V., DeLong, K. W., Stegeman, G. I., Saifi, M. A., and Andrejco, M. J., *Opt. Lett.*, *14*, 1140 (1989).
71. Kirkby, C. J. G., Cush, R., and Bennion, I., *Optical Bistability III* (H. M. Gibbs, P. Mandel, N. Peyghambarian, and S. D., Smith, eds.), Springer-Verlag, New York, 1985, p. 165.
72. Goodman, J. W., *Opt. Photon. News*, *2*(2), 11 (1991).
73. Ozktas, H. M., and Miller, D. A. B., *Appl. Opt.*, *35*, 1212 (1996).
74. Huignard, J. P., and Herriau, J. P., *Appl. Opt.*, *17*, 2671 (1978).
75. Chang, T. Y., Hong, J. H., and Yeh, P., *Opt. Lett.*, *15*, 743 (1990).
76. Thoma, R., Hampp, N., Brauchle, C., and Oesterhelt, D., *Opt. Lett.*, *16*, 651 (1991).
77. Gross, R. B., Can Izgi, K., Birge, R. R., Jamberdino, A. A., and Niblack, W., *Proc. SPIE*, *1662*, 186 (1992).
78. Downie, J. D., *Appl. Opt.*, *33*, 4353 (1994).
79. Imam, H., Lindvold, L. R., and Ramanujam, P. S., *Opt. Lett.*, *20*, 225 (1995).
80. Downie, J. D., *Appl. Opt.*, *84*, 5210 (1995).
81. Takei, H., and Shimizu, N., *Appl. Opt.*, *36*, 1848 (1996).
82. Burykin, N. M., Korchemskaya, E. Ya., Soskin, M. S., Taranenko, V. B., Dukova, T. V., and Vsevolodov, N. N., *Opt. Commun.*, *64*, 68 (1985).

17

Microgravity Processing and Photonic Applications of Organic and Polymeric Materials

Donald O. Frazier, Benjamin G. Penn, David D. Smith, and William K. Witherow
NASA Marshall Space Flight Center
Huntsville, Alabama

Mark S. Paley and Hossin A. Abdeldayem
Universities Space Research Association
NASA Marshall Space Flight Center
Huntsville, Alabama

I. INTRODUCTION

In recent years, a great deal of interest has been directed toward the use of organic materials in the development of high-efficiency optoelectronic and photonic devices. There is a myriad of possibilities among organics which allow flexibility in the design of unique structures with a variety of functional groups. The use of nonlinear optical (NLO) organic materials such as thin-film waveguides allows full exploitation of their desirable qualities by permitting long interaction lengths and large susceptibilities allowing modest power input [1]. There are several methods in use to prepare thin films, such as Langmuir–Blodgett (LB) and self-assembly techniques [2–4], vapor deposition [5–7], growth from sheared solution or melt [8,9], and melt growth between glass plates [10]. Organics have many features that make them desirable for use in optical devices such as high second- and third-order nonlinearities, flexibility of molecular design, and damage resistance to optical radiation. However, their use in devices has been hindered by processing difficulties for crystals and thin films.

In this chapter, we discuss photonic and optoelectronic applications of a few organic materials and the potential role of microgravity on processing these materials. It is of interest to note how materials with second- and third-order nonlinear optical behavior may be improved in a diffusion-limited environment and ways in which convection may be detrimental to these materials. We focus our discussion on third-order materials for all-optical switching, and second-order materials for frequency conversion and electrooptics.

A. Third-Order Materials for Optical Switching

Optical-fiber communication systems have undergone stunning growth over the past decade. The technologies that have arisen in support of these systems have been incredibly fortuitous. The operating wavelength of erbium-doped amplifiers, for instance, serendipitously coincides with the minimum loss wavelength of fused silica fibers. But, even as fiber-optic networks have been implemented on a universal scale, electronic switching is still the main routing method. Although fibers have dramatically increased node-to-node network speeds, electronic switching will limit network speeds to about 50 Gb/s. Already, it is apparent that terabit-rate speeds will soon be needed to accommodate the 10–15%/month growth rate of the Internet and the increasing demand for bandwidth-intensive data such as digital video [11].

All-optical switching using nonlinear optical materials can relieve the escalating problem of bandwidth limitations imposed by electronics. Several important limitations need to be overcome, such as the need for high $\chi^{(3)}$ materials with fast response and minimum absorption (both linear and nonlinear), development of compact laser sources, and reduction of the switching energy. The goal of minimizing optical loss obviously depends on processing methods. For solution-based processes, such as solution crystal growth, electrodeposition, and solution photopolymerization, it is well known that thermal and solutal density gradients can initiate buoyancy-driven convection. Resultant fluid flows can affect transport of material to and from growth interfaces and become manifest in the morphology and homogeneity of the growing film or crystal. Likewise, buoyancy-driven convection can hinder production of defect-free, high-quality crystals or films during crystal and film growth by vapor deposition.

The guided-wave materials used most commonly have been inorganic fibers and semiconductors. Less developed but highly promising are organic materials such as conjugated polymers which possess large nonlinearity, with fast response times, are more easily tailored at a molecular level and more malleable than their inorganic counterparts. One of the major challenges for proponents of organic materials is to find cheap, reliable methods of waveguide fabrication that takes advantage of existing technologies. Processing techniques and choices of materials should result in a minimization of both scattering and absorption losses. One of the main reasons that organic and polymeric materials are not more strongly competitive with silica fibers for switching applications is due to the maturity of silica fiber processing.

B. Second-Order Materials for Electrooptic Applications

Applications of materials with second-order nonlinearity include frequency conversion, high-density data storage, and electrooptic modulators and switches. The first demonstration of second-harmonic generation was in quartz [12] and it has traditionally been observed in inorganic crystals. A decade later it was demonstrated that the second-order nonlinearity may be several orders of magnitude larger in organic crystals possessing delocalized π-electron systems in which intramolecular charge transfer occurs between electron donor and acceptor substituents [13]. Although organic materials may offer larger nonlinearities than inorganic crystals, the utilization of organic crystals is limited by the small number of molecules with large hyperpolarizabilities that have a noncentrosymmetric crystalline state (11 of the 32 crystal classes possess inversion symmetry and cannot be used as $\chi^{(2)}$ materials). Also, the maturity of inorganic crystal growth is relatively advanced, whereas that of organic crystal growth has not had the necessary time for comparable development.

Whereas the second-order nonlinearity in inorganic systems is a bulk effect ascribable to crystalline structure, the primary contribution to bulk nonlinearity for organic systems is due to the ensemble of nonlinearly responding molecules. Van der Waals forces between mers (molecular units) are small and the induced dipoles result most directly from the external field. This provides an added degree of flexibility for organic materials because the required asymmetry does not require crystalline structure but, instead, may be achieved in amorphous geometries. For example, instead of relying on the art of crystal growth, electric field poling of polymers containing the nonlinear chromophore may be used to induce macroscopic asymmetry. Alternatively, self-assembly or liquid-crystal ordering can achieve required asymmetry, possibly encouraged in environmental quiescence such as that offered by reduced convection during microgravity processing.

In order to take advantage of the large nonlinearities in organic materials resulting from π-electron mobility while also utilizing the mechanical and thermal properties of inorganic crystals, some researchers have turned to a semiorganic approach in which organic molecules are bound to an inorganic host by complexation or salt formation. For example, semiorganic single crystals of L-histidine tetrafluoroborate (HFB) have demonstrated five times the effective second-order nonlinearity of potassium dihydrogen phosphate (KDP) [14]. Single crystals are easily obtained from solution and crystals are thermally stable with a decomposition temperature of 205°C. Solution crystal growth has been performed repeatedly in microgravity, particularly for the study of protein crystal growth. Many organics, also good NLO materials, are amenable to solution growth and are ideal candidates for studies of the kinetics and fluid dynamics of solution growth processes. Knowledge of such processes can lead to significant improvement in crystals grown in space or on Earth. Several research tasks are underway to study the growth of bulk single crystals of important materials, such as HFB, L-arginine phosphate (LAP), and several other organic and semiorganic molecules. There is certainly evidence that gravity could play a role in the growth of macromolecular crystals (e.g., protein crystals) as well as on solution polymerization processes. Rosenberger et al. studied the temperature dependence of protein solubility and applications to crystallization and growth kinetics. Within these studies, they observed relative interfacial kinetics and bulk transport as functions of supersaturation and resultant effects on the growth morphology of lysozyme and horse serum albumin [15]. Supersaturations driven by thermal fluctuations on the order of 1–2°C during growth lead to significant optical and structural nonuniformity in a growing crystal. Pusey et al., using tetragonal lysozyme, showed that forced fluid flow rates of 30–40 μm/s slow and eventually stop the growth of 10-μm crystals [16,17]. Cessation of growth occurred at flow rates as low as 2.5 μm/s in some crystals, but growth persisted over a longer period of time than at higher fluid flow rates. The conclusion is that for some crystals at least, even small convective flows are deleterious for growth of the crystal. The precise mechanism for this is not yet understood, but, clearly, the case is quite strong for studying the growth of certain organic and semiorganic crystals in microgravity environments.

It is therefore reasonable to consider that microgravity can play an important role in the formation of organic and semiorganic crystals for second-order applications. In the diffusion-limited regime of space, larger and more defect-free crystals may be grown, hence improving optical transparency and conversion efficiency. By different processing techniques, improved optical quality and molecular alignment have been observed in polydiacetylene (PDA) and phthalocyanine (Pc) thin films processed in microgravity or reduced convection environments. Perhaps of more immediate relevance, knowledge gained from low-gravity (low-**g**) experiments may enable the optimization of growth conditions on Earth.

C. Material Limitations

1. Dispersion and Loss

One limitation governing the propagation of pulses in a waveguide is the reshaping of the pulse due to group velocity dispersion (GVD). Each frequency component in the pulse sees a different refractive index and thus travels at a different speed through the medium, leading to a spreading or reshaping of the pulse. GVD is more important for short pulses, as they contain more frequency components. This pulse spreading reduces the amount of information that can be carried and destroys the cascadability (the ability of the output of a device to reenter an identical device and behave similarly) of the system. It is quite obvious that losses due to film imperfections must contribute significantly to this limitation, and processing methods must strive to ensure minimization of material defects. However, a pulse also consists of different intensities. Each intensity can also see a different refractive index due to the nonlinearity of the medium and so travels at a different speed. The ability of a pulse to become reshaped due to the nonlinearity of the medium is known as self-phase modulation (SPM). For certain pulse shapes, GVD is exactly balanced by SPM and the pulse retains its shape over an infinite propagation distance in the absence of loss. GVD pulses that are balanced by SPM are known as solitons. In practice, however, solitons cannot propagate infinitely. The effect of dissipation decreases the peak power in the pulse, hence decreasing SPM. The result is that the soliton widens as it propagates. Two main sources of loss are operative in silica fibers: an intrinsic loss due to Rayleigh scattering and, superimposed on this, absorption due to hydroxide ion impurities in the melt [18]. Due to the maturity of silica fiber fabrication technology, impurities in the melt can be kept quite low. Microgravity processing may enable polymeric materials to become more competitive with silica fibers by the reduction of similar types of scattering loss mechanisms.

2. Two-Photon Absorption

A major setback to the development of an all-optical switch is the existence of nonlinear loss manifested by multiphoton absorption. The current strategy is to search for a truly nonresonant operating wavelength. To demonstrate the deleterious effects of multiphoton absorption, consider its effect on a simple device. A nonlinear directional coupler (NLDC) is a device in which two waveguides or fibers are brought close together such that their overlapping evanescent attenuations periodically transfer power back and forth between the guides as light propagates through the device [19]. The periodicity of phase-matched power transfer depends on channel separation and refractive index. Full power transfer occurs after one coupling length, L_c. When the length of the overlapping region equals the coupling length, all the power will be transferred to the neighboring guide. If the guide material has a large cubic nonlinearity (and hence nonlinear refractive index n_2), the coupling will be perturbed at high intensities due to an intensity-dependent phase mismatch in the guides, and at the switching power, the majority of the transmission will switch from one guide to the other. The majority of the coupler's output may then be toggled between the guides simply by raising and lowering the intensity.

Multiphoton absorption occurs when the intensity of the radiation causes virtual levels to become accessible. Transition rules prohibit transitions between same-parity states unless virtual levels are involved. Because two-photon absorption (TPA) is intensity dependent, it varies with propagation distance. The result is that the power transfer is accompanied by increasing TPA in the low-intensity guide and decreasing TPA in the high-intensity guide such that complete switching does not occur or is suppressed (Fig. 1). Mizrahi et al. [20] have shown that TPA places a fundamental constraint on the usefulness of any high-$\chi^{(3)}$ material. Semiconductors, for example, are constrained to energies below half the band gap to avoid TPA [21,22].

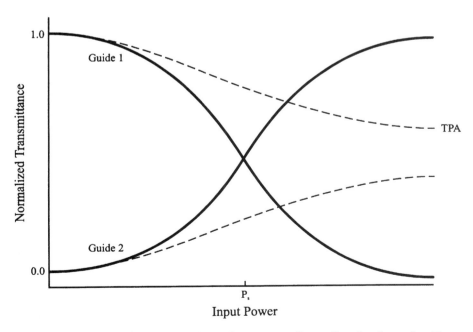

FIGURE 1 Effect of two-photon absorption on a nonlinear directional coupler. The coupler is initially in a crossed state. Increasing input power perturbs the coupling which causes switching, suppressed by the presence of TPA.

An alternative approach based on materials architecture is elimination of induced absorption. Specifically, the nonlinear absorption in a material may be canceled by the addition of small metal particles. This approach is possible because of a counterintuitive consequence of local field effects that was first recognized by Hache et al. [23]. They found that for gold nanoparticles in glass, the colloid as a whole demonstrated saturable absorption. However, the isolated metal behaved quite differently, demonstrating induced absorption. Embedding the particles in a glassy matrix altered the sign of the nonlinear absorption as a result of the local field correction. The implication of Hache's finding is that there is a concentration somewhere between pure gold and the colloidal gold glass at which the imaginary part of the cubic susceptibility goes to zero. In fact, if the sign of Im $\chi^{(3)}$ is the same for each component, then, by necessity, there will actually be two concentrations at which Im $\chi^{(3)}=0$. The smaller concentration crossing point is obviously more useful because it entails a lower amount of linear absorption and poses less of a challenge to fabricate. The significance and specificity of particle concentration demands careful control of metal particle dispersion. This requirement is also prevalent in the crystal growth of certain semiconducting alloys where the optimum band gap is determined by alloy stoichiometry. Lehoczky et al. directionally solidified $Hg_{0.84}Zn_{0.16}Te$ alloy in microgravity with some success at purely diffusion-controlled growth to achieve the desired stoichiometry throughout the space-grown crystal. However, unexpected transverse residual accelerations from the shuttle environment resulted in some radial compositional variations [24]. The result stresses the difficulty in achieving precise concentrations during materials processing where thermal or concentration gradients arise and could be fatally disruptive in terms of achieving a defined concentration objective. Considering difficulties in achieving diffusion-controlled growth and other desirable gravitationally sensitive results during microgravity processing, ground-based attempts at overcoming these effects may prove futile. Where homogeneous dispersion and concentration are requisite in materials of interest, processing on Earth-orbiting platforms could offer distinct advantages.

To demonstrate the cancellation of absorptive nonlinearity of a composite aqueous system, Smith [25] performed open-aperature Z-scan measurements on a gold colloid prepared by the recipe of Turkevich [26] at various particle concentrations. A frequency-doubled ND : YAG laser at 532 nm provided 30-ps mode-locked pulses at a repetition rate of 10 Hz. Each concentration was placed in a 1-cm optical-path-length quartz cuvette on a track near the focus of the beam. The focal length of the lens was 33 cm and the full width at half-maximum (FWHM) beam diameter was 2.5 mm. The beam waist was measured to be $w_0=70$ μm, corresponding to a Rayleigh diffraction length of $z_0=2.9$ cm.

Instead of two-photon absorption, however, the nonlinear mechanism was reverse saturable absorption. The system consisted of a known reverse saturable absorber 1,1',3,3,3',3'-hexamethylindotricarbocyanine iodine (HITCI) in methanol and water. The proportions of dye, methanol, and water were held fixed. Because these chemicals form a solution, they may be considered a single component, although their chemical association should be accounted for by a modification of the local field factor as described by Fröhlich [27]. The other component was the gold. The proportions of the various components are illustrated in Table 1. Note that the

TABLE 1 Component Properties of HITCI/Au Composite

CURVE No.	GOLD COLLOID (ml)	WATER (ml)	123 μM HITCI (ml)	METHANOL (ml)
1	0	2.75	0.4	0.6
2	0.5	2.25	0.4	0.6
3	1.0	1.75	0.4	0.6
4	1.5	1.25	0.4	0.6
5	1.9	0.85	0.4	0.6
6	2.0	0.75	0.4	0.6
7	2.2	0.55	0.4	0.6
8	2.5	0.25	0.4	0.6
9	2.75	0	0.4	0.6

same amount of dye, methanol, and water is used in each case and that the only variable is the concentration of gold. The Z-scans for various concentrations of gold are displayed in Fig. 2. It can be clearly seen that the nonlinear absorption changes sign near curve 6. The experiment was repeated several times and the sign reversal was obvious each time. Note that curves 1–5 display a valley indicative of reverse saturable absorption. At the focus, however, a small secondary peak is observed. This peak corresponds to a saturation of the nonlinear absorption and was first reported by Swatton et al., [28]. A four-level semiclassical model was used to describe the absorption of the dye.

D. Candidate Materials

1. Fused Silica Fibers Versus Conjugated Organics

Figures of merit have been proposed to describe the linear and nonlinear losses and the dispersion in candidate materials [20,22,25,29–32]. Fused silica fibers are attractive for optical communications systems because, although they have a relatively weak nonlinearity, they have among the highest figures of merit due to their low loss in the region—1.3–1.6 μm, known as the telecommunications window. Also, silica fibers exhibit anomalous (negative) dispersion for wavelengths greater than 1.3 μm, allowing solitons to propagate. The two wavelengths most commonly used for communications are the zero dispersion wavelength of 1.3 μm and the low-loss wavelength of 1.55 μm. The wavelength of choice has become 1.55 μm due to the availability of multiple-clad dispersion-shifted fibers.

A rule of thumb for all-optical switches is that $\chi^{(3)}LI$ (path length and intensity) is constant [11]. So, although silica fibers have a much lower cubic nonlinearity, $\chi^{(3)}$, the switching power can be much lower than that of other materials because there is almost no length limitation on a fiber; whereas, typically for a highly nonlinear material, the loss is much greater, imposing a length limitation. But simply increasing the switch's operating length does not take into account the dispersion or spreading of the pulse, which becomes more important for very short pulses. For fused silica fibers, the dispersion can be tuned negative to support solitons, but most highly nonlinear materials have positive GVD in the telecommunications win-

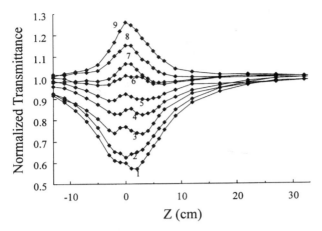

FIGURE 2 Elimination of induced absorption in HITCI by the addition of small Au particles. For each Z-scan, the peak power was $P_i = 0.16$ MW and the on-axis peak irradiance at focus was $I_0 = 2.1$ GW/cm^2.

dow and thus cannot support solitons for these wavelengths. One way to decrease the switching energy without the switching power is to use shorter pulses; however, this also increases dispersion. Thus, for long pulses, fibers are not constrained by pulse spreading and can operate at lower powers, but for short pulses, materials with higher nonlinearities may switch at lower powers due to GVD considerations. Asobe et al. have shown that for very short pulses, when group velocity dispersion is taken into account, highly nonlinear conjugated polymers will operate at a lower switching energy than positive dispersion silica fibers [32]. As we will show, microgravity may be exploited to reduce the length limitation in polymeric waveguides.

An additional reason for choosing highly nonlinear conjugated polymers over silica fibers is because of limitations which arise from the propagation design of the device. Switching in silica fibers requires very long lengths of fiber which can seriously reduce the switching time and result in cumbersome setups.

2. Gravitational Effects in Processing Organic and Polymeric Films

Two promising classes of organic compounds for optical thin films and waveguides are polydiacetylenes (PDAs), which are conjugated zigzag polymers, and phthalocyanines, which are large ring-structured porphyrins. Epitaxial growth on ordered organic and inorganic substrates under various processing conditions have been useful for preparing highly oriented polydiacetylene (PDA) and phthalocyanine (Pc) films [33–35]. The degree of significance relating processing conditions to uniformity in thickness, degree of orientation, and optical properties for a specific processing technique is the general focus of work in this area. A study on the effect of processing conditions relevant to thin-film deposition by various techniques is particularly difficult because of the possibility that convection may play a major role. It is a goal of some researchers to produce good quality anisotropic films; therefore, an important yet understudied requirement should be to assess the role of gravity during processing. This may be particularly true for the vapor deposition of diacetylenes where subsequent polymerization in the crystal is topochemical and occurs readily only when neighboring monomer molecules are sufficiently close and suitably oriented [29]. Likewise, this requirement is equally viable for the vapor deposition of Pcs in view of the results of microgravity experiments by 3M Corporation involving the preparation of thin films of copper Pc (CuPc) [34–44]. Indeed, a variety of microstructural forms was obtained in thin films of CuPc dependent on processing methods and conditions. Small changes in processing parameters caused large changes in molecular orientation within the film. Microgravity grown CuPc had several desirable features which indicate that the growth of organic films in low **g** may result in better quality films for optical and electrical applications [43,44]. The dramatic 3M result was very encouraging and has been one source of optimism toward considering the microgravity environment of value for processing high-quality organic films.

One goal of microgravity research on vapor-deposited organic films is to understand factors for improving film quality and optical properties. Important aspects of any study involving fluids, as in vapor transport, are driving mechanisms for heat transfer with natural convection and diffusion processes which determine flow profiles and temperature distributions.

A novel technique, recently discovered, for growing polydiacetylene thin films involves exposing a transparent substrate, in contact with diacetylene monomer solution, to ultraviolet (UV) light [45]. A polymer film deposits on the side of the substrate in contact with monomer in solution, and there are distinct gravitational effects which influence film quality. Good quality thin films elude growth from solutions absent of uniform flow fields and homogeneous temperature distributions near the substrate surfaces. The flow fields and temperature distributions during the polymerization process by exposure to UV light details the nature of gravitational influences on this process.

From a device perspective, the UV technique makes construction of extremely complex waveguides possible. Utilizing a computer-controlled x–y translation stage, programmed to trace out a desired pattern, researchers demonstrated that UV radiation (364 nm) from an argon ion laser could trace out a test pattern [46] (Fig. 3a). It is possible to construct a Mach–Zehnder interferometer with an optimized curvature using this technique [47]. After mounting a test cell containing diacetylene monomer solution on a translation table, a focused UV laser beam passing through the UV transparent surface of the test cell traced the desired paths to form the polymer-based optimized Mach–Zehnder waveguide (Fig. 3b).

It has been proposed that NLO thin-film properties may be improved by low-gravity processing using electrodeposition. Strong candidates for NLO thin-film applications are the polythiophenes. Polymeric thiophenes are attractive materials due to their ease of preparation, stability, and high third-order susceptabilities [48–52]. A simple and convenient method for preparation of polythiophenes is electrochemical oxidation. Earlier microgravity experimentation [53,54] on metal,metal/cermet electrodeposition of Ni provides some microgravity electrodeposition background and raises the possibility of application to improving the quality of polythiophene thin films in low gravity. Electrodeposition of Ni on a Au substrate in low gravity often results in the production of an x-ray nondiffracting surface. Similarly deposited cobalt metal does not give this result nor does Ni on a glassy carbon substrate. Further, the Co/Ni alloy composition variance during electrodep-

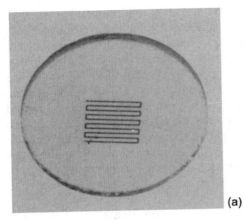

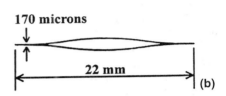

170 microns

22 mm

(a)   (b)

FIGURE 3 A PDA film derived from 2-methyl-4-nitroaniline (MNA) circuit photodeposited onto UV transparent substrates using the radiation from an argon ion laser: (a) demonstration pattern on a quartz disk; (b) enlarged image of an actual Mach–Zehnder interferometer on a glass microscope slide.

osition is strongly dependent on the amount of convection. Similar sensitivities to gravitational influences apparent in inorganics during electrodeposition should arise during electrodeposition of organics.

Electrochemical polymerization deserves thorough investigation for use in fabricating thin-film waveguides. NLO films might be prepared on the surface of various substrates during polymerization for the fabrication of devices. This method has been useful in the synthesis of several polymers, in addition to the polythiophenes, such as polypyrrole and polyazulene. The probable existence of thermal and concentration gradients in such dynamic processes suggest an assessment of gravitational influences on film morphologies is justified.

II. GROWTH OF THIN FILMS BY VAPOR DEPOSITION

A. Phthalocyanine Thin Films

This class of materials is an excellent candidate for use in developing NLO devices because of their two-dimensional planar π-conjugation, better chemical and thermal stability than most other organic materials, and ease of derivatizing through peripheral and axial positions (Fig. 4). Large [55–64] and ultra fast [52,65,66] third-order nonlinearities have been demonstrated for phthalocyanines (Pc). The nature of the central metal atom strongly influences the value of $\chi^{(3)}$. Shirk et al. [56] measured the third-order susceptibility of tetrakis(cumyl phenoxy) phthalocyanines by degenerate four-wave mixing at 1.064 μm. The $\chi^{(3)}_{xxxx}$ values for Pt–Pc (2 $\times$ 10^{-10} esu) and Pb–Pc (2 $\times$ 10^{-11} esu) were found to be approximately 45 and 5 times that of the metal-free phthalocyanine form (4 $\times$ 10^{-12} esu), respectively. Other researchers have shown that $\chi^{(3)}$ of Pc's increase about 15 times by changing the central metal atom from silicon to vanadium [67] and there is also an increase when the central atom is a heavy metal atom such as lead [68]. These positive

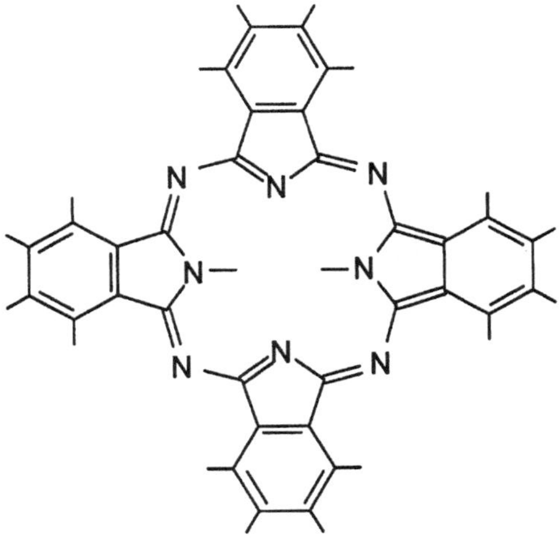

FIGURE 4 Metal-free phthalocyanine (H_2Pc).

influences of metal substitution on the third-order optical nonlinearity are attributed to the introduction of low-lying energy states derived from metal-to-ligand and ligand-to-metal charge transfer [56].

Recently, phthalocyanine-related compounds have emerged as novel candidates for second-order nonlinear optical applications. Sastre et al. [69] measured a β value of 2000×10^{-30} for the trinitro-substituted boron sub-phthalocyanine (SubPc) whose structure is shown in Fig. 5. SubPc's are two-dimensional 14 π-electron-conjugated macrocycles that are composed of three isoindole units containing boron inside. The reported second-order hyperpolarizability for the trinitro derivative is comparable to that of the most efficient linear or dipolar compound.

Difunctional tetraarylporphyrins with nitro groups as electron acceptors and amino groups as electron donors were observed to have a substantial hyperpolarizability by Suslick et al. [70]. The β values for *cis*-diamino–dinitro- and triamino–mononitro-substituted tetraarylporphyrins were 30×10^{-30} esu and 20×10^{-30} esu, respectively. These derivatives were prepared by the partial reduction of the nitro groups of 5, 10, 15, 20-tetrakis-(p-nitrophenyl) porphyrin. Li et al. [71] measured a relatively large second-harmonic generation (SHG: $\chi^{(2)}_{zzz} \sim 2 \times 10^{-8}$ esu) for covalently bound self-assembled monolayer thin films of 5, 10, 15, 20-tetra (4-pyridyl) 21 H, 23 H–porphine and its derivatives on quartz and silicon $\langle 100 \rangle$ substrate having a native oxide layer.

Thin films of Pc for fabrication of waveguides can be obtained by physical vapor transport because of their exceptional thermal stability and ease in sublimation. Matsuda et al. [58] observed that Pc with axial ligands (e.g., vanadyl phthalocyanine) have higher $\chi^{(3)}$ values than most usual unsubstituted Pc. The $\chi^{(3)}$ values of thin films of Pc vacuum deposited at 10^{-4} Pa onto fused quartz are shown in Table 2. The maximum values for the unsubstituted Pc's at 1.9 μm were 1.5×10^{-12} esu for CuPc and 0.8×10^{-12} esu for NiPc. In comparison, chloro-indium Pc and vanadyl Pc had $\chi^{(3)}$ values of 1.3×10^{-10} esu and 3×10^{-11} esu, respectively. Ho et al. [55] grew films of chloro-gallium (GaPc–Cl) and fluoro-

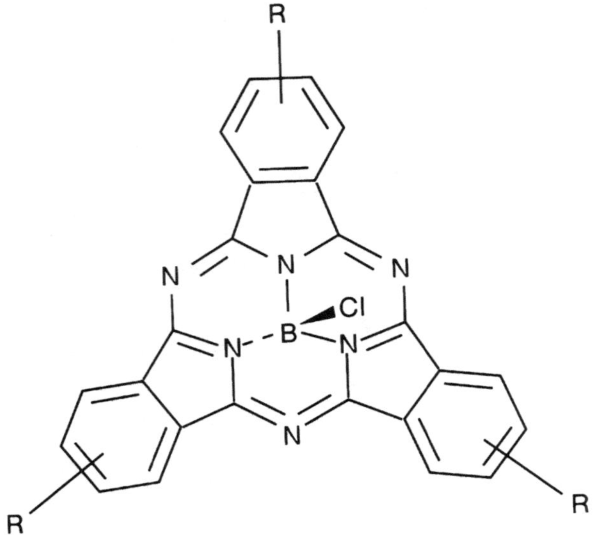

R = NO₂, or H, or *tert*-Bu

FIGURE 5 Trinitro-substituted boron sub-phthalocyanine (SubPc).

aluminum (AlPc–F) phthalocyanines onto fused silica flats at 150°C and 10^{-6} Torr. The $\chi^{(3)}$ values for GaPc–Cl and AlPc–F were 5×10^{-11} and 2.5×10^{-11} esu at 1064 nm for thickness of 1.2 μm and 0.8 μm, respectively.

Wada, et. al. [72] measured a $\chi^{(3)}$ of 1.85×10^{-10} esu at 1.907 μm for a 51.4-nm-thick film of vanadyl phthalocyanine (VOPc) vacuum deposited onto quartz.

TABLE 2

Compound	Film thickness (μm)	$\chi^3 \times 10^{-12}$ esu (1.9 μm)
Unsubstituted phthalocyanine		
Copper phthalocyanine (CuPc)	0.53	1.5
Cobalt phthalocyanine (CoPc)	0.22	0.76
Nickel phthalocyanine (NiPc)	0.35	0.80
Platinum phthalocyanine (PtPc)	0.41	0.60
Unsubstituted phthalocyanine with axial ligands		
Vanadyl phthalocyanine (VOPc)	0.28	30
Titanyl phthalocyanine (TiOPc)	0.26	27
Chloro-aluminum phthalocyanine (ClAlPc)	0.26	15
Chloro-indium phthalocyanine (ClInPc)	0.14	130

The $\chi^{(3)}$ values for two different phases in vanadyl and titanyl (TiOPc) phthalo-cyanines were measured by optical third-harmonic generation at wavelengths of 1543 nm and 1907 nm by Hosoda et al. [73]. The transformation of as-prepared Pc's films from phase I to phase II was performed by thermal annealing and was accompanied by a red shift in absorption spectra and an increase in $\chi^{(3)}$ values of two to three times. $\chi^{(3)}$ values for as-prepared films of VOPc and TiOPc were 3.8 $\times$ 10^{-11} esu and 10^{-11} esu, whereas the annealed films had values of 8.1×10^{-11} esu and 4.6×10^{-11} esu, respectively.

Recently, a relatively strong SHG was reported for vacuum-deposited copper phthalocyanine (CuPc) films which possess inversion symmetry. Chollet et al. [74] measured a $d_{eff} \approx 2 \times 10^{-19}$ esu at 1.064 μm fundamental wavelength for films with thickness ranging from 50 to 500 nm that were prepared at a pressure of 10^{-6} Torr and source temperature of 120°C. The films were homogeneous and partly oriented with a relatively large distribution of molecular axes oriented almost perpendicular to the substrate. This order was confirmed by SHG measurements. SHG in the films was attributed to quadrupolar or dipolar origins. Kamagai et al. [75] prepared 40–2000-Å thick films at a pressure of 5×10^{-6} Torr and obtained $\chi^{(2)}_{zyy}$ = 4×10^{-8} esu at a 1.06 μm fundamental wavelength, which is one-fourth the value for LiNbO$_3$. They proposed a mechanism in which an asymmetric crystal field acting perpendicular to the surface makes each CuPc molecule capable of SHG. Yamada et al. [76] performed in situ observation of SHG from CuPc films during vacuum evaporation on glass at a pressure of 1.5×10^{-5} Torr and a rate of 1.2 nm/min. Thickness dependence on SHG was compared with calculations based on electric dipole, electric quadrupole, and magnetic dipole mechanisms to clarify the origin of SH activity. Based on the comparison, SHG was ascribed to electric quadrupolar or, preferably, magnetic dipolar origin.

The research of Debe et al. [77–79] indicates that better quality organic thin films for use in NLO devices might be obtained by closed-cell physical vapor transport (PVT) in microgravity. In the PVT process, the source material is sublimed in an inert gas and allowed to convect or diffuse down a thermal gradient and to ultimately condense at a crystal or thin-film growth interface [77]. The advantage of thin-film growth in microgravity is that it provides the opportunity to eliminate buoyancy-driven convection. Recent reports [77–79] of the Space Shuttle mission STS-51 of August/September 1985 include results of experiments in which CuPc was epitaxially deposited, by PVT, onto highly oriented seed films of metal-free phthalocyanine (H$_2$Pc). The substrate was a 1.4-cm-diameter solid copper disk.

The PVT of organic solids (PVTOS) apparatus used to grow CuPc thin films consisted of nine identical metal/Pyrex ampoules housed within its own heater assembly and vacuum insulation cell [77,78,80]. Films were grown in 1.7-cm-diameter by 7.5-cm-long Pyrex tubes that were placed within resistance-wire-wound heaters which induced a nonlinear axial thermal gradient. The growth ampoules were filled primarily with CO$_2$, H$_2$, the buffer gas Xe, or He, and then either N$_2$ or CO as the next most abundant component. A computer-controlled heater maintained the hot end of the ampoule, containing the source, at 400°C for 4 h after the cruise temperature was reached. The substrate was maintained at a temperature of 70°C by a heat pipe.

The substrate seed film was prepared by vacuum sublimation of H$_2$Pc onto a copper disk at a temperature range known to produce highly oriented films. A metal-free phthalocyanine film with a thickness of 1100 ± 50 was grown at a deposition rate of 70 Å was grown at a deposition rate of 70 Å/min. The substrate was held at a temperature of 5–10°C and the source-to-substrate distance was 16 cm. Debe has shown that highly oriented films of H_2Pc are obtained when the substrate temperature range is approximately 5 ± 5°C [80]. Before starting the growth process, the source material was out-gassed by slowly increasing the temperature of the hot zone. After ∼ 2 h of "soaking," the temperature was increased to cause the H_2Pc material to deposit on the copper substrate at a pressure of about 5×10^{-6} Torr.

Microgravity grown CuPc films had several desirable features which indicate that the growth of organic films in low g may result in better quality films for NLO applications. For example, results of analysis by visual photography, bright-field and differential interference contrast microscopy, scanning ellipsometry, visible reflection spectroscopy, and direct interferometric phase-contrast microscopy imply that the Space grown films were radially more uniform and homogeneous, and an order of magnitude smoother over the submillimeter to submicron scale range [77]. Results of analysis involving the use of external reflection–absorption infrared (IR) spectroscopy, grazing incidence x-ray diffraction, and visible–near-IR reflection–absorption spectroscopy infer that the microgravity grown films are more highly uniaxially oriented and the films were found to consist predominantly of crystalline domains of a previously unknown polymorphic form of CuPc [78]. In addition, scanning electron microscopy analysis revealed that there was a distinctly different microstructure in the center of the Space grown films and that the circular perimeters of the microgravity grown films had microstructure much like that of the ground control films in both their center and edge regions [79].

As stated earlier in the chapter, electric field poling of polymers containing the nonlinear chromophore may be used to induce macroscopic asymmetry. However, in the case of the ring-structured macromolecule, phthalocyanine, poling is not generally an option to achieve ordering in deposited films. It is, therefore, beneficial to achieve, if possible, self-assembly to induce required asymmetry by other means. Whenever self-assembly might occur in molecules not prone to poling, exploitation of conditions favorable toward asymmetry could prove beneficial. In the case of Pc, for example, $\chi^{(3)}$ enhancement and possible $\chi^{(2)}$ inducement might result from self-assembly. Vapor-deposited H_2Pc have demonstrated some potentially interesting nonlinear optical properties. Researchers report these films to be randomly oriented when processing occurs in 1 **g** (Fig. 6a) [43,44,77–80]. From microgravity processing, CuPc films epitaxially deposited onto H_2Pc films are highly oriented and densely packed (Fig. 6b). Abdeldayem et al. [81] recently observed intrinsic optical bistability in vapor-deposited thin films of metal-free Pc, ranging in thickness from 40 to 800 nm, using continuous-wave (CW) and chopped He–Ne lasers at 633 nm. Source and substrate temperatures were maintained at 300°C and 5°C, respectively, whereas vapor vacuum deposition occurred at 10^{-6} Torr onto quartz disks. Bistability in the film was attributed to changes in the level of absorption and refractive index caused by thermal excitation. This nonlinear effect could improve dramatically in highly oriented microgravity processed films.

For a discussion of optical bistability, we first recognize that the absorption spectrum in the visible region shows a strong maximum at 626 nm. In one experiment [81], the CW He–Ne (632.8 nm) laser at a fixed input power of ∼30 mW

(a)

(b)

FIGURE 6 Copper phthalocyanine films epitaxially vapor deposited onto copper substrate a) μ-**g** deposition of CuPc epilayer; (b) 1-**g** deposition of CuPc epilayer. (Courtesy of 3M Corporation.)

was focused on a 230-nm-thick-film. The beam transmission increased temporally over a period of nearly 12 h, as shown in Fig. 7a. The nearly straight line of Fig. 7b is a fraction of the input power for monitoring laser stability. Fitting the transmission data to a single exponential (the solid line) gave a rise time of ~2.2 h to reach a steady state.

The temporal transmission effect in the film can be explained by the following sequence: (a) The initial low transmission of the beam through the film can be attributed to a strong absorption of the beam which generates free electrons and holes; (b) these free charges relax to excitonic states [82] and release their excess energy as heat to the system at the focal point; (c) the buildup of localized heating at the focus reduces absorption of the He–Ne radiation causing saturation of absorption. This factor is in agreement with the results of a separate experiment to measure the sample's absorption at a temperature elevated above room temperature by a water bath. Such reduction in absorption allows more transmission to occur with time.

Investigation of the bistability of metal-free Pc films of 833 nm thickness used a chopped He–Ne 632.8-nm laser beam at frequencies ranging from 100 to 750 Hz. The film was positioned on a micrometer stage, at the lens focus, and transversely translated in and out of the beam alternately to record intensity input and film transmittance. A Hewlett-Packard (HP) digitizing oscilloscope model 54120B recorded the input and the transmitted pulse with a HP plotter model 7470A (Fig. 8). The nonsymmetrical shape of the transmitted pulses (Fig. 8b) indicated the

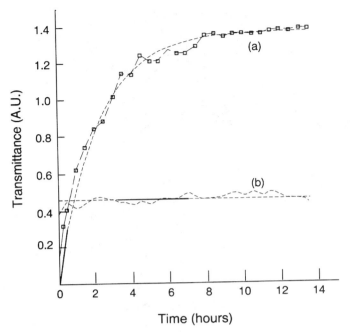

FIGURE 7 (a)Time-dependent transmittance through a metal-free phthalocyanine film (230 nm thick), using a CW He–Ne laser at 632.8 nm and 30 mW power. The dotted curve represents the experimental data, whereas the solid line represents a single exponential theoretical fit. (b) The dotted line represents laser stability throughout the experiment, and the solid line is the corresponding straight-line fit.

presence of intrinsic bistability in metal-free phthalocyanine. Figure 8c depicts typical bistable switching, constructed from the transmitted pulse. The switching power of ~0.33 mW per pulse in combination with a pulse duration of 1.37 ms recovery time yields a very low switching energy of ~0.45 nJ. Observation of bistability was repetitive in the same film using a CW He–Ne laser as shown in Fig. 9 for different time spans between successive points.

A thinner film, 230 nm, also demonstrated saturation of absorption at the same He–Ne laser frequency. Figure 10 illustrates the experimental data recording absorption by a 230-nm-thick film and the theoretical curve fit assuming Bloch-type saturable absorption with negligible scattering losses [83]:

$$a(I)L = \left(\frac{a_0 L}{1 + I/I_s} \right) \tag{1}$$

where I_s is the threshold power of the saturation, L is the thickness of the sample, and a_0 is the linear absorption coefficient. The saturation intensity, estimated from the theoretical fit, is

$$I_s \sim 2.0 \times 10^4 \quad \frac{W}{cm^2}$$

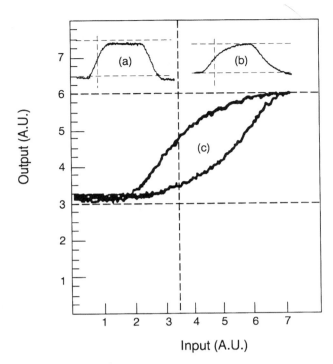

FIGURE 8 The bistability loop of a 833-nm metal-free phthalocyanine film using a chopped CW He–Ne laser at 632.8 nm: (a) the input pulse, (b) the transmitted pulse, (c) hysteresis switching constructed from (b).

Modeling phthalocyanine as a three-level system, a molecule in the ground state at saturation absorbs light at a rate

$$\frac{1}{t} = s_0 \, I_s / hf \tag{2}$$

where t is the decay time of the excited triplet state, I_s is the saturation intensity, s_0 is the absorption cross section of the ground state, and hf is the energy of the incident photon. From the measurements of the fall time of 1.0756 ms at 245 Hz in Fig. 8b, the absorption cross section was estimated [84] to be on the order of $\sim\!2.4 \times 10^{-17}$ cm^2.

The estimated third-order nonlinear susceptibility measurements by four-wave mixing using a pulsed Nd : YAG laser at 532 nm was on the order of 10^{-8} esu. This relatively large value is attributed to both resonant as well as thermal mechanisms that might be present in the system at this wavelength.

B. Polydiacetylene Thin Films

1. Second- and Third-Order Nonlinear Optical Properties of Polydiacetylenes

Polydiacetylenes (Fig. 11) are highly conjugated organic polymers that are of considerable interest because of their unique chemical, optical, and electronic prop-

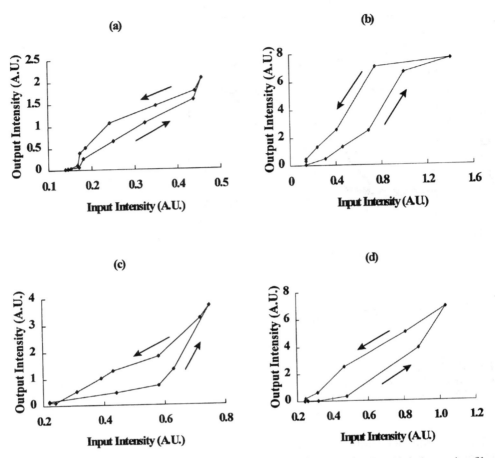

FIGURE 9 The bistability loops for different time spans of a metal-free phthalocyanine film of thickness 232.5 nm using a CW He–Ne laser. Time spans between successive points are (a) 3.84 s, (b) 10 s, (c) 342 s, and (d) 1800s. Part (a) shows the least prominent bistability loop; parts (b), (c), and (d) show a minimal effect of the time span between points.

erties [85–88]. This class of polymers has received extensive attention as organic conductors and semiconductors, as well as NLO materials. The high mobility of the π electrons in the polymer backbone allows them to have large optical/electrical susceptibilities with fast response times; they can be highly ordered, even crystalline, which is important for optimizing their electronic and optical properties, and they can readily be formed into thin films, which is the preferred form for many applications. The physical, chemical, and mechanical properties of polydiacetylenes can be varied by varying the functionality of the side groups, thereby making it possible to tailor their properties to meet specific needs. Thus, there is a great deal of interest in the use of these polymers for technological applications.

Polydiacetylenes are among the best known third-order NLO materials, and there has been considerable investigation over the past 20 years into their properties. Single crystals of poly(2,4-hexadiyne-1,6-ditosylate), also known as PTS, pos-

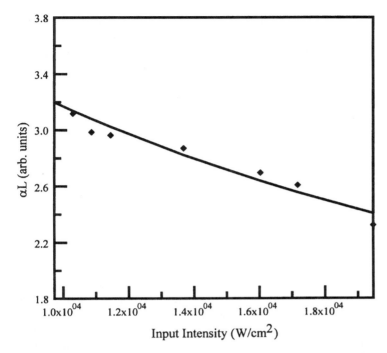

FIGURE 10 Experimental and theoretical fitting of saturable absorption of a metal-free phthalocyanine film (232.5 nm thickness) at 632.8 nm from a CW He–Ne laser. The corresponding saturation intensity is 19.8×10^3 W/cm^3.

sess one of the largest (possibly the largest) nonresonant third-order optical nonlinearities ever measured, on the order of 10^{-9} esu [89]. The third-order NLO properties of numerous other polydiacetylene crystals and thin films have also been determined. Typical $\chi^{(3)}$ values for polydiacetylenes range from 10^{-7} to 10^{-12} esu, depending on the degree of resonance enhancement and other factors. Both theoretical and experimental results have determined that the $\chi^{(3)}$ value is approximately 100 times greater along the polydiacetylene backbone than perpendicular to it, demonstrating the effect of the conjugated π-electron system. Because the origin of the nonlinearity is electronic, they can have very fast response times, on the order of femtoseconds.

More recently, polydiacetylenes have been investigated as potential second-order NLO materials. Theoretical calculations have indicated that certain polydiacetylenes could possess extremely high second-order NLO susceptibilities (e.g.,

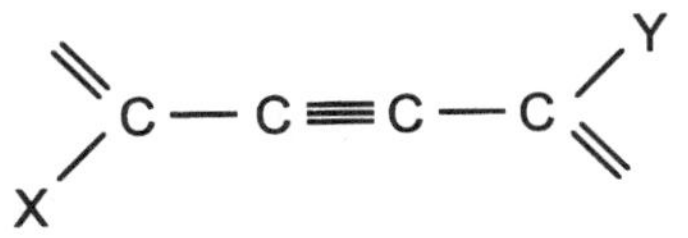

FIGURE 11 Structure of polydiacetylene repeat unit.

molecular hyperpolarizabilities on the order of 10^{-27} esu [90]. In order to make use of this second-order nonlinearity, it is necessary to orient the polymers into acentric structures, either crystals or thin films. This is not trivial; many compounds which have desirable properties at the molecular level tend to orient themselves centrosymmetrically in the bulk to minimize electrostatic interactions.

Second-harmonic generation (SHG) has been observed from certain asymmetrical liquid-crystalline diacetylene monomers (although, interestingly, not from the corresponding polymers) [91], from both LB and self-assembled polydiacetylene monolayers and multilayers [92,93] and even from a spin-coated polydiacetylene film [94]. Finally, powder SHG efficiences comparable to that of MNA (2-methyl-4-nitroaniline) have been obtained from vapor-deposited polycrystalline films of a polydiacetylene possesing MNA as a side group [90]. If the crystallites are partially aligned by growing the films quasi-epitaxially onto prealigned poly(tetrafluoroethylene) substrates, the SHG efficiency increases by almost one order of magnitude.

2. Potential Benefits of Microgravity Processing

Optical applications require the formation of high-quality thin polydiacetylene films (i.e., films possessing minimal defects such as impurities, inhomogeneities, light scattering centers, etc.). The standard techniques for obtaining polydiacetylene thin films involve the growth of crystalline diacetylene monomer films or the deposition of LB films, followed by topochemical polymerization of these films in the solid state to yield ordered polydiacetylene films. This ability to undergo solid-state polymerization is a very intriguing property of diacetylenes; in principle, one can start with a single-crystal monomer and obtain a single-crystal polymer [7,95]. However, the process is not trivial; the formation of high quality crystalline diacetylene monomer films or LB films can be very tedious and difficult, and, furthermore, by no means do all monomers polymerize readily in the solid state [8]. Achieving high-quality polydiacetylene films requires the growth of high-quality diacetylene monomer films, which are then topochemically polymerized. A commonly employed technique for obtaining diacetylene monomer films is vapor deposition. One of the chief limitations to vapor deposition of high-quality monomer films (e.g., single-crystalline films with good molecular orientation and few defects) has been a lack of understanding of how the processing conditions affect monomer film growth. It is certainly well known that parameters such as temperature, pressure, concentration, and so forth can affect vapor transport processes. One parameter which is often tacitly ignored is the influence of gravity. However, the effects of gravity, such as buoyancy-driven convection, can greatly influence heat and mass transport during the growth process, and thereby influence all of the aforementioned growth parameters. Thus, discerning the effects of gravity (or the lack thereof) could play a critical role in optimizing the growth of high-quality polydiacetylene films by vapor deposition.

One method of assessing convection in a gas phase is to perform the computation at low pressure to relieve the need for specific materials constants. It is important to note that buoyancy effects are possible only if the molecular mean free path is short enough relative to cell dimensions such that molecular flows are not in the free molecular flow regime. A mathematical model has been developed to determine buoyancy-driven heat transfer in an ideal gas under a variety of ori-

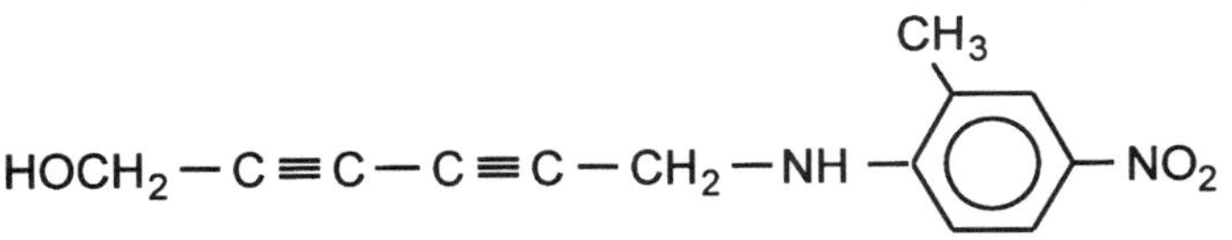

$$HOCH_2-C\equiv C-C\equiv C-CH_2-NH-\overset{\displaystyle CH_3}{\underset{\displaystyle}{\bigcirc}}-NO_2$$

FIGURE 12 Diacetylene monomer (DAMNA).

entations relative to gravitational accelerations [97]. The model demonstrates that convection can occur at total pressures as low as 10^{-2} mm Hg in cells having relatively high length-to-width ratios. A preliminary experimental test of the model involved deposition of the diacetylene monomer, DAMNA (Fig. 12), at an evacuation pressure of 10^{-2} mm Hg.

The deposition of DAMNA by physical vapor transport was of 30 min duration. Cell dimensions were the same as those depicted in Fig. 13 with source temperature of 120°C and sink temperature of 30°C. The evacuation pressure reached a minimum in the presence of DAMNA (10^{-2} mm Hg) and was significantly lower in the absence of DAMNA. It is probable that the DAMNA vapor pressure is equal to, or greater than, the evacuation pressure of 10^{-2} mm-Hg. From a different study, the measured vapor pressure of 4-*N*,*N*-dimethylamino-4′-nitrostilbene (DANS) at 120°C was reportedly 0.374 mm [98]. This is a relatively large organic molecule

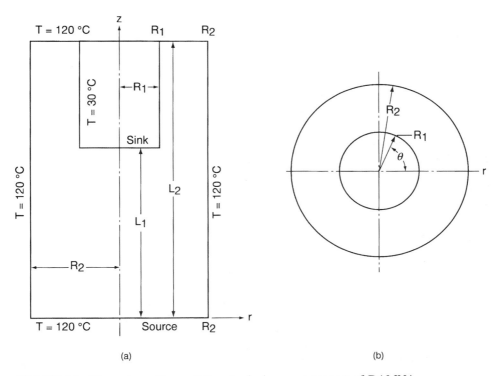

(a)   (b)

FIGURE 13 Vapor-deposition cell for physical vapor transport of DAMNA.

having a molecular weight of 211 g/mol as compared to DAMNA with a molecular weight of 247 g/mol. Considering structural and size similarities, we may approximate similarities in vapor pressure. Indeed, heptadecanol (molecular weight = 256.5 g/mol), also an alcohol with possible hydrogen bonding in its condensed phases such as expected of DAMNA, has a vapor pressure of 10^{-1} mm at 120°C [99]. There is no reason to expect DAMNA to differ drastically from these measured vapor pressures, and we may approximate that the vessel evacuation pressure minimum is largely due to the vapor pressure of DAMNA.

Using physical material parameters of air, a series of time steps demonstrates the development of flow and temperature profiles in an ideal gas. These profiles are driven by the specified temperatures with no mass fluxes (there are no subliming or condensing masses in this model) in a vessel as specified in Fig. 13. Computations show that in unit gravity, it is possibly that vapor deposition occurs by transport through an axisymmetric circulating flow pattern when applying heat to the bottom of a vertically positioned vessel. In the case where heating of the reaction vessel occurs from the top, deposition of vapor does not normally occur by convection due to a stable stratified medium. When vapor deposition occurs in vessels heated at the bottom, but oriented relative to the gravity vector between these two extremes, horizontal thermal gradients induce a complicated asymmetric flow pattern.

Comparison of Figs. 14 and 15 and Figs. 14 and 16 show the differences in induced flow fields in the $\theta = 0°-180°$ and $\theta = 90°-270°$ planes, respectively, driven by vertical and tilted cavities. There are two recirculation flows generated with centers at $(r_c, z_c) = (\pm 1.90$ cm 2.7 cm) and at $(r_c, z_c) = (\pm 0.8$cm, 3.2 cm) for flows induced by vertical and tilted orientations, respectively. For the vertical orientation, the flow depicted in Fig. 14 is axisymmetric, hence also representative of flows in the $\theta = 90°-270°$ plane. The flows are upward along the surface of the outer cylinder and downward from the central column. These are Bernard-type flows driven by counterrotating cells (clockwise on the left and counterclockwise on the right) due to an incipient instability in the narrow cylindrical cell cavity. There is no critical Rayleigh number for convection which assumes an infinite extent of the cell width. For the tilted cell, a resultant asymmetric flow in the $\theta = 0°-180°$ plane approaches an antisymmetric flow profile resulting from differential heating between vertical and opposite walls of a cavity; that is, an incipient instability causes heat to flow along the bottom surface (now tilted upward), upward along the side wall, along the cold surface (downward), and downward along the other side wall. The two recirculation flows for this orientation appear in the $\theta = 90°-270°$ plane (Fig. 16). There is an upward flow from the bottom surface of the outer cylinder transporting heat toward the bottom surface of the central column with downward flows on the outside walls (clockwise on the right and counterclockwise on the left). This asymmetric three-dimensional flow profile is quite complex and represents a greater degree of convection in the cell cavity for the tilted orientation than for the vertical orientation. The model material parameters for air predict the effect of kinematic viscosity to be of the same order as thermal diffusivity, which is the case for Prandtl number (Pr) ~1 fluids [100]. The classical heat capacity for a gaseous molecule, such as DAMNA, with its large number of vibrational degrees of freedom, is only achieved at high temperatures; that is, the heat capacity of DAMNA will approach that of a diatomic molecule, such as the

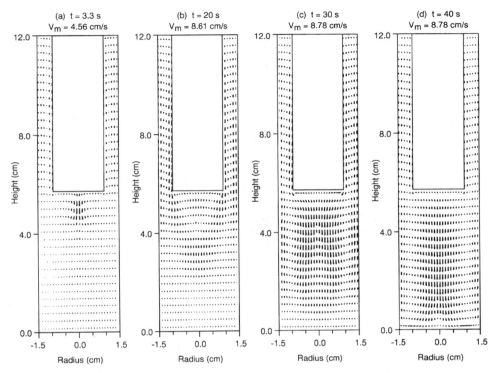

FIGURE 14 Axisymmetric flow in the $\theta = 0°-180°$ plane in a cell in which deposition occurs with cell oriented vertically.

major components of air, because its vibrational modes are unable to store energy at these relatively low temperatures. The thermal diffusivity approximation using air data is, therefore, a fair one, assuming similar heat conduction coefficients. In the classical limit, the higher heat capacity would yield $Pr \gg 1$. Because an ideal heat capacity approximation is an overestimate at the operating temperatures, deviation from ideal behavior causes $Pr \sim 1$, closer to that of air. Without actual data for DAMNA, we are left with experimental data from the low operating pressures and temperatures to compare with modeling estimates. Close agreement between experiment and the model would indicate that DAMNA at these pressures indeed approximates an ideal gas with the appropriate deviations which tend toward validation of the use of air physical constants. Furthermore, the constant heat flux provided by the circulating bath is a factor in neutralizing thermal effects specific to individual molecules.

Experimentally, it is helpful to utilize Beer–Lambert's relationship for transmission of radiation through a medium to test gravitationally sensitive flow pattern predictions [97]. We would expect that the flow patterns in Figs. 15 and 16 (cell tilted at 45° relative to gravity vector) would affect film quality over most of the deposition surface differently than the flow pattern depicted in Fig. 14 (cell oriented vertically). Preliminary experiment suggests this to be the case with respect to film thickness. The Beer–Lambert law relative to film thickness can be written

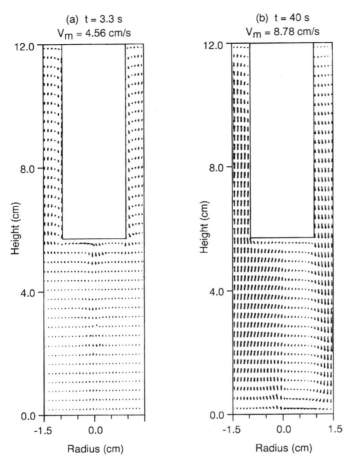

FIGURE 15 Flow in the $\theta = 0°-180°$ plane in a cell in which deposition occurs with the cell tilted 45° to the vertical axis.

$$I = I_0 e^{-\alpha l} \tag{3}$$

where I_0 is incident radiation intensity, I is the reduced radiation intensity after passing through the film, α is a proportionality constant containing the molar absorption coefficient, ε, and concentration of absorber, and l is the film thickness. The molar absorption coefficient represents a molar cross section for absorption. The greater the cross section of the molecule for absorption and absorber concentration, the greater the attenuation of the intensity of the beam. Likewise, film thickness, l also attenuates the beam intensity accordingly. The ratio I/I_0 is a measure of beam transmittance, T. If increased convection yields greater film thickness, we would expect a larger intensity in the absorption bands of films deposited in cells tilted at 45° over those where deposition occurred in cells positioned vertically. We may define the dimensionless product $A = -\alpha l$ as absorbance which incorporates all of the contributors to beam attenuation. Because the nature of the material, DAMNA, and concentration (pure material) are identical in both cases, ab-

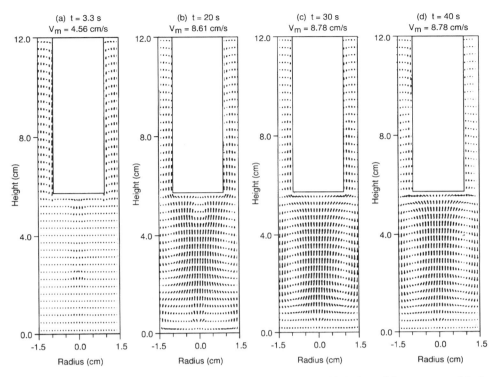

FIGURE 16 Flow in the $\theta = 90°–270°$ plane in a cell in which deposition occurs with the cell tilted 45° to the vertical axis.

sorbance is only a function of film thickness. Figure 17 is a wavelength scan in the range 190–820 nm of a vertically deposited film, and Figure 17b is that of an obliquely deposited film. Table 3 contains representative absorbance intensities from scanning an approximately 1-cm-diameter spot in similar vicinities of films formed during vapor deposition in vertically and obliquely oriented cells. Visually, the films from the obliquely oriented cells are a deeper yellow color than those from vertically oriented cells. We qualitatively assume generally thicker films from the relative appearances. The surfaces of these monomer films are translucent and microcrystalline. At the present time, it is also preferable for us to discuss relative film thickness from beam attenuation in qualitative terms (although more quantitative than visual observation) while observing that the spectroscopic irradiation spot sizes are large enough to average over about 50% of the film surfaces. The same relative result occurred, in that beam attenuation was greater at each maximum absorbance wavelength repeatedly. We may consider Table 2 a qualitative representation of a consistent result.

In keeping with the assumptions, the mathematical model correctly predicts qualitative differences in film properties between vertically and obliquely oriented cells; that is, more convection in cells having the tilted orientation is apparently responsible for correspondingly greater film thickness.

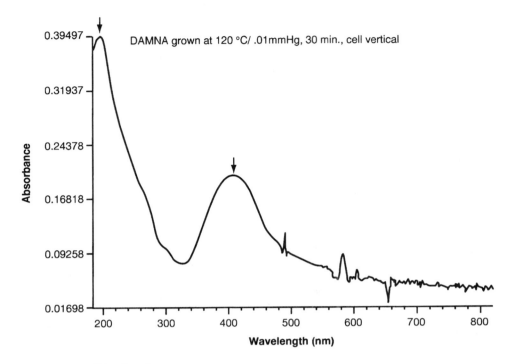

Annotated Wavelength:

(a)   1: Wavelength = 422   Result = 0.200165
       2: Wavelength = 214   Result = 0.394974

FIGURE 17 Absorbance as function of wavelength for vapor-deposited film after 30 min of deposition for (a) a vertically oriented cell and (b) a tilted cell.

III. GROWTH OF THIN FILMS BY SOLUTION PROCESSES

A. Ultraviolet Solution Polymerization

1. Polydiacetylene Films

Recently, a novel process has been discovered for the formation of thin amorphous polydiacetylene films using photodeposition from monomer solutions onto transparent substrates [45,46,90,101]. Specifically, polymeric films were directly synthesized from a diacetylene monomer, DAMNA (Fig. 12), derived from 2-methyl-4-nitroaniline (MNA) that only sluggishly polymerizes when the crystalline monomer is irradiated [102]. This compound was one of several asymmetric diacetylenes that were first studied extensively for their optical and electronic properties by Garito and co-workers in the late 1970s; however, their investigations did not include behavior in solution [103,104]. It was found that thin polymerized DAMNA (PDAMNA) films can be obtained readily from solutions of DAMNA in 1,2-dichloroethane by irradiation with long-wavelength UV light through a quartz or glass window, which serves as the substrate. This simple straightforward process yields transparent films with thickness on the order of 1 μm.

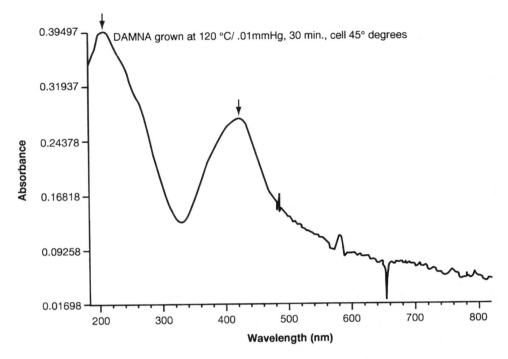

Annotated Wavelength:

    1: Wavelength = 422 Result = 0.385361
    2: Wavelength = 214 Result = 0.545670

(b)

FIGURE 17 Continued

Despite the considerable volume of literature available on diacetylenes and polydiacetylenes, this solution-state photodeposition reaction has never been reported. Thus, many of the parameters controlling the efficacy of the process are not yet known. The basic idea is quite straightforward; the diacetylene monomer solution is irradiated through a UV transparent substrate and a thin polydiacetylene film results. To date, several diacetylene monomers have been tested and found to be capable of photodeposition of polymeric films from solution.

TABLE 3 Representative Absorbance Intensities from Scanning an Approximately 1-cm-diameter Spot in Similar Vicinities of DAMNA Films Formed During Vapor Deposition

λ (nm)	A_{max} (cell vertical film)	A_{max} (cell tilted film)
422	0.200	0.385
214	0.395	0.546

Special chambers were constructed for carrying out the reaction and obtaining thin films on small round substrate disks (Fig. 18). To obtain a PDAMNA thin film, a solution of DAMNA in 1,2-dichloroethane (approximately 2.5 mg/ml = 0.01 mol/l) is placed inside the chamber, and the chamber is then irradiated through the substrate with long-wavelength UV light (365 nm). The thickness of the resulting PDAMNA film depends on the duration of exposure and the intensity of the UV source. After the photodeposition is complete, the monomer solution, which now also contains suspended particles of precipitated polymer, is removed from the growth chamber. The substrate, now coated on one side with the PDAMNA film, is then removed, washed with 1,2-dichloroethane, and dried.

Thin PDAMNA films obtained in this manner are transparent, glassy yellow–orange in appearance, suggesting an amorphous nature. Both refractive index measurements and electron beam diffraction studies indeed indicate that the films are amorphous. The films are insoluble in organic solvents, although solvents such as acetone can cause them to peel off of the substrate. Even concentrated sulfuric acid does not dissolve the films; they turn brown, shrivel, and peel off of the substrate, but do not dissolve, even after several weeks. In contrast, the PDAMNA powder precipitated from the bulk solution is soluble in solvents such as acetone and dimethylsulfoxide (DMSO).

The exact role that the substrate plays in photodeposition of polydiacetylene films from solution is not yet fully understood. Apparently, any substrate which is sufficiently transparent to UV light can be used; thus far, we have grown PDAMNA films onto glass, quartz, mica, indium–tin-oxide-coated glass, polyethylene teraphthalate, KBr, and NaCl. In order to gain some insight into the process that occurs at the surface of the substrate during photodeposition, masking experiments were conducted in which a portion of the substrate is blocked from exposure to the UV light during film deposition. The mask is placed on the *exterior* surface of the substrate (opposite the side on which the film is grown) and thus, is not in contact with the solution; it serves merely to protect part of the substrate from the light. Interestingly, the result is that film deposition occurs only where the substrate is

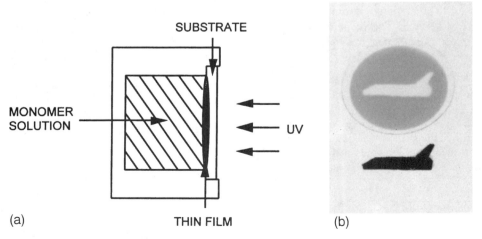

FIGURE 18 (a) Polydiacetylene thin-film growth chamber; (b) masked PDAMNA film on glass.

directly exposed to the light (see Fig. 18b); absolutely no film deposition occurs behind the mask, even though polymerization takes place throughout the bulk solution. This clearly indicates that polymerization is occurring at (or very near) the surface. If this were simply a case of bulk solution polymerization, followed by adsorption of the polymer onto the substrate, the mask, because it is on the outside, should have no effect; film deposition would be expected to occur over the entire substrate.

2. Third-Order NLO Properties of Films

Based on optical microscopy and on refractive index measurements using waveguide mode analysis, PDAMNA thin films obtained via photodeposition from solution have good optical quality, superior to that of films grown using conventional crystal growth techniques. Considering the simplicity of photodeposition, this technique could make the production of polydiacetylene thin films for applications such as nonlinear optical devices technologically feasible. Hence the nonlinear optical properties of the PDAMNA films were investigated—specifically, their third-order NLO susceptibilities.

Degenerate four-wave mixing experiments carried out at 532 nm on PDAMNA films obtained by photodeposition from solution (thickness around 1.0 μm) yield $\chi^{(3)}$ values on the order of 10^{-8}–10^{-7} esu. Qualitative measurements indicate that the response time is on the order of picoseconds, which is consistent with an electronic mechanism for the nonlinearity. It should be pointed out that because 532 nm is in the absorption edge of the polymer, the $\chi^{(3)}$ values obtained are resonance enhanced. Typically, $\chi^{(3)}$ values for polydiacetylenes can vary by several orders of magnitude, depending on the degree of resonance enhancement and other factors [87]. The largest reported nonresonant (purely electronic) $\chi^{(3)}$ value for a polydiacetylene is on the order of 10^{-9}–10^{-10} esu for PTS single crystals [105]. In order to obtain a valid measure of the inherent nonlinearity of the PDAMNA films, experiments need to be conducted at longer wavelengths, where the polymer does not absorb (in the case of PDAMNA, above 700 nm). Preliminary measurements with the PDAMNA films using a Ti–Sapphire laser at 810 nm give $\chi^{(3)}$ values on the order of 10^{-11} esu, with response times on the order of femtoseconds [106]. There are no indications of either one- or two-photon absorption at this wavelength. To ascertain the true potential for device applications (the figures of merit), thorough measurements of light scattering, linear absorption, two- and three-photon absorption, damage thresholds, and so forth must be performed [103]. Such experiments are underway and will be the subject of future publications.

Thus far, these films have not been studied for second-order NLO properties because they are amorphous, although studies (atomic force microscopy, scanning electron microscopy, and UV–visible spectroscopy) do indicate that there is partial chain alignment in the direction normal to the substrate. At present, the degree of orientation is too low to exploit any potential second-order nonlinearity. However, it may be possible to improve the orientation in the films by means such as electric field poling, surface modification of the substrate, or even modifying the polymer structure (e.g., attaching a liquid-crystal moeity). Ordered polydiacetylene films would not only be capable of second-order nonlinearity but should also exhibit increased third-order nonlinearity. Additionally, electronic applications such as one-dimensional conductors require films with aligned polymer chains.

3. Fluid Dynamic Analysis

It is well known that gravitational effects, such as buoyancy-driven convection, can affect heat and mass transport processes in solution [108]. Photodeposition of polydiacetylene films from solution is no exception. We shall first discuss how buoyancy-driven convection can arise during photodeposition of PDAMNA films from solution, and then describe how this convection can affect the morphology, microstructure, and properties of the films obtained. Both the monomer solution and the film generate heat due to absorption of UV radiation. The radiative heating, along with the thermal boundary conditions of the walls of the thin-film growth chamber, will give rise to a complex temperature pattern in the solution. Due to the lack of thermodynamic equilibrium, the solution will possess temperature and concentration gradients, and, therefore, density gradients. These gradients, under the influence of gravity, can induce convective fluid flows in the solution (buoyancy-driven convection).

The onset of thermal convection is determined by a stability parameter known as the Rayleigh number, Ra, defined as [108]

$$\mathrm{Ra} = \frac{\alpha\, g\, d^3\, \Delta T}{\nu\, \kappa} \tag{4}$$

where α is the coefficient of thermal expansion of the solution, g is the acceleration due to gravity, ΔT is the temperature difference across distance d in the solution, ν is the kinematic viscosity, and κ is the thermal diffusivity. For photodeposition of PDAMNA films, the value of ΔT (over a distance of less than 1 mm) can vary from a only a few tenths of a degree to several degrees, depending on the intensity of the UV radiation. In order to grow thicker films (>1 μm), higher-intensity radiation is necessary, making large temperature gradients unavoidable. The intensity and flow pattern of convection can be predicted when the Rayleigh number is known. For instance, for an infinite fluid layer in the horizontal direction with a temperature gradient in the vertical direction (colinear with gravity), convective motion will occur in the form of rolls with axes aligned horizontal when Ra > 1708 (the critical Rayleigh number), whereas no convection will occur if Ra < 1708 [109]. The exact value can only be determined by numerical solution of the fluid flow in the chamber. In the case of horizontal temperature gradients (orthogonal to gravity), all values of the Rayleigh number lead to convection, and the magnitude of the velocity of the fluid flow is proportional to the square root of the Rayleigh number.

Density gradients can also arise in the solution due to variations in the concentrations of the chemical species present in the solution. Variations in the monomer concentration are caused by depletion of monomer from the solution at the surface of the growing film and in the bulk. Also, generation of dimers, trimers, and other soluble by-products in the bulk solution may result in additional concentration density gradients. Such solutal gradients, along with the temperature gradients, can give rise to double-diffusive convection. This complicated convective motion is usually analyzed with the aid of the solutal Rayleigh number, in addition to the thermal Rayleigh number [108]. The solutal Rayleigh number, Ra_s, is defined as

$$\text{Ra}_s = \frac{\beta \, g \, d^3 \Delta C}{\nu \, D} \tag{5}$$

where β is the coefficient of concentration expansion, ΔC is the concentration difference across distance d in the solution, and D is the diffusion coefficient. Double-diffusive convection flows can be far more complex than simple thermal convection flows.

Hence, we see that convection can arise by several means during polydiacetylene film photodeposition from solution. The extent of convection and its intensity and structure can only be understood through accurate numerical modeling of the fluid motion and thermodynamic state of the system.

4. Transport of Particles from Bulk Solution

One significant effect of convection can be seen when PDAMNA films grown in 1-g are viewed under an optical microscope: They exhibit small particles of solid polymer embedded throughout. These form when polymer chains in the bulk solution collide due to convection and coalesce into small solid particles, on the order of a few hundredths of a micrometer in size. Because these particles are so small, almost colloidal in nature, they do not sediment out readily and, thus, remain suspended in the bulk solution. Convection then transports these particles to the surface of the growing film, where they become embedded. These particles are defects that can scatter light and, thus, lower the optical quality of the films.

To study the effects of convection on the occurrence of these particles in the films, the growth chamber was placed in different orientations with respect to gravity in order to vary the fluid flow pattern [110]. PDAMNA films were grown with the chamber vertical (irradiating from the top) and with the chamber horizontal (irradiating from the side). In the case when the chamber is vertical and the solution is irradiated from the top, the axial temperature gradient is vertical with respect to gravity, and the bulk solution is stably stratified because warmer, less dense solution is above cooler, more dense solution. Thus, in this orientation, convection should be minimized. In the case when the chamber is horizontal and the solution is irradiated from the side, the axial temperature gradient is horizontal with respect to gravity, which makes the density gradients less stable. Hence, convection should be much more pronounced in this orientation. Numerical simulations of the fluid flow are consistent with these expectations [110].

This is reflected in the distribution of solid particles observed in the PDAMNA films grown in the two different orientations. Films grown with the chamber horizontal clearly contain a greater concentration of particles than films grown with the chamber vertical (Fig. 19). This is consistent with expectations based on the relative amounts of convection in the two orientations; films grown under increased convection contain more particles than those grown under less convection. Also, waveguiding experiments with these films demonstrate that the films containing more particles exhibit greater light scattering than those containing fewer particles.

Note that even the film grown in the vertical orientation, where convection is minimized, still contains particles. Thus, although convection is lessened in this case, it is not eliminated. There are two reasons for this. First, there are still radial thermal density gradients in the horizontal direction, even when the chamber is

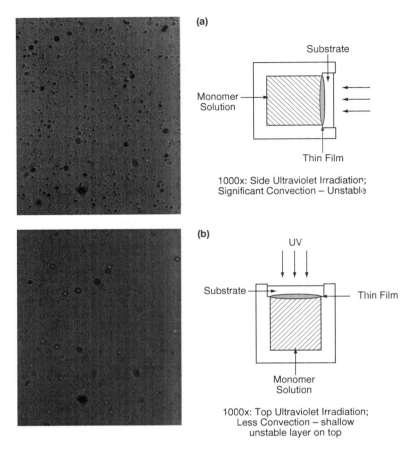

(a)

Substrate

Monomer
Solution

Thin Film

1000x: Side Ultraviolet Irradiation;
Significant Convection – Unstable

(b)

UV

Substrate

Thin Film

Monomer
Solution

1000x: Top Ultraviolet Irradiation;
Less Convection – shallow
unstable layer on top

FIGURE 19 PDAMNA films grown in two different orientations: (a) films grown with the chamber horizontal; (b) films grown with the chamber vertical.

vertical because the solution near the side walls is cooler than that near the center; these can give rise to convection. Also, because the substrate is transparent to UV light, it is not heated directly by the radiation; the only means by which it receives heat is via conduction. Thus, initially, there will be some heat flow from the warm solution to the cooler substrate, producing a very shallow unstable thermal density gradient in the immediate vicinity of the substrate–solution interface, which sits above the stably stratified bulk solution. Any convection initiated in this unstable layer may penetrate deeper into the stable layer below, giving rise to the phenomenon of penetrative convection [111]. The bottom line is that even under optimum conditions in 1 **g**, convection is still present during polydiacetylene thin film photodeposition from solution, causing particles in the films.

Not only can convection affect the transport of particles which are polymerized from the bulk solution, it can also transport colloidal particles which are purposely introduced into the system to alter the optical properties. The study of the effect of composite geometry on the nonlinear optical properties of materials is an active area of research. One of the most commonly studied systems, discussed earlier, involves small metal inclusion particles surrounded by a continuous host medium. The linear optical properties of these composites are described by the theory of Maxwell Garnett [112]. In recent years, the model has been extended to include nonlinear materials [113,114]. Unfortunately, assimilation of metal particles into highly nonlinear solid-state materials is difficult using traditional chemical techniques because most highly nonlinear polymeric host candidates require organic solvents which are incompatible with the colloid. Ion implantation is a realistic alternative but is not readily available in most laboratories. Moreover, the distribution of metal in the axial direction follows a Gaussian profile. Recently, however, Brust et al. [115], functionalized gold particles with thiol groups which serve to protect the particles from solvent degradation and attack. Interestingly, the colloidal metal can actually be dried and stored without agglomeration. The thiol groups in this case have little effect on the optical properties of the colloid. These thiol-capped metal particles can then be resuspended in many organic solvents and incorporated into the PDAMNA films through photopolymerization. Hence, this recipe offers a simple way to introduce small metal particles into highly nonlinear polymers. The resulting films, however, suffer from gradients in the metal concentration which are clearly visible under reflected room light. Moreover, the films have a much higher concentration of metal than would be expected from a diffusional process. Hence, these gradients most likely arise from the process of convection. Convection often tends to destroy the homogeneity of particle-doped systems, whether it be colloidal metal films or doped porous glass. In this case, the metal particle dopants serve not only to modify the nonlinear optical properties of the system but also to elucidate the role of convection in the formation of polymeric thin films.

5. Effects of Convection of Kinetics, Morphology, and Microstructure

We have discussed how convection can transport particles of solid polymer precipitated from the bulk solution into the films. However, convection can also affect film deposition at the molecular level. To gain some insight into these effects, it is necessary to understand the kinetics of film deposition.

The rate of polydiacetylene film photodeposition from solution can be given by the expression [116]

$$\frac{dl}{dt} = kI^m C^n \tag{6}$$

where l is film thickness, t is time, I is the intensity of the UV radiation, C is monomer concentration, m and n are the orders of the reaction in radiation intensity and monomer concentration, respectively, and k is the rate constant. Initial results indicate that for photodeposition (using 365-nm-wavelength light) of PDAMNA films from 1,2-dichloroethane at ambient temperature (25°C), $m = 1.0$, $n = 0.5$, and $k = 3.2 \times 10^{-7}$ (MKS units) [117].

Additionally, it can be shown from the principles of chemical kinetics (Arrhenius equation) [116] that the rate constant (k) is given by

$$k = Ae^{-E/k_bT} \tag{7}$$

where E is the activation energy of the reaction, k_b is Boltzman's constant, T is temperature, and A is a preexponential factor related to the frequency of collisions of molecules with the surface of the growing film.

The equations above clearly show how the rate of polydiacetylene film photodeposition from solution depends on variables such as temperature and monomer concentration. The effects of convection can also be gleaned from these equations. We know that convection affects heat and mass transport to and from the surface of the growing film, which is reflected in the temperature and concentration profiles along the surface. Variations in these parameters along the surface of the film, in accordance with the above equations, will cause variations in the rate of film deposition, leading to uneven film growth. This will be especially pronounced if the fluid flow along the surface varies drastically or is turbulent. Thus, we see how convection can directly affect the kinetics of polydiacetylene film photodeposition from solution, and thereby affect the morphology (thickness and surface roughness) of the films.

Finally, convection may also play a role in affecting the microstructure of the films, specifically the molecular orientation of the polydiacetylene chains. Preliminary studies conducted using atomic force microscopy and x-ray photoelectron spectroscopy indicate that films photodeposited onto quartz (in 1 **g**) for a very short duration of time (a few minutes) exhibit good polymer-chain alignment in the direction normal to the substrate, whereas films grown for longer duration show significantly poorer chain alignment. Thus, in the early stages of deposition, there appears to be some tendency for orientation, which lessens as the reaction proceeds and the polymer chains grow. This could be due to the fact that as the film grows, any influence that the substrate may have on molecular orientation at the film surface decreases. However, in this case, the substrate is amorphous quartz; hence, ordering of the polymer chains by the substrate (i.e., epitaxy) is not expected. Therefore, another possibility for the decrease in order is convection. The turbulent and chaotic molecular motions that occur during convection may cause the chains to become entangled and matted around each other as they grow longer. Also, variations in temperature and monomer concentration along the film surface, influenced by convection, can effect molecular orientation, and possibly even polymer-chain packing densities. Discerning the role that convection plays in affecting molecular orientation is an essential part of any fundamental study.

6. Growth of Films in Microgravity

An experiment was recently conducted aboard the Space Shuttle Endeavor (CON-CAP-IV) in which photodeposition of PDAMNA films from solution was carried out in microgravity [114]. In this environment, buoyancy-driven convection can essentially be eliminated. Because of unplanned orbiter maneuvers during the mission, leading to extraneous accelerations and limitations of the flight hardware, results varied somewhat among samples. However, the best space grown film clearly exhibits fewer particles than the best ground-based films (Fig. 20). These few particles may have resulted from slight mixing in the solution caused by the orbiter motions, or, possibly, they may have nucleated on the surface of the film itself. Nonetheless, the initial results are very encouraging; it appears that the lack of convection can, indeed, lead to PDAMNA films with significantly fewer defects,

100 μm

FIGURE 20 PDAMNA films grown in space. The best space grown films clearly exhibit fewer particles than the best ground-based films.

and thus greater optical quality. Further characterizations of the space grown films are currently underway.

This study clearly shows that photodeposition of polydiacetylene thin films in unit gravity occurs in highly convective environments and that this convection can influence the morphology and quality of the films. Indeed, even when irradiation occurs from the top of the cell, the most stable stratified cell orientation, defects remain in the films due to the persistence of buoyancy-driven convection. To achieve homogeneity, minimal scattering centers, and possible molecular-order photodeposition of polymer films by UV light exposure must proceed in a microgravity environment. Fluid mechanics simulations are useful for establishing gravitational sensitivity to this recently discovered process [45] for preparing thin films having quite promising nonlinear optical characteristics.

B. Polymer Thin Films by Electrochemical Polymerization

Electrochemical polymerization is a method that should be thoroughly investigated for use in fabricating thin-film waveguides. The procedure is relatively straightforward and films can be fabricated using many commercially available materials. Another desirable feature is that NLO films might be prepared on the surface of various substrates during polymerization for the fabrication of devices. In addition, this method has been used to synthesize several polymers that include polythiophenes, polypyrrole, polazulene, and polypyrrole.

Polythiphenes (pT) are promising materials for NLO applications because of their large third-order optical nonlinearities, environmental stability, and structural versatility. Their potential suitability for devices was demonstrated by Dorsinville et al. [118], who measured the NLO properties of thin films of pT, polythieno(3,2-b), thiophene (pTT), polydithieno(3,2-b,2′,3′-d)thiophene (pDTT). The $\chi^{(3)}$ values at 532 nm for pT, pTT, and pDTT were 6.6×10^{-9} esu, 5.9×10^{-9} esu, and 11.3×10^{-9} esu, respectively. Electropolymerization of the monomers was performed in a two-compartment cell with indium–tin oxide electrodes. Typical thickness of films were 0.5–2 μm. In addition, Logsdon et al. [119] obtained a $\chi^{(3)}$ value of 10^{-9} esu for LB films of poly(3-dodecylthiophene). The polymer was first prepared by electrochemical polymerization. This was then used to prepare LB films. The electrochemical process was carried out at 5°C using nitrobenzene as the solvent, tetra-*n*-butylammonium hexaflurophosphate as the electrolyte, indium–tin oxide as the cathode, and a platinum coil as the cathode.

The prospect for using microgravity processing to produce better polymer films by electrochemical reactions or learn principles that might be used to improve processing on Earth should be explored, as there is evidence that electrodeposition of metals in low gravity results in deposits that have significant differences relative to those prepared on Earth. Ehrhard [120,121] found that nickel deposited at high rates in microgravity during a suborbital rocket flight produced an amorphous or nanocrystalline film with grains so small that x-ray diffraction peaks associated with the crystalline structures were not seen. Further studies involving the electrodeposition of metals in microgravity have been performed by Riley and co-workers [122–124].

IV. PROTOTYPE DEVICES BASED ON SECOND- AND THIRD-ORDER NLO ORGANICS AND POLYMERS

A. Electrooptic Polymer Advances

It is an accepted conclusion that the development of all-optical devices will be gradual and will include hybrid devices in which electrons interface with photons. The advent of electrooptics occurred several years ago when researchers realized that photons could respond to electrons through certain media such as lithium niobate (LiNbO$_3$). This route toward the development of all-optical devices demands immediate and intensive searches for the ideal hybrid, which requires large second-order characteristics in the materials of interest and appropriate processing techniques. It is this technology requirement that enhances the criticality of asymmetry in crystal packing and heightens the potential value of microgravity processing. The goal is to exploit maximum and optimal directionality of hyperpolarizabilities in inorganic or organic crystals interacting with electric fields, and their capacity to modulate light. Alignment of polymers and macromolecules to take advantage of predicted second-order superiority in many of these molecules could provide the desired effects.

There have been several significant advances in the development of electrooptic devices based on polymers. One of the problems associated with organic systems is the inherent "brittle" nature of pure materials as a result of weak van der Waals intermolecular interactions in contrast to strong ionic and covalent bonds characteristic of inorganics. Dalton et al. [125] have addressed this problem by preparing a variety of highly cross-linked polymers based on spin-casting and poling a polymethacrylate with "dangling chromophores." The resultant cross-linked polymer films are "hard" and exhibit very high electrooptic coefficients. These materials retain their optical nonlinearity for several thousand hours at relatively high temperatures (90–150°C). Optoelectronic modulators developed by this group based on these and other materials and processes are available as prototypes and are capable of efficiently converting electronic signals into optical signals. These researchers and some others in the field believe that polymer-based modulators will outperform conventional inorganic devices and should be significantly less expensive to prepare. Marder and co-workers [126] have played a key role in these developments, including the development of a heterocyclic molecule with a strong electron donor and strong acceptors. This molecule is highly soluble in a polycarbonate matrix, which, after poling, yields a material with very high second-order coefficients and good mechanical strength. Although this material does not have good thermal stability, the contribution is a significant one toward the search for the ideal electrooptic material based on organic and polymeric materials.

B. All-Optical Switching

A Mach–Zehnder (MZ) interferometer (Fig. 21a) is an example of an on–off switch. Switching occurs when the phase of one arm is shifted by π relative to the phase of the other arm. For all-optical switching, the phase shift may be achieved by introducing a NLO material into one of the arms of the interferometer or by using a nonlinear material for the entire interferometer and making one arm longer than the other—an asymmetric MZ. For fused silica fibers, long lengths of fiber are required to reduce the switching energy due to the weak nonlinearity, and this makes it difficult to maintain the stability of the interferometer. Thus, for fibers, a very ubiquitous approach is to use a nonlinear optical loop mirror (Fig. 21b). The loop mirror is very stable, as it is a common path interferometer. The required phase shift is achieved through cross-phase modulation (XPM) of the signal by an orthogonal polarized control pulse. Because the control pulse propagates in one direction, the index change occurs only for the copropagating signal pulse. The bottom coupler extracts the control pulse from the loop.

Although NLO loop mirrors provide a variety of controls and options for the user, more compact devices are required for integrated optics applications and these require materials with higher nonlinearity to keep the switching energy low enough to be practically useful. The nonlinear directional coupler described earlier (Fig. 21c) or the Mach–Zehnder interferometer are better suited for these applications. Efficient femtosecond all-optical switching has been demonstrated in GaAlAs passive semiconductor systems at energies below half the band gap in a number of different devices [127–130]. Although TPA is no longer a problem for energies below half the band gap, three-photon absorption can still impede (degrade) device performance [131]. To avoid three-photon absorption, Stegeman et al. employed a longer interaction length in an NLDC to reduce the switching power, thus avoiding higher-order nonlinear absorption [132]. A switching energy of 65 pJ was achieved. GVD had little effect for 6.6-ps pulses but severely affected device operation for 430-fs pulses. An important trait of these semiconductor systems is that by varying the alloy composition, the half-band-gap energy can be tuned anywhere within the telecommunications window of 1.3–1.6 μm. An alternative approach to working

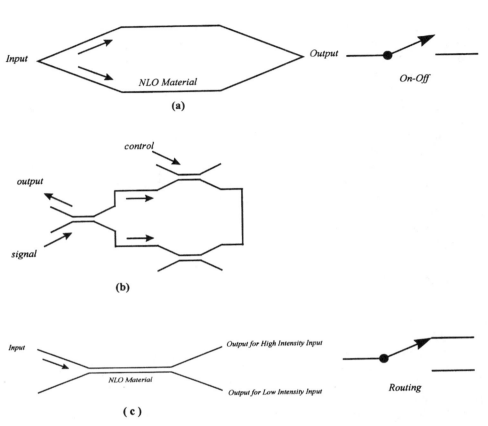

FIGURE 21 (a) Mach–Zehnder interferometer; (b) nonlinear optical loop mirror; (c) nonlinear directional coupler and their electronic analogs.

below the half-band-gap presented is to utilize the large, fast nonlinearity at the transparency point of an active semiconductor. Lee et al. have reported subpicosecond 10-pJ switching at 880 nm in a current injected GaAs/AlGaAs multiple-quantum-well NLDC [133].

Organic materials such as conjugated polymers, although less developed than semiconductor systems, are highly promising because they generally have larger nonlinearities, are more facile for molecular tailoring, and more malleable then their inorganic counterparts. Many conjugated polymers are difficult to pattern into waveguides, however, and it was nearly a decade before switching in an NLDC made from a soluble polydiacetylene commonly known as poly(4-BCMU) (poly-[5,7-dodecadiyne-1,1,12-diolbis(*n*-butoxycarbonylmethylurethane)]) was assessed by Townsend et al. [134]. An interesting fabrication technique was utilized in which the substrate was first patterned with high-index ion-exchanged channels before spin-coating. The underlying channel provides lateral optical confinement, requiring no patterning of the polymer itself. However, because refractive index of the polymer is higher than that of the channels, the polymer contains the majority of the field and, hence, provides the necessary nonlinear optical interactions. Although a slower thermooptical switching was attained, incomplete ultrafast switching at 1.06 μm was observed due to TPA. Irreversible photo-induced bleaching has also been shown to produce index changes in poly(4-BCMU) that are large enough to fabricate channel waveguides [135]. In addition to expected enhancement of $\chi^{(3)}$ and possible $\chi^{(2)}$ inducement through microgravity processing by certain techniques as discussed, it is viable to encourage self-ordering characteristics of liquid-crystal molecules into the polydiacetylene structure. Such would facilitate the formation of ordered thin films, particularly during microgravity processing by, for example, the photodeposition process. Thakur et al. have demonstrated all-optical phase modulation in a polydiacetylene by using a photolithographic technique to fabricate channel waveguides in single-crystal films of PTS prepared by a shear method [104]. The figures of merit of PTS at the communications wavelengths of 1.3 μm and 1.55 μm have been measured by Kim et al. and indicate its promise for all-optical switching applications [136]. More recently, it has been demonstrated that PDAMNA films photodeposited from solution (described earlier) may be exploited to produce integrated optical structures [45,46].

With only marginal improvements in the GaAs/AlGaAs passive semiconductors expected in the near future, attention has turned to the polymeric systems. Although the technology of fabricating third-order polymeric integrated optics is still maturing, the figures of merit are very promising. We believe that, because of the uniqueness of polymeric materials, simpler and less costly techniques such as those presented here can develop, eventually complementing standard lithographic techniques.

ACKNOWLEDGMENTS

The authors are grateful for support from NASA's Office of Life and Microgravity Science and to Hari Sunkara for helpful suggestions after reviewing the manuscript.

REFERENCES

1. Nayar, B. K., and Winter, C. S., *Opt. Quantum Electron.*, 22, 297 (1990).
2. Carter, G. M., Chen, Y. J., and Tripathy, S. K., *Appl. Phys. Lett.*, 43, 891 (1988).
3. Kajzar, F., Meissier, J., Zyss, J., and Ledoux, I., *Opt. Commun.*, 45, 133 (1983).
4. Kajzar, F., and Messler, J., *Thin Solid Films*, 11, 132 (1988).
5. Debe, M. K., and Kam, K. K., *Thin Solid Films*, 816, 289 (1990).
6. Frazier, D. O., Penn, B. G., Witherow, W. K., and Paley, M. S., *Crystal Growth in Space and Related Diagnostics*, 1557, 86 (1991). San Diego, CA, 22–23 July 1991.
7. Wegner, G. Z., *Naturforschung*, 246, 824 (1969).
8. Thakur, M., and Meyler, S., *Macromolecules*, 18, 2341 (1985).
9. Thakur, M., Carter, G. M., Meyler, S., and Hryniewicz, H., *Polym. Prepr.*, 27(1), 49 (1986).
10. Ledoux, I., Josse, D., Vidakovic, P., and Zyss, J., *Opt. Eng.*, 27(1), 49 (1986).
11. Islam, M. N., *Phys. Today*, 47(5), 34 (May 1994).
12. Franken, P. A., Hill, A. E., Peters, C. W., and Weinreich, G., *Phys. Rev. Lett.*, 7, 118 (1961).
13. Davydov, L. D., Derkacheva, V. V., Dunina, M. E., Zhabotinskii, V. F., Zolin, L. G., Koreneva, and Samokhina, M. A., *Opt. Spectrosc.*, 30, 274 (1971).

14. Marcy, Rosker, M. J., Warren, L. F., Cunningham, P. H., Thomas, C. A., Deloach, L. A., Velsko, S. P., Ebbers, C. A., Liao, J. H., and Kanatzidis, M. G., *Opt. Lett. 20*, 252 (1995).

15. Rosenberger, F., Howard, S. B., Sowers, J. W., and Nyce, T. A., *J. Crystal Growth, 129*, 1 (1993).

16. Pusey, M., Witherow, W., and Naumann, R., *J. Crystal Growth, 90*, 105 (1988).

17. Pusey, M., in *Fourth International Conference on Crystallization of Biological Macromolecules*, Freiburg, Germany, Aug. 18–23 1991.

18. Agrawal, G. P., *Nonlinear Fiber Optics*, Academic Press, San Diego, CA 1995.

19. Jensen, S. M., *IEEE J. Quantum Electron., QE-18*, 1580 (1982).

20. Mizrahi, V. K., Delong, K. W., Stegeman, G. I., Saifi, M. A., and Andrejco, M. J., *Opt. Lett., 14*, 1140 (1989).

21. Sheik-Bahae, M., Hagan, D. J., and Van Stryland, E. W., *Phys. Rev. Lett., 65*, 96 (1990).

22. Delong, K. W., and Stegeman, G. I., *Appl. Phys. Lett., 57*, 2063 (1990).

23. Hache, F., Richard, D., Flytzanis, C., and Kriebig, U., *Appl. Phys. A47*, 47 (1988).

24. Lehoczky, S. L., Szofran, F. R., Gillies, D. C., Cobb, S. D., Su, C. H., Sha, Y. G., and Andrews, R. N., *NASA Conf. Publi., 3272* (1994).

25. Smith, D. D., *Ph.D. thesis*, University of Alabama in Huntsville (1996).

26. Turkevich, J., Stevenson, P. C., and Hillier, H., The Size and Shape Factor in Colloidal Systems, *Disc. Faraday Soc., 11*, 55 (1951).

27. Fröhlich, H., *Theory of Dielectrics*, Oxford University Press, London 1958.

28. Swatton, S. N. R., Welford, K. R., Till, S. J., and Sambles, J. R., *Appl. Phys. Lett., 66*, 1868 (1995).

29. Stegeman, G. I., and Miller, A., in *Photonics and Switching*, (J. E. Midwinter, ed.), Academic Press, London, 1994, p. 81.

30. Delong, K. W., Rochford, K. B., and Stegeman, G. I., *Appl. Phys. Lett., 55*, 1823 (1989).

31. Yang, C. C., Villeneuve, A., Stegeman, G. I., and Aitchison, J. S., *Opt. Lett., 17*, 710 (1992).

32. Asobe, M., Naganuma, K., Kaino, T., Kanomori, T., Tomaru, S., and Kurihara, T., *Appl. Phys. Lett., 64*(22), 2922 (1994).

33. Paley, M. S., Frazier, D. O., McManus, S. P., Zutaut, S. E., and Sangahadasa, M., *Chem. Mater. 5*, 1641 (1993).

34. Debe, M. K., *Prog. Surf. Sci., 24*(1–4), 1 (1987).

35. Liu, C. J., Debe, M. K., Leung, P. C., and Francis, C. V., *Appl. Phys. Commun. 11*(2&3), 151 (1992).

36. Kam, K. K., Debe, M. K., Poirer, R. J. and Drube, A. R., *J. Vac. Sci. Technol., A5*(4), 1914 (1987).

37. Debe, M. K., Kam, K. K., Liu, C. J., and Poirier, R. J., *J. Vac. Sci. Technol., A6*, 1907 (1988).

38. Debe, M. K., *J. Appl. Phys., 55*, 3354 (1984); Debe, M. K., and Tommet, T. N., *J. Appl. Phys., 62*, 1546 (1987).

39. Debe, M. K., Poirier, R. J., and Kam, K. K., *Thin Solid Films, 197*, 335 (1991).

40. Debe, M. K., and Field, D. R., *J. Vac. Sci. Tehcnol., A9*, 1265 (1991).

41. Debe, M. K., *J. Vac. Sci. Technol., A10*(4), 2816 (1992).

42. Debe, M. K., *J. Vac. Sci. Technol., 21*, 74 (1992).

43. Debe, M. K., Poirier, R. J., Erickson, D. D., Tommet, T. N., Field, D. R., and White, K. M., *Thin Solid Films, 186*, 257 (1990).

44. Debe, M. K., and Poirier, R. J., *Thin Solid Films, 186*, 327 (1990).

45. Paley, M. S., Frazier, D. O., McManus, S. P., and Donovan, D. N., U.S. Patent 5,451,433 (September 19, 1995).

46. Paley, M. S., Frazier, D. O., Abdeldeyem, H. A., Armstrong, S., McManus, S. P., *J. Am. Chem. Soc.*, *117*(17), 4775 (1995).

47. Pearson, E., Witherow, W. K., and Penn, B. G., personal communication.

48. Hotta, S., Hosaka, T., Soga, M., and Shimotsuma, W., *Synth. Met.*, *9*, 381 (1984).

49. Tanaka, S. M., Sato, A., and Kaeriyama, K., *Makromol. Chem. 185*, 1295 (1984).

50. Waltman, R. J., and Bargon, J., *Can. J. Chem.*, *64*, (1986).

51. Yang, L., Dorsinville, R., Wang, Q. Z., Zou, W. K., Ho, P. O., Yang, N. L., Alfano, R. R., Zamboni, Z., Danieli, R., Ruani, G., and Taliani, C., *J. Opt. Soc. Am.*, *6*(4), 753 (1989).

52. Dorsinville, R., Yang, L., Alfano, R. R., Zanboni, Z., Daniel, R., and Taliani, C., *Opt. Lett.*, *14*(23), 1321 (1989).

53. Abi-Akar, H. M., Ph.D. thesis, University of Alabama in Huntsville (1992).

54. Abi-Akar, H., Riley, C., Coble, H. D., and Maybee, G., (private communication).

55. Ho, Z. Z., Ju, C. Y., and Hetherington, W. M., III., *J. Appl. Phys.*, *62*, 716 (1987).

56. Shirk, J. S., Lindle, J. R., Bartoli, F. J., Hoffman, C. A., Kafafi, Z. H., and Snow, A. W., *Appl. Phys. Lett.*, *55*, 1287 (1989).

57. Wu, J. W., Heflin, J. R., Norwood, R. A., Wong, K. Y., Zamani-Khamiri, O., Garito, A. F., Kalyanaraman, R., and Sounik, J., *J. Opt. Soc. Am. B*, *6*(4), 707 (1989).

58. Matsuda, M., Okada, S., Masaki, A., Nakanishi, H., Suda, Y., Shigehara, K., and Yamada Y., Proc. SPIE, *1337*, 1995 (1990).

59. Hosoda, M., Wada, T., Yamada, A., Garito, A. F., and Sasabe, H., *Jpn. J. Appl. Phys.*, *30*, L1486 (1991).

60. Hoshi, H., Nakamura, N., and Maruyama, Y., *J. Appl. Phys.*, *70*, 7244 (1991).

61. Suda, Y., Shiegehara, K., Yamada, A., Matsuda, H., Okada, S., Masaki, A., and Nakanishi, H., Proc. SPIE, *1560*, 75 (1991).

62. Casstevens, M. K., Samoc, M., Pfleger, J., and Prasad, P. N., *J. Chem Phys.*, *92*, 2019 (1990).

63. Prasad, P. N., and Williams, D. J., *Introduction to Nonlinear Optical Effects in Molecules and Polymers*, Wiley–Interscience, New York, 1991, p. 205.

64. Roger, J. R., Powell, R. C., Chang, Y. H., Ford, W. T., and Zhu, W., *Opt. Mater.*, *5*, 43 (1996).

65. Ho, Z. Z., and Peyghambarian, N., *Chem. Phys. Lett.*, *148*, 107 (1988).

66. Williams, V. S., Mazumdar, S., Armstrong, N. R., Ho, Z. Z., and Peyghambarian, N., *J. Phys. Chem.*, *96*, 4500 (1992).

67. Wada, T., Matsuoka, Y., Shigehara, K., Yamada, A., Garito, A. F., and Sasabe, H., Mater. Res. Socl Symp. Proc., *12*, 75 (1989).

68. Shirk, J. S., Lindle, J. R., Bartoli, F. J., Hafafi, Z. H., Snow, A. W., and Boyle, M. E., *Int. J. Nonlinear Opt. Phys.*, *1*, 699 (1992).

69. Sastre, A., Torres, T., Diaz-Garcia, M. A., Aguillo-Lopez, F., Dhenqant, C., Brasselet, S., Ledoux, I., and Zyss, J., *J. Am. Chem. Soc.*, *118*, 2746 (1996).

70. Suslick, K. S., Chcn, C. T., Meredith, G. R., and Cheng, L. P., *J. Am. Chem. Soc.*, *114*, 6928 (1992).

71. Li, D., Swanson, B. I., Robinson, J. M., and Hoffbarrer, M. A., *J. Am. Chem. Soc.*, *115*, 6975 (1993).

72. Wada, T., Yamanda, S., Matsuoka, Y., Grossmn, C. H., Shigehara, K., Sasbe, H., A., Yamada, and Garito, A. F., in *Nonlinear Optics of Organics and Semiconductors* (T. Kobayashi, ed.), Springer-Verlag, Berlin, 1989, p. 292.

73. Hosada, M., Wada, T., Yamada, A., Garito, A., and Sasabe, H., *Jpn J. Appl. Phys.*, *30*(8B), L1486 (1991).

74. Chollet, P. A., Kajzar, F., and LeMoigne, J., Proc. SPIE, *1273*, 87 (1990).

75. Kumagai, K., Mitzutani, G., Tsukioka, H., Yamauchi, T., and Ushioda, S., *Phys. Rev. B*, *48*(19), 14488 (1993).

76. Yamada, T., Hoshi, H., Ishikawa, K., Takezoe, H., and Fukuda, A., *Jpn. J. Appl. Phys.*, *34*, L299 (1995).

77. Debe, M. K., and Kam, K. K., *Thin Solid Films*, *186*, 289 (1990).

78. Debe, M. K., and Poirier, R. J., *Thin Solid Films*, *186*, 327 (1990).

79. Debe, M. K., *J. Vac. Sci. Technol.*, A4(3), 273 (1986).

80. Debe, M. K., *J. Appl. Phys.*, *55*(9), 3354, 1984; Erratum: Debe, M. K., and Tommet, T. N., *J. Appl. Phys.*, *62*, 1546 (1987).

81. Abdeldayem, H., Frazier, D. O., Penn, B. G., Witherow, W. K., Banks, C., and Smith, D., *Opti. Communi.* (in press).

82. Casstevens, M. K., Samoc, M., Pfleger, J., and Prasad, P. N., *J. Chem. Phys.*, *92*(3), 2019 (1990).

83. Yariv, A., *Quantum Electronics*, 2nd ed., Wiley, New York, 1975.

84. Hercher, M., *Appl. Opt.*, *6S*(5), 947 (1967).

85. Bloor, D., and Chance, R. R. (eds.), *Polydiacetylenes*, Martinus Nijhoff, Dordrecht, 1985.

86. Chemla, D. S., and Zyss, J., (eds.), *Nonlinear Optical Properties of Organic Molecules and Crystals*, 2, Academic Press., Orlando, FL, 1987.

87. Prasad, P. N., and Williams, D. J., *Introduction to Nonlinear Optical Effects in Molecules and Polymers*, John Wiley & Sons, New York, 1991; especially p. 232.

88. Carter, G. M., Thakur, M. K., Chen, Y. J., and Hryniewicz, J. V., *Appl. Phys. Lett.*, *47*, 457 (1985).

89. Hermann, J. P., and Smith, P. W., in *Digest of Technical Papers—11th International Quantum Electronics Conference*, 1980, p. 656.

90. Paley, M. S., Frazier, D. O., Abdeldeyem, H. A., and McManus S. P., *Chem. Mater.*, *6*(12), 2213 (1994).

91. Tsiboulkis, J., Werninck, A. R., Shand, A. J., and Milburn, G. H. W., *Liq. Cryst.*, *3*(10), 1393 (1988).

92. Kim, T., Crooks, R. M., Tsen, M., and J. Sun, *Am. Chem. Soc.*, *117*, 3963 (1995).

93. Cheong, D. W., Kim, W. H., Samuelson, L. A., Kumar J., and Tripathy, S. K., *Macromolecules*, *29*, 1416 (1996).

94. Kim, W. H., Bihari, B., Moody, R., Kodali, N. B., Kumar, J., and Tripathy, S. K., *Macromolecules*, *28*, 642 (1995).

95. Sandman, D. J., (ed.), *Solid State Polymerization*, American Chemical Society, Washington, DC, 1987.

96. Thakur, M., and Meyler, S., *Macromolecules*, *18*, 2341 (1985).

97. Frazier, D. O., Hung, R. J., Paley, M. S., Penn, B. G., and Long, Y. T., *J. Crystal Growth*, *171*, 288 (1997).

98. Zugrav, M., (personal communication).

99. Reid, R. C., Prausnitz, J. M., and Poling, B. E., *The Properties of Gases and Liquids* 4th ed., McGraw-Hill Book Company (1987).

100. Markham, B. L., Greenwell, D. W., and Rosenberger, F., *J. Crystal Growth*, *51*, 426 (1981).

101. Paley, M. S., Frazier, D. O., McManus, S. P., Zutaut, S. E., and Snagahadasa, M., *Chem. Mater.*, *5*(11), 1641 (1993).

102. Horner, C. J., and Garito, A. F., *Makromol. Chem.*, *182*, 19 (1981).

103. Garito, A. F., Singer, K. D., and Teng, C. C., in *Nonlinear Optical Properties of Organic and Polymeric Materials* (D. J. Williams ed.), ACS Symposium Series 233, American Chemical Society, Washington, DC, 1983, p. 1.

104. Etemad, S., Baker, G. L., and Soos, Z. G., in *Molecular Nonlinear Optics* (J. Zyss, ed.), Academic Press, San Diego, CA 1993.

105. Thakur, M., and Krol, D. M., *Appl. Phys. Lett.*, *56*(13), 1213 (1990).

106. Samoc, M. (personal communication).

107. Stegeman, G. I., Torruellas, W., in *Electrical, Optical, and Magnetic Properties of Organic Solid State Material* (A. F. Garito, A. K. Jen, C. Y-C. Lee, and L. Dalton, eds.), MRS Symposium Proceedings, Materials Research Society, Pittsburgh, 1994, p. 397.
108. Walter, H. U., (ed.), *Fluid Sciences and Materials in Space (ESA)*, Springer-Verlag, New York, 1987.
109. Antar, B., and Nuotio-Antar, V. S., *Fundamentals of Low-Gravity Fluid Dynamics and Heat Transfer*, CRC Press, Boca Raton, FL, 1994.
110. Frazier, D. O., Hung, R. J., Paley, M. S., and Long, Y. T., *J. Crystal Growth*, *173*, 172 (1997).
111. Antar, B. N., *Phys. Fluids*, *30*(2), 322 (1987).
112. Maxwell Garnett, J. C., *Phil. Trans. Roy. Soc. London 203*, 385 (1904); *205*, 237 (1906).
113. Ricard, D., Rousignol, P., and Flytzanis, C., *Opt. Lett.*, *10*, 511 (1985).
114. Sipe, J. W., and Boyd, R. W., *Phys. Rev. A46*, 1614 (1992).
115. Brust, M., Walker, M., Bethell, D., Schiffrin, D. J., and Whyman, R., *J. Chem. Soc. Chem. Commun.*, 801 (1994).
116. Paley, M. S, Armstrong, S., Witherow, W. K., and Frazier, D. O., *Chem. Mater.*, *8*(4), 912 (1996).
117. Bromberg, J. P., *Physical Chemistry*, Allyn and Bacon, Boston, MA, 824, (1980).
118. Dorsinville, R., Yang, L., Alfano, R. R., Zamboni, R., Danieli, R., Ruani, G., and Taliani, C., *Opt. Lett.*, *14*(23), 1321 (1989).
119. Logsdon, P. B., Pfleger, J., and Prasad, P. N., *Synth. Met.*, *26*, 369 (1988).
120. Erhardt, J., *Galvanotechnik*, *72*(1), 13 (1981).
121. Naumann, R. J., *Microgravity Sci. Tehcnol.*, *VIII/4*, 204 (1995).
122. Riley, C., Coble, H. D., Loo, B., Benson, B., Abi-Akar, H., and Maybee, G., Polym. Prepri., *28*(2), 470 (1987).
123. Riley, C., Coble, D., and Maybee, G., AIAA Paper 87-0510 (1987).
124. Riley, C., Abi-Akar, H., and Benson, B., *J. Spacecraft Rockets*, *27*(4), 386 (1990).
125. Dalton, L., et al., *Adv. Mater.*, *7*, 519 (1995); *Chem. Mater.*, *7*, 1060 (1995); *Appl. Phys. Lett.*, *67*, 1806 (1995).
126. Marder, S. R., Perry, J., Staehelin, M., and Zysset, B., *Science*, *271*, 335 (1996).
127. Aitchison, J. S., Kean, A. H., Ironside, C. N., Villeneuve, A., and Stegeman, G. I., *Electron. Lett.*, *28*, 1709 (1991).
128. Villeneuve, A., Yang, C. C., Wigley, P. G. J., Stegeman, G. I., Aitchison, J. S., and Ironside, C. N., *Appl. Phys. Lett.*, *61*, 147 (1992).
129. Al-hemyari, A., Aitchison, J. S., Ironside, C. N., Kennedy, G. T., Grant, R. S., and Sibbett, W., *Electron. Lett.*, *27*, 1090 (1992).
130. Atichison, J. S., Villeneuve, A., and Stegeman, G. I., *Opt. Lett.*, *18*, 1153 (1993).
131. Yang, C. C., Villeneuve, A., Stetgeman, G. I., and Aitchison, J. S., *Opt. Lett.*, *17*, 710 (1992).
132. Al-hemyari, A., Villeneuve, A., Kang, J. U., Aitchison, J. S., Ironside, C. N., and Stegeman, G. I., *Appl. Phys. Lett.*, *63*, 3562 (1993).
133. Lee, S. G., McGinnis, B. P., Jin, R., Yumoto, J., Khitrova, G., Gibbs, H. M., Binder, R., Koch, S. W., and Peyghambarian, N., *Appl. Phys. Lett.*, *64*, 454 (1994).
134. Townsend, P. D., Jackel, J. L., Baker, G. L., Shelburne, J. A., and Eternad, S., *Appl. Phys. Lett.*, *55*, 1829 (1989).
135. Rochford, K. B., Zanoni, R., Gong, Q., and Stegeman, G. I., *Appl. Phys. Lett.*, *55*, 1161 (1989).
136. Kim, D. Y., Lawrence, B. L., Torroellas, W. E., Stegeman, G. I., Baker, G., and Meth, J., *Appl. Phys. Lett.*, *65*, 1742 (1994).

18
Optical and Electrical Properties of Protein and Biopolymer Thin Films

Thomas M. Cooper
Air Force Research Laboratory
Wright-Patterson Air Force Base, Ohio

Lalgudi V. Natarajan
Science Applications International Corporation
Dayton, Ohio

I. INTRODUCTION

The investigation of the interface between materials science and biology provides insights from nature useful in a biomimetic approach to material development (1) or in direct use of biomolecular materials (2). Biomaterials have the capacity for supramolecular self-assembly, conversion between chemical, thermal, mechanical, electromagnetic, and electrical energy, and the ability to sense and adapt to the environment (3). Synthetic materials having these characteristics are called "smart materials" (4).

Optical materials found in biology provide numerous starting places for biomimetic design or direct use of biomolecular materials. Some examples demonstrate the variety of natural optical materials. Visual perception begins when the rhodopsin in the rod outer segment converts light energy into an electrical signal (5). The protein bacteriorhodopsin converts light energy into chemical energy (6). The color of butterfly wing scales results from optical interferences associated with ridges on the wing's outer scales (7). The iridescent cuticle of Ruteline scarab beetles selectively reflects circularly polarized light (8). The high transparency of the lens protein crystallin results from water-like short-range order (9) and "molecular chaperone" behavior inhibiting formation of light-scattering centers (10).

In this chapter, we will discuss electrooptic applications of biopolymer and protein thin films. We emphasize thin-film preparation techniques, nonlinear optics, and holography. The materials classes we will describe (biopolymers, gelatin, and bacteriorhodopsin) have demonstrated electrooptic applications or serve as useful model systems. Biopolymers are synthetic amino acid polymers which have both polymer and protein properties (11,12). They form protein-like (i.e., α-helix) conformations, self-assemble into ordered structures, form liquid-crystalline phases, and can be processed into optical thin films necessary for optics applications. Gelatins, formed from hydrolysis of collagen, have applications as photographic, holographic, electrooptic, and waveguiding materials (13). Bacteriorhodopsin (6) is a photochromic protein with numerous potential applications. These three materials classes demonstrate applications and serve as starting-off places for development of new optical biomolecular materials.

II. BIOPOLYMER THIN FILMS

There are numerous methods varying in cost and complexity for preparing biopolymer thin films. Important film properties include thickness, orientation, the ability to control composition along the z axis and in the xy plane.

A. Spin-Coating and Solvent-Casting

Spin-coating involves applying a solution of film material to a rapidly spinning disk. A uniform fluid film forms, becoming a solid film after evaporation of the solvent. To prepare films of known thickness, the researcher needs to measure film thickness as a function of biopolymer solution concentration, disk spin speed, and spin time (14,15). For example, the thickness of spin-cast thin films of copper–phthalocyanine-doped gelatin fits to a simple function of solution concentration and spin speed (16). The film reaches a steady state for spin times greater than 20 s. For a spin time of 30 s, the data fits to a product of power laws:

$$d = \frac{kc^\alpha}{\Omega^\beta} \tag{1}$$

d being the film thickness in micrometers, c the gelatin concentration in weight percent, and Ω the spin speed in revolutions per minute. Within a range of concentration (5–40%) and spin speeds (25–250 rpm), the film thicknesses range from 1 to 30 μm with fitting coefficients $k = 0.182$, $\alpha = 2.96$, and $\beta = 1.26$. A recent study (17) describes macroscopic poly (γ-benzyl-L-glutamate) (PBLG) structures prepared by spin-coating. By varying solvent, concentration, and temperature, Maltese cross shapes and left- or right-handed spirals form, independent of spin direction. To prepare a solvent-cast film, place a biopolymer solution in a well and evaporate to dryness. Film morphology, a function of molecular weight, initial concentration, solvent and thermal history varies from a poorly ordered (Form A) to a well-ordered (Form B) array of packed helices (18–20) or contains cholesteric regions (Form C) (18). Solvent-casting in magnetic (21) or electric (22,23) fields yields oriented thin films.

B. Covalent Attachment to a Substrate

A strategy for preparing an oriented film required for nonlinear and liquid-crystal optics applications involves covalent attachment of a helical polypeptide to a surface. The methods attempted include direct polymerization of the N-carboxy andydride monomer (NCA) on an aminosilanized surface and covalent attachment of the polypeptide at either the N- or C-terminus. Through the use of a suitably designed initiator covalently attached to the substrate, alanine NCAs have been polymerized onto gold and ITO-coated glass (24). Polarized infrared spectra provides evidence for alignment. The intensity ratio of methyl group C–H stretch to a α-helix amide I band changes with intensity, suggesting a preferred helix orientation. A similar study (25) describes films polymerized onto silanized silica, implying the polypeptide rods orient parallel to the surface. Surface graft polymerization of γ-methyl-L-glutamate NCA melt onto a silanized silicon wafer has been performed (26). The authors spin-coat an NCA solution onto a silanized silicon wafer. Following evaporation of solvent, the samples anneal at the NCA melting temperature. Fourier-transform infrared (FTIR) evidence demonstrates formation of grafted α-helical polypeptide films but no clear evidence of film orientation.

The electric field enhances placement of α-helical polypeptides onto a gold surface (27). The authors prepare PBLG with disulfide labeling at the N-terminus. An applied electric field influences orientation and causes enhanced film coverage with the disulfide moiety preferentially binding to the negative electrode. Despite increased attachment, the films show poor orientation as the helix axes orient in a Gaussian distribution perpendicular to the surface. Covalent attachment of the C-terminal carboxylate residue of PBLG to an aminosilanized substrate by carbodiimide coupling has been shown (28). The authors conclude the polypeptide rods orient at a tilt angle of 57° with respect to the surface normal. The thin film behaves as a chiral surface creating a spiral texture in a liquid-crystal phase interacting with the film (29).

C. Electrostatic Self-Assembly

Biopolymer thin films can be prepared by the electrostatic self-assembly technique (30–32). The technique uses electrostatic attraction between oppositely charged species and requires a charged substrate and solutions of positively or negatively charged materials. Film structure consists of alternating positively and negatively charged monolayers. The researcher can construct films with complete control of composition along the axis perpendicular to the plane of the film. The technique requires minimum investment in equipment and yields films with quality comparable to those made by the Langmuir–Blodgett technique. Glass slides that have been cleaned, silanized, and protonated can be used as substrates (32). Films are prepared by dipping the charged substrate into an aqueous solution of anionic material. The polyanion adsorbs onto the charged substrate. After rinsing in deionized water and drying with nitrogen gas, the slide is dipped into an aqueous solution of cationic material, followed by a rinsing and drying step. By this method, a multilayer of oppositely charged species can be obtained. Monolayer thickness is a function of solution concentration and dipping time. In our laboratory, we have made films composed of the oppositely charged polypeptides poly(L-lysine) and poly(L-glutamate) (PGA) (33), poly(L-lysine), and oppositely charged dyes (32) and films composed of only oppositely charged dyes (32). An example of a polymer/biopolymer hybrid prepared by this method is a film containing alternate layers of DNA and poly(allylamine) (34). The author also prepares films containing a virus (31). He first applies six alternating poly(styrenesulfonate)/poly(allylamine) layers onto the substrate. Then, a layer of Carnation Mottle Virus is adsorbed, followed by further deposition of six poly(styrenesulfonate)/poly(allylamine) bilayers. The electrostatic self-assembly technique has been expanded to include specific ligand–protein interactions (35). A layer of biotinylated poly(L-lysine) adsorbed onto a poly(allylamine)/poly(styrenesulfonate) substrate has specific affinity for streptavidin, which shows affinity for other biotinylated poly(L-lysine)/streptavidin layers. By biotinylating enzymes, dyes, and so forth, it is possible to use the specific biotin–streptavidin interaction for highly specific control of the film structure. More complex protein–polymer hybrid films have been prepared, an example being {poly(ethylenimine)/poly(styrenesulfonate) / poly(ethylenimine)+[poly(styrenesulfonate)/lysozyme)]$_2$+poly(styrenesulfonate)/poly(ethylenimine)   +   [glucose oxidase/poly(ethylenimine)]$_6$} (30).

D. Langmuir–Blodgett Films

The Langmuir–Blodgett (LB) film-preparation technique is used to prepare multilayers of amphiphilic molecules (36). In the techniques, a monolayer spreads on a water subphase. A barrier compresses the monolayer into a condensed two-dimensional solid at the air–water interface, followed by transfer to a substrate via mechanical dipping. Along with transfer to a surface, chemical synthesis can be performed at the air–water interface. For example, an amphiphilic polypeptide has been synthesized by hydrolyzing a poly(γ-methyl-L-glutamate) monolayer spread on water (37). Only those methyl groups exposed to the aqueous solution hydrolyze, producing an α-helix with a hydrophobic face exposed to the air and a hydrophilic face exposed to the water. A successful polypeptide that has been used in LB film assembly is the "hairy rod" (38). These polypeptides are random copolymers poly[(γ-methyl-L-glutamate)-*co*-(γ-*n*-octadecyl-L-glutamate)] having a 30% octadecyl side chain randomly placed around the α-helix. Waveguide properties of these films have been determined. Because the polypeptides have optical transitions in the ultraviolet (UV), the low-refractive-index dispersion leads to high Abbe numbers. Photochemically active LB films have been prepared containing a ternary copolymer of γ-methyl-, γ-octadecyl-, and γ-(11-cinnamoylundec-1-yl)-L-glutamate (39,40). The cinnamoyl group is used for photochemical cross-linking of polymers. By irradiating with UV light through a photoresist mask, a waveguide pattern has been written into a 20-layer film.

E. Combinatorial Chemistry

The thin-film-preparation methods described above all provide z-axis structure control, with the films being isotropic in the xy plane. Photolithographic techniques make possible control of composition in the xy plane. The technique combines solid-phase peptide synthesis with semiconductor-based photolithography (41,42). Synthesis requires an aminated surface with N-terminus-blocked amino acids covalently attached to the substrate. The blocking group is light sensitive. Upon illumination through a mask, photodeprotection occurs, exposing N-termini through the mask. Subsequent coupling takes place only at the photodeprotected amino acids. An example is the synthesis of a 20 × 20 array of all 400 possible dipeptides (42). Optical thin films, as well as waveguides and holograms, could be prepared by this method. A recent example (43) demonstrates preparation of streptavidin–biotin arrays on a surface. Similarly, photochemical functionalization of a polymer surface and production of biomolecule-carrying micrometer-scale structures by deep UV lithography has been accomplished (44). Through the use of genetic engineering, cytochrome c has been modified to have a unique sulfhydryl group on its surface (45). It attaches to a sulfhydryl-specific silane layer, forming an oriented film.

III. LIQUID-CRYSTALLINE PROPERTIES

A. Lyotropic and Thermotropic Systems

An intensively studied aspect of biopolymers is their liquid-crystalline properties. Figure 1 shows an example of a lyotropic liquid-crystal biopolymer concentration–temperature phase diagram (46). The phase diagram is composed of a curve for isotropic solutions at low concentration, and another curve for the liquid-crystal phase at high concentration, separated by a biphasic "chimney" region. Transitions between phases can be observed under crossed polarizers. The transition between isotropic and biphasic solution is termed concentration point A. Spherulites containing the "fingerprint" texture appear against the dark isotropic background at point A. The transition between the biphasic solution and liquid-crystalline phase is concentration point B. Because of the chiral α-helix conformation, the liquid-crystal phase is cholesteric, where a planar section through the solution shows nematic orientational order, with a superhelix twist having a pitch p between planar layers. Both the magnitude and sign of the helical pitch is a function of molecular weight, solvent, concentration, and temperature. A simple expression relates these (47).

$$\frac{M_A(1-w_A)}{2pdw_A} = (N\beta_A\frac{M_B}{M_A} - 2N\beta_{AB})w_A + 2N\beta_{AB} \tag{2}$$

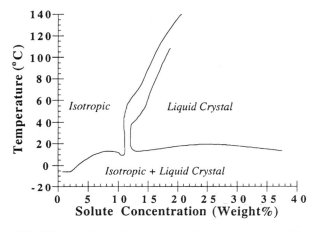

FIGURE 1 Phase diagram for lyotropic polypeptide poly(γ-benzyl-L-glutamate) (MW 310,000) dissolved in dimethylformamide. (Adapted with permission from Ref. 46, copyright 1971 American Chemical Society.)

where M_A is the polypeptide molecular weight, M_B is the solvent molecular weight, p is the cholesteric pitch, d is the solution density, w_A is the weight fraction of the cholesteric polypeptide component, N is Avagadro's number, β_A is the molecular twisting power of the polypeptide, and β_{AB} is the molecular twisting power resulting from the solvent–polypeptide interaction. Molecular twisting power is also a linear function of temperature, where the superhelix unwinds with increasing temperature, converting to a nematic phase and then rewinds to a superhelix of opposite handedness.

Polypeptides with thermotropic liquid-crystalline phases have also been synthesized. They have mesogenic side chains which liquify upon heating and form cholesteric phases (12,48). Figure 2 gives an example of a phase diagram for the thermotropic polypeptide poly(γ-benzyl-L-glutamate-*co*-γ-dodecyl-L-glutamate) (49). The solid to liquid-crystal transition temperature decreases from 100°C to 50°C as the dodecyl content increases from 30% to 100%. At high dodecyl content, interpolymer interaction between dodecyl chains leads to a low transition temperature and ease of thin-film processing.

B. Selective Reflection

When light propagates along the cholesteric axis, polypeptides in the liquid-crystal phase exhibit selective reflection according to the relationship (50)

$$\lambda_0 = np \tag{3}$$

where n is the average refractive index and λ_0 is the wavelength of maximum reflection. The film reflects circularly polarized light having the same handedness and transmits circularly polarized light having opposite handedness. The wavelength of maximal reflection is a function of temperature, composition, and processing conditions. A mixture of poly(γ-butyl-L-glutamate) dissolved in triethylene glycol dimethacrylate has a cholesteric liquid-crystal phase (51). Thermally stable

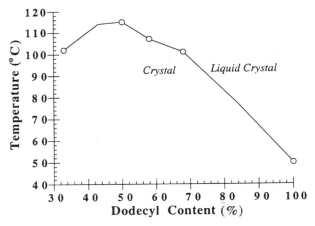

FIGURE 2 Phase diagram for thermotropic polypeptide poly(γ-benzyl-L-glutamate-*co*-γ-dodecyl-L-glutamate). Degree of polymerization = 780. (Data obtained from Ref. 49.)

polymer films having selective reflection form upon photopolymerization in the presence of a photosensitizer. The position of λ_0 is a function of prepolymer syrup composition and photopolymerization temperature. Cholesteric phases of thermotropic polypeptides based on poly(γ-methyl-L-glutamate-*co*-γ-alkyl-L-glutamate) have also been investigated (49,52). The wavelength of maximum reflection is a function of degree of polymerization, alkyl content, alkyl chain length, and temperature. Solidified films containing the cholesteric mesophases retain the solution optical properties. Film deformation and temperature influence optical pitch (53). A polypeptide-containing optical notch filter capable of rejecting radiation in a narrow wavelength band has been described (54). Thin films of right-handed poly(γ-benzyl-L-glutamate-*co*-γ-dodecyl-L-glutamate) and left-handed poly(siloxane), when assembled in series and properly tuned, show high optical density at the rejection wavelength. Cholesteric polypeptide films doped with an azobenze derivative show intense, induced-circular dichroism with potential for chirooptical-recording applications (55).

C. Liquid-Crystalline Biopolymer Gels

Mechanically and optically anisotropic cross-linked polypeptide gels have properties between solution and solid. A concentrated PBLG solution annealed 7–10 days in the presence of a diaminoalkane cross-linker converts to a gel with cholesteric order (56). Cross-linking under a magnetic field forms gels have nematic liquid-crystalline order (57). Anisotropic gel swelling and shrinking occurs. Dyes doped into these gels have induced-circular dichroism, suggesting orientation of the dye along the helix axis (58). PGA hydrogels with cholesteric order have been prepared from hydrolysis of PBLG gels (59). The PGA hydrogels hold liquid-crystalline order at neutral pH but become isotropic at high pH. When dissolved in a helix-promoting solvent, cholesteric structure appears under crossed polarizers. In a coil-forming solvent, the gel becomes isotropic. All these effects are reported to be reversible.

IV. PHOTOCHROMIC BIOPOLYMERS

Photochromic biopolymers have light-sensitive dyes attached to the side chains by a carbodiimide-mediated coupling, transesterification, or direct polymerization of the dye-modified amino acid. Examples include PBLG, PGA (60–65), poly(L-aspartic acid) (66), poly(L-lysine) (67,68), poly(L-α,γ-diaminobutyric acid) (69), poly(L-α,β-diaminopropionic acid) (70), and poly(L-ornithine) (71). Most investigations focus on polypeptides modified with either azobenzene or spiropyran dyes. The chromophores photoconvert between polar(merocyanine or *cis*-azobenzene) and nonpolar (spiropyran or *trans*-azobenzene) isomers.

Photochromic dye-modified polypeptides with greater than 20% modification show significant photo-induced changes in side-chain state, conformation, membrane permeability, and solubility. The experimental observations can be interpreted in terms of the theory of cooperative transitions. Helix content (θ) data are a function of the Zimm–Bragg initiation (σ) and propagation(s) parameters (72):

$$\theta = 1 - \frac{1}{2}\left(1 + \frac{s-1}{\sqrt{(s-1)^2 + 4\sigma s}}\right) \tag{4}$$

The σ parameter ($\sigma \approx 10^{-4}$–10^{-3}) describes low-probability initiation of a phase transition like conformation change or precipitation. The propagation constant s, a function of temperature, pH, solvent composition, and so forth, varies from 0 to 1000 or larger. In solvent-effect studies, s is directly proportional to solvent–peptide CO/NH group binding constant K and solvent volume fraction x (73) (Fig. 3). The experiment requires measurement of helix content in a series of helix- and coil-forming solvent mixtures under dark and light adaptation. At the transition midpoint $x_{1/2}$, $s = 1$ and $\theta = 0.5$. Photo-induced conformation changes occur in a narrow solvent-composition window near the midpoint of the solvent-induced helix-to-coil transition, called a "gated photoresponse" (74). The light-induced free-energy change per monomer is

$$\Delta G_{\mathrm{LA}} - \Delta G_{\mathrm{DA}} = -RT \ln\left(\frac{x_{\mathrm{LA}}}{x_{\mathrm{DA}}}\right) \tag{5}$$

where LA and DA stand for light and dark adapted, respectively, and x is the volume fraction of the coil-forming solvent giving 50% helix content. As an example, the light-induced free-energy change per monomer has been measured in spiropyran-modified succinylated poly(L-lysine) dissolved in trifluorethanol/hexafluoroisopropanol (75). The measured photo-induced solvent–monomer binding energy change is $\Delta G_{\mathrm{LA}} - \Delta G_{\mathrm{DA}} = 71$ J/mol/monomer unit. The light-induced free-energy change per monomer is less than RT, where RT (= 2.5 kJ/mole) is the Maxwell–Boltzmann thermal energy. Near the helix-to-coil transition midpoint, the helix content changes by 20% upon illumination. The photo-induced free-energy change per macromolecule is $\Delta G = 0.2 \times 1000$ monomer units $\times$ 71 J/mol/monomer unit = 14.2 kJ/mol polypeptide, a quantity greater than the thermal energy. Collective interactions between monomer units and solvent make possible

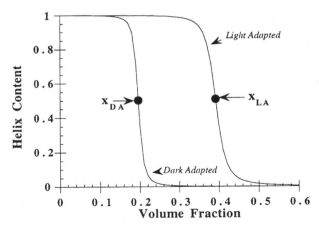

FIGURE 3 Hypothetical plot of helix content versus solvent composition for light- and dark-adapted polypeptides.

the translation of a small light-induced energy difference ($< RT$ per monomer) into large macromolecular conformation changes ($> RT$ per macromolecule). A molecular dynamics study of light-induced conformation changes (64) gives insight into the molecular mechanism of the light-adaptation state on polypeptide conformation. Measures of conformation fluctuation, including mean hydrogen-bond lengths and distribution of backbone dihedral angles, are consistent with the polar merocyanine group destabilizing the polypeptide backbone, thereby facilitating coil formation.

The pH and surfactants also influence side-chain state and conformation. Azobenzene-modified PGA shows an order–disorder transition occurring with different pK_a values for the dark adapted (trans) ($pK = 6.8$) and irradiated (cis) ($pK = 6.3$) states (60). At acidic pH, the polypeptide has a β-sheet conformation which converts to a coil at basic pH. In the presence of dodecylammonium chloride, the polypeptide undergoes a coil-to-helix transition upon UV irradiation which reverses by visible irradiation (62). Azobenzene-modified poly(L-lysine) dissolved in hexafluoroisopropanol is helical under all light/dark-adaptation conditions (67). In a hexafluoroisopropanol/water/sodium dodecyl sulfate mixture, the polypeptide is a β-sheet when dark adapted and an α-helix when light adapted. Pararosaniline-modified PGA (76) shows conformation changes that are a function of light adaptation and pH. The polypeptide is a coil at low and high pHs and an α-helix at weakly alkaline pH. Light adaptation causes a coil-to-helix or helix-to-coil transition depending on the pH. Measurements of photo-induced solubility changes in azobenzene-modified PGA as a function of cis/trans isomeric composition reveals an abrupt increase in solubility in a narrow concentration range, suggesting a cooperative transition (77).

Langmuir–Blodgett films of "hairy rod" polyglutamates containing photochromic azobenzene dyes have been prepared (78). When no solvent is present, the polypeptides adopt the α-helical conformation. The authors use surface plasmon spectroscopy to measure the two in-plane (n_x and n_y) and out-of-plane (n_z) refractive index components of a 0.37-μm (156 monolayers) film at 632.8 nm. There is a small in-plane ($n_y - n_x$) anisotropy parallel (y) versus perpendicular (x) to the dipping direction. In contrast, a large out-of-plane anisotropy ($n_z - n_{x,y}$) shows high azobenzene side-chain orientation perpendicular to the plane of the film. The azobenzene dye undergoes photo-induced interconversion between the trans and cis forms. Upon exposure to UV light, n_z decreases significantly ($\Delta n = -0.1$), whereas n_x and n_y increases by nearly the same amount. The effect is photoreversible with no evidence of photofatigue. This study gives an example of a material showing a high change in the real refractive index component that is coupled with large imaginary index component.

V. HOLOGRAPHY IN PHOTOCHROMIC BIOPOLYMER THIN FILMS

Biomolecular materials have applications in holography. Investigations of the holography of thin films of the photochromic polypeptide poly(spiropyran-L-glutamate) (PSLG) illustrate the principles of holography in a simple system and demonstrates the feasibility of writing gratings on a synthetic photochromic polypeptide thin film. (Fig. 4) (65,79). The photochromic dye exists in two possible forms: spiropyran (SP) and merocyanine (MC) that interconvert by either photochemical or thermal pathways. The PSLG thin-film absorption spectrum is a

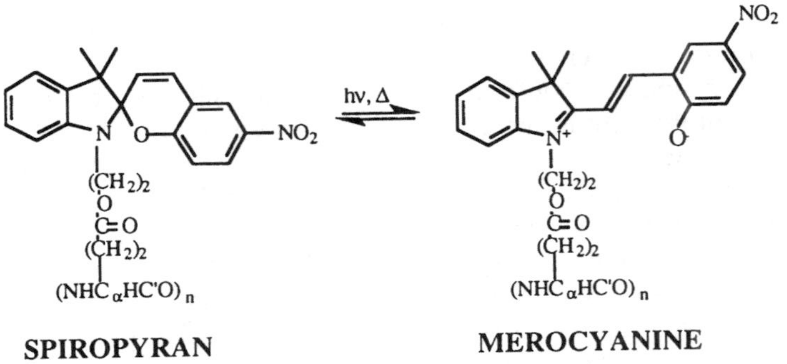

SPIROPYRAN   **MEROCYANINE**

FIGURE 4 Chemical formula of PSLG. The photochemical state of PSLG is an equilibrium between spiropyran (SP) and merocyanine (MC) moieties.

combination of the spectra of SP ($\lambda_{max} = 0.343$ μm) and MC ($\lambda_{max} = 0.555$ μm). Gratings written at 0.3638 μm result from SP-to-MC photoconversion, whereas the reverse (MC-to-SP conversion) occurs upon writing at 0.488 μm. Spin-cast films of PSLG have a thickness range of 1–10 μm and are light sensitive. Unlike the solution properties described above, the polypeptide film retains the α-helix conformation under both light and dark adaptation. We utilize the spiropyran photochemistry to write gratings with an argon ion laser at 0.3638 μm or 0.488 μm at a fringe spacing ranging from 0.39 to 100 μm.

The relation between grating spacing and interference angle is

$$\Lambda = \frac{\lambda_w}{2 \sin (\alpha/2)} \tag{6}$$

where λ_w is the hologram writing wavelength, Λ is the grating spacing, and α is the interference angle. Coarse grating spacing is measured with a microscope and reticle. Fine grating spacing is measured from the first-order Bragg reflection angle θ of a He–Ne laser probe beam. As the transmittance of the difference spectrum at 0.6328 μm is 90%, the hologram behaves as a lossless dielectric grating when probed by a He–Ne laser. A He–Ne laser to probes first-order diffraction efficiency, defined as

$$\eta_1 = \frac{I_1}{I_0} \tag{7}$$

where I_0 is the zeroth-order and I_1 is the first-order intensity of the diffracted beam. To measure angular sensitivity of a fine grating, a 50% beam splitter and energy meters is used to monitor the incident, transmitted and reflected energy of the laser beam while the sample was rotated through a $\Delta\theta = \pm$ 3° range around the Bragg angle. Equation (7) describes the behavior of a grating

$$Q = \frac{2\pi d\lambda}{\Lambda^2 n} \tag{8}$$

where d is the thin-film thickness and n is the refractive index. When $Q \ll 1$, gratings behave according to Raman–Nath theory, in which the first-order diffraction efficiency includes both phase and amplitude contributions (80):

$$\eta_1 = \left(\frac{\pi d \Delta n}{\lambda}\right)^2 + \left(\frac{d \Delta \alpha_\lambda}{4}\right)^2 \tag{9}$$

where $\Delta \alpha_\lambda$ is the absorption coefficient modulation and Δn is the index-of-refraction modulation. When $Q >> 1$, coupled-wave theory (81,82) describes grating properties with the first-order diffraction efficiency for lossless dielectric gratings given by

$$\eta = \frac{\sin^2(\nu^2 + \xi^2)^{1/2}}{1 + (\xi^2/\nu^2)} \tag{10}$$

where $\nu = \pi \Delta n d / \lambda$, $\xi = 2\pi d \Delta \theta / \Lambda$, and $\Delta \theta$ is a small deviation from the Bragg angle.

By writing gratings at increasing exposure energy, the data (Fig. 5) fit to a model (83) assuming the development of a photostationary equilibrium between the two species SP and MC (Fig. 4). At short times, a quadratic dependence of grating efficiency over time occurs:

$$\eta_1(t_w) \sim (I_w \phi \varepsilon C_0 t_w)^2 \tag{11}$$

where I_w is the irradiance of the writing beam, ϕ is the quantum yield of the photoreaction, ε is the molar absorptivity at the writing wavelength, C_0 is the dye concentration, and t_w is the writing time. An example of this analysis uses the data from Fig. 5. A plot of $\eta_1^{1/2}$ versus time for short times gives a line with slope $I_w \phi \varepsilon C_0$. Combining data from holograms written at 0.3638 and 0.488 μm gives a slope ratio $\varepsilon_{0.488} \phi_{0.488} / \varepsilon_{0.3638} \phi_{0.3638} = 0.18$. From known spiropyran and merocyanine extinction coefficients, $\varepsilon_{0.488} = 20{,}000\ M^{-1}\ \text{cm}^{-1}$ and $\varepsilon_{0.3638} = 11{,}000\ M^{-1}\ \text{cm}^{-1}$, we calculate the quantum yield ratio to be $\phi_{0.488} / \phi_{0.3638} = 0.10$. This ratio compares favorably with published quantum yield ratios (84) ranging from 0.01 to 0.1 and shows the observed efficiency differences result from PSLG photochemistry. Color decay of a UV-cured film follows a biexponential function with a minor fast component ($t_{1/2} \sim 8$ min) and a major slow component ($t_{1/2} \sim 40$ h). The decay of diffraction efficiency with time behaves similarly with decay of η_1, following a biexponential function with a minor component ($t_{1/2} \sim 17$ min) a major slow component ($t_{1/2} \sim 6$ h). Photofatigue occurs in spiropyran-containing films. A repeated write–erase cycle causes a 3–6% decrease in grating efficiency per cycle, consistent with systematic studies of spiropyran photofatigue (84).

A fine grating written on a dark-adapted PSLG film gives a grating efficiency ($\sim$0.1%) with sufficient signal-to-noise ratio to measure angular sensitivity (Fig. 6). The Bragg reflection signal could be discerned above noise +1.5° around the Bragg angle. We fit the data to Eq. (10) plus a sloping baseline. The narrow angular sensitivity (1°) is required for writing multiplexed gratings into the films. This study on PSLG demonstrates that photochromic polypeptides can function as holographic materials, and optical-quality thin films can be made by spin-casting. For photochromic polypeptides to be useful in holographic devices, high diffraction efficiency (>90%) has to be achieved by having a refractive index modulation of the order of $\sim$0.1. Photochromic materials like spiropyran suffer from photofatigue, reducing

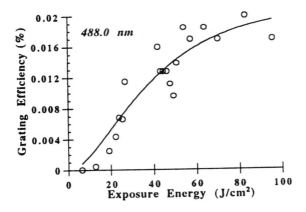

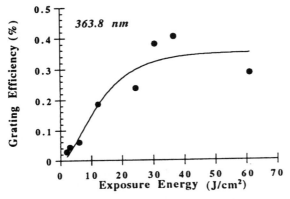

FIGURE 5 Grating efficiency as a function of exposure energy. The abscissa is given in energy $(I_w t_w)$ units per cm². Top: Gratings were written ($\lambda_w = 0.488$ μm, power = 200 mW, spot size = 0.2 cm, $\Lambda = 25$ μm) on films previously UV cured for 5 min. Bottom: Gratings were written ($\lambda_w = 0.3638$ μm, power = 40 mW, spot size = 0.5 cm, $\Lambda = 70$ μm) on dark-adapted films. (Reprinted from Ref. 65, copyright Optical Society of America.)

the number of reversible cycles. There is a need to improve the photostability of the photochromics in order for the polypeptides to be very useful as holographic materials. A recent study demonstrates high-efficiency holograms can be written onto azobenzene-containing photochromic oligopeptide films (85).

VI. GELATINS

Gelatin forms from the hydrolytic degradation of collagen (86–88). It is a high-molecular-weight protein (MW-300,000) and contains high percentages of glycine (35%), proline and hydroxyproline (12%), and alanine (11%), as well as smaller percentages of the polar amino acids lysine (2.7%), arginine (4.8%), aspartic acid (4.4%) and glutamic acid (7.4%). Gelatins are classified according to Bloom strength (89), where a higher Bloom number corresponds to a harder gel. Two

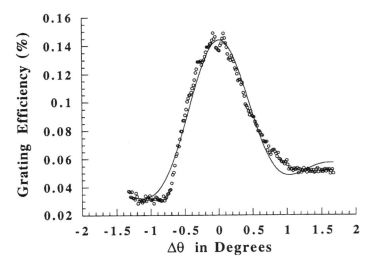

FIGURE 6 Angular sensitivity curve fro PSLG. A hologram is written on a PSLG thin film ($\lambda_w = 0.3638$ μm, power = 4.2 mW, $t_w = 2$ min, spot size = 1.5 cm, $\Lambda = 0.39$ μm). The solid line is the calculated curve assuming a lossless dielectric grating with index modulation $\Delta n = 6 \times 10^{-4}$ and film thickness $d = 10$ μm. (Reprinted from Ref. 65, copyright Optical Society of America.)

types of gelatins are commercially available. Type A gelatin is made from acid-cured tissues, whereas type B gelatin is made from lime-cured tissues. The isoelectric point of type A gelatin is 9.0, whereas that for type B gelatin is 4.9 (90). The protein consists of three helical chains in parallel association. The protein denatures to a coil conformation upon heating above 40°C. With cooling the proteins return to the helical conformation and reassociate into monomers, dimers, and trimers. A network of interhelix junctions forms, eventually leading to a gel network (91). When doped with ammonium dichromate [$(NH_4)_2Cr_2O_7$], the material, called dichromated gelatin (DCG), becomes light sensitive due to the photoreduction of Cr^{6+} to Cr^{3+}. The Cr^{3+} ions form cross-links between carboxylate groups of neighboring polypeptide strands. When exposed to interfering beams of laser radiation, the photoreduction causes the recording of a hologram (92–94). Photosensitizing dyes extend the wavelength range in which holograms can be written (95,96). There are four steps in the processing procedure for preparing hardened DCG holograms (94,97). The first step involves the preparation of a clear gel and sensitizing it with ammonium dichromate. Second, the film is exposed to laser radiation, creating a grating composed of Cr^{3+}–polypeptide cross-links. Third, the film is washed to remove residual chemicals and to create an initial refractive index modulation by swelling. The fourth step involves washing the film with isopropanol and drying, causing amplification of the initial refractive index modulation. The isopropanol-induced index modulation amplification is key to the formation of high-efficiency holograms. The amplification results from microphase separation which converts to differences in micropore density between exposed and unexposed regions (98). The alcohol-drying procedure influences the final color of the DCG hologram (99). If several alcohol baths are used, with the alcohol concentration rising from the first to the last, the reconstructed image is nearly monochromatic. If the hologram is immersed directly into alcohol, the image has more color due to the formation of secondary reflection bands. Numerous detailed descriptions of the four-step processing procedure have appeared in the literature (100–103), as well as investigations of environmental stability (104). Reproducible hologram development requires precise control of temperature, humidity, and time during the swelling and drying steps of hologram development.

Holography has been the most widely used DCG application. DCG holograms have been developed for use in holographic head-up display combiners (105). A recent success is the storage of 2000 images having a resolution 512 × 512 pixels in a DCG emulsion without cross-talk effect (106). The authors suggest the technology could be used to store 160,000 images on a 3.5-in. floppy disk. Gelatins have many properties giving them usefulness for other optical applications. A recent review (107) describes guided-wave devices demonstrated using DCG. A gelatin film shows nearly 100% transmittance from 300 nm to 2700 nm and low waveguide propagation loss (<0.1 db/cm). Waveguide devices demonstrated include a graded-index waveguide on a silicon substrate (108), a locally sensitized waveguide containing a multiplexed holographic phase grating, a rare-earth-ion-doped waveguide laser and an electrooptical waveguide modulator. Gelatin doped with an NLO chromophore and poled has a high electrooptic coefficient (109). A recent degenerate four-wave mixing experiment demonstrated generation of phase-conjugate waves in dichromated gelatin (110). A gelatin film that has been cross-linked and stretched shows high birefringence, suggesting a high degree of chain alignment (111).

VII. NONLINEAR OPTICAL BIOPOLYMER THIN FILMS

Inorganic, organic, and polymeric materials have been intensively investigated for electrooptic and photonic applications (112–114). Because biomolecular materials can be processed into optically clear thin films containing nonlinear chromophores, study of their nonlinear optical properties complements traditional optical materials research. This section discusses the nonlinear optical properties of biopolymer thin films.

The polarization of a material by a time-varying electromagnetic field is given by the expression (115)

$$P_I(\omega) = \chi_{IJ}^{(1)}(\omega)E_J(\omega) + \chi_{IJK}^{(2)}(\omega_3; \omega_1, \omega_2)E_J(\omega_1)E_K(\omega_2)$$
$$+ \chi_{IJKL}^{(3)}(\omega_4; \omega_1, \omega_2, \omega_3)E_J(\omega_1)E_K(\omega_2)E_L(\omega_3) \qquad (12)$$

The nonlinear susceptibilities $\chi^{(j)}$ are tensor quantities relating the amplitude of the nonlinear polarization to the product of field amplitudes E at frequency ω_i. The first term describes the linear processes absorption and refraction. The second term describes second-order processes like the linear electrooptic effect, second-harmonic generation, sum-frequency generation, and difference-frequency generation. The third term describes third-order processes like phase conjugation, two-photon absorption, self-focusing, and optical bistability. The frequencies ω_3 and ω_4 are the output frequencies for second- and third-order processes, respectively. For simplicity, frequencies are not included in the following equations. Older literature tends to report nonlinear susceptibility in esu units. To convert to SI units, use the conversion equations (115):

$$\chi^{(2)} \text{ (SI)} = 4.189 \times 10^{-4} \chi^{(2)} \text{ (esu)} \tag{13}$$

and

$$\chi^{(3)} \text{ (SI)} = 1.400 \times 10^{-8} \chi^{(3)} \text{ (esu)} \tag{14}$$

A. Second-Harmonic Generation

The macroscopic quantity $\chi^{(2)}_{IJK}$ relates to molecular quantities by the expression

$$\chi^{(2)}_{IJK} = N f_I f_J f_K \langle \beta_{ijk} \rangle_{IJK} \tag{15}$$

where N is the chromophore number density and the product $f_I f_J f_K$ describes the influence of the local field experienced by the nonlinear chromophore in the presence of a polarizable polymer environment.

Assuming the chromophore resides in a spherical cavity, the Onsager theory gives the local field factor

$$f^S_O = \frac{\varepsilon(\epsilon_\infty + 2)}{2\varepsilon + \varepsilon_\infty} \tag{16}$$

where ε is the dielectric constant and $\varepsilon_\infty = n^2$. More complex expressions have been developed for elliptical cavities. The quantity $\langle \beta_{ijk} \rangle_{IJK}$ is the ijk component of the molecular polarizability tensor in the molecular reference frame averaged over all orientations in the laboratory IJK reference frame. The sum-over-states expression for β_{ijk} simplifies if the chromophore has one symmetry axis, a highly allowed transition dominates its magnitude, and the exciting frequency is off-resonance. The molecular hyperpolarizability tensor then reduces to one term (116):

$$\beta_{zzz} \approx (\mu_{ee} - \mu_{gg}) \frac{\mu^2_{ge}}{E^2_{ge}} \tag{17}$$

where μ_{ee} and μ_{gg} are the excited- and ground-state dipole moments, respectively, μ^2_{ge} is the transition dipole moment for the $g \rightarrow e$ transition and E_{ge} is the transition energy. With modest degrees of electric-field-induced orientation, the expression for $\chi^{(2)}_{IJK}$ becomes

$$\chi^{(2)}_{333} = N\beta^x_{zzz} \frac{\mu F}{5kT} \tag{18}$$

The superscript x indicates the local field factor product rolls into the expression for molecular hyperpolarizability. Another variable used for describing material nonlinear susceptibility is the d tensor, related to $\chi^{(2)}_{IJK}$ by

$$\frac{1}{2}\chi^{(2)}_{IJK} = d_{IJK} \tag{19}$$

The 27-element d tensor reduces to 18 elements by collapsing subscripts: $d_{IJK} \rightarrow d_{I\mu}$; $11 \rightarrow 1$, $22 \rightarrow 2$, $33 \rightarrow 3$, $23,32 \rightarrow 4$, $13,31 \rightarrow 5$, $12,21 \rightarrow 6$. In a thin film, the system is isotropic in the 1,2 plane, giving the symmetry as ∞mm. The only nonzero tensor elements are d_{33}, d_{31} and d_{15}. When all frequencies are off-resonance, Kleinmann symmetry holds and $d_{15} = d_{31}$. Furthermore, when liquid-crystalline phases do not influence orientation, a further simplification occurs:

$$\frac{d_{31}}{d_{33}} \cong \frac{1}{3} \qquad (20)$$

Electric-field-induced alignment (poling) has been performed to prepare nonlinear polymer films (117,118). A needle oriented perpendicular to the plane of the film charges until breakdown of the surrounding atmosphere causes deposition of ions on the polymer film surface. The resulting electric field causes the chromophores to orient perpendicular to the film. While poling, heating the thin film to above its glass transition temperature gives enhanced chromophore mobility in the electric field. Subsequent cooling to below the glass transition temperature with the electric field applied maximizes chromophore orientation and film nonlinearity. When the applied field is turned off, the chromophores relax to their previous randomly oriented state. Real-time monitoring of orientation during and after poling can be performed by measuring the intensity of second-harmonic radiation at 532 nm produced upon excitation at 1064 nm. The second-order tensor parameter d_{33} is measured by the Maker fringe technique, where second-harmonic intensity is measured as a function of the angle between the surface normal and the k vector of the incident radiation at various polarizations.

A series of biopolymers with pendent nonlinear chromophores has been synthesized, put into thin-film form, and given second-order nonlinearity by corona poling (118). In light of investigations of other polymer systems (117), nonlinear biopolymers provide further insights into the behavior of nonlinear polymers. Figures 7 and 8 illustrate typical results from a corona-poling experiment. The data show real-time second-harmonic signal with the field applied, followed by decay when the field switches off. Also during the experiment, the film heats to above its glass transition temperature, followed by cooling to room temperature. Figure 7 shows data for PGA 8% modified by Disperse Red 1. The glass transition temperature for this polymer is 80–85°C. Upon heating, the second-harmonic signal increases and orientation remains after cooling with the field switched off. The measured $d_{33} = 1.8$ pm/V favorably compares to other polymer systems (117). Figure 8 shows data for poly(γ-methyl-L-glutamate) 10% modified by Disperse Red 1. The glass transition temperature for this polymer is 40–45°C. During heating, the second-harmonic signal reaches a maximum and then decreases. During cooling, the signal rises to a maximum. With the field removed, the orientation decays. Comparison of the two systems shows a higher glass transition temperature and the presence of hydrogen-bonding groups stabilizes orientation. Decay of orientation fits to a biexponential function

$$\frac{d(t)}{d(0)} = A \exp\left(\frac{-t}{\tau_1}\right) + (1 - A) \exp\left(\frac{-t}{\tau_2}\right) \qquad (21)$$

where τ_1 and τ_2 are fast and slow decay times, respectively. For the biopolymer systems investigated, both lifetimes increase with the glass transition temperature, similar to other polymer systems (Fig. 9).

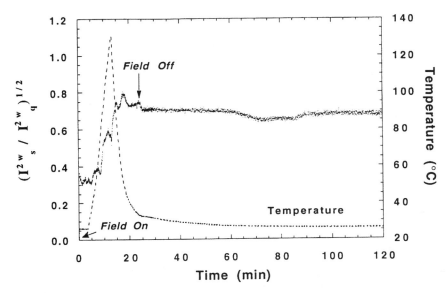

FIGURE 7 Second-harmonic signal as a function of time for a corona-poled thin film of Disperse-Red-1-modified PGA. Percent dye modification is 8%. Film is heated and cooled while poled. (Reprinted from Ref. 118, copyright 1994 American Chemical Society.)

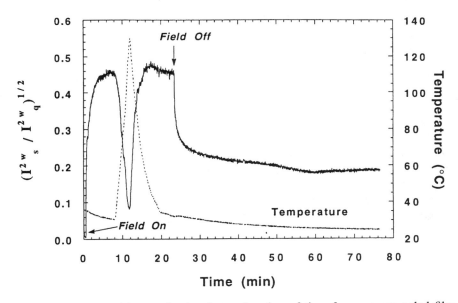

FIGURE 8 Second-harmonic signal as a function of time for a corona-poled film of Disperse-Red-1-modified poly(γ-methyl-L-glutamate). Percent dye modification is 10%. Film is heated and cooled while poled. (Reprinted from Ref. 118, copyright 1994 American Chemical Society.)

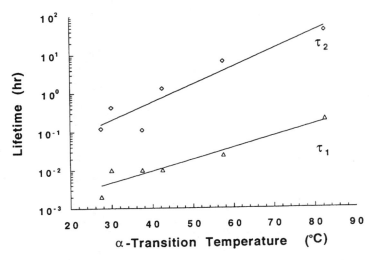

FIGURE 9   Second-harmonic lifetime of the fast (τ_1) and slow (τ_2) decay-time constants as a function of the glass transition temperature of chromophore-modified polypeptides. (Reprinted from Ref. 118, copyright 1994 American Chemical Society.)

Another method for inducing nonlinearity is main-chain poling. In this case, the electric field orients the entire polypeptide chain (119). Rod-shaped polypeptides have a large dipole moment and hyperpolarizability proportional to the degree of polymerization, giving a material nonlinear susceptibility of the form (117)

$$\chi^{(2)} \sim \frac{N^2 \mu \beta}{V(N)} \tag{22}$$

where $V(N)$ is the specific volume of the N-unit chromophore. Large nonlinear susceptibilities have been reported for electric-field-oriented PBLG solution (119,120). Large-surface-area-oriented PBLG films have been prepared by electric-field-enhanced solvent-casting (23). The authors describe a cell consisting of a rectangular teflon reservoir with embedded electrodes all contained in a glass housing. They prepare films by slow solvent evaporation in the applied field and measure film thickness and orientation factor as a function of molecular weight, initial solution concentration, and field strength. For example, a nonlinear film with an order parameter $S = 0.8$ is obtained for a PBLG of molecular weight 118,000 Da and an initial concentration of 2.5% (w/w).

B.   Linear Electrooptic Effect

A related second-order phenomenon is the linear electrooptic effect, where refractive index changes result from an applied electric field, given by

$$\Delta\left(\frac{1}{n_{ij}^2}\right) = \sum_{k}^{3} r_{ijk} E_k \tag{23}$$

where r is the linear electrooptic coefficient. With an off-resonance-exciting frequency, the r and d coefficients relate according to

$$r_{IJ} = \frac{-4d_{IJ}}{n_{II}^2 n_{JJ}^2} \tag{24}$$

For symmetry ∞mm, the nonzero elements are r_{13}, r_{33}, and r_{15}. When all frequencies are off resonance, $r_{15} = r_{31}$. Devices using the linear electrooptic effect cause an electric-field-induced change in refractive index proportional to $n^3 r$ and require a small low-frequency dielectric constant. A figure of merit rates different materials (117):

$$F_m = \frac{n^3 r}{\varepsilon} \tag{25}$$

A recent study describes measurements of the electrooptical coefficient r_{33} and the refractive index of PBLG films made nonlinear by contact poling (121). Because of the low poling voltage, the measured electrooptic coefficients are small ($r_{33} \sim$ 0.05–0.1 pm/V). The refractive indices range from 1.6 to 1.7. The protein dielectric constant, between 2.5 and 4, allows for enhanced values of the figure of merit (117,122). Poled gelatin doped with chlorophenol red and other nonlinear dyes exhibits an electrooptic coefficient $r_{33} = 28$ pm/V and a relaxation time constant of 820 h (109,123). The figure of merit is estimated to be 29, comparable to polymeric systems and larger than lithium niobate ($F_m = 4.2$) (117).

C. Third-Order Processes

Third-order nonlinear processes manifest through the intensity-dependent refractive index (114,115,124),

$$n = n_0 + n_2 I \tag{26}$$

where n_0 represents the weakfield refractive index, I is the time-averaged intensity of the optical field, given by

$$I = \frac{n_0 c}{2\pi} |E(\omega)|^2 \tag{27}$$

and n_2 is directly proportional to the third-order nonlinear susceptibility,

$$n_2 = \frac{12\pi^2}{n_0^2 c} \chi^{(3)} \tag{28}$$

In the low-frequency limit, $\chi^{(3)}$ is of the form

$$\chi^{(3)} \cong \frac{32\pi^3 N \mu_{ge}^4}{h^3 \omega_0^3} \tag{29}$$

The factors influencing $\chi^{(3)}$ include the number density N, the resonance frequency ω_0, and the transition moment μ_{ge}. Unlike second-order phenomena, third-order processes can occur in centrosymmetric systems. The intensity-dependent refractive index contains both real and imaginary components and results from electronic polarization, molecular orientation, electrostriction, thermal effects, photochemistry, and excited-state electronic processes. The third-order susceptibility is a tensor quantity, so material orientation and the geometry of the experiment will influence the components of n_2.

Some of the third-order experiments performed on biomolecular materials include four-wave mixing and Z-scan. Four-wave mixing involves interaction of four beams in a centrosymmetric medium. Two counterpropagating high-intensity pump beams, labeled as the forward and backward pumps, interact with the medium. A weak probe beam incident at some angle to the forward pump beam produces a fourth beam, the conjugate beam, which propagates counter to the probe beam. The Z-scan experiment measures the real and imaginary components of the intensity-dependent refractive index. For a constant input, the transmittance is measured as the sample is scanned along the z direction through the focus of the lens, giving the refractive index components.

VIII. BACTERIORHODOPSIN

Bacteriorhodopsin is a membrane-bound protein (MW-26,000) found in the bacterium *Halobacterium halobium* (6). The bacterium grows in highly illuminated salt marshes with NaCl concentration approaching 4 M. It functions as a photosynthetic protein which undergoes a photocycle accompanied by formation of a transmembrane pH gradient used in the synthesis of ATP. It is composed of seven α-helical segments which transverse the cell membrane. The light-sensitive chromophore all-trans-retinal attaches to lys_{216} via a protonated Schiff-base linkage. The protein aggregates into sheet-like purple membrane patches on the membrane. The isolated purple membrane has a net negative charge and dipole moment and undergoes oriented adsorption to a cationic surface (125). Central to applications is the bacteriorhodopsin photocycle. A simplified version of the cycle is given in Fig. 10.

The chromophore of dark-adapted bacteriorhodopsin (bR_{570}) has an absorption maximum at 570 nm. Upon absorption of light, it goes through a series of photointermediates ($J_{600} \rightarrow K_{590} \rightarrow L_{500}$) and converts to the M_{412} state, having an absorption maximum at 412 nm. The chromophore of the M_{412} state is 13-cis-retinal with an unprotonated Schiff-base linkage. The photocycle completes itself through the dark conversion ($M_{412} \rightarrow N_{560} \rightarrow O_{640}$), returning to all-trans-retinal protonated Schiff base. Following absorption of a photon, the time for conversion of bR_{570} to M_{412} is 50 μs. The dark conversion time of the $M_{412} \rightarrow bR_{570}$ reaction is 10 ms. Direct photoconversion from M_{412} to bR_{570} also occurs. The quantum yield for photoconversion for both the forward ($\phi_{bR \rightarrow M} = 0.65$) and reverse ($\phi_{M \rightarrow bR} = 0.95$) reactions is high (126). The forward and reverse photocycles can also be

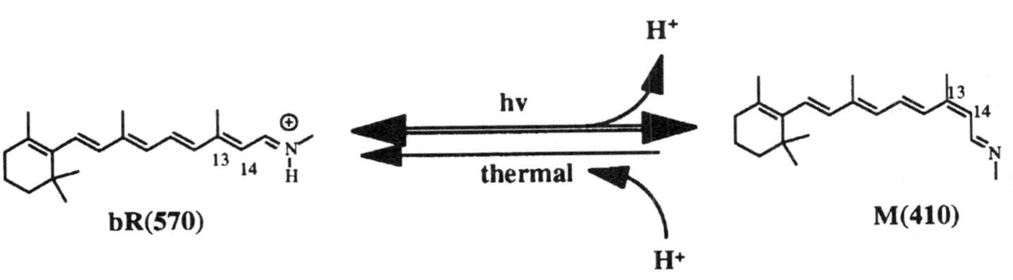

FIGURE 10 Simplified bacteriorhodopsin photocycle.

activated by two-photon excitation at 1140 nm and 820 nm, respectively (125). Because of the photoprotective properties of the chromophore binding site, the photocycle can be repeated indefinitely with no photofatigue. When prepared as oriented thin films, illumination of bR causes a photovoltage to appear (127).

A. Preparation of Thin Films and Control of M-State Lifetime

Bacteriorhodopsin films have been prepared by doping into a carrier polymer, sol–gel processing, electrostatic adsorption, and the Langmuir–Blodgett technique. Examples of polymer matrices in which bacteriorhodopsin has been incorporated include poly(vinyl alcohol) (128) and gelatin (129,130). Films are made by solvent-casting or spin-coating with thicknesses ranging from 30 to 300 μm, an optical density from 0 to 10, and high optical quality (131). Because bR has a net charge and dipole moment, oriented films showing a photoelectric response have been prepared by electrodeposition onto a poly(vinyl alcohol) film (132). Bacteriorhodopsin has been encapsulated into a silica matrix using a sol–gel technique (133). The encapsulation process does not affect the absorption spectrum and photocycle. Bacteriorhodopsin films have been prepared by the Langmuir–Blodgett technique (134,135). Films prepared by this method tend to have poor optical quality, low absorbance, and high sensitivity to moisture. Despite these problems, the films are oriented and show photovoltage signal when irradiated (136).

Photoisomerizations in photochromic dyes occur approximately 100–1000 times slower in the solid phase (137). In the case of bR, the M-state lifetime increases from 10 ms to a range of seconds to minutes when doped into a polymer film (131). Control of the M-state lifetime enables the researcher to tailor-make a film for specific applications. Several methods to alter the M-state lifetime have been described. A dehydrated poly(vinyl alcohol) film containing the denaturant guanidine hydrochloride, the base 1,4-diaminobutane, and bR has an M-state lifetime of 15 min (131). The pH of the casting solution has a pronounced effect. Films cast at pH 9.3 have an M-state lifetime of 47 s, whereas at a pH 10.3, the lifetime increases to 600 s (138). Changing amino acids at specific locations by genetic engineering causes significant effects on the photocycle. Site-directed mutagenesis experiments have substituted asp_{96} to asn_{96} (139). Asp_{96} is an acidic amino acid involved in the protonation/deprotonation of the retinal Schiff base during the photocycle. Substitution by the basic amino acid asn_{96} increases the M-state lifetime from 290 to 1050 ms in their thin-film preparations. As well as changing the amino acid sequence, substitution of retinal analogs for the native all-trans-retinal alters the characteristics of the photocycle. For example, substitution of all-trans-4-keto-retinal causes the M-state lifetime to increase from seconds to minutes (140).

B. Electrical and Optical Applications of Bacteriorhodopsin Films

Applications of bR films are based on the properties of the retinal Schiff base and its photointermediates. At any region of a film, the conversion between bR and the M state is given by (138)

$$\frac{d[\text{bR}]}{dt} = -\sigma_{\text{bR}}\phi_{\text{bR}\rightarrow\text{M}}I_1[\text{bR}] + (k_{\text{M}\rightarrow\text{bR}} + \sigma_{\text{M}}\Phi_{\text{M}\rightarrow\text{bR}}I_2)[\text{M}] \tag{30}$$

where σ_{bR} and σ_M are the absorption cross sections of the bR and M states at the write-and-read wavelength, $\Phi_{bR \rightarrow M}$ and $\Phi_{M \rightarrow bR}$ are the quantum efficiencies of the photoreactions, $k_{M \rightarrow bR}$ is the rate constant of the thermal relaxation of M to bR, and I_1 and I_2 are the intensities of the write-and-read beams at that location. This equation is valid for an optically thin layer of absorbing material. Description of a thick film requires numerical simulation of multiple thin layers that accounts for transmission changes through the optical path (141).

Numerous holographic devices have been demonstrated that utilize bR photochemistry. All rely on a distribution of bR and M in a film created by illumination. Holograms can be written from the bR state (B-type holograms) or from the M state (M-type holograms) (142). The holograms can be read nondestructively at a nonresonant wavelength or destructively at a resonant wavelength. Also, the high quantum yields of the bR/M photointerconversion and low photofatigue make feasible multiple write–erase cycles (6). The ability to control the M-state lifetime makes static or transient holograms written in a bR film possible. Device performance depends on the magnitude of real refractive index modulation (Δn) created by the photochemically induced interference pattern. Estimates of Δn, measured by Z-scan, range from 10^{-3} to 10^{-2} (142–144), with reported hologram efficiencies up to 10% (131,142). Examples of applications include nonlinear holographic correlation (142), real-time interferometric monitoring of Na vapor in a flame (145), and imaging by a phase-conjugate mirror (146).

Photovoltages arising from illumination of oriented bR films have sensor applications (147). These devices require oriented bR films, prepared by the Langmuir–Blodgett technique (148) or electrophoresis (149). Thin films of bR have a photovoltage of several hundred millivolts (150,151). A 16×16-pixel bR array has been prepared by coating the pixel-bearing side of an ITO substrate with a 10–14-layer Langmuir–Blodgett film (150,152). Photocurrents from each pixel show a differential response to light intensity, allowing for detection of images in real time and edge detection. The dynamic range of the sensor is 10^{-4}–10^{-1} W/cm^2, giving good sensitivity under variable light conditions. A bR sensor mimicking mammalian photoreceptor cells has been described (151). It consists of two etched oppositely oriented bR-coated ITO plates facing each other. The center of the pixel contains bR molecules oriented in a direction opposite to those in the periphery. The bipolar photoresponse from these films gives enhanced edge detection mimicking the properties of the retina.

IX. CONCLUSION

In this chapter, we have described three biomolecular materials and their optics applications: biopolymers, bacteriorhodopsin, and gelatin. These materials compete with inorganic, organic, and polymeric optical materials. Important factors include cost, environmental stability, and performance. Inorganic crystals like LiNbO$_3$ have long-term stability and excellent performance as holographic and nonlinear optical materials (153). However, a high-quality single crystal can cost several thousand dollars. For this reason, lower-cost organic and polymeric optical materials have been investigated. Numerous photocrosslinkable and photopolymerizable polymeric holographic materials have been developed (154). These materials have high resolution and diffraction efficiency in direct competition with DCG-based holograms.

However, the photopolymerization process continues after exposure, so photo-polymers are not useful for multiplexing, whereas multiplexed holograms are straightforward to prepare in DCG and other photo-cross-link-based holographic materials (106). Photopolymer-based single holograms have higher environmental stability than DCG holograms. Environmental sensitivity and difficulty in processing are disadvantages of DCG holograms, contrasting with simpler and more robust photopolymer-based holograms (155). Numerous DCG waveguide devices have been demonstrated (107). Current microelectronic chips required short-term manufacturing temperatures in excess of 300°C as well as environmental stability for years at temperatures between 80°C and 120°C (117). Ultrastable chromophores doped in cross-linked polymide have thermal stabilities approaching these requirements (156). High-performance DCG waveguide devices will only be used in applications having more benign manufacturing, performance, and environmental stability requirements. Biopolymer thin films have been proposed as optical storage and display materials (12). Langmuir–Blodgett films of polyglutamates containing azobenzene show large birefringence changes upon illumination and high-diffraction-efficiency holograms have been described (78,85). Similar polymeric systems have been developed (157). Polymer requirements for optical data storage include high transparency, low scatter, low water sensitivity, low noise, thermal stability, high mechanical stability, and fast manufacture (158). The chiral structure and surface-anchoring properties of biopolymer films makes possible liquid-crystal optical elements like circular polarizers, notch filters, and liquid-crystal display elements (50).

Of the three types of materials described in this chapter, bacteriorhodopsin films have the largest number of potential applications. Bacteriorhodopsin's utility arises from its multifunctionality, including large changes in spectra during the photocycle and the formation of a photovoltage. The protein can be easily processed into oriented thin films possessing dichroism. Because of the ability to control the M-state lifetime and the absence of photofatigue, bacteriorhodopsin has dynamic holography applications. The photovoltage allows for development of artificial retina and sensor devices. It would be straightforward to prepare hybrid DCG waveguide/bacteriorhodopsin materials. In conclusion, biomolecular optical materials have demonstrated applications. Because of its multifunctionality, bacterio-rhodopsin has the greatest potential for future new technologies.

ACKNOWLEDGMENTS

The authors thank Bob Crane, Wade Adams, Zbigniew Tokarski, Angela Campbell, Vince Tondiglia, Bob Epling, Weijie Su, Morley Stone, Karen Hussong, Keith Obermeier, Timothy Grinstead, Dave Zelmon, Ruth Pachter, and Steve Cline for their support and collaboration in biomolecular materials research at Air Force Research Laboratory.

REFERENCES

1. Heuer, A. H., Fink, D. J., Laraia, V. J., Arias, J. L., Calvert, P. D., et al., Innovative materials processing strategies: A biomimetic approach, *Science, 255*, 1098–1105 (1992).

2. Tirrell, J. G., Fournier, M., Mason, T. L., and Tirrell, D. A., Biomolecular materials, *Chem. Eng. News*, *72*, 40–51 (1994).

3. Alper, M., Bayley, H., Kaplan, D., and Navia, M. (eds.), Biomolecular Materials by Design, Material Research Society Proceedings, 330, Materials Research Society, Pittsburg, 1994.

4. Thompson, B.S., Biomimetic materials: Was Leonardo mistaken?— PartI, *SAMPE J.*, *32*, 38–43 (1996).

5. Ottolenghi, M., The photochemistry of rhodopsins, *Adv. Photochem.*, *12*, 97–200 (1980).

6. Birge, R. R., Photophysics and molecular electronic applications of the rhodopsins, *Ann. Rev. Phys. Chem.*, *41*, 683–733 (1990).

7. Ghiradella, H., Light and color on the wing: Structural colors in butterflies and moths, *Appl. Opt.*, *30*, 3492–3500 (1991).

8. Caveney, S., Cuticle reflectivity and optical activity in scarab beetles: The role of uric acid, *Proc. Roy. Soc. London B.*, *178*, 205–225 (1971).

9. Delaye, M., and Tardieu, A., Short-range order of crystallin proteins accounts for eye lens transparency, *Nature*, *302*, 415–417 (1983).

10. Borkman, R. F., Knight, G., and Obi, B., The molecular chaperone α-crystallin inhibits UV-induced protein aggregation, *Exp. Eye Res.*, *62*, 141–148 (1996).

11. Block, H., *Poly(γ-benzyl-L-glutamate) and other Glutamic Acids Containing Polymers*, Gordon and Breach, New York, 1983.

12. Daly, W. H., Poche, D., and Negulscu, I. I., Poly(γ-alkyl-α,L-glutamates) derived from long chain paraffinic alcohols, *Prog. Polym. Sci.*, *19*, 79–135 (1994).

13. Ammann-Brass, H., and Pouradier, J., Photographic gelatin, in *Proceedings of the Fourth IAG Conference*, Fribourg, 1983.

14. Meyerhofer, D., Characteristics of resist films produced by spinning, *J. Appl. Phys.*, *49*, 3993–3997 (1978).

15. Levinson, W. A., Arnold, A., and Dehodgins, O., Spin coating behavior of polyimide precursor solutions, *Polym. Eng. Sci.*, *33*, 980–988 (1993).

16. Cooper, T. M., Campbell, A. L., Su, W., Obermeier, K., Natarajan, K., et al., Optical limiting in chromophore-doped gelatin thin films, *MRS Symp. Proc.*, *374*, 99–104 (1995).

17. Schreckenbach, A., and Wunsche, P., Preparation of structures in liquid crystal polymer systems (PBLG) with spin coating, *Polymer*, *35*, 5611–5617 (1994).

18. McKinnon, A. J., and Tobolsky, A. V., Structure and transition in the solid state of a helical macromolecule, *J. Phys. Chem.*, *70*, 1453–1456 (1966).

19. McKinnon, A. J., and Tobolsky, A. V., Structure and properties of poly-γ-benzyl-L-glutamate cast from dimethylformamide, *J. Phys. Chem.*, *72*, 1157–1161 (1968).

20. Tobolsky, A V., and Samulski, E. T., Solid "liquid-crystalline" films of synthetic polypeptides: A new state of matter, *Adv. Chem. Phys.*, *21*, 529–535 (1971).

21. Samulski, E. T., and Tobolsky, A. V., Some unusual properties of poly(γ-benzyl-L-glutamate) films cast in strong magnetic fields, *Macromolecules*, *1*, 555–557 (1968).

22. Iizuka, E., Orientation of poly(γ-benzyl-L-glutamate) in a very low electric field, *Biochim. Biophys. Acta*, *175*, 457–459 (1969).

23. Trantolo, D. J., Gresser, J. D., Wise, D. L., Mogul, M. G., Cooper, T. M., et al., Electric field enhanced alignment of biopolymers for nonlinear applications, *SPIE Proc.*, *2528*, 219–229 (1995).

24. Whitesell, J. K., Chang, H. K., Directionally aligned helical peptides on surfaces, *Science*, *261*, 73–76 (1993).

25. Oosterling, M. L. C. M., Willems, E., and Schouten, A. J., End grafting of (co)polyglutamates and (co)polyaspartates onto Si–OH containing surfaces, *Polymer*, *36*, 4463–4470 (1995).

26. Wieringa, R. H., and Schouten, A. J., Oriented thin film formation by surface graft polymerization of γ-methyl-L-glutamate N-carboxyanhydride in the melt, *Macromolecules*, 29, 3032–3034 (1996).

27. Worley, C. G., Linton, R. W., and Samulski, E. T., Electric-field-enhanced self-assembly of α-helical polypeptides, *Langmuir*, 11, 3805–3810 (1995).

28. Machida, S., Urano, T. I., Sano, K., Kawata, Y., Sunohara, K., et al., A chiral director field in the nematic liquid crystal phase induced by a poly(γ-benzyl glutamate) chemical reaction alignment film, *Langmuir*, 11, 4838–4843 (1995).

29. Urano, T. I., Machida, S., and Sano, K., Dynamics of a nematic liquid crystal on a PBLG–CRA film: Time-resolved infrared spectroscopic study, *Chem. Phys. Lett.*, 242, 471–477 (1995).

30. Lvov, Y., Ariga, K., Ichinose, I. and Kunitake, T., Assembly of multicomponent protein films by means of electrostatic layer-by-layer adsorption, *J. Am. Chem. Soc.*, 117, 6117–6123 (1995).

31. Lvov, Y. M., and Decher, G., Assembly of multilayer ordered films by alternating adsorption of oppositely charged macromolecules, *Crystallogr. Rep.*, 39, 696–716 (1994).

32. Cooper, T. M., Campbell, A. L., and Crane, R. L., Formation of polypeptide-dye multilayers by an electrostatic self-assembly technique, *Langmuir*, 11, 2713–2718 (1995).

33. Cooper, T. M., Campbell, A. L., Noffsinger, C., Gunther-Greer, J., Crane, R. L., et al., Preparations of polypeptide-dye multilayers by an electrostatic assembly process, *MRS Proc. 351*, 239–245 (1994).

34. Lyov, Y., Decher, G., and Sukhorukov, G., Assembly of thin films by means of successive deposition of alternate layers of DNA and poly(allylamine), *Macromolecules*, 26, 5396–5399 (1993).

35. Lvov, Y., Haas, H., and Decher, G., Assembly of polyelectrolyte molecular films onto plasma-treated glass, *J. Phys. Chem.*, 97, 12385 (1993).

36. Fuchs, H., Ohst, H., and Prass, W., Ultrathin organic films: Molecular architectures for advanced optical, electronic and bio-related systems, *Adv. Mater.*, 3, 10–18 (1991).

37. Higuchi, M., Minoura, N., and Kinoshita, T., Photocontrol of micellar structure containing amphiphilic sequential polypeptide, *Chem. Lett.*, 227–230 (1994).

38. Mathy, A., Mathauer, K., Wegner, G., and Bubeck, C., Preparation and waveguide properties of polyglutamate Langmuir–Blodgett films, *Thin Solid Films*, 215, 98–102 (1992).

39. Mathauer, K., Mathy, A., Bubeck, C. and Wegner, G., Pattern formation and photophysical applications of Langmuir–Blodgett multilayers based on polyglutamates, *Thin Solid Films*, 210, 449–451 (1992).

40. Mathauer, K., Schmidt, A., Knoll, W. and Wegner, G., Synthesis and Langmuir–Blodgett multilayer-forming properties of photo-cross-linkable polyglutamates, *Macromolecules*, 28, 1214–1220 (1995).

41. Fodor, S. P. A., Read, J. L., Pirrung, M. C., Stryer, L., Lu, A. T., et al., Light-directed, spatially-addressable parallel chemical synthesis, *Science*, 25., 767–773 (1991).

42. Jacobs, J. W., and Fodor, S. P. A., Combinatorial chemistry—applications of light-directed chemical synthesis, *Trends Biotech.*, 12, 19–26 (1994).

43. Mazzola, L. T., and Fodor, S. P. A., Imaging biomolecule arrays by atomic force microscopy, *Biophys. J.*, 68, 1653–1660 (1995).

44. Yan, M., Cai, S. X., Wybourne, M. N., and Keana, J. F., Photochemical functionalization of polymer surface and the production of biomolecule-carrying micrometer scale structures by deep UV lithography using 4-substituted perfluorophenyl azides, *J. Am. Chem. Soc.*, 115, 814–816 (1993).

45. Stayton, P. S., Olinger, J. M., Jiang, M., Bohn, P. W., and Sligar, S. G., Genetic engineering of surface attachment sites yields oriented protein monolayers, *J. Am. Chem. Soc.*, *114*, 9298–9299 (1992).

46. Wee, E. L., and Miller, W. G., Liquid crystal-isotropic phase equilibria in the system poly(γ-benzyl-L-glutamate-dimethylformamide), *J. Phys. Chem.*, *75*, 1446–1452 (1971).

47. Schiau, C. C., and Labes, M. M., Correlation of pitch with concentration and molecular weight in poly(γ-benzyl-L-glutamate) lyophases, *Macromolecules*, *22*, 328–332 (1989).

48. Watanabe, J., and Tominaga, T., Thermotropic liquid crystals in polypeptides with mesogenic side chains. 1. *Macromolecules*, *26*, 4032–4036 (1993).

49. Watanabe, J., Goto, M., and Nagase, T., Thermotropic polypeptides 3. Investigation of cholesteric mesophase properties of poly(γ-benzyl-L-glutamate-co-γ-dodecyl-L-glutamates) by circular dichroic measurements, *Macromolecules*, *20*, 298–304 (1987).

50. Jacobs, S. D., Cerqua, K. A., Marshall, K. L., Schmid, A., Guardalben, M. J., et al., Liquid-crystal laser optics: Design, fabrication and performance, *J. Opt. Soc. Am. B*, *5*, 1962–1978 (1988).

51. Tsutsui, T., and Tanaka, R., Solid cholesteric films for optical applications, *Polymer*, *21*, 1351–1352 (1980).

52. Watanabe, J., and Nagase, T., Thermotropic polypeptides IV. Thermotropic cholesteric mesophase of copolyglutamates based on γ-benzyl-L-glutamate and γ-alkyl L-glutamate, *Polym. J.*, *19*, 781–784 (1987).

53. Wantanabe, J., and Krigbaum, W. R., Films solidified from a lyotropic mesophase which retain the cholesteric structure, *J. Polym. Sci. B: Polym. Phys.*, *25*, 173–184 (1987).

54. Tsai, M. L., Chen, S. H., and Jacobe, S. D., Optical notch filter using thermotropic liquid crystalline polymers, *Appl. Phys. Lett.*, *54*, 2395–2397 (1989).

55. Sisido, M., Narisawa, H., Kishi, R., and Watanabe, J., Induced circular dichroism from cholesteric polypeptide films doped with an axobenzene derivative, *Macromolecules*, *26*, 1424–1428 (1993).

56. Kishi, R., Sisido, M., and Tazuke, S., Liquid-crystalline polymer gels 1. Cross-linking of poly(γ-benzyl-L-glutamate) in cholesteric liquid crystalline state, *Macromolecules*, *23*, 3779–3784 (1990).

57. Kishi, R., Sisido, M., and Tazuke, S., Liquid-crystalline polymer gels 2. Anisotropic swelling of poly(γ-benzyl-L-glutamate) gel cross-linked under a magnetic field, *Macromolecules*, *23*, 3868–3870 (1990).

58. Sisido, M., and Kishi, R., Liquid-crystalline polymer gels 3. Facile and reversible cholesteric ordering of dye molecules doped in poly(γ-benzyl-L-glutamate) gels, *Macromolecules*, *24*, 4110–4114 (1991).

59. Matsuoka, Y., Kishi, R., and Sisido, M., Liquid-crystalline polymer gels VI. Preparation and swelling behavior of cross-linked hydrogels of poly(L-glutamic acid) possessing cholesteric order, *Polym. J.*, *25*, 919–927 (1993).

60. Pieroni, O., Houben, J., Fissi, A., Costantino, P., and Ciardelli, F., Possible conformation changes induced by light in poly(L-glutamic acid) with photochromic side chains, *J. Am. Chem. Soc.*, *102*, 5913–5915 (1980).

61. Ciardelli, F., Fabbri, D., Pieroni, O., and Fissi, A., Photomodulation of polypeptide conformation by sunlight in spiropyran-containing poly(L-glutamic acid), *J. Am. Chem. Soc.*, *111*, 3470–3472 (1989).

62. Fabbri, D., Pieroni, O., Fissi, A., and Ciardelli, F., Photochromic polymers: Effects of surfactants and side chain electrostatic charge on photocontrol of polypeptides conformation, *Chem. Ind. (Milan)*, *72*, 115–123 (1990).

63. Cooper, T., Obermeier, K., Natarajan, L., and Crane, R., Kinetic study of the helix to coil dark reaction of poly(spiropyran-L-glutamate), *Photochem. Photobiol.*, *55*, 1–7 (1992).

64. Pachter, R., Cooper, T., Natarajan, L., Obermeier, K., Crane, R., et al., Molecular dynamics simulation of poly(spiropyran-L-glutamate): Influence of chromophore conformation, *Biopolymers*, *32*, 1129–1140 (1992).

65. Cooper, T. M., Tondiglia, V., Natarajan, L. V., Shapiro, M., Obermeier, K., et al., Holographic grating formation in poly(spiropyran L-glutamate), *Appl. Opt.*, *32*, 674–677 (1993).

66. Ueno, A., Takahashi, K., Anzai, J., and Oso, T., Photocontrol of polypeptide helix sense by *cis–trans* isomerism of side chain azobenzene moieties, *J. Am. Chem. Soc.*, *103*, 6410–6415 (1981).

67. Fissi, A., Pieroni, O., and Ciardelli, F., Photoresponsive polymers: Azobenzene-containing poly(L-lysine), *Biopolymers*, *26*, 1993–2007 (1987).

68. Pieroni, O., Fissi, A., Viegi, A., Fabbri, D., and Ciardelli, F., Modulation of chain conformation of spiropyran-containing poly(L-Lysine) by combined action of visible light and solvent, *J. Am. Chem. Soc.*, *114*, 2734–2736 (1992).

69. Yamamoto, H., and Nishida, A., Photoresponsive peptide and polypeptide systems. Part 9. Synthesis and reversible photochromism of azo aromatic poly(L-α,γ-diaminobutyric acid), *Polym. Int.*, *24*, 145–148 (1991).

70. Yamamoto, H., Nishida, A., and Kawaura, T., Photoresponsive peptide and polypeptide systems: 10. Synthesis and reversible photochromism of azo aromatic poly(L-α, β-diaminopropionic acid), *Int. J. Biol. Macromol.*, *12*, 257–262 (1990).

71. Yamamoto, H., Nishida, A., Takimoto, T., and Nagai, A., Photoresponsive peptide and polypeptide systems. VIII. Synthesis and reversible photochromism of azo aromatic poly(L-ornithine), *J. Polym. Sci., Part A: Polym. Chem.*, *28*, 67–74 (1990).

72. Davidson, N., *Statistical Mechanics*. McGraw Hill, New York, (1962).

73. Hoppe, W., Lohmann, W., Markl, H., and Ziegler, H., *Biophysics*, Springer-Verlag, Berlin, (1983).

74. Fissi, A., Pieroni, O., Balestreri, E., and Amato, C., Photoresponsive polypeptides. Photomodulation of the macromolecular structure in poly(N-ε-((phenylazophenyl) sulfonyl)-L-lysine), *Macromolecules*, *29*, 4680–4685 (1996).

75. Cooper, T. M., Stone, M. O., Natarajan, L. V., and Crane, R. L., Investigation of light-induced conformation changes in spiropyran-modified succinylated poly(L-lysine), *Photochem. Photobiol.*, *62*, 258–262 (1995).

76. Sato, M., Kinoshita, T., Takizawa, A., and Tsujita, Y., Photoinduced conformational transition of polypeptide membrane composed of poly(L-glutamic acid) containing parasaniline groups in the side chains, *Macromolecules*, *21*, 3419–3424 (1988).

77. Fissi, A., and Pieroni, O., Photoresponsive polymers. Photostimulated aggregation-disaggregation changes and photocontrol of solubility in azo-modified poly(L-glutamic acid), *Macromolecules*, *22*, 1115–1120 (1989).

78. Buchel, M., Sekkat, Z., Paul, S., Weichart, B., Menzel, H., et al., Langmuir–Blodgett–Kuhn multilayers of polyglutamates with azobenzene moieties: investigations of photoinduced changes in the optical properties and structure of the films, *Langmuir*, *11*, 4460–4466 (1995).

79. Cooper, T. M., Natarajan, L. V., and Crane, R. L. Light-sensitive polypeptides, *Trends Polym. Sci.*, *1*, 400–405 (1993).

80. Yariv, A., and Yeh, P., *Optical Waves in Crystals: Propagation and Control of Laser Radiation*, Wiley–Interscience, New York, 1984.

81. Kogelnick, H., Coupled wave theory for thick hologram gratings, *Bell Syst. Tech. J.*, *48*, 2909–2947 (1969).

82. Couture, J., and Lessard, R., Effective thickness determination for volume transmission multiplex holograms, *Can. J. Phys.*, *64*, 553–557 (1986).

83. Eich, M., and Wendorff, J., Laser-induced gratings and spectroscopy in monodomains of liquid crystalline polymers, *J. Opt. Soc. Am. B*, *7*, 1428–1436 (1990).

84. Bertelson, R., Photochromic processes involving heterolytic cleavage, in *Techniques of Chemistry, Photochromism III.* G. Brown (ed.), Wiley–Interscience, New York, 1971, pp. 45–431.

85. Berg, R. H., Hvilsted, S., and Ramanujam, P. S., Peptide oligomers for holographic data storage, *Nature*, *383*, 505–508 (1996).

86. Meyerhofer, D., Dichromated gelatin, in *Holographic Recording Materials*, H. M. Smith (ed.), Springer-Verlag, Berlin, 1977, pp. 75–99.

87. Ross Murphy, S. B., Structure and rheology of gelatin gels: Recent progress, *Polymer*, *33*, 2622–2627 (1992).

88. Sturmer, D. M., and Marchetti, A. P., Silver halide imaging, in *Imaging Processes and Materials*, J. Sturge, V. Walworth and A. Shepp, (eds.), Van Nostrand Reinhold, New York, 1989, pp. 71–109.

89. Wainewright, F. W., Physical tests for gelatin and gelatin products, in *The Science and Technology of Gelatin*, A. G. Ward and A. Courts, (eds.), Academic Press, London, 1977, pp. 507–534.

90. Peng, B., and Chen, L., Isoelectric point distribution of modified gelatins, *Chem. Abstr.*, *103*, 30194 (1984).

91. Ross Murphy, S. B., Incipient behaviour of gelatin gels, *Rheol. Acta*, *30*, 401–411 (1991).

92. Shankoff, T. A., Phase holograms in dichromated gelatin, *Appl. Opt.*, *7*, 2101–2105 (1968).

93. Brandes, R. G., Francois, E. E., and Shankoff, T. A., Preparation of dichromated gelatin films for holography, *Appl. Opt.*, *8*, 2346–2348 (1969).

94. Chang, B. J., Dichromated gelatin holograms and their applications, *Opt. Eng.*, *19*, 642–648 (1980).

95. Kubota, T., Ose, T., Saaki, M., and Honda, K., Hologram formation with red light in methylene blue sensitized dichromated gelatin, *Appl. Opt.*, *15*, 556–558 (1976).

96. Blyth, J., Methylene blue sensitized dichromated gelatin holograms: A new electron donor for their improved photosensitivity, *Appl. Opt.*, *30*, 1598–1602 (1991).

97. Saxby, G., Fred Unterseher on dichromated gelatin holograms, *Br. J. Photogr.*, *131*, 176–177 (1984).

98. Mel'nichenko, Y. B., Gomza, Y. P., Shilov, V. V., and Kuzilin, Y. E., Mechanism of index modulation in developed dichromated gelatin films, *J. Photogr. Sci.*, *39*, 133–138 (1991).

99. McGrew, S., Color control in dichromated gelatin holograms, *SPIE Proc.*, *215*, 24–31 (1980).

100. Jeong, M. H., Song, J. B., and Lee, I. W., Simplified processing of dichromated gelatin holographic recording material, *Appl. Opt.*, *30*, 4172–4173 (1991).

101. Bahuguna, R. D., Beaulieu, J., and Arteaga, H., Reflection display holograms on dichromated gelatin, *Appl. Opt.*, *31*, 6181–6182 (1992).

102. Stojanoff, C. G., Brasseur, O., Tropartz, S., and Schutte, H., Conceptual design and practical implementation of dichromated gelatin films as an optimal holographic recording medium for large format holograms, *SPIE Proc.*, *2402*, 301–311 (1994).

103. Boj, P. G., Crespo, J., and Quintana, J. A., Broadband reflection holograms in dichromated gelatin, *Appl. Opt.*, *31*, 3302–3305 (1992).

104. Naik, G. M., Mathur, A., and Pappu, S. V., Dichromated gelatin holograms: An investigation of their environmental stability, *Appl. Opt.*, *29*, 5292–5297 (1990).

105. Wood, R. B., and Thomas, M. A., Holographic head-up display combiners with optimal photometric efficiency and uniformity, *SPIE Proc.*, *1289*, 50–62 (1990).

106. Ramenah, H. K., Bertrand, P., and Meyruels, P., Using dichromated gelatin as a diffractive optical storage medium and illustration of multiple applications, *Opt. Eng.*, *35*, 1407–1412 (1996).

107. Chen, R. T., Polymer-based photonic integrated circuits, *Opt. Laser Tech.*, *25*, 347–365 (1993).

108. Gerold, D., and Chen, R. T., Vacuum-tuned graded-index polymer waveguides on silicon substrates, *Appl. Opt.*, *35*, 400–403 (1996).

109. Gerold, D., Chen, R. T., Farone, W. A., and Pelka, D., Poled electro-optic photoline gel polymer doped with chlorophenol red and bromomethyl blue chromophores, *Appl. Phys. Lett.*, *66*, 2631–2633 (1995).

110. Naik, G., and Pappu, S., Generation of phase conjugate waves in dichromated gelatin and related materials, *Appl. Opt.*, *30*, 1890–1892 (1991).

111. Zhao, W., Kloczkowski, A., Mark, J. E., Erman, B., and Bahar, I., Make tough plastic films from gelatin, *Chemtech*, *26*, 32–38 (1996).

112. Ashwell, G. J., and Bloor, D., *Organic Materials for Nonlinear Optics III*, Royal Society of Chemistry, Cambridge, 1993.

113. Hann, R. A., and Bloor, D., *Organic Materials for Non-Linear Optics II*, Royal Society of Chemistry, Cambridge, 1991.

114. Sutherland, R. L., *Handbook of Nonlinear Optics*, Marcel Dekker, Inc., New York, 1996.

115. Boyd, R. W., *Nonlinear Optics*, Academic Press, Boston, 1992.

116. Kanis, D. R., Ratner, M. A., and Marks, T. J., Design and construction of molecular assemblies with large second order optical nonlinearities. Quantum chemical aspects, *Chem. Rev.*, *94*, 195–242 (1994).

117. Burland, D. M., Miller, R. D., and Walsh, C. A., Second-order nonlinearity in poled-polymer systems, *Chem. Rev.*, *94*, 31–75 (1994).

118. Tokarski, Z., Natarajan, L. V., Epling, B. L., Cooper, T. M., Hussong, K. L., et al., Nonlinear optical characterization of chromophore-modified poly(L-glutamate) thin films, *Chem. Mater*, *6*, 2063–2069 (1994).

119. Levine, B., and Bethea, C., Second order hyperpolarizability of a polypeptide α-helix: poly-γ-benzyl-L-glutamate, *J. Chem. Phys.*, *65*, 1988–1993 (1976).

120. Ishii, T., Wada, T., Garito, A., Sasabe, H., and Yamada, A., Molecular design of synthetic polypeptides for nonlinear optics, *Mat. Res. Soc. Symp. Proc.*, *175*, 129–134 (1990).

121. Choi, D., and Zand, R., Determination of the electro-optical coefficient r_{33} and refractive index of para-substituted-poly(γ-benzyl-L-glutamate) derivatives, *Polymer*, *35*, 23–29 (1994).

122. Gilson, M. K., and Honig, B. H., The dielectric constant of a folded protein, *Biopolymers*, *25*, 2097–2119 (1986).

123. Ho, Z. Z., Chen, R. T., and Shih, R., Electro-optic phenomena in gelatin-based poled polymer, *Appl. Phys. Lett.*, *61*, 4–6 (1992).

124. Bredas, J. L., Adant, C., Tackx, P., and Persoons, A., Third-order nonlinear optical response in organic materials: theoretical and experimental aspects, *Chem. Rev.*, *84*, 243–278 (1994).

125. Birge, R. R., Protein-based optical computing and memories, *Computer*, *25*, 56–67 (1992).

126. Birge, R. R., Gross, R. B., Masthay, M. B., Stuart, J. A., Tallent, J. R., et al., Nonlinear optical properties of bacteriorhodopsin and protein-based two-photon three-dimensional memories, *Mol. Cryst. Liq. Cryst. Sci. Technol. B: Nonlinear Opt.*, *3*, 133–147 (1992).

127. Groma, G. I., Szabo, G., and Varo, G., Direct measurement of picosecond charge separation in bacteriorhodopsin, *Nature, 308,* 557–558 (1984).

128. Chen, Z., Lewis, A., Takei, H., and Nebenzahl, I., Bacteriorhodopsin oriented in polyvinyl alcohol films as an erasable optical storage medium, *Appl. Opt., 30,* 5188–5196 (1991).

129. Zhang, C., Song, W. Q., Ku, C.-Y., Gross, R. B., and Birge, R. R., Determination of the refractive index of a bacteriorhodopsin film, *Opt. Lett., 19,* 1409–1411 (1994).

130. Downie, J. D., and Smithey, D. T., Red-shifted photochromic behavior of a bacteriorhodopsin film made from the L93T genetic variant, *Opt. Lett., 21,* 680–682 (1996).

131. Gross, R. B., Izgi, K. C., and Birge, R. R., Holographic thin films, spatial light modulators and optical associative memories based on bacteriorhodopsin, *SPIE Proc., 1662,* 186–196 (1992).

132. Uehara, K., Kawai, K., and Kouyama, T., Photoelectric response of oriented purple membrane electrodeposited onto poly(vinyl alcohol) film, *Thin Solid Films, 232,* 271–277 (1993).

133. Wu, S., Ellerby, L. M., Cohan, J. S., Dunn, B., El-Sayed, M. A., et al., Bacteriorhodopsin encapsulated in transparent sol–gel glass: A new biomaterial, *Chem. Mater., 5,* 115–120 (1993).

134. Ikonen, M., Peltonen, J., Vourimaa, E., Lemmetyinen, H., Study of photocycle and spectral properties of bacteriorhodopsin in Langmuir–Blodgett films, *Thin Solid Films, 213,* 277–284 (1992).

135. Furuno, T., Takimoto, K., Kouyama, T., Ikegami, A., and Sasabe, H., Photovoltaic properties of purple membrane Langmuir–Blodgett films, *Thin Solid Films, 160,* 145–151 (1989).

136. Ikonen, M., Sharonov, A., Tsachenko, N., and Lemmatyinan, H., The photovoltage signals of bacteriorhodopsin in Langmuir–Blodgett films with different molecular orientations, *Adv. Mater. Opt. Electron., 2,* 115–122 (1993).

137. Smets, G., Photochromic phenomena in the solid phase, *Adv. Polym. Sci., 50,* 17–44 (1983).

138. Lindvold, L. R., Imam, H., and Ramanujam, P. S., The sensitometric properties of chemically modified bacteriorhodopsin films, *SPIE Proc., 2429,* 22–33 (1994).

139. Hampp, N., Popp, A., Brauchle, C., and Oesterhelt, Diffraction efficiency of bacteriorhodopsin films for holography containing bacteriorhodopsin wildtype BR(WT) and its variants BR(D58E) and BR(D96N), *J. Phys. Chem., 96,* 4679–4685 (1992).

140. Vsevolodov, N. N., and Dyukovna, T. V., Retinal-protein complexes as optoelectronic components, *Trends Biotech., 12,* 81–88 (1994).

141. Song, Q. W., Zhang, C., Blumer, R., Gross, R. B., Chen, Z., et al., Chemically enhanced bacteriorhodopsin thin-film spatial light modulator, *Opt. Lett., 18,* 1373–1375 (1993).

142. Thoma, R., Dratz, M., and N. H., All-optical nonlinear holographic correlation using bacteriorhodopsin films, *Opt. Eng., 34,* 1345–1351 (1995).

143. Song, Q. W., Zhang, C., Gross, R. B., and Birge, R. R.., The intensity-dependent refractive index of chemically enhanced bacteriorhodopsin, *Opt. Commun., 112,* 296–301 (1994).

144. Rao, D. V. G. L. N., Aranda, F. J., Chen, Z., Addara, J. A., Kaplan, D. L., et al., Nonlinear optical studies of bacteriorhodopsin, *J. Nonlinear Opt. Phys. Mater., 5,* 331–349 (1996).

145. Millerd, J. E., Brock, N. J., Brown, M. S., and DeBarber, P. A., Real-time resonant holography using bacteriorhodopsin thin films, *Opt. Lett., 20,* 626–628 (1995).

146. Zhang, Y., Song, Q. W., Tseronis, C., and Birge, R. R., Real-time holographic imaging with a bacteriorhodopsin films, *Opt. Lett., 20,* 2429–2431 (1995).

147. Hong, F. T., Fundamentals of photoelectric effects in molecular electronic thin film devices: Applications to bacteriorhodopsin-based devices, *BioSystems*, *35*, 117–121 (1995).

148. Wang, J. P., Li, J. R., Tao, P. D., Li, X. C., and Jiang, L., Photoswitch based on bacteriorhodopsin Langmuir-Blodgett films, *Adv. Mater. Opt. Electron.*, *4*, 219–224 (1994).

149. Takei, H., Lewis, A., Chen, Z., and Nebenzahhl, I., Implementing receptive fields with excitatory and inhibitory optoelectrical responses of bacteriorhodopsin films, *Appl. Opt.*, *30*, 500–509 (1991).

150. Miyasaka, T., and Koyama, K., Image sensing and processing by a bacteriorhodopsin-based artificial photoreceptor, *Appl. Opt.*, *32*, 6371–6379 (1993).

151. Chen, Z., and Birge, R. R., Protein-based artificial retinas, *Trends Biotech.*, *11*, 292–300 (1993).

152. Koyama, K., Yamaguchi, N., and Miyasaka, T., Molecular organization of bacteriorhodopsin films in optoelectronic devices, *Adv. Mater.*, *7*, 590–594 (1995.)

153. Gunter, P., Holography, coherent light amplification and optical phase conjugation with photorefractive materials, *Phys. Rep.*, *93*, 199–299 (1982).

154. Manivannan, G., and Lessard, R. A., Trends in holographic recording materials, *Trends Polym. Sci.*, *2*, 282–290 (1994).

155. Natarajan, L. V., Sutherland, R. L., Tondiglia, V., Bunning, T. J., and Adams, W. W., Photopolymer materials (development of holographic gratings), in *Polymeric Materials Encyclopedia* (J. C. Salamone ed.), CRC Press, Boca Raton, FL, 1996, pp. 5230–5237.

156. Kowalczyk, T. C., Kosc, T. Z., and Ermer, S., Crosslinked polyimide electrooptic materials, *J. Appl. Phys.*, *78*, 5876 (1995).

157. Schmidt, H., Dichroic dyes and liquid crystalline chain polymers, *Angew. Chem. Adv. Mater.*, *101*, 964–970 (1989).

158. Abbott, S., Polymers for optical data storage, in *Special Polymers for Electronics and Optoelectronics* (J. A. Chilton and M. T. Goosey, eds.), Chapman & Hall, London, 1995, pp. 315–341.

19
Polymer Composites of Quantum Dots

Greg Carlson and Kenneth E. Gonsalves
University of Connecticut
Storrs, Connecticut

I. INTRODUCTION

The usual description [1] of polarization, **P**, arising in a material as a result of incident polarized light is a vectorial power series in the incident optical electric vector, **E**:

$$\mathbf{P} = \mathbf{P}_0 + \chi \cdot \mathbf{E} + \chi^{(2)} \cdot \mathbf{E} \cdot \mathbf{E} + \chi^{(3)} \cdot \mathbf{E} \cdot \mathbf{E} \cdot \mathbf{E} + \cdots \quad (= \chi_{\text{eff}} \cdot \mathbf{E}) \tag{1}$$

The constant $\mathbf{P}_0$ is the permanent, zero-field polarization. The linear term χ, a tensor, is the conventional susceptibility, which is proportional to intensity ($|\mathbf{E}|^2$). The main nonlinear optical susceptibilities of a material are its second- and third-order terms, $\chi^{(2)}$ and $\chi^{(3)}$, respectively. Although there are systems which exhibit higher-order dependence on the optical field (e.g., $\chi^{(5)}$), these four terms are sufficient to describe most materials.

$\chi^{(2)}$ is a 3×3 tensor $\{d_{11}, d_{12}, d_{13}, \ldots, d_{33}\}$ and $\chi^{(3)}$ is a 4×4 tensor $\{\gamma_{11}, \gamma_{12}, \gamma_{13}, \gamma_{14}, \ldots, \gamma_{44}\}$. Different experimental techniques and geometries measure different components of these tensors. The situation is complicated by the use of several different sets of units [2]. Also, some investigators use an alternate version of Eq. (1)—a Taylor expansion in the field, **E**. For example, two accurate measurements of $\chi^{(3)}$ on the same material might be reported as differing by a factor of 6. In general, nonlinear optical properties are also wavelength dependent. Thus, it is important to be sure the measured quantity has been fully specified when studying and comparing published values of nonlinear optical susceptibilities.

Some consequences of this nonlinearity are that the material's refractive index, n, and dielectric constant, ε, also depend on the strength (intensity) and frequency of the optical electric field. This dependence,

$$n^2(\omega) = \varepsilon(\omega) = 1 + 4\pi\chi_{\text{eff}}(\omega) \tag{2}$$

results in a linear dependence of n on the intensity for third-order materials.

769

Knowledge of these properties can be used to produce devices which can use light to modulate current, and current to modulate light. Also, light may be used to switch light. Because light (unlike current) does not generate heat and can be pulsed very rapidly, architectures [3] have already been designed to use this phenomenon to greatly increase the speed and capacity [4] of communications and computers.

Important properties of materials which have a significant $\chi^{(3)}$ include generation of a third-harmonic signal (an input at ω results in an output signal at 3ω) and four-wave mixing, which can generate a phase conjugate signal.

II. MATERIALS REQUIREMENTS

In general, structural requirements are different for second- and third-order nonlinearities. Because $\chi^{(2)}$ is a third-rank tensor, most second-order materials are anisotropic. [Exceptions to this rule [5] are centrosymmetric materials with nonzero values of the off-diagonal terms of the $\chi^{(2)}$ tensor.] Large second-order nonlinear optical susceptibilities have been measured in asymmetric crystals [6] and Langmuir–Blodgett–Kuhn films [7].

In polymeric materials, the anisotropy necessary for a significant second-order nonlinear optical susceptibility ($\chi^{(2)}$) is introduced by adding a strongly dipolar moiety (called a chromophore, or "dye"), which usually consists of a strong electron donor and a strong electron acceptor connected by a conjugated system, such as a phenyl ring (Fig. 1). The relative ease of the HOMO–LUMO (highest occupied molecular orbital–lowest unoccupied molecular orbital) transition and asymmetry of the excited state imparts nonlinear polarizability to such a molecule. These groups can be incorporated into a polymer matrix as a dopant, a side chain, or in the polymer backbone. However, the presence of these dipoles is not sufficient to produce a $\chi^{(2)}$ material. The random orientation of the groups results in a zero net anisotropy. The dye groups must be aligned in a principle direction. This is achieved through a process called poling [8]. Although many refinements have been made [9] and there have been some notable successes [10], the poled-polymer process is still complex and difficult to optimize [11].

Because the third-order susceptibility, $\chi^{(3)}$, is a fourth-rank tensor, there is no material requirement of asymmetry [12]. This property allows isotropic materials to be considered for optical devices. Currently, most third-order nonlinear optical materials are inorganic crystals, such as lithium niobate ($LiNbO_3$), potassium dideuterium phosphate (KDP), and barium titanate ($BaTiO_3$). These crystals have

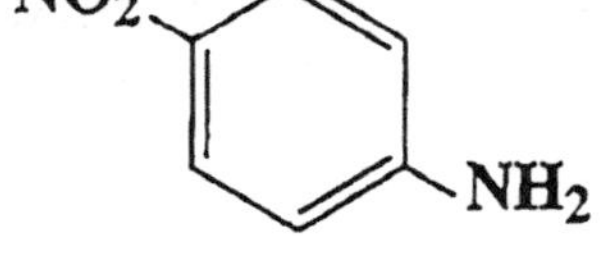

FIGURE 1 Chromophore structure.

relatively large values of $\chi^{(3)}$, because of periodic lattice vacancies. Some of their drawbacks are that they are expensive and difficult to deposit into layers of well-controlled thickness and they are very intolerant of impurities or structural defects.

Some attempts have been made to produce third-order susceptibility in polymers by doping. These chromophoric groups do not require poling, but they generally have only moderate nonlinear susceptibilities and if the loading is too high, plasticization can occur.

Certain classes of polymers have considerable $\chi^{(3)}$ in the undoped state. This is because they have charge delocalization (conjugation) along their backbone. This ability to transport charge also makes many of these polymers conductive. However, conduction in polymers involves "hopping," or charge transfer between chains. Optical nonlinearity is thought to occur by an intrachain rearrangement of charge [13]. Some examples of polymers (Fig. 2) with π-conjugated repeat units are polyacetylenes, poly(p-phenylene vinylene), and polythiophenes. Polysilanes, with their all-silicon backbone, exhibit conjugation through their σ electrons. Studies of varying-length oligomers [14] have shown that the extent of conjugation (and the $\chi^{(3)}$ measured) increases with increasing chain length, reaching its full value by degree of polymerization of about 50–60 carbons [15], 30 silicones [16], or 8 thiophene rings [17].

Another type of material which can exhibit $\chi^{(3)}$ is a nanostructured composite containing small particles [13] of a metal or semiconductor. The nonlinearity arises, in this case, from the large percentage and irregular structure of surface atoms. The effect can be enhanced by using one of the $\chi^{(3)}$ polymers as the continuous phase.

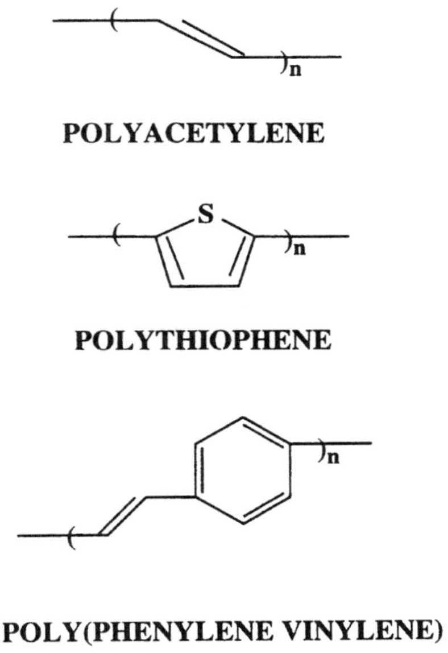

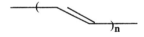

POLYACETYLENE

POLYTHIOPHENE

POLY(PHENYLENE VINYLENE)

FIGURE 2 Polymer structures.

III. $\chi^{(3)}$ MEASUREMENT TECHNIQUES

The different methods used to probe various aspects of third-order nonlinear optical susceptibilities have several things in common. It is important to remember that nonlinear optical phenomena are wavelength dependent. Also, most experiments are performed using pulsed lasers. Different mechanisms may contribute to $\chi^{(3)}$, operating on different time scales [18]. Thus, the pulse duration and repetition rate become additional variables. Also, values of the susceptibility are obtained relative to some reference material. Different reference materials are used for different techniques and even the reported $\chi^{(3)}$ of these materials can vary among techniques [19,20], so it is necessary to state the $\chi^{(3)}$ value of the reference material when reporting results.

Degenerate four-wave mixing (DFWM) is one technique for measuring $\chi^{(3)}$. It uses the interaction of three pulsed laser beams with the same frequency to create a fourth beam whose intensity is proportional to the third-order susceptibility of the material. The characteristics of the grating, such as diffraction efficiency and decay time, together with the dependence of FWM signal on parameters such as beam intensities, crossing angles, and polarization, provide information about the nonlinear optical properties of the material.

In Fig. 3, the BOXCARS geometry [21] of a DFWM experiment is shown (there are other DFWM setups, but this one is illustrative). Usually, a thin-film sample is used, although $\chi^{(3)}$ measurements have also been obtained on solutions [22] and in the vapor phase [23]. Two strong "pump" beams and a weaker "probe beam" are pulsed (~50-ps pulses repeated at 10 Hz) simultaneously. Their paths intersect within the sample. The pump beams, through their nonlinear interaction with the material, create ("write") a grating. The interaction of the probe beam with this grating causes interference: The probe beam is diffracted. A fourth "signal" beam is created. By measuring the intensity of the "signal" beam and comparing to a known standard, such as CS_2, the value of this nonlinearity is obtained using [24]

$$\frac{\chi^{(3)}}{\chi^{(3)}_R} = \left(\frac{\eta}{\eta_R}\right)^{1/2} \left(\frac{n}{n_R}\right)^2 \frac{L}{L_R} \tag{3}$$

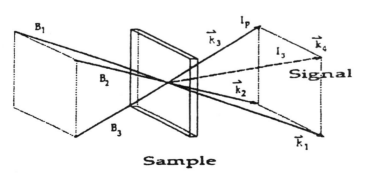

FIGURE 3 Degenerate four-wave mixing (DFWM) beam diagram, BOXCARS geometry.

where the subscript R refers to parameters pertaining to the reference, L is the effective interaction length, n is the refractive index, and

$$\eta = \frac{I_s}{I_p} \qquad (4)$$

is the diffraction efficiency. I_s and I_p are the intensities of the signal and probe beams, respectively. [Usually, the intensities of the probe beams, I_1 and I_2, are equal. Otherwise, a correction factor $(I_{1R}I_{2R}/I_1I_2)$ is used.] In $\chi^{(3)}$ experiments, the beam frequencies and geometry is usually reported as a subscript. For example, values obtained by this DFWM method would be designated as $\chi^{(3)}_{-\omega,\omega,\omega,-\omega}$.

Another consequence of the tensorial nature of $\chi^{(3)}$ is that, unlike $\chi^{(2)}$, the third-order nonlinearity has a sign; it can be positive or negative. To determine this sign, a technique called a Z-scan [25] is used. A collimated beam is focused into the sample, usually a thin film (Fig. 4). The sample is then translated along the optical, or Z, axis. The third-order nonlinearity causes an intensity-dependent lensing effect in the medium, which leads to a self-defocusing of the beam on one side of the original focus ($Z = 0$) and a self-focusing on the other. This can be measured by an aperture detector (a defocused beam has lower intensity). By convention, the $\chi^{(3)}$ of the standard carbon disulfide is assigned a positive sign. The Z-scan for CS_2 exhibits an intensity minimum as the sample is translated toward the source ($Z < 0$) and a maximum as it is moved past the focus, away from the source ($Z > 0$). Materials which behave in the opposite manner have a negative $\chi^{(3)}$.

Third-harmonic generation [26] results from the nonlinear interaction of the material with three incident photons, resulting in an output signal at three times the input frequency ($\lambda/3$). This ability to generate new frequencies has many signal processing applications. At sufficiently high intensity, the third-harmonic is observed. The measurement is made in a collinear geometry, with the fundamental frequency (ω) filtered out by using a solution which absorbs strongly at this frequency. The remaining signal at 3ω is measured by a photodetector. The nonlinear susceptibility measured in this way is called $\chi^{(3)}_{\omega,\omega,\omega,3\omega}$.

The time dependence of the third-order nonlinearity is an important characteristic for device applications and may also indicate the nature of the process causing

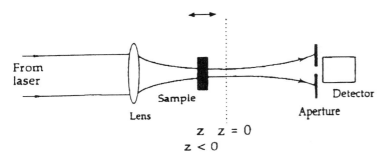

FIGURE 4 Z-scan apparatus.

the nonlinearity [27]. For optical memory applications, a long decay time is preferable because it allows less frequent refresher signals and more time between these signals in which to read the grating. However, for signal processing, a fast response is preferable because it makes very fast switching possible. This is also the case for time-domain multiplexing applications, in which several information streams may pass simultaneously through the same device. [Materials requirements for wavelength-domain multiplexing, although they are outside the scope of this chapter, are technologically important [28].]

Time-resolved measurement of the nonlinearity are performed in order to probe the temporal behavior of the grating. Consider the DFWM apparatus described above. To alter this experiment so that it measures the time dependence of $\chi^{(3)}$, a change is made to the timing of the pump and probe beams. The pump beams are still synchronized with each other. The pump and probe pulses are still 50 ps long, repeating at 10 Hz. But a lag time is introduced in the probe, so that it arrives at the material slightly after the pump beams have written the grating. If this lag time is long enough, the grating will have decayed completely and no signal beam will be generated. If this lag time is zero, there will be no decay of the grating and the signal beam will have its full intensity. To make time-resolved measurements of this sort, the lag time is usually a few tens or hundreds of picoseconds. By varying the lag time, a complete picture of the decay of the grating is obtained. It is also possible to use a negative lag time (i.e., the probe arrives before the pump) of less than the pulse duration to study the evolution of the nonlinearity.

IV. NANOSTRUCTURED MATERIALS

In recent years, nanostructured materials (Fig. 5), those with an average domain size below 100 nm (0.1 μm), have attracted considerable attention [29]. Because of their high surface-to-volume ratio, they exhibit behavior quite different from their coarser-grained counterparts [30] of the same composition. Atoms within a

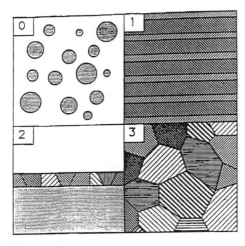

FIGURE 5   Different geometries of nanostructured materials.

bulk material (assuming a simple cubic structure) are bonded on six sides: left, right, front, back, top, bottom. Atoms at the surface are bonded on five sides; not on top. Atoms at a nanostructured particle surface are even more loosely bound. The increasing ratio [31] of surface atoms to those "inside" the particle (with six nearest neighbors) makes it necessary to relax the assumption that all surface atoms are bonded from below as strongly as they would be in a bulk material. The more loosely bound surface atoms constitute a significant fraction of the sample (Fig. 6) and their properties influence its behavior. For example, the melting point of gold is dramatically reduced [32] when the particle diameter drops below 5 nm. Great improvements have also been made in the mechanical [33] and magnetic [34] properties of metals when they are prepared with nanostructured grain size. Unfired ("green") nanostructured ceramics [35] can be shaped and molded to an extent not possible with conventional ceramics.

At a sufficiently small domain size, the particle roughness [36] increases and becomes comparable to the radius of curvature. This reduces the strength of bonding to adjacent atoms and often results in surface atoms with increased reactivity. Also, the abundance of particle edges and corners can create nonequivalent reactive sites on a single particle. The unique chemical properties of nanostructured materials has led to their use as sensor materials and catalysts [37].

The first nanostructured materials [38] were prepared by physical methods, such as inert-gas evaporation. A vapor of the material is generated by heating it above the boiling point, then condensed rapidly onto the walls of a reactor by introducing a cool inert gas, such as helium. Another physical method being studied is high-energy ball milling [39], or mechanosynthesis. These processes are both very energy intensive, and high-energy ball milling has the additional disadvantage of contamination by attrition of the grinding medium.

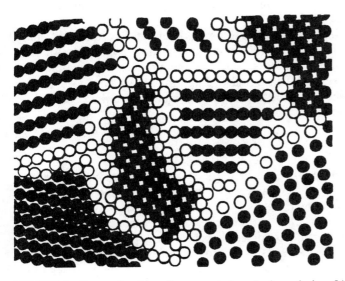

FIGURE 6 Large fraction of atoms at domain boundaries. [After Gleiter, H., *Nanostruct. Mater.*, *1*, 1 (1992).]

Chemical methods for producing nanostructured materials have been studied intensively [40]. Because these materials are built up to nanometer size from molecules, rather than broken down from the bulk, there is the potential for much better homogeneity and control of composition. Chemical vapor deposition (CVD) and sol–gel synthesis have produced some nanostructured materials. Recently, nanometric semiconductor and metal particles of controllable size have also been synthesized by the reaction of chemical precursors such as metal salts and organometallics.

Nanostructured metals and semiconductors exhibit quantum confinement, a property which is particularly important in determining their electronic and optical characteristics. This is a size-dependent change in the properties of the material's free electrons.

Atoms and molecules have discrete orbitals. Bulk materials have valence and conducting bands, separated by a band gap in the case of semiconductors. In a crystal with fewer atoms, the electronic environment is more like that in free space. In bulk, charge is transported easily. The effective mass of an electron is small, and its de Broglie wavelength is long. As the particle size decreases, it approaches the scale of this wavelength [41]. Simultaneously, the de Broglie wavelength is decreasing (i.e., electron effective mass increases).

When the domain size becomes comparable to the de Broglie wavelength [42], the valence and electron bands begin to change (Fig. 7) from their pseudocontinuous structure to a more discrete set of quantized [43] bands. Taken to its extreme, this size reduction leads to the smallest particle, the atom, in which the energy levels are completely discrete. This effect becomes significant below approximately 50 nm for metals and 20 nm for semiconductors. At low temperature, single "quantum dots," or "artificial atoms," have been shown to exhibit step-like current–voltage behavior [44]. Each of these steps corresponds to a single addition to the free-electron cloud of the particle. The altered electronic environment causes such electronic effects as third-harmonic generation and the optical phase conjugation [45] measured by DFWM.

At sufficiently small size, the surface atoms (and surface free electrons) have a distribution of internuclear distances, rather than a repeating crystalline structure. This results in many surface processes, such as the creation of excitons and electron–hole pairs, and excited surface states. Experimentally, enhanced Raman scattering [46] and increased $\chi^{(3)}$ near the resonant frequency can be observed.

V. NANOCOMPOSITES

Composites containing a phase with one or more average dimension smaller than 100 nm may occur with any of several geometries. These include multilayer or monolayer films [47], nanograined metals [48], and clays with polymers intercalated [49] into their structures. One particularly versatile nanocomposite is the quantum dot geometry, with small particles dispersed in a polymer matrix. The particles are also referred to as a "zero dimensional" and such systems are often modeled using the Maxwell–Garnet geometry, which considers the particles to be spherical. The particles, being small, have a high fraction of surface atoms. This causes the system to have a very large surface energy.

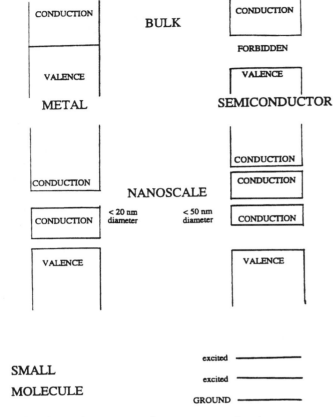

FIGURE 7 Quantum confinement energy bands.

There is a natural tendency for the particles to combine, reducing the amount of surface and, thus, the total surface energy. However, these smaller particles can be stabilized through a suitable treatment of their surfaces. Stabilization may be achieved through steric or electrostatic mechanisms, or some combination of the two.

Steric stabilization results from the chemical functionalization of the surface to create steric repulsion [50] between particles. Metal particles are formed by chemical reaction in the presence of molecules containing a functional group which coordinates with the metal atoms. These functionalized atoms become part of the particle surface, with the remainder of the stabilizing molecule extending outward. These groups cover the particle surface and introduce steric repulsion between particles (i.e., the metal atoms cannot get close enough together to agglomerate because the surface molecules get in the way). Nanometric particles of several different metals have been prepared in this way [51]. If the stabilizing molecule is chosen correctly, it can also make the particle soluble in the matrix.

One synthesis of this type uses 12-carbon alkanethiol molecules ($C_{12}H_{25}SH$) to stabilize gold nanoparticles. In the absence of the thiol, zero-valent gold would combine rapidly into macroscopic particles of gold. However, the natural affinity of gold for thiol [52–56] causes a competing process whereby gold atoms (or clusters of several gold atoms) are coordinated to a thiol group. In the resulting proposed structure (see Fig. 8), gold atoms which combine with each other form the particle core, and thiol-functionalized gold forms the surface.

A. Synthesis

The particles were synthesized using a phase transfer reaction [57]. Gold salt ($HAuCl_4$) was dissolved in water (30 ml, 0.030 M). Normally, the resulting $AuCl_4$ anion would be insoluble in organic solvents [58]. However, when the aqueous solution is stirred for 1 h with a toluene solution (80 ml, 0.050 M) of the phase transfer catalyst, tetraoctylammonium bromide ($N[C_8H_{17}]_4Br$), the gold species is transferred into the toluene.

Next, the surface-functionalizing reagent, dodecanethiol (0.2 ml), was added. Then, sodium borohydride solution (25 ml, 0.40 M) was dropped in gradually to reduce the gold from Au^{3+} to Au^0. The reaction was complete 3 h after the borohydride was added.

The product (organic phase) was decanted. Approximately 90% of the toluene was removed by evaporation; then, the remaining material was precipitated into ethanol. After filtration, the product was purified by redissolving in toluene and reprecipitating with ethanol. The resulting material was a waxy, purple solid which was stable over a period of months.

B. Characterization

The functionalized particles were dissolved in toluene and optical absorbance was measured versus a toluene blank. The spectrum showed a shoulder peak between

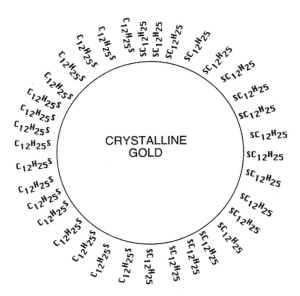

FIGURE 8 Surface-functionalized gold particle, proposed structure.

520 and 530 nm. This agreement with previous studies [59] signifies that the gold is nanostructured.

Isolated particles will rearrange their surfaces to reach an energetic minimum [60]. In particles stabilized by chemical coordination, these processes are suppressed, leading to metastable structures. These surfaces cannot be probed by many traditional "surface" techniques [e.g., X-ray photoelectron spectroscopy (XPS), with a penetration depth of 70 Å, cannot distinguish between surface and bulk atoms of particles smaller than 14 nm]. Numerical simulation and improvements in electron microscope technology may yield some insight into structure–property relationships in nanocomposites. Electron microscopy can yield virtually atomic resolution of structure.

High-resolution transmission electron microscopy (HRTEM) was used to examine the gold particles directly. Figure 9 shows a HRTEM image of the particles, which exhibit faceting as shown by the arrows. All particles display the characteristic lattice fringes of the common HRTEM images. Some of the gold particles clearly have faceting. The smallest gold particles usually show perfect crystalline structure. However, in some cases, the presence of twin boundaries are evident. It is also interesting to point out that, from these HRTEM images, there is no evidence of crystalline defects such as stacking faults or dislocations. The particle size distribution (Fig. 10) shows that most of the particle diameters fall in the range 1.0–3.4 nm.

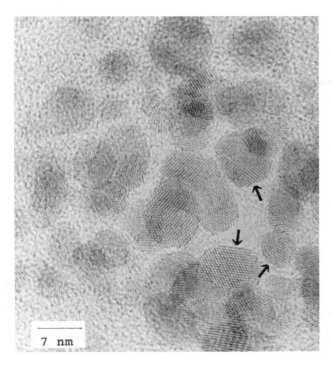

FIGURE 9 HRTEM image of gold particles.

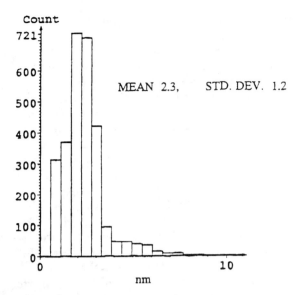

FIGURE 10 Gold particle size distribution.

C. Composite Film Preparation

Composites of the gold nanoparticles in polymer were obtained easily by dissolving the particles in a suitable monomer (methyl methacrylate or styrene) and initiating free-radical polymerization using azobisisobutylnitrile under thermal conditions (70°C). Polymerization proceeded more slowly in the particle-containing reactions than in the control reactions. This may indicate the presence of a chain-terminating side reaction. Polymerization was terminated by cooling when the mixture reached a viscosity suitable for spin-coating.

Films were cast from solution onto clean quartz substrates. Relatively slow spinning (~250 rpm) was found to produce the best films. At high speeds, centrifugation effects cause the gold concentration to vary radially. After 3 min of spinning, the solvent (toluene or monomer) evaporated, leaving a flat film 5–10 μm thick.

These free-radical polymers are relatively tractable and provide good, clear films. However, better nonlinear optical properties may be obtained using polymers which are, themselves, nonlinear. Polysilane–matrix composites were prepared to increase the $\chi^{(3)}$ of the resulting material.

D. Polysilane Synthesis

The synthesis of poly(phenylmethylsilane) was carried out by the Wurtz coupling of phenylmethyldichlorosilane assisted by ultrasonication [61]. First, sonication at 40% amplitude for 20 min was used to disperse sodium metal in toluene. Then, the silane monomer was added to this dispersion over a 30-min period. The reaction continued with sonication at 20% amplitude for 1 h. A 50/50 ethanol/water mixture was used to quench the reaction.

The polymer was precipitated into isopropanol, filtered, redissolved in toluene, reprecipitated, filtered again, and vacuum dried. The reaction yielded 17% soluble polymer. There was also a significant amount of an insoluble side product, probably cross-linked material.

The in situ method described above could not be used to form composites with a polysilane matrix because of the harsher polymerization conditions. In this case, the polymer and the nanoparticles were both dissolved in toluene and codeposited during spin-coating.

E. Film Characterization

The films were examined by UV–visible spectroscopy (Figs. 11 and 12) and showed similar peak positions to those for the solutions (plain polymer films were used as blanks). The linear absorption coefficient at 532 nm (where the NLO characterization was performed) was approximately 0.6 cm^{-1}. Also, TEM images of thin (8 nm) sections of the composite material showed that the particles were incorporated into the films without agglomeration.

VI. NONLINEAR OPTICAL MEASUREMENTS

A. Z-Scan

By convention, the standard carbon disulfide is assigned a positive sign. The gold particles were found to exhibit the opposite behavior (a prefocal minimum at $Z < 0$ and postfocal maximum intensity), thus having a negative sign (see Fig. 13). This is also known to be the case for polysilanes. Thus, their nonlinearities should add to create a larger, negative $\chi^{(3)}$ in the composite.

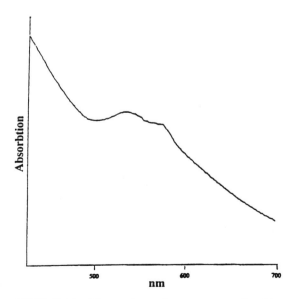

FIGURE 11 Linear absorption spectrum of gold particle/PMMA composite.

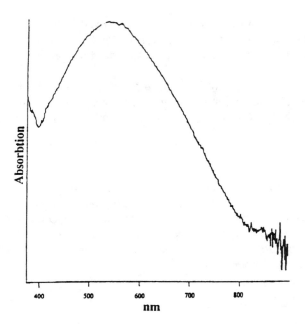

FIGURE 12 Linear absorption spectrum of particle/polysilane composite.

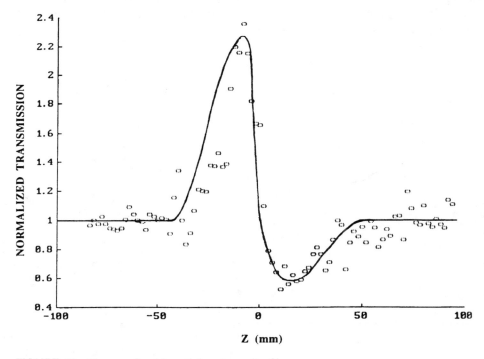

FIGURE 13 Z-scan of gold particle composite film.

B. Degenerate Four-Wave Mixing

The geometry used for the measurements in this study is the counterpropagating pump geometry, rather than the simpler BOXCARS. The two "write" beams are incident on the medium from opposite sides. These are examined by a weaker "read" beam. When the three beams are incident on a nonlinear optical medium they generate a fourth beam, the phase conjugate, which is counterpropagating to the probe.

The magnitude of the third-order nonlinear susceptibility, $\chi^{(3)}$, was estimated from a measurement of the intensity of the phase conjugate beam relative to that of a reference sample of CS_2 placed in a quartz cell of 2-mm path length. The value of $\chi^{(3)}$ was obtained from Eq. (3). The measured values of $\chi^{(3)}$ are given in Table 1. Theoretically, $\chi^{(3)}$ is expected to be proportional to particle concentration. However, the proportionality is not exact. Some possible sources of this error are dispersity of particle size and nonuniformity of the films.

C. Time-Resolved Measurements

The temporal behavior of the phase conjugate signal in the gold-containing films was monitored by delaying the probe beam with respect to the pump. A 30-ps pulse duration was used for the pump and probe beams.

Experimental data (see Fig. 14) show that the grating is composed of two components: a fast process (which has a shorter duration than the laser pulse) and a slower one (which decays exponentially). The fast process can be attributed to the electronic response of the medium. The slow process may be the result of a physical phenomenon, such as plasmon resonance or lattice heating. This slow process has a longer decay time than in previous studies of nanostructured gold in liquid solution. This may be evidence of slower diffusion of, or heat dissipation from, the particles when embedded in a polymer matrix.

D. Experimental Details

Tetrachloroauric acid ($HAuCl_4$), tetraoctylammonium bromide ($N[C_8H_{17}]_4Br$), dodecanethiol, sodium borohydride, and sodium metal were used as received from

TABLE 1 Third-Order Nonlinear Optical Susceptibilities of Composite Films

Matrix	Gold concentration (mg/ml)	$-\chi^{(3)}$ (esu)
PMMA	0	not measurable
PMMA	1.3	1.6×10^{-11}
PMMA	4.4	1.0×10^{-10}
poly(phenylmethylsilane)	0	6.1×10^{-11}
poly(phenylmethylsilane)	5.1	1.5×10^{-10}

Note: Measured by DFWM, counterpropagating pump beams, 532 nm.

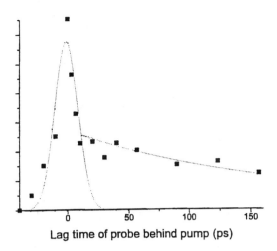

Lag time of probe behind pump (ps)

FIGURE 14   Time dependence of $\chi^{(3)}$ in gold particle/PMMA composite.

Sigma Chemical Co. Water and toluene were deoxygenated by reflux and distillation under an inert atmosphere. The monomers phenylmethyldichlorosilane (Gelest), methyl methacrylate, and styrene (Aldrich) were purified by fractional distillation under vacuum. The initiator, azobisisobutylnitrile (AIBN), was recrystallized from methanol before use.

Ultrasonic treatment was performed using a high-intensity ultrasonic probe (Sonic and Materials VC-600, 1/2-in.-titanium horn, 20 kHz, 100 W/cm²). Films were cast using a Specialty Coatings Systems P-6204-A spin-coater. Film thickness was measured using a Tencor Alpha-Step 200 profilometer. UV–visible spectra were measured on a Perkin-Elmer Lambda 6 spectrophotometer from 190 to 900 nm using a 1-nm slit at a scan rate of 120 nm/min.

High-resolution TEM samples were prepared from methanol dispersions of the particles and deposited on copper grids with a carbon film. The instrument was a JEOL-4000EX with a point-to-point resolution of approximately 1.7 A. Transmission electron microscopy was also performed on poly(methyl methacrylate) (PMMA)/gold composite samples using a Phillips EM 300 instrument. The samples were prepared by microtoming to a thickness of 80 Å.

The nonlinear optical properties were measured using 532-nm, 30-ps pulses from a frequency-doubled, *Q*-switched neodymium-doped yttrium–aluminum–garnet (Nd : YAG) laser (Quantel). The laser operated at a 10-Hz repetition rate. Average pulse energy was 25 mJ. A neutral density filter was placed in the probe beam to reduce its intensity to about 1% of that of the pump beams. The phase conjugate beam was separated from the signal using a beam splitter. The crossing angle was 6°. Fast silicon photodiodes calibrated against a laser-energy meter were used to monitor the signal, the probe, and the pump pulse energies.

VII.   NANOSTRUCTURED SEMICONDUCTORS

Semiconductor nanoparticles have several interesting size-dependent properties. The optical absorption maximum is blue-shifted from the value for the bulk material. Theory also predicts that composites of these particles will have an appreciable $\chi^{(3)}$, which can be measured by DFWM.

Syntheses of several nanostructured II–VI semiconductors have been reported. By reacting cadmium ions with $Se[Si(CH_3)]_2$ in inverse micelles, CdSe particles of approximately 4.5 nm diameter [62] were prepared. These were sterically stabilized by capping with phenyl-$Se[Si(CH_3)]$. Cadmium-based semiconductor particles (CdSe, CdS, CdTe) have also been grown by injection of organometallic reagents into hot, (~300°C) coordinating solvents, such as trioctylphosphine [63]. Through selective precipitation, relatively monodisperse fractions of these materials were separated, with diameters ranging from 2 to 11 nm. Using copolymers with zinc-containing blocks, a polymer containing 3-nm zinc particles was prepared. The particles were subsequently converted to ZnS by treatment with H_2S gas [64]. The similarity between these methods is that precursors to II–VI semiconductors decompose readily under relatively mild conditions.

Because they have a wider band gap (3.1–3.8 eV), III–V semiconductors have attracted significant research interest recently. This higher energy corresponds to light of shorter wavelength and thus the potential for greater information transfer in the same time. However, precursors to these materials do not decompose readily and are generally more difficult to synthesize than II–VI precursors. By decomposing cyclotrigallazane at 600°C, nanostructured, wurtzite gallium nitride [65] was obtained. Also, several kinds of binary and ternary III–V nanocrystals have been synthesized using dehalosilylation reactions of group III halides with $P[Si(CH_3)]_3$ or $As[Si(CH_3)]_3$ in organic solvents. A sodium–potassium alloy (Na/K) has been used to produce other nanostructured III–V's [66]. Another synthetic route has led to nanostructured, zinc blende gallium nitride of 3–8 nm diameter [67].

A. Synthesis

The dimeric precursor [68], $Ga_2[N(CH_3)_2]_6$ (Fig. 15), was synthesized by the reaction of gallium chloride, $GaCl_3$, with lithium dimethylamide, $LiN(CH_3)_2$. A slight excess (6.44 g) of $LiNMe_2$ was reacted with 7.17 g $GaCl_3$ at room temperature for 2 days in dried, distilled hexane under an inert atmosphere. After stirring, the solution was filtered through celite and the volatiles were removed under vacuum to yield a colorless solid. LiCl by-product remained on the frit and was discarded.

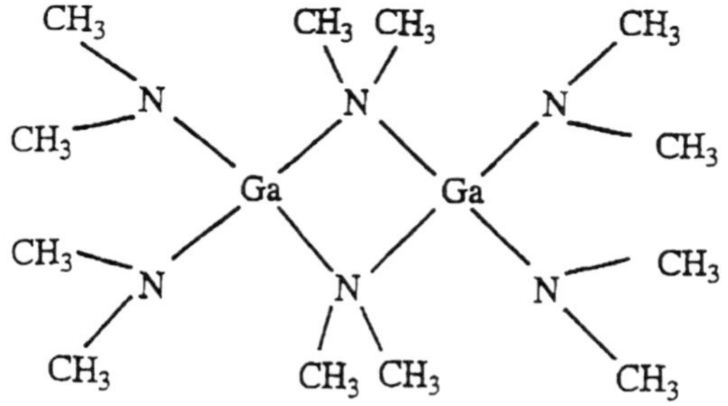

FIGURE 15 Structure of dimeric precursor, $Ga_2[N(CH_3)_2]_6$.

The pale yellow pure dimer was isolated by vacuum sublimation at 110°C and 10^{-3} Torr. Yield was 3.82 g (71%).

The precursor dimer was placed in an alumina boat and decomposed [69] using a 1-in.-inner diameter (ID) quartz tube placed in a Thermolyne 21100 furnace with a 12-in. heating zone. The material was heated for 4 h at 600°C. Ammonia flow at ambient pressure was maintained during heat treatment and while the reaction cooled to room temperature over several hours. The product, a grayish-green powder, was handled only under an inert atmosphere in order to prevent oxidation.

To form the nanostructured composite, 60 mg of this powder and 1 ml of methyl methacrylate were charged to a flask under argon. The mixture was sonicated in a cleaning bath for 2 h. The liquid was then decanted and placed in a sealed vial with a small amount of initiator (azobisisobutylnitrile, ~5 mg). This mixture was polymerized thermally at 72°C for 50 min. From the resulting solution, films were spin-cast onto clean quartz plates. The speed was 225 rpm and the time was 5 min. Film thicknesses, measured by profilometry, were ~5 μm.

B. Characterization

Powdered samples of GaN were examined under mineral oil to inhibit oxidation. The analysis was performed on a Siemens D-5000 diffractometer equipped with a CuK_α radiation source and a graphite monochromator. The intensity was determined by step scanning in the 10°–70° range, with a 2θ step of 0.02° every 3 s. The crystalline structure was using the Rietveld technique as described in the new version of the DBWS-9411 program [70]. The profile-breadth fitting was determined by considering the averaged crystallite size and crystal-microstrain parameters using a modified (71) pseudo-Voight function.

The x-ray diffraction pattern (Fig. 16) clearly exhibits the {111}, {220}, and {311} reflections of the metastable, face-centered cubic (zinc blende) GaN phase at angles 2θ of 35.5°, 58°, and 69°, respectively. The asymmetry of the first reflection can be interpreted as an overlapping of the {111} and a confined {002} of weak intensity. X-ray refinement using the Rietveld method shows that the best match with experiment is obtained by considering a structure that is a mixture of zinc blende and wurtzite (hcp) GaN phases, with a particle size determined from the coherence length calculated from the Debye-Scherrer formula of 2.8 ± 0.3 nm. The simulation considers that each particle consists of a majority (76.7%) of zinc blende GaN mixed with a smaller amount (23.3%) of wurtzite GaN in such a way that the $\langle 111 \rangle$ zinc blende crystallographic axis is parallel to the $\langle 0001 \rangle$ wurtzite axis. This result suggests that the nanostructured GaN crystallizes mainly in the zinc blende lattice with some disorder along the $\langle 111 \rangle$ axis. The lattice constant deduced from our x-ray Rietveld refinement is 0.4500 nm, which is consistent with measurements reported previously.

Duplicate chemical analyses of the as-synthesized GaN were performed (Galbraith Labs) to determine gallium and nitrogen. A separate test (Microlytics) was run to measure hydrocarbon contamination. The tests showed a nitrogen-poor material with empirical formula $GaN_{0.86}$, that contains traces of carbon (0.8 wt%) and hydrogen (0.5 wt%). Thermodynamically, a Ga interstitial should be much more difficult to form in zinc blende GaN than a N vacancy. Thus, it can be assumed that the GaN nonstoichiometry is due to N vacancies.

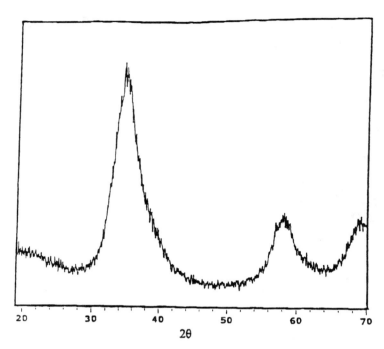

FIGURE 16 X-ray spectrum of GaN powder.

High-resolution transmission electron microscopy (HRTEM) was used to examine the powder sample and the composite. The instrument used was a JEOL-4000EX with an accelerating voltage of 400 keV and a point-to point resolution of approximately 1.7 Å. The powder sample was prepared by grinding the powder between two glass plates and bringing the fine powder into contact with a carbon-coated copper grid under nitrogen. The composite sample was embedded in epoxy resin and sliced down to a thickness of about 80 nm using a LKB Ultratome V equipped with a diamond knife. Slices were then picked up onto a TEM grid. HRTEM images were obtained at optimum (Scherzer) defocus.

A typical electron micrograph (Fig. 17) shows that powdered GaN is composed of porous particles of relatively large size (around 50 nm major axis). Examination of the particles at higher magnification (Fig. 18) indicates that each of these large particles is composed of an agglomeration of smaller particles with nanostructured (~3 nm) domains, in accordance with the value deduced from the x-ray reflections. The TEM image of the composite shows that the GaN is uniformly dispersed in the PMMA matrix. The particle size distribution (Fig. 19), determined by analysis of these images, had a mean of 5.6 nm and a standard deviation of 2.6. This is slightly larger than in the assynthesized powder; some particles may contain two or more of the primary domains. A higher-magnification image of the composite confirms that the dispersed phase is crystalline GaN.

Examinations of crystallite structure in the as-synthesized powder and in the composite show that defects are abundant. Close analysis of the atomic structure in the powder indicates very short-range order and a considerable number of stack-

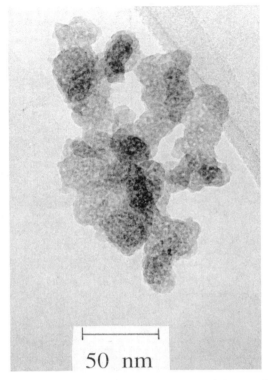

FIGURE 17   Nanostructured GaN at high magnification.

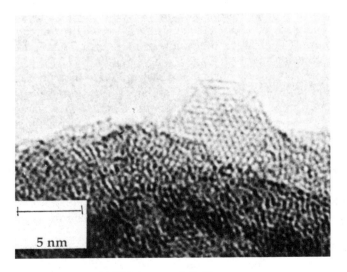

FIGURE 18   HRTEM image of GaN.

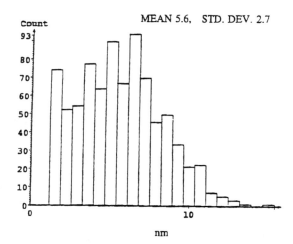

MEAN 5.6, STD. DEV. 2.7

FIGURE 19 GaN particle size distribution.

ing faults. Also, one can notice that the order range is higher in the composite; the incorporation of nanocrystalline GaN into the polymer remarkably reduces the stacking-disorder density. It is not clear whether the increase in order is due to the larger particle size or to the surrounding polymer's interaction with the surface, or both.

C. Optical Measurements

The absorption of the composite film was examined using a Perkin-Elmer Lambda 6 spectrophotometer. The sample was scanned at a rate of 2 nm/s and the slit width was 0.5 nm. The resulting spectrum had a strong, symmetric peak between 315 and 340 nm.

Photoluminescence was measured on a Perkin-Elmer LS 50B fluorescence spectrophotometer. An emission spectrum was run with 310 nm excitation, 2.5 nm slit width, scanning at 120 nm/min. A single broad peak centered at 370 nm was observed.

Third-order nonlinear optical properties of nanoparticles in PMMA were studied using degenerate four-wave mixing (DFWM). The backward wave geometry was used, where the signal is a phase conjugate replica of the probe, counterpropagating in the direction of the probe. The 532-nm, 30-ps pulses from a frequency-doubled, Q-switched neodynium-doped yttrium–aluminum–garnet (Nd : YAG) laser (Quantel) were used. The laser operated at a 10-Hz repetition rate. Average pulse energy was 25 mJ. A neutral density filter was placed in the probe beam to reduce its intensity to about 1% that of the pump beams. The phase conjugate beam was separated from the signal using a beam splitter. The crossing angle was 6°. Fast silicon photodiodes calibrated against a laser-energy meter were used to monitor the signal, the probe, and the pump pulse energies. By comparison with a CS_2 standard, the $\chi^{(3)}$ was 2.6×10^{-11} esu, which is relatively low, compared to the values measured for the gold composites. However, this is expected because of the wavelength dependence of $\chi^{(3)}$. The 532-nm pulses used in both experiments fall at an absorption maximum for gold (on resonance) and away from this maximum for gallium nitride.

VIII. CONCLUSIONS

The fact that optical nonlinearity has been known since the last century [72,73] speaks highly of the abilities of early investigators, and of the sensitivity possible in interferometric and polarization measurements. However, the nonlinearity was difficult to detect using conventional light sources.

With the advent of lasers, nonlinear optics has become an important technology. The ability to run experiments to fully characterize nonlinear optical materials will enable their incorporation into integrated photonic and electrooptical systems. Applications for $\chi^{(3)}$ materials exploit the properties measured by these experiments. The defocusing properties underlying the Z-scan can be used to protect sensors and, more importantly, human eyes from high-intensity light. The protection device (or pair of glasses) constructed with a third-order material passes light at normal intensity, but defocuses it at high intensity. The optical phase conjugation [74] demonstrated in the DFWM experiment can be used to create holograms. The fast response of the third-order nonlinearity also makes it useful for all-optical modulation, which could be used to build much faster computers [75]. Also, in certain applications, $\chi^{(3)}$ materials can have the same effect as a $\chi^{(2)}$ material when used under an applied DC voltage [76].

To achieve the best electronic and optical properties in the composite, the particles must be small enough to exhibit quantum confinement; generally, the smaller the better. Also, a narrow distribution of sizes is desirable. Particles of different sizes have slightly different forbidden energy levels. Thus, dispersion of the size results in a more diffuse band-gap architecture and devices with diminished or ill-defined properties. Because the interaction of light or current with the composite takes place mostly at particle surfaces, higher particle loading will result in a stronger interaction, allowing the device to be designed with smaller dimensions. Another feature which could potentially lead to much faster, denser circuits is the periodic placement of the particles within the composite to form an ordered array [77].

Quantum-confined, polymer–matrix composites represent one class of $\chi^{(3)}$ materials capable of great flexibility in design. The size, morphology, composition, and loading of the dispersed phase, as well as matrix–polymer properties, may be varied to yield a wide variety of optical and optoelectronic properties. Progress in this area has been quite rapid and the large number of new materials produced recently is encouraging. However, efforts to devise composites with new and better properties should be undertaken to take advantage of this class of advanced materials.

ACKNOWLEDGMENTS

K. E. Gonsalves acknowledges partial support of this work by the ONR grant #N0014-94-1-0833 and NSF grant INT-9503854. Significant contributions to this work were also made by Professor M. Benaissa, Professor R. Perez, and Professor

M. José-Yacamán, Universidad Nacional Autonóma de México (UNAM) and National Institute for Nuclear Research (ININ); Professor J. Kumar, University of Massachusetts, Lowell; Dr. F. Aranda, University of Massachusetts, Boston; the U.S. Army Soldier Systems Command at Natick, MA; and Dr. S. P. Rangarajan and Dr. L. Khairallah, University of Connecticut.

REFERENCES

1. Kaminow, I., *An Introduction to Electro-Optic Devices*, Academic Press, New York, 1974, p. 56.
2. Skinner, I., and Garth, S., *Appl. Phys. Lett.*, 58, 177 (1990).
3. Stegeman, G. I., and Miller, A., in *Physics of All-Optical Switching Devices* (J. E. Midwinter, ed.), Academic, Orlando, FL, 1994; Bowden, C. M., Giftan, M., and Robl, H. R., *Optical Bistability*, Plenum, New York, 1981.
4. de Melo, C. P., and Sibley, R., *J. Chem. Phys.*, 88, 2567 (1988).
5. Zyss, J., LeDoux, I., and Nicoud, J. F., *Molecular Nonlinear Optics*, Academic Press, New York, 1994, p. 180.
6. Willand, C. S., and Albrecht, A. C., *Opt. Commun.*, 28, 235 (1985).
7. Neal, D. B., Petty, M. C., Roberts, G. G., Ahmad, M. M., Feast, W. J., Girling, I. R., Cade, N. A., Kolinsky, P. V., and Peterson, I. R., *Electron. Lett.*, 22, 460 (1986).
8. Singer, K. D., Kuzyk, M. G., and Sohn, J. E., *J. Opt. Soc. Am. B*4, 968 (1987).
9. Barry, S., and Soane, D., *Appl. Phys. Lett.*, 58, 1134 (1991).
10. *SPIE Proceedings on Optical Computing*, Vol. 1560, SPIE, K. Singer, ed. Bellingham, WA (1991); Chen, M., Dalton, L. R., Yu, L. P., Shi, Y. Q., and Steier, W. H., *Macromolecules* 25, 4032 (1992).
11. Page, R. H., Jurich, M. C., Reck, B., Sen, A., Tweig, R. J., Swalen, J. D., Bjorklund, G. C., and Wilson, C. G., *J. Opt. Soc. Am. B*7, 1239 (1990).
12. Heeger, A. J., et. al. (eds.), *Nonlinear Optical Properties of Polymers*, MRS Symp. Proc. No. 109, Materials Research Society, Pittsburgh, Pennsylvania 1987.
13. Prasad, P. N., in *Materials for Nonlinear Optics* S. R. Marder, J. E. Sohn, and G. D. Stucky, eds.), ACS Symp. Series 455, American Chemical Society, Washington, DC, 1991, p. 50.
14. Zhao, M. T., Samoc, M., Singh, B. P., and Prasad, P. N., *J. Phys. Chem.*, 93, 7916 (1989); Daniel, C., and Dupuis, M., *Chem. Phys. Lett.*, 17, 209 (1990); Bosma, W. B., Mukamel, S., Greene, B. I., and Schmitt-Rink, S., *Phys. Rev. Lett.*, 68, 2456 (1992).
15. Shuai, Z., and Brédas, J. L., *Phys. Rev. B*,. 46, 4395 (1992).
16. Klingensmith, K., Downing, J. W., Michl, J., and Miller, R. D., *J. Am. Chem. Soc.*, 108, 7438 (1993).
17. Beljonne, D., Shuai, Z., and Brédas, J. L., *J. Chem. Phys.*, 98, 8819 (1988).
18. Wu, C. K., Agostini, P., Petite, G., and Fabre F., *Opt. Lett.*, 8, 67 (1983).
19. Ho, P. P., and Alfano, R. R., *Phys. Rev. A*, 20, 2170 (1979).
20. Guha, S., Kang, K., Porter, P., Roach, J. F., Remy, D. E., Aranda, F. J., and Rao, D. V., *Opt. Lett.*, 17, 264 (1992).
21. Eckbreth, A. C., *Appl. Phys. Lett.*, 32(7), 421 (1978).
22. Gonsalves, K. E., Carlson, G., Chen, X., Gayen, S. K., Perez, R., and Jose-Yacaman, M., *Nanostruct. Mater.*, 7(3), 293 (1996).
23. Ai, B., Glassner, D. S., Knize, R. J., and Partanen, J. P., *Appl. Phys. Lett.*, 64(8), 951 (1994).
24. Bogdan, A. R., Prior, Y., and Bloembergen, N., *Opt. Lett.*, 6, 82 (1981).
25. Cheung, Y. M., and Gayen, S. K., *J. Opt. Soc. Am.*, B11, 636 (1994).

26.   Terhune, R. W., Maker, P. D., and Savage, C. M., *Phys. Rev. Lett.*, *8*, 404 (1962).
27.   Bloemer, M. J., Haus, J. W., and Ashley, P. R., *J. Opt. Soc. Am.*, *B7*, 790 (1990).
28.   Dentai, A., Stone, J., Burrows, E. C., Burrus, C. A., Stulz, L. W., and Zirngibl, M., *IEEE Phot. Tech. Lett.*, *6*(5), 629 (1994).
29.   Dubin, P., and Tong, P., *Colloid-Polymer Interactions*, ACS Symp. Series 532, American Chemical Society, Washington, DC, 1993.
30.   Chakravorty, D., and Giri, A. K., Nanomaterials, in *Chemistry for the 21st Century*: *Chemistry of Advanced Materials* (C. N. R. Rao, ed.), Blackwell Scientific Publications, London, 1993.
31.   Stwalley, W. C., and de Llano, M., *Z. Phys. D*, *2*, 153 (1986).
32.   Ichinose, N., Ozaki, Y., and Kashu, S., *Superfine Particle Technology*, Springer-Verlag, London, 1992 (translated from Japanese).
33.   Kear, B. H., and McCandlish, L. F., *Adv. Mater.*, *10*, 11 (1993).
34.   Shull, R. D., et al., *Nanostruct. Mater.*, *2*, 205 (1993).
35.   (B. I. Lee and E. J. A. Pope, eds.), *Chemical Processing of Ceramics*, Marcel Dekker, Inc., New York, 1994.
36.   Bovin, J. O., and Melm, J. O., *Z. Phys. D*, *19*, 293 (1991).
37.   Boakye, E., et al., *J. Colloid Interf. Sci.*, *163*, 120, 1994.
38.   a) Hahn, H., and Averback, R. S., *J. Appl. Phys.*, *67*, 1113 (1990); R. Uyeda, *Prog. Mater. Sci.*, *35*, 1 (1990).
39.   Koch, C. C., *Nanostruct. Mater.*, *2*, 109 (1993).
40.   Gonsalves, K. E., U.S. Patent 4,842,641 (1989); Rivas, J., et. al., *J. Magnet. Magnet. Mater.*, *122*(1–5), 1 (1993).
41.   Brus, L. E., *J. Chem. Phys.*, *79*, 5566 (1983).
42.   Wang, Y., et. al., *J. Chem. Phys.*, *87*, 7315 (1987).
43.   Wang, Y., and Herron, N., *J. Chem. Phys.*, *95*, 525 (1991).
44.   Kastner, M. A., *Phys. Today*, *46*, 24 (January 1993).
45.   Ricard, D., Roussignol, P., and Flytzanis, C., *Opt. Lett.*, *10*, 511 (1985).
46.   Wokaun, A., *Molec. Phys.*, *56*(1), 1 (1985).
47.   Ulman, A., Williams, D. J., Penner, T. L., Robello, D. R., Schildkraut, J. S., Scozzafava, M., and Willand, C. S., U.S. Patent 4,792,208 (1988).
48.   Gonsalves, K. E., Rangarajan, S. P., Law, C. C., Feng, C. R., Chow, G.-M., and Garcia-Ruiz, A., in *Nanotechnology*: *Molecularly Designed Materials*, (G.-M. Chow and K. E. Gonsalves, *eds.*), ACS Symp. Series 622, American Chemical Society, Washington, DC, 1995, p. 220.
49.   Pinnavin, T. J., Lan, T., Wang, Z., Shi, H., and P. D. Kaviratna, in *Nanotechnology*: *Molecularly Designed Materials*, G.-M. Chow and K. E. Gonsalves, eds.), ACS Symp. Series 622, American Chemical Society, Washington, DC, 1995, p. 250.
50.   Verwey, E. J. W., *Chem. Weekbl.*, 39:563 (1942); Derjaguin, B. V. and Landau, L., *Acta Physicochim. URSS*, *14*, 633 (1941).
51.   Duteil, A., Schmid, G., and Meyer-Zaika, W., *J. Chem. Soc., Chem. Commun.*, 31 (1995); Bonnemann, H., Brinkman, R., Koppler, R. Neiteler, P., and Richter, J., *Adv. Mater.*, *4*, 804 (1992).
52.   Nuzzo, R. G., and Allara, D. L., *J. Am. Chem. Soc.*, *105*, 4481 (1983).
53.   Nuzzo, R. G., Fusco, F. A., and Allara, D. L., *J. Am. Chem. Soc.*, *109*, 2358 (1987).
54.   Li, T. T.-T., and Weaver, M. J., *J. Am. Chem. Soc.*, *106*, 6107 (1984).
55.   Finklea, H. O., Avery, S., Lynch, M., and Furtsch, T., *Langmuir*, *3*, 409 (1987).
56.   Porter, M. D., *J. Am. Chem. Soc.*, *109*, 3559 (1987).
57.   Brust, M., Walker, M., Bethell, D., Schiffrin, D. J., and Whyman, R., *J. Chem. Soc., Chem. Comun.*, 801 (1994).
58.   Dehmlow, E. V., *Phase Transfer Catalysis*, Weinheim, New York, 1993.
59.   Olsen, A. W., and Kafafi, Z. H., *J. Am. Chem. Soc.*, *113*, 7758 (1991).

60. Bovin, J. O., and Melm, J. O., *Z. Phys. D*, *19*, 293 (1991).
61. Kim, H. K., and Matyjasewski, K., *J. Am. Chem. Soc.*, *110*, 3323 (1988).
62. Steigerwald, M. L., Alivasatos, A. P., Gibson, J. M., Harris, T. D., Kortan, R., Muller, A. J., Thayer, A. M., Duncan, T. M., Douglass, D. C., and Brus, L. E., *J. Am. Chem. Soc.*, *115*, 8706 (1987).
63. Murray, C. B., Norris, D. J., and Bawendi, M. G, *J. Am. Chem. Soc.*, *115*, 8706 (1993).
64. Sankaran, V., Yue, J., Cohen, R. E., Schrock, R. R., and Silbey, R. J., *Chem. Mater.*, *5*, 1133 (1993).
65. Hwang, J.-W., Campbell, J. P., Kozubowski, J., Hanson, S. A., Evans, J. F., and Gladfelter, W. L., *Chem. Mater.*, *7*, 517 (1995).
66. Halaoui, L. I., Kher, S. S., Lube, M. S., Aubuchon, S. R., Hagan, C. R. S., Wells, R. L., and Coury, L. A., in *Nanotechnology: Molecularly Designed Materials*, G.-M. Chow and K. E. Gonsalves, eds.), ACS Symp. Series 622, American Chemical Society, Washington, DC, 1996, p. 178, and references therein.
67. Gonsalves, K. E., Carlson, G., Rangarajan, S. P., Benaissa, M., and Jose-Yacaman, M., *J. Mater. Chem.*, *6*(8), 1451 (1996).
68. Noth, H., and Konrad, P., *Z. Naturf.*, *30b*, 681 (1975).
69. Waggoner, K. M., Olmstead, M. M., and Power, P. P., *Polyhedron*, *9*, 257 (1990).
70. Young, R. A., Sakthivel, A., Moss, T. S., and Paiva-Santos, C. O., *J. Appl. Crystallogr.*, *28*, 366 (1995).
71. Thompson, P., Cox, D. E., and Hastings, J. B., *J. Appl. Crystallogr.*, *20*, 445 (1974).
72. Kerr, J., *Phil. Mag.*, *1*, 337 (1875).
73. Lord Rayleigh, *Phil. Mag.*, *XLI*, 274 (1871).
74. O'Meara, T. R., and Pepper, D. M., in *Optical Phase Conjugation* (R. A. Fischer, ed.), Academic, New York, 1983, p. 537.
75. *SPIE Proceedings on Optical Computing*, Vol. 963, SPIE, Bellingham, WA, 1988.
76. Stamatoff, J., Demartino, R., Haas, D., Khanarian, G., Man, H. T., Yoon, H. N., and Norwood, R., *Ang. Makr. Chem.*, *183*, 151 (1990).
77. Lent, C. S., Tougaw, P. D., Porod, W., and Bernstein, G. H., *Nanotechnology*, *4*, 49 (1993).

20
Molecular Self-Assemblies as Advanced Materials

DeQuan Li and Octavio Ramos, Jr.
Los Alamos National Laboratory
Los Alamos, New Mexico

I. AN OVERVIEW OF SELF-ASSEMBLED MONOLAYERS

A. Introduction

Cross-cutting chemistry, physics, biology, and materials science, molecular self-assemblies, or supramolecular systems have become one of the most rapidly emerging fields in science (Fig. 1). The fundamental ideas of this field began with ancient philosophers in Asia and Europe who inadvertently formulated a general definition: Only the mutuality of the parts creates the whole and its ability to function—the whole is much more greater than the sum of its parts.

Molecular/macromolecular assemblies commonly occur in nature; some examples include lipid self-organization, protein folding, and biomolecular replication/fabrication. At the cell membrane level, glycocalyx contributes to molecular recognition, lipids self-organize into double layers, cytoskeleton proteins stabilize the cell, and channel-forming proteins provide mass and information exchanges between the interior and exterior of a cell.

In 1894, Fischer first postulated the concept of supramolecular interaction. It was not until approximately 1985, however, that it was generally realized that molecular assemblies (which mimic natural systems) offer a rational link between structures and properties. The field's initial focus was on guest–host systems but has expanded to studies on self-organization, mesoscale structures, regulations, pattern formation, molecular recognition, and novel electrical and optical materials such as second-order nonlinear optical properties.

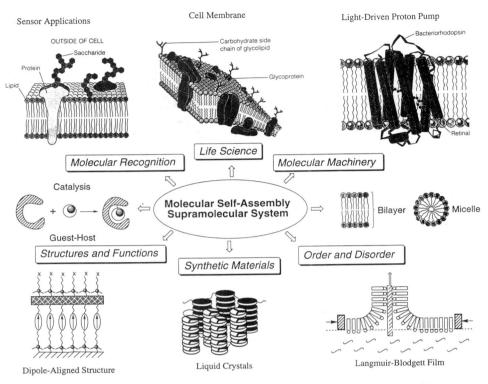

FIGURE 1 Illustration of the relationship between molecular self-assembly systems and several disciplines, such as life sciences, materials synthesis, and supramolecular chemistry.

B. Self-Assembled Monolayers: A Definition and Their Characteristics and Behavior

In essence, molecular self-assemblies are phenomena in which a hierarchical organization or ordering is spontaneously established in a complex system without external intervention. The driven "forces" of these spontaneous processes are the chemistry of intermolecular interactions. A self-assembled monolayer (SAM) is typically formed by a chemical reaction between surface-active sites (silanol for SiO_2 surfaces; aluminum for the Al_2O_3 surfaces) and the corresponding functional groups of the molecules in solution (gas or liquid phase). For example, competitive adsorption of short- and long-chain alkanethiols onto gold surfaces was conducted by Folkers et al. [1]. In this experiment, the composition of the mixed SAM was not equal to the composition of the solution because of kinetics. Typically, ionic interactions have faster kinetics than covalent bond formations. For example, let us consider two amphiphile–solid systems: ionic binding arachidic acid on ZnSe and covalent-binding *n*-octadecyltrichlorosilane (OTS) on silicon [2]. In the ionic system, complete monolayers were formed with immersion times no longer than 15 s. On the other hand, several minutes were required to form a complete monolayer of OTS on silicon. Both systems formed highly crystalline SAMs, with $\nu_a(CH_2) =$

2918 cm^{-1} and $\nu_s(CH_2) = 2850$ cm^{-1} in Fourier transform infrared (FTIR) spectroscopy measurements.

There are two principal types of SAMs: alkanethiols on noble metals (e.g., gold, silver, and copper) [3] and organosilane on oxides [SiO$_2$, ITO indium–tin oxide, and superconductors]. A third SAM type (which has not been extensively studied) consists of fattic acids on metals or metal oxides (Al$_2$O$_3$) [4]. The bonding nature of the silane–oxide systems is regarded as a covalent bond. Studies [5,6] have shown that there is still an incomplete understanding of the reactions involved in the interactions of sulfur with gold. Presumably, the nature of sulfur and gold binding is ionic and coordinative. The sulfur atoms are located in threefold hollow sites of gold lattices, a position rarely observed for covalent bonds (i.e., three equivalent covalent S–Au bonds). But it is quite possible for sulfur in this position to coordinate the gold atoms as a bridging ligand.

Although thiol–gold interactions are probably ionic and coordinative in nature, thiols such as HS(CH$_2$)$_{10}$Y, where Y = CH$_3$ or CH$_2$OH and HS(CH$_2$)$_{21}$X, where X = CH$_3$ or CH$_2$OH do bind gold and silver strong enough that they do not exchange with other thiol molecules in solution, as monitored by the contact-angle measurement.

C. Formation and Properties of Self-Assembled Monolayers

In general, SAMS are formed simply by dipping a substrate into a dilute solution that contains the desired species to be assembled. OTS has been studied extensively as a model system for the formation of SAMs on silica-based substrates (see Fig. 2). The hydration state of the oxide greatly influences the SAM formation of OTS and dimethyloctadecylchlorosilane (DMOCS). An OTS on a well-hydrated SiO$_2$ surface yields tightly packed and ordered SAMs [$\nu_a(CH_2) \approx 2919$ cm; narrow-band full width at half-maximum (FWHM) ≈ 17], whereas OTS on dry SiO$_2$ surface and DMOCS on either surface exhibit much lower coverage and more disorders [$\nu_a(CH_2) \approx 2926$–28 cm^{-1}; narrow-band FWHM ≈ 17] as a result of bulky dimethyl groups [7]. In all cases, there was no indication of islands of tightly packed molecular structures.

Upon OTS deposition, the isolated silanols ($\nu_{Si-O-H} = 3720$ cm^{-1}) decrease, whereas hydrogen-bonded silanols ($\nu = 3600$ cm^{-1}) and silanols bonded to phys-

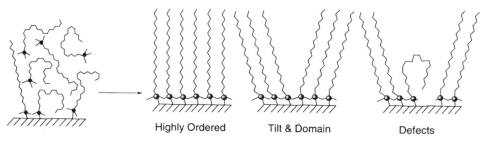

Highly Ordered   Tilt & Domain   Defects

FIGURE 2 Formation of an *n*-octadecyltrichlorosilane SAM on an oxide surface and potential mesoscale structures.

isorbed water ($\nu = 3260$ cm^{-1}) increase. These results do not unambiguously prove the formation of covalent bonds between organosilane and SiO$_2$ surfaces.

Despite the hydrophobicity observed macroscopically, water vapor can penetrate even a tightly packed, fully covered OTS and hydrogen-bond to the free silanol groups at the organic–inorganic interfaces. Curing at 150°C cross-links the silanol groups and reduces water penetration. This effect is minimal for the DMOCS systems because no additional silanols are available to induce hydrogen-bonding.

The water surface-wetting phenomenon is governed mostly by intermolecular hydrogen-binding. Surface wetting using water was found to be a sensitive function of depth of polar or nonpolar functionality also. This effect began to disappear when the functionality was three to six carbon chain lengths away from the surface. Wetting was also affected by pH and acidic or basic functional groups on surfaces.

Furthermore, the wetting studies indicate that the hysteresis of the advancing and receding contact angles depend strongly on the nanometer-scale structures of the surface domain of functional groups.

For organic thiol on gold systems [8], the adsorption kinetics of alkanethiolate on microcrystalline gold were found to follow the Langmuir isotherm over a limited concentration range. This study further found that the alkanethiolate is at an equilibrium with the free gold sites, with $K_{eq} = 1.5 \times 10^4$ M^{-1} and $K_{eq} = 1.9 \times 10^3$ M^{-1} for n-C$_{18}$H$_{37}$SH and n-C$_8$H$_{17}$SH in hexane solution.

X-ray photoelectron spectroscopy was used to study the thermal stability of a dodecanethiol SAM on Au(111) [9]. It was determined that domains grow with different sulfur lattice positions and chain-tilt azimuths. As temperature increased (50°C), larger domains formed first. The top layer of gold and the SAM (which became mobile at 100°C) then began to reorganize, maintaining a crystalline-like structure as the thiols slowly evaporated (before the SAM collapsed at 130°C). In this experiment, the collective interactions among molecules accommodated missing molecules and maintained ordered structures.

Electrochemistry of SAMs on metallic electrodes is a large field. In this chapter, we will merely mention a few examples to illustrate its applications in the studies of SAMs on electrodes. For example, a SAM of 4-(4-mercaptobutyl)-4'-ferrocenylazobenzene was formed on the surface of Au(111) or Pt supported by mica [10]. Cyclic voltammetry in 0.2 M NaClO$_4$ showed the presence of a ferrocenyl group on the Au and Pt surfaces. The Au/mica surfaces were found to enhance the surface Raman scattering by a factor of $(2–6) \times 10^3$ without additional treatment.

When combined, real-time ellipsometry and electrochemistry are potent tools to investigate SAMs such as 11-ferrocenyl-1-undecanethiol (F$_c$C$_{11}$H$_{22}$SH) on a gold electrode [11]. The thickness of the F$_c$C$_{11}$SH layer was estimated to be 2.9 nm, with a complex refractive index of $n = 1.464 - 0.074i$ at a wavelength of 632.8 nm. The electrochemical oxidation of the ferrocenyl tail yields an increased thickness of 0.3 nm. Moreover, the simultaneous uptake of anions from the electrode (1 M HClO$_4$ or 0.5 M H$_2$SO$_4$) caused an absorption increase of 10%. Electrochemistry also can be successfully combined with other techniques. For instance, surface-enhanced infrared and Raman were used to monitor the electrochemical reduction of p-nitrothiolphenol on a silver film (approximately 20 nm thick) on a Ge hemisphere prism: The disappearance of the symmetric NO$_2$ stretching mode

$[\nu_s(NO_2) = 1345 \text{ cm}^{-1}]$ and the appearance of the NH_2 deformation mode correspond to the chemical conversion [12].

For organosilane–oxide systems, there have been many attempts to chemically attach functional moieties to surface functional groups. For instance, a number of N-containing heterocycles (4-methylpyridine, 5,6-dimethyl-1,10-phenanthroline, 4,4′-dimethyl-2,2′-bipyridine) were anchored to the surface with Br-terminated alkyl silane chains [13] (see Section II.C.1).

Conducting an S_{N2}-type nucleophilic substitution of a benzyl chloride SAM with iodides was found to be very difficult for densely packed monolayers; attaining a 50% conversion took place at a 10-fold slower rate than the same reaction in solution because of steric (restrain) requirements [14]. On the other hand, the benzyl halide SAM was found to be reactive with strong nucleophiles, such as lithium salts of ethylenediamine and 3-pyridine.

SAMs do have complex behaviors; film preparation protocols play a crucial role in their successful formation [15]. Temperature-dependent electrochemistry of a redox pair, such as $[Fe(CN)_6]^{3-}$ at a gold electrode, had yielded the following order–disorder transitions in alkanethiol monolayers: The peak of this transition takes place at $T = 55°C$ with a shoulder at $T = 46°C$. This effect is not caused by electron tunneling but by the monolayer's permeability to the redox species.

In a similar experiment, a transverse shear mode (TSM) device was used to detect a secondary phase transition (which was accompanied by a change in mechanical and viscous properties) at 18–20°C for a monolayer of octadecanethiol on gold [16]. These mechanical and viscous property changes cause a negative frequency shift in the acoustic device. This secondary phase transition is reversible and repeatable, as long as the film is held to less than 50°C. This order–disorder transition is distinct from the bulk melting point. If the molecules are only physically adsorbed, only their melting points will be observed. For tetradecanol, this phase transition takes place at 38–40°C; for octadecylamine, it takes place at 55–56°C.

Brittain [17] monitored the formation and the phase change of silver *n*-octadecanethiolates at an air–water interface by using an in situ real-time ellipsometer ($d\Delta = 0.07$ rad). It was found that the interaction between the silver and organic functional group decreased in the order of RSH > RCOOH > ROH.

The studies of molecular self-assemblies can lead us to an understanding of biological systems and the mimicking of nature's strategy. For example, the orientation of negatively charged porphyrins in matrices of ammonium bilayers is determined by optimized electrostatic interactions. Electron paramagnetic resonance shows that the porphyrins were found distributed horizontally when the negative charges were evenly distributed, and were vertically incorporated into the membrane when the charges are localized on a single point [18]. The orientations of myoglobin and other heme proteins in bilayers such as phosphate amphiphile were controlled in a similar manner.

Molecular self-assemblies can also be used to isolate inorganic nanoparticles or connect inorganic nanoparticles together. For example, a bifunctional molecule (16-mercaptohexadecanoic acid) was attached to nano-sized particles (γ-Fe_2O_3) through a carboxylate head group [19]. These surface-activated magnetic particles potentially can be used to bond antibody specific cells, which then can be separated using magnetic methods. These nano-sized particles could also serve as a data storage unit when linked into an ordered three-dimensional lattice. The nanoparticles could serve as a reservoir of electrons and the data storage mechanism could be a capacitor effect by removing an electron from the nanoparticles. The choice of ultrafine particles, rather than single molecules as storage units, is because these particles are large enough so that the data stored can be easily retrieved.

D. The Structure of Self-Assembled Monolayers

The structure of a SAM is dictated by the interplay of intermolecular and molecular substrate interactions. The structure can be further divided into the structure of the head group, the chain structure, and the terminal group structure. Many techniques, which yield a subset of information of these structures, have been used to investigate these SAMs. In this section, we organize them in such a way that we will discuss vibration spectroscopy, diffraction techniques, and microscopy. Finally, we will discuss the perturbation of these structures when a chemical functional group is introduced into the system.

Of the many self-assembled systems, a well-characterized system is alkanethiols on gold. The fundamental structure of this system consists of a commensurate $(\sqrt{3} \times \sqrt{3})R30°$ overlayer (see Fig. 3), which balances the interactions among sulfur–gold, sulfur–sulfur, and chain–chain [20].

The exception is CH_3SH monolayers on Ag(111), which were found to be $(\sqrt{7} \times \sqrt{7})R10.9°$ overlayers. On the average, alkanethiols have a chain that is tilted 33° from the surface normal (α) in an all-trans conformation that is 14° away from the nearest neighbor (NN direction). It is also rotated, or "twisted," 55° about the chain (β). Figure 4 shows these characteristics in graphic form.

Studies have also revealed the orientation of alkanethiols at gold and silver surfaces. For example, Ulman used free-energy calculations to show a minima at a chain tilt of approximately 30°, with a rotation of approximately 55°, which was based on van der Waals interactions between adjacent molecules [21a]. Molecular dynamics simulations have revealed an average tilt of approximately 30°, with a nearest-neighbor distance of 5.0 Å [21b]. Nuzzo et al. [21c] estimated a tilt angle of approximately 40° for long-chain alkanethiols on gold, with a rotation about the chain of 50° based on infrared measurements. Porter et al. proposed the chain tilt to be approximately 20°–30°; these calculations were derived from infrared and ellipsometry data [21d]. Strong and Whitesides estimated the tilt angle at 25°–35° based on their pioneer work on electron diffraction studies (see Ref. 33).

1. Surface Vibration Spectroscopy

Surface vibration spectroscopy is a sensitive technique to characterize SAMs. Figure 5 shows an experimental p-polarized, infrared external reflection spectra at 86° incident angle in the 2750–3050-cm^{-1} frequency region for single monolayers of a series of n-alkanethiols [$CH_3(CH_2)_{n-1}SH$, C_n; $n = 16–19$] on polycrystalline Au(111). Vibrational mode assignments for the dominate CH_2 and CH_3 stretching vibrations are indicated as follows: d^+, CH_2 sym.; r^+, CH_3 sym.; d^-, CH_2 antisym.; r_a^- and r_b^-, CH_3 asym. (in- and out-of-plane of CCC backbone). The intensities of the methyl-stretching mode—$\nu_s(CH_3) = 2879$ cm^{-1} and $\nu_a(CH_3) = 2963$ cm^{-1} for n-alkanethiols [$(CH_3(CH_2)_nSH$] on gold—show an even–odd behavior (see Fig. 5); ν_a is strong for odd chains (where $n = 15, 17, 19, 21$) and ν_s is strong for even

(a) (√3×√3)R30°

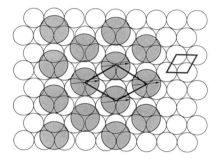

(b) (2√3×√3)R30°

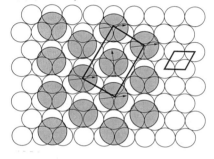

FIGURE 3 This SAM consists of an alkyl thiolate monolayer on Au(111). Arrows represent an uncanted orientation pointing toward the polymethylene chains. The large solid circles are sulfur atoms, which are located in the threefold hollow sites of gold atoms. The nearest sulfur–sulfur distance is $\sqrt{3}$ times the gold–gold distance; the lattice of sulfur atoms is rotated 30° with respect to the gold lattice [i.e., $(\sqrt{3} \times \sqrt{3})R30°$]. (a) A single chain model; (b) a two-chain model.

chains ($n = 16, 18$). This even–odd effect is caused by the terminal methyl group orientation; the methyl group points toward the direction of the surface normal for even chains.

Surface vibration spectroscopic techniques have revealed that the formation of 1-alkanethiol monolayers on gold substrates might involve the cleavage of S–H or S–S bonds [21]. However, the formation of *covalent* Au–S bonds has not been proven; this suggests that the Au–S bond might be largely ionic and coordinative in nature. Another study showed that the spectra obtained from electrochemically roughened and mechanically polished polycrystalline silver electrodes with SAMs (1-butanethiol, 1-dodecanethiol, and 1-octadecanethiol) were similar in all spectral regions [22]. These monolayer films are most ordered in the case of 1-butanethiol and 1-octadecanethiol.

Tao et al. [23] studied the self-assembling structure of a series of acid derivatives on the surfaces of silver and copper (see Fig. 6); these derivatives included 4-alkoxybiphenyl-4′-carboxylic acids (ABCA), $CH_3(CH_2)_m OC_6H_4COOH$, and 6-alkoxy-2-naphthoic acid (ANA), $CH_3(CH_2)_m OC_{10}H_6COOH$. These derivatives were

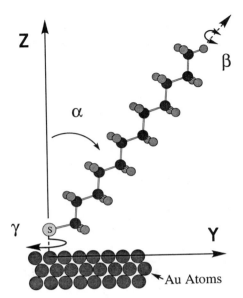

FIGURE 4 Side view of the orientation of a single, all-trans, alkyl-chain thiol on gold. The tilt angle (α) corresponds to the surface normal, whereas the twist angle (β) corresponds to a plane established by the alkyl chain axis and the surface normal vector. The angle of precession is the rotation around the surface normal (γ), away from the nearest-neighbor (NN) direction.

compared to simple *n*-alkonic acids, such as $CH_3(CH_2)_{m-1}COOH$, and *n*-alkanethiols, such as $CH_3(CH_2)_mSH$ (where $m = 15–19$), both on silver. The disappearance of the out-of-plane aromatic C–H deformation, which has a transition dipole perpendicular to the aromatic ring at 837 and 772 cm^{-1} for ABCA and at 861, 824, and 809 cm^{-1} for ANA, reveals that the aromatic plane aligns almost perpendicular to the surface. Methylene stretches $\nu_a(CH_2)$ and $\nu_s(CH_2)$ at 2918 and 2849 cm^{-1}, respectively; this indicates a crystalline-like monolayer, with its intensity in correlation with chain length.

This series of experiments also showed that strong odd–even carbon-chain effects take place in the intensity of both $\nu_a(CH_3)$ and $\nu_s(CH_3)$. Also observed was the contact angle in bicyclohexyl and hexadecane, both suggesting a chain tilt configuration. As the chain starts to tilt, the top surface will start to diffuse between odd and even chain-derivatized monolayers. The critical surface tensions of wetting, γ_c, on a monolayer of ABCA with hydrocarbon liquid yield 19 dyn/cm for even chains and 16 dyn/cm for odd chains, whereas H_2O wetting is insensitive to even and odd chains. The large tilt angle is caused by the large head groups (–COO⁻, –ArCOO⁻), which create more free volume in the carbon-chain region. As a result, the hydrocarbon chain is tilted toward its *second* nearest neighbor rather than the nearest neighbor, as in the case of *n*-alkanoic acid, thereby maximizing van der Waals interactions.

In short, on surfaces in which binding interaction is dominant and binding sites are specific, the chain tilts in accordance with binding geometry (e.g., thiol and

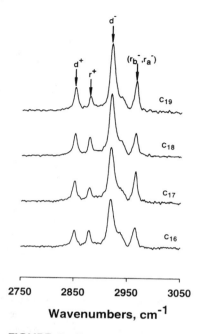

FIGURE 5 Experimental *p*-polarized, infrared external reflection spectra at 86° incident angle in the 2750–3050-cm^{-1} frequency region for single monolayers of a series of *n*-alkanethiols [CH$_3$(CH$_2$)$_{n-1}$SH, C$_n$; n = 16–19] on polycrystalline Au(111).

carboxylic acid on silver) and binding lattice. On surfaces in which binding interaction is weak or nonspecific, however, normal orientation is favored because of the chain–chain interactions. The following two examples of thiol on copper and thiol on gold further illustrate this hypothesis.

Monolayers (C$_{18}$H$_{37}$SH) on copper are typically of poor quality and are structurally ill-defined with respect to their gauche conformations. The structure of the substrate–sulfur interaction is believed to control the molecular orientations of the alkyl chains in these films.

Both H$_2$S and dimethyl disulfide dissociatively chemisorb on Au(111) and are predominantly formed with the following structure: ($\sqrt{7} \times \sqrt{7}$)R10.9°, with a nearest-neighbor distance of approximately 4.41 Å. This structure is approximately 8% denser (approximately 20.3 Å^2/RS) than those typically observed, such as the ($\sqrt{3} \times \sqrt{3}$)R30° overlayer of alkylthiol on gold. The chain–chain interactions are sufficient to outweigh any preferences caused by the sulfur–gold interactions, thereby supporting the hypothesis that surface sulfur–gold bonds have largely ionic and coordinative characteristics.

Conducting a quantitative analysis of infrared data (C$_{18}$H$_{37}$SH) that uses numerical simulation (based on an average single-chain model) revealed that the alkyl chain in monolayers on silver is all-trans, zigzag, and canted ±12° from the surface normal; these chains have a 45° twist of the plane containing carbon chains, which is defined by the tilt and surface normal [24]. However, alkanethiols on gold "fit" better on two-chain models; these possess a tilt of α = 26° and two possible

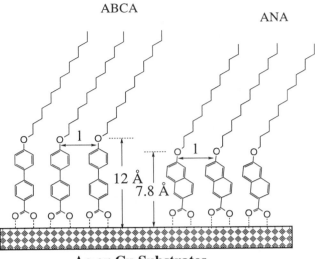

Ag or Cu Substrates

FIGURE 6 Proposed structures for ABCA and ANA on silver; note that the alkane chains are more disordered on copper. The alkane chains are in a crystalline environment. The odd–even effect of methyl vibrational modes $\nu_a(CH_3)$ and $\nu_s(CH_3)$ was clearly observed in these systems, both on silver and copper. On silver, the distance to the second nearest neighbor (diagonal direction) was $l = 2a \cos 30° = 10$ Å ($l < 10$ Å on Cu), which is the direction of alkyl chain tilt caused by the favored conformation of the O–C$_{alkyl}$ bond eclipsing the aromatic plane. ANA is in tight contact with its nearest neighbor because of the large van der Waals contour (7.6 Å); on the other hand, ABCA (van der Waals contour ~6.3 Å) can comfortably fit into the same lattice of *n*-alkanoic acid (lattice constant of $a = b = 5.78$ Å).

twists—$\beta = 50$ and 48—and the gauche configuration is allowed at the terminal methylene group (see Fig. 3).

Pemberton used surface Raman scattering to study SAMs formed from a series of 1-alkanethiols [$CH_3(CH_2)_nSH$, where $n = 3$–5, 7, 8, 11, and 17] at mechanically polished and electrochemically roughed gold surfaces [21e,22]. The C–S bond was found to be perpendicular to the silver surface. On a gold surface, the C–S bond's intensity diminished, a phenomenon which implies that the C–S bond is mostly parallel to the surface.

The ratio of $\nu(C–C)/\nu(C–H)$ values for surface-adsorbed SAMs to bulk liquids, referred to as the surface/liquid ratio, is greater than 1 for butanethiol and pentanethiol; this suggests that the carbon backbone is more perpendicular to the surface than the C–S bond (with a chain tilt of approximately 30° and a twist of 45°). As shown in Fig. 7, the methyl group is largely perpendicular to the surface for butanethiol and hexanethiol (even chains) and largely parallel to the surface of pentanethiol (odd chain). Therefore, the intensity of $\nu_{asym}(CH_3)$ mode—orthogonal to the carbon backbone—is stronger for odd chains on gold and weaker for even chains on gold, whereas the $\nu_{sym}(CH_3)$ band is strong for even chains and weak for odd chains.

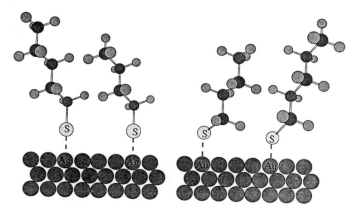

FIGURE 7 The orientation of alkanethiols on silver (left) and gold (right). The S–C bond is more parallel to the gold surfaces and approximately perpendicular to the silver surfaces. As a result, the odd–even effect on gold and silver is reversed.

On silver surfaces, however, the ν(C–S)/ν(C–H) surface/liquid ratio is greater than 1; this suggests that the C–S bond is largely perpendicular to the silver surface. Moreover, the ν(C–C)/ν(C–H) surface/liquid ratio is greater than that of those on gold, thereby suggesting a smaller chain tilt angle of approximately −15°, with a twist of 45°. The odd–even behavior of the methyl groups on silver was also observed; it was found to be the opposite of that observed on gold because of the orientation of the C–S bond. Raman studies have shown that the alkyl chain is mostly an all-trans conformation with some gauche configurations.

2. Diffraction Techniques

The first surface x-ray (synchrotron) study was conducted for docosyl mercaptan on Au(111) at the synchrotron light source [25]. This study revealed an interrow spacing of 4.32 Å (q_{xy} = 1.45 Å^{-1}). This result is in good agreement with the 4.33 Å expected, given the ($\sqrt{3} \times \sqrt{3}$)R30° structure with a domain size of approximately 70 Å (FWHM Δq_{xy} = 0.088 Å^{-1}). The tilt angle of 12° ± 1° to its nearest neighbor was estimated using the q_z scan, in which a maximum in intensity was found at q_z = 0.267 Å^{-1}. This tilt angle consists exclusively of the coherent components of the total tilt; spectroscopic techniques can measure both the ordered and random domains and hence yield a higher tilt angle.

Low-angle x-ray reflectivity (R) was measured using Eq. (1) from alkylsiloxane monolayers on the native oxide of silicon as a function of q_z, which represents the momentum change of the photon upon reflection.

$$R = R_{\mathrm{F}} \left| \rho_\infty^{-1} \int_{-\infty}^{\infty} \left\langle \frac{d\rho_{\mathrm{el}}}{dz} \right\rangle \exp(iq_z z) \, dz \right|^2 \tag{1}$$

where q_z = $4\pi\lambda^{-1} \sin \theta$, ρ_∞ is the electron density of the bulk substrate, and R_F is the Fresnel reflectivity.

Fitting of the observed interference pattern to a structure model will enable the calculation of film thickness and surface roughness. Moreover, it projects an area of each alkylsiloxane group in the plane of the surface of 22.5 ± 2.5 Å^2; this packing density is equivalent to the $(\sqrt{3} \times \sqrt{3})R30°$ of alkanethiols on gold and is similar to close-packed Langmuir–Blodgett films. The advantages of x-ray reflectivity—when compared to ellipsometry—are (1) that it does not require preassumption of an optical constant refractive index and (2) it can differentiate island or "uniform" incomplete layers.

X-ray diffraction studies (synchrotron light source) of docosane–selenol ($C_{22}H_{45}SeH$) on Au(111) showed an in-plane structure of an incommensurate structure with an oblique unit cell ($a = 5.204$ Å, $b = 4.897$ Å, and $\gamma = 120°$)—a distortion of 3% from a perfect hexagonal $(\sqrt{3} \times \sqrt{3})R30°$ lattice [26a]. The tilt is measured to be 15° in the direction of Au[$\bar{2}02$] or R30°. The oblique structure of the unit cell represents a distorted hexagonal close-packed lattice.

A recent study of $CH_3(CH_2)_9SH$ on Au(111) that used grazing incidence x-ray diffraction showed a nearly hexagonal structure of the hydrocarbon chains with dimerization of the sulfur head groups [26b]. The sulfur–sulfur distance was 2.2 Å, accommodated through a gauche bond. These results demonstrate the complexity of molecular interactions in self-assembled systems.

Another experiment [27] demonstrated that a SAM containing eight perfluorinated carbon segments [$CF_3(CF_2)_7CONHCH_2CH_2SH$] exhibited good wetting behavior [$\Theta_{adv}(H_2O) = 114°$]. NEXAFS (near-edge x-ray absorption fine structure) measurements further showed that the chain axis was nearly normal to the gold surface. In the polarized NEXAFS spectra, glancing incidence takes place when the electric field vector (of the x-ray) is nearly perpendicular to the gold surface. This effect enhanced the adsorption at 295 eV, which is assigned to a transition from the C_{1s} orbital to the C–C σ^* orbital. The absorptions at 292 and 299 eV (assigned to transitions from the C_{1s} orbital to the C–F σ^* orbital) are enhanced at normal incidence when the electric field vector is parallel to the gold surface.

Neutron reflection studies have demonstrated that OTS—or concentrated OTS—forms a SAM with the following characteristics: a thickness of 24 ± 2 Å and a coverage that ranges from 4.7 molecules/nm^2 (a crystal-like film) to 3.3 molecules/nm^2 (a liquid-like film) [28a]. The presence of an oxide and an organic layer on a substrate greatly complicates the analysis of neutron reflectivity data. These problems are similar to other techniques, such as the problems encountered when using an ellipsometer. Water always successfully penetrates the OTS hydrophobic layer and water inside of the film readily exchanges with water in bulk. This effect is consistent with other experiments (7).

Helium diffraction measurements on octadecanethiol SAMs showed a rectangular primitive unit of 8.68 Å $\times$ 10.02 Å containing four hydrocarbon chains, which were equivalent to a $c(4 \times 2)$ superlattice [28b]. Thermal annealing was found to increase the average domain area, which consists of the alignment of domain boundaries along the next-nearest-neighbor direction of the sulfur lattice and abundant superstructures superimposed on the $(\sqrt{3} \times \sqrt{3})R30°$ lattice. All domains show a surface coverage of 21.5 ± 0.5 Å^2/molecule, or 4.65 molecules/nm^2, figures consistent with the expected areal density required for close-packed chains.

3. Microscopy Techniques

Scanning tunneling microscopy (STM), which complements other techniques, reveals a superstructure of the $(\sqrt{3} \times \sqrt{3})R30°$ hexagonal lattice identified as the $c(4 \times 2)$ structure, in which $a = 3a_{Au} = 8.5$ Å, $b = 2\sqrt{3}a_{Au} = 10.1$ Å, where a_{Au} is the interatomic spacing of Au, $a_{Au} = 2.88$ Å. These twists of the trans hydrocarbon chains cause a slight difference of the chain height of film thickness, which is caused by the chain orientation of the topmost C–C bond. Several molecular conformations can coexist in SAMs.

Both STM and atomic force microscopy (AFM) have been extensively used to characterize the surface topology of organic thiols on gold. Work by Widrig et al. [29] has shown that most of these alkanethiols form $(\sqrt{3} \times \sqrt{3})R30°$ structure, which is consistent with infrared and x-ray diffraction results.

More recently, STM and AFM have been used to study the interaction between two layers of molecules. The tunneling current probability between an STM tip and Au(111) is increased by about a factor of 5 per carbon unit in alkyl chains for a given distance [30]. The tunneling current was also predicated to increase (with a large tilt angle), whereas the alkyl chain tilts away from the surface normal. While oscillating near its resonance frequency in a combined STM and AFM, the amplitude and phase of oscillation changed rapidly when the tip approached within a few nanometers of the surface. When the tunneling current made contact with the alkyl SAMs [HS(CH$_2$)$_n$CH$_3$, where $n = 12$, 11], the current was found to be in accord with a 10–20% decrease in oscillation amplitude.

The STM studies, in conjunction with molecular dynamics simulations, effectively focused on the effect of hydrogen-bonding in two polar end groups of OH (in mercaptoundecanol) and NH$_2$ (in mercaptododecylamine) [31]. This amino-terminated SAM was found to have mostly parallel striped patterns with a width of 7.5 Å, which consisted of "tails" of dimerized molecules (a result of hydrogen-bonding). This spacing between strips is almost twice as large when compared to the hydroxyl-terminated SAM whose width is 4.2 Å. The spacing also is in agreement with the configuration of chains of dimerized molecules that were spontaneously formed in the Monte-Carlo Dynamics simulation. This SAM reconstruction (which was induced by surface hydrogen-bonding) was not stable—wetting can completely dissolve it.

To examine the surface interactions between an AFM tip and a substrate (in contact mode), sharp silicon AFM tips (with a 5-nm nominal radius) and silicon substrates were used with or without monolayer modifications. These were chemically treated to yield either hydrophobic or hydrophilic surfaces [32]. Uncoated Si tips (Si–OH) with hydrophilic surfaces (Si–OH) yielded high adhesive forces between the tip and the surface (6.9 ± 0.2 nN), thereby creating spurious trench artifacts and/or artificially higher asperities. The latter effect is dependent on the scanning direction. The artifacts were suppressed by using hydrophobic surfaces, or tips, to weaken the adhesive forces. OTS created AFM tips with hydrophobic surfaces (Si–R), which yielded very low adhesive forces of approximately 1 nN. These results yielded markedly enhanced surface topography images, especially on rough surfaces.

Strong and Whitesides have used transmission electron microscopy and diffraction to study the docosylmercaptan [CH$_3$(CH$_2$)$_{21}$SH] and didocosyldisulfide

$[CH_3(CH_2)_{21}S-S(CH_2)_{21}CH_3]$ overlayer on Au(100) and Au(111) [33]. A hexagonal structure of $c(7 \times 7)$ with a short interchain distance of 4.97 Å was observed for both docosylmercaptan and didocosyldisulfide on Au(111); this result is different from the typical $(\sqrt{3} \times \sqrt{3})R30°$ structure. On the Au(100) surface, the disulfide exhibits an additional phase, such as $c((7/2)\sqrt{2} \times (7/2)\sqrt{2})$ and $c(5 \times 5)$, both of which are commensurate with the predominant subsurface (100) lattice. Both systems exhibit an in-plane structure $c(10 \times 10)$, which consists of a base-centered square array with a nearest-neighbor distance of 4.54 Å. The tilt angles for these structures are 25°–35° for $c(7 \times 7)$ hexagonal close packed, 6°–12° for $c(10 \times 10)$ square overlayer, and less than 6° for $c((7/2)\sqrt{2} \times (7/2)\sqrt{2})$ and $c(5 \times 5)$ square structures.

4. Perturbation of Basic SAM Structures

Incorporating aromatic rings alters the mesoscale structures of resultant SAMs. For instance, wetting and electrochemistry studies show that *p*-biphenylmercaptan in EtOH and *p*-terphenyl mercaptan in EtOH form stable and better organized SAMs, but not for the thiophenol in H_2O system [34].

Studies showed that introducing a sulfone group results in a preferential tilt ($40° \pm 8°$) in SAMs of $CH_3(CH_2)_{n-1}SO_2(CH_2)_mSH$ [35]. This phenomenon is a result of $SO_2 \cdots SO_2$ interactions. The disappearance of the S–O symmetric stretching mode in the SAM suggests that the SO_2 group is parallel to the gold surface. This change in molecular conformation within monolayers was induced by large in-plane dipole moments [36].

Furthermore, flat aromatic rings were incorporated into long alkyl chain SAMs [37]. These experiments synthesized two series of thiols: 1-alkylthio-4-(ω-thiolalkyl)benzene and 1-alkylsulfonyl-4-(ω-thiolalkyl)benzene. The chemical structures of these thiols appear in Fig. 8.

High-quality SAMS can be formed if the aliphatic chain above the aromatic group is greater than eight carbons, which is indicated by a narrow peak width for

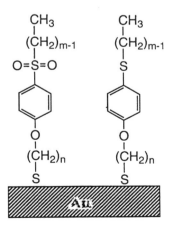

FIGURE 8 The structure of 1-alkylthio-4-(ω-thiolalkyl)benzene (left) and 1-alkylsulfonyl-4-(ω-thiolalkyl)benzene (right) on a gold substrate. An ordered methylene chain was observed when there are eight or more carbons on top of the phenyl rings ($m > 7$).

$\nu_{as}(CH_3)$, $\nu_s(CH_3)$, and $\nu_s(CH_2)$, as well as smaller hysteresis in their water contact angles [$\Delta\theta(H_2O) \approx 10°$]. The chain between the aromatic group and the gold surface appears to have little, if any, effect on the surface ordering. Indeed, the more polar sulfone series was found to be less ordered than their sulfide counterparts. The large $\nu_s(CH_3)/\nu_{as}(CH_3)$ ratio suggests that the methyl group is largely perpendicular to the surface. The chain tilt and twist are estimated to be $\alpha = 40°$ and $\beta = 43°$, respectively. To arrive at these results, the following formula was used: $\cos^2\alpha - I_{obs}/3I_{calc}$, where α is the angle between the transition moment and the surface normal for a given mode.

Similar studies were also carried out on the incorporation of a phenoxy group into 11-(*p-n*-nonylphenoxy)undecanyltrichlorosilane [$CH_3(CH_2)_8–C_6H_4O–(CH_2)_{11}SiCl_3$] and focused on the perturbation effect upon alkylsilane [38]. The SAMs were almost crystalline-like, with $\nu_a(CH_2) = 2920$ cm^{-1} and $\nu_s = 2851$ cm^{-1}, results that are approximately 2 cm^{-1} away from a perfectly ordered structure. The phenoxy possessed a tilt of approximately 20°, whereas the alkyl chain was tilted by an additional 10°, when compared to OTS. The shorter the alkyl chains on either side of the phenyl ring, the more disorder or liquid-like monolayers are formed.

X-ray interferometry/holography proved an effective technique in the investigation of profile structures of *n*-hexadecyltrichlorosilane SAMs that were chemisorbed onto the SiO$_x$ surface layer of Ge/Si multilayer substrates [39]. The investigation concentrated on the SAM's initial "as-deposited" form and its form throughout the annealing process. The results revealed that these forms consisted of small domains of highly tilted chains (relative to the normal and monolayer plane) within a positionally disordered and distorted hexagonal in-plane lattice. The annealed form of the SAM consisted of larger domains of much less tilted chains within a positionally disordered hexagonal in-plane lattice. This SAM was structurally stable over a range of 293 K to 363 K.

E. Binary Self-Assembled Monolayers

1. Mixed Self-Assembled Monolayers

When more than two components of SAM precursors were exposed to a substrate, the kinetic competition of various precursors will result in the formation of mixed SAMs. Depending on the intermolecular forces among individual components, the resulting SAMs may be mixed at a molecular level if they are compatible or they may form into domains of various sizes. For example, hydrophobic surfactants tend to separate themselves from hydrophilic molecules and each form their own domains. The final structure is determined by competition of the chemical interactions among the molecules and the molecular bonding affinity to the substrate.

By exposing an isooctane solution containing both a thiol [$L_1 = Cl(CH_2)_{11}SH$] and a carboxylic acid [$L_2 = CF_3(CF_2)_8COOH$], a simultaneous formation of orthogonal SAMs on gold and alumina (oxidized from aluminum spontaneously in air) surfaces occurred. As illustrated in Fig. 9, thiols were only anchored to gold, and carboxylic acids were only absorbed by alumina; no cross-reaction took place. The formation of these orthogonal SAMs reenforced that the driving force for SAMs consists of selective chemical interactions, such as ionic attractions, coordination, hydrogen-bonding, and covalent bonds.

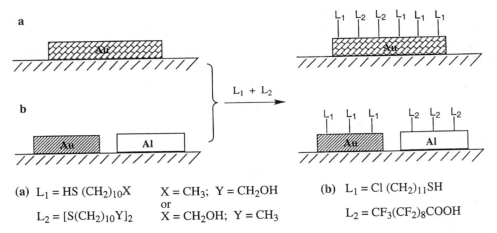

(a) $L_1 = HS(CH_2)_{10}X$ $X = CH_3$; $Y = CH_2OH$ (b) $L_1 = Cl(CH_2)_{11}SH$

or

$L_2 = [S(CH_2)_{10}Y]_2$ $X = CH_2OH$; $Y = CH_3$ $L_2 = CF_3(CF_2)_8COOH$

FIGURE 9 Chemically favorable interactions and competition govern the final structure of a SAM. (a) Competitive absorption of two species on a single substrate may result in either domains or a system that is mixed at the molecular level. (b) Selective chemical interactions will "dictate" the self-assembly process and generate orthogonol SAMs.

These studies were extended to mixed (solution) systems of disulfide and thiols, such as $HS(CH_2)_{10}CH_2OH$ + $[S(CH_2)_{10}CH_3]_2$ or $HS(CH_2)_{10}CH_3$ + $[S(CH_2)_{10}CH_2OH]_2$ [40]. The binary mixtures of these molecules were selected such that one component was terminated by a hydrophobic methyl group and one by a hydrophilic alcohol group. In the solution of mixtures of a thiol and a disulfide, adsorption of thiol was strongly preferred (~75:1).

The coadsorption of $HS(CH_2)_{11}OH$ and $HS(CH_2)_{19}OH$ revealed that these compounds disperse at the molecular level, rather than phase segregate into macroscopic islands [41]. The maximum hydrophobic surface formed as a result of this dispersion is a ratio of $R = [HS(CH_2)_{11}OH]/[HS(CH_2)_{19}OH] = 6$ in solution.

A binary SAM that consisted of OTS and 11-(2-naphthyl)undecyltrichlorosilane (2-Np) was prepared by backfilling a partial monolayer of one component with the solution of another [42]. Both methods produce monolayers with similar molecular orientation and tilting of naphthyl groups. However, this ordering is significantly different from partial SAMs of either component, which possess more liquid characters. These orderings were verified by using an ultraviolet–visible naphthyl tag that was perpendicular to the surface. This process revealed a shift of CH_2 stretching to a lower frequency, as well as the disappearance of the full excimer.

2. Patterned Self-Assembled Monolayers

Patterned SAMs were an extreme situation of mixed SAMs when they were purposely controlled in such a way that they formed into large localized domains. A number of techniques can be used to create patterned SAMs of thiols and silanes on substrates such as gold and oxide. These techniques include microwriting, micromachining, stamping, and ultraviolet microlithography.

Self-assembled monolayers of *n*-octadecanethiol on GaAs and *n*-OTS on SiO_2 were patterned with electron beams at a dose rate of 100–160 mC/cm^2 to generate a line pattern that consisted of a 50-nm period. AFM effectively removed all or part of the OTS layer after it was exposed to electron beams. The pattern was then transferred onto GaAs and SiO_2 by using an ammonium hydroxide (aq.) and buffered HF etch, respectively. AFM measurements of the depths of these etches were determined to be approximately 30 nm [43].

A more selective etchant was used [44] to transfer patterns of alkanethiolate SAMs onto gold substrates. Known as $S_2O_3^{2-}$/ferri/ferrocyanide, this etchant completely removes bare gold within a shorter time interval with a higher edge resolution. It also yields fewer defects in the SAM, as compared to CN^-/O_2. Copper and silver can be patterned in a similar fashion by using this methodology. The etching solution uses thiosulfate to coordinate the metal ions and ferricyanide as the oxidant [44]. This etchant is less toxic, less hazardous, and has a smaller environmental impact than traditional etchants.

Ultraviolet light radiation also can be used for patterning SAMs, especially on halogen-terminated SAMs. For example, ultraviolet light induced the loss of a halogen atom (chlorine or iodine) from benzyl halide SAMs [45]. The subsequent nucleophilic reaction led to the development of a highly general and efficient process that produces patterned SAMs with specific chemical functionalities. As illustrated in Fig. 10, this chemical template's effects can be used in microlithography, multianalyte microsensors, and molecular devices.

A more straightforward way of generating a pattern on surfaces is by using a poly(dimethylsiloxane) (PDMS) stamp. The stamp can be created by cross-linking PDMS polymer on a photolithographically patterned surface. This process transfers the photolithography on the surface to the PDMS stamp. The PDMS stamp can then be used to transfer organic "ink" to the substrates. For example, patterns of 60-μm OTS parallel lines on silicon surfaces were generated. The potential viability of this technique for integrated microelectronics was explored by patterning (Pb,

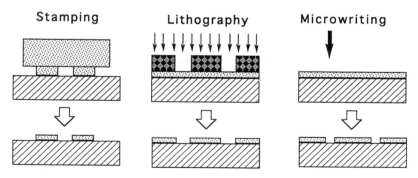

FIGURE 10 Techniques used to pattern SAMs include stamping and printing, which transfer organic molecules to the substrate (left); lithography, which involves the irradiation of the exposed SAMs through either patterned photoresist or mask (center); or microwriting, which could be either positive writing (pen filled with thiols) or negative writing (an AFM tip scratch at the surface) (right).

La)TiO$_3$ and LiNbO$_3$ layers (100-μm feature) on sapphire, silicon, and indium–tin oxide [46].

Similarly, a PDMS stamp was used to transfer patterns onto a gold surface, followed by an immersion in ethanolic solution of 50 mM [HOOC(CH$_2$)$_{15}$S]$_2$ to fill the unpatterned surface area [47]. Subsequent wetting (H$_2$O) generated self-organized liquid structures down to 10 μm^2 in size. The dimensions of the patterned structure were approximately 2 nm perpendicular to the plane of the SAM. This strategy for assembling liquid features could potentially prove useful in building and controlling three-dimensional structures near the surface of SAM films.

Patterned SAMs have a number of applications. First, to modify the hydrophilicity of surfaces, mixed SAMs [48] were generated by incorporating oligomers of ethylene glycol [HS(CH$_2$)$_{11}$(OCH$_2$CH$_2$)$_n$OH, where n = (3–7)] into 1-dodecanethiol [HS(CH$_2$)$_{11}$CH$_3$]. Collective evidence from x-ray photoelectron spectroscopy, measurements obtained from the contact angles, and ellipsometry suggest that there is substantial disorder in the oligo(ethylene glycol)-containing segment. Moreover, it was found that SAMs that terminated with oligo(ethylene glycol) were resistant to protein adhesion [R = –(OCH$_2$CH$_2$)$_6$OH], whereas SAMs that terminated with nonpolar (R = –CH$_3$) and ionic (R = CO$_3^{2-}$, –PO$_3$H$^-$, 2-imidazolo) groups promoted protein adsorptions [48]. Proteins that underwent evaluation included pyruvate kinase, bovine carbonic anhydrase, RNaseA, fibrinogen, fibronectin, streptavidin, bovine IgG, and bovine serum. This feature suggests that these SAMs may prove a useful model system for studying the adsorption of proteins onto organic surfaces.

Second, to generate mixed, two-component SAMs with feature sizes of approximately 12 μm, SAMs of *N*-(2-aminoethyl-3-aminopropyl)trimethoxysilane (EDA) were patterned with deep ultraviolet light at 193 nm (photolithography) [49]. This procedure was followed by the treatment of tridecafluoro-1,1,2,2-tetrahydro-octyl-dimethylchlorosilane. Neuroblastoma cells, explanted rat hippocampal cells, and human umbilical vein endothelial cells were found to selectively adsorb and grow on the EDA region of the mixed SAMs. This phenomenon might provide a new methodology for cell rearrangement, prosthetic implants, and tissue repair.

Finally, a monolayer of bifunctional silane [(CH$_3$O)$_3$Si(CH$_2$)$_3$X, X = –CH$_3$, –SH, and –NH$_2$] was used as a silica surface modifier to control the vapor deposition of gold films [50]. Thiol (X = SH) and amine-terminated propylsilanes induced the formation of similarly sized gold crystallites—the gold films on the amine-terminated surfaces were substantially more conductive than those on the thiol-terminated surfaces (seven orders of magnitude more for 8.5-nm-thick gold film). For these short-chain systems, well-organized structures are not expected and these results should not be generalized to long alkyl chain systems.

Because the feature sizes of many patterned SAMs are on the scale of 1 μm, many techniques can be used to verify the formation of patterned structures. These techniques include scanning electron microscopy, atomic force microscopy, and scanning tunneling microscopy. Spectroscopic techniques were also used to image the pattern formation of SAMs; these included second-harmonic techniques, x-ray photoelectron spectroscopy, and a secondary ion mass spectrometry (SIMS). For example, a secondary ion mass technique was used to image SAMs of (11-mercaptoundercanoyl)ferrocene and deuterated ferrocene [51]. SIMS maps of Fe$^+$,

$[(C_5H_5)Fe]^+$, Au^+, and Si^+ indicate that the above-mentioned thiol molecules only absorbed on gold, not on the surface of Si_3N_4.

II. CHROMOPHORIC SELF-ASSEMBLED MULTILAYERS FOR NONLINEAR OPTICAL APPLICATIONS

A. Introduction

For second-order nonlinear optical applications to function successfully, the materials must be noncentrosymmetric, with all the dipolar molecules aligned in the same direction. This requirement presents an extremely challenging situation for a chemist: Although it is usually possible to synthesize the desired chemical structures, it is impossible to align them in planned mesoscale or macroscale structures. Poled polymers [52–54] represent promising approaches to high-efficiency second-harmonic generation (SHG) materials. Nevertheless, net achievable nonlinear optical (NLO) chromophore alignments, number densities, and alignment temporal stabilities are still significant issues in these poled polymeric materials [55]. Molecular self-assemblies offer a new and innovative approach to dipole-aligned structures at the molecular level that apply the sequential construction of covalently bound multilayers containing NLO-active chromophores [56].

Designing and constructing artificial supermolecular self-assembly arrays with a planned structure and desired physical properties have attracted growing interest during the past few years [57]. Dramatic advances made in superconducting materials during the past decade have led to an increased awareness of the importance of the art of synthetic strategy and methodology. For example, both NLO materials and high-T_c superconductors [58] achieve their unusual properties by the structures they adopt. Superconductor structures can be neither adequately predicted nor controlled with ease, thereby posing a difficult challenge to the successful design and synthesis of new materials.

On the other hand, there are methods that can design and control the structure of NLO materials; successfully applying these methods will yield many new NLO materials. Although scientists are presently investigating a variety of organized molecular systems in bulk phase and at interfaces, ordered monolayer films transferred onto solid supports from an air–water interface by the Langmuir–Blodgett (LB) method seem to be favored as one of the most attractive candidates for constructing ordered molecular assemblies [59].

Langmuir–Blodgett films [60,61] offer unique opportunities for direct external handling and control because the films are insoluble and float at the air–water interface. Unfortunately, traditional LB films suffer from several drawbacks [62]:

- Weak physical adsorption or noncovalent interactions between interfaces
- The large number of mechanical manipulations required to produce complex structures
- The presence of scattering microdomains and structure irregularities

These drawbacks are caused by two factors: (1) LB films are metastable structures that can be easily disrupted and (2) the films are prepared via the transfer of preassembled (not covalently linked) monolayers onto the substrate. Therefore, LB films tend to be fragile and short-lived.

The LB methodology does provide a convenient starting point for dipole-aligned molecular assemblies that yield controlled molecular organization and device applications. Molecular self-assembly technology has been recognized as a simple albeit superior path to nanostructured materials and "molecular" devices [63]. Self-assembled monolayers [64] or multilayers [65] with unsophisticated structures, such as alkylsilane on glass or alkylthiol on gold (discussed in previous sections), provide some knowledge to fabricate highly aligned NLO thin films.

Self-assembled monolayers can overcome the problem of weak physical interaction (such as dipole–dipole) among themselves and between interfaces, provided that they do not introduce strong steric interactions to prevent them from forming ordered structures. It is generally true, however, that molecules with high-even-order nonlinearities usually require a large ground-state dipole moment [66]. Therefore, it would be an instructive and unique challenge to sequentially react and grow these dipolar chromophores from monolayers to multilayers on a defined surface in a self-assembled manner, with a general alignment along the surface normal in structurally interlocked configurations.

Direct adsorption on (then bonding) solids may be regarded as an optimized route to molecular organization that combines desirable features of "highly artificial" nanostructures with those of the "natural" self-association of molecules occurring in biological systems [67,68]. A solid actively participating in forming the film assists molecular organization, orients reactants into molecular assemblies of arrays, and fabricates the film in a more facile and controllable manner (as compared to when self-association takes place in bulk liquids). Using primary covalent forces in self-assembled films on solid surfaces results in structures of greater stabilities than those structures arising from LB methods or self-assembly in a bulk liquid phase.

It is extremely important to understand the interactions on molecular assemblies because such an understanding would open the way to developing an array of new applications, such as two-dimensional chemistry [69], monolayer catalysts [70], nonlinear optical materials [71], cell membranes, and molecular electronics.

This section addresses the building of organic polar superlattices of self-assembled, highly aligned multilayers of high-β molecular chromophores as thin-film NLO materials. Noteworthy features include a stilbazole chromophore precursor [72] in which layer-building quaternization affords both an anchored strong electron acceptor [73] and diagnostic changes in the optical spectrum. In addition, polymeric layers or inorganic silica layers are introduced transverse to the stacking direction as a cross-linking layer to promote structural stability. Similarly, Katz has used an azo dye chromophore sandwiched between zirconium phosphate/phosphonate layers [73c].

B. Strategy for Covalently Linked Polar Self-Assembled Multilayers

A counterpart to genetic engineering, molecular engineering is aimed at constructing artificial systems by using cooperative molecular components as functionalized entities. In the context of nonlinear optics, photons and electrons are the only "mobile" parts of these systems functioning as designed, built-in components. The unique feature of the molecular architecture is that the molecular organizates or self-assemblies are not just monolayers, micelles, or aggregates, but molecular and supermolecular systems with built-in donor and acceptor pairs oriented in the surface normal direction. These donors and acceptors are linked by an excellent charge transfer medium—organic π conjugation. Upon excitation, large induced polarization will take place, yielding ultrahigh even–order nonlinearities. In this system, chromophores are infrastructures and electrons are the "machinery," with photons as the raw materials and the SHG light wave as the ultimate product.

Crucial for SHG efficiency are noncentrosymmetric structures that have a very high degree of chromophore alignment. In this section, multilayer structures are built on the strategy that a defined surface is acentric by definition. This novel approach enables the chromophore molecules in solution to couple their molecular functional groups with active functional sites on the surface.

Inorganic oxides (in the form of crystals or thin films) were selected as the desired substrates because oxides yield satisfactory hydrophilic surfaces with a high density of hydroxyl groups and possess a clear window in the visible region [74]. These hydroxyl groups react readily with a variety of silyl coupling reagents to introduce desired coupling functionalities that, in turn, facilitate the introduction of organic high-β NLO chromophores.

This sequential procedure consists of two steps. In step 1, the reactive silyl groups attach to the hydroxyl surface; in step 2, the chromophore attaches to the silylated surface. This process effectively isolates the two chemical functional groups, which otherwise will polymerize before surface reaction, unless the chemical functional groups are designed such that surface reactions can only be photoinduced or initiated electrochemically.

Figure 11 summarizes this two-step strategy, where Cp and Ch represent coupling spacers and high-β chromophores, respectively. Cp has a great degree of flexibility, whereas Ch is rather limited because the most effective charge transfer chromophores are donor–acceptor pairs that have amino or substituted amino groups combined with cyano or nitro groups. Amino groups allow the introduction of other functionalizations, but cyano and nitro groups are essentially a "dead end" from a synthetic point of view (i.e., they have no available reactive functionality). Therefore, the challenge was to identify a group or structure that has the desired electron–acceptor properties; it also must have the synthetic flexibility to propagate multilayer growth.

The pyridinium structure possesses both characteristics. The nonaromatic analog to the pyridinium cation, trimethylammonium ($-N^+Me_3$ group), has an electron-withdrawing parameter, $\sigma_p = +0.82–0.96$, derived from a Hammett free-energy relationship [75], which is even larger than that of the nitro group ($\sigma_p = +0.81$), the best simple acceptor presently used in organic NLO materials. Although the nitro and cyano groups are monofunctional groups, the pyridine structure has bifunctional properties (electron-withdrawing and synthetic bonding abilities). On the other hand, the phosphonates and sulfonates are moderate electron-withdrawing groups and can also be used for multilayer constructions because of their available synthetic functionalities. These characteristics enable the incorporation of these structures onto the surface without blocking subsequent layer formation and hence allow great flexibility in the molecular architectural design.

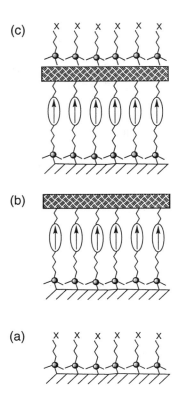

FIGURE 11 Schematic showing the sequential growth of self-assembled chromophoric multilayers: (a) the introduction of a coupling layer on defined substrates (SiO_2); (b) construction of an NLO active chromophore layer on top of a coupling layer, followed by cross-linking to reconstruct/regenerate an active layer for the coupling layer; (c) repeat of above processes of (a) and (b) to achieve multilayer systems.

Coupling molecular functionalities into purposely oriented organized entities provides synthetic steric interlocking of the rod-like chromophore, but such interlocking is subject to dipole–dipole interactions that tend to lead to randomization of orientation. To keep the chromophores "standing up" on the surface of the substrate, further assistance is preferred. This assistance is provided by structural interlocking or cross-linking of the chromophore arrays into a three-dimensional network in a step subsequent to the formation of the chromophore layer (see Fig. 11). This structural interlocking proved to be very effective in terms of preventing the dipole randomizations. The structurally interlocking layer can be inorganic, or organic, or polymeric.

C. Self-Assembling Chemistry on Surfaces

1. Coupling Layer Formation on Surfaces

In previous sections, we have illustrated how hydroxyl groups on an oxide surface will anchor and orient any molecules with functionalities that can couple or bond to them. Bifunctional silane-coupling reagents, which have a general formula of $YRSiX_3$ (X = Cl, Br, I, and OMe), are ideal anchors. The halogen (X) is involved in hydrolysis to form a covalent linkage with the inorganic substrate; the nonhydrolyzable organic group (Y = Cl, Br, I, and NH_2) possesses a latent functionality that will enable the coupling agent to covalently bond to an NLO active chromophore in subsequent steps. There are numerous suitable organosilane coupling agents, with a few examples shown in Fig. 12.

The formation of these coupling layers is similar to those of alkylsilane on glass; the formation can be prepared by simply immersing the freshly cleaned substrates into an organic solution containing various concentrations of organosilanes. The completion of this coupling reaction can be verified by the water contact angle, $\theta_a(H_2O)$, because halogenated hydrocarbons are very hydrophobic. The change of $\theta_a(H_2O)$ is remarkable from a cleaned smooth substrate contact angle of $\theta_a(H_2O) < 15°$, to a silylated surface with a very high contact angle (typically $>82°$). The reactions are also typically verified with polarized, surface FTIR; in most cases, various internal reflection techniques are chosen to study SAMs on silicon wafers.

2. Synthesis and Characterization of Chromophore Layer

Forming chromophore layers required a specific chemical reaction. For example, a stilbazole chromophore layer was generated by refluxing the silylated substrates in a solution of *n*-propanol containing 1 m*M* *p*-bis(3-hydroxylpropyl)aminostyrylpyridine (see Fig. 13). The reason that *p*-bis(3-hydroxylpropyl)aminostyrylpyridine was selected as the desired chromophore is that the pyridine group is bifunctional and has highly nonlinear optical properties [73].

Although pyridine derivatives generally react rapidly with various halogenated organics in solution, surface-bound organic halogens were found to be less reactive than their solution counterparts because of surface steric reasons. Therefore, the choice of more reactive halogenated hydrocarbons is desired; sample choices include alkyl bromoacetate, benzyl iodide, or iodoalkane.

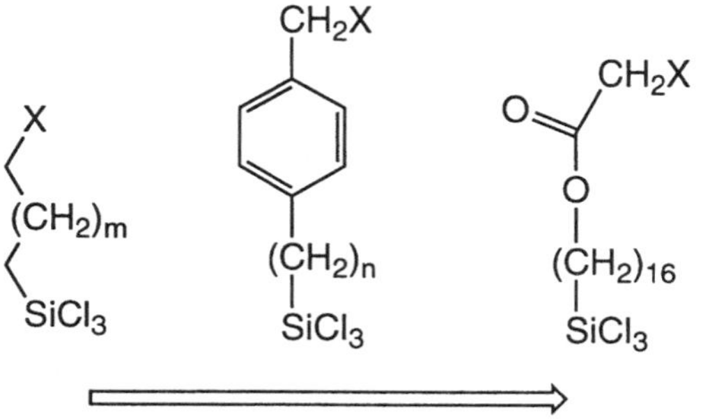

Increase in reactivity towards S_{N2} reactions

X = Cl, Br, or I; m = 1, or 15; n = 0, 2, or 12

FIGURE 12 Molecular structures of coupling layers. For S_{N_2} reactions, the reactivities decrease in the following sequence: bromoacetate > benzyl bromide > bromoalkane.

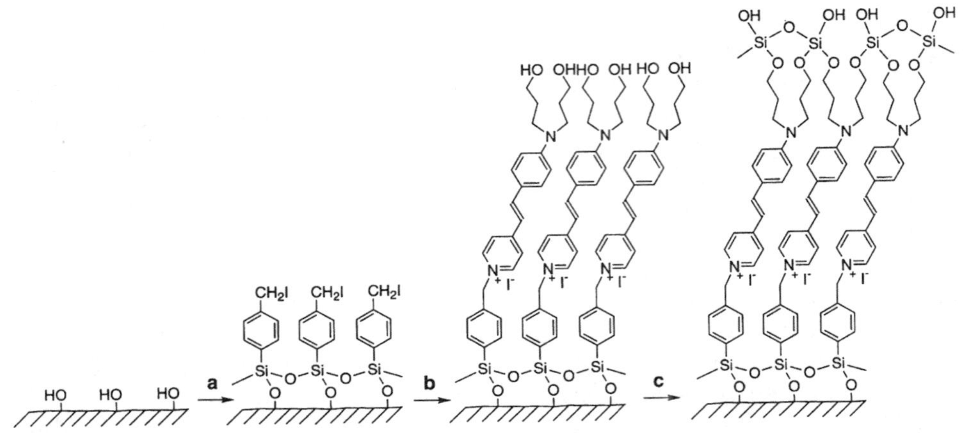

FIGURE 13 Construction of a self-assembled, chromophoric, multilayer, NLO thin films. (a) *p*-Iodomethylphenyltriiodosilane in benzene at 25°C; (b) reflux in *n*-PrOH containing 5 m*M* *p*-bis(3-hydroxypropyl)aminostyrylpyridine; (c) Cl₃SiOSiCl₂SiCl₃ in tetrahydrofuran.

In situ SHG experiments were used to monitor the reaction kinetics of the benzyl halide monolayers with stilbazole. Because the SHG only probes the dipole-aligned molecules on surfaces and neglects the bulk solution during in situ measurements, SHG is an ideal method for monitoring the growth of ordered chromophoric self-assemblies. At 60°C, these results show that the initial quaternization reaction is extremely rapid during the first 10 h; it then slows down as the majority of the surface sites are occupied and eventually reaches a steady state at approximately 75 h. This characteristic growth of chromophoric SAMs with 50% surface coverage (~2 molecules/nm^2) fit well into a biexponential with a fast growth phase, followed by a slow growth phase.

Katz selected a moderate NLO acceptor, phosphonate, as an anchoring group [73c]. These phosphonate groups react readily with inorganic metal ions such as $ZrOCl_2$. The inorganic $ZrOCl_2$ layer was introduced by first treating silica surfaces with 3-aminopropyltriethoxysilane, followed by phosphorylation with $POCl_3$. Zhang et al. [76], however, reacted the surface-terminated amino group with a series of 4-substituted benzaldehyde to generate monolayers and multilayers for SHG measurements. Alternatively, the amino-terminated surfaces were treated with glyoxal, followed by a reaction with a series of 4-substituted anilines. Moderate SHG [$\chi^{(2)} = 7.9 \times 10^{-9}$ esu] was observed; the thermal stability of the films, measured by second-harmonic reflection, was found to be stable until 130°C. Choi et al. prepared chromophoric SAMs with ionic bondings and observed moderate SHG signal caused by low surface coverage and poor orientation [77]. Page and Neff have incorporated NLO chromophore between a phosphonic acid moiety and a phosphonate ester group in an attempt to prepare polar structures [78].

Yokoyama et al. used a polyimide with a polar azobenzene pendant and a surfactant alkylamine salt from a hybrid LB film [79]. The LB film was immersed in an OTS to replace the alkylamine surfactant and to form a hybrid SAM. In the hybrid monolayers, the hydrocarbon chain tilt angles relative to the surface normal direction were estimated at approximately 16° by using FTIR. SHG measurements yielded a tilt angle of 39° for the azobenzene unit.

D. Temporal Stability and Cross-linking of Chromophoric SAMs

Second-harmonic generation measurements revealed that all films are metastable because the intensity of SHG decays with time. Decaying SHG intensity can be slowed down if the substrates are immersed in solvents such as water or alcohol. Such a procedure helps solvate the cation–anion pairs and fills in voids between chromophore units.

The decay of SHG intensity could be caused by (1) chemical decomposition of the monolayer species or (2) reorientation of the rod-like dipole chromophores. No disappearance or shift in peak position took place in the ultraviolet–visible spectra; rather, the intensity of each absorption peak increased with time at first, then leveled off. This phenomenon suggests that the chromophores are reorienting themselves or undergoing a photo-induced chemical reaction to minimize the dipole–dipole repulsions.

In submonolayer systems, the dipolar molecules lower their potential interaction energy by falling to the surface, thereby reaching more stable configurations because the dipoles are close to the paired-up configurations and they become more densely packed. Meanwhile, their transition moments (most of the charge transfer transition moments are along the dipole direction) become more and more parallel to the incoming ultraviolet–visible light polarization directions. Hence, the absorptions are more pronounced while these rod-like molecules fall to the surface.

The final polar structure is determined by the balance of attractive van der Waals forces, dipole–dipole repulsions, and chromophore–surface interactions. These metastable structures can be disrupted if no attractive forces are introduced; this is done by cross-linking the chromophores with various linker molecules. These linkers include oligomers such as octachlorotrisiloxane, polymers such as polyvinyl alcohol, and multivalent metal ions such as Zr phosphonate. These cross-linking procedures can be repeated until dipole–dipole interactions in the chromophore layer are stabilized.

Cross-linking can be used not only to strengthen the multilayer structure but also to explore the distance between two nearest neighbors on the surface by monitoring the point in which effective cross-linking is evident. Designing a series of cross-linkers with various lengths and studying their reactivity with hydroxyl groups on the surface of the chromophore layer enable the probing of the distance between the two adjacent hydroxyl groups in the chromophore layer. This may be extrapolated to give the approximate distance between two adjacent chromophore molecules on the surface. Furthermore, cross-linking not only prevents randomization on the surface but also "stretches" the rod-like chromophores and causes them to be more aligned on the surface. These orientational effects were detected by SHG techniques. The SHG increase caused by cross-linking varies from sample to sample; poorly oriented, low-chromophore-density samples usually exhibit a large increase—as high as 260%, whereas high-chromophore-density samples exhibit little effect [80]. This concept was repeated by Kakkar et al. 3 years later [81].

E. Synthesizing Self-Assembled Multilayers

Multilayers are prepared by repeating the basic sequential procedure for each layer. The general strategy in this procedure is to minimize harsh reaction conditions (high temperature), reduce reaction time, and maximize chromophore density and structure regularities. In the stilbazole system, organic multilayers were built up in the following sequence (see Fig. 13): silane-coupling (Cp) layer, NLO active chromophore (Ch) layer, and cross-linking (Csl) layer (Si and PVA or Si only) for each cycle. For the stilbazole system, values of $d_{33} = 200 \times 10^{-9}$ esu and $d_{33} = 700 \times 10^{-9}$ esu were found for multilayer and monolayer materials, respectively.

Similarly, Katz employed a procedure consisting of a repeating cycle of phosphorylation ($POCl_3$) on an amino/hydroxyl-terminated surface, followed by a zirconation with $ZrOCl_2$, and finally bound to the phosphonate of NLO active azo benzene chromophores (see Fig. 14). This cycle was repeated successfully 34 times; moreover, good structural regularity was indicated by the linear growth of optical absorption, as well as thickness as a function of the number of layers (15.7 Å per layer). The d_{33} value was determined to be 25×10^{-9} esu. The SHG signal was found to be stable up to 150°C, at which point the chromophore decomposed.

One critical concern is whether each layer generates enough functionalities to ensure the formation of the next high-quality layer. The number of layers is unlimited by regenerating the surface and "repairing" the defect of unreacted sites through cross-linking (if the cross-linking layer can successfully repair all the defect sites). A defect will grow exponentially if no repairing layer is introduced between each cycle.

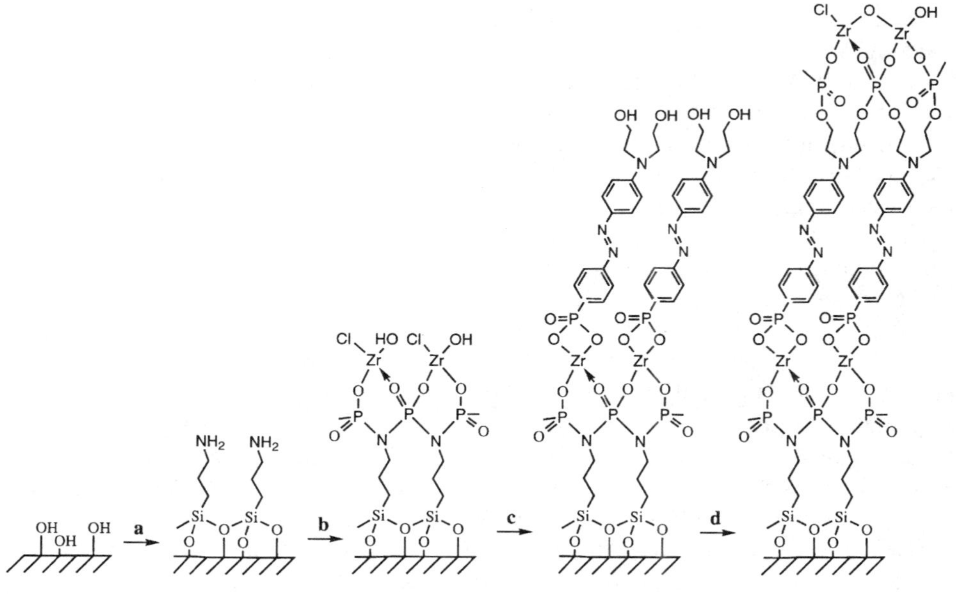

FIGURE 14 Idealized azo dye multilayer structures: (a) aminopropyltrimethoxysilane; (b) 0.2 *M* $POCl_3$—0.2 *M* collidine in acetonitrile with 5 m*M* $ZrOCl_2$ (aq.); (c) 1.5 m*M* 4-{4-[*N,N*-bis(2-hydroxyethyl)amino]phenylazo}phenylphosphonic acid; (d) same as (b).

F. Synthesizing Supermolecular Self-Assembled Monolayers

Tethering these linear molecular units through methylene bridges yields a calixarene-based supermolecule that consists of four π-conjugated molecules with a D-π-A structure, where D and A are the electron donor and acceptor, respectively, and π is a π-conjugated segment or sensitizer [81]. To achieve optimum molecular β, the orientation of each D-π-A unit should be fixed in nearly the same direction, thereby forming a molecular "pyramid." The calix[4]arene was synthesized with a frozen pyramid conformation by attaching ethoxyethyl groups to the phenol groups (see Fig. 15).

A ^{1}H nuclear magnetic resonance (NMR) spectrum in CDCl$_3$ confirmed a rigid "cone" conformation in which the D-π-A dipolar units were all aligned in the same direction; two doublets ($J = 12$ Hz) corresponding to equatorial and axial methylene protons, respectively, were observed [81]. Long ethoxyethyl groups prevent flip-over of the individual NLO units in the imine derivative of calix[4]stilbazole, and the "cone" conformation is virtually frozen. With a molecular building block that has a rigid structure and a fixed dipole orientation, it is expected that a SAM of calix[4]stilbazole derivative will have higher alignment and better thermal and structural stability than SAMs using a single chromophoric unit.

The aforementioned methodologies serve as a foundation for constructing covalently bonded SAMs of this molecular pyramid. The formation of these pyramids is similar to those of chromophoric SAMs. The silica surfaces were first treated with *p*-chloromethylphenyltrichlorosilane, followed by the quaternization reaction with the imine derivative of calix[4]stilbazole in toluene at 110°C.

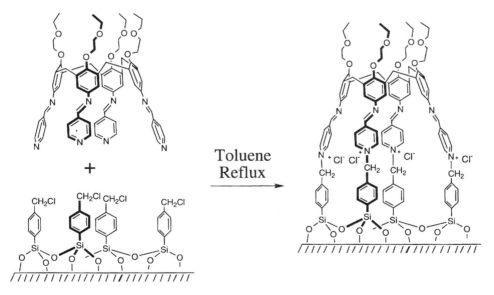

FIGURE 15 Synthetic strategy for constructing polar monolayers that consist of molecular pyramids on silica surfaces.

No pyridine mode at 1597 cm^{-1} was observed, indicating that the surface quaternization was complete—it left no dangling stilbazole imine structures in the SAM. The advantages of four-point C–N attachment are that the molecular orientation angle is fixed by the cone configuration and that it does not depend on the surface coverage. The surface coverage of this calix[4]stilbazole was found to be 0.92 molecules/nm^2, which corresponds to a densely packed monolayer consisting of calix[4]stilbazole pyramids with bases of ~100 Å^2, as derived from ultraviolet and visible optical spectra and a space-filling model. An average molecular orientation of $\psi \sim 35° \pm 5°$ was deduced by modeling the angle-dependent SHG response ($n_{\text{film}} = 1.7$); an absolute magnitude of d_{33} value of 191×10^{-9} esu was derived from these calixarene-based monolayers at 862 nm by calibrating to reference Y-cut quartz.

G. Characterization of Chromophoric Self-Assembled Monolayers

1. Surface Wetting and Contact-Angle Measurements

Measuring the contact angle is one of the most sensitive techniques [82] used to examine the nature of surface functional groups and hence the monitoring of alternative multilayer deposition procedures. Surface silanol groups effectively bind to water through hydrogen-bonding and yield excellent hydrophilic surfaces and, hence, lower water contact angles [low $\theta_a(\text{H}_2\text{O})$ values], whereas organic halogens are typically hydrophobic and consequently result in higher contact angles [high $\theta_a(\text{H}_2\text{O})$ values].

The bishydroxylpropyl groups on the outer surface of the chromophoric layers process both characteristics and yield moderate $\theta_a(\text{H}_2\text{O})$ values. As the organic multilayers are built up, cyclic change behavior of water contact angle was observed, verifying that the growth of the multilayer structure proceeds smoothly. The contact angles are 15°, 82°, 55°, 17°, and 84° for the cycle of glass surface, Cp layer, Ch layer, Csl layer, and again the Cp layer (on the Csl layer), respectively.

2. Surface-Polarized Internal-Reflection Infrared Spectroscopy

Surface infrared (IR) spectroscopy is a powerful tool used to characterize monolayer thin films. The grazing-angle IR (discussed in Section I) for organic molecules on metal surfaces cannot be applied to chromophoric SAMs because these SAMs are typically grown on oxide substrates that have ideal optical transparency. To solve this problem, a new technique, internal-attenuated reflection (IAR), was introduced (see Fig. 16). Chromophoric SAMs are deposited on silicon wafers, which are pressed against internal reflection hemisphere crystals, typically Ge or ZnSe. Surface infrared spectra with monolayer sensitivity were obtained from a single reflection at the Ge–SAM–Si interface. This technique allows the incidence angle to be adjusted from 5° to 85° in either *p*- or *s*-polarized mode, whereas the grazing-angle technique used for SAMs on metal surfaces is restricted to a high incident angle and *p*-polarization.

The formation of covalent bonds between self-assembled layers was directly proved using IAR techniques. For chromophores containing pyridyl groups, a strong IR mode is located at ~1595 cm^{-1}; this mode will shift to a higher frequency, ~1635 cm^{-1}, upon quaternization of the pyridyl group. Figure 17 shows the formation of covalent C–N bonds between the pyridyl group in the

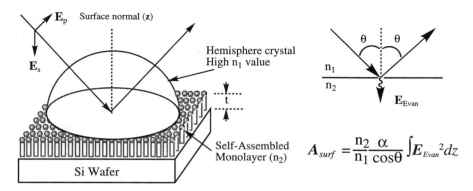

$$A_{surf} = \frac{n_2}{n_1} \frac{\alpha}{\cos\theta} \int E_{Evan}^2 dz$$

FIGURE 16 Polarized, variable-angle internal reflection setup. A monolayer on a silicon wafer, pressed against the internal reflection hemisphere crystal, is detected by the evanescent wave of the infrared beam (E_{Evan}). Both *p*- and *s*-polarized spectra can be obtained because no conductive substrate was used. The angle can be adjusted from 5° to 85°; the highest sensitivity was observed at the internal total reflection angle.

calix[4]stilbazole imine and the coupling silane-anchoring layers, as revealed by the presence pyridinium species.

This technique also was used to characterize the formation of covalent C–N bonds between porphyrins and the benzyl chloride layer. After introducing 5,10,15,20-tetra(4-pyridyl)-porphine (TPyP), two strong vibrations were observed at 1635 cm^{-1} and 1593 cm^{-1}; these vibrations are associated with two weak vibrations (at 1716 cm^{-1} and 1407 cm^{-1}) for both *p*- and *s*-incident polarizations. The two strong characteristic bands at 1635 cm^{-1} and 1593 cm^{-1} were assigned to

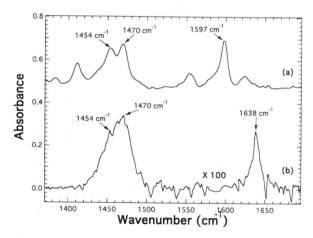

FIGURE 17 FTIR spectra of (a) calix[4]stilbazole imine in bulk (KBr disk) and (b) a monolayer of calix[4]stilbazole imine anchored on the coupling layer of *p*-chloromethyl-phenyltrichlorosilane. The geometry used for the surface IR consists of *p*-polarized light incident at 45°.

ν_a, the N-substituted pyridinium (a_1), and ν_a modes of pyridyl groups [83], respectively. The simultaneous appearance of these two bands is the signature of an asymmetric structure (i.e., free pyridyl at the outer surface and the pyridinium linkage of the porphyrin to the coupling layer). Therefore, the porphyrin monolayer is polar.

The intensity of the pyridyl mode at 1593 cm^{-1} can be used to diagnose the quaternization of the outer pyridyl groups. For example, after reacting the porphyrin monolayer with iodomethane, the strong band at 1593 cm^{-1} that corresponds to the mode of pyridyl groups disappeared completely. Conversely, the N-substituted pyridinium mode increased in intensity because of the conversion of pyridyl groups into pyridinium moieties. These results suggest that quaternization reaction was complete, including the formation of the C–N covalent bonds.

Ultraviolet–visible spectroscopy proved another sensitive technique in detecting surface monolayer formation because the chromophore possesses highly delocalized π electrons. The quaternization of stilbazole during the formation of the chromophore layer involved π electrons, and, as a result, caused a diagnostic redshift in λ_{max} from 390 to 510 nm. The growth of chromophoric multilayers can be monitored by the increase in absorbance as the number of layers increases. Absorption at 510 nm is caused by the charge transfer from the donor (dimethylamino group) to the acceptor (pyridinium ring). The growth of the Ch layer increases the absorption linearly.

X-ray photoelectron spectroscopy (XPS) and SIMS provide surface microanalysis to characterize multilayer growth. High-resolution XPS spectra of the Cp and CpCh layers on glass substrates was carried out to identify the iodine element in monolayers with peaks at 622 eV and 634 eV binding energy, which correspond to electrons ejected from iodine $3d^{5/2}$ and $3d^{3/2}$ configurations, respectively.

In the case of reacting stilbazole with a bromopropyl group, the shift of the bromine $3d$ peak from 70 to 68 eV was also observed (84). XPS was also used to monitor the growth of incoming anion signals, such as iodide, and the diminishing of chloride signal in the exchange of monolayer of stilbazole chloride with iodide, p-aminobenzenesulfonate, and ethyl orange (85,86). After the anion exchange, the intensity of SHG from these chromophoric SAMs was increased by as much as a factor of 2. This effect was caused by two principal factors: (1) the incoming anions changed the polarizability and hyperpolarizability of the monolayers and (2) the incoming anions disturbed the molecular orientation of the monolayers.

In the bromine-doped self-assembled porphyrin monolayers, XPS was used to estimate the amount of bromine in the system by monitoring the photoelectrons of Br($3d$) at 70.0 eV, which was consistent with the formation of CpTPyPBr$_x$ monolayers. The amount of Br$^-$ estimated from the XPS spectra was $x = 0.78$–1.85. The presence of Br$^-$ in the covalently bound SAM film of CpTPyPBr$_x$ was also confirmed by the observation of the negative ions ^{79}Br$^-$ and ^{81}Br$^-$ in the static SIMS spectra.

X-ray reflectivity (XRR) has proven reliable in determining the monolayer thin-film thickness. Collected on the periodical multilayer structure that consists of silane layer, XRR data show a smooth, monotonic increase of film thickness as a function of each repeating CpChCsl cycle. A Bragg peak caused by scattering from each individual layer was also observed at $K_z = 0.17$ Å^{-1} (maximum), which corresponds to an interlayer spacing of 36 ± 1 Å—in good agreement with molecular models (84). Typical XRR experiments show regular-spaced minima in the data that are generated by the reflection from the air–SAM and SAM–substrate interfaces. The spacing between these fringes depends on film thickness.

3. Chromophoric Orientation and Self-Assembled Monolayer Surface Coverage

To determine the molecular orientation (the direction of electronic optical transition moment) in chromophoric SAMs, the following are measured: the optical absorption of the oriented chromophores on surfaces and the optical absorption of the exact amount of randomized chromophores in solution (87). These measurements were conducted using azobenzene sulfonate dyes as optical indicators (Fig. 18). Because the sulfonate in the zirconium–sulfonate bond can be replaced by a stronger binder—phosphate buffers (HPO_3^{2-}/PO_3^{3-}; pH = 11), the following experiments were designed: Azobenzene chromophores first underwent self-organization from solution onto the zirconium ion surface and subsequently returned to solution by *quantitative* exchange with a phosphate salt.

By examining the starting solution, surface-bound state, and end solution, the following data were obtained: (1) the azobenzene dyes were intact in the process of self-assembling, as well as in disassembling; (2) surface coverage was determined accurately using the end solution and the Beer–Lambert law ($\Gamma_{surf} = A\varepsilon^{-1}$); and (3) molecular orientation was experimentally deduced by comparing the absorbance of the surface-bound species to the absorbance of the same amount of that species in the end solution.

These azobenzene chromophores can be considered rigid rod-like sensitizers with a charge transfer excitation in the visible region of the optical spectrum. The optical transition dipole moment is approximately parallel to the rigid-rod direction and can be excited only when the external electromagnetic radiation has a nonzero **E** field along the molecular axis. The average angle $\langle \psi \rangle$ of the transition moments, with respect to the surface normal, can be derived by comparing the absorbance of the thin film for light polarized parallel to the surface (A_{surf}) against that for the same number of identical species in solution (A_{soln}). Assuming a narrow distribution

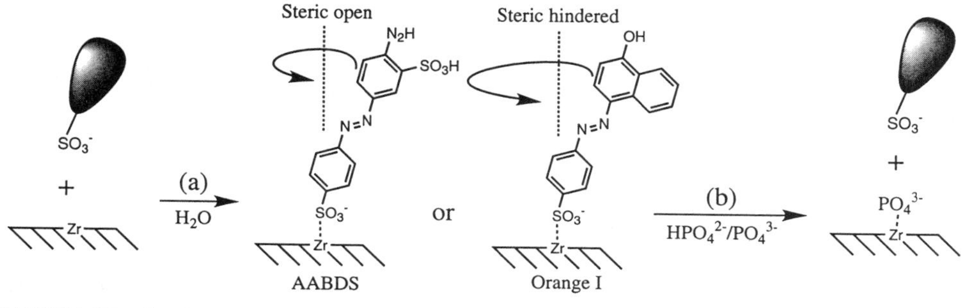

FIGURE 18 Azobenzene chromophores underwent self-organization (step a) from solution onto the surface and subsequently returned to solution (step b) by quantitative exchange with phosphate salts. These experiments revealed that molecular orientation (ψ) is a function of the surface coverage (Γ_{surf}).

around $\langle\psi\rangle$ in the film, the formula is given in Eq. (2) for an unpolarized light geometry.

$$\sin^2\langle\psi\rangle = \frac{2A_{surf}}{3A_{soln}} \qquad (2)$$

Figure 19 shows the molecular orientation deduced from Eq. (2) versus surface coverage for the chromophoric SAMs. The formation of SAMs can be divided into three categories: partially ordered submonolayers, well-ordered monolayers, and randomized aggregates. When the surface coverage was low, submonolayers were formed with some net alignment. As the surface coverage increased, highly ordered mesoscale structures started to organize, and the optimized alignment, $\langle\psi\rangle = 32°\pm 3°$, was achieved while the surface coverage approached a close-packed state of ~2.0–4.0 molecules/nm². Aggregation formed on surfaces when the surface coverage was greater than a monolayer and the orientation was random ($\sin^2\langle\psi\rangle = 2/3$). This shows that the average molecular orientation is a complex function of the surface coverage and the resulting steric interactions of adjacent rod-like molecules.

These studies further show that chlorosilane forms better organized thin films than trimethoxysilanes (higher probability of smaller $\langle\psi\rangle$ at optimum coverage), which reinforces the idea that hydrolysis of chlorosilanes with silanols are more selective than that of methoxysilanes. The optimum surface coverage (minimum points in Fig. 19) is governed by the cross-sectional area of the molecule on surfaces. For example, a steric bulky naphthanyl group occupies more surface area and hence has lower coverage [2.3 molecules/nm² for Orange I versus 3.7

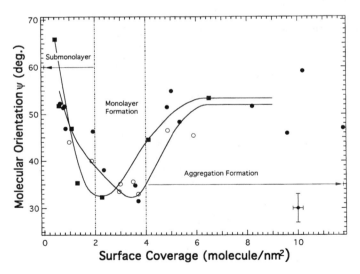

FIGURE 19 The plot of molecular orientation against surface coverage shows that the formation of chromophoric SAMs can be divided into three categories: partially ordered submonolayers, well-ordered monolayers, and randomized aggregates. The minimum molecular orientation angle is $32°\pm 3°$, with optimized surface coverage of 2–4 molecules/nm².

molecules/nm² for 4-amino-1,1′-azobenzene-3,4′-disulfonic acid, sodium salt (AABDS), and Sulfarsazene].

H.   Second-Harmonic Generation from Self-Assembled Multilayers

1.   *Transmitted Second-Harmonic Generation*

Second-harmonic-generation experiments were performed on these chromophoric self-assembled thin films to determine their second-order nonlinearity, as well as to obtain additional structural information. Single-wavelength-transmission SHG measurements were conducted in a *p*-polarized geometry (see Fig. 20) using the 1064-nm output of a *Q*-switched Nd:YAG laser. No in-plane anisotropy in the SHG signal was detected as the glass substrate (coated with polar SAMs) was rotated about the film normal. This indicates that these films possess uniaxial symmetry about the film normal and that the distribution of molecular orientation of chromophores does not have an azimuthal dependence.

Figure 20 shows SHG intensity as a function of fundamental beam incident angle from a glass slide having a covalently bonded self-assembled CpCh monolayer on either side. The figure also illustrates a nearly complete destructive interference of SHG waves from the monolayer on both sides of the glass substrate. These results indicate that the quality of the monolayers on the two sides of the

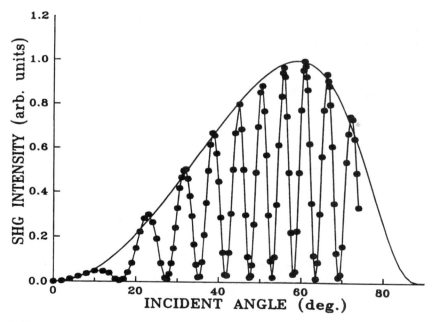

FIGURE 20   SHG intensity as a function of fundamental beam incident angle from a glass slide having a covalently bonded self-assembled CpCh monolayer on either side. The interference pattern arises from the phase difference between the SHG waves generated at either side of the substrate during propagation of the fundamental wave. The solid envelope is a theoretical curve generated for $\chi^{(2)}_{zzz}/\chi^{(2)}_{zxx} = 2\langle \mathrm{ctg}^2\psi\rangle$, where $\langle\psi\rangle = 39°$.

glass slide are nearly identical and quite uniform, suggesting that the present self-assembly method could generate excellent quality monolayers in a reproducible way. An excellent fit to the envelope of the data in Fig. 20 is obtained with a ratio of $\chi^{(2)}_{zzz}/\chi^{(2)}_{zxx} = 2\langle \text{ctg}^2\psi \rangle$, where $\langle \psi \rangle$ is the average polar angle between the surface normal, and the molecular dipole vector orientation is in the range 35°–39° from the transmission SHG measurements. The coordinates are defined in such a way that the z direction is along the surface normal and the y axis is the intersect of the incident plane and the surface.

The $\chi^{(2)}_{zzz}/\chi^{(2)}_{zxx}$ ratio also can be determined by measuring p-polarized SHG from both p- or s-polarized excitation, as described in Eqs. (3) and (4). The first subscript refers to the fundamental polarization and the second subscript refers to the polarization of the second-harmonic light:

$$P^{2\omega}_{p,p} = \chi_{zzz}E_zE_z + \chi_{zyy}E_yE_y + \chi_{yzy}E_zE_y + \chi_{yyz}E_yE_z \tag{3}$$

$$P^{2\omega}_{s,p} = \chi_{zxx}E_xE_x \tag{4}$$

Within the Kleinman symmetry, the last three terms in Eq. (3) combine into one (i.e., $\chi_{zyy} = \chi_{yzy} = \chi_{yyz}$). Based on C_∞ uniaxial symmetry in these chromophoric SAMs, these tensors can be further simplified into only two independent tensors: $\chi^{(2)}_{zzz}$, and $\chi^{(2)}_{zxx}$. Following from the derivations of Bloembergen and Pershan [88] and Zhang et al. [89], the p-polarized SHG wave output profile from a p-polarized fundamental wave is given in Eq. (3). The theoretical envelope in Fig. 20 was generated by Eq. (3) by adjusting the ratio of $\chi^{(2)}_{zzz}/\chi^{(2)}_{zxx} = 2\langle \text{ctg}^2\psi \rangle$. Once a theoretical fit was obtained, the average molecular orientation angle $\langle \psi \rangle$ was determined. We should emphasize that this orientation angle is the angle between the surface normal and the principal molecular tensor β_{111} (the only molecular tensor assumed to be nonzero in this derivation), rather than exact orientation of the molecules.

The interference pattern in Fig. 20 is caused by the phase mismatch of the two SHG waves generated at both sides of the substrate during the propagation of the fundamental wave. This interferogram disappears and becomes a complete envelope when one side of the film is removed. The SHG intensity of this envelope is one-fourth of those SHG generated from two films on both sides of the substrate. Partial interference will appear if the two sides of the substrate consist of different components; one example of incomplete destructive interference is observed when the chromophoric SAMs are prepared on a glass substrate with ITO (indium–tin oxide) coating on one side.

In the infrared region ($\lambda = 1900$ nm), we observed a $\chi^{(2)}_{zzz} = 130 \times 10^{-9}$ esu for the CpCh multilayer thin films. These values of $\chi^{(2)}_{zzz}$ are extremely large compared to those observed in typical poled polymer films. The multilayers are still extremely thin compared to the wavelength of the 1064-nm laser light radiation. Hence, they are also much smaller than the expected coherence length in which the homogeneous wave and the inhomogeneous wave went through a period of constructive and destructive interaction. As a result, we would expect the SHG intensity to increase quadratically as the film thickness or the number of chromophore layers. A precise quadratic enhancement in the SHG intensity as a function of chromophore layer number was observed in multilayer films (see Fig. 21). These results indicate that it is possible to maintain the same degree of noncentrosym-

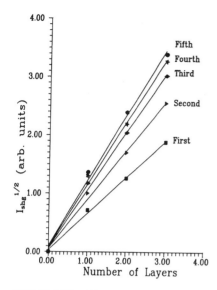

FIGURE 21 Plot of the square root of SHG intensity versus the number of chromophore layers in multilayer NLO materials after cross-linking. The straight lines are linear least-square fits to the experimental data. The line labels correspond to the SHG interference maxima, counting from the zero angle.

metric ordering as the first monolayer in the subsequent layers during the synthesis of a multilayer structure.

2. Reflected Second-Harmonic Wave

There are many advantages to performing reflection SHG measurements [90]. For example, there are no phase-mismatch problems because only the homogeneous wave is present in the reflection. Moreover, more information about the molecular orientation and the dielectric constants ($\varepsilon_{2\omega}/\varepsilon_{\omega}$) can be obtained; the latter are not available from transmission measurements.

For p-polarized, reflected second-harmonic waves, one observes the following relationship between SHG intensity with an arbitrarily polarized, incoming electric field angle (α):

$$I_p(2\omega) = (A_p \cos^2\alpha + B_p \sin^2\alpha)^2 \tag{5}$$

The limiting situation of $A_p = E_r(p - p)$ and $B_p = E_r(p - s)$ represent p-polarized SHG from p-polarized fundamental and p-polarized SHG from s-polarized fundamental, respectively. Specifically, A_p is defined as the intensity of p-polarized second-harmonic light, whereas the incident light is also p-polarized and B_p is defined as the intensity of p-polarized second-harmonic light while the incident light is s-polarized.

The ratio of A_p/B_p yields a function with one variable $2\langle \text{ctg}^2\psi \rangle = \chi_{zzz}^{(2)}/\chi_{zxx}^{(2)}$, and by measuring the ratio of A_p/B_p, one can calculate the average polar angle.

For s-polarized reflected second-harmonic wave, one finds that

$$I_s(2\omega) = (C_s \sin 2\alpha)^2 \tag{6}$$

C_s is defined as the intensity of *s*-polarized second-harmonic light when the incident polarization is 45°. The value of C_s/B_p gives the ratio of dielectric constants $\varepsilon_{2\omega}$ and ε_ω for the NLO medium.

Figure 22 shows the experimental SHG profile for both *p*- and *s*-polarized waves (data marked) along with the theoretical fit by using Eqs. (5) and (6). Averaged over a few chromophoric SAMs with film on only one side of the substrate, the polar orientation angle of the chromophore was determined to be 34°–35°, with the dielectric constant ratio $\varepsilon_{2\omega}/\varepsilon_\omega = 1.12$. This is in good agreement with the transmission measurement in which we observed a polar angle $\langle\psi\rangle$ in the range of 35°–30° (*vide supra*). The reflection measurements yield relatively more reliable molecular orientation angle because there are fewer parameters to adjust.

I. Spectroscopic Second-Harmonic Generation from Supermolecular Self-Assembled Monolayers

Recently, ultrafast spectroscopic SHG was available for measuring the frequency-dependence behaviors of first-order hyperpolarizabilities. These measurements were typically performed using a mode-locked Ti:sapphire laser in either femtosecond (150 fs autocorrelation width) or picosecond (2 ps width) mode for various fundamental wavelengths between 890 nm and 750 nm.

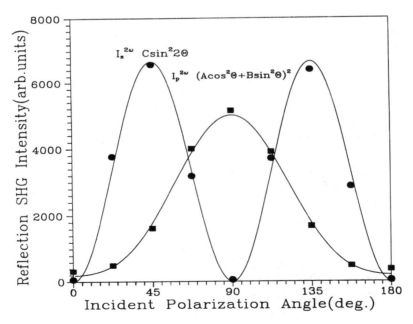

FIGURE 22 Response of reflected *p*- and *s*-polarized second-harmonic waves from a self-assembled (CpCh) monolayer on one side of a glass substrate, which was plotted as a function of the fundamental wave polarization angle (α). For the extreme situation of *p*- and *s*-polarized fundamental waves, the polarization angles (α) are 90° and 0°, respectively. The solid lines represent a theoretical fit to the experimental data.

Figure 23 shows results obtained by using femtosecond pulses to generate *p*-polarized SHG from both *p*- and *s*-polarized fundamental wavelengths of 862 nm. The solid lines in Fig. 23 are the best fits to an expression [34], modulated by an interference term, which has the same physical origin and angular dependence as the well-known Maker fringes in bulk samples [91]. However, Fig. 23 reveals that the amplitude of the fringes is much smaller, which is caused by the increased spectral width of the femtosecond pulses; complete destructive (or constructive) interference cannot be achieved across the entire pulse spectrum. For the 1-mm fused silica samples considered here, effects of pulse walkoff caused by mismatched group velocities of the fundamental and SH pulses can be neglected.

Figure 24 shows the measured values of d_{33} versus fundamental wavelength for monolayer films of calix[4]stilbazole imine, together with the linear absorption spectrum αd taken from the same sample. In this case, α and d are the linear absorption coefficient and film thickness, respectively. The nonlinear optical spectrum closely resembles the linear optical absorption spectrum, which is fairly featureless. The wavelength dependence (fundamental $\lambda = 890$–750 nm; harmonic $\lambda = 445$–375 nm) indicates a gradual increase in d_{33} as the fundamental moves to a shorter wavelength, which is consistent with normal refractive index dispersion in this spectral region approaching the charge transfer resonance at 390 nm. From Fig. 24, the charge transfer resonance is fairly broad; the measured value for d_{33} at the longest wavelength (60 pm/V at 890 nm) is still somewhat resonantly enhanced. We find a value of $d_{33} \sim 110$ pm/V at $\lambda = 775$ nm, indicating a nearly twofold two-photon resonant enhancement at the peak of the charge transfer band.

In essence, ultrahigh nonlinearities, $\chi_{zzz} = 6 \times 10^{-7}$ esu, can be grown on various inorganic surfaces in a covalently bonded self-assembled manner. This exceptionally large second-order susceptibility is one of the highest second-order nonlinearities observed to date. Chromophoric SAMs offer tremendous promise in

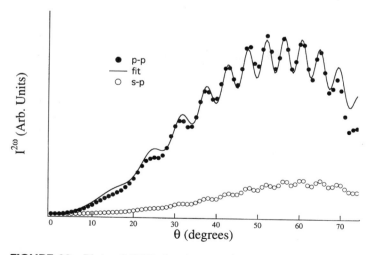

FIGURE 23 Plots of SHG signal versus incident angle for *p*-polarized SHG from both *p*-(dots) and *s*-polarized (circles) fundamental at a wavelength of 862 nm and 150 fs pulse width.

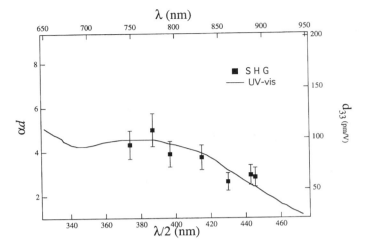

FIGURE 24 Plots of second-order nonlinear susceptibility d_{33} (in pm/V) (solid squares) versus fundamental wavelength (top and right axes) or second-harmonic wavelength (bottom and right axes) along with the linear absorption spectrum (solid line) versus wavelength (bottom and left axes) for the SAM of calix[4]stilbazole imine.

the arena of nonlinear optical thin films and integrated optical devices. Coupled with the knowledge of patterned SAMs, integrated optical devices can be potentially fabricated using self-assembling chemistry. Key issues that require additional study include the efficiency of generating multilayers in a relatively short time and a multilayer's long-term stabilities (such as structural, chemical, mechanical, and thermal stabilities). The construction of supramolecular architecture, with pyramid-like molecular building blocks, represents a significant step forward in the direction of solving thin-film stability for chromophoric SAMs.

III. MOLECULAR SELF-ASSEMBLIES FOR SENSOR APPLICATIONS

A. Sensor Concepts Based on Self-Assembled Monolayers

Advances made in the area of molecular self-assemblies, especially their ability to mimic biological membranes (e.g., lipid bilayers), have promoted studies in surface molecular recognition and sensors. Sensor applications consist of two general types: (1) chemical sensors based on molecular interactions and (2) biosensors based on biological interactions.

A critical component for sensors is the interaction between the analytes and the active surface. Molecular self-assemblies offer a new method that can tune the surface properties schematically according to specific applications. The resulting surface possesses new chemical and physical properties such as wetting phenomena. It is also possible to build in molecular recognition features that function as a smart "skin" that can detect the presence of certain chemical compounds. Such molecular recognition at the surface offers the promise of developing chemical and biological sensors. The molecular interaction information at the surface must be translated to an electronic device that can further process the information and display it. In other words, a transducer is required to interpret the chemical or biological interactions at the surface. This interfacial process typically consists of information carriers such as photons, electrons, and sound waves that correspond to optical, electrical, and frequency measurements.

In essence, a sensor consists of two key components: a sensitive/specific sensing layer and a transducer (see Fig. 25). We will discuss the sensing layers in Sections III.B and III.C; we will briefly discuss in this section a few common transduction mechanisms. Optical waveguides can be used as information carriers. For example, the evanescent wave of a guided wave can pick up molecular interaction (changes in optical spectra) at or near the surface; sensors based on this mechanism are typically optical fibers. Another optical technique utilizes the refractive index change to the guided wave; this change can be easily measured through a device configuration such as a Mach–Zender device.

Electrochemistry is another interfacial probe. One can either use a reference redox pair to monitor the changes in the environment or analyze the electrochemically active reagents (metal ions and aromatic species) directly. Although a sensitive method, electrochemistry has a drawback: the analytes must be electrochemically active or interact strongly with a well-behaved redox couple.

Surface acoustic wave (SAW) resonators can be used to measure mass changes on surfaces; i.e., surface coverage and sensor response to analytes (92). The phase velocity of this acoustic wave—and therefore the resonant frequency of the device—is a sensitive function of the physical properties of any contacting materials. In these nonconductive thin films having no lateral connectivity, electrical and viscoelastic effects are minimal and the resonant frequency shift (Δf) [93] depends mainly on the mass loading per unit area ($\Delta m/A$) of the sensing thin film [94] and any adsorbed vapors, according to Eq. (7):

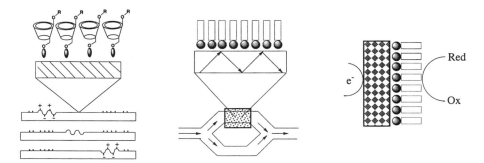

FIGURE 25 Illustration showing selected information transduction mechanisms. Surface acoustic wave devices are based on the mass change on the surfaces (left); an optical waveguide employs the change in optical properties such as absorptions or refractive index (center); the electrochemical method utilizes the redox properties of analytes or sensing layers (right).

More specifically, techniques such as SAW, electrochemical analysis, optical waveguide, and surface plasma resonance are all used in information transduction. SAW resonators, which can detect down to 10 pg of surface-bound materials, are commonly employed as chemical sensors for volatile organic compounds (VOCs). A less sensitive acoustic device, quartz crystal microbalance, is also frequently used (nanogram sensitivity).

Chemical microsensors are expected to play a growing role in cost-effective environmental monitoring, site remediation, and industrial process characterization. The analyses of chemical compounds are presently performed with standard analytical instruments such as NMR, FTIR, gas chromatography, and mass spectrometry. By integrating a physical measurement platform with a selective sensing layer, a desired chemical microsensor is constructed that provides some of the functionality of analytical instrumentation, but with drastically reduced cost, size, and power consumption.

B. Chemical Microsensors

Advanced work continues in constructing superior chemical sensors based on tailoring the surface chemistry of the sensing layer on SAW devices. Such sensors are desired for cost-effective environmental monitoring, site remediation, and industrial process characterization [95].

Two versatile families of organic "bucket" molecules known as cyclodextrins [96] and calix[n]arenes [97] have attracted a great attention recently because of their inclusion chemistry (see Fig. 26). These receptors can be further functionalized and then reacted with the transducer surfaces or organic functionalized transducer surfaces [98]. The surface-attached nanometer-sized host molecules can be predominantly aligned upward and endowed with a locally modified chemical environment to complex VOCs. The inside of the host cavity exhibits hydrophobic properties and, depending on the variety and functionalization, possesses the size and chemical environment to readily incorporate specific VOCs via host–guest interactions. The sensor mechanism in this example is based on a novel concept of noncovalent interactions—hydrophobic affinity, hydrogen-bonding, and inclusion force. As a result of these weak interactions, the resulting sensor is completely reversible.

Schierbaum et al. used modified resorcin[4]arenes as a synthetic receptor on the Au surface of a quartz microbalance to detect VOCs such as perchloroethylene and toluene [99]. Formed by four aromatic resorcinol units bridged by the methylene groups at the adjacent oxygen positions, the resorcin[4]arene cavity is more selective to perchloroethylene, with selectivity of ~8 and ~16 times over toluene and chloroform, respectively. The thermal deposition temperature for C_2Cl_4 is higher for resorcin[4]arene ($T = 286$ K) than didecylsulfide ($T = 210$ K), indicating stronger host–guest interactions. The detection limit for this sensor configuration of resorcin[4]arene and quartz microbalance is ~34 ppm, based on the noise of 3 Hz.

Alternatively, Moore and co-workers [98] employed a stronger siloxane linkage appropriate for attachment to oxide surfaces terminated with hydroxyl groups. The asymmetric host molecules used in this study include the following:

- α-Cyclodextrin dodeca(2*O*,3*O*)benzoate (CDCB)

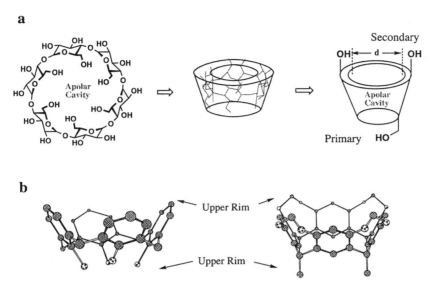

FIGURE 26 Host compounds possessing well-defined cavities are employed as surface receptors for various sensor applications. (a) Cyclodextrins, which contain 6–12 glucose units bonded together through α-linkage, are enzyme products of hydrolyzed starch (α-cyclodextrin with $n = 6$ is shown). The primary and secondary hydroxy groups can be further functionalized to enhance sensor specificity. (b) Synthetic macrocycle arene compounds: calix[n]arene ($n = 4$–11) (left) and resorcin[4]arene (right). The upper and lower rims provide further functionalization.

- β-Cyclodextrin tetradeca($2O,3O$)benzoate (CDTB)
- β-Cyclodextrin tetradeca($2O,3O$)acetate (CDTA)
- c-Undecylcalix[4]resorcinarene (UCR)
- 37,38,39,40,41,42-Hexahydroxycalix[6]arene (HCA)

Calixarenes and resorcinarene are synthetic, whereas cyclodextrins are chiral, toroidal-shaped enzymatic products formed under the catalyses of enzyme cyclodextrin transglycosylase on hydrolyzed starch (see Fig. 26). The molecular-engineered cyclodextrin or calixarene host monolayers have a sensitivity of 5 ppm to PCE, based on a 5-Hz noise, which is comparable to that of the aforementioned resorcin[4]arene. The optimized interactions between hosts and guests are attributable to proper alignment, lipophilic cavity, and functionalization of the upper rim. Monolayers of β-cyclodextrin derivatives are also more sensitive to PCE; the response to PCE is ~5 and ~20 times over toluene and chloroform, respectively.

The SAW sensor responses of these acoustic devices are typically real time, reversible, and rapid (~1–5 sec). For sensitive thin-film coatings, the SAW sensors can show a saturation behavior at high-organic-vapor partial pressures. However, the sensor also exhibited a linear response within its dynamic concentration range, especially at low concentrations. Typical cyclodextrin-monolayer sensor responses to VOC partial pressures at 15.0 ± 0.1°C [100] gave frequency shifts of 8.0 kHz or less, corresponding to approximately a monolayer or smaller amounts of the VOC. This suggests an average of one or less analyte molecule per surface-attached sensing molecule. As expected, the sensor responses (Δf) increased almost linearly with the cyclodextrin thin-film thickness; ultrahigh, sensitive VOC sensors were successfully demonstrated in this manner.

In the host–guest detection scheme, sensor selectivity depends on optimum chemical or physical interactions between the analyte and the host cavities; such interactions include mutual matching of polarity, size, and structural properties.

The cyclodextrin monolayers show appreciable selectivities within a group of organic template analytes, which are chosen for their variety of structures and polarities to probe the cavities of the host compounds (Fig. 27). The observed selectivities are expected—the inclusion complexes of phenyl units into α-cyclodextrin and the optimum fit of PCE into β-cyclodextrins are well documented by ultraviolet–visible spectroscopy [101], NMR [102], and x-ray crystallography [103].

If a pure adsorption into interstitials took place, one would expect that the response of α-CDBD versus β-CDTB should be nearly equal because they have the same chemical composition, monomeric structure, and polarity. Furthermore, the sensors each show a distinct set of relative responses among a group of analytes, attributable to inherently different local host chemical environments.

Conversely, a judicious assortment of sensing layers in an array of microsensors can be used to provide a unique characteristic pattern of responses for each compound. For example, the response of PCE (as illustrated in Fig. 27) to an array of microsensors coated with CDCB, CDTB, CDTA, UCR, and HCA is a broad peak. The same sensor array responds to toluene with a "doublet," and each compound has its own characteristic response to this sensor array. As expected, the "spectroscopic" patterns of such microsensor arrays share common characteristics—but are

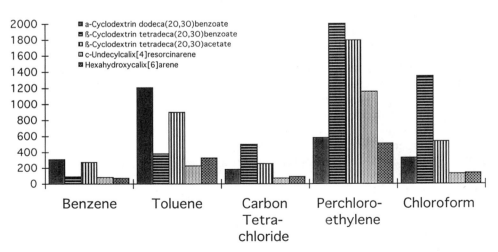

FIGURE 27 Comparison of the responses of 200-MHz SAW resonators coated with self-assembled cyclodextrins and calixarenes to a variety of analytes at unit organic vapor concentration (Hz/Torr). For each analyte (from left to right), the microsensor array coatings are CDBD, CDTB, CDTA, UCR, and HCA. Note that each organic compound gives a characteristic set of responses.

nonetheless distinguishable—when the structures of two compounds resemble one another closely, such as benzene and toluene, or chloroform and carbon tetrachloride.

Host receptors based on cyclobis(paraquat-*p*-phenylene) were functionalized with disulfide moieties and formed mixed SAMs with $C_{10}H_{21}SH$ (104). The paraquat moiety included π organics, such as indole, catechol, benzonitrile, and nitrobenzene, at micromolar concentrations. These receptors caused a shift in the formal potential for the monoelectronic reduction of the surface-confined paraquat groups. The paraquat host and aromatic guests have an equilibrium constant, K, in the range $(1–2) \times 10^5$. Similarly, Wrighton used electrochemistry and mixed self-assembly of a ferrocenyl thiol $(F_cCO(CH_2)_{10}SH)$ and a quinone (Q) thiol $(H_2Q(CH_2)_8SH)$ as a one of many ongoing pH sensors. The detection mechanism is based on the measurements of the potential difference of current peaks for oxidation and reduction of reference ferrocene and indicator (quinone) in a two-terminal linear sweep voltammogram with a large counterelectrode. The half-wave potential, $E_{1/2}$, of ferrocene is insensitive to pH; the quinone has an $E_{1/2}$ that is sensitive to pH and shifts from ~0.5 V negative (pH = 11) to ~0.5 V positive (10 M $HClO_4$) of ferrocene. The response to pH changes is basically linear within this concentration region.

Rubinstein et al. used mixed SAMs of 2,2′-thiobisethyl acetoacetate $[S(CH_2OCOCH_2COCH_3)]$ or TBEA and *n*-octadecyl mercaptan $[CH_3(CH_2)_{17}SH]$ to selectively detect the presence of Cu^{2+} ions against Fe^{3+} metal ions [105]. The divalent metal Cu^{2+} formed a square planar structure with TBEA, whereas the trivalent Fe^{3+} ion required octahedral coordination, which cannot be provided by the surface-confined TBEA. These studies were carried out with cyclic voltammograms in 0.1 M H_2SO_4, which contained 1.0 mM of Cu^{2+} and 2.0 mM of Fe^{3+} (based on the observation of redox potential of Cu^{2+}/Cu at $-0.40–0.55$ V and the absence of Fe^{3+}/Fe^{2+} at $0–0.10$ V).

The *n*-octadecyl mercaptan served as an electrochemically inert blocking element that prevented leakage of Fe^{3+} through the uncovered portion of the electrode. The electron tunneling from Cu^{2+} to the electrode is believed to be the charge transfer mechanism; this was demonstrated with a consistent negative shift of the Cu^{2+} reduction peak potential, because the Cu^{2+} is located further away from the electrode surfaces. This system was also used to effectively detect Pb^{2+} directly and Zn^{2+}, indirectly, by stripping Cu^{2+} ions [106].

Currently, molecular self-assemblies are competing with other coating technologies in sensor applications, such as polymeric thin films and sol–gel technologies. The great advantage for molecular self-assemblies is their ability to tailor material properties at the molecular level.

C. Biological Sensors

At present, biosensing mechanisms consist of immunosensors (based on immunochemistry) or biomimetic interactions. In the field of immunosensors, highly specific antibody–antigen interactions yield precise detection of the unknown antigen. There are a few general disadvantages of using this method. For example, the data acquisition and analysis are still expensive and time-consuming. Moreover, the detection limits are typically low because of the small signal generated by the binding of an antibody to the surface of a transducer. Reversibility is another critical problem in immunological reactions. Some immunological reactions can be reversed with high salt concentration or pH changes, but, unfortunately, these conditions also cause denaturing of the corresponding sensing enzyme. Some immunoreactions are simply irreversible. Nonetheless, the field of biosensors offers a bright future for highly specific detections, as response (bioreaction) time and longevity of the enzyme are being improved.

One recent advance in this area applies only the active part of the antigen (e.g., a viral protein) instead of the whole antigen protein. Known as epitopes, these active sites are recognized by specific antibodies. Epitopes consist of a small peptide sequence that makes them more robust and less likely to be denatured. Knichel and co-workers chose an epitope as a model system that consists of amino acids 135–154 of the capsid protein VP1, which causes the foot-and-mouth disease virus (107). The 20-amino-acid strand was then coupled to ω-hydroxyundecanethiol $HS(CH_2)_{11}OH$ via a succinic acid. Specific antibody at the concentration of 17.4 $\mu g/ml$ recognize the epitope, regardless of using the C- or N-terminal as an anchor site to the SAM of ω-hydroxyundecanethiol on gold electrodes.

Surface plasma resonance spectroscopy is an ideal method to conduct in situ and real-time measurements of the nonspecific adsorption of proteins to a mixed SAM, which consists of $S(CH_2)_{10}CH_3$ and $S(CH_2)_{11}(OCH_2CH_2)_6OH$ on gold (108). Mixed SAMs that possessed less than 50% of hexalethylene glycol irreversibly adsorbed proteins, such as RNaseA, lysozyme, fibrinogen, and pyruvate kinase.

Mrksich [109] employed an enzyme substrate to recognize the presence of that particular enzyme; a surface-bound benzenesulfonamide group in a mixed SAM was used to interact with bovine carbonic anhydrase. The interactions between *p*-substituted benzene sulfonamide and bovine carbonic anhydrase are well characterized; the equilibrium dissociation constants (K_d) were approximately 10^{-6}–10^{-9} *M*. Monitored by surface plasma resonance, the sensor response to 0.3 mg/ml of carbonic anhydrase increased with the benzenesulfonamide component in the mixed SAM up to 8% of surface coverage. The mixed SAM resists nonspecific adsorption of proteins; 4-carboxybenzene-sulfonamide inhibits specific surface attachment of carbonic anhydrase. Unfortunately, this sensor is not completely reversible.

A series of molecules, which terminated with biotin and hydroxyl groups, were anchored to gold surfaces via a Au–S linkage [110–113]. The recognition ability of biotin to streptavidin was studied in situ by using a surface plasmon resonance technique. This study revealed that thiols (R–SH) and disulfides (R–S–S– R) form similar SAMs, whereas sulfides (R–S–R) form very different films—less closely packed, with a high degree of unspecified streptavidin binding. A high specific-binding capacity for streptavidin can be obtained by dilution of biotinylated thiols with hydroxy-terminated thiols, as well as the inclusion of a spacer (optimized concentration $\sim$10%) in the biotinylated thiols.

Monitored by surface plasmon resonance spectroscopy and microscopy [114], thiollipid and palmitic acid (1:4) were used to form the mixed SAMs on gold; the fluorescence micrograph indicated that palmitic acid isolated into domains $\sim$14 μm in size. The round palmitic acid domains can be easily rinsed off with ethanol; a mixture of thioglucose and a peptide with a terminal cysteine (–SH group) was used to fill this region. The resulting patterned SAM showed different binding properties toward bovine serum albumin and monoclonal antibody against the cysteine-terminated peptide. The thiolipid selectively bound with bovine serum albumin, whereas the monoclonal antibody specifically bound to the peptide.

The above mixed SAM approach can be expanded to a multilayer system, which sequentially consists of biotinylated thiol, streptavidin, biotinylated Fab, human chorionic gonadotrophin, and monoclonal antibody [111]. This system's sensitivity to human chorionic gonadotrophin yields a detection limit of approximately 10^{-8} M, which is two orders of magnitude away from commercial usage (e.g., pregnancy tests).

One of the most explored types of biosensors are glucose sensors. For example, glucose oxidase was anchored on a SAM of cysteamine via a bifunctional coupler—*trans*-stilbene-(4,4′-diisothiocyanate)-2,2′-disulfonic acid [115]. The resulting enzyme network immobilized onto the electrode was modified by *N*-(2-methylferrocene)-caproic acid. This enzyme electrode shows a nonlinear response to glucose from concentration of ~2 to ~20 mM, with ferrocene as the electron transfer mediator.

IV. CONCLUDING REMARKS

We have reviewed simple SAMs (thiols on gold) and functional SAMs (chromophoric multilayers for nonlinear optical materials and surface-bound receptors for sensor applications). SAMs provide a technique that enables us to control the microstructures of organic thin films. It is these microstructures that will determine the desired physical properties. Molecular self-assemblies are now branching into multidisciplinary fields. For example, patterned SAMs have led to lithography and micromachining. Mimicking cell membrane and lipid bilayers, SAMs consisting of molecular recognition or surface receptors have led to the development of chemical or biological sensors. Using a self-assembly approach, we will learn to use the smallest units—molecular building blocks as the information carrier and exploit chemical and physical interactions at the molecular level.

Multilayer polar structures of molecular self-assemblies have been demonstrated without the requirement of the electric field poling to achieve highly aligned microstructures. We have discussed several systems that exhibit second-order nonlinearities larger than poled polymeric materials. Key issues being addressed are chemical, mechanical, and thermal stabilities of these microstructured systems. Optical devices can be fabricated through molecular engineering of chromophoric SAMs and polar SAMs once efficient routes are discovered to achieve stable systems with required film thickness. In closing, we would emphasize that the construction of polar SAMs is a paradigm on controlled nanostructures and desired physical properties for contemporary materials chemistry.

REFERENCES

1. Folkers, J. P., Laibinis, P. E., and Whitesides, G. M., Self-assembled monolayers of alkanethiols on gold: Comparisons of monolayers containing mixtures of short-chain and long-chain constituents with CH$_3$ and CH$_2$OH terminal groups, *Langmuir*, *8*(5), 1330–1341 (1992).

2. Gun, J., and Sagiv, J., On the formation and structure of self-assembling monolayers.
    3. Time of formation, solvent retention, and release, *J. Colloid Interf. Sci.*, *112*(2),
    457–472 (1986).

3. Laibinis, P. E., Fox, M. A., Folkers, J. P., and Whitesides, G. M., Comparisons of self-
    assembled monolayers on silver and gold: Mixed monolayers derived from $HS(CH_2)_{21}X$
    and $HS(CH_2)_{10}Y$ (X, Y = CH_3, CH_2OH) have similar properties, *Langmuir*, *7*(12),
    3167–3173 (1991).

4. Tour, J. M., Jones, L., Pearson, D. L., Lamba, J. J. S., Burgin, T. P., Whitesides, G.
    M., Allara, D. L., Parikh, A. N., and Atre, S. V., Self-assembled monolayers and mul-
    tilayers of conjugated thiols, alpha, omega-dithiols, and thioacetyl-containing adsor-
    bates: Understanding attachments between potential molecular wires and gold surfaces,
    J. Am. Chem. Soci., *117*, 9529–9534 (1995).

5. Whitesides, G. M., and Laibinis, P. E., Wet chemical approaches to the characterization
    of organic-surfaces: Self-assembled monolayers, wetting, and the physical organic-
    chemistry of the solid liquid interface, *Langmuir*, *6*(1), 87–96 (1990).

6. Laibinis, P. E., Hickman, J. J., Wrighton, M. S., and Whitesides, G. M., Orthogonal
    self-assembled monolayers: Alkanethiols on gold and alkane carboxylic-acids on alu-
    mina, *Science*, *245*, 845–847 (1989).

7. Angst, D. L., and Simmons, G. W., Moisture absorption characteristics of organosilox-
    ane self-assembled monolayers, *Langmuir*, *7*(10), 2236–2242 (1991).

8. Karpovich, D. S., and Blanchard, G. J., Direct measurement of the adsorption-kinetics
    of alkanethiolate self-assembled monolayers on a microcrystalline gold surface, *Lang-
    muir*, *10*(9), 3315–3322 (1994).

9. Delamarche, E., Michel, B., Kang, H., and Gerber, C., Thermal-stability of self-
    assembled monolayers, *Langmuir*, *10*(11), 4103–4108 (1994).

10. Caldwell, W. B., Chen, K. M., Herr, B. R., Mirkin, C. A., Hulteem, J. C., and Van-
    duyne, R. P., Self-assembled monolayers of ferrocenylazobenzenes on Au(111), *Lang-
    muir*, *10*(11), 4109–4115 (1994).

11. Ohtsuka, T., Sato, Y., and Uosaki, K., Dynamic ellipsometry of a self-assembled
    monolayer of a ferrocenylalkanethiol during oxidation–reduction cycles, *Langmuir*,
    10(10), 3658–3662 (1994).

12. Matsuda, N., Yoshii, K., Ataka, K., Osawa, M., Matsue, T., and Uchida, I., Surface-
    enhanced infrared and Raman studies of electrochemical reduction of self-assembled
    monolayers formed from para-nitrohiophenol at silver, *Chem. Lett.*, *7*, 1385–1388
    (1992).

13. Paulson, S., Morris, K., and Sullivan, B. P., A general preparative route to self-
    assembled monolayer surfaces of polypyridine ligands and their metal-complexes, *J.
    Chem. Soc., Chem. Commun.*, *21*, 1615–1617 (1992).

14. Koloski, T. S., Dulcey, C. S., Haralson, Q. J., and Calvert, J. M., Nucleophilic
    displacement—reactions at benzyl halide self-assembled monolayer film surfaces,
    Langmuir, *10*(9), 3122–3133 (1994).

15. Badia, A., Back, R., and Lennox, R. B., Phase-transitions in self-assembled monolayers
    directed by electrochemistry, *Angew. Chem. Int. Ed. Engl.*, *33*(22), 2332–2335 (1994).

16. Garrell, R. L., and Chadwick, J. E., Structure, reactivity and microrheology in self-
    assembled monolayers, *Colloids Surf. A—Physicochem. Eng. Aspects*, *93*, 59–72
    (1994).

17. Zhao, W. Z., Kim, M. W., Wurm, D. B., Brittain, S. T., and Kim, Y. T., Silver *n*-
    octadecanethiolate Langmuir monolayers mimicking self-assembled monolayers on sil-
    ver, *Langmuir*, *12*(2), 386–391 (1996).

18. Kunitake, T., Supermolecular engineering based on self-assembling monolayers and
    bilayers, *Thin Solid Films*, *210*(1–2), 48–50 (1992).

19. Liu, Q. X., and Xu, Z. H., Self-assembled monolayer coatings on nanosized magnetic particles using 16-marcapthohexadecanoic acid, *Langmuir, 11*(12), 4617–4622 (1995).

20. Delamarche, E., Michel, B., Gerber, C., Anselmetti, D., Guntherodt, H. J., Wolf, H., and Ringsdorf, H., Real-space observation of nanoscale molecular domains in self-assembled monolayers, *Langmuir, 10*(9), 2869–2871 (1994).

21. (a) Ulman, A., Eilers, J. G., and Tillman, N., *Langmuir, 5*(9), 1147 (1989); (b) Bareman, J. P., and Klein, M. L., *J. Phys. Chem.*, *94*, 5202 (1990); (c) Nuzzo, R. G., Dubois, L. H., and Allara, D. L., *J. Am. Chem. Soc.*, *112*, 558 (1990); (d) Porter, M. D., Bright, T. B., Allara, D. L., and Chidsey, C. E. D., *J. Am. Chem. Soc.*, *109*, 3559 (1987); (e) Bryant, M. A., and Pemberton, J. E., *J. Am. Chem. Soc.*, *113*(22), 8284–8293 (1991).

22. Bryant, M. A., and Pemberton, J. E., Surface Raman-scattering of self-assembled monolayers formed from 1-alkanethiols at silver, *J. Am. Chem. Soc.*, *113*(10), 3629–3637 (1991).

23. Tao, Y. T., Lee, M. T., and Chang, S. C., Effect of biphenyl and naphthyl groups on the structure of self-assembled monolayers: Packing, orientation, and wetting properties, *J. Am. Chem. Soc.*, 115(21), 9547–9555 (1993).

24. Laibinis, P. E., Whitesides, G. M., Allara, D. L., Tao, Y. T., Parikh, A. N., and Nuzzo, R. G., Comparison of the structures and wetting properties of self-assembled monolayers of normal-alkanethiols on the coinage metal-surfaces, Cu, Ag, Au, *J. Am. Chem. Soc.*, *113*(19), 7152–7167 (1991).

25. Samant, M. G., Brown, C. A., and Gordon, J. G., Structure of an ordered self-assembled monolayer of docosyl mercaptan on gold(111) by surface x-ray diffraction, *Langmuir*, *7*(3), 437–439 (1991).

26. (a) Samant, M. G., Brown, C. A., and Gordon, J. G., Formation of an ordered self-assembled monolayer of docosaneselenol on gold(111): Structure by surface x-ray diffraction, *Langmuir*, *8*(6), 1615–1618 (1992); (b) Fenter, P., Eberhardt, A., and Eisengerger, P., Self-assembly of *n*-alkyl thiols as disulfides on Au(111), *Science*, *266*, 1216–1218 (1994).

27. Lenk, T. J., Hallmark, V. M., Hoffmann, C. L., Rabolt, J. F., Castner, D. G., Erdelen, C., and Ringsdorf, H., Structural investigation of molecular-organization in self-assembled monolayers of a semifluorinated amidethiol, *Langmuir*, *10*(12), 4610–4617 (1994).

28. (a) Fragneto, G., Lu, J. R., McDermott, D. C., Thomas, R. K., Rennie, A. R., Gallagher, P. D., and Satija, S. K., Structure of monolayers of tetraethylene glycol monododecyl ether adsorbed on self-assembled monolayers on silicon: A neutron reflectivity study, *Langmuir*, *12*(2), 477–486 (1996). (b) Chidsey, C. E. D., Liu, G. Y., Rowntree, P., and Scoles, G., Molecular order at the surface of an organic monolayer studied by low energy helium diffraction, *J. Chem. Phys.*, *91*, 4421–4423 (1989).

29. Widrig, C. A., Alves, C. A., and Porter, M. D., Scanning tunneling microscopy of ethanethiolate and normal-octadecanethiolate monolayers spontaneously adsorbed at gold surfaces, *J. Am. Chem. Soc.*, *113*(8), 2805–2810 (1991).

30. Salmeron, M., Neubauer, G., Folch, A., Tomitori, M., Ogletree, D. F., and Sautet, P., Viscoelastic and electrical-properties of self-assembled monolayers on Au(111) films, *Langmuir*, *9*(12), 3600–3611 (1993).

31. Sprik, M., Delamarche, E., Michel, B., Rothlisberger, U., Klein, M. L., Wolf, H., and Ringsdorf, H., Structure of hydrophilic self-assembled monolayers: A combined scanning-tunneling-microscopy and computer-simulation study, *Langmuir*, *10*(11), 4116–4130 (1994).

32. Alley, R. L., Komvopoulos, K., and Howe, R. T., Self-assembled monolayer film for enhanced imaging of rough surfaces with atomic-force microscopy, *J. Appl. Phys.*, *76*(10), 5731–5737 (1994).

33. Strong, L., and Whitesides, G. M., Structures of self-assembled monolayer films of organosulfur compounds adsorbed on gold single-crystals: Electron-diffraction studies, *Langmuir*, *4*(3), 546–558 (1988).

34. Sabatini, E., Cohenboulakia, J., Bruening, M., and Rubinstein, I., Thioaromatic monolayers on gold: A new family of self-assembling monolayers, *Langmuir*, *9*(11), 2974–2981 (1993).

35. Ulman, A., Evans, S. D., and Snyder, R. G., Self-assembled monlayers of alkanethiols on gold: Sulfone groups enhancing two-dimensional organization, *Thin Solid Films*, *210*(1–2), 806–809 (1992).

36. Evans, S. D., Goppertberarducci, K. E., Urankar, E., Gerenser, L. J., and Ulman, A., Monolayers having large in-plane dipole moments: Characterization of sulfone-containing self-assembled monolayers of alkanethiols on gold by Fourier-transform infrared-spectroscopy, x-ray photoelectron-spectroscopy, and wetting, *Langmuir*, *7*(11), 2700–2709 (1991).

37. Evans, S. D., Uranker, E., Ulman, A., and Ferris, N., Self-assembled monolayers of alkanethiols containing a polar aromatic group: Effects of the dipole position on molecular packing, orientation, and surface wetting properties, *J. Am. Chem. Soc.*, *113*(11), 4121–4131 (1991).

38. Tillman, N., Ulman, A., Schildkraut, J. S., and Penner, T. L., Incorporation of phenoxy groups in self-assembled monolayers of trichlorosilane derivatives: Effects on film thickness, wettability, and molecular-orientation, *J. Am. Chem. Soc.*, *110*(18), 6136–6144 (1988).

39. Murphy, M. A., Nordgren, C. E., Fischetti, R. F., Blaise, J. K., Peticolas, J. J., and Bean, J. C., Structural study of the annealing of alkylsiloxane self-assembled monolayers on silicon by high-resolution x-ray diffraction, *J. Phys. Chem.*, *99*(38), 14039–14051 (1995).

40. Bain, C. D., Biebuyck, A., and Whitesides, G. M., Comparison of self-assembled monolayers on gold: Coadsorption of thiols and disulfides, *Langmuir*, *5*(3), 723–727 (1989).

41. Bain, C. D., and Whitesides, G. M., Molecular-level control over surface order in self-assembled monolayer films of thiols on gold, *Science*, *240*, 62–63 (1988).

42. Mathauer, K., and Frank, C. W., Binary self-assembled monolayers as prepared by successive adsorption of alkyltrichlorosilanes, *Langmuir*, *9*(12), 3446–3451 (1993).

43. Lercel, M. J., Tiberio, R. C., Chapman, P. F., Craighead, H. G., Sheen, C. W., Parikh, A. N., and Allara, D. L., Self-assembled monolayer electron-beam resists on GAAS and SiO$_2$, *J. Vac. Sci. Technol. B*, *11*(6), 2823–2828 (1993).

44. Xia, Y. N., Zhao, X. M., Kim, E., and Whitesides, G. M., A selective etching solution for use with patterned self-assembled monolayers of alkanethiolates on gold, *Chem. Mater.*, *7*(12), 2332–2337 (1995).

45. Calvert, J. M., Lithographic patterning of self-assembled films, *J. Vac. Sci. Technol. B*, *11*(6), 2155–2163 91993).

46. Jeon, N. L., Clem, P. G., Nuzzo, R. G., and Payne, D. A., Patterning of dielectric oxide thin-layers by microcontact printing of self-assembled monolayers, *J. Mater. Res.*, *10*(12), 2996–2999 (1995).

47. Biebuyck, H. A., and Whitesides, G. M., Self-organization of organic liquids on patterned self-assembled monolayers of alkanethiolates on gold, *Langmuir*, *10*(8), 2790–2793 (1994).

48. Palegrosdemange, C., Simon, E. S., Prime, K. L., and Whitesides, G. M., Formation of self-assembled monolayers by chemisorption of derivatives of oligo(ethylene glycol) of structure HS(CH$_2$)$_{11}$(OCH$_2$CH$_2$)meta-OH on gold, *J. Am. Chem. Soc.*, *113*(1), 12–20 (1991).

49. Lopez, G. P., Biebuyck, H. A., Harter, R., Kumar, A., and Whitesides, G. M., Fabrication and imaging of two-dimensional patterns of proteins on self-assembled mono-

layers by scanning electron-microscopy, *J. Am. Chem. Soci.*, *115*(23), 10774–10781 (1993).

50. Dunaway, D. J., and McCarley, R. L., Scanning force microscopy studies of enhanced metal nucleation: Au vapor-deposited on self-assembled monolayers of substituted silanes, *Langmuir*, *10*(10), 3598–3606 (1994).

51. Frisbie, C. D., Martin, J. R., Duff, R. R., and Wrighton, M. S., Use of high lateral resolution secondary ion mass-spectrometry to characterize self-assembled monolayers on microfabricated structures, *J. Am. Chem. Soc.*, *114*(18), 7142–7145 (1992).

52. (a) Singer, K. D., Sohn, J. E., and Lalama, S. J., *Appl. Phys. Lett.*, *49*, 248–250 (1986); (b) Singer, K. D., Kuzyk, M. G., and Sohn, J. E., *J. Opt. Soc. Am. B*, *4*, 968–975 (1987).

53. (a) Ye, C., Marks, T. J., Yang, Y., and Wong, G. K., *Macromolecules*, *20*, 2322–2324 (1987); (b) Ye, C., Minami, N., Marks, T. J., Yang, J., and Wong, G. K., *Macromolecules*, *21*, 2901–2904 (1988); (c) Dai, D. R., Marks, T. J., Yang, J., Lundquist, P. M., and Wong, G. K., *Macromolecules*, *23*, 1894–1896 (1990); (d) Park, J., Marks, T. J., Yang, J., and Wong, G. K., *Chem. Mater.*, *2*, 229–231 (1990).

54. (a) Singer, K. D., Kuzyk, M. G., Holland, W. R., Sohn, J. E., Lalama, S. J., Commizzoli, R. B., Katz, H. E., and Schilling, M. L., *Appl. Phys. Lett.*, *53*, 1800–1802 (1988); (b) Eich, M., Sen, A., Looser, H., Bjorklund, G. C., Swalen, J. D., Twieg, R., and Yoon, D. Y., *J. Appl. Phys.*, *66*, 2559–2567 (1989).

55 (a) Hampsch, H. L., Yang, J., Wong, G. K., and Torkelson, J. M., *Macromolecules*, *21*, 526–528 (1988); (b) Hampsch, H. L., Yang, J., Wong, G. K., and Torkelson, J. M., *Poly. Commun.*, *30*, 40–43 (1989).

56. (a) Li, D., Ratner, M. A., Marks, T. J., Yang, J., Zhang, C. H., and Wong, G. K., preliminary communication, in *Abstracts, American Chemical Soc. Nat. Mtg.*, Boston, MA, April 23–27, 1990, ORG 288; (b) Li, D., Ratner, M. A., Marks, T. J., Yang, J., Zhang, C. H., and Wong, G. K., *J. Am. Chem. Soc.*, *112*, 7389–7390 (1990).

57. (a) Lee, H., Kepley, L. J., Hong, H. G., and Mallouk, T. E., *J. Am. Chem. Soc.*, *110*, 618–620 (1988); (b) Lee, H., Kepley, L. J., Hong, H. G., Akhter, S., and Mallouk, T. E., *J. Phys. Chem.*, *92*, 2597–2601 (1988); (c) Ulman, A., and Tillman, N., *Langmuir*, *5*, 1418–1420 (1989); (d) Tillman, N., Ulman, A., and Penner, T. L., *Langmuir*, *5*, 101–111 (1989); (e) Tillman, N., Ulman, A., and Elman, J. F., *Langmuir*, *5*, 1020–1026 (1989).

58. (a) Wu, M. K., Ashburn, J. R., Torng, C. J., Hor, P. H., Meng, Y. Q., and Chu, C. W., *Phys. Rev. Lett.*, *58*, 908–910 (1987); (b) Tarascon, J. M., Greene, L. H., McKinnon, W. R., and Hull, C. W., *Phys. Rev.*, *B35*, 7115–7118 (1987); (c) Hwu, S., Song, S. N., Thiel, J., Poeppelmeir, K. R., Ketterson, J. B., and Freeman, A. J., *Phys. Rev.*, *B35*, 7119–7121 (1987).

59. (a) Kuhn, H., Mobius, D., and Bucher, H., *Techniques of Chemistry* (A. Weissberger and B. W. Rossiter, eds.), Wiley, New York, 1972, Vol. 1, Part III B, pp. 577–702; (b) *Proceedings of the First International Conference on Langmuir–Blodgett Films*, Durham, 1982; (c) Fendler, J. H., *Membrane Mimetic Chemistry*, Wiley–Interscience, New York, 1982.

60. (a) Bosshard, C. H., Kupfer, M., Gunter, P., Pasquier, C., Zahir, S., and Seifert, M., *Appl. Phys. Lett.*, *56*, 1204–1206 (1990); (b) Popovitz-Biro, R., Hill, K., Landau, E. M., Lahav, M., Leiserowitz, L., Sagiv, J., Hsiung, H., Meredith, G. R., and Vanherzeele, H., *J. Am. Chem. Soc.*, *110*, 2672–2674 (1988); (c) Cross, G. H., Peterson, I. R., Girling, I. R., Cade, N. A., Goodwin, M. J., Carr, N., Sethi, R. S., Marsden, R., Gary, G. W., Lacey, D., McRoberts, A. M., Scrowston, R. M., and Toyne, K. J., *Thin Solids Films*, *156*, 39–52 (1988).

61. (a) Messier, J., Kajar, F., Prasad, P., and Ulrich, D. (eds.) *Nonlinear Optical Effects in Organic Polymers*, Kluwer Academic Publishers, Dordrecht, 1989; (b) Khanarian, G.

(ed.) *Nonlinear Optical Properties of Organic Materials*, Bellingham, Washington, SPIE, 1988; (c) Heeger, A. J., Orenstein, J., Ulrich, D. R. (eds.), *Nonlinear Optical Properties of Polymers*, Mats. Res. Soc. Symp. Proc., 109, MRS, Pittsburgh, 1988; (d) Chemla, D. S., and Zyss, J. (eds.), *Nonlinear Optical Properties of Organic Molecules and Crystals*, Academic Press, New York, 1987, Vols. 1 and 2; (e) Zyss, J. *J. Mol. Electron.*, *1*, 25–56 (1985); (f) Williams, D., *J. Angew, J. Chem. Int. Ed. Engl.*, *23*, 690–703 (1984).

62. (a) Schildkraut, J. S., Penner, R. L., Willand, C. S., and Ulman, A., *Opt. Lett.*, *13*, 134–136 (1988); (b) Lupo, D., and Ledoux, I., *J. Opt. Soc. Am. B*, *5*, 300–308 (1988); (c) Ledoux, I., Josse, D., Vidakovic, P., Zyss, J., Hann, R. A., Gordon, P. F., Bothwell, B. D., Gupta, S. K., Allen, S., Robin, P., Chastaing, E., and Dubois, J. C., *Europhys. Lett.*, *3*, 803–809 (1987); (d) Hayden, L. M., Kowel, S. T., and Srinivasan, M. P., *Opt. Commun.*, *61*, 351–356 (1987).

63. (a) Kuhn, H., *J. Photochem.*, *10*, 111–132 (1979); (b) Kuhn, H., *Thin Solid Films*, *99*, 1–16 (1983).

64. (a) Wasserman, S. R., Tao, Y. T., and Whitesides, G. M., *Langmuir*, *5*, 1074–1087 (1989); (b) Murphy, M. A., Nordgren, C. E., Fischetti, R. F., Blaise, J. K., Peticolas, J. J., and Bean, J. C., Structural study of the annealing of alkylsiloxane self-assembled monolayers on silicon by high-resolution x-ray diffraction, *J. Physi. Chem.*, *99*(38), 14039–14051 (1995); (c) Troughton, E. B., Bain, C. D., Whitesides, G. M., Nuzzo, R. G., Allara, D. L., and Porter, M. D., *Langmuir*, *4*, 365–385 (1988); (d) Randall, S., Farley, H., and Whitesides, G. M., *Langmuir*, *3*, 62–76 (1987).

65. (a) Maoz, R., and Sagiv, J., *J. Colloid Interf. Sci.*, *100*(2), 465–496 (1984); (b) Netzer, L., Iscovici, R., and Sagiv, J., *Thin Solid Film*, *99*, 235–241 (1983); (c) Sagiv, J., *Israel J. Chem.*, *18*, 339–345 (1979).

66. (a) Li, D., Minami, N., Ratner, M. A., Ye, C., Marks, T. J., Yang, J., and Wong, G. K., *Synth. Met.*, *28*, D585–D593 (1989); (b) Li, D., Marks, T. J., and Ratner, M. A., *Mater. Res. Soc. Symp. Proc.*, *109*, 149–155 (1988); (c) Li, D., Yang, J., Ye, C., Ratner, M. A., Wong, G. K., and Marks, T. J., in *Nonlinear Optical and Electroactive Polymers*, (P. N. Prasad and D. R. Ulrich, eds.), Plenum Press, New York, 1988, pp. 217–228.

67. (a) Ringdorf, H., Venzmer, J., *Angew. Chem. Int. Ed. Engl.*, *27*, 113–158 (1988); (b) Luisi, P. L., *Agnew. Chem. Int. Ed. Engl.*, *24*, 439–450 (1985); (c) Mittal, K. L. (ed.), *Colloid Dispersions and Micellar Behavior*, ACS Symp. Ser. 9., American Chemical Society, Washington, DC, (1975); (d) Luisi, P. L., and Straub, B. E. (eds.), *Reverse Micelles*, Plenum Press, New York, 1982.

68. (a) Darszon, A., Vandenberg, C. A., Ellisman, M. H., and Montal, M., *J. Cell Biol.*, *81*, 446–452 (1979); (b) Kornberg, R. D., and McConnell, H. M., *Biochemistry*, *10* 111–1120 (1971).

69. Somorjai, G. A., *Chemistry in Two Dimensions*, Cornell University Press, Utica, NY, 1981.

70. (a) Zental, R., *Angew, Chem. Int. Ed., Engl.*, *28*, 1406–1407 (1989); (b) Weiss, K., and Lossel, G., *Agnew. Chem. Int. Ed. Engl.*, *28*, 62–64 (1989); (c) Iwasawa, Y., *Adv. Catal.*, *35*, 187–264 (1987); (d) Asakura, K., and Iwasawa, Y., *Chem. Lett.*, 859–862 (1986).

71. Yang, X., McBranch, D., and Li, D. Q., *Angew. Chem. Int. Ed. Engl.*, *35*(5), 538–540 (1996).

72. (a) Tournilhac, F., Micond, J. F., Simon, J., Weber, P., Guillon, D., and Skoulios, A., *Mol. Cryst. Liq. Cryst.*, *2*, 55–61 (1987); (b) Lu, T. H., Lee, T. J., Wong, C., and Kuo, K. T., *J. Chin. Chem. Soc.*, *25*, 131–139 (1979); (c) Kuo, K. L., *J. Chin. Chem. Soc.*, *25*, 131–139 (1978).

73. (a) Li, D., Marks, T. J., and Ratner, M. A., *Chem. Phys. Lett.*, *131*, 370–375 (1986); (b) Li, D., Ratner, M. A., and Marks, T. J., *J. Am. Chem. Soc.*, *110*, 1707–1715 (1988);

(c) Katz, H. E., Scheller, G., Putvinski, F. M., et al., C. E. D., *Science*, *254*, 1485–1487 (1991).

74. (a) Below, N. V., *Crystal Chemistry of Large Cation Silicates*, Consultants Bureau, New York, 1963; (b) Hodgson, A. A., *Fibrous Silicates*, Royal Institute of Chemistry, London, 1964; (c) Britten, M. D., *Angew. Chem. Int. Ed.*, *15*, 346–354 (1976); (d) Iler, R. K., *The Chemistry of Silica: Solubility, Polymerisation, Colloid and Surface Properties and Biochemistry*, Wiley, New York, 1979; (e) Unger, K. K., *Porous Silica*, Elsevier, Amsterdam, 1979.

75. (a) Carey, F. A., and Sundberg, R. J., *Advanced Organic Chemistry*, 2nd ed., Plenum Press, New York, 1984, part A, pp. 179–190; (b) March J., *Advanced Organic Chemistry*, 3rd ed., Wiley–Interscience, New York, 1985, pp. 237–250; (c) Taft, R. W., and Topsom, R. D., *Prog. Phys. Org. Chem.*, *16*, 1–84 (1987); (d) Topsom, R. D., *Prog. Org. Chem.*, *16*, 85–124 (1987).

76. Zhang, Z. Q., You, X. Z., Ma, S. H., and Wei, Y., *J. Mater. Chem.*, *5*(4), 643–647 (1995).

77. Choi, J. U., Lim, C. B., Kim, J. H., Chung, T. Y., Hahn, J. H., Kim, S. B., and Park, J. W., *Synth. Met.*, *71*(1–3), 1729–1730 (1995).

78. Neff, G. A., and Page, C. J., *Langmuir*, *12*(2), 238–242 (1996).

79. Yokoyama, S., Yamada, T., Kajikawa, K., Kakimoto, M., Imai, Y., Takezoe, H., and Fukuda, A., Novel orientation of azobenzene pendent group in hybrid monolayers composed of polyimide Langmuir–Blodgett–Film and alkyl polysiloxane self-assembled monolayer, *Langmuir*, *10*(12), 4599–4605 (1994).

80. Li, D. Q., Synthesis and Characterization of Chromophoric Self-Assembled Multilayers as Nonlinear Optical Materials, Ph.D thesis, Northwestern University, 1990.

81. Kakkar, A. K., Yitzchaik, S., Roscoe, S. B., Hubota, F., Allan, A. S., and Marks, T. J., *Langmuir*, *9*(2), 388–390 (1993).

82. (a) deGennes, P. G., *Rev. Mod. Phys.*, *57*, 827–863 (1985); (b) Pomeau, Y., and Vannimenus, J., *J. Colloid Interf. Sci.*, *10*, 477 (1985); (c) Schwartz, L. W., and Garoff, S., *Langmuir*, *1*, 219–230 (1985).

83. (a) Bunding, K. A., Bell, M. I., and Durst, R. A., *Chem. Phys. Lett.*, *89*, 54–58 (1982); (b) Kobayashi, Y., and Itoh, K., *J. Phys. Chem.*, *89*, 5174–5178 (1985); (c) Long, D. A. and George, W. O., *Spectrochimica Acta*, *19*, 1777–1790 (1963); (d) Spinner, E., *Aust. J. Chem.*, *20*, 1805–1813 (1967).

84. Lin, W. B., Yitzchaik, S., Lin, W. P., Malik, A., Durbin, M. K., Richter, A. G., Wong, G. K., Dutta, P., and Marks, T. J., *Angew. Chem. Int. Ed.*, *34*(13–14), 1497–1499 (1995).

85. Roscoe, S. B., Yitzchaik, S., Kakkar, A. K., Marks, T. J., Lin, W. P., and Wong, K., *Langmuir*, *10*(5), 1337–1339 (1994).

86. Bella, S. D., Fragala, I., Ratner, M. A., Marks, T. J., *Chem. Mater.*, *7*(2), 400–404 (1995).

87. Buscher, C. T., McBranch, D., and Li, D. Q., *J. Am. Chem. Soc.*, *118*(12), 2950–2953 (1996).

88. Bolembergen, N., and Pershan, P. S., *Phys. Rev.*, *128*(2), 606–622 (1962).

89. Zhang, T. G., Zhang, C. H., and Wong, G. K., *J. Am. Opt. Soc.*, *7*, 902–907 (1990).

90. Ye, C., Minami, N., Marks, T. J., Yang, J., and Wong, G. K., in *Nonlinear Optical Effects in Organic Polymers* (J. Messier, F. Kajzar, P. N. Prasad, and D. R. Ulrich, eds.), NATO ASI Series, pp. 173–183. Kluwer Academic, Dordrecht (1989).

91. Maker, P. D., Terhune, R. W., Nisenoff, M., and Savage, C. M., *Phys. Rev. Lett.*, *8*, 21 (1962).

92. (a) Campbell, C., *Surface Acoustic Wave Devices and Their Signal Processing Applications*, Academic Press, Boston, 1989; (b) Feldmann, M., and Henaff, J., *Surface Acoustic Wave for Signal Processing*, Artech House, Boston, 1989; (c) Grate, J. W.,

Snow, A., Ballantine, D. S., Jr., Wohltjen, H., Abraham, M. H., McGill, R. A., and Sasson, P., *Anal. Chem.*, *60*(17), 869–875 (1988).

93. Grate, J. W., and Klusty, M., *Anal. Chem.*, *63*(17), 1719–1727 (1991).

94. Grate, J. W., Klusty, M., McGill, R. A., Abraham, M. H., Whiting, G., and Andonian-Haftvan, J., *Anal. Chem.*, *64*, 610 (1992).

95. (a) Cornell, F. W., in *Proceedings of the Natl. Symp. on Measuring and Interpreting VOCs in Soil: State of the Art and Research Needs, Las Vegas, NV*, Environmental Liability Management, Inc., Princeton, NJ, 1993; (b) Henricks, A. D., and Grant, D. E., in *The Cost Effectiveness of Field Screening for VOCs, Emerging Technology Symposium*, Los Alamos National Laboratory, Los Alamos, NM, 1993.

96. Croft, A. P., and Bartsch, R. A., *Tetrahedron*, *39*(9), 1417 (1983).

97. Gutsche, C. D., *Calixarenes*, Royal Society of Chemistry, Cambridge, 1989.

98. Moore, L. W., Springer, K. N., Shi, J.-X., Yang, X., Swanson, B. I., and Li, D., Surface acoustic wave chemical microsensors based on covalently bound self-assembled host monolayers, *Adv. Mater.*, *7*(8), 729–731 (1995).

99. Schierbaum, K. D., Weiss, T., Vanvelzen, E. U. T., Engbersen, J. F. J., Reinhoudt, D. N., Gopel, W., *Science*, *265*, 1413–1415 (1994).

100. Dean, J. A., *Lange's Handbook of Chemistry*, 13th ed., McGraw-Hill Book Co., New York, 1985, Chap. 10, pp. 28–54.

101. Cramer, F., and Spatz, H.-Ch., *J. Am. Chem. Soc.*, *89*, 14 (1967).

102. (a) Yonemura, H., Kasahara, M., Saito, H., Nakamura, H., and Matsuo, T., *J. Phys. Chem.*, *96*, 5765 (1992); (b) Teiichi, M., Kazuaki, H., and Satoshi, M., *Chem. Express.*, *4*, 645–548 (1989).

103. Harata, K., *Bull. Chem. Soc. Jpn.*, *48*, 2049 (1975).

104. Rojas, M. T., and Kaifer, A. E., *J. Am. Chem. Soc.*, *117*(21), 5883–5894 (1995).

105. Rubinstein, I., Steinberg, S., Tor, Y., Shanzer, A., and Sagiv, J., *Nature*, *331*, 426–429 (1988).

106. Steinberg, S., and Rubinstein, I., *Langmuir*, *8*, 1183–1187 (1992).

107. Knichel, M., Heiduschka, P., Beck, W., Jung, G., and Gopel, W., *Sensors Actuators B—Chemical*, *28*(2), 85–94 (1995).

108. Mrksich, M., Sigal, G. B., and Whitesides, G. M., Surface-plasmon rsonance permits *in-situ* [sic] measurement of protein adsorption on self-assembled monolayers of alkanethiolates on gold, *Langmuir*, *11*(11), 4383–4385 (1995).

109. Mrksich, M., Grunwell, J. R., and Whitesides, G. M., *J. Am. Chem. Soc.*, *117*(48), 12009–12010 (1995).

110. Spinke, J., Liley, M., Schmitt, F. J., Guder, H. J., Angermaier, L., and Knoll, W., *Chem. Phys.*, *99*(9), 7012–7019 (1993).

111. Spinke, J., Liley, M., Schmitt, F. J., Guder, H. J., Angermaier, L., and Knoll, W., *Langmuir*, *9*(7), 1821–1825 (1993).

112. Schmitt, F. J., Haussling, L., Ringsdorf, H., and Knoll, W., *Thin Solid Films*, *210*(1–2), 815–817 (1992).

113. Haussling, L., Ringsdorf, H., Schmitt, F. J., and Knoll, W., *Langmuir*, *7*(9), 1837–1840 (1991).

114. Duschl, C., Liley, M., Corradin, G., and Vogel, H., *Biophys. J.*, *67*(3), 1229–1237 (1994).

115. Willner, I., Riklin, A., Shoham, B., Rivenson, D., and Katz, E., *Adv. Mater.*, *5*(12), 912–915 (1993).

21

Electrooptic Polymers and Applications: Materials Based on Heteroaromatic Nonlinear Optical Chromophores

Alex K-Y. Jen and Yue Zhang*
Northeastern University
Boston, Massachusetts

I. INTRODUCTION

Organic and polymeric electrooptic materials have received considerable attention in recent years due to their large nonlinearities, low dielectric constants, and ease of fabrication [1–7]. Great effort has been made in both understanding the basics of nonlinear optical effects in organics and in the application of these materials to practical devices. Many materials with electrooptic coefficients exceeding that of $LiNbO_3$, a standard electrooptic crystal, have been developed [8,9]. The use of these materials in the fabrication of high-speed (~ 100 GHz) electrooptic devices has been demonstrated [10–12].

The electrooptic effect in polymeric materials arises from the field-induced alignment [13] of second-order nonlinear optical (NLO) chromophores which normally consists of a π-electron conjugated structure terminated by an electron-withdrawing group on one end and an electron-donating group on the other [6,7]. These chromophores are often incorporated into amorphous polymers either as a dopant in guest–host systems or as a side chain covalently attached to the polymer backbone. Electric field poling is performed by heating the polymer to a softened state, normally around its glass transition temperature (T_g), followed by the application of a strong electric field. After the material is cooled to room temperature, the electric field is removed and the chromophores are locked into the aligned state. Such an electric-field-aligned arrangement is a quasi-steady-state; the chromophores with large dipole moments tend to slowly relax back to a randomly oriented state. To achieve both thermally and temporally stable electrooptic coefficient, polymers with sufficiently high glass transition temperature must be used [14]. Alternatively, cross-linking, either by a thermally induced or chemically induced process,

---

*Current affiliation: Lightwave Microsystems Corporation, Santa Clara, California

may result in more rigid structures which prevent NLO chromophores from relaxing back to random orientation [15,16].

Traditionally, second-order chromophores have been based on such conjugation paths as phenyl, stilbene, and polyenes, and electron donors/acceptors as NH$_2$, N(Me)$_2$, N(Et)$_2$, NO$_2$, dicyanovinyl, and so forth [1,6]. The first molecular hyper-polarizabilities of second-order NLO chromophores strongly depend on the strengths of the donors and acceptors. Many electron-donating groups have been tested in the design of such chromophores. Common electron-donating groups include –OCH$_3$, –OH, –NR$_2$ (R = –H, –CH$_3$, –CH$_2$CH$_3$, etc.) with increasing electronegativity. Commonly employed electron-withdrawing groups are –COH, –SO$_3$H, –CN, –NO$_2$, –C=C(CN)$_2$ and –C(CN)=C(CN)$_2$, with the tricyanovinyl group demonstrating the strongest electron-withdrawing strength. Although these moieties provide reasonable electron delocalization, there exist several deficiencies such as poor solubility, low thermal stability, and poor chemical stability. Recently, it has been demonstrated that large molecular nonlinearities can be achieved in donor–acceptor-substituted stilbenes by replacing phenyl moieties with easily polarizable heteroaromatic moieties, partially due to either the electron richness or deficiency of the five-membered heteroaromatic rings [17,18,50]. Further studies showed that by replacing the stilbene connection with a bi-thiophene or a fused-thiophene ring structure, higher thermal stability can be achieved [19]. The use of heteroaromatic moieties as electron donors to enhance optical nonlinearity and thermal stability has also been demonstrated [20–23].

This chapter discusses electrooptic polymers that are based on the incorporation of heteroaromatic nonlinear optical chromophores. Section II presents a brief introduction to the nonlinear optical and electrooptic effects in organic polymers. Material preparations and various techniques that are employed to characterize nonlinear optical chromophores and electrooptic polymers are presented in Section III. In Section IV, we will discuss various classes of NLO chromophores that are based on heteroaromatic structures, their synthesis, and optical and nonlinear optical properties. Section V discusses high-performance electrooptic polymers and their properties. Finally, the application of these materials in the fabrication of integrated optical devices such as interferometers and electrooptic switches are presented in Section VI.

II. SECOND-ORDER NONLINEAR OPTICAL EFFECTS IN POLED POLYMERS

When a material is subject to an optical field, a dipole polarization is induced which can be described as [1,2]

$$P_I(\omega) = \chi^{(1)}_{IJ}(\omega)E_J(\omega) + \chi^{(2)}_{IJK}(-\omega; \omega_1, \omega_2)E_J(\omega_1)E_K(\omega_2)$$
$$+ \chi^{(3)}_{IJKL}(-\omega; \omega_1, \omega_2, \omega_3)E_J(\omega_1)E_K(\omega_2)E_L(\omega_3) + \cdots \quad (1)$$

where $\chi^{(1)}_{IJ}(\omega)$ is the linear susceptibility, $\chi^{(2)}_{IJK}$ and $\chi^{(3)}_{IJKL}$ are the second- and third-order nonlinear susceptibilities, E_J, E_K, and so forth are the components of the electric field. Because nonlinear optical effects in polymers arise from the optical nonlinearities of constituent molecular components, the dipole polarization induced by light on the molecular level can be expressed by a similar expansion:

$$p_i(\omega) = \alpha_{ij}(\omega)E_j(\omega) + \beta_{ijk}(-\omega; \omega_1, \omega_2)E_j(\omega_1)E_k(\omega_2)$$
$$+ \gamma_{ijkl}(-\omega; \omega_1, \omega_2, \omega_3)E_j(\omega_1)E_k(\omega_2)E_l(\omega_3) + \cdots \tag{2}$$

where $\alpha_{ij}(\omega)$ is the linear polarizability, β_{ijk} and γ_{ijkl} are the first and second hyperpolarizabilities, respectively, and E_j, E_k, and so on are the components of the applied field in the molecular coordinate frame.

This chapter deals with second-order NLO effects and we will, therefore, focus on the second-order nonlinear susceptibility, $\chi^{(2)}_{IJK}$, which is related to the molecular first hyperpolarizability, β_{ijk}, by [1,2]

$$\chi^{(2)}_{IJK}(-\omega; \omega_1, \omega_2) = Nf_I(\omega)f_J(\omega_1)f_K(\omega_2)\langle\beta_{ijk}(-\omega; \omega_1, \omega_2)\rangle_{IJK} \tag{3}$$

where $f_I(\omega)$, $f_J(\omega_1)$, and $f_K(\omega_2)$ are the local field factors which describe the relationship between applied electric field and the local field felt by an individual molecule. The local field factors consist of both the Lorentz–Lorenz correction [24], f_∞, and the Onsager local field correction [25], f_0, given by

$$f_\infty = \frac{\varepsilon_\infty + 2}{3} \quad \text{and} \quad f_0 = \frac{\varepsilon(\varepsilon_\infty + 2)}{2\varepsilon + \varepsilon_\infty} \tag{4}$$

respectively. In Eq. (4), $\varepsilon_\infty = n^2$ and ε are the optical-frequency and low-frequency dielectric constants, respectively.

For a material to exhibit second-order optical nonlinearity, it must not possess a center of symmetry. On the molecular level, nonlinear optical molecules with noncentrosymmetry is obtained by using donor- and acceptor-substituted conjugated structures. On the bulk level, second-order NLO materials can be processed by growing crystals with proper symmetries [1,26], forming Langmuir–Blodgett films [27], use of self-assembled thin layers [28,29], use of chiral chromophores [30], or poling the NLO molecules with a strong electric field [13], which is the main subject of this chapter.

When a DC electric field is applied across a medium containing molecules with a nonzero dipole moment, the interaction between the field and the molecular dipoles leads to a reorientation of the dipoles toward the direction of the applied field. This reorientation gives rise to a nonzero average of $\langle\beta_{ijk}(-\omega; \omega_1, \omega_2)\rangle_{IJK}$, which, under the assumption of rigid-rod molecules ($\beta = \beta_{zzz}$) and weak interaction, lead to the only nonvanishing components of the second-order nonlinear susceptibility as [31]

$$\chi^{(2)}_{ZZZ} = N\beta FL_3(p) \tag{5}$$
$$\chi^{(2)}_{ZXX} = N\beta F\{L_1(p) - L_3(p)\}$$

where $L_1(p)$ and $L_3(p)$ are the first- and third-order Langevin functions given by

$$L_1(p) = \frac{1}{3}p - \frac{1}{45}p^3 + \frac{2}{945}p^5 - \frac{2}{9450}p^7 + \cdots \tag{6}$$
$$L_3(p) = \frac{1}{5}p - \frac{4}{185}p^3 + \frac{8}{9450}p^5 + \cdots$$

with $p = \mu E$, where μ is the molecular dipole moment and E is the strength of the applied electric field.

The second-order susceptibility, $\chi^{(2)}_{IJK}$, gives rise to a number of processes, including second-harmonic generation, sum- and difference-frequency generation, optical rectification, and electrooptic effect. The electrooptic effect, or the Pockels effect, results from an electric-field-induced change in the material's refractive index [32],

$$\Delta n = -\frac{1}{2} n^3 r E \tag{7}$$

where the electrooptic coefficient, r, is related to the second-order susceptibility, $\chi^{(2)}_{IJK}$, by

$$r_{IJK}(-\omega; \omega, 0) = \frac{8\pi}{n^4} \chi^{(2)}_{IJK}(-\omega; \omega, 0) \tag{8}$$

The linear electrooptic effect represents a second-order nonlinear optical process where one of the interacting fields is a low-frequency electric field, and the nonlinear susceptibility $\chi^{(2)}_{IJK}(-\omega; \omega, 0)$ can be obtained from $\chi^{(2)}_{IJK}(-2\omega'; \omega', \omega')$ [often measured by second-harmonic generation (SHG)] using the two-level model [33]

$$\chi^{(2)}(-\omega; \omega, 0) = \frac{\chi^{(2)}(-2\omega; \omega, \omega)}{n^4(\omega)} \frac{(f^\omega)^2 f^0}{f^{2\omega'}(f^{\omega'})^2}$$
$$\times \frac{(3\omega_0^2 - \omega^2)(\omega_0^2 - \omega'^2)(\omega_0^2 - 4\omega'^2)}{3\omega_0^2(\omega_0^2 - \omega^2)^2} \tag{9}$$

where ω_0 is the angular frequency for the absorption maximum, ω' is the frequency of the fundamental optical field used in SHG measurement, and ω is the frequency at which the electrooptic modulation is performed.

III. MATERIAL PROCESSING AND MEASUREMENT TECHNIQUES

The development of electrooptic (EO) polymers consists of several important steps. The first step is the design, synthesis, and characterization of second-order NLO chromophores. The next step is to incorporate the chromophore into a suitable polymer matrix either as a dopant or as a side-chain group. The polymer is then processed into thin films, normally by spin-coating from solutions, and baked or cured to remove residual solvents. This section discusses the techniques employed to characterize electrooptic polymers from the molecular level to thin films. These include the measurement of molecular nonlinearity, thermal stability of EO chromophores, electrical properties, electrooptic coefficient, optical losses, and temporal stabilities of poling-induced chromophore alignment.

A. Molecular Hyperpolarizabilities

1. Electric-Field-Induced Second-Harmonic Generation

The molecular properties of a second-order chromophore are often studied in the solution phase by dissolving the chromophore into a suitable solvent. If the solution is dilute enough, the interaction between molecules can often be neglected. The electric-field-induced second-harmonic (EFISH) generation technique [34] is most commonly employed to measure the molecular first hyperpolarizability. A typical EFISH setup is presented in Fig. 1. The fundamental output at 1.064 μm from a

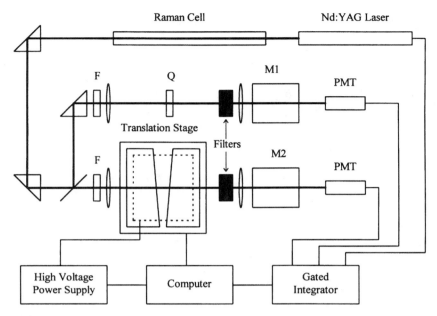

FIGURE 1 Experimental setup for second-harmonic generation. M1 and M2: monochromators; PMT: photomultiplier tube; F: filters; Q: quartz crystal.

pulsed Nd:YAG laser is used to pump a high-pressure hydrogen-filled Raman cell, and the anti-Stokes line from the Raman cell at 1.907 μm is used as the fundamental frequency for second-harmonic generation.

The fundamental beam is split to two arms. A Y-cut quartz crystal is placed in one arm to serve as a reference for two purposes: (1) to compare the SHG signal from the EFISH cell and (2) to compensate for laser power fluctuation. A set of filters is placed in each beam to allow the pass of only the 1.907-μm light which is focused onto the quartz crystal and the EFISH cell, respectively. A computer-controlled, motor-driven translation stage moves the EFISH cell across the laser beam in the signal arm. The SHG signals pass through a second set of colored glass filters and dielectric interference filters, and a monochromator ensures only lights at the second-harmonic frequency reaches the detectors which, in many cases, are photomultiplier tubes (PMT). Signals from PMT are fed into gated integrators and processed by personal computers.

The intensity of the second-harmonic light emerging from the EFISH cell is given by [35]

$$I_{2\omega} = AE_0^2 I_\omega^2 (T_1 \Gamma_G l_c^G - T_2 \Gamma_L l_c^L)^2 F(l) \tag{10}$$

where E_0 is the DC electric field, I_ω is the fundamental laser intensity, Γ_G and Γ_L are the macroscopic susceptibilities of the window material and that of the solution, respectively, and the constant A and the transmission factors T_1 and T_2 are given by

$$A = \frac{8\pi}{c}\left(\frac{8\omega}{c}\right)^2$$

$$T_1 = t_{2\omega}^G t_\omega^2 \frac{1}{n_{2\omega}^G + n_\omega^G} \frac{n_\omega^G + n_{2\omega}^L}{n_{2\omega}^G + n_\omega^L} \tag{11}$$

$$T_2 = t_{2\omega}^G t_\omega^2 t_\omega'^2 \frac{1}{n_{2\omega}^L + n_\omega^L} \frac{n_\omega^L + n_{2\omega}^L}{n_{2\omega}^G + n_\omega^L} \tag{11}$$

where

$$t_{2\omega}^G = \frac{2n_{2\omega}^G}{1 + n_{2\omega}^G}$$

$$t_\omega = \frac{2}{1 + n_\omega^G} \tag{12}$$

$$t_\omega' = \frac{2n_{2\omega}^G}{n_\omega^G + n_\omega^L}$$

and the coherence length of the solution is defined as

$$l_c = \frac{\lambda}{4(n_{2\omega}^L - n_\omega^L)} \tag{13}$$

The factor $F(l)$ describes the dependence of the SH intensity on the optical path inside the EFISH cell and is given by

$$F(l) = 2 \exp\left[-\left(\alpha_\omega + \frac{\alpha_{2\omega}}{2}\right)l\right]\left\{\cosh\left[\left(\alpha_\omega - \frac{\alpha_{2\omega}}{2}\right)l\right] - \cos\left(\frac{\pi l}{l_c}\right)\right\} \tag{14}$$

where α_ω and $\alpha_{2\omega}$ are the absorption coefficients at the fundamental and second-harmonic frequencies, respectively.

The macroscopic susceptibility of the solution, Γ_L, which consists of those of the solvent and the chromophore molecules, can be expressed in terms of the effective EFISH hyperpolarizability, γ^{eff}, as

$$\Gamma_L = f(N_S \gamma_S^{\text{eff}} + N_C \gamma_C^{\text{eff}}) \tag{15}$$

where N_S and N_C are the number density of solvent and chromophore molecules, respectively, and f is the local field factor. Therefore, the measured signal contains the second-harmonic signal from both the solvent and the chromophore molecules. After obtaining the effective Γ values, the chromophore nonlinearity is extracted by fitting the chromophore concentration dependence of Γ_L according to Eq. (15).

3. Hyper-Rayleigh Scattering

The electric-field-induced second-harmonic generation technique is a well-established method in determining molecular hyperpolarizabilities. Because an effective hyperpolarizability, γ^{eff}, is obtained from the EFISH measurement, one needs to determine the dipole moment and the second hyperpolarizability of a molecule before the first hyperpolarizability, β, can be extracted. A recently reported, yet quickly accepted, the hyper-Rayleigh scattering technique [36,37] measures the first hyperpolarizability directly from the concentration dependence of the intensity of the scattered light at the second-harmonic frequency, without requiring the knowledge of dipole moment or the second hyperpolarizability. The experimental setup is much simpler than that used for EFISH, eliminating the need for a high-voltage power supply and cell translation. The intensity of the scattered light is given by

$$I_{2\omega} = GI_0^2(N_S\beta_S^2 + N_C\beta_C^2) \tag{16}$$

where I_0 is the laser beam intensity, N_S and N_C are the number densities of the solute and solvent, respectively, and β_S and β_C are the first hyperpolarizabilities of the solute and solvent, respectively.

B. Thermal Analysis

Thermal analysis of a material provides information on its thermal stabilities which are important in determining the usefulness of the material. Differential scanning calorimetry (DSC) is often used to determine the decomposition temperature of a second-order chromophore as well as the glass transition temperature (T_g) of a polymer, which determines the temperature at which the polymer can be effectively poled.

Ultraviolet–visible (UV–vis) absorption spectroscopy can be an effective tool in determining the thermal stabilities of electrooptic polymers. Besides the loss of chromophore alignment, the instability is almost always associated with the loss of chromophore content in a polymer at elevated temperatures which can occur in several different fashions, namely (1) sublimation, which often occurs in guest–host systems, (2) chromophore degradation due to decomposition and isomerization, and (3) chromophore degradation due to chemical reaction with polymer matrices or processing solvents. In UV–vis absorption spectroscopy, thermal degradation is observed either as a shift of absorption peak, λ_{max}, in the spectrum or a combination of λ_{max} shift and reduction in absorbance. On the other hand, sublimation only results in a reduction of absorbance. Figure 2 shows the absorption spectra of a side-chain polymer after several cure/aging processes. The inset shows the maximum absorbance as a function of cure temperature. Notice that, in this case, λ_{max} remains unchanged after the polymer undergoes a complete cure cycle necessary for the conversion from polyamic acid to polyimide.

C. Electrical Properties

Determination of electrical properties of an EO polymer is important, especially if the material is to be used in multilayer structures such as a channel waveguide device. The conductivity of an EO polymer directly affects the efficiency of electric field poling in multilayer devices when at least one cladding polymer layer is employed. The general rule of thumb is that the conductivity of the active EO polymer must not be higher than that of the cladding layer. Otherwise, more field will be dropped on the cladding layer instead of on the active EO polymer, resulting in lowered poling efficiency.

To study the conductivity, an EO polymer is normally deposited by spin-coating on a substrate covered with a conductive material [e.g., indium–tin oxide (ITO)]. After necessary baking and curing, a second metal electrode is deposited on top of the polymer, either by vacuum evaporation or by plasma sputtering. The resistance

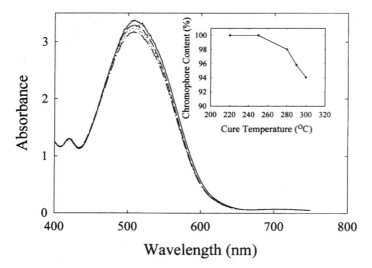

FIGURE 2 UV–vis absorption spectroscopy as a tool of examining the thermal stability of NLO chromophores and electrooptic polymers. Solid curve: 250°C/30 min; dashed curve: 280°C/15 min; dotted curve: 290°C/10 min; dash–dot curve: 300°C/10 min. Inset: Absorbance as a function of cure temperature.

of the polymer film can be measured by a digital multimeter while the film temperature is raised. The conductivity is normally measured by applying a DC voltage across the sample while monitoring the current flow through it.

A simple resistivity or conductivity measurement can also quickly provide valuable information in determining an appropriate poling temperature of an EO polymer. An EO polymer normally has very high resistivity or low conductivity at temperatures well below its glass transition temperature. As the polymer is heated to the vicinity of its T_g, the molecular segment motion sharply increases [38]. As a result, the polymer's electric resistivity sharply decreases while the conductivity increases as a result of the increased mobility of trace trapped ionic species. Figure 3 shows a typical temperature dependence of resistivity and conductivity from an EO polymer which has a T_g of 250°C.

D. Electric Field Poling

Second-order nonlinear optical effects, including the linear electrooptic effect, require that the materials possess no center of symmetry. Second-order NLO chromophores inherently possess large dipole moments whose interaction tend to make the chromophores' orientation randomly distributed, making the material centrosymmetric. Many approaches have been proposed and demonstrated to break this center of symmetry. These include (1) growing crystals with proper symmetries [1,26], (2) the formation of Langmuir–Blodgett films [27], (3) the use of self-assembled thin layers [28,29], and (4) the use of chiral chromophores [30]. However, the most commonly used method to achieve a noncentrosymmetric chromophore alignment is the electric-field-poling technique [13]. In this technique, a

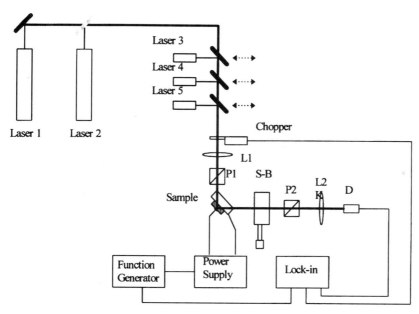

FIGURE 3 Conductivity and resistivity measurement. (a) Sample geometry and experimental setup. (b) A typical temperature dependence of resistivity and conductivity obtained from a side-chain electrooptic polymer with a glass transition temperature of 250°C.

material is first brought to a state, often by heating the sample to its glass transition temperature, where the chromophores are free to rotate, followed by the application of a strong electric field across the material to align the chromophores in the direction of the field. Cooling the sample to a temperature considerably below its T_g locks the chromophores in the aligned direction, preserving the second-order nonlinearity.

A strong electric field can be achieved by several methods. In the simplest case of contact poling, a DC voltage applied to the top electrode as shown in Fig. 3a produces a controlled electric field across the EO polymer film. The chromophores will be aligned perpendicular to the sample surface plane. Alternatively, parallel electrodes deposited on the substrate or the polymer surface can be used to achieve in-plane poling where chromophores will be aligned parallel to the surface plane [39]. It is somewhat difficult, however, to obtain uniform field distribution. The strength of the electric field achievable in contact poling is normally limited to less than 100 V/μm due to charge injection or the presence of defects or pinholes. A much stronger field can be achieved by the corona-discharge technique which has been widely used in the photographic industry [40]. In this technique, a high voltage is applied to a thin needle or wire which is placed several millimeters above the EO polymer; see Fig. 4. This high voltage causes a breakdown of the air surrounding the corona wire or needle, producing charges having the same polarity as the high voltage. These charges are driven to the surface of the EO polymer by the potential gradient between the wire and ground. Sample breakdown associated with charge injection and the presence of defects or pinholes can be effectively avoided,

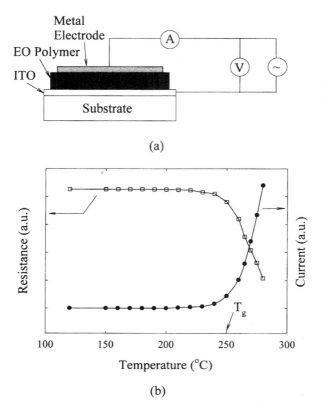

(a)

(b)

FIGURE 4 Experimental setup for corona poling; HV = high-voltage power supply.

resulting in much higher electric field. However, care must be taken in such experiment, as the charged particles may cause surface damage as they strike the sample. Such damage can be avoided with the use of a grid or mash placed between the corona wire and the sample surface. A lower voltage with the same polarity as that of the corona voltage can reduce the speed of the charged particles. A moving grid or mash also helps achieve more uniform field distribution. The use of a removable protective layer above the EO polymer can also help reduce poling-induced surface damage [41]. The protective layer, often soluble in a solvent which does not dissolve the EO polymer, is removed after poling.

E. Electrooptic Coefficient Measurement

The application of an electric field to an EO material causes a change in the material's refractive index. Thus, the characteristics of a light beam can be modulated upon traveling through the material by the application of such a field. The most commonly employed is the ellipsometric technique using an experimental configuration shown in Fig. 5 [42]. A thin film of polymer is spin-coated onto ITO covered glass substrate. The conductive ITO serves as one electrode used for poling and EO modulation. The second electrode is fabricated by sputtering gold on top

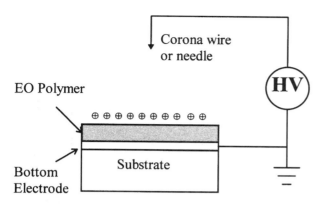

FIGURE 5 Experimental setup for the measurement of electrooptic coefficient by the ellipsometric technique. L1 and L2: focusing lenses; P1 and P2: polarizers; S-B: Soleil–Barbinet compensator; D: photodiode.

of the EO polymer. A laser beam is incident on the back of the glass substrate at an angle θ. It propagates through the substrate, the ITO, the polymer layer, and is then reflected back onto air by the top gold electrode. The polarization of the input laser beam is set at 45° with respect to the plane of incidence so that the parallel (*p* wave) and the perpendicular (*s* wave) components of the optical field are equal in amplitude. The reflected beam propagates through a Soleil–Babinet compensator, an analyzer, and into a detector. In the measurement, the compensator is adjusted such that the output DC signal at the intensity is at the half-intensity point, I_0, as shown in Fig. 6.

Assuming that $r_{33} = 3r_{13}$, the EO coefficient is then calculated by (43)

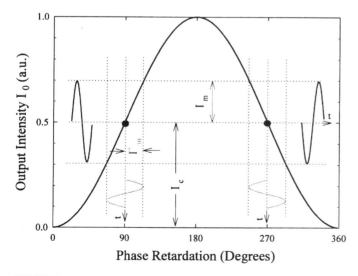

FIGURE 6 Light intensity as a function of phase retardation.

$$r_{33} = \frac{3\lambda I_m}{4\pi V_m I_c n^2} \frac{(n^2 - \sin^2 \theta)^{1/2}}{\sin^2 \theta} \tag{17}$$

where I_m is the amplitude of modulation, I_c is the half DC intensity, θ is the angle of incidence, and V_m is the amplitude of the modulating voltage.

A revised version of this technique, utilizing transparent ITO films for both the bottom and top electrodes [44], can improve measurement accuracy and reduce the measurement time. The two components of EO coefficient, r_{33} and r_{13}, can be independently measured by the use of Fabry–Perot devices fabricated from an electrooptic polymer sandwiched between two semitransparent metal (preferably gold) electrodes [45]. Refractive index modulation due to the electrooptic effect causes changes in the transmission characteristics of the devices. Care must be exercised in this measurement because the applied electric field also causes a change in the film thickness due to converse piezoelectric effects, whose contribution to the change in transmission characteristics can be measured by repeating the measurement on the base polymer of which the EO polymer is made. Alternatively, a free-space interferometric technique [46] has also been used for EO coefficient measurement.

F. Waveguide Loss and Refractive Index

Waveguide-integrated optical circuits is an area in which electrooptic polymers find their ultimate applications with great advantages over inorganic crystals. Important materials properties such as absorption and scattering losses, refractive indices, electrical conductivity, thermal coefficient of expansion (CTE), and chemical resistance need to be determined before a material can be considered for such applications. Optical losses in waveguides made of electrooptic polymers can be measured by several different methods. The most commonly employed technique uses a CCD camera to measure the intensity of a light streak launched into a slab waveguide by prism coupling (Fig. 7). The output from the camera is fed into a personal computer and the intensity data are analyzed. Alternatively, optical losses can be determined by a two-prism technique by measuring the intensities of the incident beam, I_i, the reflected beam, I_R, and the output beam, I_o. Optical losses are calculated based on these intensities and the distance over which the light travels inside the waveguide. The same setup can be used to measure the refractive index and thickness of polymer thin films by determining the angles at which light is coupled [47].

IV. CHROMOPHORES

During the past decade, considerable progress has been made in understanding the factors that affect the nonlinear optical molecular properties [1,2]. Molecular nonlinear optical response (β, first hyperpolarizability) is generally observed in donor–acceptor-substituted π-conjugated molecules containing built-in dipole moment. In such molecules, the π conjugation provides a pathway for the redistribution of electronic charge under the influence of an electric field, whereas the donor and acceptor substituents provide the required ground-state charge asymmetry. Established structure–property relationships indicate that β increases with

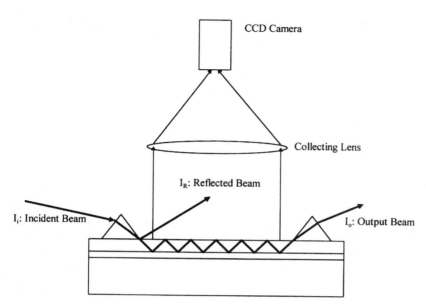

FIGURE 7 Optical loss measurement setup.

increasing donor and acceptor strengths and with increasing π-conjugation length [3,4]. Polyenes are often used as π-conjugating units as they provide the most effective pathway for efficient charge transfer between the donor and acceptor groups [48]. Incorporation of benzene rings into the push–pull polyenes is observed to limit or saturate molecular nonlinearity but enhance thermal stability [35]. The barrier due to the aromatic delocalization energy of the benzene ring is believed to be responsible for the reduced or saturated β values. To overcome the problem of saturation of molecular nonlinearity, several groups have recently developed systems that contain easily delocalizable five-membered heteroaromatic rings [17,21,49–51].

A. Push–Pull Heteroaromatic Stilbenes

1. Effect of Heteroaromatics as Conjugation Moieties

Using thiophene (in place of benzene) as a conjugating segment, several classes of donor–acceptor compounds have been studied [52,53]. Initial studies were focused on the replacement of the benzene rings of the well-known *N,N*-diethylamino-4-nitrostilbene (DANS, **1**) with thiophene moieties [52]. The rationale was that the lower resonance energy of thiophene relative to benzene should reduce the aromatic delocalization and thereby increase the electronic transmission between donor and acceptor substituents, resulting in an increase in β. Indeed, experimental EFISH studies (1.907 μm, dioxane solvent) revealed that the $\beta\mu$ value of **2** was significantly higher than that of **1** (Table 1). This result in conjunction with the comparable dipole moments of **1** (7.3 Debye) and **2** (7.1 Debye) demonstrates that the increase in $\beta\mu$ upon thiophene substitution is solely due to the increase in first hyperpolarizability, β.

TABLE 1 Experimental Linear and Nonliner Optical Properties (EFISH Data Measured at 1.907 μm) of a Donor-Acceptor Substituted Stilbene and Its Thiophene Analog

Compound	λ_{max}^{a} (nm)	μ^{b} (Debye)	$\beta\mu^{c}$ (10^{-48} esu)	β (10^{-30} esu)	β_{0}^{d} (10^{-30} esu)
1	424	7.3	580	79.5	60.6
2	516	7.1	1040	146.5	95.6

[a] All UV–vis spectra cited in this chapter measured in 1,4-dioxane unless otherwise noted.
[b] All dipole moment data cited in this chapter measured in 1,4-dioxane unless otherwise noted.
[c] All EFISH $\beta\mu$ data cited in this chapter measured at a fundamental wavelength of 1.907 μm in 1,4-dioxane.
[d] The dispersion-free value of first hyperpolarizability.

In view of the results obtained with thiophene stilbene derivatives such as **2**, it would be of interest to understand the role of other common five-membered heteroaromatic rings such as furan and pyrrole in influencing the molecular NLO properties. Although there are some scattered reports on furan-substituted NLO compounds [21,51], pyrrole-substituted NLO compounds had never before been reported. In order to gain some insight into how pyrrole and furan affect the molecular nonlinearity with respect to thiophene, theoretical investigations of the second-order NLO responses of two classes (**3–5** and **6–8**) of nitro-amino-substituted heteroaromatic stilbenes were performed (Table 2) [54]. These structures differ only in the nature of heteroaromatic rings and their positions in the molecular framework. In stilbenes **3–5**, the ring attached to the donor moiety is the heteroaromatic ring, whereas in structures **6–8**, the acceptor-substituted ring is the heteroaromatic ring.

First, geometry optimizations were performed at the MNDO level [55a] assuming planar geometries for all the structures considered. Then, complete geometry optimizations (which also include dihedral angles) were carried out using the AM1 method [55b]. Hyperpolarizability calculations were performed on planar as well as fully optimized geometries. The commonly used finite field and sum-over-states methods were employed for the computation of the first hyperpolarizability [56]. In the finite field method, the molecular hyperpolarizabilities are obtained from the energy derivatives of the ground-state energy. The alternative sum-over-states approach involves the use of a perturbation expression in terms of the excited states of the molecule. Both methods have been extensively used for the calculation of β in conjunction with semiempirical as well as *ab initio* molecular orbital methods. Calculations were performed both at the *ab initio* (4–31G) [57] and semiempirical (AM1/CI) [58] levels. In the latter, single and pair excitations over a space of 18 orbitals were considered. Contributions about 150 singlet states were included. The calculated β values presented in Table 2, represent β_{vec}'s as defined in Ref. 1.

TABLE 2 Calculated Hyperpolarizabilities (β_0), Transition Energies (λ) and Change in Dipole Moments ($\Delta\mu$) for Several Amino-Nitro Substituted Heteroaromatic Stilbenes

| | | *ab initio* AM1 | AM1 (fully optimized geometry) | | |
Compound		β_0 (10^{-30} esu)	β_0 (10^{-30} esu)	λ (nm)	$\Delta\mu$ (Debye)
3	H$_2$N— (thiophene) —CH=CH— (benzene) —NO$_2$	62.5	161.8	142.9	376.6	28.59
4	H$_2$N— (furan) —CH=CH— (benzene) —NO$_2$	69.4	174.1	153.1	379.5	32.11
5	H$_2$N— (pyrrole) —CH=CH— (benzene) —NO$_2$	90.4	212.6	182.9	384.2	35.54
6	H$_2$N— (benzene) —CH=CH— (thiophene) —NO$_2$	58.7	183.1	165.1	380.5	30.62
7	H$_2$N— (benzene) —CH=CH— (furan) —NO$_2$	51.5	153.9	138.4	373.4	27.44
8	H$_2$N— (benzene) —CH=CH— (pyrrole) —NO$_2$	43.2	132.7	105.1	365.0	22.17

However, as the first excited state is believed to be the dominant excited state that contributes to the first hyperpolarizability in many molecular systems [6], some of the relevant excited-state parameters such as the transition energy (λ) and the change in dipole moments ($\Delta\mu$) between the ground and the first excited states from sum-over-states calculations were also highlighted.

Three sets of calculated hyperpolarizability values are reported here. Using planar geometries, the first two sets of β values were computed at *ab initio* and AM1/CI levels. Additional AM1/CI calculations were performed on fully optimized geometries. In all the structures examined, the β values obtained from AM1 calculations are higher (by a factor of 2–3) than those obtained from *ab initio* calculations. However, the trends in β values are virtually identical in all the three sets of calculations. The β values obtained for fully optimized geometries are lower than those obtained for planar geometries and this emphasizes the relationship between molecular planarity and effective charge transfer in donor–acceptor-substituted stilbenes [59]. Comparison of the computed β values obtained for 3–5 indicates that pyrrole and furan placed on the donor end are more effective in enhancing the molecular nonlinear response than thiophene (3 < 4 < 5). On the other hand, β values obtained for structures 6–8 suggest that the reverse is true (6 > 7 > 8) when the substituted heteroaromatic rings act as acceptors.

These trends suggest that heteroaromatics play a subtle role in influencing the molecular nonlinear optical properties of donor–acceptor compounds. Because the second-order nonlinear optical responses depend on the extent of intramolecular charge transfer in the donor–acceptor systems, one tends to correlate the observed trends in β to a few relevant parameters of heteroaromatic rings. These parameters include the resonance energy as well as the electron-rich or electron-deficient characteristics of heteroaromatic rings. Whereas the resonance energy of heteroaromatic rings affects the electronic transmission between donor and acceptor substituents, their electron-rich or electron-deficient characteristics affect the overall electron-donating and electron-accepting effects. For the heteroaromatic rings considered here, the experimental resonance energies decrease in the order: thiophene (28 kcal/mol) > pyrrole (21 kcal/mol) > furan (16 kcal/mol); their electron-rich character (assessed from chemical reactivity) increases in the order: thiophene $<<$ furan < pyrrole (60). If the aromatic delocalization is a dominant factor in determining β, one would expect much higher calculated β values for furan derivatives **4** and **7**, with respect to the corresponding pyrrole and thiophene derivatives (**5** and **8**, **3** and **6**, respectively). The calculated β values indicate that it may not be the case; instead, they suggest that furan and pyrrole either enhance or decrease the molecular nonlinearity relative to thiophene, depending on where they are located in the molecular framework. The highly electron-rich nature of pyrrole is due to the ring nitrogen which serves as an additional electron donor and assists the amino group to produce a more powerful electron-donating effect in compound **5**, whereas in compound **8**, the opposite effect occurs because the ring nitrogen counteracts the electron-withdrawing effect of the nitro group. The resulting electron-donating effect in **5** may account for its higher β value, whereas the lower β value of **8** can be attributed to its weaker electron-withdrawing effect. One would expect a similar trend with furan-containing compounds **4** and **7**, as furan is also very electron rich in nature.

Another important piece of information which can be discerned from calculations is the relationship between β and the first excited-state properties (e.g., the change in dipole moment between the ground and excited states, $\Delta\mu$, and the transition wavelength, λ) in this class of compounds. Based on the perturbation expression, larger values of λ, $\Delta\mu$, and transition dipole matrix element r would lead to an increased β. Within each of the two series of heteroaromatic compounds, r values are roughly constant, whereas transition wavelengths show a small variation consistent with the trends in β. The dominant factor determining the variations in β appears to be μ.

The effectiveness of the thiophene ring as a conjugation path is further demonstrated experimentally in a series of compounds shown in Table 3. Experimental results show that the molecular hyperpolarizability grows with the increasing number of thiophene rings in the conjugation path with a dependence shown in Fig. 8, where the measured $\beta\mu$ values are plotted on a log–log scale against the molecular length between donor and acceptor [61]. In both cases for the nitro and dicyanovinyl acceptors, the data fit reasonably well to straight lines, corresponding to a power-law dependence of the molecular nonlinearity on the molecular length. A least-squares fit of experimental data results in a power-law exponent of 2.08 and 1.91 for the nitro and dicyanovinyl acceptors, respectively, indicating an intrinsic exponent of 2 for conjugating chromophores. Furthermore, the smaller increase in dipole moment, as calculated using molecular orbital theory and shown in Table 3, further proves that the increase in the measured $\beta\mu$ values is attributable to the

TABLE 3 Effect of Thiophene Rings as Conjugation Path

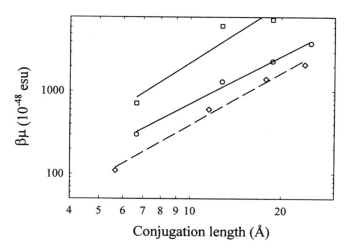

Compound		n	μ (Debye)	λ_{max} (nm)	$\beta\mu$ (10^{-48} esu)
	9	0	7.64	385	110
	10	1	8.85	478	600
	11	2	9.68	506	1400
	12	3	9.83	518	2100
	13	0	7.43	419	300
	14	1	8.66	513	1300
	15	2	8.55	547	2300
	16	3	9.26	556	3800
	17	0	—	505	710
	18	1	—	640	6200
	19	2	—	653	7400

FIGURE 8 Log-log plot of experimental $\beta\mu$ values versus molecular length between donor and acceptor for a series of compounds. The acceptor groups used are nitro (diamonds), dicyanovinyl (circles), and tricyanovinyl (squares), respectively. Straight lines are a least-squares fit of experimental data to a linear function.

increase in molecular first hyperpolarizability, β. In the case of the tricyanovinyl acceptor, no meaningful conclusion can be drawn due to the limited number of data points available. However, the trend does show a tendency of saturation.

2. *Effect of Acceptor Strength*

The dependence of molecular nonlinearity on the acceptor strength of a series of stilbene-like compounds is listed in Table 4. The electronic absorption data of these compounds clearly show the difference of charge transfer properties as a function of acceptor strength. In the comparison of dicyanovinyl-substituted compound **14** and tricyanovinyl-substituted compound **18**, a dramatic red shift (~ 120 nm) of the absorption peak and a large $\beta\mu$ enhancement ($\sim$ a factor of 5) clearly indicate that the tricyanovinyl group is a more effective electron acceptor. The dipole moments of compounds **14** and **18** were not measured. However, the reported values for single-ring compounds *N,N*-dialkylamino-4-dicyanovinyl benzene and *N,N*-dialkyl-amino-4-tricyanovinyl benzene are quite similar. Thus, one may infer that the observed enhancement of $\beta\mu$ values in the tricyanovinyl derivatives here may be attributed mainly to the increase in β (52). A comparison of dicyanovinyl-substituted compound **14** and thiobarbituric-acid-substituted compound **21** also shows a very large absorption peak shift (56 nm) and $\beta\mu$ value enhancement ($\sim$ a factor

TABLE 4   Effect of Acceptor Strengths on Molecular Nonlinearity

Compound		λ_{max} (nm)	$\beta\mu$ $(10^{-48}$ esu)
10		478	710
14		513	1300
20		578	1760
21		578	2400
18		640	6200

of 2). These phenomena may be explained by the contribution of the gain of partial aromaticity of this barbituric acid acceptor in the resonance charge separated state.

3. *Effect of Donor Strength*

The dialkylamino functional group is one of the most commonly used efficient electron-donating groups for NLO chromophores. In a search of more efficient NLO materials containing substituents whose electron-donating ability exceeds that of the amino group, a series of "unconventional" electron-donating functional groups based on the derivatives of the dithiolyldinemethyl group has been proposed and demonstrated [62,63]. The studies of Katz et al. on dithiolyldinemethyl-4-nitrobenzene ($\beta\mu = 359 \times 10^{-48}$ esu, measured at 1.3 μm) and dithiolyldine-methyl-4-tricyanovinylbenzene ($\beta\mu = 1200 \times 10^{-48}$ esu, measured at 1.58 μm) reveal that the dithiolyldinemethyl donor is far superior to amino donors (a factor of 2.4 greater than the dialkylamino group). However, this functional group is very sensitive to high temperature, especially under very polar conditions. The presence of this donor in an NLO chromophore severely limits the usefulness in many practical applications.

A series of dithiolyldinemethyl derivatives was later developed by Jen et al. focusing on improving the electron-donating property and thermal stability [63]. The nonlinear optical property and absorption spectra of these derivatives are presented in Table 5. Compound **22** has the lowest $\beta\mu$ value and absorption peak when compared with dithiolyldinemethyl model compound **24**. This may be due to fewer π electrons (4) on the saturated cyclic dithiane ring which cannot form the stabilized dithiolium aromatic structure like compounds **23**, **24**, and **25**. On the contrary, compound **25**, which possesses two more electron-rich sulfur atoms fused on the double bond linkage of the dithiolyldinemethyl structure, further enhance the stability of the resonance thiolium structure. This contribution causes a dramatic red shift of absorption peak and the enhancement of molecular nonlinearity.

TABLE 5   Effect of Donor Strengths on Molecular Nonlinearity

		λ_{max} (nm)	$\beta\mu$ (10^{-48} esu)
22		556	940
23		604	1350
24		625	1600
25		638	2400

In addition to the significant enhancement of $\beta\mu$ values, the derivatives of modified dithiolyldinemethyl donors further improve the thermal stability of these highly charge transfer compounds. Compounds **22**, **23**, and **25** demonstrated high thermal stability in several preimidized polyimides at temperatures between 275°C and 300°C. This improvement greatly enhances the usefulness of these chromophores in high-temperature polymer matrices for achieving large electrooptic responses.

B. Trade-offs Among Molecular Nonlinearity, and Chemical and Thermal Stabilities

For the organic NLO chromophores to be useful for electrooptic device applications, the chromophores must possess large nonlinear optical coefficients as well as high thermal stability. The family of tricyanovinylthiophene derivatives (e.g., **18**) possess dramatically enhanced molecular nonlinearities. However, their limited thermal stabilities prevent them from being used in high-temperature applications. These instabilities arise from two separate sites: the olefinic linkage and the tricyanovinyl acceptor. At high temperatures, the olefinic linkages undergo cis–trans isomerization which leads to a significant reduction of molecular nonlinearity. Furthermore, this cis–trans isomerization results in a local concentration of π-electron density, which, in turn, favors a reaction with electrophiles such as singlet oxygen. The tricyanovinyl group is also unstable, especially under acidic conditions.

Two approaches have been proposed for the improvement of the thermal stabilities of such compounds. First, to eliminate the thermal instability associated with olefinic linkages, a new class of materials lacking olefinic bonds, such as **22**, and a set of compounds based on bithiophenes [20] and fused-thiophenes [64] where the olefinic bond has been replaced have been prepared, as shown in Table 6.

In these compounds, the ketendithioacetal group serves as the electron donor and the tricyanovinyl group as the electron acceptor. The molecular nonlinear optical properties obtained from EFISH experiments and the thermal stability data obtained from DSC measurement (20°C/min) are presented in Table 6. The experimental $\beta\mu$ values indicate that the thiophene–stilbene **28** possesses higher molecular nonlinearity than the fused thiophene **27**, which, in turn, possesses higher activity than the bithiophene **26** derivative. The molecular nonlinearity of **28** can be attributed to its longer and effective π conjugation offered by the thiophene–stilbene unit. The higher $\beta\mu$ value obtained for **27** in comparison with **26** can be attributed to the better planarity of the fused thiophene compared to the bithiophene.

Inherent thermal stability of these compounds were obtained by DSC. All the samples were heated in a sealed DSC pan at the heating rate of 20°C/min, and the decomposition temperatures (T_d) were estimated from the intercept of the leading edge of the decomposition exotherm with the baseline. All of the four compounds (**22**, **26–28**) possess high thermal stability (>250°C). The bithiophene and fused thiophene derivatives **26** and **27**, respectively, were more stable than the thiophene–stilbene derivative **28**. Because the T_d's vary with heating rate, care must be taken when interpreting the results obtained from this study. For this reason, the stabilities of these compounds were studied by isothermally heating the samples

TABLE 6 Experimental Linear and Nonlinear Optical (EFISH Data, Measured at 1.907 mm) and Thermal Properties of Donor–Acceptor Substituted Thiophene, Bithiophene, Fused-Thiophene and Thiophene–Stilbenes

Compound	λ_{max} (nm)	$\beta\mu$ (10^{-48} esu)	T_m (°C)	T_d (°C)
22	556	940	230	290
26	594	1500	215	305
27	570	2200	270	310
28	612	3420	250	250

at two different temperatures, 250°C and 275°C, respectively, for half an hour. At 250°C, all of the four compounds studied were stable, whereas only compounds **26** and **27** were stable at 275°C. These thermal stabilities may be further improved by functionalizing the bithiophene and the fused thiophene with donor and acceptor groups possessing much higher thermal stabilities.

The second approach is aimed at improving the thermal stabilities of compounds, such as **18** possessing a dialkylamino group which has been proven to be an effective site for connection to polymer backbones. First, compound **29** was developed, which is an analog of **18** but without the olefinic bridge. Removing the olefinic bridge led to a lowered molecular nonlinearity, as anticipated. However, a much improved thermal stability with T_d = 296°C was also obtained.

An attempt to incorporate compound **29** into high-temperature polyimides revealed that the tricyanovinyl group is extremely sensitive to the polyamic acid curing conditions. Jen and co-workers showed that heating compound **29** in polyamic processing solvent such as N-methylpyrolidone (NMP) and dimethylacetamide (DMAc) results in ready decoloration at the solvent's boiling temperatures. It is believed that the cyano group at the vinylcarbond-2 position is most reactive due to the ease of attacking by a nucleophile.

In the next step, the most reactive cyano group was replaced by a bulky aryl group to further enhance thermal stability (compound **30**, T_d = 346°C). The replacement also resulted in a lowered $\beta\mu$ value as well as a significant blue shift in the absorption spectrum, because the introduction of aryl group lowers the electron deficiency of the acceptor. Attaching an electron-donating group on the phenyl ring and extending its conjugation path further increase the molecular nonlinearity while preserving the already enhanced thermal stability (Table 7) [65]. Compounds **30–32**, with T_d well above 300°C, have been tested in polyamic acid processing solvent, including NMP, DMAc, and cyclohexanone, and are shown to be stable. UV–vis absorption spectroscopy studies on thin films of guest–host systems containing polyimide and compounds **30–32** showed less than 10% loss of chromophore at temperatures up to 300°C after the samples have been fully cured. Compound **33** is listed in Table 7 for the sake of comparison.

C. Configuration-Locked Polyenes for Improved Thermal Stability

It has been shown that molecular nonlinearities increase with the length of conjugation path and polyenes provide effective charge delocalization [51,57]. However, these polyene moieties are inherently thermally unstable due to thermally induced isomerization. Shue et al. [66] recently developed a novel method to utilize configuration-locked polyenes to improve molecular nonlinearity and thermal stability simultaneously. This synthetic method combines the advantage of using thiophene ring and triene as efficient conjugating moieties for easier charge separation, and a 2,2′-dimethyl-propyl-group-connected six-membered ring system that provides a configuration-locked geometry for trans-triene to prevent the thermally induced cis–trans isomerization. This allowed the synthesis of a series of NLO chromophores with broad variation of electron acceptors for fine-tuning the linear absorption property (λ_{max}) and $\beta\mu$ (Table 8).

Electronic absorption spectra of these compounds were obtained to compare their intramolecular charge transfer properties. The triene compounds **35, 37,** and **39** have substantially red-shifted charge transfer bands when compared to the stilbene compounds **14, 21,** and **18** and diene compounds **34, 36,** and **38**, indicating more efficient electron delocalization nature of the triene bridge. Comparison of the $\beta\mu$ values of the compounds in Table 8 reveals that the combination of the configuration-locked trans-triene bridge with tricyanovinyl acceptor provides a very efficient mechanism to enhance nonlinearity. The absorption maximum λ_{max} for compound **39** is 684 nm; and thus correcting this value for dispersive enhancement using the two-level model correction gave a zero-frequency value; $\beta(0)$, of 5480×10^{-48} esu, which is 15 times greater than that for the commonly employed NLO chromophore 4,4′-diethylamino-nitrostilbene [DANS, $\beta(0) = 370 \times 10^{-48}$ esu]. Thermal stability studies were performed by dissolving **38** and **39** in polyquinoline PQ-100 (Maxdem), and isothermally heating the thin film samples at 175, 200, and 225°C for 20 min at each temperature. The $\pi \rightarrow \pi^*$ charge transfer absorption band was used to monitor the decomposition temperature. Compound **39** with a configuration-locked trans-triene bridge possesses much better thermal stability at 225°C (>91%) than compound **38**, which has a trans-diene bridge (<30%) (Fig. 9). It is well known that thermal stability of the polyene NLO chromophores decreases with the increase of numbers of double-bond linkage. However, by using this configuration-locked triene approach, it significantly enhances the thermal stability of compounds with even longer chain lengths. Furthermore, when **39** was poled as a guest–host system in PQ-100 (210°C) at 10 wt% loading level and 1.4

TABLE 7 Trade-Off Between Molecular Nonlinearity and Thermal and Chemical Stabilities

Compound	λ_{max} (nm)	$\beta\mu$ (10^{-48} esu)	T_d (°C)
29	607	2700	296
30	514	480	346
31	467	840	369
32	513	1300	354
33	545	2450	307

TABLE 8 Electronic Absorption and Molecular First Hyperpolarizabilities of Configuration-Locked Polyenes

Compound		λ_{max} (nm)	$\beta\mu$ $(10^{-48}$ esu)
	14	513	1,300
	34	545	2,700
	35	559	3,450
	21	578	2,400
	36	597	4,200
	37	611	5,380
	18	640	6,200
	38	662	9,800
	39	684	13,000

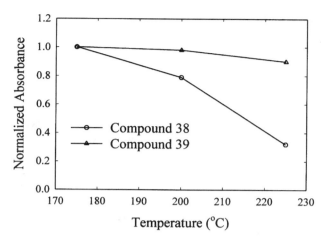

FIGURE 9 Comparison of thermal stabilities between chromophore with and without configuration locking. Samples of thin polyquinoline films containing the chromophores were isothermally heated at 175, 200, and 225°C, respectively, for 20 min. Measured by following the change in absorbance of a 20 wt% chromophore-doped polyquinoline as a function of temperature.

MV/cm poling field, the resulting polymer had an electrooptic coefficient, r_{33}, of 20 pm/V measured at 1.3 μm, consistent with the large molecular nonlinearity.

V. HETEROAROMATIC CHROMOPHORE-BASED ELECTROOPTIC POLYMERS

Over the past decade, electrooptic polymers have evolved through many stages, from simple guest–host systems to cross-linked high-temperature side-chain polymers. The earliest electrooptic polymers were systems based on readily processible poly(methyl methacrylate) (PMMA), polystyrenes, polyurethanes, and so on. Various second-order NLO chromophores have been either doped as guest molecules or covalently attached as side-chain groups to form electrooptic polymers, taking advantage of PMMA's excellent optical quality and processability. Electrooptic polymers based on many other polymer structures have also been demonstrated. However, the poor thermal and chemical stability of these polymers severely limit their use in practical devices.

Polyimides and polyquinolines have been found to be attractive candidates for polymeric EO materials particularly because of their low dielectric constants, high glass transition temperatures (T_g) and compatibility with semiconductor process. Many systems that utilize polyimides as the host for NLO chromophores have been demonstrated, although limited attention has been paid to the recently developed polyquinoline systems. This section discusses various electrooptic polymers based on high-temperature polyimides and polyquinolines and heteroaromatic NLO chromophores. Different approaches toward the preparation of highly active and highly thermally stable side-chain polyimides are presented along with a brief summary of their electrooptic properties.

A. Guest–Host Systems

1. *Polyimide-Based Guest–Host Systems*

A guest–host EO polymer system can be prepared relatively easily by blending the chromophore and the polymer together under proper conditions. Many guest–host electrooptic polymers have been made over the years which include polymers ranging from simple methacrylates to more complicated polyimides and polyquinolines, and chromophores such as PNA, DANS, DR-1, and so on. Recently, several classes of NLO chromophore-doped high-temperature polyimides have been examined [14,67–69] to achieve long-term thermal stability of EO coefficients. Among many examples, the first guest–host system that combines a high-temperature polyimide and a highly efficient and thermally stable heteroaromatic chromophore (**18**) was prepared by Wong and Jen [70]. A polyamic acid (PIQ-2200), used as the precursor polymer, was blended with **18** with concentrations ranging from 5 to 15 wt%. Higher chromophore concentrations resulted in phase separation. Curing at 200°C for 30 min removes any residual solvent or by-product such as water and completes the imidization process. Poling a sample containing 12 wt% of **18** at 220°C with 100 V/μm resulted in an EO coefficient of $r_{33} = 10.8$ pm/V at a wavelength of 1.52 μm.

The rate of relaxation of chromophore alignment in poled polymers depends on the mobility of the polymer-chain segmental motion and is generally believed to be related to the free volume within the polymer. It is well known in the field of polymer science that annealing of a polymer below its glass transition temperature can effectively reduce the polymer-free volume and segmental motion, thus resulting in the higher stability of the chromophore alignment [71]. The effect of annealing was demonstrated by monitoring the relaxation of EO coefficient of two samples: one annealed and one directly quenched. The annealed sample showed much slower relaxation in EO coefficient at both 120 and 150°C, as shown in Fig. 10a. In Fig. 10b, the thermal decay of EO coefficient at 120 and 150°C is presented. Over 85% and 65% of the original EO coefficient was retained at these two temperatures, respectively, for more than 700 h. Thus, through the use of high-glass-transition-temperature polyimides and physical aging process, a significant improvement in the thermal stability of poled structure is achieved.

2. *Polyquinoline-Based System*

Polyquinolines have emerged in the last few years as a new class of materials for applications in electronic packaging due to their superior mechanical and electrical properties and thermal stabilities [72,73]. In addition, their mild processing conditions make them ideal hosts for highly active NLO chromophores which cannot survive the harsh processing conditions of polyamic acids. This was demonstrated by Cai and Jen [9], using a highly efficient N,N'-dibutylamino-tricyanovinyl cinamyl thiophene stilbene (**40**, $\beta\mu = 9800 \times 10^{-48}$ esu) and Maxdem PQ-100 (**41**) with a weight percentage of 20%. The glass transition temperature dropped from the original 265°C (pure PQ-100) to 180°C due to plasticization by the chromophores. After poling at 180°C with 80 V/μm for 5 min, they obtained an EO coefficient of 45 pm/V at a wavelength of 1.3 μm, considerably higher than that of LiNbO$_3$. The temporal stability of the EO coefficient was monitored at 80°C. After an initial quick relaxation, the EO coefficient retained over 60% of its original

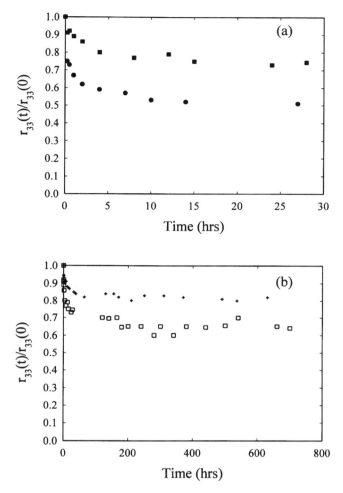

FIGURE 10 Temporal stabilities of the electrooptic coefficient. (a) Effect of physical aging. Filled circles: quenched sample; filled squares: sample physically aged at 150°C for 7 h. (b) EO coefficient relaxation at 120°C (cross) and 150°C (open squares).

value (Fig. 12). Fitting of experimental data to a Kohlrausch–Williams–Watts (KKW) stretched exponential function [74],

$$r_{33}(t) = r_{33}(0) \exp\left[-\left(\frac{t}{\tau}\right)^{\beta}\right] \tag{18}$$

gives a characteristic time constant of $\tau = 1 \times 10^6$ with $\beta = 0.085$. Alternatively, the data were fitted to a bi-exponential function [75]

$$r_{33}(t) = r_{33}(0)(r_1 e^{-t/\tau_1} + r_2 e^{-t/\tau_2}) \tag{19}$$

where $r_1 + r_2 = 1$ and τ_1 and τ_2 are the short-term and long-term relaxation time constants, respectively. Values obtained for τ_1 and τ_2 are 20 and 40,000 h, respectively.

(a)

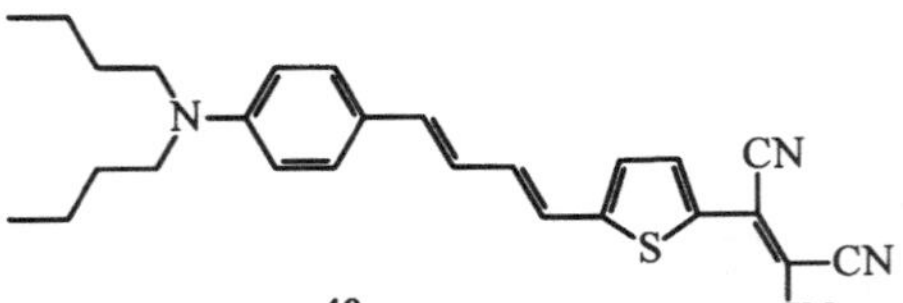

40

(b)

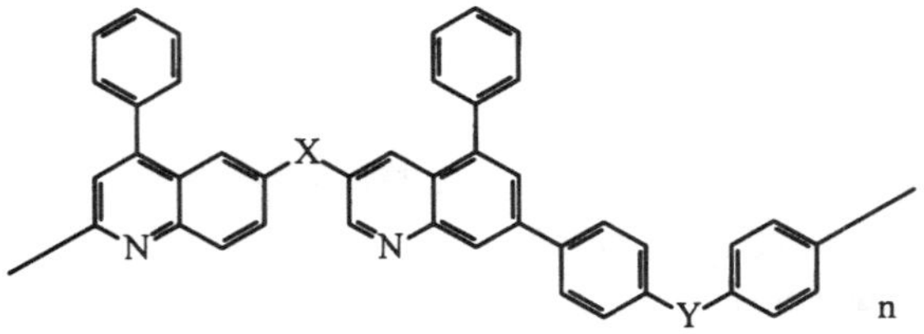

41

FIGURE 11 Chemical structures of (a) RT9800 and (b) PQ-100.

B. Side-Chain Aliphatic Polyimides

Despite the improvement in long-term stability of the electrooptic activity compared to the previous results on side-chain polyacrylates, the polyimide-based guest–host polymer approaches have certain disadvantages. These include (a) sublimation of NLO chromophore from the polymeric films at device fabrication temperatures (>275°C), (b) plasticization effects due to higher chromophore loading levels, thus resulting in lower T_g's and lower long-term thermal stabilities, and (c) difficulty in

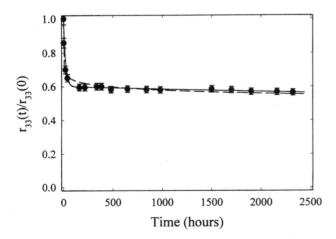

FIGURE 12 Normalized electrooptic coefficients of RT9800/PQ-100 at 80°C. The data points are the experimental values. The dotted line is a "KKW" stretched exponential relaxation function fitting and the solid line is a bi-exponential function fitting.

fabricating channel waveguides caused by chromophore erosion during the multi-layer process. To overcome these deficiencies associated with guest–host materials, various approaches have been developed to covalently bond the NLO chromophores to these high-temperature polymer matrices [6]. However, very few reports on NLO-functionalized polyimides or polyquinolines had appeared, perhaps due to synthetic difficulties. Dalton et al.'s [76] and Yu and Peng's [77] groups have developed synthetic schemes for aliphatic polyimides containing Disperse Red and DANS as side chains, respectively. However, many known chromophores which possess inherent high thermal stability are unstable in the polyamic acid to polyimide curing process. This may be attributed to the sensitivity of NLO chromophores to (1) the very acidic environment of the medium, (2) the polar processing solvents, and/or (3) the by-products formed at high curing temperatures (200–300°C). In this connection, Jen et al. have developed a two-stage process to obtain side-chain aliphatic polyimides containing efficient tricyanovinyl substituted NLO chromophores (Scheme 1) [78]. By functionalizing the chromophore at the last stage, the harsh environment during the imidization process is avoided. Polyimide prepolymer **45** has a lower T_g (158°C) and is readily soluble in common organic solvents. After tricyanovinylation, the resulting polyimide **46** has a much higher T_g (189°C) and is soluble only in polar solvents such as DMSO, DMF, and cyclohexanone. The large change in T_g has been observed in side-chain polyacrylates systems and may be attributed to the influence of the highly dipolar nature of the NLO chromophores on the polymer-chain motions. The T_g of these side-chain polyimides can easily be fine-tuned by reacting the dianhydride monomer with more rigid diamine monomers. This approach provides almost unlimited variations of monomers to fine-tune the structures to achieve desired optical, mechanical, and electrical properties of the EO polymers. The EO activity of the poled polymer film was measured at 0.83 μm and an r_{33} value of 16 pm/V was obtained. The thermal stability of the poled polyimide was demonstrated by heating the poled samples in an oven at 90°C for over 1000 h with negligible change of the sample's EO coefficient [78].

C. Side-Chain Aromatic EO Polyimides

1. *Chromophore-Containing Diamines*

Although the above-described aliphatic polyimides provide fairly promising results, there is a strong need to further improve their mechanical properties (for multilayer integration) and thermal stability in order to accommodate the stringent processing requirements of electrooptic devices. Aromatic polyimides possess much higher thermal stability (higher glass transition temperature) and provide very broad variation of monomers for fine-tuning the structural and electrical properties [79]. In order to achieve such desirable properties, the IBM group [80,81], Yu et al. [82,83] and Jen et al. [84] have developed several synthetic methodologies for the incorporation of NLO chromophores into aromatic polyimides. Examples are shown in Schemes 2–4.

In Scheme 2, the chromophore-containing diamine (**48**) was reacted with 2,2′-bis-(3,4-dicarboxyphenyl)hexafluoropropane dianhydride (6FDA) or 4,4′-oxydiphthalic dianhydride (ODPA) to obtain polyamic acids which were then thermally imidized to produce corresponding polyimides. Polyimides **50(a)**–**50(d)** have

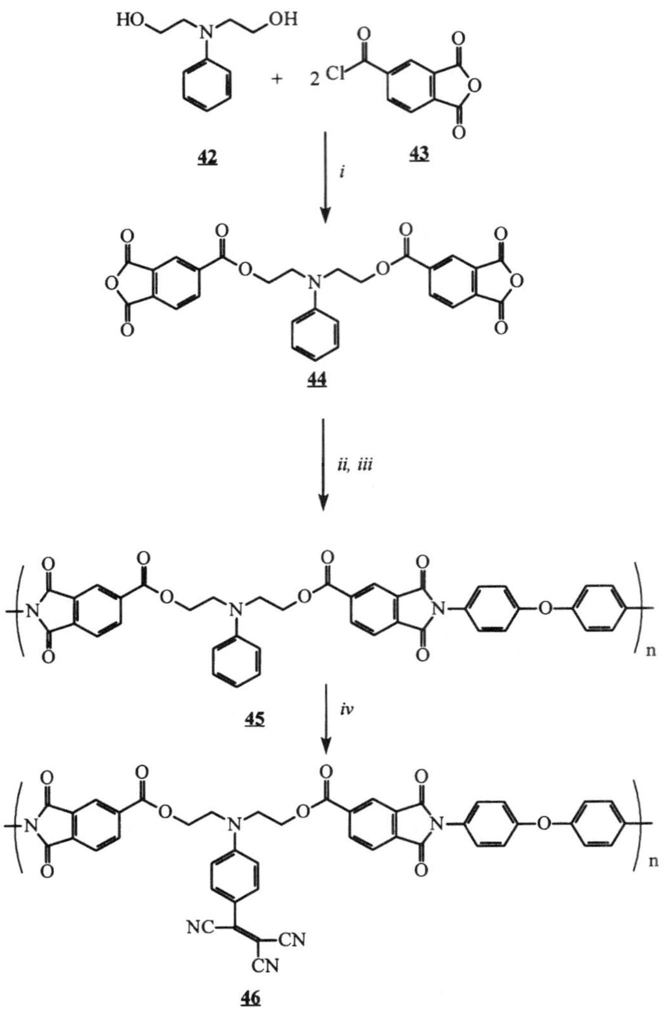

SCHEME 1 Reagents and conditions: *i*, pyridine, CH$_2$Cl$_2$, 0 → 25°C; *ii*, 4,4'-oxydianiline, NMP, N$_2$, 0 → 25°C; *iii*, 1,2-dichlorobenzene, N$_2$, heat, 180°C; *iv*, TCNE, DMF, N$_2$, heat, 70°C.

glass transition temperatures ranging from 210°C to 230°C, and electric field poling at temperatures close to T_g resulted in electrooptic coefficients from 2.5 to 13 pm/V at 1.3 μm. Temporal stabilities of poled polymers were studied at an elevated temperature (100°C), with over 90% of the original EO coefficients retained for over 1000 h [81].

In Scheme 3, compound **53** is obtained by reacting compound **51** and the hydroxy-containing chromophore **52** in the presence of DEAD and Ph$_3$P. The dinitro compound **53** is then reduced using standard chloride in hydrochloric acid to produce diamine **54**, which was then reacted with pyromellitic dianhydride **55** in NMP to obtain a viscous polyamic acid solution **56**. After thermal treatment, a thin

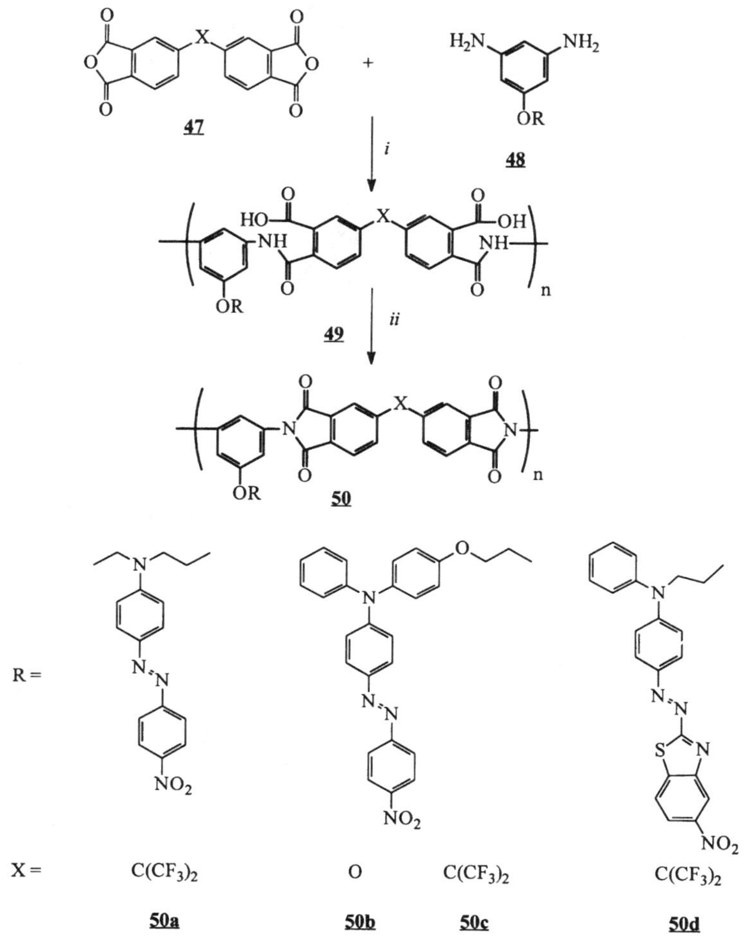

SCHEME 2 Reagents and conditions: *i*, NMP, 25°C; *ii*, heat, 250°C, or Ac$_2$-pyridine.

film of **56** was converted to polyimide **57** through ring-closure cyclization. Performed during the imidization process, electric field (corona) poling yielded a second-harmonic generation coefficient of d_{33} = 51 pm/V at a fundamental wavelength of 1.064 μm. After correcting for dispersion, the authors calculated the disperseless d_{33} value to be 18 pm/V. Temporal stability study shown that 85% and 60% of the initial d_{33} value was retained after the samples were heated for 150 h at 150°C and 170°C, respectively [82,83].

2. Post-tricyanovinylation

We showed in Section IV that the tricyanovinyl group exhibits excellent electron-withdrawing properties and that extremely large $\beta\mu$ values can be obtained from chromophores containing such an electron acceptor. It was also shown that the same group is highly susceptible to the harsh environment in which polyamic acid to polyimide conversion is performed. A posttricyanovinylation approach was re-

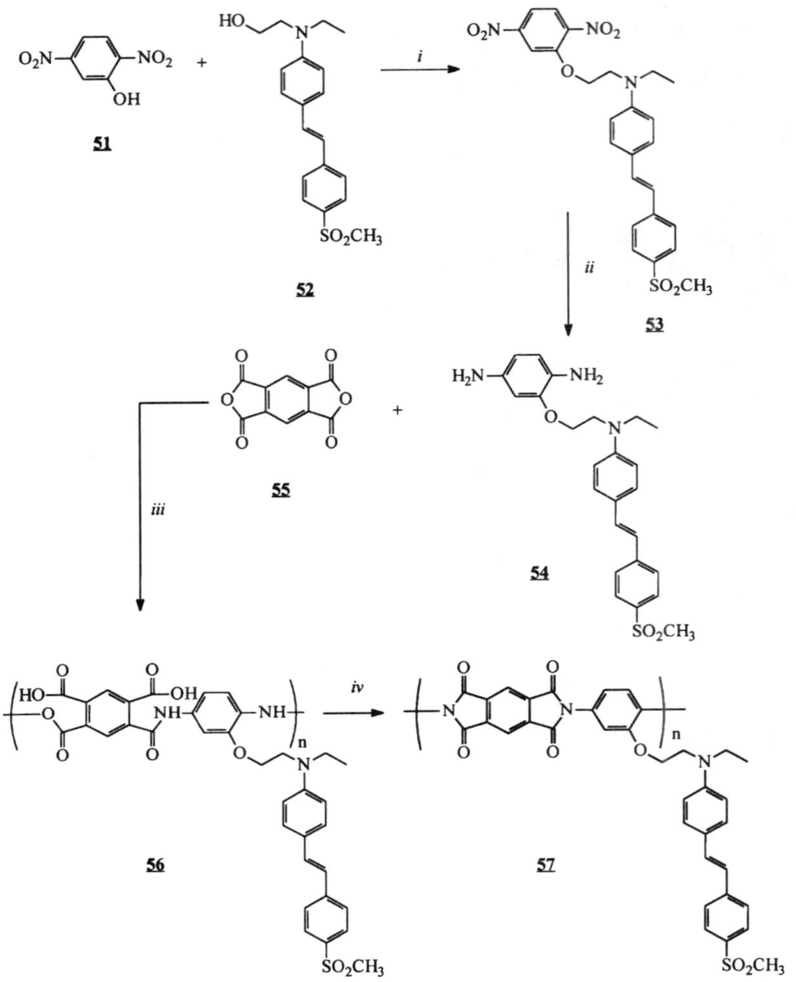

SCHEME 3 Reagents and conditions: *i*, DEAD, Ph₃P, THF; *ii*, SnCl₂, HCl; *iii*, NMP; *iv*, thermal treatment.

cently developed by Jen et al. to prepare aromatic polyimides which contain only part of the NLO chromophore. The tricyanovinyl electron acceptor is attached to the fully imidized polyimide, avoiding the imidization environment.

The synthetic approach is shown in Scheme 4. Compound **59** was obtained in 92% yield by condensing 2,5-dinitrophenol **51** with 2-(N-ethylamino) ethanol **58** under a reaction condition using the Mitsunobu reaction [7]. The dinitro compound **59** was then reduced by hydrogenation in dimethylformamide (DMF) with a catalytic amount of palladium/carbon to give diamine **60** as a viscous liquid (80% yield). Diamine **60** was reacted with 2,2′-bis-(3,4-dicarboxyphenyl)hexa-fluoropropane dianhydride (6FDA) **61** (1 equiv.) in NMP at 0°C (N₂). After approximately 12 h, 1,2-dichlorobenzene (same amount as original NMP) was added

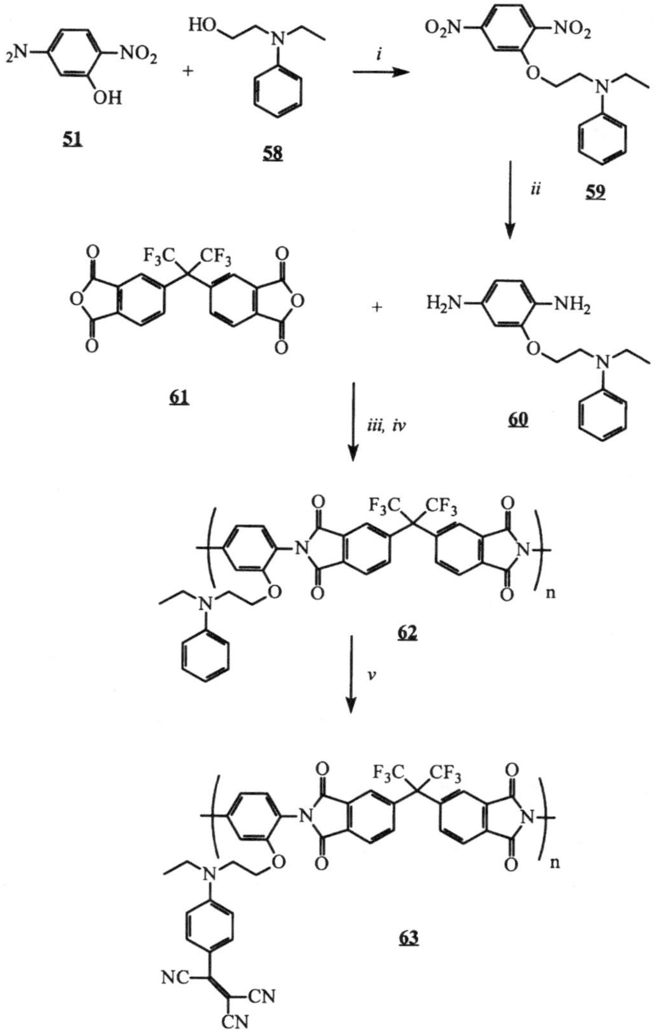

SCHEME 4 Reagents and conditions: *i*, DMAc, 0°C → RT; *ii*, xylenes, 160°C; *iii*, PPh$_3$, DEAD; *iv*, THF, RT.

to the solution as a cosolvent to aid in the removal of water formed during the imidization process. Posttricyanovinylation of polyimide **62** was carried out by dissolving **62** and tetracyanoethylene (TCNE) (1.1 equiv.) in DMF and heating at 70°C (N$_2$) for 24 h (92% yield).

Polyimide **63** (32% m/m of the NLO chromophore *N*,*N*-diethylamino-4-tricyanovinylbenzene) is readily soluble in such polar solvents as cyclohexanone, DMSO, and DMAc. The T_g of this compound (224°C) is 35°C higher than that of the aliphatic polyimide containing the same chromophore and the same loading level. This increase in T_g may be attributed to the hindrance of the polymer chain motion by the more rigid aromatic structure of this polyimide. From TGA studies, polyimide **63** is thermally stable up to 310°C under nitrogen atmosphere. These studies were based on dynamic heating at a relatively faster rate (20°C/min) and cause some ambiguity in determining the inherent thermal stability of the polymer. For this reason, thin polymer films from polyimide **63** were heated on a hot stage isothermally at 250°C (0.5 h) and 275°C (0.5 h). The $\pi-\pi^*$ charge transfer absorption band of the chromophore was used to monitor the decomposition temperature. After 275°C, there was less than 5% change in the absorption intensity, indicating the high thermal stability of the polymer.

Poling thin-film samples of **63** at 215°C with a field of 50 V/μm yielded an EO coefficient of 15 pm/V at a wavelength of 830 nm. The thermal stability of the poled polyimide **63** was demonstrated by heating the poled samples in an oven (air atmosphere) at 100°C. After an initial drop to 90% of its original value, the EO coefficient remained unchanged for over 1000 h.

3. Post-Mitsunobu Reaction

Although the above-discussed side-chain polyimide systems provide encouraging results, all these synthetic methods for aromatic side-chain polyimides involve a tedious procedure for the synthesis of the chromophore-containing diamine monomers. Moreover, the fact that few chromophores can survive the relatively harsh chemical environment of the monomer synthesis and the imidization of the polymers severely limits the application of the methodologies. A facile and generally applicable two-step approach for the synthesis of NLO side-chain aromatic polyimides was recently developed by Chen et al. [85]. This is a one-pot preparation of the preimidized, hydroxy-containing polyimide [86], followed by the covalent bonding of a chromophore onto the polyimide backbone via a post-Mitsunobu reaction [87]. By introducing the chromophores at the last stage through the very mild Mitsunobu condensation, the harsh imidization process of the polyamic acid is avoided and the tedious synthesis of chromophore-containing diamine monomers is also eliminated. A series of aromatic polyimides with broad variation of polymer backbone and side-chain NLO groups were synthesized using this method (Schemes 5 and 6).

Step 1 in Scheme 5 is for the synthesis of hydroxy-containing aromatic polyimides. The hydroxy-containing diamine, 3,3′-dihydroxy-4,4′-diamino-biphenyl **64** was reacted with the stoichiometric amount of the dianhydride, 2,2′-bis-(3,4-dicarboxyphenyl) hexafluoropropane dianhydride **61**. The viscosity of the solution was dramatically reduced during this period. It was then obtained in the polyamic acid solution. Dry xylenes were added into the flask and the polyamic acid was thermally cyclized at 160°C for 3 h. Water eliminated by the ring-closure reaction was removed as a xylenes azeotrope at the same time. Step 2 of the synthesis is for the covalent bonding of the hydroxy-containing chromophore onto the polyimide **65** backbone. Disperse Red 1 (0.8 equiv., relative to the equivalent of the hydroxy groups of polymer **3**) was reacted with polymer **65** via Mitsunobu condensation to afford NLO side-chain aromatic polyimide **67** in 92% yield.

Proton NMR spectroscopy monitoring of the polymerization process of hydroxy-containing preimidized polyimides and the Mitsunobu condensation show that post-Mitsunobu condensation between hydroxy polyimides such as **3** and hydroxy chromophores such as Disperse Red 1 are quantitative. This allows the pre-

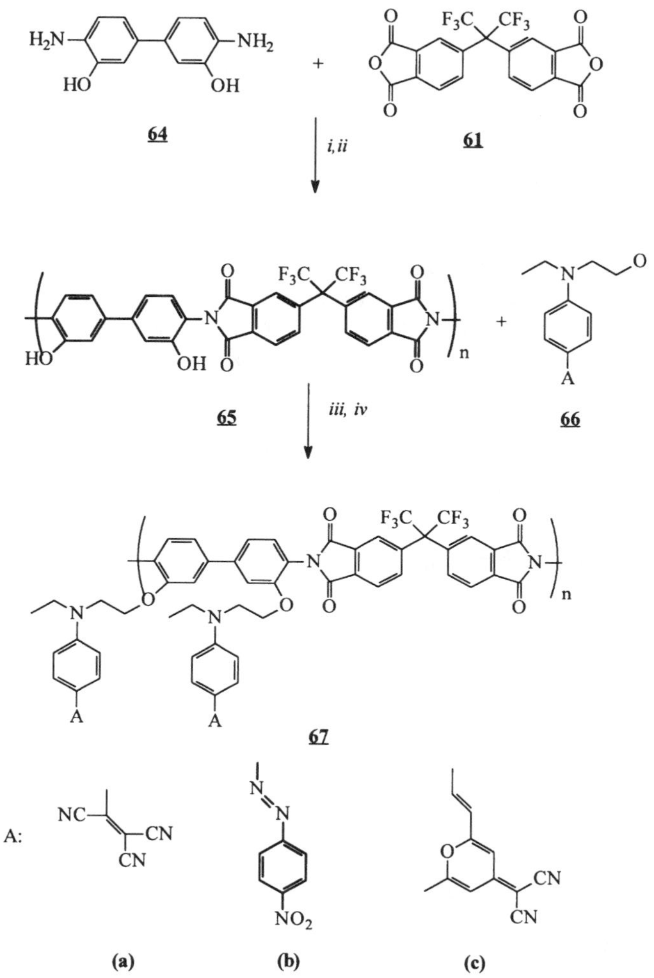

SCHEME 5 Reagents and conditions: *i*, PPh$_3$, DEAD, THF; *ii*, Pd/C, [H]; *iii*, NMP, 0 → 25°C; *iv*, dichlorobenzene, 180°C, 3 h; *v*, TCNE, DMF, 70°C.

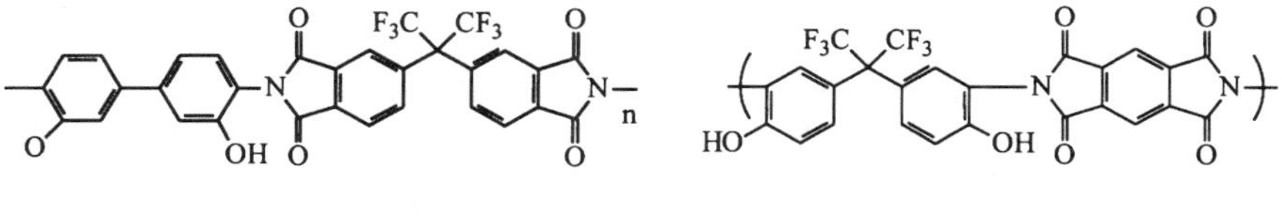

SCHEME 6

cise adjustment of chromophore loading levels and the density of the intact OH groups. The hydroxy-containing side-chain polyimides can further be cross-linked to a different extent to further improve the materials' mechanical properties, solvent resistance, and thermal stability of the chromophore alignment [88].

The NLO polyimide is soluble in polar solvents such as cyclohexanone, DMAc, and THF. The molecular weights of the polymers were measured by gel permeation chromatography (GPC). Polymer **67(a)** has a weight-averaged molecular weight (relative to polystyrene standard) M_w of 57,000 with a polydispersity index of 1.51. All the resulting side-chain polyimides have high glass transition temperatures and good thermal stability. Polyimide **67(a)** (with a chromophore loading level of up to 45%), for example, has a T_g of 222°C (by DSC) and a thermal stability of less than 1% weight loss up to 300°C (by TGA). The UV–vis spectrum of thin films of **67(a)** exhibited a strong absorption pattern (λ_{max} = 520 nm) due to the π–π^* charge transfer band of the chromophore. Electric field poling at 80 V/μm led to an EO coefficient (r_{33}) of 11 pm/V at a wavelength of 830 nm. More than 90% of the original EO coefficient was retained after heating at 100°C for 400 h.

D. High-Performance Electrooptic Polymer Systems

The application of electrooptic polymers to integrated optical devices poses a series of stringent requirements on the polymers. These include large and stable electrooptic coefficients, low optical losses, low electrical conductivity, and good mechanical properties to prevent thermally induced cracking and solvent erosion. In addition, it is equally important to identify and develop polymers with matched properties such as refractive index and electrical conductivity.

Using the approaches described in this section, a series of high-temperature EO polyimides have recently been developed by incorporating highly thermally and chemically stable NLO chromophores onto the high-temperature polymer backbones (Schemes 5 and 6). These high-performance EO polyimides also demonstrated excellent mechanical properties and both long-term (at 100°C) and short-term (at 240°C) thermal stabilities. For example, an electrooptic polyimide system, Optimer™, combining EO active polyimide (with a general structure shown in Fig. 13) and refractive index and electrical conductivity matched passive cladding polyimides, was recently developed [89].

Electrooptic polymers with various chromophore loading levels were formulated. Thin films of Optimer™ RT-2700 are spin-coated onto an oxide-covered silicon wafer. After soft baking at 100°C for 2 min, the films undergo a series of curing temperatures (e.g., 250°C, 280°C, 290°C, and 300°C). The polymer's refractive index and optical losses are measured after each cure step. Results shown in Fig. 14 clearly show that the optical losses increase with cure temperature and the refractive index decreases with cure temperature. DSC spectra (20°C/min) show that Optimer™ RT-2700 with 15 and 20 wt% loading levels have glass transition temperatures of 280°C and 275°C, respectively. Electric field poling at temperatures close to their T_g's resulted in electrooptic coefficients of 6 and 10 pm/V at a poling field of 100 V/μm.

Zhang et al. studied the stability of the poling-induced electrooptic coefficient at various temperatures up to 240°C [90]. The relaxation of the EO coefficient is shown in Fig. 15. Experimentally obtained data were fitted to a stretched exponen-

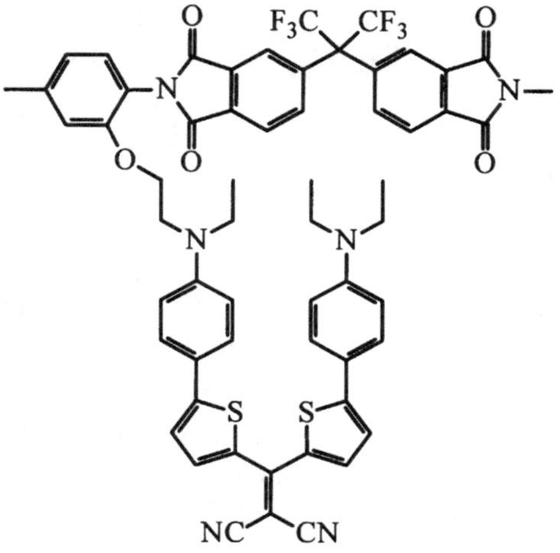

FIGURE 13 General structure of Optimer™.

tial [Eq. (18)] and relaxation time constants were obtained. These time constants are then plotted versus temperature and can be fitted to an Arrhenius equation [91].

$$\tau^{-1} = \tau^{-1}(\infty)e^{-A/kT} \tag{20}$$

where τ is the relaxation time constant, $\tau(\infty)$ is the relaxation time constant at infinitely high temperature, A is the activation energy, and kT is the Boltzmann energy. An activation energy of 36 kcal/mol has been obtained.

VI. APPLICATIONS OF ELECTROOPTIC POLYMERS IN INTEGRATED OPTICS

The driving force behind the research and development effort on EO polymers is their advantage over their inorganic counterpart in the application in integrated optical devices. These applications require materials with a large electrooptic coefficient, preferably in excess of 30 pm/V, and certain thermal stabilities [92]. In general, a material must demonstrate thermal stability at high temperatures (e.g., 250°C) for a short period (30 min) in order to survive the packaging process. It must also exhibit long-term stability at moderate temperatures (e.g., 100°C) for at least 10,000 h. Because electrooptic activities in amorphous polymers arise from an electric-field-induced alignment of the nonlinear optical chromophores, these stability requirements translate into either high glass transition temperature or some type of cross-linking to achieve rigid structures.

In addition, the fabrication of integrated electrooptic devices poses a series of stringent requirements on materials. This may be better understood with the help of an illustration of a typical fabrication process, as shown in Fig. 16. The process starts with plasma sputtering deposition followed by lithographic definition of a bottom electrode. A bottom cladding polymer is then deposited by spin-coating and then cured. The next step of the process involves the formation of trenches by means of plasma etching. An active electrooptic polymer is then spin-coated and

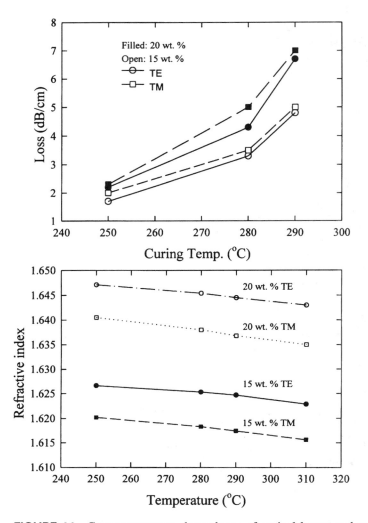

FIGURE 14 Cure temperature dependence of optical losses and refractive index.

cured, followed by the coating and curing of a top cladding layer onto which a top metal electrode is finally deposited. It is clear that the following requirements must be met: (1) Both the bottom cladding layer and the active EO polymer must be robust enough to survive the coating of subsequent layers without corrosion and interlayer diffusion. (2) The refractive index of each of the cladding polymer layer must be slightly lower than that of the active EO polymer; the optimum refractive index difference for single mode waveguide of 3 μm in dimension is $0.005 < \Delta n < 0.015$. (3) The cladding layer and the active EO polymer must have comparable mechanical properties (e.g., thermal expansion coefficient) to minimize mechanical cracking. (4) The conductivity of the cladding polymer must be equal to or higher than that of the active EO polymer to achieve effective electric field poling.

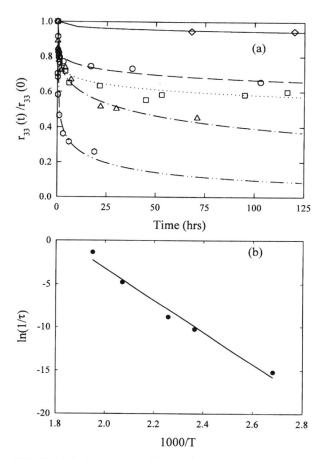

FIGURE 15 Thermal stability of electrooptic coefficient. (a) Relaxation of EO coefficient at elevated temperatures. Diamonds: 100°C; circles: 150°C; squares: 170°C; triangles: 210°C; hexagons: 240°C. Curves represent theoretical fit to Eq. (18). (b) Plot of $\ln(1/\tau)$ versus temperature. Filled circles: values obtained by fitting data in (a) to Eq. (18); solid line: least-squares fit to the Arrhenius equation.

Many materials or material systems have been developed over the past years aimed at these applications. However, only a few have been tested in the fabrication of waveguide electrooptic devices such as EO modulators and switches. A Mach–Zehnder modulator was constructed by Girton et al. [10] using DANS as the side-chain group of a polymer with a T_g of 140°C and electrooptic modulation at 20 GHz was demonstrated. Thackara et al. also reported polymeric electrooptic Mach–Zehnder switches which operate at 1.32 μm and up to 2 GHz [93]. Van Eck et al. fabricated an electrooptic polymer waveguide switch by uniform poling and selective photobleaching [94]. The switch taps off a fraction of the optical power from a main rail into two complementary outputs, passing on the rest of the power to additional switches downstream. A traveling-wave electrooptic phase modulator

(1) Base electrode   (4) Active core fill

(2) Bottom clad   (5) Top clad

(6) Top electrode

(3) Form trench

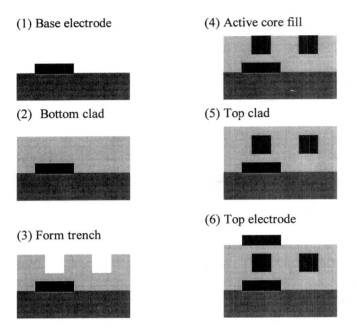

FIGURE 16 Fabrication procedure for polymeric integrated electrooptic devices using a trench-and-fill technique. (1) Deposition of base metal electrode; (2) spin-coating and cure of bottom cladding layer; (3) trench formation by plasma etching; (4) spin-coating and cure of core active electrooptic polymer; (5) spin-coating and cure of top cladding layer; (6) deposition of top metal electrode.

was fabricated by Wang et al. using cross-linked polyurethane–Disperse Red 19 [11,12]. In their case, a commercial polyurethane was used as a cladding material. Modulation was demonstrated up to 18 GHz with a half-voltage of 35 V. Later experiments have shown that this device can function up to 60 GHz [95].

A highly active and thermally stable EO polymer has recently been developed by ROITech for the fabrication of integrated optical devices [89]. The polymer, Optimer™ RT2700, incorporates a highly active NLO chromophore, RT2700, into an electronic-grade polyimide to form a side-chain polymer with a glass transition temperature of 250°C. A series of integrated devices were designed and fabricated including Mach–Zehnder EO modulator, directional coupler, Mach–Zehnder switch, and HEOS™, a high-speed electrooptic switch (see Fig. 17). The properties of these devices are shown in Table 9 with HEOS™ having the advantage of digital operation of an abrupt modulation function as well as small size and lower switching voltage. The small size of HEOS™ translates into higher level of integration. HEOS™ functions by modulating the propagation constant of the light in the switch with a voltage-induced index change of the nonlinear optical polymer. The modulation changes the evanescent coupling between the waveguides leading to a change in the power transfer.

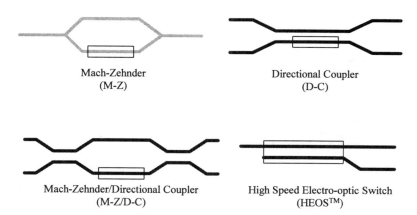

FIGURE 17 Structures of switches, couplers, and modulators used in integrated electrooptic devices.

A 1 × 4 integrated optical digital switch was fabricated using Optimer™ and HEOS™ (Fig. 18). The switch consists of all the basic circuit elements that are necessary for the fabrication of other electrooptic devices including optical multichip modules (MCMs). The whole device occupies an area less than 0.4 cm² and 122 of them were fabricated on a single 6-in. wafer using standard integrated circuit (IC) processes. Digital optical switching at 2 GHz has been demonstrated with an applied voltage of 10 V.

VII. OUTLOOK

This chapter has discussed the development of high-temperature electrooptic polymers based on heteroaromatic nonlinear optical chromophores. In addition, several important techniques for the characterization of materials and process for the fabrication of integrated multilayer channel waveguide devices were briefly described. There remains much interesting and critical work in the areas of developing materials with superior properties, processing techniques that are more suitable for

TABLE 9 Comparison Between Various Integrated Electro-Optic Devices

Type	Device area (mm²)	Devices per cm²	Switching voltage	Transfer function	Fabrication tolerance
M-Z (modulator)	5.6	17	$V_\pi = \dfrac{\lambda d}{2\Gamma n^3 r_{33} L}$	Cosine	Moderate
D-C (switch)	5.6	17	$1.3\sqrt{3}V_\pi$	Cosine	Moderate
M-Z/D-C (switch)	10.2	9	V_π	Cosine	Difficult
HEOS™ (switch/modulator)	1.9	50	$<1.3\sqrt{3}V_\pi$	Digital	Easy

1 to 2 splitter

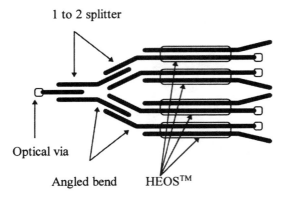

Optical via

Angled bend   HEOS™

FIGURE 18   A 1 × 4 integrated optical switch.

electrooptic polymers, and device designs that truly demonstrate the advantage of polymeric electrooptic materials.

The use of electrooptic polymers for device applications awaits solutions to a number of key issues. For example, simultaneous optimization of electrooptic coefficient and poled-order stability, processes to achieve efficient poling and to overcome the strong dipole–dipole repulsion and phase segregation which occurs at high-chromophore concentrations, the precise measurements of relaxation of EO coefficient and refractive index to assure efficient light coupling, switching, and modulation, and the integration of a suitable material system including core/clad that can fulfill the requirements of multilayer process without optical and mechanical failure. Fortunately, these problems are slowly being recognized by academic and industrial scientists in this field, and several prototype EO devices have attracted the attention of users. It will take the combined attention from chemists, physicists, and material and device engineers to continue the development and turn from original scientific curiosity to practical applications.

Another area where high-performance electrooptic materials can find important application is in the development of photorefractive polymers which have emerged in the last few years [96–99] as a new class of materials for such applications as erasable optical holographic storage and optical phase conjugation. Photorefractivity is a combined effect of photoconductivity and electrooptic response simultaneously present in a material [100]. Therefore, the development of photorefractive polymers is closely related to that of electrooptic polymers and the methodologies employed for EO polymers can be extended to photorefractive polymers as well. In particular, many photorefractive polymers developed thus far have very low glass transition temperatures [101]. These low-T_g materials potentially suffer such problems as phase segregation and loss of optical quality. The use of high-T_g and multifunctional polymers can reduce or even eliminate these problems and, thus, improve the performance of photorefractive polymers [102].

ACKNOWLEDGMENTS

We wish to thank our colleagues at ROITech and Enichem America for their excellent contribution to the work which has been cited in this chapter. Especially, we are indebted to Drs. T. Chen, Y. Liu, V. P. Rao, K. J. Drost, Y. Cai, J. T. Kenney, E. S. Binkley, K. Y. Wong, X. Zhang, and M. Hess. Partial financial support from AFOSR (F49620-95-C-0069) and ONR (N00014-95-1-1319 and N00014-95-C-0321) is highly acknowledged.

REFERENCES

1. Chemla, D. S., and Zyss, J. (eds.), *Nonlinear Optical Properties of Organic Molecules and Crystals*, Academic Press, Orlando, FL, 1987.
2. Prasad, P. N., and Williams, D. J., *Introduction to Nonlinear Optical Effects in Molecules and Polymers*, Wiley, New York, 1991.
3. Lindsay, G. A., and Singer, K. D. (eds.), *Polymers for Second-Order Nonlinear Optics*, American Chemical Society, Washington, DC, 1994.
4. Wessels, B. W., Marder, S. R., and Walba, D. M. (eds.), *Thin Films for Integrated Optics Applications*, Mater. Res. Soc. Symp. Proc. 392, Materials Research Society, Pittsburgh, 1995.
5. Jen, A. K-Y., Lee, C. Y-C., Dalton, L. R., Wnek, G. E., and Chiang, L. Y., *Electrical, Optical and Magnetic Properties of Organic Solid State Materials III*, Mater. Res. Soc. Symp. Proc. 413. Materials Research Society, Pittsburgh, 1996.
6. Burland, D. M., Miller, R. D., and Walsh, C. A., Second-order nonlinearity in poled-polymer systems, *Chem. Rev.*, *94*, 31–75 (1994).
7. Dalton, L. R., Harper, A. W., Ghosn, R., Steier, W. H., Ziari, M., Shi, Y., Mustacich, R. V., Jen, A. K.-Y., and Shea, K. J., Synthesis and processing of improved organic second-order nonlinear optical materials for applications in photonics, *Chem. Mater.*, *7*, 1060–1081 (1995).
8. Ahlheim, M., Barzoukas, M., Bedworth, P. V., Blanchard-Desce, M., Fort, A., Hu, Z.-Y., Marder, S. R., Perry, J. W., Runser, C., Staehelin, M., and Zysset, B., Chromophores with strong heterocyclic acceptors: A poled polymer with a large electro-optic coefficient, *Science*, *271*, 335–337 (1996).
9. Cai, Y., and Jen, A. K.-Y., Thermally stable poled polyquinoline thin film with very large electro-optic response, *Appl. Phys. Lett.*, *67*, 299–301 (1995).
10. Girton, D. G., Kwiatkowski, S. L., Lipscomb, G. F., and Lytel, R. S., 20 GHz electro-optic polymer Mach–Zehnder modulator, *Appl. Phys. Lett.*, *58*, 1730–1732 (1991).
11. Wang, W., Chen, D., Fetterman, H. R., Shi, Y., Steier, W. H., and Dalton, L. R., Traveling wave electro-optic phase modulator using cross-linked nonlinear optical polymer, *Appl. Phys. Lett.*, *65*, 929–931 (1994).
12. Shi, Y., Steier, W. H., Chem, M., Yu, L., and Dalton, L. R., Thermosetting nonlinear optical polymer: polyurethane with disperse red 19 side groups, *Appl. Phys. Lett.*, *60*, 2577–2579 (1992).
13. Meredith, J., VanDusen, J. G., and Williams, D. J., Optical and nonlinear optical characterization of molecularly doped thermotropic liquid crystalline polymers, *Macromolecules*, *15*, 1385–1389 (1982).
14. Wu, J. W., Valley, J. F., Ermer, S., Binkley, E. S., Kenney, J. T., Lipscomb, G. F., and Lytel, R., Thermal stability of electro-optic response in poled polyimide systems, *Appl. Phys. Lett.*, *58*, 225–227 (1991).
15. Hubbard, M. A., Marks, T. J., Yang, J., and Wong, G. K., Poled polymeric nonlinear optical materials: Enhanced second-harmonic generation stability of cross-linkable matrix/chromophore ensembles, *Chem. Mater.*, *1*, 167 (1989).
16. Eich, M., Bjorklund, G. C., and Yoon, D. Y., Poled amorphous polymers for second-order nonlinear optics, *Polym. Adv. Technol.*, *1*, 189–197 (1990).

17. Jen, A. K-Y., Rao, V. P., Wong, K. Y., and Drost, K. J., Functionalized thiophenes: Second-order nonlinear optical materials, *J. Chem. Soc., Chem. Commun.*, 90–92 (1993).

18. Wong, K. Y., Jen, A. K-Y., Rao, V. P., and Drost, K. J., Theoretical and experimental studies of the molecular second-order nonlinear optical responses of heteroaromatic compounds, *J. Chem. Phys.*, *100*, 6818–6825 (1994).

19. Drost, K. J., Jen, A. K-Y., and Rao, V. P., Designing organic NLO materials, *ChemTech* 16–25 (September 1995).

20. Jen, A. K-Y., Rao, V. P., Drost, K. J., Cai, Y. M., Minini, R. M., Kenney, J. T., Binkley, E. S., Dalton, L. R., and Marder, S. R., Progress on heteroaromatic chromophores in high temperature polymers for electro-optic applications, in *Organic, Metallo-Organic, and Polymeric Materials for Nonlinear Optical Application*, Proc. SPIE 2143, 30–40 (1994).

21. Rao, V. P., Cai, Y. M., and Jen, A. K-Y., Ketene dithioacetal as a π-electron donor in second-order nonlinear optical chromophores, *J. Chem. Soc., Chem. Commun.*, 1689–1691 (1994).

22. Karna, S. P., Zhang, Y., Samoc, M., Prasad, P. N., Reinhardt, B. A., and Dillard, A. G., Nonlinear optical properties of novel thiophene derivatives: experimental and ab initio time-dependent coupled perturbed Hartree–Fock studies, *J. Chem. Phys.*, *99*, 9984–9993 (1993).

23. Pan, H., Gao, X., Zhang, Y., Prasad, P. N., Reinhardt, B., and Kannan, R., A new class of heterocyclic compounds for nonlinear optics, *Chem. Mater.*, *7*, 816–821 (1995).

24. Bittcher, C. J. F., *Theory of Dielectric Polarization*, 2nd ed., Elsevier, Amsterdam, 1973.

25. Onsager, L., Electric moments of molecules in liquids, *J. Am. Chem. Soc.*, *58*, 1486–1493 (1936).

26. Badan, J., Hierle, R., Perigaud, A., and Vadakovic, P., Growth and characterization of molecular crystals, in Nonlinear Optical Properties of Organic Molecules and Crystals (D. S. Chemla and J. Zyss, eds.), Academic Press, Orlando, FL, 1987, pp. 297–356.

27. Roberts, G. G., in *Electronic and Photonic Applications of Polymers*, (M. J. Browden and S. R. Turner, eds.), Advances in Chemistry Series Vol. 218, American Chemical Society, Washington, DC, 1988, pp. 225–270.

28. Ulman, A., Williams, D. J., Penner, T. L., Robello, D. R., Schildkraut, J. S., Scozzafava, M., and Willand, C. S., U.S. Patent 4,792,208 (1988).

29. Kakkar, A. K., Yitzchaik, S., Roscoe, S. B., Kubota, F., Allan, D. S., Marks, T. J., Lin, W., and Wong, G. K., Chromophoric self-assembled nonlinear optical multilayer materials. Synthesis, properties, and structural interconversions of assemblies with rodlike alkynyl chromophores, *Langmuir*, *9*, 388–390 (1993).

30. Kauranen, M., Verbiest, T., Boutton, C., Teerenstra, M. N., Clays, K., Schouten, A. J., Nolte, R. J. M., and Persoons, A., Supramolecular second-order nonlinearity of polymers with orientationally correlated chromophores, *Science*, *270*, 966–969 (1995).

31. Williams, D. J., Nonlinear optical properties of guest-host polymer materials, in Ref. 1, pp. 405–435.

32. Yariv, A., and Yeh, P. C., *Optical Waves in Crystals*, Wiley, New York, 1984.

33. Singer, K. D., Kuzyk, M. G., and Sohn, J. E., Second-order nonlinear optical processes in orientationally ordered materials: Relationship between molecular and macroscopic properties, *J. Opt. Soc. Am. B*, *4*, 968–976 (1987).

34. Bethea, C. G., Experimental technique of the dc induced SHG in liquids: measurement of the nonlinearity of CH_2I_2, *Appl. Opt.*, *14*, 1447–1451 (1975).

35. Singer, K. D., Sohn, J. E., King, L. A., Gordon, H. M., Katz, H. E., and Dirk, C. W., Second-order nonlinear optical properties of donor- and acceptor-substituted aromatic compounds, *J. Opt. Soc. Am. B*, *7*, 1339–1349 (1989).

36. Clays, K., and Persoons, A., Hyper-Rayleigh scattering in solution, *Phys. Rev. Lett.*, *66*, 2980–2983 (1991).

37. Clays, K., and Persoons, A., Hyper-Rayleigh scattering in solution. *Rev. Sci. Instrum.*, *63*, 3285–3289 (1992).

38. Roe, R.-J., Glass transition, in *Encyclopedia of Polymer Science and Engineering, Vol. 7*, Wiley, New York. 1987, pp. 531–544.

39. Willand, C. S., and Williams, D. J., Nonlinear optical properties of polymeric materials, *Ber. Bunsenges. Phys. Chem.*, *91*, 1304–1310 (1987).

40. Pai, D. M., and Pringett, B. E., 1993. Physics of electrophotography, *Rev. Mod. Phys.*, *65*, 163–211 (1993).

41. Hill, R. A., Knoesen, A., and Mortazavi, M. A., Corona poling of nonlinear polymer thin films for electro-optic modulators, *Appl. Phys. Lett.*, *65*, 1733–1735 (1994).

42. Teng, C. C., and Mann, H. T., Simple reflection technique for the measuring the electro-optic coefficient of poled polymers, *Appl. Phys. Lett.*, *56*, 1734–1736 (1990).

43. Mortazavi, M. A., Knoesen, A., Kowel, S. T., Henry, R. A., Hoover, J. M., and Lindsay, G. A., Second-order nonlinear optical properties of poled coumaromethacrylate copolymers, *Appl. Phys. B*, *53*, 287–295 (1991).

44. Lundquist, P. M., Jurich, M., Wang, J.-F., Zhou, H., Marks, T. J., and Wong, G. K., Electro-optic characterization of poled-polymer films in transmission, *Appl. Phys. Lett.*, *69*, 901–903 (1996).

45. Eldering, C. A., Knoesen, A., and Kowel, S. T., Use of Fabry–Perot devices for the measurement of polymeric electro-optic films, *J. Appl. Phys.*, *69*, 3676–3686 (1991).

46. Hayden, L. M., Sauter, G. F., Ore, F. R., and Pasillas, P. L., Second-order nonlinear optical measurements in guest-host and side-chain polymers, *J. Appl. Phys.*, *68*, 456–465 (1990).

47. Metricon® Model 2010 Prism Coupler Thin Film Thickness/Refractive Index Measurement System Operating and Maintenance Guide, Rev. 9/91. Metricon, Pennington, N.J., 1991.

48. Marder, S. R., Cheng, L.-T., Tiemann, B. G., Friedli, A. C., Blanchard-Desce, M., Perry, J. W., and Skindhoj, J., Large first hyperpolarizability in push–pull polyenes by tuning of the bond length alternation and aromaticity, *Science*, *263*, 511–514.

49. Rao, V. P., Jen, A. K-Y., Wong, K. Y., and Drost, K. J., Novel push–pull thiophenes for second-order nonlinear optical applications, *Tetrahedron Lett.*, *34*, 1747–1750 (1993).

50. Dirk, C. W., Katz, H. E., Schilling, M. L., and King, L. A., Use of thiazole rings to enhance molecular second-order nonlinear optical susceptibilities, *Chem. Mater.*, *2*, 700–705 (1990).

51. (a) Cheng, L.-T., Tam, W., Stevenson, S. H., Meredith, G. R., Rikken, G., and Marder, S. R., Experimental investigation of prganic molecular nonlinear optical polarizabilities. 1. Methods and results on benzene and stilbene derivatives, *J. Phys. Chem.*, *95*, 10631–10643 (1991); (b) Cheng, L.-T., Tam, W., Marder, S. R., Stiegman, A. E., Rikken, G., and Spangler, C. W., Experimental investigation of prganic molecular nonlinear optical polarizabilities. 2. A study of conjugation dependences, *J. Phys. Chem.*, *95*, 10643–10652 (1991).

52. Rao, V. P., Jen, A. K-Y., Wong, K. Y., and Drost, K. J., Dramatically enhanced second-order nonlinear optical susceptibilities in tricyanovinylthiophene derivatives, *J. Chem. Soc., Chem. Commun.*, 1118–1120 (1994).

53. Jen, A. K-Y., Wong, K. Y., Rao, V. P., Drost, K. J., and Cai, Y., Thermally stable polar polymers: Highly efficient heteroaromatic chromophores in high temperature polyimides, *J. Electron. Mater.*, *23*, 653–657 (1994).

54. Jen, A. K-Y., Rao, V. P., and Chandrasekhar, J., Highly efficient and thermally stable second-order optical chromophores and electro-optic polymers, in Ref. 3, pp. 147–157.

55. (a) Dewar, M. J. S., and Thiel, W. J., Ground states of molecules. 38. The MNDO method. Approximations and parameters, *J. Am. Chem. Soc.*, *99*, 4899–4907 (1977); (b) Dewar, M. J. S., Zoebisch, E. G., Healy, E. F., and Stewart, J. J. P., AM1: A new general purpose quantum mechanical molecular model, *J. Am. Chem. Soc.*, *107*, 3902–3909 (1985).

56. Kanis, D. R., Ratner, M. A., and Marks, T. J., Design and construction of molecular assemblies with large second-order optical nonlinearities. Quantum chemical aspects, *Chem. Rev.*, *94*, 195–242 (1994).

57. Jain, M., and Chanchasekhar, J., Comparative theoretical evaluation of hyperpolarizabilities of push–pull polyenes and polyynes. The important role of configuration mixing in the excited states, *J. Phys. Chem.*, *97*, 4044–4049 (1993).

58. Clark, T., and Chanchasekhar, J., NDDO-based CI methods for the prediction of electronic spectra and sum-over-states molecular hyperpolarization, *Israel J. Chem.*, *33*, 435–448 (1993).

59. Kodaka, M., Fukaya, T., Yonemoto, K., and Shibuya, I., Theoretical study on the unusual effect of phenyl substituent on second-order hyperpolarizability, *J. Chem. Soc., Chem. Commun.*, 1096–1098 (1990).

60. Gilchrist, T. L., *Heterocyclic Chemistry*, Wiley, New York, 1965.

61. Wong, K. Y., Jen, A. K-Y., and Rao, V. P., Experimental studies of the length dependence of second-order nonlinear optical responses of conjugated molecules, *Phys. Rev. A*, *49*, 3077–3080 (1994).

62. Katz, H. E., Singer, K. D., Sohn, J. E., Dirk, C. W., King, L. A., and Gordon, H. M., Greatly enhanced second-order nonlinear optical susceptibilities in donor–acceptor organic molecules, *J. Am. Chem. Soc.*, *109*, 6561–6563 (1987).

63. Jen, A. K-Y., Rao, V. P., Drost, K. J., Wong, K. Y., and Cava, M. P., Optimization of thermal stability and second-order nonlinear optical properties of thiophene derived chromophores, *J. Chem. Soc., Chem. Commun.*, 2057–2058 (1994).

64. Rao, V. P., Wong, K. Y., Jen, A. K-Y., and Drost, K. J., Functionalized fused thiophenes: A new class of thermally stable and efficient second-order nonlinear optical chromophores, *Chem. Mater.*, *6*, 2210–2212 (1994).

65. Rao, V. P., Jen, A. K-Y., and Cai, Y., Achieving excellent trade-offs among optical, chemical and thermal properties in second-order nonlinear optical chromophores, *J. Chem. Soc., Chem. Commun.*, 1237–1238 (1994).

66. Shu, C-F., Tsai, W. J., Chen, J.-Y., Jen, A. K-Y., Zhang, Y., and Chen, T-A., Synthesis of second-order nonlinear optical chromophore with enhanced thermal stability and nonlinearity: Using configuration-locked trans-polyene approach, *J. Chem. Soc., Chem. Commun.* (in press).

67. Wu, J. W., Valley, J. F., Ermer, S., Binkley, E. S., Kenney, J. T., and Lytel, R. S., Chemical imidization for enhanced thermal stability of poled electro-optic response in polyimide guest–host systems, *Appl. Phys. Lett.*, *59*, 2213–2215 (1991).

68. Hubbard, S. F., Singer, K. D., Li, F., Cheng, S. Z. D., and Harris, F. W., Nonlinear optical studies of a fluorinated poled polyimide guest–host system, *Appl. Phys. Lett.*, *65*, 265–267 (1994).

69. Kowalczyk, T. C., Kosc, T. Z., Singer, K. D., Beuhler, A. J., Wargowski, D. A., Cahill, P. A., Seager, C. H., Meinhardt, M. B., and Ermer, S., Crosslinked polyimide electro-optic materials, *J. Appl. Phys.*, *78*, 5876–5883 (1995).

70. Wong, K. Y., and Jen, A. K-Y., Thermally stable poled polyimides using heteroaromatic chromophores, *J. Appl. Phys.*, *75*, 3308–3310 (1994).

71. Hampsch, H. L., Yang, J., Wong, G. K., and Torkelson, J. M., Dopant orientation dynamics in doped second-order nonlinear optical amorphous polymers. 2. Effects of physical aging on poled films, *Macromolecules*, *23*, 3648–3654 (1990).

72. Sybert, P. D., Beever, W. H., and Stille, J. K., Synthesis and properties of rigid-rod polyquinolines, *Macromolecules*, *14*, 493–502 (1981).

73. Stille, J. K., Polyquinolines, *Macromolecules*, *14*, 870–880 (1981).

74. Teraoka, I., Jungbauer, D., Reck, R., Yoon, D., Twieg, R., and Willson, C. J., Stability of nonlinear optical characteristics and dielectric relaxations of poled amorphous polymers with main-chain chromophores, *J. Appl. Phys.*, *69*, 2568–2576 (1991).

75. Lindsay, G. A., Henry, R., Hoover, J., Knoesen, A., and Mortazavi, M., Sub-T_g relaxation behavior of corona-poled nonlinear optical polymer films and views on physical aging, *Macromolecules*, *25*, 4888–4894 (1992).

76. Becker, M., Sapochak, L., Ghosen, R., Xu, C., Dalton, L. R., Steier, W. H., Shi, Y., and Jen, A. K-Y., Large and stable nonlinear optical effects observed for a polyimide covalently incorporating a nonlinear optical chromophore, *Chem. Mater.*, *6*, 104–106 (1994).

77. Peng, Z., and Yu, L., Second-order nonlinear optical polyimide with high-temperature stability, *Macromolecules*, *27*, 2638–2640 (1994).

78. Jen, A. K-Y., Drost, K. J., Cai, Y., Rao, v. P., and Dalton, L. R., Thermally stable nonlinear optical polyimides: synthesis and electro-optic properties, *J. Chem. Soc., Chem. Commun.*, 965–966 (1994).

79. Hergenrother, D. M. (ed.), in *The Interdisciplinary Symposium on Recent Advances in Polyimides and Other High Performance Polymers*, Sparks, NV, 1993.

80. Moylan, C. R., Tweig, R. J., Lee, V. Y., Miller, R. D., Volksen, W., Thackara, J. I., and Walsh, C. A., Synthesis and characterization of thermally robust electro-optic polymers, *Proc. SPIE*, *2285*, 17–30 (1994).

81. Miller, R. D., Burland, D. M., Jurich, M., Lee, V. Y., Moylan, C. R., Tweig, R. J., Thackara, J. I., Volksen, W., and Walsh, C. A., High-temperature nonlinear polyimides for $\chi^{(2)}$ applications, in Ref. 3, pp. 131–145.

82. Yu, D., and Yu, L., Design and synthesis of functionalized polyimides for second-order nonlinear optics, *Macromolecules*, *27*, 6718–6721 (1994).

83. Yu, D., Peng, Z., Gharavi, A., and Yu, L. Development of functionalized polyimides for second-order nonlinear optics, in Ref. 3, pp. 173–180.

84. Jen, A. K-Y., Liu, Y., Cai, Y., Rao, V. P., and Dalton, L. R., Design and synthesis of thermally stable side-chain polyimides for second-order nonlinear optical applications, *J. Chem. Soc., Chem. Commun.*, 2711–2712 (1994).

85. Chen, T-A., Jen, A. K-Y., and Cai, Y., Facile approach to nonlinear optical side-chain aromatic polyimides with large second-order nonlinearity and thermal stabilities, *J. Am. Chem. Soc.*, *117*, 7295–7296 (1995).

86. Ho, B.-C., Liu, Y.-S., and Lee, Y.-D., Synthesis and characterization of organic soluble photoactive polyimides, *J. Appl. Polym. Sci.*, *53*, 1513–1524 (1994).

87. Mitsunobu, O., The use of diethyl azodicarboxylate and triphenylphosphine in synthesis and transformation of natural products, *Synthesis*, *1*, 1–28 (1981).

88. Marks, T. J., and Ratner, M. A., Design, synthesis and properties of molecule-based assemblies with large second-order nonlinearities, *Angew. Chem. Int. Ed. Engl.*, *34*, 155–173 (1995).

89. Chon, J. C., Stiller, M. A., Ball, D., Nurse, J., Binkley, E. S., Sherman, R., Kenney, J. T., and Jen, A. K-Y., High thermal stability electro-optic polymer 1 × 4 switch, *SPIE Critical Reviews of Optoelectronics*, San Jose, CA, February 1996.

90. Zhang, Y., Jen, A. K-Y., Chen, T.-A., Liu, Y.-J., Zhang, X.-Q., and Kenney, J. T., High performance electro-optic polymers and their applications in high speed electro-optic switches and modulators, *Proc. SPIE, 3006* (in press).

91. Nozaki, R., and Mashimo, S., Dielectric relaxation measurements of poly(vinyl acetate) in glassy state in the frequency range 10^{-6}–10^6 Hz, *J. Chem. Phys., 87,* 2271–2277 (1987).

92. Lytel, R. S., Lipscomb, G. F., Kenney, J. T., and Binkley, E. S., Large-scale integration of electro-optic polymer waveguides, in *Polymers for Lightwave and Integrated Optics*, Marcel Dekker, Inc., New York, 1993.

93. Thackara, J. I., Chon, J. C., Bjorklund, G. C., Volksen, W., and Burland, D. M., Polymeric electro-optic Mach–Zehnder switches, *Appl. Phys. Lett., 67,* 3874–3876 (1995).

94. Van Eck, T. E., Tickner, A. J., Lytel, R. S., and Lipscomb, G. F., Complementary optical tap fabricated in an electro-optic polymer waveguide, *Appl. Phys. Lett., 58,* 1588–1590 (1991).

95. Wang, W., Chen, D., Fetterman, H. R., Shi, Y., Steier, W. H., and Dalton, L. R., Optical heterodyne detection of 60 GHz electro-optic modulation from polymer waveguide modulators, *Appl. Phys. Lett., 67,* 1806–1808 (1995).

96. Ducharme, S., Scott, J. C., Twieg, R. J., and Moerner, W. E., Observation of the photorefractive effect in a polymer, *Phys. Rev. Lett., 66,* 1846–1849 (1991).

97. Zhang, Y., Cui, Y., and Prasad, P. N., Observation of photorefractivity in a fullerene-doped polymer composite, *Phys. Rev. B, 46,* 9900–9902 (1992).

98. Moerner, W. E., and Silence, S. M., Polymeric photorefractive materials, *Chem. Rev., 94,* 127–155 (1994).

99. Zhang, Y., Burzynski, R., Casstevens, M. K., and Ghosal, S., Photorefractive polymers and composites, *Adv. Mater., 8,* 111–125 (1996).

100. Yeh, P. C., *Introduction to Photorefractive Nonlinear Optics*, Wiley, New York, 1992.

101. Meerholz, K., Volodin, B. L., Sandalphon, Kippelen, B., and Peyghambarian, A photorefractive polymer with high optical gain and diffraction efficiency near 100%, *Nature, 371,* 497–500 (1994).

102. Peng, Z., Bao, Z., and Yu, L., Large photorefractivity in an exceptionally thermostable multifunctional polyimide, *Macromolecules,* 6003–6004 (1994).

22
Organic Complex Thin Films for Ultrahigh Density Data Storage

Z. Q. Xue
Peking University
Beijing, China

H.-J. Gao and S. J. Pang
Chinese Academy of Sciences
Beijing, China

I. INTRODUCTION

The microelectronics industry has pursued a path of aggressively reducing feature sizes in integrated circuits. There is every reason to expect this trend to continue, as there seems no limit to our demand for more computer processing power and memory. Manufacturing of integrated circuits with nanometer feature size is now projected. Today, even in the research laboratory, fabrication of structures and electronics devices with these dimensions presents formidable challenges.

In spite of the tremendous progress, recent studies revealed that quite a few crucial limitations exist in fabricating integrated circuits with nanometer feature sizes using inorganic semiconductor materials such as Si or GaAs wafers. A possible way to overcome these limitations is to develop organic or biological semiconductor materials for the purposes of fabricating new type of electronics and optic electronics devices with nanometer- or even atomic-scale feature sizes.

A nanometer-scale electronics component should have less than 10^9 atoms, indicating that the corresponding data storage density could be as high as 10^{12} bits/cm^2, "ultrahigh data storage density." On a nanometer or atomic scale, various quantum effects and statistical fluctuations prevail over classical phenomena and become the predominant factor which requires serious consideration. Along with high-purity materials, low signal power, and short response time, developing reliable methods of fabrication for nanometer-scale electronics involves new theory, novel techniques, and new materials.

Because of many interesting features, organic polymers have become a subject of basic and technological interest. Recent research efforts have focused on developing the high-quality functional polymers which should have almost perfect crystal structures with much less impurity and less defects both on surface and at the interfaces. Novel techniques developed were used to prepare some functional polymers with fairly satisfactory results. These functional polymers were also characterized with transmission electron microscope (TEM), x-ray diffraction (XRD), ultraviolet–visible (UV) absorption spectra, and scanning tunneling microscope (STM).

II. TECHNOLOGICAL LIMITATIONS OF MICROELECTRONICS ELEMENT BASED ON Si AND GaAs WAFERS

A. Reduction of Electronics Element Sizes

Roberts [1] reported a statistics of the electronics device development in the past 50 years, and gave the road map of element sizes in Fig. 1. It showed vacuum tubes (in cm) in the 1940s; solid electronic devices (in mm) in the 1950s; very large-scale integration (VLSI) (in a few tens μm) in the 1970s; ultralarge-scale integration (ULSI) (in μm) in the 1990s. Technological development is oriented toward fabricating smaller devices to enable them to be packed more closely on chips. Future development will be a giga-scale integration (GSI). At that size, the devices still obey classical operation principles.

The rapid reduction of feature sizes of electronic devices accelerates the development of the computer's processing power and memory. Recently, the best information processing technique is digitalization and the simplest method is the binary system. Its base is "0" state and "1" state, which is the unity of opposites in philosophy, such as positive and negative, cation and anion, up and down, right and left, yes and no, insulator and conductor. Both "0" and "1" states are called one bit. A density of bits is determined by image factors and the resolution of

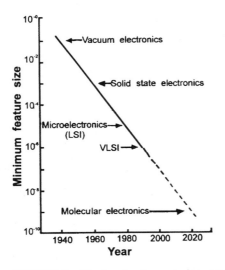

FIGURE 1   Electronic element size versus years. (From Ref. 1.)

patterns, so that the higher density means the stronger capability for possessing information. The decrease in size of electronics devices also considerably reduces the size of all electronics instruments. The rapid growth of personal computers results from the development of integrated circuit technology. Forty years ago, much less powerful computers used thousands of vacuum tubes, which were large and required enormous power consumption. The information society will build on the base of the personal computers and the high-speed computers. Today, the computer plays an important role in the rapid development of science and technology.

B. What Is the Next Generation of Electronic Devices?

The integrability on a wafer continually increases one step per 1.8 months. Up to now, the element size was reaching its upper physical limit. Once devices and circuits begin to work, distortion and electronic waves can affect adjacent elements; conventional design and analysis of integration circuits are not valid. Quantum effects and statistic fluctuations become prominent factors; therefore, one has to take serious consideration.

Chiabrea et al. [2] discussed the physical limits on integration of classical electronic devices. They investigated these limits from the collective behavior of atoms, the power dissipation, the voltage- or current-induced breakdown, the background white noise, and Heisenberg's uncertainty principle. The results show that nanometer-scale devices do not obey classic rules, but demonstrate new effects, and may work on unknown principles. New materials and fabrication techniques on an atomic or molecular scale should be developed to make these devices. In order to go beyond these limits, research efforts have been taken to develop information processing systems that operate at the molecular level.

Sugano [3] reported on a perspective on next-generation silicon devices. Besides the classical scale-down of conventional silicon devices, utilization of quantum-mechanical phenomena such as the modification of the silicon energy band with mechanical stress, Brillouin zone folding, and the reduction of electron degrees of freedom are expected to improve electronic properties of silicon and, consequently, the characteristics of devices, so as to invent new devices. The possible quantum-mechanical engineering of electronic properties of silicon will be addressed later. The heat transmission of a chip is well known to limit the maximum density of devices integrated on the chip. To achieve higher-density devices, the standby power consumption in the circuit must be negligible, and the signal energy representing one bit of information must be reduced as well. Effects of the statistical fluctuation and doping distribution may also affect minimum feature sizes.

Meindl et al. [4] discussed the physical limits on giga-scale integration. They defined the chip performance index (CPI) as the number (N) of transistors per chip and the associated power–delay product N/Pt_d. The three most important fundamental limits on giga-scale integration (GSI) were derived from thermodynamics, quantum mechanics, and electromagnetic theory. The thermal energy of an electron gas requires that the power–delay products Pt_d of a switching transition be greater than about $4kT$, were k is Boltzmann's constant and T is absolute temperature. The Heisenberg uncertainty principle of quantum mechanics requires that the average power transfer during a measurable energy change associated with a switching transition of duration t_d be greater than h/t_d^2, where h is Planck's constant. The CPI

has increased by about 12 degrees since 1960 and is projected to increase by about another 6 degrees before 2020.

Kaga et al. [5] investigated the technological issues involving the miniaturization of dynamic random-access memory into the gigabit era. Ever-smaller giga-generation dynamic random-access memory cells require three-dimensional high-charge-density capacities with insulating films, which leads to the need for further improvements in lithographic resolution for ever-smaller and higher-ratio memory cells, and planarization technologies for reducing the memory-cell height.

According to the discussions mentioned above, an element size decrease in a large integration circuit will lead to a new era of science and technology development. As shown in Fig. 2 [5], the element size will be from millimeters to micrometers, and traditional science and technology will still be useful. But from micrometers to nanometers, new effects, new concepts, and new technology will be studied.

For the fabrication of a new generation of devices, there are three major aspects: (1) theory based on quantum effect and statistics fluctuations; (2) materials, nanometer-scale materials consisting of finite atoms; the surface and interface atomic and electronic structures play key roles; (3) assembly technique, to assemble functional molecular, ultramolecular and clusters on atomic and molecular scales.

There are too many problems to be solved, such as characterization of materials on a molecular scale. Electronic materials will probably change from inorganic materials to organic and biology protein materials. In this transition period, we will have three stages: (1) study of nanometer-scale electronics devices based on inorganic materials, such as investigation of new development of GSI and UGSI, device operating principles and process techniques on which quantum effect and statistic fluctuation play prominent roles; (2) study of organic or metal–organic materials; (3) study of nanometer-scale electronic devices based on organic or biology protein materials.

Optical and electrical switching devices based on organic molecules, polymer, and biology protein materials are the main functional devices possessing "0" and "1" states. These kind of devices are different from vacuum electronic devices. The electron devices based on organic materials possess low-dimension characteristics, such as a one-dimensional molecular chain and van der Waals force inter-

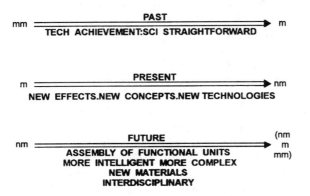

FIGURE 2 Challenge in miniaturization. (From Ref. 3.)

action between molecular chains, so that their functional elements possess less volume and larger integrability in organic molecular chips. The functional devices based on organic materials also require assembling techniques on a molecular level. These kinds of electrical and optical functional devices based on assembled carbon–hydrogen molecules are called molecular electronic devices.

Molecular electronic devices will take the place of the current micrometer-scale electronics devices, and the advantages are less size, rich materials, easy process, low cost, and direct assembled function devices by carbon–hydrogen molecules, which is called the molecular engineering or molecular architecture. They possess a larger connection rate, faster response, and higher computing and processing speeds.

III. ULTRAHIGH DENSITY DATA STORAGE TECHNOLOGY

In recent years, much attention has been focused on ultrahigh density data storage. Some materials and writing–reading methods have been developed.

A. Scanning Capacity Microscope

Barrett et al. [6] reported ultrahigh density data storage using a scanning capacity microscope (SCM). A schematic diagram of their function element is shown in Fig. 3. A SiO_2 thin film of 5 nm was formed on a p-Si substrate, and then a Si_3N_4 layer of 50 nm was deposited by the chemical evaporation method. A metal electrode was deposited on top of the Si_3N_4 layers. The metal electrode might be fixed or mobile. When -40 V was applied between the electrode and the p-Si substrate, electron tunneling was created from the Si_3N_4 layer to the p-Si substrate. As a result, a spatial charge region was built, and the capacity property was changed. Charges could be stored for a long period. When a STM tip replaced a metal electrode, the signal writing and reading could be performed with STM. The operating principle of the SCM writing and reading is shown in Fig. 4. An AFM with a conductive tip was used, and the signal could be written in and read out, and the topography of the sample could be observed. Barrett et al. recorded the independence declaration of Swiss Confederation in 1291 on an area of 120 μm $\times$ 120 μm, and the school badge of Stanford University in 1891 on 90 μm $\times$ 90 μm across. Each of these patterns contains 256 kilobits of information. The storage density was close to 10^9 bits/cm^2, and the writing rate was 10^6 bits/s.

Wiesendanger [7] studied the SCM images of charge patterns written with the voltage pulse method. The stored charge was found to be stable for several days.

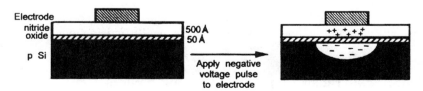

FIGURE 3 The NOS structure and the charge profile of the SCM storage medium. (From Ref. 6.)

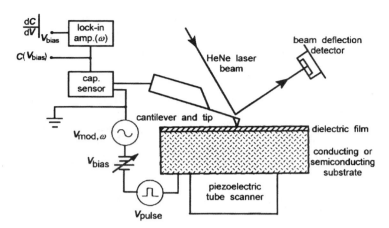

FIGURE 4 Schematic of the SCM. The instrument is an optical beam deflection atomic force microscope (AFM) with an added capacitance sensor. (From Ref. 6.)

They studied the nitride–oxide–silicon (NOS) devices with varying oxide film thickness. By applying voltage pulses between a conducting AFM cantilever in contact with the top nitride layer and the silicon substrate, it was possible to store charge in the NOS heterostructure. The writing was achieved by applying a voltage pulse of −35 V and 60 ms. Further reducing the size of the written charge bits seem possible by improving the detection sensitivity. As compared with other nano-fabrication methods by scanning probe techniques, charge storage based on voltage pulses applied to NOS structures seems to be fast, highly reliable, and stable.

B. $Na_2O–V_2O_5–P_2O_5$ Medium Film for Data Storage

Sato and Tsukamoto [8] studied nanometer-scale recordings with STM. Their recording material was a $Na_2O–V_2O_5–P_2O_5$ thin film. The glass composition was 14 mol% Na_2O–71 mol% V_2O_5–15 mol% P_2O_5. The V_2O_5 is electrical conduction and the mechanism of electrical conduction depends on the Na_2O concentration. The P_2O_5 was added to stabilize the amorphous form. Marks were formed by applying voltage pulse of +4 V (the polarity refers to the sample) for 1 ms width and the tunneling current was about 50 nA. The recording rate was about 1 ms. Marks could be erased by applying reverse-polarity pulses of −5 V for 10 ms. This reversible transformation could be repeated more than 10 times. Any recording mark subjected to writing–erasing cycles of more than 10 times became impossible to be erased. The reproducible signal was the change in tunneling current measured with the STM in the constant-height mode. Figure 5a and 5b showed reproducible waveforms after recording five marks and subsequently erasing one mark, respectively.

C. Nanometer-Scale Recording on Chalcogenide Films

Kado and Tohda [9] used amorphous chalcogenide films of $GeSb_2Te_4$ as the storage medium. The 20-nm-thick $GeSb_2Te_4$ film was deposited by sputtering on a bottom

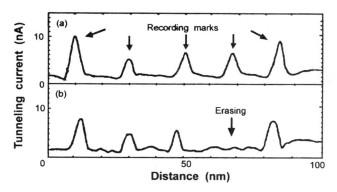

FIGURE 5 Tunneling current waveforms after recording (a) and erasing (b). (From Ref. 8.)

electrode. The bottom electrode was made of 300-nm-thick Pt film deposited on a Si substrate by the sputtering method. They used atomic force microscopy (AFM) to measure the topography and the electrical conductance of a sample simultaneously. Figure 6 showed the experimental system. The AFM was operated in air under a constant-repulsive-force mode. The probe was prepared by vacuum evaporation of 15 nm of Cr and 100 nm of Au over the Si_3N_4 pyramidal tip. The electrically conductive probe was used to apply a voltage to the $GeSb_2Te_4$ film. The conduction images were obtained by monitoring the current when a bias voltage between the probe and the sample was applied. In the recording process, a pulse voltage was applied to the probe that touched the sample surface.

The recording experiment was carried out by applying a pulse voltage of +3 V for 5 ms while the probe scanned the sample surface. The current for recording was 10–20 nA. The size of the recorded regions depended on the pulse width. The shortest pulse width with which the data could be recorded so far was 10 ms. However, the probability of recording was low with the 10-ms-wide pulse voltage. Figure 7 demonstrated a few tens of nanometer scale recording by applying a pulse voltage of +3 V for 0.5 ms. The conductance image shown in Fig. 7a was obtained

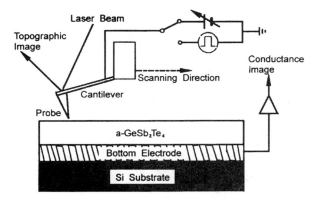

FIGURE 6 Schematic diagram of AFM system used recording. (From Ref. 9.)

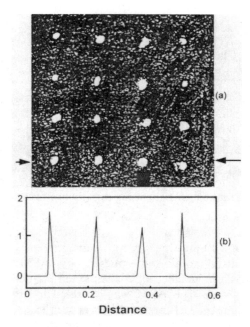

FIGURE 7 An example of a few nanometer-scale recording by applying pulse voltages of +3 V for 0.5 ms, scan area 0.6×0.6 mm^2; (a) conductance image, (b) line scan profile of the conductance image. (From Ref. 9.)

with a bias voltage of 0.5 V. The scanned area was 0.6×0.6 μm^2. The smallest recorded region was 10 nm in diameter, corresponding to a data storage density of 1 Tbit/cm^2. Figure 7b showed a line scan profile of the conductance image. The current of the recorded and the unrecorded region was about 1 nA and less than 10 pA, respectively, indicating that the technique is capable of reading at a high signal-to-noise (SN) ratio.

D. Data Storage Used Cis–Trans Isomerization [10]

Liu et al. [11] reported a stable recording medium based on the two-step photo-electrochemical reduction of 4-octyl-4'-(5-carboxy pentamethyleneoxy) azobenzene Langmuir–Blodget (LB) film. The photochromic properties of the azobenzenes were well established; isomerization occurred when they were irradiated with an ultraviolet (UV) source, whereas reversal processes can occur either by thermal treatment or by visible-light irradiation. Liu et al. demonstrated that the less stable cis form might be electrochemically reduced to hydrazobenzene, which was stable in an inert atmosphere but might be electrochemically oxidized to *trans*-azobenzene (see Fig. 8). This photoelectrochemical process was stable in several hundred write–read cycles.

Although the preliminary progress revealed that the azobenzene can be used as erasable optical data storage media, with the limit being 10^8 bits/cm^2, one-step writing process would be preferable. Moreover, the number of write–read cycles

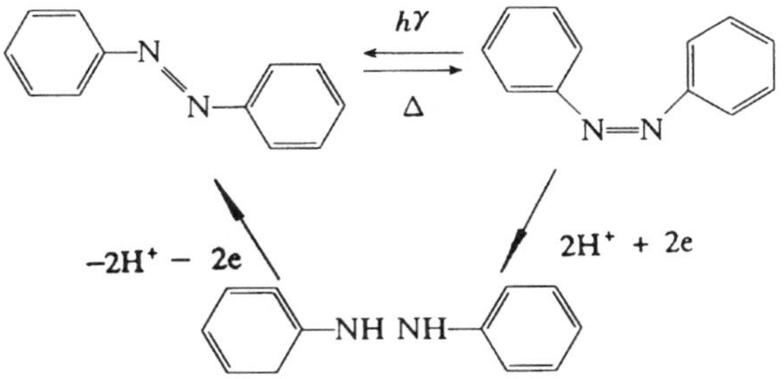

FIGURE 8 Photoelectron switching of 4-octyl-4'-(5-carboxypentame thyleneoxy). (From Ref. 10.)

on this kind of film was much less than that of magnetic tape. Liu et al. have suggested that uniformly irradiating the film may convert it to the cis form, and storage densities as high as 10^{12} bits/cm^2 may be realized with an electrochemical scanning tunneling probe to write on the *cis*-azobenzene film.

E. Scanning Near-Field Optical Data Storage

Terris et al. [12] reported their near-field optical data storage. Because of circumventing the diffraction limit, the near-field optical technique can have a very high resolution by using a submicron aperture and operating in the near-field regime. The aperture is produced by imaging through a tapered metallic optical fiber, at the end of which is a small pinhole. Using such a system, the optical resolution can be in the 10-nm range. The other method able to obtain near-field optical resolution is to use an immersion lens made of a high index of refraction material, the solid immersion lens (SIL). The lens is formed by placing a truncated sphere between a focusing objective and the sample of interest. The wavelength inside is reduced by the high index of the glass, leading to a reduction in the diffraction limited spot size.

To realize the near-field data storage, SIL was used to reduce the minimum achievable spot size in air. Figure 9 showed their flying SIL, which incorporated an air-bearing surface as an integral part of the lens design [13]. To form an optical spot at the air-bearing surface, an incident collimated laser beam was focused by using an auxiliary objective lens toward a point nr below the center of curvature of the SIL, where n is the index of refraction and r is the radius of curvature of the SIL. The incident rays were refracted at the curved SIL surface and were focused to the air-bearing surface, a distance r/n below the center of curvature.

A scanning near-field optical microscope (SNOM) generally uses a tapered optical fiber to form a subwavelength-sized light source. The aperture light source consists of an optical fiber with one tapered end that has a tip size of 50–100 nm. A schematic diagram of the SNOM is shown in Fig. 10 [14]. The 488-nm light from an Ar ion laser is sent to the SNOM fiber probe via a single-mode fiber coupler. The alignment of the fiber coupler is optimized for maximum optical throughput. The output intensity from the probe is typically several nanowatts. Light emitted is collected through an objective lens. In this case of reflection (the

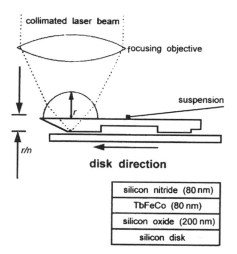

silicon nitride (80 nm)
TbFeCo (80 nm)
silicon oxide (200 nm)
silicon disk

FIGURE 9   Schematic diagram of the flying solid immersion lens and disk structure. (From Ref. 13.)

scattered light from the surface), a 20 × objective lens is used for collection and is positioned off-axis from the sample's normal. The collected light is then sent to a spectrometer for analysis. If samples are data storage media, the SNOM can be used to write and read signals.

F.   Ultrahigh Density Data Storage on Magnetic Ultrathin Films

Metallic ferromagnetic ultrathin films and multilayers exhibit new outstanding properties, such as perpendicular magnetic easy axes, enhanced magnetooptical Kerr rotations, and giant magnetoresistances (GMR). It is possible to use these

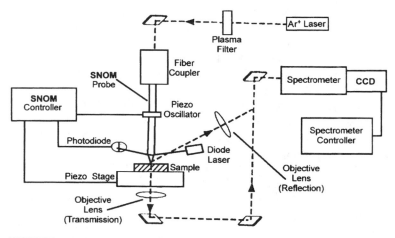

FIGURE 10   Schematic diagram of the SNOM setup. (From Ref. 14.)

films for the ultrahigh density data storage. For instance, the submicron-patterned magnetic media can be used as high density data storage or magnetooptical recording. To increase recording density, Roussaux et al. [15] studied a large area of dot arrays in Au/Co/Au(111) sandwiches with an ultrathin Co layer (0.6–2 nm) which exhibits interface-induced perpendicular anisotropy. Hosaka et al. [16] reported a nanometer-sized structure fabrication on an insulator and a fine magnetic domain formation in a magnetic material. Application of negative voltage to the gold-coated AFM probe could make gold lines 40 nm wide and dots 20 nm in diameter on SiO_2/Si by field evaporation. In the case of a positive voltage, thermal processes were dominant. For example, nanometer-sized magnetic domains can be produced by electron-induced local heating by using STM and magnetic force microscopy (MFM). The proposed MFM recording is shown in Fig. 11. The probe used the birdbeak-type cantilever coated with double-layered films of magnetic PtCoCr (30 nm thick) and carbon (20 nm thick). The MFM recording could form 60×240-nm^2 domains in Pt/Co multilayer magnetic films.

The flying SIL was incorporated into a magnetooptical storage test system (12). Included in the system were a diode laser, a 0.5 NA (the numerical aperture of the focusing lens) focusing objective, a focus servo system, and a differential detector to sense the Kerr rotation of light reflected from the magnetooptical (MO) disk. The disk was mounted on an air-bearing spindle, which was stable enough so that the focused spot remained on a written track without the need for a tracking servo.

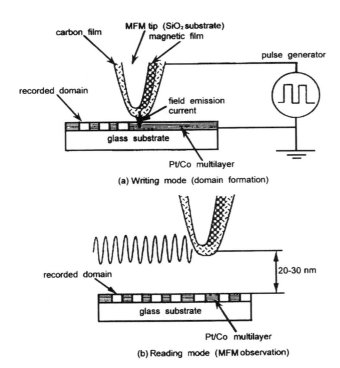

FIGURE 11 Schematic diagram of the MFM recording: (a) writing mode, (b) reading mode. (From Ref. 16.)

The focus servo maintained focus by moving only the focusing objective, with the SIL height held constant by the air-bearing spindle. Data was written by the laser pulse to an applied magnetic field and could be erased by reversing the direction of the magnetic field. The disk structure was similar to that used in conventional MO recording, except that the beam incidents from the air side instead of the substrate side of the disk.

They demonstrated a linear density of 26.7 kbits/cm and a linear velocity of 1.25 m/s translating to data for data storage, which is considerably higher than any other near-field or scanned-probe data storage demonstrations. The data rate is not close to any fundamental limit, but is limited by the low linear velocity needed to maintain a low fly height. By improving the air-bearing design, which has been shown to be possible in magnetic recording, it may be possible to increase both the linear velocity and data rates by at least a factor of 5, while maintaining a small spot size and high data storage density. By combining a new near-field technique, the SIL, with conventional magnetooptical storage technology, they achieved a realistic demonstration of near-field optical data recording. With the track pitch of 0.7 mm and linear density of 26.7 kbits/cm, the real density was 3.8×10^8 bits/ cm^2. Moreover, the SIL can be combined with a short-wavelength source, with the data density scaling as the wavelength squared. This, together with some improvement in performance gained by lowering the fly height, should enable the density to reach at least 1.5×10^9 bits/cm^2.

A SNOM, a variation of the STM, uses the laser and the optical fiber [17]. Data storage on optical functional thin films can be obtained by the SNOM. The magnetic scanning near-field optical microscope (MSNOM) is a variation of the SNOM. By scanning a small collecting aperture very close (about 10 nm) to the sample surface, one can obtain an optical image with spatial resolution close to $\lambda/40$. The MSNOPM uses the magnetooptic effect—changes in the polarization of the reflected light—for magnetic imaging. The ultrahigh density data storage may be realized if the magnetic sensor is small enough.

Giant magneto-resistance (GMR) is a term coined to describe the behavior of materials consisting of alternating layers of ferromagnetic and nonmagnetic metals deposited on an insulating substrate [18]. The resistance is the greatest when the magnetic moments in the alternating layers are oppositely aligned and the smallest when they are all parallel. Berkowitz [19] and Xiao et al. [20] reported Co–Cu particle systems which possessed GMR properties. Advances in GMR materials have led the development of memory elements to current applications. The recording data density of new magnetic material thin films may achieve 10^{10} bits/cm^2.

IV. REVIEW OF ORGANIC POLYMER THIN FILMS USED FOR DATA STORAGE

In the past 20 years, many research workers have studied organic polymer thin films for data storage. The main results are as follows.

A. Playing the Role of Metal Filaments in Thin Films

Shortly after discovering switching phenomena of the inorganic semiconductor thin films, Bui et al. [21] discovered similar effect on a Au/PS(polystyrene)/Au device produced by glow-discharge deposition. Later, several research groups studied switching phenomena and mechanism of polyethylene, polystyrene, and polystyleneamino thin films [22,23]. Two distinctive forms of memory switching were observed in thin films of glow-discharge polymerized styrene, acetylene, benzene, and aniline. The occurrence of these phenomena depends on three parameters, namely the electrode's thickness, the film's thickness, and the nature of the "forming" atmosphere, but it is independent of electrode materials, such as Au, Al, Cr, and Pb. Switching through voltage was proportional to film thickness. Switching time was proportional to electrode thickness. The state of low resistance was relevant to highly conductive filaments fusing from the two electrodes. Couch et al. [24] reported that metallic conduction occurs through Langmuir–Blodget (LB) films. The systems investigated were Ag/ω-tricosenic acid/Ag and Au–Pd/cadmium stearate/Au. The conduction gave rise to high electrical conductivity through a Langmuir–Blodget film in a metal/LB film/metal sandwich. Metallic pathways were found to exist through the LB films, even for relatively thick multilayers ($N = 40$ layers). They were estimated to cover an area of 6×10^{-3} mm^2. Switching effects were observed in these filaments with the application of voltage pulse. For the thickest LB films (about 100 layers), there was evidence that the conduction was predominantly through the organic regions and was of the form log I–$V^{1/2}$.

Godehardt and Heydenreich [25] reported that an inorganic Ag/SiO$_x$/Ge sandwich system can also record data, which was an insulating layer embedded between a metallic electrode and a Ge (36 nm) semiconducting top electrode, prepared on glass substrates by vacuum evaporation. Local conducting channels form at weaker points of the insulating layer by applying a voltage between the electrodes. On the sample surface, these current filaments produced corresponding electric microfields of radial symmetry owing to variations of the surface potential. The microfields could be detected with scanning electron microscopy (SEM) by revealing characteristically bright- or dark-image structures, depending on the polarity of the voltage applied. For some samples, the filament current was evaluated to be 2.9 nA, taking into account the measured 144 MΩ of the sheet resistance of the semiconducting top electrode. When there were filaments in organic or organic insulator thin films between metal electrodes, electric bistable phenomena could occur, but relating parameters were unstable and less reproducible.

B. Charge Transfer Complex Thin Films

In 1979, Potember et al. [26] found that the TCNQ–M (M = Ag, Cu) thin films possessed stable and reproducible current-controlled bistable electrical switching properties. For the Cu/Cu–TCNQ/Al, the I–V characteristics revealed an abrupt decrease in resistance from 2 MΩ to less than 200 Ω at a field strength of 4×10^3 V/cm. The transition from the high- to low-resistance state occurred with delay, with switching times of approximately 15 ns and 10 ns, respectively. Switching with high-power dissipation yields a low-resistance memory state that could be erased by application of a short current pulse. After that, the electrical, optical, and optoelectronic switching properties of this kind of material devices have been studied widely [27–29]. Generally, when an electrical field of sufficient strength is applied, electrical switching phenomena can be observed [i.e., the thin film will switch to a lower resistance state (several hundred ohm)].

C.   Phase Change Complex Thin Films

Machida et al. [30] reported a switching phenomenon in lead phthalocyanine (PbPc) films. PbPc have two types of crystal structures: monoclinic structure and triclinic structure. The switching effect was only observed in the film consisting of a mixture of monoclinic grains and the amorphous phase. Both the switching threshold voltage and the OFF/ON ratio of resistance became large when the evaporation rate increased. Their sample device was of the Au/PbPc/Au sandwich structure. Its ON state remained even after the applied voltage was removed. The largest OFF/ON ratio of resistance obtained was 4.5×10^4, and the deposition rate was 1 nm/s. The switching time was at the order of 10–100 ns, depending on the applied voltage.

Sakai et al. [31] reported a metal/LB film/metal (M/LB/M) sandwich-like structure with a noble-metal base electrode. The base electrode, consisting of 30 nm of Au on 5 nm Cr, was vacuum evaporated on the substrate. The LB films prepared were squarylium dye (SOAZ) and polyimide (PI) and head-to-head (Y type) assembly films. After the fabrication process, the devices required a preliminary forming procedure, in which a sufficiently high voltage (such as 10 V or less) was applied to each M/LB/M junction to create a permanent change in the conductivity of the devices. Afterward, the anticipated switching behavior was observed. The voltage was applied across the device in series with a 100-Ω-load resistor. In the initial state, the formed device was in a very high resistance state (OFF state, higher than 10^6 Ω). When a certain voltage was applied to the device, it switched from the OFF state to a high-conductance state (ON state, 20–100 Ω) via an intermediate state. The conduction was non-Ohmic in the OFF state, but ohmic in the ON state. The ON state could be returned to the OFF state by applying a voltage of 1–2 V above the threshold value between the top and base electrodes; when the voltage was subsequently reduced, both OFF and ON states remained and were stable even in the air. The switching and memory phenomenon occurred reproducibly. Similarly, they have studied Al/PI/Au devices.

D.   Nanometer-Scale Data Storage Thin Films

Organic complex switching thin films have already found their application for data storage, but the quality of these thin films prepared by chemical methods are far from perfect. In addition, analysis and measure methods are not precise, so that the experiment data are random in distribution and the obtained results are different from each other. In recent years, efforts have been paid to better the quality of these films by preparing them with the ultrahigh-vacuum deposition or the molecular beam epitaxy methods. We have built the ionized cluster beam deposition (ICBD) system [32]. Many kinds of organic complex thin films have been studied. Several of them possessed the electrical and optical bistabilities [33]. We will describe these thin films in Sections V–VII.

V.   SWITCHING PROPERTIES OF Ag(Cu)–TCNQ THIN FILMS

Among the organic materials so far reported, the most attractive and well-investigated thin films were made of either copper or silver charge transfer complexes of TCNQ and its congeners [26,29,34]. Many important applications of two-terminal bistable thresholds and memory logic elements, optoelectronic switches, erasable optical recording media, fast photochromic filters, and ultrahigh density information storage addressed by a tip of a STM have been proposed. As a result, these kinds of thin films are receiving increasing interest and have become one of the most important institution parts of molecular electronics [35]. Hence, the developments of new materials, fabrication, and performance of the devices are worthwhile to explore and will also provide important insight into these kinds of thin film devices [33].

General organic crystalline are electrical insulators. These molecules are held together by van der Waals forces, and the interaction between them is too weak to be appreciable for π-electron overlap. Ashwell [36] demonstrated that conductivity enhancement can be achieved by turning to charge transfer complexes that may be classified into separate groups based on the degree of charge transfer. Nonbonding complexes have low conductivity, and their structures are characterized by mixed stacks in which the donor and acceptor moieties alternate, whereas dative complexes have intermediate to high conductivity of $10^{-8} < \sigma < 10^4$ S/cm. In this category, the less conductive examples exhibit complete charge transfer and/or mixed stacking, whereas for metallic behavior, the criteria are nonintegral charge transfer and segregated stacking with a uniform spacing and favorable π-overlap; see Fig. 12.

A. Charge Transfer Model

Molecular electronic devices should be assembled on the molecular level; stacked molecular (parallel or series) thin films can be realized. Now, let us explore the operating mechanism of electric element on this base.

Rombidi et al. [37] have proposed an electric conduction process between a donor and an acceptor, and its structure is D–σ–A, where D represents the donor molecule, A represents the acceptor molecule, and σ is the bridge connecting D and A. In the donor and the acceptor, there are the highest occupied molecular orbital (HOMO) of donor (D_H) and acceptor (A_H), and the lowest unoccupied molecular orbital (LUMO) of donor (D_L) and acceptor (A_L). When a strong donor D and a strong acceptor A are connected, the lower unoccupied orbitals are occupied, so that the donor is charged positive (D^+) and the acceptor is charged negative

$D°$ $A°$ $D°$ $A°$	D^+ A^-	D^+ A^-	$D^{\delta+}$ $A^{\delta-}$ $D^{\delta+}$ $A^{\delta-}$	
$A°$ $D°$ $A°$ $D°$	A^- D^+	D^+ A^-	$D^{\delta+}$ $A^{\delta-}$ $D^{\delta+}$ $A^{\delta-}$	
$D°$ $A°$ $D°$ $A°$	D^+ A^-	D^+ A^-	$D^{\delta+}$ $A^{\delta-}$ $D^{\delta+}$ $A^{\delta-}$	
$A°$ $D°$ $A°$ $D°$	A^- D^+	D^+ A^-	$D^{\delta+}$ $A^{\delta-}$ $D^{\delta+}$ $A^{\delta-}$	
$D°$ $A°$ $D°$ $A°$	D^- A^+	D^+ A^-	$D^{\delta+}$ $A^{\delta-}$ $D^{\delta+}$ $A^{\delta-}$	
$A°$ $D°$ $A°$ $D°$	A^- D^+	D^+ A^-	$D^{\delta+}$ $A^{\delta-}$ $D^{\delta+}$ $A^{\delta-}$	
Low—σ	Intermediate—σ		High—σ	

FIGURE 12 Staking and electrical characteristics of charge transfer complexes. (From Ref. 36.)

(A$^-$). For added metal electrodes (M), the molecular element can be shown as M$_1$/D$^+$–σ–A$^-$/M$_2$, and its energy diagram schematic shown in Fig. 13, where $\emptyset$ is the work function. The M$_1$/D$^+$–σ–A$^-$/M$_1$ device operating mechanism is as follows: when the applied suitable bias between electrode M$_1$ and M$_2$, Fermi energy $E_F(M_2)$ becomes higher than A$_L$, D$_H$ becomes higher than Fermi energy $E_F(M_1)$, and the electron tunneling σ bridge trends upward from A$_L$ to D$_H$, so that electrons run from M$_2$ to M$_1$ (see Fig 13b). If the out circuit is connected, there is electric current. If applying reverse-bias voltage (see Fig. 13c), different current values appear. So the M$_1$/D$^+$–σ–A$^-$/M$_2$ device can possess rectifying properties.

When applying voltage to this system, and the voltage changes from lower to higher, electrons hardly transmit at the beginning, and the thin film appears to be an insulator; when the voltage is higher than a threshold voltage, so that $E_F(M_2)$ is higher than the A$_L$ and the D$_H$ higher than $E_F(M_1)$, charge transfer occurs, and the thin film changes to an electric conductor state. So if there is a threshold voltage, when the applied voltage is beyond the threshold, phase change of the device can occur, and the device maintains at this phase (e.g., the M$_1$/D$^+$–σ–A$^-$/M$_2$ device can be a functional element possessing electronic bistability).

B. Organic Materials Possessing Switching Property

However, the bridge is not necessary, as is discussed above and is shown in the example of M-TCNQ (M = Ag, Cu), which possess bistable properties, and can be used for electrical and optical as well as photoelectronic switching devices.

M and TCNQ are a nonstoichiometric complex in the M-TCNQ thin film, and the donor and acceptor show charge states at ambient temperature. Under the laser beam or electric field, charge transfer can happen and some neutral metal and TCNQ will occur; the film can return to its original state when heated. So this reaction can be shown as follows:

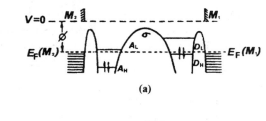

(a)

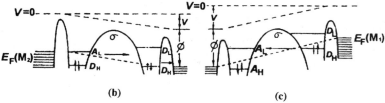

(b)   (c)

FIGURE 13 The simplified model of electron tunneling in a molecular rectifier. (From Ref. 37.)

$$n[M^+(TCNQ)^-] \rightleftharpoons xM^0 + (n - x)[M^+(TCNQ)^-] + x(TCNQ)^0$$

Various analysis techniques such as UV-absorbed spectra, XRD, XPS, AES, and Raman spectra have been taken. For example, the Raman spectra of the Cu–TCNQ thin film in Fig. 14 are (a) Raman spectrum of neutral TCNQ, (b) Raman spectrum of Cu–TCNQ film, and (c) Raman spectrum of Cu–TCNQ after exposure to the Ar$^+$ laser [42]. The strong Raman bands observed in the switched and unswitched Cu–TCNQ film can be used to differentiate the two states. The TCNQ ν_4 (C=C stretching) Raman bands are strongly affected by the π-electronic structure of TCNQ. The fully charged charge transfer species (b) has a maximum at 1451 cm^{-1}, whereas in the neutral TCNQ species (c), the band is shifted to 1375 cm^{-1}. So when neutral Cu and TCNQ are produced in the Cu–TCNQ film by light, the C=C bonds and index n will change comparatively.

The optical bistable state depends on the irradiated power (W/cm^2) of the incident light beam. At low irradiated levels, an optoelectric switching from a high to a low electrical resistance state can be induced; whereas at increased irradiated levels, high optical contrast patterns can be generated directly on the thin-film material. In the first case, small amounts of neutral donor and acceptor molecules are produced during the irradiation. They remain locked in the crystal structure of

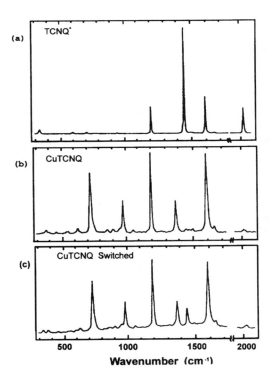

FIGURE 14 Raman bands: (a) Raman spectrum of neutral TCNQ; (b) Raman spectrum of Cu–TCNQ film; (c) Raman spectrum of Cu–TCNQ film after exposure to Ar$^+$ laser. (From Ref. 42.)

the thin film and are ready to recombine if heated. In the second case, macroscopic amounts of neutral products appear and the patterns are visible by the unaided eye.

C. Erasable Optical Recording [43]

Both Ag and Cu complexes with the electron acceptors TCNQ, tetracyanoethylene (TCNE), tetracyanonaphthoquinodimethane (TNAP), or other derivatives of TCNQ can be used for erasable optical recording media through the electric-field-induced switching effect. The switching in these materials is reversible and fast, with switching times of less than 10 ns, as is observed in static switching experiments. Writing can be fulfilled by laser. Write-in threshold power is relatively independent of wavelength throughout the visible and infrared regions of the spectrum but varies between 3 and 150 mW depending on the specific material choice of metal and acceptor complex. These power levels are well within the capabilities of most moderately powered commercial laser systems.

D. Writing and Reading on the M–TCNQ Thin Films by STM

Matsumoto et al. [44] reported the switching and memory effects of Cu–TCNQ thin film by applying voltage to the film with STM tips. The switching and memory effect of the films were examined by changing the stimulus with the variation of the applied voltage and the duration period of voltage. When the stimulus is 1.5 V and 1 min, a convex image appeared. The protuberance in the image, ~120 nm in diameter and ~40 nm in height, is probably due either to the increased conductivity of the triggered portion of the film or to the rising of the particular portion of the surface. The reversibility of the switching phenomena was confirmed by a heat treatment at 330 K for less than 1 min.

Yamaguchi et al. [39] reported field-induced switching phenomenon of M–TCNQ (M = Ag, Cu) using STM. Their preliminary STM study on polycrystalline Cu–TCNQ showed clear surface topography without field-induced switching at a lower tunneling voltage ($V_t = 0.55$ V, $I_t = 0.9$ nA) than Ag–TCNQ. This result is important in investigation of structural changes before and after switching. Their experimental images reveal the continual switching of Ag–TCNQ with nanometer-scale resolution and can clearly discern the neutral reaction products of a segregate as a hole forms and white streaks appear across the image. The more conductive regions (white streaks) are due to the formation of neutral silver and the less conductive area (widening hole) is due to the production of $TCNQ^0$. These experiments suggest that the STM may have potential as a molecular-based information storage system when used in combination with the M–TCNQ system.

VI. STRUCTURES AND PROPERTIES OF C_{60}–ORGANIC COMPOUND THIN FILMS

Polymers are applied in some microelectronics devices based on their thermal, mechanical, and dielectric properties. A C_{60} is a cluster of 60 carbon atoms arranging on a spherical shell with a football-like geometry, and its diameter is 0.71 nm [45]. The solid C_{60} is an insulator, but when doped with K, Rb, Cs, or other metals, it may become the conductor, or the semiconductor, or even the superconductor. They possess very important and potential applications.

Xue et al. [46] prepared C_{60}–PE (polyethylene) thin films by the ionized cluster beam–time-of-flight mass spectrometer (ICB–TOFMS) deposition system and obtained the crystal thin film. Its crystal structure was not only different from the C_{60} crystal film, but also different from the polyethylene thin film. The C_{60}–PE thin film was a good insulator. Morita et al. [47] reported on the C_{60}-doped PDAF-6 [poly(9,9-dihexylfluorene)] conducting polymer system and obtained the enhancement of photoconductivity. Gao et al. [48] prepared the C_{60}–TCNQ multilayer thin films by the ICB method and observed its fractal patterns. Their experimental results demonstrated the great potential of the ICB deposition method for the investigation of fractal patterns.

Because the ICB deposition is carried out in a high vacuum and the structures and properties of the deposited films can be controlled by adjusting parameters of the ICB system, the ICB deposition system is a powerful tool to fabricate desired films with certain properties and few impurities and defects. In the past decades, many kinds of inorganic material thin films with high quality have been successfully fabricated by the ICB deposition method. It is now a promising method for preparation of organic and metal–organic complex thin films with high quality.

The deposition of a C_{60}–TCNQ thin film was fulfilled by the ICB method with the TOFMS to inspect the conditions of the controlled clusters in a high-vacuum chamber. The C_{60} and the TCNQ were deposited alternatively to form about 100-nm-thick C_{60}–TCNQ films sandwiched between two Au electrode layers for current–voltage (I–V) measurement. For absorption spectrum measurement, the films were prepared on quartz or CaF_2 plates. The background vacuum was better than 8.5×10^{-6} Pa and the vacuum level during the deposition was 4.0×10^{-5} Pa. All substrates for the I–V measurement, and TEM, XRD, and STM analyses were simultaneously put on the holder of the ICB–TOFMS system during depositing films.

Figure 15 shows a TEM microstructure of a typical crystal C_{60}–TCNQ thin film fabricated by the ICB method. In this investigation, different crystal C_{60}–TCNQ thin films can be produced by adjusting the parameters of the ICB–TOFMS system. From the electron and the x-ray diffraction data of the C_{60}–TCNQ thin films, we have determined the crystal structures of the films. The surface structures of films were also imaged by STM.

The I–V characteristics of the C_{60}–TCNQ thin films were measured in air at room temperature. The film showed reproducible electrical bistability (Fig. 16). When the electric field was applied at the start, the film was an insulator with a resistance of about 10^9 Ω, namely the "0" state; but at the threshold voltage of 1.75 V, the film abruptly entered a low-resistance region of about 500 Ω, namely the "1" state.

Xue et al. [33] reported a novel polymer, toluylene 2,4-dicarbamidonitrile (TDCN) with carbon–nitrogen-conjugated backbones. The TDCN easily formed a homogenous nanometer thin film by the ICB deposition method. The C_{60}–TDCN thin films were prepared by the ICB–TOFMS deposition system; see Fig. 17a. The top and the bottom Au thin films were electrodes for measuring the I–V property of the films. The result is shown in Fig. 17b. The film possessed the electric bistable

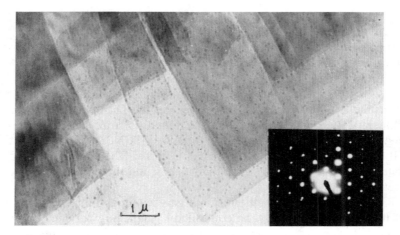

FIGURE 15 The C_{60}–TCNQ thin film: TEM micrograph and electron diffraction pattern.

property, and its threshold voltage was about 1.56 V. Below the threshold, the film was an insulator with a resistance of about 10^9 Ω cm; over the threshold, the thin film possessed conducting features with a resistance of 1.4 10^2 Ω cm. The C_{60}–TDCN thin film was analyzed by the UV absorption spectra; see Fig. 18. When the C_{60} was doped into the TDCN thin film, three peaks at 220, 260, and 340 nm wavelength appeared. The peak intensity increasing with the C_{60} concentration indicated that the C_{60}–TDCN thin film was a complex film, in which charge transfer might occur.

We have synthesized a novel polynitrile π acceptor 1,4-bis(2,2-dicyano-vinyl)benzene (BDCB) [49] with molecular structure shown in Fig. 19. The C_{60}-

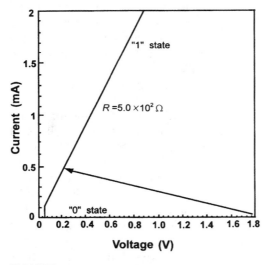

FIGURE 16 *V* curve of the C_{60}–TCNQ thin film.

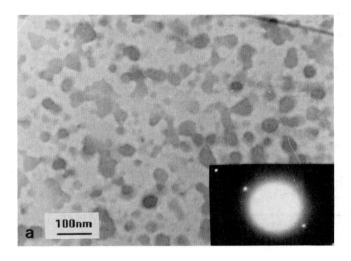

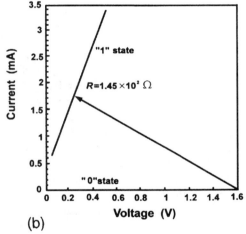

(b)

FIGURE 17 The C_{60}–TDCN thin film: (a) TEM micrograph and electron diffraction pattern; (b) *I–V* curve.

BDCB thin film was prepared by vacuum coevaporation of isomolar BDCB and C_{60} from two separate sources set just below one of the arms of an additional U-type tube (1.5-cm inner diameter, 4.5-cm height, 2.5 cm arm-to-arm distance). The tube was heated so that the C_{60} and BDCB vapors could go through it and were simultaneously deposited onto a Au-coated glass substrate set just below the other arm. A current–voltage (*I–V*) curve of the Au/C_{60}–BDCB/Au devices was measured in series with a 10^4-Ω load resistor, as shown in Fig. 20. When the DC voltage applied to the device increased to a threshold of 4.7 V (field strength 7.8 × 10^3 V/cm), the device switched abruptly from the higher resistance "0" state (10^9 Ω) to the lower resistance "1" state (160 Ω). Once entering the conductive state, the device remained at that state even though the voltage was removed, unless it was heated, in which condition it could switch back to the original highly resistive

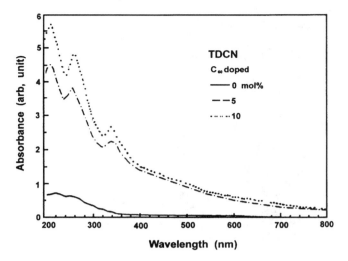

FIGURE 18   UV absorption spectra of the C_{60}–TDCN thin film.

state. The switching process (interchange between "0" and "1" state) was very stable and reproducible. As shown in Fig. 21, the XRD pattern of the C_{60}–BDCB thin films showed significant differences from those of both BDCB and C_{60} thin films. The BDCB thin film had only one sharp peak at $2\theta = 27.0°$, corresponding to a plane spacing of 0.330 nm. The C_{60} thin film exhibited three peaks at $2\theta =$ 20.8°, 28.0°, and 38.3°, representing spacings of 0.427, 0.319, and 0.235 nm, respectively. With reference to the literature [50], the above diffraction of C_{60} thin film can be indexed as (112), (310), and (304) crystal planes of the hexagonal system with cell constants $a = 1.005$ nm and $c = 1.641$ nm. It is interesting that the diffraction characteristics of BDCB and C_{60} thin film disappeared in the diffraction pattern of C_{60}–BDCB thin film, which possessed three weak peaks at $2\theta =$ 27.1°, 28.2°, and 38.4°, and a strong peak at $2\theta = 46.9°$ with a totally different set of plane spacings of 0.523, 0.317, 0.234, 0.194, and 0.186 nm, respectively. Obviously, a new phase formed due to the reaction in the simultaneously sublimated C_{60} and BDCB thin film, although the reaction mechanism is still unclear.

VII. STRUCTURES AND PROPERTIES OF NOVEL ORGANIC COMPLEX THIN FILMS

Electronic switching thin films paid more and more attention due to their potential application in future electronic industry [51] and ultrahigh density information stor-

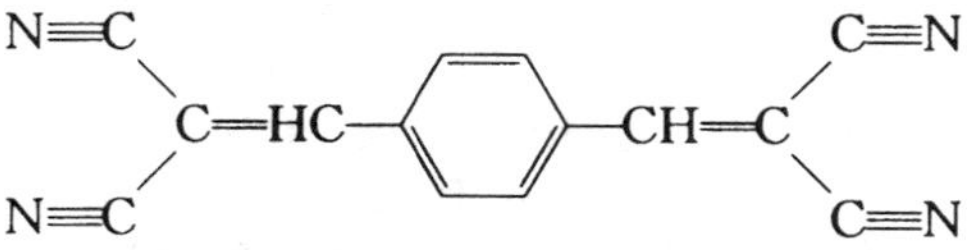

N≡C   C≡N
   \   /
    C=HC—⟨ ⟩—CH=C
   /   \
N≡C   C≡N

FIGURE 19   Molecular structure of the BDCB.

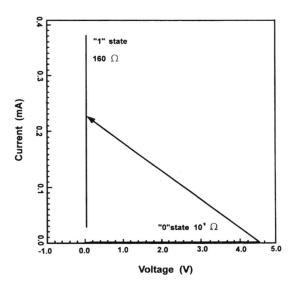

FIGURE 20 *I–V* characteristic of the Au/C$_{60}$–TDCN/Au.

age [39,41,44]. Organic or metal-organic compounds for electronic switching devices are superior to inorganic because of their extremely small size, diversity of compositions and structures, ease of fabrication, potential low cost, and superfast response. These make them promising candidates for replacing conventional inorganics and thus revolutionizing electronic and computer technology in speed and capacity.

Current interests have been focused on conjugated organic and polymeric system with large nonlinear optical and electrical active properties for a variety of high technique applications. We have synthesized novel organic materials and have investigated fabricating technique of novel complex function thin films. Some complex thin films are described as follows.

A. Ag–BDCB Complex Thin Films

The BDCB (as mentioned earlier) easily formed a homogenous nanometer-scale thin film by vacuum deposition technique in pressure of 1×10^{-4} Pa. A scanning electron microscope (SEM) micrography of the thin film was taken on an AMRAY 1910 field emission microscope; see Fig. 22. In Fig. 22a, the structure of the BDCB film consisted of long flake-like crystals with very large sizes. The flakes oriented randomly and were very loosely packed. However, in Fig. 22b, the Ag–BDCB film appears to be more uniformly and tightly packed.

The Ag–BDCB complex thin film about 1 μm thickness was prepared by vacuum coevaporation of the Ag and BDCB from separate sources. The BDCB source was set just below one of the arms of an additional U-type tube. The tube was heated so that the BDCB vapor could go through it and deposited simultaneously with Ag onto a Au-coated glass substrate set just below the other arm. Then another Au film was deposited as the top electrode. Thus, the sandwiched device Au/Ag–BDCB/Au was fabricated.

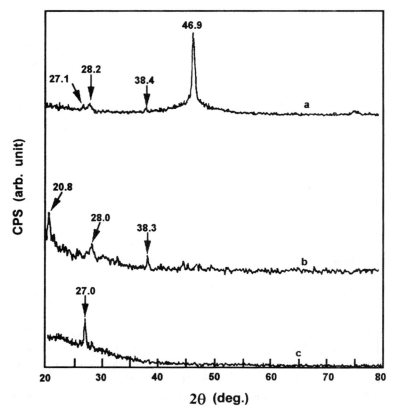

FIGURE 21   X-ray diffraction pattern of (a) C_{60}–BDCB, (b) C_{60}, and (c) BDCB thin films.

The *I–V* properties were measured in series with a 10^4-Ω load resistor and the result is shown in Fig. 23. The Ag–BDCB thin film possessed the electrical bistable property. When the applied voltage was beyond the threshold of 0.37 V, the film was switched from its original high-resistance state ("0" state) to a conductive state ("1" state). Once switched to the "1" state, the film remained at that state even though the voltage was removed, until it was heated to the "0" state. The process was stable and reproducible.

We have tried to substitute the Ag with Au, Pt, Pd, Al, and Sb, but those complex thin films did not possess electrical bistable property.

In order to derive the switching mechanism, the IR spectra of the films of BDCB, the KBr pellet of BDCB, and Ag–BDCB were compared in Fig. 24. As seen from Fig. 24, the peak at 2229.6 cm^{-1} due to C=N stretching vibration for the spectrum of BDCB film was nearly identical to that for the KBr pellet of BDCB. For the Ag–BDCB film, however, this peak was characterized as a shoulder centered at 2153.6 cm^{-1}. The red-shift of 76 cm^{-1} compared with silver-free BDCB film, was about twice as large as those for simple 1:1 charge transfer salts of TCNQ. This indicated that the charge transfer between Ag and BDCB occurred in the film of Ag–BDCB we studied, and BDCB might behave as a two-electron acceptor,

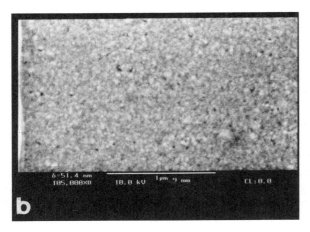

FIGURE 22 SEM micrographs of (a) BDCB thin film and (b) Ag–BDCB thin film.

unlike TCNQ and its congeners in their charge transfer complexes with Ag. Thus, the observed switching was probably associated with the charge transfer induced by the applied electric field, as shown by

$$[(Ag^+)_2BDCB^{2-}]_n \rightleftharpoons Ag_{2x}^{\cdot} + (BDCB^0)_x + [(Ag^+)_2BDCB^{2-}]_{n-x}$$

The UV spectra of the films of BDCB, Ag–BDCB, and Ag are shown in Fig. 25. The BDCB film exhibited a peak at ~320 nm and two weak peaks at 555 and 590 nm, respectively. The latter two still appeared in the spectrum of the Ag–BDCB film, but the former one disappeared and an intense and wide peak centered at 460 nm appeared. This peak may be responsible for the charge transfer between Ag and BDCB and may be compared with BDCB-free Ag film, for which strong absorption peak due to silver ultrafine particles is centered at 660 nm. According to Mulliken's charge transfer theory [52], the charge transfer energy is given by

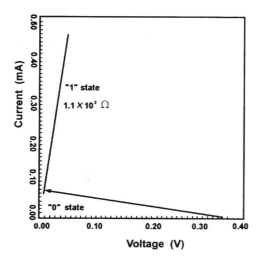

FIGURE 23 The *I–V* properties of the Ag–BDCB thin film.

$$E_{CT} = E_I + E_A - E_C + \frac{2t^2}{E_I + E_A - E_C}$$

where E_I and E_A are the ionization energy of donor Ag and the affinity energy of acceptor BDCB, respectively; E_C is the Coulomb energy and t is the transfer integral. The latter terms of the equation, $2t^2/(E_I + E_A - E_C)$ is usually trivial compared with the former unless, by chance, E_D and E_A are closely matched [53]. In principle, the electron affinity ability of BDCB and TCNQ could be compared based on the charge transfer spectra of Ag–TCNQ and Ag–BDCB. However, this

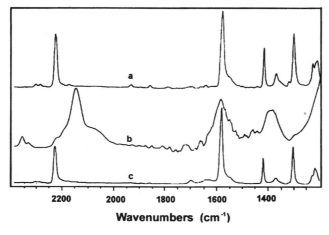

FIGURE 24 Infrared spectra (a) BDCB film, (b) Ag–BDCB film, and (c) KBr pellet of BDCB.

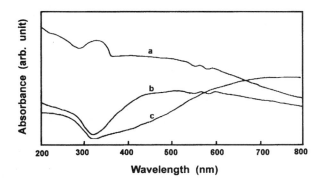

FIGURE 25 Ultraviolet–visible spectra (a) BDCB film, (b) Ag–BDCB film, and (c) Ag film.

is complicated by different type of their charge transfer salt, 1:1 for Ag–TCNQ, probably 2:1 for Ag–BDCB as discussed above, namely the Coulomb energy terms for these two systems differ considerably.

B. Ag–DDME Complex Thin Films

The polymer π acceptor with simple molecular structure shown in Fig. 26 [41], 1,1-dicyano-2,2-(4-dimethylaminophenyl)ethylene (DDME), has been synthesized and well characterized by elemental analysis, ^{1}H nuclear resonance spectroscopy (NMR), XRD, IR, and UV spectroscopy.

An electric bistable property similar to the above-mentioned Au/Ag–BDCB/ Au device was also found in a Ag/Ag–DDME/Ag device which was prepared by a method similar to the Au/Ag–BDCB/Au device. The *I–V* curve is shown in Fig. 27.

As shown in Fig. 28, the IR spectrum of DDME film was similar to that of the KBr pellet of DDME. On the contrary, the positions and strengths of some peaks in the spectrum of Ag–DDME film were evidently changed relative to those for the spectra of DDME film or KBr pellet. The peak at 2210.2 cm^{-1} for the spectrum of DDME, typical of the C$\equiv$N stretching vibration, was red-shifted to 2150.2 cm^{-1} in the spectrum of Ag–DDME; it should be pointed out that the red-shift of 60 m^{-1} was about 1.5 times as large as those for simple 1:1 charge transfer salts of TCNQ. In addition, the doublet at 1388.7 and 1363 cm^{-1} due to the δ_{-CH_3} vibration in the spectrum of DDME film was changed to a singlet at 1377 cm^{-1} in the spectrum of Ag–DDME. Two other peaks that might be assigned to

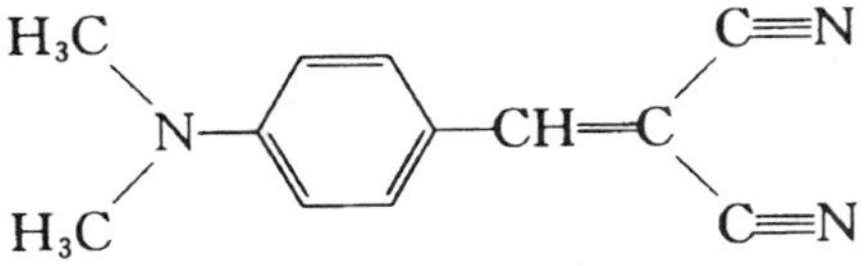

$$\underset{H_3C}{\overset{H_3C}{>}}N-\!\!\!\left\langle\;\;\right\rangle\!\!\!-CH\!\!=\!\!C\underset{C\equiv N}{\overset{C\equiv N}{<}}$$

FIGURE 26 The molecular structure of DDME.

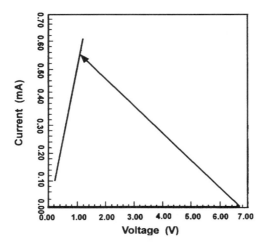

FIGURE 27 *I–V* characteristic of Ag–DDME film.

C=C (wing) stretching modes were blue-shifted to 1609.1 and 1668.2 cm^{-1}, respectively. However, peak positions at 1523.8 and 1544.2 cm^{-1} due to phynyl ring vibration basically remained uncharged, but greatly decreased in intensity. Thus, the observed switching was probably associated with the charge transfer induced by the applied electric field.

The UV spectra of the film of DDME showed three peaks at 290, 564, and 600 nm. They disappeared in the spectrum of Ag–DDME, but an intense and wide peak centered at 460 nm appeared, which might be due to the charge transfer absorption of Ag–DDME and which was different from those for Ag–BDCB and Ag–TCNQ. This can be easily understood by Mulliken charge transfer theory [52]; thus, the charge transfer spectra for different thin films may vary considerably. It is noteworthy that the charge transfer absorption of the Ag–DDME thin film ex-

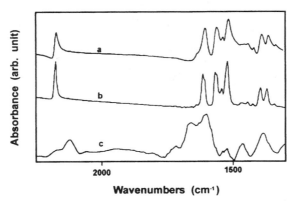

FIGURE 28 Infrared spectra (a) KBr pellet of DDME, (b) DDME film, and (c) Ag–DDME film.

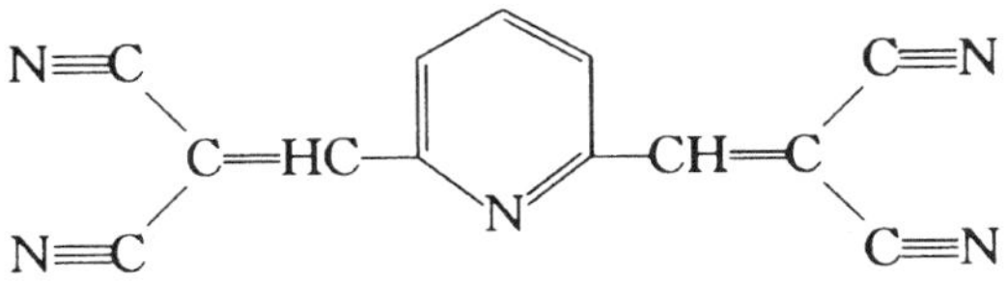

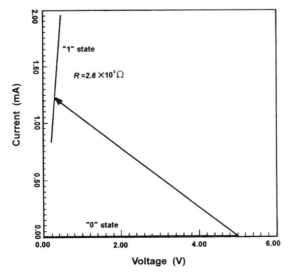

FIGURE 29 The molecular structure of BDCP.

pended to long wavelengths matchable to the semiconductor laser (e.g., GaAsAl diode laser). This feature may be attractive for optical storage media.

C. Ag–BDCP Complex Thin Films

A new organic material 2,6-bis(2,2-bicyanovinyl)pyridine (BDCP) with carbon–nitrogen-conjugated backbone was synthesized in our laboratory. The BDCP molecular structure is shown in Fig. 29.

Electric bistable properties similar to the above-mentioned Au/Ag–BDCB/Au and Ag/Ag–DDME/Ag devices was also found in the Ag/Ag–BDCP/Ag device prepared by a method similar to the two devices. The *I–V* curve is shown in Fig. 30.

The chemical bond configuration in the thin films deposited on the CaF$_2$ substrate were identified by the UV spectra. The UV transmission spectra of the Ag–BDCP, the BDCP, and KBr pellet of BDCP are shown in Fig. 31. The Ag–BDCP complex thin films showed very different spectral features when compared with the BDCP thin films and the Ag thin films. Pure BDCP film had five absorption peaks at 222 nm, 263 nm, 318 nm, 570 nm, and 600 nm, respectively; the Ag thin film absorption peak was at 780 nm. Nevertheless, the Ag–BDCP

FIGURE 30 *I–V* curve of Ag–BDCP thin film.

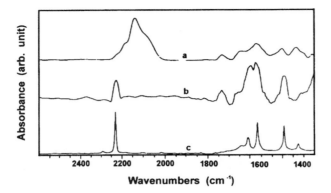

FIGURE 31 Infrared spectra: (a) Ag–BDCP thin film, (b) BDCP thin film, and (c) KBr pellet of BDCP.

complexes exhibited only one wide band centered at 525 nm. This new peak might be attributed to the charge transfer between Ag and BDCP. Unfortunately, the crystal structure of Ag–BDCP is not exactly known yet; thus, it is difficult to give a more exact interpretation of the UV spectrum of the Ag–BDCP thin films. The IR spectra of the Ag–BDCP thin film, the BDCP film, and the KBr pellet of BDCP are shown in Fig. 32. BDCP thin film had a $-C\equiv N$ band at 2234.7 cm^{-1} nearly identical to that of the KBr pellet of BDCP; the corresponding band in the Ag–BDCP thin film was at 2145.7 cm^{-1} with a shoulder at around 2170 cm^{-1} and was considerably broadened, so the bathochromic shift was 89 cm^{-1}. The bathochromic shift revealed that the charge transfer might have occurred in the Ag–

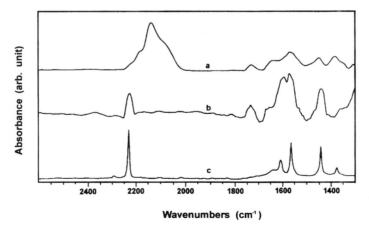

FIGURE 32 Infrared spectra: (a) Ag–BDCP thin film, (b) BDCP thin film, and (c) KBr pellet of BDCP.

BDCP complex thin film. Furthermore, it is worth noting that the bathochromic shift of 89 cm^{-1} in Ag–BDCP is about twice as large as that of simple 1:1 charge transfer salts of TCNQ [54], indicating that BDCP may behave as a two-electron acceptor:

$$[(Ag^+)_2BDCP^{2-}]_n \rightleftharpoons 2x(Ag^0) + x(BDCP^0) + x[(Ag^+)_2BDCP^{2-}]_{n-x}$$

D. Decacyelene–BDCB Complex Thin Films

Decacyelene (DC) was recrystallized twice from nitrobenzene before being used. Vacuum deposition was made at a pressure of 1×10^{-3} Pa. The DC–BDCB film was deposited by coevaporation of DC and BDCB from two evaporation sources through an additional U-tube.

In Fig. 33, the XRD patterns of BDCB and DC films were compared with that of DC–BDCB. The pattern of DC film showed three weak peaks at $2\theta = 26.4°$, 38.35°, and 44.55°, representing periodic spacings of 3.37, 2.35, and 2.03 nm, respectively, and two strong peaks at 28.20° and 28.25° representing periodic spacings of 3.17 and 3.15 nm, respectively. Based on the crystal parameters [40], the orthorhombic system has $a = 1.277$, $b = 2.070$, $c = 0.395$ nm, and $\beta = 99°$, and following crystallographic formula and Bragg equation,

$$(d_{hkl})^{-2} = h^2a^{-2} + k^2b^{-2} + l^2c^{-2}$$

$$2d_{hkl}\sin\theta = 1$$

So the three weak peaks are assigned to (113), (068), and (069) diffraction and the two strong peaks to (133) and (081) diffraction. In other words, we have obtained single-crystal thin films of DC. The pattern of BDCB and DC exhibited two weak peaks at $2\theta = 28.15°$ and 75.4° representing periodic spacings of 3.17 and 1.26 nm, respectively, and one strong peak at $2\theta = 46.9°$ with a periodic spacing of

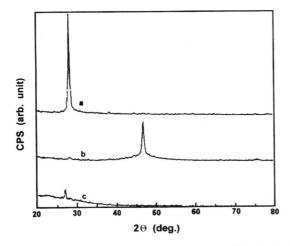

FIGURE 33 X-ray diffraction patterns of (a) DC film, (b) DC–BDCB film, and (c) BDCB film.

1.94 nm. However, these peaks cannot be definitely assigned because of the unavailable crystal structural data. Considering the significant differences of DC–BDCB film with DC and BDCB films, we can conclude that DC–BDCB films are a new phase rather than just a mixture of DC and BDCB.

Figure 34 shows the *I–V* characteristic of the device Ag/DC–BDCB/Ag. When the voltage applied to the device was increased to a threshold of 7.9 V, the device switched from the high-resistance state (10^9 Ω) to the lower one (45 Ω). No preliminary forming was required for the device to be able to switch. Once switched to the "1" state, the film remained at that state even though the voltage was removed, until it was heated to the "0" state. The process was stable, reproducible, and independent to the polarity.

So far, three switching mechanisms of the electric conductive filament [25], the chemical phase changes of charge transfer complexes, and the change in molecular or crystal structures [55,56] have been reported. Usually, the films switching as an electric conductive filament have appreciable physical damage on the surface; we have never found any damage on DC–BDCB film under a microscope. Because both the films of pure DC and BDCB are highly insulating, the switching mechanism of the chemical phase change of the charge transfer complex can also be ruled out in the DC–BDCB film. So the switching mechanism of the thin film we studied is most probably due to the change in molecular or crystal structures.

VIII. WRITING AND READING ON ORGANIC COMPLEX THIN FILM WITH SCANNING TUNNELING MICROSCOPE

The scanning probe microscope (SPM), such as the STM, the AFM, and the SNOM have been widely used for developing ultrahigh density data storage devices. Some

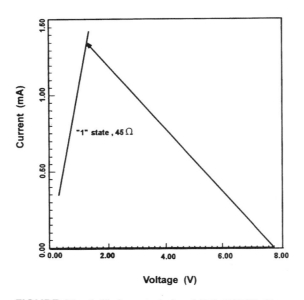

FIGURE 34 *I–V* characteristic of DC–BDCB film.

inorganic compound thin films have been studied for nanometer-scale recording with SPM [9,57]. Recently, many scholars have studied the organic polymers used in solid electron devices. We have studied several organic complex thin films used for ultrahigh density data storage—for instance, the Ag–TDCN, the Ag–CPU (*N*-cyno-*N'*-phyenylurea), and the nitrobenzal malononitrile–diamine benzene (*m*-NBMN–DAB) thin films. These films possess electrical and optical bistable properties, so we may write and read data on these thin films by the STM or the laser.

A. Metal–Organic Complex Thin Films Used for Data Storage by STM

We have synthesized novel organic material, *N*-cyno-*N'*-phyenylures (CPU); its molecular structure is shown in Fig. 35. Data of the molecular structures and the crystalline have been given in Ref. 50.

CPU easily formed a homogenous nanometer-scale thin film by the ICB deposition method. CPU and Ag were deposited, in turn, on a Au-coated glass substrate by the ICB deposition method and the PVD method, respectively. The conditions for deposition of CPU are the crucible temperature of 200°C, the substrate temperature of 40°C, the accelerate voltage of -600 V, the ion current of 20 μA, and the chamber pressure of 3×10^{-5} Pa. Then the film was heated to accelerate the reaction between CPU and Ag. The obtained Ag–CPU thin film was about 100 nm in thickness. At last an upper electrode of Au was deposited to form a Au/Ag–CPU/Au device. The *I–V* properties were measured and the results are shown in Fig. 36. The device possessed electric bistable properties. When the applied voltage was increased from 0 V to a threshold of 0.65 V, the device turned from its high-resistance state of about 10^9 Ω cm to conductive state of 71 Ω cm; after heating at 150°C for 10 min, the sample returned to the high-resistance state. This process was stable and reproducible.

We prepared the samples of the Ag–CPU thin films with the ICB–TOFMS system for the STM analysis. The substrate was a piece of highly oriented pyrolitic graphite (HOPG). The tunneling condition was $V_b = 0.16$ V and $I_t = 0.44$ nA. The write-in voltage pulses were 4 V and 2 ms. Two pulses made two marks successively. The result is shown in Fig. 37. As shown in Fig. 37a, when the STM worked

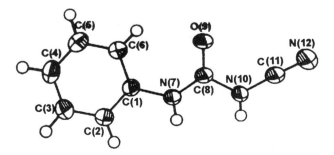

FIGURE 35 Molecular structure of a CPU and atom-numbering system of compound. Displacement ellipsoids are plotted at the 50% probability level. (From Ref. 58.)

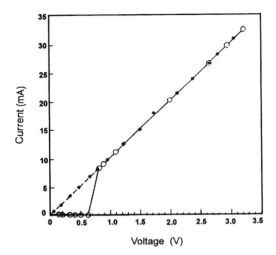

FIGURE 36 *I–V* properties of Ag–CPU thin film: the open circle is from insulator to conductor with a voltage increase; the solid circle is the conductor after conduction.

at the constant height mode, the two marks appeared to be two holes 50 nm in diameter; however, they appeared to be two hills when the STM worked at the constant current mode, as was shown in Fig. 37b. This phenomenon indicated that the write-in pulses had changed the electrical property rather than the atomic structures of the Ag–CPU thin film [59]. Reading remained stable under STM work conditions for several days. Figure 38 showed scanning tunneling spectroscopy (STS) of a writing point (i.e., *I–V* properties of the writing point); 5 nA was set to be the saturation current. We can find that the conduction through voltage was 0.8 V. Experimental results showed that the write-in process was reproducible.

Optically written and read samples were prepared by methods mentioned earlier. Ag–CPU was sandwiched between a glass substrate and a reflection layer of Ag. Writing and reading experiments were carried out on a signal writing and reading instrument with a high-density laser. The power and wavelength of the laser were 50 mW and 780 nm, respectively. Light incident from glass into the Ag–CPU thin film, and then was reflected by the Ag layer. The intensity of the reflection light was recorded under some certain incident light intensity. The write-in pulses were 10 mW and 10 μs. The read-out power was 1.1 mW. The erasing pulse was 8 mW and 1 ms. Signal writing, reading, and erasing could be reproduced many times. Figure 39 shows the writing and erasing of signals.

As is shown in these results, the Ag–CPU thin films possess excellent electrical and optical bistable properties for data storage devices. The switching mechanism is not clear yet.

B. Organic Donor and Acceptor Complex Thin Films Used for Data Storage by STM

Our group formed two kinds of organic material: nitrobenzal malononitrile (*m*-NBMN) and diamine benzene (DAB). Their molecular structures are shown in Fig.

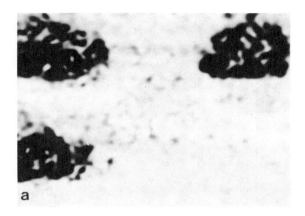

FIGURE 37 Writing STM graph of the Ag–CPU thin films, $V_b = 0.16$ V, $I_t = 0.44$ nA, $V_p = 4$V, $t_w = 2$ ms: (a) constant height mode, (b) constant current mode.

40. The mixture of *m*-NBCN and DAB in a 1:1 molar ratio was finely abraded before being evaporated to deposit on a freshly cleaved HOPG substrate by the vacuum evaporation method. In this complex thin film about 50 nm thick, the *m*-NBMN acted as an acceptor, and the DAB as a donor.

Experiments were performed with a homemade STM under ambient conditions. The STM tips were 0.25-mm-diameter Pt/Ir (80/20) wire snipped by a cutter. The STM operated in constant height mode. Different tips and samples were used for checking the certainty of the experiments. The *I–V* curves were obtained using the STS. Recording experiments were carried out by applying voltage pulses between the tip and the sample. The tunneling conditions were $V_b = 0.98$ V and $I_t = 0.17$ nA, and recording marks were made by applying a pulse voltage of 4 V for 10 ms. A typical 30×30 nm^2 STM image of the insulating *m*-NBMN/DAB film with recorded marks is shown in Fig. 41. The marks remained stable for at least several days. The probability of recording was over 90%. The finest mark size obtained

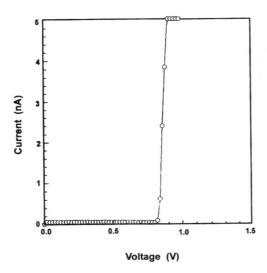

FIGURE 38 *I–V* properties of the Ag–CPU thin films with STS.

up to now is 1.3 nm in diameter, corresponding to a data storage density of about 10^{13} bits/cm^2. To analyze the recording mechanism, we have measured *I–V* curves before and after recording, as shown in Fig. 42. Curve a is the one before recording, and curve b is the one after recording. It can be concluded from Fig. 42 that the unrecorded regions showed an insulator behavior with a critical voltage of 2.4 V whereas the recorded regions showed a conductor behavior.

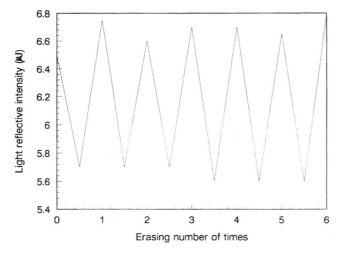

FIGURE 39 Writing and erasing of light signals for the Ag–CPU thin films.

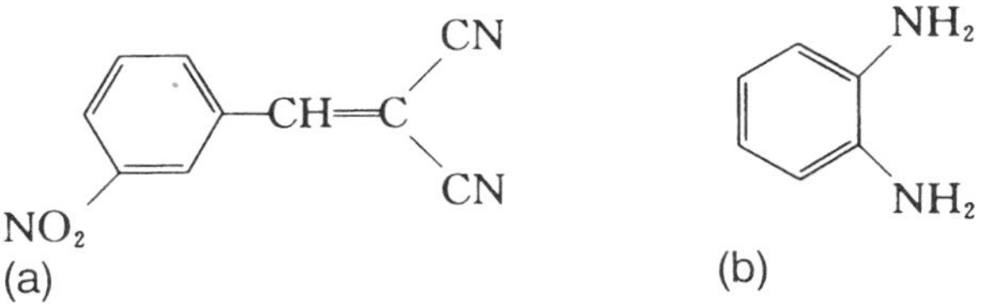

FIGURE 40 Molecular structures: (a) *m*-NBMN and (b) DAB.

IX. CONCLUSIONS

A few electronic organic materials and Ag (or C_{60}, DC)–organic acceptor complex thin films have been successfully developed by the ICBD method and coevaporation deposition in vacuum. We found that some complex thin films, for example, Ag–CPU, Ag–BDCB, Ag–DDME, Ag–BDCP, and C_{60}–TDCN, possess electrical bistable states. Signal writing and reading on the Ag–CPU and *m*–NBMN/DAB thin films have been performed by laser or STM. Recording spots 1.3 nm in diameter have been realized, corresponding to data storage density up to 10^{13} bits/cm^2. Our preliminary results demonstrate that these novel organic complex thin films can be promising data storage materials used for future nanoelectronic device fabrication.

ACKNOWLEDGMENTS

The authors thank all the colleagues in our interdisciplinary group. This project is supported in part by the National Natural Science Foundation of China (NSFC) and the Doctoral Program Foundation of Institution of Higher Education.

FIGURE 41 STM image of the *m*-NBMN/DAB thin film with recording marks.

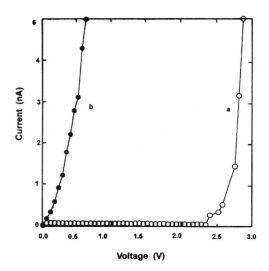

FIGURE 42 *I–V* curves of the *m*-NBMN/DAB thin film obtained by STS (a) before recording and (b) after recording.

REFERENCES

1. Roberts, G. G., An applied science perspective of Langmuir–Blodgett films, *Adv. Phys.*, *34*(4), 475–512 (1985).
2. Chiabrera, A., Zitti, E. D., Costa, F., and Bisio, G. M., Physical limits of integration and information processing in molecular systems, *J. Phys. D: Appl. Phys.*, *22*, 1571–1579 (1989).
3. Sugano, T., A perspective on next-generation silicon cevices, *Jpn. J. Appl. Phys.*, *32*, 261–265 (1993).
4. Meindl, J. D., Physical limits on gigascale integration, *J. Vac. Sci. Technol.*, *B14*(1), 192–195 (1996).
5. Kaga, T., Ohkura, M., Murai, F., Yokoyama, N., and Takeda, K., Process and device technologies for 1 Gbit dynamic random-access memory cells, *J. Vac. Sci. Technol.*, *B13*(6), 2329–2334 (1995).
6. Barrett, R. C., and Quate, C. F., Large-scale charge by scanning capacitance microscopy, *Ultramicroscopy*, *42*, 262–267 (1992).
7. Wiesendanger, R., Recent advances in nanostructural investigations and modifications of solid surfaces by scanning probe methods, *Jpn. J. Appl. Phys.*, *34*, 3388–3395 (1995).
8. Sato, A., and Tsukamoto, Y., Nanometre-scale recording and erasing with the scanning tunnelling microscope, *Nature*, *363*, 431–432 (1993).
9. Kado, H., and Tohda, T., Nanometer-scale reading on chalcageninde films with an atomic aforce microscope, *Appl. Phys. Lett.*, *66*(22), 2961–2962 (1995).
10. Ashwell, G. J., Sage, I., and Trundle, C., in *Molecular Electronics* (G. J. Ashwell, ed.), Research Studies Press Inc. New York, 1991, p. 1.
11. Liu, Z. F., Hashimoto, K., and Fujishima, A., Photoelectrochemical information storage on azobenzene derivative, *Nature*, *347*, 658–659 (1990).
12. Terris, B. D., Mamin, H. J., Rugar, D., Studenmund, W. R., and Kino, G. S., Near-field optical data storage using a solid immersion lens, *Appl. Phys. Lett.*, *65*(4), 388–390 (1994).

13. Terris, B. D., Mamin, H. J., and Rugar, D., Near-field optical data storage, *Appl. Phys. Lett.*, *68*(2), 141–143 (1996).

14. Nagahara, L. A., Scanning near-field optical microscopy/spectroscopy of thin organic films, *J. Vac. Sci. Technol.*, *B14*(2), 800–803 (1996).

15. Rousseaux, F., Decanini, D., Carcenac, F., Cambril, E., Ravet, M. F., Cappert. C., Bardou, N., Bontenlian, B., and Veillet, P., Study of large area high density magnetic dot arrays fabricated using synchrotron radiation based X-ray lithography, *J. Vac. Sci. Technol.*, *B13*(6), 2787–2781 (1995).

16. Hosaka, S., Koyanagi H., Kikukawa, A., Miyamoto, M., Ryo Imura, and Ushiyama, J., Fabrication of nanometer-scale structures on insulators and inmagnetic materials using a scanning probe microscope, *J. Vac. Sci. Technol.*, *B13*(3); 1307–1311 (1995).

17. Dahlberg, E. D., and Zhu, J. J., Micromagnetic microscopy and modeling, *Phys. Today*, 34–40 (April 1995).

18. Baibich, M. N., Broto, J. M., Fert, A., Nguyen van Dau, F., Petroff, F., Etienne, P., Crreuzt, G., Friedrich, A., and Chazeles, J., Giant magnetoresistance of Fe(001)/Cr(001) magnetic superlattices, *Phys. Rev. Lett.*, *61*, 2472–2475 (1988).

19. Berkowitz, A. E., Mitchell, J. R., Carey, M. J., Young, A. P., Zhang, S., Spada, F. E., Parker, F. T., Hutter, A., and Thomas, D., Giant magnetoresistance inheterogeneous Cu–Co alloys, *Phys. Rev. Lett.*, *68*(25), 3745–3748 (1992).

20. Xiao, J. Q., Jiang, J. S., and Chen, C. L., Giant magnetoresistance in nonmultilayer magnetic systems, *Phys. Rev. Lett.*, *68*(25), 749–3752, (1992).

21. Bui, A., and Carchano, H., French Patent 7113321 (1971).

22. Pender, L. F., and Fleming, R. J., Memory switching in glow discharge polymerized films, *J. Appl. Phys.*, *46*, 3426–3431 (1975).

23. Segui, Y., Bui, A., and Carchano, H., Switching in polystyrene films: transition on to off state, *J. Appl. Phys.*, *47*, 140–143 (1976).

24. Couch, N. R., Montgomery, C. M., and Jones, R., Metallic conduction through Langmuir–Blodgett films, *Thin Solid Films*, *135*, 173–182 (1986).

25. Godehardt, R., and Heydenreich, J., Detection and evaluation of current filaments in MIS sandwich systems by applying mirror electron microscopy (MEM) and scanning tunneling microscopy (STM), *Int. J. Electron.*, *73*, 1093–1094 (1992).

26. Potember, R. S., Poehler, T. O., and Cowan, D. O., Electrical switching and memory phenomena in Cu–TCNQ thin films, *Appl. Phys. Lett.*, *334*, 405–407 (1979).

27. Benson, R. C., Hoffman, R. C., Potember, R. S., Bourkoff, E., and Poehler, T. O., Spectral dependence of reversible optically induced transition in organometallic compounds, *Appl. Phys. Lett.*, *42*, 855–857 (1983).

28. Hottman, R. C., and Potember, R. S., Organometallic materials for erasable optical storage, *Appl. Opt.*, *28*, 1417–1421 (1989).

29. Sato, C., Wakamatsu, S., and Tadokoro, K., Polarized memory effect in the device including the organic charge-transfer complex, copper–tetracynoquinodimethane, *J. Appl. Phys.*, *68*, 6535–6537 (1990).

30. Machida, Y., Saito, Y., Taomoto, A., Waragai, K., and Asakawa, S., Electrical switching in evaporated lead phthalocyanine films, *Jpn. J. Appl. Phys.*, *28*(3), 297–298 (1989).

31. Sakai, K., Kawoda, H., Takatsu, O., Matsuda, H., Eguehi, K., and Nakagiri, T., Electrical memory switching in Langmuir–Blodgett films, *Thin Solid Films*, *178*, 137–142 (1989).

32. Xue, Z. Q., Liu, W. M., Zhao, X. Y., Gao, H. J., Xu, Y. H., Zhu, C. X., Ma, Z. L., and Pang, S. J., Study of the organic films with STM, *Thin Films Beam-Solid Interact.* *5*, 229–232 (1992).

33. Xue, Z. Q., Gao, H. J., Liu, W. M., Wu, Q. D., Chen, H. Y., Qiang, D., Pang, S. J., and Liu, N., Study of C_{60}–TDCN nanometer scale thin films, *Jpn. J. Appl. Phys.*, *34*, 197–199 (1995).

34. Hoffman, R. S., and Potember, R. S., Organometallic materials for erasable optical storage, *Appl. Opt.*, *28*, 1417–1421 (1989).
35. Potember, R. S., *Molecular Electronic Devices* (F. L. Carter, ed.), Marcel Dekker, New York, 1992, p. 73.
36. Ashwell, G. J., *Molecular Electronics*, Wiley, New York, 1992, p. 18.
37. Rombidi, N. G., Chernavsk, D. S., and Krinsky, V. I., *Information Processing and Computing Devices Based on Biomolecular Nonlinear Dynamic Systems* (K. Sienicki, ed.), CRC Press, Boca Raton, FL, 1993, p. 85.
38. Ashwell, G. J., Sage, I, and Trundle, C., in *Molecular Electronics* (G. J. Ashwell, ed.), Wiley, New York, 1992, p. 1.
39. Yamaguchi, S., Viands, C. A., and Potember, R. S., Imaging of silver and copper tetracynoquinodimethane using a scanning tunneling microscope and an atomic force microscope, *J. Vac. Sci. Technol.*, *B9*, 1129–1132 (1991).
40. Wang, K. Z., Xue, Z. Q., Ouyang, M., Zhang, H. X, and Huang, C. H., Electronic memory switching in a new charge transfer-complex thin films, *Solid State Commun.*, *96*(7), 481–484 (1995).
41. Wang, K. Z., Xue, Z. Q., Ouyang, M., Wang, D. W., Zhang, H. X., and Huang, C. H., A new polynitrile p acceptor for electronic switching devices, *Chem. Phys. Lett.*, *243*, 217–221 (1995).
42. Potember, R. S., Poehler, T. O., and Benson, R. C., Optical switching in semiconductor organic thin films, *Appl. Phys. Lett.*, *41*(6), 548–550 (1982); Takenaka, T., *Spectrchim. Acta*, *27A*, 1735 (1971).
43. Potember, R. S., Poehler, T. O., Hoffman, R. C., Speck, K. R., and Bensen, R. C., in *Molecular Electronic Devices II.* (F. L. Carter, ed.), Marcel Dekker, Inc., New York, 1987, p. 91.
44. Matsumoto, M., Nishio, Y., Tachibana, H., Nakamura, T., Kamabata, Y., and Samura, H., Switching and memory phenomena of Cu–TCNQ films triggered by a stimulus with an STM tip, *Chem. Lett.*, 1021–1024 (1991).
45. Kroto, H. W., Heath, J. R., O'Brien, S. C., Curl, R. F., and Smally, R. E., C_{60}: Buckminsterfullerene, *Nature*, *318*, 162–163 (1985).
46. Xue, Z. Q., Liu, Y. W., Gao, H. J., Liu, W. M., Wu, Q. D., Qiang, D., Gu, N. Z., Zhou, H. X., Pang, S., Zhu, C., Ma, Z., and Shen, J., The study of the C_{60}–polyethylene thin films, in *Proceedings of the Third China–Japan Symposium on Thin Films*, 1992, Vol. 2, pp. 162–165.
47. Morita, S., Kiyomatsu, S., Fukuda, M., Zakhidov, A. A., Yoshino, K., Kikuchi, K., and Achiba, Y., Effective photoresponse in C_{60}-doped conducting polymer due to forbidden trandition in C_{60}, *Jpn. J. Appl. Phys.*, *32*, L1173–1175 (1993).
48. Gao, H. J., Xue, Z. Q., Wu, Q. D., and Pang, S., Observation of fractal patterns in C_{60}-polymer thin films, *J. Mater. Res.*, *9*(9), 2216–2218 (1994).
49. Wang, K. Z., Studies on electronic switching and electroluminescent thin films made of organic and metal–organic compounds, Postdoctoral dissertation, Peking University, 1995.
50. Gu, Z. N., Qian, J. X., Zhou, X. H., Wu, Y. Q., Zhu, X., Feng, S. Q., and Gan, Z. H., Buckminsterfullerene C_{60}: Synthesis, spectroscopic characterization, and structure analysis, *J. Phys. Chem.*, *95*, 9615–9618 (1991).
51. Potember, R. S., in *Molecular Electronic Devices* (F. L. Carter, ed.), Marcel Dekker, New York, 1992, p. 73.
52. Mulliken, R. S., and Person, W. B., *Molecular Complexes*, Wiley, New York, 1969.
53. Ashwell, G. J., Dawney, E. C., Kuczynski, A. D., Sandy, I. M., Bryce, M. R., Grainger, A. M., and Hasan, M., Langmuir–Blodgett alignment of witterionic optically non-linear D-σ-A materials, *J. Chem. Soc. Faraday Trans.*, *86*, 1117–1120 (1990).

54. Jamada, M., and Omichi, H., Determinaiton of composition of TEM(TCNQ)$_2$ thin films prepared, *Thin Solid Films*, *232*, 13–15 (1993).
55. Sakai, K., Kawada, H., Eguchi, K., and Nakagiri, T., Switching and memory phenomena in Langmuir–Blodgett films, *Appl. Phys. Lett.*, *53*, 1274–1276 (1988).
56. Hamann, C., Switching behaviour of lead phthalacyanine thin films, *Int. J. Electron.*, *73*, 1039–1040 (1992).
57. Sato, A., and Tsukamoto, Y., Nanometer-scale recording and erasing with the scanning tunnelling microscope, *Nature*, *363*, 431–432 (1993).
58. Yang, Q. C., Huang, D. M., Chen, H. Y., and Tang, Q. Y., 1:1 Molecular complex ω,ω'-diphenyl-biuret and phyenyl carbamidonitrile, *Acta Crystallogr.*, *C51*, 1412–1413 (1995).
59. Gao, H. J., Wang, D. W., Liu, N., Xue, Z. Q., and Pang, S. J., Ultrahigh density data storage on Ag–TDCN thin films by scanning tunneling microscopy, *J. Vac. Sci. Technol.*, *B14*(2), 1349–1352 (1996).

Index